CONSUMER ELECTRONICS TROUBLESHOOTING & REPAIR HANDBOOK

Other McGraw-Hill Books of Interest by Homer L. Davidson

Troubleshooting and Repairing Audio Equipment
Troubleshooting and Repairing Compact Disc Players
Troubleshooting and Repairing Camcorders
Troubleshooting and Repairing Consumer Electronics Without a Schematic
Troubleshooting and Repairing Microwave Ovens
Troubleshooting and Repairing Solid-State TVs
TV Repair for Beginners
Consumer Electronics Component Handbook

CONSUMER ELECTRONICS TROUBLESHOOTING & REPAIR HANDBOOK

HOMER L. DAVIDSON

McGraw-Hill
New York San Francisco Washington, D.C. Auckland Bogotá
Caracas Lisbon London Madrid Mexico City Milan
Montreal New Delhi San Juan Singapore
Sydney Tokyo Toronto

Library of Congress Cataloging-in-Publication Data

Davidson, Homer L.
Consumer electronics troubleshooting and repair handbook / Homer Davidson.
p. cm.
ISBN 0-07-015809-6
1. Electronic apparatus and appliances—Maintenance and repair—Handbooks, manuals, etc. I. Title.
TK7870.2.D375 1999
621.3815'4—dc21 99-19660
CIP

McGraw-Hill

A Division of The ***McGraw-Hill*** *Companies*

1 2 3 4 5 6 7 8 9 0 DOC/DOC 9 0 4 3 2 1 0 9

ISBN 0-07-015809-6

The sponsoring editor for this book was Scott Grillo, the editing supervisor was Andrew Yoder, and the production supervisor was Tina Cameron. It was set in Times New Roman by Lisa M. Mellott through the services of Barry E. Brown (Broker—Editing, Design and Production).

Printed and bound by R. R. Donnelley & Sons Company.

This book was printed on acid-free paper.

Dedication

I dedicate this book to the thousands of electronic technicians who troubleshoot and repair millions of consumer electronic products every year. A great deal of thanks goes to those electronic manufacturers who provide the world with electronic entertainment throughout every day of our lives. I salute those who place the electronic product in the hands of the consumer, and the electronic technician who stands at the service bench each day and brings the product back to life.

CONTENTS

INTRODUCTION

Servicing today's TV, audio and video, cassette and CD players, and camcorders requires more knowledge, constant use of the oscilloscope, and crucial test instruments. Most of the latest consumer electronic products have crucial circuits with miniature surface-mounted components (SMD), requiring dedicated troubleshooting. The components might be crammed together and require removing different sections to access the defective component. Each section must be isolated, tested, and repaired. This book is crammed with service data, electronic tips, and how to make those difficult repairs within its 1200 pages.

The purpose of this book is to provide practical service data on consumer audio/video equipment. It can help the beginning college student, intermediate, and experienced electronic technician further their knowledge with practical service applications of test equipment. Examples are taken from many case histories in the electronic field. Different methods of servicing the electronic chassis are given because each year many changes and modifications are included in TVs, audio equipment, cassette and CD players, and camcorders.

This large book is broken down into five different parts with an introduction to basic and general troubleshooting, and a safety chapter. Part I explains "Getting Started." Part II is "Troubleshooting and Repairing Solid-State TVs." Part III provides service data in "Troubleshooting And Repairing Audio And Video Cassette Players." Part IV explains how to "Troubleshoot And Repair CD Players." Last but not least, Part V winds up the book with "Troubleshooting And Repairing Camcorders."

How to troubleshoot and service the circuits in the modern TV, such as SAW filters, comb filter, high-definition circuitry, chopper and switched-mode power supplies, start-up circuits, scan-derived voltage, and stereo sound are included in "Troubleshooting And Repairing The Solid-State TV Chassis."

A wealth of detailed illustrations and photographs help you learn how to service defective horizontal and vertical sweep circuits, diagnose high-voltage circuit problems, identify and cure tuner malfunctions, perform AGC and sync circuit tests, maintain brightness and picture tube problems, how to service remote-control transmitter and receiver circuits. This incredibly complete workbench reference gives you practical information on how to troubleshoot and repair the latest solid-state circuitry used by all major TV manufacturers.

The cassette player is used in every room of the house or office, outdoors, in the car, and while running for more exercise. Inside the section "Troubleshooting and Repairing Audio and Video Cassette Players," you will find detailed coverage of a wide range of electronic audio and video components, which includes personal and portable CD players, boom-box and double cassette decks, home and car stereo cassette players, professional portable cassette and microcassette recorders, and VCRs.

How to troubleshoot the cassette player without the exact schematic, and how to perform speed adjustments, fix tape and erase head problems, demagnetize the tape heads, the erase head, and repair erratic switching, accidental erasing, defective recording circuits, nonfunctional fast forwarding, weak stereo circuits, garbled recording, and noisy stereo channels are only a few of the symptoms included in this section.

Just about anyone can make the most simple audio and cassette repairs in these chapters. Although certain cassette problems are provided in a given chapter, the same problem might be included in another chapter and related to the cassette player you are now servicing. Cassette troubleshooting and servicing methods are all here, so you can bring that cassette player and recorder back to life.

The compact disc (CD) player has come a long way in the past 15 years. The little rainbow-reflecting disc has brought clear, crisp, noise-free reproduction of music to our ears. The digital music source can now be reproduced with more depth, greater detail, and more imaging than ever before. This new digital-to-analog technology has brought the ultimate in glorious sound.

"Troubleshooting and Repairing CD Players" covers the newest makes and models, addresses technological advance, with troubleshooting charts throughout. You will find hands-on instructions on how to service servo systems, remove and replace the defective laser head, how to locate the defective optical assembly, how to service the signal and system-control circuits, and how to locate and replace the defective slide, load, spindle up-down, magazine, chucking, elevator, and changer motors.

Compact disc players are loaded with special components, such as surface-mounted parts (SMD), crucial large-scale integration (LSI) and integrated circuits (ICs), optical lens-and-laser assemblies, and many different motor circuits. You should obtain these parts through the CD distributor, manufacturer, manufacture service depot, or a dealer that handles CD players. Very few universal components are used in compact disc players. Always replace these special parts with those that have the original part number.

All formats of today's camcorders (8 mm, Beta, VHS, and VHS-C) are covered in the section "Troubleshooting And Repairing Camcorders." The various chapters include practical service techniques, video circuits, servo circuits, control systems, motor circuits, mechanical tape operations, mechanical adjustments, electrical adjustments, and camera pickup circuits.

Like today's compact disc player, the camcorder is loaded with special components, such as SMDs, integrated ICs, main MI-COM processors, COD and MOS image devices, and special optical components. Besides correct test equipment, the schematic diagram and service literature is a must-have item. Many manufacturers have special test equipment and jigs to provide quicker camcorder service.

In the many pages of this book, you will find how to troubleshoot and repair 90% of the electronic products in the consumer electronics field. Actually, this is four complete books in one large volume. Servicing electronic products can be a learning experience, a lifetime job, and can be a lot of fun. This volume contains more than 40 years of practical experience with repairing data makes an excellent reference for anyone involved in consumer electronic repair, professionally, or as a hobby.

PART 1

GETTING STARTED

1

PRACTICAL TROUBLESHOOTING TECHNIQUES

CONTENTS AT A GLANCE

The experienced electronic technician finds a symptom and ends up replacing the defective component. Just by looking at the TV screen or listening to a speaker, you can isolate the stage where the trouble appears (Fig. 1-1). By using the three senses, you can sometimes pinpoint the possible stage on the block diagram or schematic. Knowing how to locate, remove, and replace the defective part saves service time.

Precious Time

Time is one of the electronic technician's greatest factors. You can waste it or make money with it. The time servicing the difficult or intermittent problem will determine if the repair job is profitable. Any time that the experienced technician spends more than one hour on a given electronics problem without locating the defective component, time is lost. Lost time can also occur when a coffee break extends from 15 to 30 minutes (or even longer).

Call backs or repeated repairs cost the electronic technician extra money. Doing a thorough repair job at the beginning eliminates repeated calls. Preventive maintenance can reduce call backs. Of course, I must admit that the electronic chassis does produce a lot of service problems that can happen after a repair. Always charge for repeated service when the original repair has nothing to do with the present problem. Remember, the doctor or auto mechanic charges for the additional call.

FIGURE 1-1 By checking the TV screen, you can note the symptom and decide the location of the trouble.

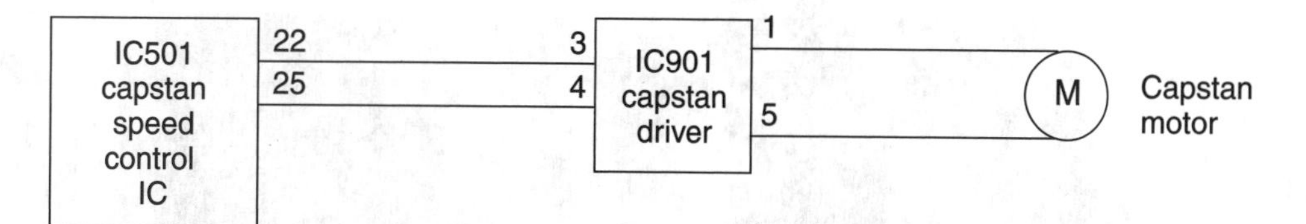

FIGURE 1-2 **The slow-speed symptom of a camcorder can point to a defective speed-circuit capstan driver IC motor or belt.**

The Symptom

On a TV, a distorted picture and mushy sound point to a defective IF or tuner within the chassis. If the volume control is turned completely on and only weak sound within the speaker of an audio amplifier can indicate that the transistor, IC, or coupling capacitor is defective. Distorted and weak sound might indicate a dirty tape head or output audio stage within the cassette player.

A camcorder with slow speed might have a defective speed circuit or motor (Fig. 1-2). Improper or intermittent loading of the CD player or camcorder might result in a dirty or defective switch, loose motor belt, dry gearbox, or a defective motor. The tape might not rotate in the camcorder or cassette player if the motor is defective, capstan is dirty or dry, or if the motor belt is loose. Just remember to watch, listen, and smell for the various symptoms, before tearing into the electronic chassis.

The Three S's

The three S's might be your most important tool. Sight, sound, and smell can help you to solve a lot of TV, cassette, camcorder, CD player, and microwave oven problems. You can see burned resistors, fried flyback or power transformers, and lightning damage. A leaky electrolytic capacitor might have a black or white substance oozing out at the bottom connections. Poor board connections might indicate an overheated solder connection. Cracked or overheated connections of a high-power resistor might indicate possible trouble. Definitely, you can see a spark gap, tripler unit, or high-voltage lead arc over. Above all, your eyes identify the trouble symptoms from the front of a picture tube, overheated IC, or transistor on the chassis.

Insufficient or distorted audio sound might be traced to the sound stages. Arcover at the picture tube or high-voltage transformer within the TV chassis might be heard. The ear might pick up the tic-tic sound of the flyback transformer with high-voltage or chassis shutdown. Some TV technicians can hear the 15,750-Hz sound of the horizontal output transformer, indicating that the horizontal oscillator stages are performing (Fig. 1-3). Noisy sounds from a interstage or output transformer might indicate loose particles within the component.

FIGURE 1-3 **The 15,750-Hz sound from the flyback can be heard by some people.**

You might smell an overheated voltage-dropping resistor or degaussing thermistor within the TV chassis. The ozone smell might be traced to the flyback transformer, tripler unit, screen-focus assembly, focus controls, or the anode connection of the picture tube. The sweet smell of the overheated power transformer in the cassette player might indicate lightning or power line outage damage. Not only can you smell an overheated transistor or diode, but you can feel them. Overheated components within the electronic chassis are always a source of trouble.

Isolation

Isolate the symptom to a block diagram. Begin troubleshooting procedures with the block diagram. Upon checking the block diagram, you can quickly isolate the defective sections in the TV, cassette player, audio amplifier, and CD player. You can see where the signal path goes from section to section (Fig. 1-4). The block diagram can further be broken down to show several blocks of one particular section. Not only does the block diagram illustrate how the different sections are tied together, but it can be used to show how the circuits operate in a given chassis.

After locating the defective section or circuits, locate the various components upon a schematic diagram that might cause the electronic chassis to not operate. The first look at a schematic of a TV or CD player might appear complicated, but if you break the schematic down into various sections, servicing becomes more systematic. For instance, if the loading motor in the camcorder is not operating, go directly to the servo or system control IC and trace the signal back to the driver and loading motor (Fig. 1-5). Each functional problem can be circuit-traced using the same logic.

A lot of the manufacturers have the different circuits drawn in various colors, so they stand out. Others use a variation of dotted lines to separate the various circuits. Most CD schematics have arrows or a thick color path of arrows, indicating the signal path throughout the various stages. Crucial voltages are found on the schematic or listed separately in a voltage chart. Some schematics have voltages listed in red or green. If not, mark them directly on the diagram before you start servicing procedures. After locating a defective component, circle it, and draw a line out to the side to the margin area to record the service problem.

Locating the suspected component on the PC board chassis might be difficult if a parts layout diagram is not handy. Sometimes the components are labeled and others are not. The electronic components are often mounted on one board and the mechanical parts are located on another assembly. Small components, such as transistors, capacitors, resistors, and diodes might be difficult to locate because many are not individually marked. If you do not have a service manual or schematic, you must trace out wiring and components, which takes up a lot of valuable time.

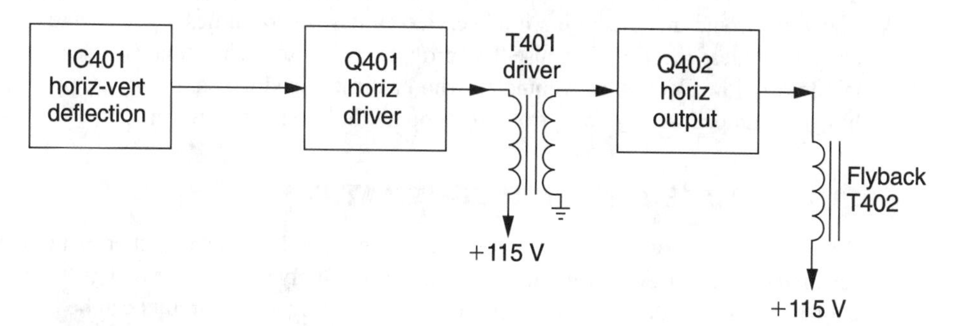

FIGURE 1-4 **Isolate the symptom to a block diagram and then use the circuit diagram.**

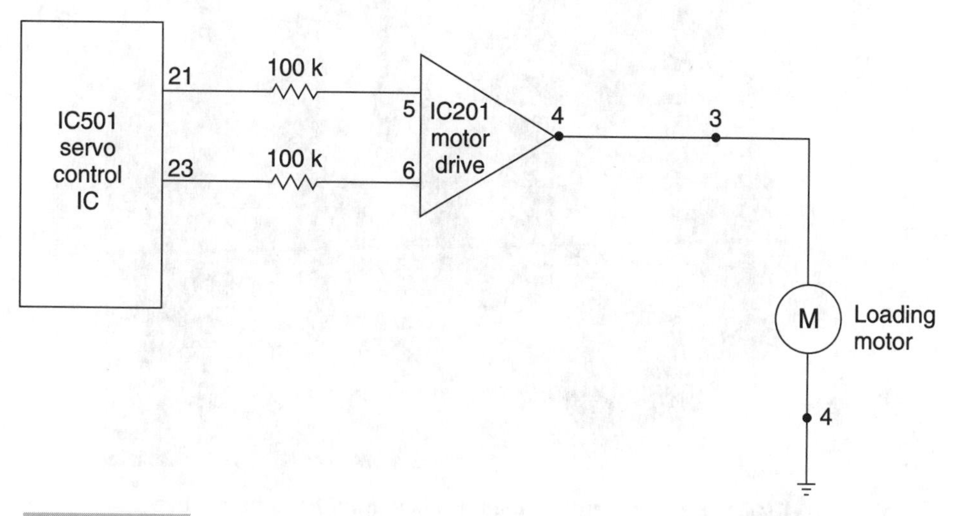

FIGURE 1-5 **Trace the dead loading motor symptom to the system control IC, driver IC, and loading motor.**

Service Literature

Schematics and service literature are a must for the electronic technician. Although many TV technicians might service a TV or camcorder without a schematic, it is not practical over the long run (Fig. 1-6). You can lose a lot of valuable service time without a good electronic schematic. Several years back, the Japanese TV chassis was difficult to service because wiring diagrams and parts were difficult to obtain.

This is not so today because most foreign imports are covered by schematics and service literature. Howard Sams Photofacts cover most TV chassis. Manufacturers Profax schematics are found in the pages of *Electronic Servicing & Technology* magazine.

Service literature, including detailed diagrams, is published by each electronics manufacturer. Here, additional data as to how each stage operates might clear up the tough job. This service literature can be purchased yearly or for each individual electronic chassis. You will also receive important production and modification information for each chassis. Many electronic manufacturers hold service clinics several times a year. Take a day off and attend these meetings because they provide crucial servicing data. Besides, you might talk to a fellow electronics technician who has just licked the same service problem that has you stumped. A day away from the shop might bring future rewards.

SUBSTITUTE ANOTHER SCHEMATIC

When the exact service manual or schematic is not available, use another manufacturer's schematic diagram. Although the substitute is not exactly the same, it will give you different test points to troubleshoot the circuit. Signal tracing audio circuits can be checked by

FIGURE 1-6 **A schematic diagram is a must-have item when troubleshooting a difficult problem.**

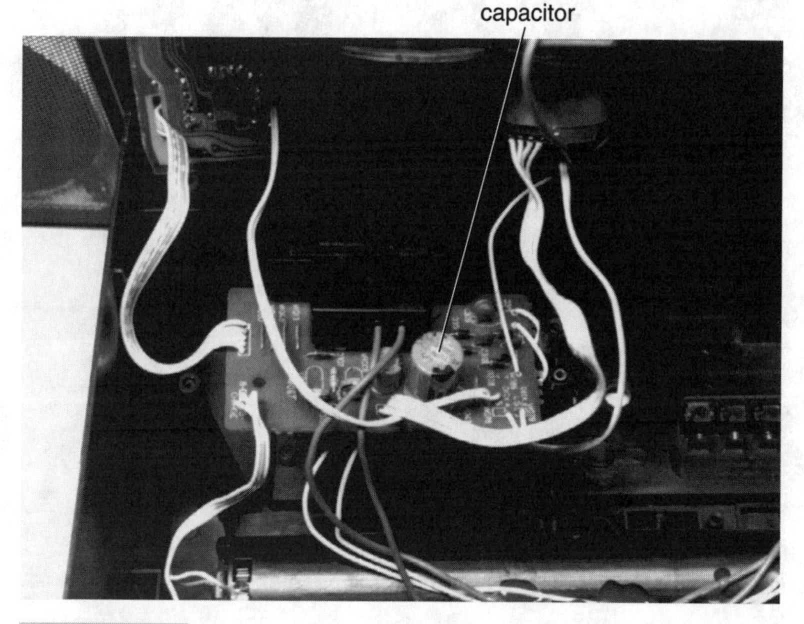

FIGURE 1-7 **Locate the low-voltage circuits by identifying the large filter capacitor.**

starting at the tape head winding and going from base to base of each transistor. The output stages can be checked by starting at the volume control.

Usually, cassette audio stereo circuits are laid out on the PC board with the left channel on the left and the right channel on the right. You can start at the speaker and trace the circuit back to locate the output transistors or ICs. Some of these power output transistors and ICs are located on heatsinks. The supply lead from the IC or transistor can be traced back to the power supply. The power supply can be identified with large filter capacitors, diodes, and bridge rectifiers (Fig. 1-7). The suspected part can be checked against the identical one found in the good stereo channel.

Often the horizontal and vertical circuits within the TV chassis can be compared to another schematic with very few changes. The different stages can be scoped with the same type of waveforms as another chassis. Likewise, the voltage and resistance measurements are quite common in the horizontal and vertical circuits.

Chassis Comparison

Many times the same chassis might be located in more than one different TV. This is especially true if you are servicing chain-store merchandise. It's also possible to have trouble with a new TV right out of the box. The new TV chassis might also be compared with those in the showroom if you carry that certain brand of TV.

FIGURE 1-8 **If the correct schematic is not handy, compare the defective circuits with those in a comparable chassis.**

If, by chance, you are stumped on a new circuit or do not have a schematic handy, compare the voltage and resistance measurements of the working chassis can help solve the problem (Fig. 1-8). Many chain-store TVs are manufactured for them by another source, and many TV brands actually have the same equipment. Of course, comparison checks take a lot of time and should only be made when you have no other choice.

Spilled Liquid

After removing the back cover of a TV, you can quickly decide if some kind of liquid has spilled down on the chassis. Besides being a big cleaning mess, the chassis might not be repairable if liquid has been spilled inside the back of the TV. It is very difficult to clean small resistors and capacitors that are soldered flat against the PC board. Often, the chassis is turned on, causing arcing in the horizontal and high-voltage circuits.

Sometimes when a voltage arc between components occurs, the arc breaks down the PC board and can cause burning of the board. The board can be saved by cutting out the burned area. Use a pocket knife or drill out the burned area, then replace burned components across the area. Place hookup wires around burned-out PC wiring connections. The wiring schematic and component replacement charts can be handy with this procedure.

If a beverage is spilled in the chassis, the PC board and components become sticky and can be difficult to repair. Place a fan near the chassis to dry the areas if water has been spilled inside the TV. After cleaning and repairing the board, spray the entire area with clear high-voltage insolation spray after the chassis operates for several hours. The chassis should then be operated for several days to be sure that no further breakdowns occur. Sometimes you will find that the TV cannot be repaired and must be replaced.

Basic Test Equipment

Back in the vacuum tube days, many TVs were repaired with a tube tester, a VTVM or VOM, and an oscilloscope (Fig. 1-9). Today, a digital multimeter (DMM), oscilloscope, CRT checker, color-dot-bar generator, and tube-test jig are the basic test equipment. Crucial voltage and resistance measurements can be made with the DMM. The DMM can even check diodes and transistors in or out of the circuit. The oscilloscope is required to look at waveforms in the various circuits. Most waveforms are checked with the color-dot-bar generator connected to the antenna terminals to provide a signal. A tube test jig is ideal to monitor when servicing the pulled chassis. A good CRT tester eliminates guessing about the status of the picture tube.

As new circuits arrive and old test equipment is replaced, an older oscilloscope should be replaced with a dual-trace 40- or 100-MHz unit. An isolation variable line transformer is a must item when servicing a new integrated flyback chassis. A capacitance meter and CRT restorer-analyzer are handy test instruments to have around. A frequency counter and sweep-marker generator are required for TV-alignment purposes. To check high voltage at the CRT, select a good high-voltage probe (Fig. 1-10).

With the latest portable and console TVs, the chassis can be loosened, slid backwards, or tilted on edge or side for easy servicing. Often, the connecting lead wires and cables from the TV chassis to other components mounted within the plastic portable case can be untied and extended so that the chassis can be slid backwards or actually removed from the cabinet (Fig. 1-11). Likewise, remove chassis bolts and screws so that the small TV chassis inside the large TV console can be moved to test most parts within the wooden cabinet.

You might find that the busy electronics technician uses only three or four test instruments in his daily service routine. Many other test instruments are collecting dust on the

FIGURE 1-9 A dual-trace scope, DMM, FET-VOM, and isolation transformer are basic test instruments for an electronic technician.

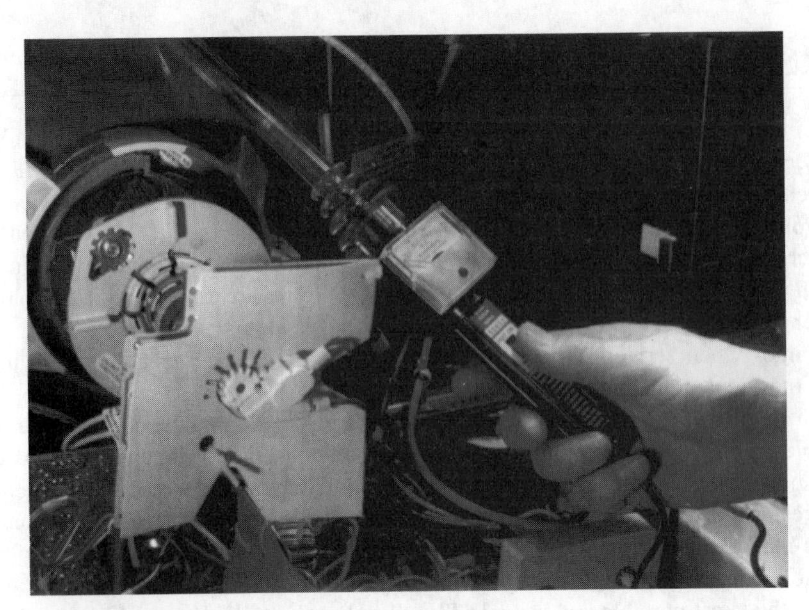

FIGURE 1-10 Check the high voltage at the picture tube with a high-voltage probe.

FIGURE 1-11 Pull the portable TV chassis back or remove it for easy access.

service bench and not being used. Knowing how to use the basic test equipment is the secret to quick, efficient, and practical electronics servicing. One quick way to learn how the test instruments operate is to attend the classes sponsored by many test instrument distributors and manufacturers. Check the list that follows for required basic test instruments.

BASIC TV TEST INSTRUMENTS

1 Digital multimeter (DMM)
2 VOM, VTVM, or FET voltmeter
3 Oscilloscope (at least 40 MHz)
4 Variable isolation transformer
5 CRT tester and rejuvenator
6 Diode, SCR, semiconductor tester
7 Capacitance tester
8 High-voltage probe
9 Color-dot-bar generator

REQUIRED AUDIO TEST INSTRUMENTS

Besides several screwdrivers, a pair of long-nose pliers and side cutters do the bulk of audio work, combined with a VOM and DMM. The digital multimeter (DMM) can check voltage, resistance, current, capacitance, diodes, and transistors; some testers even have a frequency counter. A VOM or FET-VOM is ideal in audio alignment (Fig. 1-12). It's much easier to see the meter hand respond to the audio signal than the DMM. You will find that the same TV test equipment can be used to service audio equipment.

BASIC AUDIO TEST EQUIPMENT

1 VOM and DMM
2 FET-VOM
3 Dual-trace oscilloscope
4 Noise signal generator
5 Audio generator
6 Sine-squarewave generator

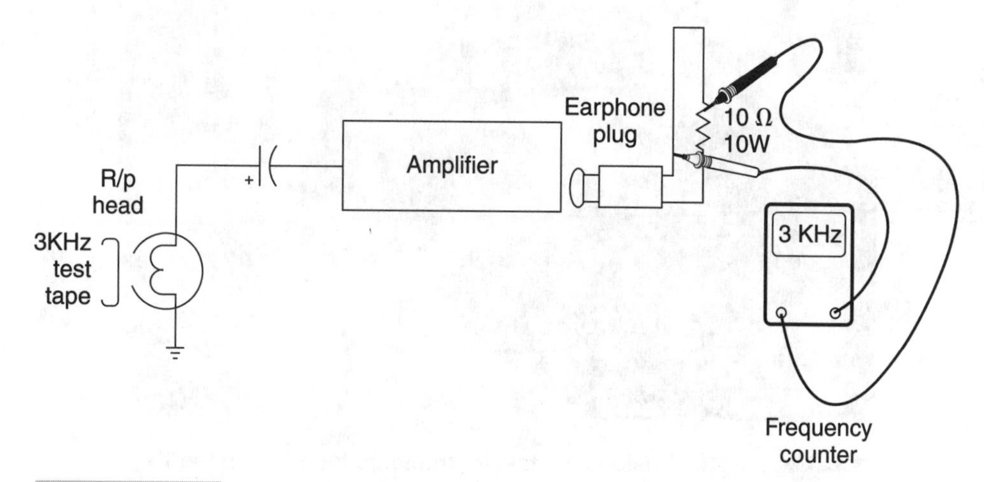

FIGURE 1-12 **The VOM, FET-VOM, and frequency counter are ideal for tape-head and audio-alignment procedures in cassette players.**

7 Semiconductor tester
8 Capacitance meter
9 Audio signal tracer
10 Test speaker
11 Speaker loads
12 Test discs and cassettes

For those technicians who specialize in troubleshooting audio equipment, the additional test equipment might speed up making crucial repairs and alignment:

1 Frequency counter
2 Function generator with counter
3 Audio oscillator and frequency counter
4 Distortion meter or analyzer
5 Wow and flutter meter

REQUIRED CD TEST EQUIPMENT

You need several test instruments to troubleshoot and make the necessary alignment adjustments in the compact disc player. Most of these test instruments are already common to the average electronics technician's service bench (Fig. 1-13).

1 Dual-trace oscilloscope
2 Optical power meter

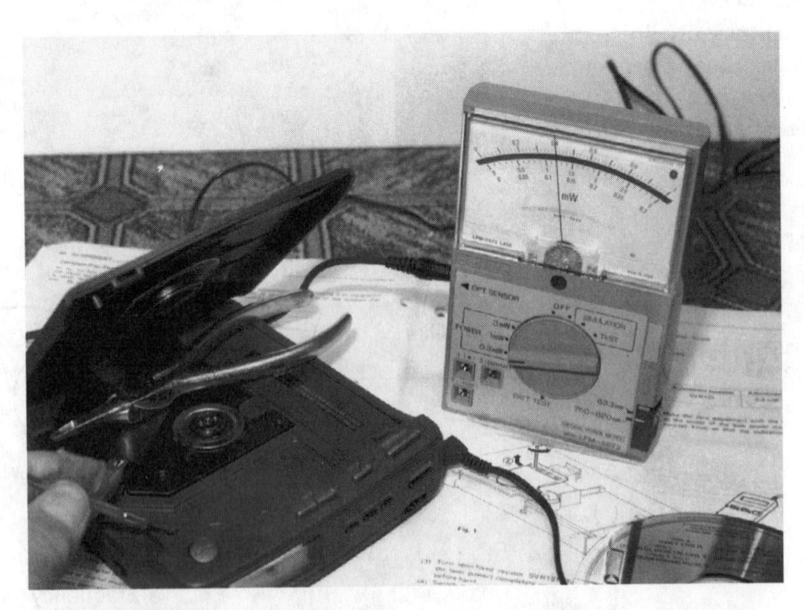

FIGURE 1-13 Besides the test instruments found upon the TV bench, the optical power meter, frequency counter, and AF signal generator are essential for compact disc (CD) servicing.

FIGURE 1-14 **The optical power meter and special adjustment tools are necessary for CD servicing.**

3 Digital Multimeter (DMM)
4 Low-frequency AF oscillator
5 Signal generator
6 Capacitance meter
7 Frequency counter
8 Test discs
9 Special tools, filter adjustment circuits, wrist trap, manufacturer special jigs, etc.

The optical power meter, test discs, special tools, filter circuits and special jigs purchased from the manufacturer for special adjustments are probably the only devices that the established repair shop might not have (Fig. 1-14).

Special tools, such as a grating tool, feeler gauge, or special screwdrivers might be needed for some adjustments. The focus and tracking, loop-gain harness, and special manufacturer circuits can be handmade. Special manufacturers jigs, monitor devices, and test cables can be purchased for certain tests required by the manufacturer. Although, each manufacturer might require a certain test disc, the most common are the YEDS7, YEDS18, and SZZP1014F.

The laser power meter is used to measure the laser diode output and infrared remote-control units. The laser meter is particularly suitable for the service of compact disc and laser disc players because of its narrow, tillable probe. The meter can be used to check the function of cassette compartment LED in video recorders and transmitting diode in infrared 04 remote controls (Fig. 1-15). With the three measuring ranges of 0.3 mW, 1 mW, and 3 mW, all laser light sources found in many CD players can be checked. The switchable wavelength of 633 nm, 750 to 820 nm, has an accuracy of 5%.

When measuring the laser light beam, do not look directly at the laser light. Remember that CD and LD players emit invisible, infrared light. You simply cannot see the infrared beam. Keep your eyes at least 1½ feet away from the laser beam.

FIGURE 1-15 Besides checking the laser beam, the optical power meter can help you to check the LED, transmitting diode, and infrared remote-control units.

REQUIRED MICROWAVE OVEN TEST INSTRUMENTS

Several different test equipment is required to service and maintain microwave ovens. Besides regular hand tools, torque and star screwdrivers, a microwave leakage tester and magnameter are must-have test instruments (Fig. 1-16). Many of the basic tools you'll need are found in the shop or on a TV repair bench.

1. Torque and star screwdrivers
2. DMM
3. Microwave leakage tester
4. Triac and SCR tester
5. Magnameter
6. Fuse saver
7. Test bulb

CRUCIAL VOLTAGE TESTS

No doubt, crucial voltage and resistance tests have been used to find more defective components than any other tests. Most defective solid-state components, such as transistors, diodes, and ICs can be located with crucial voltage tests. A quick voltage check across the main filter capacitor can help you to determine if the power supply of any electronic product is normal or has an improper voltage source (Fig. 1-17). A leaky or shorted audio output transistor or IC can be spotted with a crucial voltage test.

TRANSISTOR VOLTAGES AND RESISTANCE TESTS

Taking crucial voltage measurements on the transistor element can determine if the transistor is normal. Suspect that the transistor is open if the collector voltage is much higher than normal, with no voltage on the emitter terminal. An open emitter resistor can produce the same voltage readings (Fig. 1-18).

FIGURE 1-16 **The Magnameter can check the current, high-voltage, and magnetron operation within the microwave oven.**

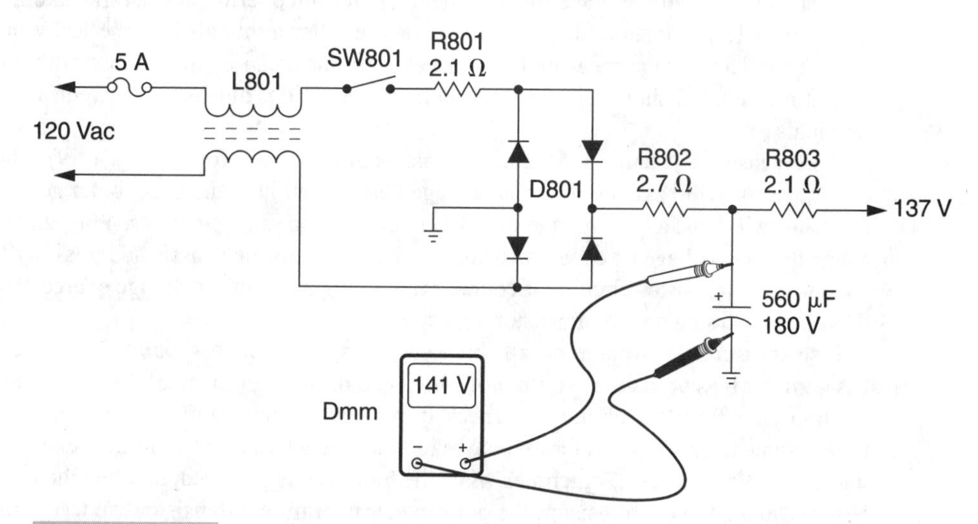

FIGURE 1-17 **A quick voltage test of the large filter capacitors can indicate if the low-voltage source is normal.**

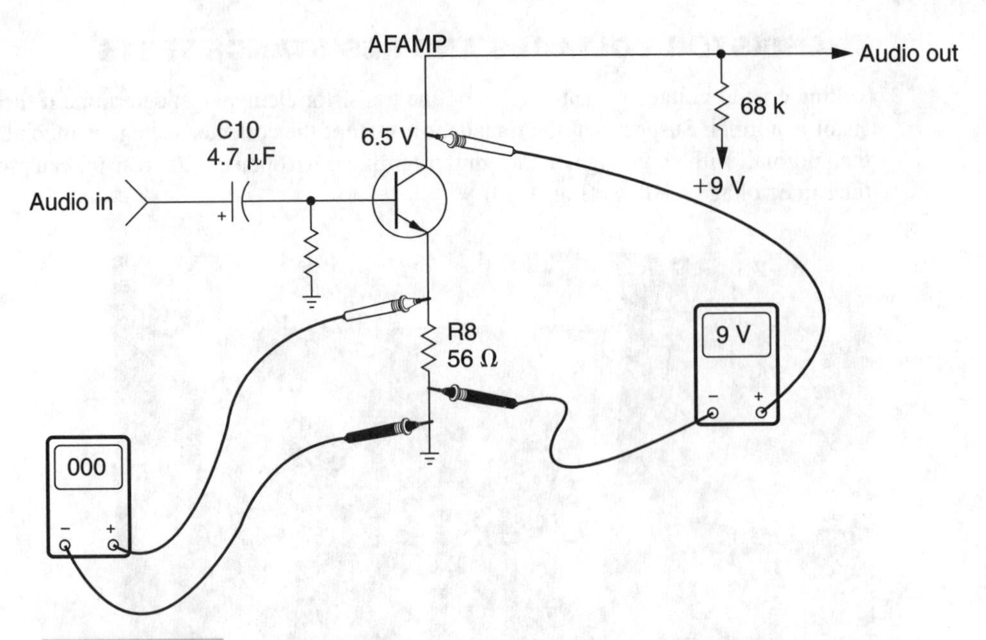

FIGURE 1-18 **An open emitter resistor can increase the voltage at the collector terminal.**

Connect the voltmeter across the collector and emitter terminals and mark down the voltage measurements. Remember, the DMM can read these low-voltage measurements. The collector terminal of a NPN transistor is positive, but a PNP transistor is negative. Now short the base to emitter terminals together and the voltage should increase if the transistor is normal.

Practically the same voltage measurements on all three terminals can indicate that the transistor is leaky or shorted. The collector and emitter terminals become leaky in most transistors. The voltage measurement on both terminals can be quite close with a direct collector-to-emitter short. Discard transistors with any signs of leakage between any of the terminals.

The transistor is usually good if normal base-emitter bias is found (Fig. 1-19). The silicon transistor will have a 0.6-V bias voltage between emitter and base, but a germanium transistor will have a bias voltage of 0.3V between these same elements. You can quickly check the bias voltage of each transistor within the electronic chassis and possibly locate the defective transistor circuit. Of course, with an improper or no voltage source, very little voltage is found on any transistor terminals.

First, measure the voltage on all three elements to common ground. Then check the base-emitter bias voltage. If you're not convinced of the legitimacy of the results, remove the transistor from the circuit and check it out of the circuit with a DMM or transistor tester. Some technicians prefer to remove only the base terminal from the circuit and give it another in-circuit test. Sometimes when the transistor is removed, or when the in-circuit tester is applied to the transistor, the defective or intermittent transistor can test normal.

Besides using the voltage and resistance methods, the transistor can be checked in or out of the circuit with any of the many different transistor testers on the market. The suspected transistor can be checked out of the circuit with resistance measurements of the VOM or DMM (Fig. 1-20). Leaky or open transistors can be located with the ohmmeter scale.

A quick method to check a transistor in or out of the circuit is with the diode or transistor test of the digital multimeter (DMM) (Fig. 1-21). Comparable resistance measurements of the diode junction from base to collector and base to emitter will identify an NPN or PNP transistor, and will indicate if a leakage or open junction occurs between the elements. The leaky transistor resistance might be low between collector and emitter or base

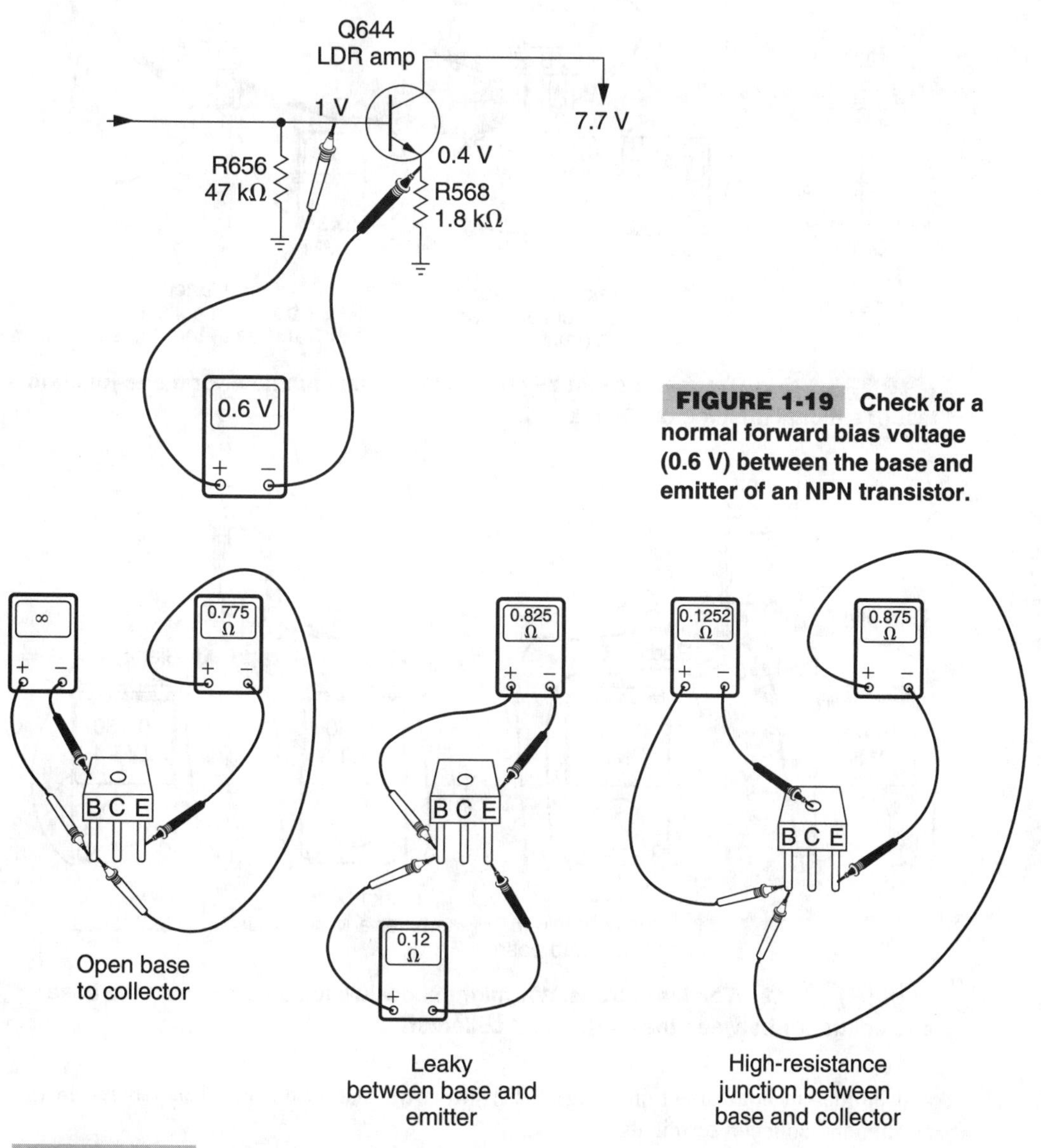

FIGURE 1-19 **Check for a normal forward bias voltage (0.6 V) between the base and emitter of an NPN transistor.**

FIGURE 1-20 **Check the suspected transistor with a diode-junction transistor test of DMM.**

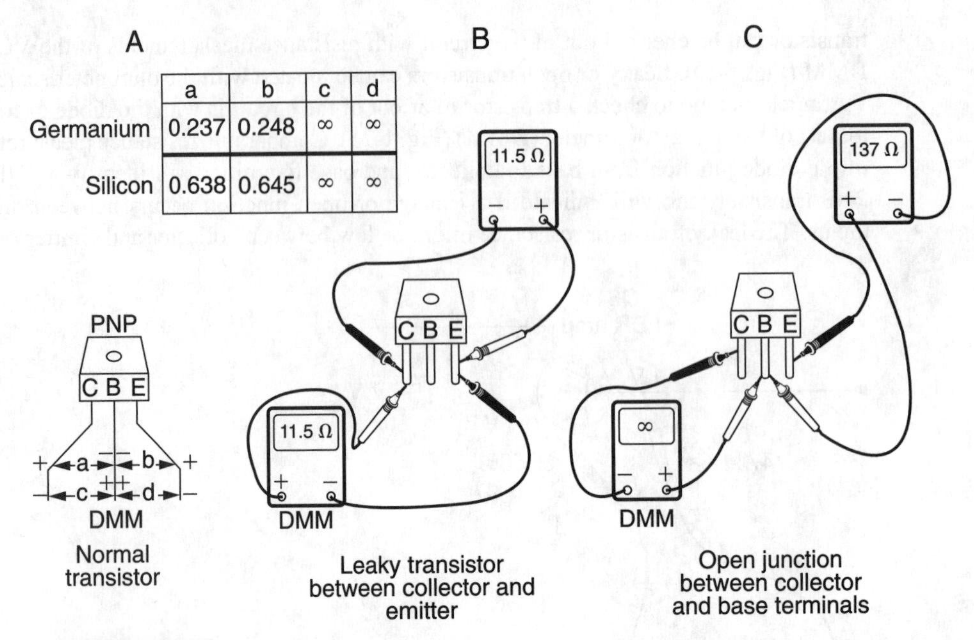

	a	b	c	d
Germanium	0.237	0.248	∞	∞
Silicon	0.638	0.645	∞	∞

FIGURE 1-21 **Transistors can be checked with the ohmmeter or diode-junction test of a digital multimeter (DMM).**

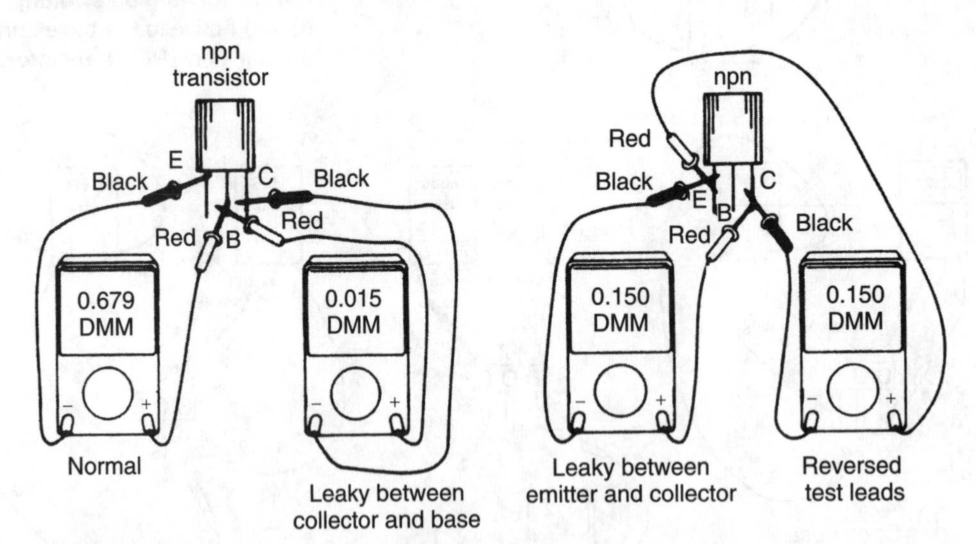

FIGURE 1-22 **The leaky transistor might show low leakage between the base and emitter or between the emitter and collector.**

and emitter in both directions (Fig. 1-22). Most transistors become leaky between the emitter and collector terminals.

The transistor might be open between base and collector or base and emitter terminals, with no measurement on the DMM. A transistor with a high-resistance joint can cause a weak or dead response when the leakage is greater with one set of transistor elements than the other set. A high-resistance joint exists when the measurement is different by several hundred ohms between two elements and not the other two. Transistors can be quickly checked with the DMM diode or transistor tests.

DIODE TESTS

A quick voltage test with the positive terminal of voltmeter at the positive terminal of suspected diode and the negative terminal to ground can help you to locate a defective diode. Likewise, a zener diode can be located in the same manner. The zener diode is used in a circuit to regulate the voltage at a certain level. Really low voltages at the positive terminal indicate that the diode is leaky or shorted.

The defective diode can be located with a low-resistance measurement. Low-voltage diodes can be checked for open or leaky conditions with the diode test of a DMM. Be sure that a power transformer winding, low-resistance resistor, or transistor is not across the path of the diode when making in-circuit tests. It's best to remove one end of the diode for accurate leakage test (Fig. 1-23).

Check the suspected diode with the red probe of DMM to the anode terminal of the diode and the black probe to the cathode terminal. You should get a reading. Now reverse the procedure. No measurement or an infinite measurement is noted on a normal diode. When making a diode test with the FET-VOM, start with the red or positive probe to the cathode and

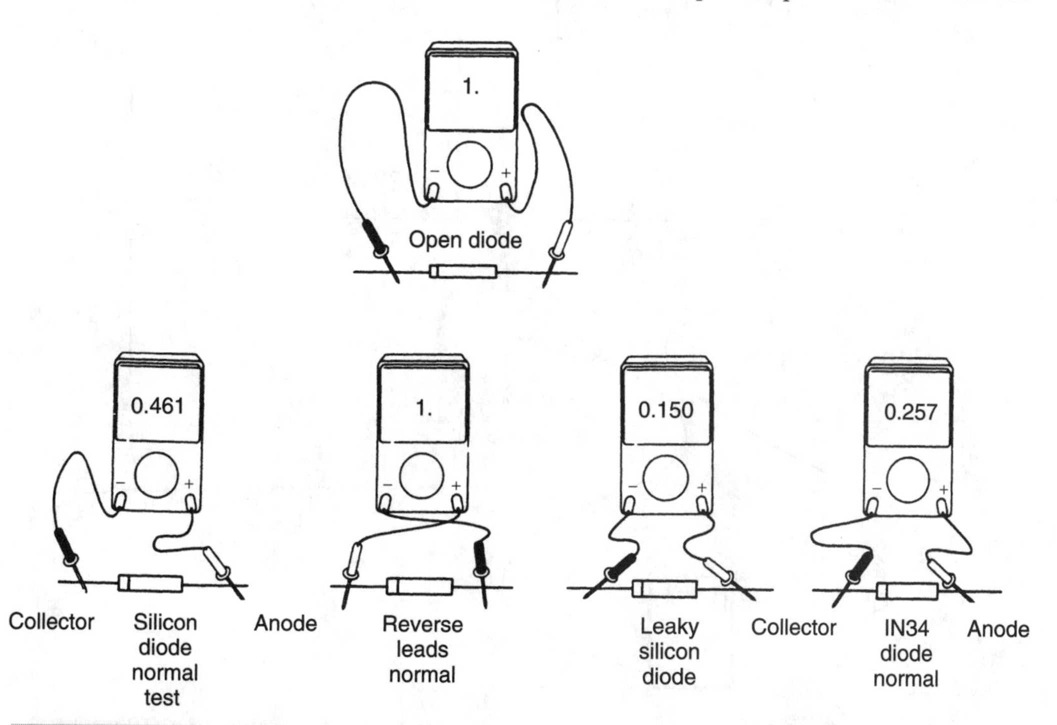

FIGURE 1-23 **Remove one end of diode from the circuit for accurate open or leaky tests.**

black probe to the anode terminal. The normal diode will show a resistance measurement. If a low-resistance measurement is found in any direction, the diode is leaky or shorted.

Most low-voltage and damper diodes can be checked for open or leaky conditions with an ohmmeter or DMM. The RF, video, and audio detector diodes can be checked with an ohmmeter and DMM, but will have a higher resistance measurement. Some older boost rectifiers within the TV can only be checked for heavy leakage. The high-voltage rectifier stick found in the black-and-white TV chassis and microwave oven cannot successfully be measured with the VOM or DMM tests.

IC VOLTAGE AND RESISTANCE TESTS

The defective IC can be traced with voltage and resistance measurements. Check the signal in and out of the IC with the scope. The audio IC can be signal traced with audio in and out tests of the suspected IC with an external audio amplifier. Very low voltage on several terminals of the IC can indicate a leaky component. A low voltage at the IC terminal supplied directly from the low-power supply can indicate a leaky IC (Fig. 1-24). Remove the

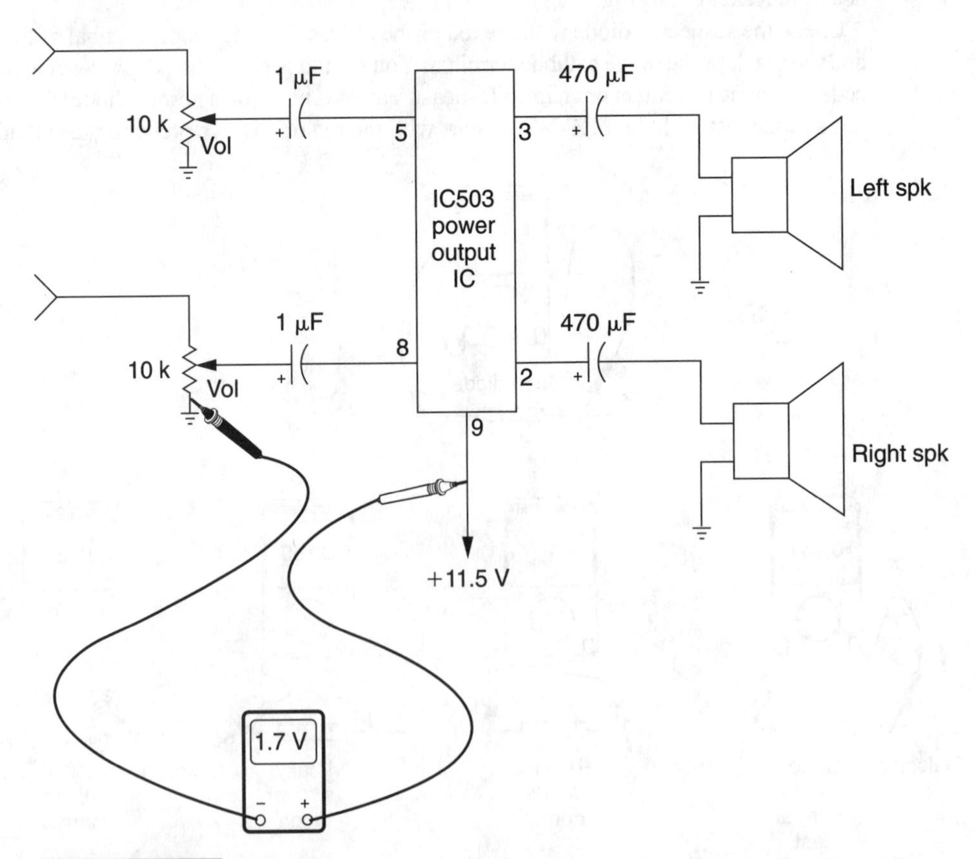

FIGURE 1-24 **A low resistance at the voltage-supply terminal of an IC might indicate that it is leaky.**

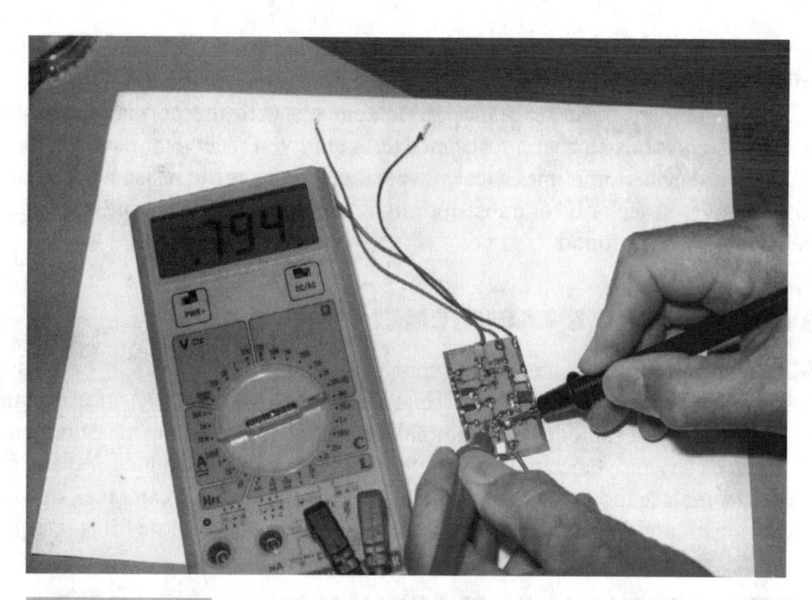

FIGURE 1-25 **A leaky or open transistor might appear to operate properly after removing it from the chassis.**

terminal pin from the circuit and take another resistance measurement between the pin and the chassis ground. Apply the soldering iron tip to the mesh material and suck up all solder around the voltage-supply pin. Flick the pin with a pocket knife or small screwdriver to be sure that the pin is free. Replace any IC that has a low-resistance measurement.

Internal open connections inside of the IC can cause it to be dead or intermittent. If the signal is applied to the input and no signal or a low signal is at the output terminal, suspect a defective IC. Take accurate voltage measurements and check them against the schematic to determine if the IC is defective. Sometimes voltage and resistance measurements are normal compared to the schematic; replacing the IC is the only way to solve the problem. Replace the suspected IC at once if it is a plug-in socket type.

Sometimes, with ohmmeter tests, you can locate a change in a resistor or leaky capacitor from one pin to common ground. Be sure that IC is defective by taking signal-in and signal-out tests, with crucial voltage and resistance measurements. Always compare these measurements with the good channel in the stereo audio amplifier. Replacing the IC requires a little more time than replacing a transistor.

CRUCIAL RESISTANCE AMPLIFIER TESTS

If the audio amplifier keeps blowing a fuse and the chassis shuts down, sometimes it is difficult to take accurate voltage tests. Sometimes an IC or transistor will break down under load and test normal within the transistor tester (Fig. 1-25). You might find that in-circuit transistor tests with the DMM indicates that all transistors are normal. It is difficult in directly driven transistor or IC output circuits to locate the defective component without accurate voltage measurements.

The best method is to take in-circuit resistance measurements of the transistors and ICs. Start at the speaker output terminals and take crucial resistance measurements to common ground. Now compare these resistance measurements with the normal stereo channel. Check each IC and transistor with resistance tests until you receive or have a different resistance measurement. Sometimes a leaky capacitor or a change in resistance can cause the amp to shut down. A leaky IC or transistor might be located with accurate resistance measurements to common ground.

SCR RESISTANCE MEASUREMENTS

The SCR found in low-voltage power supply and high-voltage rectifier circuits can be checked with resistance measurements (Fig. 1-26). A low-resistance measurement between the gate (G) and cathode (K) is normal. Replace the SCR if the measurement is below 50 ohms. If any resistance measurement is found between the anode (A) and cathode (K) terminals, replace the leaky SCR. Like the transistor, the SCR can break down or become intermittent under load and should be replaced if you suspect that it is faulty.

TRANSISTOR AND IC REPLACEMENTS

Most transistors and ICs can be replaced with original parts or universal replacements. Use original transistors within the TV tuner, control, and IF stages. Large, crucial ICs should be replaced with those that have the original part numbers (Fig. 1-27). Foreign transistors and ICs are found at most wholesale or mail-order TV part establishments.

After locating the defective transistor in the audio amplifier, the transistor must be removed and replaced. Most transistors found in audio and radio circuits can be replaced with universal replacements if the original part is not available. For instance, the common AF transistor (2SC374) can be replaced with universal RCA SK3124A or ECG289A. The AF transistor (2N3904) can be replaced with a universal RCA SK3854 or ECG123AP.

Look up the transistor number within the RCA SK series, NTE, or with Sylvania's ECG series replacement guide book (Fig. 1-28). Most universal solid-state transistors and ICs can be replaced with RCA, GE, Motorola, NTE, Sylvania, Workman, or Zenith replacements. Simply look up the part number and replace it with an universal replacement. Test the new transistor before installing it.

After obtaining the correct replacement, remove the old transistor with iron and solder wick from the PC board. Remove the mounting screws on the power output transistors, then unsolder the emitter and base terminals. Be sure that you have the correct terminals in

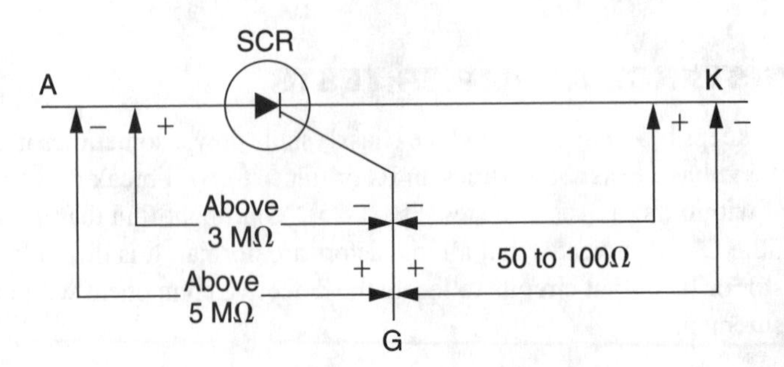

FIGURE 1-26 Check the suspected SCR with resistance tests or upon an SCR tester.

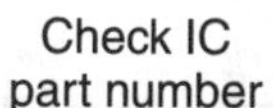

FIGURE 1-27 Replace large processors and ICs with those that have the original part number.

FIGURE 1-28 Replace the defective transistors or IC with universal replacements found in the RCA, GE, and NTE universal solid-state replacement manuals.

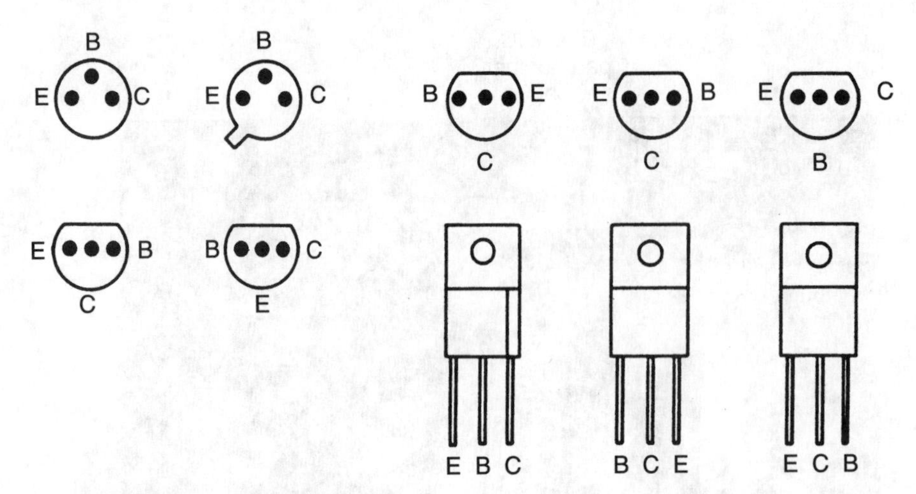

FIGURE 1-29 **There are many different transistor terminal connections found in signal and audio transistors.**

the right PC board holes. Doublecheck the transistor wire terminals and the bottom base diagram (Fig. 1-29). Do not leave the soldering iron on the transistor terminals too long. Use long-nose pliers or heatsink tool to dissipate the heat, so as not to ruin the new replacement. Place silicon grease between the transistor, insulator, and heatsink before mounting.

Look in the replacement guide for a universal IC replacement if the original one is not available. Handle it with care; the IC cannot be tested before installation. Leave it in the envelope until it's ready to be installed.

Remove the old IC with solder-wick and soldering iron. Start down the outside row of contacts or pins and keep the iron on the mesh at all times. Move the mesh down as it picks up solder. Then, go back and be sure that each contact is unsoldered from each pin to PC board wiring. Flick the pin with a pocket blade or with a small screwdriver. Mark pin 1 on the PC board with a felt pen. Lift the defective IC out by prying underneath the component. Be careful not to damage other components nearby or break the PC wiring.

Check for the terminal 1 dot, while the line or indexes determine how to correctly mount the IC (Fig. 1-30). Sometimes, only an index identification is found on the IC. When looking down on top of the IC, pin 1 is to the left of the index. Pin 1 is indicated with a white dot; on other PC boards, the number one and last terminal numbers are marked. Correct mounting of IC is found in the solid-state replacement guide. Place silicon grease and metal heatsink behind large ICs.

CRUCIAL WAVEFORMS

The waveforms taken by the oscilloscope within the TV, camcorder, CD player, and audio amplifier can quickly locate the defective stage and component. Crucial waveforms taken within the horizontal, vertical, and color circuits of the TV can pin point a defective circuit (Fig. 1-31). The horizontal drive waveform from the oscillator stage can be traced through the driver transistor to the final flyback output circuits. If one of the waveforms is missing or if there is incorrect polarity and improper waveform, suspect trouble within the horizontal circuits. Likewise, the TV vertical circuits can be signal-traced with waveforms in the same manner. Usually, these vertical waveforms are very unsteady.

A crucial waveform taken at the RF amplifier within the compact disc player can indicate the correct RF or EFM waveform. The CD player will shut down if no EFM waveform is found at the RF amplifier output circuits (Fig. 1-32). Crucial waveforms taken in the signal and servo circuits might indicate these circuits are functioning. A dead focus or tracking coil circuit can be located with a correct waveform. The defective motor circuits in the CD player and camcorder can be located with the oscilloscope.

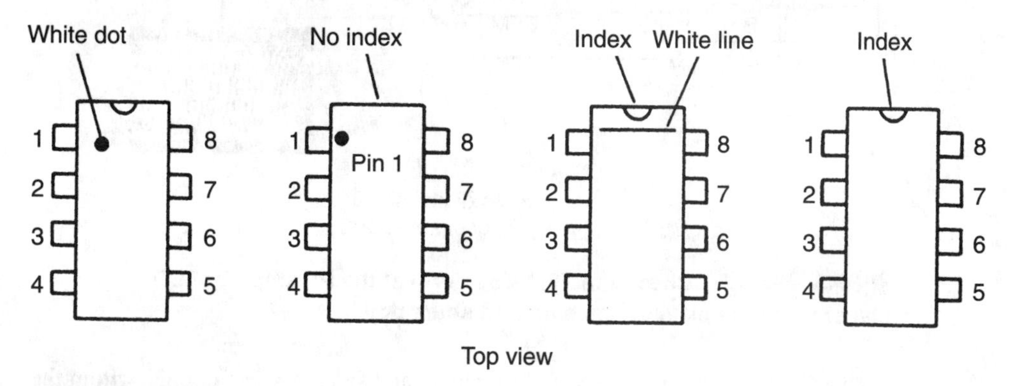

FIGURE 1-30 Check for a white dot, index, or white line to locate terminal 1 of suspected IC.

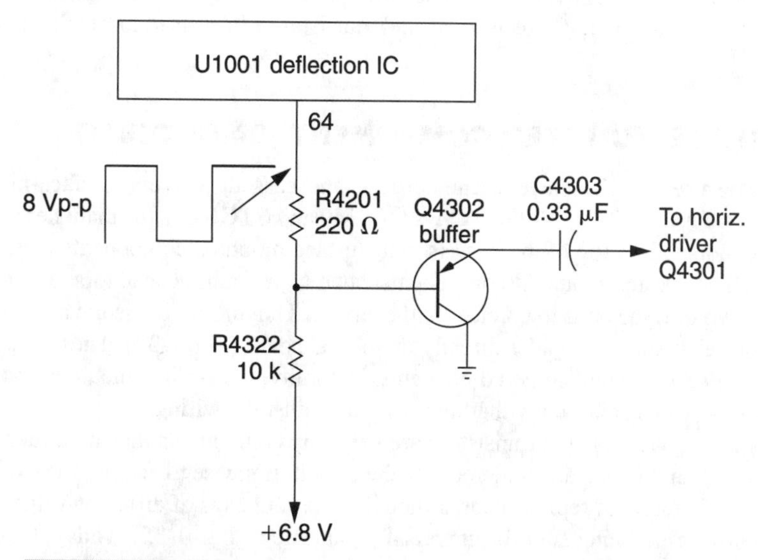

FIGURE 1-31 The missing or improper horizontal waveform can indicate a defective horizontal circuit within the TV chassis.

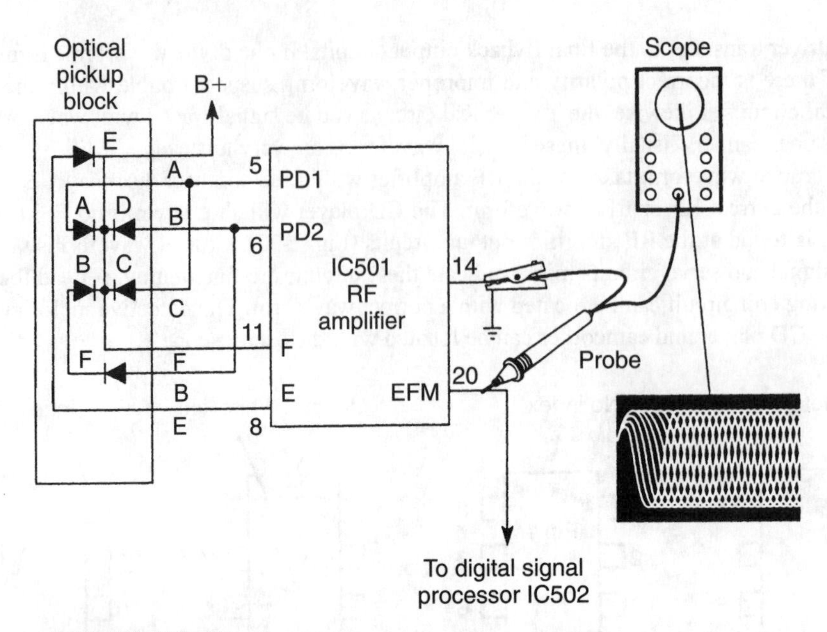

FIGURE 1-32 Check the EFM waveform at the RF amp IC of CD player when the player comes on and shuts down.

The oscilloscope can be used for alignment and signal tracing circuits within the audio circuits. The scope waveform can locate a weak, intermittent, or dead channel within the stereo amplifier with a dual-trace oscilloscope. Just compare the audio signals in both channels at a given point. If the waveform is improper or missing, you have located the defective stage. Now take voltage, resistance, and transistor or IC tests to locate the defective component.

TV SURFACE-MOUNTED COMPONENT (SMD) TESTS

Besides being used in CD players, camcorders, and cassette players, surface-mounted components are used in TVs. Within RCA's CTC140 and CTC145E, the main heavy components are mounted on top of the chassis with surface-mounted components underneath (Fig. 1-33). The surface-mounted components, such as resistors, capacitors, and diodes, might have two or three ends to connect to the circuit. The microprocessor can have gull-type wings or elements that solder directly on the PC board (Fig. 1-34). These ICs might have more than 80 different soldered elements. Naturally, the surface-mounted transistor has three leads, with a small body that mounts flat against the wiring.

The resistors, capacitors, and transistors are very tiny components that are either black, brown, or white, and which are soldered into the circuit. It's wise to acquire the exact part from the manufacturer for replacement, although cards and bags of different-value capacitors and resistors are being sold for universal replacement (Fig. 1-35). Today, these tiny parts can be bought by the tens, hundreds, or thousands.

You might find a part that looks like a resistor, but is actually a solid-tie feed-through component. A double resistor chip contains two resistance elements of a different value in

FIGURE 1-33 In the RCA portable TV, SMD components are mounted upon the PC wiring underneath the PC board.

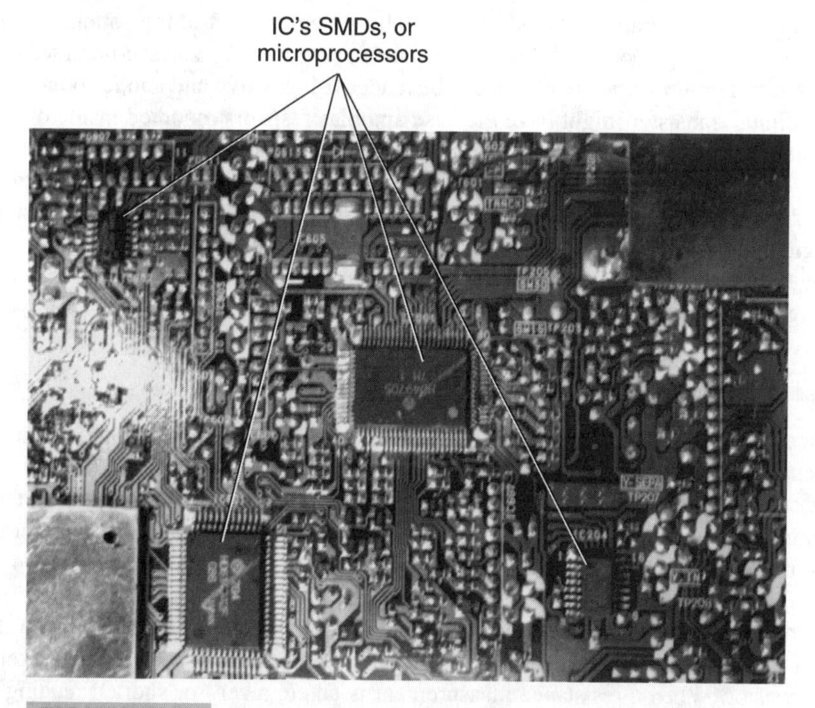

FIGURE 1-34 The microprocessor might have many gull-type wings or elements that solder to the PC wiring.

FIGURE 1-35 **Small SMD replacement capacitors, resistors, diodes, and transistors might be mounted on cardboard or found in plastic bags of 10 or 100 components.**

one chip. The small transistor might have all three elements out of the bottom, or the collector out of the top. You might find more than one diode in the surface-mounted component. A different hookup connection can be made with positive and anode connections.

The digital transistor might have the base and bias resistors mounted inside of the surface-mounted component, but the standard resistor has leads directly to the internal elements. Notice that the PNP digital transistor has the base and bias resistors tied to the emitter terminal. Likewise, the NPN transistor is the same. When making transistor tests on digital transistors, allow for the resistor in series with the base terminal and a resistance leakage between base and emitter terminals. Determine whether the transistor is a standard or a digital type before testing (Fig. 1-36).

HOW TO READ SMD VALUES

Surface-mounted parts (such as resistors, capacitors, and transistors) might have numbers and letters stamped upon the top side of the component. Fixed resistors can be marked with white stamped numbers. A resistor with the numbers 123 has a value of 12 kΩ. The first two numbers indicate the resistance value, with the last number indicating the number of zeros that follow. In the last position, the number three equals 000, and a number 4 equals 0000 (Fig. 1-37). For instance, the number 692 equals 6.9 kΩ.

Fixed capacitors and transistors might contain stamped letters and numbers on the top side of the component. Notice that a feed-through SMD part can be easily mistaken for a fixed resistor. When a resistance measurement is taken, a zero or shorted reading is obtained. These chip components take the place of wire jumpers to tie two different circuits together on the regular PC board. You might find round capacitors and resistors instead of flat-chip components in some TVs.

CD SURFACE-MOUNTED IC CHIPS

In addition to the optical pickup assembly, many IC processors, CPUs, and LSIs must be handled with care, not only in testing, but also during removal and replacement of the many terminal leads attached to the chip. Some have from 40 to 80 different terminal connections found in the CD player (Fig. 1-38). After the defective IC has been located, be careful not to damage the printed wiring while removing it.

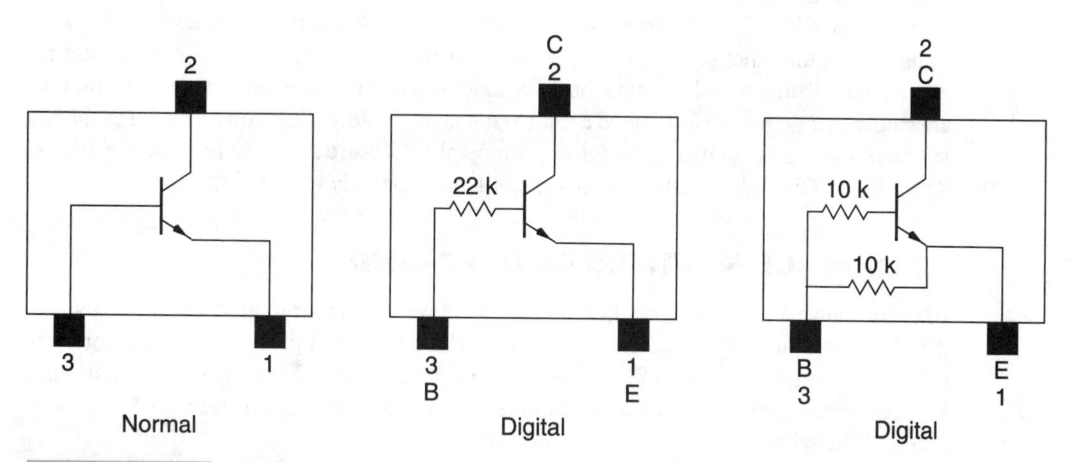

FIGURE 1-36 Determine if the transistor is a standard or digital type found in CD decks, camcorders, and VCRs.

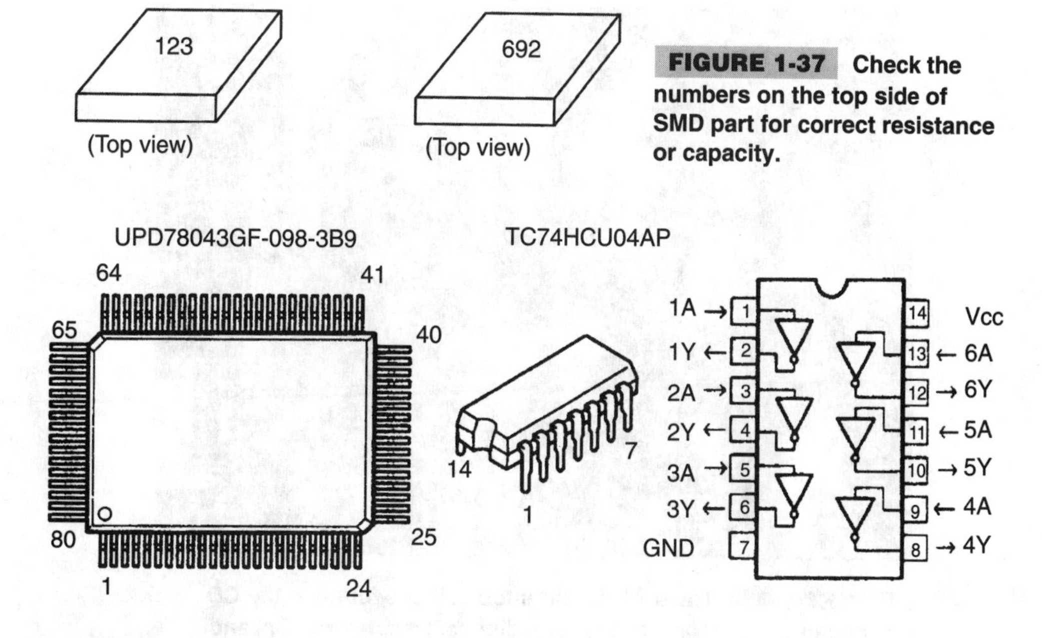

FIGURE 1-37 Check the numbers on the top side of SMD part for correct resistance or capacity.

FIGURE 1-38 Some microprocessors or ICs might have more than 80 terminal connections.

The LSI and ICs perform many CD player operations. The digital signal processing LSIs can integrate up to 13 or more different digital functions in a single chip. ICs handle many other functions, such as optical focusing, transport servo control, tracking, pulse width modulation, and digital-to-analog conversion.

The large-scale integrated (LSI) and control-processing unit (CPU) components should be handled with kid gloves, so to speak. Remove the solder around the leads with a solder sucker or solder wick. Heat the area by laying the wick material along the row of terminals. Slowly slide the wick along as the solder is melted. Now, remove the solder from the same row of the other side. A solder sucker that fits over an IC terminal can quickly remove solder from around the pins, but it will not remove excess solder where several pins are soldered together at a grounded area. Do not keep the soldering iron on one area for any length of time to avoid lifting the wiring. Remember, ICs are sensitive to static electricity. Keep them in the foil or static containers until you are ready to install them.

CD SURFACE-MOUNTED DEVICES (SMD)

Like the latest TVs, some of the CD chassis have surface-mounted components soldered directly upon the board wiring. The surface-mounted parts can be resistors, capacitors, transistors, or ICs (Fig. 1-39). You might also see LEDs with a compact, thin, leadless type of structure. These components are mounted flat on the board, then soldered to the wiring.

FIGURE 1-39 The surface-mounted devices found in the CD player can be resistors, capacitors, diodes, transistors, ICs, and microprocessors.

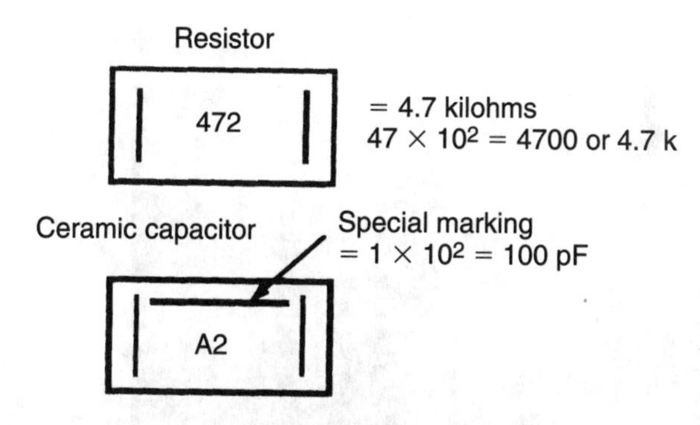

FIGURE 1-40 Check the numbers and letters for the exact value of capacitor chip in the CD player.

Many different surface-mounted devices are used in CD and cassette players. Some have more than five different kinds of ICs. The SMD ICs might have gull-type flat terminals with over 80 connecting flat terminals. Be very careful not to short them out when you take voltage measurements on these terminals with test probes. Use small-pointed test probes when making tests on SMD components. Often, terminal number one is indicated with a dot on top or numbers stamped on the PC board.

The surface-mounted transistor might appear in a chip form with flat contacts at one side, top and bottom, or on both sides. You might find more than one transistor inside one chip. The same applies to fixed diodes and LED SMD parts. Two or more fixed diodes might be found in one component. These transistors and diodes can be tested like any regular diode or transistor. The quickest and easiest method is with the diode test of a DMM.

Because many surface-mounted devices have similar shapes and sizes, it is quite difficult to identify them at a glance. The resistors appear as round, flat, leadless devices. The ceramic capacitor is a flat, solid device with the terminal connections at the outside, tinned ends. Always obtain these parts from the manufacturer because they mount perfectly. If they are not obtainable, you might be able to secure these universal components from large electronic wholesales or mail-order firms. The resistor has a number for identification with lines at the ends, and the ceramic capacitor has a line at the top with a letter of the alphabet and number (Fig. 1-40). Transistors and diodes are often identified with two letters.

REMOVING SURFACE-MOUNTED COMPONENTS

To remove a defective resistor or capacitor, remove the solder at one end. Grasp the end of the component or lead and melt the solder from the other end. Hold the part with a pair of tweezers and twist as the component is removed. Throw away all surface-mounted components removed from the chassis. They should not be used after removal.

To remove a surface-mounted transistor or diode, melt the solder at one end and lift the lead upward with a pair of tweezers. Do this to each terminal until all are removed. Some larger components might be glued underneath. Remove jagged or overlapping solder from the PC wiring with solder mesh material. Clean off lumps of solder from the wiring with the soldering iron and solder wick.

Do not use a large soldering gun and solder wick to remove small surface-mounted components. Use a 30-W pencil or regulated soldering iron (Fig. 1-41). Use extreme care so as not to damage the PC wiring. Be sure that all solder is removed before lifting small components.

FIGURE 1-41 Use a soldering pencil or regulated soldering iron to replace SMD components.

REPLACING SURFACE-MOUNTED COMPONENTS

Do not apply heat for more than four seconds; you might destroy the new chip or PC wiring. Avoid using a rubbing stroke when soldering. Do not bend or apply pressure to the transistors or IC terminals in the replacement. Use extreme care not to damage the new chip. Do not drag the TV chassis over the bench because the surface-mounted components can be damaged, mounted underneath a large PC board.

Replace all chip components, except semiconductors and ICs, by preheating them with a hair dryer for approximately two minutes. Tin or presolder the contact points on the circuit pattern or wiring. Press down the component with tweezers and apply the soldering iron to the terminal connections (Fig. 1-42). Apply more solder over the terminal if it is needed for a good bond. Do not leave the soldering iron on the transistor or IC terminals for too long; you might damage the internal junctions of the semiconductor. The large IC or processor should be flat and should mount right over each correct terminal connection.

Remove only one SMD component from the package because you might confuse the values. Place it upon a piece of white paper so that you can see it. Check the new component before installation. Use the smallest diameter of solder available for SMD connections. Lay the component over the end pads with a pair of tweezers. Be careful because these tiny devices can flip out of sight. Tack in one end with solder to hold it in place. Now solder the other end. Touch up both ends of the flat component with solder. Inspect the soldered joint with a magnifying glass. Recheck the SMD after installation.

DOUBLE-SIDED BOARDS

The PC board with wiring on both sides can cause a lot of service problems if the board is large and becomes warped. The PC wiring and SMD chip joints can pop or break, producing a dead or intermittent chassis. A hair-line crack is difficult to see on a PC wiring connection. Double-sided PC boards are used in small portable radios, camcorders, TVs, VCRs, CD players, and auto receivers (Fig. 1-43).

In many of the compact CD players with only one large PC board, you might find wiring and components on both sides of the board. Usually, the LSIs, CPUs, or ICs are surface-

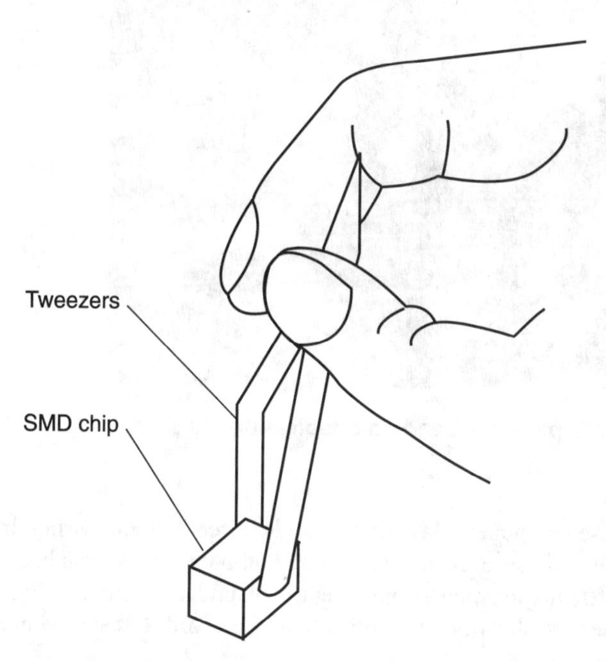

FIGURE 1-42 Hold the SMD part with tweezers and solder the ends with a low-wattage soldering iron.

FIGURE 1-43 The double-sided board is used in portable radios, camcorders, VCRs, CD players, and auto receivers.

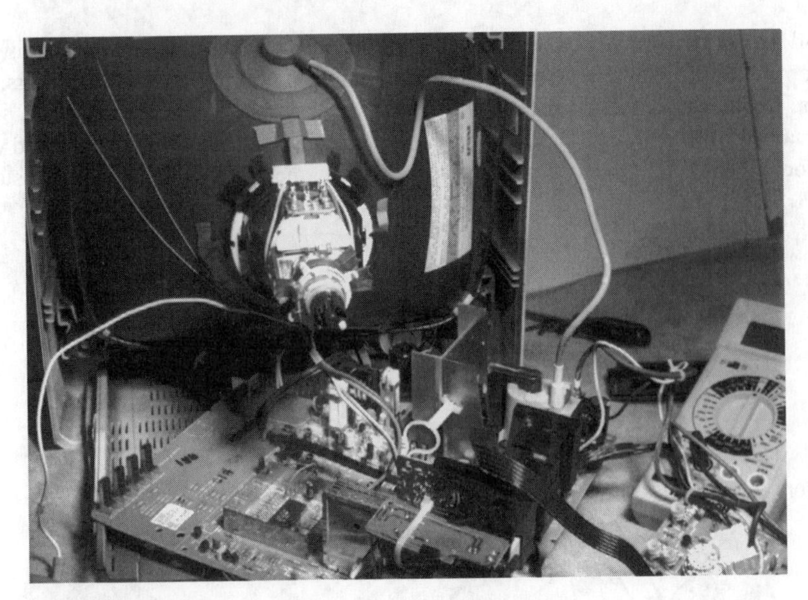

FIGURE 1-44 **The SMD parts are used on double-sided boards in the TV chassis.**

mounted underneath. These components lay flat and solder directly to the wiring. In some chassis, the single-sided board has components mounted on both sides of the board, so if you are having special difficulty locating a component, look under the chassis (Fig. 1-44). However, most IC outlines are shown on the top side of the board. Locate the manufacturer's parts layout section.

Double-sided wiring presents various service problems when troubleshooting the chassis. If test points are not provided on the top side of the chassis, the board must be removed and turned over to access the various surface-mounted components. Cracked wiring and surface-mounted element connections can break loose if the board has the opportunity to warp or sag. You might need to make various voltage and signal tests on components on the top side that feed through to the suspected component underneath.

Read and Study

With the electronic service business changing from year to year, the electronic technicians must keep abreast of each new development. Besides attending the manufacturer's service meetings, they must read books and magazines on different electronic products. Howard W. Sams, Parker, Hayden, McGraw-Hill, and TAB Books (an imprint of McGraw-Hill), cover many subjects related to solid-state servicing. Subscribe to *Electronic Servicing & Technology* and *Electronics Now* magazines for additional electronics servicing information.

Set aside one hour a day for study. This hour a day can keep you abreast of new developments and provide more efficient use of your time in electronic servicing. Fifty years ago, an old-time motor rewinder would study one hour each day. He would go to work at 7 a.m. and study from 7 to 8 a.m. Motors were brought to him from 100 miles around, he turned down more work than he could handle and he was successful. He was efficient, furthered his knowledge about motors, and made money.

Case Histories

The tough or unusual repairs on TV, camcorder, CD, audio, and microwave ovens will not be forgotten if they are recorded. Some technicians place the repairs in a 3"-×-5" file-card system. Today, the computer can keep case histories for easy reference. One quick method is to list the problem along the side of the schematic diagram. Each part can be identified on the schematic for easy reference. Long troubleshooting techniques can also be stapled to the service literature.

One case history might save hours of time in the busy world of electronic servicing. The same service problem might occur over and over again in another chassis. You cannot possibly remember all of the various problems related to a special TV chassis, so it's wise to keep a file or mark the case history on the schematic. Additional case histories can be found in *Electronics Now* and *Electronic Servicing & Technology* magazines (Fig. 1-45).

Preventative Maintenance

Repeated call backs can be eliminated if preventive maintenance procedures are followed in servicing the electronic chassis. Often, it takes just a few minutes to check various items on the chassis after the original repair is made. Check for poor or burned connections of

FIGURE 1-45 TV and electronic case histories are listed in *Electronics Now* and *Electronic Servicing & Technology* magazines.

components on the PC board or chassis. Check for good soldered connections of high-wattage resistors. Check for poorly soldered terminals and hookup wires around high-wattage resistors. Repair or replace burned terminal strips connected to various heated components.

Always check each dial light and see if it is lit. Replace any dial light that comes on, but remains black. This indicates that the bulb will not last very long. The same applies to neon lights or neon bulbs. Check for a loose dial bulb assembly or poorly soldered connections if the lamp is intermittent.

Do not forget to clean the TV tuner assembly after each repair job. Most rotary tuners need to be cleaned once per year. Squirt an approved cleaning fluid inside of the volume, tint, and contrast controls. Lubricate the tuner bearing areas. If a small control is frozen, squirt silicone lubricant inside of the shaft area and work the control back and forth to free it. Check each knob. Be sure that it fits properly and is not loose or falls off. Replace the spring or the entire knob, if necessary.

Take a peak at each large electrolytic filter capacitor. If a white or black substance is oozing out around the terminal connections, replace the capacitor. Enlarged areas on the body of the capacitor might indicate the need for replacement. Check the ends of small filter or bypass capacitors for signs of deformed or cracked ends, where each terminal is inserted. Sometimes the capacitor can explode, leaving blown pieces laying around. Usually, the ends come loose and make an intermittent connection.

Always clean the tape heads, capstan and tape areas after making repairs upon the cassette player. Clean the capstan bearing and apply a drop of light grease on bearing area. Recheck the tape speed. Likewise, clean the tape head, capstan, and all tape spindles after servicing the VCR recorder. Check the motor drive belts for slippage and replace them, if necessary. Last, but not least, polish and clean the entire television cabinet, picture tube, control knobs and dial assemblies.

2 PLAY IT SAFE

CONTENTS AT A GLANCE

The electronic technician servicing the electronic chassis can have fun, earn money, and gain valuable knowledge within the electronic consumer entertainment field. Besides making a good living, the electronic technician can live and work with danger. As in every job, danger lurks just around the corner.

Remember, the electronic technician works on electronic equipment that is powered by 120-Vac power (Fig. 2-1). Although the high voltage generated in the TV chassis might be above 32 kV, it might not kill, but it could cause severe shock hazards that might eventually cause a heart attack. The 3.6-kV peak voltage in microwave ovens produces heavy current and it is very dangerous if you come in contact with it. Low- and high-voltage

FIGURE 2-1 **Be careful while working on electronic products that operate from the ac power line.**

sources found in the electronic chassis can be dangerous, just respect it, as you work around it. Safety rules followed everyday can prevent numerous injuries. Get in the habit of playing safe.

Safety Lessons

Besides having each service bench fused with grounded polarity outlets, the service technician must follow safety procedures when repairing the TV or microwave oven. Use an isolation variable power transformer with all TVs. Today, most TV chassis are ac-grounded without a power transformer. Keep the chassis away from grounded pipes or metal bench posts. Work on a rubber mat to protect yourself not only from electrical shock, but also to save those tired feet.

Think before you leap when working in the ac power line or high-voltage circuits. Be sure that voltage test instrument cables are good, with no bare wires or exposed meter terminals. Discharge the picture tube before attempting to remove it. Wear safety goggles when replacing a picture tube. It takes only one picture tube explosion to damage your eyes for life. Discharge filter capacitors before checking voltages in the power-supply circuits.

Be careful when replacing parts. A replacement electrolytic capacitor with lower voltage than required might blow up in your face. The same can occur if the capacitor is installed backwards. Be sure that the power cord is disconnected if you touch a component (such as the flyback transformer, tripler, or power-output transistor). Sometimes a leaky horizontal or power regulator transistor can be red hot, and you will come away with a burned finger. Look and think before placing a meter probe or your finger into the TV chassis.

Before buttoning up the rear cover, be sure no leads or cables are laying on parts of TV chassis. Tie cables up by folding long wires, then wrap them up with plastic ties and rubber bands. Be sure that the speaker wires or cables are not pinched between chassis and bottom side areas. Keep the high-voltage lead tied up away from components on the chassis.

Take the antenna cold check and leakage hot check. For the antenna cold check, remove the ac power cord, turn the ac switch on, connect one lead of the ohmmeter to the ac plug prongs, and touch the metal tuner with the other probe. If the resistance measurement is less that 1.0 MΩ or greater than 5.2 MΩ, something is wrong, and the problem should be corrected before the TV is returned to the customer.

Make the hot current test as follows: plug the TV into the ac outlet, turn the power on, connect the leakage current meter probe to the tuner, screw leads, or metal control shafts, and then connect the other probe to an earth ground or water pipe (Fig. 2-2). This current measurement check should be less than 0.5 mA. Now reverse the power cord in the outlet and repeat the test. If the current reading is higher, check the component leakage to the power-line components.

TV SAFETY CAPACITORS

Bypass and hold-down capacitors in the horizontal output transistor, yoke, and flyback circuits should be replaced with factory-specified parts (Fig. 2-3). The safety capacitors in the collector circuit of the horizontal output transistor prevents the HV from increasing to a dangerously high voltage. Always replace with exact capacity and working voltage. If components with the manufacturer's exact part number are not available, use those with exact capacity and high-operating-voltage. Remember that the safety capacitors in the horizontal output transistor circuit hold down the high voltage at the flyback and anode terminals of the picture tube. Excessive HV can cause sharp cracking of arcover voltage to the chassis, spark gaps, or to the picture-tube grounding strap. If the exact capacity is not used, the high voltage can be raised extremely high or lowered unacceptably.

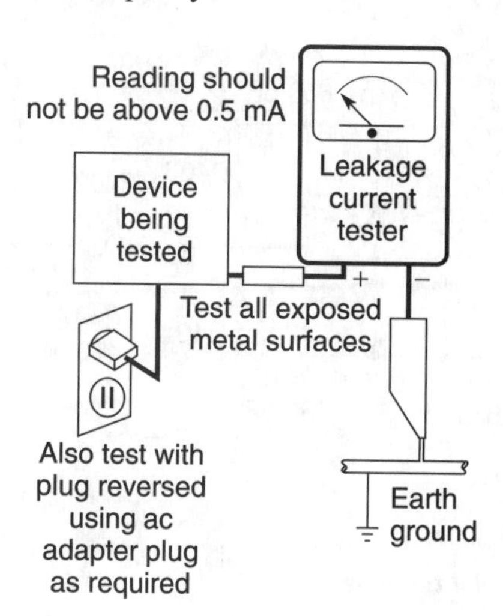

FIGURE 2-2 Take a hot current test of the TV from the chassis to the earth ground.

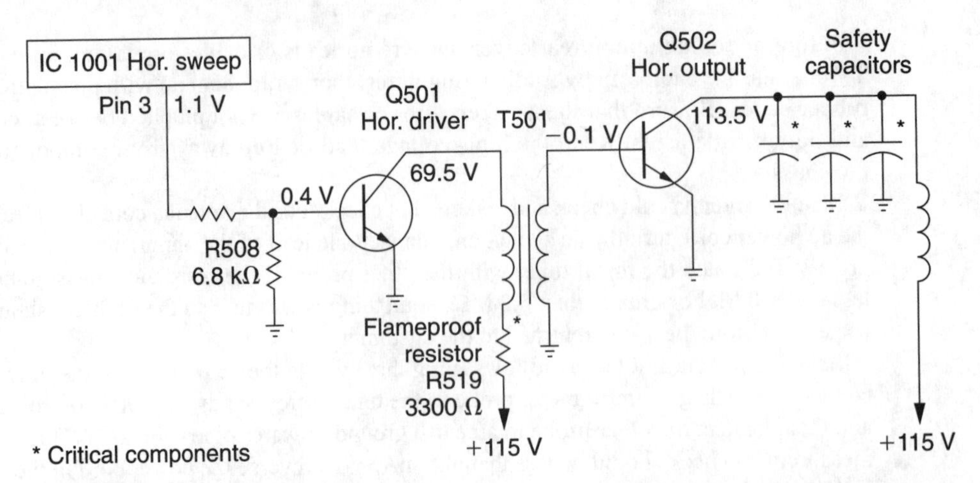

FIGURE 2-3 **Check the safety or hold-down capacitors in the collector circuit or check the horizontal output transistor for open conditions.**

FIRE DAMAGE

Always replace crucial components with those that have exact part numbers that have shaded or star indicators. Some crucial parts are identified by a black triangle symbol in some schematics ▲. These crucial components are used in the low-voltage power supply, regulator, high-voltage, sweep, and horizontal output circuits of the TV circuits.

Safety components in the low-voltage power supply include capacitors, fuses, silicon diodes, and power resistors. You can find oxide metal-film type resistors in voltage-divider networks. Replace crucial capacitors and resistors in the horizontal output circuits with the exact type of part, or one with equivalent universal replacements (Fig. 2-4). High-voltage fixed resistors should be replaced with those that have the same

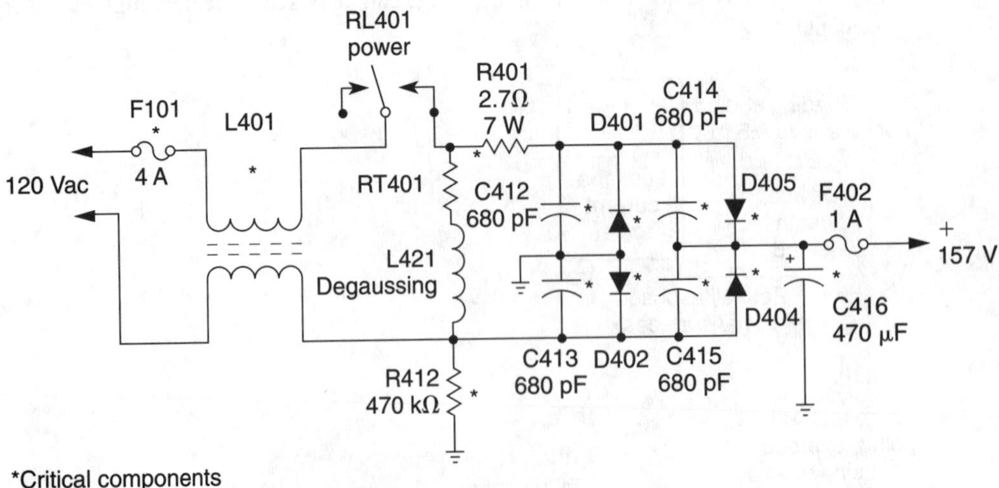

FIGURE 2-4 **Replace all safety parts in the TV chassis.**

voltage and resistance ratings. Replace wire-wound resistors with those that have the same or higher-wattage parts.

Route wires covered with PVC tubing, double insulated wires and high-voltage leads away from components that operate fairly warm. Use specified insulated materials for hazardous live parts; such materials include insulation tape, PVC tubing, spacers, and insulators for transistors. Keep all loose wire and cables away from overheated components, such as power transistors, heatsinks, oxide metal-film resistors, fuseable resistors, and power resistors. Replace the new component in the exact space as the original. Do not leave long leads on capacitors and resistors that can flop around. Make clean soldered joints or bonds. Take pride when replacing those new components. Sloppy part replacement can cause flames, resulting in extensive fires and even death.

HOT TV CHASSIS

Many components within the TV run warm in normal operation. Overheated horizontal output and regulator transistors might be too hot to touch. Do not touch the metal body of the horizontal output transistor while the chassis is operating. Although, the dc voltage might not be greater than 150 Vdc, the peak voltage can cause shock hazards and make the technician drop tools or break costly test instruments.

The TV chassis might be called a *hot chassis*, in references with a hot or cold ground (Fig. 2-5). The new ac/dc TV chassis can have a "hot" and a "cold" chassis within the large overall

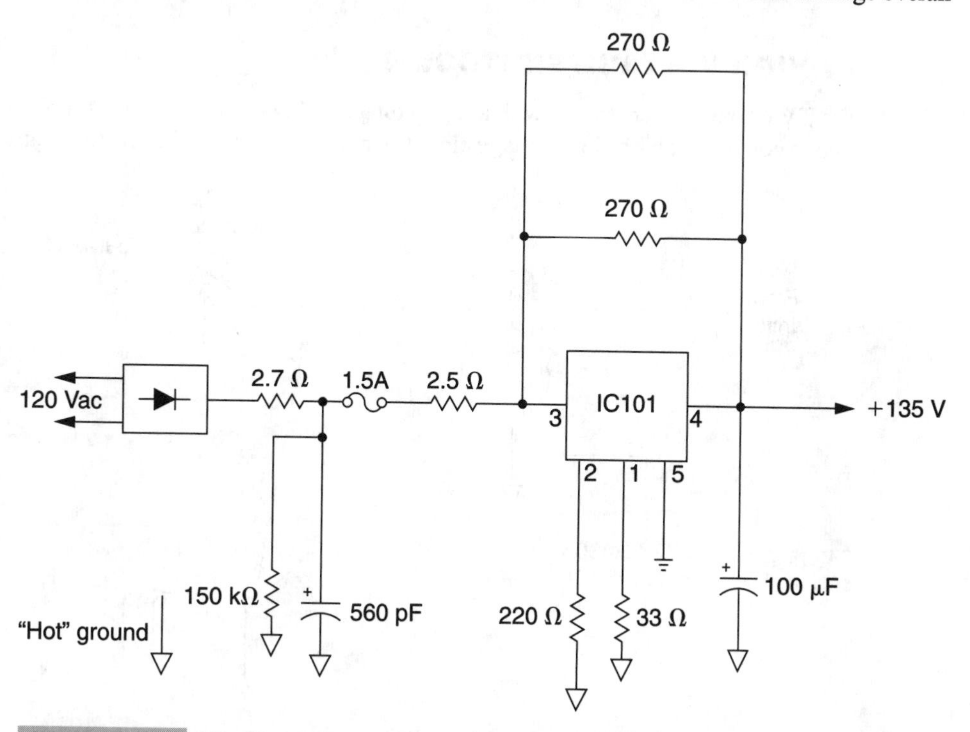

FIGURE 2-5 **The "hot" TV chassis might include the low-voltage power supply and horizontal output circuits.**

chassis. Usually, the "hot" ground is insulated away from the metal chassis ground. The hot ground is used in the ac power line, low-voltage power supply, switching and switched-mode power supplies. The horizontal circuits within the TV chassis might operate with a hot ground. Always use the "hot" ground as instrument ground when taking voltage measurements within these circuits or the voltage measurements will be inaccurate. Suspect that the HV shutdown circuits are defective if the chassis shuts down before the correct HV has been reached.

KEEPS BLOWING THE FUSE

Suspect an overloaded condition or a leaky component if the blown fuse has been replaced and it opens again, at once. The blown fuse in the TV can be caused by leaky silicon diodes, the filter capacitor, regulator, and horizontal output transistors. If both the line fuse and secondary fuse is blown, suspect a shorted or leaky horizontal output transistor or flyback components. An improper drive voltage at the base of the horizontal output transistor can blow both fuses (Fig. 2-6).

The primary line fuse in the ac TV power supply might be a 1.5-, 4-, 5-, or 7-A fuse. Most TV chassis contain a 4- or 5-A fuse. If the glow area of the fuse is black, check for a shorted overload condition. A small break in the fuse internal wire might indicate a flash-over component or power-line outage. Suspect lightning damage if the fuse holder is blown apart and the PC wiring is stripped. Always replace the line fuse with one that has the exact ratings. Never replace the fuse with a high rating because it could later on cause a fire.

TV HIGH-VOLTAGE SHUTDOWN

The TV chassis can shutdown with a high-voltage or low-voltage service problem. A defective component in the low-voltage circuits can operate for a while and then cause chas-

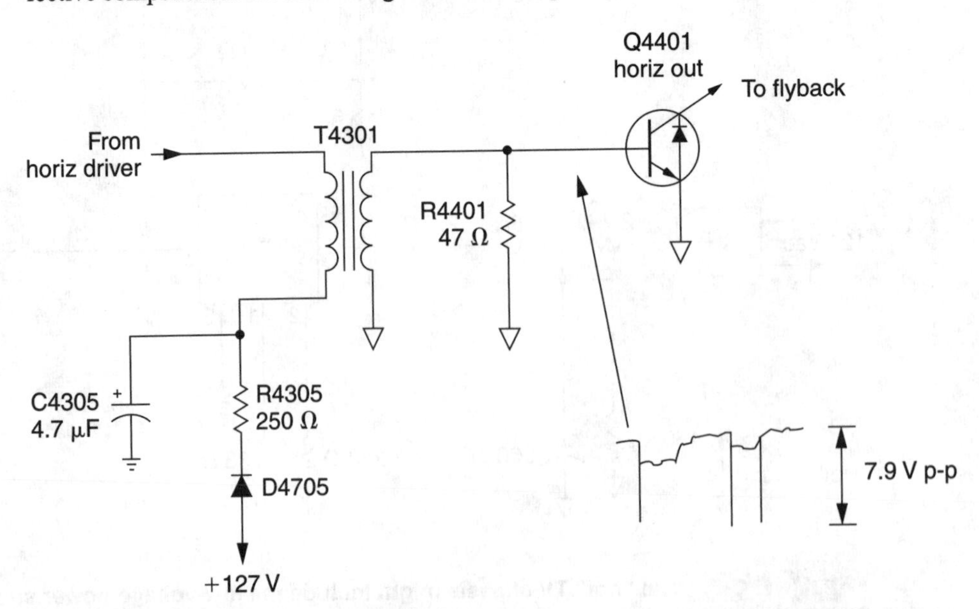

FIGURE 2-6 Check for improper drive voltage or waveforms at the horizontal output transistor if the fuse keeps blowing.

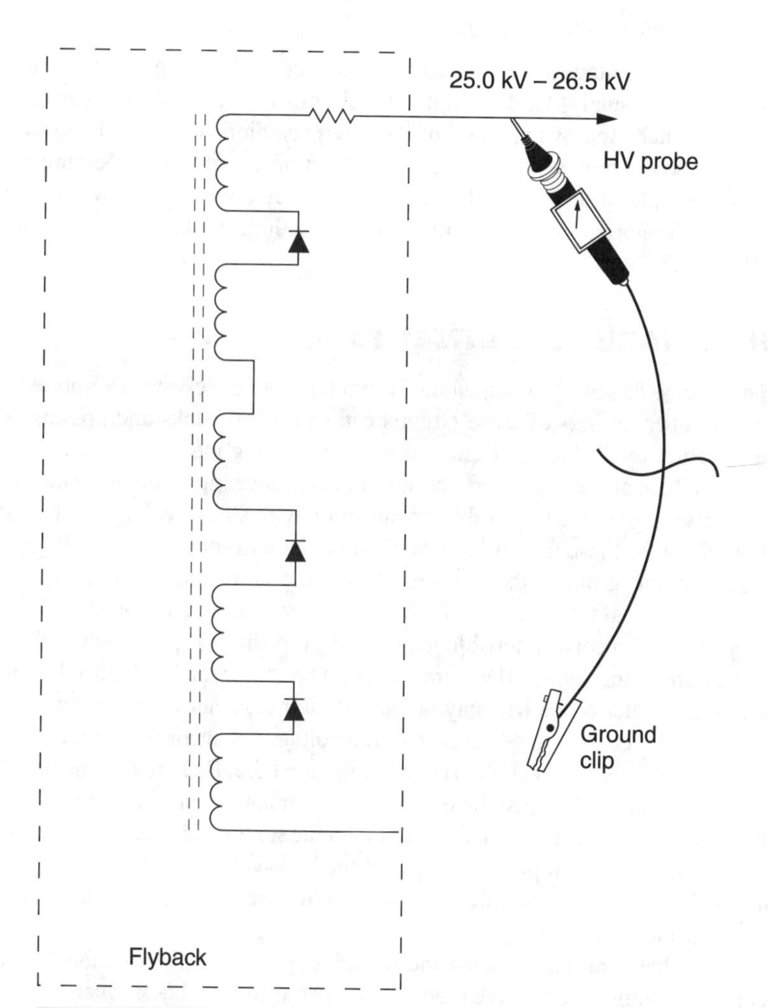

FIGURE 2-7 **Measure the high voltage at the TV CRT.**

sis shutdown. Excessive high voltage can cause the chassis to shutdown when the internal HV shutdown circuits are operating. If loud cracking and arc-over sounds can be heard around the HV cable, flyback, and CRT anode socket, suspect excessive high voltage. If the HV shutdown circuits are functioning, the chassis should shutdown at once.

Improper adjustment of the low-voltage power source or B+ adjustment might cause HV shutdown. If the B+ control is set too high and the power-line voltage increases by a few volts, the high voltage will increase, shutting down the chassis. Readjust the B+ control for the correct voltage applied to the horizontal output transistor. Locate in the manufacturer's service literature the correct low-voltage (B+) adjustment.

A defective HV shutdown circuit might cause the chassis to shut down at once or it might not shut down the chassis at all. You would now hear a cracking and arcover symptom with excessive high voltage. Be sure that the B+ voltage is correctly applied to the horizontal output transistor. Measure the HV at the CRT anode button (Fig. 2-7). Suspect a defective B+ circuit if high voltage exists and the B+ circuit voltage cannot be adjusted. Repair the low-voltage regulator circuits.

Suspect that the high-voltage shutdown circuit is defective if the high-voltage shuts the chassis down at once. Usually, the HV shutdown circuits are coupled to a winding from the flyback transformer. Remove one end of the reference diode or zener diode from the circuits. Slowly raise the power-line voltage with isolation transformer. Measure the HV at the CRT. Suspect a hold-down or safety capacitor is defective or open within the collector terminal of the horizontal output transistor, if the high-voltage raises beyond the critical level.

TV HIGH-VOLTAGE ADJUSTMENTS

After making TV chassis repairs, readjust the B+ voltage and check the HV applied to the picture tube. Improper settings of these circuits can cause call backs and possible shock hazards. Never return the TV to the customer without making these adjustments. Excessive power-line voltage at the customer's home can also cause high-voltage shutdown.

Check the high-voltage at the CRT rubber connection with a high-voltage probe. Do not use a VOM or DMM for these measurements. Ground the wire-clip from the HV meter to the outside ground that grounds the picture tube. The picture tube ground wire is flat against the black outside glass area of CRT. If the meter ground wire is not grounded to the picture tube, you could receive a terrible shock and drop the HV probe when the sharp probe is inserted under the rubber HV wire. Often, when the probe is inserted under the rubber insulator, the meter probe will stay put and B+ adjustments can be made.

Adjust the B+ or HV control with the correct B+ voltage set according to the manufacturer's service literature. Now check the HV at the picture tube. Be sure that the brightness control is set at a minimum. Readjust the B+ or HV adjustment. Recheck the HV at the picture tube and compare it to the manufacturer's HV on the schematic. Return the brightness and contrast control to a normal picture. Notice if the high-voltage measurement is below the original. The high voltage at the picture tube should never go over the manufacturer's listed high-voltage measurement.

Be very careful when making HV tests and working around the picture tube. Discharge the CRT before removing the defective tube. Carefully remove the back cover. Loosen the top screws and then the bottom. Hold the back so that it will not fall down against the CRT and possibly break or crack the neck of the tube. Remove the chassis.

Place the front of the tube or cabinet face down on a rubber mat or rug. Notice the mounting of each component from the neck of the CRT and around it. Remove the picture tube and lay it on its face on a rug or carpet so as not to scratch the screen area. Do not carry the picture tube against your stomach. Two service technicians might be needed to remove and replace the very large picture tubes. Always wear safety glasses or goggles when removing and installing a new CRT.

TV STAR COMPONENTS

Look at the schematic to see if a star is shown beside component numbers. If so, only replace these components with exact replacements (Fig. 2-8). Often, these special parts work in crucial voltage circuits, such as the power sources, horizontal circuits, and flyback circuits. Sometimes universal parts replacement of crucial circuits will not operate within certain TVs. Try to replace crucial safety-marked components with originals when possible.

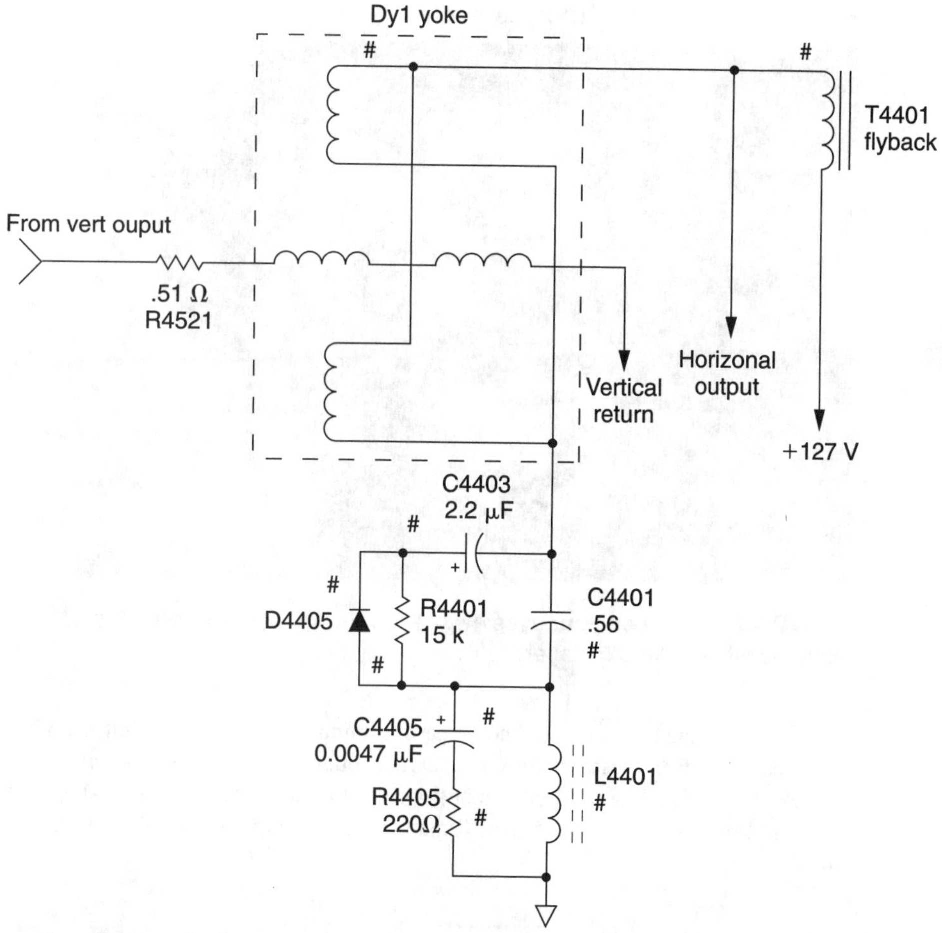

FIGURE 2-8 **Replace crucial components with originals when marked with a star or triangle.**

CD PLAYER SAFETY PRECAUTIONS

For the continued protection of the customer and service technician, several safety precautions must be followed. Always make a leakage current check after repairs. Keep your eyes at least one foot away from the optical laser beam (Fig. 2-9). Replace the crucial safety parts identified by the safety symbol ▲ with originals. Do not alter the design or circuitry of the CD player to where it could result in injury or property damage. Remove all test clips and shorting devices around interlocks after repairs. Keep a conductive mat under the test equipment and the CD player while servicing it.

With the ac plug removed from any source, connect an electrical jumper across the two ac plug prongs. Place the instrument ac switch in the ON position. Connect one lead of an ohmmeter lead in turn to each pushbutton control, exposed metal screw, metalized overlay, and each cable connector. If the measured resistance is less than 1.0 mW or greater than 5.2 mW, an abnormality exists that should be corrected before the instrument is returned to service.

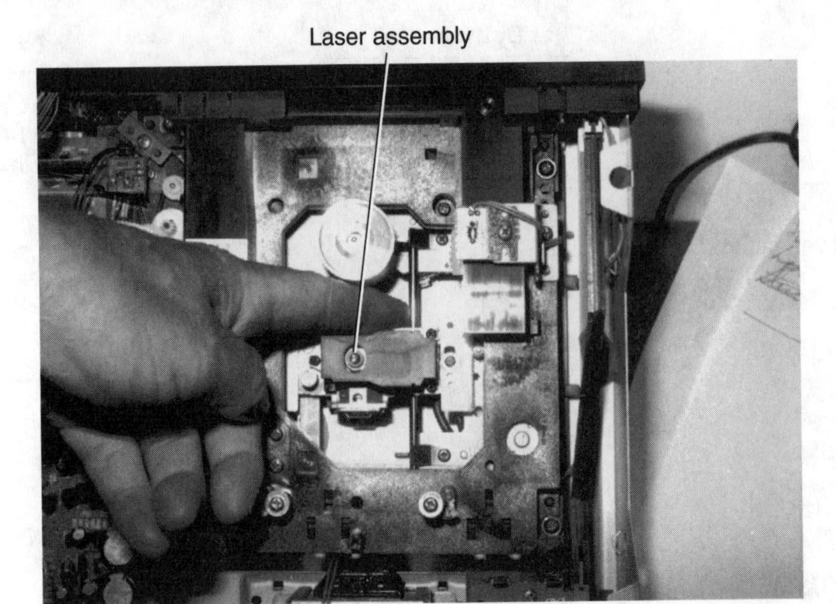

FIGURE 2-9 **Keep your eyes away from the CD optical assembly while servicing the CD player.**

Measure leakage current to a known earth ground (water pipe, conduit, etc.) by connecting a leakage current tester between the earth ground and all exposed metal parts of the CD player. Plug the ac line cord into ac outlet and turn the switch on (Fig. 2-10). Any current measured must not exceed 0.5 mA. Repair the electronic product if more current is measured.

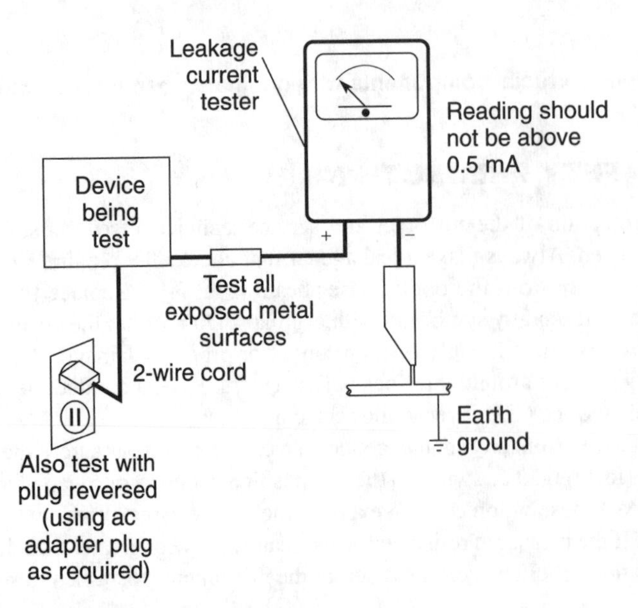

FIGURE 2-10 **The current measurement of a CD player leakage should not be greater than 0.5 mA.**

CD LASER SAFETY

Be careful when working around the CD player to avoid laser-beam radiation exposure. You can damage your eyes if you stare at the bare optical lens assembly while the player is operating, so keep your eyes away from this area. The laser-beam warning label is usually fastened to the back of the laser optical assembly citing danger. Remember, the laser beam is not visible like that of an LED or pilot light.

HANDLE WITH CARE

Many ICs and especially the laser diode assembly are crucial components that are sensitive to static electricity or high voltage, so be careful around them. Remove excess solder with an anti-static suction-type solder iron or braid. Replace the sensitive devices and solder them in with a low-wattage iron; the battery-operated solder iron is ideal. Clip the metal part of the iron to the CD chassis to ground it. Wear a protective arm band and also clip it to the chassis (Fig. 2-11). Use a conductive metal sheet or mat under the CD player and test equipment to keep all units at the same ground potential.

CD TRANSPORT SCREW REMOVAL

The CD player optical mechanism is protected from excessive vibration or rough handling during shipment with one or more transit or transport screws. These screws hold the pickup assembly in position; if they are not removed, the player might shut off and not operate. Look for these screws (and any labels identifying them) on the bottom chassis. Simply loosen the screw so that the mechanism is free. Some screws fall out, and others cannot be completely removed. If the screws are removed, tape them to the back of the chassis with masking tape. You should replace these screws (or tighten them if they were not completely removed) when you take the player in for repair or when you move it.

Check for tight transport screws if the player won't operate. A lot of CD players have been brought into the shop or taken back to the dealer for repairs with the transport screws

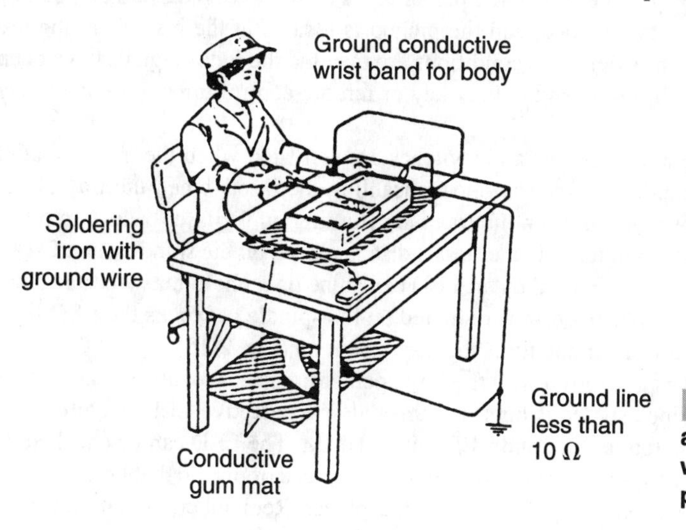

FIGURE 2-11 Wear a protective arm band while servicing CD players.

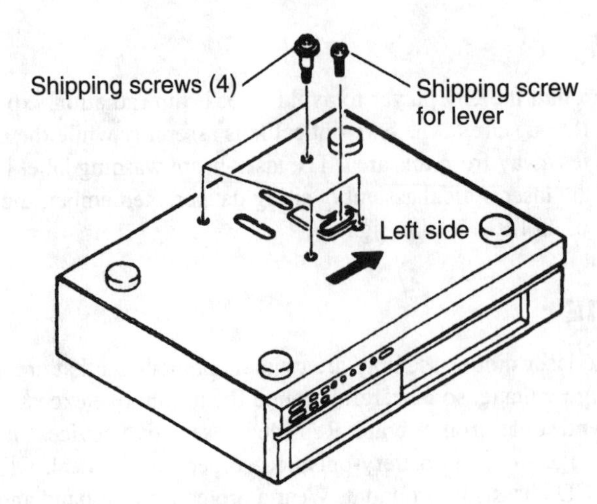

FIGURE 2-12 Remove the transport screws at the bottom side of the CD player before trying to operate.

still in place. A Phillips or straight screwdriver can usually be used to loosen these screws (Fig. 2-12).

SAFETY CD INTERLOCKS

The safety interlocks are provided so that the operator or service personnel are not exposed to any laser radiation that can damage your eyesight. Never look directly at the laser lens assembly while servicing the CD player. Keep your eyes away from that area. Also be sure that a disc is in the tray at all times.

The laser outputs are controlled by the injection or cut off of the constant voltage source to the laser diode at pin 30 of IC501 (Fig. 2-13), and also by the automatic laser power-control circuit. When pin 30 is in "H" (high) level, the laser emits the beam. When pin 30 is in "L" (low) level, the laser does not emit the beam.

Pin 30 is set in high level when the unit is loaded with the disc and is reading the index signals or in the Play mode. When the player reads the index signals and the loading limit switch has the disc tray loaded, and the pickup is located at the inside starting area, the laser emits the beam. After these conditions are met and the index signals have been read, the laser emits the beam when the Play key or remote-control transmitter is pressed, and when the display is on.

Other interlock safety circuits and switches are activated when the tray is loading and unloading. The interlock switch is open (disabling the laser) when the tray is out; it is closed with the disc loaded when the tray is in (turning on the laser beam). In other players, the laser beam is not turned on unless a disc is placed on the spindle (Fig. 2-14). With the tray empty, the light emitted by an LED can shine on a phototransistor and shut down the lower assembly. When a disc is mounted on the spindle, it blocks the LED light from striking the phototransistor and turns the laser beam on (Fig. 2-15).

If the disc motor starts to run with the tray open or no disc is loaded on the spindle, suspect that the interlock switch is bent or damaged or a defective LED/phototransistor circuit. Check the phototransistor and LED with a DMM. The LED can be checked like any ordinary diode. Always remove the shorting pin or piece of plastic that holds the interlock switch engaged after servicing the portable CD player. Remember, do not look at the lens assembly if the disc spindle is rotating with no disc in position.

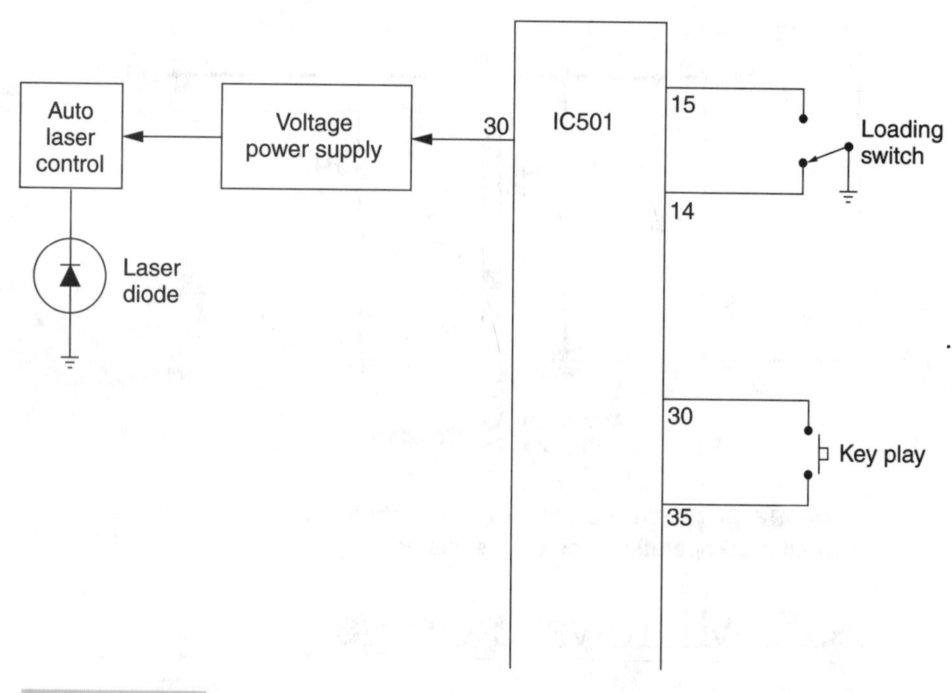

FIGURE 2-13 The voltage source to the laser diode is cut off by the interlock of an unloaded disc.

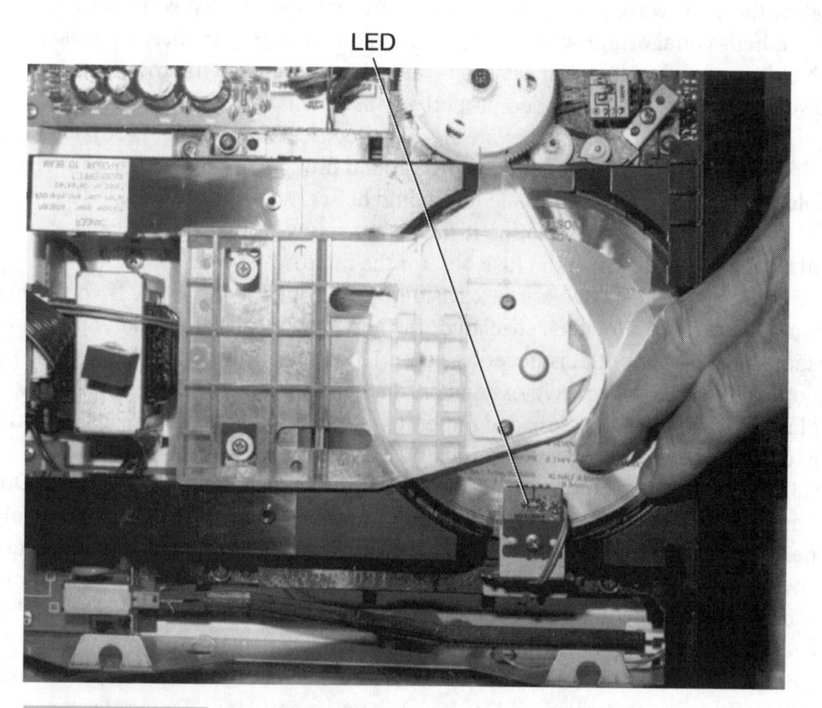

FIGURE 2-14 The laser assembly will not operate until a disc is loaded on the spindle or disc.

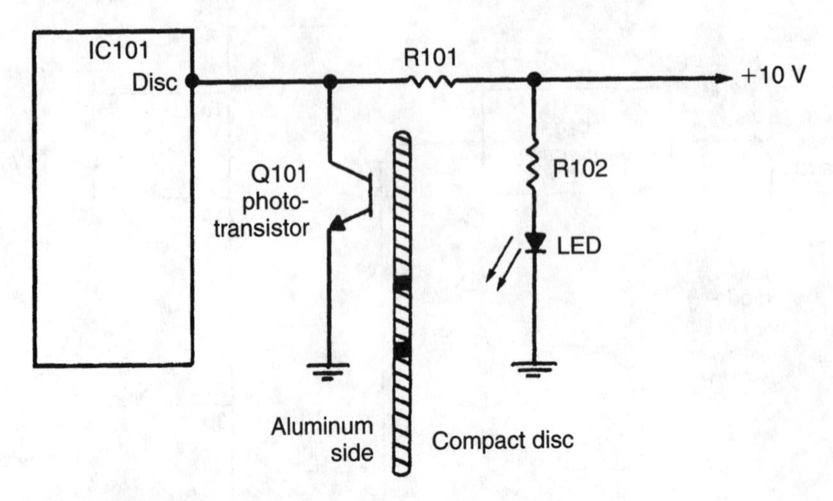

FIGURE 2-15 **The LED shines on the phototransistor and shuts down the CD operation if no disc is loaded.**

Safe Microwave Usage

To prevent the risk of burns, fires, and electrical shock, the owner or operator should observe the following guidelines:

Follow the microwave oven instructions before attempting to cook. Don't forget that the oven is a high-voltage, high-current appliance. You must be extremely careful at all times, but this does not mean that you have to be afraid to operate the microwave oven.

Be sure that the oven is installed properly. Most ovens include a three-prong plug. You can pick up one at your local hardware store or electrical dealer (Fig. 2-16). Do not, under any circumstances, cut or remove the third ground prong from the power plug. The small, flexible grounding wire with a spade lug should be screwed under the plate screw of the ac outlet. Use a voltmeter to determine if the center screw of the ac receptacle is grounded. Install a separate outlet from the fuse box for the microwave oven.

Do not use an ordinary two-wire extension cord to operate the oven. If you need an extension cord, contact a local electrician or the appliance dealer that sold you the oven. Be sure that the ac power outlet is never less than 105 volts or more than 125 volts for proper oven operation. The microwave oven is designed to operate from 115 to 120 Vac. Don't just plug the oven into an overloaded outlet if several appliances are already tapped into it. If you do, the oven might operate erratically and never cook properly.

Install or locate the oven only in accordance with the installation instructions. Don't install the oven where the side or top opening might be blocked. The hot cooking air must escape, and fresh air must be pulled into the oven for normal operation. Don't operate the microwave oven if the power cord or plug is damaged. Repair the three-wire cord or install a new cord, and be sure that the ground wire is intact. Check the continuity from the metal cabinet to the ground wire terminal with the low range of an ohmmeter (Fig. 2-17).

Supervise children who are operating the oven. Most microwave oven problems are related to improper oven operation. Don't use the oven outdoors. Do not let the cord hang over edges of table or cooking counter. Keep the ac cord away from heated surfaces or a nearby cooking stove. Don't use metallic cooking containers. Use only cooking utensils or accessories of the type recommended by the manufacturer or microwave cookbooks.

Don't use any material that might explode in the oven. Some sealed glass containers or plastic jars could build up pressure and explode. A regular paper sack with food inside might explode and cause a fire if steam or air holes are not punched in the top of the sack.

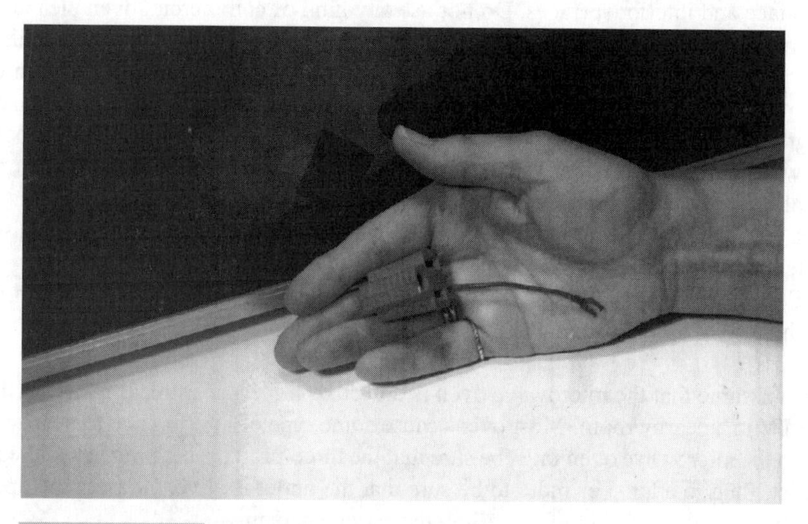

FIGURE 2-16 Ground the microwave oven with a three-prong plug and ground wire.

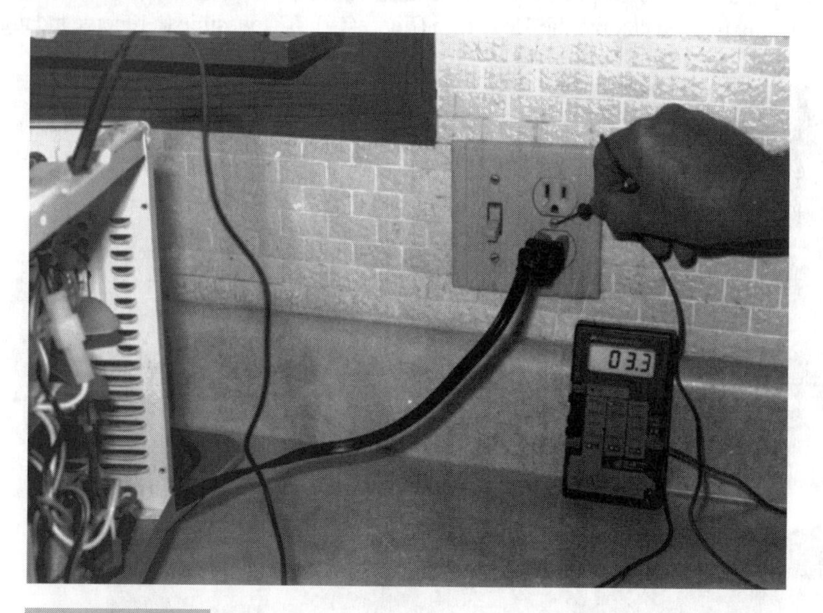

FIGURE 2-17 Measure the resistance between the oven's metal case and the ground-wire terminal.

A fire inside the oven might cause the plastic heat shields and front oven coverings to melt. This could result in a fairly expensive oven repair. Keep the oven cavity spotlessly clean. Cooking fatty items, such as bacon, might, in time, cause grease to collect behind the plastic shelf supports on plastic microwave guide cover, resulting in excessive arcing and damage. Do not operate the oven with nothing in it.

Use liquid window cleaner or a very mild detergent with a soft, clean cloth over the oven's face and interior surfaces. Do not use any kind of commercial oven cleaner inside of the oven area. Use a paper towel or sponge to blot up and remove spills while the oven is still warm. Pull the oven's power cord out if fire is discovered inside the oven cavity. Keep the oven door closed to help smother the fire. Clean out the top area or exhaust areas for signs of food pulled up by the fan (Fig. 2-18). Have the service person brush out these areas when the oven is serviced.

A collection of food particles might in time cause a persistent odor. Some odors can be eliminated by boiling a one-cup solution of several tablespoons of lemon juice dissolved in water in the oven cabinet.

Don't change the operation of the oven while it is in operation. This is a good way to blow the fuse of a microwave oven. Touch the Stop button and start the operation all over again.

Don't assume that the microwave oven is defective if lines are streaking across the face of the TV in a nearby room. Most ovens cause some type of interference to nearby TVs.

When the microwave oven fails, be sure that the three-prong cable plug is installed in the ac outlet. Plug in a lamp or radio to be sure that the outlet is alive. Go over the operation instructions very carefully before calling the service person.

Don't' remove the back cover of the microwave oven unless you know what you are doing. Keep the power cord pulled when the back cover is off. Discharge the HV capacitor before trying to even change the 15-A fuse (Fig. 2-19). It's possible to receive a dangerous

FIGURE 2-18 Clean food crumbs out of the oven exhaust area.

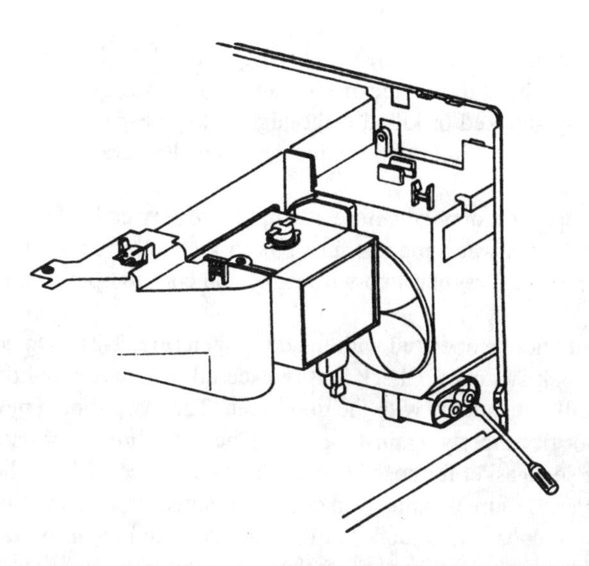

FIGURE 2-19 Discharge the HV capacitor before even trying to change the open fuse.

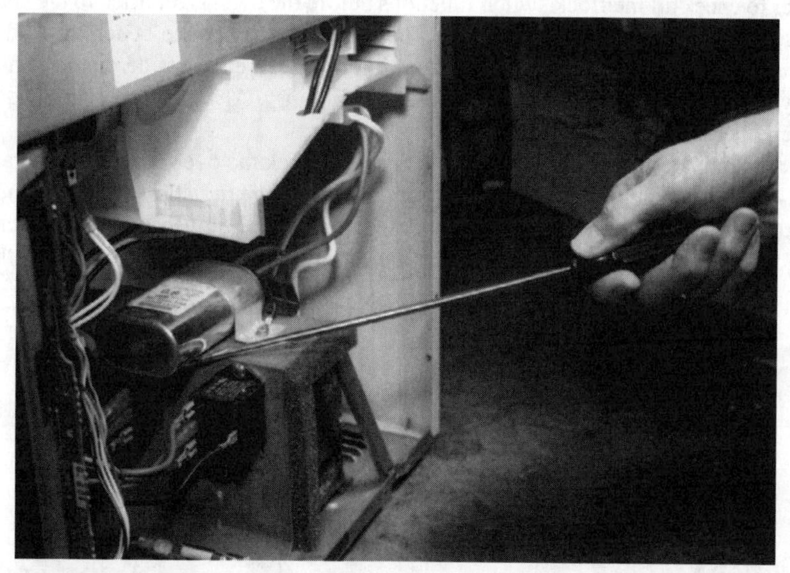

FIGURE 2-20 Do not stick a tool or connect test equipment in the oven while it is operating.

shock or possibly be killed because your hands are in the oven operation area. This capacitor can hold a dangerous charge for weeks after the oven has been shut off.

BASIC MICROWAVE REPAIR SAFETY TIPS

Before taking off the back cover or outside wrap, remove your wristwatch. Never stick a tool or your hand inside the oven area while the oven is operating (Fig. 2-20). Before checking any component or wiring in the oven, discharge the high-voltage capacitor, just like you discharge the CRT anode connection before working around the HV section of a TV. The big difference between the high-voltage of a TV and a microwave oven is the amperage; you might be severely shocked or killed if the high-voltage capacitor is not discharged. Always discharge the capacitor with an insulated-handle screwdriver before working in the high-voltage area.

Be sure that the oven is properly grounded before attempting to service it. If you're not certain, clip a flexible wire from a water pipe or fuse box to a metal screw on the back cover or metal chassis. Do not use a regular two-wire extension cord to operate the oven while servicing it.

The microwave oven should not be operated with the door open (Fig. 2-21). Do not, for any reason, defeat the interlock switches. Check and replace all defective monitors and latch switches if the oven will not shut off with the door open. The oven should never be operated if the door does not fit properly against the seal. Check for broken or damaged hinges. Visually inspect the seal gasket for possible cut or missing pieces. Check the gasket seal area for foreign matter. Be sure that the oven door is mounted straight and fits snug against the oven. Readjust the door if it is loose between the door and oven. Always remember to check all interlock switch functions before the oven is returned to the customer.

While servicing a microwave oven, it's best to have the service bench away from the customer and not talk to anyone while working on the oven. Outside distractions might cause you to make a mistake that could result in an injury. Always keep your hands out of the oven operation area when the oven is operating.

After making any repairs or service adjustments, check the oven for excessive radiation (Fig. 2-22). To ensure that the oven does not emit excessive radiation and that it meets the Department of Health and Human Services guidelines, the oven must be checked for leakage with an approved radiation meter. Especially check for radiation around the door

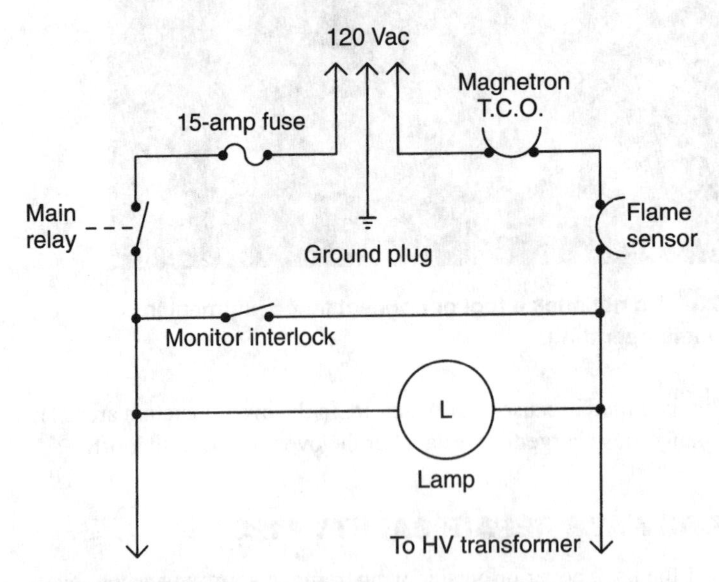

FIGURE 2-21 **Check the interlock switches if the oven operates with the door open.**

FIGURE 2-22 After repairs, check the oven door for possible radiation leakage.

and vent areas. If by chance the radiation leakage is greater than 5 mW/cm^2 or if someone has been hurt pertaining to microwave radiation leakage, report this to the oven manufacturer.

Although a few components can be substituted, most microwave oven parts should be replaced with those that have the original part number. These can be obtained from the manufacturer or oven distributors. When replacing control-panel circuits, do not touch any part of the board because static electricity might damage the control. Usually these controls arrive packed in a static-free wrap and carton (Fig. 2-23).

BASIC MICROWAVE OVEN PRECAUTIONS

When servicing an microwave oven, follow these basic safety precautions:

Do not try to operate the oven with the door open, an improper seal, and/or damaged door hinges. If the oven has been dropped, inspect the door seal and hinges before you do any other servicing on it. Be sure that the door closes properly and that no foreign matter is inside the door. Check the door for leakage with a radiation leakage meter (Fig. 2-24).

Always be sure that the oven is grounded with a three-prong wall receptacle. The oven should operate from a wall receptacle directly fused by the fuse box. It is best to have a competent electrician install the three-prong wall receptacle. Use only a three-wire extension cord between the oven and the wall outlet.

Never reach into the oven compartment area while the oven is operating. When making ohmmeter or continuity tests, unplug the oven from the ac power line. If ac voltage measurements are to be made, clip in the voltmeter while the oven is off. Do not touch the

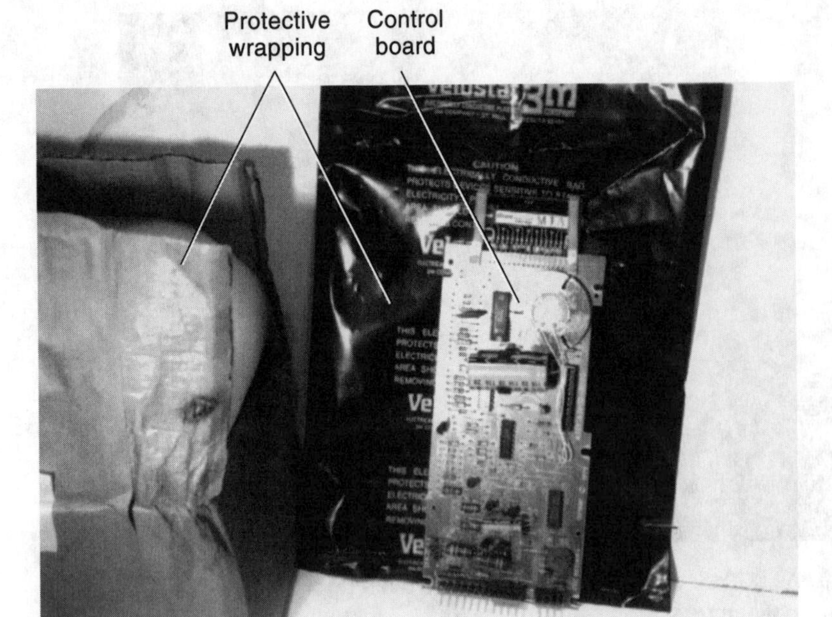

FIGURE 2-23 Most control panels are packed in static-free wrapping and carton.

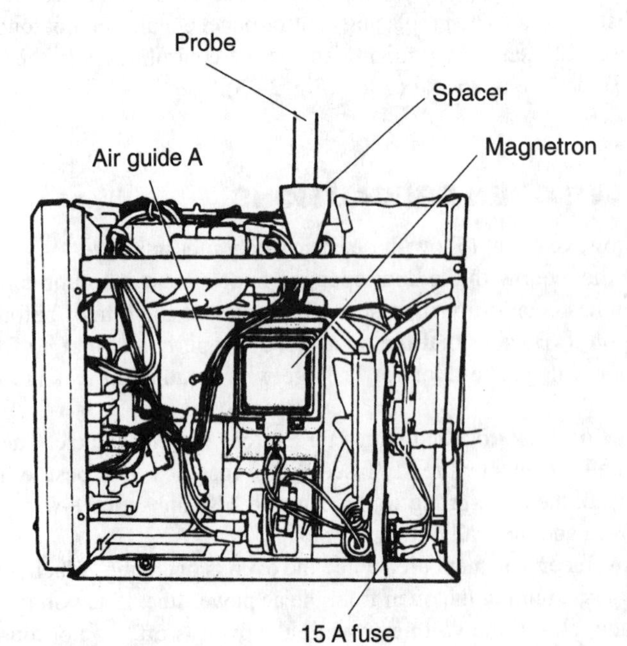

FIGURE 2-24 Check the oven for radiation leakage with an approved leakage survey meter.

voltmeter test probes while the oven is plugged in. It's best to keep small VOM or DMM testers out of the oven.

Before checking any components in the oven, short out the two capacitor terminals. Always discharge the HV capacitor each time that the oven is energized (Fig. 2-25). Failure to discharge the capacitor could result in a severe electrical shock.

Do not defeat or bypass the interlock door switches. They are there to protect the operator. Each interlock should be checked for proper operation. Check each interlock with an ohmmeter and the oven power cord disconnected from the wall socket. Discharge the HV capacitor before checking any oven component.

The power transformer, HV capacitor, HV diode, and magnetron tube contain high voltage (HV) when the oven is operating (Fig. 2-26). Do not touch any component in the oven while it is operating. When the oven shuts off, discharge the HV capacitor before attempting to perform any type of service. Be very careful when taking high voltages in the HV section. Connect a Magnameter to the HV section for high-voltage and current measurements.

Check the magnetron level and secure bolts to eliminate hot spots or radio frequency (RF) arcing. Always replace the RF gasket around the magnetron tube antenna when replacing the magnetron. Do not use metal tools near the magnetron because it can jerk them out of your hands, sometimes breaking off the glass-enclosed antenna assembly (Fig. 2-27).

Always make a microwave leakage test with the approved radiation meter after all repairs are made. This should be done before the oven front and side cleanup procedures.

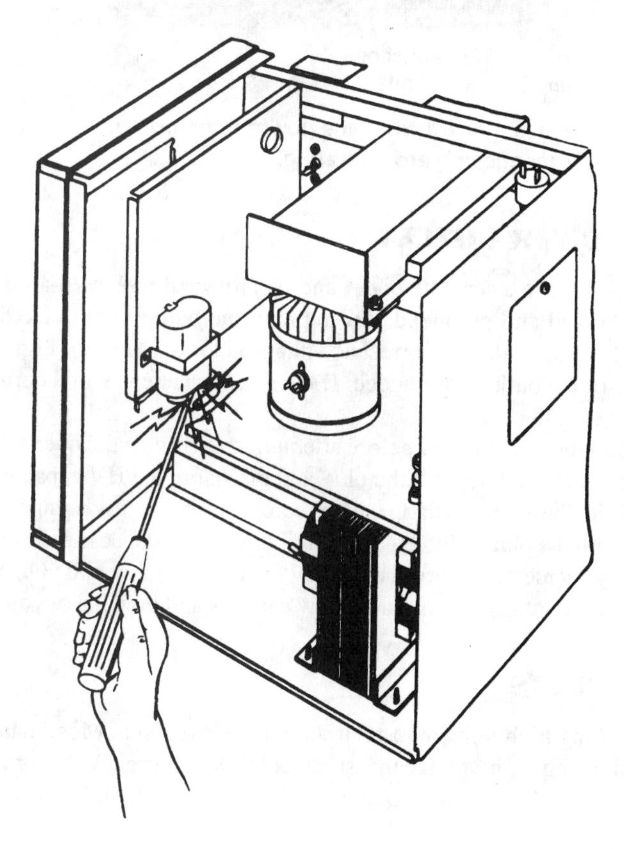

FIGURE 2-25 Discharge the HV capacitor each time that the oven is turned on and off.

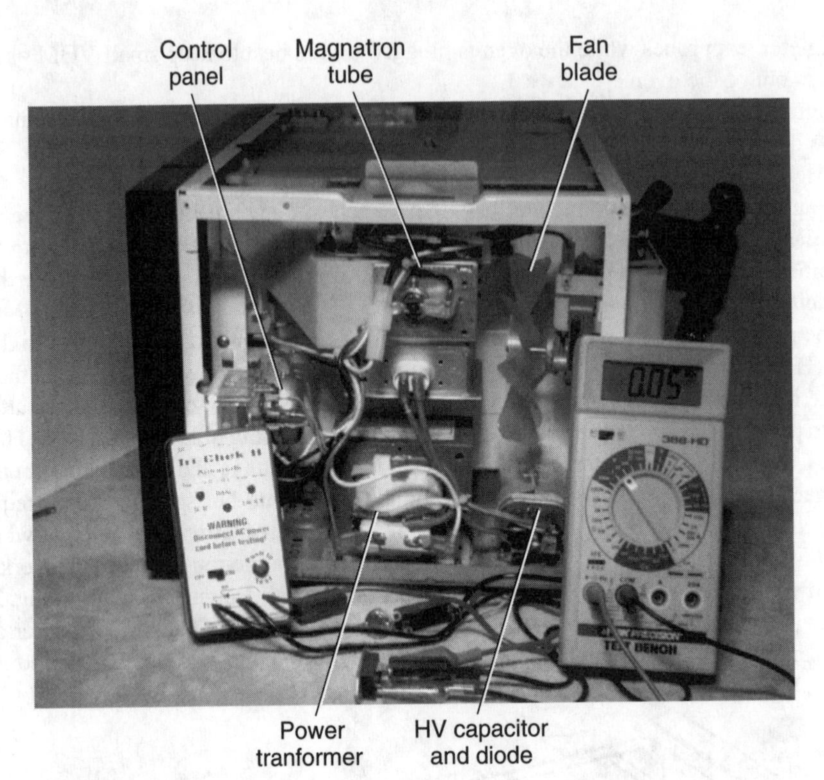

FIGURE 2-26 **High-voltage is present when the power transformer, HV diode and capacitor, and magnetron are operating.**

CHECKING THE POWER OUTLET

The microwave oven should have a separate outlet and be protected with a 20-amp fuse. The outlet should be polarized and grounded. An improper ac outlet might electrically shock the operator. Check the ac outlet for correct ac voltage (117 to 120 Vac) (Fig. 2-28). Determine if the left side of the outlet is grounded. If the ac receptacle is wired correctly, the right side of the outlet should be hot.

Always remember to pull the plug or cord before attempting to service the oven (Fig. 2-29). Each time that the oven is fired up, pull the plug and discharge the HV capacitor before putting your hands in the oven. Pull the plug before clipping test equipment to components in the oven. Pull the plug before taking continuity or resistance measurements of oven components. If, by accident, you are shocked by the HV capacitor and you end up with a metallic taste in your mouth, you will know how careless and how lucky you are.

DANGEROUS HV TESTS

Be very careful when making high-voltage tests in the microwave oven. Most manufacturers do not recommend taking high-voltage measurements. Do not use a VOM or DMM voltmeter that can measure 3000 volts; the voltmeter can be damaged. Only use a Magnameter test instrument to take voltage and current measurements across the magnetron, HV capacitor, and diode.

OVEN LEAKAGE TESTS

After the oven has been repaired, make a leakage test to protect the oven operator. Every oven should be checked with a U.S. Government leakage standard survey meter (Fig. 2-30). Most people are afraid of microwave radiation. So, check for leakage around the front door and gap areas. When testing near a corner of the door, keep the probe perpendicular to the surface, being sure that the probe end at the base of the cone does not get closer than two inches from any metal. Slightly pull on the door while the oven is operating and notice if there is leakage. The door might have some play that lets the microwave energy escape.

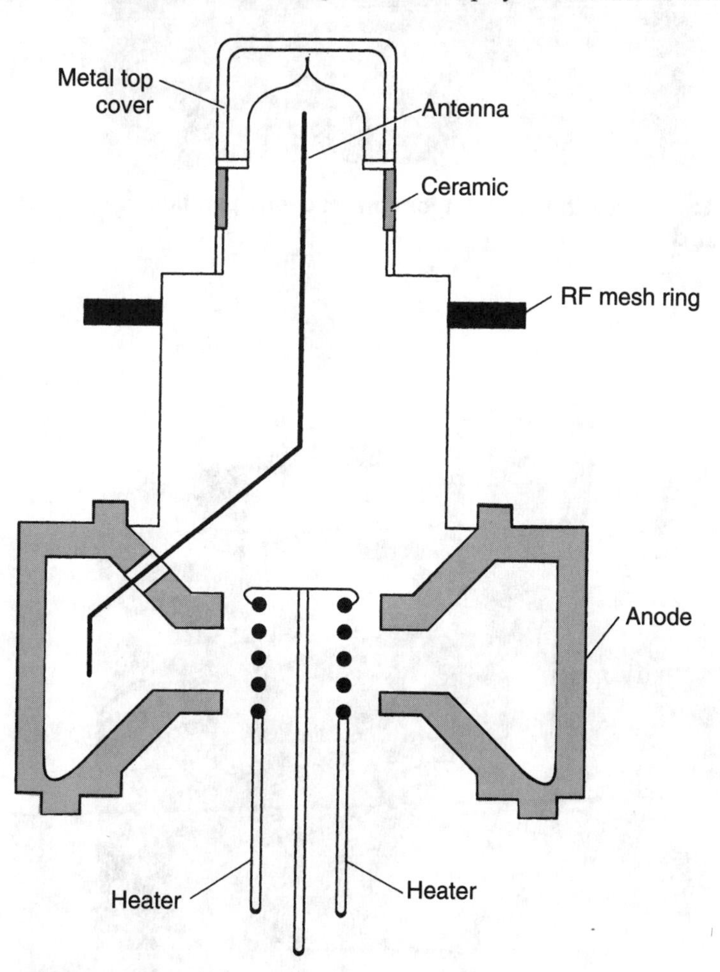

FIGURE 2-27 **Do not place metal tools near the glass antenna envelope and magnet, to prevent breaking the glass antenna assembly.**

FIGURE 2-28 Check the ac outlet for correct oven operation (117 to 120 Vac).

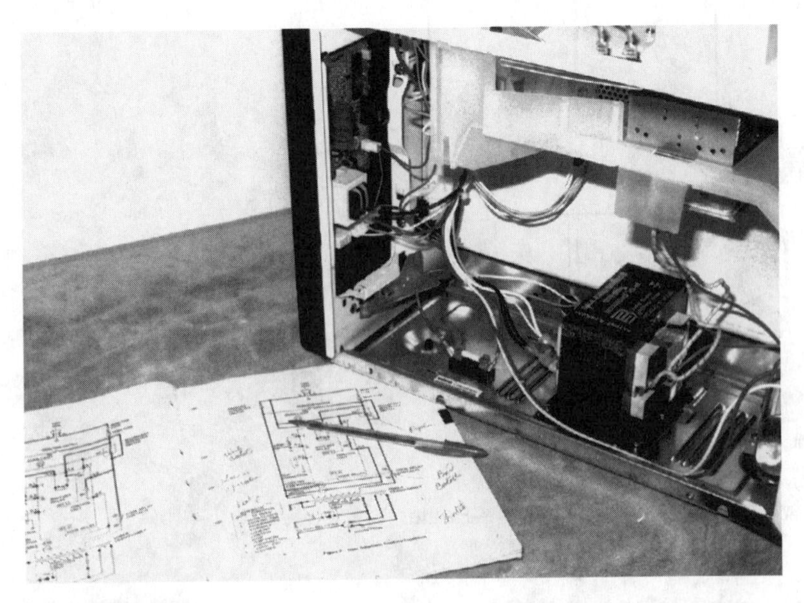

FIGURE 2-29 Pull the ac plug before attempting to repair the microwave oven.

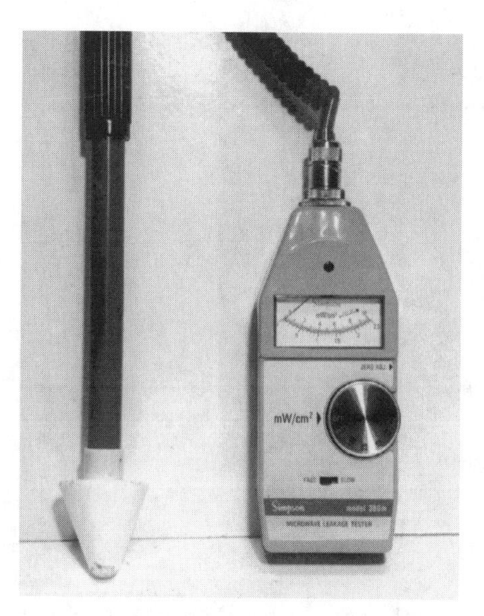

FIGURE 2-30 Check the oven for radiation with a U.S. government leakage standard survey meter.

Conclusion

Servicing electronic products within the consumer electronic field can be quite rewarding and profitable. Remember, the danger that lurks in corners, over a hot soldering iron, in high-voltage areas, and within heavy-current components. Have fun and play it safe out there.

PART 2

TROUBLESHOOTING AND REPAIRING SOLID-STATE TVS

3

SERVICING THE LOW-VOLTAGE POWER SUPPLY

CONTENTS AT A GLANCE

The low-voltage power supply of the solid-state TV receiver supplies voltage to the transistors, ICs, and filament voltage to the picture tube (in some cases). In the latest solid-state TV chassis, the filament of the picture tube and other circuits might be powered with voltage developed from the flyback or horizontal output transformer. A power transformer

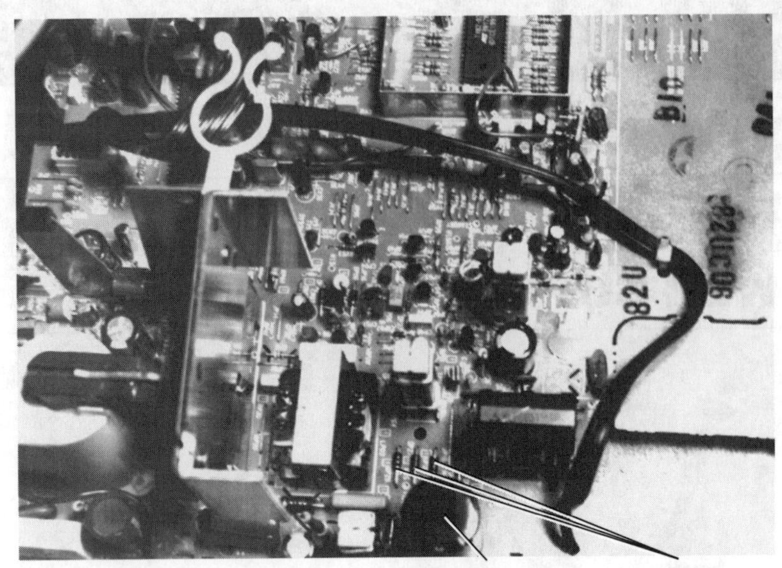

FIGURE 3-1 Locate the low-voltage circuits by finding the large filter capacitor and parts nearby.

was found in earlier chassis, and most of today's low-voltage power supplies operate directly from the power line (Fig. 3-1).

Most line-connected low-voltage power supply circuits consist of either a half-wave silicon rectifier or a full-wave bridge circuit that uses four different silicon diodes. Either circuit can have a zener diode, transistor, and SCRs in a filter regulator circuit. You might find very large filter capacitors in line-connected power supplies.

The integrated or flyback low-voltage power supplies are rectified and filtered from a separate winding of the horizontal output transformer. A single silicon diode rectifies the high ripple voltage from the flyback winding. The horizontal circuits must be operating before any voltage is developed by the output transformer.

The following test instruments are used for servicing the low-voltage power supply:

- VTVM, either VOM or DMM.
- Oscilloscope.
- Rectifier tester (could be included in DMM).
- Capacity tester.
- Isolation transformer.

Low-Voltage Power-Supply Symptoms

The low-voltage power supply is the heart of any consumer electronic product. Nothing operates without power or voltage furnished from the low-voltage source. When trouble occurs in the TV chassis, most electronic technicians first check the low-voltage source. A defective power supply can cause a dead, intermittent, hum and improper TV symptom.

The low-voltage power source can cause chassis shutdown because of excessively high or low output voltage. Poor or narrow width can be caused with low power-supply voltage, and high-voltage output can cause excessive width and high voltage at the CRT. Defective or dried-up filter capacitors can produce hum in the sound and crawling pictures in vertical circuits. Chassis shutdown can result from a defective component in the low-voltage power supply.

ISOLATION TRANSFORMER

The isolation transformer is a must-have item when servicing the transformerless ac/dc chassis. Besides providing direct ac voltage on the TV chassis, it will prevent fuse and low-voltage circuit damage when attaching the test instrument (Fig. 3-2). If you do not use the isolation transformer, ac-operated test equipment cannot be used directly on the live chassis.

Choose an isolation transformer with variable voltage taps or continuous voltage control. The isolation transformer can vary from 0 to 150 Vac with an isolated outlet. Some of the transformers are fused with a 10-A fuse. Select an isolation transformer that has a 1.5- to 5-A outlet. You might find that the isolated transformer is contained with the low-voltage power supply.

You can make your own isolation transformer by ordering a regular line isolation transformer (115 Vac to 115 Vac). These transformers might have to be ordered through an electronics wholesale distributor. Of course, you might have to wait 30 days for delivery, but they are worth it. Stancor and Thordorson sell line-isolation transformers. Remember, these transformers weigh between 4 and 10 pounds.

FIGURE 3-2 **Always use a variable isolation transformer when servicing today's TV.**

Choose a line transformer from 115- to 125-Vac output for ordinary ac line operations. Although this type of isolation transformer is not variable, it will isolate the ac/dc TV chassis from the power line. Isolation is required when connecting test equipment to the ac/dc TV chassis. If not, you might damage components in the TV.

The isolation transformer might be included in the same cabinet as the other power supplies. A separate ac outlet is placed on the outside of the cabinet to plug in the TV chassis or other equipment. With this method, the isolation transformer can be used without cranking up the other power supplies.

THE HALF-WAVE RECTIFIER

The half-wave rectifier circuit was once found only in the portable TV. Today, the half-wave circuit, such as the one in Fig. 3-3, can be found in both console and portable TVs. In fact, a modern TV chassis is rather small compared to the early solid-state chassis. The power-line operated power supply does not contain a heavy power transformer, large choke filter, or several large filter cans (electrolytic capacitors).

The circuits of the entire receiver are usually protected by a 5-A fuse located on the main chassis board. Most ac and dc circuits are fused instead of using circuit breakers. The power plug is polarized to prevent a "hot" metal chassis. Always use an isolation power transformer when servicing a line-connected TV to prevent injury from the power line.

L900 and C900 (Fig. 3-3) prevent spurious signals from entering the power supply from the power line. Power is applied to the TV chassis when the on/off switch (SW1) is closed. The silicon rectifier (D900) rectifies one-half of the full cycle, and the ripple dc voltage is filtered by L101, C901, and C902. Because silicon diodes have a tendency to radiate spurious signals, C903 keeps the diode from radiating noise lines into the picture. The half-wave circuits have RC (resistance-capacitance) or LC (inductance-capacitance) filter networks.

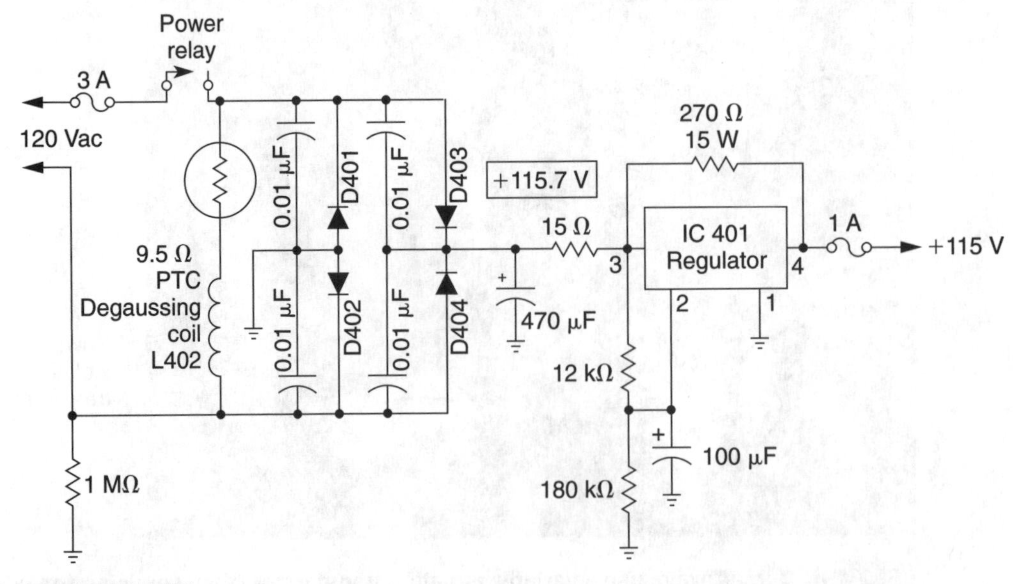

FIGURE 3-3 The low-voltage power supply might have one large power-line IC regulator.

FIGURE 3-4 **Hum bars in the sound circuits and crawling pictures can result from defective filter capacitors.**

The degaussing coil is energized through the R900 (3-Ω cold thermistor). After a few seconds, R900 heats up and increases in value to prevent current from entering the degaussing coils. Thus, the face plate inside of the picture tube is degaussed for a few seconds only when the receiver is turned on. This prevents spotted or magnetically contaminated areas on the face of the picture tube.

F800 protects the power supply circuit if the horizontal output, damper, and flyback circuits become leaky. Besides feeding the horizontal circuits, the 140-Vdc supply can be tied into transistorized or SCR-regulated low-voltage circuits. The low-voltage regulated circuits can furnish voltage-to-sound, IF, video, sync, horizontal, and vertical oscillator countdown stages. The power supply circuits are easily serviced by making voltage and resistance measurements.

Fuse and silicon diode replacement are the most commonly needed fixes within low-voltage power supplies. Always replace the blown fuse with another of the same current rating. D900 can be replaced with a 2.5-A silicon diode. Some half-wave rectifiers are replaced with 3-A types. The fuse and diode can open because of overloaded components in the power supply or in connecting circuits. The fuse and silicone diode can also be damaged by lightning or power-line spikes.

The defective silicon diode can become leaky, shorted, or open. A shorted diode can blow the line fuse and open the line-isolation surge resistor. The leaky diode with high resistance might not blow the line fuse at once. Look for a shorted diode or electrolytic capacitor if the fuse is exploded, seared, or black. Besides a shorted silicon diode, the fuse might open with a 1-watt voltage dc line regulator.

Hum bars or audio hum can mean bad filter capacitors (Fig. 3-4). Defective filter capacitors can cause unstable vertical rolling and horizontal sync problems. Chattering relays

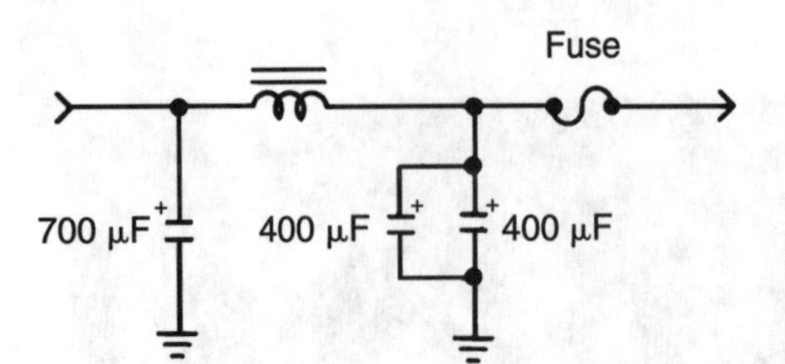

FIGURE 3-5 Connect two electrolytic capacitors in parallel to obtain the correct capacity.

and improper operation of remote-control systems can also be caused by defective filter capacitors.

The filter capacitors (C901 and C902) in Fig. 3-3 can be replaced with the same (or higher) capacitance and voltage. Notice how large the filters are in the half-wave circuit (700 μF/800 μF). If they are replaced with units of lower operating voltage, the capacitors could run warm and blow up in your face. Replacing a filter capacitor with one of a lower capacity rating can produce more hum and an unstable picture. Check the hum bar waveform with the scope if a larger-value capacitor is added to the filter circuit.

Filter capacitors can be paralleled to provide greater filtering action in most power supplies. For instance, if you have a capacitor rated at 400 μF, 400 μF, 700 μF at 200 V, the capacitor could be substituted for C901 and C902. Just connect the two 400-μF sections together as a 800-μF capacitor (C902) replacement and use the 700-μF section to replace C901 (Fig. 3-5). The original capacitors (C901 and C902) were rated at 700 μF and 800 μF with a 185-V operating voltage. Sometimes you might have to add more capacity to the power filter circuits to eliminate the hum.

The defective filter capacitor can be located by checking the waveform with a scope or by shunting another capacitor across the suspected one. Choose a capacitor with the same or higher capacity and voltage. Most filter replacements are of a higher operating voltage than the original one.

Use clip leads to clip a new filter capacitor across the suspected one. Observe correct capacitor polarity. Disconnect the power cord and discharge the shunted capacitor each time that a different capacitor is shunted to prevent damage to transistor and ICs. Replace the capacitor when hum bars or instability stops altering the picture.

FULL-WAVE RECTIFIER CIRCUITS

Although power transformers were used in the early solid-state chassis, very few were replaced unless damaged by lightning. In some chassis, a small power transformer might be used to supply low-voltage circuits and a power line full-wave rectifier circuit is then used to provide a higher voltage to the horizontal output transistor and the sound circuits. The full-wave rectifier can use single silicon diodes or a bridge-type rectifier. Figure 3-6 shows a typical full-wave rectifier circuit.

A 3-A fuse (F2) protects the entire TV receiver from the power line. L101 and L102 prevent noise signals from entering the TV from the power line. In many of the latest TVs, a line-operated varistor (V201) protects the receiver from lightning or high-voltage surges from the power line. If excessive voltage is found across V201, the varistor shorts and arcs over, causing the 3-A line fuse to open. This prevents the high voltage from entering the primary winding of the power transformer. Capacitors CPR201 and CPR202 prevent static charges from nearby metal objects to enter via the ground when the receiver is turned off or on.

The power transformer (T201) has two separate full-wave power circuits supplied by the power transformer in this special circuit. Terminals 1 and 5, with center tap 3, supply high ac voltage to D201 and D202. Terminals 2 and 4, with center tap ground 3, provide another full-wave rectifier circuit to D203 and D204 with a lower-voltage source and output. Each single silicon rectifier can be replaced with a 2.5-A diode. Capacitors C110 through C113 prevent stray voltages from entering the TV chassis via the low-voltage power supply. The four diodes can be checked within seconds with the diode tester of a DMM.

The full-wave 120-Hz ripple voltage is filtered out by capacitors C201A and C201B. The filter choke L202, C201A, and C201B provide an LC filter network. F3 provides protection to the low-voltage circuits. The 40-V output can have zener diode or transistor regulator circuits attached to provide regulated voltage to crucial receiver stages.

The power transformer power supply can be serviced like any power supply circuit. If the pilot light and the end of the picture tube light, you can assume that the power transformer is normal. Check and replace the 3-A fuse if there is no pilot or dial light. Suspect a problem with the primary winding, on/off switch (SW1), or chassis wiring if replacing the fuse does not solve the problem. Disconnect the power cord and measure for a low resistance (50 W or less) across the power plug. This indicates that the primary winding is normal.

Measure the output voltage at capacitor terminals C201A and F3. An ac voltage measured at the anode terminal of each diode will indicate that the power transformer is normal.

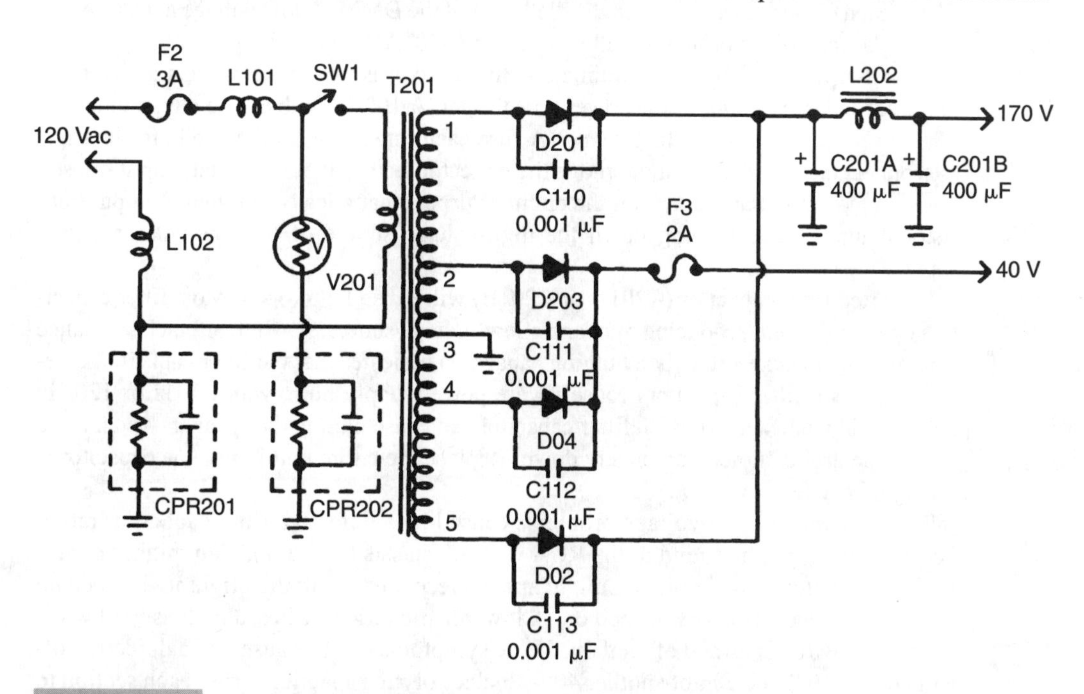

FIGURE 3-6 **Full-wave rectification can use individual silicon diodes or one bridge rectifier component.**

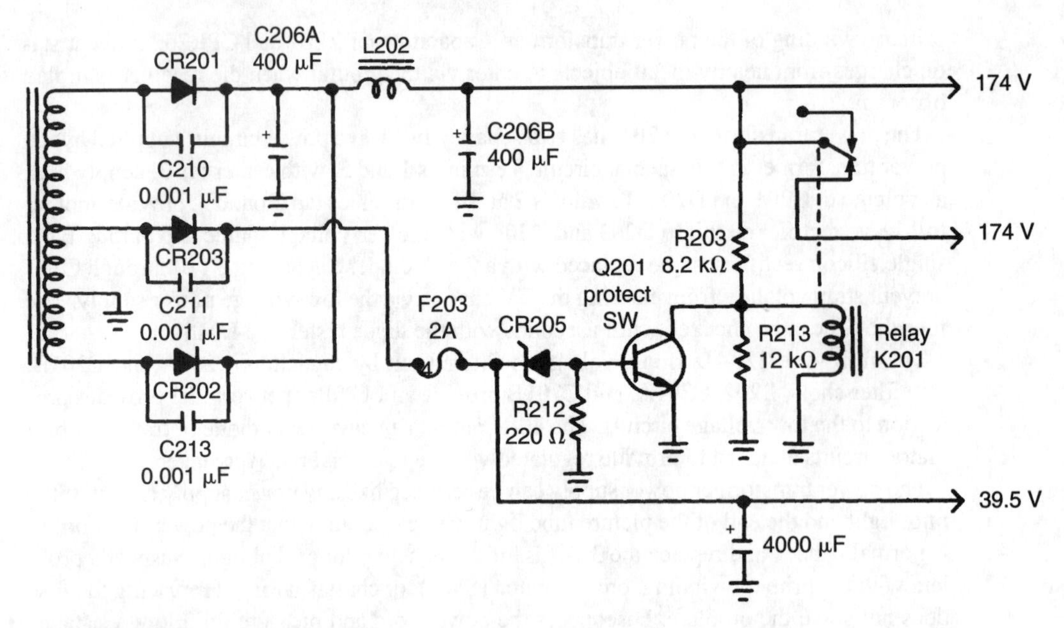

FIGURE 3-7 **A defective 400-µF electrolytic capacitor caused the relay to chatter or shut down the TV chassis.**

Check each diode for leakage with the diode test of the DMM. A low voltage at C201A can indicate a leaky diode or that the filter capacitor (C201A) is open.

If the 3-A fuse (F2) opens each time that the receiver is turned on, suspect an overload in the secondary winding of the power transformer. A 100-W bulb can be clipped in place of F2 to prevent replacing the blown 3-A fuse each time. Check each diode for leakage. Disconnect the power cord and remove the collector lead of the horizontal output transistor, or remove the transistor from the circuit. Suspect that a leaky horizontal output transistor damper diode loading down the high-voltage source of the low-voltage power supply.

A shorted filter capacitor (C201A or C201B) will cause F2 to open. Most filter capacitors open or dry out, producing hum and a low-voltage source. A white or black substance leaking from the capacitor (or a bulging capacitor) indicates that the filter capacitor is defective. Leaky filter capacitors cause severe power supply hum, with 60-Hz or 120-Hz hum bars in the raster. An open filter capacitor can cause hum in the speaker—even if the volume control is turned completely down. Replace the entire unit if any one capacitor is found to be defective.

Besides hum and low-voltage problems caused by a defective filter capacitor, rather odd symptoms can be found in the RCA CTC47 chassis (Fig. 3-7). You might hear relay K201 chatter, and the picture might appear very noisy with the brightness turned up or when the receiver is first turned on. A low whistle might be heard in the sound when the on/off switch is turned off and on. These symptoms can be caused by a defective filter capacitor (C206). Shunt another 400-µF electrolytic capacitor across each section to locate the defective capacitor. Remember, defective filter capacitors can produce odd symptoms.

BRIDGE RECTIFIER CIRCUITS

The bridge rectifier circuit can be operated from a full-wave power transformer or directly from the power line (Fig. 3-8). The bridge rectifier can consist of four separate silicon diodes or of four diodes molded into one component. Sometimes two silicon diodes are found in one envelope, and it takes two separate units to complete the bridge circuit (Fig. 3-9). Replace the whole unit—even if it has only one leaky diode.

A 3-A fuse offers power supply protection of the full-wave line-operated power supply in Fig. 3-10. Line filter L101 prevents noise from entering the power supply of the power line. Thermistor T701 provides low resistance to the ac line voltage to energize the degaussing coil for a few seconds. This happens whenever SW1 is turned on; then T701 increases in resistance to shut off any current flow to the degaussing coils.

The ac input voltage is rectified by a bridge rectifier circuit (D401, D402, D403, and D404). Capacitors C402, C403, and C404 bypass the silicon diodes to keep them from radiating noise into the dc voltage source. C405 is a capacitor filter at the circuit output.

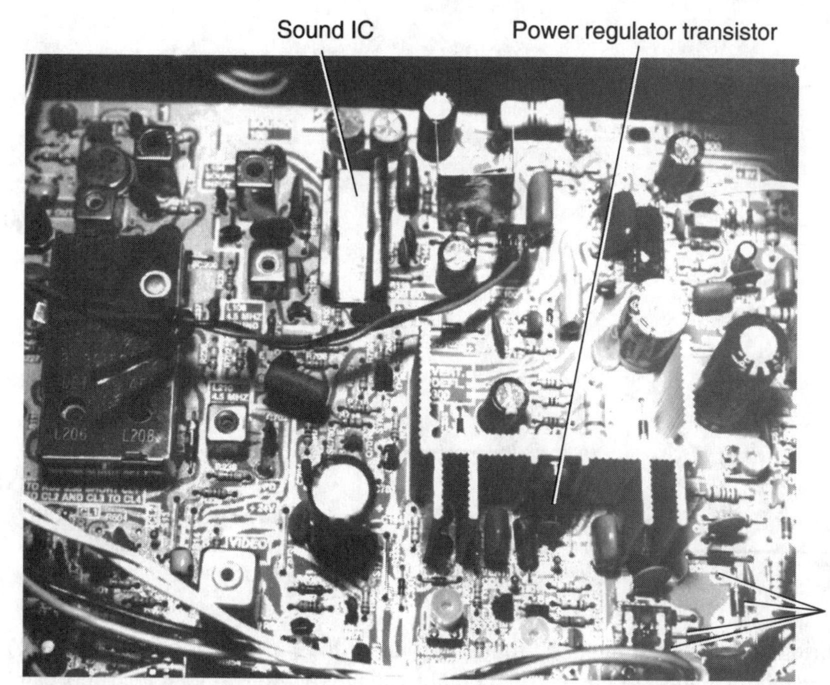

FIGURE 3-8 You might find four individual silicon diodes or a bridge rectifier in the low-voltage power supply.

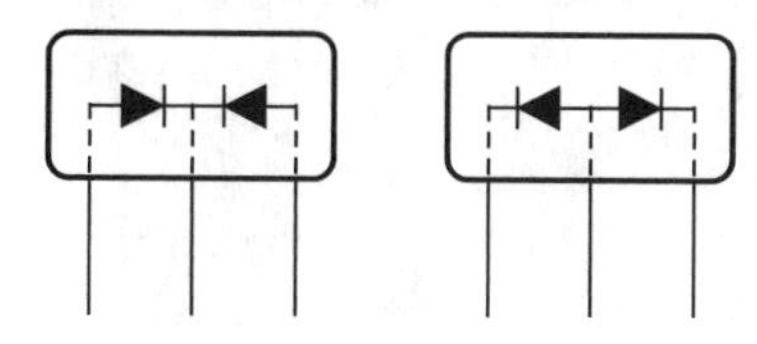

FIGURE 3-9 You might find two silicon diodes in one molded body of a bridge-type rectifier. Two such rectifiers are necessary for a full-wave bridge circuit.

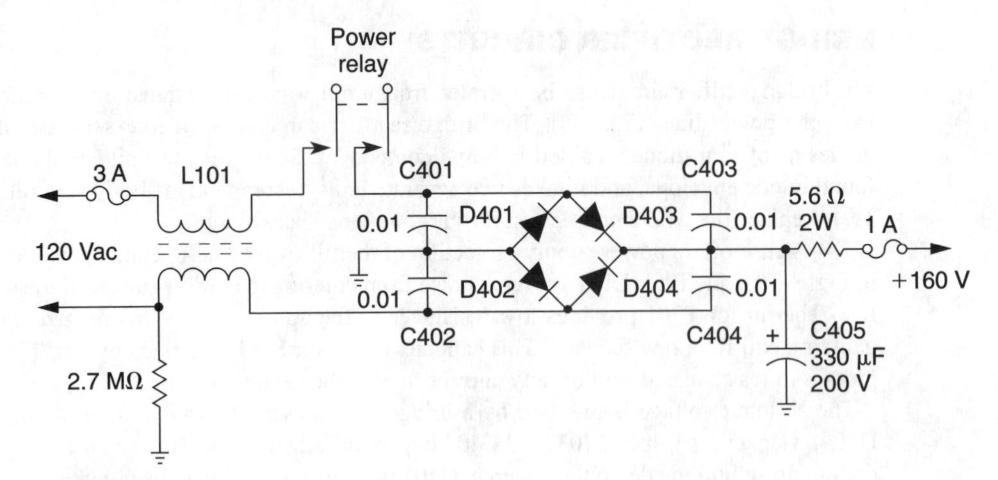

FIGURE 3-10 A typical remote low-voltage power supply with separate silicon diodes.

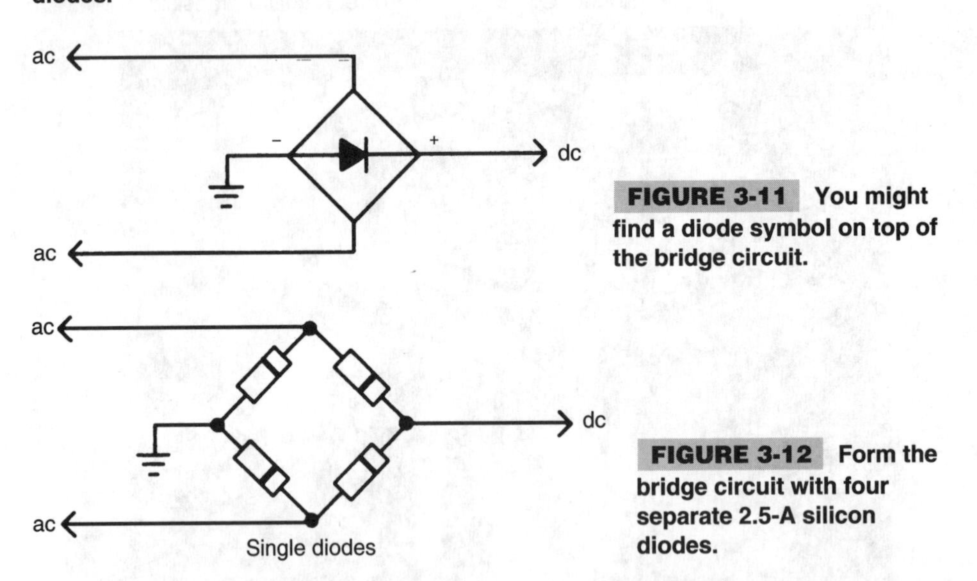

FIGURE 3-11 You might find a diode symbol on top of the bridge circuit.

FIGURE 3-12 Form the bridge circuit with four separate 2.5-A silicon diodes.

A bridge-type symbol might be found in some of the Japanese models (Fig. 3-11). The diode symbol "points" to the rectified output voltage. All four silicon rectifiers can be replaced with single 2.5-A diodes if one whole bridge rectifier component is not readily available. Simply connect the diodes (as shown in Fig. 3-12) and solder them into the respective holes of the PC board. Likewise, both rectifiers can be replaced (in the half-molded bridge unit mentioned earlier) with two separate 2.5-A diodes.

HOT CHASSIS

The new ac/dc TV chassis can have a "hot" and a "cold" chassis within the large overall chassis. Be careful when taking voltage measurements around these chassis. Often the "hot" chassis is above regular chassis ground and has a different ground marking (Fig. 3-13). If the "hot" ground is not used and you use common chassis ground, the voltage measurement is incorrect. Always use the "hot" ground when making voltage measurements on components in that area.

THE VARIABLE-FREQUENCY SWITCHING AND SWITCHED MODE POWER SUPPLIES

The variable-frequency switching power supply (VIPUR) is used in the RCA and some other TVs, but the switched-mode power supply (SMPS) is found in the Sylvania, Philips, and some other TVs. The VIPUR circuits in most chassis are located on a separate PC board. The variable-frequency switching power supply provides hot-to-cold ground isolation between the primary TV power source (hot ground) and the video-audio input/output circuits (cold ground).

The raw 150 Vdc (hot) is fed from the bridge rectifier circuits to the switching transformer (T1) and power MOSFET transistor (Q1). Regulator IC U2 received the on and off commands from the remote amplifier circuit or on and off switch. These regulators provide a drive signal to the gate of power MOSFET Q1 (Fig. 3-14). Likewise, this signal switches Q1 off and on, causing energy to be stored in the primary winding of T1. The secondary windings of T1 develop a regulated secondary voltage of 20, 25 and 150 Vdc. Notice that the secondary windings have a cold ground.

Another small step-down transformer from the power line provides a standby and run voltage to operate the remote-control circuits. The run voltage is fed to the on/off switch (Q2), U1 on/off optoisolator, and to the U2 regulator IC. IC U2 controls the overcurrent-protection circuit that can shut down the regulator circuit, turning off all power to the TV.

Remember to use an isolation transformer while working on this chassis. If not, damage can occur to the chassis or test equipment, or the technician could be injured.

Check for a shorted or leaky MOSFET output transistor (Q1) when the chassis is dead. Measure the raw 150 volts applied to TV and Q1. Remove the MOSFET transistor from

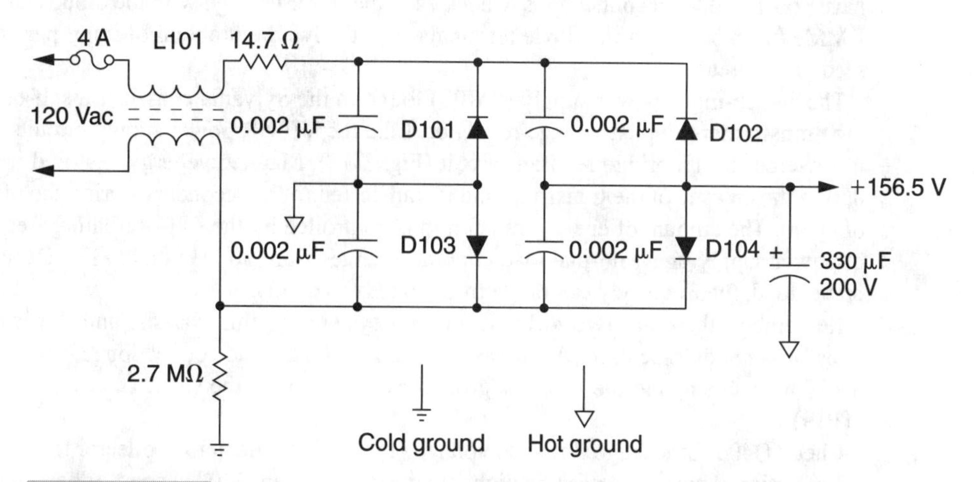

FIGURE 3-13 Notice that the low-voltage power supply has both "cold" and "hot" grounds.

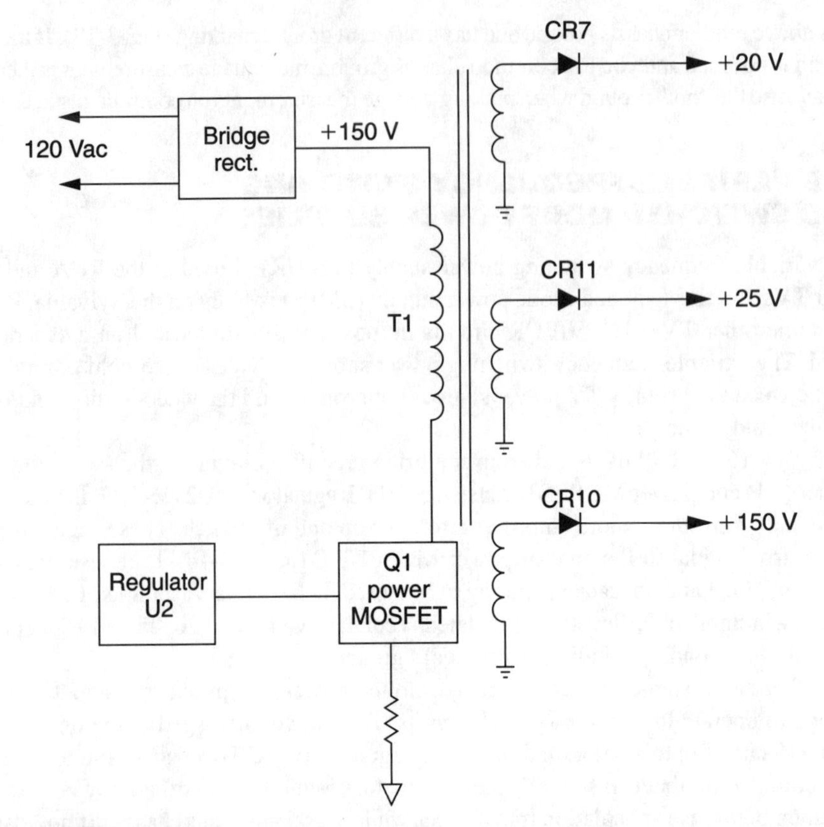

FIGURE 3-14 The block diagram of the RCA VIPUR power supply.

the circuit and test for leakage. In the early VIPUR chassis, CR7, CR10 and CR11 might have poorly soldered connections. Check each diode for leakage with the diode test of the DMM. Just resoldering the diode terminals might solve the problem of improper voltage secondary circuits.

The switch-mode power supply (SMPS) found in the Sylvania chassis is a self-oscillating, transformer-coupled voltage regulator. Like the VIPUR power supply circuits, T401 is powered from a bridge rectifier circuit (Fig. 3-15). Magnetic energy is stored in T401 during the on time of the transistor, and is transferred to the secondary during the off time of Q400. The amount of energy transferred is controlled by the self-oscillating frequency of power supply under normal load condition ranges from 20 kHz to 40 kHz. Depending on the load, the frequency can range from 10 kHz to 60 kHz.

Remember, there are two different ground systems in this chassis, and the isolation transformer should be used. A chassis common signal ground connection can be made to the IF area shield, and the "hot" ac ground can be made at the emitter terminal of Q400 (TP19).

Check Q400 for leakage if raw dc voltage (155 V) is found at the collector terminal and no operation. This voltage can be higher than normal when Q400 is open. When Q400 appears shorted or leaky, check D425 and D421 for burned or shorted conditions. Replace

R422 (562) and R430 (56 ohm) for a change in resistance. Check all transistors and diodes within the SMPS circuits. A blown fuse will almost always point to a leaky switched-mode regulator transistor (Q400).

A shorted Q400 can cause D404 through D407 of bridge diodes to become leaky or open. Always check Q403 and Q402 when Q400 fails. Test D422, D424, and D434 for leakage or opens if one of the secondary voltages is improper or missing. Notice that the hot ground is on the primary side of T401 and the common ground is on the secondary side.

REGULATOR CIRCUITS

A defective regulator circuit can cause low (or no) voltage, hum bars, intermittent operation, and chassis and high-voltage shutdown. The regulator circuit can consist of transistors, SCRs, zener diodes, ICs, or a combination of these, to provide accurate voltage regulation, regardless of the fluctuations of the power line. Most regulator circuits have a B+ control that the technician can use to adjust the operating voltage applied to the horizontal output and low-voltage circuits.

A typical B+ regulated circuit is shown in Fig. 3-16. Here a B+ regulator and error amp transistor provides 115-V, 114-V, and 12-V regulated voltages for the TV chassis. Adjustment of R812 (115 V) places the exact operating voltage to the collector terminal of the horizontal output transistor. The F1 fuse protects the power supply from an overloaded horizontal circuit.

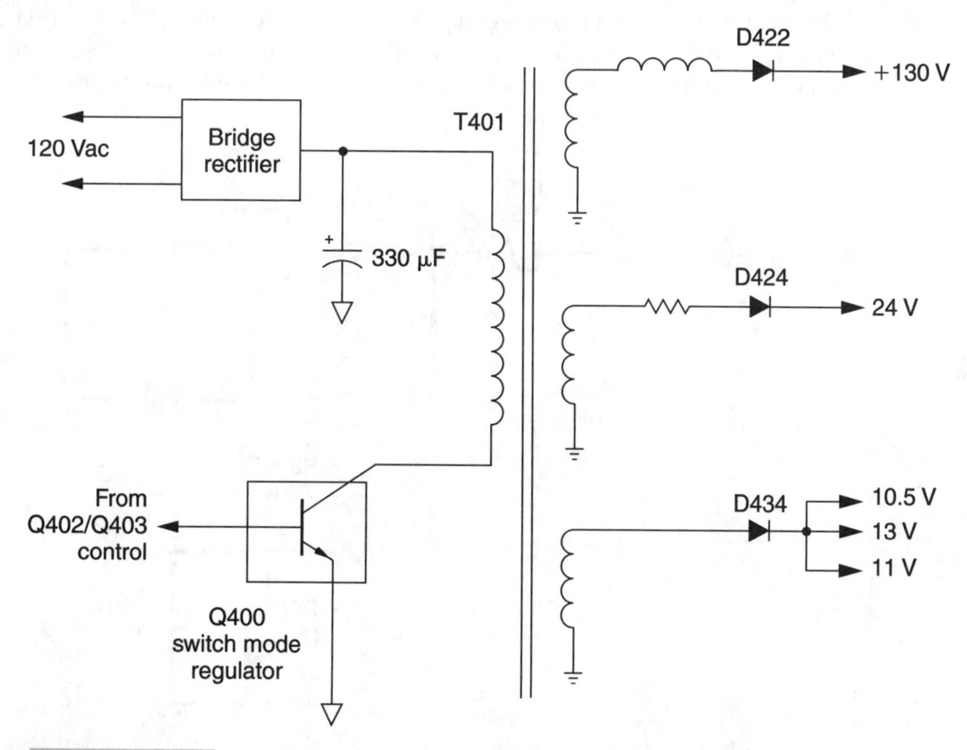

FIGURE 3-15 The block diagram of the Sylvania switch-mode power supply (SMPS).

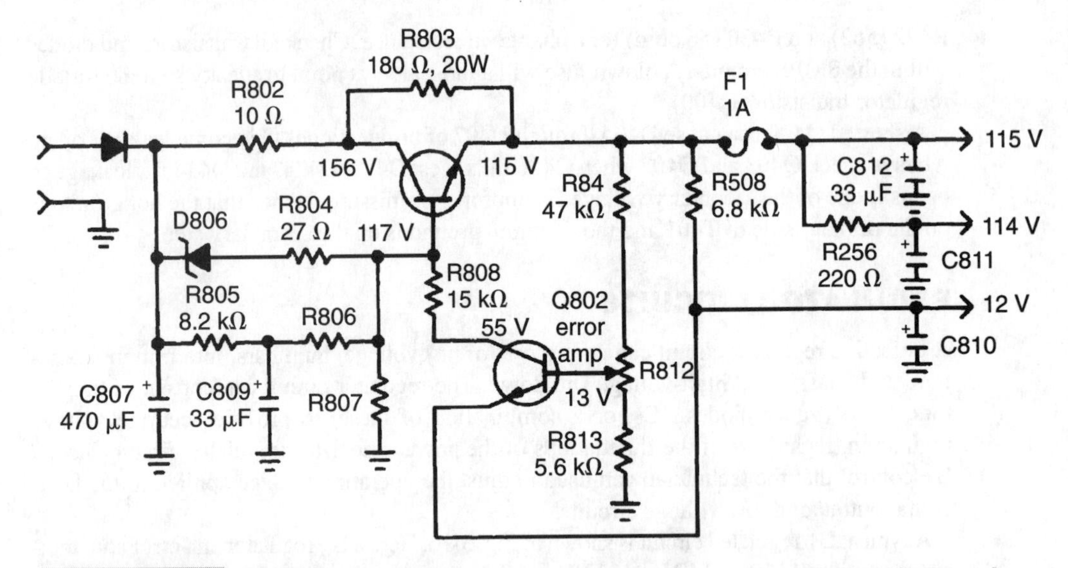

FIGURE 3-16 A typical low-voltage regulator supply circuit with a B+ regulator and error amp transistor.

The line-operated low-voltage power supply can be regulated with an IC circuit (Fig. 3-17). A half-wave diode rectifier (D803) provides 135-Vdc to the regulator system, which consists of IC801 and associated components. The regulated 115-Vdc voltage provides power to the horizontal output transistor.

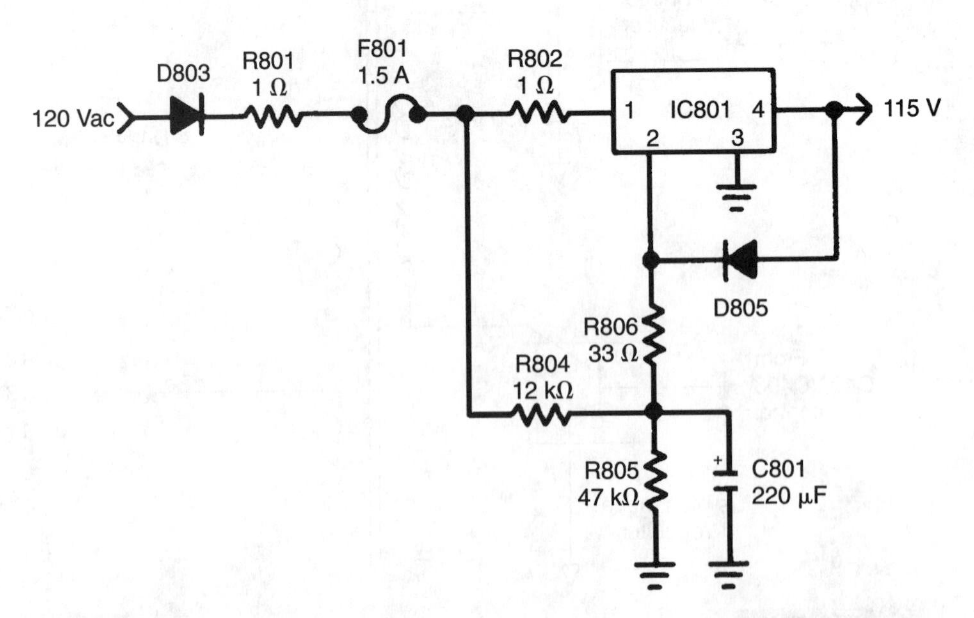

FIGURE 3-17 IC801 regulates the output voltage in this line-operated power-supply.

To quickly check the IC regulator system, measure the voltages at the input terminal (1) and at the output terminal (4) of IC801. Suspect a leaky IC801 if there is low output voltage. Excessive high voltage at pin 1 and 4 of IC801 can indicate a leaky IC regulator. Higher than normal output voltage on pin 4 can produce high-voltage shutdown. Check for a leaky D805 diode with one terminal removed from the circuit board. If the chassis is hit by lightning, inspect D803 and the IC for visible component body damage.

The defective line-operated regulator can become shorted, leaky, or open. A leaky or shorted IC regulator to common ground can blow the line fuse, open the surge resistor, and destroy silicon diodes in the power supply. Open or shorted components tied to the body of the regulator might cause the TV to appear "dead." The open regulator can cause improper voltages to the horizontal output transistor.

The defective line voltage regulator might destroy the horizontal output transistor. These regulators can cause chassis shutdown. A shorted or leaky regulator from input to output terminals can place a high voltage on the collector terminal of the horizontal output transistor and destroy the transistor. Overloaded components or an overloaded flyback, yoke, and horizontal output transistor can destroy the line voltage regulator. Intermittent operation and chassis shutdown can result from a defective regulator. Suspect that the flyback is leaky if the line-voltage regulator is leaky and the output transistor is shorted.

CHOPPER POWER SUPPLIES

The chopper power supply was first used in Japanese TVs and now is used in many American TV sets. The main components of the chopper power supply are the chopper regulator transistor, chopper transformer, regulator control transistor, x-ray latch, and overcurrent shutdown transformer (Fig. 3-18).

FIGURE 3-18 **The chopper transformer located in RCA's CTC131 chassis.**

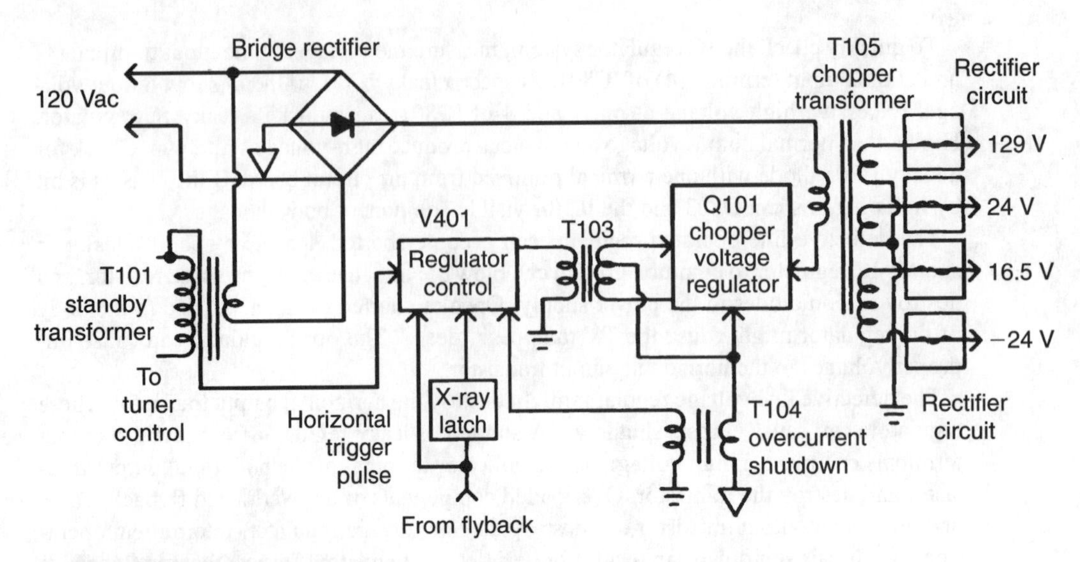

FIGURE 3-19 A block diagram of RCA's CTC131 chopper circuit.

A chopper-type regulator circuit is used in RCA's CTC131 and CTC132 chassis (Fig. 3-19). Line (ac) power is supplied to a full-wave bridge rectifier circuit and a standby power transformer (T101). It is active at all times when ac power is applied.

The bridge rectifier circuit supplies an unregulated B+ of 150 Vdc to the chopper output circuit through chopper transformer T105. The standby transformer (T101) supplies power to the tuner-control module and standby startup power to the chopper regulator IC (U401).

The pulse-width-modulated (PWM) chopper-regulated power supply is similar to the horizontal deflection system found in many TVs. The regulator free-running oscillator frequency of approximately 15 kHz is triggered by a horizontal pulse derived from the flyback, thereby locking it to the horizontal scan frequency (15734 Hz).

The output of the regulator circuit is applied to regulator drive transformer T103. Increased pulses on the secondary of T103, in turn, control the on/off state of the chopper output circuit. The chopper output circuit is powered by the unregulated ("raw") B+ (150 Vdc) developed by the full-wave bridge rectifier circuit. The on/off state of the chopper output circuit causes a pulsating dc action to occur in the primary winding of the chopper transformer (T105). Increased pulses in the secondary windings of T105 are then rectified to produce a number of dc voltages that power the TV. The 129-Vdc source is fed back to the regulator control circuit, where it is applied to a voltage-comparator circuit. The voltage comparator output is applied to an error amp, which is used to control the duty cycle of the pulse width modulator, thereby providing regulation for the B+ sources produced by the secondary windings of chopper transformer T105 (Fig. 3-20).

Suspect an open input fuse or a leaky chopper output transistor (Q101) with a no sound/no picture/no control symptom in the RCA CTC131 chassis. Sometimes a voltage isolation resistor ahead of the bridge-rectifier diodes will open. Check U401 for poor voltage regulation.

FLYBACK TRANSFORMER LOW-VOLTAGE CIRCUITS

Today, many of a TV's low-voltage circuits are driven from separate windings on the flyback or horizontal output transformer. Several advantages are gained by tapping the flyback winding for a low-voltage source. A positive or negative voltage can be obtained by reversing the low-voltage diode polarity. Voltages can be produced without large voltage-dropping resistors in a regulator voltage network; thus, less heat is generated. Filtering is less crucial because of the much-higher ripple frequency. Often, small filter capacitors (less than 470 µF) are included in the flyback voltage circuits. You might find transistors and zener diode regulators in some low-voltage circuits of the flyback winding. Remember that the horizontal deflection and horizontal output circuits must function before any voltages are found at the flyback voltage power source.

A typical low-voltage flyback circuit has a half-wave rectifier diode with a capacitor filter network (Fig. 3-21). Various voltages are tapped off with small resistor-capacitor filter components. A zener diode can regulate the voltage source (6.8 V) that feeds the luminance and chrome circuits. In other circuits, a transistor or SCR will regulate the low-voltage source.

Another voltage source from the flyback winding can supply voltage to the screen or color output circuits (Fig. 3-22). Usually, this voltage is more than 200 V with a single half-wave rectifier (D701). R702 and C111 form an RC filter network. A negative or low brightness is a possibility if the voltage is missing at the color output stages. Sometimes only a faint color can be seen in the raster. Check both D701 and R702 for leaky or burned conditions.

Remember, when secondary flyback voltage sources feed voltage to the horizontal oscillator IC, the horizontal circuits must operate before any of these voltages are produced. If no or improperly low voltages are found in the secondary sources, the horizontal oscillator sweep IC, driver, horizontal output transistor, and flyback circuits might be defective.

FIGURE 3-20 A close-up view of the chopper components in the RCA CTC131 chassis.

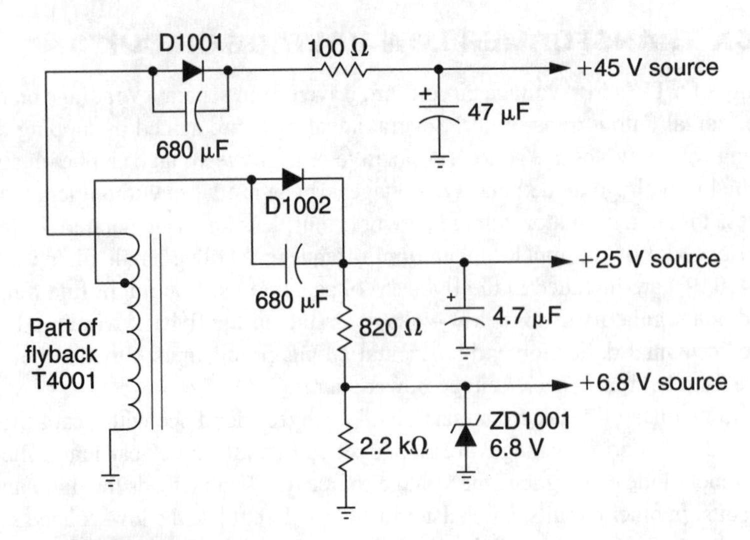

FIGURE 3-21 Check the secondary circuits of the flyback transformer for voltage sources feeding other TV circuits.

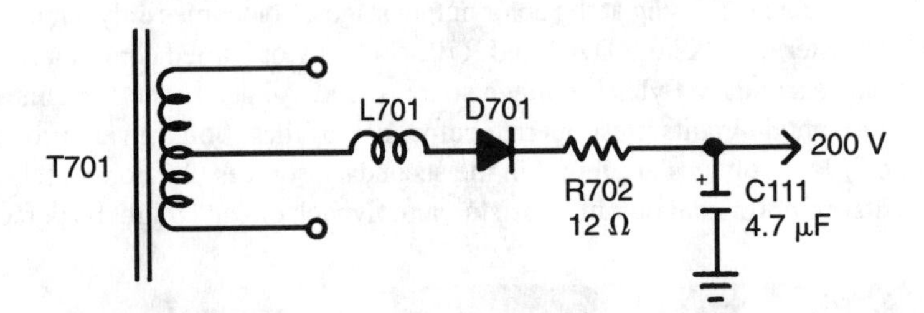

FIGURE 3-22 Voltages feeding the color output transistors and screen grid circuits of the picture tube are developed in the secondary circuits.

Any one of these circuits or a combination of faulty components can result in no secondary voltage source.

SERVICING THE LOW-VOLTAGE CIRCUITS

Three quick tests of the low-voltage power supply can locate the defective component or dead chassis within seconds (Fig. 3-23).

- Take a peek at the fuse.
- Measure the voltage on the case of the horizontal output transistor.
- Measure the voltage at the output of the low-voltage power supply.

The fuse can open because of an overloaded condition inside or outside of the power supply. A power-line outage or lightning can also open the suspected fuse. The fuse might open with a flash or arc over a component, and when replaced, restore the set to normal operation.

After checking the fuse, test to see if a low dc voltage is present from the low-voltage power supply. Quickly check fuseable resistor RF201 (3.9 Ω) for an open condition if there is no output voltage. With the digital multimeter (DMM) set at diode test, check all diodes in the low-voltage circuit. A leaky diode will show leakage in both directions. The conventional bridge, half-wave, or full-wave rectifier power supply can be checked within two minutes.

In J.C. Penney model 685-2033 (shown in Fig. 3-23), the 7-A fuse and RF201 were open. Diodes D3 and D4 were shorted. The two diodes were replaced with 2.5-A silicon units.

Besides fuses and low-value resistors and diodes, look for an open switch or poor wiring connections. If the on/off switch appears to be erratic, replace it. Disconnect the power cord and measure it for continuity with the switch in the on position. Another quick method is to plug in the power cord and measure the 120-V power line voltage across the switch terminals. Replace the defective switch if the full power line voltage is measured across the switch terminals in both on and off positions. A quick ac voltage measurement across the bridge rectifier will help you to identify problems with an open fuse, low-ohm isolation resistor, and power cord circuits.

You can locate a separate winding of the flyback transformer in the ac circuits of the low-voltage power supply. The horizontal circuits are "kicked on" by ac applied to this winding (Fig. 3-24). If the ac voltage cannot be measured across the diodes at points 1 and 2, suspect that a winding of T702 is open or that the board connections to the flyback winding and R705 are bad.

BLOWING FUSES

Suspect overloaded conditions in the low-voltage power supply or connecting circuits if the fuse blows immediately after replacing it. The most commonly leaky components in

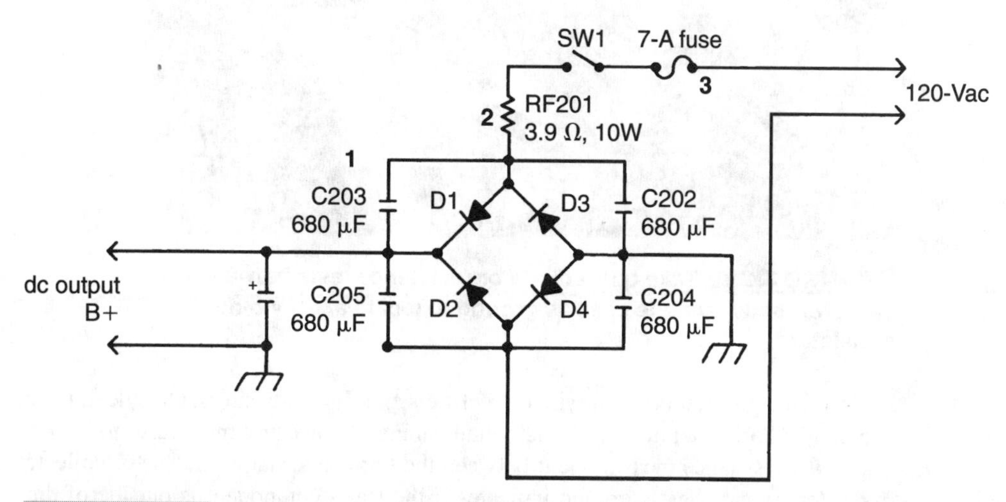

FIGURE 3-23 Check the low-voltage circuits by the numbers 1, 2, and 3 in a J.C. Penney 685-2033.

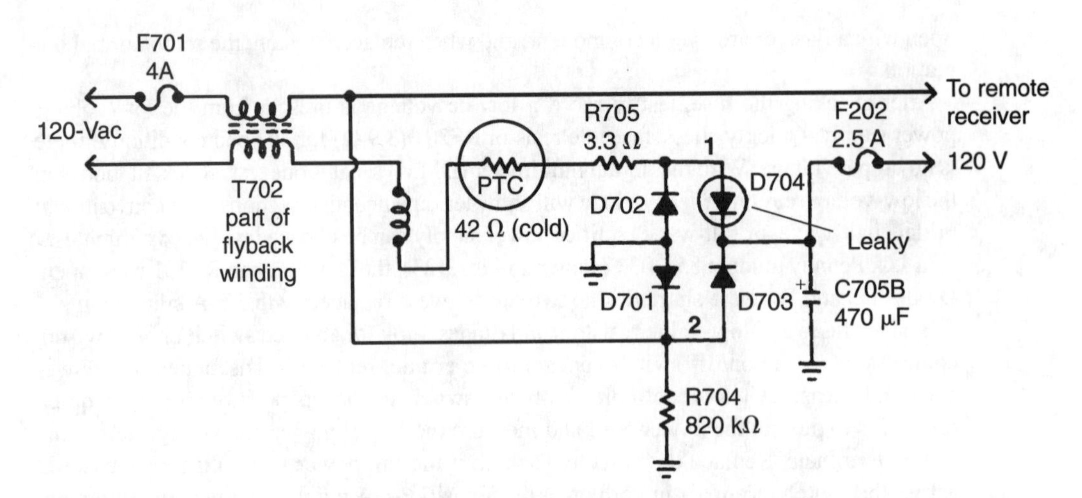

FIGURE 3-24 Suspect an open winding in T702, poor board connections, or an open R705 if the ac voltage cannot be measured at numbers 1 and 2.

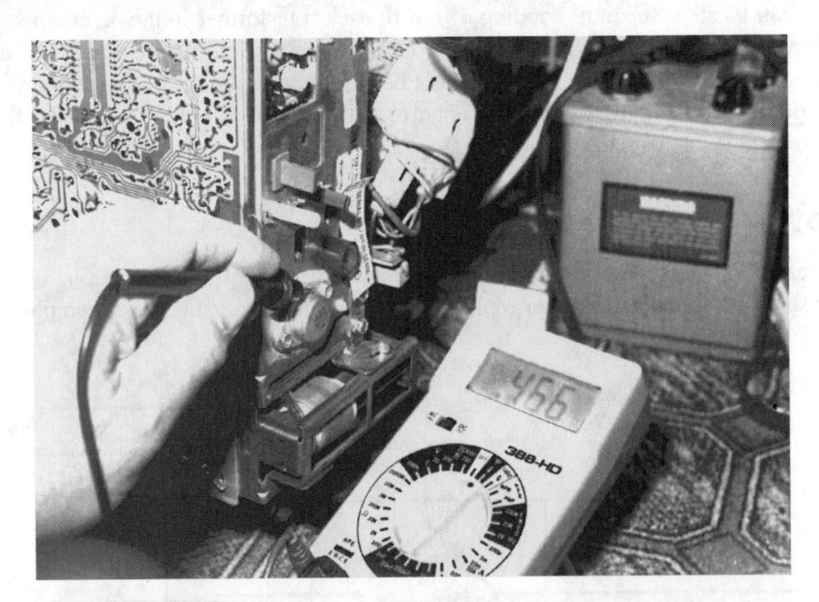

FIGURE 3-25 Take quick diode or resistance tests between the collector (body) and the chassis ground to locate a leaky output transistor.

the low-voltage circuits are the silicon rectifiers and filter capacitors. Outside of the power supply, the horizontal output transistor and damper diode cause most fuses to open.

A quick resistance measurement between the horizontal output transistor collector terminal (case) and chassis ground indicates if the leaky component is outside of the low-voltage power supply and in the horizontal section (Fig. 3-25). Set the DMM at the diode test and check the transistor case and chassis ground. A low-resistance reading indicates a leaky output transistor. Any resistance measurement below 500 Ω indicates that the transistor or damper diode is leaky.

Suspect that the diodes in the power supply are leaky if the resistance measurement is above 1 kΩ across the transistor and chassis ground. Check each diode for low resistance in both directions. A leaky input filter capacitor can cause the fuse to open.

Another method to determine if the overloaded condition is outside the power supply is to clip a 100-W bulb across the fuse terminal (Fig. 3-26). The light bulb will be bright if a leaky component is inside or outside of the power supply. Under normal conditions, at first the light bulb will appear bright and slowly go out because of the degaussing circuit action. Remove the horizontal output transistor and notice if the bright light goes out. This indicates an overloaded condition in the horizontal output circuit. Each circuit tied to the low-voltage power supply can be cut loose. When the light goes out, you have located the overloaded circuit outside of the low-voltage power supply. Even a shorted picture tube can blow the main power fuse.

A really dark area inside of the glass fuse indicates that a direct short is ahead of the fuse. If both the line fuse and secondary fuse open, suspect a defective component in the horizontal output transistor circuits. You might find an open fuse and after replacement, the TV chassis operates without any other problems. The loose fuse in the clip-type socket can cause an intermittent or shutdown symptom.

INTERMITTENT POWER SUPPLY

Most intermittent problems in the low-voltage power supply are caused by improper component connections, poor wiring connections, and defective transistors. Monitor the ac voltage applied to the rectifier at point 1 and the dc voltage output of the low-voltage power supply at point 2 (Fig. 3-27). If the voltage is missing from either or both circuits, you know in what section the defective component is located.

Sometimes by just moving or tapping the TV chassis, the intermittent condition occurs. Do not overlook a loose fuse holder. Suspect a defective relay or relay contact in TVs with remote or relay-controlled ac circuits. Poor ac-cord interlock contacts can also produce intermittent low-voltage conditions.

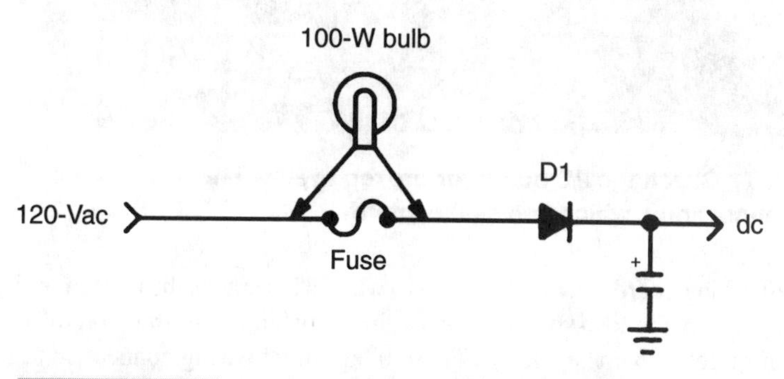

FIGURE 3-26 **Clip a 100-W bulb across the fuse slips to determine if an overload condition exists, if the line fuse keeps blowing.**

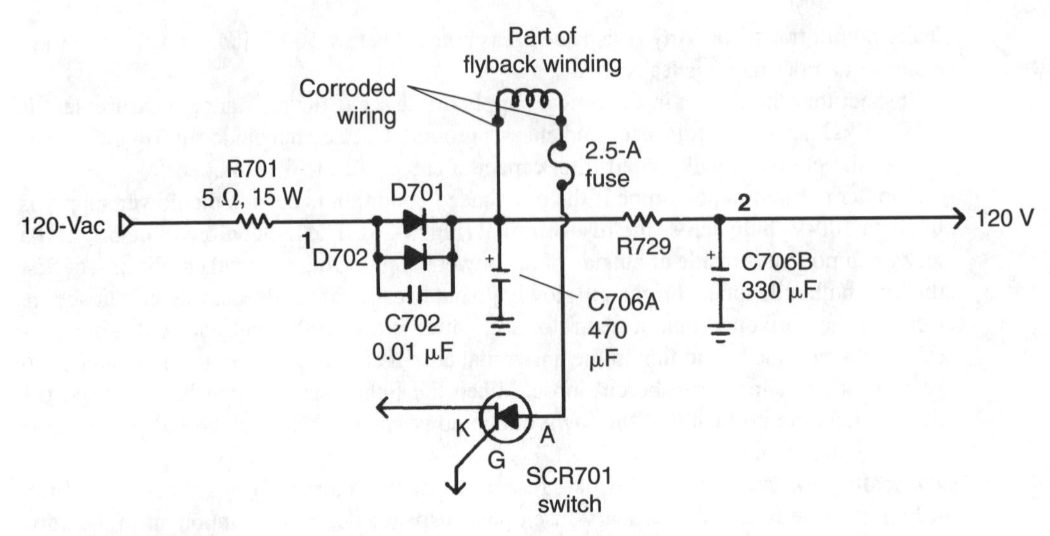

FIGURE 3-27 **Poor or corroded terminal connections to the flyback winding caused intermittent reception in a Sharp C1950.**

FIGURE 3-28 **Check the PC board for broken eyelets and component connections, which can cause intermittents.**

Double-sided wiring boards have a tendency to warp. This cracks the printed wiring at the feed-through eyelets in the low-voltage sources. Just touching or pushing on the board can cause the raster or color to come and go. These intermittent wiring connections can be repaired with lengths of hookup wire fed through the eyelet and connected to the nearest corresponding component connection of the same broken circuit (Fig. 3-28). Low-resistance tests between broken wiring connections can help you to locate the intermittent connection.

B+ VOLTAGE ADJUSTMENT PROBLEMS

If adjusting the B+ control has no effect on the output voltage from the low-voltage power supply, suspect component breakdown within the power supply circuits. The B+ control will have no effect if the load is removed from the power supply. This is especially important to remember with the horizontal output transistor removed or if chassis or high-voltage shutdown occurs.

Improper adjustment of the B+ control can cause hum bars in the raster or insufficient image width. Incorrect setting of the B+ control can cause a very high voltage to be applied to the horizontal output transistor, resulting in high-voltage shutdown. In some chassis, an incorrect B+ control setting can cause the horizontal section to go out of sync (diagonal bars) instead of causing high-voltage shutdown.

Check for leaky or open regulator transistors in the regulator circuit of the low-voltage power supply with no B+ adjustment. Look for overheated zener diodes, which produce a lower regulated voltage. Measure the resistance of each resistor in crucial low-voltage regulator circuits. Always adjust the B+ control at the manufacturer's required operating voltage.

LIGHTNING DAMAGE

Lightning or high-voltage power-line surges can destroy several components in the low-voltage power supply. Usually, lightning damage occurs in the antenna lead or in the power-line circuits. A heavy lightning strike can damage several sections of the TV. Sections of printed wiring might be stripped as a result of lightning damage.

Check for burn or smoke marks on the chassis and around the ac wiring connected to the TV. Look for blown-apart components, such as transistors, ICs, and resistors. Burned silicon diodes and connecting board wiring can be signs of lightning damage. The receiver can be repaired if there is only a little damage to the power supply or antenna terminals. Total the cost of the TV if lightning damage is excessive and if the set cannot successfully be serviced.

HUM BARS

Hum bars of 60 Hz or 120 Hz can be caused by defective filter capacitors, leaky regulator transistors, or by an improperly adjusted B+ voltage. One dark bar moving up the raster represents 60 Hz, and two hum bars is a 120-Hz problem. Try to adjust the B+ control to the proper voltage setting before tearing into the chassis. Suspect that the large filter capacitors have dried up in the half- or full-wave rectifier circuits.

Check each regulator transistor for leaks if either 60-Hz or 120-Hz hum bars are present. Often the leakage is between the emitter and collector terminals of the regulator or APF amp in voltage-regulator circuits (Fig. 3-29). After locating one transistor with leakage, remove it and test it out of the circuit. Then test the other transistors for leakage with in-circuit tests. A leaky zener diode can also cause hum bars in the raster.

Shunt large filter capacitors with those that have the same capacity and voltage rating if hum bars are found in the raster. Always shut the chassis down and clip the filter capacitor in place to prevent damage to transistors or ICs. Clip across each filter capacitor until the defective section is located. Replace the entire filter can if more than one filter capacitor is in the same container—even if only one capacitor is defective.

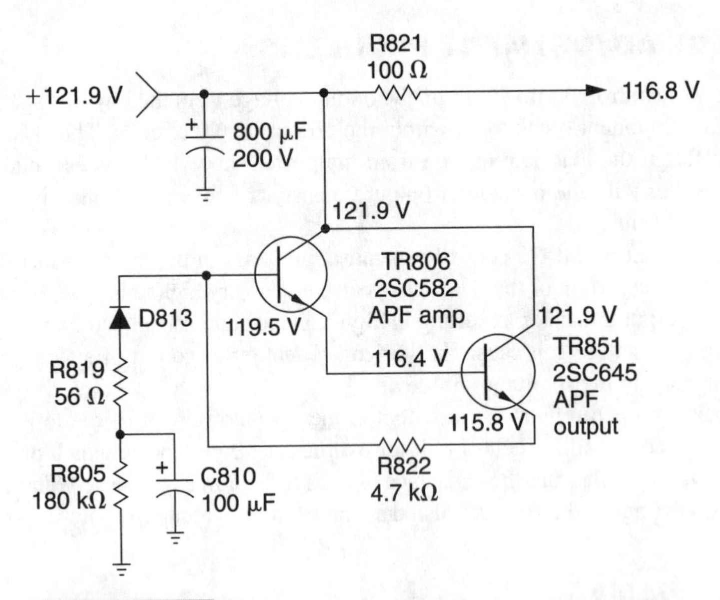

FIGURE 3-29 Hum bars can result from leaky regulator transistors in the low-voltage circuits.

CHASSIS AND HIGH-VOLTAGE SHUTDOWN

High-voltage shutdown can occur with too much voltage from the low-voltage power supply fed to the collector terminal of the horizontal output transistor. The chassis might shut down instantly or operate for a few minutes. Check the B+ voltage adjustment. Sometimes, if the B+ adjustment is too high and the line voltage increases, the higher dc voltage can cause the chassis to shut down.

Monitor the dc voltage at the collector terminal of the horizontal output transistor. Often a 10-V increase in voltage at the collector can cause high-voltage shutdown. Lower the dc voltage with a variable ac power transformer at the power cord. Set the transformer for correct voltage at the collector terminal. If the chassis shuts down, lower the ac voltage by about 10 V. Keep lowering the ac voltage as the dc voltage creeps up. If the chassis operates for 30 minutes without shutdown, notice the ac setting. If the chassis shuts down with a low ac voltage, suspect a defective horizontal output or shutdown circuit.

In an Admiral 4M10 chassis, the raster would collapse after operating for about 10 minutes. The B+ adjustment had no control over the high dc voltage (Fig. 3-30). The chassis would shut down at 131 Vdc and operate normally at 120 Vdc. Voltages were high at the collector and base terminals of the regulator transistor. After operating a few minutes, D902 (125-V zener diode) should become warm and allow the voltage increase to 130 V. Replacing D901 solved the shutdown problem. After replacing D901, the B+ adjustment would vary the dc voltage from 114 V to 130 V.

Leaky regulator transistors can cause chassis shutdown. Be sure that regular or zener diodes in the regulator circuits are not damaged. Check for a leaky regulator IC in many of the later TV regulator circuits. While the transistor is out of the circuit, check each resistor for correct resistance.

SHUTDOWN HORIZONTAL LINES

Suspect that the voltage is higher than normal at the collector terminal of the horizontal output transistor and low-voltage power supply if the raster displays horizontal lines. Measure the dc voltage at the low-voltage source (Fig. 3-31). In some early solid-state TVs, to prevent excessive high voltage at the picture tube, a shutdown circuit would make the pic-

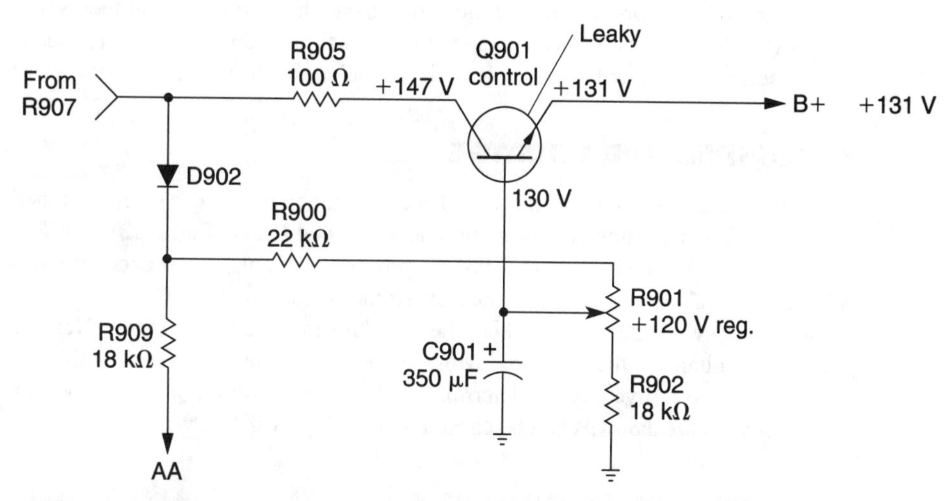

FIGURE 3-30 **A leaky Q901 caused the raster to collapse after 10 minutes of operation.**

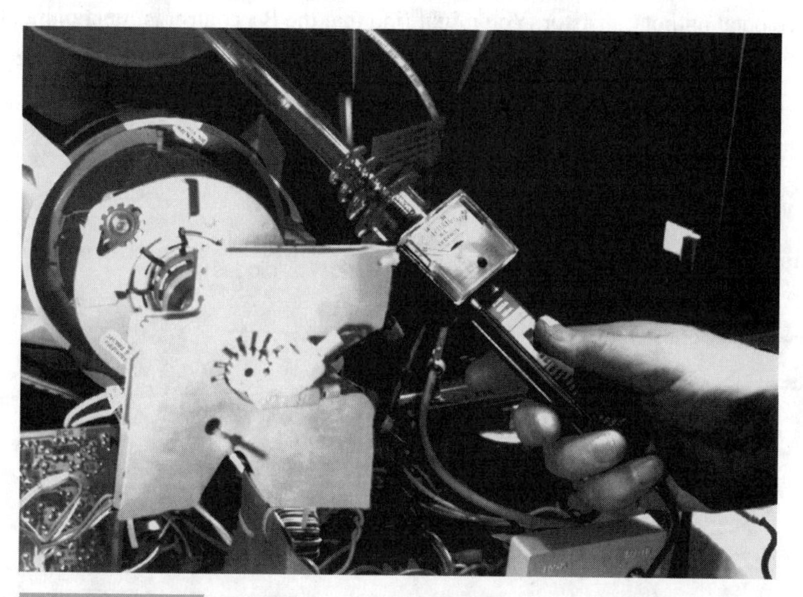

FIGURE 3-31 **Check the high voltage at the anode of the CRT if horizontal lines are present, indicating high-voltage shutdown.**

ture go into horizontal lines instead of complete chassis shutdown. These horizontal lines cannot be removed with the horizontal hold control as long as the chassis is in shutdown.

To determine if the horizontal oscillator circuits are defective or if the chassis is in shutdown mode, check the dc voltage applied to the horizontal output transistor. If the B+ adjustment will not turn the voltage down, suspect that the low-voltage regulator circuit is defective. First, check all transistors and diodes for leakage. Next, measure the resistance of each crucial resistor—especially those that have changed color and feel warm. Check electrolytic filter capacitors with a capacitor tester for a change in capacity. Suspect horizontal oscillator trouble if the dc voltage is normal at the horizontal output transistor.

EXCESSIVE LINE VOLTAGE

An excessive power-line voltage can destroy components and cause continued breakdowns. The ac power-line voltage should never exceed 125 V (Fig. 3-32). The TV was designed to work on a 117-Vac to 120-Vac power line. Usually, an excessive power-line voltage is found where the power line transformer is located nearby.

Check the power-line voltage where the TV plugs into the ac receptacle. Next, check the power-line voltage at the fuse box. Most power companies will monitor the power-line voltage for three to five days to determine if the voltage is high or low. Repeated service calls or chassis breakdowns can be caused by a high power-line voltage.

INSUFFICIENT IMAGE WIDTH

Often, insufficient width is caused by component breakdown in the horizontal output circuits. The raster can pull in from the sides if insufficient low-voltage power is applied to the horizontal output transistor. You might find that the B+ control is functioning, but the dc voltage is lower than normal and cannot be raised above the operating voltage.

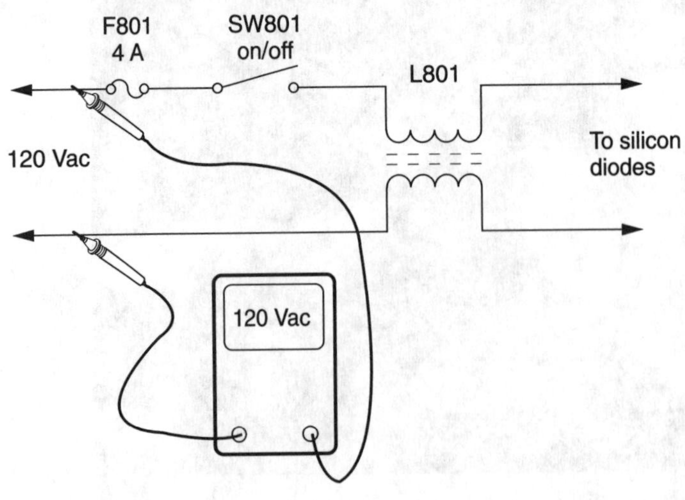

FIGURE 3-32 Check the ac voltage at the line input for normal 120 Vac.

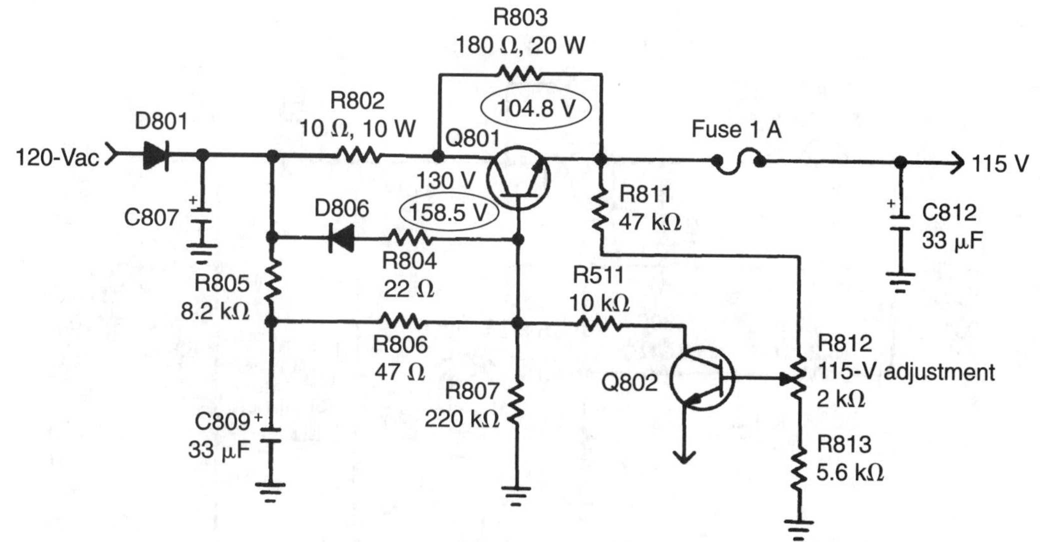

FIGURE 3-33 An open filter capacitor or regulator transistor can produce a low output voltage.

A leaky regulator transistor can cause an increase in the dc voltage source. Although an open transistor can lower the dc voltage, an open zener diode will let the voltage increase. Suspect that a regulator transistor is open or a filter capacitor is defective if the dc voltage cannot be returned to normal operation (Fig. 3-33). Sometimes these regulator transistors will test normal, but will open under actual working conditions.

UNUSUALLY LOW VOLTAGES

The 1.25-A fuse would blow in a K-Mart KMC1311G portable TV when the ac cord was plugged in. A voltage check at the horizontal output transistor indicated low voltage, then shutdown. A 100-W bulb was clipped across the fuse terminals. It operated with low sound and no raster. This indicated that the horizontal and low-voltage circuits were functioning, but without high voltage. A 2-A fuse was inserted and it blew at once. After two hours, a 3-A fuse was inserted. Right away, the picture tube started arcing in the gun assembly.

Although fuses with larger amperage should never be left in a TV chassis, the symptoms here indicated a defective component in the high-voltage section. The larger fuse was tried as a last resort. If left operating too long, the horizontal output transistor could have been damaged. Use this method only to smoke out a defective component and only for a few seconds. Installing a new picture tube solved the repeated blown-fuse problem.

"TOUGH DOG" POWER SUPPLY

Anyone who has spent hours trying to find the defective component in a TV will not forget the problem by the next time the same type of chassis comes across the service bench.

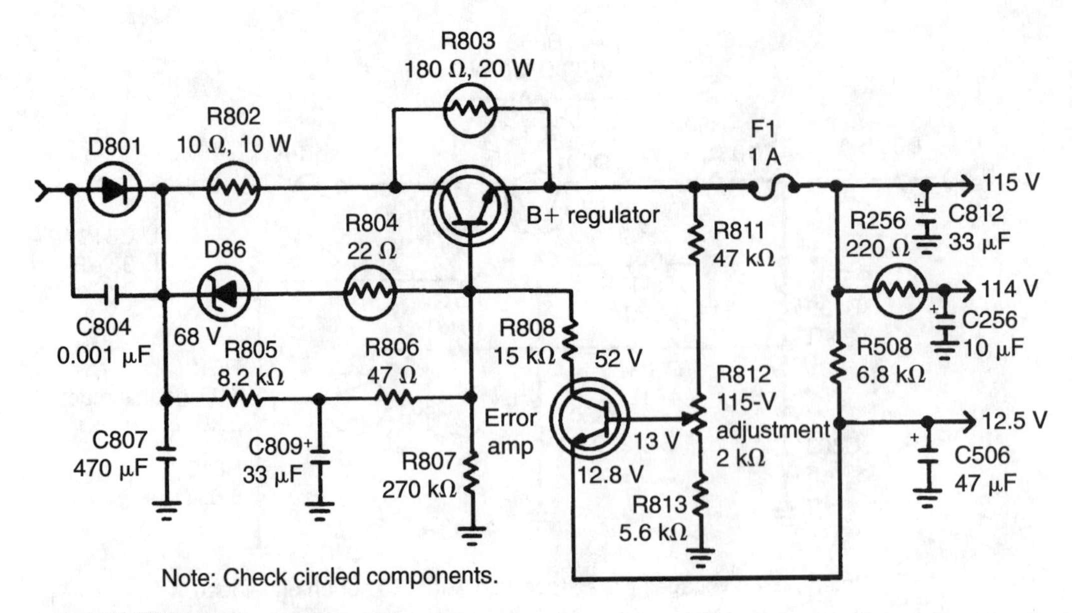

FIGURE 3-34 **This Goldstar CNR-405 chassis required a lot of service time to locate all of the defective components in the low-voltage circuits.**

You might have one set that bothers you, but it might not be a service problem for the next technician. Do not blame the manufacturer for a these problems.

The first time that a Goldstar CNR-405 chassis appeared on the service bench, it was classified as a "tough dog" repair of the power supply circuit. Not only was servicing the chassis without a schematic a great problem, but several different components were found to be defective (Fig. 3-34). Now, servicing the same power supply circuit is a cinch and it takes very little service time.

Check the B+ regulator (2SC1629) and error amp (2SC1573P) for leakage conditions. Doublecheck for leakage after removing them from the circuit. The B+ regulator can be replaced with a universal ECG86 or an SK3563.

Zener diode D86 can become leaky and produce a higher-than-normal output dc voltage. When both D86 and the regulator transistor become leaky, the dc source can go to 143 V, producing high-voltage shutdown. Check fuseable resistor R804. Other crucial power resistors in the circuit of the low-voltage regulator transistor are R802 and R803.

STANDBY POWER SUPPLIES

If the remote transmitter will not turn on the TV and the remote is normal, suspect that the standby power supply is defective. The standby power supply is found in many of the new chassis to provide some means of turning the TV off and on by remote control.

In older remote TVs, the standby power was turned on with relays from a separate power supply. Today, the standby circuits are switched on with switching transistors or ICs (Fig. 3-35). The IR remote transmitter signal is picked up by a photo transistor and fed to a receiver circuit within the mechanism IC.

SERVICING STANDBY CIRCUITS

First, take crucial voltage measurements and observe the waveform of each component in the standby circuitry. Place a jumper across the emitter and collector terminals of the mode switch and notice if the chassis turns on. Test each transistor and diode with in-circuit tests. If you notice leaky or open measurements, remove the collector terminal of each transistor from the circuit. Likewise, check the suspected leaky diodes in the same manner. Remember, most of the stages in the standby power circuits are either on or off.

FIVE NEW LOW-VOLTAGE POWER SUPPLY PROBLEMS

Here are five problems that occur in some common TVs.

Fuse keeps blowing Fuse F802 (1.2 A) would keep blowing after replacement in a Samsung TC-2540 TV. All components in the 130-volt source tested normal. No B+ voltage was noted at the collector terminal of the horizontal output transistor (Fig. 3-36). High leakage was found from the collector terminal of output transistor. Q402 was tested out of the circuit and was normal.

Capacitor C414 (0.36 μF) was leaky. Replace both C414 and C415 if either one is shorted or leaky. In another chassis, Q402 and C414 were replaced to prevent F802 from opening.

Intermittent RCA FPR2722T TV Sometimes the chassis would operate for days before quitting; when the chassis was touched or bumped, it came back on. The raw ac voltage

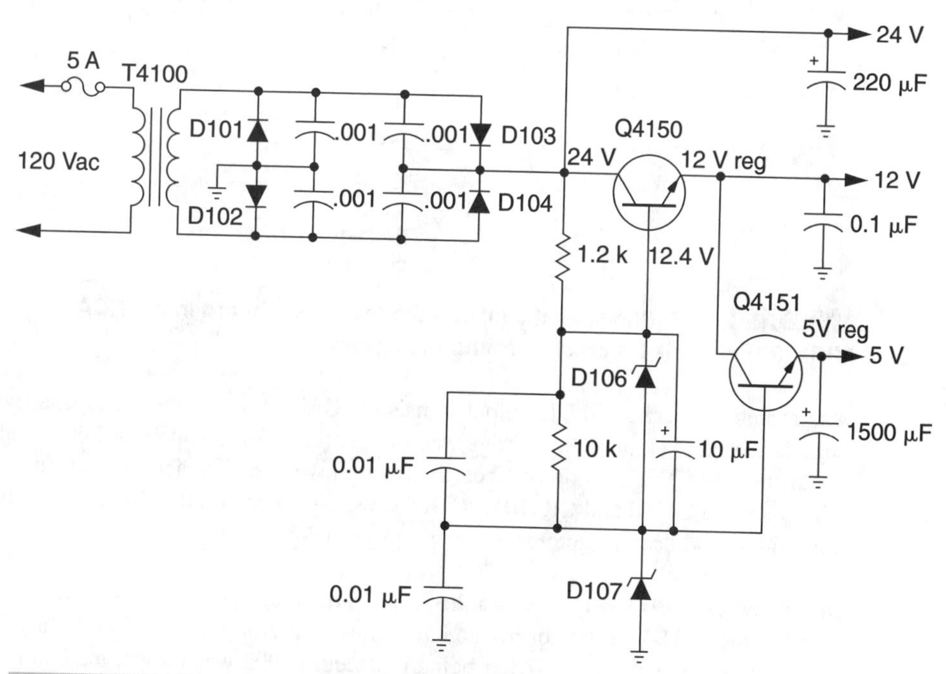

FIGURE 3-35 A typical standby power supply in the latest TV chassis.

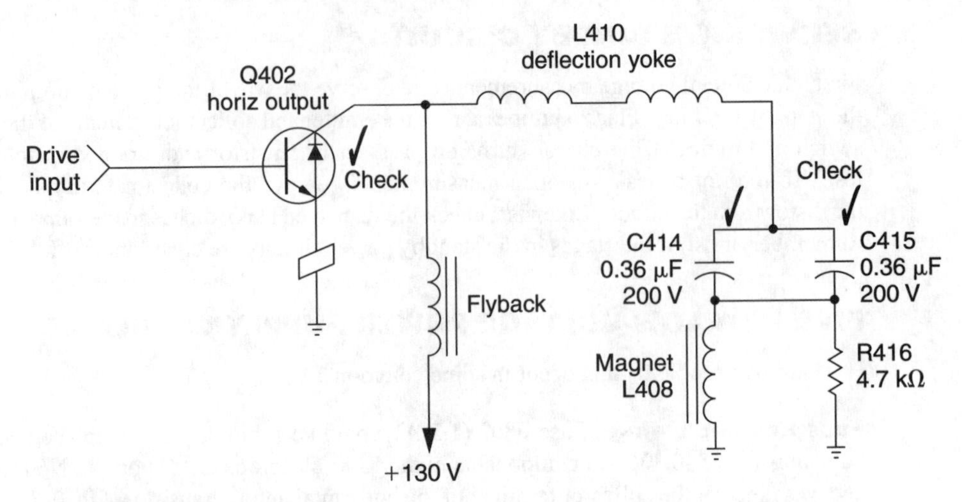

FIGURE 3-36 Check capacitors C414 and C415 for leakage if the main fuse blows and the relay clicks on.

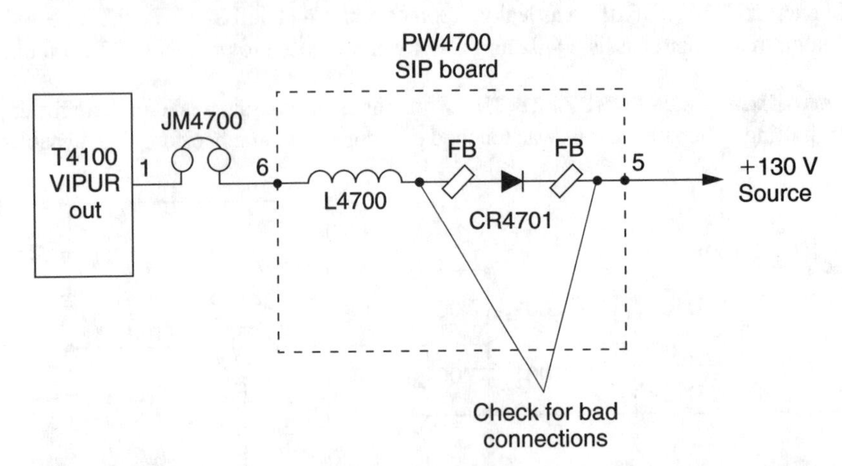

FIGURE 3-37 Check all the silicon diodes on SIP board in the RCA FPR2722T chassis for cracks around the terminals.

was monitored at the VIPUR output transistor (Q4100) and appeared normal when the chassis ceased operation. No voltage was found at the 130-V source or horizontal output transistor. PW4700 (a diode SIP board) was removed and a crack was found in the PC wiring around L4700 and CR4701. All leads were soldered on the SIP board, solving the intermittent and dead symptoms in several places (Fig. 3-37).

Dead Quasar ALCD177 chassis After replacing the main fuse, no voltage was found at the +131-V source or horizontal output transistor. R801 and F002 (a 1.5-A secondary fuse) were also open. After being replaced, F1002 was blown, indicating that the IC regulator or horizontal output transistor was overloaded. The horizontal output transistor tested normal with a leaky IC801. Replacing the STR30130 line-voltage regulator solved the dead-chassis problem (Fig. 3-38).

A blown fuse in a dead Sony SCC-548D portable F001 (a 5-A fuse) opened again after replacement, indicating that the low-voltage power supply was defective or that the horizontal output transistor was leaky. Q502 (2SD1555) was shorted in the horizontal output circuits. The secondary fuse F602 (1.25 A) was replaced with the same results (Fig. 3-39). Low voltage was measured at filter capacitor C602 (560 µF, 200 V), but not on the

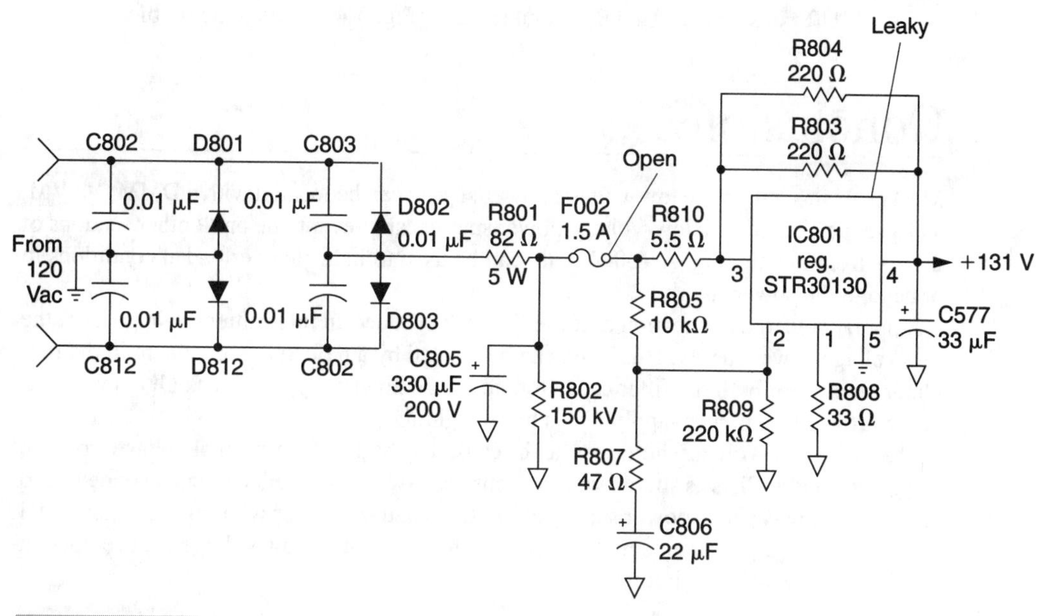

FIGURE 3-38 R801 and F002 were found open in a dead Quasar ALCD177 chassis. The cause was a leaky IC801 line-voltage regulator.

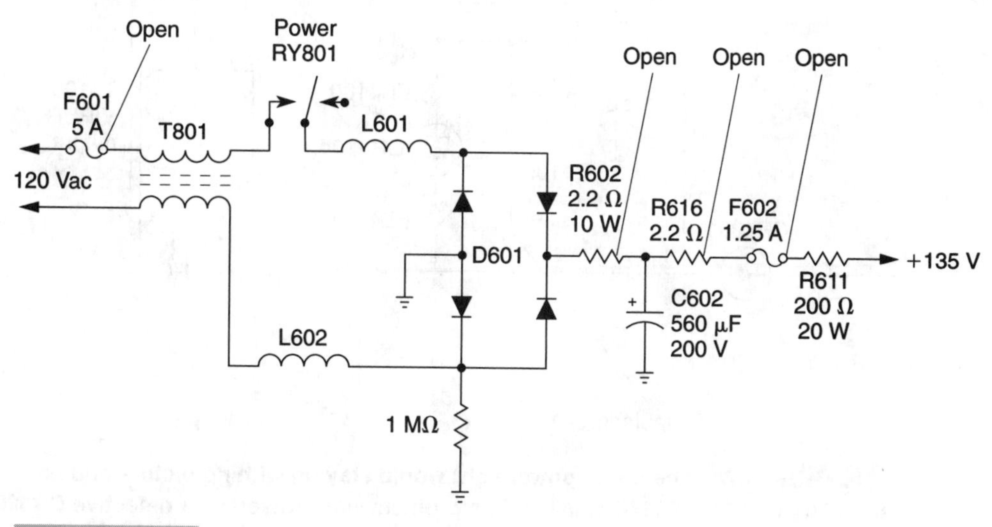

FIGURE 3-39 A shorted horizontal output transistor (Q502) in a Sony SCC548D chassis opened resistors R616, R602, and R603 in the low-voltage circuits.

fuse terminal of F602. R616 (2.2 Ω) was also found open. Replacing F601, F602, R616, and Q502 put the TV back into operation.

Green light on—No picture In an RCA CTC140 chassis, the green power light would not turn off, with a no-sound/no-picture/no-HV symptom. Voltage measurements in the VIPUR circuits were fairly normal. Replacing both CR4104 and CR4201 in the leg of the VIPUR transformer (T4100) output circuits (Fig. 3-40) solved the problem.

Conclusion

Most problems within the low-voltage power supply can be located with a DMM, VTVM, and a scope. A defective low-voltage power supply can prevent one or all other sections of the TV from functioning. A 100-W bulb can be used to indicate overloaded conditions in or beyond the power supply.

A poorly soldered connection or loose fuse can produce an intermittent voltage from the low-voltage power supply. Hum bars can be created by a defective component in the regulator circuits, or by large filter capacitors in the dc power supply. Check all components in the low-voltage power supply if it is hit by lightning.

Chassis or high-voltage shutdown can be caused by higher-than-normal voltage from the dc power supply. Chassis shutdown can occur because of overloaded circuits connected to the rectifier low-voltage power supply of the flyback transformer winding. The horizontal and horizontal output circuits must function before secondary low voltage is developed in the flyback low-voltage circuits.

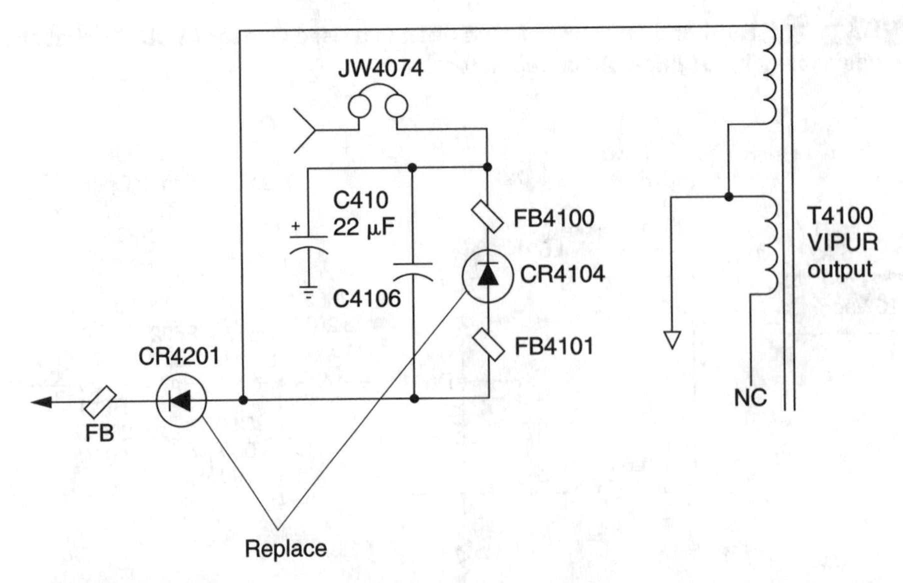

FIGURE 3-40 **The green power light would stay on with no picture and no sound in an RCA CTC140 chassis. The problem was caused by a defective CR4201 and CR4104.**

TABLE 3-1 USE THIS POWER SUPPLY TROUBLESHOOTING CHART FOR EASY SERVICING

WHAT TO CHECK	HOW TO CHECK IT
Inspect line fuse.	Test in circuit with ohmmeter.
Check voltage across filter capacitor.	Test silicon diodes and isolation resistors.
Check output voltage at regulators.	Check transistor regulator and zener diodes.
Check horizontal fuse to horizontal output transistor	Check fuse with ohmmeter if soldered in circuit.
Check voltage at horizontal output transistor.	Open isolation resistor or coil and flyback winding.
Overloaded circuits.	Remove horizontal output transistor from chassis. (Voltage should be normal or a little higher with no output transistor in circuit.)

Always remember that defective components in the power-supply circuits can cause different symptoms in other connecting circuits. Do not overlook the possibility that a filter capacitor is defective in the power supply. These capacitors cause strange things to happen in other circuits. If in doubt, shunt or bypass each filter capacitor to help isolate power-supply problems. Carefully visually inspect the components and wiring to locate the defective part. Check Table 3-1 for a power-supply troubleshooting chart that enables easy servicing.

4

TROUBLESHOOTING THE HORIZONTAL SWEEP CIRCUITS

CONTENTS AT A GLANCE

CONTENTS AT A GLANCE *(Continued)*

The most troublesome section of the solid-state TV chassis is the horizontal sweep section; it also has the most interesting circuits (Fig. 4-1). After mastering horizontal sweep circuits, most other electronic circuits are a cinch. New service techniques must be used to ensure quick and proficient servicing methods. Servicing the horizontal section can be fun—and very frustrating, at times.

Although many of the regular horizontal oscillator and output symptoms are the same, several new circuits (which have the same symptoms) are used in today's solid-state chassis. Horizontal drifting and off-frequency symptoms can be caused by the horizontal oscillator circuits. The horizontal oscillator can be contained in one IC with both a horizontal driver and vertical circuits. You might find a horizontal and vertical frequency-countdown circuit in one IC.

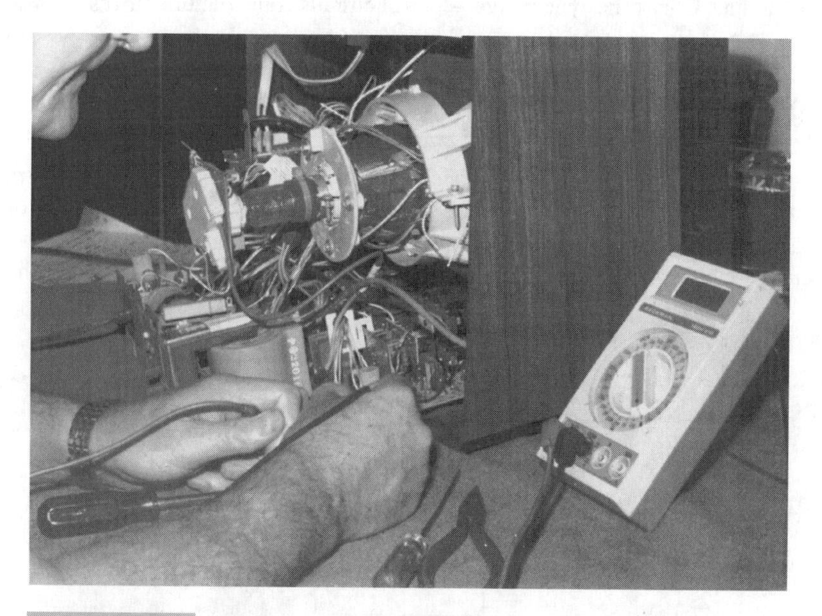

FIGURE 4-1 Eighty-five percent of TV chassis problems are found in the horizontal and vertical circuits.

The no-raster/no-sound/no-high-voltage symptom can be caused by just about any component in the horizontal sweep or low-voltage stages. The no-raster, good-sound symptom can be caused by a defective component in the high-voltage circuits. High-voltage shutdown can be caused by the HV shutdown circuits, excessive low voltage, and defective components in the horizontal output circuits. Chassis shutdown can be caused by a breakdown in the low-voltage or horizontal-sweep sections.

The integrated flyback can produce arcing when self-enclosed high-voltage diodes shut down the chassis. The horizontal output transistor can be quickly destroyed by the flyback before the fuse opens. With the new integrated flyback transformer, a breakdown in one of the low-voltage circuits tied to one of the extra windings of the flyback can load down the transformer, causing chassis shutdown. Today, the flyback transformer provides voltage to many separate circuits and additional waveforms to other circuits.

Horizontal Oscillator Problems

A dead TV chassis can be caused by an open or leaky horizontal oscillator transistor or IC. High-pitched sound can indicate that the horizontal oscillator is off frequency. After several seconds, the horizontal oscillator might drift off frequency because of a defective transistor, IC, or corresponding components. If the TV comes on with horizontal lines, the horizontal hold control might need adjusting, the horizontal oscillator circuit might be defective, or the horizontal stage has been shut down by a high-voltage shutdown circuit (Fig. 4-2). Excessive pulling at the top of the picture can be caused by improper sync or filtering of the low-voltage circuits. Check the sync or sync-control circuits if the picture moves sideways.

Horizontal Oscillator Circuits

The horizontal oscillator stage can be checked by looking at the sawtooth waveform at the output terminal. The circuit might be one transistor in a Colpitts oscillator circuit or an IC with oscillator and driver stage inside one component. In the latest IC circuits, a vertical-horizontal countdown IC can contain a horizontal sync separator, AFC detector, horizon-

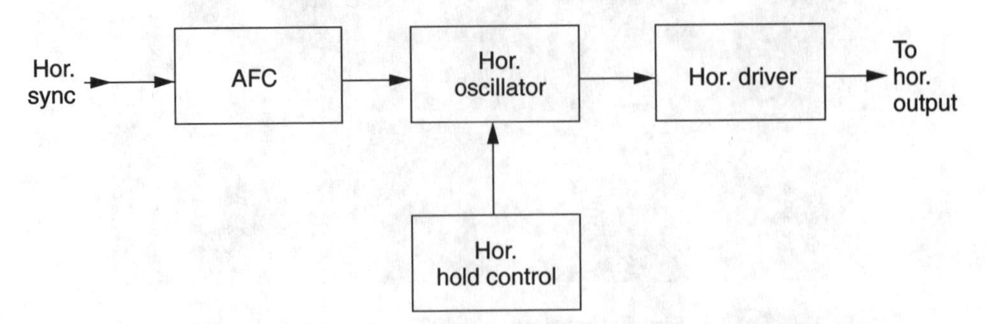

FIGURE 4-2 The horizontal circuits consist of the AFC, horizontal oscillator or countdown, horizontal driver, and horizontal output stages.

tal oscillator flip-flop, horizontal amp, and x-ray high-voltage protection, as well as vertical, luminance, and color circuits.

The horizontal frequency in a simple oscillator circuit (Fig. 4-3) is controlled by L601, C609, and C610. L601 is the horizontal hold control. If the horizontal oscillator drifts off frequency, suspect Q601, C609, C610, and L601. Spray Q601, C609, and C610 with coolant and notice if the horizontal lines flop into a picture. Replace C609, C610, and L601 with original replacement parts.

During the past 10 years, most horizontal oscillator transistors have been replaced by ICs (Fig. 4-4). Here C605 is charged and discharged repeatedly to produce the horizontal oscillator frequency. The charge time constant is controlled by R601, R603, R607, and R609. A feedback pulse from the flyback transformer passes through R605 and C605 and is turned into a sawtooth waveform that feeds the AFC circuits at pin 35. The sawtooth signal is compared to the horizontal AFC circuit output and kept in phase with the horizontal pulse waveform at pin 24, feeding the horizontal drive transistor (Fig. 4-5).

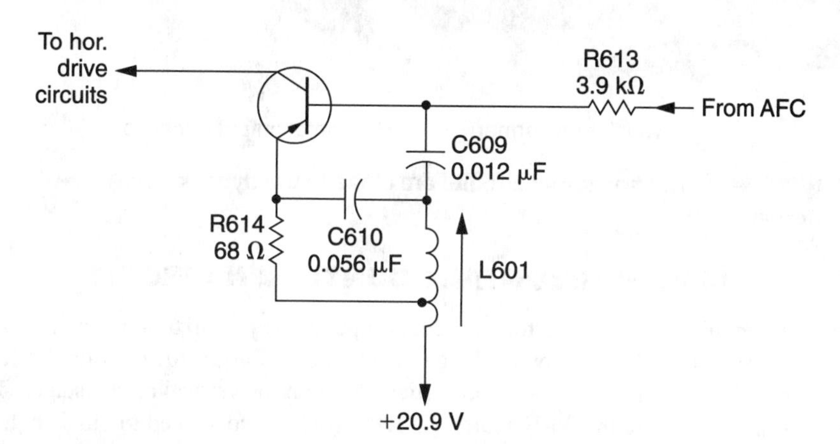

FIGURE 4-3 **In early low-priced TVs, a simple horizontal oscillator circuit incorporated a plastic shaft as a horizontal hold control.**

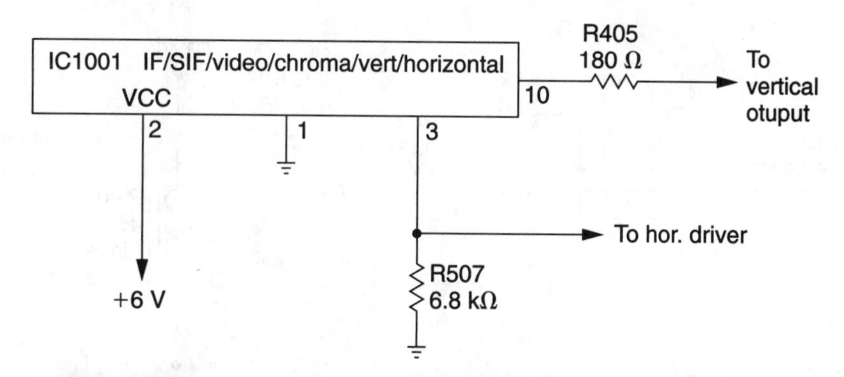

FIGURE 4-4 **Today's horizontal oscillator or deflection circuits are often located in one large IC with IF/video/chroma and vertical/ horizontal control circuits.**

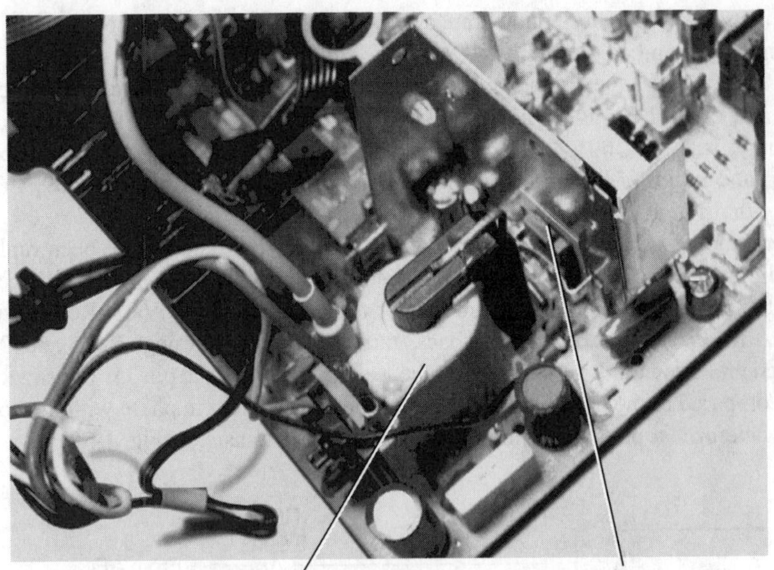

FIGURE 4-5 The horizontal circuits are close to the flyback transformer.

RCA CTC130C HORIZONTAL COUNTDOWN CIRCUIT

Integrated circuit (IC) U401 performs the dual operations of horizontal synchronization (oscillator) and vertical countdown. Along with other functions, U401 provides horizontal drive for the horizontal output stage and vertical drive for the vertical output stage. The IC (U401) is powered from the 26-V source via R116 (680 Ω) connected to pin 5 (Fig. 4-6). Internal to the IC, pin 5 is connected to a 10-V shunt regulator.

To determine if the horizontal oscillator is working, scope pin 16 of U401. If the chassis is shut down and will not start, the horizontal oscillator should be checked by injecting

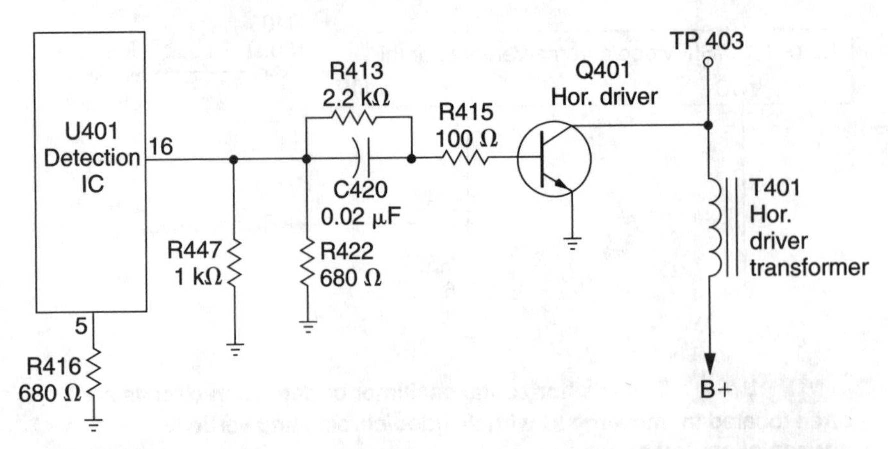

FIGURE 4-6 The horizontal countdown circuit in the RCA CTC130C chassis.

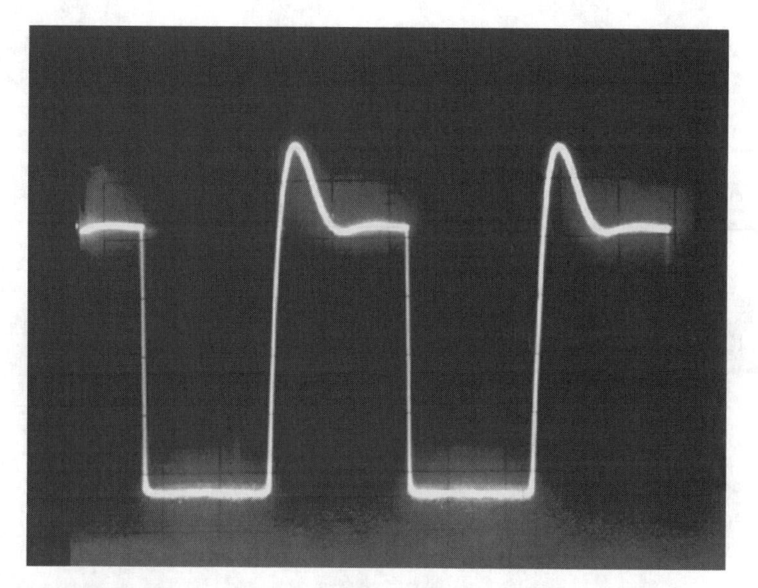

FIGURE 4-7 The horizontal output waveform at pin 16 of U401 in the RCA CTC130C chassis.

26 V at pin 5 and scoping the horizontal drive at pin 16 (Fig. 4-7). Check the horizontal waveform with that shown on the schematic. If there is no output waveform, replace U401.

HORIZONTAL OSCILLATOR IC TESTS

Take crucial waveform test measurements of the horizontal IC oscillator, sync, and output pulse waveforms. Check pin 17 for the horizontal oscillator drive pulse to the horizontal driver transistor (Q601) (Fig. 4-8). Check for proper horizontal sync pulse from the sync separator to pin 16. Check the AFC waveform at pin 14. Improper waveforms at any of these pin numbers can indicate a defective IC.

Go a step further and take crucial voltage measurements on the horizontal oscillator (pin 19) (12.5 V). A very low voltage at pin 19 can indicate that the IC is leaky. Check crucial voltage test points at pins 17, 19, 15, and 18. If any pin has a low-voltage measurement, disconnect the pin or component tied to that pin number to see if it is loading down the suspected IC. Sometimes voltage measurements can be quite close—even if the IC is defective. Replacing the IC is the only answer.

POOR HORIZONTAL SYNC

Although horizontal and vertical sync problems are covered in Chapter 7, the horizontal sync covered here determines if a horizontal hold problem is found in the horizontal circuits. Check the AFC diodes by removing one end from the circuit and looking for leakage (Fig. 4-9). Remove one end of each resistor above 50 kΩ in the AFC circuit to make accurate resistance measurements. These large resistors have a tendency to increase in resistance or go open. Check each small capacitor with a capacitance meter.

Short out the AFC diode to ground to determine if the AFC circuits or the oscillator is causing the sync or off-frequency problem. Readjust the horizontal hold control to try and

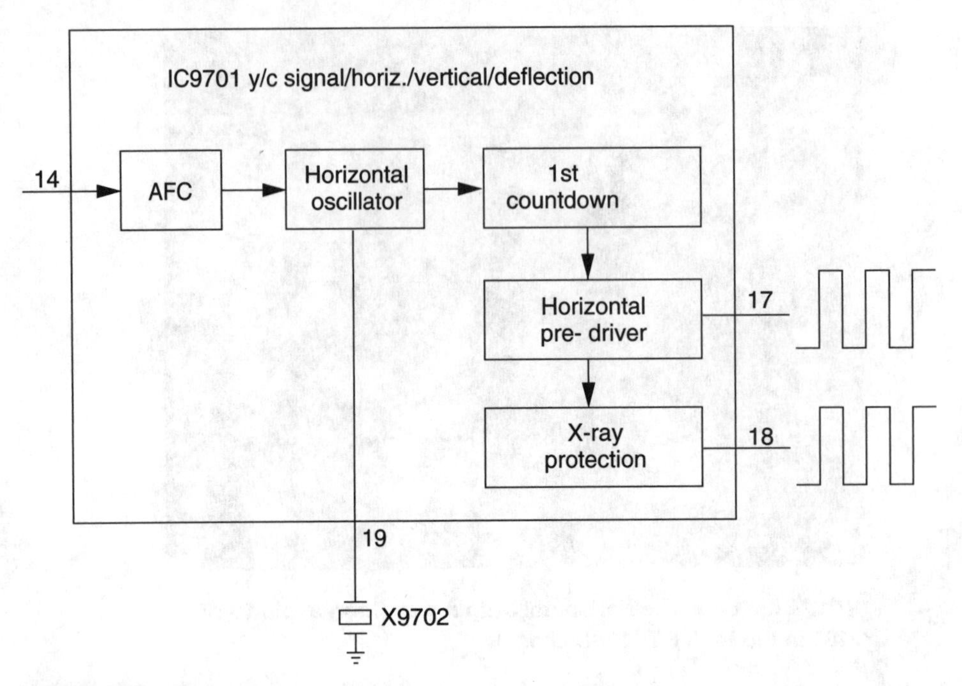

FIGURE 4-8 **A waveform at pin 17 and 18 can indicate if the oscillator section is working.**

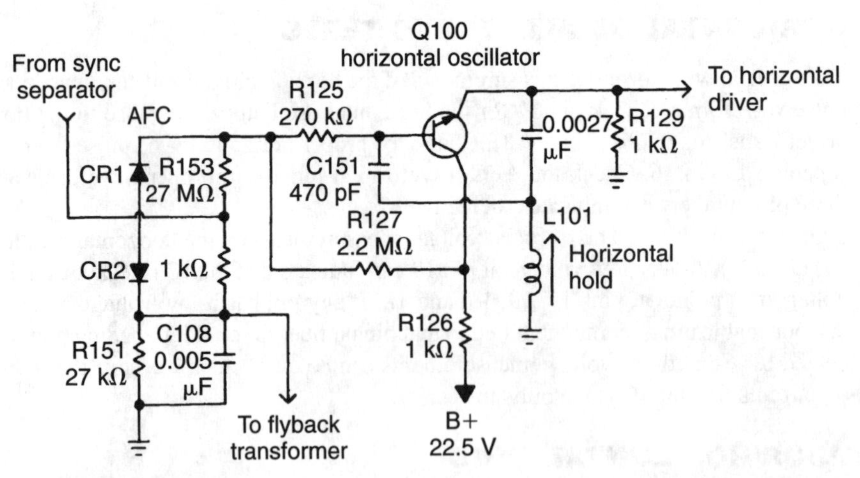

FIGURE 4-9 **Poor horizontal sync can be caused by AFC diodes CR1 and CR2.**

lock in a picture. If the picture is fairly stable, the oscillator circuits are good. Now check the AFC and sync input circuits for lack of horizontal sync. Scope the flyback waveform fed back to the horizontal oscillator in some chassis (Fig. 4-10).

In today's TV, horizontal and vertical sync circuits are included inside one large IC with video, brightness, color, tint, horizontal and vertical sweep, and x-ray protector circuits. Check the input waveform of the sync separator terminal on the IC. If the sync is normal here, suspect that an IC is defective or that the voltage source at the supply pin is improper. This signal-processor IC can cause poor vertical and horizontal sync while other circuits are normal. Often, poor sync is found with a poor video picture.

CRUCIAL HORIZONTAL WAVEFORMS

The horizontal sweep waveform can be traced from horizontal oscillator or countdown IC to the collector output terminal of a horizontal output transistor with the oscilloscope. Figures 4-11A to 4-11D, represent the sweep output waveform from countdown sweep IC. The drive waveform is applied to the base of driver transistor, then is amplified and applied to the primary winding of the driver transformer. This drive signal is applied to the

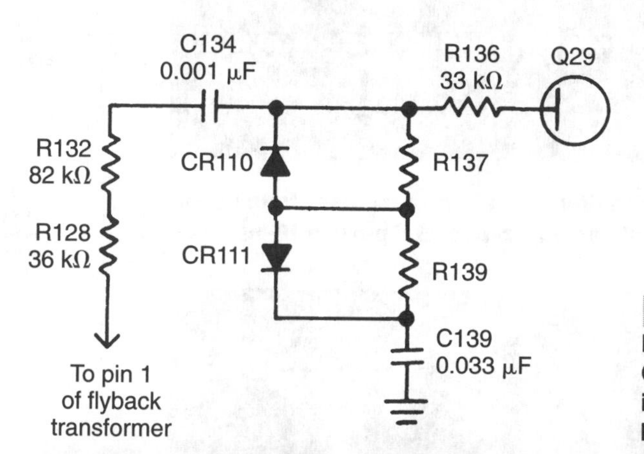

FIGURE 4-10 Poor horizontal sync in the RCA CTC97 chassis resulted in insufficient waveform at the base of Q29.

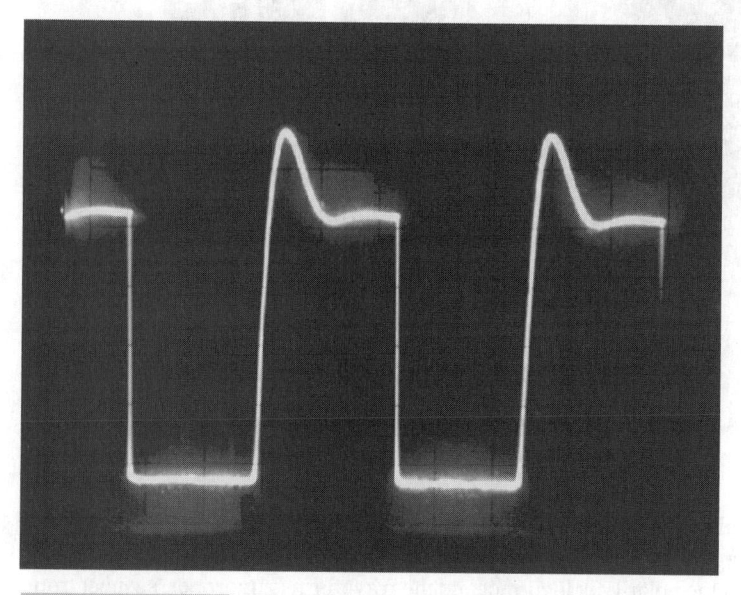

FIGURE 4-11A The horizontal pulse at the base of the horizontal driver transistor (1 V p-p to 4.5 V p-p in a 19-inch TV).

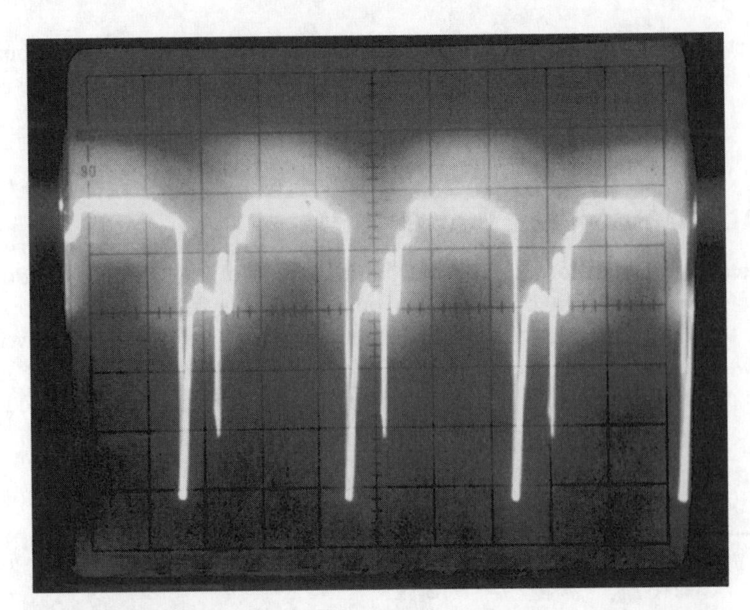

FIGURE 4-11B The drive waveform at the base terminal of horizontal output transistor (14 V p-p to 25 V p-p in a 19-inch TV).

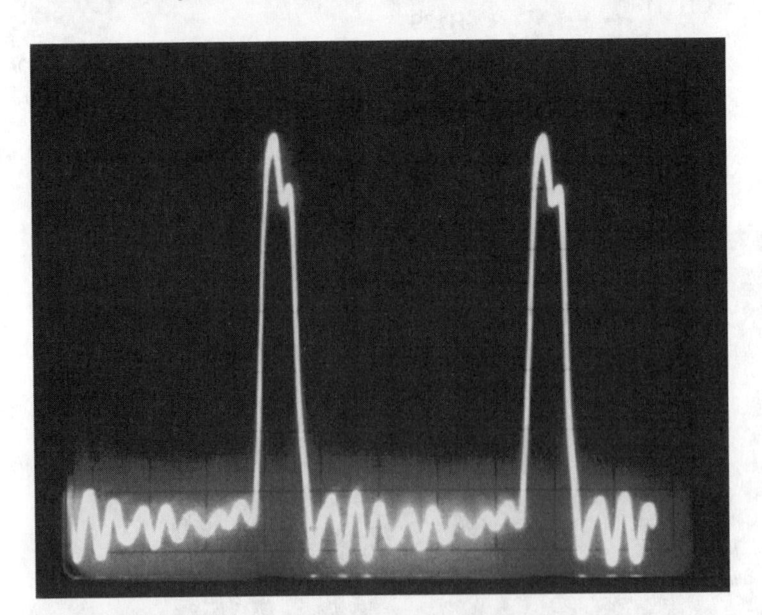

FIGURE 4-11C The high-voltage waveform found by placing scope probe near the flyback transformer.

base terminal of the horizontal output transistor, then is amplified and applied to the primary winding of the horizontal output transformer or flyback. If one of the waveforms is missing, or if the polarity is incorrect and the waveform is improper, suspect trouble within that horizontal circuit.

HORIZONTAL OFF-FREQUENCY PROBLEMS

Most TV technicians can hear the horizontal oscillator when it has drifted off frequency. Additionally, horizontal lines appear on the screen and there is no horizontal hold control. Sometimes the oscillator will drift off after operating for a few minutes. This indicates that a component is getting warm. Check the horizontal transistor or IC for leaky conditions. Replace the resonating and coupling capacitors in the emitter-tapped oscillator coil circuit (Fig. 4-12).

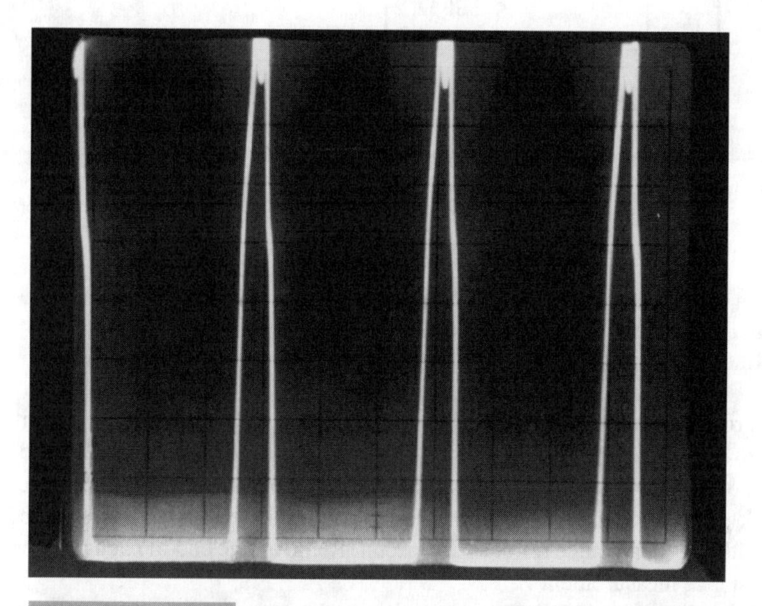

FIGURE 4-11D The horizontal output waveform applied to the primary winding of the horizontal output transformer (70 V to 250 V p-p in a 19-inch TV).

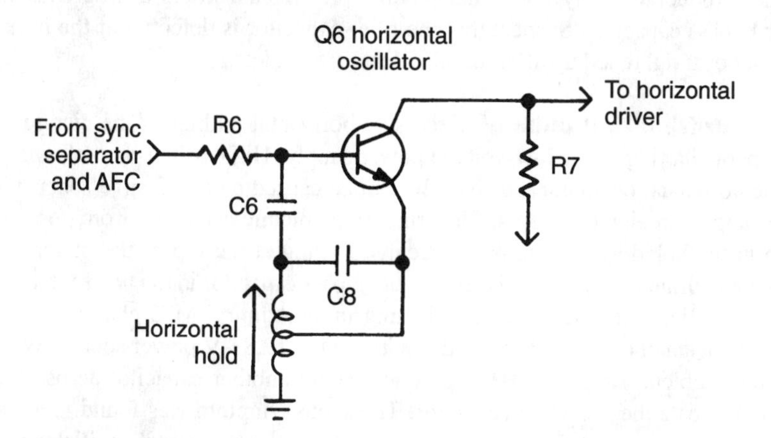

FIGURE 4-12 Check capacitors C6 and C8 in the horizontal coil circuit if the circuit drifts off the horizontal frequency.

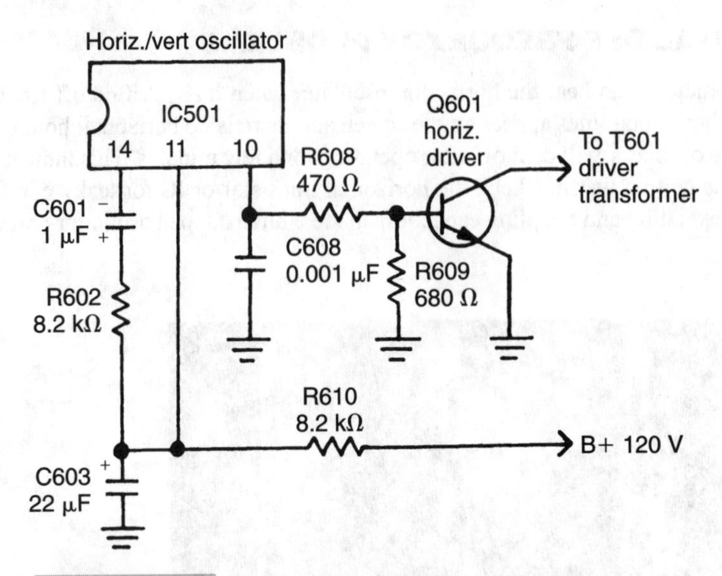

FIGURE 4-13 **Clip another capacitor across the one that you suspect and notice if the picture returns in a Sharp C1355A chassis.**

Suspect problems in the IC, small filter capacitor, and RC time-constant circuits if the horizontal oscillator is off frequency. First, adjust the horizontal hold control. Take a quick voltage test on the supply voltage and horizontal oscillator pins of the IC. Replace the IC. Next, check each small electrolytic capacitor in the circuit (Fig. 4-13). Test each bypass capacitor with the capacitance tester in the circuit. Remove one end of each resistor for the correct resistance measurement.

A dried-up or open electrolytic filter capacitor in the power source feeding the horizontal circuits can cause poor horizontal lock or an off-frequency condition. Sometimes the electrolytic capacitors tied to the horizontal oscillator circuit do not filter out the sync or hash signal, producing funny horizontal pictures. Horizontal jitters can be caused by a poor filter bypass capacitor. Suspect that the filter capacitor is defective if the horizontal lines cannot be straightened up after all other tests are made.

Horizontal drifting and pulling Excessive horizontal pulling of the picture can be caused by poor filtering in the low-voltage power supply. Higher-than-normal voltage applied to the horizontal oscillator transistor IC can be caused by a defective zener diode in the power supply circuits. Horizontal flickering of the picture can result from poor voltage regulation in the holddown or regulator circuits. Pulling at the top of the picture can be caused by the malfunction of a small electrolytic bypass capacitor in the horizontal circuits.

Horizontal pulling with the frequency drifting in an Admiral M25 chassis (Fig. 4-14) was caused by a small electrolytic filter capacitor in the 19.83-V power source. When addressing this problem, scope the B+ source and shunt another capacitor across the suspected one to locate the defective capacitor. The same symptom was found in a General Electric CD chassis, except that the voltage applied to the horizontal oscillator was too high (Fig. 4-15). Often, with dried-up filter capacitors, the voltage source is too low. Here

34.1 V was found at the collector terminal of the horizontal oscillator transistor (Q502). Replacing the defective 22-V zener diode with a 5-W type solved the horizontal pulling symptom. Do not overlook the possibility that horizontal pulling and lines is caused by excessive high voltage at the picture tube.

Can't tune out the lines Suspect that the horizontal oscillator is off frequency or that there is poor filtering in the low-voltage source if the horizontal lines cannot be adjusted out of the picture. The picture might come on with uncontrollable lines or, after several minutes, go into horizontal lines that cannot be straightened up. Determine if the horizontal line problem is in the oscillator circuits, power supply, or other sources. Monitor the voltage at the power source. If the voltage is either higher or lower than normal, check the low-voltage source that feeds the horizontal oscillator circuits (Fig. 4-16). Check the low-voltage circuits for poor voltage regulation.

Check the high-voltage and shutdown circuits for excessive high voltage. Simply measure the voltage at the picture-tube anode terminal. In the early high-voltage shutdown circuits, if the high voltage went too high, the horizontal circuit was shut down and all you could see were horizontal lines. An improper setting of the HV or B+ control can cause the same condition. Raise and lower the brightness control and notice if the picture goes into horizontal lines, indicating that too much high voltage is at the picture tube.

In the RCA CTC46 and CTC48 chassis (Fig. 4-17), high voltage (greater than 30 kV) at the picture tube can cause the shutdown circuits lines in the picture. Sometimes adjusting the brightness control can raise or lower the high voltage, causing the raster to bloom with a picture and then go into horizontal lines. Often the horizontal-line symptom is caused by a leaky zener diode (CR5) or HV regulation transistor. Poor regulation transformer board connections can cause horizontal lines with excessively high voltage.

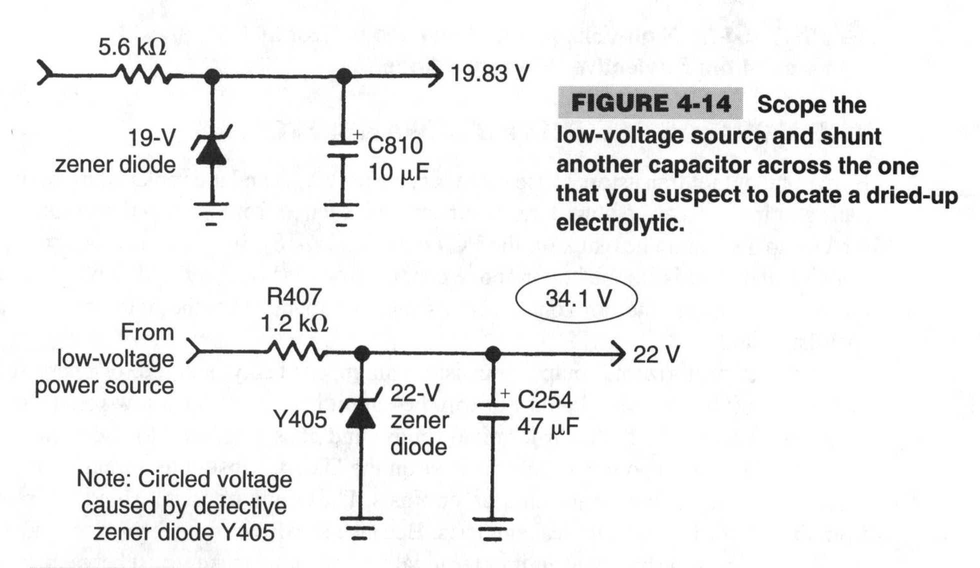

FIGURE 4-14 Scope the low-voltage source and shunt another capacitor across the one that you suspect to locate a dried-up electrolytic.

FIGURE 4-15 A defective Y405 caused a higher dc voltage than specs, resulting in horizontal lines in the raster.

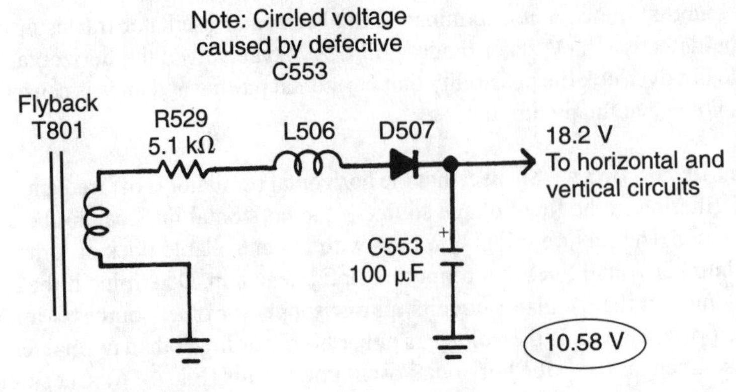

FIGURE 4-16 Low voltage in the scan-derived flyback source produced horizontal lines in a Sony 1500.

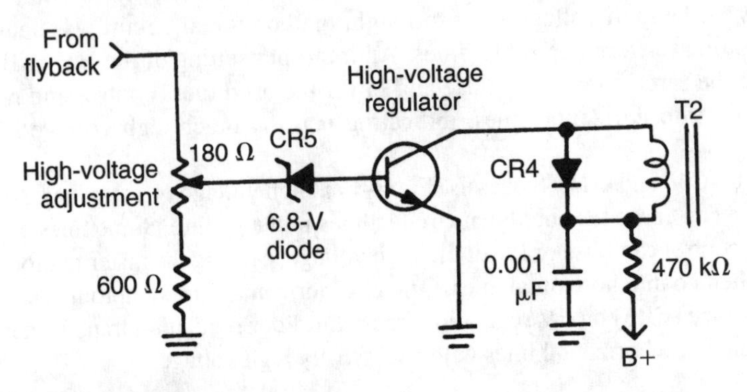

FIGURE 4-17 High-voltage shutdown and horizontal lines result from a defective CR5 zener diode.

THE HORIZONTAL OUTPUT TRANSISTOR

Horizontal output transistors cause more service problems and are replaced more than any transistor in the TV. The output transistor can be insulated from the metal chassis or bolted directly to a separate heatsink on the PC board (Fig. 4-18). In some chassis, the separate heatsink and chassis can be hot or above chassis ground (Fig. 4-19). All voltage measurements made around the horizontal circuits must be made from the hot chassis—not the chassis ground.

The defective horizontal output transistor can appear leaky, shorted, or open. A leaky output transistor can blow the fuse or trip the circuit breaker. Take a low-resistance measurement between the collector terminal (body) and chassis ground to check for a leaky transistor. Only a few ohms should register on the DMM. Most output transistors leak or short between the collector and emitter terminals. The output transistor should be removed from the circuit for accurate leakage tests. Because the driver transformer secondary terminals are across the base and emitter terminals, an accurate leakage test between base and emitter cannot be made (Fig. 4-20).

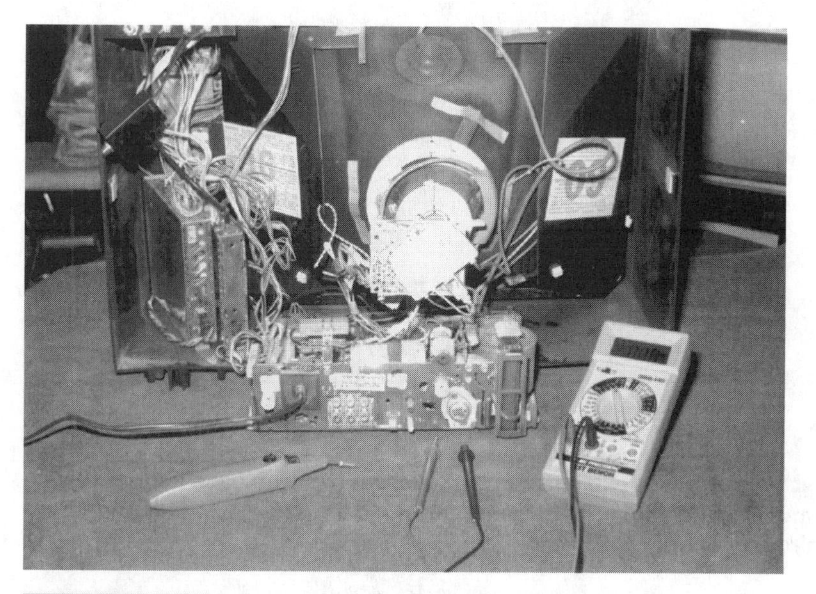

FIGURE 4-18 The horizontal output transistor is bolted to the metal chassis or heatsink in the RCA CTC108 chassis.

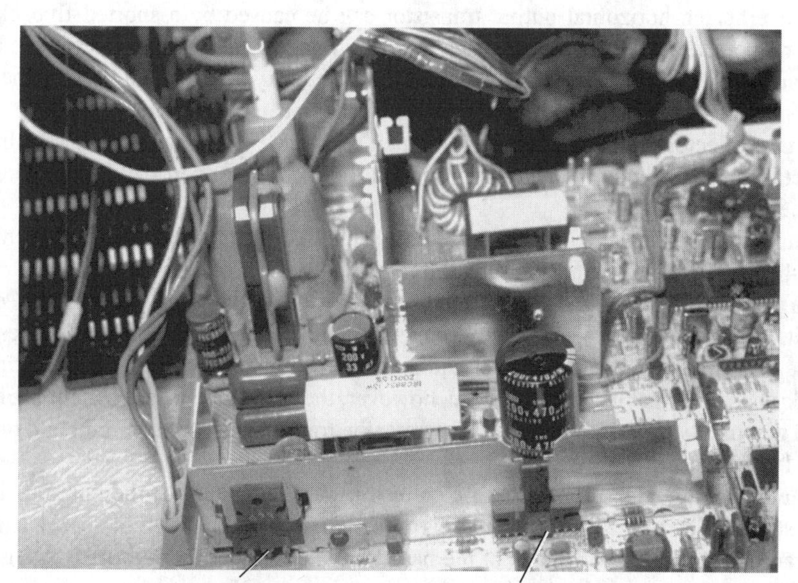

FIGURE 4-19 The horizontal output transistor is insulated above the heatsink and might have a "hot" ground.

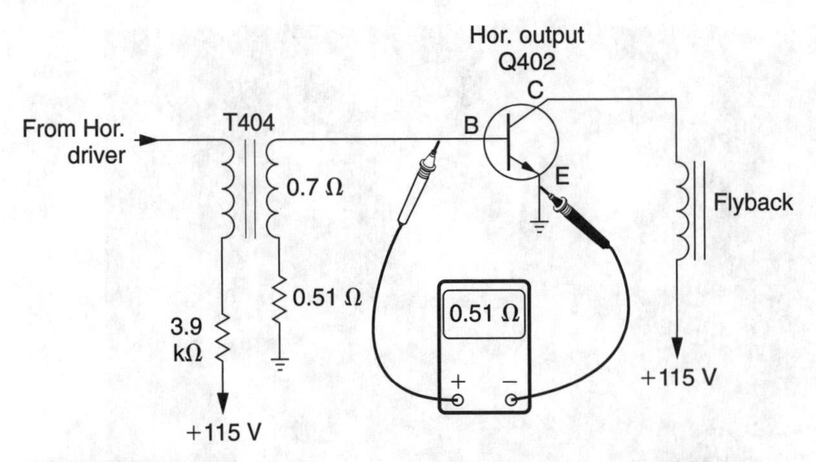

FIGURE 4-20 **A resistance or leakage test between base and common ground might not be accurate with a transformer winding in the circuit.**

Open output transistor tests between the base and collector terminals can be accurate, but not between the base and emitter terminals while the transistor is in the circuit. Remove the horizontal output transistor from its socket to test for leaks or opens. In addition to the secondary driver winding in the base circuits, the damper diode in the collector circuit can produce inaccurate measurements.

An overheated horizontal output transistor can be caused by a shorted flyback transformer, open damper diode, or insufficient drive voltage. Suspect that a flyback is leaky if the transistor heats up at once and shuts down. The leaky flyback can destroy the output transistor each time that the transistor is replaced. Because the damper diode is shunted from the collector terminal to ground, a quick resistance check with the DMM from the collector terminal (body) to chassis ground can indicate an open if the ohmmeter measurement is low (Fig. 4-21).

You might find the damper diode inside the same case as the horizontal output transistor (Fig. 4-22). This type of transistor is used in the latest TVs. Check the schematic diagram if you find that the output transistor has some leakage. If the transistor is checked in the circuit, a low resistance can exist between the emitter and base terminals. A low resistance can be measured between the emitter and collector terminals, indicating that the transistor is leaky under normal tests. In this case, however, the resistance measurement is of the internal damper diode in one direction. Remove the transistor from the circuit and make another leakage test.

A quick voltage measurement at the collector terminal (body) of the horizontal output transistor can determine if the transistor is defective. Low voltage can indicate a leaky output transistor or no drive voltage at the base terminal. Higher-than-normal voltage at the collector terminal can indicate that the transistor is open. No voltage measurement at the collector terminal might indicate that the horizontal protection fuse, isolation resistor, or flyback primary winding are open.

Always place silicone grease on both sides of the insulator and transistor when installing a new output transistor. Wipe off all old grease and dust from the metal heatsink. Snug up the mounting screens, but not too tight, because sometimes the corner of the insulator can break through, causing voltage leakage between the transistor and heatsink.

Testing horizontal output transistors with damper diodes In many of the new TVs, the damper diode is inside of the output transistor. Do not be alarmed if you have a low-resistance measurement between the emitter and collector terminals. Usually, when a low resistance is found between the emitter and collector, the transistor is discarded (Fig. 4-23). However, the SK9119/ECG89 horizontal output transistor has a 480-Ω resistance between the emitter and collector terminals, with the positive lead at the emitter terminal.

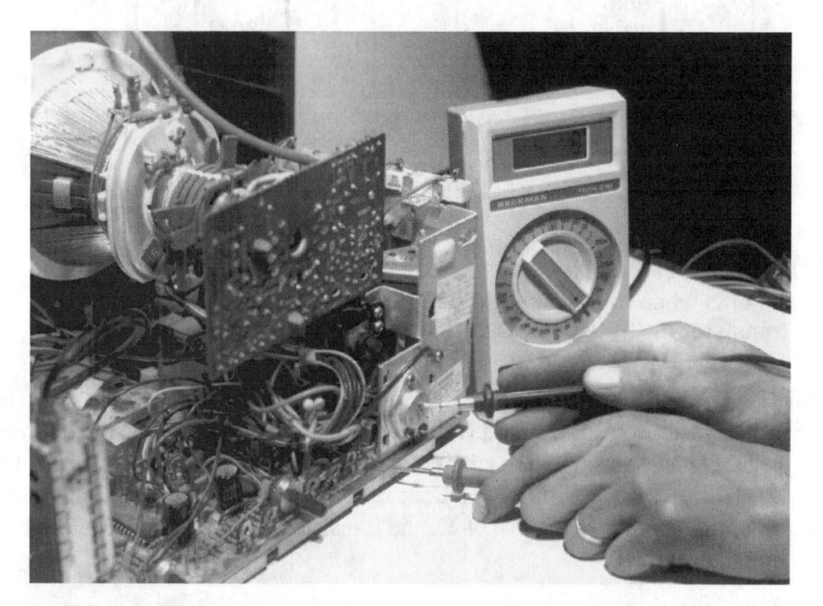

FIGURE 4-21 Check the horizontal output transistor from the collector (leaky) to common ground with the diode test of the DMM.

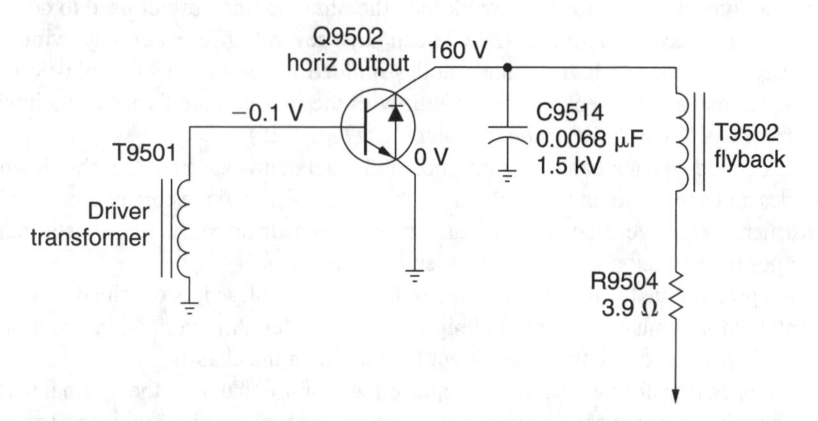

FIGURE 4-22 Notice that the damper diode is located inside the horizontal output transistor (Q9502) 25D2331.

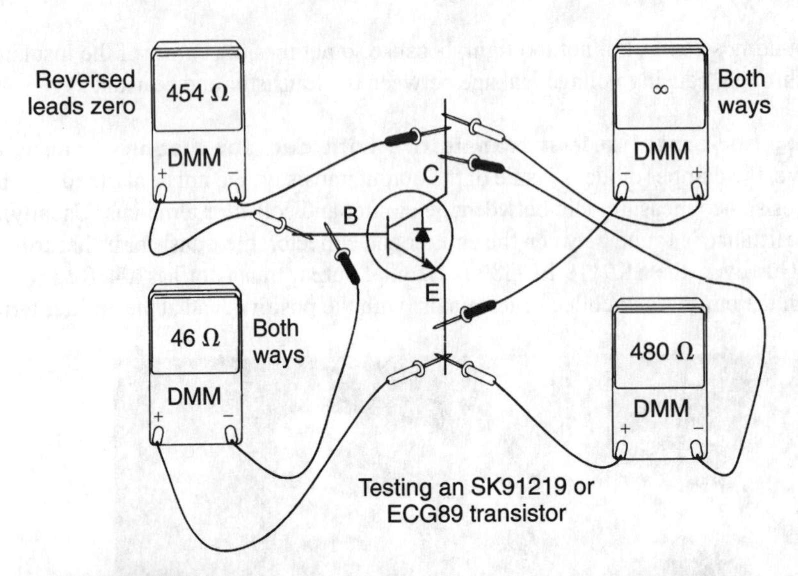

FIGURE 4-23 **Typical ohmmeter measurements across the three elements when the damper diode is located inside of the output transistor (SK9119 or ECG89).**

Actually, you are measuring the resistance of a silicon diode inside of the transistor. If a low-resistance measurement is found with reverse test leads, replace the output transistor. Notice the 50-Ω resistance from the emitter to the base terminal. This measurement will be found in both directions with the DMM diode test. Do not throw away the transistor. If the reading in both directions is below 11 Ω, the transistor is shorted between the base and emitter terminals.

THE FLYBACK TRANSFORMER

Today, the flyback does a lot more work than the old tube transformer used to do. The integrated high-voltage transformer (IHVT) can have several different voltage windings for low-voltage sources and high-voltage diodes molded inside of the same plastic housing (Fig. 4-24). Instead of a separate tripler unit with the low-voltage flyback, the integrated transformer has diodes built inside the windings (Fig. 4-25).

The integrated flyback can run warm, pop and crack, and cause chassis shutdown. Like the diodes and capacitors in the tripler unit, the high-voltage diodes break down inside the transformer. Excessive arcover can cause overheating or arcing of the plastic material. Sometimes the flyback arcs over to the metal core area.

The integrated transformer should operate fairly cool, unless it is overloaded by a leaky horizontal output transistor or internal high-voltage diodes. An overload in one of the secondary voltages can cause the transformer to shut down the chassis.

Most flyback transformers should be replaced with those that have the original part number. Although the American-made transformer can be replaced with universal types, imported transformers must be replaced with the original replacement transformer (Fig. 4-26). The Japanese component might include the screen and focus controls as one unit.

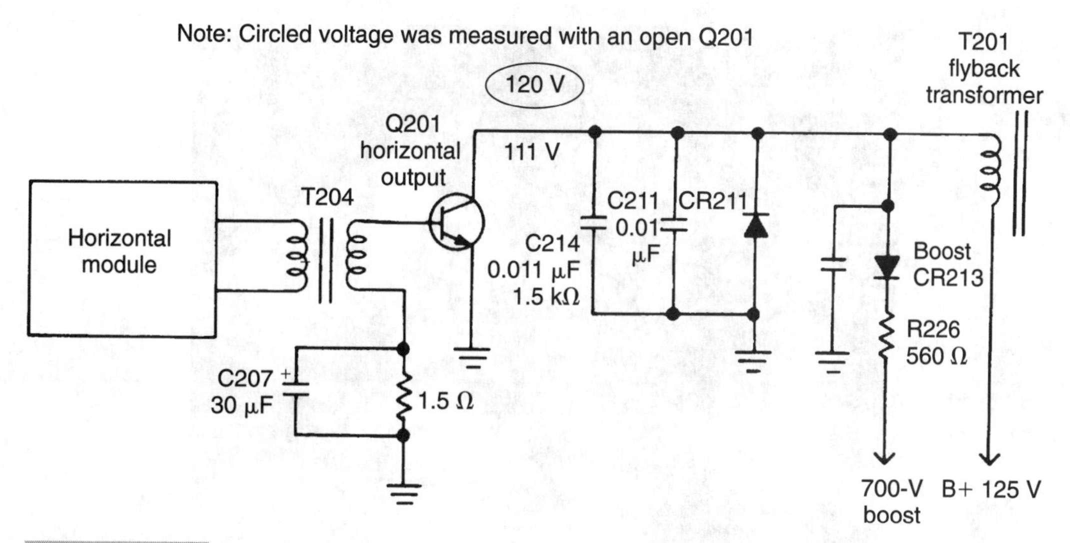

FIGURE 4-24 Collector voltage that is very low can indicate that the output transistor is leaky. A very high voltage can indicate an open output transistor.

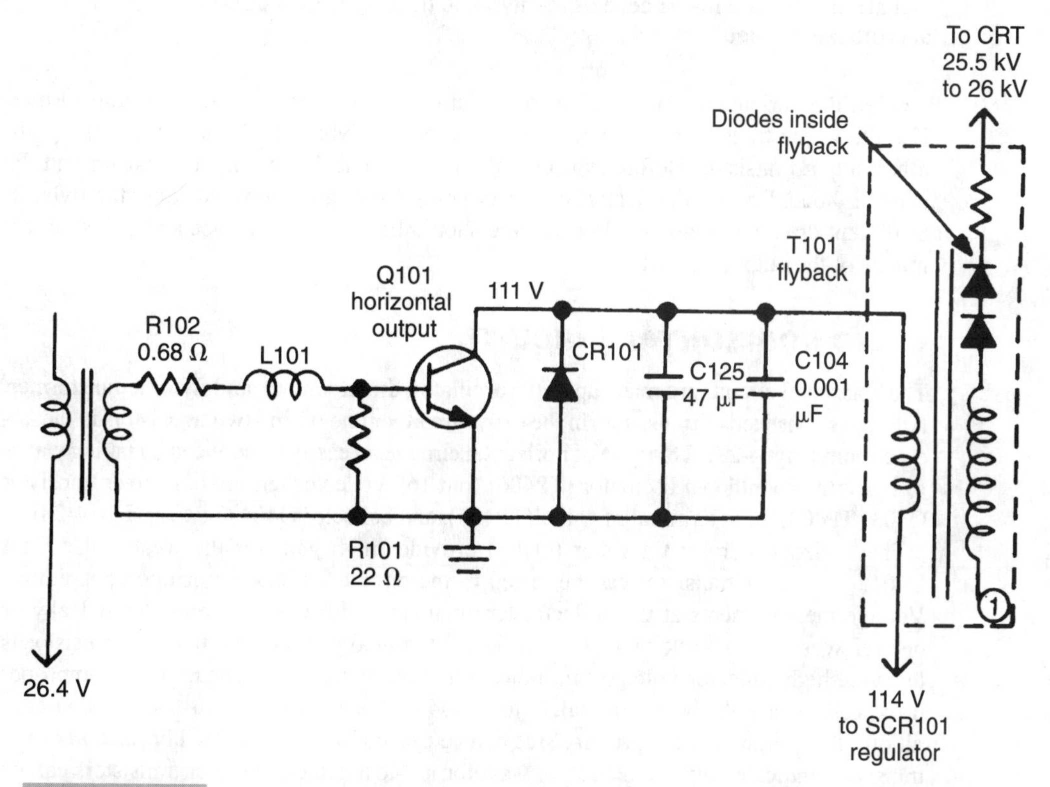

FIGURE 4-25 No resistance can be measured in the secondary winding of the IHVT flyback with high-voltage diodes in the circuits.

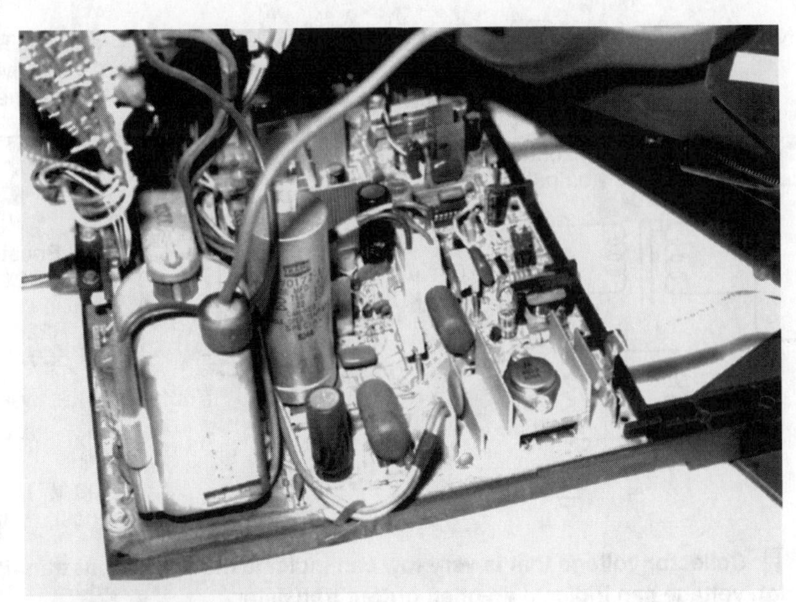

FIGURE 4-26 The high-voltage diodes or secondary winding can arc over to the metal core of the flyback, making a loud popping and cracking noise.

Often the horizontal output transistor is damaged by a defective flyback transformer. Use the variac transformer tests to locate a defective flyback and prevent damaging another output transistor. Before replacing the flyback, check the output transistor and the damper diode. Be sure that leaky secondary components are not overloading the flyback. Sufficient drive and regulated low voltage should be found at the base and collector terminals of the output transistor.

BASIC HORIZONTAL CIRCUITS

The horizontal circuits are made up of the oscillator, driver, output, and flyback transformer. All stages consisted of transistors in the early circuit, or one IC and two transistors in the latest circuits (Fig. 4-27). This type of horizontal circuit can easily be serviced by taking waveforms at the countdown oscillator (TP406) (pin 16), collector terminal of driver transistor Q403 (TP408), base terminal of Q404 (TP404), and collector Q404 (TP407) (Fig. 4-28).

The horizontal driver transistor (Q403) provides high gain for the weak pulse from U401. The driver transistor can be tested in the circuit for leakage or open conditions. Voltage measurements at the collector terminal can indicate if the transistor is leaky or open. Excessively low voltage at the collector terminal can indicate that the transistor is leaky. A high collector voltage can indicate that the transistor is open, or if an improper drive voltage is at the base terminal. Often R434 will become quite warm with either condition. Shorted turns in the primary side can be caused by an overheated horizontal driver transistor, indicated by a weak output waveform. Most defective driver transistors can be located with crucial voltage and waveform tests. The universal transistor can be used effectively in the driver circuits.

The peak-to-peak waveform tests at the base and collector terminals of the horizontal output transistor can indicate if the stage is weak or dead. Remember, when taking a base waveform with the transistor out of the socket, the waveform will look different from that of a normally mounted transistor. Often, the leaky horizontal output transistor or flyback

FIGURE 4-27 The screen and focus controls might be part of the output-transformer replacement.

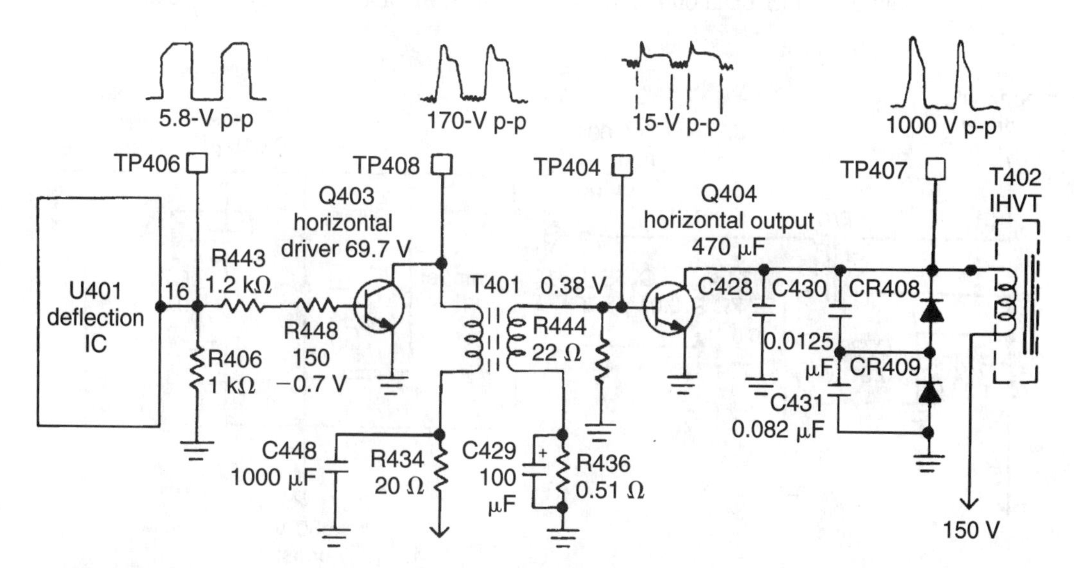

FIGURE 4-28 Crucial waveforms and voltage measurements at the various test points identify most horizontal circuit problems in the RCA CTC121 chassis.

can load down the base waveform with the transistor in the circuit. At least with the transistor out, the horizontal driver output and base waveforms should be close to normal if the preceding circuits are working.

HOLDDOWN CAPACITORS

In the early Zenith, Admiral, and Magnavox chassis, a four-legged collector-shunt capacitor could open, producing excessive high voltage or no voltage applied to the horizontal output transistor (Fig. 4-29). These tuning capacitors had a tendency to open at one end. Often, the horizontal output transistor was destroyed in the process. Always replace this holddown capacitor if you find a leaky output transistor.

If the capacitor does open, the high voltage will increase, and it might shut down the chassis. Besides the higher-than-normal B+ voltage applied to the horizontal output transistor, check the holddown capacitor for excessive high voltage. High-voltage arcover can occur at the picture tube or in the flyback circuits when the shutdown high-voltage circuits do not function because of a defective capacitor. This capacitor should be replaced with one that has the exact capacitance and operating voltage. The voltage can be from 1.2 kV to 2 kV (Fig. 4-30).

In some chassis, the tuning capacitor can open and destroy the flyback and the output transistor. It's best, when replacing a flyback transformer (in the older models), to replace the holddown capacitor and output transistor. Often, the fuse or isolation resistor opens because of a leaky transformer and output transistor (Fig. 4-31). Check the damper diode and boost diodes when you find a leaky output transistor.

THE DAMPER DIODE

A leaky damper diode can blow the fuse or trip the circuit breaker. The open damper diode can cause the horizontal output transistor to run warm and be destroyed. The diode can be

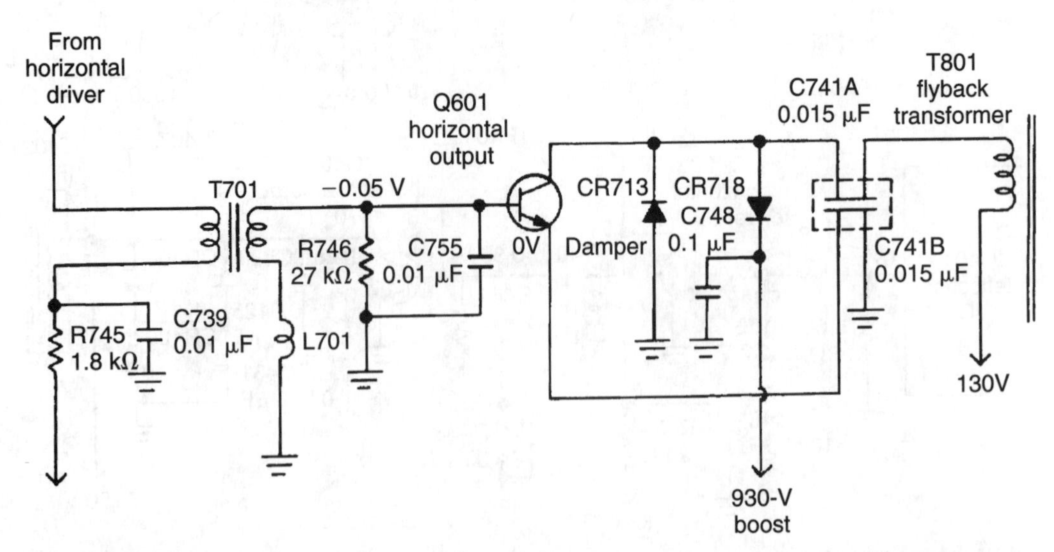

FIGURE 4-29 **A typical output safety circuit using the four-legged collector-shunt capacitor.**

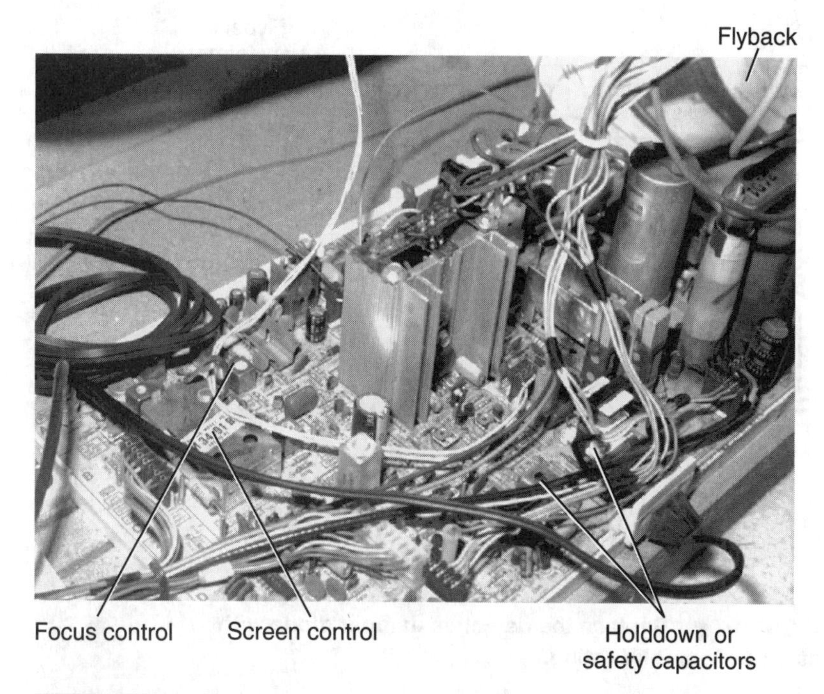

FIGURE 4-30 The voltage of the holddown or safety capacitor can range from 1.2 kV to 2 kV with a 5% tolerance.

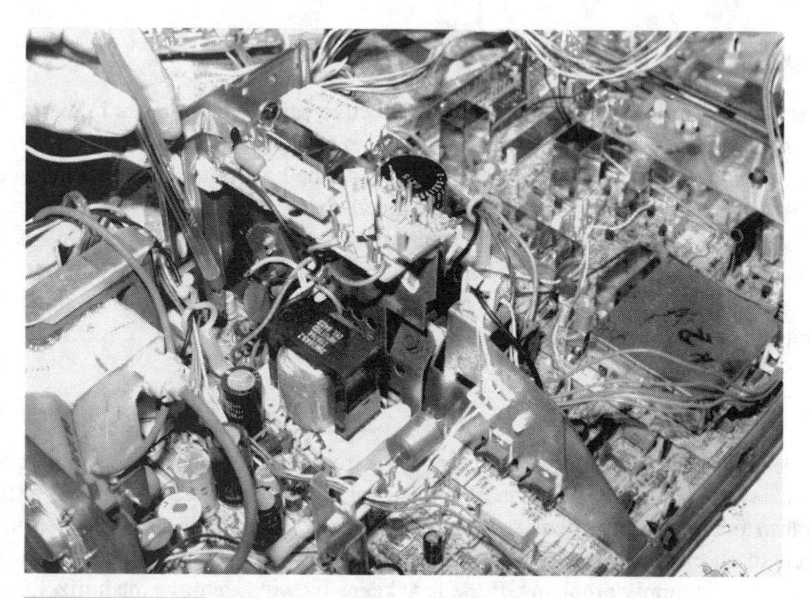

FIGURE 4-31 Check the safety capacitor if the output transistors keep shorting out.

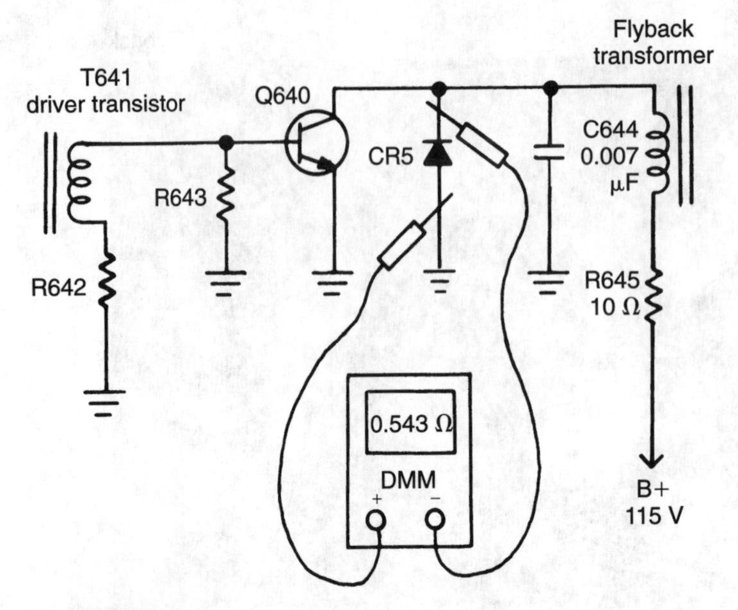

FIGURE 4-32 **Replace the defective damper diode with one of a 1.8-kV or 2-kV rating.**

checked with the diode test of the DMM or VOM (Fig. 4-32). Low resistance measurements should be found in one direction; infinite resistance should be found with reversed test leads. A leaky damper diode will have a low resistance in both directions. Replace the damper diode with a 1.8-kV to 2-kV rated unit. Do not replace it with an ordinary power diode.

CRUCIAL SAFETY COMPONENTS

Replace (with original replacement parts) defective safety components in the horizontal circuits. Check the manufacturer's schematic for a star, diamond, or letters alongside each safety component. Important components, such as the horizontal driver transformer, flyback, safety capacitors, bypass capacitors, metal-oxide resistors, the yoke, and the horizontal output transistor should be replaced with exact replacement parts. Always replace holddown capacitors and capacitors in the flyback and yoke circuits with exact replacements. These crucial capacitors have a higher voltage measurement than ordinary bypass capacitors.

THE DEAD CHASSIS

Most dead-chassis symptoms occur because of trouble in the horizontal output and low-voltage power supply circuits. Check for low voltage at the collector terminal of the output transistor and the low-voltage supply. Measure the low voltage at the fuse terminal that protects the horizontal circuits. If you find no voltage at the fuse or collector terminal, there are power supply problems. If the fuse keeps blowing, remove the horizontal output transistor to determine if the horizontal circuits are defective.

Measure the high voltage at the anode connection of the CRT if you find a normal voltage at the output transistor. Suspect that the yoke is defective if the high-voltage current at the picture tube is low. Check for an open yoke-return capacitor if there is no high voltage. An open flyback transformer or capacitor in the flyback circuits can cause a dead chassis. Usually, a leaky flyback or capacitor will lower the output transistor collector voltage and can open the fuse.

Suspect an open horizontal output transistor if the voltage at the collector terminal is higher than normal. The transistor should be removed and tested out of the circuit. Low voltage at the collector terminal can indicate that the output transistor is leaky or that the drive voltage is insufficient. Scope the base terminal of the output transistor. If you find an insufficient waveform at the base terminal, check the horizontal drive and oscillator circuits for correct waveforms.

RCA CTC121 chassis The most common components that break down in the CTC121 chassis are shown in Fig. 4-33. Check Q404 and CR408 for leaky conditions. Low voltage at Q404 or chassis shutdown can be caused by a leaky T402. Improper drive voltage at the base terminal can be caused by a leaky Q403, an open T401, or a burned R434. After replacing a leaky Q404, be sure that the base resistor (R436) is not open. Sometimes Q404 can have a dead short, taking out R436, which is in series with T401. Check for open horizontal winding of the deflection yoke or capacitor C447 if the chassis is dead.

Montgomery Ward GGY16215A The dead chassis can be caused by high-voltage shutdown. Sometimes the chassis shuts down at once and the technician does not hear the high voltage come up. In a Montgomery Ward GGY16215A, the chassis was dead with

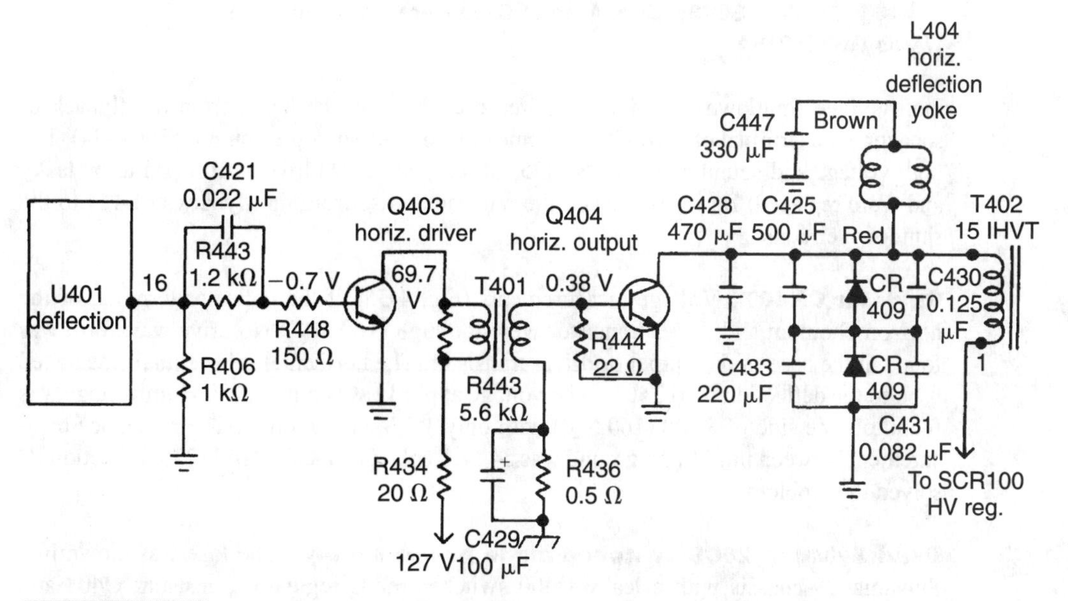

FIGURE 4-33 The horizontal output circuit of the RCA CTC121 chassis.

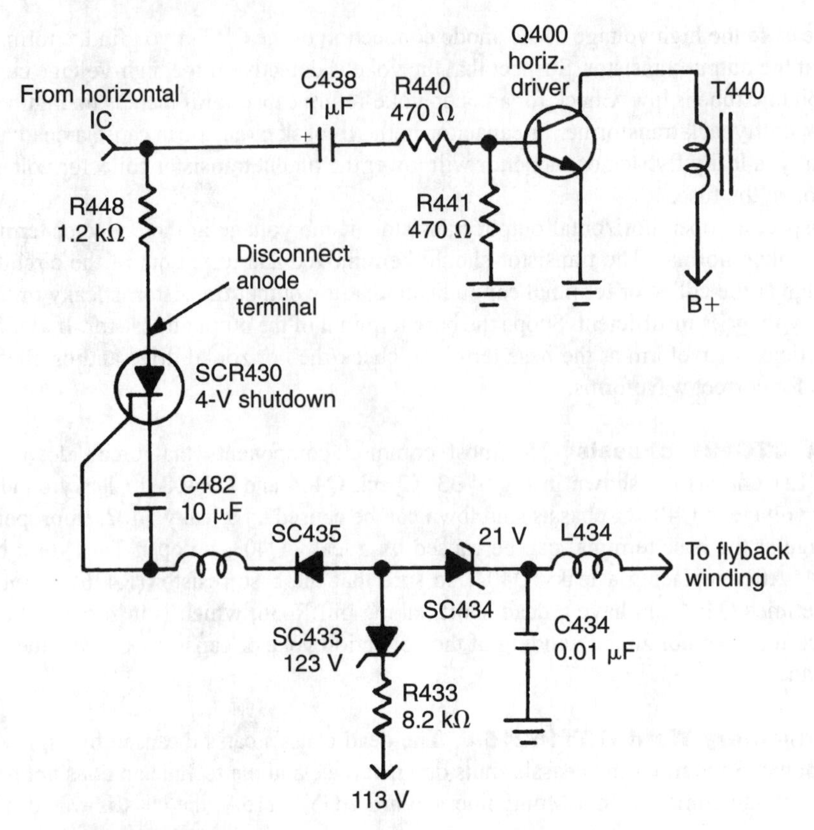

FIGURE 4-34 **SC435, SC434, and SC433 were suspected in a Wards GGY16216A.**

high-voltage shutdown. Nothing, not even a click, could be heard from the flyback or speaker. One terminal of SCR430 was removed from the shutdown circuit (Fig. 4-34). The high voltage and sound came on. SC435, SC433, and SC434 were all found to be leaky and were replaced. Be sure to check the feedback pulse from the flyback at L434 to the shutdown circuits.

Goldstar CR407 Voltage measurements (Fig. 4-35) at the collector terminal of the horizontal output and driver transistor were too high (160.5 V). No drive waveform was found at the base of the output or driver transistor. The horizontal drive signal was traced back to the deflection IC (IC401). The voltage at pin 11 was only 1.5 V. High voltage was found on one side of R508 (160.5 V) with only 1.5 V on the other. The resistance measurement between pin 11 and ground was 54 Ω. Replacing a leaky HA11235 deflection IC solved the problem.

Dead Sylvania 26C9 switched-mode power supply The fuse was blown in a Sylvania C9 chassis with a leaky Q400 switched-mode regulator transistor. Q403 and Q402 were also found to be leaky. After checking the diodes and small resistors in the primary circuit, D425 and D421 were found to be shorted. R430 and R432 were burned, so they were replaced (Fig. 4-36).

The chassis was fired up, but had no output voltage (130 V). R531 was removed and a 75-W bulb was used as the load on the power supply. All components in the secondary were checked for open or leaky diodes and resistors. R434 (1 Ω) was open and D434 was leaky. The voltages on Q406 and Q407 were found to be normal when 24 V were injected at the collector of D434. All other components were good.

Q400 was replaced again with a new fuse. The chassis was normal. The switched-mode transistor was fine after replacing defective components R434 and D434 in the 24-V power source.

Miscellaneous dead checks Poor pincushion and regulator transformer terminals can cause a dead chassis. If the connections are intermittent, the raster will narrow down with low high voltage at the picture tube. Improper connections between the col-

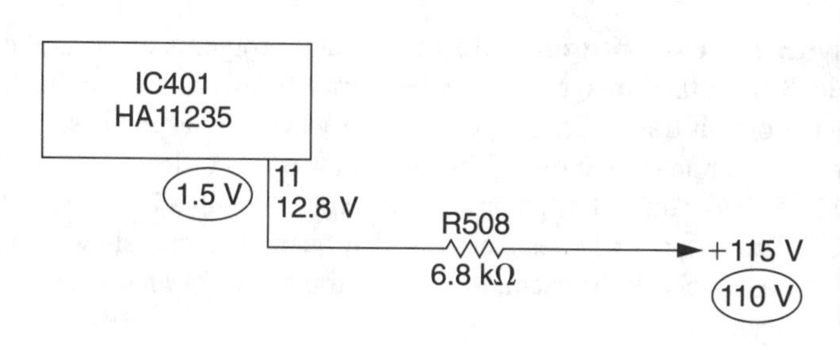

FIGURE 4-35 A leaky deflection IC (IC401) in a Goldstar CR-407 caused the dead-chassis symptom. A 54-ohm resistance between pin 11 and chassis ground indicates that the IC is leaky.

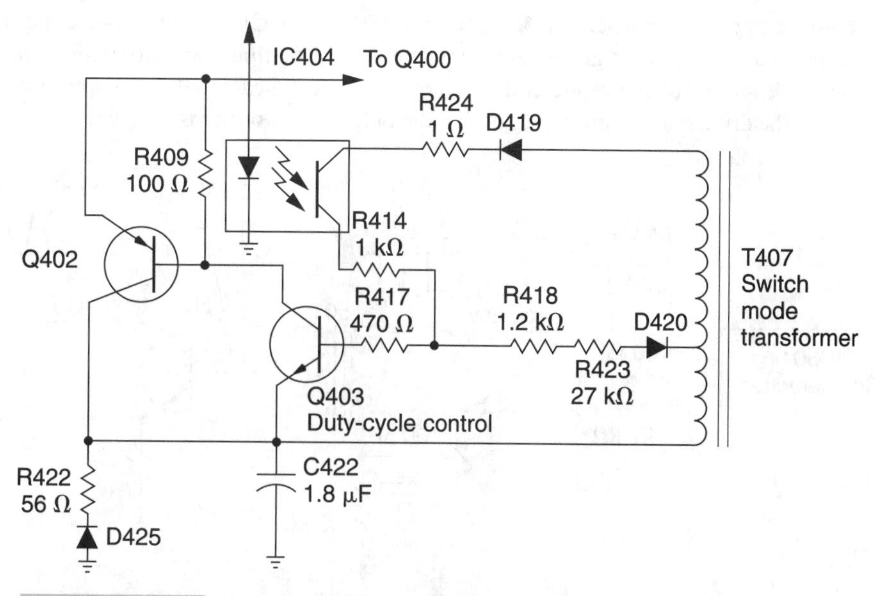

FIGURE 4-36 Q403 and Q402 were found leaky in the switched-mode power supply (SMPS) of the Sylvania 26C9 chassis.

lector bolt and metal body of the output transistor can cause a dead raster with no voltage at the collector terminal. Sometimes, if the mounting bolts are too tight, the transistor can short through the insulation and ground out the collector voltage. Be careful when taking voltage measurements at the horizontal output circuits in some chassis, where the output circuits are above ground. Always measure the collector voltage at the isolated ground terminal or chassis. Otherwise, low-voltage measurements will not be accurate.

No raster, no sound, no high voltage With a no-raster, no-sound symptom, check the high voltage at the picture tube with a high-voltage probe. If the high voltage is present without a raster, troubleshoot the video and picture tube circuits. Go directly to the horizontal output stages if there is no high voltage. Quickly check each horizontal stage with the scope. Start at the base of the output transistor and go backwards toward the horizontal IC to locate the lost signal (Fig. 4-37). When the signal is located, troubleshoot the preceding stage with voltage and resistance measurements.

Sharp C1335A with shorted output Most horizontal problems occur in the horizontal output circuits. Check the voltage at the flyback transformer and the collector terminal of the horizontal output transistor. Scope the input waveform at the base terminal. Remove the output transistor and test it out of the circuit for opens or leaks.

In Fig. 4-38, Q602 was leaky on all terminals and was replaced with a 2SD869, 2SD870, or a universal ECG89 transistor. R613 resistor was also open. The chassis will "motorboat" with R613 open. Also, check the secondary driver transistor for an open winding if the output transistor is leaky.

Open output transistor in Magnavox T995-02 Often, a higher-than-normal low voltage can be measured at the collector terminal of the output transistor with an open transistor. A lower voltage at the collector terminal can indicate that the output transistor is leaky, the drive pulse is improper, or that the output transformer is leaky. Here (Fig. 4-39),

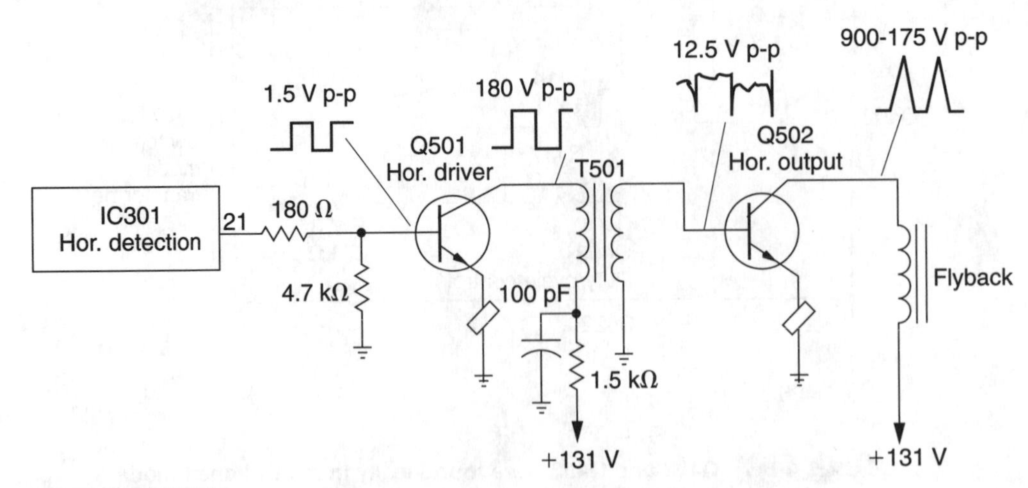

FIGURE 4-37 Typical waveforms found in the horizontal circuit.

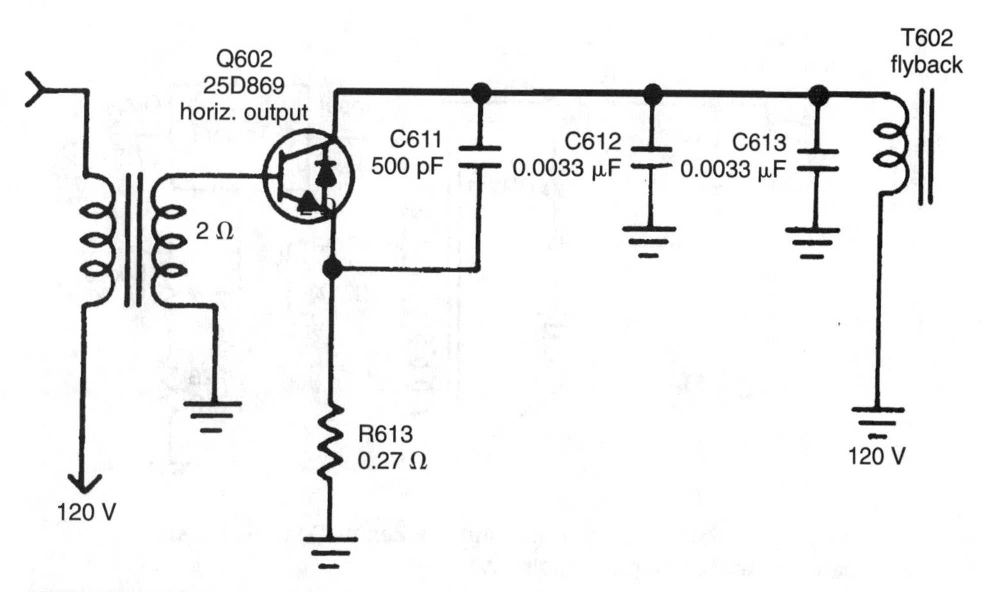

FIGURE 4-38 Q602 was found leaky in a Sharp C1335A chassis caused by a shorted internal damper diode.

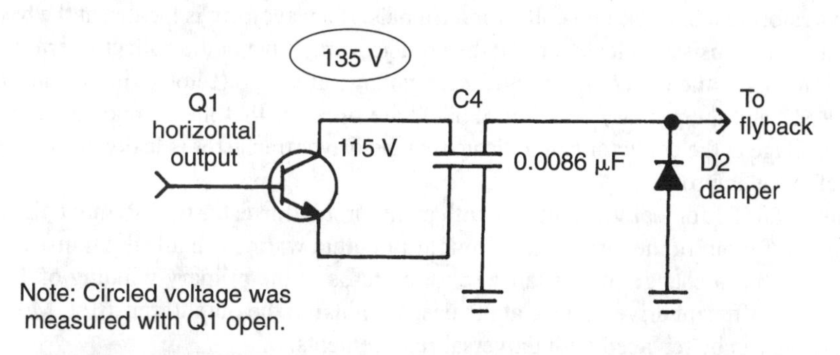

FIGURE 4-39 Replace C4 a four-legged safety capacitor and check damper diode D2 if you find a shorted output transistor (Q1) in a Magnavox T995-02.

the low voltage had increased to 135 V, indicating that output Q1 is leaky. Q1 was replaced with a GE-259 transistor. Always replace C4 if the output transistor is leaky. Check the condition of the damper diode D2 before replacing Q1.

Leaky tripler in Zenith 17FC45 With a leaky flyback or tripler unit, the voltage at the collector terminal of the output transistor will be lower than normal (Fig. 4-40). First, check the output transistor for leaky conditions. Be sure that a drive waveform is found on the base terminal with the scope. Remove the input lead to the tripler unit. If the collector voltage goes back to normal (144 V), suspect that the tripler unit is leaky. Check the flyback transformer if the voltage remains the same after removing the tripler input lead.

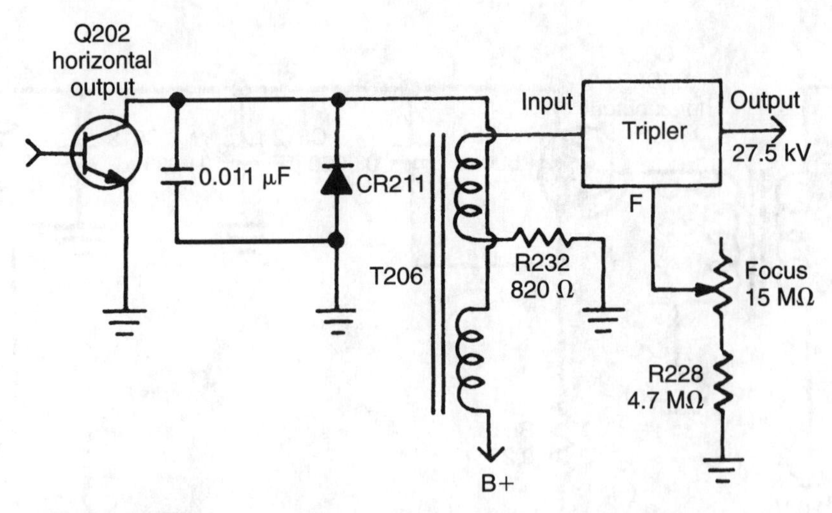

FIGURE 4-40 A leaky tripler unit in a Zenith 17FC45 chassis loaded down the output transistor.

No drive pulse in J.C. Penney 685-2012 Go directly to the collector terminal of the driver transistor if no drive pulse is at the base terminal of the output transistor. Check the waveform on the base and collector terminals. If a waveform is located at the base, suspect that the transistor is leaky or that the voltage is improper at the collector terminal (Fig. 4-41). Measure the collector voltage. Zero voltage at the collector terminal can indicate that R518 and the primary winding of T502 are open, or that there is no supply voltage. Low voltage at the collector can indicate that the driver transistor is leaky or that the drive waveform is improper.

Check TR503 for leaky conditions with an in-circuit transistor test. Remove the transistor and test it out of the circuit. R518 might run quite warm with a leaky transistor. Measure R518 for a change in resistance. Shorted turns in the primary winding of T502 can provide insufficient drive voltage at the base terminal of the output transistor. Most driver transistors can be replaced with universal replacements.

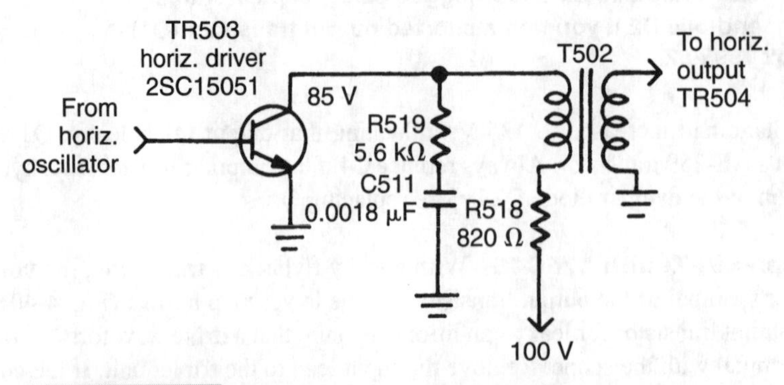

FIGURE 4-41 TR503 was found leaky in a J.C. Penney 685-2012 with R518 (820 ohms) burned open.

Realistic 16-410 improper drive pulse Check the horizontal circuits by taking quick waveforms of each stage. Start at the horizontal oscillator, sweep, or countdown IC terminal that contains the drive signal fed to the horizontal driver transistor. Sometimes a large IC might contain a countdown horizontal signal with a drive amplifier inside of the large IC.

In the Memorex 16-410 color TV/VCR portable, IC9701 contains the horizontal amp pre-driver, which feeds a drive signal to Q1001. The drive signal is taken from pin 17 of IC9701 to driver transistor Q9501 and driver transformer (T9501) to the base of the horizontal output transistor (Fig. 4-42). No waveform was found at pin 17, indicating no horizontal drive voltage. Low supply voltage (V_{CC}) indicated that IC9701 was leaky.

No oscillator waveform in RCA CTC111A chassis Check the output of the oscillator transistor or IC with the scope. No waveform or an improper waveform can indicate that a transistor, IC, or surrounding component is defective. Measure the collector voltage of the transistor and the supply voltage terminal of the IC. Low voltage can indicate that a transistor or IC is leaky.

Check all voltages tied to the IC deflection circuit (U401). Notice if any voltages are very far off from normal (Fig. 4-43). In this RCA chassis, zero voltage was found at terminal 5 of U401. Terminal 5 was unsoldered from the PC wiring. A low resistance measurement was not found from the terminal to chassis ground. A resistance measurement across CR401 indicated that a diode was leaky. The leaky CR401 prevented Q100 from starting and prevented IC401 from applying a sawtooth voltage to the driver circuits.

Bad connection in a GE AB-B chassis Besides locating defective components in the horizontal section, do not overlook the possibility of bad board connections. A poorly soldered connection where the component ties into the PC wiring or a broken wiring connection

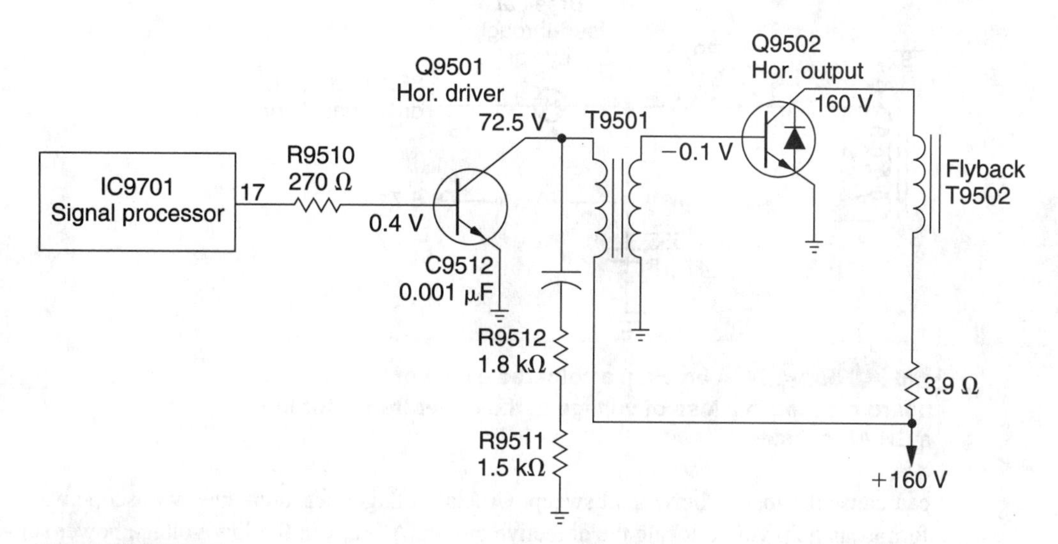

FIGURE 4-42 A schematic diagram of the horizontal output circuits within the Realistic 16-410 portable TV/VCR.

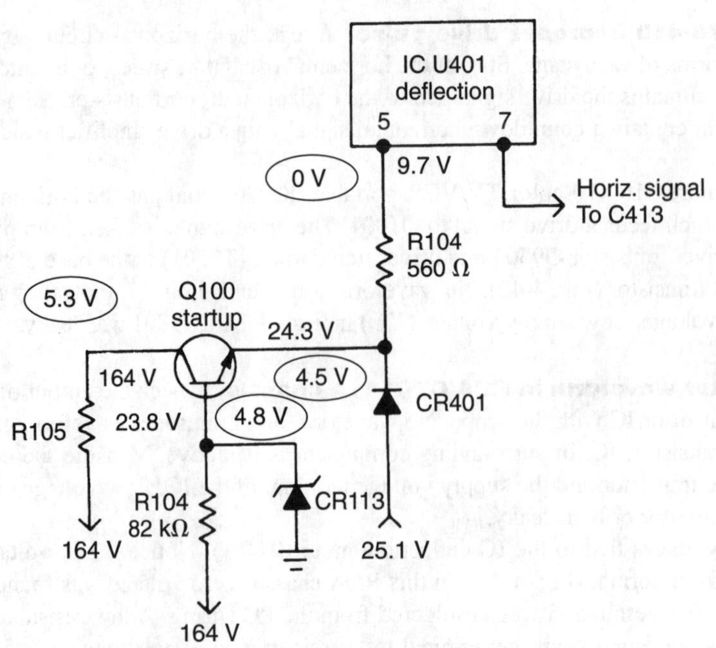

FIGURE 4-43 **No horizontal drive signal at pin 7 was found in an RCA CTC111 chassis. The problem was caused by a leaky diode (CR401).**

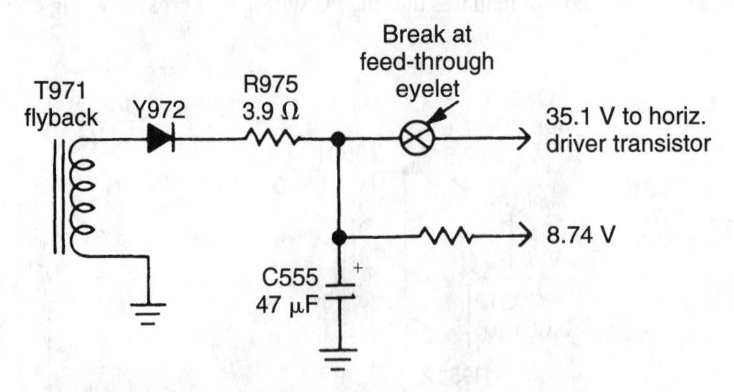

FIGURE 4-44 **A break in a soldered eyelet of the PC board resulted in a loss of voltage at the driver transistor in a GE AB chassis.**

can cause the loss of horizontal sweep. Crucial voltage measurements with scope waveforms can help you to locate the defective circuit. A break in the low-voltage power supply (35.1 V) of the General Electric chassis (Fig. 4-44) was located when no voltage was at the collector terminal of the horizontal driver transistor.

No raster, normal sound In older TVs, you could have normal sound with a defective horizontal section. Today, with the horizontal circuits and the sound powered from a low-voltage source connected to the secondary windings of the flyback transformer, the horizontal circuits must function to receive good sound. Check the high-voltage circuits for possible trouble if the sound is normal, but there is no raster. A defective picture tube or its circuit can prevent a normal raster. If the high voltage and picture voltages are normal, suspect a problem with the video circuits.

INTERMITTENT RASTER

An intermittent raster can originate in the horizontal or high-voltage circuits. A defective gun assembly of the CRT or picture-tube circuit can produce an intermittent picture. The intermittent raster can be caused by a poor regulated low-voltage power supply. Although high voltage might be present, intermittent video components can cause an intermittent raster. Poor board or wiring connections cause many intermittent problems.

An intermittent raster that changes constantly is fairly easy to troubleshoot and locate the defective component. A TV that might take hours to quit working or only "dies" once a week is very difficult to diagnose. Besides voltage and waveform monitoring, the intermittent component might be found with coolant and heat applications. Moving the board or prodding components on the board might turn up a poor board connection.

Transistors and ICs produce most of the intermittent problems in the TV. The horizontal output transistor and the high-voltage regulator are likely suspects (Fig. 4-45). Of course, any transistor in the horizontal circuits can break down. The suspected transistor can be

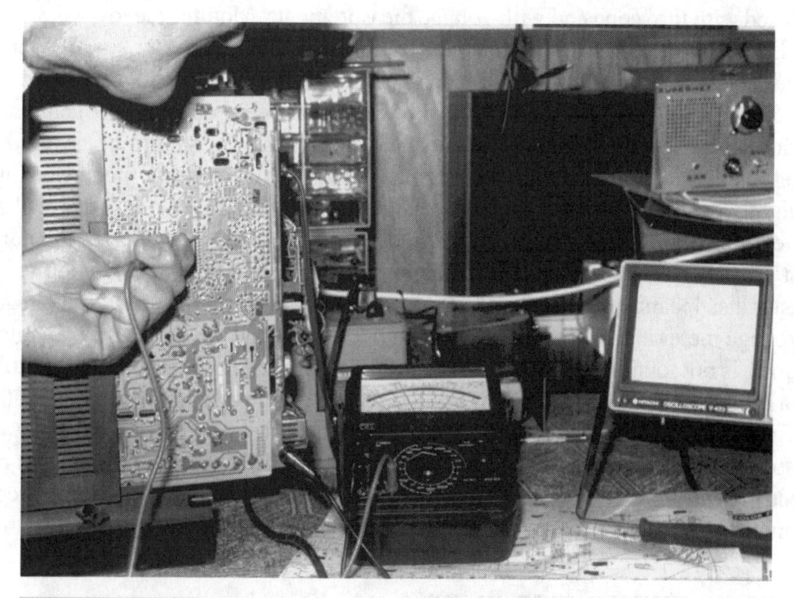

FIGURE 4-45 Check the waveform at the base and collector terminals of the horizontal output transistor with a scope to determine if the output transistor is leaky or open.

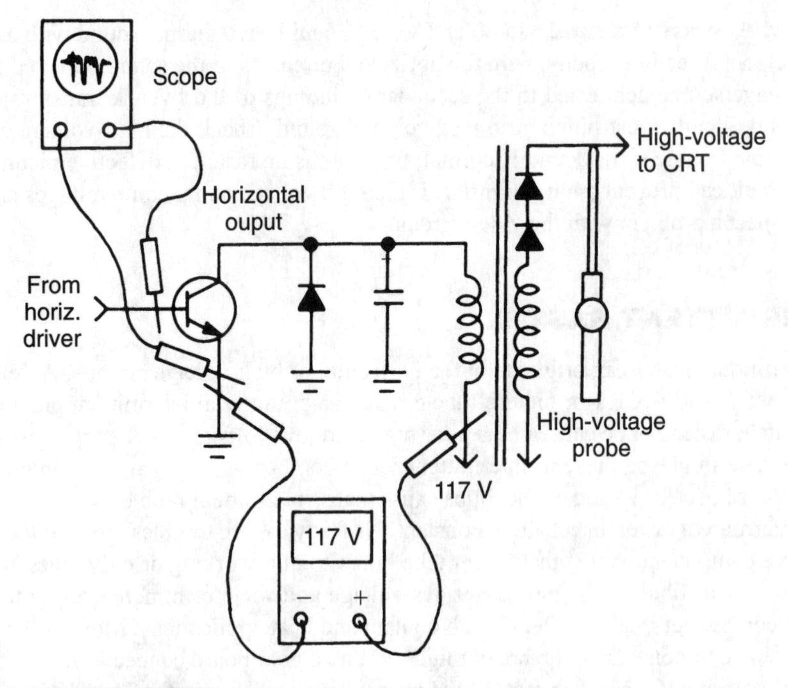

FIGURE 4-46 **Monitor the intermittent chassis with the scope and take crucial voltage measurements.**

monitored with the scope and with voltage measurements. Monitor the low-voltage supply at the horizontal fuse for intermittent power problems. Poor transistor socket connections can cause intermittent problems. Bad component-to-board connections or bad double-board feed-through eyelets can be located by pressing up and down on the chassis.

With difficult intermittents, the whole back side of the TV should be covered with a blanket to add more heat so that the intermittent will act up. Raising the power-line voltage slightly can help. Monitor the low-voltage supply at the fuse or where dc voltage enters the flyback transformer. Connect the scope lead to the base terminal of the horizontal output transistor (Fig. 4-46). Insert the high-voltage probe to monitor the high voltage. If the raster disappears, note which monitor indicates a poor measurement. No voltage at the low-voltage meter indicates that trouble is in the low-voltage power-supply circuits. If the waveform is not found at the base terminal of the horizontal output transistor, troubleshoot the horizontal stages. The high voltage will disappear with horizontal problems. If the base waveform is normal and the high voltage is low or absent, suspect a defective picture tube or suspect the video and high-voltage circuits. In case the high voltage increases rapidly, suspect high-voltage shutdown. Concentrate on locating the defective component after isolating the intermittent to a given section of the chassis.

KEEPS BLOWING THE FUSE

Suspect that a component in the horizontal or low-voltage circuits is defective if the fuse blows at once. If the fuse takes a few seconds to open, suspect a problem with tripler unit or a component in the high-voltage circuits. A leaky power supply diode or filter capacitor can cause the fuse to repeatedly open. Remove the horizontal circuit fuse or output transistor to determine if the component is in the low-voltage or horizontal circuits.

Check the damper diode and horizontal output transistor if the horizontal fuse will not hold. Measure the resistance between the collector (body) terminal of the output transistor to chassis. Remove the transistor with a low measurement and test it out of the circuit. A leaky flyback transformer can cause the case of the horizontal output transistor to run warm before the fuse is blown. The leaky flyback test should be made with the universal power-line transformer to prevent damaging a new output transistor. Isolate a possible shorted yoke by removing the plug or the red wire connected to the yoke plug. It's best to remove the red lead if the yoke plug has a shorting wire for the B+ circuits.

Suspect that the tripler unit is leaky if the fuse does not blow until after four or five seconds have passed. Simply disconnect the input lead from the tripler unit. If the fuse does not open now, suspect that the tripler is leaky. In an Admiral 2M10C chassis, fuse F102 (1 A) would open after four seconds. A voltage check at the collector terminal of the horizontal output transistor was normal until the fuse opened (Fig. 4-47). A 100-W bulb was clipped across the open fuse and it appeared fairly bright with the overloaded component. When the output transistor and tripler lead were removed, the light went out. Replacing the tripler with a GE-537 replacement solved the fuse-blowing problem.

J.C. Penney 682-2114 open fuse The F910 (1 A) fuse would constantly open when a new fuse was inserted. The leaky component was beyond the horizontal output transistor with the 100-W bulb test. At first, the flyback transformer was suspected. But when the red lead of the yoke was removed from the circuit, the light bulb brightness dimmed (Fig. 4-48). A quick measurement of the horizontal and vertical yoke connections indicated a 2.7-Ω short between the two windings. Replace the deflection yoke with the original replacement part.

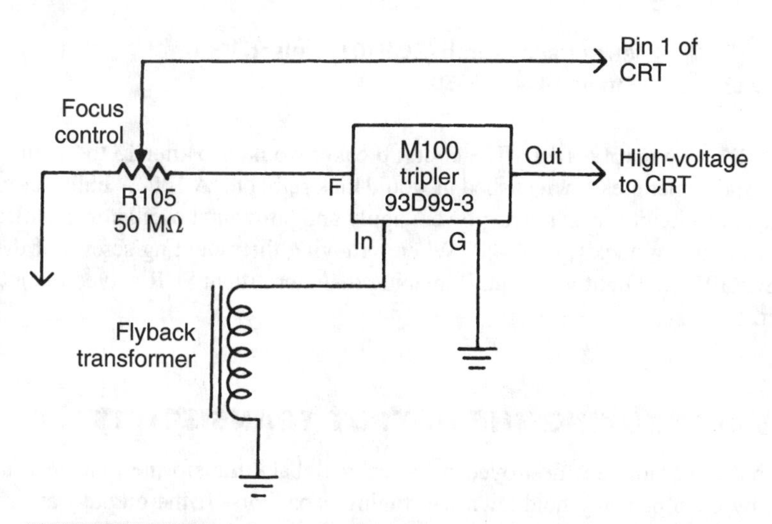

FIGURE 4-47 After a while the fuse would blow if there is a leaky tripler unit in the horizontal output circuits.

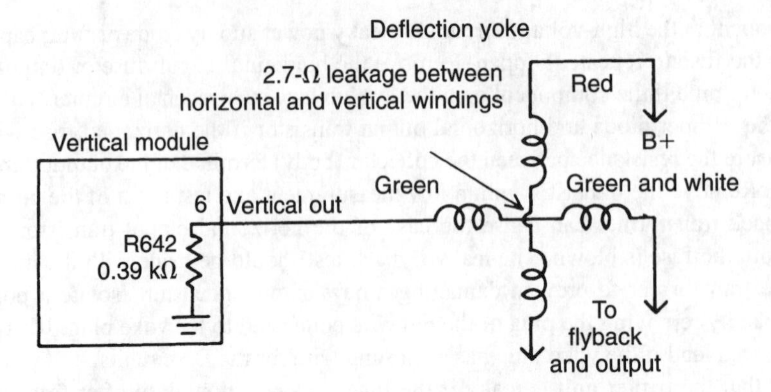

FIGURE 4-48 **The deflection yoke leakage in a J.C. Penney 682-2114 chassis caused the line fuse to blow.**

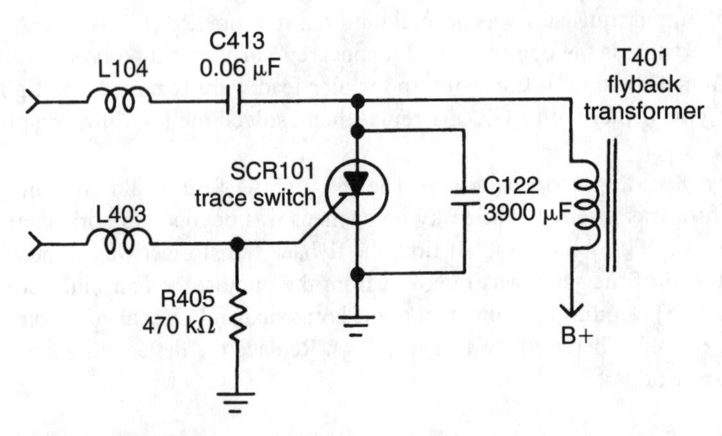

FIGURE 4-49 **A leaky trace switch (SCR101) caused the chassis to shut down in an RCA CTC59.**

RCA CTC59 15-minute trip The circuit breaker would trip after 15 to 20 minutes of operation, and sometimes when pushed in would kick right out. A 100-W bulb was clipped across the open circuit breaker. The power supply and horizontal oscillator modules were replaced without any luck (Fig. 4-49). When removing the mounting screws of the trace switch (SCR101), the light went out. Replacing the intermittent SCR solved the intermittent raster.

KEEPS DESTROYING THE OUTPUT TRANSISTORS

Most output transistors are destroyed by a leaky flyback transformer, an open damper diode, or by open or leaky holddown and tuning capacitors. If the output transistor becomes warm without any high voltage present, suspect a problem with either component.

Remove one end of the damper diode and test for open conditions. Replace the tuning capacitor if the output transistor is leaky or shorted.

Besides the most common components, do not overlook the possibility that the deflection yoke is leaky. Check the yoke-return capacitor for leaky conditions. Check the low-voltage source. If it is quite high, the output transistor could be damaged by a defective high-voltage regulator circuit. Disconnect the boost diode and check for possible leakage. Check the diodes in the secondary voltage sources for leakage before inserting a new output transistor.

The new output transistor might be defective. Before installation, check each transistor or diode. The output transistor can be damaged because of lower breakdown voltage. If the original transistor is not available, install one with a 1200-V or 1500-V rating. Many service technicians use the higher-voltage-rating output transistors so that they will last longer. If the case is warm when the transistor is operating, either insufficient drive or too high of an operating voltage is applied to the output transistor. Checking the leaky output transformer takes a little longer.

To prevent damaging another output transistor, use a variable line transformer at the power cord (Fig. 4-50). Bring the line voltage up from 60 V to 80 V and check the voltage at the output transistor. Because only two-thirds of the line voltage is used, the dc voltage should be down one-third from normal. Suspect that the flyback is leaky if the dc voltage is low and under 400 peak-to-peak waveforms at the collector terminal of the output transistor. Remove possible leaky secondary components on the transformer winding with an integrated flyback. Simply remove each lead of the diodes that furnish power to other circuits on the secondary windings and notice if the dc voltage returns.

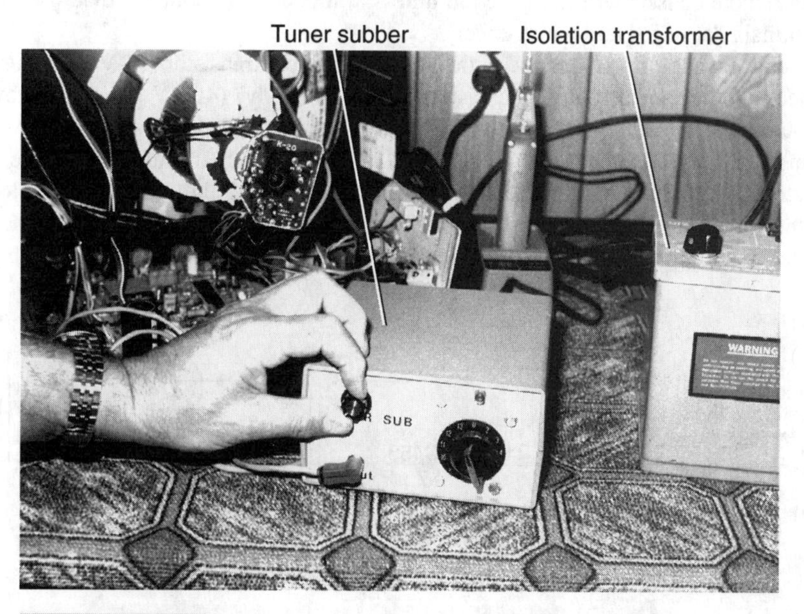

FIGURE 4-50 Connect the TV chassis to variable line transformer and vary the voltage to about 60 to 80 Vac to check horizontal circuit before it shuts down.

RED-HOT OUTPUT TRANSISTORS

If the horizontal output transistor runs red hot and sometimes ends up with a dead short, suspect that the drive waveform is improper, a driver transformer is faulty, or that a small filter capacitor is bad. Determine if the waveform drive voltage is the same as shown upon the schematic. Solder all terminals on the driver transistor. Check the primary resistance to see if it is the same as in the schematic. The Sams Photofacts lists the primary and secondary resistance, but some manufacturers do not.

If the output transistor is still hot, replace the small electrolytic capacitor (usually a 1 μF, 4.7 μF, and 10 μF at 250 V) located on the primary of the horizontal driver transformer (Fig. 4-51). Also, replace the filter capacitor connected to the primary of the flyback in the B+ source. Replace the driver transformer after all other parts have been replaced. Do not overlook the possibility that a flyback transformer is leaky.

CHECKING THE FLYBACK IN AN RCA CTC111L CHASSIS

To determine if the flyback transformer is defective in almost any RCA chassis with an integrated transformer, the following method can be used:

1 Short capacitor C113 with test leads (Fig. 4-52).
2 Connect a jumper across the gate and cathode terminals of SCR100.
3 Adjust the variable line transformer to 60 Vdc at the cathode terminal of SCR100.
4 The remote chassis will not function unless a jumper lead is attached across the relay terminals (large red and white tracer).
5 Check the waveform at the base of the horizontal output transistor (15-V p-p) (Fig. 4-53).
6 Check the horizontal output transistor waveform (body) (450-V p-p). The waveform should be clean and without ringing. If ringing is found, suspect a shorted high-voltage transistor. Defective high-voltage diodes (inside transformer) can cause noise spikes. Distortion or retrace spikes denote mistuning or a cracked high-voltage transistor core.
7 If the waveform is clean, raise the ac voltage to a normal B+ 130 V.
8 If heavy current is pulled, suspect a leaky horizontal output transformer.

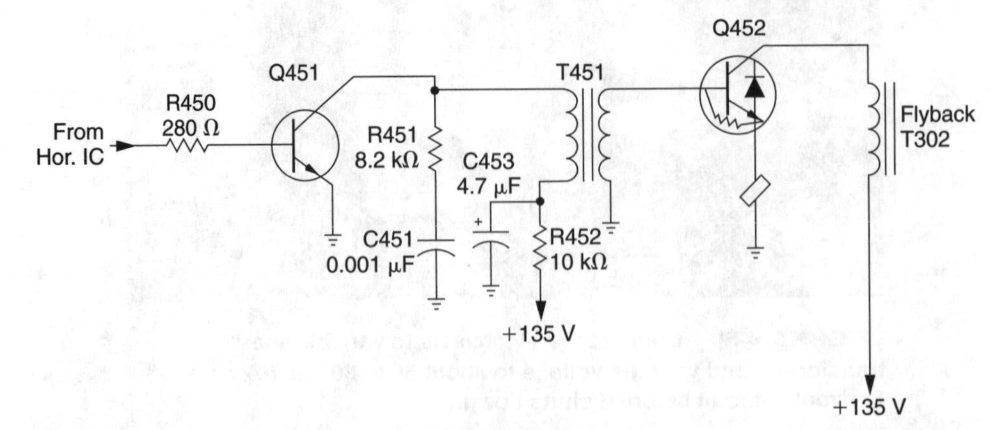

FIGURE 4-51 Check small electrolytic capacitors (1 μF to 10 μF) in the horizontal driver circuits if the horizontal output transistor runs red hot.

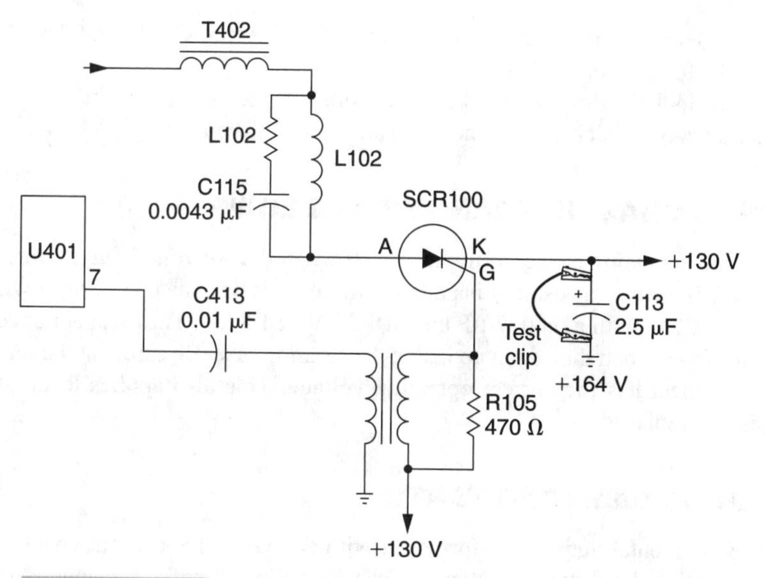

FIGURE 4-52 Testing the flyback in an RCA CTC111L chassis.

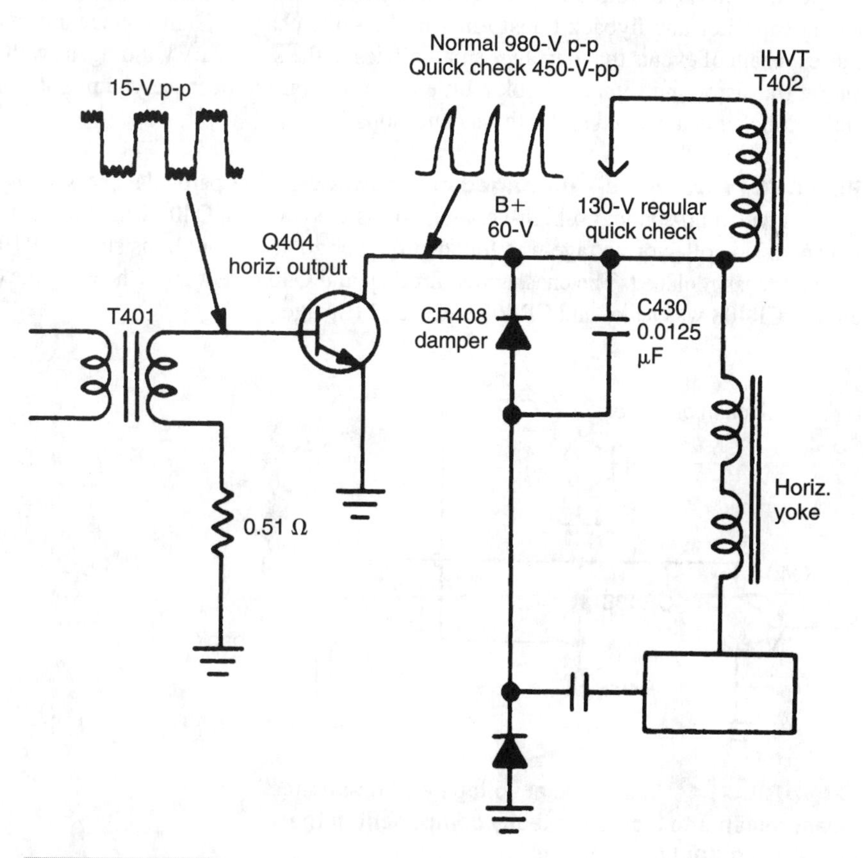

FIGURE 4-53 Check the waveforms at the base and collector terminals with the scope of the horizontal output transistor.

9 Check the flyback secondary voltages and diodes CR106, CR107, CR109, CR110, and CR112 for possible leakage.
10 Do not overlook the deflection yoke, pincushion, and centering circuits.
11 Replace a leaky flyback transformer with an exact replacement.

A GE 19QB CHASSIS TRANSISTOR BLOWS

Always replace the tuning capacitor in the chassis when the horizontal output transistor is found shorted. If a new transistor is installed without replacing the capacitor, it will blow again. Replace C234 with a 0.0047-µF unit at 1.2 kV or 1.8 kV. These capacitors can appear normal—even when they open up inside the ceramic case. Be sure that the tuning capacitor replacement has the correct operating voltage. This also applies to the damper diode when it is replaced.

ADDITIONAL DAMAGED PARTS

With a leaky horizontal output transformer or with arcover in the chassis, check for several parts that might be damaged. Often the chassis comes in with a no-sound, no-raster symptom. You might find a leaky horizontal output transistor, damper diode, or an open tuning capacitor and flyback transformer in the same chassis. High-voltage arc-over can cause a chain of events that damages small diodes in the secondary winding, as well as the output transistor and flyback. Look a little closer than usual because you might find several components damaged under these conditions.

RCA CTC111H chassis damaged components This particular chassis (Fig. 4-54) came in with a no-sound, no-high-voltage, no-raster symptom. Q404 was shorted and was replaced. No collector voltage was found at the horizontal output transistor. SCR100 was open and was replaced. The chassis was fired up and Q404 went out. Checking the output circuit, CR408 was leaky and CR409 had a 2-Ω leakage.

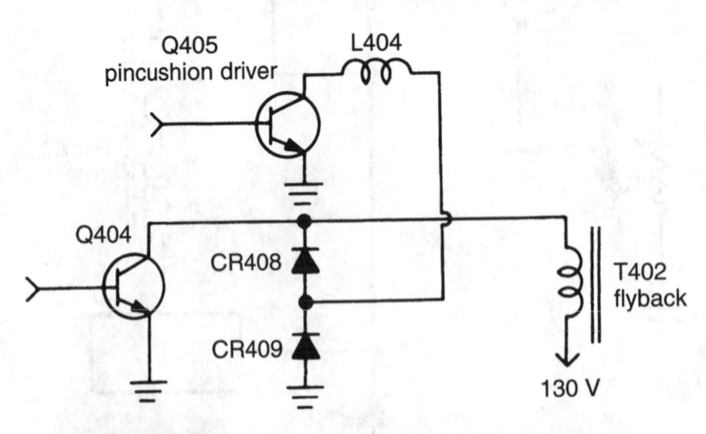

FIGURE 4-54 Take crucial voltage and resistance measurements to locate the leaky component in the horizontal output circuits.

After replacing both diodes with original parts, a line-voltage transformer was inserted in the line. As the line voltage was raised, Q401 was pulling heavy current. Right away, the flyback transformer was suspected (T402). The yoke assembly was disconnected without any results. Next, the pincushion circuits were checked and transistor Q405 was found to be leaky. The dc voltage came up at the output transistor, indicating that the overload circuit was lifted. The blown 5-A fuse, Q404, SCR100, Q405, CR408, and CR409 were replaced to put the chassis in operation.

Variac Transformer Tests

The variac step-up (or down) power-line transformer is the ideal test instrument for locating intermittents with defective flyback transformers and in high-voltage shutdown cases. With intermittents, the voltage can be raised above the normal powerline voltage to apply more voltage or heat to the intermittent component. To prevent damage to another horizontal output transistor, the dc voltage can be lowered until a leaky flyback or excessively high low-voltage source is located. The high-voltage shutdown circuits or the other components that are causing the shutdown can be found by applying lower voltage to the horizontal circuits.

Sometimes, when raising the variac line voltage to around 90 volts, the chassis might operate with narrow width, indicating that a line voltage regulator is defective. Monitor the voltage at the output of the line voltage regulator as the line voltage is raised. Monitor the base terminal of the output transistor with the oscilloscope. Suspect the line voltage regulator when voltage is raised higher and the chassis shuts down. Also, suspect the regulator if the chassis is dead and will only operate with about 87 to 90 volts applied from the variable line transformer.

HIGH VOLTAGE TOO HIGH

Excessive high voltage in the TV can cause the chassis to shut down, turn the raster into horizontal lines, arcover, and destroy components in the horizontal and high-voltage circuits. In the early HEW holddown circuits, the excessive high voltage would affect the horizontal oscillator circuit and cause the raster to go into horizontal lines or off frequency. Today, if the high voltage is too high, the high-voltage shutdown circuits shut the chassis down.

A higher-than-normal high voltage can cause the raster to expand or go out when the brightness control is raised in some TVs. The chassis might intermittently go into horizontal lines when adjusting the brightness control. First check the high voltage at the picture tube with the high-voltage probe. Adjust the B+ source or high-voltage control for lower voltage. The chassis might be shutting down with too high of a low-voltage adjustment.

Some TV chassis shut down quicker than others. Connect the chassis to a universal line transformer. Adjust the line voltage until the chassis shuts down. Under normal operation, you might find a few chassis will shut down at 122 or 125 Vac, and others can operate at up to 140 Vac. Monitor the high voltage with the high-voltage probe. Suspect a defective shutdown circuit if the high voltage shuts the chassis down too early. Check the power-line voltage at the customer's house for a higher-than-normal line voltage, which might shut the chassis down at certain times of the day.

HIGH-VOLTAGE SHUTDOWN

The chassis might shut down with a high power-line voltage, with a defective component in the chassis, or by action of the high-voltage shutdown circuits. Use a universal power-line transformer to help solve the shutdown problem. Slowly raise the ac voltage and notice at what voltage the chassis shuts down. Start all over and keep the line voltage lower than shutdown to determine what section is causing the chassis to shut down. A 100-W bulb in series with the open fuse can help you to locate the shutdown component.

To determine if the chassis is shut down by another component or if it is actually high-voltage shutdown, use the universal power-line transformer or remove a diode in the shutdown circuit. Often, one end of a diode or transistor can be easily removed from the shutdown circuit (Fig. 4-55). Monitor the high voltage with a probe and monitor the dc voltage source to the horizontal output transistor with a voltmeter. Be very careful at this point. Hold on to the ac plug as the chassis is plugged in. Do not leave the chassis on too long, so as to prevent damage to the picture tube, high-voltage, and horizontal circuits. Notice the high-voltage and low-voltage source measurements. If high-voltage shutdown is causing the problem, the chassis will remain on during this test. After making repairs, do not forget to replace the diode or transistor lead to the shutdown circuit.

Suspect that a component in the low-voltage source is defective if the dc voltage is higher than normal. If the low-voltage source is normal and the high voltage is too high, suspect a problem with the high-voltage regulator SCR, or the tuning or holddown capac-

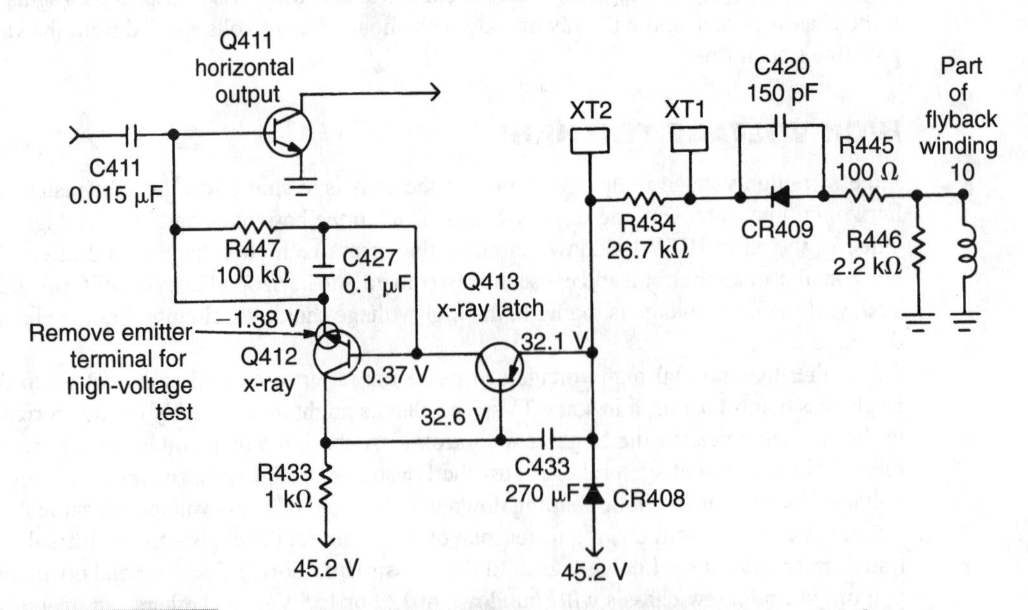

FIGURE 4-55 Remove one end of the diode or transistor from the circuit to determine if the high-voltage shutdown circuits are shutting down the chassis.

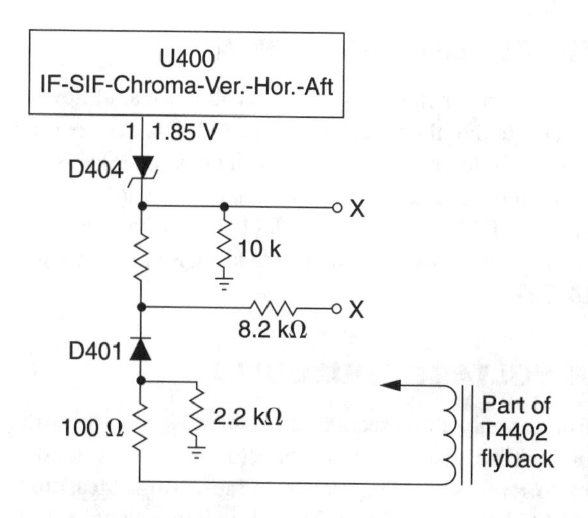

FIGURE 4-56 Short the two X connections to see if the shutdown circuits will shut down the chassis.

itor in the horizontal circuits. Look for a defect in the high-voltage shutdown circuits if the high voltage is normal and the chassis shuts down.

HIGH-VOLTAGE SHUTDOWN CIRCUITS

Most high-voltage shutdown circuits operate the same way, although some can be quite complicated and use several transistors and diodes. If the high voltage becomes too high, the shutdown circuit turns the chassis off. If the high-voltage shutdown circuits have a defective component, causing overloading, the chassis might shut down by itself. Disconnect the shutdown circuits momentarily to see if the high voltage at the picture tube is normal.

Some high-voltage shutdown circuits use a silicon-controlled rectifier (SCR) to kill the horizontal oscillator stage. Other circuits can kill the drive pulse at the horizontal drive transistor or at the defective IC. The primary concern of the high-voltage shutdown circuits is to protect the operator from radiation at the picture tube because of excessive high voltage.

Excessive high voltage can be caused by higher-than-normal low voltage at the horizontal output transistor. A leaky high-voltage SCR can apply excessive voltage to the output transistor. If the tuning or holddown capacitor goes open, the high voltage will increase. Use a universal line-voltage transformer to determine if the high voltage is shutting the chassis down or if the problem is a defective chassis component. A leaky or defective integrated horizontal output transformer can cause high-voltage shutdown.

Reconnect the diode or transistor terminal to the shutdown circuit after making repairs. A defeated high-voltage shutdown circuit is too dangerous to leave unattended. Be sure that the high-voltage shutdown circuit is working by shorting out a component. In Fig. 4-56, the shutdown circuits can be tested by shorting test points XT2 and XT1 together. Most chassis have an easy method to check the high-voltage shutdown circuits. The chassis should not operate with these two test points shorted. If the chassis does not shut down, suspect a faulty high-voltage shutdown circuit.

RCA CTC101A HIGH-VOLTAGE SHUTDOWN

With the variac transformer in the power line, this chassis would not shut down until 30 kV was measured at the CRT. One end of the holddown diode (CR419) was removed from the circuit to verify high-voltage shutdown. The collector voltage at Q100 was 139 V and should be about 121.5 V. In this model, a new regulator resistor trim component kit was installed to lower the dc voltage. CR115, R113, R114, R115, R116, and R117 were removed from the circuit. New components from the regulator kit (CR115, R113, and R117) were installed (kit number 146399)

SANYO 91C64 HIGH-VOLTAGE SHUTDOWN

The chassis and sound came on for only a few seconds and shut down. To determine if the chassis was shut down by high voltage, disconnect the collector terminal of shutdown transistor Q403. Use solder wick to remove the terminal from the wiring. Measure the high voltage if the chassis stays on (25 kV to 27 kV). If the high voltage is above 28 kV, suspect the high-voltage shutdown circuit.

Check the voltage at the horizontal output transistor (normally 114 V). Try to lower the dc voltage with low-voltage control. If the dc voltage is above 120 V to 150 V, the chassis will shut down. Suspect that a low-voltage regulator circuit is faulty if the low voltage cannot be adjusted or is too high. Notice if the high voltage varies with the low-voltage regulator control.

If the low voltage and high voltage can be adjusted and are normal with a 120-Vac line, suspect trouble in the high-voltage shutdown circuit (Fig. 4-57). Check for zero voltage at pin 8 of LA1461. Any voltage measurement at pin 8 indicates a defective high-voltage shutdown circuit.

Test Q403 in the circuit and replace it if it is leaky. Check all diodes—especially D404 (33-V zener), D405, D415, and D406. Remove one end of each diode for accurate leakage tests. Be sure that 12 V is found on the anode side of D405. Check all 1% resistors in the shutdown circuits. Doublecheck the emitter voltage on Q403. It should be lower than the base terminal.

RCA CTC97 BLAST-ON SHUTDOWN

Suspect that the flyback transformer is leaky if the sound blasts on, then the set shuts down. Often the horizontal output transistor is shorted and SCR101, the high-voltage regulator, is leaky. Remove the collector lead of Q38 to see if the chassis has high-voltage shutdown. If the high voltage does not come up and the chassis shuts down, insert a universal transformer in the power line. Slowly increase the ac voltage to 60-V or 80-Vac. Feel the case of the horizontal output transistor. Shut the chassis down if it gets warm. Replace the leaky flyback with a 145722.

TIC-TIC NOISE

If, with one ear close to the flyback transformer, a tic-tic noise is heard, the horizontal circuit might not have sufficient drive voltage to the output transistor, or it could be a defective flyback transformer, leaky tripler unit, or a loading down of the flyback in secondary circuits. In

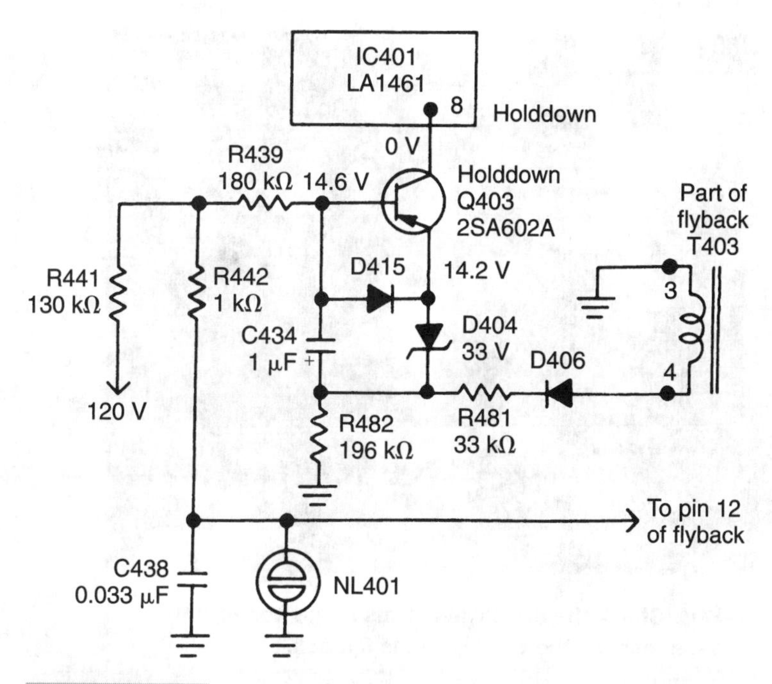

FIGURE 4-57 **Check components C434, IC401, D415, D404, and D406 if chassis shutdown occurs in a Sanyo 91C64 chassis.**

the modular chassis, remove each module, except the horizontal oscillator, to locate the defective module that is loading down the low-voltage circuits in the secondary voltage source of the flyback. Remove the transformer input lead to the tripler unit with a tic-tic noise to determine if the tripler is loading down the output transformer. Check the waveform at the base terminal of the output transistor to determine if horizontal circuits are normal (Fig. 4-58).

Take an in-circuit diode test with the DMM of each low-voltage diode tied to the secondary low-voltage source before suspecting the flyback transformer. A leaky diode or component tied to one of the low-voltage sources can load down the secondary circuits. Often, with a leaky secondary component, the high voltage will come up with sound, then die down in a few seconds.

DEFECTIVE DEFLECTION YOKE

A leaky horizontal yoke winding to the vertical winding can cause the fuse to blow in older solid-state TVs. Shorted turns in the horizontal deflection yoke can cause shutdown in the latest TV. Often, the chassis might operate for a few seconds before the defective yoke shuts down the chassis. The raster might come up, or only a quick flash might be in the picture tube before it shuts down.

Usually, the yoke loads down the flyback and horizontal output transistor with some of the high voltage with the variac transformer in the power line. The output transistor might run warm with an overloaded, leaky yoke assembly. Remove the red lead from the yoke assembly. The voltage will increase right away if the yoke is defective. Of course, the high voltage will be quite low with the yoke out of the circuit. Remove the yoke assembly and

FIGURE 4-58 **Check the drive waveforms in the horizontal circuits with the scope, all the way up to the flyback.**

check for leakage between the horizontal and vertical windings. It's very difficult to locate a shorted turn or two with the ohmmeter because the horizontal windings have a very low resistance. If the yoke loads down the voltage at the collector terminal of the horizontal output transistor, replace it. Do not overlook the possibility that a pincushion circuit, which is tied to the same spot as the yoke assembly, is leaky.

CHASSIS SHUTDOWN

Practically any component in the horizontal circuit can produce chassis shutdown. The horizontal output transistor, high-voltage regulator, flyback, and IC oscillator are the most likely to fail. The chassis might shut down if the component terminal connections are poor or if the transistor sockets are loose. Chassis shutdown can occur at once or several hours later. Check for arcing sounds in the flyback if the chassis shuts down. The defective component might be intermittent.

Approach the shutdown problem as an intermittent or high-voltage problem. Use a variable line-voltage transformer in IHVT circuits. Take scope waveforms with crucial voltage measurements. Isolate the high-voltage shutdown circuits temporarily to a defective chassis component by removing the lead of the transistor or diode from the high-voltage shutdown circuits.

RCA CTC93 chassis shutdown After an hour or so, this RCA chassis would quit operating (Fig. 4-59). At the collector terminal of the horizontal output transistor, no voltage was measured when the chassis shut down. The open high-voltage regulator (SCR600) was located with voltage tests at the collector and anode terminals when the chassis shut down.

Poor socket connections in an RCA CTC87A chassis would intermittently shut it down. Either replace the socket or solder the SCR terminal pins to the socket temporarily.

POOR FOCUS

Poor focus can result from a low high voltage at the CRT. Check the anode and focus voltage at the picture tube with the high-voltage probe. Intermittent focus problems can be caused by poor picture tube pin connections. A defective focus control can produce erratic focus problems (Fig. 4-60). Isolate the focus problem to the focus or horizontal circuit with low high voltage. A defective IHVT can cause poor focus problems.

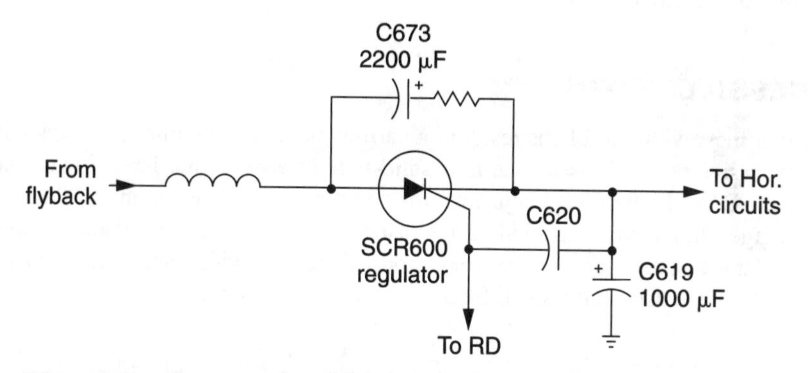

FIGURE 4-59 Regulator SCR600 caused the chassis to be intermittent in an RCA CTC93 chassis.

FIGURE 4-60 Check the focus control for erratic focus and replace it with a new one, if necessary.

POOR WIDTH

Check the high-voltage regulator circuits for poor picture width in the latest TV chassis. The high-voltage regulator transistor, SCRs, and zener diodes in the regulator circuits can produce insufficient width. Poor solder connections at pincushion, regulator, and driver transformers can result in poor width. Low voltage applied to the horizontal output circuits can cause poor width. Open bypass or coupling capacitors in the horizontal and high-voltage circuits also can cause poor width.

EXCESSIVE WIDTH

Although most width problems result in a narrow picture, sometimes a chassis comes in with excessive width. Most of the new solid-state chassis do not have a width control. Check the B+ or high-voltage adjustment for correct high voltage at the CRT. Excessive low voltage can increase the width of the raster. This can be caused by a leaky low-voltage regulator transistor or an open zener diode. Excessive width can be caused by a high B+ boost voltage. Check the small filter capacitor in the boost circuit.

POOR HIGH-VOLTAGE REGULATOR

An improper voltage applied to the horizontal output transistor can result from a poor high-voltage regulation circuit. Leaky or intermittent high-voltage regulator SCRs can produce poor regulation (Fig. 4-61). A high regulated voltage applied to the horizontal output transistor can cause chassis shutdown in the IHVT chassis. Check the high-voltage regulation circuits for poor voltage fed to the horizontal output stage. Replace the high-voltage regulator module for poor regulation in a modular chassis (Fig. 4-62). A regulator transistor or zener diode in the regulator circuit can cause poor high-voltage regulation.

RCA CTC68 cannot adjust high voltage The horizontal frequency was off and the raster would go into horizontal lines intermittently in this chassis. The high-voltage control could change the high-voltage setting, but the high voltage would constantly shift high and low. The open Q401 was replaced and it immediately burned out (Fig. 4-63). Replacing the open CR404 and Q401 solved the poor high-voltage regulation problem.

IMPROPER B+ OR HIGH-VOLTAGE ADJUSTMENTS

You might find both a B+ and a high-voltage adjustment control in early TV sets. Some of the latest TVs do not have either control. Adjust for the correct voltage at the horizontal fuse or collector terminal of the horizontal output transistor with the B+ control. Set the high-voltage control with correct voltage at the anode socket of the picture tube. These voltages are listed on the schematic.

Lower the B+ voltage if the line voltage is high at the customer's home. This often happens with rural power lines. Now readjust the high-voltage control. Suspect that the high-voltage regulator circuit is defective if the B+ voltage has to be lowered with higher-than-normal

FIGURE 4-61 **Suspect a high-voltage regulator if the voltage is missing at the horizontal output transistor.**

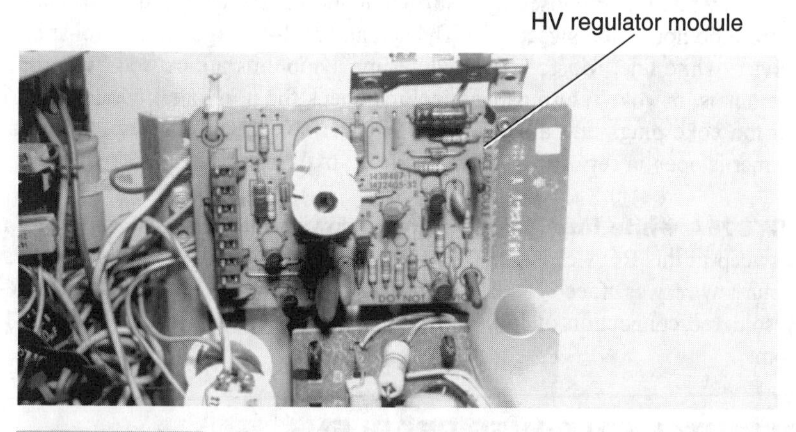

FIGURE 4-62 **Replace the high-voltage regulator module if no voltage is at the horizontal output transistor.**

voltage at the collector terminal of the horizontal output transistor. The wrong zener diode replaced in a high-voltage regulator circuit can produce improper high voltage. Doublecheck the voltage of a replaced zener diode. Lower the B+ voltage as far as possible (while retaining adequate raster width) in extreme high-voltage areas.

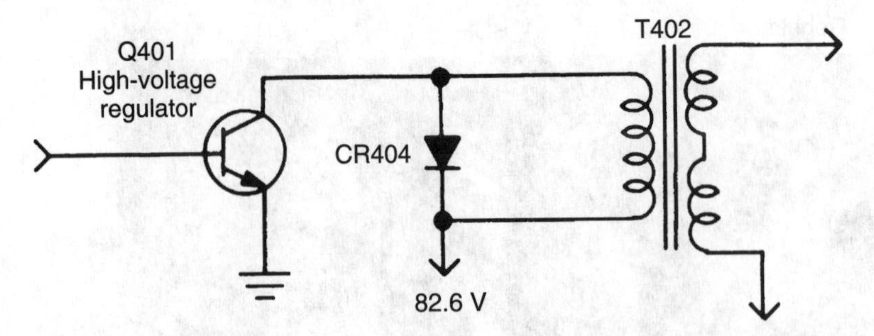

FIGURE 4-63 The high-voltage adjustment will not vary the voltage if CR404 is leaky or shorted.

FLASHING HORIZONTAL PICTURES

Check the flyback transformer for poorly soldered connections or loose wires if the picture is flashing. Often the whole raster will flash off and on. The raster might flash across the screen and go dead. Poor PC wiring or board connections can cause a flashing picture. Inspect the yoke socket for poorly soldered or loose wire connections. A poorly soldered input terminal connection from the flyback to the tripler can cause flashing in the horizontal circuits. Discharge the CRT and refasten the high-voltage cable to the picture tube. Sometimes the high-voltage cable is not correctly inserted into the CRT socket.

VERTICAL WHITE LINE

A horizontal white line is caused by insufficient vertical sweep, but the vertical white line results from no horizontal sweep. The flyback and high-voltage circuits must be working to receive a white line. Most vertical white-line symptoms are caused by an open yoke, yoke terminals, or yoke return capacitor. First check the horizontal winding and connections of the yoke plugs into a socket. Do not overlook the possibility that a pincushion transformer is open in series with the horizontal yoke return winding.

RCA CTC76A white line Only the up and down white line was found with no horizontal sweep in this RCA chassis. After the yoke winding and terminals were tested, the yoke return wire was traced to the pin-phase transformer, R5, or L401 (Fig. 4-64). A poorly soldered connection of L401 to the PC wiring resulted in the vertical white-line symptom.

HORIZONTAL FOLDOVER PROBLEMS

Horizontal foldover can be caused by a defective damper diode, safety or holddown capacitors, capacitors in pincushion circuits, and the pincushion transformer. A leaky horizontal output transistor, trace or retrace SCR, or flyback can cause foldover. Open bypass capacitors in the high-voltage regulator or SCR horizontal circuits can cause some type of horizontal linearity or foldover. Open or poorly soldered terminals of reactors and the regulation transformer can cause horizontal foldover (Fig. 4-65).

PECULIAR HORIZONTAL NOISES

Many different horizontal noises are caused by a component in the horizontal circuits. A high-pitched, high-voltage squeal or a ringing noise can be caused by loose particles or by vibration in the flyback transformer. Sometimes the flyback bolts need to be tightened to eliminate the noise. Pinning the flyback winding core with toothpicks and glue might solve the vibration or singing problem. Replace an IHVT flyback that suffers from loose particles (Fig. 4-66).

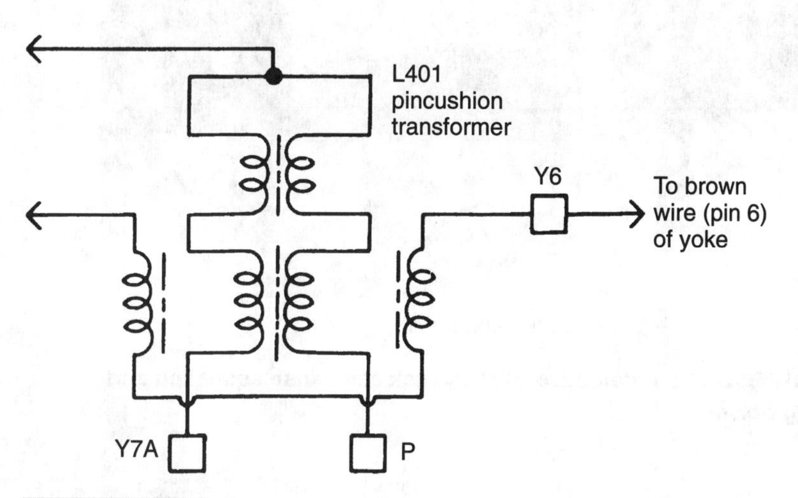

FIGURE 4-64 **Check the pincushion terminal connections for bowing or a white line in the picture.**

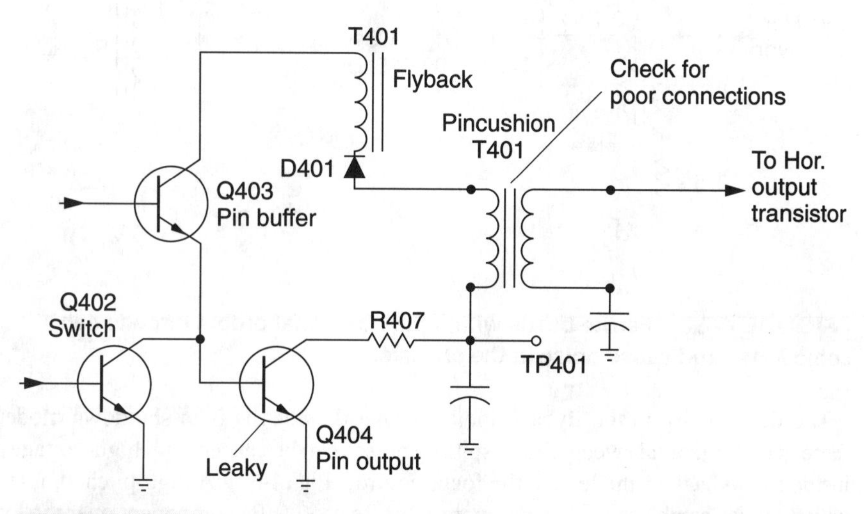

FIGURE 4-65 **Solder all pincushion transformer terminals and check the pin output transistor for leakage in late-model pincushion circuits.**

FIGURE 4-66 **A defective IHVT flyback can cause squealing and ringing noises.**

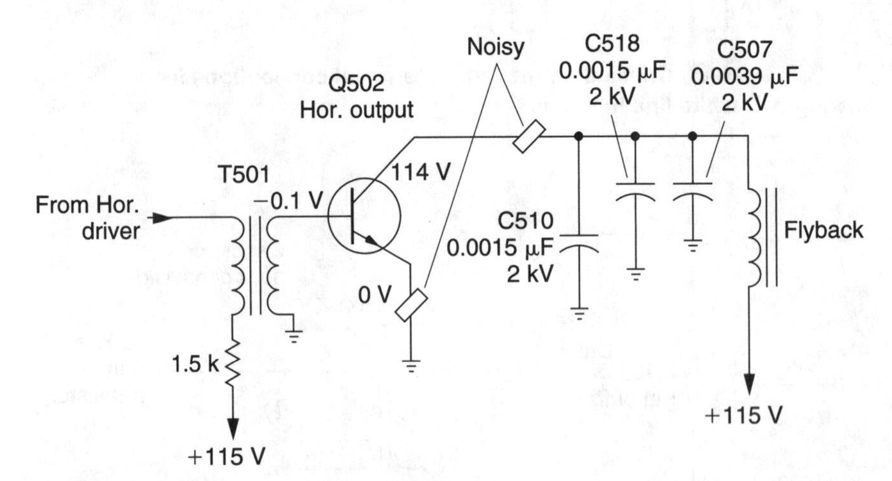

FIGURE 4-67 **Ferrite beads within the horizontal output circuits can come loose and cause noise in the chassis.**

The tic-tic noise in the flyback indicates that the chassis is in shutdown mode or that there is no horizontal sweep. Small spitting noises can be caused with high-voltage arcover inside the flyback or the lead to the focus control (Fig. 4-67). A high-pitched, intermittent whistle noise can be caused by a vibrating ferrite bead (FB) component over the leads of a diode or transistor element in the horizontal circuits. The "firing" noise with a high-pitched squeal can be caused by arcover in a focus control.

NOISE LINES IN THE RASTER

A firing flyback transformer can cause noise lines in the picture—even if you cannot hear the internal arcing noise. Poorly soldered connections on the flyback terminals can cause line noise in the raster. Be sure that the picture tube shield is grounded to the metal TV chassis. Check the high-voltage cable to the CRT for breaks or for a loose screw in the anode clip. Notice if the high-voltage lead is clipped properly into the picture tube.

Excessive arcing inside the focus control can cause firing lines in the raster. Simply rotate the focus control and notice if the lines disappear. Replace the defective focus assembly. Horizontal lines that look like auto ignition noise can be caused by open filter capacitors in the B+ voltage-regulation circuits. Remove the antenna lead to determine if the noise is picked up or if it occurs within the TV.

ARCOVER AND FIRING

High-voltage arcover can occur at the CRT, flyback, or tripler unit. Often, you can see it arcing over. Sometimes the high-voltage arcover at the picture tube anode will show lines of firing down to the banded area of the CRT. Replace arcing flyback and tripler units. Arcover can occur at spark-gap assemblies, which indicates excessive voltage or a defective component nearby.

Usually, arcover inside the picture tube gun assembly indicates that a tube filament is open or that an assembly is open. After replacing the defective component that is causing the arcover, check the B+ and high voltage. Service the high-voltage regulator or low-voltage circuits that produce excessive voltage at the picture tube.

RCA CTC92L DARK LINES IN PICTURE

Two dark lines appeared at the top and bottom of the screen and rolled upward. It looked like 120-Hz hum. All filter capacitors were shunted and the results were the same. CR305 was found to be leaky in the startup diode circuit at T201 (Fig. 4-68). Any one of the

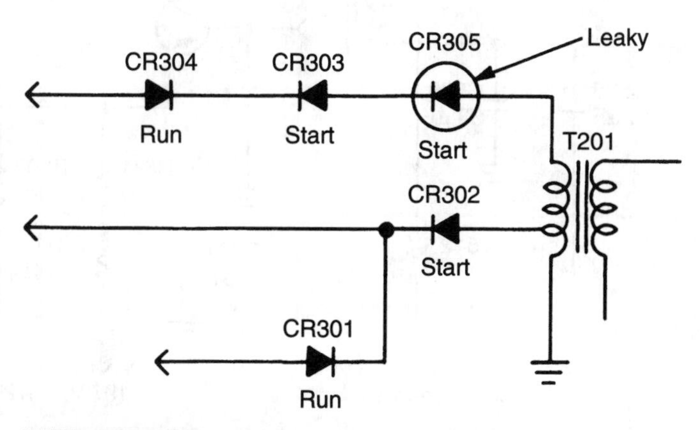

FIGURE 4-68 Check for a leaky startup diode, which will cause two dark lines to roll up the screen.

startup diodes can cause the dark bars in the raster. Replacing the start diode solved the noisy line problem.

JAIL BARS

In the tube chassis, vertical bars to the left or right of the raster indicate Barkhausen oscillations. A defective flyback or horizontal output transistor can cause vertical firing lines in the raster. Check the horizontal output and high-voltage components for vertical lines in the raster.

Sony KV-2643R jail bars Several dark vertical bars to the left of the raster were visible on this TV (Fig. 4-69). Sometimes, if the contrast and brightness were adjusted, the bars could barely be seen. Resistor R812 was found burned in the pincushion circuits. Replacing both electrolytic capacitors (C535 and C539) in the horizontal output circuit solved the problem. Both of these capacitors are mounted close to the flyback transformer area.

PIE-CRUST LINES

The pie-crust lines often appear to the left side of the picture. These lines can be intermittent and can vary with the brightness control. The pie-crust effect can occur with open resistors across the clamp diodes or poor filtering in the horizontal stages.

J.C. Penney 685-2041 horizontal pie-crust lines These unusual pie-crust horizontal lines were caused by a defective electrolytic capacitor in the horizontal oscillator circuits (Fig. 4-70). The 3.3-μF capacitor was replaced by one with a 250-V rating.

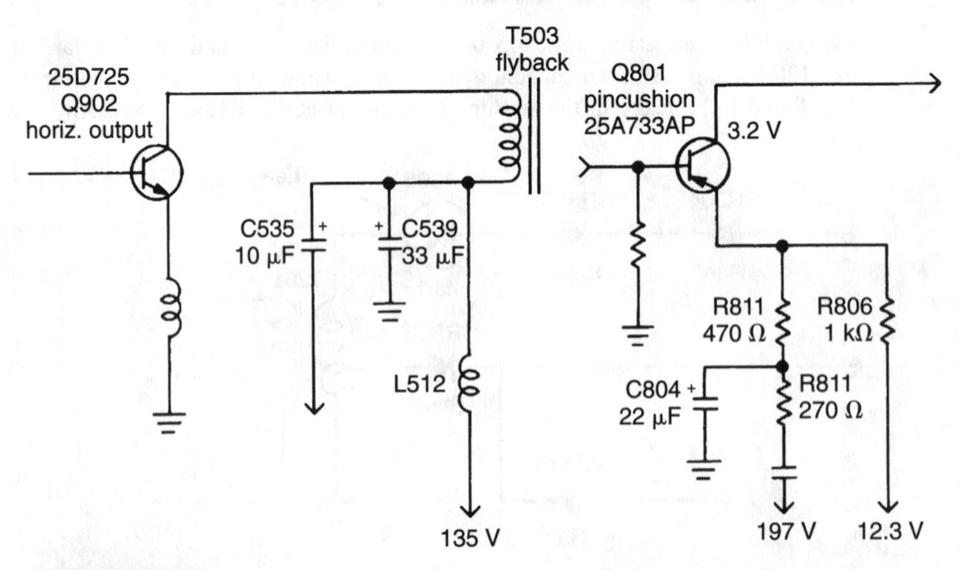

FIGURE 4-69 Open electrolytic capacitors cause "jail bars" to the left side of the picture.

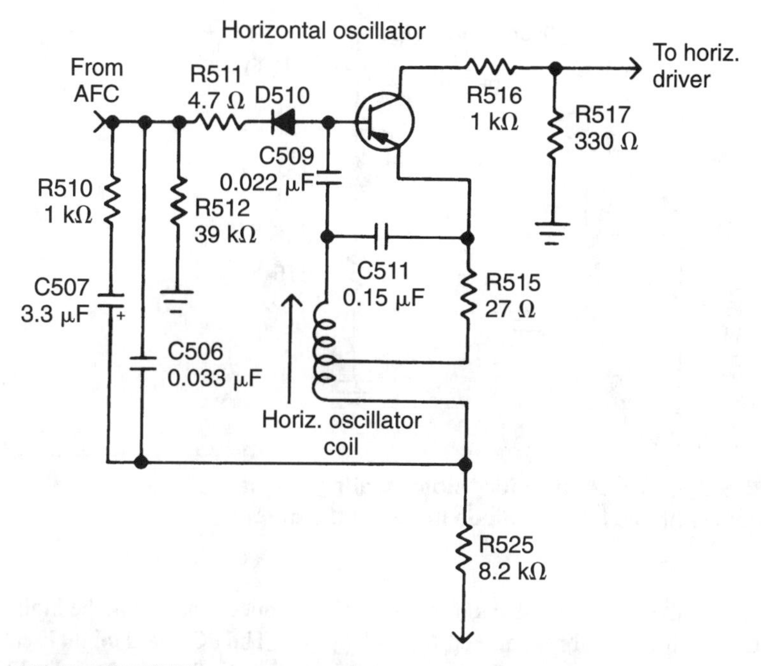

FIGURE 4-70 Replace C507 in a J.C. Penney 685-2041 that shows horizontal "pie-crust" lines.

HORIZONTAL MOTORBOATING

A putt-putt noise in the horizontal output circuit can be caused by a defective output transistor or by a corresponding component. Check for an open emitter or base resistor in the horizontal output transistor circuit. Replace the output transistor when no emitter resistors are located in the circuit. These transistors might test normal and still oscillate under load.

Montgomery Ward C1935 motorboating The neon dial lights would blink, demonstrating that the B+ voltage was unstable. At first, the trouble suspected was poor low-voltage regulation. When Q602 was removed from the circuit, the motorboating sound quit. Replacing the horizontal output transistor (2SD870) did not solve the problem (Fig. 4-71). Installing a new emitter resistor (R615) solved the motorboating condition. Evidently, Q602 had shorted, taking out the emitter resistor.

POOR BOARD CONNECTIONS

Poor board or component terminal connections can produce a dead chassis, intermittent horizontal section, and high-voltage troubles. A poorly soldered connection around a regulator or pincushion transformer can cause the raster to narrow. The poorly soldered terminals of any horizontal transistor can cause an intermittent picture. Check all eyelet feed-through terminals that can cause improper voltage to a horizontal circuit.

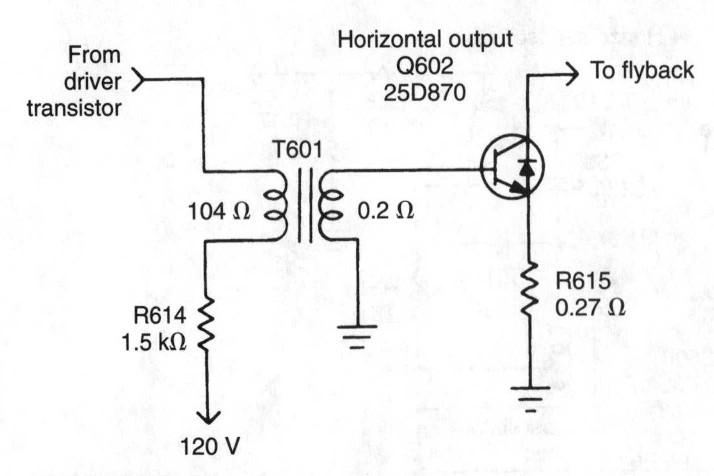

FIGURE 4-71 Check R615 for "motorboating" in the output circuit with the damper diode inside of the output transistor.

PC wiring might break at a flyback transformer lug connection, killing the high voltage. Sometimes the wiring will break at feed-through eyelets. The PC board might break where the flyback transformer is mounted, if the chassis is dropped or knocked off the TV stand. Often, heavy components, such as transformers will cause the board to crack when the TV is dropped. Repair the PC wiring with short lengths of hookup wire across each broken area. Check for broken PC wiring where the wiring joins a large terminal post. Check for PC wiring breaks at the various controls if the set is dropped on the front side.

SOLDERING IC TERMINALS

Look for cracked board areas around large IC processors. Repair the cracked wiring with hookup wire. Sometimes the intermittent symptom might be caused by poor pin connections of the IC. If the chassis is carried or moved around, the contacts might break.

Solder each pin with a small soldering iron. The battery-operated iron is ideal for small, delicate connections (Fig. 4-72). Use very small solder (0.015) for IC work. Do not hold the iron on too long, just long enough to make a good, flowing joint.

SCR HORIZONTAL CIRCUITS

In the early modular chassis, two SCRs were used in the sweep circuit in place of a regular horizontal output transistor. The same SCR components were used in the RCA chassis, starting with the CTC40, CTC46, CTC48, CTC58, and CTC68 chassis (Fig. 4-73). You will find the same sweep circuits in Philco, Coronado, J.C. Penney, and other models during this same period. Although the SCR sweep circuits were fairly reliable, most defective components were common to all chassis.

Retrace and trace switch SCRs provide a horizontal sweep pulse to the flyback transformer. Often SCR101 is open and SCR102 becomes leaky. CR402 and CR401 provide damping action for SCR102 and SCR101 respectively (Fig. 4-74). If CR402 goes open,

SCR102 will become shorted. Replace the small insulator with silicone grease when installing a new SCR.

Take a voltage check at the case of each SCR with a dead symptom. Normal voltage at SCR102 and no or low voltage at SCR101 can indicate a normal retrace switch with

FIGURE 4-72 Solder the pins around the IC terminals with a low-wattage soldering iron.

FIGURE 4-73 Two SCRs developed the drive to the flyback transformer in the early TV chassis.

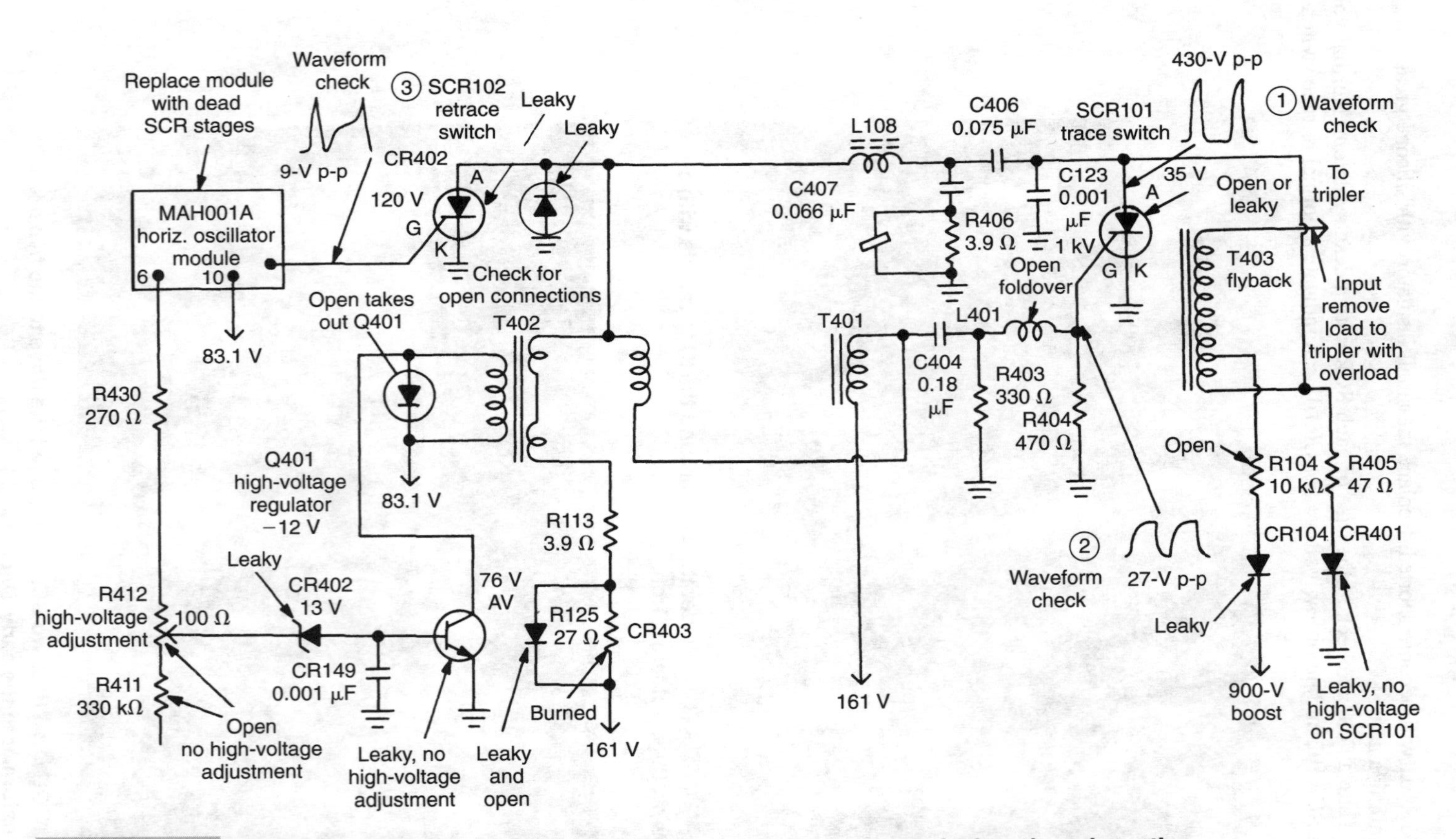

FIGURE 4-74 The various problems caused by defective components are marked on the schematic.

a defective trace component. Low voltage at SCR101 can be caused by a leaky tripler unit. Remove the tripler input lead for further tests. A quick scope test at SCR101 will indicate if the normal drive pulse is applied to the flyback. Remove both SCRs and test them out of the circuit with no drive pulse. Disconnect one end of CR402 and CR401 and test for leakage.

HIGH-VOLTAGE PROBLEMS

Excessive high voltage will cause these chassis (Fig. 4-75) to go into horizontal lines. The horizontal might sync in when the brightness is raised and lowered. Try to adjust the high voltage with R412. If not, troubleshoot the high-voltage regulator circuits. Check Q401, CR402, and CR404 for leakage. Remove one lead of each diode for accurate leakage tests. Check the primary of T402 for a correct resistance of 37 Ω.

Poor horizontal and vertical sync can occur with 32- to 35-kV high voltage at the picture tube. Excessive high voltage can be caused by poorly soldered terminals on T402. Poor terminal connections on T402 and T401 can produce intermittent and high-voltage problems. Replace CR402 with a 13-V, 1-W zener, diode. If CR404 opens, Q401 is destroyed. Q401 can be replaced with a universal SK3024 transistor if the original cannot be found.

Servicing the RCA CTC110A chassis Many of the latest RCA chassis can be serviced in the same manner as the CTC110A chassis. For that matter, several different brands of TVs can be serviced with the following service procedures. After blown fuses have been replaced in sets with a no-high-voltage/ no-raster/ no-sound symptom, determine if the chassis has high-voltage shutdown or a defective component. With this particular chassis, the flyback circuit must perform or no voltage is applied to the horizontal oscillator and other circuits.

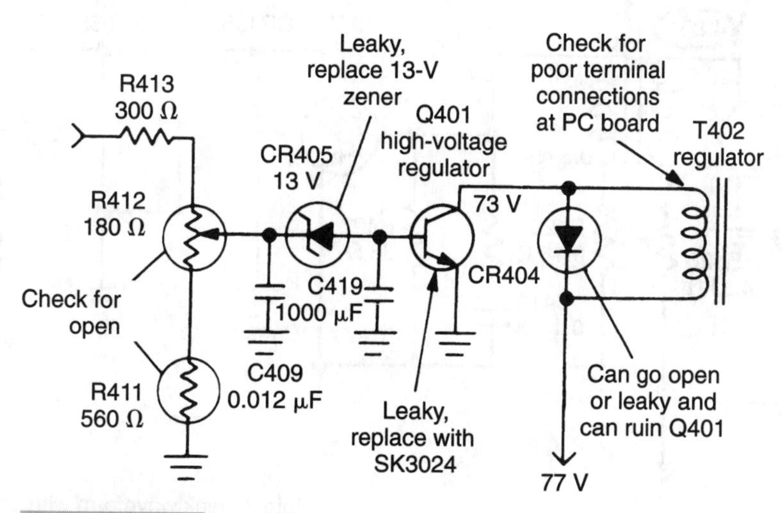

FIGURE 4-75 Check the various components in the high-voltage regulator circuits.

Use a universal power-line transformer to set the voltage at about 60 to 80 Vac. Notice if the chassis has high-voltage shutdown. Another method is to remove the emitter terminal of the x-ray latch transistor (Q414) to prevent high-voltage shutdown. If the high voltage does not come up, suspect that a horizontal circuit is defective.

Take a waveform check at the output transistor. Without a drive pulse, measure the voltage at the collector terminal of the output transistor (Q412). A normal voltage should be around 121 V (Fig. 4-76). A lower voltage can indicate an output transistor or damper diode is leaky. Remove Q412 from the circuit and test both components for leakage. If the voltage is high, suspect that the drive signal is improper or that the high-voltage regulator circuit is defective. In this particular case, 159.7 V was found at the regulator with output of 132.5 V at the collector terminal of the output transistor, indicating that the high-voltage regulator circuits were working.

A low collector voltage was found at Q411 (4.8 V). This indicated that no drive signal was at the base of the horizontal driver or that Q411 was leaky. Because the flyback circuits were not working, the 19.9 V feeding the horizontal was missing. To determine if the horizontal or output circuits were defective, the external voltage supply must be connected to the 19.9-V supply. The low-voltage source must function to provide drive signal to the horizontal output circuits. A universal power source is ideal in this situation.

Remove the power cord and connect the external voltage source to the 19.9-V source in the horizontal circuits (Fig. 4-77). Adjust the external voltage to 19 V or 20 V. The horizontal output transistor (Q412) should be removed for these tests. Check the waveform at the base of the horizontal driver transistor (Q411) with the scope. Now plug the chassis into the power line. Signaltrace the waveform from the horizontal switch through to the

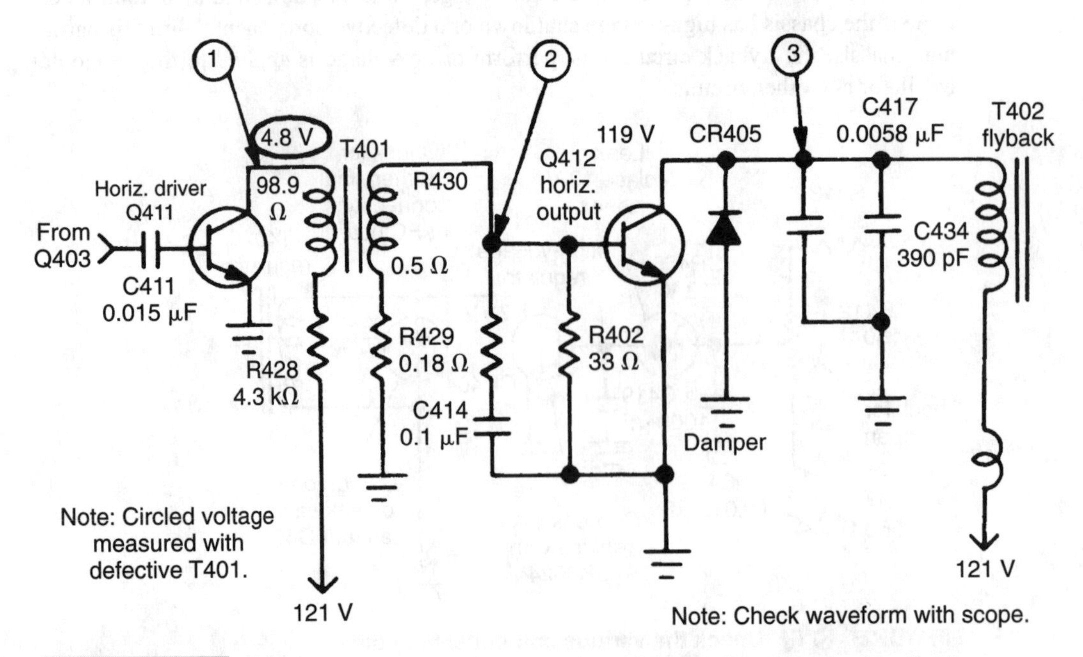

FIGURE 4-76 Check the voltage source and waveform at the collector terminals of Q412.

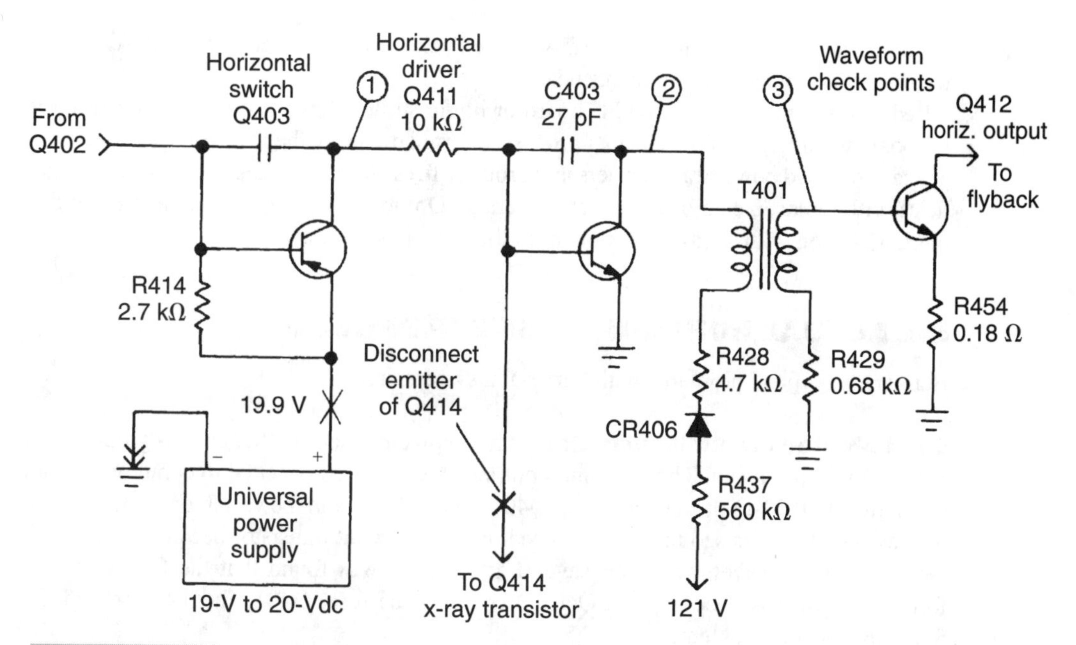

FIGURE 4-77 **Connect an external power supply to determine if the horizontal circuits are working.**

base of the output transistor. Suspect problems in the output circuits if a normal waveform is found at the base of Q412.

In this particular case, the drive signal was good at the base of Q411 and no signal was found at the collector terminal. Q411 was checked for leakage out of the circuit. Q411 was replaced and still no waveform was found. Both R428 and the primary winding of T401 were checked for the correct resistance. When the chassis was reconnected, the waveforms were good at the base of the output transistor and collapsed in 30 minutes. The primary winding of T401 had an internal short, yielding a resistance of 0.05 Ω, instead of 98.9 Ω. Replacing the shorted driver transformer solved the no-raster/no-high-voltage symptom.

The TV chassis with a flyback transformer (IHVT) must be serviced in a slightly different manner than the standard output circuit. First, determine if the chassis is shut down with high voltage or a bad component in the chassis. Use a universal power-line transformer or remove a leg of a transistor or diode in high-voltage shutdown circuit to determine if high-voltage shutdown is the problem. Next, determine if the horizontal circuits are functioning with an external power source and monitor the circuits with the scope.

FIRE DAMAGE

Components that can cause fires in the TV chassis are the deflection yoke, flyback transformer, degaussing coils, overheated resistors, large power resistors, and transformers. Replace these defective components with parts that have the exact part number as the originals. A stuck relay or relay transistor circuit can cause the relay to stay on and apply power to the TV. A defective relay with a weak release spring or broken leaf assembly might let the relay on and not shut off. A shorted yoke or flyback transformer can cause the windings to burn other components nearby.

Red-hot resistors close to the PC board or liquid spilled down inside TV can cause the PC board to undergo a slow burning process. Shorted diodes in the low-voltage power supply can overload power transformers and produce fires. Always replace line and secondary fuses with replacements of the exact amperage. Do not place larger-amperage fuses that might not open with a leaky component in the voltage sources.

SIX ACTUAL HORIZONTAL SWEEP PROBLEMS

Here are six typical problems with horizontal sweep circuits.

Dead-shorted output transistors After replacing the fuse, the chassis groaned and out went the fuse again. The horizontal output transistor was checked to common ground and showed a 0.13-Ω short (Fig. 4-78). Q401 was replaced and the variable isolation transformer was slowly raised to 65 volts. The horizontal output transistor became warm, with very little drive waveform. After several attempts, it was found that the flyback transformer (T408) was knocking out Q401. Replacing the fuse, Q401, and T408 solved the Sylvania 25C504 problem.

Destroys output transistor in a few hours After the horizontal output transistor was replaced in an Emerson MS1980 TV, the chassis shut down in about three hours. Again, Q401 was checked, and it was found to be shorted. Replacing the transistor once again and slowly raising the line voltage with the variable isolation transformer did not cause the chassis to shut down.

Crucial voltage and waveform tests were made on Q401 and the horizontal driver transistor. The collector voltage upon the driver transistor was too high, according to the schematic. C435 (1 μF, 160 V), an electrolytic capacitor, was replaced, which eliminated

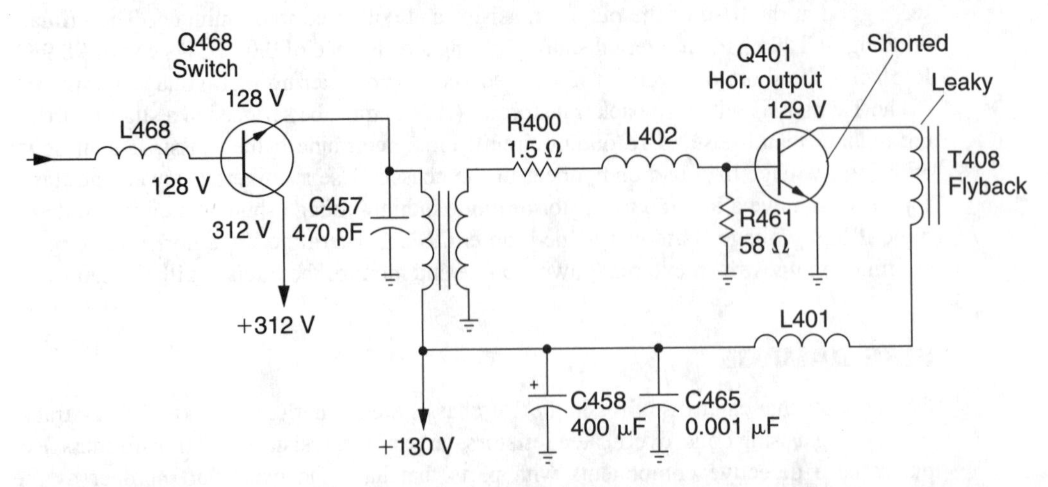

FIGURE 4-78 A shorted Q401 was caused by a defective flyback (T408) in a Sylvania 25C504 chassis.

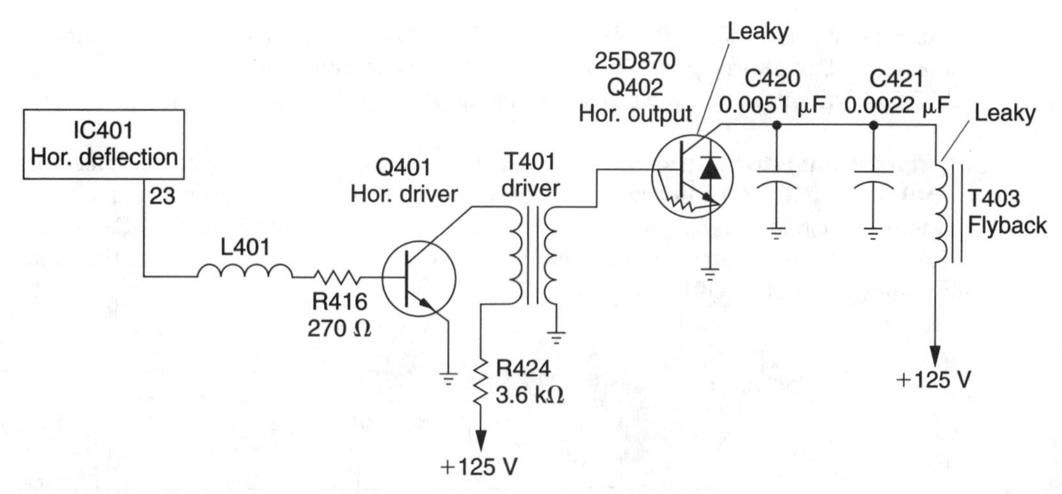

FIGURE 4-79 Replacing Q402, C420, and T403 solved the horizontal problems in a Samsung K-25 chassis.

the damaging of the high-output transistor. Replace small-capacity capacitors in the collector circuit if they keep destroying the output transistor

Blown fuse—shorted output transistor The main line fuse in a Strung K-25 chassis would open after replacement, indicating an overload in the power supply or horizontal output circuits. Q402 (25D870), a horizontal output transistor was found to be leaky. After replacing Q402, the chassis still blew the main fuse. A resistance measurement to collector of Q412 and common ground indicated 0.27 Ω (Fig. 4-79).

The output transistor was removed and it tested normal. Again, a resistance test was made from the flyback and collector terminal with the same resistance. C420 and C421 checked normal. T403 was found to be shorted. Before the repair was finished, the main fuse, line-voltage regulator (IC801), Q402, and T403 were replaced.

Dead, no start—RCA CTC109A The TV chassis (Fig. 4-80) seemed to want to start, but only a buzz could be heard in the flyback transformer. High voltage was found on the collector terminal of Q412 (137 V). The high-voltage regulator was checked, with 178 V at the anode terminal of SCR101 and 137 V at the gate and cathode terminals. Although the voltage was high, the high-voltage regulator SCR was functioning with a voltage drop of 41 V.

Often, a higher voltage found at the output collector terminal is caused by a defective high-voltage regulator circuit or by open output transistors. Because the regulator circuit was working, Q412 was checked for opens and leaks. Remove Q412 from the chassis because T401's secondary winding will show a short between the base and emitter terminals. Replacing the open output transistor (Q412) between the emitter and base terminals cured the dead chassis.

No regulator voltage—RCA CTC115A No voltage was measured at the output transistor in a 9-inch portable color TV. Voltage measurements of 117.6 V were found at the gate and cathode terminals of the high-voltage regulator (SCR101), and no voltage was found at the anode terminal. Because U401 triggers SCR101, voltage measurements were found low at pins 5 and 17 (0.5 V). After pins 5 and 17 were removed from the PC wiring, a leakage of 54 Ω was found from pin 5 to the chassis ground. Replacing R401 with a 149253 IC kit solved the no-picture/no-sound regulator problem (Fig. 4-81).

Output transistor keeps blowing After replacing the horizontal output transistor (Q402) in the RCA CTC149 chassis, the fuse would blow and the horizontal output transistor was shorted. After applying an external voltage source to the deflection IC, no vertical or horizontal output pulses were noted. After several hours of frustration, the countdown crystal (Y3301) was found to be defective.

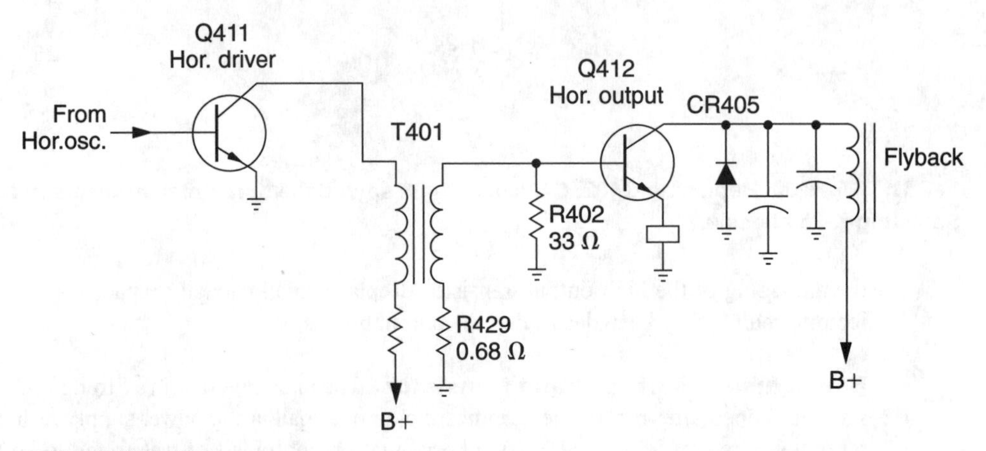

FIGURE 4-80 **Remove Q412 from the chassis for an accurate open test.**

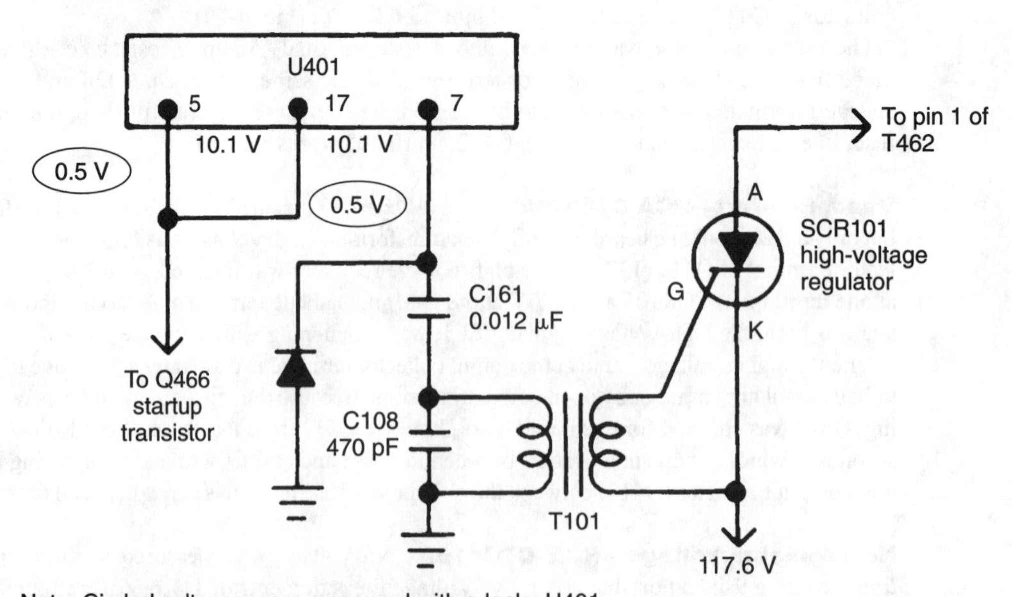

FIGURE 4-81 **A leaky ICU401 caused chassis shutdown in the RCA CTC115A chassis.**

FIGURE 4-82 The new horizontal output transistors might look the same as the originals on the schematic, but they have a TO-218 body.

TABLE 4-1 CHECK THE HORIZONTAL SWEEP CIRCUITS TROUBLESHOOTING CHART

WHAT TO CHECK	HOW TO CHECK IT
No horizontal sweep.	Check voltage to flyback and horizontal output transistor.
Voltage normal, no sweep.	Check waveform on base of output transistor.
No waveform on horizontal driver transistor.	Check output waveform on sweep IC.
Check voltage on horizontal sweep IC.	Check low voltage source (V_{CC}) to sweep IC.
Is sweep IC defective?	Inject dc voltage to IC sweep and monitor output with a scope.
Check circuit diagram for correct voltage applied to sweep IC.	Unplug ac cord for this test.

The new horizontal output transistor The new horizontal output transistor from the latest TV circuits resemble the same type on the schematic, except that it has a TO-218 mounting. This output transistor looks like the TO vertical transistors. Q4400 (179743) found in the RCA CTC140 chassis is an NPN, silicon, 8-A, 1500-V replacement (Fig. 4-82). Although the universal replacements can dissipate 125 watts, they seem to run warmer than the older output transistors. See Table 4-1 for the horizontal troubleshooting chart.

5

HIGH-VOLTAGE TESTS

CONTENTS AT A GLANCE

Problems with HV Circuits

Although the high-voltage circuits do not produce quite as many problems as in the old tube chassis, some new and old service problems still arise. Because the tripler unit replaces the high-voltage rectifier, fewer flyback transformers need to be replaced in the new TVs. The tripler unit, which had many service problems at first, was replaced with the integrated flyback transformer with high-voltage diodes molded inside (Fig. 5-1). These transformers have produced various service problems.

Besides the high-voltage rectifier molded inside the flyback transformer, additional windings are wound on the same core, providing low voltage to many dc circuits (Fig. 5-2).

FIGURE 5-1 The modern integrated horizontal output transformer (IHVT) contains high-voltage diodes molded inside the flyback.

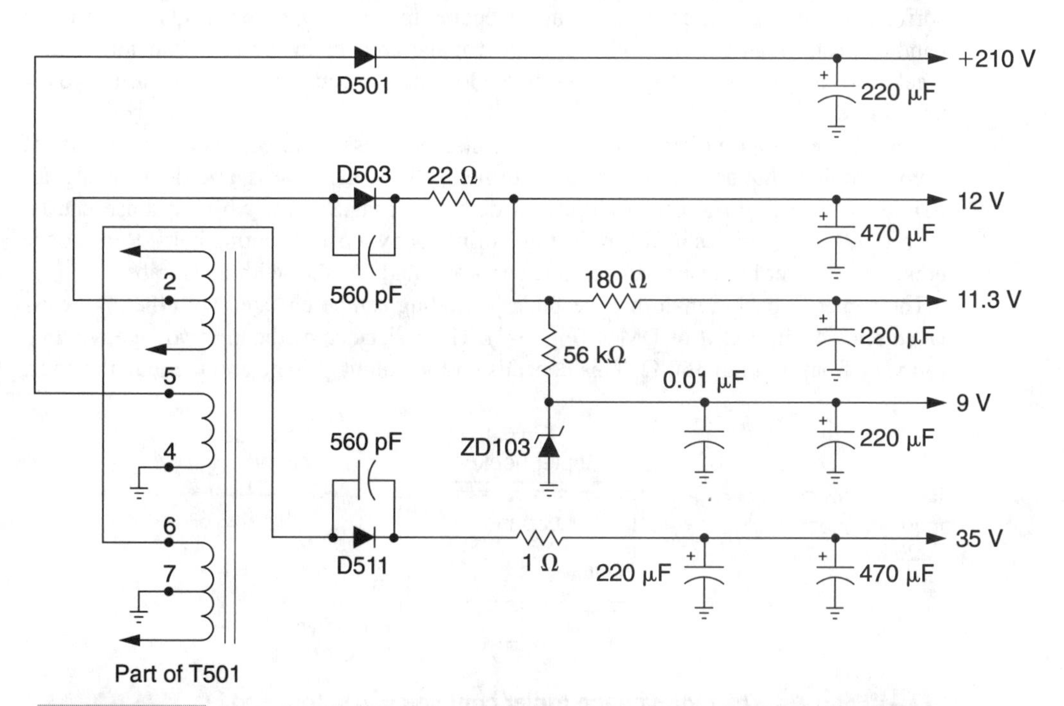

FIGURE 5-2 The IHVT can supply voltages to the screen grid and focus grid, and provide several different low-voltage sources from separate windings on the secondary of the flyback.

You might find a separate filament winding for the CRT on the flyback transformer. These additional windings shutdown the chassis if the circuits are overloaded by a leaky component in the connecting circuits. Do not overlook the possibility of a high-voltage shutdown problem in the flyback secondary circuits.

A defective tripler unit and integrated horizontal output transformer can destroy horizontal output transistors. The tripler or flyback can arc over, producing a loud arcing noise. Sometimes the defective integrated flyback only makes a tic-tic noise with chassis shutdown. Poor focusing of the CRT can be caused by a defective flyback or tripler unit. Servicing the new high-voltage circuits might require different troubleshooting techniques, but they are rather easy to service after a few have crossed your service bench.

FLYBACK TRANSFORMER PROBLEMS

The horizontal output or flyback transformer is physically small, compared to the old tube TV transformers. Very few service problems are found with the flyback transformer tied into a tripler unit (Fig. 5-3). These transformers operate at a very low voltage, compared to the integrated flyback. This reduces breakdown problems. The RF output voltage is usually under 10 kV, but the integrated transformer can be over 30 kV. The tripler unit breaks down more than the transformer.

The integrated transformer with several low-voltage windings and molded high-voltage diodes can cause a lot of service problems. Suspect that the flyback transformer when the horizontal output transformer is arcing or begins to run warm. The insulation between windings can break down, producing high-voltage arcover in the new transformers. A breakdown of the high-voltage diodes molded inside can cause arcing or a warm flyback transformer.

With the new high-voltage shutdown circuits, the chassis can come on with a blast of sound and then shut down. In some transformers, all you can hear is a tic-tic noise. An intermittent squealing noise can indicate a defective flyback or a whining noise can be caused by loose particles inside of the transformer or by loose mounting bolts. Replace the noisy transformer because most ceramic cores are held together with metal tabs.

The tripler flyback transformer secondary winding can be checked with the low-resistance scale of the VOM or DMM (Fig. 5-4). The resistance of the high-voltage winding can vary from 55 Ω to 350 Ω. This depends on the output voltage and the manufacturer.

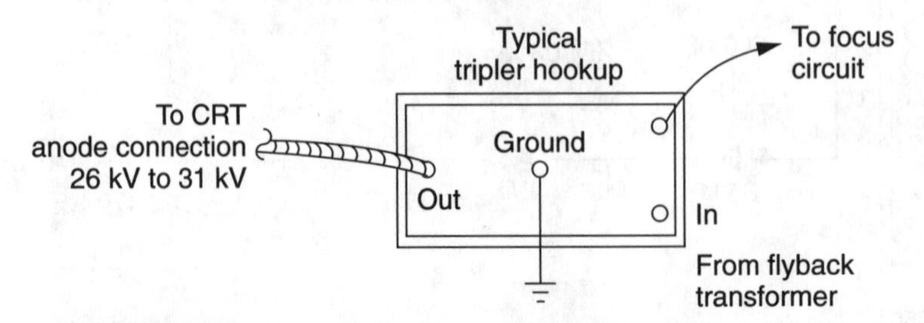

FIGURE 5-3 **The high-voltage tripler contains capacitors and diodes to provide a higher ac voltage for the anode terminals of the CRT.**

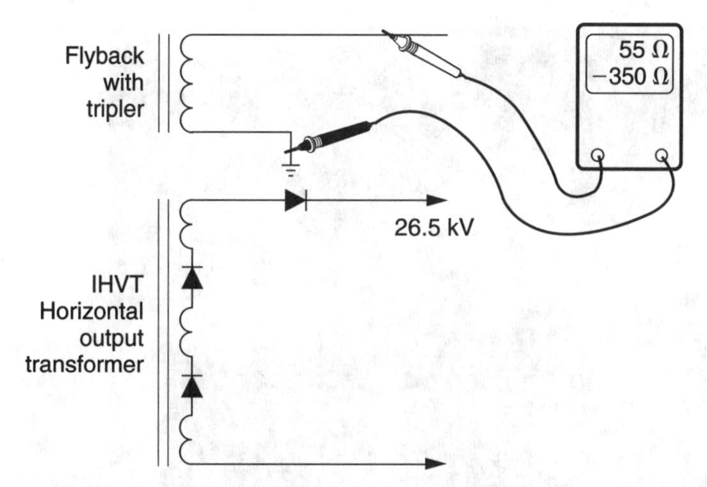

FIGURE 5-4 **Disconnect the input lead from the tripler and measure the secondary HV winding to see if it's open.**

Although the resistance measurement does not indicate a shorted turn, the measurement does show us if the winding is open. In the integrated transformer, winding continuity cannot be measured because of high-voltage diodes in the circuit.

The output lead of the tripler transformer can be checked by creating a 0.25-inch arc between a screwdriver blade and the input lead removed from the tripler unit. A high-voltage measurement at the anode CRT lead can indicate high voltage with the integrated transformer. If the integrated transformer is in shutdown, the transformer's condition must be checked with a universal line voltage or variac transformer. All circuits in the secondary winding of the horizontal output transformer are covered in this chapter.

A defective flyback can destroy the horizontal output transistor or produce a high-voltage shutdown. The shorted flyback can cause an excessively bright raster with heavy retrace lines because of a leaky path to the core of the transformer. Because there is no high voltage, the defective integrated flyback can cause pulsating blooming of the raster.

A defective flyback in an RCA CTC97A chassis produced a pulsating and blooming raster that was out of focus. Only 12 kV were measured with the high-voltage probe. A normal voltage was measured on the horizontal output transistor, indicating that the high voltage at the picture tube was improper.

THE HIGH-VOLTAGE PROBE

A high-voltage measurement at the anode connection of the CRT indicates that the high-voltage and horizontal section is working. The ground lead of the high-voltage probe must be grounded to the chassis or you will receive a jolt. A high-voltage shock can injure you. Never use a low-voltage VOM or DMM in the high-voltage tests. Be extra careful when working around high-voltage circuits (Fig. 5-5).

The VTVM can be used as a high-voltage meter when a special high-voltage probe is attached, instead of the regular voltage probe. Again, clip the ground wire to the metal TV

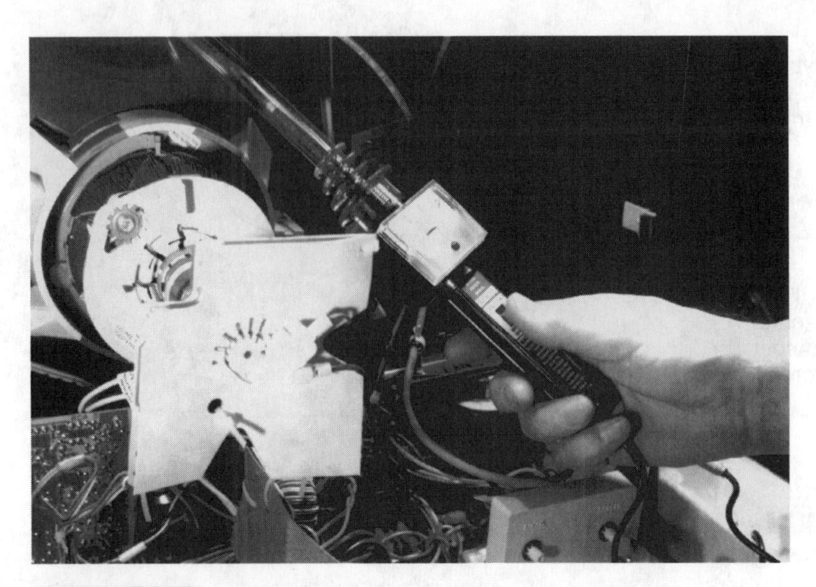

FIGURE 5-5 **Monitor the high voltage with the high-voltage probe at the anode terminal of the CRT.**

TABLE 5-1 THE HIGH VOLTAGE LEVEL, PICTURE TUBE, AND MODEL NUMBERS OF SEVERAL DIFFERENT TV CHASSIS

MAKE	MODEL OR CHASSIS	HIGH VOLTAGE
Bohsei	13B081R	19.9 kV
J.C. Penney	685-2048	22.5 kV
Panasonic	GXLHM	29.5 kV
RCA	CTC130C	25.8 kV
RCA	CTC117	24.2 kV
Zenith	L1740W	26.5 kV

chassis. Always choose a high-voltage probe that will measure up to 35 kV because many of the new TVs generate about 31 kV. Discharge the CRT at the anode terminal before working in the high-voltage circuits. Remove the anode lead. Isolate it on heavy insulation or on a cardboard box to prevent arcing when the high-voltage measurement is low. A leaky picture tube or circuits might be pulling down the high-voltage measurement (Table 5-1)

REMOVING AND REPLACING THE FLYBACK

Be sure that the flyback transformer is defective before installing a new one because replacements take time to remove and install. Mark down all color-coded leads to each terminal on a separate piece of paper. List each component and the tie points that solder directly to the transformer terminals. Many integrated transformers solder directly into the circuit of the PC board.

The flyback transformer in the latest RCA chassis mounts into soldering eyelets on the fiber chassis (Fig. 5-6). Long lugs from the transformer fit into the eyelet and then are soldered. The best method is to use a solderwick material with a 150-W (or higher wattage) soldering gun. If the solderwick mesh does not have enough soldering paste, apply extra soldering paste to help suck up the melted solder. A desoldering tool can be used to pick up the melted solder.

Start about 0.5 inches down the braid of the solderwick and apply heat to the first eyelet. Slowly go around the terminal with the iron tip pressed tightly against the wick material. Make a complete turn around the eyelet and pick up excess solder. If the wick material is full, start down another inch and pick up the remaining solder from the eyelet. Keep the heat applied as the solderwick goes around the terminal. If excess solder is left in the hole, the terminal is hard to remove and it might crack the board or wiring when it is removed. Disconnect each terminal with the same method.

Be careful not to misplace or break components on the top side of the board when removing the transformer. A screwdriver blade slipped under the metal or plastic body of the transformer can help to loosen each terminal. Most transformers will pop right up if all excess solder is removed. Suck out or remove all excess solder from each eyelet before installing the new transformer. Cut off the focus and screen cables, as well as any other wires found on top of the chassis connected to the transformer. Leave the leads long enough to identify the color code of each wire.

Route the filament wires down at the bottom side of the transformer and through the back metal brace area. An improper voltage might be placed on the filament wire if it is left high around the plastic belt of the flyback. Sometimes this can burn out the picture tube. Solder the filament wires before soldering the bottom transformer lugs.

Keep the transformer tight against the PC board while soldering the eyelet terminals. Be sure that each eyelet is full of solder all around the terminal. Now check each terminal to

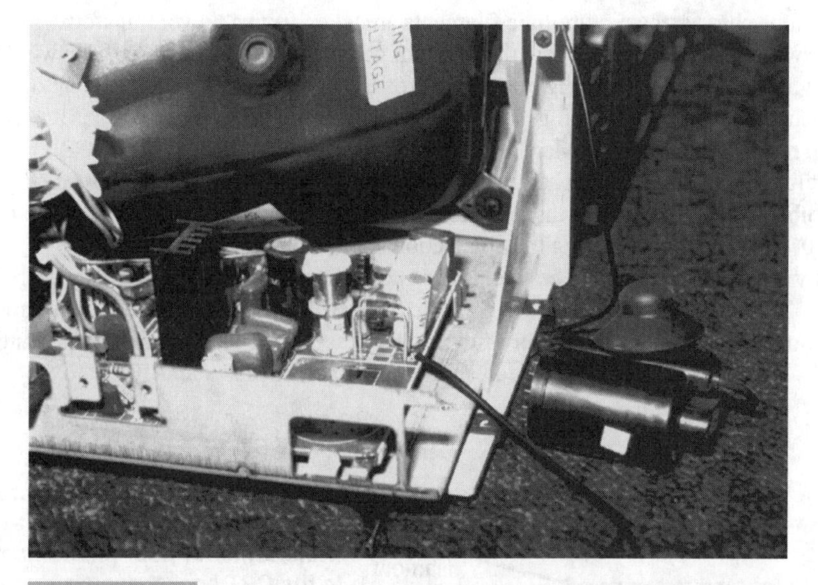

FIGURE 5-6 Remove excess solder from the output transformer eyelet or pin connection with a large iron and solder wick.

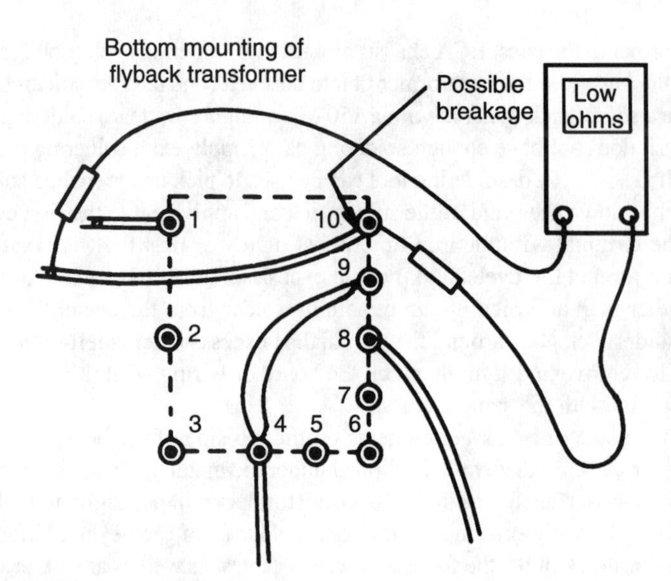

FIGURE 5-7 After installing a new flyback, check each terminal with the corresponding foil wire to be sure that no foil is broken at the transformer connection.

the PC wiring for a broken connection. Sometimes the wiring will break right at the eyelet. Take a continuity check with the low-resistance scale of a VOM or DMM from the transformer terminal to the nearest component tied to each terminal of the transformer (Fig. 5-7). This check can save a few hours of troubleshooting if the transformer does not work after installation.

PICTURE-TUBE FILAMENT CIRCUIT

In the tube chassis, the picture-tube filaments are taken from a power transformer or are in series with the power line. Today, the filament voltage is taken from a separate winding on the flyback transformer. No light at the end of the picture tube can indicate an open tube filament or transformer winding.

The raster was out in a K-Mart SK1310A with no light at the end of the CRT. The picture-tube filaments were checked across pins 6 and 7 with the socket removed (Fig. 5-8). The tube was normal, except that transformer and lead continuity was found to be open at pin 10. Cleaning and soldering the connection solved the problem.

Only a very dim light could be seen in the end of the picture tube in a Montgomery Ward GA1-129940 with normal high voltage (Fig. 5-9). Because RF filament voltage cannot be measured accurately on the picture-tube socket, the socket was removed and continuity checks were made. Practically a dead short existed across pins 9 and 10.

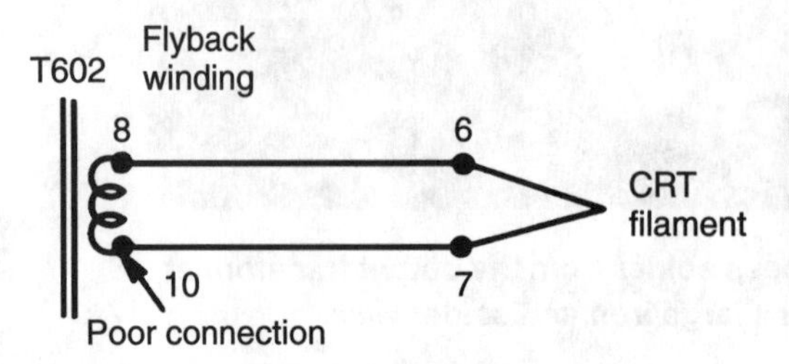

FIGURE 5-8 Poorly soldered connections at the flyback transformer or poor pin contacts in the CRT socket can produce intermittent or no filament voltage in the CRT.

A normal filament winding measurement should be around 2.8 Ω with voltage-dropping resistor R115 in the circuit. Diodes D102 and D103 were replaced because they contained white burn marks on the body and measured a dead short across each diode.

TRIPLER PROBLEMS

The defective high-voltage tripler unit (Fig. 5-10) can cause many problems within the TV. A leaky tripler can keep blowing fuses or it might cause the fuse to open after several seconds of operation and keep tripping the circuit breakers. Very little or low high voltage can result from a defective tripler unit. A defective tripler can load down the horizontal circuits, destroying the horizontal output transistor or causing a low voltage at the collector terminal.

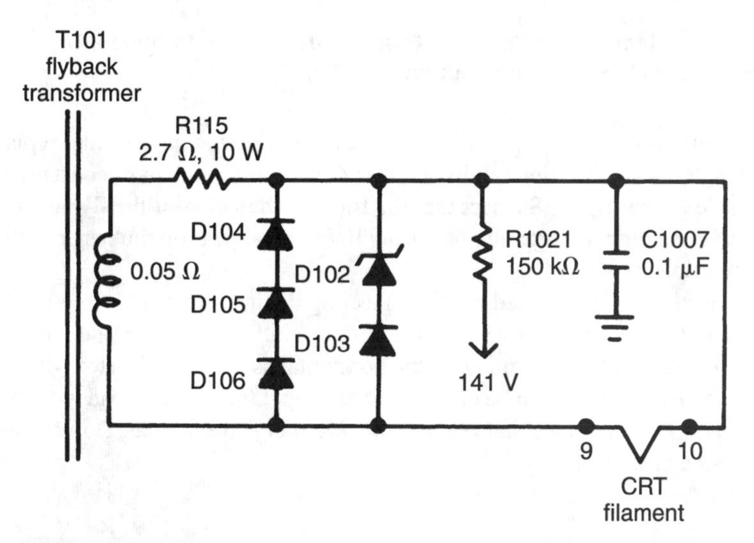

FIGURE 5-9 Overheated diodes in a Wards GA1-12994C caused a very low filament light.

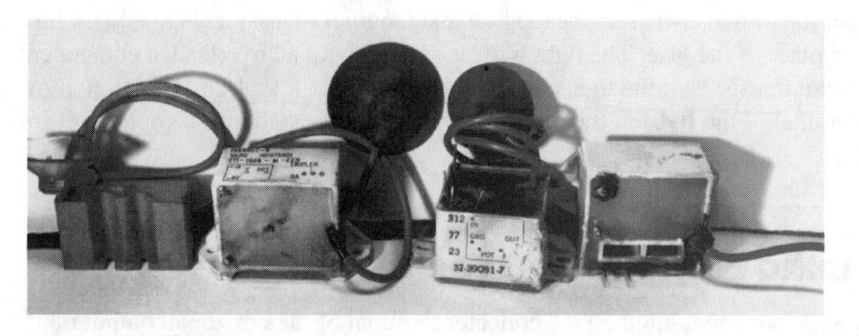

FIGURE 5-10 Several different types of tripler units are found in the various high-voltage chassis.

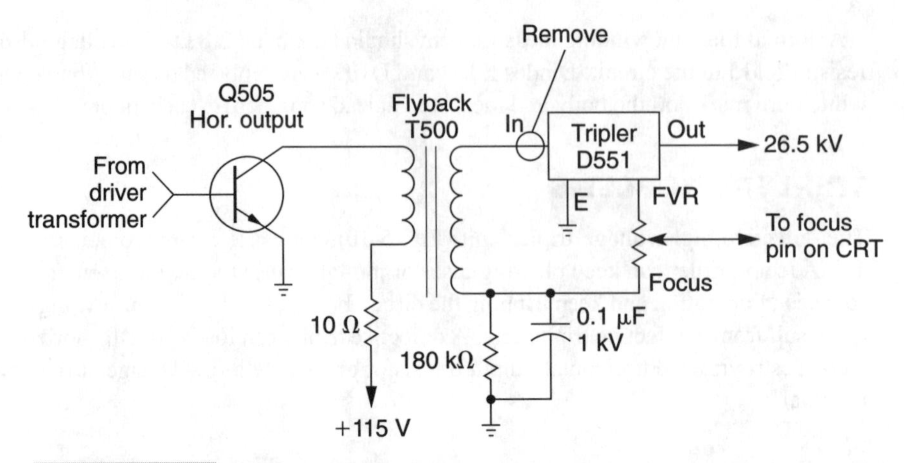

FIGURE 5-11 **Remove the "in" lead from the tripler to determine if the unit is loading down the horizontal output circuits.**

Feel the outside case of the suspected tripler. After pulling the power plug, replace it if it is running quite warm. All tripler units should run cool. Intermittent or constant arcover can occur inside of the tripler. Suspect that the tripler is defective if the flyback makes a tic-tic noise with no high-voltage output. Firing lines can appear on the raster with an internally arcing tripler unit.

A defective tripler can be isolated by disconnecting the input terminal from the flyback (Fig. 5-11). If the tripler unit was causing the fuse or circuit breaker to open, the voltage will be restored at the collector terminal of the horizontal oscillator transistor with no shutdown. A normal 0.25-inch arc can be drawn from the input transformer lead with the blade of a screwdriver. This indicates that the horizontal circuits are normal. Do not touch the metal blade to the chassis.

KEEPS BLOWING FUSES

The horizontal output transistor and fuse was found blown in an RCA CTC92. After replacing both components, the fuse blew again. To determine if the horizontal circuits are causing the fuse to open, remove the horizontal output transistor (Fig. 5-12). Clip a 100-W bulb in place of the fuse. The light will stay bright with an overloaded component. Check the output transistor. If the light comes on bright, check for a leaky tripler. Remove the input terminal of the flyback transformer to the tripler. Install a new tripler unit if the light goes out.

LOADING DOWN—MAGNAVOX T995-02

Only 87 V were measured on the collector terminal of the horizontal output transistor before the circuit breaker tripped in a Magnavox TV. The leaky horizontal output transistor was replaced with the same results. After the flyback input lead was removed from the tripler, the voltage was normal at 114 V without shutdown. Replace the tripler component with one that has the original part number or use the correct universal replacement.

POOR FOCUS

A defective tripler or components in the high-voltage circuit can produce poor focus problems. The leaky flyback transformer with enclosed high-voltage capacitors can have low high voltage, resulting in a poor focus voltage. A defective focus control or bad high-resistance dropping resistors can cause poor focusing. Check the focus pin at the tube socket for poor connections.

First, measure the high voltage at the anode terminal of the CRT. Poor focus will result if the high voltage is low. Normal high voltage with poor focus can be caused by a poor focus assembly or picture tube. Check the focus voltage at the focus pin of the CRT socket. Place the end of a paper clip inside of the focus socket and measure the focus voltage from it. A low focus voltage can be caused by a poor socket or spark-gap assembly inside of the CRT socket (Fig. 5-13).

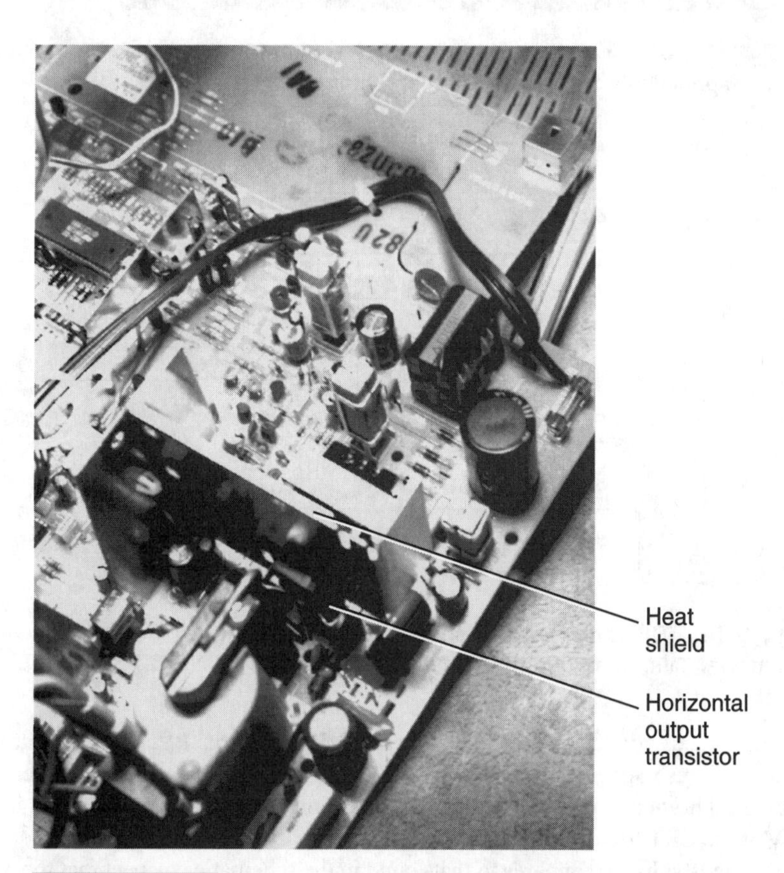

FIGURE 5-12 Remove the horizontal output transistor to determine if the horizontal circuits are blowing the fuse.

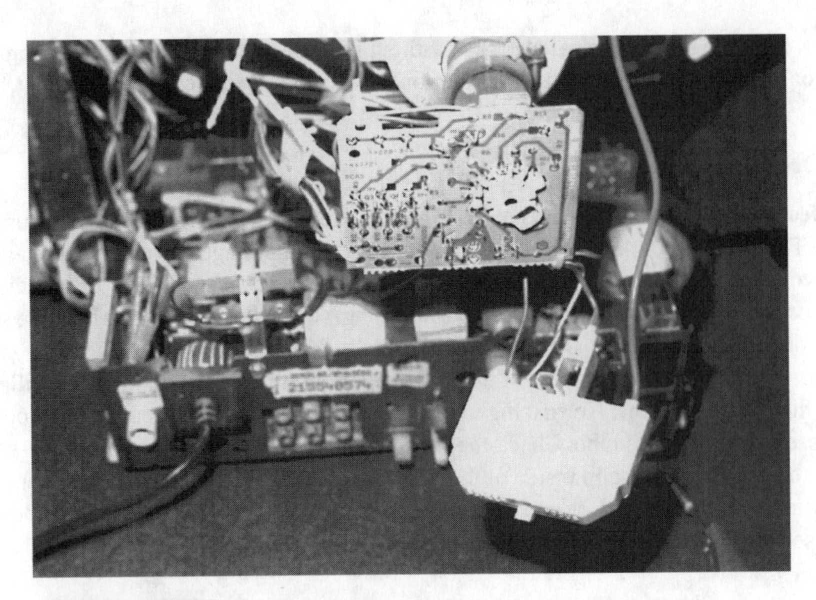

FIGURE 5-13 Check all the pins in the CRT socket for corrosion—especially in the focus and filament terminals.

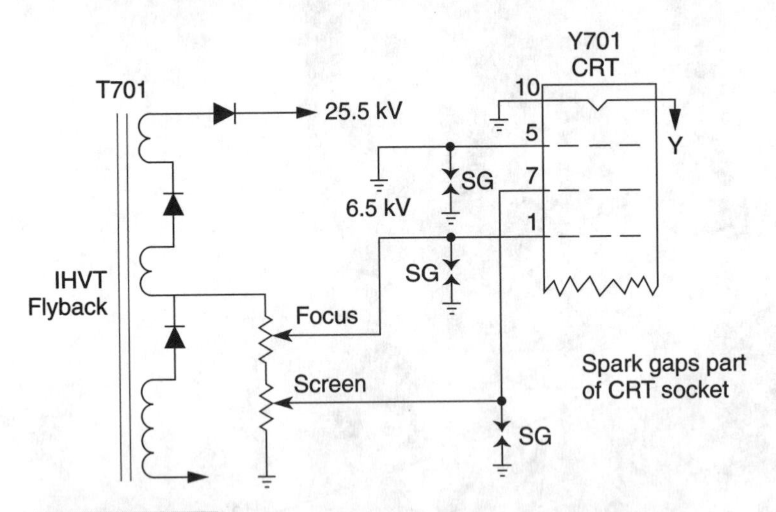

FIGURE 5-14 The focus voltage should vary from 3.5 kV to 6.5 kV; in newer TVs, this voltage should be as high as 9 kV.

The average focus voltage should vary between 3.5 kV and 6.5 kV. With the latest picture tubes, the focus voltage can vary from 6 kV to 9 kV (Fig. 5-14). Check the focus voltage with the high-voltage probe. The normal focus control should vary the voltage by at least 1 kV at the CRT focus terminal.

When the focus voltage compares to that found in the schematic, suspect that the picture tube is defective. Check the picture tube in the tube checker. Measure the crucial voltages at the picture tube terminals. A weak or gassy picture tube can cause poor focus. Suspect a problem with the focus spark gap or focus control if the focus voltage at the CRT socket will not change the focus. Blow out the socket and the focus spark-gap assembly. Do not overlook the possibility of poor heater terminals.

HIGH-VOLTAGE ARCING

Excessive high-voltage arcing can cause high-voltage shutdown with excessive high voltage at the picture tube. Use the variable line transformer to help locate the arcing shutdown problem. The leaky tripler can arc between the plastic case and chassis. Insulate the tripler with a pasteboard box or tube carton to see if the chassis is operating normally. Do not try to apply putty or high-voltage insulation. Simply replace the leaky tripler. It might break down later under hot and humid weather conditions if it is repaired with putty or insulation spray.

Poor focus with high-voltage arcover can occur in a defective focus control or spark gap. Remove one terminal of the spark gap to determine if it is defective. In a Philco C-2518KW, excessive arcing could be heard across the spark gap mounted on top of the tripler unit (Fig. 5-15). The spark gap was removed and cleaned. The tripler unit was replaced. The integrated horizontal output transformer can arcover internally, with firing lines in the raster or with a no-high-voltage symptom. Check the outside case of the transformer for signs of arcing or melted plastic (Fig. 5-16). The integrated transformer can run warm, with high-voltage arcover of the molded high-voltage diodes. Sometimes tapping the transformer with the insulated handle of a screwdriver will make it act up. If the arcing can be heard but not seen, darken the area around the flyback and focus-screen units so that the defective component can be identified.

FIGURE 5-15 **Check for arcing in the flyback, tripler, high-voltage lead, and picture-tube anode terminal.**

FIGURE 5-16 Check for signs of arcover on the flyback transformer mold case area.

ARCING IN THE FOCUS CONTROL

Thin white lines or dark zigging lines going across the screen horizontally can be caused by high-voltage arcing or poor connections to the focus controls. Some of the RCA focus and screen controls are located in one component that mounts on the TV chassis. Improper soldering of the focus control leads or a cracked ceramic or plastic base that the printed control is placed on can cause fine-line arcing (Fig. 5-17). Carefully inspect the control and feel it for warm sections. The inside elements of both controls can be ordered and replaced.

SECONDARY FLYBACK CIRCUITS

In the latest horizontal circuits, startup and secondary power supplies taken from the flyback transformer are found in practically every manufacturer's TV chassis. This means that the horizontal circuits must function before voltage is applied to the various circuits (Fig. 5-18). A breakdown of components in any power source can cause the chassis to shut down. In most cases, secondary voltage-source shutdown occurs a little later than high-voltage shutdown. Usually, there are a few minutes of lag, with the sound blasting on and high voltage coming up before shutdown.

Try to isolate the defective power source before it shuts down, if possible. If the picture and raster are normal for a few seconds, but there is no sound, suspect the sound circuits are loading down the flyback transformer. A blast on of sound with normal high voltage and no picture before shutdown can be caused by the luminance or vertical circuits. Check for high-voltage shutdown if the chassis immediately shuts down.

Isolate the shutdown problem with a variable line or variac transformer. Slowly bring up the line voltage and notice what circuits are working. Check the horizontal and low-volt-

FIGURE 5-17 Check the focus and screen control assembly for cracks and corroded terminals that could arc over.

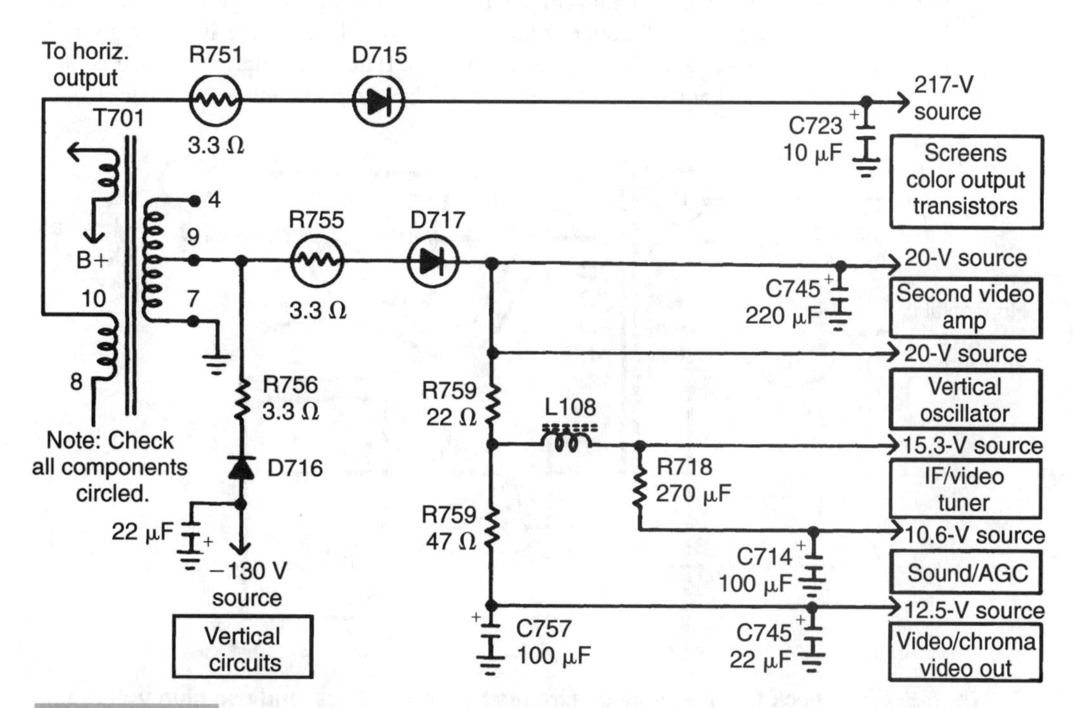

FIGURE 5-18 The horizontal output transistor could be overloaded if a shorted or leaky component is in the secondary voltage sources.

age circuits for no high voltage. All horizontal and high-voltage circuits must be functioning before any other voltage source is developed. Often, the high voltage might be lower, with improper width and an overloaded secondary voltage source.

Go directly to the secondary power source if the circuits are not operating or appear to be dead. Be careful not to cut off any dc voltage connected to the horizontal circuits. Measure each voltage source. A low measurement can turn up the defective or overloaded circuit. Shut the chassis down and remove this circuit from the flyback circuit. Simply remove one end of the diode rectifier to disconnect it from the horizontal output transformer. Remember, these voltages are lower than normal because the universal line transformer is adjusted lower before chassis shutdown.

You have located the overloaded circuit if voltage on the circuits drastically increases. Check for an increase in voltage at the collector terminal of the horizontal output transistor. Now raise the line voltage to 120 Vac. Locate the defective component in the overloaded circuit before connecting into the circuit.

Only a tic-tic noise could be heard at the flyback transformer in an RCA CTC85 chassis. The chassis would come on with sound and then shut down. Only a horizontal white line could be seen before shutdown. Because a vertical sweep collapse was noticed, the vertical module was replaced. With the module out of the circuit, the horizontal and high voltage were normal. Replacing the leaky vertical module fixed the problem.

QUICK HIGH-VOLTAGE CIRCUIT CHECKS

With a no-raster, no-sound symptom (Fig. 5-19), check the high voltage at the picture tube (1). If high voltage is normal, notice if the picture-tube filaments are lit. Now go to the video circuits if the CRT and high voltage are normal. If no high voltage is measured, take a voltage measurement at the collector terminal of the horizontal output transistor (2).

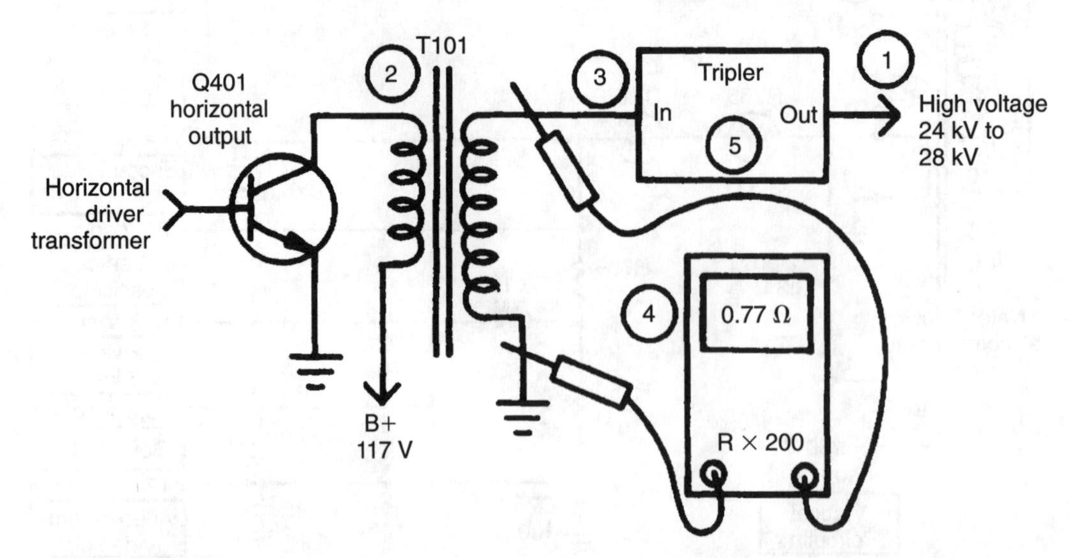

FIGURE 5-19 **Check the high-voltage circuits by the numbers, and use high-voltage and resistance measurements.**

FIGURE 5-20 **Plug the chassis into a variable power transformer to determine what is shutting down the chassis.**

Check the tripler circuits if the horizontal voltage at the output transistor is a little low or normal. Remove the tripler input wire from the flyback if you suspected that it is loading down the secondary circuits (3). Fire up the chassis and draw a small 0.25-inch arc from the flyback input cable. Take a resistance continuity check of the flyback winding if there is no voltage (4). Replace the defective tripler if the flyback seems normal (5).

If the line fuse or circuit breaker keeps tripping, remove the horizontal circuit fuse or horizontal output transistor. Now if the fuse or circuit breaker holds, suspect that the horizontal or high-voltage circuits are defective. Replace the horizontal output transistor and remove the input terminal to the tripler unit. If the fuse or circuit breaker is normal, replace the leaky tripler unit. Always use a variable power transformer as B+ for these tests (Fig. 5-20).

Check the high-voltage circuits with the integrated horizontal output transformer in the very same manner (Fig. 5-21) (1). If the transformer has a tic-tic noise and is in chassis shutdown, connect a variable line transformer to the chassis (1). Slowly bring up the ac voltage and notice at what voltage shutdown occurs. Check the chassis for high-voltage shutdown with a higher-than-normal low or high voltage. Check the voltage at the collector terminal of the horizontal output transistor (3). Determine if the secondary voltage circuits are overloading the flyback in chassis shutdown. Disconnect each voltage source until the overloaded circuit responds with normal voltage at the horizontal output transistor (4). Do not overlook the possibility that the horizontal output transformer is defective.

VOLTAGE SUPPLIES TO CRT

High voltage is supplied to the anode terminal of the CRT from the horizontal output transformer. Focus and screen voltages are fed to the focus and screen grids of picture tube from an HV divider network, consisting of separate focus and screen controls. In some

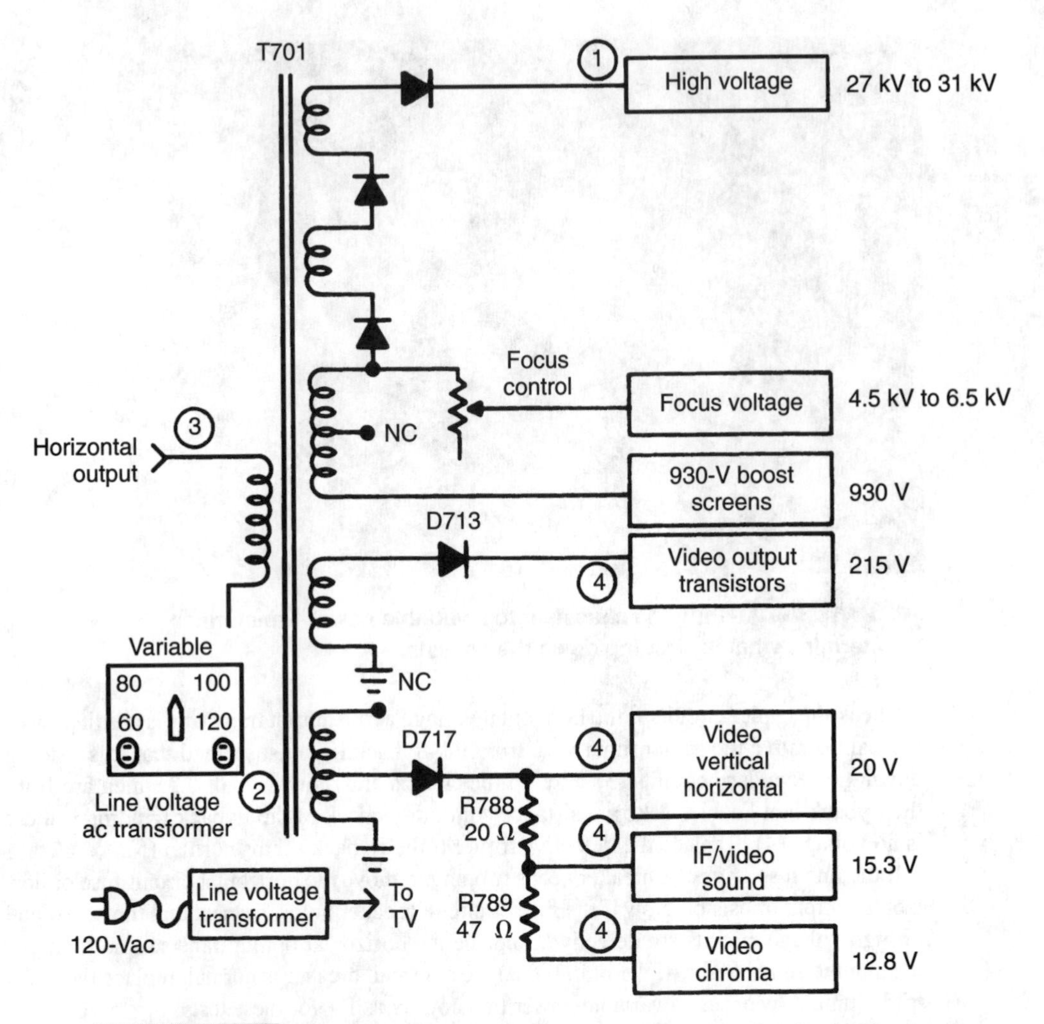

FIGURE 5-21 **Check the horizontal output transformer circuits by the number to locate a defective component.**

TVs, high boost voltage is supplied to the CRT from a lower winding of the flyback. One or two turns of large heater wire is wound around the horizontal output transformer core to develop heater or filament voltage, which is then fed to the heater pins of the CRT. Be sure that all these voltages are found on the picture tube before you replace anything.

CHECK COMPONENTS BEFORE REPLACING FLYBACK

Before replacing the horizontal output transformer, be sure that proper driver voltage is applied to the base of the horizontal output transistor. Test and, if necessary, replace the output transistor. Check for proper collector voltage at the output transistor. Remove the red lead of the yoke assembly to determine if the yoke is defective. Check the continuity of the primary winding or flyback (Fig. 5-22). Determine with the variable power-line transformer if the output transistor and flyback are loading down the horizontal output circuits. Feel the horizontal output transistors and flyback and see if they run warm at 60 to 90 volts applied to the collector of the output transistor. Remove each secondary-derived voltage from the flyback to determine if a low-voltage component is loading down the flyback voltage source. Replace a defective flyback with a part that has the correct part number.

FIVE ACTUAL CASE HISTORIES

The following are five actual problems and their solutions.

Pulses off and on The RCA CTC167 chassis would pulse off and on and then shutdown when first turned on. This problem occurred in many of the RCA TVs. Check all diodes within the sawtooth generator circuits. CR4101 caused the chassis to pulse off and on; in another chassis, CR4103 and C4108 became leaky (Fig. 5-23). Check diodes CR4103 and CR4101 if the chassis pulses on, then shuts down. C4108 has caused shutdown problems in a few other RCA chassis.

Intermittent width—shutdown In an RCA CTC145 chassis, the picture width would pull in sometimes, and would remain normal at other times. Finally, the chassis shut down with a blown fuse. No dc voltage was found upon the collector terminal of the horizontal output. Sometimes resistor R4413 (680 Ω) would run red hot. In checking over the flyback winding and yoke circuits, capacitor C4415 (0.25 μF) was found to be intermittent and leaky (Fig. 5-24).

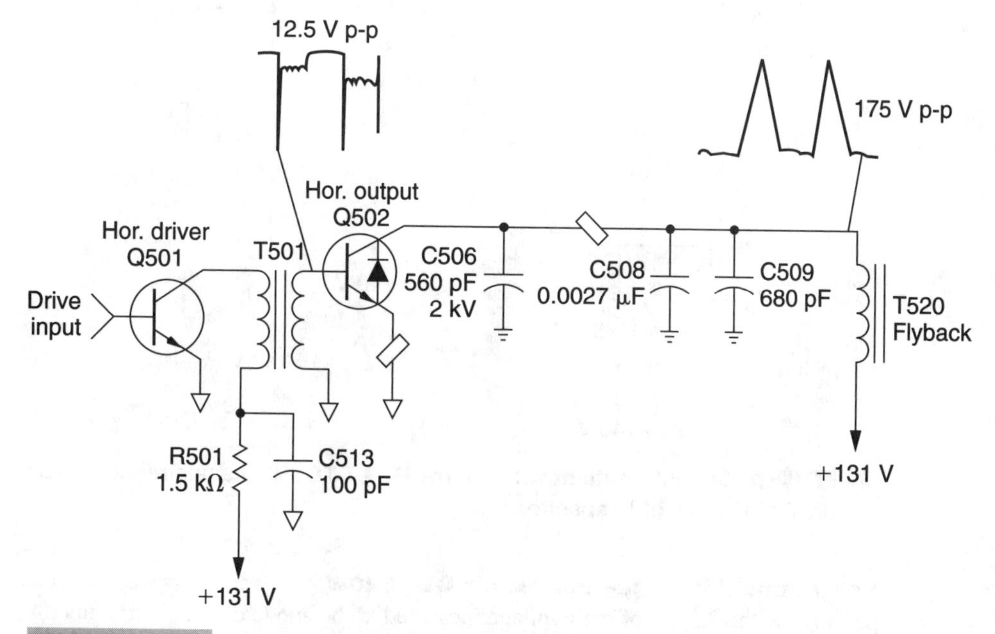

FIGURE 5-22 Check waveforms, voltages, resistances, and all components in the horizontal circuits before replacing the horizontal output transformer.

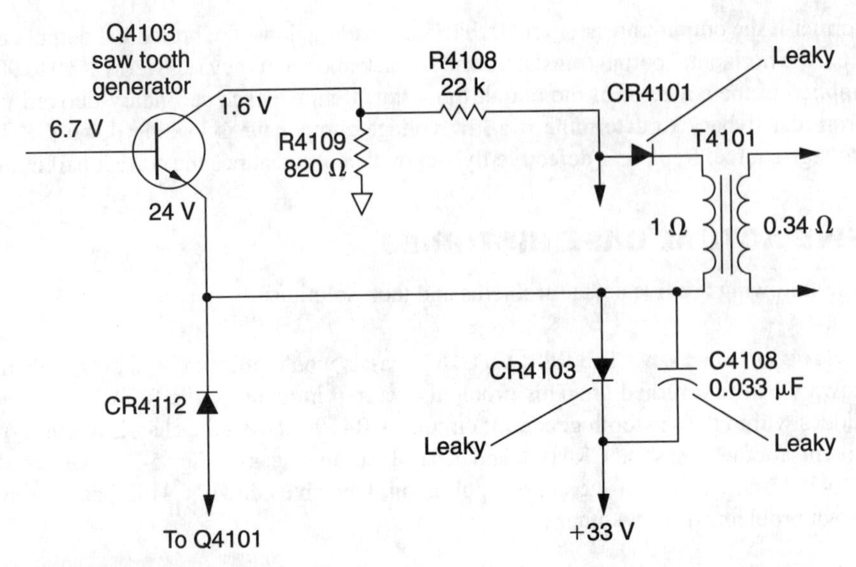

FIGURE 5-23 Check CR4101, CR4103, and C4108 if an RCA CTC167 TV pulses off and on, then shuts down.

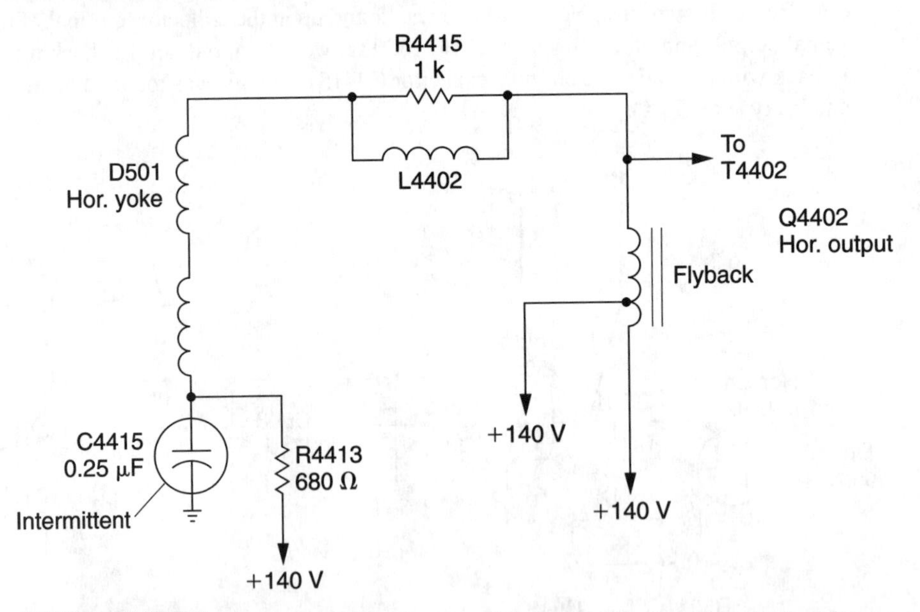

FIGURE 5-24 Intermittent width in the RCA CTC145 chassis resulted from a leaky C4415 (0.25 µF) capacitor.

Low filament voltage—Panasonic CT-1320W No brightness was indicated in this portable, with 22.3 kV of high voltage measured at the anode of the picture tube. After the picture tube was checked and found to be good, voltage measurements were fairly normal on all elements of the CRT. In fact, no filament light could be seen at the glass of picture tube.

A continuity check was made across the filament of the CRT and was normal. Removing the CRT socket and taking another measurement across the filament winding yielded a measurement of 210 Ω. Resistor R372 (2.4 Ω, 2 W) was found corroded and was replaced, solving the no-filament voltage problems (Fig. 5-25).

Chassis normal—HV shutdown The chassis in a Quasar NTS-989 chassis would come up and look normal, then shut down. A quick HV test and scope waveform test at horizontal output transistor Q501 was made and the voltage was monitored at the line-voltage regulator STR380 (IC801) to determine what section caused the shutdown. Slowly, the line voltage was varied by the isolation line transformer; when it was brought up to 90 volts, the chassis shut down.

To eliminate the voltage shutdown circuits, one end of D513 was removed from the circuit (Fig. 5-26). Again, the line voltage was raised and the chassis operated at 120 Vac. This indicated problems within the high-voltage protection section. Q554 was tested and proved to be leaky. Q553 tested normal, but both Q554 and Q553 were replaced to cure the HV shutdown problem.

No HV—no picture, no raster Although low voltage was found at the collector terminal of output transistor Q602, no HV was measured at the picture tube. Q602 was found to be open and was replaced with a universal ECG89 transistor. Now the chassis blew the main fuse F701 (4 A). All safety components were checked in the output collector circuits and all tested good.

Q602 was replaced again. Slowly, the line voltage was raised to 60 Vac. Q602 began to get hot (Fig. 5-27). The drive waveform was fairly normal, indicating an overload in the flyback circuits. The red lead of yoke DY601 was removed from the circuit and the results were the same. Replacing flyback T602 with the exact replacement in the Sharp 19F90 model solved the no-HV problem.

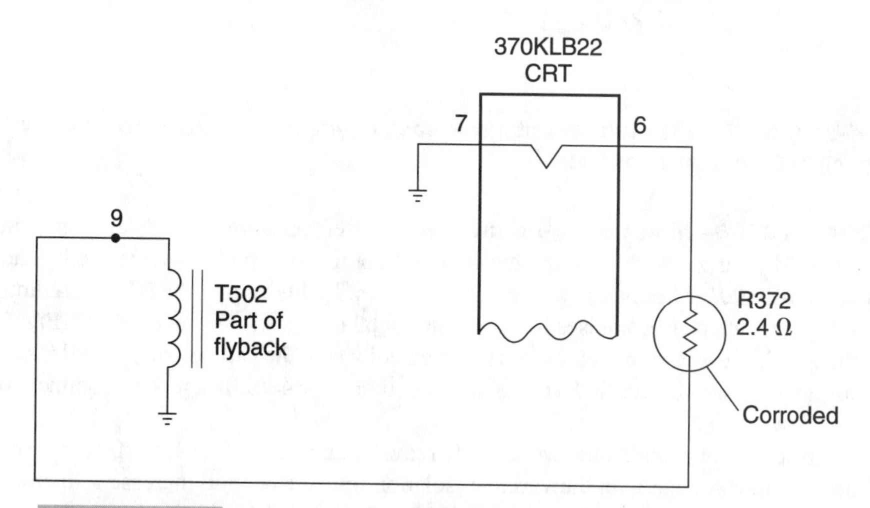

FIGURE 5-25 **Very low filament voltage in a Panasonic CT-1320W was caused by corroded resistor R372.**

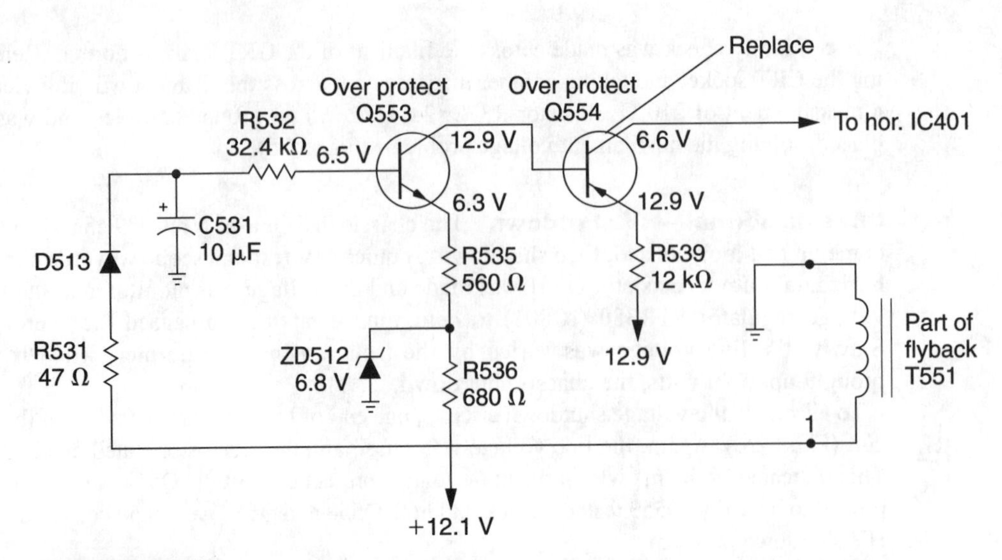

FIGURE 5-26 High-voltage shutdown was eliminated by replacing Q554 and Q553 in the overvoltage-protection circuits in a Quasar NTS-989.

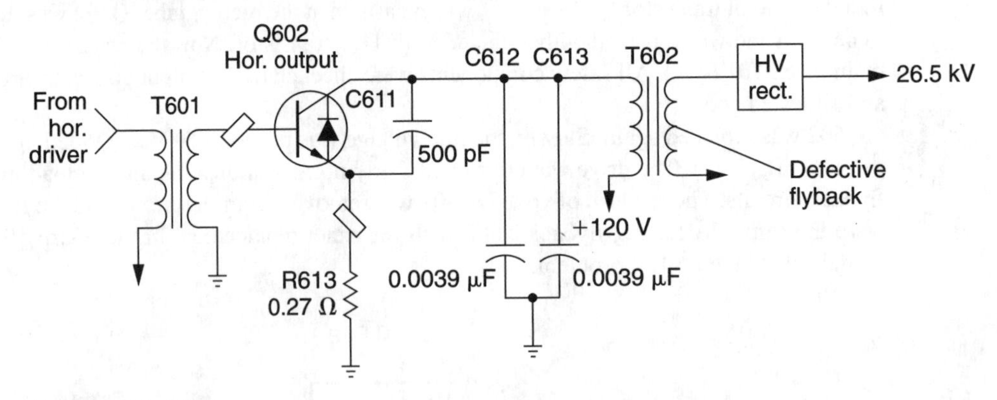

FIGURE 5-27 The no-high-voltage/no-raster symptom resulted from a leaky flyback T602 in a Sharp 19F90.

RCA CTC166—high-voltage shutdown After replacing horizontal output transistor Q4401 in an RCA chassis, the high-voltage shutdown circuits were tested by shorting 1 and 2 of J4901. The chassis would not shut down. In this circuit, the HV is monitored by CR4901, which rectifies pulses from the horizontal output transformer, T4401 (Fig. 5-28). When the high voltage increases, the rectified voltage at the cathode of CR4901 increases; with an increase of rectified voltages, zener diode CR4902 triggers to conduct, which shuts down the set.

To troubleshoot this shutdown circuit, remove one end of CR4901 from the circuit. Raise the line voltage with the variable isolation transformer and increase it to around 85 to 90 V. Repair the shutdown circuits. Both CR4902 and CR4901 were replaced. Now raise the line voltage up to 125 to 130 Vac, and see if the shutdown circuits are working. Test the shutdown circuits by shorting 1 and 2 terminals of J4901 together.

CR4902 is an 11-volt zener diode and should be replaced with original part number 159429 or NTE5019T1; SK9970 is the universal replacement. Replace CR4901 with a general-purpose 1-A silicon diode.

OVERLOADED SECONDARY CIRCUITS

Remember, overloaded circuits within the secondary winding voltage sources can shut down the chassis or damage the horizontal output transistor. Any low resistance between a leaky IC, transistor, capacitors, or regulators in the secondary circuits, such as sound, brightness, horizontal, color, IF, and video circuits, can cause chassis shutdown. Often, there will be a lag or delay in shutdown with either a no-sound or no-picture symptom.

For instance, if the picture went down to a white line before shutdown, check to see if the vertical circuits are overloading the flyback circuits. If the picture goes black with a delay in high voltage in the CRT, suspect that a video component is leaky. Likewise, in chassis that depends on a secondary voltage source to supply voltage to the deflection IC, check the IC for leakage leading to shutdown.

The sound was distorted before the chassis shut down in an Emerson MS250R portable (Fig. 5-29). Because the picture was normal for a few seconds with distorted sound, the sound voltage source was monitored. IC1203 had only 0.75 volts at pin 2 when fired up, indicating that an IC (AN5836) is leaky. The AN5836 IC was replaced with RCA SK9731, a universal replacement.

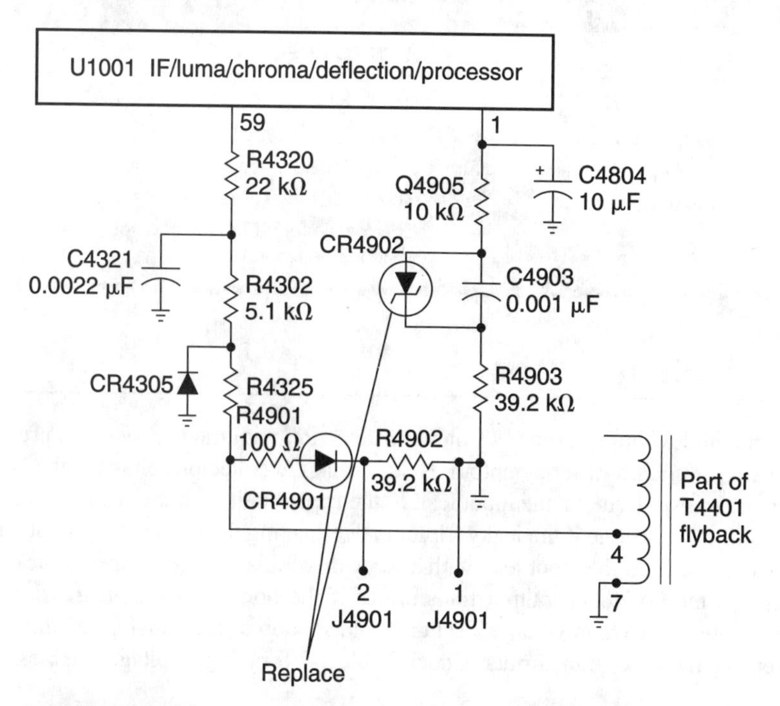

FIGURE 5-28 **High-voltage shutdown in an RCA CTC166 chassis was caused by CR4901 and CR4902.**

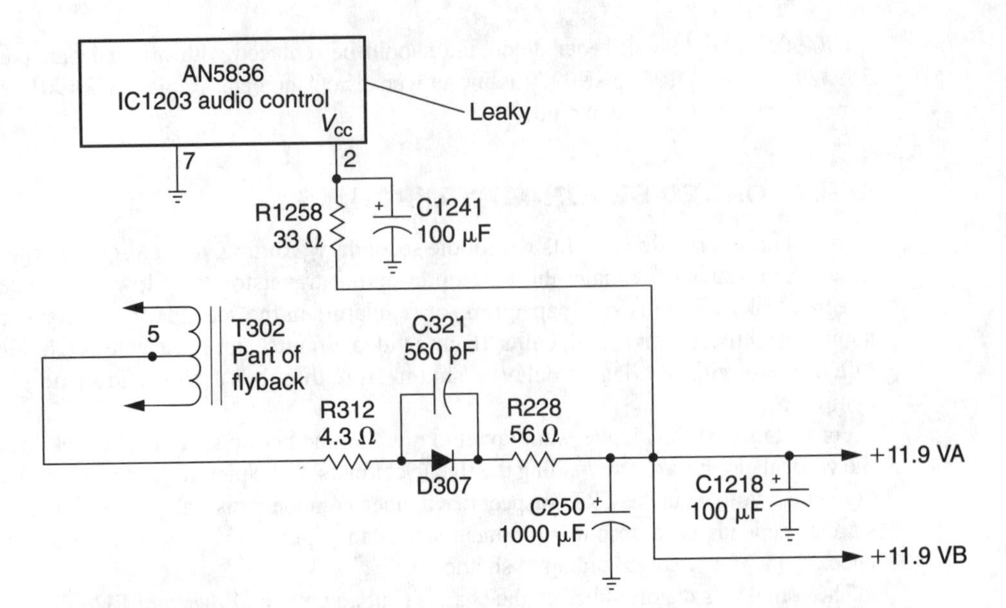

FIGURE 5-29 **A leaky IC1203 overloaded the 11.9-V source of the secondary winding of flyback T302, shutting down the chassis of an Emerson MS250R.**

TABLE 5-2 FOR EASY SERVICING, CHECK THIS TROUBLESHOOTING CHART FOR HIGH-VOLTAGE CIRCUITS

WHAT TO CHECK	HOW TO CHECK IT
Check HV at crt with high voltage probe.	Check flyback and horizontal circuits.
Check focus and grid voltages.	Check flyback circuits.
Check continuity of flyback primary winding.	Open winding or isolation resistor.
Disconnect red yoke wire or lead.	Check horizontal sweep circuits and voltages.
	Check waveform on base of output transistor. Repair horizontal circuits.

Conclusion

Measure the high voltage at the CRT anode connection with the high-voltage probe on a TV with a no-sound/no-raster symptom. Next, check the collector voltage at the horizontal output transistor. Remove the input lead to the tripled unit with low voltage at the output transistor to determine if the leaky tripler is overloading the horizontal output circuits. Draw a small arc from the input lead with a screwdriver blade to see if the flyback is normal. Replace the horizontal output transformer if the horizontal output transistor runs warm with low (or no) high voltage. Feel the molded body of the transformer and, if quite warm, replace the leaky transformer. Check Table 5-2 for a high-voltage troubleshooting chart.

6

REPAIRING THE VERTICAL CIRCUITS

CONTENTS AT A GLANCE

Vertical Circuit Problems

After the horizontal section, the vertical stages rate next as causing the most trouble in the TV. A horizontal white line and insufficient and intermittent sweep are the most common vertical problems. Vertical foldover, crawling, and rolling of the picture can occur in the vertical circuits. Black and white bars in the raster are a little more difficult to locate in the latest solid-state TV chassis.

A horizontal white line can be caused by many crucial components in the vertical circuits and is the result of no vertical sweep. Insufficient vertical sweep can appear as from 1 to 8 inches of raster. The top and bottom portion of the raster cannot be adjusted with either the vertical height or the linear controls to fill out the screen. Improper vertical sweep can be caused by a leaky output transistor, burned resistors or diodes, and improper voltage in the vertical output circuits. An intermittent vertical sweep can result from transistors, ICs, or poor board wiring connections. Check for leaky output transistors with improper bias resistance for vertical foldover. Vertical crawling and rolling can be caused by defective filter capacitors and improper sync signal.

The horizontal white line indicated by a no-vertical-sweep symptom is the easiest vertical trouble to locate. Intermittent vertical sweep and foldover are more difficult to find. Often, insufficient vertical sweep symptoms are caused by abnormal voltages and can be located with the DMM or VOM. Look for poor filtering in the vertical voltage source causing vertical crawling.

The most useful test instruments are the oscilloscope, DMM, and VOM (Fig. 6-1). The crosshatch generator is ideal to check vertical height and linearity. Crucial waveforms taken by the scope can indicate where the waveform is missing. Low voltage and resistance measurements with the DMM can quickly locate the defective component. You can test the transistors and diodes in the circuit with the DMM. Of course, the correct schematic diagram is a must when servicing the vertical circuits.

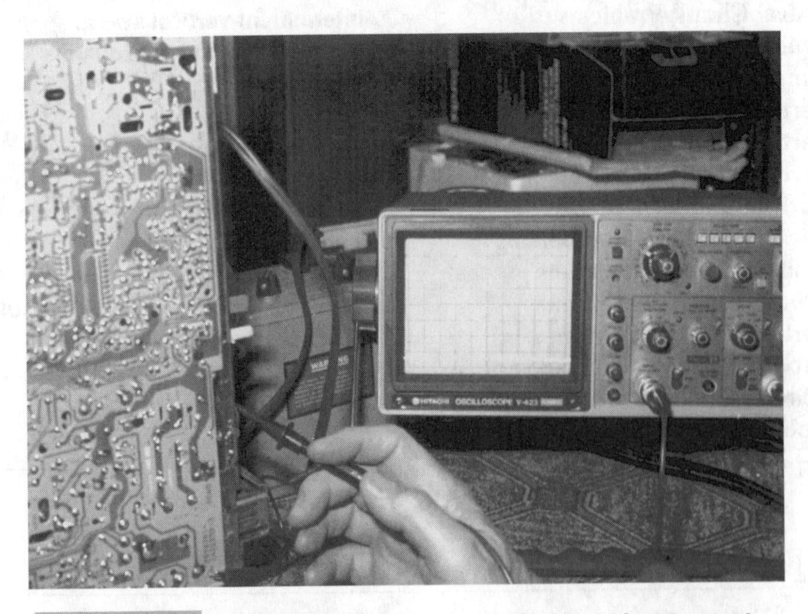

FIGURE 6-1 **The DMM and scope can help you to locate most problems in the vertical circuits.**

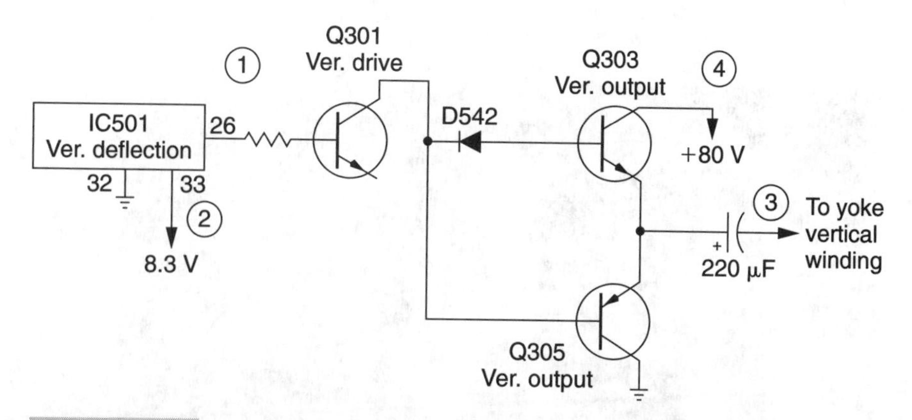

FIGURE 6-2 **Troubleshoot the vertical section by the number, take waveforms at 1 and 3, and take crucial voltage measurements at 2 and 4.**

CRUCIAL TEST POINTS

Three crucial test points within the vertical circuit can determine where and if the vertical section is functioning (Fig. 6-2). Take a scope waveform of the vertical oscillator and output stages to determine where the improper vertical sweep symptom occurs. Measure the supply voltage to the vertical circuits.

A voltage measurement on the heatsink or metal collector terminal of the vertical output transistor can indicate improper or no B+ power source with no vertical sweep. A lower-than-normal power source can produce insufficient or vertical foldover problems. No voltage at a given stage can indicate that transistor (or other component) is leaky.

The waveforms taken at the output of the oscillator transistor indicate the beginning of the sweep circuit. A waveform check at the vertical coupling capacitor shows that the vertical stages do not have yoke problems or corresponding component problems. If a waveform is found at the oscillator transistor and not at the output stages, look for the defective component between the two test points. A low-amplitude waveform can indicate that the vertical sweep is insufficient. Scope waveforms taken throughout the vertical circuits, especially with feedback circuits, are useless in these type circuits.

TYPES OF VERTICAL CIRCUITS

The vertical blocking oscillator circuits are used extensively in the early tube circuits with capacitor coupling to the final output stage. Many of the early solid-state chassis used the single transformer-coupled output transistor (Fig. 6-3). Then came the multivibrator circuits. Today, many of the vertical circuits contain a relaxation oscillator circuit, using an NPN and a PNP transistor (Fig. 6-4). In the latest solid-state chassis, you might find the ICs as both horizontal and vertical deflection in a frequency-counter circuit (Fig. 6-5).

The relaxation oscillator and frequency-counter circuits are directly coupled to the transformerless (OTL) output stages. No large output transformer is found in these vertical circuits. A PNP transistor is used as the top vertical output transistor, with an NPN transistor

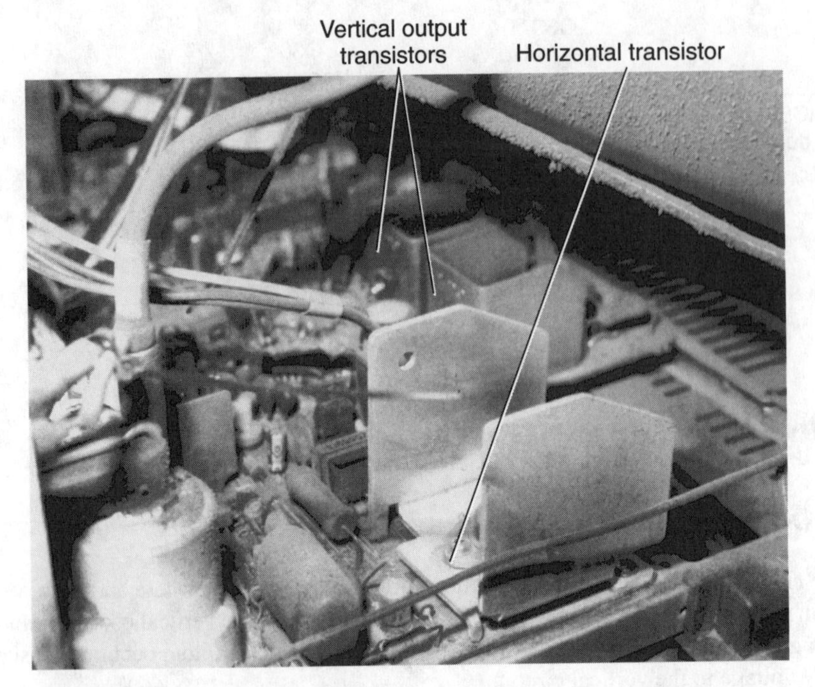

FIGURE 6-3 The vertical output transistors are mounted in separate heatsinks in the vertical section.

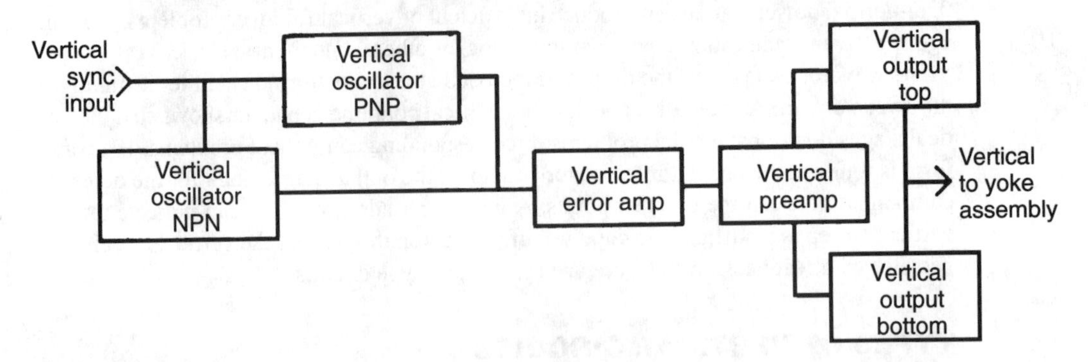

FIGURE 6-4 A block diagram of a transistorized vertical circuit, from the oscillator to the vertical output transistors.

as the bottom vertical output (Fig. 6-6). Direct or capacitor coupling to the vertical yoke section completes the vertical deflection operation.

VERTICAL FREQUENCY-COUNTER IC

The vertical output signal is developed inside an IC frequency-counter chip (Fig. 6-7). The vertical drive signal is generated by charging C702 through the vertical height control (R705) to ground. The charging and discharging of Q702 generates a sawtooth voltage.

The sawtooth voltage is amplified by a vertical amp in the IC; this voltage is fed to the base terminal of the bottom vertical output transistor (Q703).

Check the sawtooth waveform at pin 18 of IC701 (Fig. 6-8). If there is no waveform, measure the dc voltage on pin 17. Go to the 24-V voltage source in the low-voltage power supply with no or improper voltage at terminal 17. Suspect IC701 of leakage if the 24-V source is fairly normal and low voltage exists at terminal 17. With normal B+ voltage and no sawtooth signal, check the vertical sync waveform at pin 24. If the sawtooth signal is fairly normal at pin 18 and the vertical sweep is insufficient, check the vertical output stages.

VERTICAL OUTPUT CIRCUITS

In the tube and early solid-state chassis, the vertical output circuits were transformer-coupled to the yoke assembly. In the 1970s, two transistors of different polarity were incorporated in

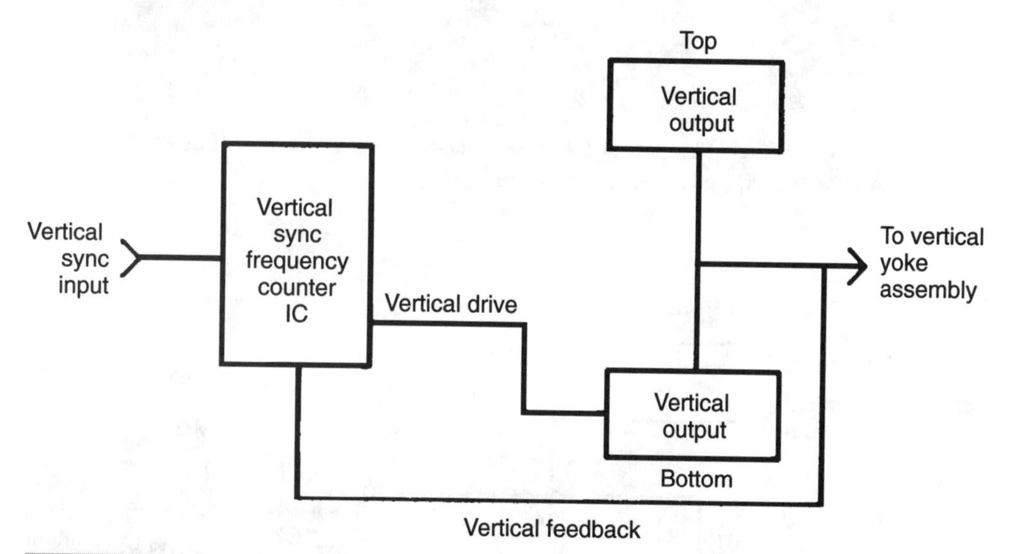

FIGURE 6-5 **A frequency-counter IC can provide signal and drive for the vertical and horizontal output transistors.**

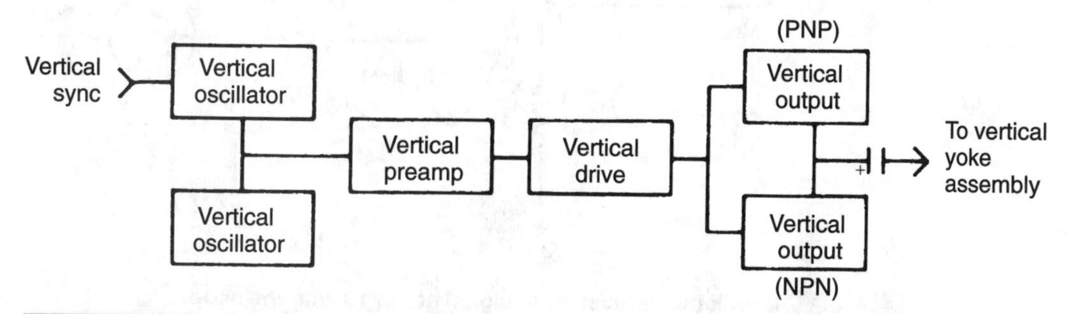

FIGURE 6-6 **The vertical output circuits are coupled directly to the vertical yoke winding or through an electrolytic capacitor.**

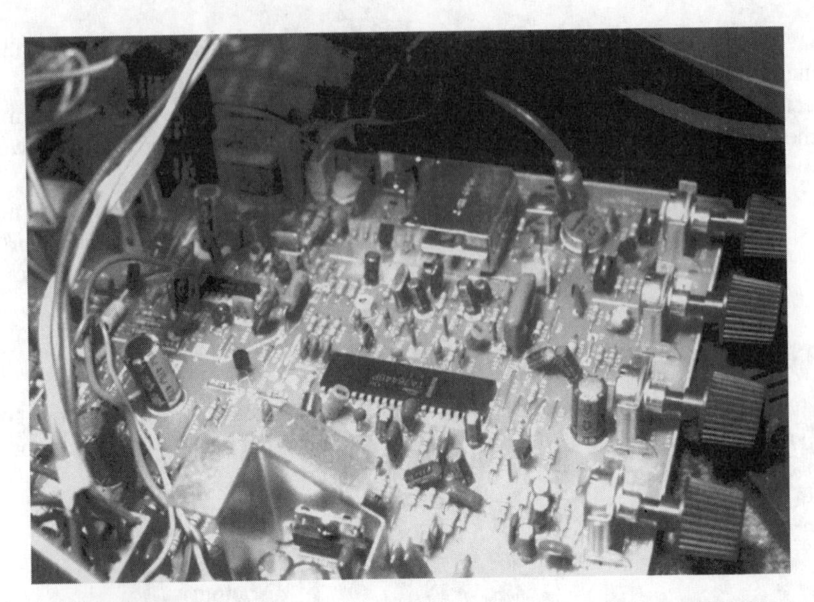

FIGURE 6-7 The latest TV chassis contains a vertical output IC on a heatsink, instead of transistors.

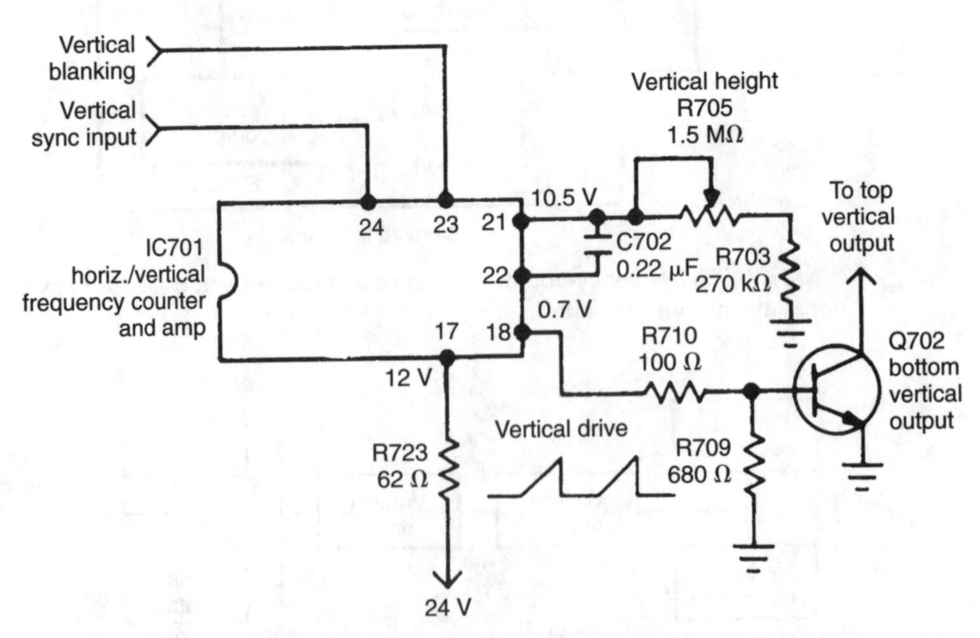

FIGURE 6-8 Check the vertical drive signal at pin 18 with the scope.

a transformerless (OTL) vertical output circuit (Fig. 6-9). Each transistor conducts only one-half of the input signal and cuts off with reverse polarity. One transistor sweeps the bottom half while the other transistor sweeps the top half of the raster. You might find one IC containing all vertical input and output circuits in some of the latest TVs (Fig. 6-10).

A separate vertical driver transistor can be found between the IC oscillator and the vertical output transistors (Fig. 6-11). In the latest vertical ICs, the vertical amp is built right in the IC. Thus, the amplified sawtooth signal applies directly to the base terminal of the vertical output transistor.

Most vertical sweep problems are related to a defective component in the vertical output circuits. A horizontal white line, insufficient vertical sweep, intermittent sweep, and foldover can be caused by a defective output transistor or component. Usually, the vertical output transistors are mounted on a separate heatsink or metal TV chassis. You might find a top-hat or flat-mounting vertical output transistor (Fig. 6-12). The flat transistors can be insulated from the heatsink with small pieces of insulation. These flat-type transistors have a tendency to break down under load and still test normal when out of the circuit.

Troubleshooting the vertical output stages is fairly easy with a known oscillator sawtooth signal. Check the input waveform and output signal with the scope. With a fairly normal sawtooth voltage at the output of the vertical IC and no waveform at the vertical output transistors, suspect problems within the vertical output stages. Measure the voltage at the

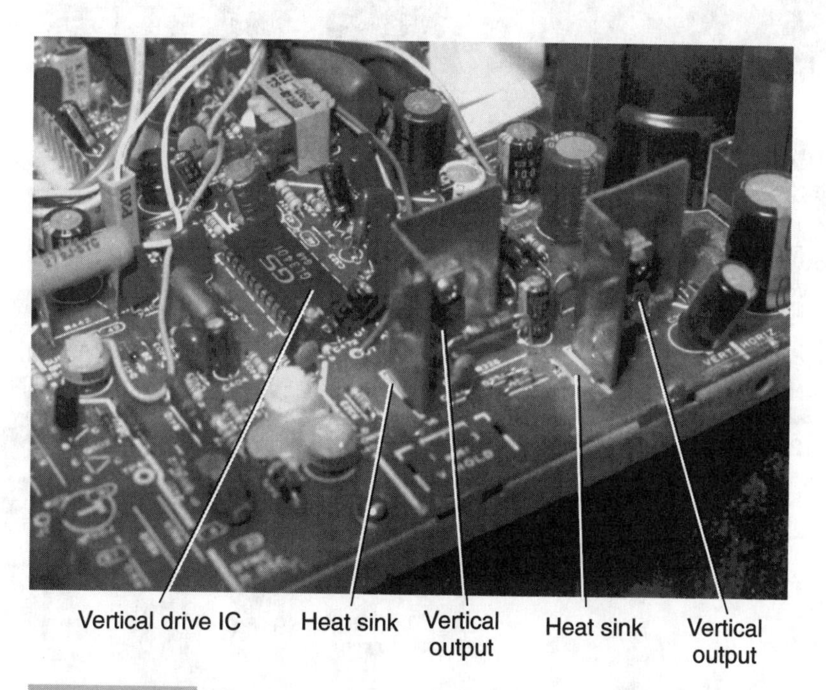

FIGURE 6-9 **The vertical output transistors on separate heatsinks with different voltages on the collector terminals.**

FIGURE 6-10 The entire vertical assembly could be included in one large IC in today's TV chassis.

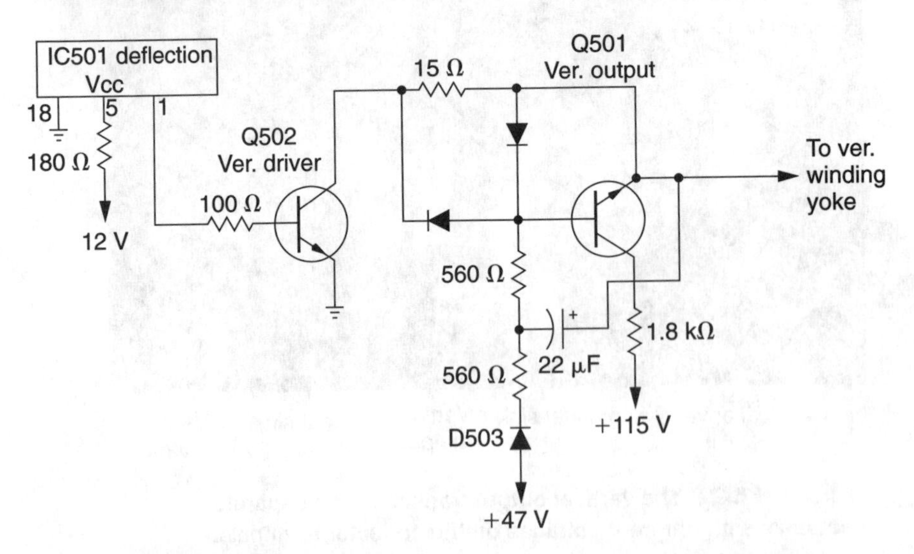

FIGURE 6-11 Q502 drives the single vertical output transistor in this vertical output circuit.

FIGURE 6-12 Older vertical circuits have flat-top and top-hat transistors, but the new chassis have one power-output IC.

metal collector terminals of each output transistor. The collector terminal of one transistor is usually grounded, with the other having a high B+ voltage. Correct voltage measurements on each transistor can determine if the transistor or surrounding components are defective.

VERTICAL OUTPUT IC

Today, most vertical output circuits consist of one fairly large power IC (Fig. 6-13). The vertical output IC is mounted to a separate heatsink on the TV chassis. The vertical IC receives a drive pulse from a vertical countdown or sweep IC and is coupled directly to the input terminal. The output terminal of the vertical power output IC is coupled to the vertical winding of the deflection yoke.

The vertical IC may have one or two different voltage sources at voltage supply pins (V_{CC}). When the supply pin voltage is low, suspect that a vertical output IC is leaky or that the voltage from the supply source is improper. Take voltage measurements at all pin terminals and compare them to the schematic. Check the drive waveforms at the input and output terminal of the vertical output IC (Fig. 6-14). If you don't find the input waveform, check the vertical sweep IC circuits. Suspect a defective vertical IC with adequate input waveform and no output waveform. Check the yoke winding with a fairly normal output waveform from the vertical output IC.

HORIZONTAL WHITE LINE

No vertical sweep is the easiest symptom to locate in the vertical circuits. Practically any defective component in the vertical circuit can produce a horizontal white line. First, measure

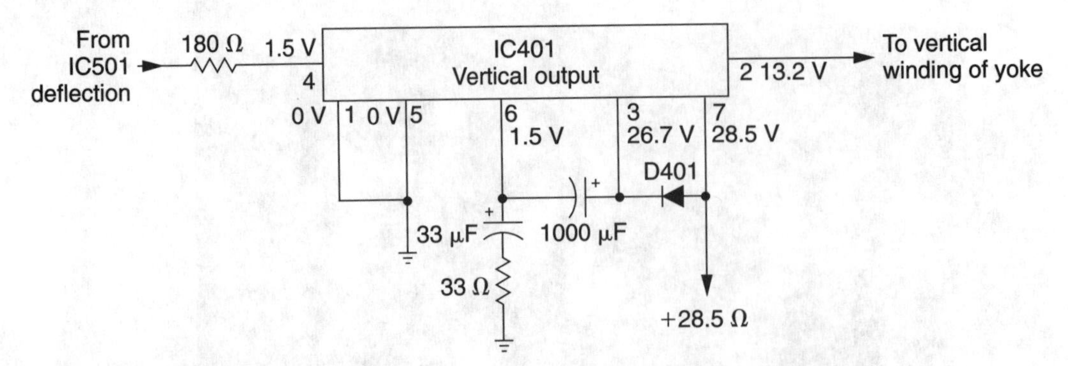

FIGURE 6-13 The vertical output circuit consists of a few capacitors, resistors, diodes, and the vertical power-output IC.

FIGURE 6-14 The vertical output pulse at the output IC and as it applied to the vertical yoke winding.

the B+ voltage that feeds the vertical circuits. You might find a different voltage supplying the vertical IC and the vertical output circuits. Improper voltage at any stage can cause no vertical sweep.

Only a horizontal white line was found in the raster of a Sharp 19D72 (Fig. 6-15). No voltage was found at either vertical output transistor. A voltage check at the 48-V voltage source indicated that R520 (a 22-Ω resistor) overheated. D503 was leaky when tested in the circuit with the diode test of the DMM. Replacing both R520 and D503 restored voltage and proper vertical sweep to the vertical circuits.

Another common problem is when no waveform is at the vertical output transistors, but there is a vertical sweep. Check the oscillator waveform at the transistor or IC oscillator stage. Check for a waveform at the base and collector terminal of each transistor within the vertical circuit.

No waveform was found at the collector terminal of TR602 in a Montgomery Ward 12926 (Fig. 6-16). A normal signal was found on the base terminal. TR602 was checked in the circuit with the diode transistor test of a DMM. High internal leakage was found between the emitter and collector terminals. TR602 was replaced with an ECG159 universal transistor.

In directly coupled vertical circuits where vertical feedback controls the vertical oscillator circuits, the scope is useless. If you find no waveforms, check each transistor in the circuit with the DMM diode test. The vertical transistors can be checked within seconds with the DMM. If leakage is noted between two elements, check the schematic for resistors or diodes across the two elements that are causing the leakage measurement. If in doubt, remove the emitter terminal with solder wick and take another measurement. The open or leaky transistors can be found within a few minutes in the vertical circuits.

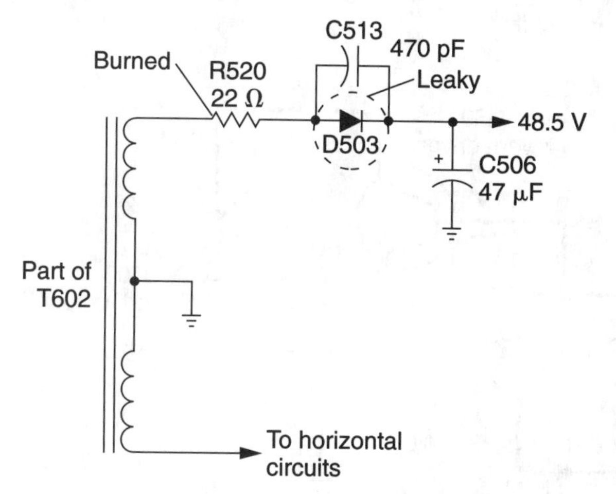

FIGURE 6-15 **Leaky diode D503 and burned resistor R520 prevented the dc voltage from reaching the vertical output circuits.**

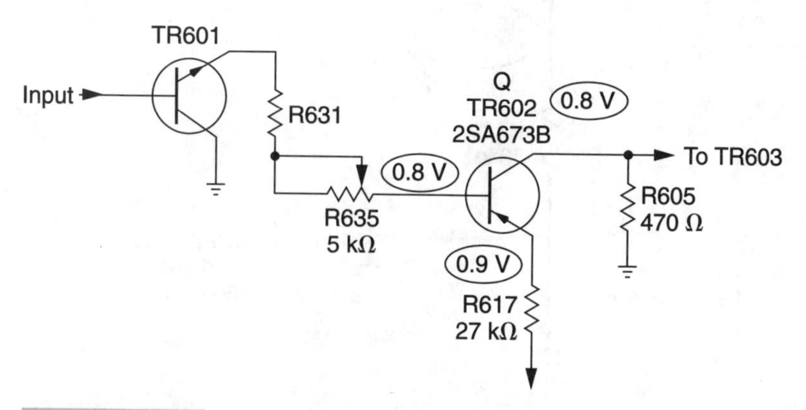

FIGURE 6-16 **The circled voltages occurred with a leaky TR602 in a Wards 12926.**

VERTICAL OSCILLATOR IC PROBLEMS

Check the vertical oscillator IC with a scope and make voltage tests. Check the vertical sawtooth waveform fed to the vertical driver or output transistor. Improper or no waveform signal can indicate that an IC is defective, surrounding components are bad, or that there is no voltage input. Next, measure the B+ voltage at the IC terminal. If the voltage is low, suspect that the IC, components, or voltage source is faulty. Remove the voltage source at the B+ pin. Suspect that the IC is leaky if the correct voltage returns.

Only a horizontal white line with no vertical waveform at pin 7 was found in a Sharp 19C81B (Fig. 6-17). Only 2.7 V was noted at the B+ voltage source (pin 11). Pin 11 was removed from the circuit with a solder wick. The supply voltage increased to 18 V, indicating that IC1501 was leaky. Extra care must be exercised when replacing the defective IC with a universal replacement. Sometimes the new replacement can be damaged during installation or the new chip can be defective. If the replacement IC does not work, try another one. Replacing IC1501 with one of the original part number (RH-1X0094CEZZ) solved the no-vertical-sweep problem.

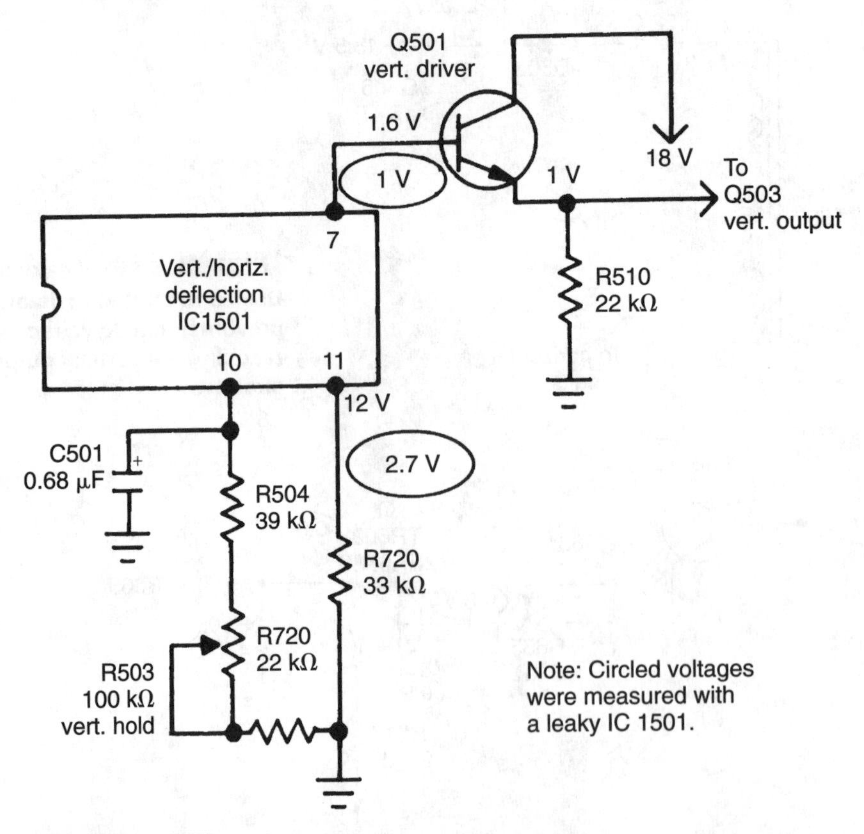

FIGURE 6-17 No vertical waveform was found at pin 7 in a Sharp 19C81B.

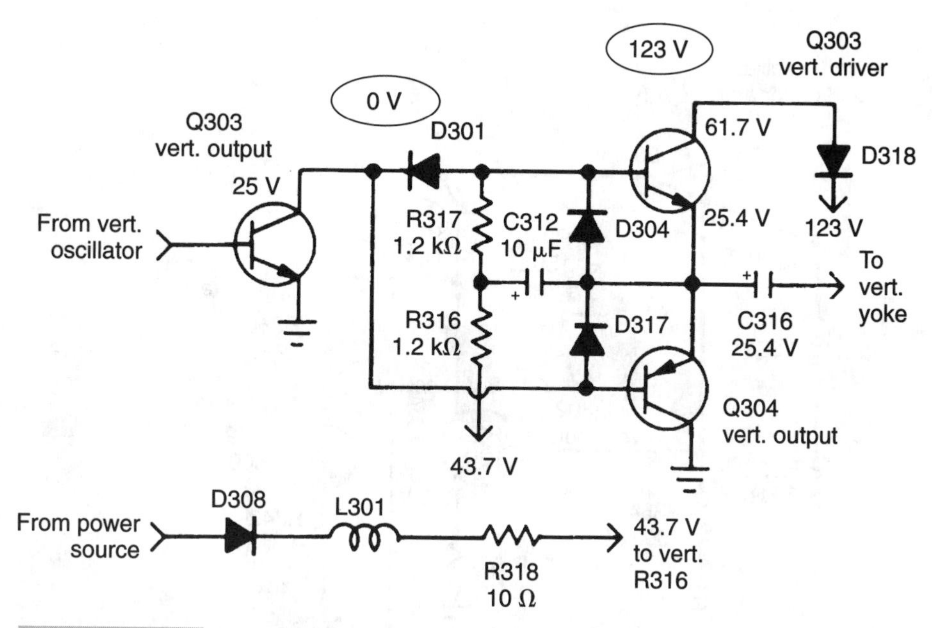

FIGURE 6-18 Check the resistor (R318) that provides voltage to the output transistor in a K-Mart KMC1921G.

WHITE LINE—OUTPUT CIRCUITS

Check the vertical output circuits if the proper vertical oscillator or driver signal is feeding the output transistors. Measure the voltage at each collector terminal. Go to the power source if no voltage is at both collector terminals of the transistor. A low collector voltage on both transistors can indicate that a transistor is leaky. Test each transistor in the circuit for an open or leaky condition with the DMM diode test.

Suspect that bias components have failed: burned resistors or leaky diodes. Remove one end of each resistor or diode to check for correct resistance. Bridge crucial electrolytic bypass and coupling capacitors in the output circuits. If trouble still exists, measure the resistance of each resistor in the vertical circuits, removing one terminal to check for correct resistance.

A horizontal white line was found in the raster of a K-Mart KMC1921G (Fig. 6-18). No voltage was found at the collector terminal of the driver transistor (Q303), and high voltage was measured at the collector terminal of Q305. The 43 Vdc was missing from the output circuits. R318 (10 Ω) in the 43-V power supply was open. Replacing the open resistor did not solve the no-sweep problem.

A quick in-circuit transistor test of both transistors revealed that Q305 was open and Q304 was leaky. Although the transistors were out of the circuit, the bias diodes and resistors were checked for leakage and correct resistance. Replacing both transistors and R318 solved the no-vertical-sweep problem. Always replace both vertical output transistors if one is defective.

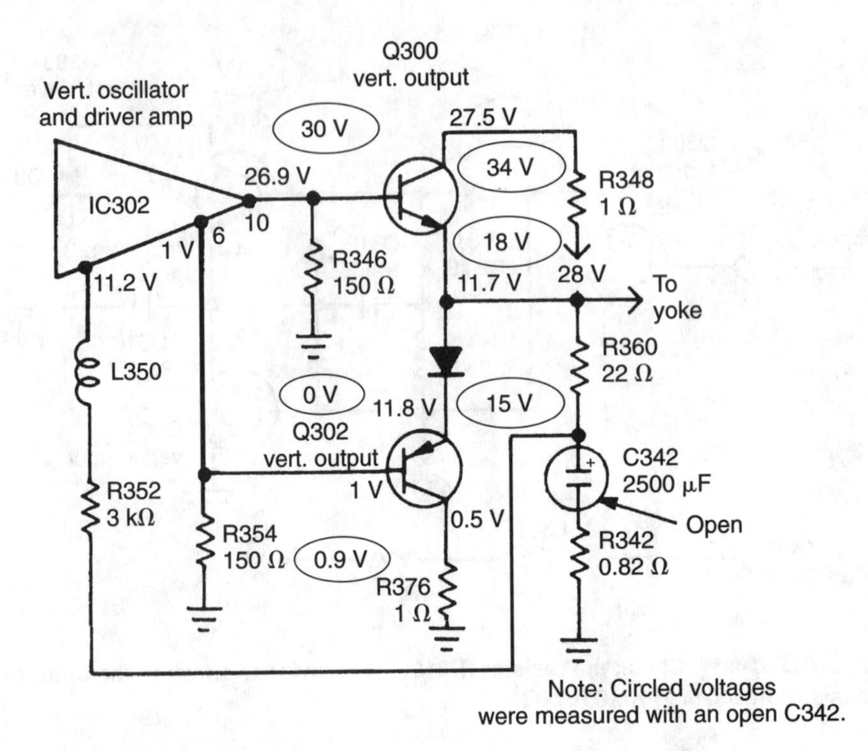

FIGURE 6-19 **Open capacitor C342 (2500 µF) produced a horizontal white line in a Sylvania E08-1 chassis.**

Although most sweep problems in the vertical output circuits are caused by the output transistors, this was not the case in a Sylvania E08-1 chassis (Fig. 6-19). Both Q300 and Q302 tested normal in the circuit, with higher-than-normal voltage on both collector terminals. A sawtooth waveform was fed to the base of Q300. All resistors within the vertical circuits were normal. The full raster was restored when C342 (2,500 µF) was bridged with a good capacitor.

CRUCIAL VERTICAL WAVEFORMS

Defective transistor output waveforms are difficult to obtain or read if a defective component is again the vertical output circuit. You can check the output waveform and see (if the waveform is visible) what is feeding to the vertical yoke winding. In the vertical output ICs, crucial waveforms can help you to determine if the vertical input or output IC circuits are defective (Fig. 6-20).

With no vertical sweep (a horizontal white line), scope the output terminal feeding to the deflection yoke. If there is no waveform or an improper waveform, attach the scope probe to the output IC's input terminal. Next, scope the vertical drive waveform from the sweep or countdown IC. Notice the peak-to-peak voltage at each waveform. Correct vertical output waveforms can help you to determine what stage or circuit is defective.

VERTICAL VOLTAGE INJECTION

The vertical sweep IC and output IC circuits can be serviced when the voltage source is taken from the flyback secondary winding by voltage injection. Check the schematic for the voltage-supply terminal of the sweep IC and vertical output IC. Inject a voltage that is shown into these V_{CC} terminals from an external universal power supply. Two different voltages must be injected at each IC. This means a dual variable power supply, or separate battery pack. Trace the scope waveform at the vertical sweep output terminal to the deflection yoke winding.

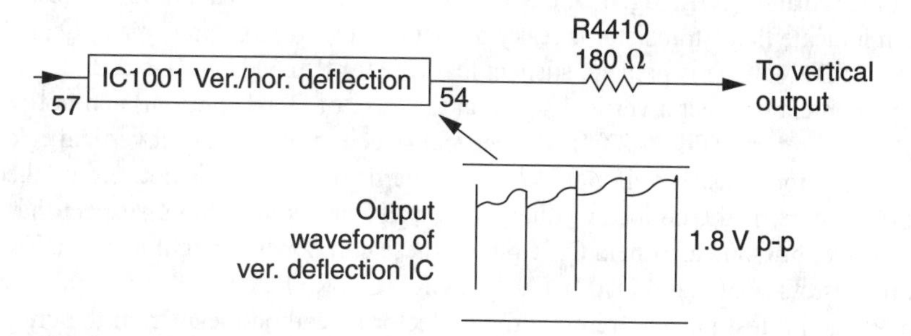

FIGURE 6-20A The output drive waveform from the deflection IC to the vertical output IC.

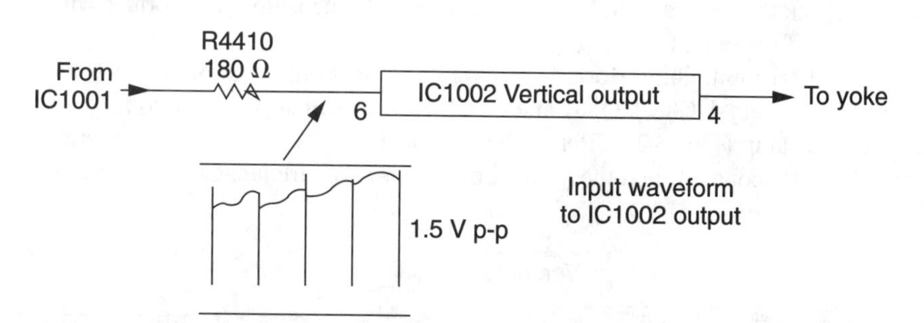

FIGURE 6-20B The input waveform at the output IC.

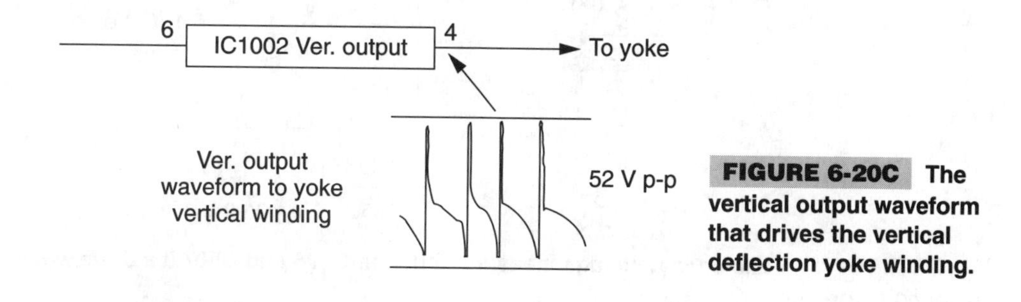

FIGURE 6-20C The vertical output waveform that drives the vertical deflection yoke winding.

SERVICING DIRECTLY COUPLED VERTICAL CIRCUITS

In many of the early TV chassis, directly coupled transistors formed the vertical circuits. The vertical oscillator might be a multivibrator type with vertical error and preamp circuits between the oscillator and output circuits. The feedback circuit from the output to the oscillator or error amp stages must be before or after vertical operation circuits. Practically any one component in the vertical circuit can keep the vertical circuit from functioning.

Scope waveforms are useless in these circuits. Only the vertical output check can indicate that vertical circuits are normal. No vertical sweep indicates yoke problems. If there is no vertical output signal, start with voltage tests. Check for positive voltage on both collectors of Q506 and Q507 (Fig. 6-21). No or low voltage on vertical output transistor Q506 (top) can indicate that a transistor is leaky or that the low-voltage power supply (116 V) is defective. If B+ voltage is present, suspect that resistor R526 (15 kΩ) is bad.

Some technicians inject a vertical signal at the base of Q507 (bottom) and notice if the vertical sweep returns (Fig. 6-22). If the vertical sweep returns, suspect vertical error, preamp, or oscillator transistor circuits. When the vertical sweep does not return, check the output transistors. If the collector voltage of Q506 (top) is very high, suspect that either Q506, CR508, or Q507 are open. Only one of these components might be open. Test both output transistors (Q506 and Q507), CR503, CR510, CR506, and CR508 in the circuit. If any one does not test normal, remove the collector or cathode lead from the circuit and check it again.

Sometimes one of the output transistors will test normal, but it then breaks down under load. Also check the deflection yoke (CY100), CR505, C424, and pincushion transformer T403. Check the vertical windings of the yoke and pincushion transformer with the low-ohm scale of the ohmmeter (Fig. 6-23).

The vertical multivibrator oscillator circuits must oscillate before there is any vertical output sweep signal. Check Q501 and Q502 when no signal is applied to the base terminal of the error amp (Fig. 6-24). Remember, vertical waveforms are not very steady—even with the best scopes. When the vertical oscillator is off frequency, check electrolytic ca-

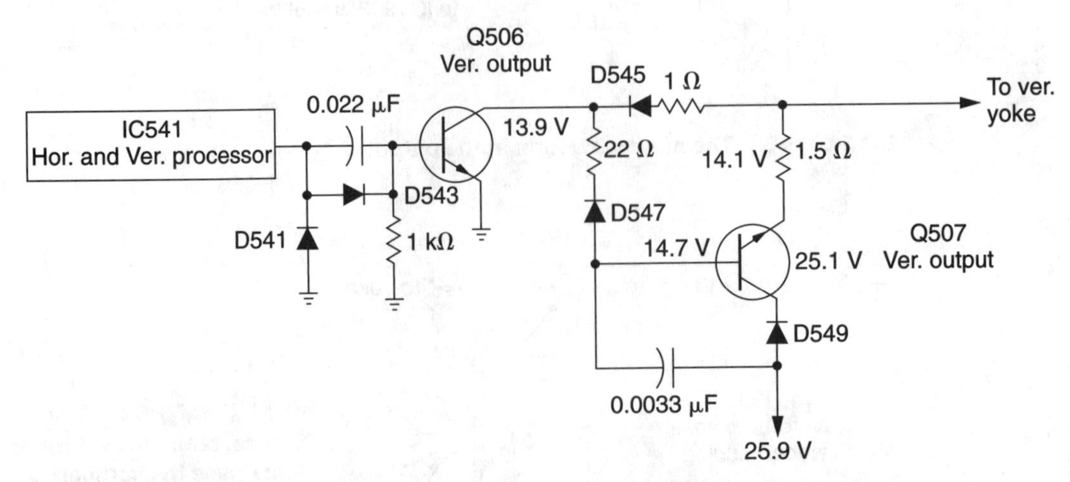

FIGURE 6-21 Take crucial voltage measurements on Q506 and Q507 if a drive waveform is found at the input.

FIGURE 6-22 The correct vertical output waveform applied to the yoke winding.

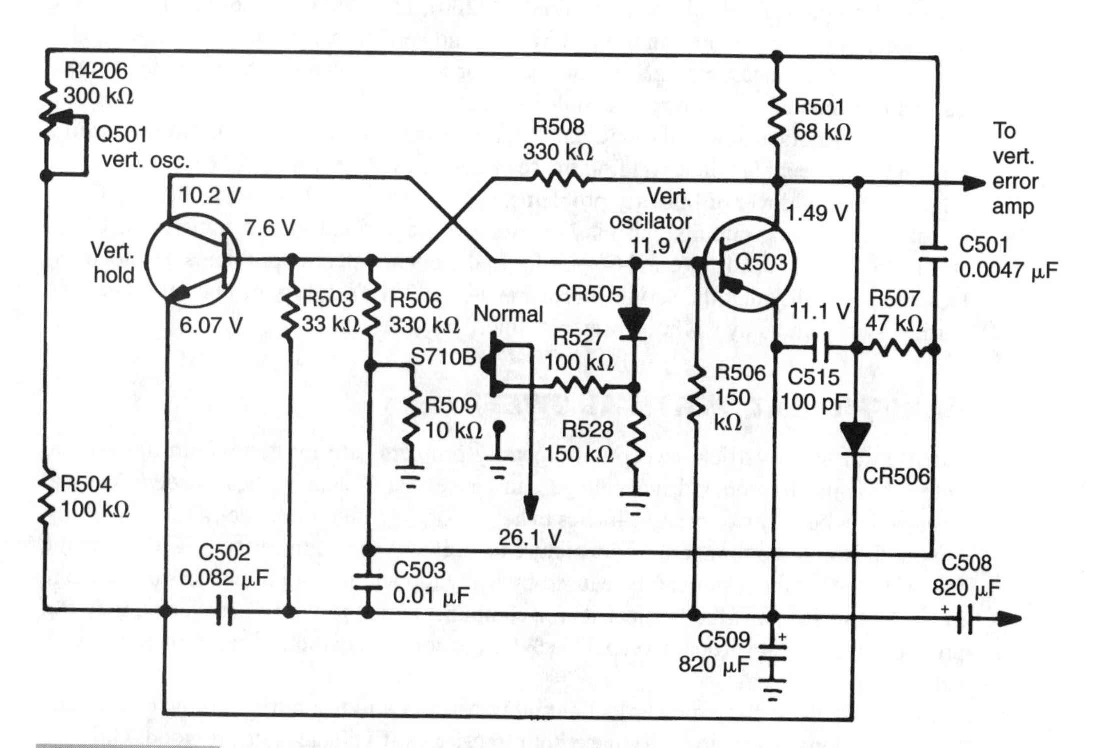

FIGURE 6-23 A multivibrator circuit with Q501 and Q503 as oscillators.

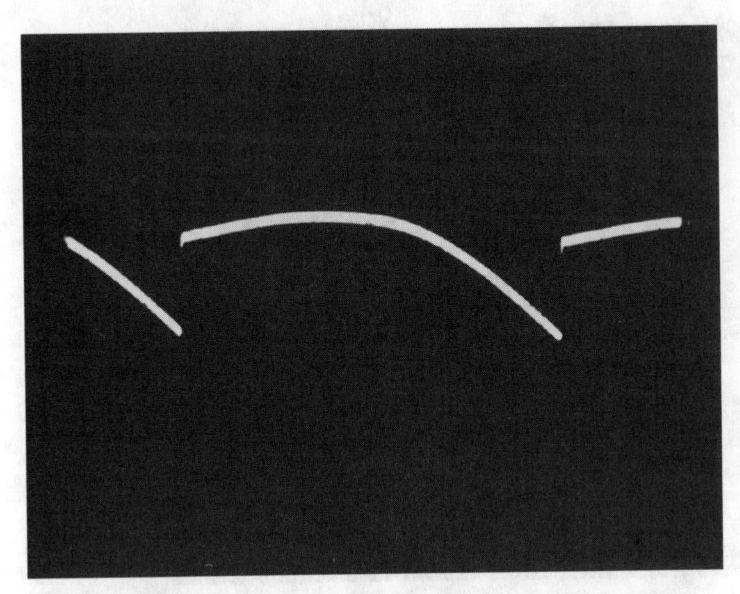

FIGURE 6-24 **Vertical drive waveform from the directly coupled oscillator transistor to the base error amp.**

pacitors C508 and C509, the vertical hold (R4206), and C424 (Fig. 6-25). Dried-up or open electrolytic capacitors can produce weak or no vertical oscillation.

Crucial voltage measurements on the transistor terminals can indicate a leaky or open transistor. Remember, in directly coupled circuits, the voltage on one transistor can directly affect another. Inject the internal voltage to the vertical oscillator circuits from a separate power supply when you find the scan-derived voltages (Fig. 6-26).

For vertical foldover or linearity problems, check C506, C507, C508, C509, and C424. Check the feedback circuits and bias resistors in the vertical output transistors. Check R521, R533, CR510, CR506, and CR508 for foldover and linearity problems. Do not overlook the possibility that the service normal switch (S701B) is dirty or open if there is no vertical sweep and only a white horizontal line.

INSUFFICIENT VERTICAL SWEEP

Often, insufficient vertical sweep and linearity problems are located within the vertical output circuits. Improper drive voltage can cause insufficient vertical sweep. Vertical sweep might be only two or three inches in height, or the raster might lack a fraction of filling out the top or bottom area of the picture tube. Poor vertical linearity (Fig. 6-27) at the top or bottom of the raster can be caused by leaky top and bottom transistors or a change in the bias resistors. If the vertical IC (or components tied to the IC) are leaky, they can produce insufficient vertical sweep. Check for the correct sawtooth waveform at the vertical IC output terminal.

Sometimes those flat-type vertical output transistors will test normal in the chassis and then break down under load. Replace both transistors if voltage tests are good. Only 1.5 inches of vertical sweep was found in a Sanyo 31C41N (Fig. 6-28). Both output transistors

FIGURE 6-25 The vertical drive waveform at the output transistors to drive the vertical yoke winding.

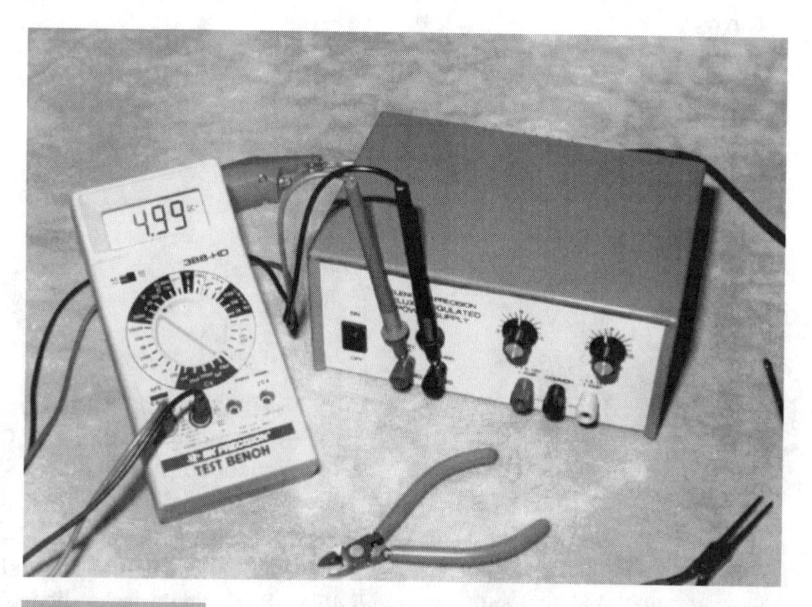

FIGURE 6-26 Inject the external dc voltage from a dual-power supply when the vertical circuits are powered by the secondary circuits of the flyback.

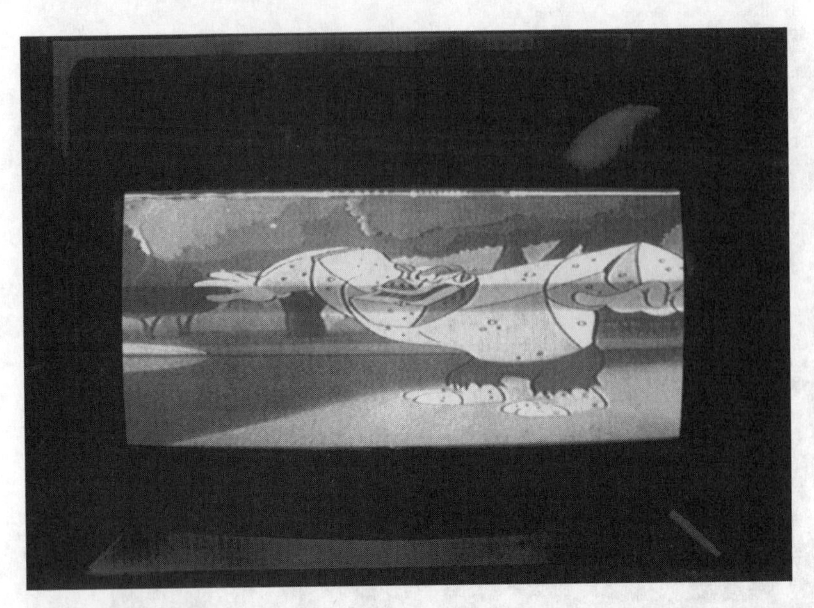

FIGURE 6-27 Poor vertical linearity can be caused by faulty output transistors, ICs, or an improper dc voltage source.

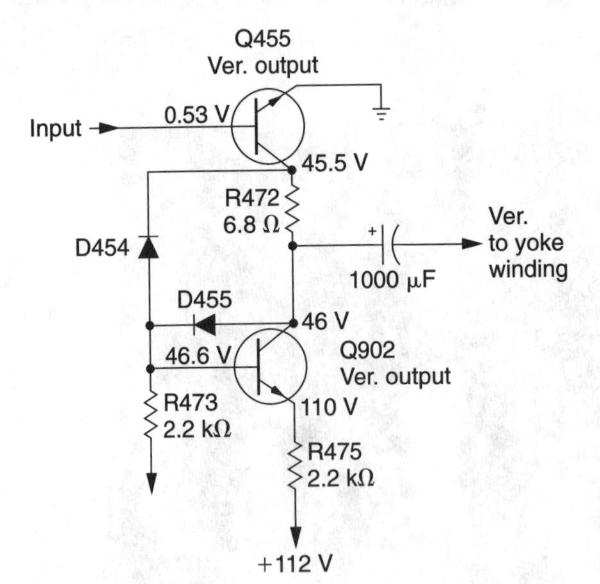

FIGURE 6-28 D454 was found leaky in a Sanyo 31C41N vertical output circuit.

(Q455 and Q902) were replaced with GE-32 universal replacement transistors. Now the height lacked 2 inches at the top of the raster. A voltage measurement indicates only 0.57 V at the base of Q902, which should be around 46 V. Replacing a leaky diode (D454) solved the insufficient-vertical-raster problem.

INTERMITTENT VERTICAL SWEEP

Of all the vertical problems, intermittent vertical sweep symptoms are the most difficult to locate and fix. The intermittent symptom can be caused by transistors breaking down, poor board connections to a loose collector transistor mounting screw, or many other things. Suspect that the board connection, component terminal, or eyelet or griplet connections are bad or loose if the raster collapses when the chassis is touched or moved. Try to narrow the intermittent to a certain section of the board. Intermittent transistors and ICs can be located by application of several coats of coolant. Monitoring the vertical circuits with the scope and DMM can help you to locate the intermittent stage and component.

Sometimes the raster would go to white line; at other times, a white horizontal line would appear in the picture with normal sweep in an RCA CTC68AF chassis (Fig. 6-29). The vertical drive signal from the MAG001B module was fairly normal when the raster went to a white horizontal line. A voltage test with the collapsed raster found higher-than-normal voltage on the base and emitter terminals of Q101. Replacing leaky Q101 and Q102 solved the intermittent raster problem.

Both vertical output transistors should be replaced when one is found to be leaky or open. The collector voltage can be quickly checked by measuring the voltage at each heatsink (Fig. 6-30). Often, heatsinks mounted in the center of a board are not grounded directly to the metal chassis. Suspect a vertical top output transistor if the top part of the raster pops down with a normal bottom half.

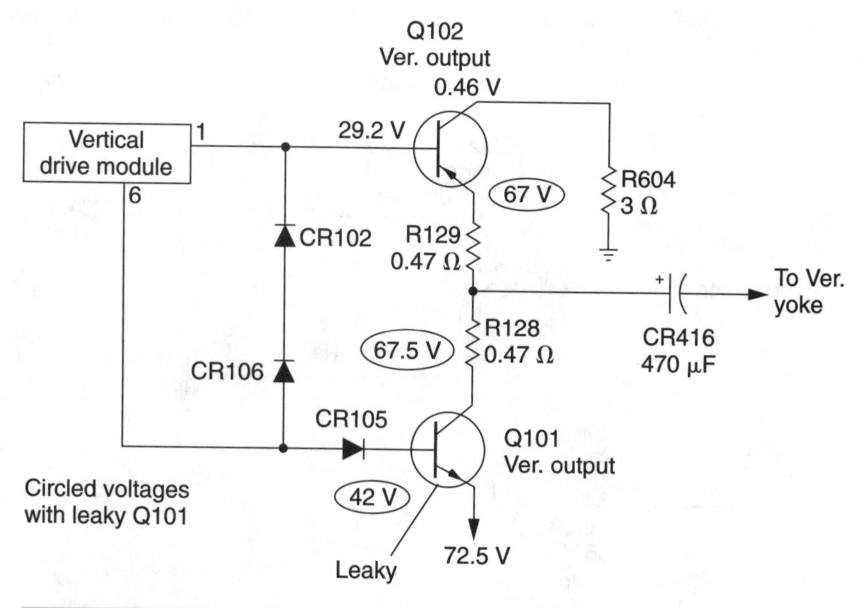

FIGURE 6-29 Check for leaky or open output transistors if there are improper voltages on all terminals.

FIGURE 6-30 **The metal heatsink of the vertical output IC is grounded.**

Intermittent vertical sweep—Sylvania E21 The picture in a Sylvania E21 chassis would operate for several hours, and then collapse. Again, the chassis might operate for days. The vertical output transistors were suspected, so both were replaced. Sometimes these output transistors test normal, but can still open up (Fig. 6-31). Q300 was replaced

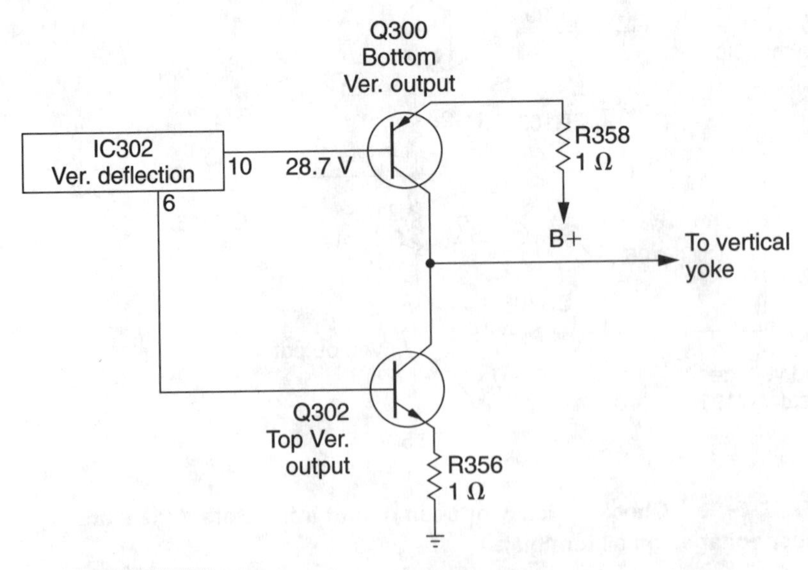

FIGURE 6-31 **IC302 was found intermittent in a Sylvania E21 chassis.**

with an RCA SK3083 and Q302 was replaced with RCA SK3054. Everything was normal for several days, but again the raster collapsed. When IC302 was in the intermittent state, several coats of cold spray were applied. The picture popped in. Replacing intermittent vertical driver IC302 restored the TV.

VERTICAL PINCUSHION PROBLEMS

Check the pincushion circuits if the sides of the picture are pulled inward. The pincushion circuits prevent TV picture distortion, in which each side of the raster can sag toward the center of the screen. Usually, pincushion problems do not cause too much trouble in small-screen TVs. Special pincushion circuits are used in 27-inch or larger screens.

Pincushion problems are caused by a defective pincushion transformer or poor solder joints. A narrow or bowed picture can result, with a shorted primary winding of the transformer. Replace any pincushion transformer that constantly emits a high-pitched squeal.

Solder the terminal connections of the transformer in all pincushion circuits. Check the primary winding resistance and compare them with the values listed on the schematic. In transistorized pincushion circuits, test the pin output transistor for leakage and improper voltages on the collector terminal (Fig. 6-32). Test all other transistors in the circuit for leakage or open conditions. Often the pin output transistor becomes leaky if the horizontal output transistor is shorted.

VERTICAL FOLDOVER

Check for leaky output transistors and improper bias resistors when vertical foldover occurs. Improper negative and positive supply voltages applied to the vertical circuits can produce vertical foldover. Most vertical foldover problems occur in the vertical output circuits. Vertical foldover can occur at the top or bottom of the raster. Usually, vertical foldover occurs with insufficient height problems.

The raster was down 5 inches at the top with vertical foldover in a J.C. Penney 685-2124. This particular chassis was made by General Electric (Fig. 6-33). The complete vertical

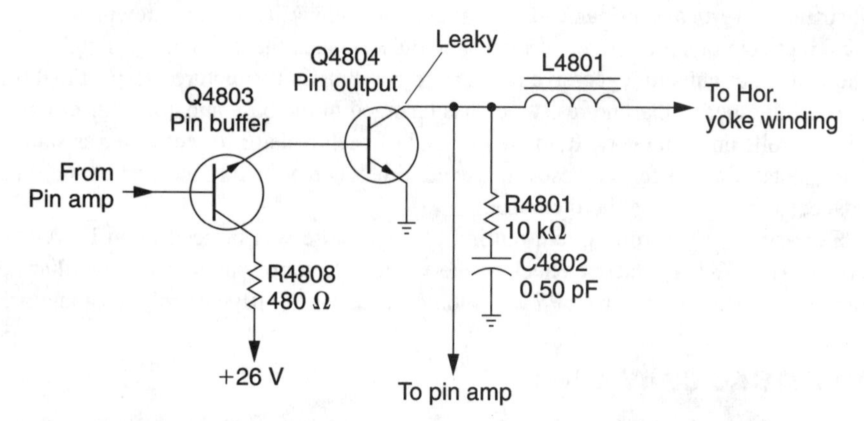

FIGURE 6-32 **Vertical blowing was caused by a leaky pin output transistor in a CTC140 RCA chassis.**

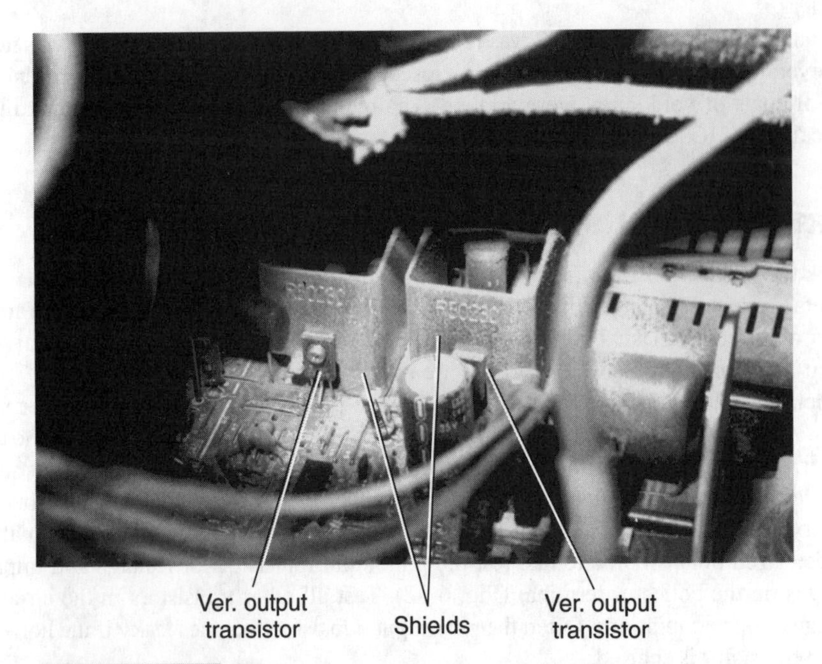

FIGURE 6-33 **Vertical foldover can be caused by a defective output transistor, bad IC, or by malfunctioning electrolytic capacitors in the output and feedback circuits.**

chassis can be removed for transistor and component tests out of the circuit. The vertical output transistor (Q267) was found to be leaky. Q267 was replaced with an SK3054 and Q268 with an SK3083 universal transistor. Again, replace both vertical output transistors if you find that one is defective.

VERTICAL ROLLING

Insufficient vertical sync can cause the picture to roll vertically up or down. Scope the vertical input circuits for correct vertical sync pulse. Check the low-voltage supply that feeds the vertical circuits for excessive rolling and crawling in the picture. Suspect that the voltage is improper or that the resistance has changed in the base and collector circuits if the picture rolls only one way. Remove one end of each resistor for accurate resistance measurements. If the raster collapses to a white line when adjusting the vertical hold control, suspect an open vertical hold control.

Excessive vertical rolling and pulling of the picture was noticed in an RCA CTC81C chassis (Fig. 6-34). A scope check of the vertical dc supply indicated poor filtering. The picture was restored when C305 was shunted with a 4000-μF electrolytic capacitor.

VERTICAL CRAWLING

Scanning lines slowly moving up the picture with a dark section is called *vertical crawling*. A dark bar might be at the top or bottom of the crawling area. Improper filtering in the low-voltage source feeding the vertical circuits produces vertical crawling. A lower dc voltage goes along with poor filtering. Check the speaker for additional hum in the sound.

Scope the low-voltage power-supply circuit to locate the defective electrolytic capacitor. Each filter capacitor can be shunted to locate the defective one. Always discharge the capacitor before clipping across the suspected one. You can damage transistors and IC circuits while shunting filter capacitors with voltage applied to the circuits. Some of these filter capacitors might have values greater than 1000 µF.

Vertical crawling was found in a Motorola D18TS chassis with the picture pulled down from the top of the screen (Fig. 6-35). The negative supply voltage feeding the vertical circuits was low. The picture returned to normal when another electrolytic capacitor was clipped across C809.

BLACK TOP HALF

You might find the top or bottom half of the picture missing in the newer vertical circuits. Go directly to the top vertical output transistor if the top section of the picture is black. Likewise, check the bottom vertical output transistor if a black area is at the bottom portion of the picture. With heavy retrace lines at the top of the raster in directly coupled vertical circuits, check the switch driver transistor.

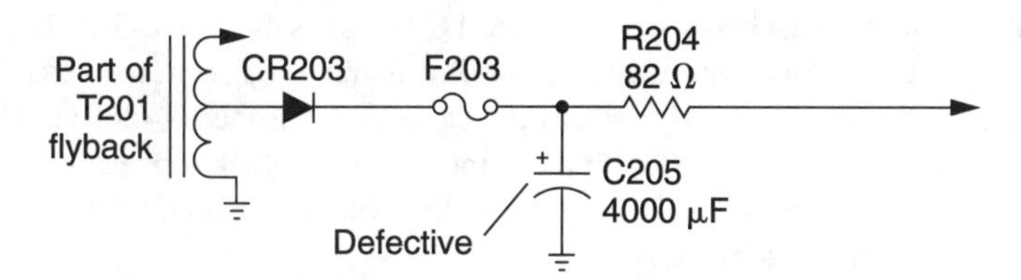

FIGURE 6-34 Dried-up filter capacitor C205 caused vertical pulling and rolling in this RCA CTC81C chassis.

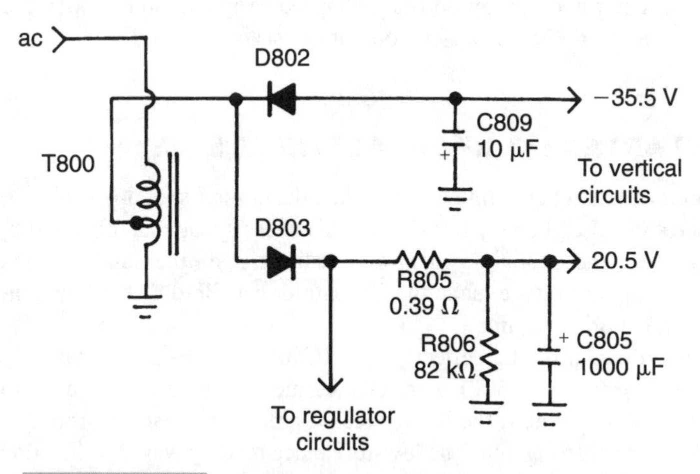

FIGURE 6-35 Vertical crawling was caused by a bad capacitor (C809) in a Motorola D18TS chassis.

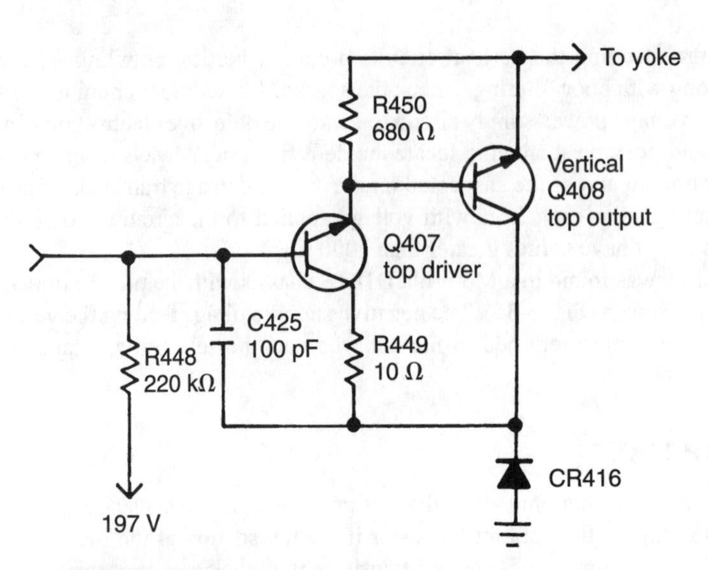

FIGURE 6-36 Leaky transistors Q408 and CR416 caused the top half of the picture to be black.

The top half of the picture was black in an RCA CTC93E chassis (Fig. 6-36). The voltages of the top output transistors were way off. Q408 was removed from the chassis and was leaky. The transistor was replaced that had the same part number (142691). The results were the same. Q407 was replaced with the same results, a black top section. An in-circuit diode test of CR416 revealed that it was leaky. Replacing both transistors and CR416 solved the black section at the top.

After a General Electric 10JA chassis was on for 10 minutes, the top part of the screen appeared black with no vertical sweep. Each vertical output transistor was sprayed with coolant. When coolant was applied to Q268, the picture was normal. Heat was applied to the top vertical transistor and the top portion collapsed. Application of coolant and heat spray helped to locate the defective top-half output transistor (Fig. 6-37).

SHUTDOWN AFTER HORIZONTAL WHITE LINE

The raster went into a horizontal white line and then the chassis shut down after being on for several minutes in a J.C. Penney 685-2084. This chassis was manufactured by Samsung Corporation. After the set shut down, a variable line transformer raised the ac voltage to 80 Vac, and any higher voltage made the TV shut down. Before shutdown, the raster would go into a white horizontal line.

The voltage measurement on the deflection IC (IC501) (Fig. 6-38) indicated the 12-V source at pin 3 was very low (5.6 V). A resistance measurement from pin 3 to chassis ground was 71 Ω. To determine if the IC was leaky, pin 3 was removed from the circuit with solder wick and a soldering gun. The low-resistance reading was still measured, indicating that IC501 was leaky.

Because ICs have a tendency to break down and become leaky, the 12-Vdc terminal (3) was removed from the circuit with the low-resistance measurement. These IC terminals can be easily removed with a mesh solder-removing material. Of course, C252 and R225 could have caused the low voltage at pin 3. The leaky vertical IC was placing a heavy load on the flyback winding of the horizontal output transformer, causing shutdown. A low-resistance measurement on any IC terminal that is not grounded can indicate that an IC is leaky.

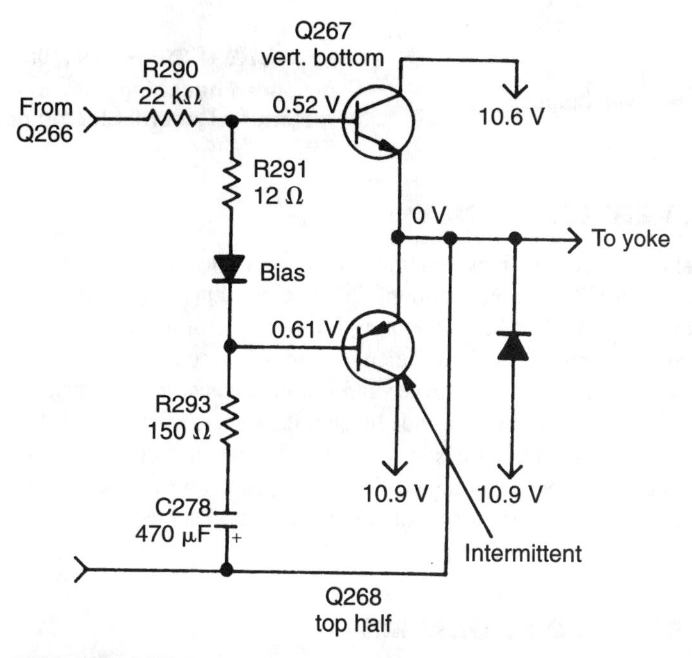

FIGURE 6-37 Q268 was found intermittent with an application of temperature extremes.

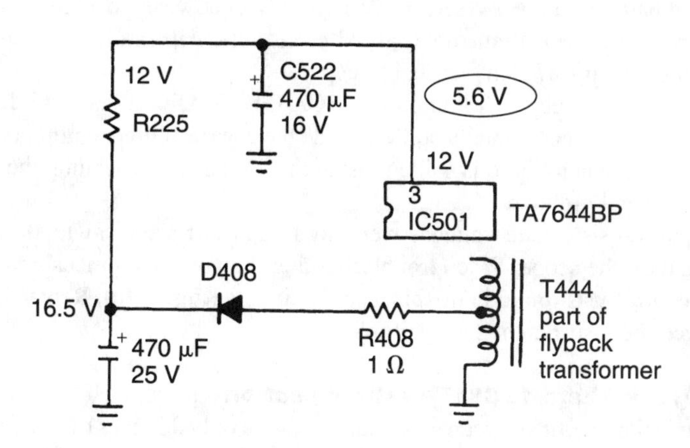

FIGURE 6-38 Improper voltages at pin 3 in IC501 caused a horizontal white line in a J.C. Penney 685-2084.

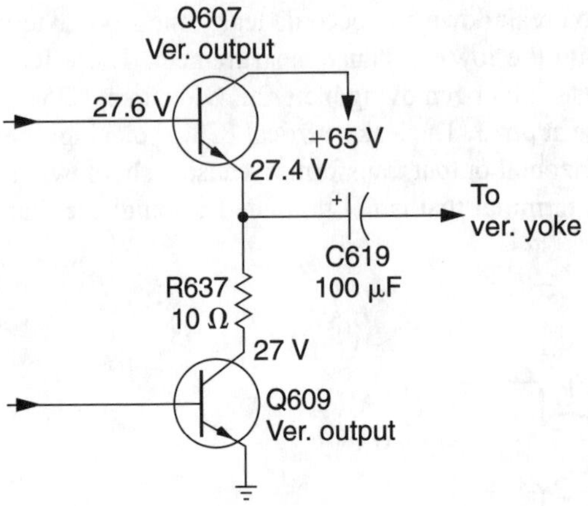

FIGURE 6-39 C619 (100 μF) produced bunching lines in the raster of a General Electric portable TV.

BUNCHING VERTICAL LINES

The symptom of heavy scanning lines was found at the top of the picture in a General Electric ECA chassis. After a 10-minute warmup, the bunching lines would appear. The picture was normal until the set had been operating for awhile. Adjusting the vertical height control would take out the lines, but insufficient height would remain.

All voltages were normal with the driver and output transistors. Because the output transistors have a tendency to cause various lines in the raster, Q607 and Q609 were replaced (Fig. 6-39). The bunching lines still remained. Bias and emitter resistors were checked for correct resistance with one terminal removed from the circuit. The vertical output coupling capacitor (C619) was shunted with another. This capacitor turned out to be defective.

VERTICAL-RELATED PROBLEMS

Sometimes insufficient or absent vertical sweep can happen outside the vertical circuits. An open yoke winding can produce a white horizontal line. A leaky or shorted yoke can look like a trapezoid pattern on the TV screen. Poor yoke socket wiring can develop into an intermittent vertical problem. Remember, a bad electrolytic capacitor in the convergence circuits can produce insufficient vertical sweep.

Only one inch of vertical sweep was found in an RCA CTC74AF chassis and it looked like a soft drink or liquor was accidentally spilled inside the chassis. Often, components or board arcing can develop when liquid is spilled inside the TV chassis. Cleaning the chassis did not cure the vertical problem.

The vertical output transistor and voltages were normal. A fair-sized sawtooth output waveform was found on the scope. When troubleshooting the yoke and pincushion components, a burned control was found in the pincushion circuits (Fig. 6-40). Replacing the 500-Ω control solved the insufficient-sweep problem.

Troubleshooting the Sharp 19C81B vertical section From the TV screen of a Sharp 19C81B insufficient vertical sweep was indicated, with only 2 inches of sweep (Fig.

6-41). Although the sawtooth waveform at the vertical amp terminal (7) of the deflection processor IC was fairly normal, very little waveform was found at the output coupling capacitor (C507). A waveform test at the driver transistor (Q501) was normal, pointing to possible trouble in the output transistor circuits.

A voltage measurement at the collector terminal indicated higher than normal voltage. Both collector terminals had about the same voltage (52 V). The voltage on Q503 should be around 20 V. Q502 was tested in the circuit with the transistor test of the DMM and was found to be open. Q503 tested good in the circuit. Both transistors were replaced with a universal ECG373 replacement, which restored the full picture.

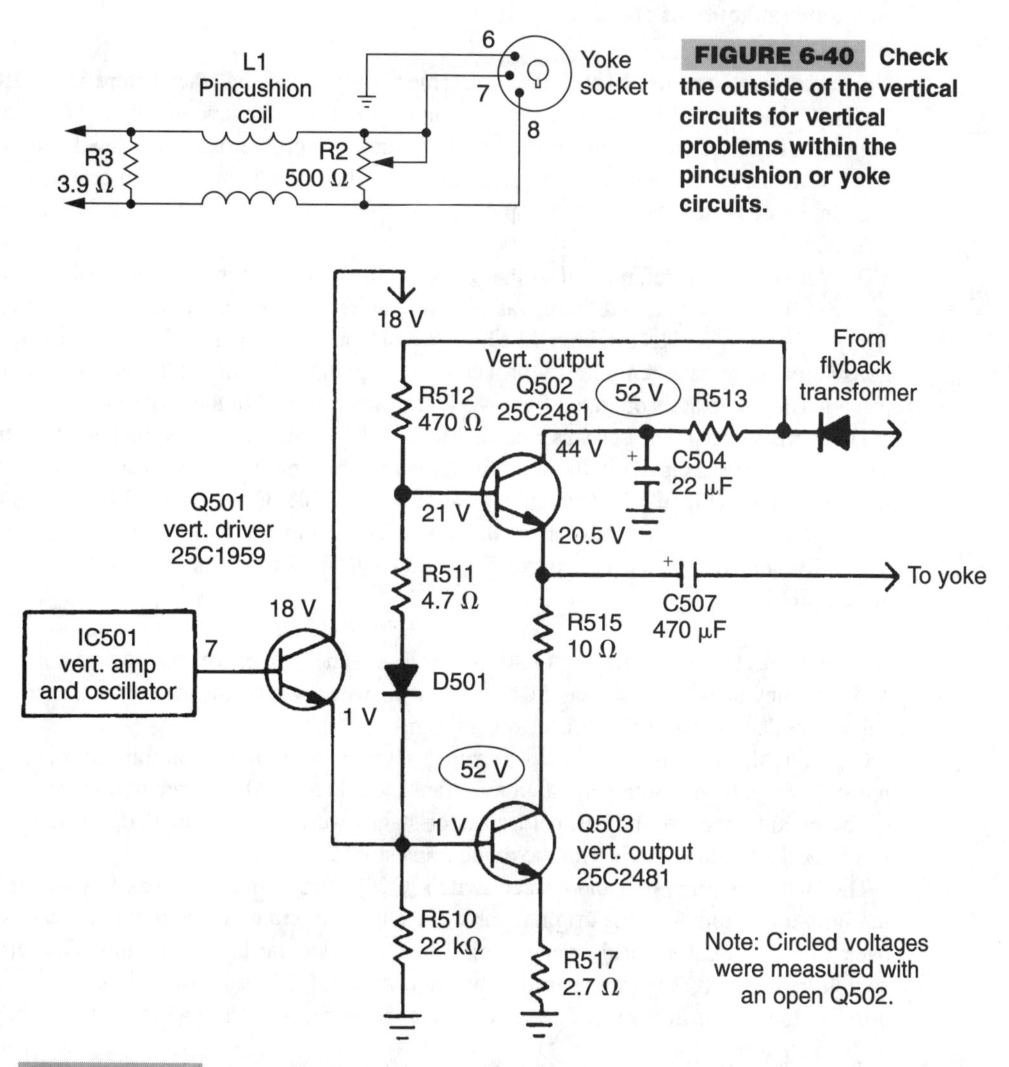

FIGURE 6-40 Check the outside of the vertical circuits for vertical problems within the pincushion or yoke circuits.

FIGURE 6-41 An open C502 produced only a two-inch vertical sweep in a Sharp 19C818 chassis.

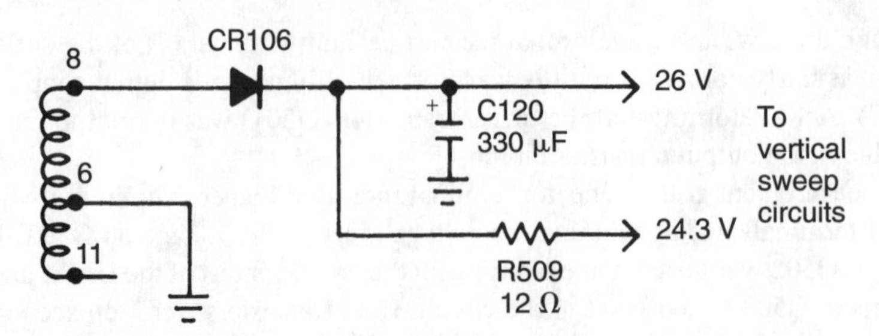

FIGURE 6-42 R509 caused intermittent vertical sweep after several hours of operation.

Unusual vertical problem The customer complained that the picture in an RCA CTC120 chassis would go down to a horizontal white line after several hours of operation. Sometimes the set would operate for days without any problems. When the TV chassis was on the bench, no vertical sweep was found on the TV screen. Pushing around on the chassis did not cause the raster to pop out. A normal vertical waveform was found at vertical IC401.

Both transistors tested normal in the circuit with the transistor test of the DMM. A dc voltage check indicated that there was no 24-V source. R509 was found to be open (Fig. 6-42). R509 was replaced with a 12-Ω, 1-W resistor. Because one of the vertical output transistors might have temporarily broken down, opening R509, both transistors were replaced. The TV chassis operated for a week and was returned to the customer.

Three weeks later, the set was returned to the shop with only three inches of vertical sweep. The chassis played in the shop for five days and finally collapsed again. R509 was running red hot (Fig. 6-43). Coolant was applied to IC401, Q502, and Q501 without any results. Next, each diode was sprayed and the raster went to a full screen. Heat was applied to CR504 and the sweep collapsed. Replacing CR504 solved the intermittent vertical sweep problem.

Difficult vertical problem Only four or five inches of vertical sweep was noticed in a J.C. Penney 685-2020. In fact, the only vertical sweep was below the center of the vertical raster. The top half of the raster was missing.

Checking the schematic, the vertical sweep section has a top and bottom transistor output stage (Fig. 6-44). Both vertical output transistor Q22 and Q24 tested normal in and out of the circuit. Because the vertical transistor stages were directly coupled, all transistors were checked in the circuit with the diode transistor test of the DMM.

The switch driver (Q19) and retrace switch (Q20) were found to be leaky. After replacing both transistors with the original parts, the chassis operated for two minutes and went back to insufficient vertical sweep, with a bright horizontal line at the top. The retrace switch transistor (Q20) was operating red hot. Again, Q20 was replaced because it was possible that it was defective. The results were the same. After a few seconds, Q20 began to get very warm.

All voltage measurements were quite close, and some were off only a volt or two. Both bias diodes in the base circuit of Q21 checked normal with one terminal lead removed. Next, PT101 (62-Ω cold) and R120 (680 Ω) were checked for correct resistance. R117 and R115 were checked in the circuit and were fairly close in resistance.

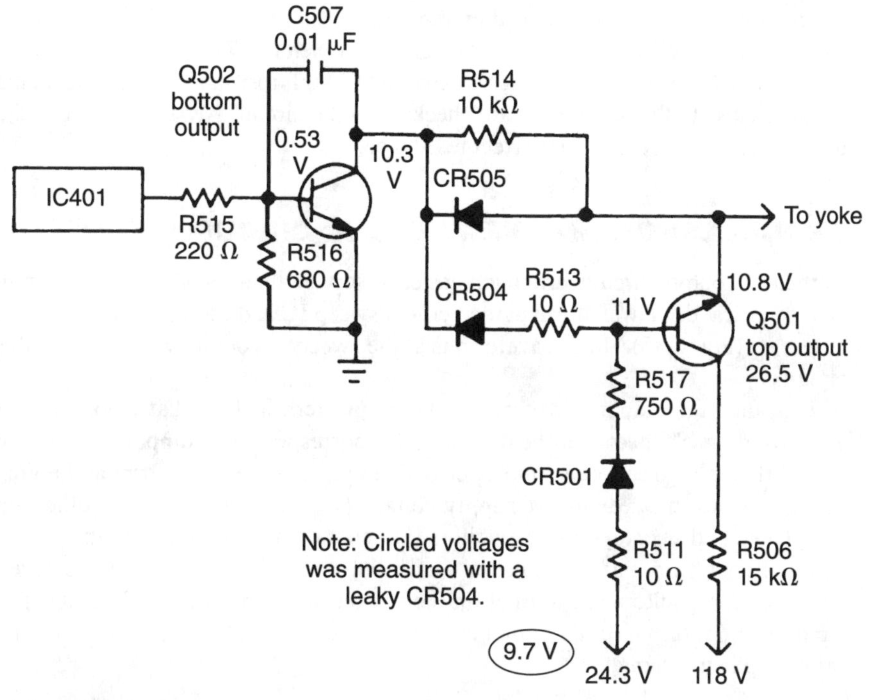

FIGURE 6-43 Coolant sprayed on capacitor CR504 caused the picture to become intermittent, then collapse.

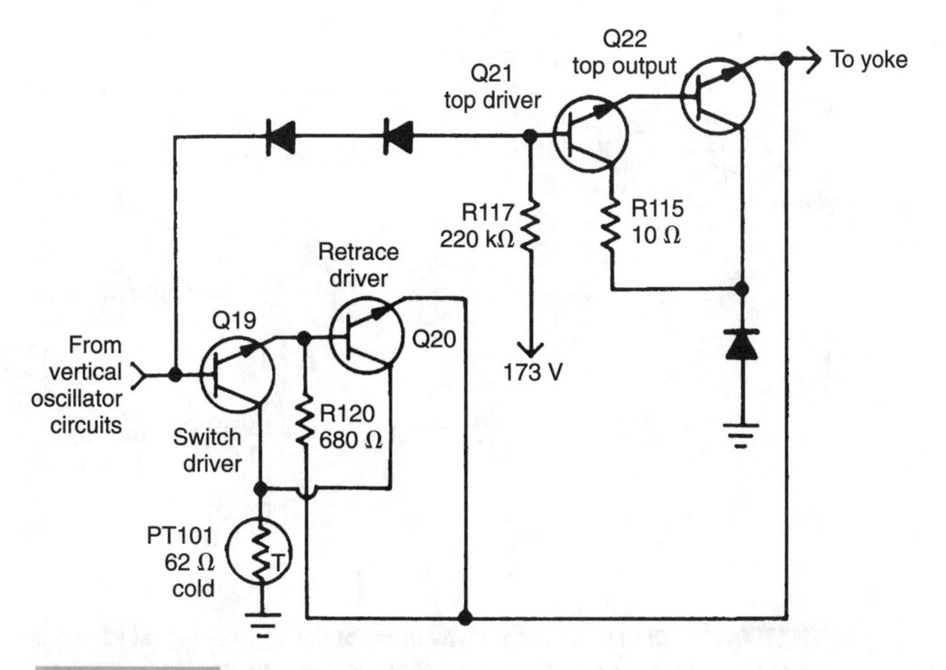

FIGURE 6-44 Replacing leaky transistors Q19, Q20, and R117 restored the missing top half of the vertical sweep.

Actually, the culprit was located by removing each resistor lead from the circuit board. Each resistor was checked for correct resistance. When R117 (220 kΩ) was checked again, it was found to be open. Sometimes we overlook a high-resistance measurement that might be quite close to the resistor we are checking in the circuit. Always remove one end of a diode or resistor to check for correct resistance.

TROUBLESHOOTING VERTICAL IC CIRCUITS

Vertical IC output circuits are much easier to service than transistor output circuits. Just take a vertical drive waveform at the vertical sweep IC and trace it to the input of vertical output IC (Fig. 6-45). If no waveform is at the sweep or countdown IC, repair this sweep circuit.

Next, take the output waveform at the output terminal IC that feeds to the vertical yoke winding. Suspect that the output IC (or corresponding components) are defective or that the voltage source is improper with a normal input waveform and no or a weak output waveform. Measure the supply voltage (V_{CC}) and voltage on all other pin terminals. Check all resistors and capacitors that connect to each pin terminal. Replace the output IC with normal components and input vertical sweep signal. Be sure that all electrolytic capacitors are normal on the pin terminals of output IC. Suspect the yoke winding and return circuit components if the output waveform is normal, but the vertical sweep is improper.

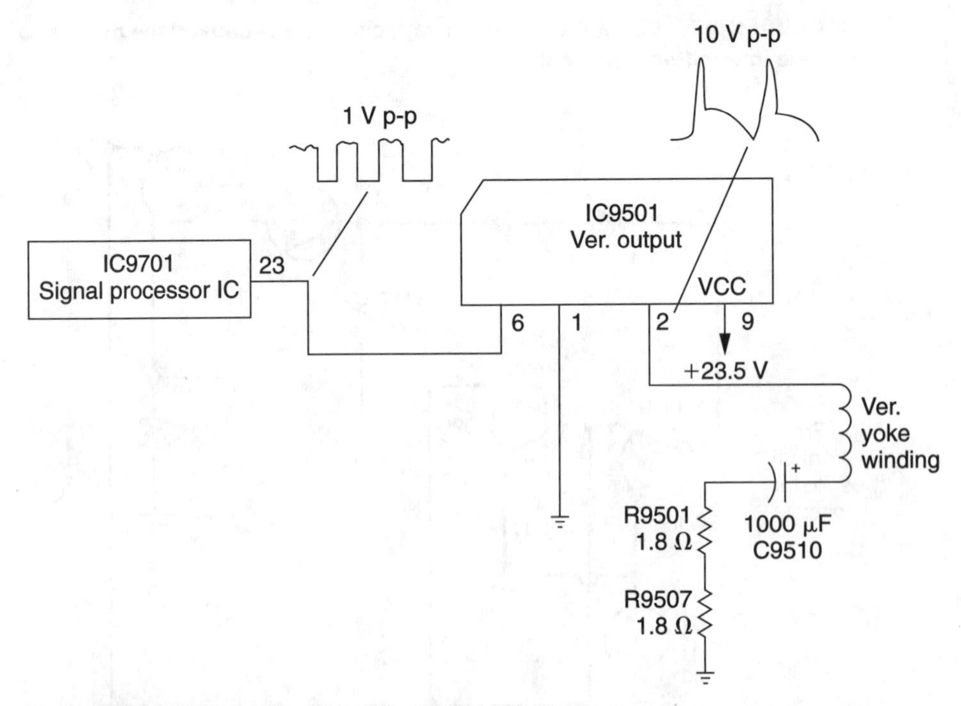

FIGURE 6-45 **Check the output waveform of the vertical signal IC and the output of the vertical power IC for correct vertical waveforms.**

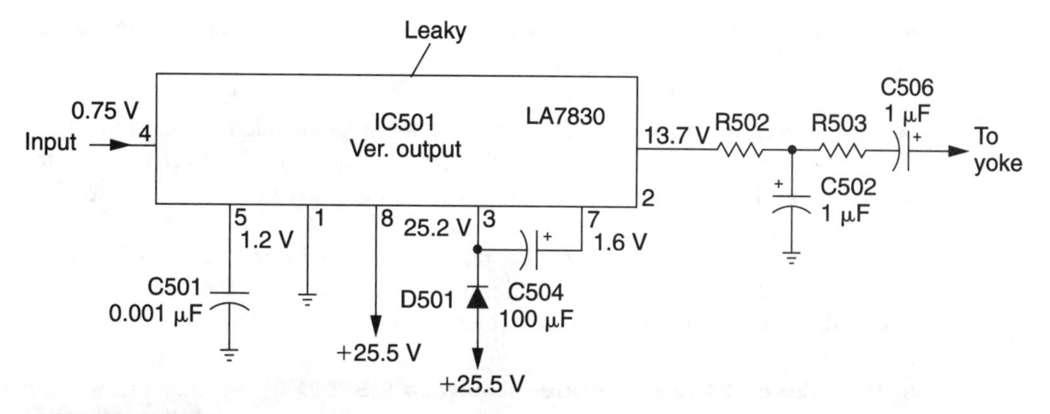

FIGURE 6-46 A leaky IC501 caused a lack of vertical sweep in an Emerson EC1OR TV.

FIVE ACTUAL CASE HISTORIES

The following case histories are actual vertical problems that have occurred in different TVchassis. Each trouble has a different symptom with quick servicing methods.

No vertical sweep—Emerson EC10R Only a horizontal white line was visible on the screen of an Emerson EC10R TV. Voltage measurements were taken upon the output terminals of vertical output IC501 after a fairly normal waveform was fed into pin 4 (Fig. 6-46). The voltage in pin 3 was very low, indicating that the output IC was leaky. When pin 3 was removed from the foil, the voltage source measured 26.7 volts. IC501 was replaced with original part number 4152078500.

Insufficient vertical sweep—Goldstar CMT 2612 The vertical drive output IC was suspected in this Goldstar chassis with insufficient sweep. The scope wave input waveform at pin 8 was fairly normal, indicating that the output IC (or corresponding components) were defective (Fig. 6-47). IC301 was replaced with an ECG1797 universal IC

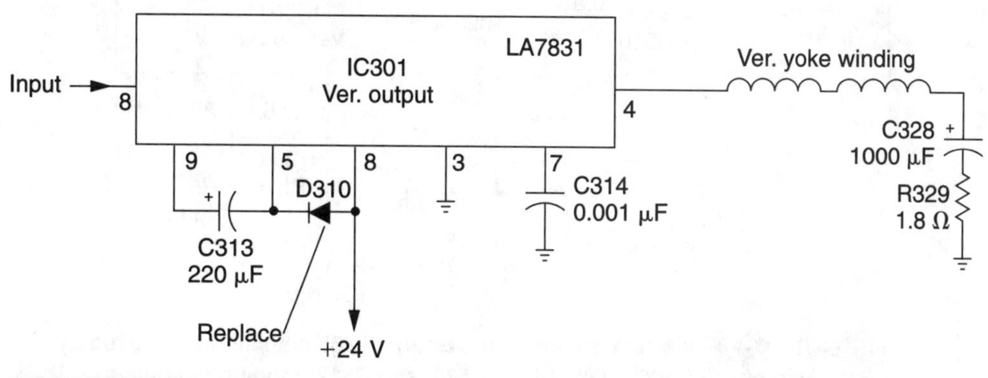

FIGURE 6-47 Replace D310 for insufficient sweep in the Goldstar CMT-2612.

and the results were the same. D310 was found to be leaky and replaced with a general-purpose 1-A silicon diode (SK3311).

Vertical foldover—Sharp 19J65 At first, a Sharp portable had very little vertical sweep with low voltage on pin 6 (1.7 V). IC501 was ordered out (X0238CE) because there was no universal replacement. The input waveform was fairly normal, with no waveform at output pin 2.

After IC501 was replaced, the vertical picture had a foldover problem. C512 was replaced, curing the foldover problem (Fig. 6-48). Check C505, C506, C509, C510, and C512 in the vertical circuits for possible foldover or poor vertical linearity.

Vertical lines at top of picture—Emerson MS250XA Several horizontal lines were found at the top of the picture in an Emerson MS250XA portable TV. All voltages were fairly normal. One end of D305 and D304 were removed and tested good (Fig. 6-49). Q303 and Q302 were tested out of the circuit. Electrolytic capacitor C305 (4.7 μF) was almost open, causing lines at the top of the picture.

Unusual vertical problem—RCA CTC145 The vertical sweep was intermittent, the picture would collapse to a white line, there was insufficient vertical sweep, and sometimes there was no sweep at all. This intermittent problem might occur after the set has run for a few hours.

U4501 was monitored with the scope at pin 4 and a voltmeter was clipped to pin 8. When the vertical section went into the intermittent state, very little output waveform was found, with practically no voltage change. Before replacing IC4501, all components were checked on each terminal. C454 is mounted quite close to the LA7831 IC, and this IC runs fairly warm (Fig. 6-50). When C454 was shunted, the picture returned. Several of these electrolytic capacitors have acted up in the recent RCA chassis and should be checked in the event of vertical problems. Replace C454 with longer leads and bend it away from U4501.

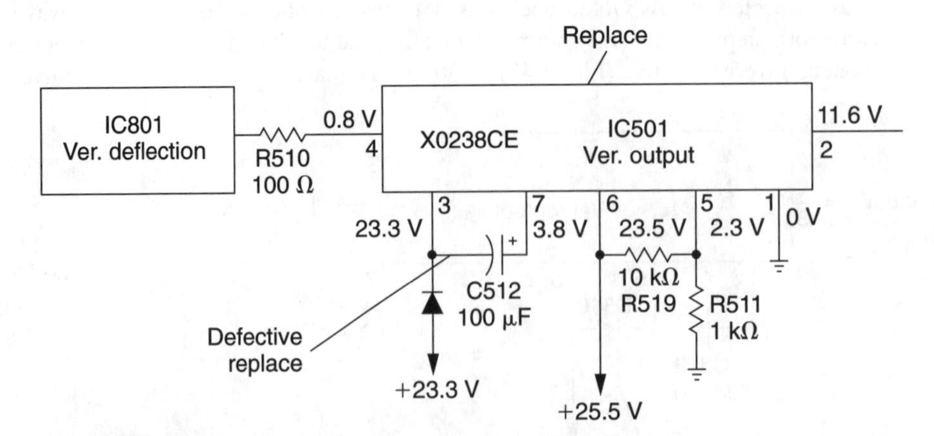

FIGURE 6-48 Vertical foldover in a Sharp 19J65 chassis was caused by C512. Also check C505, C506, C509, C510, and C512 if the unit displays vertical foldover and poor linearity.

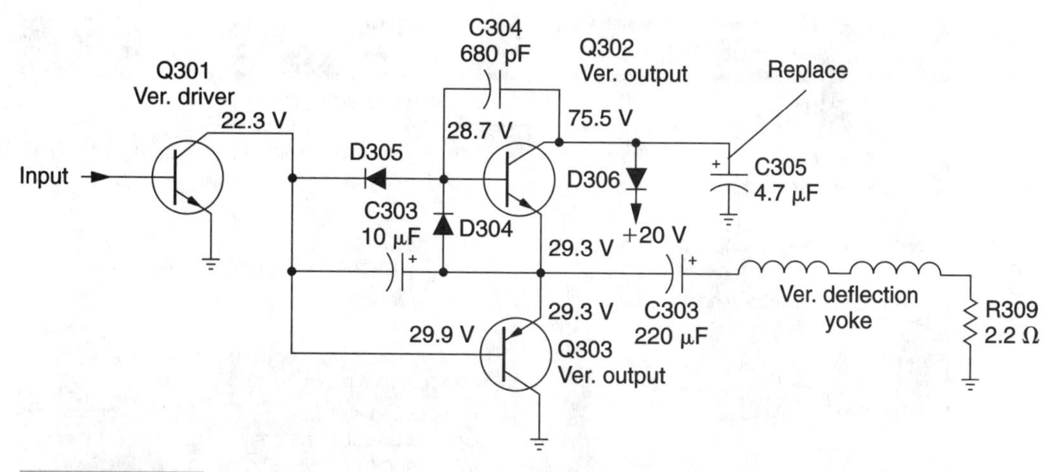

FIGURE 6-49 **Vertical lines formed at the top of the picture because of a defective C305 (4.7 μF) in an Emerson MS250VA TV.**

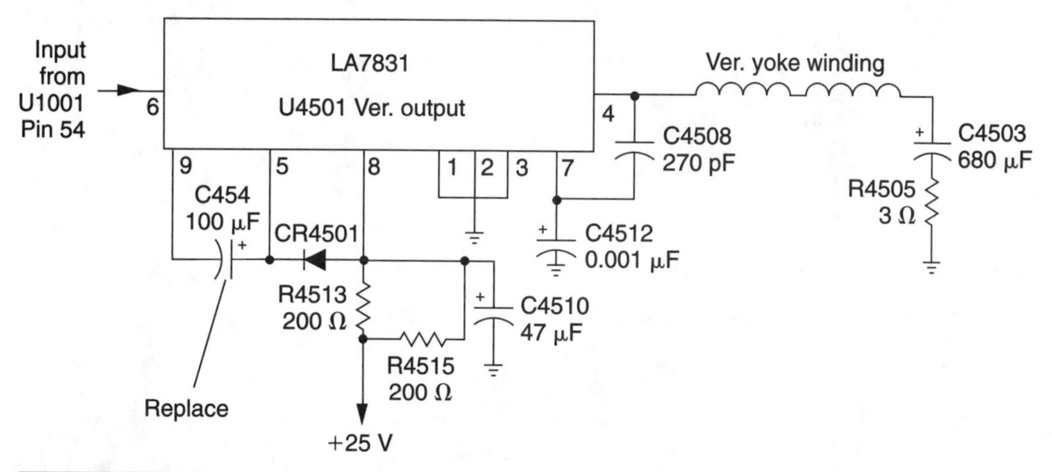

FIGURE 6-50 **Replace electrolytic capacitor C454 (100 μF) across pins 9 and 5, leave longer leads, and bend the component away from U4501.**

Conclusion

Scope the vertical oscillator and output transistors to isolate the defective stage. A voltage measurement on the collector (metal) terminals of the vertical output transistors will indicate if the transistors are defective or if there is no dc voltage source. Remember, the vertical output transistors can break down under load. Automatically replace them if all other components appear to be normal. Check outside of the vertical circuits for a defective vertical component. Do not overlook the possibility of a problem in the convergence circuit with insufficient vertical sweep or a horizontal white-line symptom. Check for easy methods of servicing the vertical circuits in troubleshooting chart Table 6-1.

TABLE 6-1 USE THIS VERTICAL TROUBLESHOOTING CHART FOR EASY SERVICING

WHAT TO CHECK	HOW TO CHECK IT
No vertical sweep.	Check low voltage source to vertical circuits.
Check waveform on base of output transistor or IC.	Check drive waveform on vertical sweep IC.
Check waveform at output of transistors or IC.	Test output transistors. Check voltages on output IC.
Check for open yoke winding.	Check for open ground return capacitor or resistor.
Vertical foldover.	Sub all electrolytic capacitors in output and feedback circuits.
Insufficient vertical sweep.	Power supply source. Check output transistors. Measure bias resistors. Check capacitors and diodes in output circuits.

7

HOW TO CHECK IF AND VIDEO CIRCUITS

CONTENTS AT A GLANCE

The biggest problem when servicing the intermediate frequency (IF) stages is just getting to these circuits. Often, the IF circuits are found directly under the bell of the picture tube and are enclosed in metal shields (Fig. 7-1). After separating the chassis from the picture tube mounting braces, the IF stages are fairly easy to service. Sometimes the chassis can be serviced from the bottom. In console models, the chassis can be pulled back or removed. Accurate in-circuit voltage and transistor tests can solve most IF repair problems.

The low voltages found in the IF transistor stages can be checked with the DMM. Often, collector voltages can change or the collectors can be difficult to access in the IF circuits. Accurate voltage measurements on the emitter terminal can indicate if a transistor is leaky or open (Fig. 7-2). These voltages are very low (1 to 3 V) across the emitter resistor. If there is no voltage, the transistor might be open or the dc voltage applied to the collector terminal might be incorrect. Check the emitter resistor for an open or incorrect resistance with the low-resistance scale. Often, the emitter resistance can vary from 1 kΩ to 47 kΩ.

Besides the DMM, a scope with a demodulator probe, sweep marker, and color-bar generator is useful when servicing IF circuits. The oscilloscope with a detector or demodulator probe can be used to signal trace each IF stage with the color broadcast signal or from the color-bar generator. The tuner-subber can be used to inject signal at the IF input and to the base of the second IF amp transistor with the picture tube as a monitor to locate a defective first or second IF stage. The sweep-marker generator can be used for complete IF alignment.

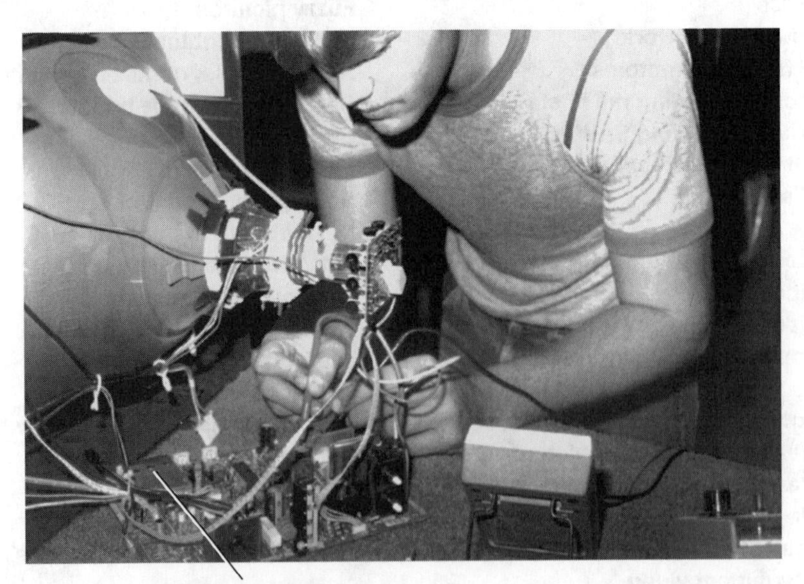

FIGURE 7-1 **Most IF stages are shielded, with components inside of the shielded area.**

FIGURE 7-2 A block diagram of the IF amplifiers and video buffer circuits.

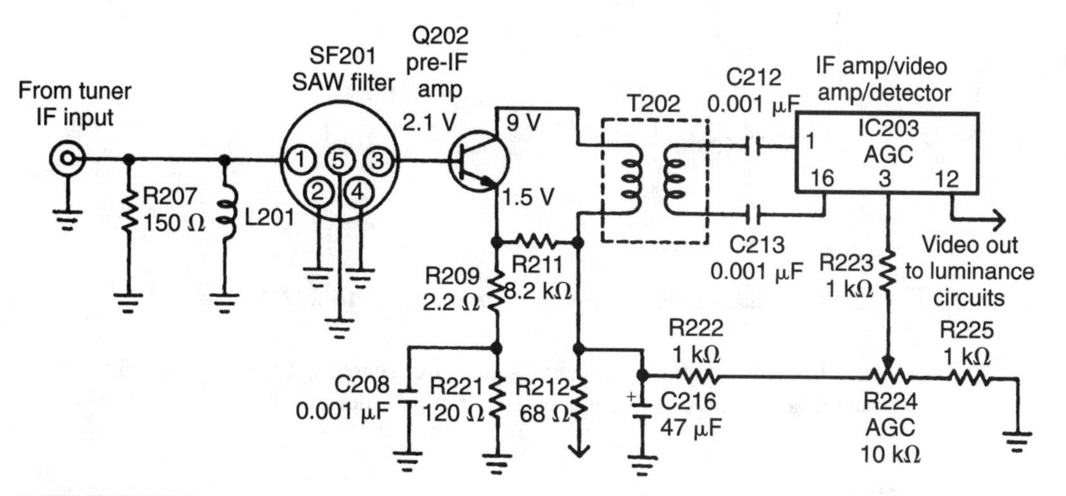

FIGURE 7-3 The SAW filter network is connected between the tuner and IF stages.

IF Circuits

The IF circuits can consist of three or four transistor stages or one IC (Fig. 7-3). The conventional IF circuits might contain a tuned output or some form of tuned input with transistor RC-coupling methods. In earlier solid-state TVs, stacked IF circuits were quite common. Today, the IC has taken over the IF circuits.

The output signal of the tuner is fed through several tuned traps and connected to the base circuit of the first IF video transistor (Fig. 7-4). The amplified IF signal is transformer-coupled to the second and third IF video stages. Notice that the second IF video stage is controlled by AGC voltage from 5.5 V to 7 V. The video detector is a fixed-diode transformer that is coupled to the first video amp. The sound take-off coil can be tapped at the collector terminal of the third IF video or video amp.

Although universal transistors can be used as replacements in the IF section, it is best to use parts with the original part numbers. In some RCA chassis, universal transistors cannot function as IF amplifiers. You will save time by using the right part because original transistors will work every time. The universal transistor terminal leads must be cut to the exact length of the original and spaced for mounting. Do not mount the IF transistors underneath the chassis. Replace each defective IF transistor inside the shielded area. They are easier to replace under the chassis, but they must be shielded. Always replace shields

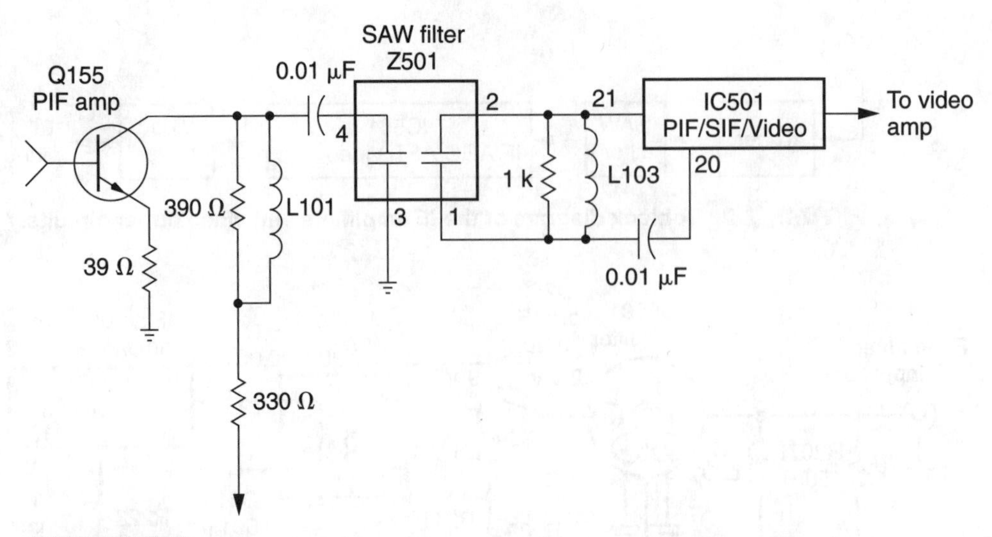

FIGURE 7-4 Today's PIF amp, SAW filter, and IC501/PIF/SIF/video IC.

and solder banded areas to prevent stray signals from entering the IF section (Fig. 7-5). IF alignment is not required with direct transistor replacement.

IF ICS

In the early solid-state IF stages, the first and second IF video circuits were contained in one IC (Fig. 7-6). Later on, the tuner was coupled to the IC IF stages with a single input transistor. Today, the IF input amp transistor is located between the tuner, SAW filter network, and IF AFT-AGC IC circuits (Fig. 7-7). The RF and IF AGC circuits are developed inside the IC.

You can use a complete plug-in IF module with the modular TV chassis. Simply remove and plug in a new IF module if there are IF trouble problems. These same IF modules can be found in the newer one-piece chassis (Fig. 7-8). Be sure that all IF terminals plug into and are firm on the TV chassis. Fasten the module to the chassis with mounting screws to prevent future intermittent IF problems. Check for a voltage adjustment on some models when replacing the IF module.

SAW FILTER NETWORK

Today, a new circuit found in the IF stages is called the surface acoustic wave SAW network. The SAW filter component is made up of a piezoelectric material with two pairs of transducer electrodes. One is the input and the other the output transducer. Voltage applied across the positive and negative terminals causes distortion and mechanical waves. The SAW filter establishes the proper IF frequency, which, in turn, eliminates IF alignment. Here, the SAW filter is located between the IF preamp and the IF processor IC (Fig. 7-9).

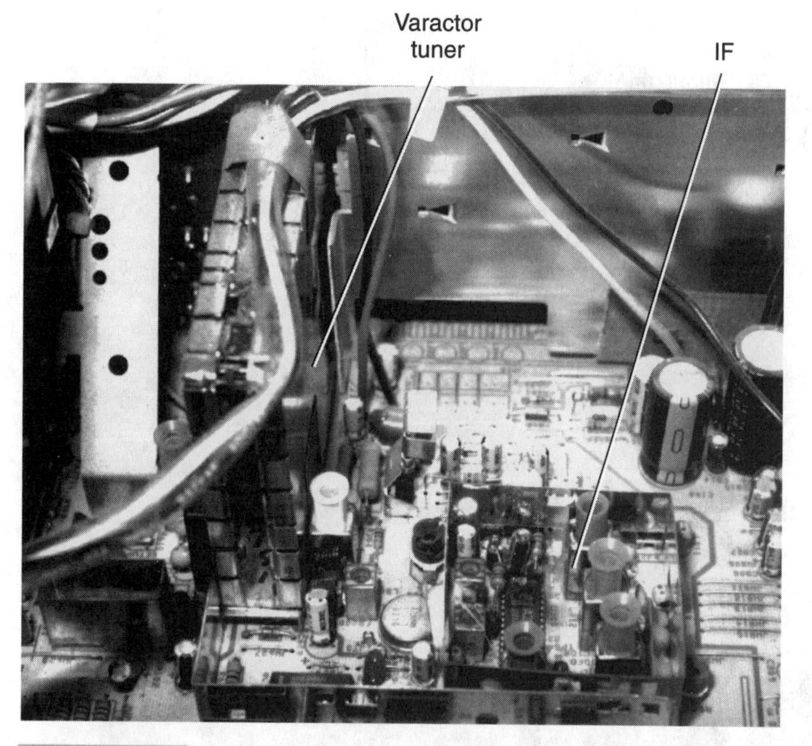

FIGURE 7-5 Notice the shielded IF with varactor tuner nearby.

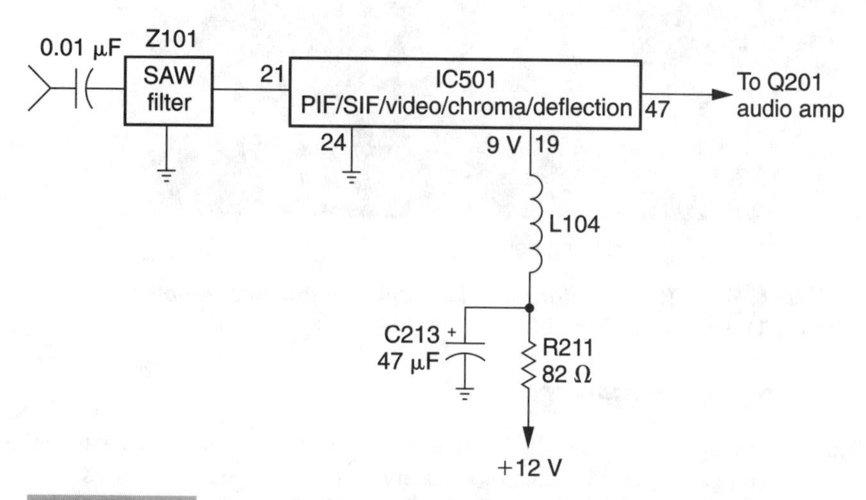

FIGURE 7-6 The IF stages might consist of a picture IF amp, SAW filter, and IC501 (which contains PIF/SIF/video), and a deflection IC.

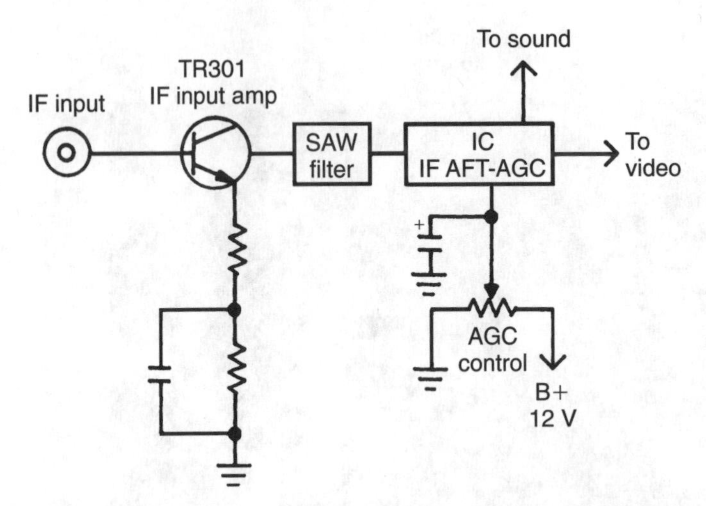

FIGURE 7-7 The SAW filter does not require alignment.

FIGURE 7-8 The varactor tuner has replaced the tuner module in many TV front-end circuits.

IF TROUBLE SYMPTOMS

The most common troubles in the IF section are no picture and no sound with white raster; a faint picture and garbled sound, or a smeary picture with multiple ghosts; and a snowy picture with fair sound. Weak or no color can result from a defective IF stage. In earlier tube and solid-state chassis, a "birdie" type sound could be heard with a defective IF stage. Most IF symptoms can be seen on the picture tube.

The same symptoms can be found with defects in the tuner, detector, video, and AGC circuits. A snowy picture can be caused by the first or second IF stage or by a defective tuner and AGC circuit. The faint and weak picture symptoms can occur in the IF picture detector and first video amp (Fig. 7-10). Similar symptoms can be found in the IF circuits. Improper IF AGC voltage can cause a snowy picture or weak, garbled sound with a poor picture, indicating the IF circuit is defective.

Before attempting to repair the IF section, be sure that the tuner and IF AGC circuits are normal. Substitute the suspected tuner with a tuner-subber. Measure the IF AGC voltage at the IF test points or in the AGC circuits. Clamp the IF AGC voltage with the internal power supply to determine if the AGC circuits are normal. After determining that the IF circuits are defective, locate the correct schematic and voltage chart. The defective IF stage can be located by scope waveforms and voltage measurements.

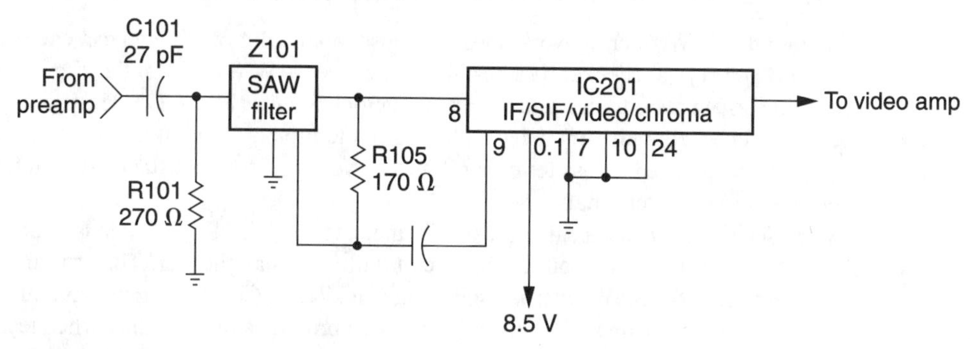

FIGURE 7-9 The SAW filter (Z101) establishes the correct frequency.

FIGURE 7-10 The SAW filter network is located close to the IF section.

TROUBLESHOOTING THE IF STAGES

A dead or weak IF stage can be located with a strong local station tuned in and scope waveforms taken at each IF stage. The oscilloscope must have a demodulator or detector probe attached. The color-bar generator is clipped to the antenna terminals and the signals are traced in the IF subber signal at the input terminal of the IC. Notice the picture on the screen (Fig. 7-11). Check the scope waveform at the output terminals of the IF section with a strong station or the color-bar generator signal. Often, IF IC terminals can easily located under the chassis with the IF section shielded and covered in a metal box. Accurate voltage measurements on the IC terminals or on the emitter terminal of the IF transistors can help you to locate the open or dead stage.

TROUBLESHOOTING SAW FILTER CIRCUITS

Although the SAW filter network does not cause too much trouble, you can check it with the VOM and crystal checker. It's best to remove the SAW filter from the circuit because low-value resistors are found in the input and output circuits (Fig. 7-12). The primary leakage can be checked with ohmmeter probes across terminals 3 and 4. Likewise, the output circuit can be checked across terminals 1 and 2. No significant resistance should be measurable across these terminals.

A resistance check does not necessarily indicate that the SAW filter is functioning. Check the SAW filter in or out of the circuit with a crystal checker. The crystal checker will determine if the SAW filter is oscillating (Fig. 7-13). Check the input and output terminals in the same manner. The crystal meter will provide a high reading when tested out of the circuit and a lower reading in the circuit. You can build your own crystal checker by following the instructions found in *Build Your Own Test Equipment* (McGraw-Hill).

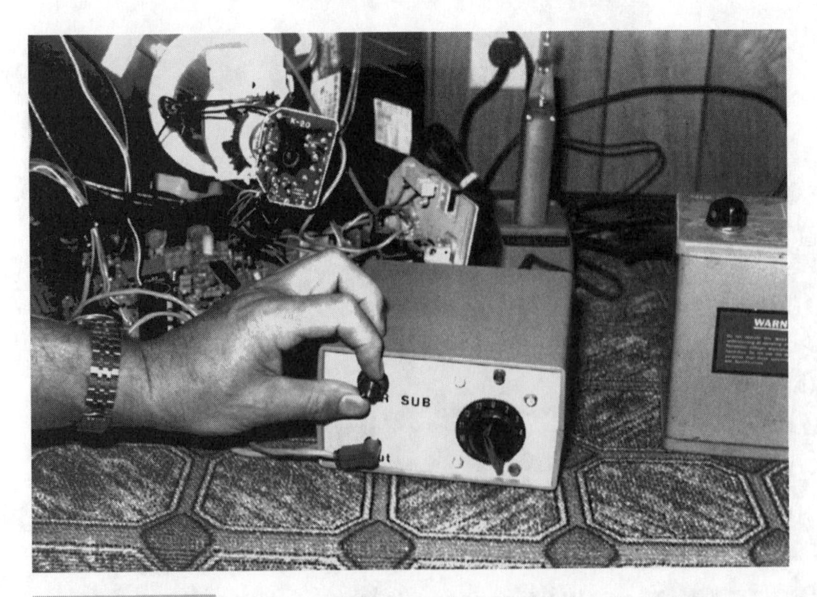

FIGURE 7-11 **A tuner subber can be used to locate a dead or open first or second IF transistor stage.**

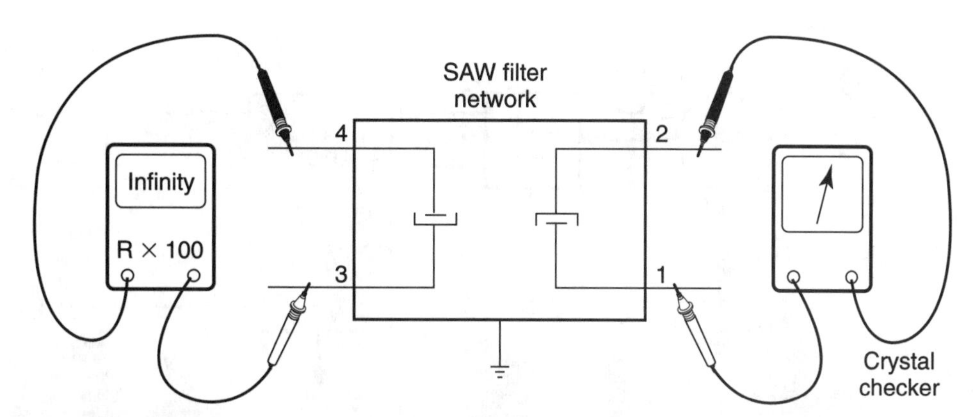

FIGURE 7-12 Test the SAW filter with resistance and a crystal checker.

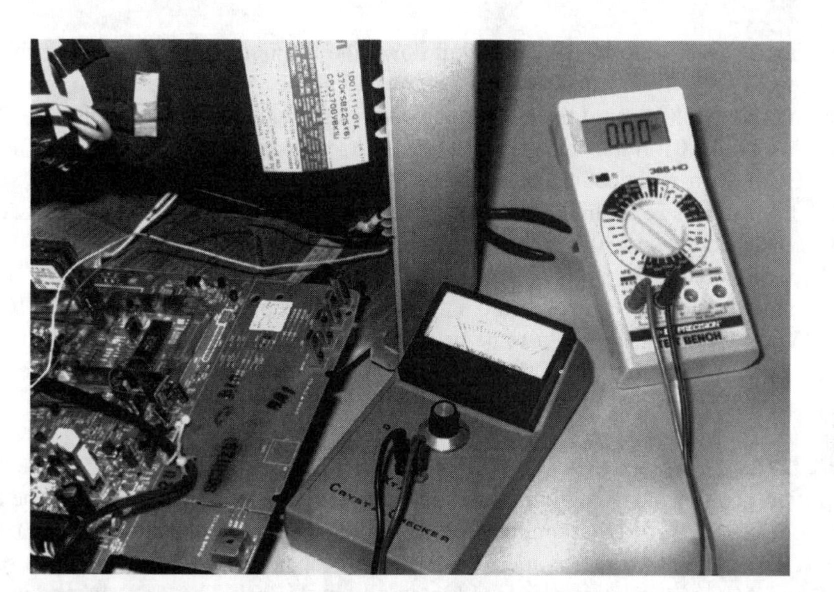

FIGURE 7-13 The crystal checker can check both input and output terminals of SAW filter.

Servicing RCA CTC130C IF circuits To determine if the tuner section or IF stages are defective, connect a tuner-subber to the IF cable. Plug the IF cable into the subber and tune in a local station. If there is no sound or picture, but normal raster, suspect problems within the IF stages. If the picture at the CRT screen is normal, repair the tuning system.

Inject an IF signal at the IF input cable and check for a picture signal at the CRT. If it is normal, check the tuner, AFT, AGC, and tuner AGC. Scope the video waveform at pin 12 of U301 (Fig. 7-14). If there is no signal, check the waveform at pin 16 of the IC. If there is still no signal, proceed to the collector of the IF preamp (Q301). Although the signal will be quite weak, you can still determine if the IF preamp and tuner circuits are normal.

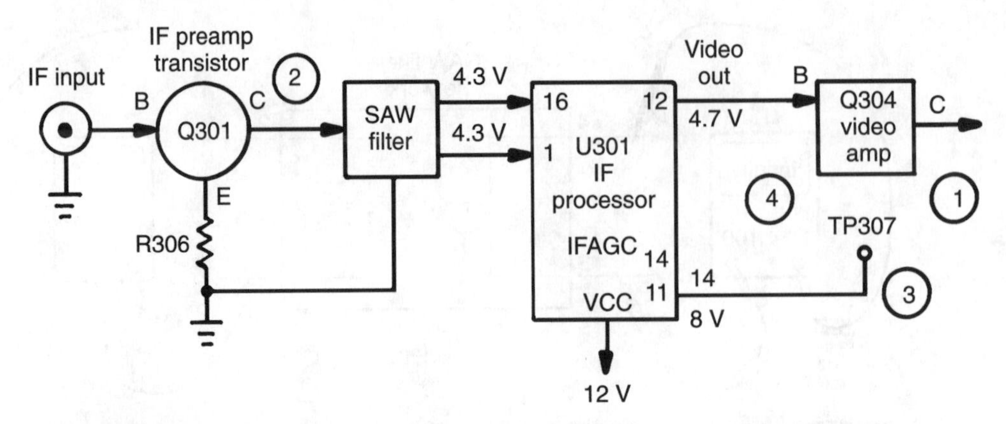

FIGURE 7-14 **Service the RCA CTC 130C IF stages by the numbers.**

An improper IF AGC voltage can shut down the IF picture stages. Measure the AGC voltage at TP307. This voltage can vary between 7 and 11 Vdc. If there is no video waveform at pin 12 (U301) or at the base of the video amp, inject AGC bias voltage at TP307 or at pin 14 of U301. If the video waveform appears with external bias voltage, suspect that the AGC system is defective. If the picture does not return with proper AGC voltage, check the voltages in preamp transistor Q301, and look for a shorted IF cable or an open L309, L302, R306, or R307.

Check all voltages on pins 1, 3, 11, 12, 14, and 16 of the IF processor (U301). If the IF signal is entering the IF processor and not coming out, suspect a problem with the IC. Regular TV broadcast signals can also be used to scope the preamp and U301 circuits with a demodulator probe and scope. Check the SAW filter (SF301) with a crystal checker, if you are in doubt.

RCA CTC157 IF circuits The IF output signal is taken from the tuner to the IF preamplifier (Q2301) and the output is applied to a SAW filter network (SF2301). The IF signal from the SAW filter is applied directly to the CTV processing IC (U1001). Inside U1001, the signal is amplified and passed on to a video detector circuit (Fig. 7-15). Both AFT and AGC voltages are developed internally by U1001. The AFT voltage is routed back to the A1U control (U3300) in the system-control circuit. This voltage determines if the tuner frequency is correct for a given channel by the A1U system. The AGC voltage is applied to the tuner to control tuner amplifier gain.

IF or tuner? The no-picture/no-sound symptom can occur if the tuner or IF video components are defective. Connect a tuner-subber to the IF cable. If the picture is normal with the subber, suspect the tuning system. If there is still no picture, troubleshoot the IF and video circuits. The tuner-subber can be used for signal injection to the IF IC.

Tuner-subbers were used constantly years ago and there were several different manufacturers. Today, it's difficult to locate a tuner-subber. You can make your own out of a transistorized VHF tuner and power source. How to build a tuner-subber is described in *Build Your Own Test Equipment* (from McGraw-Hill). The tuner-subber can save a lot of valuable service time, which means money!

Snowy pictures Besides the tuner, check the first and second IF stages for a snowy picture. Measure the IF AGC voltage without a signal to the tuner. An improper AGC voltage can produce a snowy picture. Take accurate voltages on the emitter and collector terminals. Check the dc voltage source fed to the IF section for low voltage. If possible, make an in-circuit test of each transistor with the DMM. The snowy picture can be caused by an open or leaky IF transistor.

In an RCA CTC44W chassis, all VHF stations were snowy. The local UHF station picture contained a little snow. A quick in-circuit test of each transistor located an open second IF transistor. Besides transistor tests, accurate voltage measurement can locate that leaky or open IF transistor.

Brightness, no picture, garbled sound A smooth white raster with no snow or noise ripple is caused by the IF stages. Usually, a rushing noise in the sound and some channel noise is seen without a station tuned in. Garbled sound indicates poor sound takeoff at the IF stages. The dead IF stage can be caused by an open or leaky IF transistor or IC.

The sound was garbled, with a white raster in a J.C. Penney 685-2124. The IF stages were signal-traced with the scope and demodulator probe. The video signal was found at the base and none at the collector terminal of the third IF video transistor (Fig. 7-16). Zero voltage was found at the emitter terminal, indicating that Q103 or R136 was open. The collector voltage was high, at 20 V. Q103 was found to be open and was replaced with an SK3018 universal transistor replacement.

AGC or video A pulling, washed-out picture can be caused by a defective AGC or video IF circuit. Often, the picture is out of sync and the AGC control has no effect on the

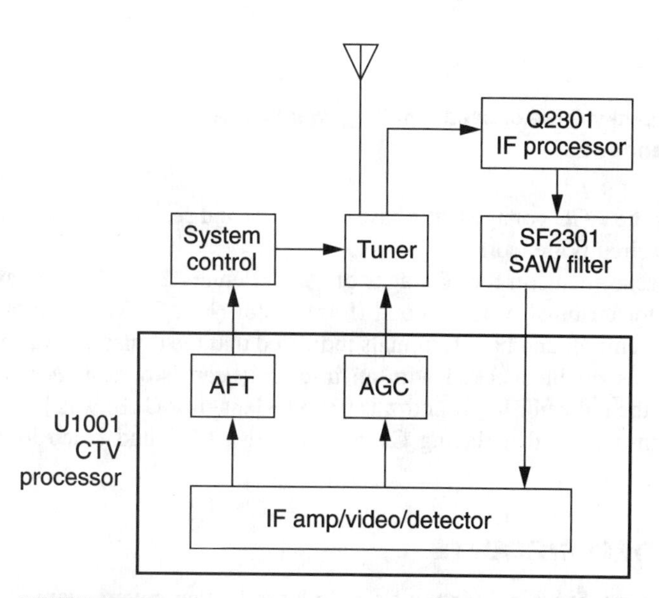

FIGURE 7-15 A block diagram of the RCA CTC157 IF circuits.

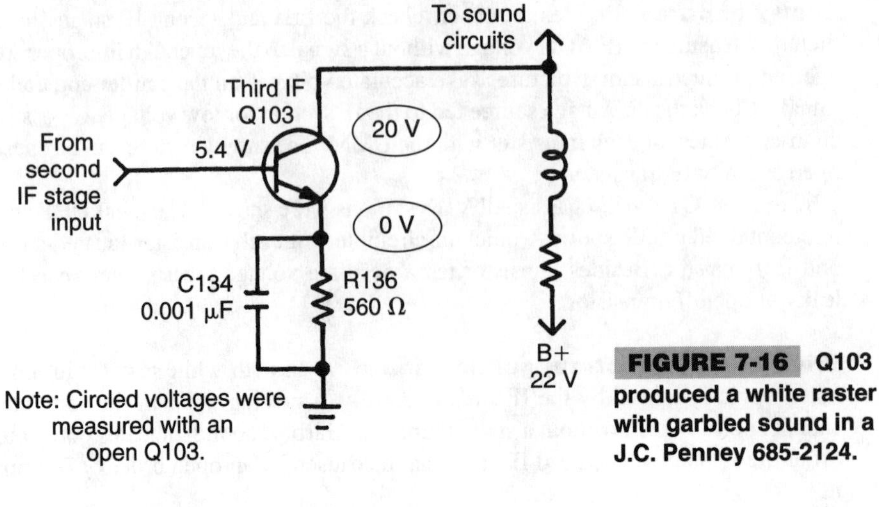

FIGURE 7-16 **Q103 produced a white raster with garbled sound in a J.C. Penney 685-2124.**

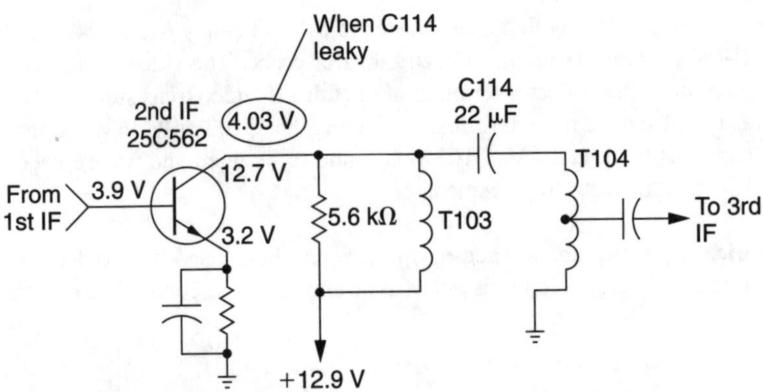

FIGURE 7-17 **Leaky C114 caused a pulling, washed-out picture in a J.C. Penney 2857.**

picture. First, check the AGC voltage at the base of the first and second IF transistors. If it is normal, the AGC circuits are working.

With voltage measurements in the IF stages of a J.C. Penney 2857, low voltage was found at the collector terminal of the second IF transistor (Fig. 7-17). A forward-bias voltage between the emitter and base terminals indicated that the transistor was normal, but the transistor was doublechecked with an in-circuit transistor test. A resistance measurement from the collector to ground was 939 Ω. Capacitor C114 was found to be leaky to T104. Removing and replacing C114 solved the AGC and video look-alike picture.

LATEST IF VIDEO CIRCUITS

The video IF circuit might be included in a single large IC that covers many different circuits in the latest TV chassis. IC1001 contains the IF/LUMA/chroma/deflection circuits in one large processor. Besides these circuits, AGC/AFT/sound IF and preamp circuits are found in one section of the 64-pin IC. This processor IC, when found defective, can be replaced with an ECG1790, SK9850, or NTE1790 universal replacement.

The large IC might have an IF preamp transistor and SAW filter within the input circuits, between tuner and IC1001 (Fig. 7-18). RF and IF AGC circuits are included with picture IF input pin 21, RF AGC out pin 46, RF AGC delay pin 18, picture AGC pin 22, RF supply voltage pin of 9 volts (19), PIF AGC TC pin 23, video detector tank pin 45, AFT output 42, and pin 47 of video output.

To troubleshoot, inject a video signal at the IF input and check for video on the face of the picture tube. If no video is at the CRT, check the video waveform at the base of the video buffer transistor. When video is missing at the base, inject AGC bias to pin 22 of IC1001. Check all voltages, waveforms, and components at IC1001, pins 18 through 47. Connect a bar sweep generator to the antenna terminals and check for correct waveforms in IF video circuits.

IF ALIGNMENT

IF alignment should be checked after all circuits are found to be normal in the IF stages when there are faint, smeary, fluttering, or oscillating pictures. Do not try to adjust IF cores or capacitors to improve a weak or faint picture. These adjustments do not change unless someone turns them. IF alignment should be done with a sweep-marker generator. If you do not have the correct alignment test equipment, it is best to take the chassis to the manufacturer or to a qualified TV alignment technician. Most alignment generators collect dust on the shelf because they are seldom used in TV servicing. IF alignment is not needed in the latest TV chassis with a SAW filter network.

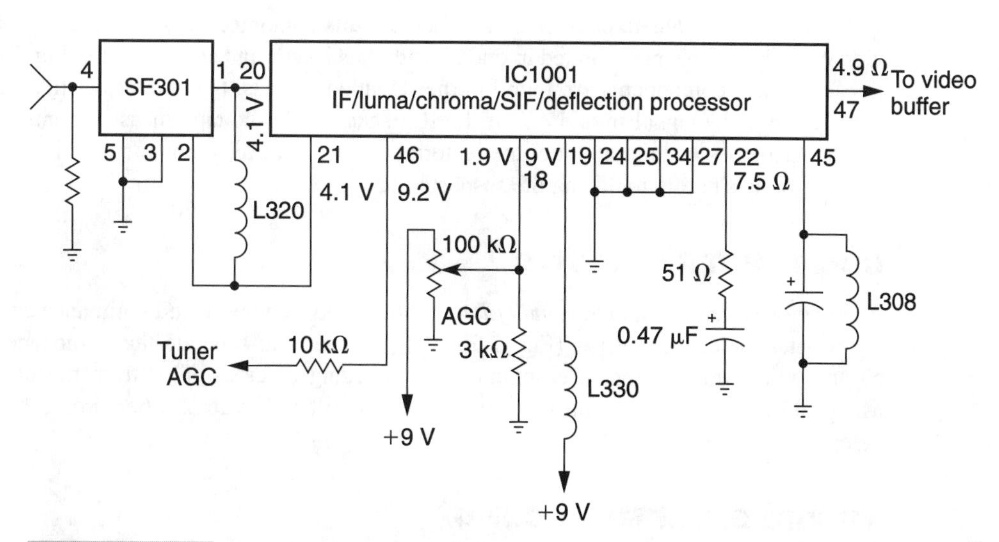

FIGURE 7-18 The latest IF circuits are included in one large IC with luma, chroma, SIF, detection, horizontal, and vertical circuits.

Video Test Equipment

The video stages can be serviced with signal tracing of IF stages with the oscilloscope. Connect a color-dot-bar generator to the TV set antenna terminals for signal tracing. Take in-circuit transistor tests of the IF and video circuits. Check crucial voltages with a DMM. A video or audio square-wave generator can be used to locate the stages, causing a smear or an oscillating picture. All IF and video signals are taken with the color dot-bar generator and scope, according to Howard Sams schematics.

VIDEO PROBLEMS

A weak or a washed-out picture can be caused by a defective video stage. Very little or no control of brightness can occur in the video circuits. Heavy scanning or retrace lines in the picture are caused by the video circuits or by improper voltages at the picture tube. Most smeary picture symptoms are caused by open peaking coils and leaky video transistors in the video circuits. A defective video output stage can cause a no-picture/no-brightness symptom.

Some video symptoms can be caused by the picture-tube circuits. Determine if the picture-tube circuits are at fault by taking crucial voltage measurements on the CRT terminals or corresponding circuits. With a low or insufficient brightness symptom, the service/normal switch can be switched to service to determine if the CRT is weak. Rotate each screen grid control to form a white line. If a bright line is noted, assume that the CRT and circuits are normal. The same method can be used on TV chassis without the service switch. Simply turn up the screen or substitute a brightness control and notice if the picture tube responds.

VIDEO IC CIRCUITS

With the early solid-state chassis, most video circuits contained only transistors. Today, video circuits can be incorporated in one IC with AGC, sync and color circuits (Fig. 7-19). Because fewer components are found in the IC video circuits, they are easier to service. Simply scope the signal in and out of the IC. Take crucial voltage measurements if the video signal goes in, but not out of the IC terminals. Check each component tied to the IC video terminals before replacing the suspected IC.

COMB FILTER CIRCUITS

A new video circuit, called the *comb filter*, is found between the IF and luminance circuits in the latest color TV chassis (Fig. 7-20). Its purpose is to separate the luminance and chroma video information to eliminate color bleeding or cross color from the picture. Many additional circuits are provided with the comb filter. The comb filter circuits are developed with an IC.

NO VIDEO, NORMAL SOUND

A normal sound and white-raster symptom without any picture occurs with a video circuit problem. The video circuits can be signal traced with a tuned station signal and scope or a color-bar generator and scope. Start at the first video amp with the scope probe and check the signal at the base and collector terminal of each video transistor until the signal is lost. Take crucial voltage measurements on the suspected transistor or IC.

Sharp 9B12

The sound was normal with no video in a Sharp 9B12 (Fig. 7-21). Because the sound was normal, the video signal was scoped after the sound takeoff point. The scoped signal stopped at the first video amp (Q202). A quick voltage check showed no positive voltage at the base and collector circuits. Although the 12-V supply source was normal, voltage was not applied to the vertical circuits. On closer inspection, the PC board was found to be cracked behind the balance control. No doubt, the TV had been dropped on the knobs, cracking the board wiring. Solid hookup wire soldered across the cracked wiring brought back the picture.

NO VIDEO, NO RASTER

Check for high voltage at the CRT anode connection with a no-video, no-raster symptom. If the high voltage is normal, suspect video problems at the last video stage or the CRT and

FIGURE 7-19 Check for one large IC with most all the TV functions in one component.

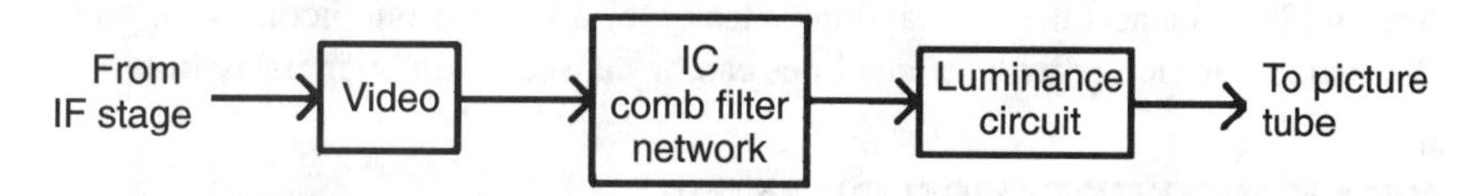

FIGURE 7-20 The comb filter is located between video and luminance circuits in the latest TV chassis.

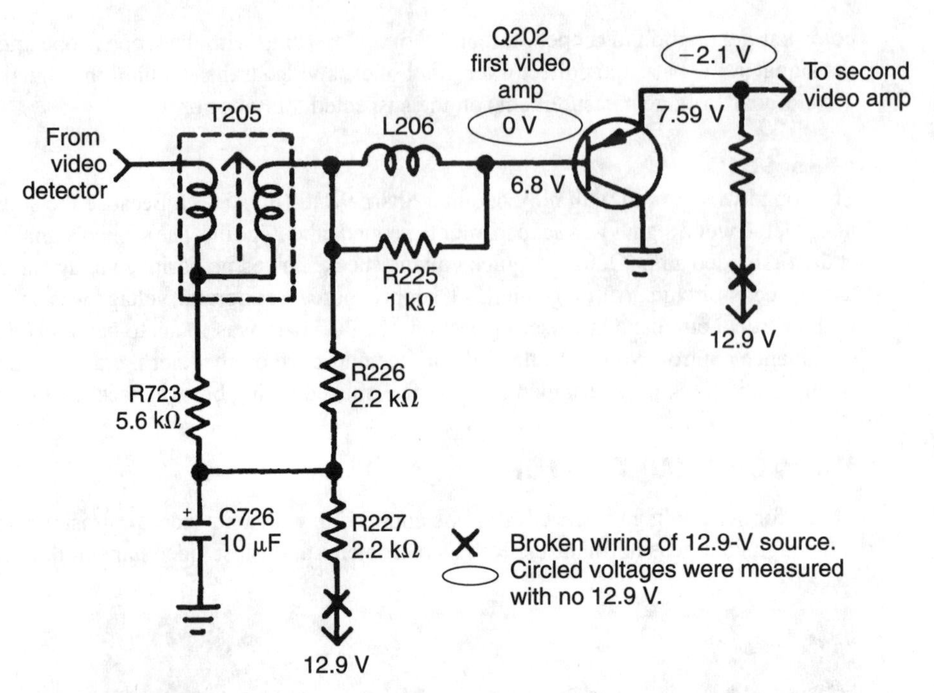

FIGURE 7-21 A broken foil wire in a Sharp 9B12 chassis caused improper voltages on Q202.

circuits. Next, check all voltages on the CRT terminals. Normal CRT voltages indicate a defective video output circuit.

The high voltage was normal with no video or raster in an RCA CTC86D chassis (Fig. 7-22). When the CRT voltages were measured, very low screen voltage was found. R435 and R441 were running red hot. A quick resistance measurement found 546 Ω of leakage on the 200-V source. After disconnecting the screen circuits from the 200-V source, other circuits tied to the same voltage source were checked. Both the luminance/sync module (MDL001B) and tuner module (MST002B) were replaced to bring back the screen grid voltage source. Remember, more than one module can be defective in a modular chassis.

WEAK OR WASHED-OUT PICTURE

A weak and/or negative picture can be caused by an open peaking coil, leaky video transistor, or a defective picture tube. Check the picture tube with a good CRT tester. Very light and weak pictures with normal sound can be caused by leaky or open video transistors and ICs. Suspect that a delay line is leaky for a washed-out picture with poor contrast. A washed-out picture with retrace lines can be caused by an open delay line.

WEAK PICTURE, RED OUTLINE

No picture was noticed until the color control was turned up in a J.C. Penney 19YC chassis (Fig. 7-23). The picture would disappear when the color was turned down. The video signal was scoped to the base terminal of the third video amp, and no signal was found at the collector. Q325 tested normal in the circuit with a DMM. When the fourth video amp was tested in the circuit, Q330 was found to be open. Replacing Q330 with a GE-18 universal transistor solved the very-weak-picture symptom.

WASHED-OUT PICTURE, RETRACE LINES

Video problems were suspected in a Panasonic CT-994 with a washed-out picture and retrace lines (Fig. 7-24). The video signal was found at the collector terminal of the first video amp (TR301) and not at the base of the second video amp (TR302). Continuity resistance measurements of L301, L302, and L303 indicated an open in the delay line.

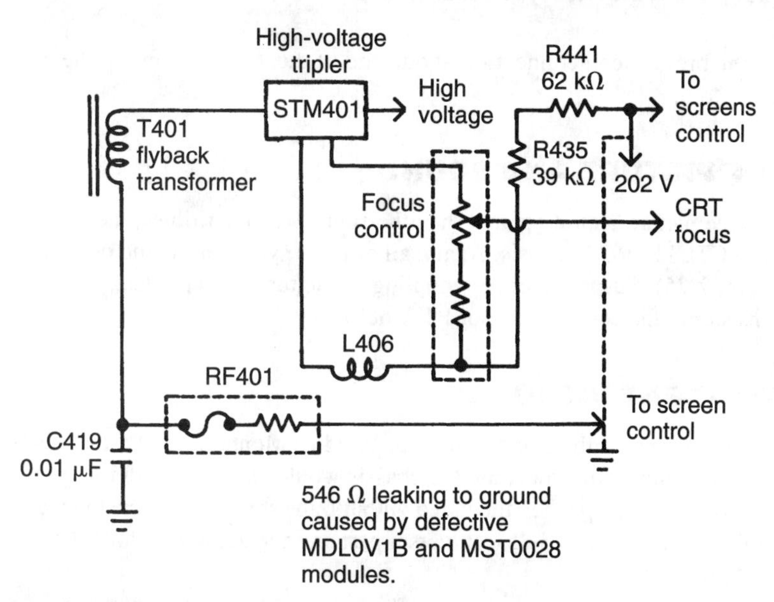

FIGURE 7-22 **Check voltages at the picture tube socket to correct a no-video/no-raster symptom.**

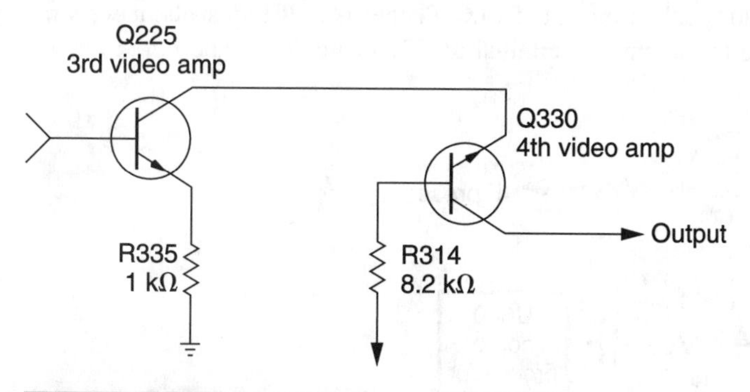

FIGURE 7-23 **An open Q330 caused a weak picture in the J.C. Penney 19YC chassis.**

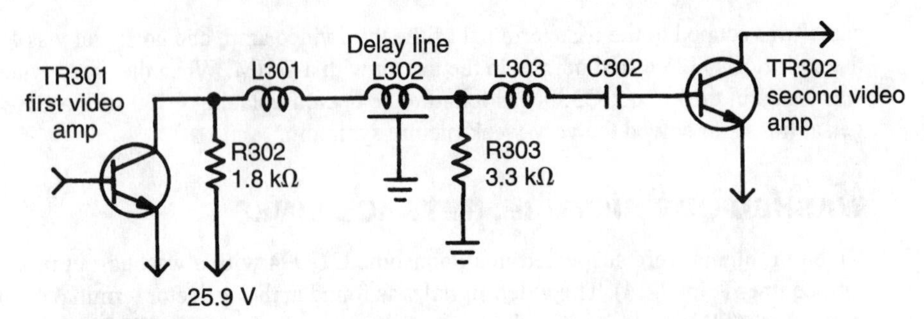

FIGURE 7-24 An open delay line in a Panasonic CT-994 caused a washed-out picture with heavy retrace lines.

Repairing the broken lead connection at one end of the delay line cured the washed-out picture symptom.

LOSS OF PICTURE AND SOUND

Sometimes the picture would go, and then the sound. At other times, the color would be missing in a CTC111 RCA chassis. Again, all of these symptoms could occur or only one of them (Fig. 7-25). Suspect that the coupling capacitor (C611) is leaky from the video buffer transistor to the comb processor IC (U600).

INTERMITTENT VIDEO

The sound was normal with an intermittent picture in a Montgomery Ward WG-17196AB (Fig. 7-26). Sometimes, just touching the chassis would cause the raster to go black. The picture would really act up when the video output transistor (Q312) was touched. Replacing transistor Q312 with a GE-251 universal part and replacing resistor R370 solved the intermittent problem.

INTERMITTENT BLACK SCREEN

The picture would "pop" in and out of a J.C. Penney 685-2014 just about every minute. The video was traced to the emitter terminal of Q225 (Fig. 7-27) and not at the video driver

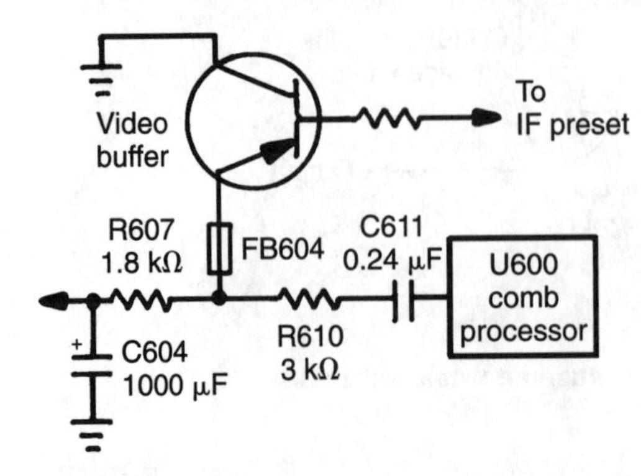

FIGURE 7-25 A leaky C611 caused a loss of color and picture in an RCA CTC111 chassis.

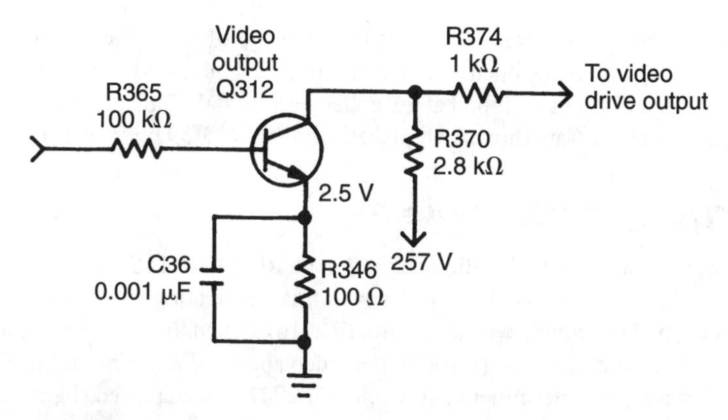

FIGURE 7-26 An intermittent Q312 in a Wards 17196AB chassis caused an intermittent picture with normal sound.

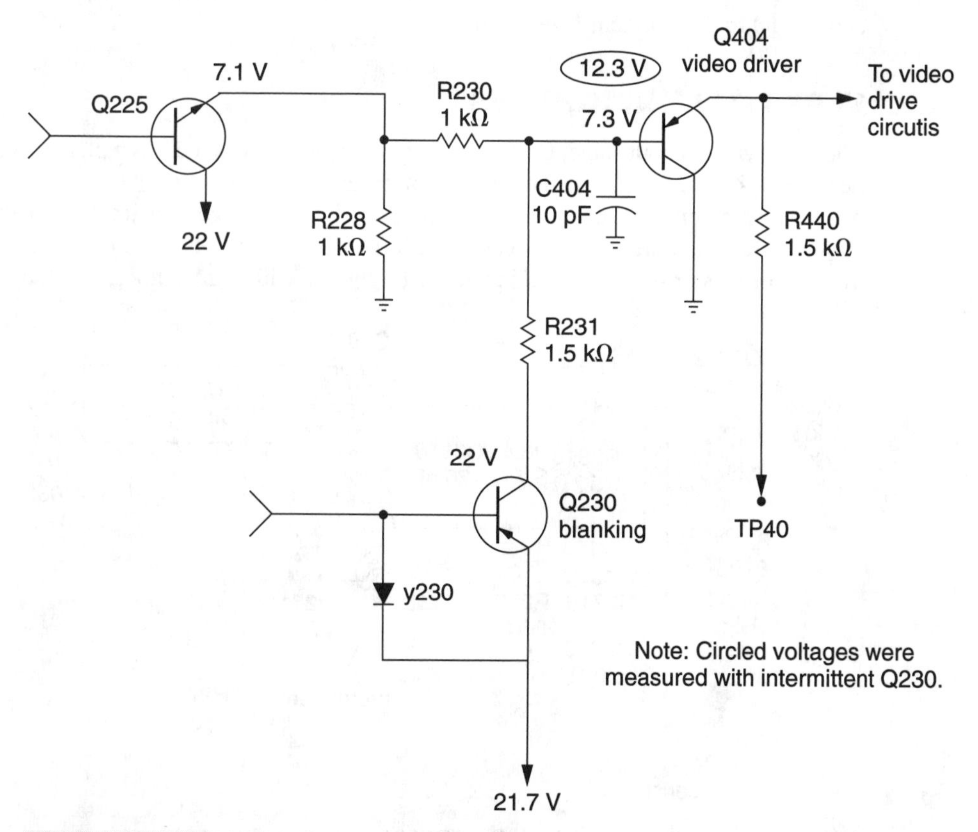

FIGURE 7-27 A high base voltage at Q404 was caused by a leaky and intermittent transistor (Q230) in a J.C. Penney 685-2014 chassis.

transistor (Q404). In the intermittent mode, the base voltage of Q404 would increase. At first, Q404 was suspected of being open. A quick in-circuit test indicated that Q404 was normal. Very high voltage (22 V) was found at the collector terminal of the blanking transistor (Q230). The leaky blanking transistor was replaced with an ECGI59 universal component.

INTERMITTENT VIDEO, AUDIO OK

The intermittent video with normal audio occurred in an RCA CTC146 chassis. The video was monitored at TP2307. The video signal was normal, indicating that the intermittent component was beyond the contrast preset control (R2716). When the video was monitored at pin 52 and 53 of the luminance IC (U1001), the video appeared to be intermittent. Upon checking the schematic, the intermittent delay line (DL2701) was replaced (Fig. 7-28).

FUZZY PICTURES

You could tap around the tuner or IF amp transistor (Q2300) in an RCA CTC140 chassis and the picture would intermittently appear fuzzy. In one model, poor pin connections were found at the IF amp transistor (Q2300). In another CTC140 chassis, poor soldering of the pins of U2300 caused the same symptom. It's wise to solder all the pins on U2300 and Q2300 for intermittent fuzzy pictures.

VERY LITTLE BRIGHTNESS

Check the video output stage, CRT, or picture-tube circuits when the symptom of very little brightness occurs. Turn up the screen controls and the service switch to determine if the picture-tube circuits are normal. Suspect that the boost voltage is inaccurate or that the CRT is defective if the brightness cannot be turned up with the screen or color output controls. If the bias or screen controls raise the brightness with retrace lines, check the video

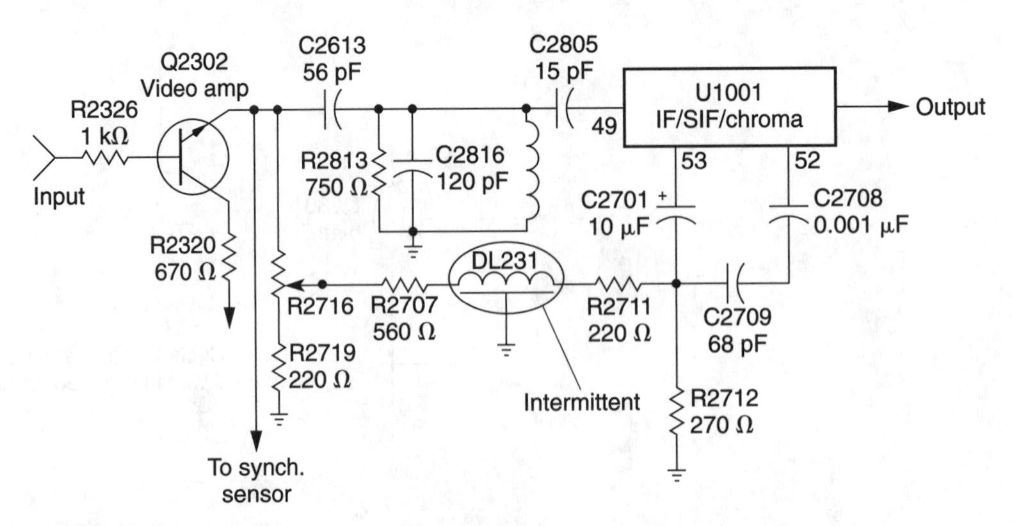

FIGURE 7-28 **Check the delay line (DL231) for intermittent video.**

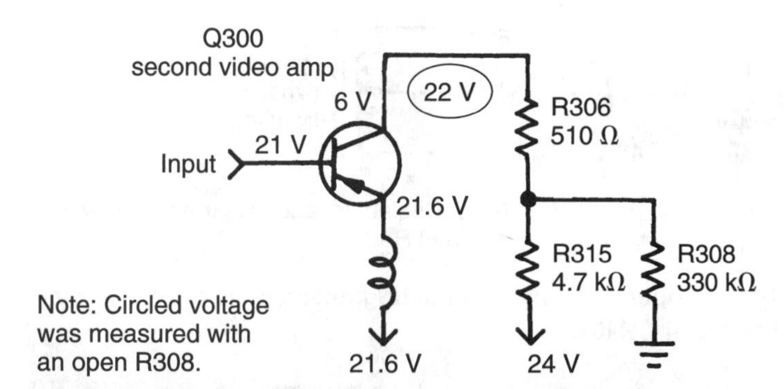

FIGURE 7-29 An open R308 in a K-Mart portable produced a black screen with normal high voltage.

output circuits. Check the automatic limiter brightness transistors and circuits. Do not overlook the possibility that a brightness control is open.

NO BRIGHTNESS, NORMAL HIGH VOLTAGE

Lack of brightness with retrace lines was noted in a K-Mart SKC1970. These occurred with the brightness and screen controls wide open (Fig. 7-29). Although the CRT tested weak, the no-brightness problem must have originated in the video circuits. A quick signal and voltage check of the second video amp (Q300) indicated that a transistor was open. Although Q300 tested normal out of the circuit, it was replaced with an SK3114 universal transistor. The voltage and brightness conditions were the same. With resistance measurements, it was found that R308 was open.

CAN'T TURN DOWN THE BRIGHTNESS

If the brightness cannot be turned down, suspect trouble in the picture-tube circuits. The picture tube is conducting too much and cannot be cut off with the controls. Check for missing boost, screen, and drive voltages at the CRT. Heavy retrace lines with no control of brightness can develop with a defective normal/service switch. Suspect a defective automatic brightness limiter (ABL) when the brightness cannot be turned down. If the picture-tube circuits are okay, check the video output circuits where the brightness control is located. Look for a sub-brightness control adjustment.

The picture-tube circuits can be double-checked with the service switch and screen controls. With many brightness-control problems, the screen controls cannot be lowered far enough to turn out the brightness. Measure the boost voltage at the screen and picture-tube circuits for missing or high boost voltage.

BRIGHTNESS WITH RETRACE LINES

The brightness could not be turned down in a Goldstar CR401 with retrace lines in the raster (Fig. 7-30). Only 48 V were found on the cathode elements of the CRT (pins 3, 8,

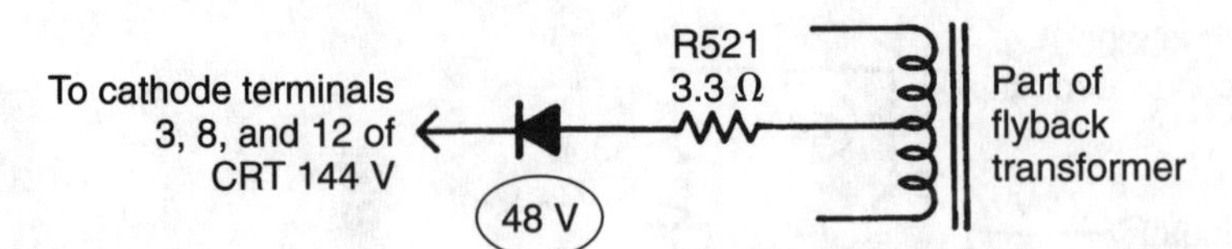

FIGURE 7-30 An open R521 caused the brightness to be fixed at maximum in a Goldstar CR401.

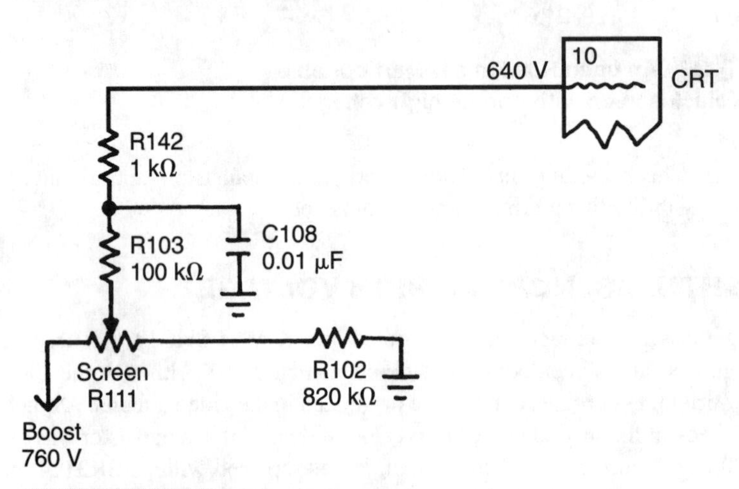

FIGURE 7-31 No brightness or screen control was found in this RCA CTC92A chassis.

and 12). The normal voltage is around 144 V. R521 in the flyback boost voltage source was open. After testing D410 for leakage, R521 was replaced, restoring the boost voltage.

In another similar symptom, the brightness or screen control had no effect in an RCA CTC92A chassis (Fig. 7-31). With the service switch at the service position, a very bright-white vertical line could not be turned down. With voltage and resistance measurements, it was found that R102 was open in the ground leg of the screen control.

CANNOT TURN DOWN BRIGHTNESS

Heavy retrace lines and no brightness control were found in an RCA CTC90D chassis. The video and other modules were replaced, with no change in the symptom. All voltages were normal on the picture-tube socket. Correct waveforms were entering the luminance/sync module (MDL002A). Voltage checks on the luminance/sync module indicated improper voltage at pin 17 (Fig. 7-32). The voltage remained at pin 17 when R4202 was rotated. Resistance measurements of the brightness control and R310 were normal. A continuity check between R309 and the service switch indicated that S301 was open. An open or corroded service switch can cause many different brightness and picture problems.

NO BRIGHTNESS CONTROL

Check the brightness and sub-brightness control adjustments if no brightness control is found in the raster. Improper brightness control can be located with a defective component in the automatic brightness limiter (ABL) circuits. A broken or open raster service switch can produce uncontrollable brightness. A very bright raster with chassis shutdown can be caused in the luminance IC. A leaky luminance IC can cause no brightness control. Check for leaky transistors or capacitors in the video circuits.

NO CONTROL OF THE LUMINANCE IC

The brightness and contrast controls had no effect on the raster in an RCA CTC109. Crucial voltage measurements on U701 turned up low voltage at pin 26 (Fig. 7-33). All resistors checked normal. A resistance of 768 Ω was found between pin 26 and the chassis ground. One lead of C710 was removed from the board. No leakage was found. When pin 26 was unsoldered from the PC wiring, 768 Ω of resistance was measured from IC pin 26 to ground. Replacing leaky U701 with original part 146858 cured the no-brightness control problem.

VIDEO IC REPLACEMENT

After locating the defective video or luminance IC, extreme care must be exercised in removing and replacing the IC. Excessive solder can be removed from the wiring side with a solder gun and solder wick. Be careful not to loosen tie points or component leads next to the IC terminals. If they are loosened, resolder them back into the circuit. Too much heat can pop or buckle the PC wiring.

Use a low-wattage or battery-operated soldering iron to solder each terminal of the new IC (Fig. 7-34). Small-diameter solder is ideal to prevent excess solder from flowing between

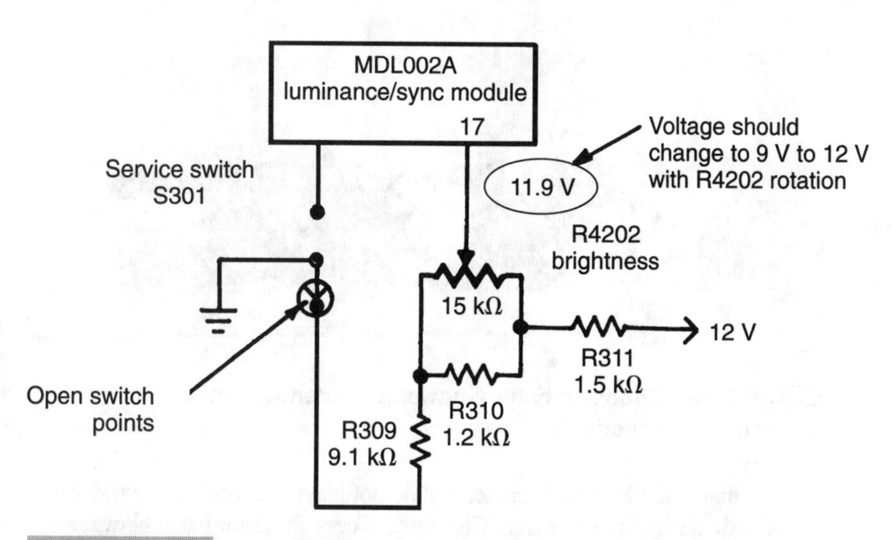

FIGURE 7-32 The poor condition of the service switch terminals resulted in no brightness control.

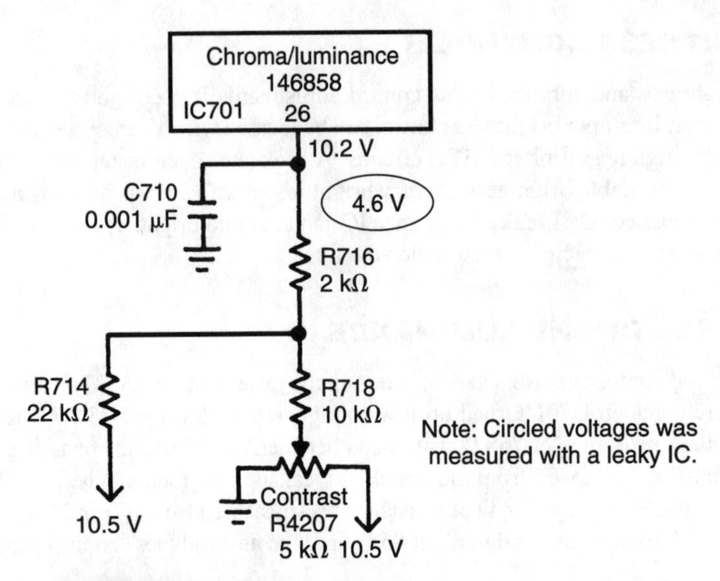

FIGURE 7-33 A leaky IC701 caused poor brightness control in this RCA CTC109 chassis.

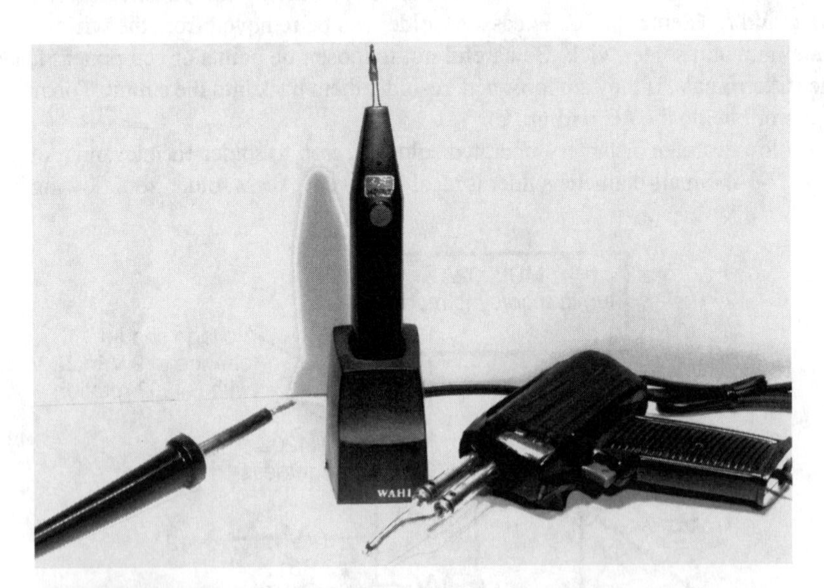

FIGURE 7-34 Choose a battery-powered soldering iron for transistor or IC connections.

the pin terminals. Make a good contact, but do not leave the iron on one pin too long. Place solder on both sides of the pin area. Clean out excess solder or paste between the pin contacts with the back edge of a pocket knife to prevent excess solder from touching the adjoining pin terminal.

VERY LOW BRIGHTNESS

No brightness control with very low brightness was found in a Sharp 19A63. All voltages were fairly normal on the picture-tube elements. The video stages were normal, except that the brightness and sub-brightness control had no effect on the raster. A voltage measurement in the ABL circuits indicated a higher-than-normal voltage on the ABL transistors (Fig. 7-35). Because the voltages were about the same on all transistor terminals, the ABL transistor was tested in the circuit and was found to be good. R431 was found to be open in the base circuits, causing high voltage at the collector terminal with no brightness control.

FAINT PICTURE WITH NO CONTRAST

The picture was very faint, without any contrast in a Montgomery Ward GEN-12907A portable. Some overscan lines were at the top of the picture without any function of the brightness or contrast controls. No video was noticed at the video amp.

After carefully checking over the video and picture-tube circuits, high voltage found on the ABL transistor focused attention on the ABL circuits (Fig. 7-36). R431 was found burned open in the base circuit of the brightness limiter transistor (Q401). Still, the picture and brightness were very faint. All resistors were checked in the brightness circuits. A short was found between R433 and ground. C614 in the flyback circuit was directly shorted, producing higher-than-normal voltage on Q401, which, in turn, left a faint and uncontrollable picture. The loss of brightness was not directly in the video circuits, but in the circuits controlling it.

INTERMITTENT BRIGHTNESS

Intermittent brightness can occur in the video circuits, ABL, picture tube, or with the voltage source feeding these circuits. Monitor the dc and boost voltage source for intermittent voltage. Check the brightness, sub-brightness, and screen controls for erratic operation. Scope the

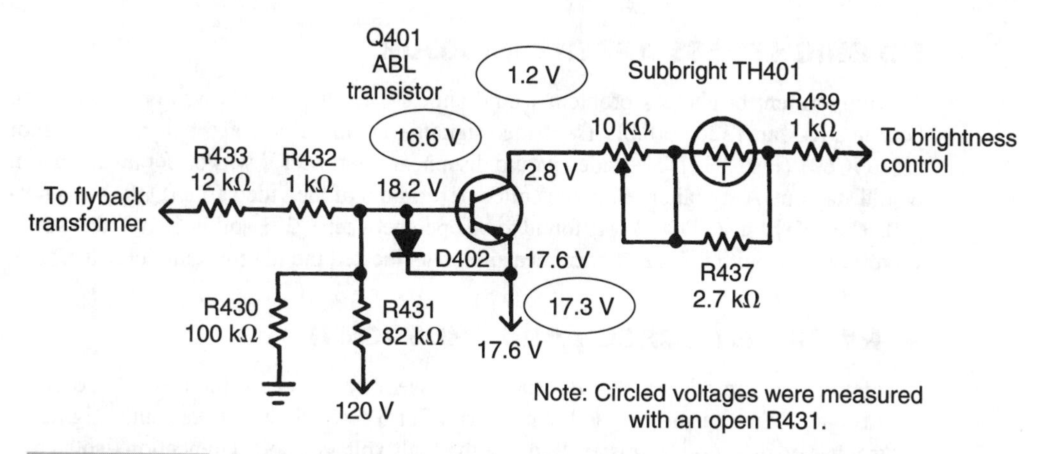

FIGURE 7-35 The very low brightness in this Sharp 19A63 was caused by an open R431.

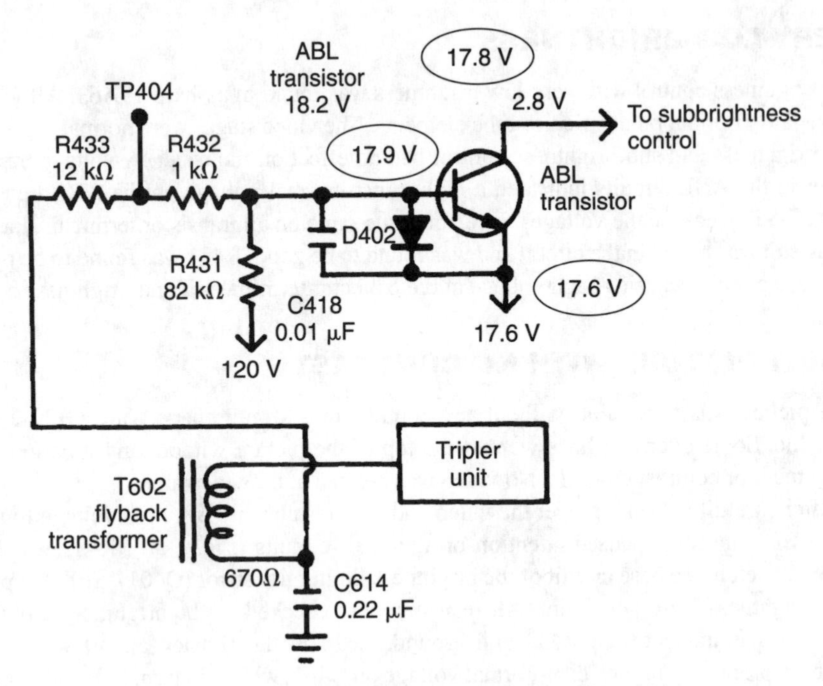

FIGURE 7-36 **An open R431 in a Wards GEN-12907A portable caused a faint picture with little contrast.**

video circuits if the chassis is intermittent. Voltage and coolant tests can locate an intermittent transistor or IC circuit. Check the delay line for an open or intermittent connection. Do not overlook possible intermittent peaking coils between the video transistors or delay lines.

NO BRIGHTNESS AFTER ONE HOUR

The intermittent brightness problem would appear after the chassis was on at least one hour in a K-Mart 1320 portable TV. Video signals were monitored at the input terminal of video IC601 (Fig. 7-37). The video would disappear after the TV was on for awhile and it would stay out. A signal check at the collector terminal of the video amp (Q301) was normal. The delay line (TD301) was found to be open between Q301 and IC601. Video scope waveform tests with low-resistance measurements located the intermittent delay line.

VERY BRIGHT SCREEN AND SHUTDOWN

A defective picture tube or shorted spark-gap assembly can cause the raster to come on very bright without a picture, and shut down after a few seconds. Often, the brightness control has no effect on the raster. Remove the high-voltage anode connection. Isolate the high-voltage lead away from the TV chassis. If the chassis turns on, but does not shut down, replace the shorted picture tube.

Sometimes the chassis will shut down before any voltages can be measured. Remove the picture tube socket and anode connection to determine if the video circuits are causing the shutdown. Check the video circuit if the chassis does not shut down with the CRT socket removed. Take voltage measurements on the video output transistor or the luminance IC.

BRIGHTNESS SHUTDOWN

The brightness-shutdown problem was isolated to the chrome/luminance IC700 circuit. Critical voltage measurements on pin 26 of U700 were fairly low (Fig. 7-38). Either U700 or some component tied to pin 26 was leaky. A resistance measurement of 1.5 kΩ was found between pin 26 and chassis ground. Pin 26 was removed from the PC wiring with no resistance between the pin and chassis. This indicated that IC700 might be good. A 1.2-kΩ leakage was found across the terminal of CR705. Replacing CR705 solved the very bright raster shutdown problem.

SMEARY PICTURES

Leaky video transistors and open peaking coils in the video circuits cause most smeary picture symptoms. Improper universal video transistor replacement can produce a smeary picture. Substitute a video transistor of another brand to see if it clears up the video problem. If not, try to locate the original. A defective video IC luminance component can cause a smeary picture.

AN UNUSUAL VIDEO PROBLEM

Heavy retrace lines with no control over the brightness occurred in a J.C. Penney 2039. Sometimes small firing lines could be seen across the raster, with no sound. The tuner control

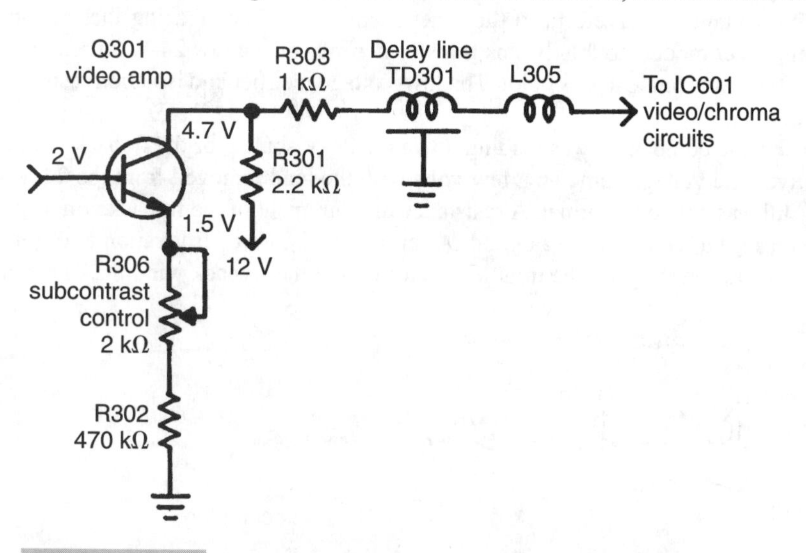

FIGURE 7-37 **An open TD301 delay line caused intermittent brightness in a K-Mart portable.**

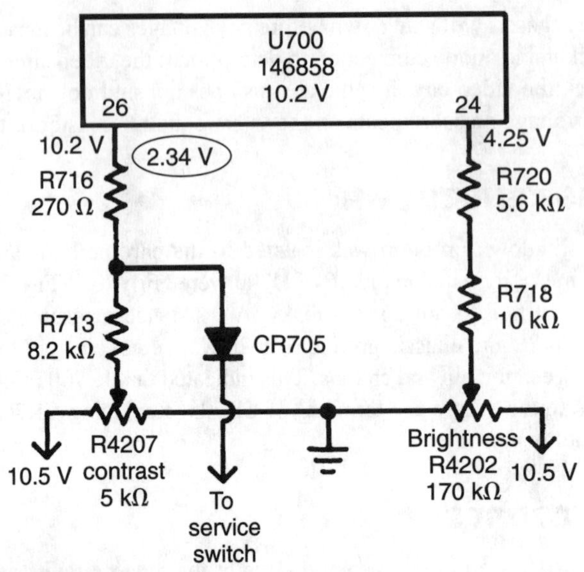

FIGURE 7-38 A leaky diode (CR705) caused chassis shutdown in an RCA CTC109 chassis.

system was dead. The tuner control module was replaced, with no change. Because the tuner would operate with no sound, no picture, and only retrace lines on the raster, the defective component must be tied to all circuits (such as a dc power source).

Voltage measurements were taken on the tuner terminals, indicating low voltage and no negative voltage or waveform to the tuner memory module. Tracing these connections from the tuner module to the flyback transformer indicated a low 24-V source (Fig. 7-39). Only 13.6 V was found at this point. The low-voltage rectifier and isolation resistor (R415) were normal.

Either some component was loading down the 24-V source, or the flyback winding was defective. The voltage came up a few volts with the load removed from the 24-V source, but it still was not up to normal. A resistance measurement of the flyback winding is difficult to take, but continuity was good. After several hours of frustration and going over tuner voltages once again, the trouble was located in the flyback winding. The continuity

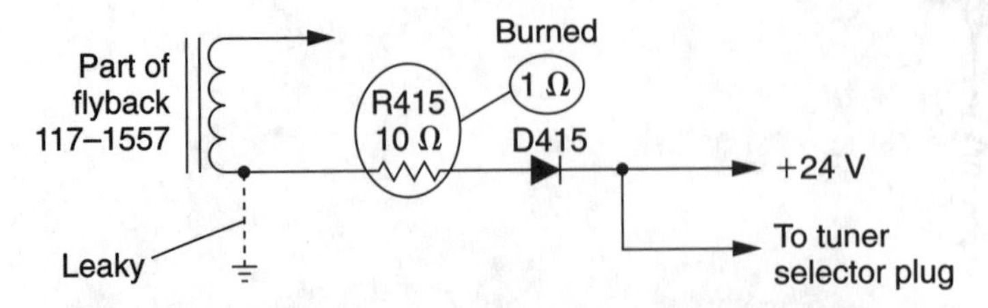

FIGURE 7-39 Heavy retrace lines were caused by a shorted winding to the core in a J.C. Penney 2039 chassis.

ohm measurement of the flyback winding was normal, but inside the transformer, the winding was grounded to the core of the transformer. Replacing the flyback transformer (117–1557) solved the unusual video problem.

FIVE ACTUAL VIDEO CASE HISTORIES

The following are five real problems related to the video raster.

Intermittent video—normal sound The picture would pop in and out of the screen in a Toshiba CT37C portable. Sometimes the TV chassis would operate properly for hours. A dot-bar generator was connected to the antenna terminals and the video stages were monitored with the scope at TP12. Very little video waveform was found at the emitter terminal of video amp Q201 when the set was intermittent. Coolant and heat applied to Q201 pointed out an intermittent transistor. Q201 (2SC1815Y) was replaced with an ECG85 universal replacement (Fig. 7-40).

Smeary picture—normal sound The picture was smeary on all channels in a Goldstar NC-07X1 chassis. Because the sound was good, the defective component had to be in the video output circuit. Waveform tests were made in the comb filter and video amp stages. These waveforms were fairly normal up to Q205. A voltage check on collector of Q205 was low and should equal the voltage source of 11.2 V (Fig. 7-41). Q205 was tested for leakage and was good. L202 was found to be open, lowering the dc collector voltage.

Black lines across screen The black lines streaking across the screen would come and go in a Sanyo AVM255 TV. The dot/bar/color generator was connected to the antenna terminals and waveforms were taken of the video circuits. The signal was normal up to the buffer amp Q313 and was very weak on the base of video amp Q315. The delay line (L304) was checked and seemed to be open. L304 was replaced with exact replacement part number LG0005KH (Fig. 7-42).

Negative picture, retrace lines The picture from a RCA CTC93E (Fig. 7-43) had a green tint with a faint, negative picture in the background and heavy retrace lines. Both

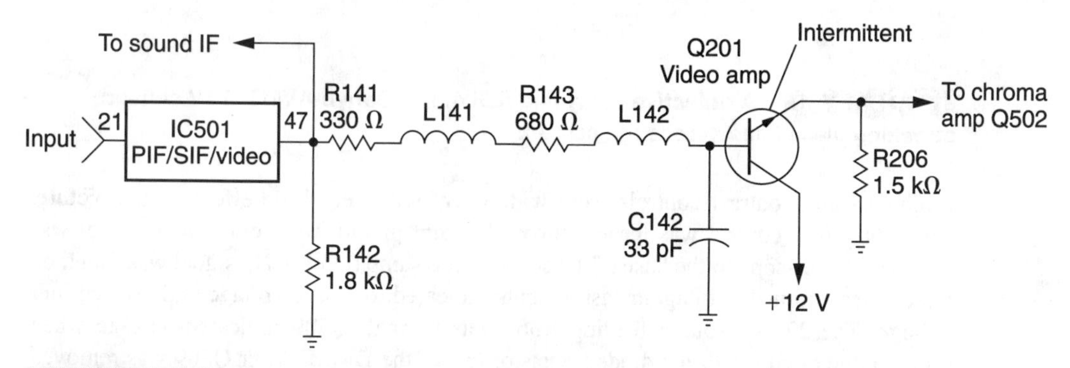

FIGURE 7-40 Q201 in a Toshiba CT37C portable caused intermittent video.

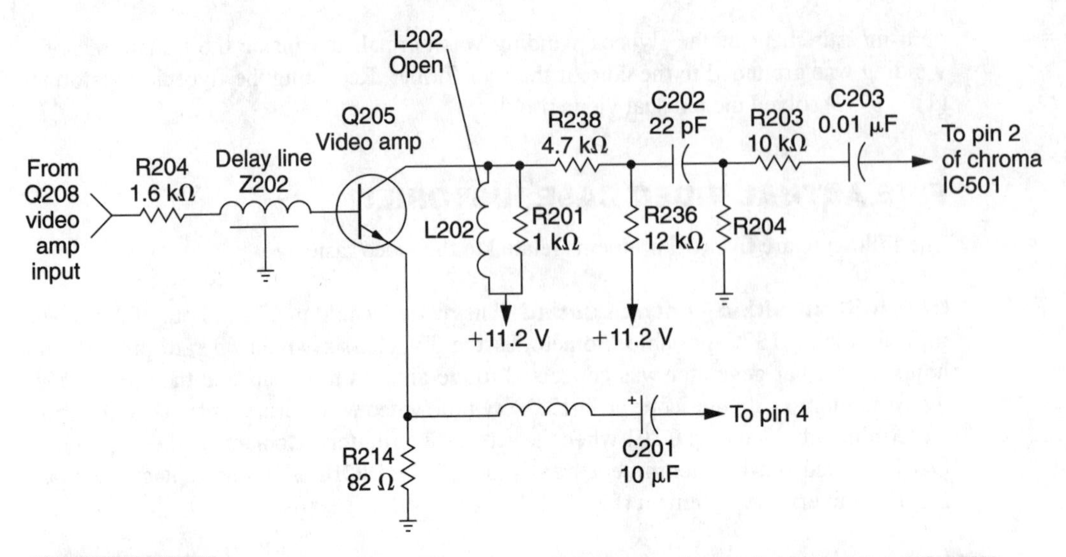

FIGURE 7-41 An open coil (L202) produced a smeary picture in a Goldstar NC-07X1 chassis.

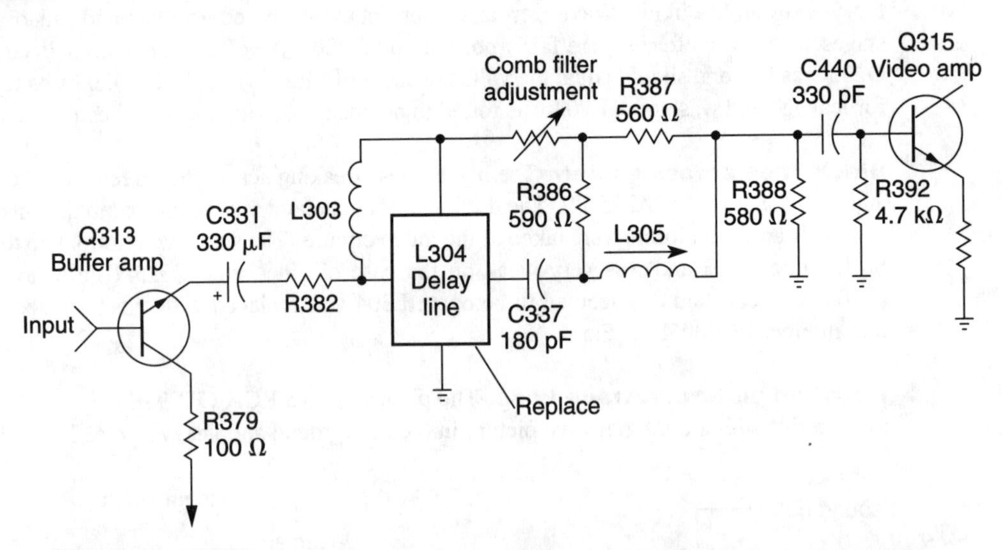

FIGURE 7-42 A defective delay line (L304) in a Sanyo AVM255 TV caused streaking black lines across the screen.

brightness and contrast controls were wide open, with very little effect on the picture. When the color control was turned down, the faint picture went out. Video signal was traced with the scope to the base of the second video amp (Q309). No signal was found on the emitter terminal. Voltage measurements indicated low base voltage and zero emitter voltage. The 27.5-V source feeding R366 was normal. Q309 indicated leakage when tested in the circuit with the diode transistor test of the DMM. After Q309 was removed from the circuit, a 78-Ω leakage was found between the collector and emitter terminals. Q309 was replaced with an original type of component.

No brightness control The brightness control had no effect on the raster in an RCA CTC111L chassis with normal sound (Fig. 7-44). The brightness reference-transistor (Q703) voltage is used to control the gain of luminance IC701 to maintain a consistent brightness level. The voltages on Q703 were changing and the low voltages were difficult

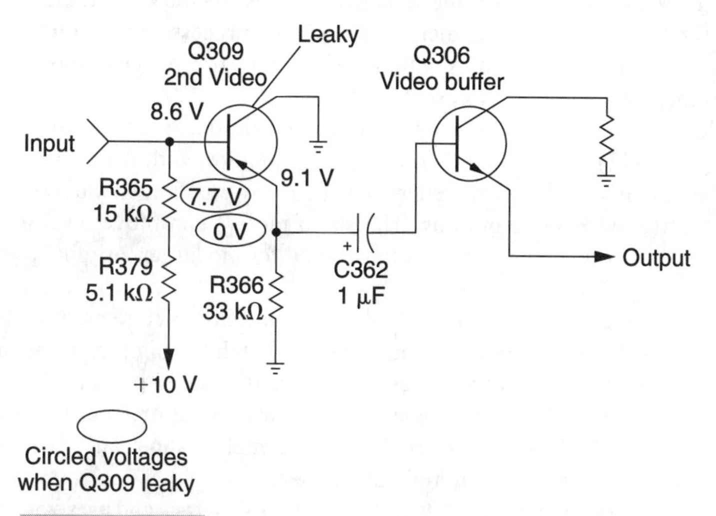

FIGURE 7-43 A leaky transistor Q309 caused a negative picture with retrace lines.

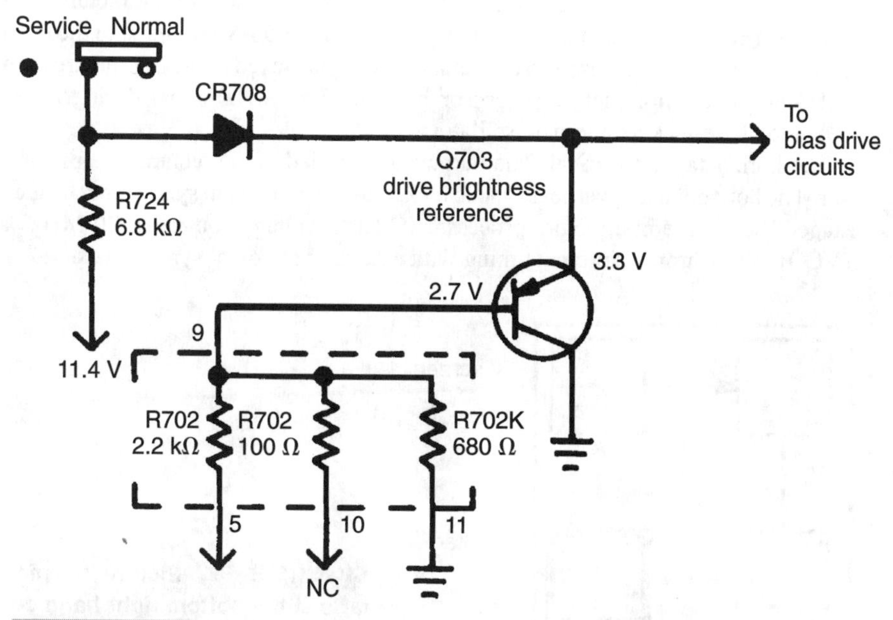

FIGURE 7-44 An open R702 in the base of Q703 resulted in no brightness control in an RCA CTC111L chassis.

to measure accurately. R702 was found to be open from base terminal to ground. The resistor network of R702 was replaced with the original replacement part because it contains many resistors in one component.

Picture in a picture You can see more than one picture on the CRT at the same time with the special feature of picture in a picture. The picture process allows you to have the large picture with several small pictures within the larger picture. You can zoom the small picture size, freeze, swap, or move it in some TVs.

When the picture is selected, the small picture appears in the lower right-hand corner of the screen (Fig. 7-45). The screen displays the big picture, with the small picture to the right. The small picture can be moved to just about any position on the screen by using four different arrow keys or buttons. The small picture can move until the arrow button is released. Of course, the sound is transmitted for the larger, original picture on the screen.

There are many different functions found with the picture-in-a-picture process. The picture can be moved with the move button and exchanged with the big picture by pressing the swap key. The freeze button can still either the small or large picture. Zoom in increases the size of the small picture each time the zoom-in button is pressed. The zoom-out button decreases the size of the small picture. Pan, strobe, multichannel, and special-effects modes can be found in some picture-in-a-picture models.

The picture-in-a-picture module plugs directly into the TV chassis and uses voltage from the power supply and flyback circuit for operation. Both horizontal and vertical sync is applied to the picture-in-a-picture module from the TV chassis.

The TV and CAV video is selected with electronic switching to the picture-in-a-picture processor. Also, the luminance and color (Y and C) are selected to the input of an R-Y and B-Y decoder (Fig. 7-46). The R-Y and B-Y signal is applied to the picture-in-a-picture processor IC. A burst oscillator provides a continuous 3.58-MHz signal to the encoder and picture-in-a-picture processor. This burst oscillator is locked to the big-picture chroma signal. The picture-in-a-picture processor uses the 3.58-MHz signal during multichannel mode to phase-lock its internal oscillator.

The horizontal and vertical sync outputs are applied to the picture-in-a-picture processor. The horizontal and vertical sync is locked to the composite sync output of the decoder stage. The picture-in-a-picture processor IC has a voltage-controlled 20-MHz oscillator (VCO) to synchronize internal timing with external horizontal sync signals.

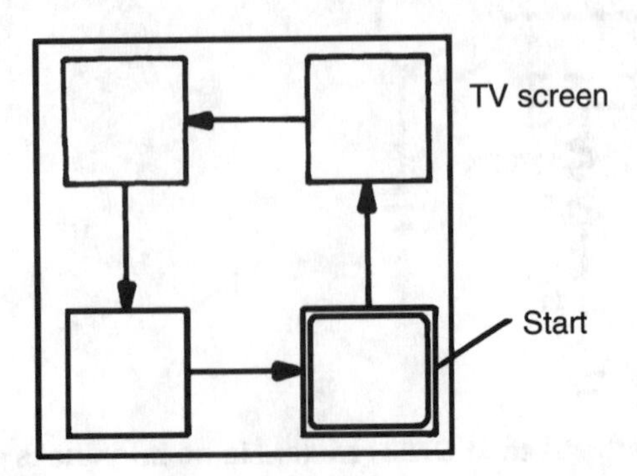

FIGURE 7-45 Picture in a picture starts at the bottom right-hand corner of the screen.

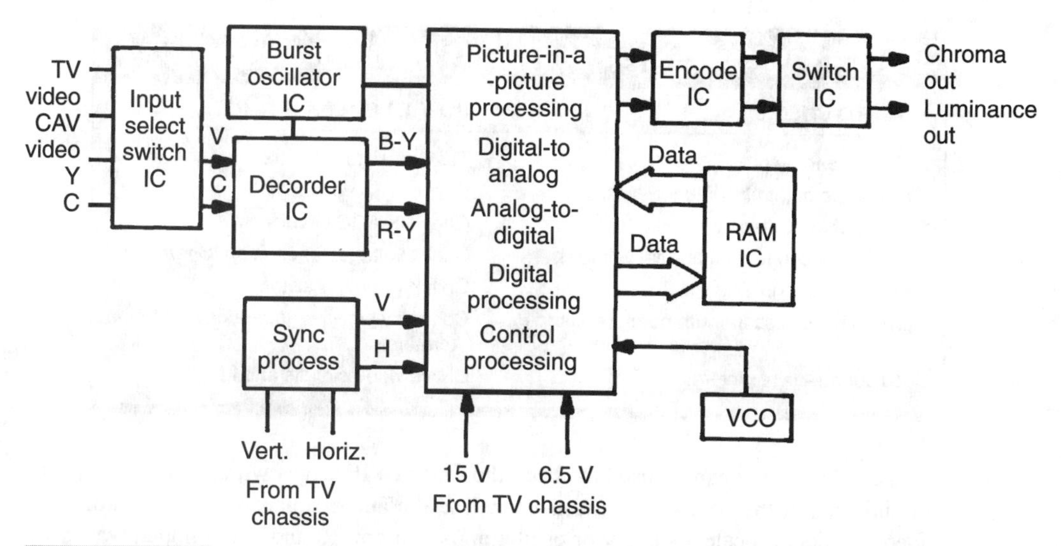

FIGURE 7-46 A block diagram of the picture-in-a-picture processing circuits.

Analog-to-digital conversion occurs in the picture-in-a-picture processor and converts analog luminance, B-Y, and R-Y signals into digital information. The digital video information is stored in the RAM IC. This digital information can be operated by the digital processing circuits.

The R-Y and B-Y and Y/C signal are fed through an encoder IC to a fast switch IC and then applied to the TV chassis. The signal information is connected to the on-screen display or line-drive circuits of the picture tube.

To troubleshoot the picture-in-a-picture circuit, check for correct voltage applied to the picture-in-a-picture module. Some chassis have an internal power supply and regulator circuits. For a no-big-or-small picture symptom, check the input TV and CAV input circuit signal. Check for luminance (Y) and chroma (C) signals at the input stage. Signaltrace the composite and component video signals throughout the input for proper switching.

For a symptom of no color in the small picture, check chroma at the input and check the B-Y and R-Y signals at the output of the decoder IC. Scope the burst oscillator IC.

If the small picture rolls vertically while the large picture is normal, scope the vertical input sync signal. Also check the vertical output of the vertical processor sync signal. If the small picture takes some time before locking in horizontally, check the adjustment of the horizontal frequency coil of the horizontal and vertical processor IC.

Conclusion

The oscilloscope can be the most useful test instrument when servicing the IF and video circuits. Locate the defective circuit with waveforms taken with the scope. Check signal waveforms at the input and output of the IC video component. Take voltage and resistance measurements within the suspected stage. Check each transistor in the circuit.

TABLE 7-1 TROUBLESHOOTING CHART FOR IF VIDEO CIRCUITS

WHAT TO CHECK	HOW TO CHECK IT
No IF or video output.	Test transistors in circuit.
Sub tuner to determine if tuner is okay.	Check IF IC voltages.
Check IF AGC voltages.	Check IF AGC circuits.
Suspect IF-video IC with normal voltages.	Check voltages upon each IC terminal.
Normal sound—no video.	Check video transistor.
Signal trace with scope with normal voltages.	Connect color-bar generator to antenna terminals.
Good sound—faint picture.	Check video circuits and IC.

A substitute tuner can be used to check the first two IF stages with signal from the station injected at the base of each transistor. Use the audio square-wave generator in the video circuits to locate a smeary or oscillating symptom. Rounded or clipped waves will indicate the defective stage.

Always replace IF shields above and below the circuit board after repairing the IF or video sections. Replace IF transistors with the original when available. Most universal transistors and ICs work very well in the video circuits. Check Table 7-1 for troubleshooting IC and video circuits.

8

AGC AND SYNC CIRCUIT PROBLEMS

CONTENTS AT A GLANCE

Servicing the AGC and sync circuits is not as complicated as it was in the past. The automatic gain control (AGC) circuit basically automatically controls the incoming signal level at the IF and RF amplifier stages. These stages are controlled with a small dc bias voltage that changes the gain of these stages. If the incoming signal becomes greater, the bias voltage rises, reducing the gain of the IF and RF stages. Likewise, if the incoming signal is weaker, the bias voltage is lower, increasing the gain of these stages (Fig. 8-1). Some of the early TV chassis had an RF AGC circuit separate from the IF AGC stage.

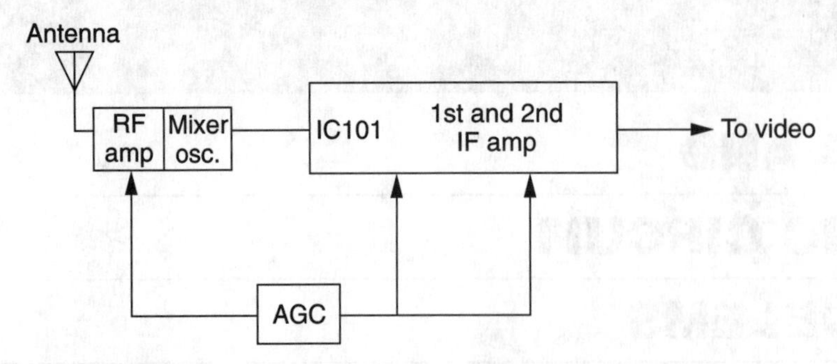

FIGURE 8-1 **A block diagram of the AGC circuits that feed the RF amp, mixer, oscillator, and the first and second IF amps.**

Today, in the keyed AGC and delayed AGC circuits, a flyback pulse is fed to the AGC circuits for a more accurate and dependable system. The incoming video signal is fed from the video stage to the AGC keyer circuit. A dc-controlled amplifier feeds the bias voltage to the first and second IF amplifier and AGC delay circuit (Fig. 8-2). The AGC delay circuit delays the bias voltage to the RF tuner stage until a fixed level of gain is obtained in the IF stage for maximum performance.

In the latest TVs, you might find an IC that internally develops the AGC action and varies the gain of the first and second IF stages. A comparison voltage is developed to control the RF AGC voltage. This type of AGC circuit is no longer keyed with a pulse from the flyback transformer circuit.

The AGC circuit can be tested by scoping the waveforms of the video input signal and flyback pulse to the AGC circuit. Crucial voltage tests on the AGC transistor or IC circuits should be made, with no signal at the tuner. Turn the manual tuner between channels or to

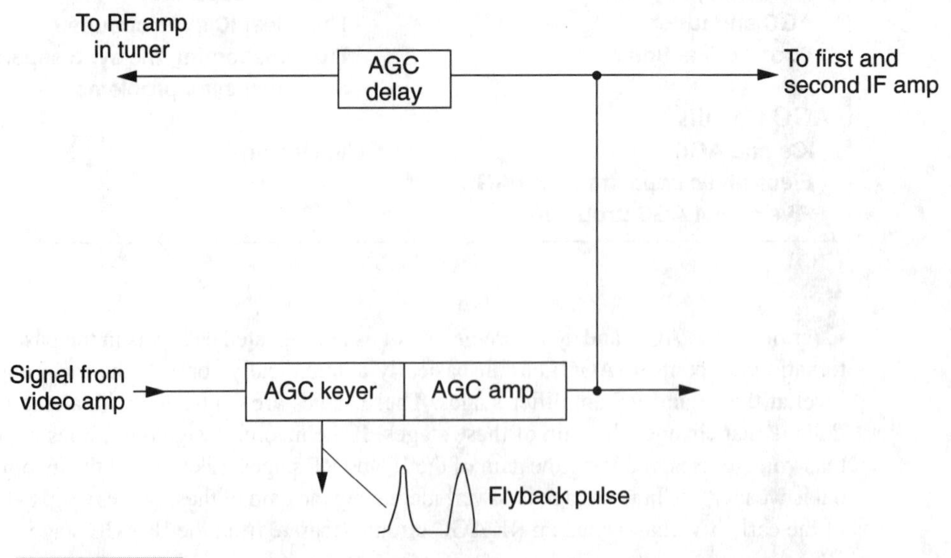

FIGURE 8-2 **The delayed AGC is taken from the AGC amp and keyer.**

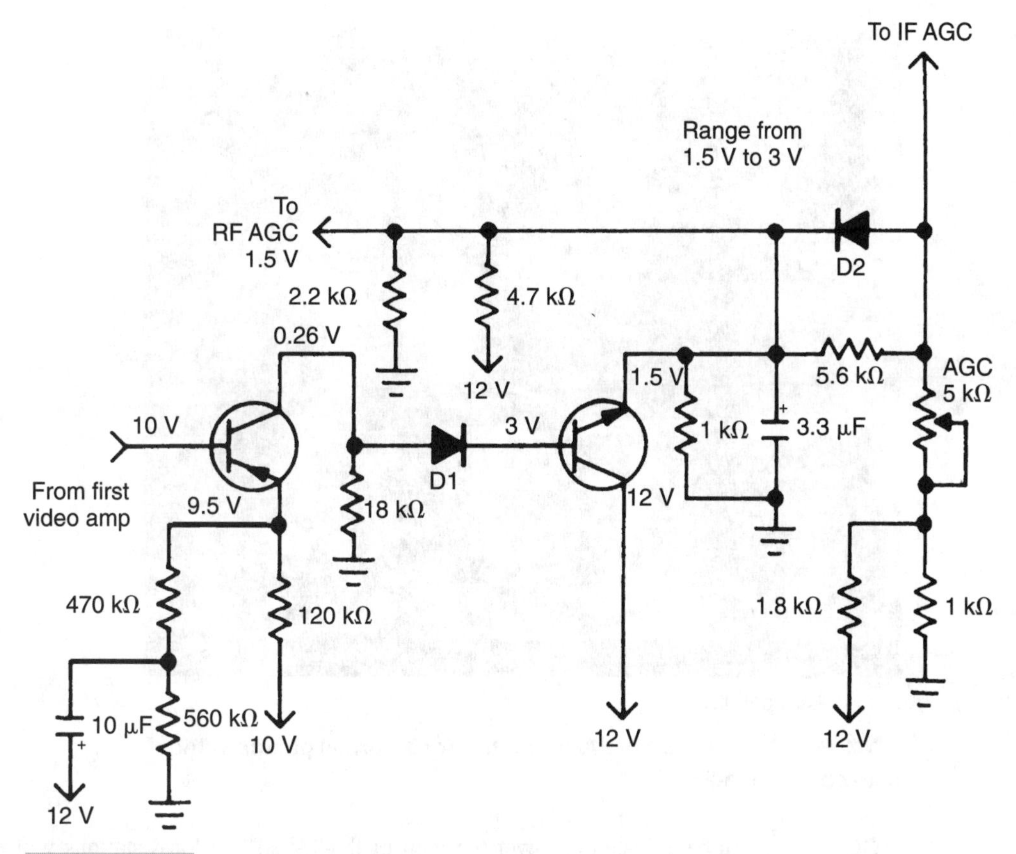

FIGURE 8-3 **Compare the AGC voltage measurements with the schematic at the first and second IF amplifiers.**

a dead channel for accurate voltage measurement. All dc voltages shown on any schematic are without any signal applied to the antenna. When signal is applied to the TV chassis, these voltages will rapidly change. If not, the AGC circuits are not functioning. One good test is to notice the change in the AGC voltage when a station is tuned in. The AGC circuits are working if the voltage changes.

Abnormal voltages taken on the AGC, transistor, or ICs can indicate that the AGC stages are defective. High voltage fed to the base circuits of the IF stages can cut off the transistor, producing a white screen. Low voltage can cause overloading and pulling of the picture. Carefully check the IF and RF bias voltage listed on the schematic for normal AGC operating conditions (Fig. 8-3).

AGC Controls

In some TVs, you might find one AGC control, a separate RF delayed and AGC control, or no control at all. Improper adjustment of either control can cause picture rolling. The

FIGURE 8-4 **Adjust the AGC control for a normal picture without snow or overloading.**

AGC control can be adjusted to lower the gain of the TV so that local stations will not overload the picture. Weaker TV stations can be received by readjusting the control to provide more gain in a given area. Some TVs might list the AGC control as the noise control (Fig. 8-4).

You can assume that the AGC circuits are functioning if rotating the AGC control makes the picture darker or appear as overloading. No reaction from adjusting the AGC control can indicate that problems are within the AGC circuits. Improper adjustment of the AGC control can cause the set to overload. The AGC control setting should always be done after repairs or when the TV is returned to the home.

AGC Problems

AGC problems can include a very dark and unstable picture. A white raster can be caused by a defective AGC tuner or video circuit. Check the AGC circuit for a "flagwaving" picture with a buzz in the sound. Low contrast and erratic pulling in the picture can be caused by improper AGC voltages. Excessive snow in the picture can be caused by a defective AGC, tuner, or IF stages. Besides the AGC circuits, these same symptoms can be caused by defects in the tuner, IF, or video stages.

AGC CLAMPING TESTS

To determine if the AGC circuits or tuner are defective, a variable positive voltage fed to the AGC terminal on the tuner can bring back a normal picture. If the picture is good, you can assume that the tuner is normal with a defective AGC system (Fig. 8-5). This test is performed with an external power supply. A tuner-subber can be plugged into the IF cable to help prove that the tuner is operating. Be sure that the dc voltages at the tuner are correct.

Although clamping tests have limitations in some circuits, check that the AGC voltage applied to the tuner and IF stages is accurate, according to the schematic. Excessive positive voltages at the tuner AGC terminal can cause snow, but when applied to the IF stages, it can produce a white screen. All AGC voltages should be taken with no signal at the TV. Simply rotate the tuning knob to a point between channels.

AGC AND TUNER

In many cases, the tuner is removed and replaced with no improvement in the picture. This can be caused by AGC defects. The tuner-subber can indicate if the tuner is defective. If the tuner-subber brings back a normal picture, you can assume that the original tuner is defective. Go a step further and measure the AGC and B+ voltages at the tuner with the tuner locked between channels or on an unused channel. Abnormal voltage at the tuner AGC terminal can indicate that an AGC circuit is defective. Replace the tuner if the AGC and B+ voltages are normal.

POOR AGC ACTION

Most AGC problems are caused by leaky or open transistors and ICs. Besides voltage and transistor tests, the defective component can be located by applying coolant or heat to the board. Sometimes, spraying each transistor or IC will help you to find the intermittent AGC symptom.

Dried-up or leaky electrolytic capacitors within the AGC circuits can produce AGC problems. Sometimes shunting a good capacitor across the suspected one can solve the

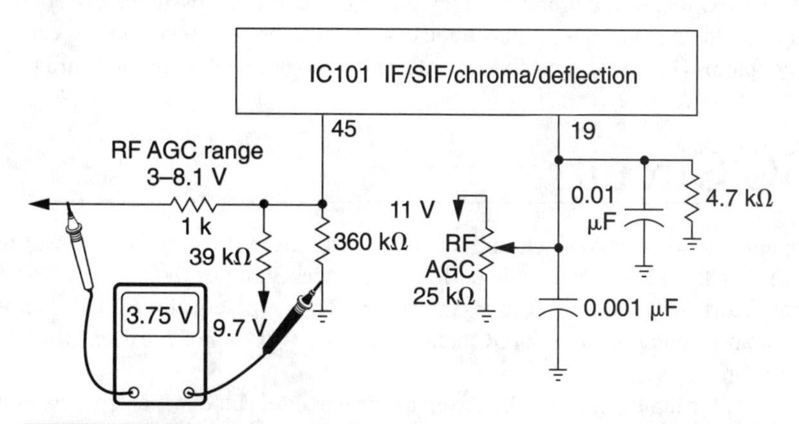

FIGURE 8-5 Check the AGC voltage to the tuner to determine if the AGC circuits are operating.

TROUBLESHOOTING AND REPAIRING SOLID-STATE TVS 2

FIGURE 8-6 **Repair the AGC PC wiring on the foil with bare hookup wire.**

AGC condition. Remove one end of the capacitor for accurate leakage tests. A good in-circuit capacitor tester is ideal in locating defective capacitors in the AGC circuits.

Take resistance measurements within the circuit when all voltages and in-circuit tests are normal. Remove one end of all resistors above 50 kΩ to make accurate resistance measurements. In many cases, an in-circuit resistor measurement can be very close to the resistor value and is passed over as good. Emitter bias and collector resistors should be checked for close resistance tolerance in AGC circuits.

Besides defective AGC components, poor IF cables and board connections can cause intermittent AGC look-alike problems. Flex the IF cable to find possible loose connections. Push up and down at various points on the chassis. Broken or cracked wiring can produce AGC symptoms (Fig. 8-6). Localize the intermittent to one section of the board (Fig. 8-7).

AGC Circuits

The input signal of any AGC circuit is usually taken from the emitter circuit of the first video amp or from a separate cathode-follower stage. Some of the earlier AGC circuits with transistors included four or more transistors. Today, practically all TV chassis use a section of an IC component for AGC control. The AGC IC is much easier to service than the transistor stages.

In Fig. 8-8, the input signal is taken from a noise inverter stage, which gets the video signal from the first video amp. The gain of the AGC keying transistor is varied with AGC control R221. The keyed pulse from the flyback transformer winding is found at the collector circuit of Q201. The AGC amp output voltage feeds both the RF AGC tuner and the

FIGURE 8-7 Try to localize the intermittent to one section of the boards.

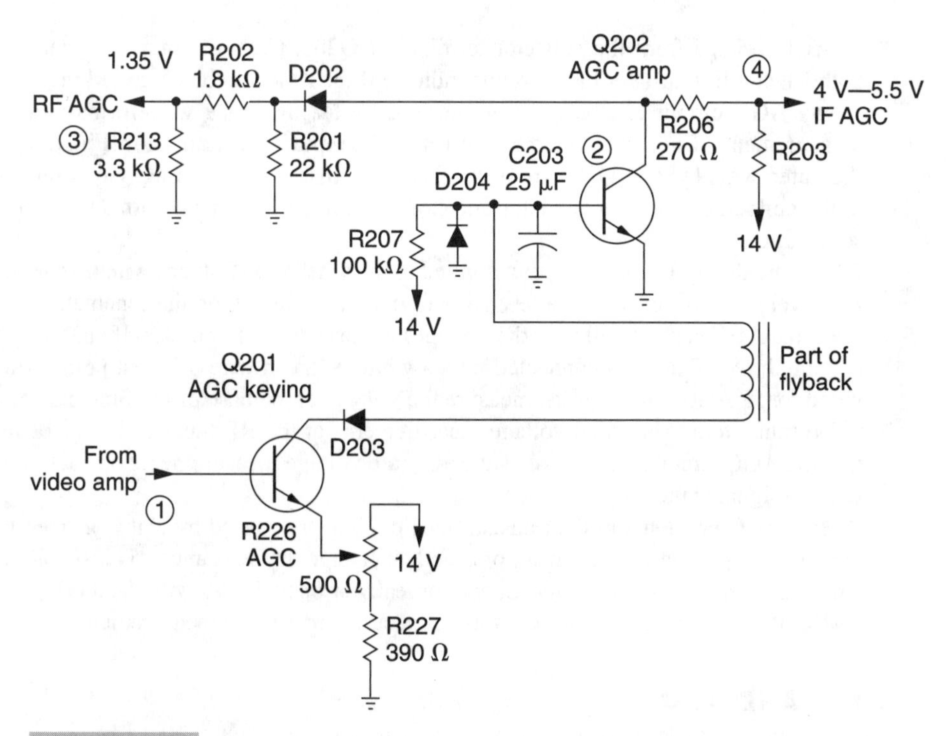

FIGURE 8-8 Check the AGC circuits by the number, with waveforms and voltage measurements.

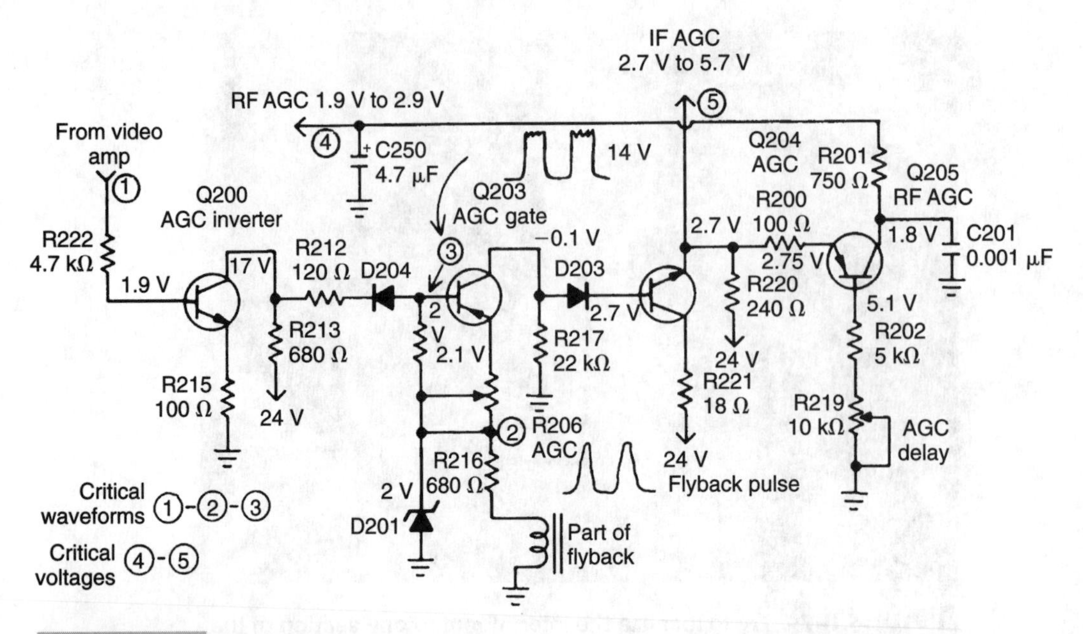

FIGURE 8-9 **Take waveforms and voltage measurements to locate the defective component in the AGC circuits.**

IF AGC voltage from the collector terminal of Q202. Crucial voltage measurements at both RF and IF transistor base circuits indicate if the AGC circuits are working.

Any AGC circuit can be quickly checked by taking scope waveforms and voltage measurements. Check the video waveform at the input AGC transistor (Fig. 8-9). Scope the gated waveform at Q303 from the flyback transformer winding. Go from the base to the collector terminals of each transistor and compare the waveforms to those on the schematic.

Measure the AGC voltage output applied to the tuner and IF stages with no signal. Most AGC voltages will only vary a few volts from what is shown on the schematic. Check to see if the AGC control will vary the voltages at these different points. Often, the RF AGC post on the VHF tuner is connected with a white or green wire. If a test point is not provided for IF AGC voltage tests, measure the voltage at the base of the first and second IF video transistors. Abnormal voltage measurements at the RF tuner or IF stages indicate that the AGC circuit is defective. Take accurate voltage measurements on each transistor with no signal at the tuner.

Most AGC problems in the transistorized circuits are caused by leaky or open transistors, dried-up electrolytic capacitors, and a change in resistance. Transistors can be checked in and out of the circuit. Check for leaky or open diodes with the diode test of the DMM. Remove one end of all resistors for accurate resistance measurements.

ICS AND AGC

Most color and black-and-white TVs manufactured today use ICs in the AGC circuits (Fig. 8-10). Locating the defective part is much easier with fewer components. Adjust the RF

AGC control, if one is found in the circuit. Suspect a defective AGC circuit if no action is noted in the picture.

Take a voltage test at the RF tuner terminal (Fig. 8-11) and IF test points (1 and 2). Scope the waveforms at 3 and 4. Check the flyback winding or wiring board connections with no keyed waveform at point 4. Check all voltages on the IC terminals related to the AGC circuits without a signal at the tuner. Low voltages can indicate a defective IC. Shunt C111 with another electrolytic capacitor. Check all resistors for accurate resistance before replacing the suspected IC.

ELECTROLYTIC CAPACITORS AND AGC

Electrolytic capacitors in the power source feeding the AGC circuits can cause some very unusual pictures. Automatically shunt each filter capacitor. Clip into the circuit with the power off. Small electrolytic bypass capacitors within the AGC circuits can cause excessive pulling and tearing of the picture. Temporarily tack another capacitor across the suspected one. Remove one end of the capacitor and check for leakage.

Unusual firing lines The picture in a Zenith 12AC10C15 chassis would pull and tear with small firing lines. Sometimes the lines were eliminated when the AGC delay control was turned down. Replacing the small polarized capacitor in the base circuit of the AGC delay transistor solved the problem (Fig. 8-12).

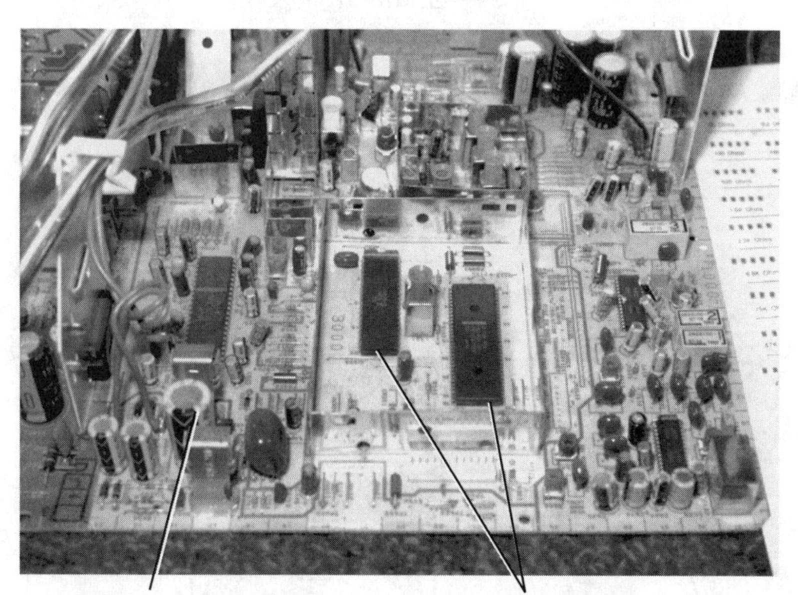

FIGURE 8-10 One large IC might contain the AGC and sync circuits, along with many others.

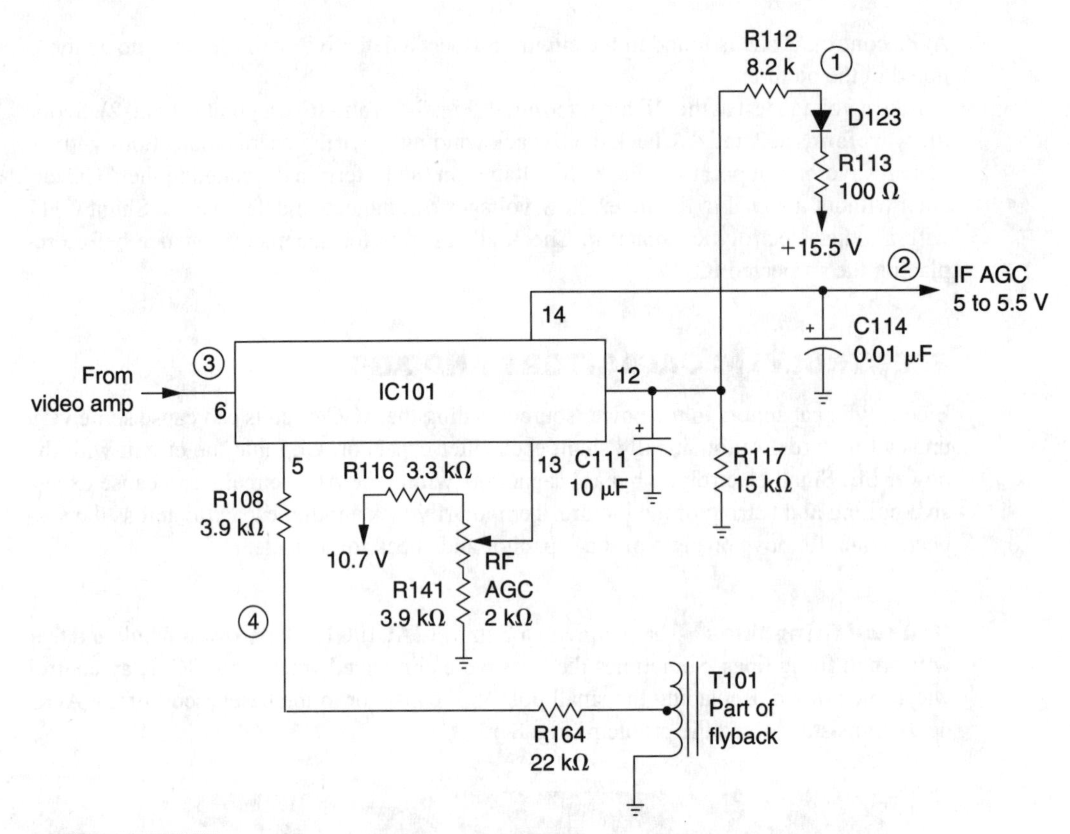

FIGURE 8-11 Low supply or pin voltages can indicate that an AGC IC is leaky.

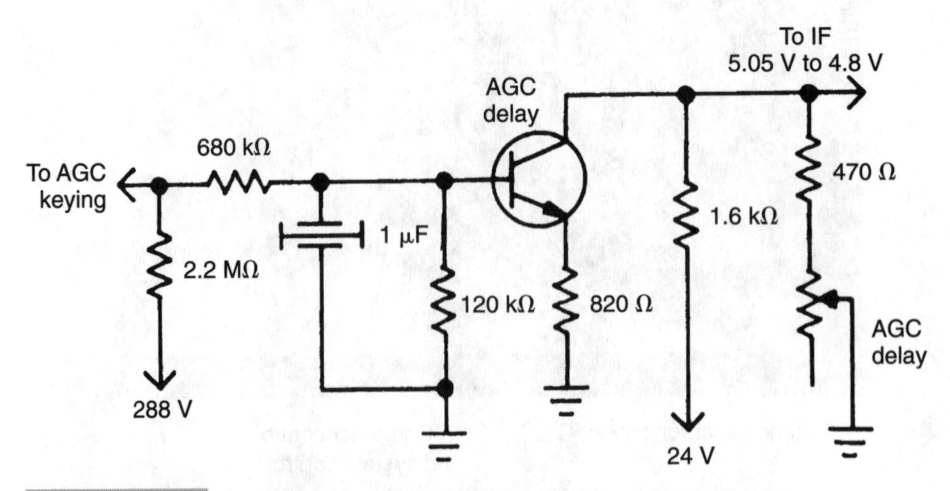

FIGURE 8-12 A twisting, pulling picture was caused by a defective nonpolarized capacitor (1 µF).

FIVE ACTUAL AGC PROBLEMS

The following items are examples of real AGC-related problems in common TV chassis.

Intermittent AGC and negative picture in a J.C. Penney 685-2520 The picture was intermittent and sometimes appeared negative, indicating AGC problems. Voltages were checked on pins 5, 6, and 7 of the VIF/SIF/AFT/Detector IC, IC101. The voltages were way off in the intermittent mode, but would not stay off long enough for a complete voltage measurement. When the probe touched pin contacts, the set would act up (Fig. 8-13). All connections were resoldered on IC101 and the problem was still the same. IC101 was replaced with AN5136KR in the J.C. Penney 685-2520 chassis, thus fixing the problem.

Poor connection—AGC problem in a General Electric 19PC-J Sometimes a General Electric 19PC-J would work perfectly; at other times, the picture would snow up. Pushing IC101 and the nearby components would make the problem act up. The snowy picture was cured by soldering the RF AGC control contacts of R116 (Fig. 8-14). Remember, the voltage measurement on pins 12 and 16 will change as the IF and RF AGC controls are adjusted.

White screen and sometimes snow in a Philco 20ST30B The screen was white and sometimes snow was found on an unused channel, indicating an AGC overload problem in a Philco 20ST30B chassis. The tuner was substituted and the AGC voltage was clamped at the tuner, indicating that problems existed within the AGC circuit. The picture would return with the tuner subbed or a clamped AGC voltage.

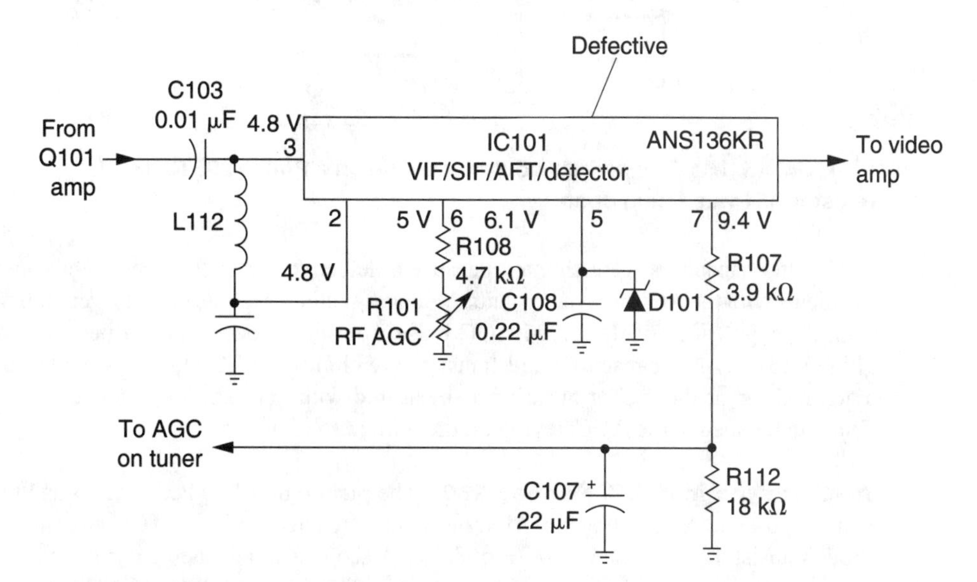

FIGURE 8-13 Intermittent AGC and a negative picture was caused by IC101 in this J.C. Penney 685-2520 chassis.

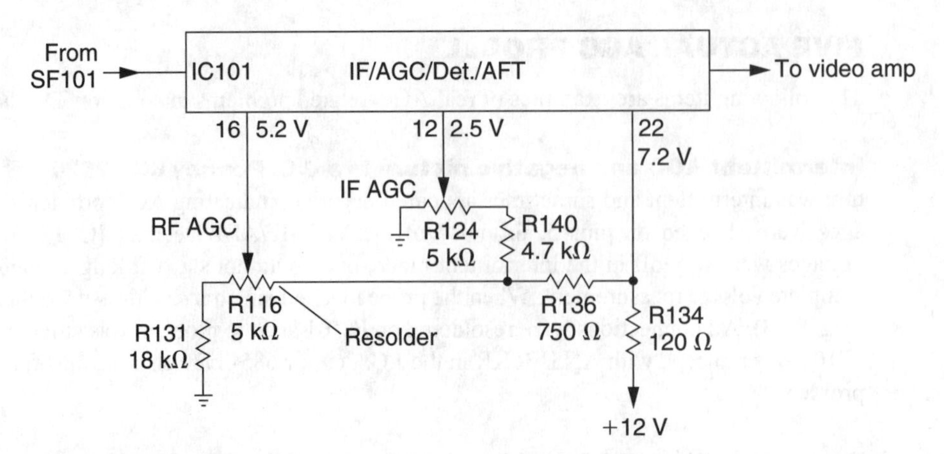

FIGURE 8-14 A poor contact on control R116 resulted in a snowy picture in a General Electric 19PC-J.

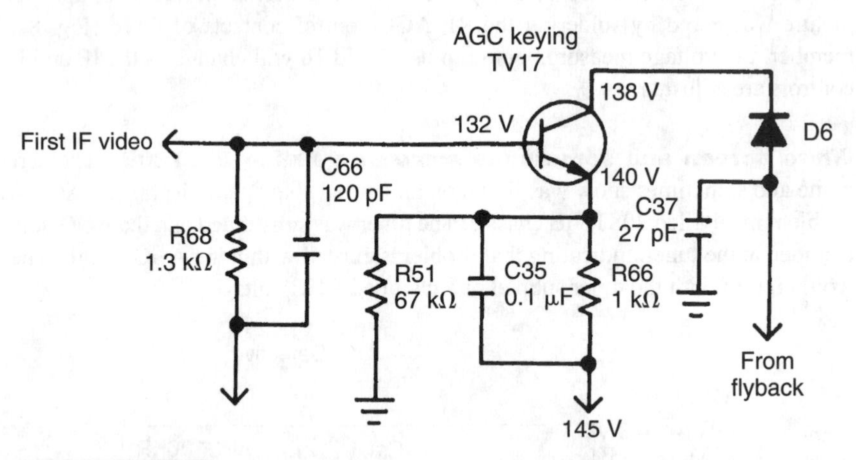

FIGURE 8-15 The screen would turn white in a Philco 20ST30B TV; resistor R51 was found open.

All filter capacitors were shunted because a defective electrolytic capacitor can cause problems with the AGC or sync circuits. Voltages within the AGC circuits were fairly normal. A scope check of the keyed AGC pulse from the flyback transformer was normal (Fig. 8-15). Bypass capacitors and transistors within the AGC circuits seemed normal. Each resistor or those over 50 kΩ was measured with one terminal removed. R51 was found to be open in the AGC keying emitter circuit to ground.

AGC or sync in a J.C. Penney 286 The picture in a J.C. Penney 286 had tearing sync or possible AGC problems. All transistors were checked in the AGC circuit with the diode transistor test of the DMM (Fig. 8-16). A scope test indicated a normal pulse from the flyback circuits. Voltage measurements were fairly normal. When C403 was shunted with another 10-µF capacitor, the picture returned to normal. Dried-up electrolytic bypass capacitors in the AGC circuit can cause the picture to appear as sync or AGC problems.

Vertical lines outside AGC circuits in a Zenith 20Y1C48 Vertical lines at the left side of the picture have been noted to be caused by defective electrolytic capacitors in the AGC circuits. Four vertical lines were found in a Zenith 20Y1C48 chassis with a very normal picture. Sometimes they would appear in the background with the brightness turned up.

Although all capacitors in the AGC circuit were shunted without any results, the problem turned out to be three blanking gate diodes in the emitter circuit of the video amp (Fig. 8-17). Replacing all three leaky diodes with 1N34As solved the vertical line problems.

AGC Circuits Conclusion

Always check the AGC control before attempting to service the AGC circuits. Momentarily short the RF AGC terminal at the tuner with a test clip or screwdriver and notice the change in the picture. If you notice a big change, perhaps the AGC is working. There will be no AGC action if the IF, detector, and video-amp sections are defective.

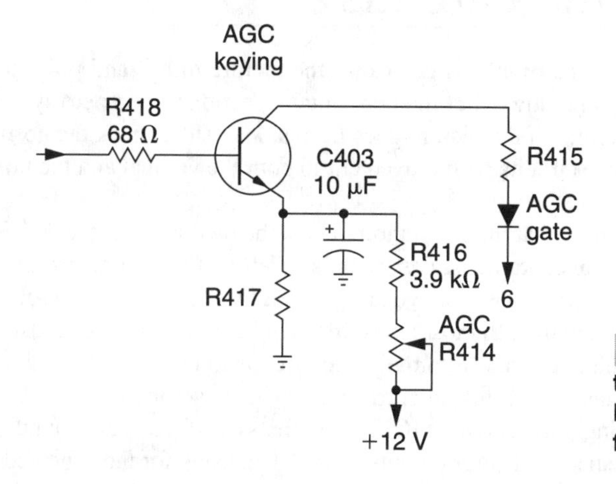

FIGURE 8-16 Excessive tearing of the picture in a J.C. Penney TV was caused by a faulty capacitor (C403).

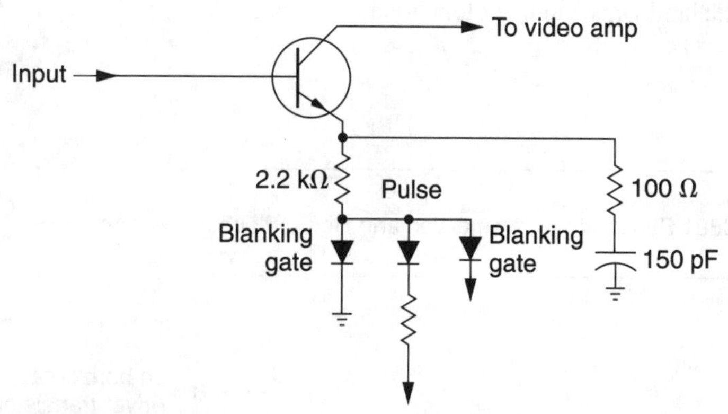

FIGURE 8-17 Firing lines in the picture were caused by leaky diodes in the emitter of the video amp.

Measure the voltage at the RF AGC tuner terminal. Turn the TV off-channel; in the case of a pushbutton or remote-control set, select an unused channel. If a big voltage change occurs, the AGC circuits are probably working. Locate the IF test point. In case of no test points, measure the dc voltage at the base terminals of the first and second IF transistors. Compare the voltage measurements for the RF tuner and IF without a signal to those on the schematic. Suspect that the AGC circuits are defective if the voltages are lower or higher than normal.

Clamp the RF AGC voltage with a variable dc voltage supply and notice if the picture returns. A fairly normal picture indicates AGC problems. Inject a variable dc voltage at the IF test point to determine if the IF AGC circuits are defective.

Scope the video input waveform at the input of the AGC circuit. Go from stage to stage and notice where the waveform is missing. Check the key waveform pulse from the flyback transformer winding. An open winding or connection can prevent the keyed AGC circuit from operating.

Servicing Sync Circuits

Poor vertical or horizontal sync problems can cause the picture to roll and pull horizontally. If the picture flips up and down, but the horizontal is stationary, suspect sync problems within the vertical circuits. If the picture goes from side to side, check the horizontal sync circuits. Check the sync circuits with movement in both the vertical and the horizontal sweep of the raster.

The sync pulses are transmitted at the TV station to lock the pictures in at the TV. Sometimes poor or snowy reception contains weak sync pulses, letting the picture intermittently roll or flip. In extreme fringe-area reception, you might find that the picture will roll or flip sideways with normal sync circuits. In most cases, the AGC control must be adjusted.

Because the two sync pulses are fed from the video circuits to the separate sync stages, the scope is the ideal test instrument to signal trace the sync waveforms (Fig. 8-18). The sync circuits might be a single transistor or IC. Today, the sync circuits are found in one large IC with AGC, luminance, and color circuits (Fig. 8-19). Look for the sync and AGC circuits within the luminance module of the modular chassis.

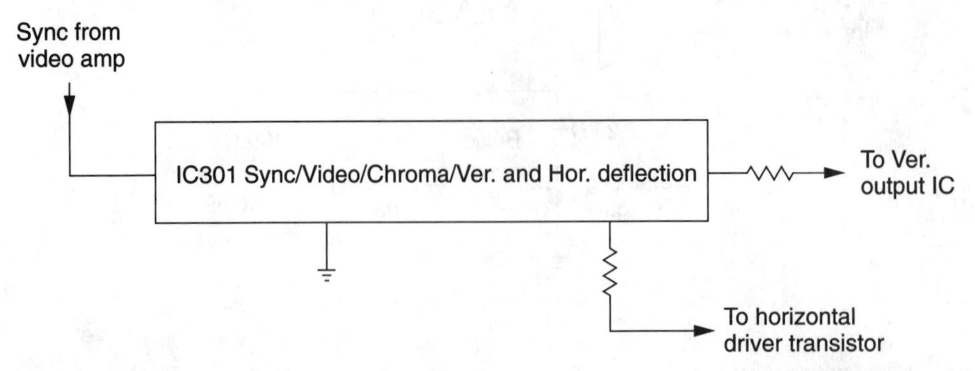

FIGURE 8-18 **The sync pulses are fed from the video amp to the input of IC301 to synchronize the vertical and horizontal circuits.**

FIGURE 8-19 Today, the sync and AGC circuits are contained in one large IC.

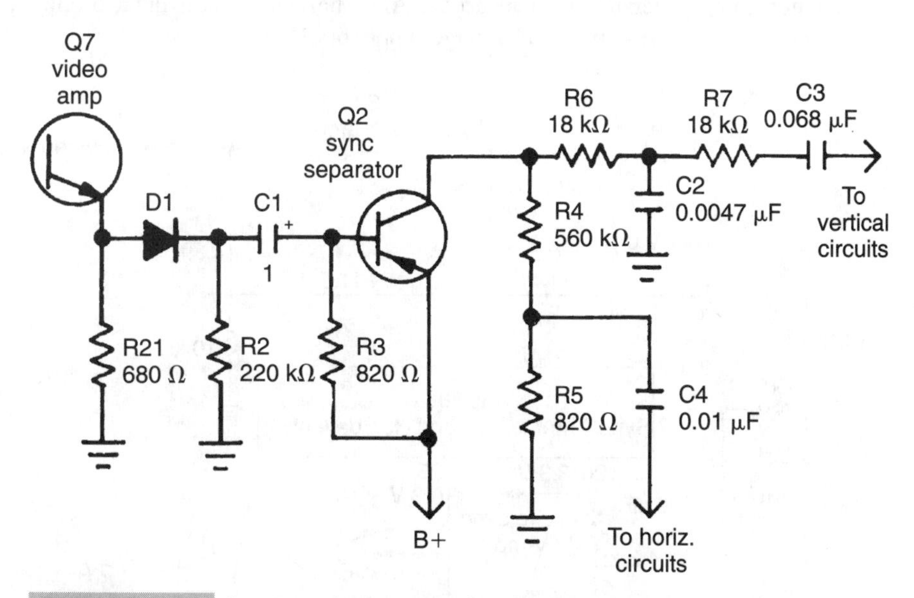

FIGURE 8-20 In a transistor stage, C3 couples to the vertical circuits and C4 couples to the horizontal circuits.

THE SYNC SEPARATOR

The transistor sync separator is a very simple circuit with the video and sync pulses fed into the base circuits (Fig. 8-20). Both vertical and sync pulses are separated from the video at the collector circuits. The vertical sync pulse goes to the vertical input circuits, but the horizontal sync feeds the AGC and horizontal circuits. Both sync circuits can be affected if trouble exists in the sync separator stage. You might find a sync amp transistor preceding the sync separator in some TV sync circuits.

A very poor signal can cause insufficient vertical and horizontal sync, including AGC problems. A defective tuner, IF, and first video amp results in weak sync signal. Scope the video waveform at the emitter terminal of the first video amp and signal trace to the base of Q101. A weak or insufficient-height waveform pulse from the separator stage can result in vertical rolling. The 60-Hz sync pulse is fed through C1 to the vertical oscillator stage. C11 couples the horizontal sync pulse to the horizontal AGC circuits.

THE LATEST IC SYNC CIRCUITS

The latest sync circuits have not changed too much with the sync taken off of the video amp (Q208) emitter terminal and fed to the sync separator stages in one large IC801. The sync signal can be checked with the scope at pin 27. This sync separator provides sync for the deflection and chroma circuits inside IC801. Pin 25 is the vertical sync separator or terminal, and the horizontal sync is provided inside of the IC. Check the waveforms and voltages at pins 27 and 25; take V_{CC} voltage measurements on pin 18 (Fig. 8-21).

IC801 provides video, contrast, pedestal clamp, brightness, R-G, Y-Y and B-Y output, color killer, tint, VCO, ACC, ground (gnd), sync separator, burst gate pulse, vertical sync separator, ramp generator, vertical driver, AFC, horizontal, horizontal oscillator, x-ray protect, horizontal pre-driver, and voltage supply pin 16 operations.

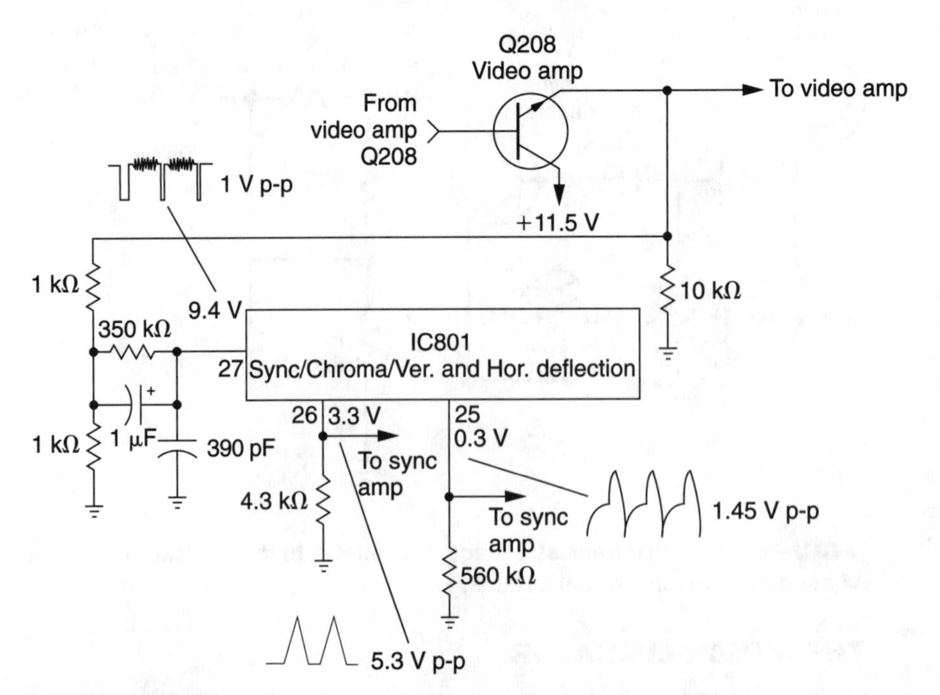

FIGURE 8-21 IC801 provides sync for both vertical and horizontal circuits.

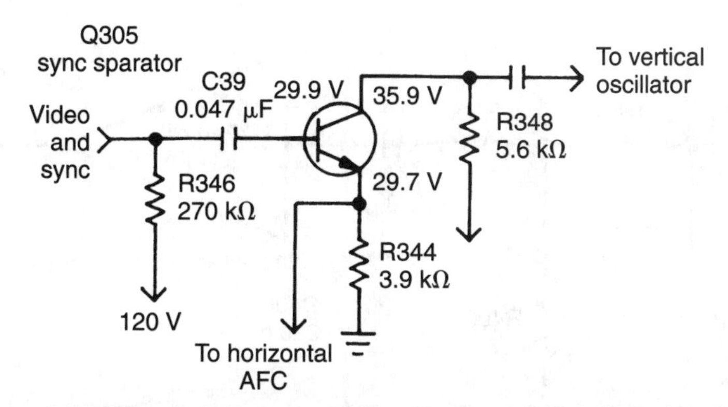

FIGURE 8-22 The picture kept rolling in a Packard Bell 98C38 chassis; the problem was burned resistors.

TROUBLESHOOTING THE SYNC SEPARATOR

Check the video composite waveform at the input of the sync separator or sync amp transistor. The TV must be connected to the antenna with a normal picture tuned in, because the sync pulses are transmitted along with the TV signal. Compare the video waveform with those listed on the schematic. Be sure that the video signal has the correct amplitude or height.

Check the waveform at the collector terminal of the sync separator. A clear-cut narrow waveform should be found here. If the height of the vertical sync pulse is low, the picture might be rolling on the screen. Horizontal and vertical jitter can be caused by a "dirty" sync pulse. Excessive hash or noise in the sync waveform can produce vertical jitters. A bending or pulling of the sync pulse can indicate that hum or 60-Hz signals are getting into the signal at the sync separator or from the vertical or horizontal circuits. If the sync pulse contains a video signal, the sync separator stage is not working properly.

Voltage measurements on the sync separator transistor can indicate that a transistor or connecting component is defective. Check the transistor in the circuit with the transistor test of the DMM. If you are not satisfied with transistor tests, remove the emitter terminal and take another test. Sometimes low emitter and bias resistors within the circuits can indicate false transistor leakage. Check each resistor with one terminal removed from the circuit.

The vertical picture would keep rolling with normal horizontal sync in a Packard Bell 98C38 chassis (Fig. 8-22). The video sync was fairly normal at the base terminal of the sync separator (Q305), with very low pips at the collector terminal. Voltage measurements were high at all transistor terminals. A careful inspection showed that the collector load resistor was burned. Replacing R348 helped some, but the picture "wanted" to roll. R344 was found to be open in the emitter circuit of the sync separator transistor. Undoubtedly Q305 had shorted and took out both resistors. In this circuit, the horizontal sync is taken from the emitter with the vertical sync at the collector terminal.

Poor video and sync were noticed in a Montgomery Ward GGY-12949A portable TV. Because both video and sync were bad, the video signal was traced through IC400 (Fig. 8-23). Pin 4 of IC400 was fairly weak. IC400 was replaced with no change. Although voltage measurements did not show anything interesting, resistance measurements did. A video peaking coil was found to be open. Replacing L408 solved the poor sync and video picture.

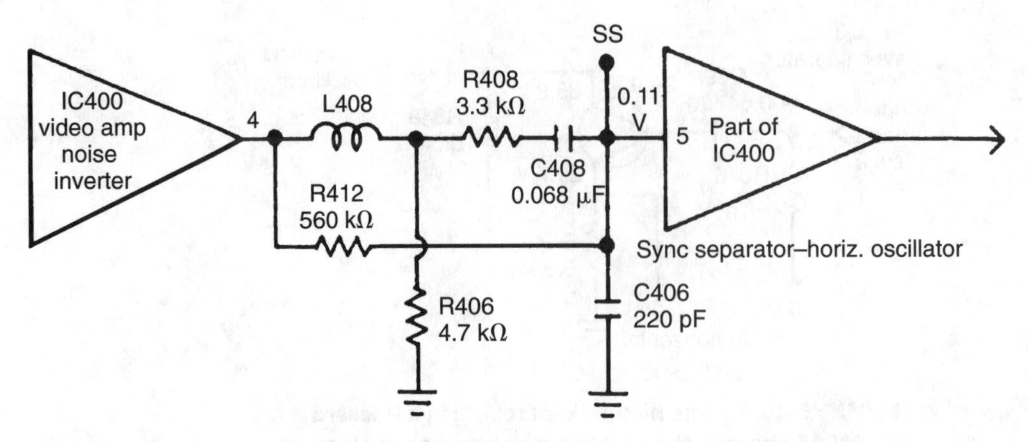

FIGURE 8-23 An open L408 caused poor video and sync in a Wards GGY-12949A portable.

SIX ACTUAL SYNC PROBLEMS

The following accounts describe six actual problems that occurred in connection with sync circuits.

Poor horizontal and vertical sync in a Panasonic PC11T30R Both the horizontal and the vertical sync were very unstable in a Panasonic PC11T30R chassis. The sync signal was normal at pin 18 of the video IF (IC101) and poor at pin 17 of the sync separator. This waveform should have a 1.4- to 1.5-V p-p sync signal. C401 (3.3 µF) was found opened (Fig. 8-24).

Poor horizontal and vertical sync in a Goldstar CMT 2612 A normal sync waveform was found at the video amp in a Goldstar CMT-2612 portable, and poor sync at pin 27 of IC301. The voltage on pin 27 was very low. A resistance measurement from pin 27 to common ground showed no signs of leakage. R302 (a 360-kΩ resistor) was found open (Fig. 8-25).

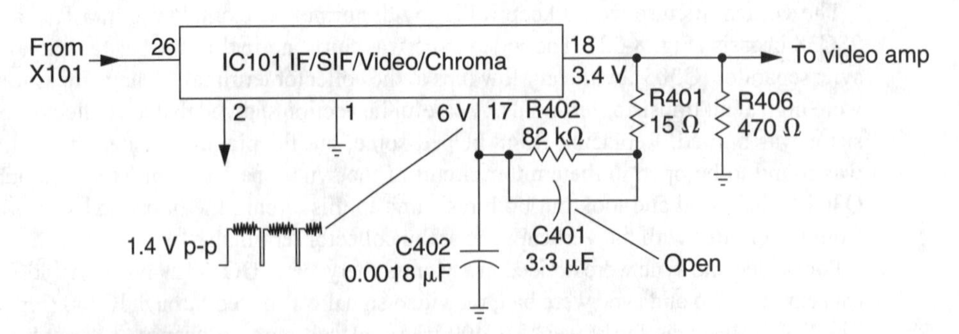

FIGURE 8-24 Poor horizontal and vertical sync were caused by an open C401 in a Panasonic PC11T30R.

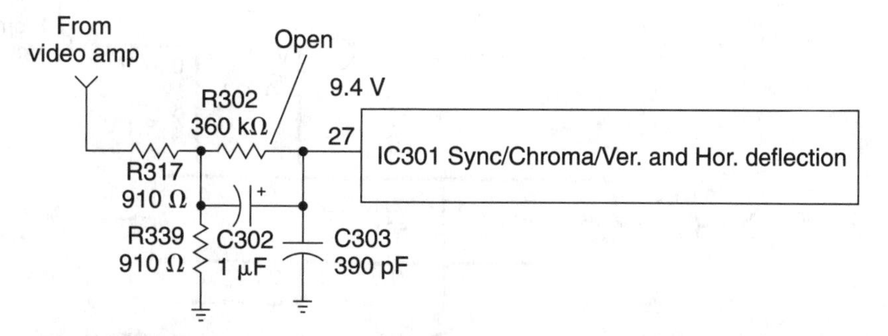

FIGURE 8-25 **Poor sync was found in a Goldstar CMT-2612 portable with open R302.**

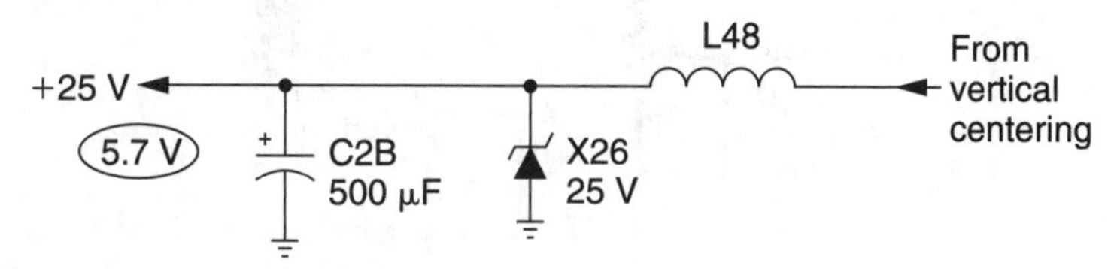

FIGURE 8-26 **Check the power voltage source for poor sync in a Zenith 4B25C19 chassis.**

Poor picture and sync in a Zenith 4B25C19 The picture in a Zenith 4B25C19 chassis was very poor with a faint outline. The picture had poor sync and color, which could have meant video trouble (Fig. 8-26). The 25-V power source measured only 5.7 V. When checking over the 25-V power-supply circuits, a leaky zener diode (X26) was located. Replacing the 25-V zener diode with a 1-W type solved the poor-picture and sync problem.

No vertical or horizontal sync in a Sharp 19D72 The picture would roll and slide sideways in a Sharp 19D72. After locating the sync circuits within IC501, voltage measurements were taken on each terminal. All voltage measurements were normal, except that higher voltages were discovered at pins 12 and 14 (Fig. 8-27). To determine if the voltage was cut down by the video amp or within IC501, pin 14 was unsoldered from the circuit. Actually, IC501 was leaky, which was causing sync problems.

The suspected IC leakage could be identified by voltage and resistance measurements. After locating the terminal with abnormal voltage, remove that terminal from the PC board. Most IC terminals are numbered or can be identified from each end (Fig. 8-28). Here, the large video AGC sync/luminance IC is numbered in several places on the wiring side of the PC board.

Poor horizontal sync in a J.C. Penney CTC97 An abnormal sync waveform at the base terminal of a J.C. Penney CTC97 chassis indicated pulling of the picture sideways (Fig. 8-29). The vertical and horizontal waveform at the sync separator were quite normal.

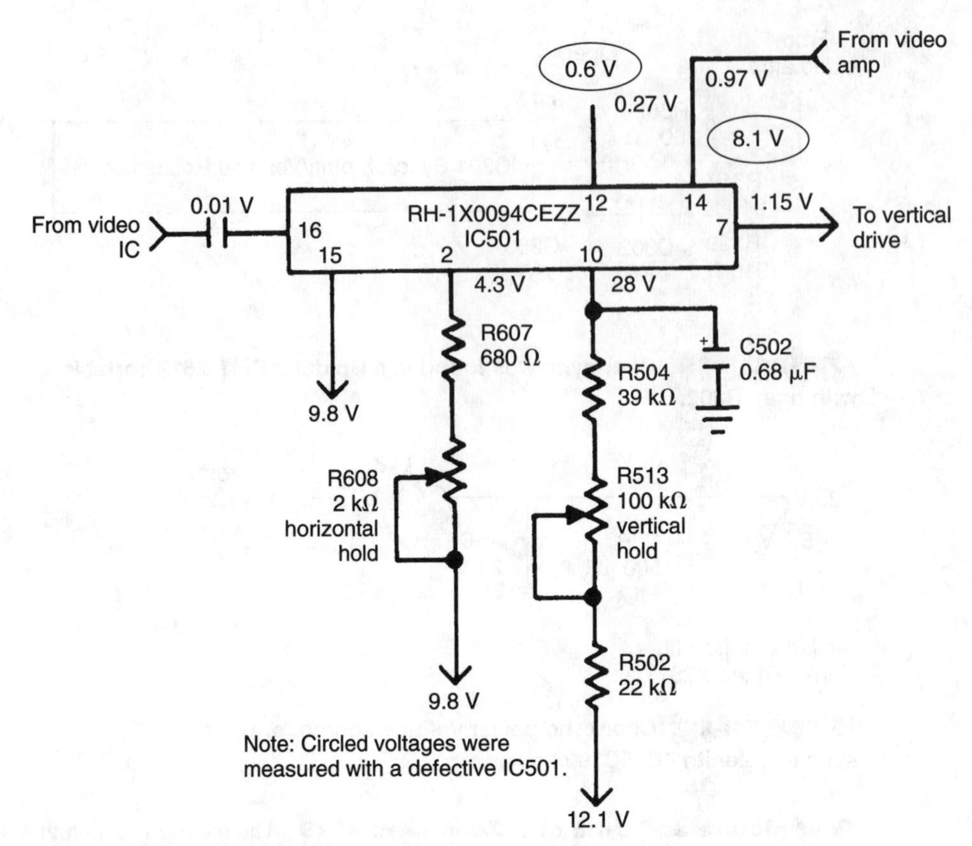

FIGURE 8-27 **Improper voltages on pins 12 and 14 indicated that IC501 was defective in a Sharp 19D72 TV.**

The keyed sync waveform from the flyback transformer was normal up to R132. The signal waveform was very low at C132. A quick resistance measurement of R132 was higher than normal. One end of R132 was removed from the circuit and the resistor was found to be open. Waveform checks with the scope and resistance measurements helped find the defective component.

No vertical or horizontal sync in a Sylvania 20B1 chassis Although the horizontal and vertical drive circuits were normal in the Sylvania B1 chassis, both pictures were moving up, down, and sideways. IC250 (TV signal processor) contains the drive pulse at output pin 24. The vertical drive pulse to the yoke at pin 21 was normal (Fig. 8-30).

The horizontal sync input waveform at pin 42 was normal and the vertical sync input at pin 44 was quite normal, but poor mixing was inside IC250. The voltages on pin 42 and 44 were off a little, but the input sync was normal. This indicated the TV signal processor (IC250) was defective. Although the sync input waveforms were fairly normal with normal sweep circuits, IC250 was replaced, solving both sync problems.

AGC delay in an RCA CTC130C The AGC delay control (R334) has been preset at the factory. Readjustment of this control is only needed if the tuner has been replaced, if

FIGURE 8-28 To check a certain pin on an IC, unsolder the pin with solder wick and an iron, then measure the pin to ground for leakage.

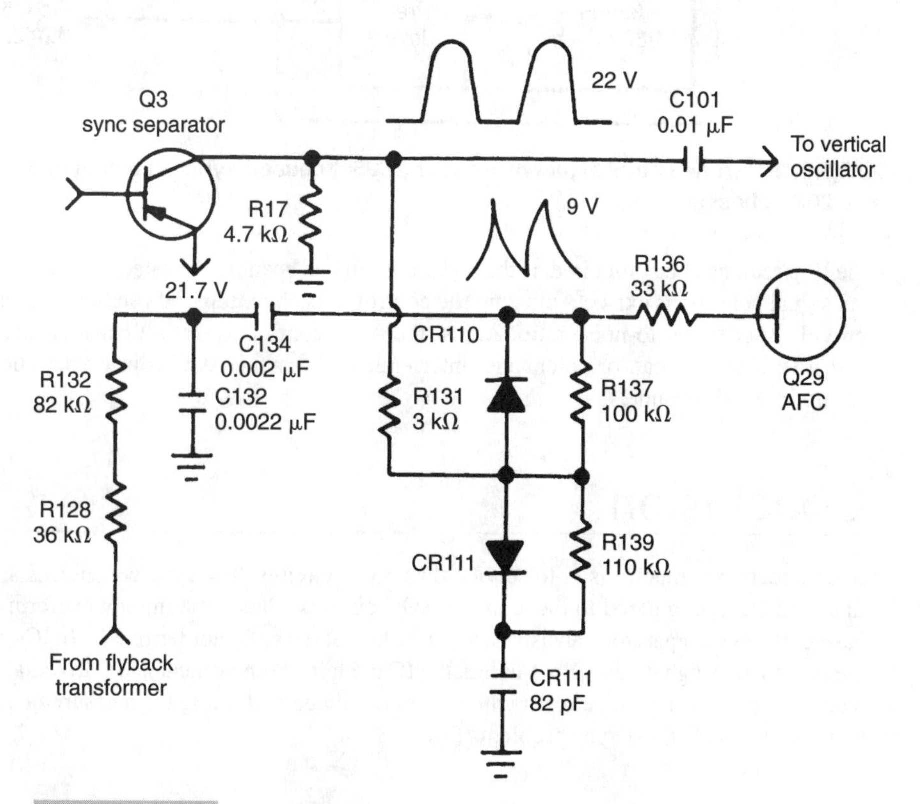

FIGURE 8-29 R132 was found open in a J.C. Penney CT97 chassis, producing poor horizontal sync.

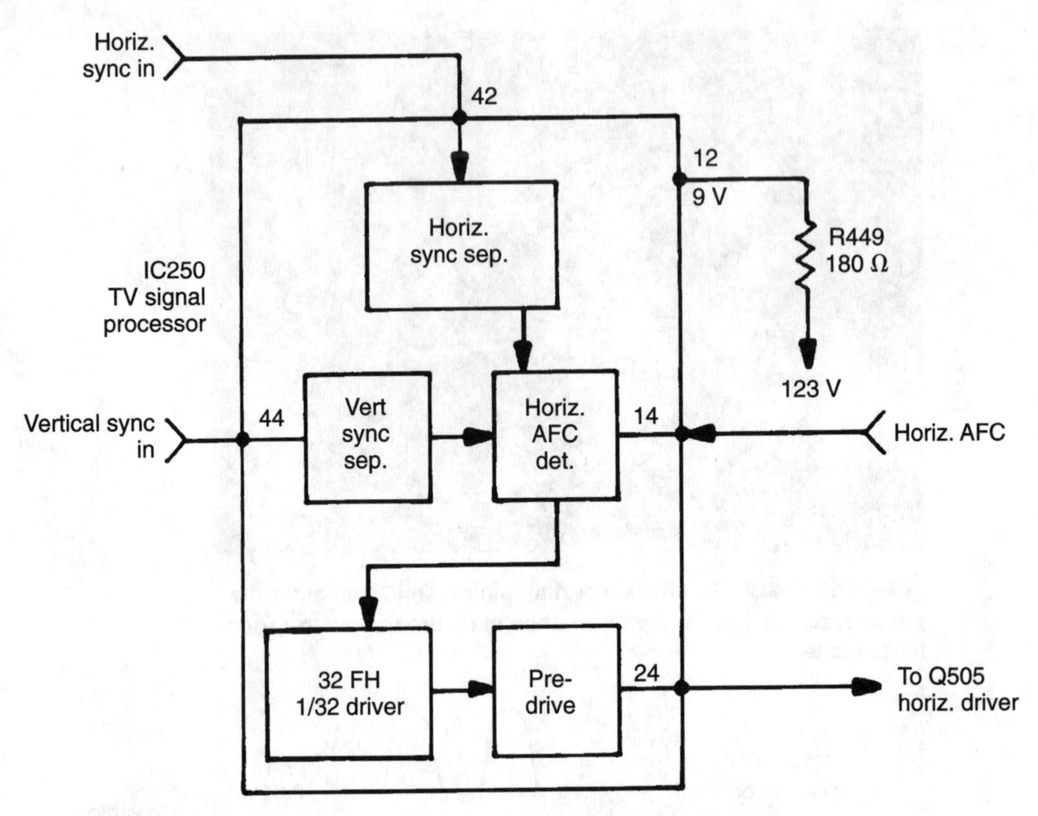

FIGURE 8-30 **A defective signal processor (IC250) caused sync problems in a Sylvania 20B1 chassis.**

the IF circuit has been repaired, if the AGC control has been misadjusted, or if unusual local signal conditions exists. Adjusting the control to each extreme of rotation will usually provide poor signal-to-noise ratio. Adjust the AGC control so that all channels are good and free of color, beat, or co-channel interference. After the AGC adjustment, check all strong local TV channels.

Conclusion

Sync circuits are much easier to service than AGC circuits. The sync waveforms are distinct and clear, compared to those in the AGC circuits. Check the input waveform at the base of the sync separator transistor and the output at the collector terminals. In ICs, check the input video signal and output unless the IC contains both vertical and horizontal oscillator circuits. Correct waveforms with accurate voltage and transistor measurements will help you to locate most sync problems (Table 8-1).

TABLE 8-1 FOR EASY TROUBLESHOOTING, CHECK THIS TROUBLESHOOTING CHART FOR AGC AND SYNC CIRCUITS

WHAT TO CHECK	HOW TO CHECK IT
White raster or snowy picture.	Sub tuner with tuner-subber.
Determine if tuner or AGC.	Measure AGC voltage to tuner and IF stages.
Check AGC circuits.	Replace AGC IC.
Poor horizontal and vertical sync.	Check sync separator waveforms on IC.
Check sync waveform from video circuits.	Check video circuits.
Replace sync IC with normal voltages.	Make sure sync input is normal.

9

TUNER REPAIRS

CONTENTS AT A GLANCE

About the Tuner

Most TV technicians admit that the tuner, IF, and AGC circuits are the most difficult stages to test and repair in the whole TV chassis. The tuner can be the most confusing, because parts are difficult to get to. New varactor tuners with control boards have many different ICs, synthesizers, and processor components. Servicing the TV tuner can be made easy by taking each unit one step at a time.

New names have been added to the tuner assembly since the old mechanical wafer tuner was first invented (Fig. 9-1). The varactor tuner can be controlled with a tuning system, control processor, initializer, presetter board, tuner control, memory, and frequency-synthesizer control units (Fig. 9-2). A tuner-control assembly can have soft-touch, keyboard, pushbutton, selector-board, auto-board, and remote-control tuning. They all do the same thing by controlling the tuner assembly, which, in turn, selects the correct channel (Fig. 9-3).

Tuner repair can be made easy with isolation and signal-injection techniques. A defective tuner can produce a snowy, weak, or erratic picture. The defective tuner-control unit can cause a dead, intermittent, or drifting varactor tuner with an improper tuning voltage. Each unit can be isolated with voltage-measurement and signal-injection methods with a tuner-subber or tuner replacement.

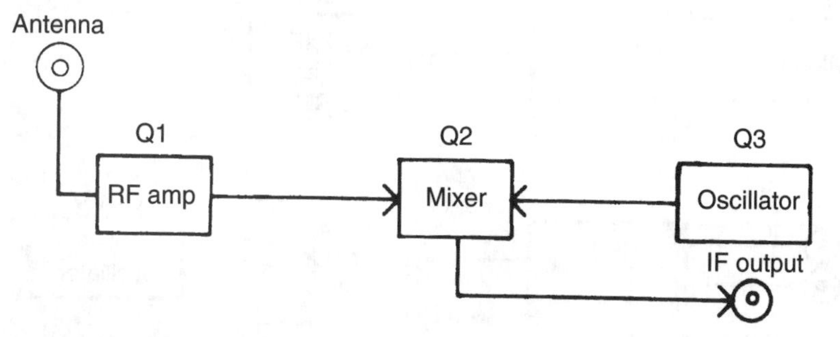

FIGURE 9-1 A typical block diagram of a mechanical VHF tuner.

FIGURE 9-2 An RCA tuner and memory module are mounted together in several different chassis.

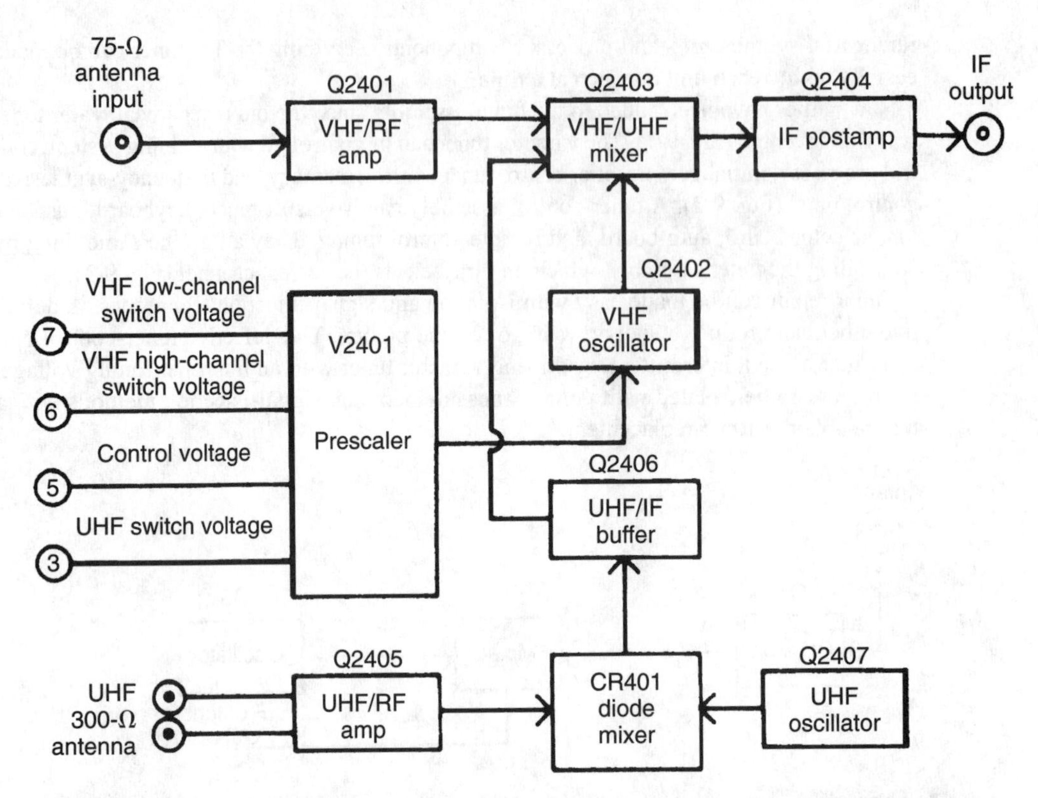

FIGURE 9-3 A block diagram of the RCA CTC120 tuner-control module (MST005A).

NO PICTURE, NO SOUND, WHITE RASTER

A defective tuner, IF section, AGC circuit, or shorted IF cable assembly can cause a white raster without picture or sound (Fig. 9-4). Because no snow is in the picture, check the mixer section and B+ voltage at the tuner terminals. An IF squeal or a birdie sound can indicate that the IF section is defective. The all-white screen can be caused by an improper

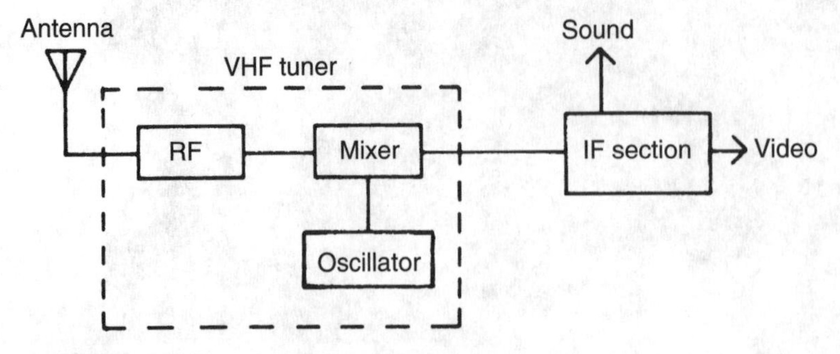

FIGURE 9-4 A defective tuner and IF section can cause a snowy picture, no picture and no sound, or only a white screen.

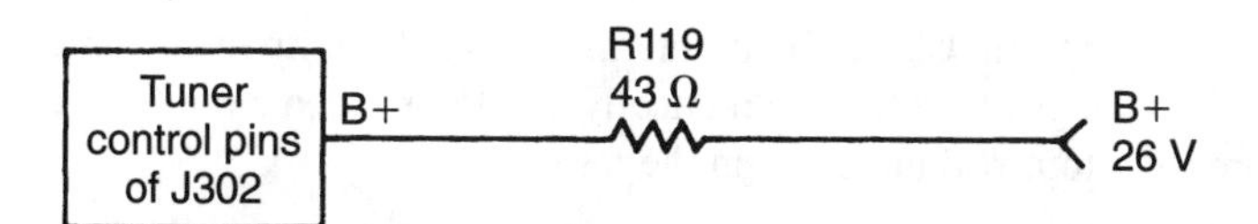

FIGURE 9-5 **Absent picture and sound was caused by a lack of B+ voltage at J302 in this RCA CTC120 chassis.**

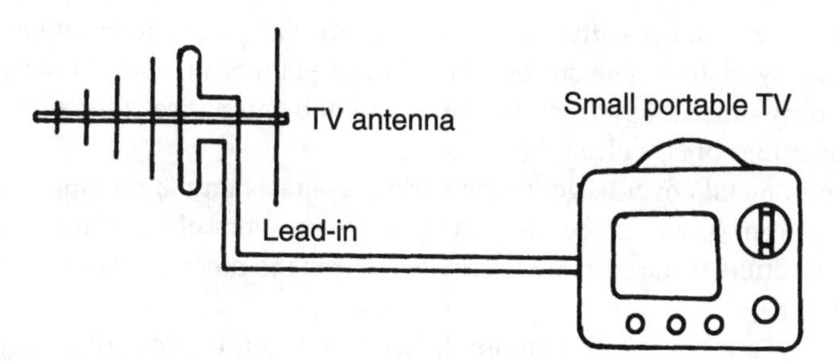

FIGURE 9-6 **Check the antenna or cable with a portable TV.**

AGC voltage applied to the VHF tuner terminals. A low-resistance continuity check of the IF cable between the tuner and chassis can help you to locate a leaky IF cable assembly.

In an RCA CTC120 chassis (Fig. 9-5), the symptoms were no picture, no sound, and normal high voltage. A B+ voltage measurement at the tuner indicated that no voltage was at the tuner control. R119 was open in the 26-V source, which supplies B+ voltage to the tuner-control unit. Replacing the burned resistor restored the picture and sound.

SNOWY PICTURE

The snowy raster can have a very faint picture with poor sound, or just a plain snowy raster. A snowy picture can result from improper signal at the antenna terminals, a defective tuner, improper AGC voltage, or a defective IF system. The poor antenna or cable system can be checked with another portable TV (Fig. 9-6). A damaged balun circuit in the front end of the tuner can cause a snowy picture. Often, the oscillator and mixer stages of a solid-state tuner are normal, with snow in the picture indicating an RF or FET transistor is leaky or open. A dirty tuner can produce a snowy picture. Check the AGC voltages at the tuner for excessively high or low AGC voltage. Leaky or open first and second IF stages can cause snow in the raster.

INTERMITTENT PICTURE

The intermittent picture can result from a broken antenna lead-in or a bad connection right at the TV. A dirty tuner can cause the picture and sound to drop out. Slightly move the tuner knob and see if the picture returns. Often, the intermittent picture with snow is caused by a problem in the front end of the TV. An intermittent transistor within the tuner can produce the same symptom. Check for an intermittent AGC component, which would cause improper AGC voltage at the tuner. Practically any IF or video circuit can cause the intermittent picture symptom without snow in the raster.

Cleaning the Tuner

A dirty mechanical tuner can cause the station to drop off-channel, or intermittently produce a snowy and noisy picture. The customer might complain of no color in the picture, which relates to a dirty tuner. Slightly move the tuner knob and notice if the picture acts up. Remove the tuner to properly clean it (Fig. 9-7).

Remove the bottom metal cover to get to the various contacts on the rotating wafers or drum area. The old drum tuner can be cleaned by applying tuner cleaner to a cloth and holding it against the drum as the assembly is rotated. Clean the tuner contact springs with a silicon spray cleaner.

The wafer switch-type tuner is a little more difficult to clean. It does not matter if the switch contacts are made of gold or silver; they still corrode. Some TV tuners have excess grease applied to the contacts when they're manufactured. The grease collects dirt and dust. Dislodge excess grease with a tuner wash spray. Select a silicone contact spray that will not destroy or damage plastic components within the tuner.

A good tuner cleanup consists of spraying the tuner, a tuner washout, and cleaning each individual contact. Some technicians use a pencil eraser, the end of a pocket knife, or a small, stiff wire brush on each contact. Always remember to replace the metal cover to prevent stray RF signals from entering the tuner or upsetting the tuner alignment. Send the

FIGURE 9-7 Clean the mechanical tuner if it drifts off frequency. Use tuner wash and spray.

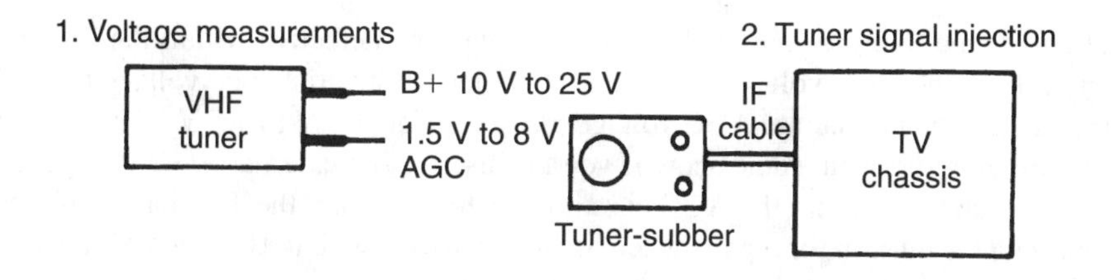

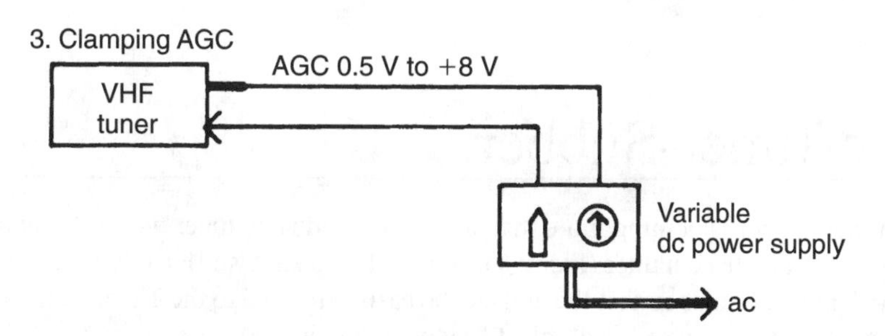

FIGURE 9-8 **Perform three different tuner tests with voltage, using an external voltage source and a tuner-subber.**

tuner into a tuner repair depot for a supersonic wash if it will not clean up or if it appears to have excessively worn switch contacts.

Three Quick Tuner Tests

Any tuner can be checked with voltage measurements on the tuner terminals, signal injection with a tuner-subber, or tuner substitution and correct AGC voltage (Fig. 9-8). Check for B+ voltage at the tuner to chassis ground. This voltage can vary from 10 V to 50 V on most solid-state tuners. The B+ voltage on varactor tuners can vary from 10 V to 30 V. Besides a B+ operating voltage, all varactor tuners have a dc tuning voltage that lies between 1.5 V and 30 V. This voltage will change each time that the channel is switched. The AGC voltage can vary between 1.5 V and 8 V on a solid-state tuner. Remember that this AGC voltage is always positive, but a negative AGC voltage is found on the tube-type tuner.

Signal injection with a TV signal generator or tuner-subber at the IF cable will quickly help you to determine if the tuner is defective or if the problem is in the TV chassis. After a voltage measurement, plug the IF cable into the tuner-subber. If there is still no picture and sound, you can assume the tuner is normal and that the signal problem lies within the TV chassis. The tuner should be repaired or sent in for repair if the picture and sound returns with the subber connected. Simply substitute another tuner module in the modular chassis to determine if the tuner is defective.

Clamping the AGC terminal at the tuner with the proper dc voltage can determine if the AGC circuits are normal. Remember that the tuner-subber does not have the AGC voltage connected to it when it is subbed into the circuit. The tuner-subber can show normal picture and sound, but the AGC voltage source might be defective within the TV chassis. Doublecheck the AGC voltage at the tuner by injecting a variable dc voltage at the AGC terminal or with an accurate AGC voltage measurement (1.5 V to 7.5 V). The picture and sound might return with some snow if you try inserting an external AGC voltage. In this case, you can assume that the AGC circuit is defective within the TV chassis. If not, you might send the tuner in for repairs when nothing is wrong with it. Besides additional tuner-repair cost, you still have the very same signal problem.

The Tuner-Subber

The tuner-subber is nothing more than a regular solid-state tuner with self-contained batteries in a separate container (Fig. 9-9). Most subbers have an IF cable jack or plug-in cable with alligator clips so that the unit can be easily attached to the TV chassis. An RF gain control should be adjusted to eliminate snow in the picture.

To operate it, simply remove the IF cable from the TV tuner and plug it into the jack at the back of the tuner-subber. Attach the external antenna wire or cable to the antenna input terminals of the subber. Turn the unit on with the RF gain control fully clockwise. Tune in a station with the tuning selector and fine tune it for the best color picture. The tuner-subber will eliminate a lot of guesswork—especially when servicing controlled varactor tuners.

You can make your own tuner-subber with a solid-state tuner out of an old TV (See *Build Your Own Test Equipment*, published by McGraw-Hill).

FIGURE 9-9 **Inject a signal at the IF cable from the tuner-subber to determine if the tuner or the first IF stage are defective.**

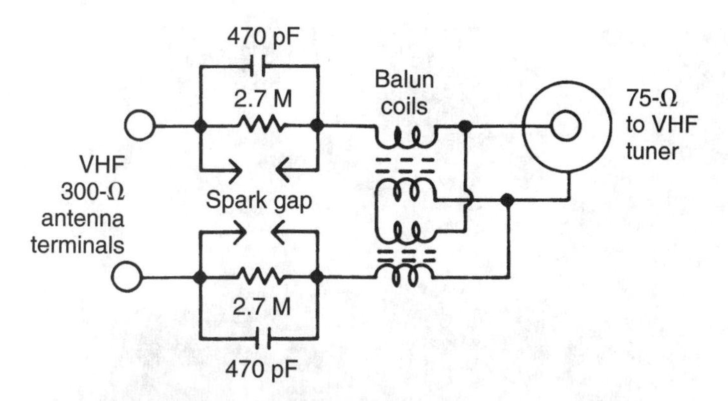

FIGURE 9-10 **Broken wires at the antenna terminal or lightning damage can cause a snowy picture.**

Defective Antenna Balun Coils

Although blocking capacitors are placed in each leg of a 300-Ω balanced antenna, a direct lightning strike can damage the entire antenna assembly. You might find small capacitors blown apart, with small wires burned off the antenna connectors (Fig. 9-10). The picture will be very snowy.

The antenna coils must be repaired or replaced. Often, the small wire ends can be unwound a turn to get enough wire to solder back to the various components. Order a replacement for the entire antenna assembly if the unit is damaged beyond repair.

An RF balun transformer that changes the 300-Ω antenna input terminals to 75 Ω can be used; just connect the balun backwards. These balun transformers are used to connect the TV receiver to a 75-Ω cable system or VCR. Sometimes these antenna-balun assemblies can be difficult to obtain.

Types of Tuners

THE CONVENTIONAL MECHANICAL TUNER

The mechanical tuner is still found in most black-and-white and some color portable TV receivers. A wafer-type switching mechanism rotates the various coils into each RF, mixer, and oscillator circuit (Fig. 9-11). The VHF solid-state tuner might have an FET in the RF stage. Usually, the mixer and oscillator transistors are located inside, beneath the wafer assemblies.

The snowy picture can be caused by a defective FET or RF transistor. In the earlier solid-state tuners, these transistors were mounted on top and were easily replaced. The FET must be handled with extreme care to prevent damage to the transistor. All tuner transistors should be replaced with a correct universal replacement or with parts that bear the original part number.

FIGURE 9-11 **A typical tuner consists of transistors in the oscillator and mixer stages.**

The absence of picture or sound in the TV receiver is the sign that an oscillator is leaky. No sound and no picture with a white raster can result from a leaky or open mixer transistor. Replacing transistors inside the tuner can pose a problem because components are tightly packed. It's difficult to get side cutters or a soldering iron tip down under the various wafer sections of the tuner.

A poorly soldered connection that causes intermittent picture and sound can be located with an insulated tool. Simply tune in a station and probe around with the plastic tool until the picture acts up. Often, you can locate a poor solder connection or a broken component lead. Suspect that an oscillator coil lead is broken or that a switch contact is dirty if only one station cannot be tuned in. Inspect the soldered coil leads where they are brought together at the front of the tuner for poorly soldered connections. The tuner assembly should be sent in for service if cleaning and simple repairs fail to fix the tuner (Fig. 9-12).

UHF TUNER

In many color TV chassis, the UHF tuner is a separate mechanical tuner, except in those with a combined VHF-UHF varactor assembly. Today, the UHF tuner can be a 70-channel, mechanically tuned type. When the VHF channel has been set for UHF operation, the B+ voltage is switched from the VHF tuner to the UHF tuner. UHF-VHF switching can be found at the rear of or inside of the VHF tuner.

Transmitted signals received by the UHF antenna are fed to a balanced 300-Ω input and coupled to L50 in a typical UHF tuner assembly (Fig. 9-13). The UHF oscillator transistor (Q50) generates an oscillator signal that is 45.75 MHz above the picture carrier frequency of the broadcast station. The incoming signal and the oscillator frequency are mixed by

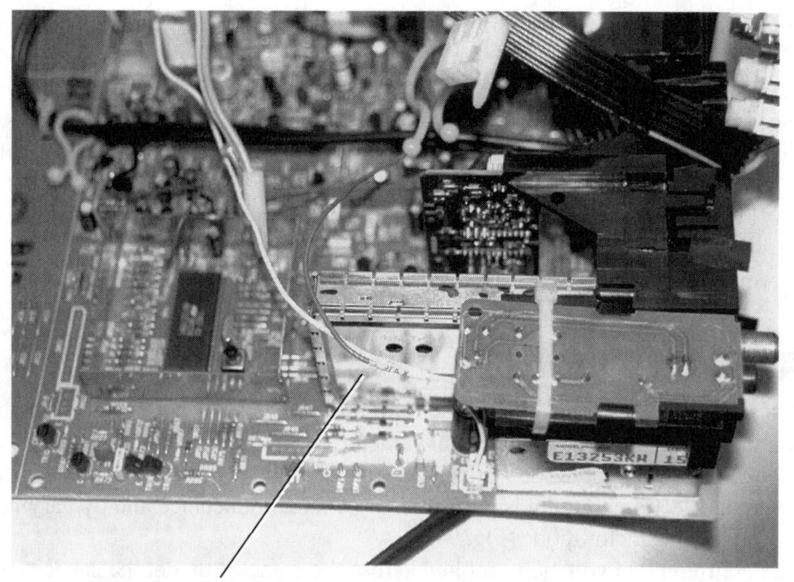

FIGURE 9-12 Today, the varactor tuner mounts directly on PC board.

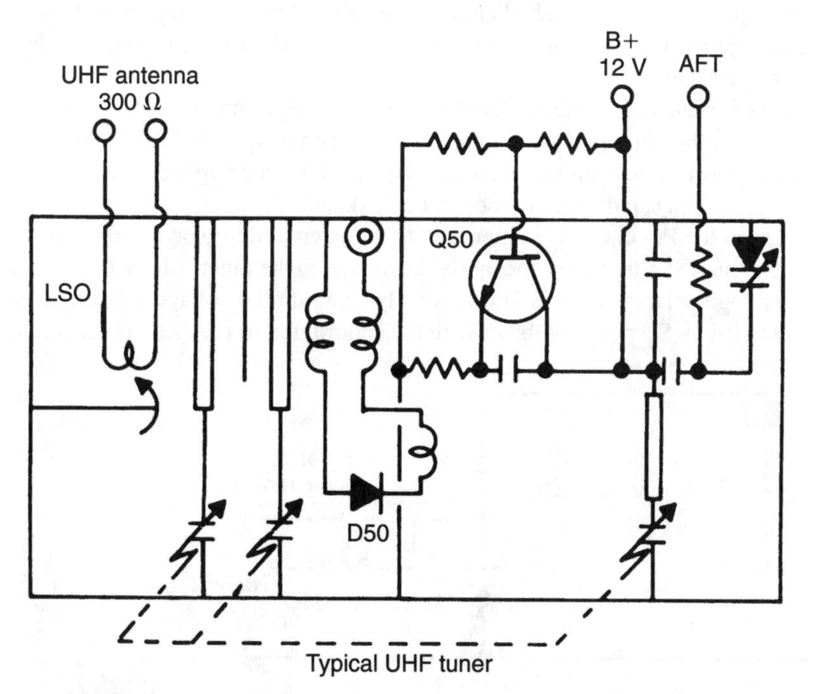

FIGURE 9-13 A typical UHF mechanical tuner schematic.

diode D50, which produces a third frequency, resulting in the IF signal. This IF signal is fed to the VHF tuner with a 75-Ω shielded cable.

Most problems in the mechanical UHF tuner are in Q50 or D50. No UHF station will be received if either the transistor or diode become leaky. Q50 should be replaced with the original-type part. UHF diode D50 can be replaced with a IN82 UHF diode. Do not replace this diode with an ordinary IN34 type. Check the rubbing of tuning capacitor plates if one station is erratic while tuning. Low voltage at the UHF tuner terminal can indicate that a transistor (Q50) is leaky or that the switching voltage is improper. Flex the UHF IF cable for intermittent or drifting UHF signals.

THE VARACTOR TUNER

Most of the present-day tuners use some form of the varactor tuner within a tuner module. The varactor tuner might contain several varactor diodes that change the frequency of each TV channel. The varactor diode works like a capacitor when a reverse bias is applied to the PN junction. Electronic tuning is performed by a fixed inductance and by varying the capacitance with a varactor diode (Fig. 9-14).

When a different dc voltage is applied across the varactor diode, the capacity changes, tuning the inductance circuit. With a different variable resistance on each fixed inductance, each channel can be tuned manually or electronically (Fig. 9-15). You can manually turn each variable resistor with a screwdriver or fingernail and tune in each channel.

Because the varactor tuner cannot cover all VHF channels, the entire band is divided into low and high bands. The low band covers channels 2 through 6 and the high band covers channels 7 through 13. A switching diode is used to switch from high to low in an electronic tuner system.

Any varactor tuner can be tested by checking the voltage applied to the tuner (Fig. 9-16). Remember, as the tuner is tuned manually or electronically to a different station, the applied voltage is different on each channel. By checking the applied voltage (variable for each channel), you can determine if the tuner is defective.

Another method to check the tuner is with an external dc voltage source. Remove the wire to the tuning voltage post and apply dc voltage to the tuner. Check the schematic for correct voltage polarity. Start at 1.5 V. Slowly bring up the voltage and notice what stations are tuned in. Suspect a problem with the control unit or the AGC if various local sta-

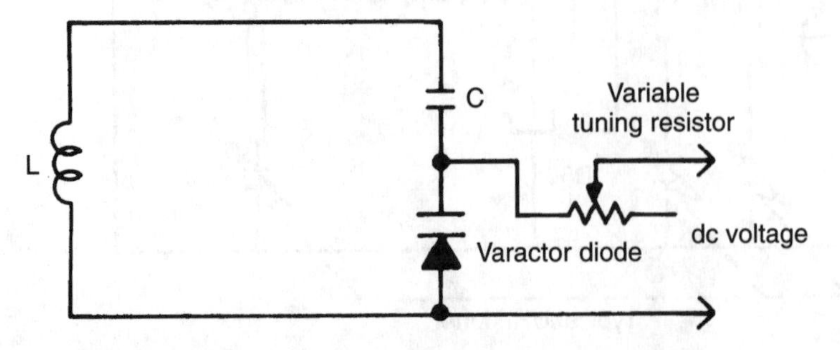

FIGURE 9-14 **Electronic tuning with the varactor tuner is performed by a fixed coil inductor and a varactor diode.**

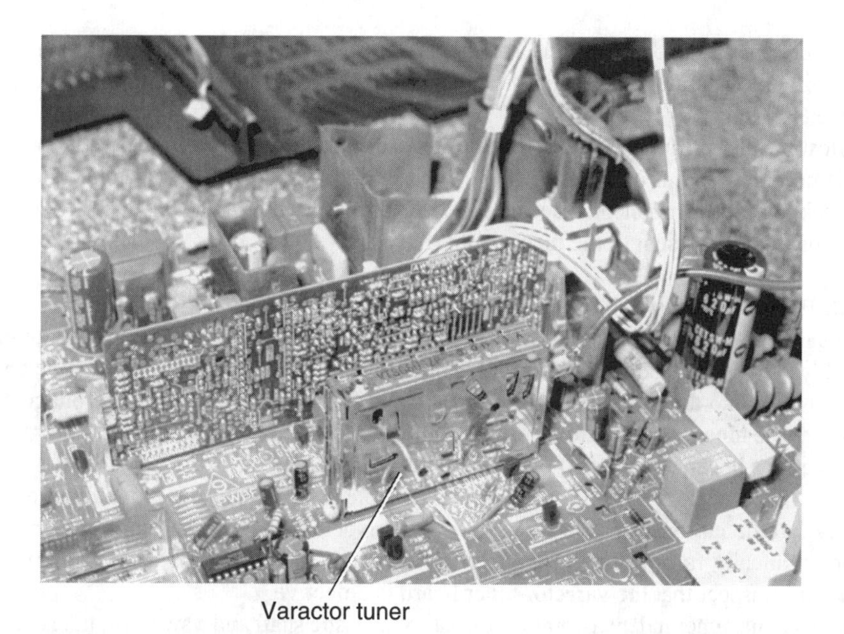

FIGURE 9-15 **Stations can be tuned in with the remote or with separate buttons on the RCA CTC146.**

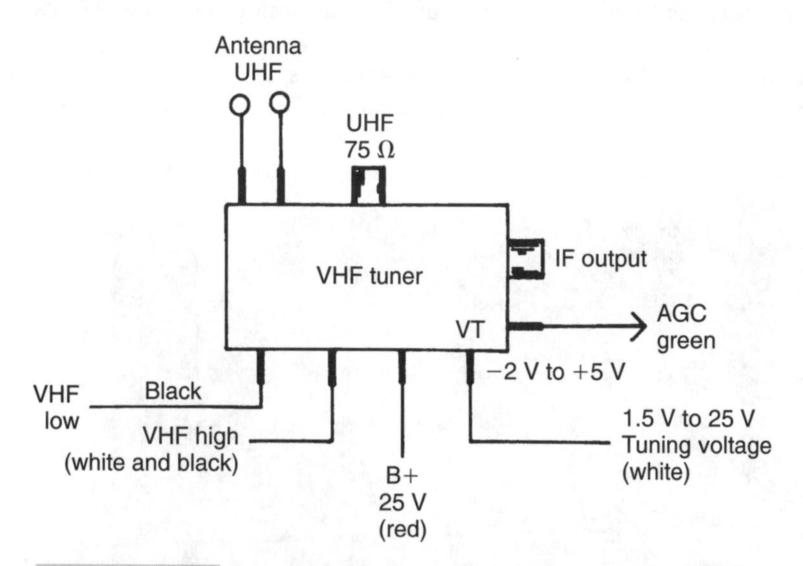

FIGURE 9-16 **Check the varactor tuner with the correct supply voltage, AGC voltage, and tuning voltage.**

tions can be tuned in with the external voltage applied. A missing channel can be found in the same manner. Compare the station program with that on another TV.

The UHF tuner can be a separate tuner or it can be combined with the VHF varactor tuner. Six or more blank channels are provided for tunable UHF stations with the combined varactor tuner. The UHF stations are tuned in with the same type of variable resistor. Often, with the separate UHF tuner, the voltage is switched from the VHF tuner and applied to the UHF tuner.

Most problems found in the varactor tuner are caused by transistor and varactor diodes. Intermittent problems or tuner drifting can be caused by poorly soldered connections or by a breakdown of components. Because these components are inside and are difficult to get to, most tuners are sent to the tuner depot for repair.

J.C. Penney manual varactor tuner In a J.C. Penney 685-2026 (made by RCA), a manual-type varactor tuner is found with a varactor tuner board and a separate tuning assembly. The tuner module can be easily changed if you suspect that it is defective because cables plug into the tuner board assembly. A check of the varactor tuning voltage can be made at the tuner assembly to determine if the varactor board is defective.

If stations began to drift or if one station cannot be tuned in, suspect a defective varactor tuning board (Fig. 9-17). Sometimes the variable resistor control is malfunctioning so badly that the station drops out when the screwdriver is taken from the control. Sometimes, when manually turning from channel to channel, the station drops out or intermittently tunes in. Suspect that the varactor tuner board is defective.

The manual tuner indent assembly consists of a plastic shaft and a switch to tune in the various stations. Erratic or intermittent reception can be caused by a dirty or defective switch. It's best to change the variable tuning board assembly. This assembly comes in two different sections; it can be ordered directly from your RCA parts distributor or from J.C. Penney Co.

Varactor tuner drifting In many of the latest TVs, varactor tuners are used in both the VHF and UHF tuners. Station drifting of either the high or low band of the VHF tuner

FIGURE 9-17 Remove the varactor tuner from the PC board by unsoldering the pins from the PC board.

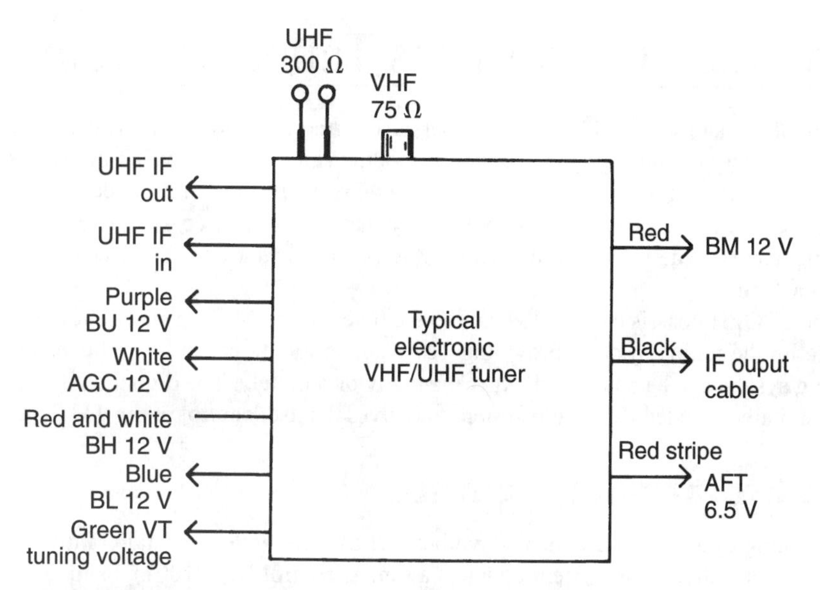

FIGURE 9-18 **Check all voltages on the varactor tuner before removing it and sending it in for repair.**

can be caused by a defective UHF tuner. A leaky capacitance diode or component within the UHF tuner can lower the tuning voltage applied to the VHF tuner. Often, this condition exists after the TV chassis has operated for several hours.

Tune in the station that drifts or fades out and leave the TV operating. If the station disappears, cut the B+ wire that feeds the UHF tuner. If the station pops back in, you can assume that the UHF tuner is defective, not the VHF tuner. Let the chassis operate for several hours; be sure that the VHF stations do not fade or drift off channel.

Mark down all color-coded leads or wires to the UHF tuner on a piece of paper (Fig. 9-18). Tape the paper to the TV chassis or cabinet for replacing the tuner after factory repair. Remove the UHF tuner and send it to the tuner repair depot. Besides the UHF tuner, a defective VHF tuner can cause stations to drift off channel.

RCA CTC145E tuner drifting In a new CTC145E chassis, the higher channels would operate for several hours, then drift off. The tuning voltage that was applied to the tuner was fairly normal. All voltages were normal at the other tuner terminals. A new tuner was ordered, solving the drifting of channels.

RCA CTC177 board-tuner With intermittent solder joints in the RCA CTC177 board-tuner, the picture would "flop," go out, become snowy, and collapse. This tuner is built on the PC board chassis. Check the shield solder joints and tuner ground connections.

Remove the bottom cover of the tuner. Solder all shields to ground. Renew the old solder joints with new solder. Do not apply too much heat or you will raise the foil PC wiring from the board. Solder all joints that lead to ground or shield them within the tuner circuits. Return the pads and be sure that the tabs in both the shields and the cover are clean and can accept solder. Reinstall the tuner shield and tuner cover. Solder all tabs with fresh solder.

Frequency-Synthesis Tuner Servicing

Many of the latest RCA TV module tuners use a frequency-synthesis system. Replacing the MST tuner module or MSC frequency module can determine which module is defective (Fig. 9-19). Most technicians substitute a new module and turn the defective module in for repair. Do not overlook the possibility that an IF cable or module connecting the wiring harness is defective. Some manufacturers do not want you to service modules, only replace them.

The CTC131 chassis has an MST multiband tuner and an MSC tuner-control module that make up the tuning system, providing 127-channel tuning capability. This channel-lock tuning system can be controlled either manually or remotely. The digital command center is a digitally encoded IR remote system that gives the user control of the TV (Fig. 9-20).

RCA CTC157 TUNER CONTROL

The tuning system is a frequency-synthesis (FS) type with a crystal-controlled phase-locked loop. The tuning system contains a tuning-control IC (U1001), band decoder and switch (U3600), and varactor diode tuning (Fig. 9-21). The tuner uses a number of varactor diodes in the local oscillator circuits. These diodes act like a variable capacitor when a different voltage is applied from the tuning-control IC.

Both the bandswitching and tuning voltage control circuits must operate correctly for proper tuning voltage to be produced. In order for the tuning-control circuit to synthesize the correct tuning control voltage, all signals used by the tuning-control circuit must be correct.

FIGURE 9-19 **The tuner and system-control modules in a CTC108, 109, and 110 chassis.**

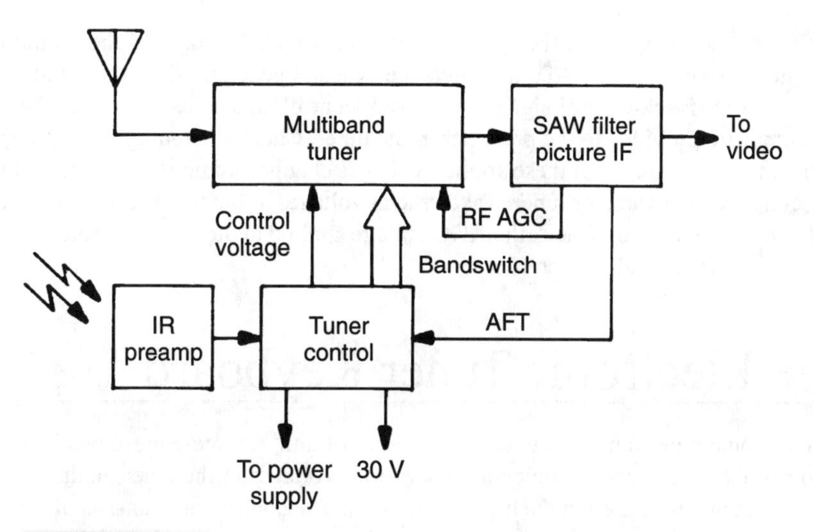

FIGURE 9-20 A block diagram of the RCA CTC131 tuner system.

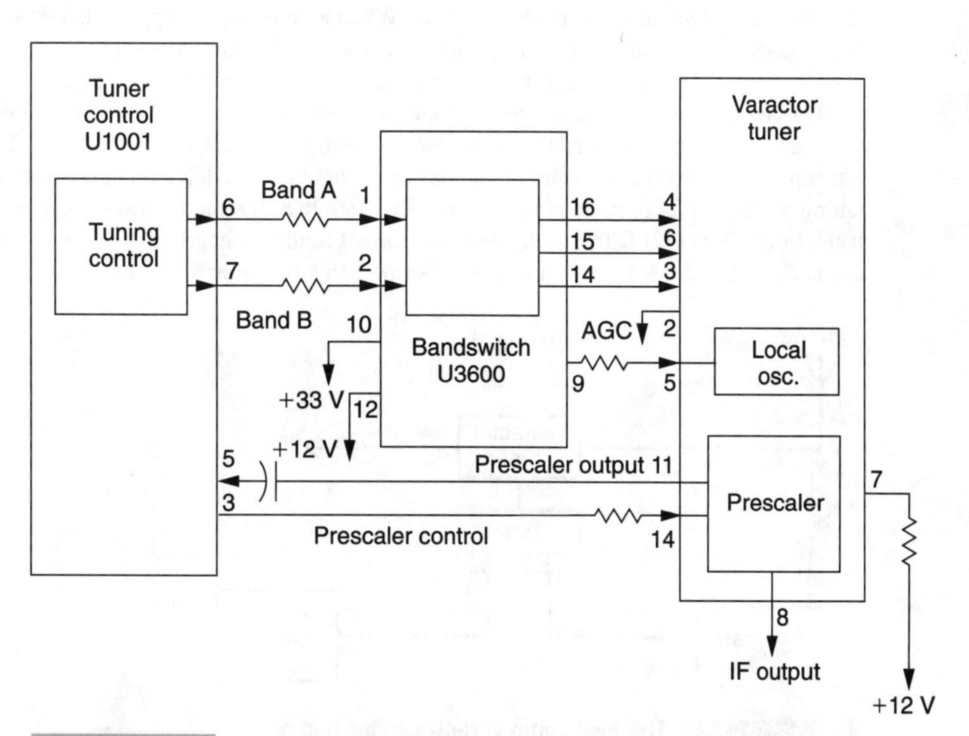

FIGURE 9-21 RCA CTC157 frequency-synthesis tuning-control circuits.

If the channel is changed, the system-control computer IC supplies the tuning control data signal to the AIV. The AIV interprets the data and generates the exact bandswitching signals and feedback-control signals. Feedback-control signals used within AIV and the PSC signal are provided to the prescaler in the tuner. Check the tuning voltage applied to the varactor tuner (pin 5). If these applied voltages change and no stations are still visible, suspect a defective varactor tuner. Take crucial voltages at the tuner, U3600 band switch, and U1001 tuner control. Check the AGC voltage applied to the tuner. Be sure that all supply voltages are normal in each IC.

The Electronic Tuner Keyboard

The electronic tuner functions with a tuner, control unit, remote-control receiver, or keyboard assembly. The control unit supplies various voltages to the tuner module or assembly. No moving parts used in the tuner (Fig. 9-22). The control unit consists of several ICs and diodes and a frequency synthesizer with microprocessor-control modules.

In the keyboard chassis, simply press the correct numbers to select the right station. With the RCA chassis, to select any station below 10, push zero and the station number. Above 10, select the correct station number. In Admiral and many Japanese keyboards, push the station numbers and push the select button, then selected station pops in.

Actually, the keyboard selects the station tied to the control unit and the control unit selects the correct voltage applied to the tuner. When a dc voltage is applied across the varactor diode, the capacitance of the diodes changes the inductance-tuned circuit to the desired station. The control unit operates the same in a remote-control receiver.

The tuner selects the channel with voltage from the control unit. Check the tuning voltage if a wrong station is tuned in, if it refuses to change channels, or if it drifts off channel. Improper or erratic voltage indicates that a control assembly is defective. Look for various tuning voltages found in the manufacturer's service literature and Howard Sam's Photofacts. Each channel is listed, along with the correct tuning voltage (Fig. 9-23). If the voltage is correct and there is no tuner action, suspect that the tuner is defective.

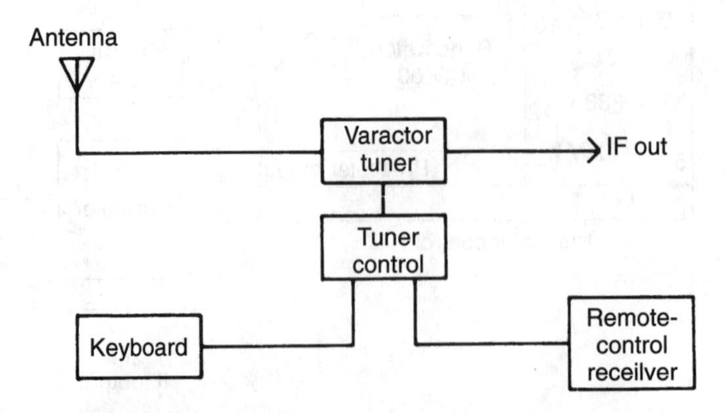

FIGURE 9-22 **The electronic varactor tuner has no moving parts; it is controlled only by a change in the tuning voltage.**

Terminal Voltage	VHF tuner	UHF tuner
B+ voltage	15 V	15 V
Tuning voltage	1.5 V to 25 V	1.5 V to 30 V
AFT voltage	0.2 V to 15 V	0.2 V to 15 V
AGC voltage	−2 V to 5 V	1.5 V to 5 V

FIGURE 9-23 Typical tuner voltages.

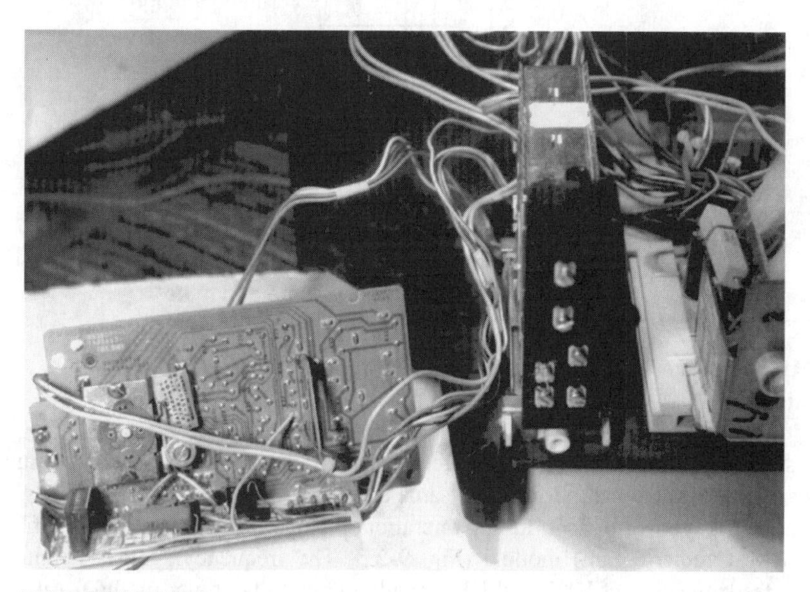

FIGURE 9-24 A pushbutton selector board for VHF and UHF tuners.

Another form of keyboard tuner control is used in Japanese and other foreign TVs. The keyboard can be mounted directly on the selector board or if there is a separate keyboard (push-button) assembly. Each channel has a pushbutton to locate the exact station (Fig. 9-24).

The keyboard selects each station with the help of the selector board. A channel control and memory IC controls the band-switching and tuning voltage connected to the VHF and UHF tuner. You might find a separate VHF tuner, or both tuners can be included in one unit. The IF signal is cabled to the TV chassis.

Some of these selector-control boards have plug-in connections for easy removal. Most of the foreign control units are wired directly into the tuner and chassis. Determine if the tuner is defective. Check for a variable tuning voltage at the tuner. Suspect that the tuner is defective if all voltages are normal, but the tuning voltage is different as each channel button is pressed.

SCAN TUNING

Scan tuning allows the operator to slowly scan up or down the channels for favorite programs. This type of tuning is especially useful when connected to a cable TV system, with

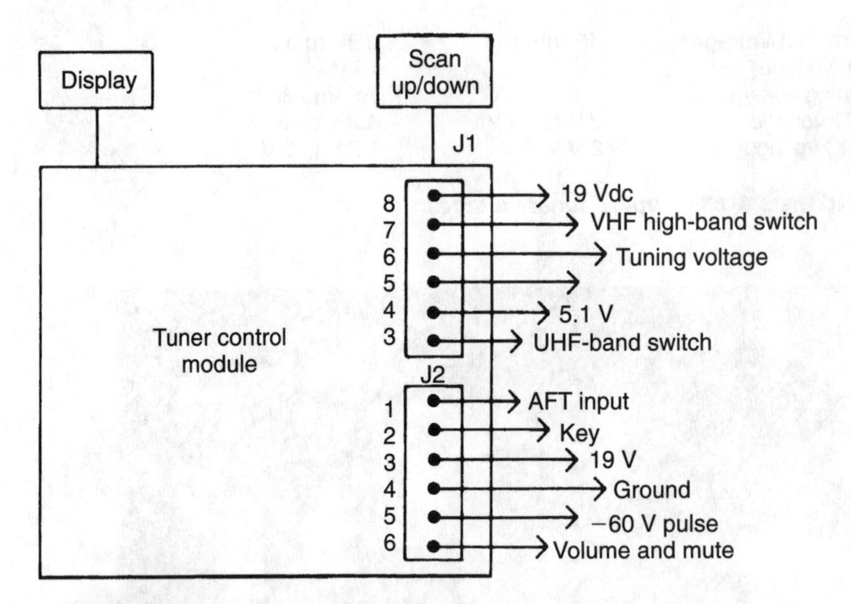

FIGURE 9-25 **An RCA scanning-control module in an RCA TV chassis.**

many stations to choose from. Many different types of scanning and pushbutton tuning devices are used, but they all perform the same task.

In an RCA keyboard or scan-tuning operation, the up-and-down switch assembly feeds into an MSC tuner-control module (Fig. 9-25). The frequency-synthesis tuner-control module feeds the correct VHF and UHF band voltage to the tuner module. An LED display indicates the channel number.

A defective MSC scan module can cause improper (or no) channel up and down action, improper addition or deletion memory of an incorrect or missing LED display, improper (or no) volume and mute control, no tuning voltage to the tuner, or improper (or no) channel change or skipping of channels. Replace the module for any of these symptoms.

If the suspected module does not restore a missing LED number, check the cable and plugs going to the LED display assembly. An intermittent display can be caused by a defective cable or wiring. Check for a poorly crimped wire at the LED socket. A poor ground connection of the MSC module can cause erratic (or no) scanning action.

A snowy raster with no picture or sound can be caused by a missing 60-V pulse at terminal 5 (Fig. 9-26). Look for broken wiring or a defective plug or connection at L103 on the flyback transformer. Check for a pulse waveform at terminal 6 on the tuner and terminal 7 of the flyback transformer.

TUNER CONTROL OR TUNER MODULE

To determine if the tuner is defective, replace the tuner in the modular chassis. Most of these tuners plug into the tuner-control or memory module. If the raster is still snowy or has a white screen, replace the control module. In the remote-control TV, replace the power-control module if the chassis is completely dead.

Be sure that all plugs are firmly pressed down on each module. If the tuner and control module plug together, be sure that the plug-in assembly inside of the module lines up properly. In some modules, you might find that the AGC wire plugs into the tuner separately. This wire can accidentally pull off when exchanging tuner modules and can cause a snowy (or no) picture. Press down on the cable wire assembly after the tuner is operating.

ELECTRONIC TUNER REPAIR

The defective electronic tuner in all TV chassis can be replaced. Some of these tuners are mounted together with a memory or bandswitching module. Simply remove the two metal screws and unplug the tuner module. Another tuner with the same part number can be substituted to cure the snowy picture. If both units are tightly plugged together, voltage measurements between the control and tuner are difficult to make (Fig. 9-27). Simply substitute each module until the correct one is found. Sometimes, if lightning strikes the TV, the tuner, memory, and control units are all damaged.

The quickest method to repair the front end of any TV is with tuner or control-module substitution. The suspected module is unplugged and another one is mounted in its place. Some manufacturers would rather if you replaced the module instead of trying to repair it (Fig. 9-28). Unless you have a lot of time, the correct test equipment, and the knowledge to repair these tuners and control modules, they should be returned to the TV distributor or manufacturer. Most tuner-repair depots repair many different American and Japanese modular tuners and control units (Fig. 9-29).

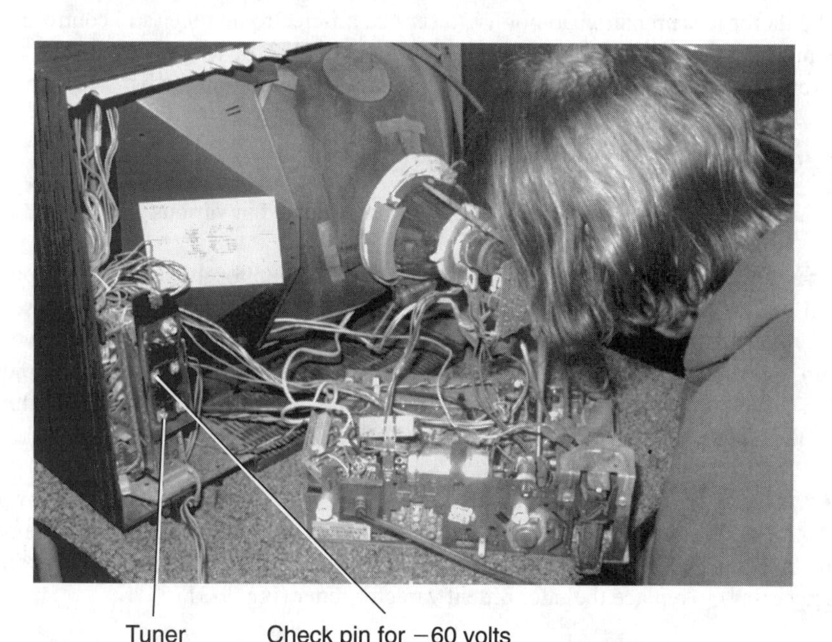

FIGURE 9-26 **A snowy raster caused by –60 volts at the memory module in an RCA chassis.**

FIGURE 9-27 **In the RCA CTC108 chassis, the tuner and memory module mount together.**

As time goes by, the electronics technician might be able to service the electronic tuner, memory, and control modules. Service data supplied by each manufacturer provides technical data for these repairs. Do not try to service all-electronic tuner and control modules without exact service data. Usually, most service establishments are swamped with many different service problems and repairing these modules is last on the list.

SEVEN ACTUAL TUNER PROBLEMS

Here are seven actual tuner-related problems that occurred in various TVs.

Tuner drifts off channel in an RCA CTC145 Monitor the waveform at output of first IF transistor (Q2301) to determine if the IF or the tuner is at fault. Check to see if the AGC voltage is from 3.1 V to 8 V with a station tuned in. Check the tuner supply voltage on pin 7; it should be 12 V (Fig. 9-30). Monitor the different tuning voltages on pins 4, 5, and 6. If the tuning voltage is applied to the varactor tuner when it drifts off channel, replace the tuner in the RCA CTC145 and CTC146 chassis.

Intermittent snowy picture in an RCA CTC140 The picture became snowy in an RCA CTC140 chassis. By tapping on the tuner, the picture might return, then fade out. All of the tuner connections were soldered around the tuner; all voltages were fairly normal. It was necessary to replace the intermittent varactor tuner (Fig. 9-31).

No channel tuning in an Emerson EC10 The tuner of an Emerson EC10 would not tune in a station. The voltage supply source was checked at the tuner. It was a normal

Tuner system	Defective component
Laser channels	Replace control module
No picture on 9 and 11	Replace control module Check AFT adjustment
Won't change channels	Replace control module
Searching on all channels	Replace control module
VHF searching	Replace tuner module
Intermittent channel 5	Replace tuner module
Keyboard malfunction	Replace control module Check plug-in connection Check keyboard asssembly
Numbers missing on display	Replace control module
Incorrect channel display	Replace control module Check plug-in connection
Intermittent display	Replace control module
No display	Replace control module
Erratic channel 5	Defective pushbutton on keyboard
Weak picture on channel 5	Replace turner module
Weak picture on channel 13	Replace turner module
No picture, no sound	Replace tuner module and control module
Flashing picture	Replace tuner module Check IF and input cables on tuner Loose AGC connection
Snowy picture	Replace tuner module Damaged balun coil AGC lead off
No UHF picture	Replace tuner module Check for UHF tuning voltage at UHF tuner
Weak channel 21	Replace tuner module Check UHF antenna cable leads

FIGURE 9-28 **A modular tuner troubleshooting chart.**

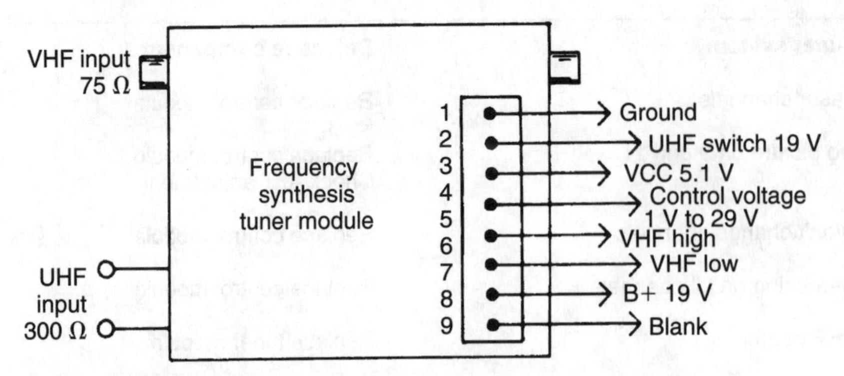

FIGURE 9-29 **Most tuner modules can be returned or exchanged at the manufacturer's distributor or be sent in to a repair center.**

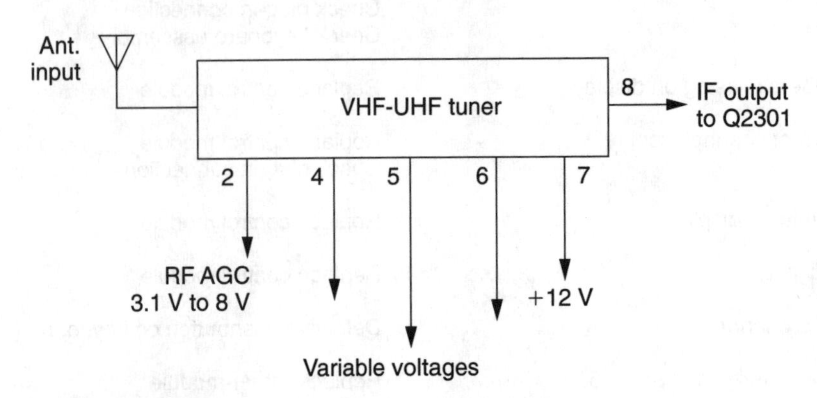

FIGURE 9-30 **Check the various voltages on the VHF-UHF varactor tuner to determine if the tuner is defective.**

11.3 volts. When the RF pin 5 was checked, no voltage was found on this RF AGC terminal. At first, IC101 was suspected, but all voltages were normal. Finally, C110 (10 µF) was found to be defective and was replaced, which solved the tuning problem (Fig. 9-32).

Flashing picture in a Sanyo 31C40A Although the outside antenna was old and could have caused the flashing problems in a Sanyo 31C40A, the set was brought into the shop for observation. Often another set in the house was used to check the outside antenna reception.

The TV was still flashing when it was connected to the shop antenna. Although the picture was fairly normal with the subber clipped to the IF cable, the picture was still flashing. One thing for sure, the picture was better with the tuner-subber.

The IF and antenna cables were flexed and held in many different positions, with the picture still flashing. A shorted or poorly connected IF cable can cause a picture to flash off and on. All wiring connections were inspected at the various tuner terminals.

Voltage measurements at the tuner would rapidly change as the picture continued to flash off and on. Even the RF AGC voltage would change rather rapidly. The AGC voltage was clamped with a variable dc voltage power supply at the tuner RF AGC terminal.

The tuner responded with 3 to 8 V applied from the variable dc source. The flashing picture was caused by improper AGC voltage to the VHF tuner (Fig. 9-33). Replacing IC101 (PIF amp/AGC/AFC sound/detector) with a direct manufacturer's replacement (M5186AP) solved the flashing picture.

Unusual tuner problem in an RCA CTC118A A no-raster/no-picture/no-sound/no-channel-display symptom was found in an RCA CTC118A. Both the memory and tuner modules were replaced, with no results. A normal high voltage was measured at the picture tube. Sometimes when channel 13 was pressed, a flash of light went across the screen. A tuner-subber on the IF cable indicated that the rest of the chassis was normal.

FIGURE 9-31 The defective tuner caused a snowy picture in the RCA CTC140 chassis.

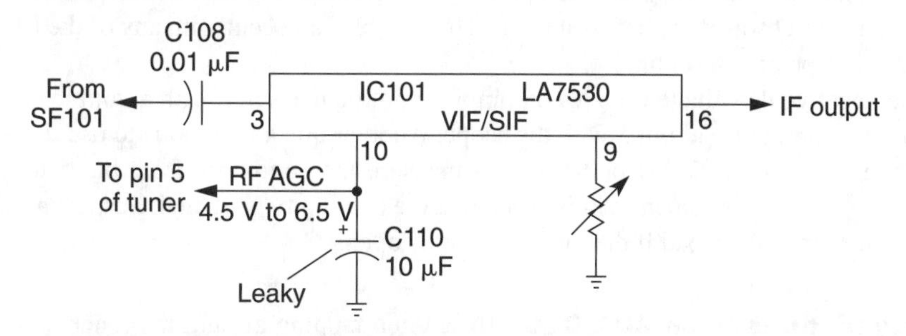

FIGURE 9-32 Defective capacitor C110 caused a lack of channel tuning in an Emerson EC10 chassis.

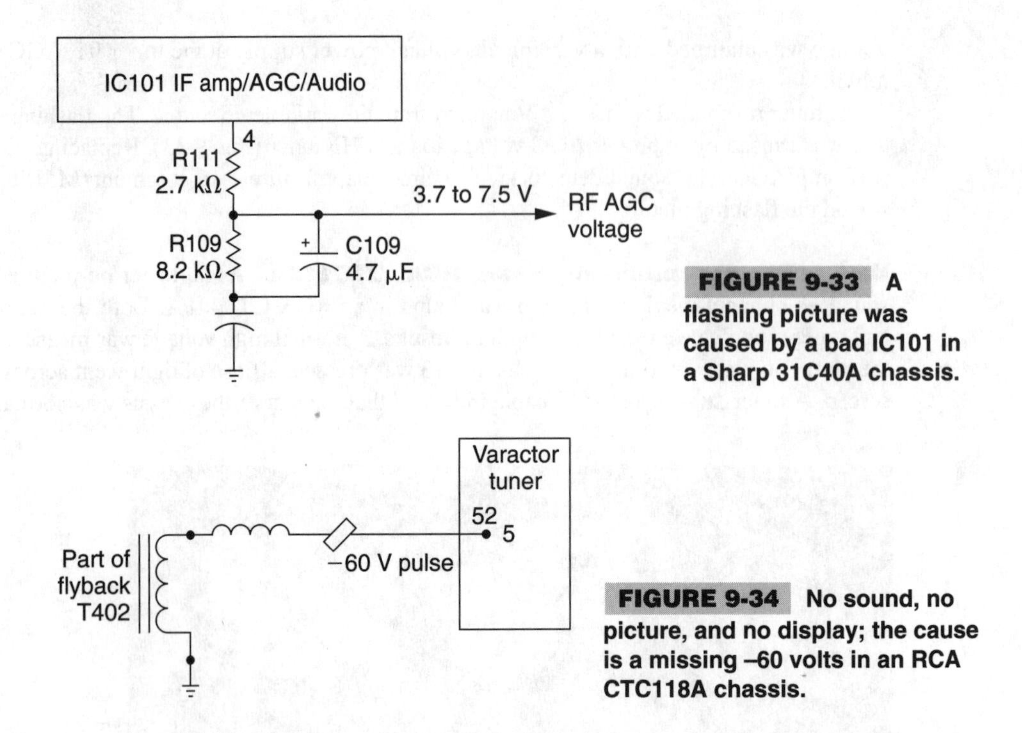

FIGURE 9-33 A flashing picture was caused by a bad IC101 in a Sharp 31C40A chassis.

FIGURE 9-34 No sound, no picture, and no display; the cause is a missing –60 volts in an RCA CTC118A chassis.

The voltage measurements on the tuner (18 V) and memory module were fairly normal. The picture and sound were brought back with an injected B+ tuning voltage of from 1 V to 27.5 V on pin 7 of the tuner. Another memory module (MSC-006A) was installed with the same results.

By rechecking the schematic and again taking voltages on the memory-control module, it was found that pin 5 of J2 should have had a –60-V pulse. After a quick check with the scope, it was found that the pulse waveform was missing from pin 5. This –60-V waveform is taken from a separate winding on the flyback transformer (Fig. 9-34). A continuity check across pin 7 and 14 of T402 was good.

Again, the chassis was fired up and pin 7 showed a good waveform. A continuity resistance reading across L104 indicated that the coil was open. The unusual problem was a poorly soldered connection at one end of L104. Resoldering the coil terminals brought back the correct waveform pulse at pin 5. This trouble can occur with any of the RCA sets using this type of control tuning.

The negative dc voltage (–60 V) on pin 5 cannot be measured with a voltmeter. Check for a waveform pulse at pin 5 with the scope. Another quick method is to use the low-resistance scale of the DMM or VOM and measure the continuity from pin 5 to chassis ground. A low measurement will indicate that the continuity is normal. Suspect a soldered connection at L104 is bad if the measurement is open.

Fuzzy pictures in an RCA CTC140 When tapping around the tuner in an RCA CTC140 chassis, the picture would fuzz up, then appear normal. Solder all tuner terminals, IF amp (Q2300) terminals, and the pins on U2300. In some CTC140 chassis, the trouble was the tuner and IF amp; in others, the IF amp and U2300 pins needed to be soldered.

Channel jump from one channel to another in an RCA CTC135D The channel would automatically jump up from one channel to another an RCA CTC135D (Fig. 9-35). When this occurred, the voltage at pin 42 of U1001 ranged from 1.2 V to around 2 V. The voltage at pin 43 was a little low, but the voltage at pin 42 should be 5.95 V. The 12-V source was normal, but it had a poor connection with L312. Although the connection looked good, soldering L312's connections solved the intermittent channel change.

Make your own tuner-subber from a used or discarded color or black-and-white TV chassis. Select a manual-type tuner. Keep the knob, etchem plate, and numbers so that you can identify the channel numbers from the old TV. Clean up the tuner and install it in a separate enclosure or cabinet.

Conclusion

No matter how complicated it is, the varactor tuner was designed to bring in the desired channel. Simply break down the tuning problem and repair it. A tuner-subber with accurate voltage measurements helps solve most tuner problems. An improper tuning voltage can indicate that a control unit is defective. Inject a dc voltage to the tuning-voltage terminal of the varactor tuner to determine if the tuner is defective.

A poor AGC system can produce symptoms like a defective tuner. Clamp the AGC terminal of the solid-state tuner with a variable dc voltage to eliminate the AGC circuits. Remember, the AGC voltage is always positive at the solid-state tuner terminal.

After locating a defective tuner, remove it and send it in for repair. Most tuners or tuner modules can be returned to the manufacturer's distributor, or they can be sent to a tuner-repair depot for repair. Leave tuner repair and alignment to the experts, unless you have the time and correct test equipment. For easy troubleshooting, check Table 9-1.

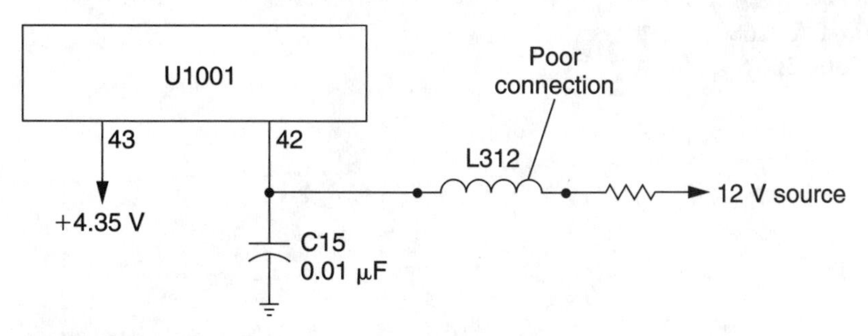

FIGURE 9-35 A poor connection at the end of L312 changed channels intermittently in an RCA CTC135D chassis.

TABLE 9-1 A TROUBLESHOOTING CHART FOR TUNER PROBLEMS

WHAT TO CHECK	HOW TO CHECK IT
No stations tuned in.	Check for dirty tuner on manual tuners.
Snowy pictures.	Check tuner voltages.
Sub tuner-subber to determine if tuner or IC circuits are defective.	Check IF circuits.
Check AGC voltages on tuner.	Check AGC circuits.
Checking tuning voltages on varactor tuner.	Check control IC and voltages.
Tuner normal.	Check IF and AGC circuits.

Tuner Repair Centers

PTS Corporate Headquarters
5233 Highway 375.
Bloomington, IN 47401
800-844-7871

PTS Corporation
4941 Allison St. #11
Arvada, CO 80002
800-331-3219

PTS Corporation
110 Mopack Rd.
Longview, TX 75602
800-264-5082

PTS Corporation
15042 Parkway Loop
Suite D
Tustin, CA 92680
800-380-2521

COLOR CIRCUIT PROBLEMS

CONTENTS AT A GLANCE

The TV must have a sharp and contrasting black-and-white picture before you attempt to tune in a color picture (Fig. 10-1). Improper tuner, IF, and video signals can result in a poor color picture. Make all color adjustments, including correct fine tuning, to improve the color picture. Check for improper color symptoms on the TV screen.

Besides defective color circuits, the color picture might appear messy, with color impurities on the raster. A defective degaussing section can foul up the color picture. Turn the color controls off and check the black-and-white picture. Degauss the CRT to eliminate

FIGURE 10-1 **A technician working on the color section of a TV chassis.**

patches of color impurities in the corners or at the bottom of the raster. Remember, if the black-and-white picture is not good, the color picture will be the same.

A weak or gassy picture tube can cause a poor color picture. To determine if the CRT is at fault, set the brightness control for average brightness. Now rotate the contrast control and notice if the picture forms deep patches of color. A blotchy-colored raster can be caused by a weak picture tube.

Using the Correct Test Equipment

The color-bar generator and scope are ideal test instruments for locating and servicing the color circuit. Attach the color-bar generator to the antenna terminals of the TV and scope each color stage according to the waveforms found in the manufacturer's literature. A dual-trace scope is ideal for troubleshooting the latest color circuits. Correct manufacturers' service literature and schematics are necessary when servicing the color circuits.

Take accurate voltage and resistance measurements within the transistor or IC circuits to quickly locate a defective color stage. An improper voltage source feeding the color circuits can cause a poor (or no) color signal. Check the low-voltage power supply if the voltages are low. Suspect that the bypass capacitor, transistor, or IC is defective if the voltage is low on a certain terminal.

Color signal injection from the test instrument can help you to locate a defective color stage. Simply inject the color signal at the various test points throughout the color stage or on the correct IC terminals. For correct color alignment, a sweep-marker generator, and scope must be used (Fig. 10-2).

COLOR-DOT-BAR GENERATOR

The color-dot-bar generator or NTSC color generator are required instruments when setting up and troubleshooting the color circuits. The color signal from the generator is used in Howard Sams Photofacts when scoping the various color circuits. Crucial waveforms are taken upon the chroma-luminance IC. Besides color tests, the color dot-bar generator can be used to check vertical and horizontal linearity, correct color bars, and pincushion circuits (Fig. 10-3).

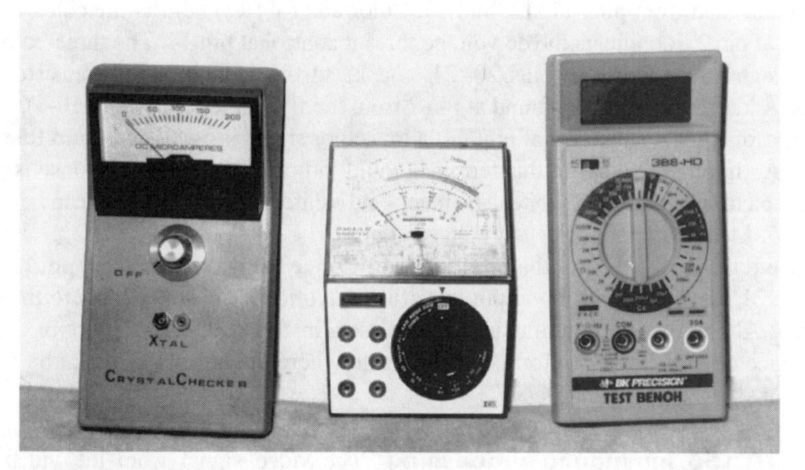

FIGURE 10-2 Besides the DMM, the crystal checker is another valuable test instrument for the color section.

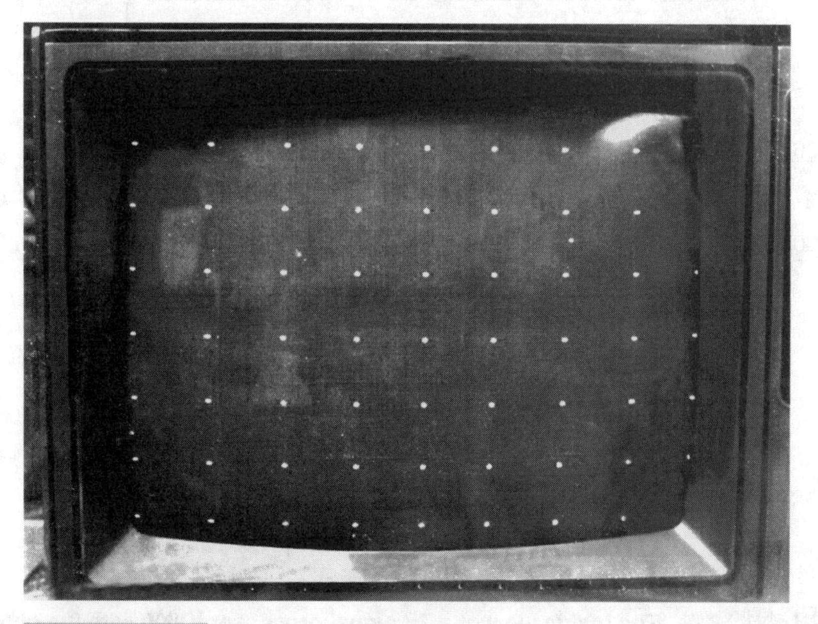

FIGURE 10-3 The color-dot-bar generator is used for setup, adjustment, and color troubleshooting.

COLOR WAVEFORM TEST POINTS

Because the IC circuits have been used in color circuits over the last 10 years, crucial color test points will help you to solve most color problems. Some manufacturers include the various block diagrams of color stages within the IC (Fig. 10-4) on the schematic diagrams, but others do not. The main thing is to know where to take the various color waveforms on the IC terminals.

The chroma input signal is applied to pin 3 of IC701 (Fig. 10-5). The 3.58-MHz oscillator is controlled from pins 11, 12, and 13. Color control (R2) adjusts the color with a dc voltage at pin 2. R1 adjusts the dc voltage for tint control at pin 14. The three-color video matrix signals are fed from pins 20, 21, and 22 to the separate color-transistor output stages. A keyed waveform is found at pin 7 from the flyback circuits (Fig. 10-6).

Check for color input signal at pin 3. If a low color signal is noted here, turn R3 to maximum. An improper signal at this terminal could indicate that the IF response is poor or that the picture is black and white. The black-and-white picture waveform can be checked at pin 27. Measure it for correct dc voltage at pin 23.

To see if the 3.58-MHz oscillator is oscillating, take the waveform from pin 13. Check pins 20, 21, and 22 for matrix output waveforms. Notice if the keyed waveform is found at pin 7. Although each manufacturer might use a different IC to take care of color and luminance circuits, the waveforms will indicate where the color signal can be found or is missing.

RCA CTC156 luminance processing The video signal from the video buffer (Q2302) is applied to the contrast preset control (R2716) (Fig. 10-7). The video signal is

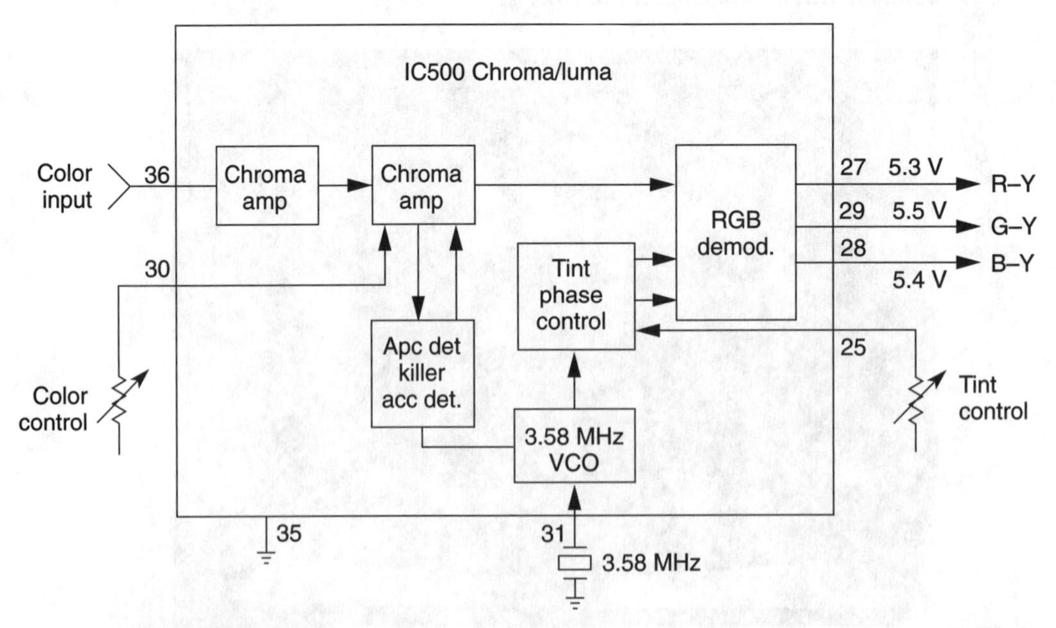

FIGURE 10-4 The inside view of a TV signal processor IC500 with the different color circuits.

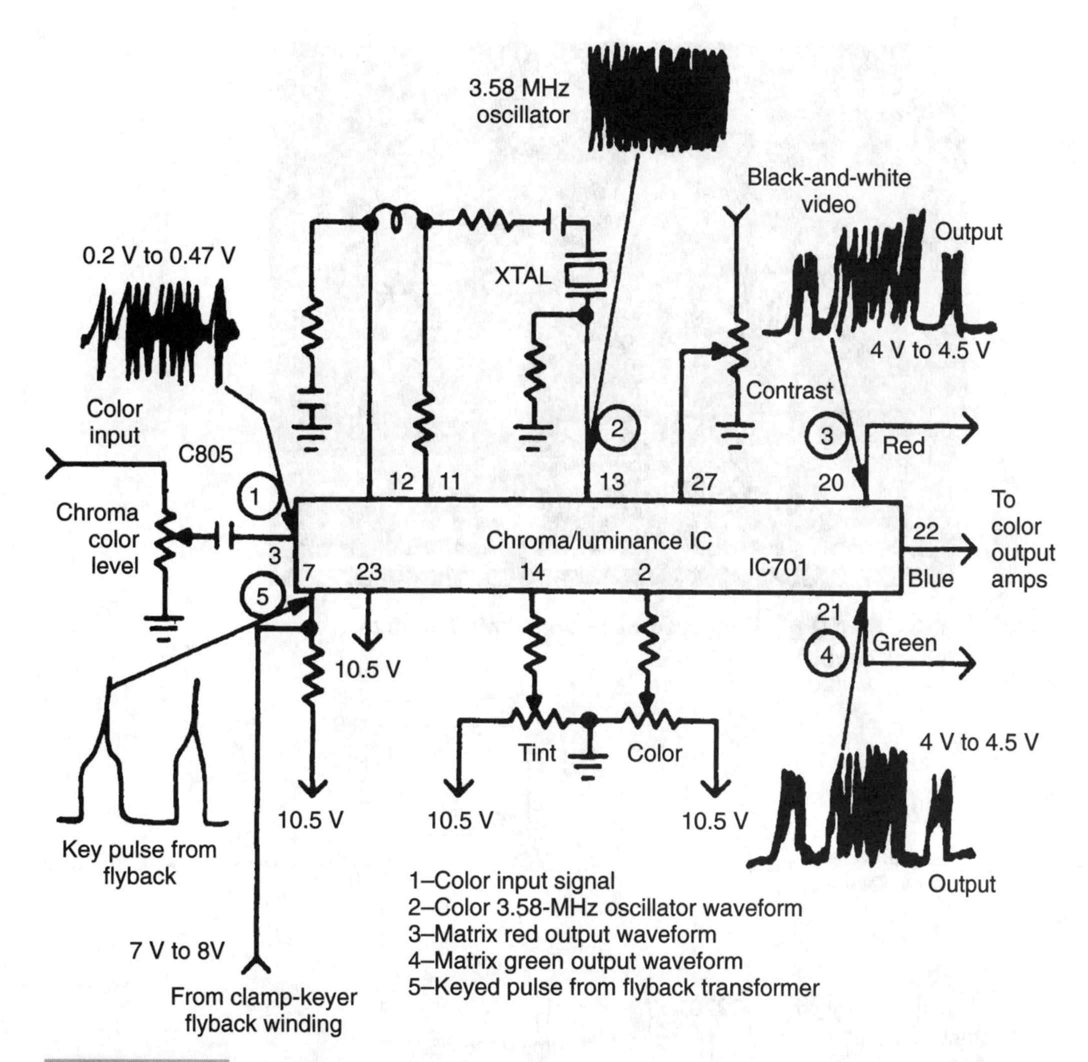

FIGURE 10-5 **The various color waveforms found on the chroma/luma IC (IC701) in the RCA CTC111 chassis.**

passed through the phase-compensation delay line (DL2701). After the delay line, the signal is divided into two paths (Fig. 10-8). The high-frequency component of the video signal goes to the sharpness circuit within U1001 at pin 52. The low-frequency portion of the signal is applied to pin 53 of the contrast circuit within the IC.

The gain of the sharpness control is controlled by dc voltage at pin 51. The dc voltage is produced by filtering the digital sharpness-control signal (BRM) output from U3300 (pin 16). The high-frequency luminance output of the sharpness circuit is input to the contrast circuit. The gain of this circuit is controlled by the dc voltage at pin 8. This contrast voltage comes from the BRM signal at U3300 (pin 20).

The output of the contrast amplifier is filtered and passed to a dc clamp circuit. The reference level for the video signal is set by the clamp circuit. Pin 19 of the AIU (U3300)

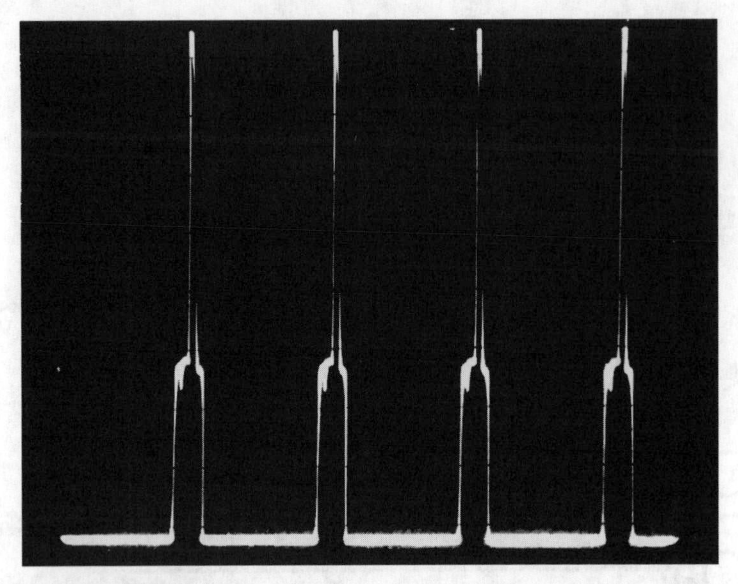

FIGURE 10-6 The keyed waveform at pin 7 from the flyback circuits.

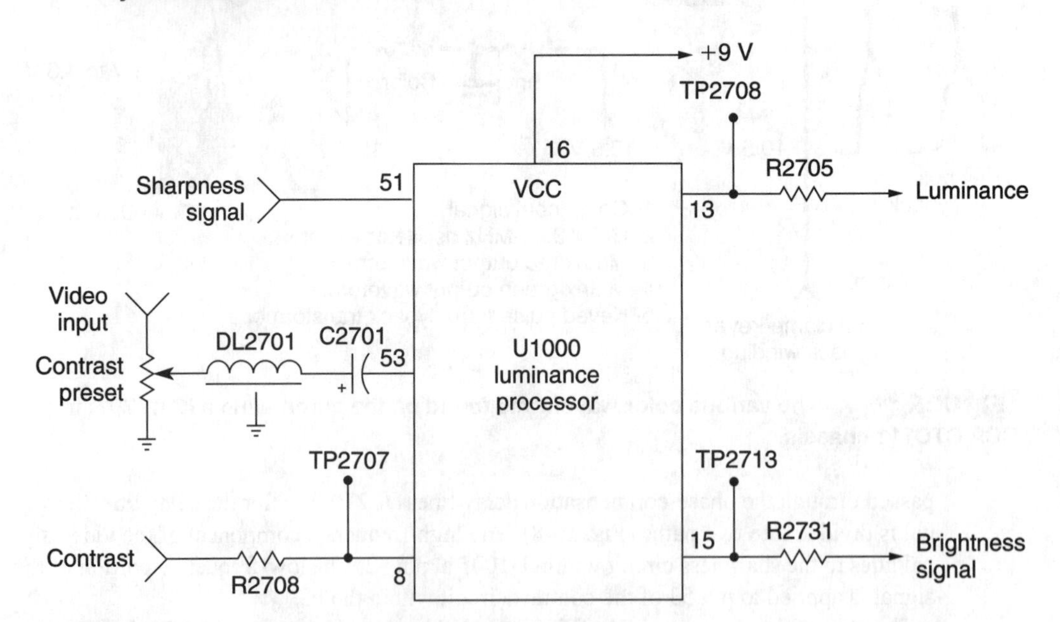

FIGURE 10-7 The block diagram of the RCA CTC156/157 luminance processing circuits.

develops the brightness control voltage. This dc level will vary as the brightness- and contrast-control voltages are varied.

The luminance signal from the dc clamp is applied to the luminance output amplifier (Y output) and it exits from pin 13 of U1001. CR2703 adds the horizontal and vertical blanking signals. Now the luminance signal is applied to pin 8 of the luminance buffer (Q2901). The luminance information is mixed with the chroma output circuit at the CRT driver/bias transistors.

The horizontal blanking signal is applied at pin 13 of U1001. This blanking signal inside of U1001 is used as a timing reference. Both chroma and luminance circuits can be affected if the horizontal blanking signal is missing or weak. Both the horizontal and vertical blanking signals are positive pulses, with 9-V peak voltage.

The deflection circuit's sync signal is developed in the luminance-processing section. Generation of the sync signal is required for proper operation of deflection and chroma circuits within U1001. Check the sync separator at the tint-control input of U1001. You should see the negative-going composite sync pulses.

The beam-limiter circuit reduces the amount of drive to the CRT when a predetermined level of beam current is exceeded. The beam-limiter circuit is tied to the brightness and contrast-control lines with CR2701 and CR2707. The beam current overrides the contrast and brightness settings generated by U3300 during periods when the beam current exceeds the predetermined level.

The high-voltage resupply voltage from pin 6 of T4402 begins to fall as the beam current rises. Now the base voltage of Q2703 goes down. If the base voltage falls, the transistor turns on and pulls down, controlling the brightness and contrast voltages at pins 8 and 15. R2701, R2718, and R2781 determine the point where the beam-limiter circuit starts to override the brightness and contrast voltages.

Troubleshooting CTC156 luminance circuits Check for video signal at the contrast preset control and pin 53 of U1001 (TP2307). Check the CRT driver/bias stages if the video signal is present. Scope for missing blanking signals at pin 13 and CR2703 (TP2705). If the video output signal is missing at pin 13 with normal input voltage at pin 53, check the

FIGURE 10-8 **The latest delay line in a 13-inch portable TV.**

brightness, contrast, and sharpness-control voltages while varying the controls. Check for dc clamp voltage at pin 14. The voltage can vary slightly because of adjustments and different test instruments. The dc clamp voltage can vary from 1.6 V to 2.6 V as the brightness control is rotated.

Check pin 16 of U1001 for correct supply voltage (9 V). Do not overlook the possibility that the ground connection at pin 12 is poor. Check the control-pulse outputs of BRM AIU (U3300). If the BRM signals are normal with abnormal brightness and contrast voltages, suspect that the problem lies in the beam-limiter circuit and Q2703. Suspect U1001 if the input signal is present, but no output signal is at pin 13.

Color IC Circuits

The IC has taken over the color and luminance circuits within the latest TV. You might find that one large IC contains the color, luminance, video, sync, and AGC circuits (Fig. 10-9) in a single device. Any breakdown in one particular circuit can destroy the IC and cause a loss of color in the chroma section.

One of the most important tests points within the color circuits is to check the power voltage source. A low voltage applied to the IC terminal can indicate that the power supply is defective or the IC is leaky. Check the schematic for other circuits that feed from the same power source. If the voltage is low at all other circuits, you can assume that the power supply is at fault. A lower-than-normal voltage at the power source of the IC can indicate that the IC is leaky (Fig. 10-10).

FIGURE 10-9 The chroma stages can be found in one large IC with other circuits in today's TV chassis.

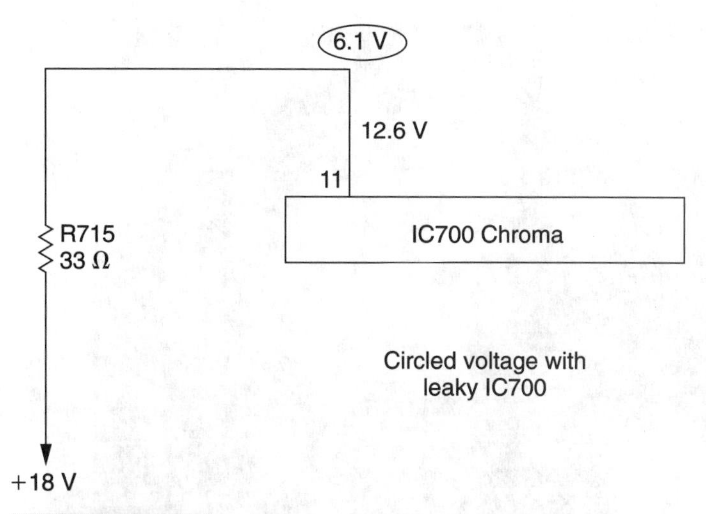

FIGURE 10-10 **Check for a leaky IC if the voltage is low at the supply pin of the IC.**

To determine if the IC is leaky or if the power supply is defective, the PC wiring can be cut at the power-source terminal or the IC pin (11) can be removed from the circuit. The quickest method is to remove pin 11 from the PC wiring. Lift the solder from around pin 11 so that the terminal is free from the board wiring. A leaky IC will measure under 500 Ω at pin 11.

Now check the dc voltage at the power source or wiring. If the voltage returns to normal, suspect that the IC is leaky. Check for a burned or open voltage-dropping resistor (R715) and power supply if the voltage source remains low. Often, if the power source is low in voltage, you will find other weak or dead circuits tied to the same power source.

A defective component within the color circuits tied to the IC can cause lower voltage at a certain terminal. Measure the voltages at each color terminal, and compare them with those on the schematic. Leaky bypass and electrolytic capacitors tied to the IC terminals can cause color problems. Open or increased resistances of fixed resistors at IC terminals can produce higher voltages.

A defective IC can cause intermittent, weak, or missing color in the picture. The wrong tint or a missing color can be caused by a leaky chroma processor IC. Color bars or colors within the black-and-white picture can result from a defective chroma IC.

Before removing the suspected chroma IC, take accurate voltage measurements on each terminal. Be sure that a leaky or open component tied to the IC is not causing the condition (Fig. 10-11). Take accurate voltage measurements at the power-source terminal. Check for correct color waveforms at the most crucial color test points. If you are still in doubt, remove and replace the color IC with one that has the correct part number or use a universal replacement. Most universal ICs and transistors work well in color circuits.

Bypass and electrolytic capacitors should be shunted if you notice color ringing in the picture. Check each capacitor for leakage. Remove one terminal for correct leakage tests after a lower-than-normal measurement is found in the circuit. An intermittent or weak color problem can result from a malfunctioning electrolytic capacitor.

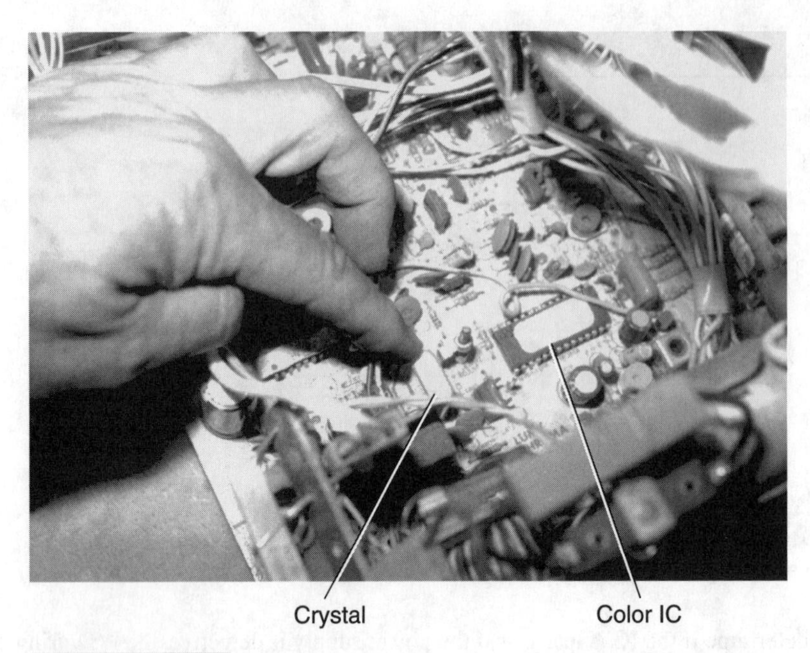

FIGURE 10-11 Look for the color crystal and color IC in the TV chassis.

COLOR MATRIX CIRCUITS

Within the tube and early transistorized color chassis, the demodulator, x, and y circuits mixed and amplified each separate color and were connected to the correct picture-tube color element. Today, the demodulator and matrix circuits are provided inside of the IC, with each color output fed to a single-color output transistor. The color-output transistors amplify the red, green, and blue signals to their respective gun assembly within the CRT.

A scope waveform at the three-color output terminals of the IC will indicate which color is missing or weak (Fig. 10-12). The waveform and amplitude of the red and green color-output signal at the IC terminals should be the same. If one color is weak or missing at the IC terminals, suspect that the chroma IC is defective.

Look for a leaky or open color output transistor if one color is missing or weak at the CRT. Be sure that the picture tube is normal. Raise and lower the screen voltage of each screen control to determine if the weak or missing color is indicated on the CRT. Check the picture tube with a good CRT tester for the missing color.

A quick voltage measurement on the collector (body) terminal of each color-output transistor can help you to locate the defective output transistor (Fig. 10-13). Normally, these collector voltages are within 5 V of each other. Go directly to the green output transistor if weak color or if no green is found in the picture. Check for low or very high voltage measurements at the collector terminal. A low collector voltage measurement can indicate that a color-output transistor is leaky or that the CRT element is shorted; a very high collector-voltage reading indicates that a color-output transistor is open. Very low collector voltage on all three color transistors can be caused by an improper boost or power-supply voltage.

Remove the color transistor that you suspect might be leaky when you test it for leakage. Sometimes these transistors can be intermittent, but when removed, they test normal. Replace the suspected color transistor if the CRT and IC are normal, but one color is weak or missing. These color transistors can be easily replaced because most of them are located on the chassis assembly of the CRT socket (Fig. 10-14).

COLOR CIRCUITS OR CRT?

To determine if the color amp or picture tube is defective, take a color waveform measurement at the collector terminal of each color-output transistor. Go directly to the color-output transistor if one color is weak or missing. Take a quick voltage measurement at the collector terminal. Raise and lower each screen control to determine if the picture-tube circuits are defective (Fig. 10-15).

Suspect that a color-output transistor is leaky if low voltage is at the collector terminal. A shorted CRT gun assembly could also be the cause. Remove the transistor and take another voltage measurement. If the voltage is still low, suspect that a spark-gap assembly or picture-tube element is leaky. Remove the CRT socket and notice if the voltage increases. Suspect that the problem is with the CRT if the voltage increases.

The screen was all blue with no picture in a Goldstar KMC1344G. The brightness control had no effect and the sound was normal (Fig. 10-16). A voltage measurement at the metal end (collector) of the blue output transistor was low (7.2 V). The normal collector voltage should be about 135 V.

Q509 was removed from the circuit board and it tested leaky. In fact, a 1152-Ω leakage measurement was found between the emitter and collector terminals. Because high leakage was found between the two terminals and the collector voltage was very low, R542 was measured for correct resistance. Sometimes these collector load resistors will become

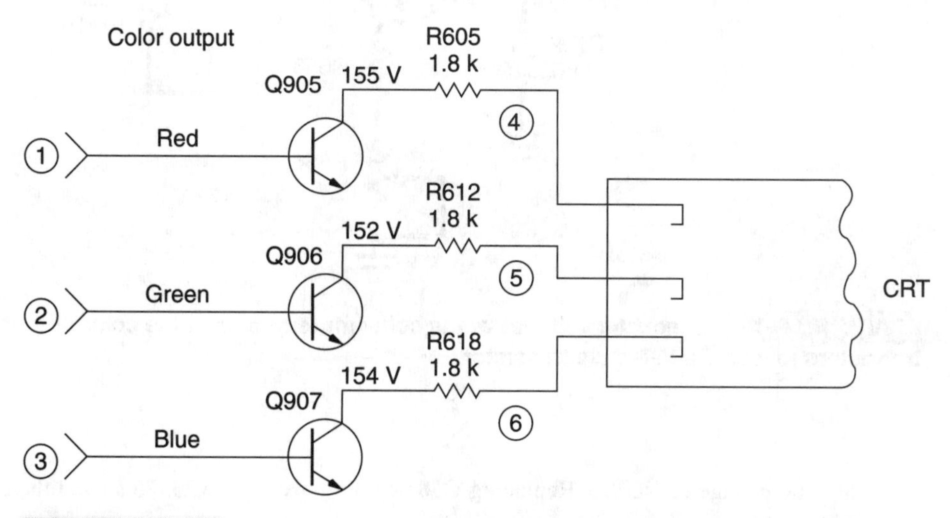

FIGURE 10-12 **Check the color output circuits by the numbers if the color is lost or weak.**

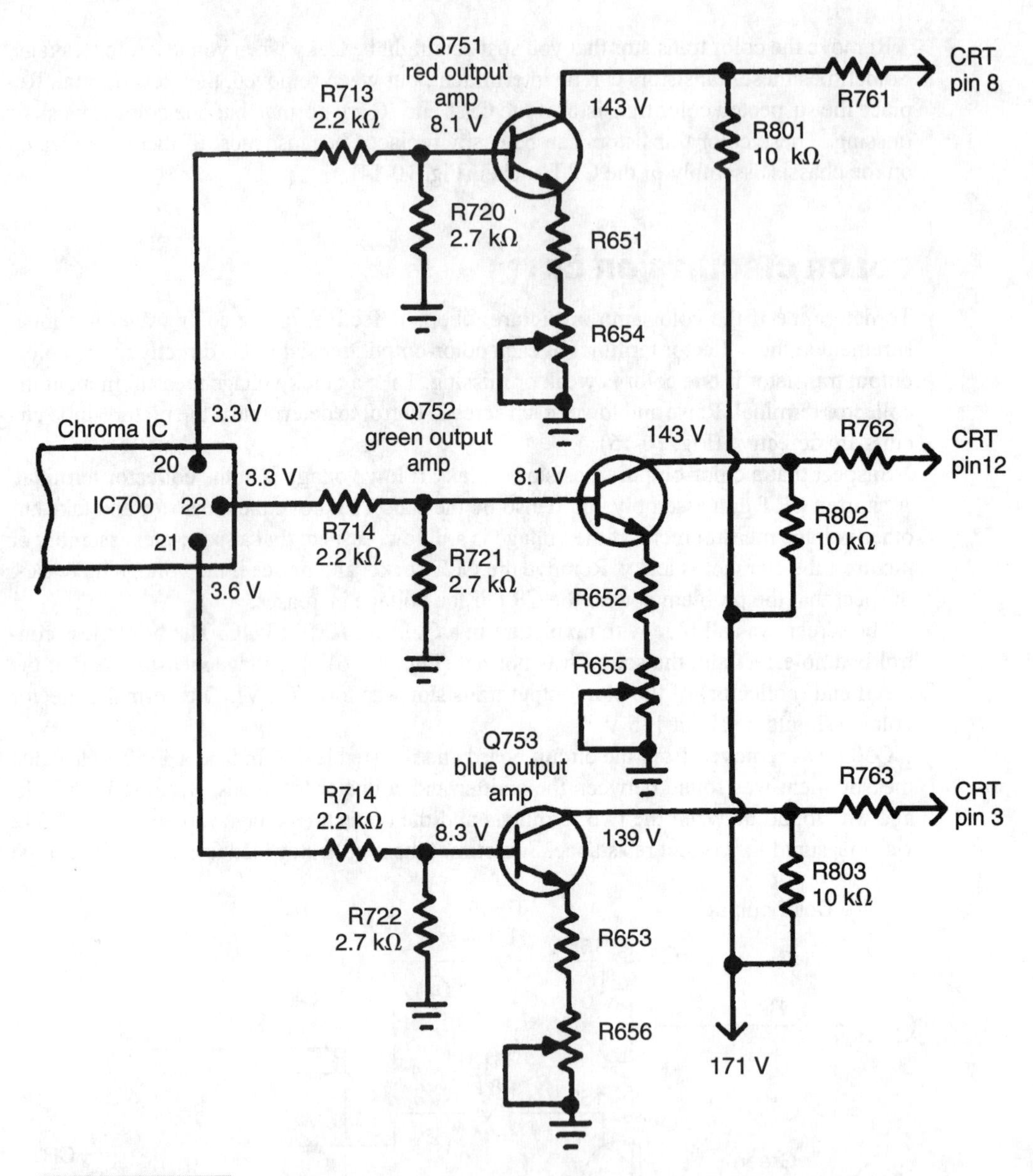

FIGURE 10-13 **Check the voltages at the collector terminals of the color output transistors to locate a defective transistor.**

hot and change resistance. Replacing Q50s with a universal ECG376 solved the all-blue raster problem.

Replace the entire color-amp module in a modular color chassis to determine if the color-amp circuits are defective (Fig. 10-17). A separate module feeds each color to the CRT, or they might be combined in one module.

FIGURE 10-14 The color output transistors are located on the CRT PC board.

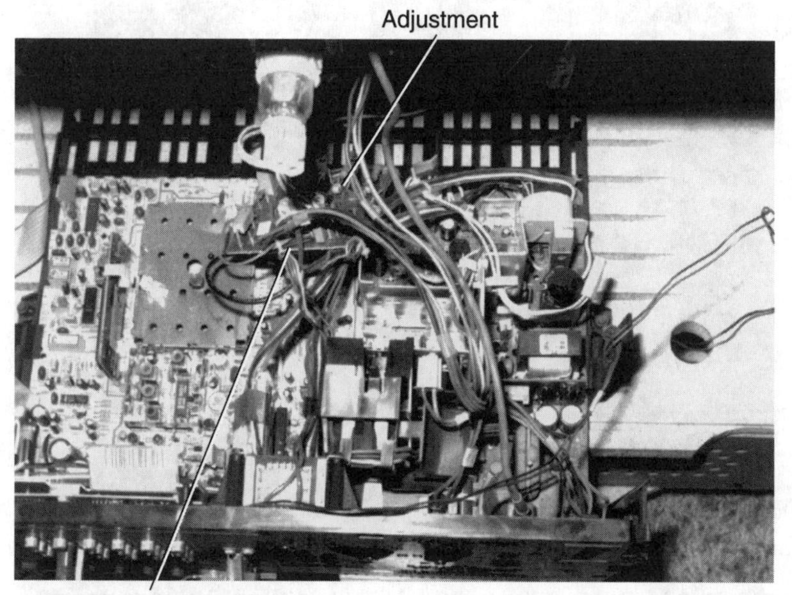

FIGURE 10-15 The color bias adjustments might be on the CRT board.

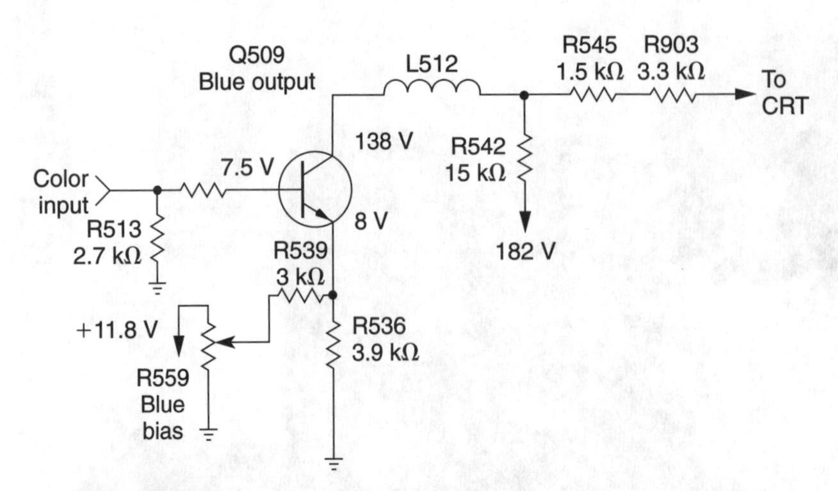

FIGURE 10-16 **The all-blue screen was caused by a leaky blue output transistor (Q509).**

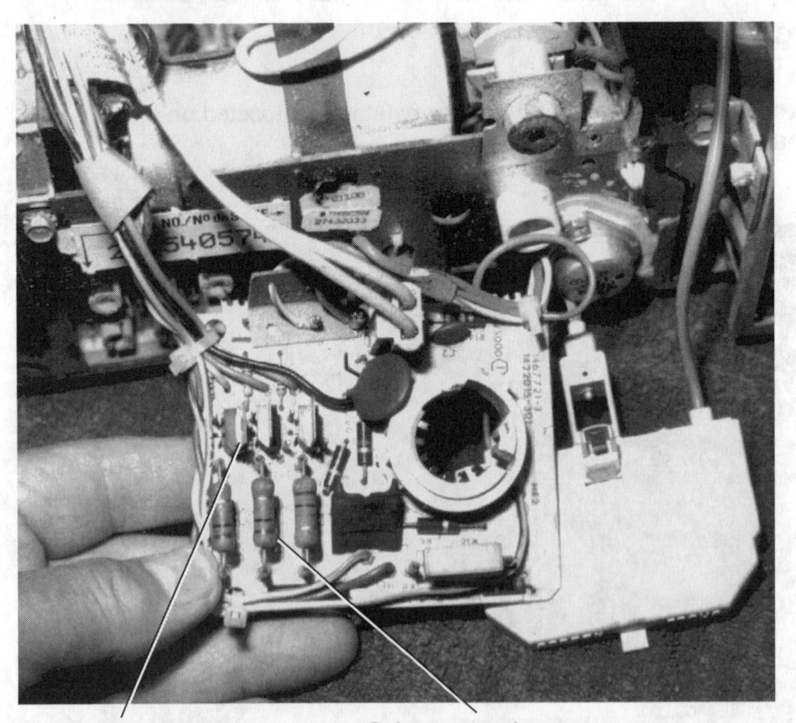

FIGURE 10-17 **Replace the leaky color output transistor and collector load resistor if they are burned or have changed in value.**

NO COLOR

Before checking the color circuits for no color, be sure that the color control is wide open with the tint control set in the middle of the tint range. Readjust the fine-tuning control behind the selector knob to adjust the color, picture, and sound. Turn up the color-killer and color-level controls for greater color in the picture. Be sure that the black-and-white picture is normal.

The no-color symptom can be caused by any component in the color circuits. If the 3.58-MHz crystal does not oscillate, no color will be in the picture (Fig. 10-18). A leaky transistor or IC can cause the no-color symptom. Faulty bypass and electrolytic capacitors tied to the transistor or IC terminals can cause no color in the picture. Improper or no low voltage from the power supply can produce weak or no color. Check the color waveforms of the 3.58-MHz oscillator. Connect the color-dot-bar generator to the antenna terminals and check for normal color waveforms at the demodulator or matrix stages. Go back to the burst and bandpass stages if no color waveform is at the color-output transistors. Check the color at the base and collector terminals of each stage. Check all waveform test points on the chroma IC. Follow the manufacturer's troubleshooting chart, which is included in most service literature.

A quick in-circuit transistor test with the diode-transistor test of the DMM located a leaky first bandpass transistor in a Panasonic CT301 (Fig. 10-19). Voltage measurements on all terminals were quite close, indicating transistor leakage. TR601 was removed from the circuit and it showed high leakage between all three terminals.

The no-color symptom was found in a Sharp 9B12B with a normal black-and-white picture. Voltage measurements on all IC801 terminals were fairly normal. No color waveforms were found on the IC terminals. IC801 was replaced with the exact replacement part

FIGURE 10-18 **Locate the color crystal and large IC to take crucial waveforms and voltage measurements.**

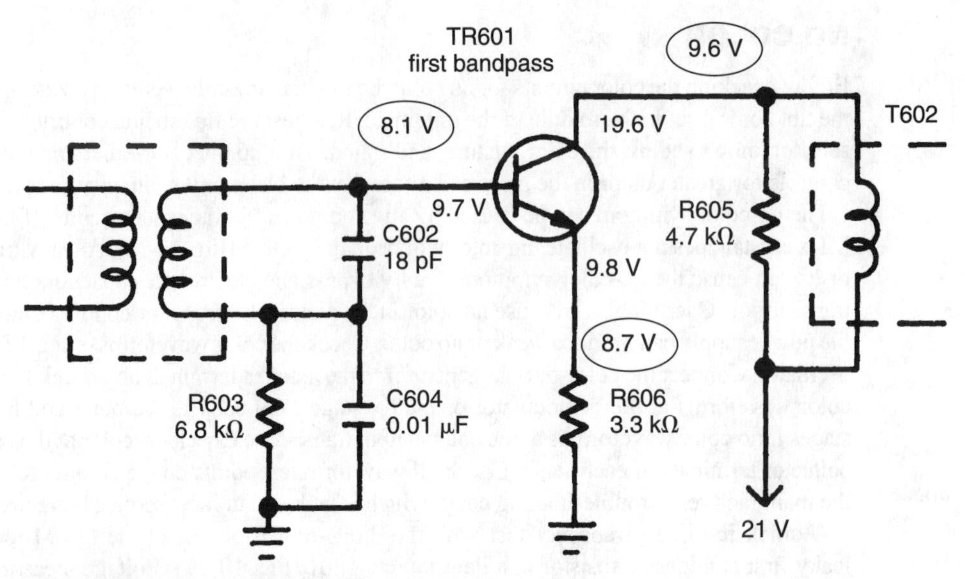

FIGURE 10-19 **In-circuit transistor tests helped to locate a leaky TR601 band-pass transistor in a Panasonic CT301.**

(part number RH-1X0093 CE22). This color IC must be replaced with an exact-type part because, at this time, there is no universal replacement.

Check component C39 and the 3.58-MHz crystal for no-color symptoms in the RCA CTRC97 chassis (Fig. 10-20). C39 has been found leaky in several chassis. All voltages are normal on IC U1, except pins 2 and 3. You might find that pin 5 has a higher voltage (8.6 V). Check C39 with a resistance measurement at pin 3 to ground. This reading is always under 5 kΩ.

WEAK COLOR

A weak-color symptom can result from a poor black-and-white picture or a defective component in the color stages. Low or improper voltages from the power source can produce weak color. A leaky transistor or IC can cause weak or no color. Check for burned resistors and open bypass capacitors in the bandpass amplifier stages. Suspect that leaky or open diodes are in the AGC color circuits.

Open coil windings can produce weak color symptoms. Improper color alignment can cause a weak-color symptom, but do not try to adjust these coils or capacitors without proper color test equipment. These adjustments do not jump out of alignment by themselves. Poor board wiring connections can cause weak color conditions.

INTERMITTENT COLOR

The intermittent color problem can take a little longer to locate than the ordinary TV symptom. First, check for a dirty tuner. Often, intermittents are caused by heat, poor transistor junctions, and poor board connections. Spray each transistor and IC in the color circuits with coolant. Push up and down and around the color board to make the color come and go. Sometimes heat applied near the component can cause it to act up. If only one color is intermittent, go to the demodulator, the x and y circuits, and the color output transistors. Also, suspect that a color-gun assembly of the picture tube is defective.

Poor eyelet or griplet connections within the General Electric ABC chassis can produce intermittent color pictures. Solder all double-sided griplet connections around the color circuits. Some technicians place a small, bare piece of hookup wire through the eyelet and solder the lead on both sides for a good connection.

Check the following components in an RCA CTC120A chassis if intermittent color conditions occur (Fig. 10-21). Spray C814 and C815 with coolant to see if the color disappears. Often, replacing both capacitors solves the intermittent color problem. Check L804 for open coil connections. Sometimes just moving C818 and Y801 will make the color come and go. You should notice very little voltage change when one of the intermittent components acts up.

Suspect that the 3.58-MHz crystal is defective if color bars are visible. Poor board or component connections within the crystal circuits can develop into visible color bars. Automatic frequency phase control (AFPC) color alignment can help prevent the color picture from drifting out of frequency and going into color bars. Also, a defective color-processor IC can produce color bars.

NO RED

Go directly to the demodulator (x) and red color-output transistors if the red color is missing or intermittent. Likewise, if only one color is missing, proceed to the color-output

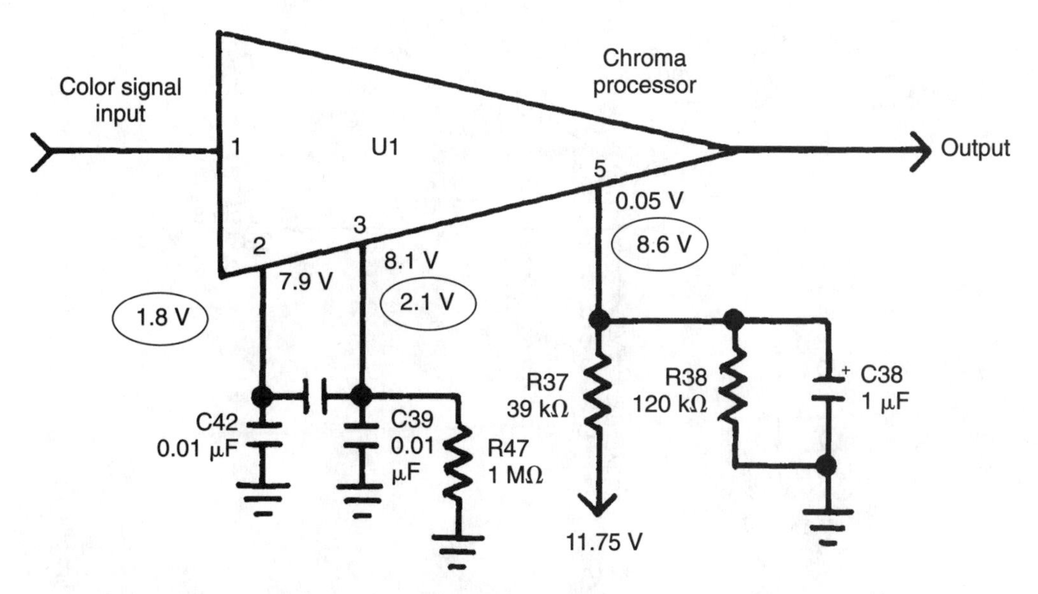

FIGURE 10-20 Leaky capacitor C39 produced a no-color symptom in this RCA CTC97 chassis.

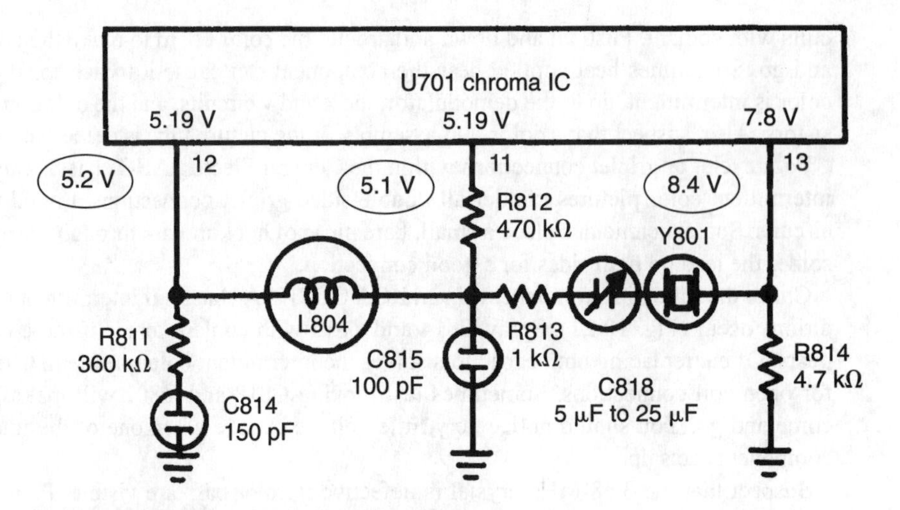

FIGURE 10-21 **Check the circled components at the chroma IC if the color is intermittent or missing.**

circuits and picture tube. Check the picture-tube circuits and the CRT for a missing color. Raise and lower the screen control to determine if the picture tube or circuits are defective.

The color-output circuits can be checked for correct waveforms or voltage measurements. In an RCA CTC87 chassis, the red color was missing from the picture (Fig. 10-22).

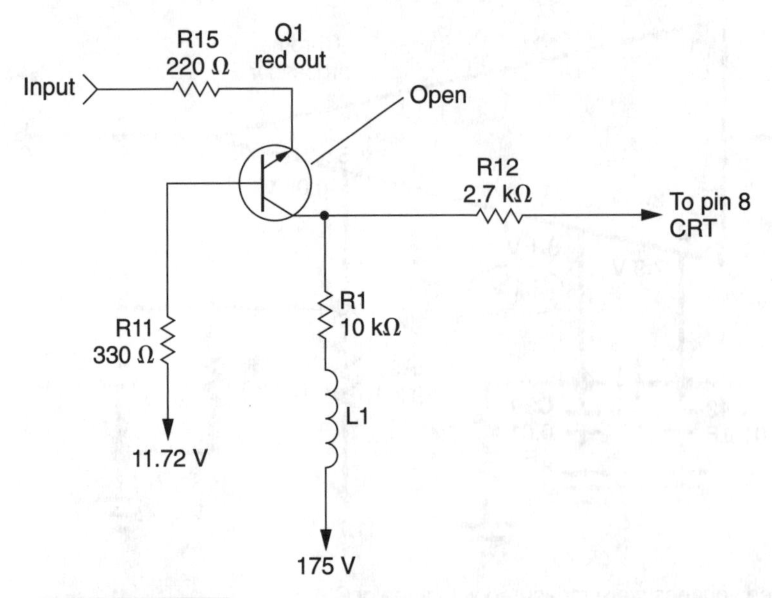

FIGURE 10-22 **Red was missing from the picture because of an open red output transistor (Q1).**

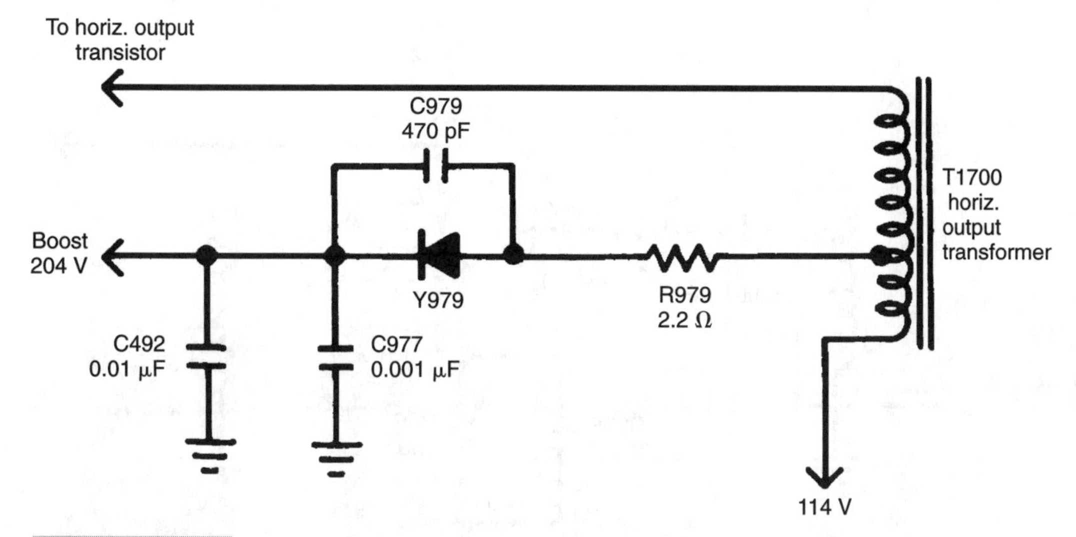

FIGURE 10-23 **Check for weak or absent color at poor boost voltage in a General Electric ECA chassis.**

A normal black-and-white picture was noted with the color control turned down. A quick voltage test at the collector terminal indicated that the voltage was high (186 V). Q1 was found to be open in the circuit with an in-circuit transistor test. Replacing Q1 with a GE-27 transistor solved the no-red symptom.

Suspect a problem in the low-voltage power supply if no color or only weak color is visible in the picture of a GE-ECA chassis (Fig. 10-23). With one particular GE-ECA, very little collector voltage was measured at Q403, Q405, and Q401. A voltage check at the boost-voltage source indicated no boost voltage. Resistor R979 in the flyback winding of the boost voltage was open. It's wise to also check Y979 for leakiness.

Too much green, just a little blue, and no red was found in a Zenith 14DC15 chassis (Fig. 10-24). The 3.5-MHz waveform was missing at the collector terminal of the 3.5-MHz amp transistor (Q206). A voltage measurement indicated that Q206 was leaky. The transistor was shorted on all elements. Replacing Q206 with an SK3122 universal replacement solved the messy color picture.

NO TINT CONTROL

Check the tint-control circuits if the tint control has no effect on the color picture. Often, one leg of the tint control is open within the tint and color circuits. Take a quick resistance measurement on both sides of the tint control. The voltage on the tint control can run high with an open ground connection to the grounded side of the control. Also, check for an open tint control.

The tint control had no effect on the color picture in an RCA CTC68L chassis (Fig. 10-25). With a resistance continuity check on each side of the control, an open was found between the tint control and the chroma module. One side of L111 was found to be open. L111 was removed and repaired. Replacing L111 solved the open connection between AF and the tint control.

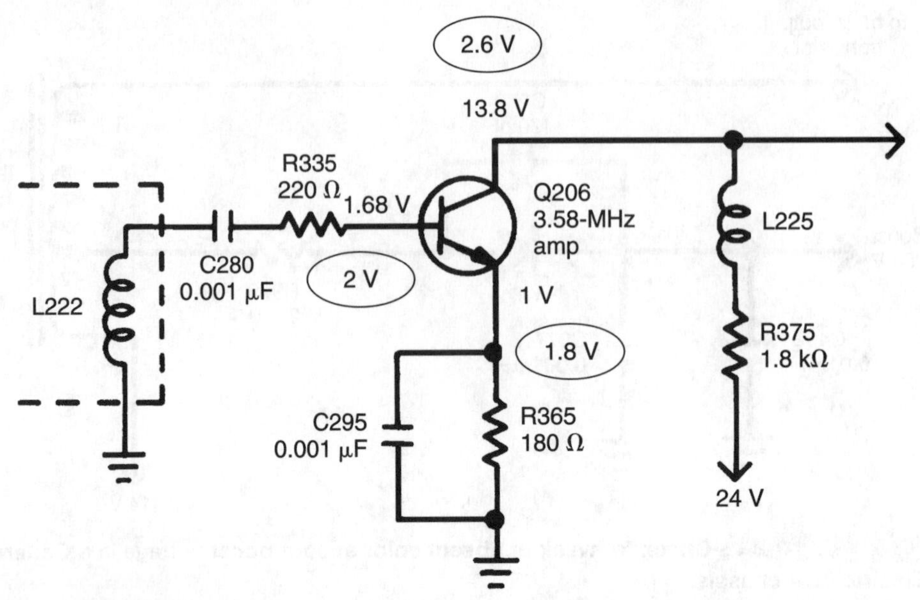

Note: Circled voltages were measured with a leaky Q206.

FIGURE 10-24 **No red was found in the picture of a Zenith 14DC15 chassis, because of a leaky Q206.**

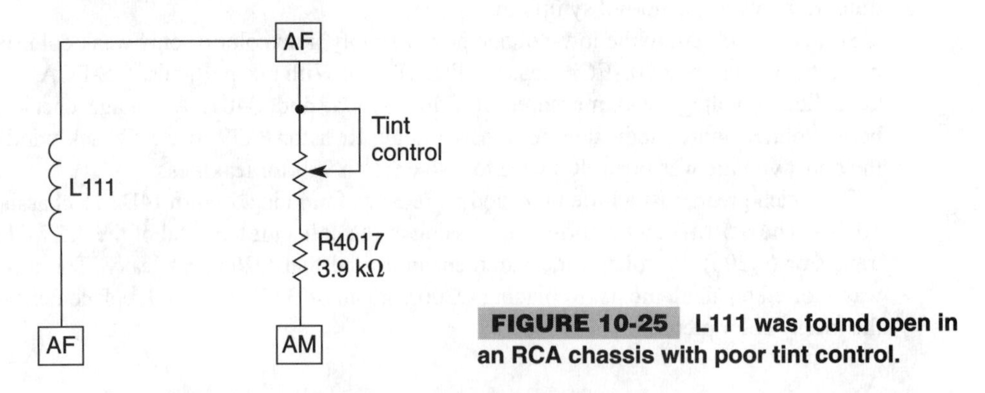

FIGURE 10-25 **L111 was found open in an RCA chassis with poor tint control.**

ALL-BLUE RASTER AND SHUTDOWN

Suspect that a picture tube or spark-gap assembly is shorted on the CRT elements if the TV comes on with a bright color, then shuts down. The raster gets extremely bright within a few seconds. Often, chassis shutdown is caused by excessive high voltage at the anode terminal of the picture tube. Test the picture tube for a shorted element. Sometimes the CRT tester might not show a leaky gun assembly. The gun assembly can become shorted only when voltage is applied to the heater and other elements. Pull off the CRT socket and fire up the chassis. If the chassis does not shut down, replace the picture tube. Check for a leaky spark-gap assembly if the chassis does not shut down with the CRT socket removed.

The raster was all blue in an RCA CTC111A chassis before shutdown. Within a few seconds, the blue raster became brighter before shutdown. The spark-gap assembly (pin 11) was shorted to chassis ground (Fig. 10-26). Replace the component with one that has the original part number (146169). Remember, in some TVs, the spark-gap assemblies can be found inside of the CRT tube socket.

CRUCIAL COLOR WAVEFORMS

Six color waveforms can be checked at the color IC processor to indicate that the IC circuits are functioning. Take a color input terminal to determine if color is at this point (Fig. 10-27). The color waveform (Fig. 10-28) will indicate if the color oscillator is functioning. The color-output signals for the three different color-output transistors are taken and all three look somewhat alike (Fig. 10-29). If the color-output signals are missing, suspect

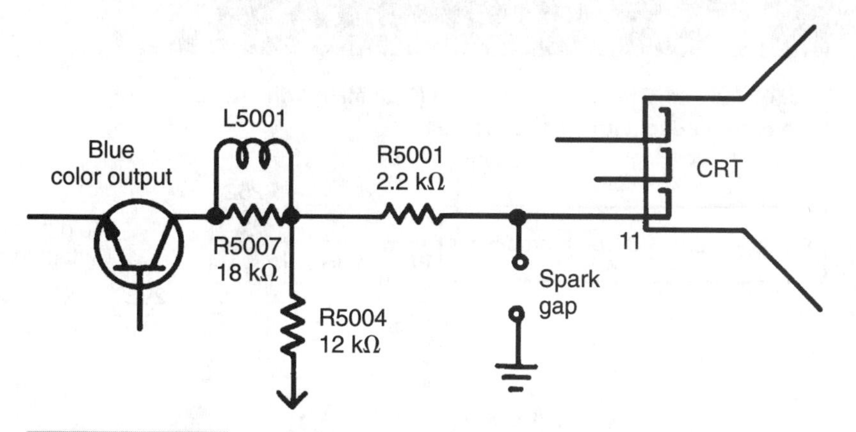

FIGURE 10-26 Check for a shorted spark gap if the all-blue screen is noted on an RCA CTC111.

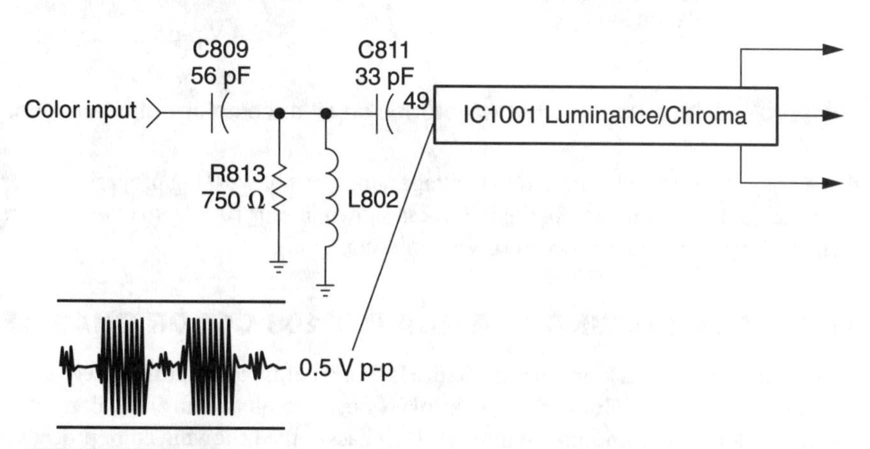

FIGURE 10-27 Check for the correct color input waveform to determine if color is present here.

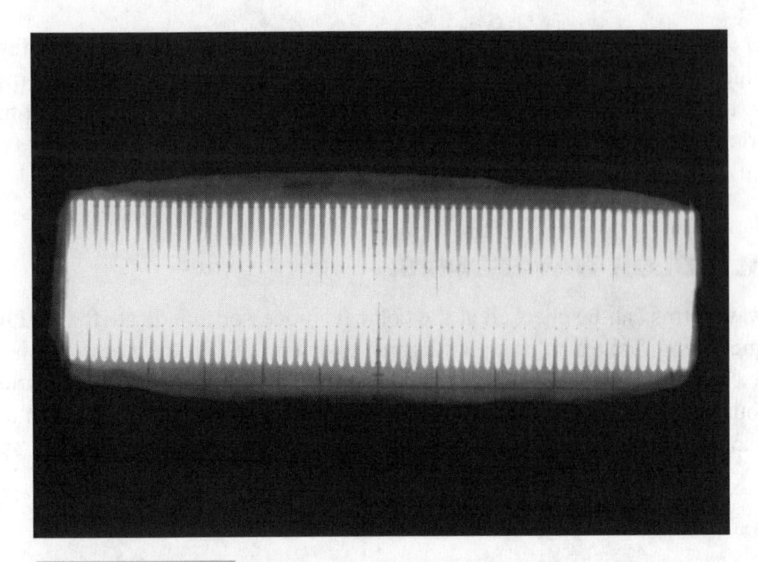

FIGURE 10-28 The color waveform (3.58 MHz) will indicate that the chroma oscillator is functioning.

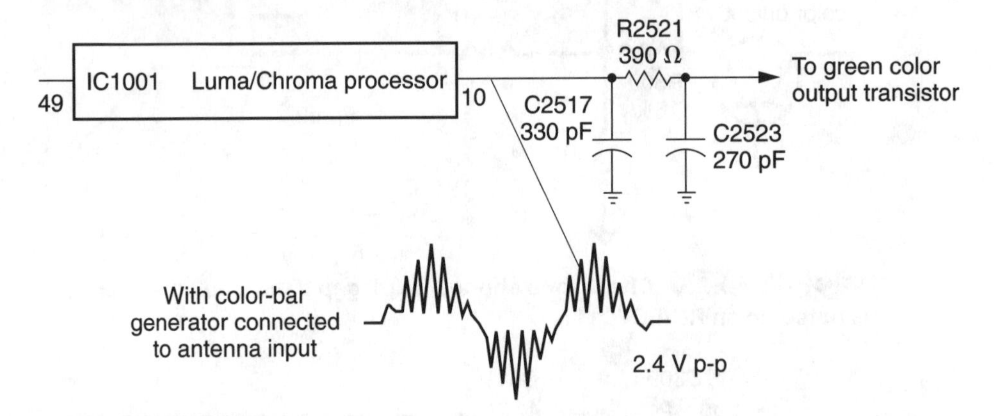

FIGURE 10-29 Color output waveforms on all three color-output circuits.

that a color IC is defective or that the voltage supply is not working properly. Last, but not least, a color trigger pulse from the flyback is shown in Fig. 10-30. No color appears in the picture if any one of these waveforms are missing.

TROUBLESHOOTING THE RCA CTC108 COLOR CHASSIS

Be sure that a good black-and-white picture is found before attempting to service the color section. Readjust the color and tint controls. Check the picture for no color, intermittent color, or color bars. In addition to the CTC108 chassis, the following color procedure is the same for CTC107, 109, 111, and 115.

Measure the dc voltage at pin 23 (10.5 V) of U701. If the voltage is between 10.5 V and 12 V, assume that the power supply voltage is normal (Fig. 10-31). A lower voltage can indicate that IC U701 is leaky or that the power supply is defective. Remove pin 23 or cut the foil to determine if the low voltage is caused by the power supply or by a leaky IC.

Now check the input color waveform at pin 3. Turn up the color-level control (R816) if the amplitude of the color waveform is low. Check the chroma signal at the chroma buffer

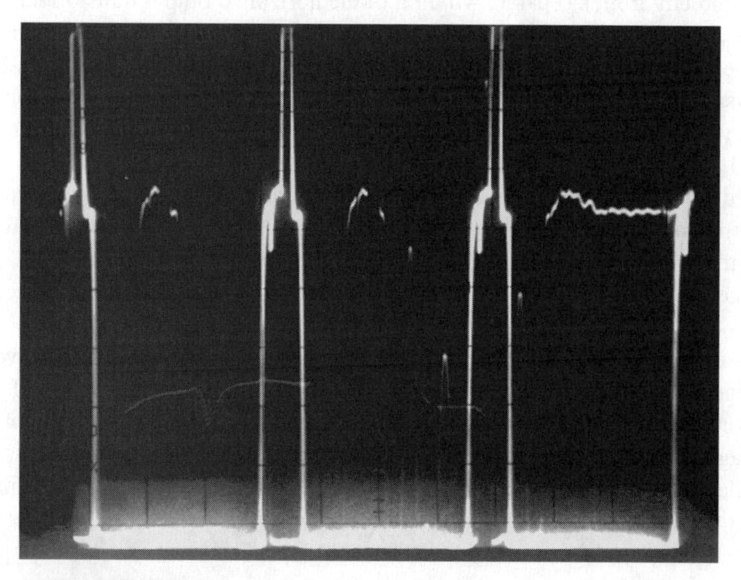

FIGURE 10-30 Trigger pulse from the flyback winding to the color IC.

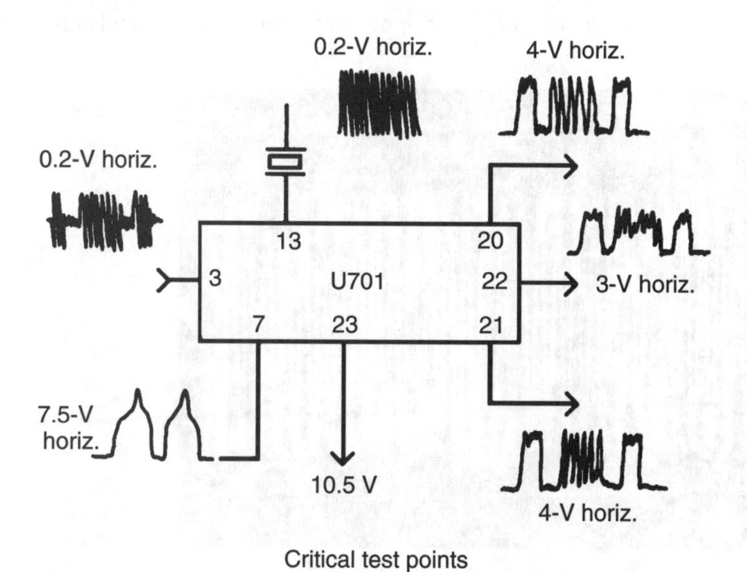

FIGURE 10-31 Check the power-supply pin (23), color input, oscillator, three color outputs, and the trigger pulse if there is loss of color.

transistor (Q800) if poor or no color signal is found at pin 3. Check the IF circuits if the color signal is lost.

Proceed to pin 7 and check the sandcastle waveform. The sandcastle input waveform must be present or U701 will have no color output. In most color circuits, the keying waveform comes directly from a separate winding of the horizontal output transformer.

Check the 3.58-MHz oscillator waveform at pin 13. No color will be present in the picture if the oscillator is not working. Measure the voltage at pin 13 (7.65 V). Substitute the crystal (Y801) if the voltage is normal without the oscillator waveform. Replace the crystal with an exact replacement part. Suspect C818 for intermittent or no color (Fig. 10-32).

A normal demodulator waveform at terminals 20, 21, and 22 indicates that the color circuits are normal at U701. Check the chroma input signal at pin 1 and 17 if the demodulator waveforms are missing or improper. Measure the voltage on pins 1 and 17. Take a resistance measurement between the terminals and chassis ground. The resistance measurement should be infinite at terminal 17 (Fig. 10-33).

Suspect that U701 is defective if there is no demodulator output, but the color voltages and the signal to the demodulator circuits are normal. If the demodulator waveforms at U701 are normal and one color is missing, suspect the color output amp or picture tube. Take a voltage test on the missing-color output-transistor collector terminal. Measure the voltage at the metal heatsink at the end of transistor. Compare this voltage measurement with those at the other two color transistors. If the collector voltage is high, suspect that an output transistor is open. A low-voltage measurement can indicate that a transistor or picture-tube element is leaky.

Check the following components (Fig. 10-34) for color problems related to the chroma processor (U701). No color is identified by number 1. For intermittent color, check number 2. For weak color, check number 3. Check number 4 for oscillating and color bars.

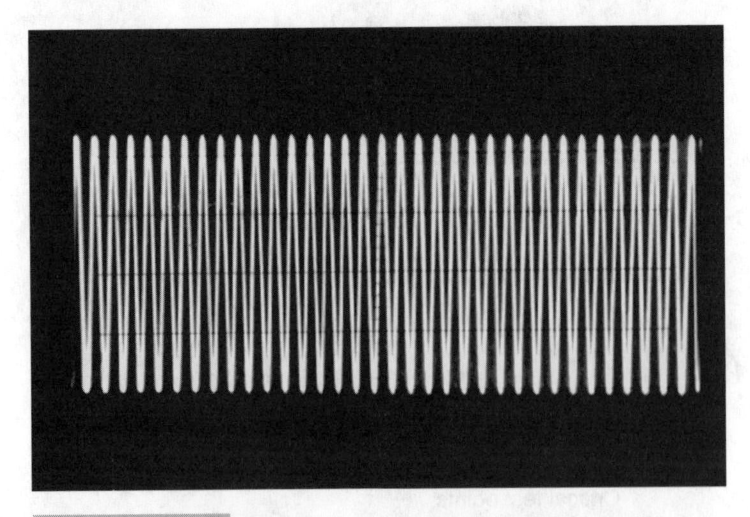

FIGURE 10-32 **The color oscillator (3.58 MHz) waveform taken from the IC of a crystal pin terminal.**

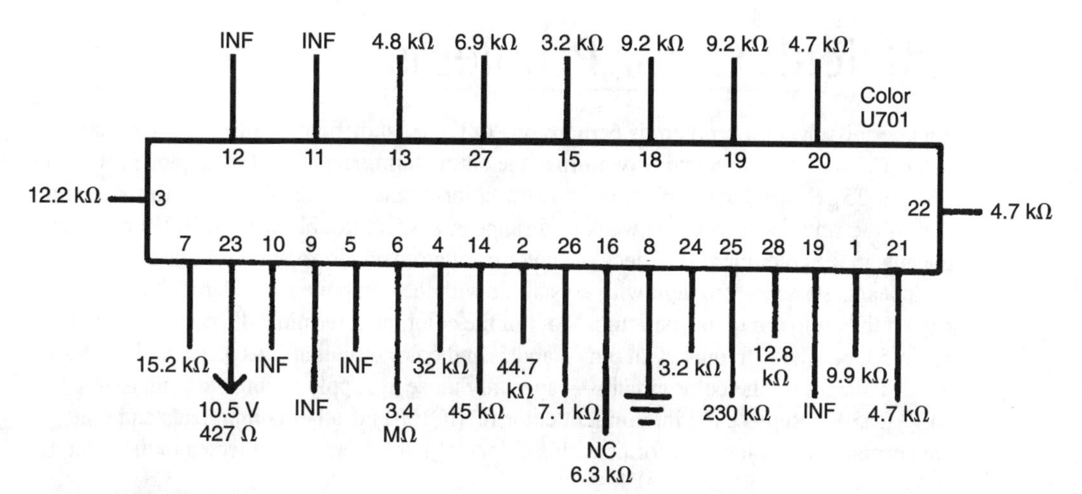

FIGURE 10-33 Normal resistance taken with a DMM in an RCA CTC107, 108, 110, 111, or 115 chassis.

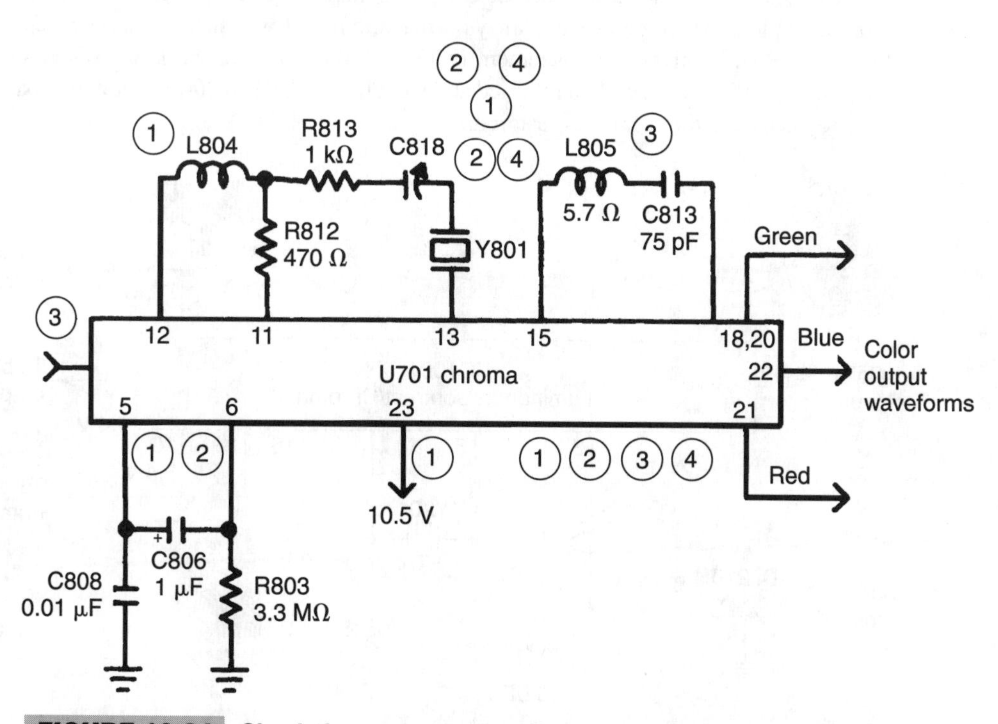

FIGURE 10-34 Check these components if the color is weak, intermittent, or missing at the IC (U701).

Difficult Color Problems

An intermittent color problem is perhaps one of the most difficult malfunctions to diagnose in the TV. Monitor the signal waveform at the output terminals of IC1001 at pins 9, 10, or 11 (Fig. 10-35). Suspect that problems are in the color-output circuits if the waveforms are normal. If the waveforms become weak or disappear, suspect trouble with IC1001 or the components in the color circuits. Check the color oscillator waveform at pins 4 and 5 if the color disappears. Suspect a problem with crystal (or with the components soldered to these terminals) if the oscillator quits operating. Monitor the color input terminal 49, contrast waveform at pin 53, oscillator frequency at pins 4 and 5, and color output at pins 9, 10, and 11. Monitor all voltages on the color circuits—especially those at supply-voltage terminals 14 (9 V) and 7 (3.5 V). Replace the intermittent color IC (IC1001) if other components and voltages are normal. Take color waveforms with a color-dot-bar generator connected to the antenna.

UNUSUAL COLOR PROBLEMS

No color was the symptom in an Emerson MS250R model. The waveforms at output pins 19, 20, and 21 were not there, indicating trouble within the IC201 chroma IC and circuits. No color waveform was found at pin 16 of the color oscillator (Fig. 10-36). The X201 color crystal was replaced, but still the color was missing. The input color waveform was normal at pin 6. All voltages were fairly normal and IC201 was suspected. After replacing TA7644BP with universal replacement ECG1547, the color was still out. When VC201 was accidentally prodded, then returned and went out. VC201 (a 20-pF trimmer capacitor) was replaced and the color was restored.

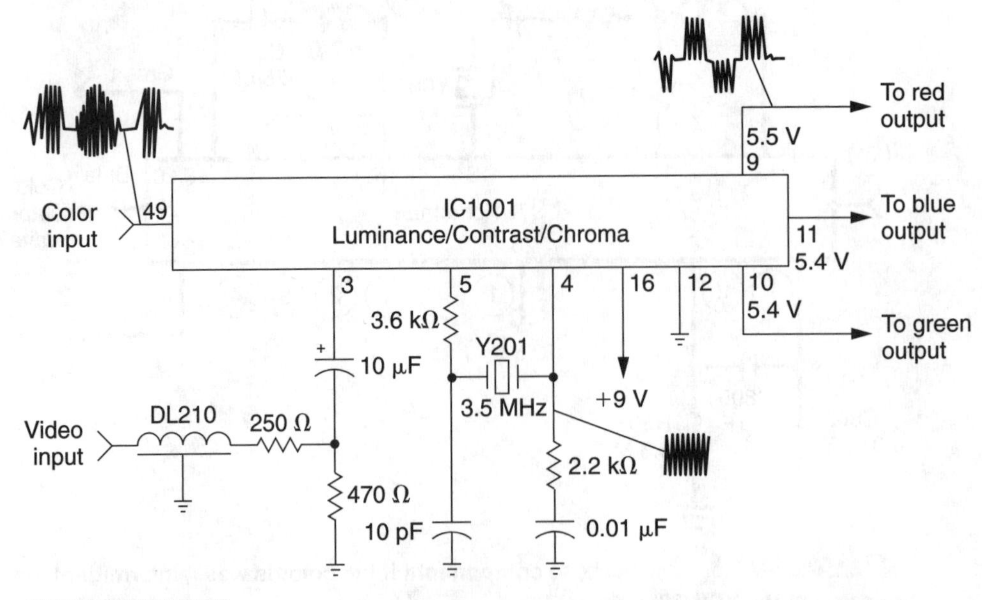

FIGURE 10-35 Monitor the intermittent colors at pins 9, 10, and 11 of IC1001.

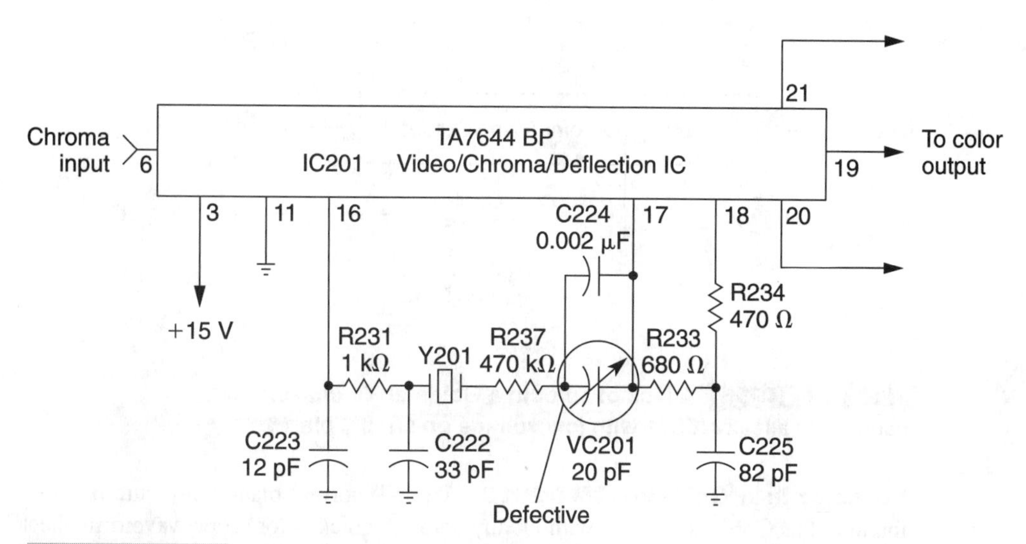

FIGURE 10-36 The unusual color problem in an Emerson MS250R was caused by trimmer capacitor VC201 (20 pF).

FIVE ACTUAL COLOR CASE HISTORIES

The following are five actual color-related TV problems.

No red in picture in a Sony SCC-548D The red output color was fairly normal on pin 18 of chroma IC301, but none was in the picture tube. A quick voltage test on red color-output transistor Q703 (located on the CRT "C" board) indicated a leaky transistor. Q703 was removed and found to be leaky (Fig. 10-37). R709 (15 kΩ) to the collector terminal of Q703 was burned. Q703 (2SC2276) was replaced with an ECG171 universal component, and R709 was also replaced, which cured the no-red symptom.

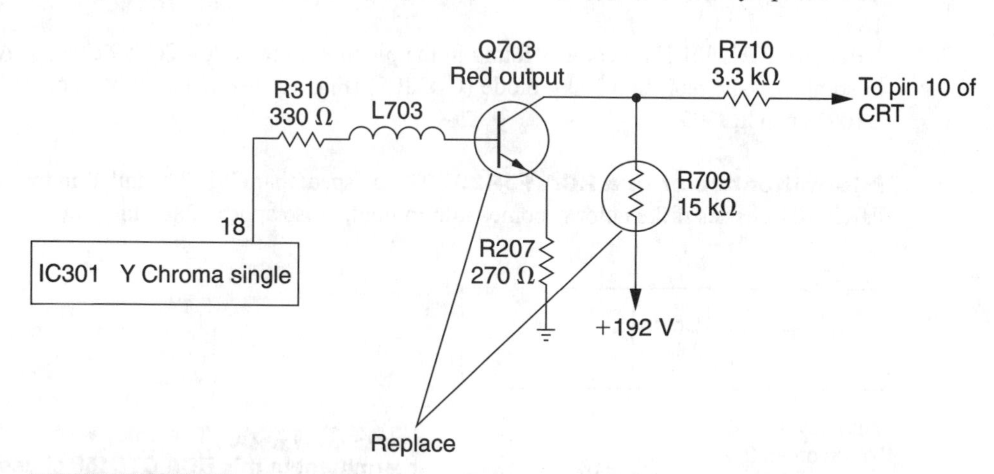

FIGURE 10-37 Leaky transistors Q703 and R709 (15KΩ) were replaced to restore the red output in a Sony TV.

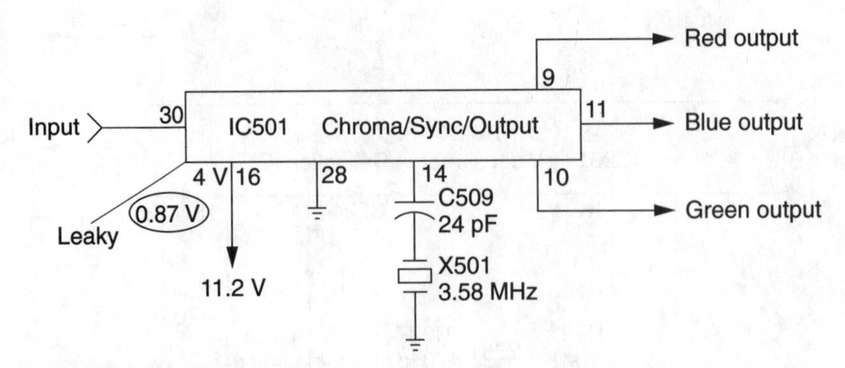

FIGURE 10-38 A lack of color in a Goldstar TV chassis was caused by a leaky IC501 with low voltage on supply pin 16.

No color in a Goldstar CMT-2612 A good, normal black-and-white picture was found in this Goldstar chassis, without any color. A quick color scope waveform check on pins 9, 10, and 11 showed no color. A normal color waveform was found at pin 30 of IC501. Pin 14 of the oscillator had no waveform. The supply voltage on pin 16 was less than 1 volt (Fig. 10-38). Pin 16 was removed from the PC board; it had a 0.15-Ω measurement to common ground. IC501 (LA7629) was replaced with an NTE7008 universal replacement. R550 and D504 were also replaced in the low-voltage (11.2-V) source of the power supply.

Intermittent or no color in an RCA CTC156 Sometimes the color in an RCA CTC156 color chassis (Fig. 10-39) would disappear and return without any remote-control transmitter adjustments. Capacitor C2810 on pin 5 of the color IC (U1001) was suspected. When the color was tuned in, the voltage was around 8 V; it should be about 5.2 V on a black-and-white or no-color picture. If the voltage is below 5 V, suspect that C2810 is leaky. This capacitor has caused a lot of different color problems on several different RCA TVs.

Replace ICU1001 if no color is found in the picture of an RCA CTC159 chassis. Also, in some chassis, replace a leaky diode (CR3305) (Fig. 10-40). If in doubt, replace both U1001 and CR3305.

Intermittent color in a RCA FJR2020T Suspect that C611 has failed in the RCA FJR2020T chassis if the color becomes intermittent, noise appears, and the picture disap-

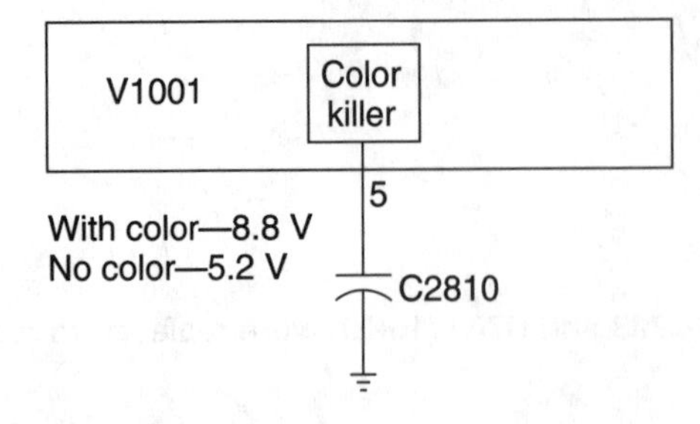

FIGURE 10-39 The color was intermittent in this RCA CTC156 chassis; it was caused by a faulty capacitor (C2810).

pears (Fig. 10-41). This capacitor is noted for developing intermittent and open conditions. C611 is located between the video buffer and comb processor (U600).

TROUBLESHOOTING WITH THE COLOR-DOT-BAR GENERATOR

The low-cost portable color-pattern generator provides a variety of test signals and patterns for TV and VCR servicing. These testers are usually crystal controlled, with outputs on channel 3 or channel 4 of the TV. The video patterns consist of a dot, crosshatch, and color bars, a gated rainbow, three bars, or ten bars. There is also a color raster, full-field color raster, user-adjustable hue, blank raster, purity, white wide bar, and half-screen white bar. The video output is 1 volt peak-to-peak (Fig. 10-42).

The color-dot-bar generator is ideal when setting up and adjusting the TV, or for making adjustments after a new picture tube has been installed. This tuner is also connected to the TV antenna for color troubleshooting. Most manufacturers (and Howard Sams Photofacts) use the color bar or NTSC color generator for color waveforms within the color circuits.

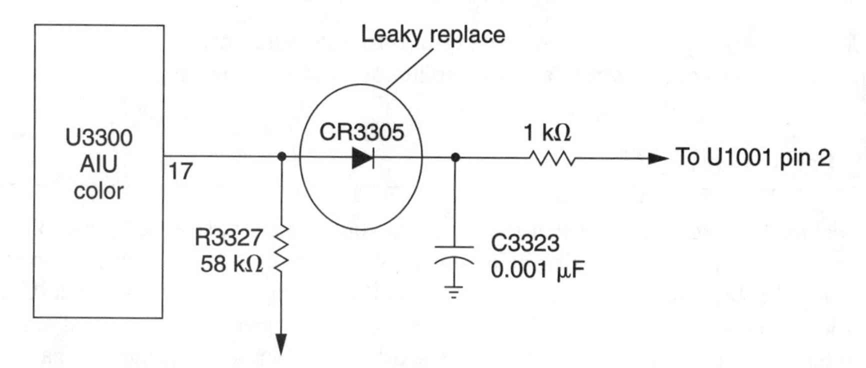

FIGURE 10-40 If no color is found in the picture, replace the leaky components (U1001 and CR3305) in the RCA CTC159 chassis.

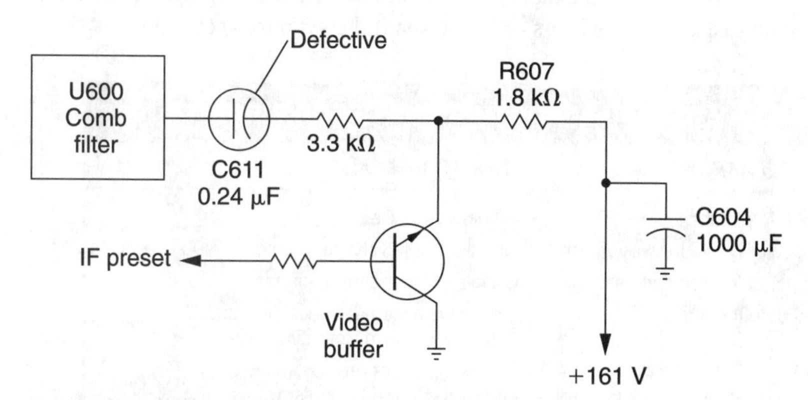

FIGURE 10-41 A defective C611 (0.24 μF) caused intermittent color in an RCA FJR2020T.

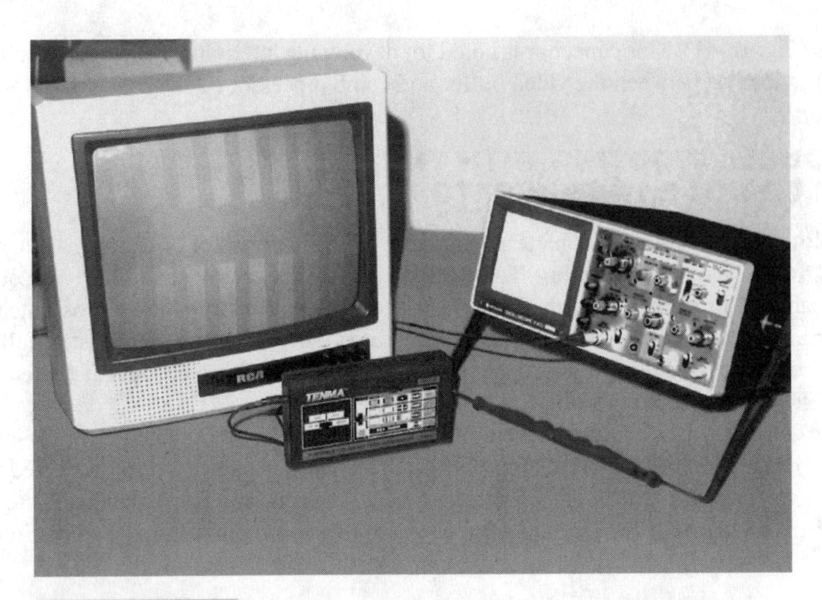

FIGURE 10-42 **The color-dot-bar generator can help you to troubleshoot the color circuits if the oscilloscope is an indicator.**

Conclusion

Checking the color-section waveforms can help you to solve the most-difficult color problems. After locating the missing color signal, take voltage and resistance measurements. Testing the suspected transistor within the circuit can help. Besides the chroma IC, check for leaky capacitors or a change in resistance between the ground and terminal pin.

Intermittent color symptoms can be caused by poor board connections. Chassis with double-sided board wiring can have intermittent feed-through connections. Do not overlook the possibility of a bad 3.58-MHz color crystal if no-color or intermittent-color symptoms occur. Always use an exact replacement part when changing crystals. Use coolant and heat on suspected transistors or ICs to see if the symptoms change.

TABLE 10-1 TROUBLESHOOTING THE COLOR CIRCUITS

WHAT TO CHECK	HOW TO CHECK IT
No color in picture.	Adjust color killer.
Check color oscillator waveform.	Check supply voltage to color IC-sub crystal.
Check for color input waveform.	Check luma and input circuits.
Still no output color.	Check waveforms on color output IC. Check waveform from flyback.
No red.	Check output of red IC. Check red output transistor. Check voltage on red output.
Weak color pictures.	Test picture tube—red, green, and blue guns.

Color alignment should not be attempted unless the required test equipment is available. Follow the manufacturer's color-alignment procedures for each TV. Do not tamper with color adjustments. The trouble might be elsewhere and the adjustments do not move by themselves. If you feel that color alignment is needed, take the chassis to a qualified TV technician or to a factory-service center that performs color alignment. For more information on troubleshooting the color circuits, see Table 10-1.

11

TESTING THE REMOTE-CONTROL CIRCUITS

CONTENTS AT A GLANCE

Servicing the remote-control units is just as easy as repairing the horizontal stages of a TV. Most TVs sold today are of the remote-control variety. With many additional cable stations, the remote receiver has moved rapidly into most homes. The remote-control circuits offer easy operation of the TV from a distance. Like any other TV component, the remote circuits break down and need repair.

FIGURE 11-1 **Many different sizes of batteries are used in remote transmitters.**

The remote circuit consists of a transmitter and receiver to operate the various functions of the color receiver. In the very early remote TVs, the mechanical transmitter operated the on/off, volume, and channel selector. The remote receiver picked up the transmitted signal, operating the various functions with several small ac motors. Later, the mechanical transmitter was replaced with a supersonic unit (Fig. 11-1). Today, the infrared transmitter controls many different operations, including scanning stations up and down, selecting individual channels, turning power on or off, selecting volume, and muting the sound. Besides controlling all TV functions, some remote transmitters that control the various operations of a VCR.

Basic Remote Transmitters

One of the first methods used to generate a supersonic wave for remote control was a mechanical or physical method of tapping a metal rod with a hammer or trip-type plunger (Fig. 11-2). To generate three different signals, each rod was of a different length. Each rod represented a different frequency to control the on/off switch and the channel up/down selector. The supersonic signal was picked up by the remote transducer of the remote receiver to control the various operational functions (Fig. 11-3).

Another transmitting method was to generate an electronic signal to control certain frequencies to be radiated through a speaker or transmitter (Fig. 11-4). The frequency zone chosen was between 44.75 kHz and 47 kHz. This frequency avoids most signals that might erroneously trigger the remote receiver. Of course, garage door openers, door and telephone bells, and other sources of supersonic signals can trigger the TV's supersonic receiver.

In a Sharp transmitter circuit, you will find that the on/off and volume are controlled at 41.5 kHz, channel-up at 40 kHz, and channel-down at 38.5 kHz. Likewise, in an RCA

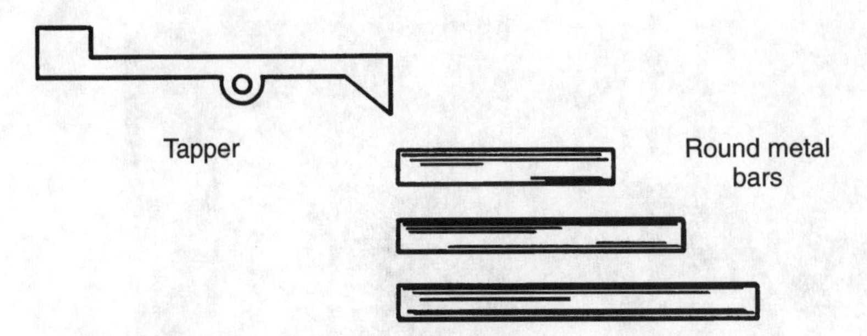

FIGURE 11-2 The mechanical plunger used in many early remote transmitters.

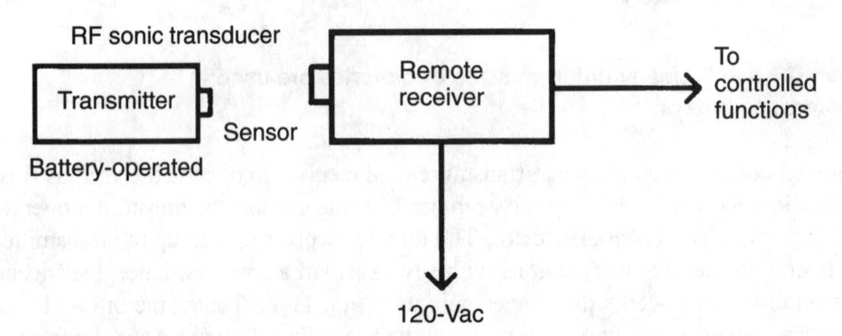

FIGURE 11-3 A block diagram of the sonic transmitter, with the transducer in the receiver.

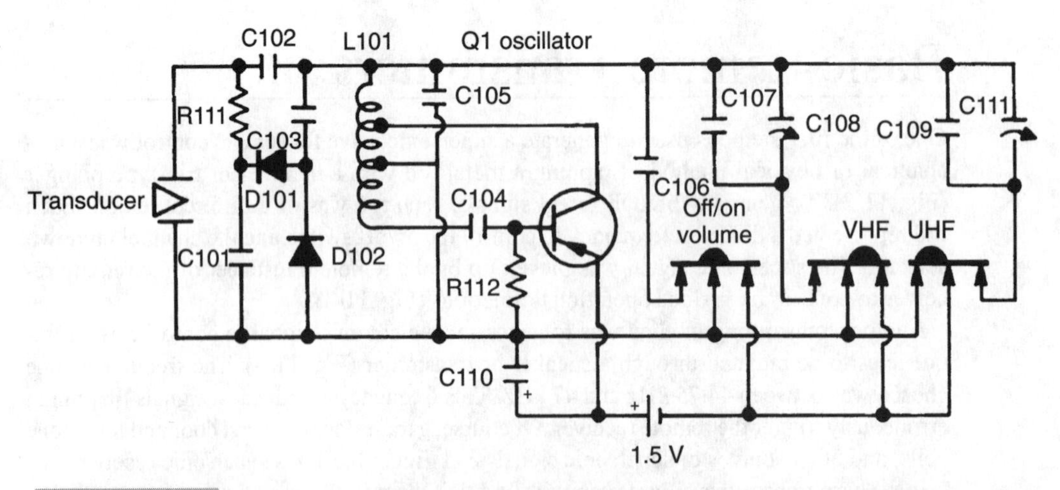

FIGURE 11-4 A typical sonic transmitter, controlling on/off/volume and VHF/UHF tuning.

CRK19A remote transmitter, the on/off and volume operate at 44.75 kHz, with channel-up and -down at 41.75 kHz. Other manufacturers use different frequencies for volume and channel selections. You might find that one frequency controls the on/off and volume in one model; in another model, the same frequency is used for channel-up or -down. This helps prevent the interaction of remote transmitters when two or more remote-control TVs are in the same home.

Although the supersonic remote transmitter was a big improvement over the mechanical transmitter, it had a few drawbacks in operation. The biggest disadvantage was when several remote-control TVs were used in a large apartment house. Your TV set might be turned on by a neighbor with another remote TV. Erroneous signals from various electrical and RF-generating devices can trigger the remote-control receiver. The design of the infrared remote transmitter has solved most of these signal problems.

TESTING SUPERSONIC REMOTE TRANSMITTERS

The small supersonic remote transmitter can be checked with another TV receiver using the same type of remote. Always take the remote to the shop when on a house call and the remote control does not function. Most remote problems involve the remote transmitter. The remote receiver in the TV set can be checked with a new remote transmitter to determine if the receiver is defective. If battery replacement does not repair the remote transmitter, the unit should be sent in for exchange or repair.

The supersonic remote transmitter can be checked with the help of the remote receiver amplifier alignment procedure (Fig. 11-5). The oscilloscope is connected to the test point of the receiver with the remote transmitter close to the receiver transducer or microphone. Each coil is peaked for maximum indication on the scope with the remote transmitter operating. Repeat the adjustments with the transmitter several feet from the TV receiver. Follow the manufacturer's alignment procedure for remote amplifier adjustments.

SERVICING THE SONIC TRANSMITTER

Most remote-control transmitters can be exchanged or serviced by the manufacturer's distributor or the factory, or they can be repaired at a tuner repair center. Simple transmitter

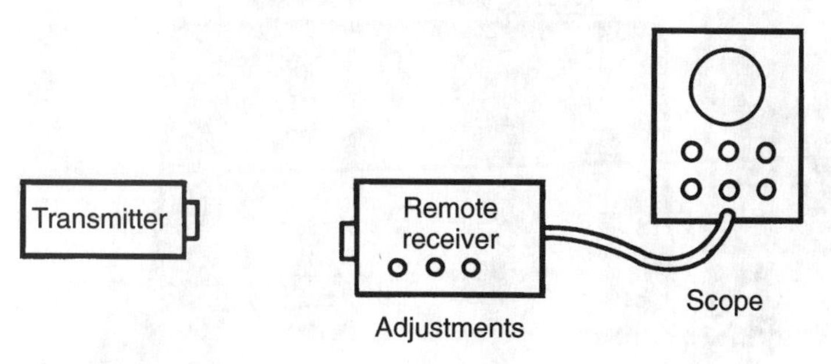

FIGURE 11-5 Check the remote transmitter signal with the remote receiver and scope as indicator.

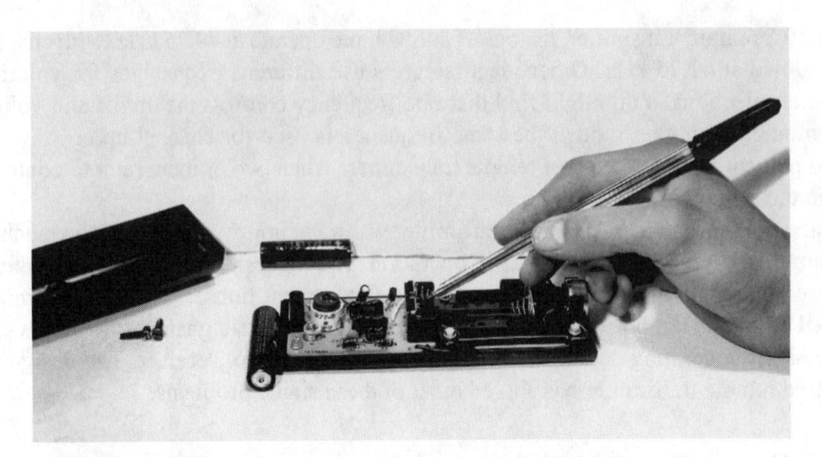

FIGURE 11-6 **Check the corroded or loose battery terminals in the remote transmitter.**

repairs can be made by the operator or service technician. Battery replacement and minor repairs should be attempted before packing up the remote-control unit.

Check the batteries with a meter or replace them. Because heavy current drain is placed on the battery, replace the batteries with heavy-duty types. Today, the remote transmitter is often used constantly. Inspect the battery terminals for corrosion (Fig. 11-6) and for broken wires.

The small transmitter has only a few working components and is easy to repair. In an RCA CRK26E, replacing a few components can solve the defective remote (Fig. 11-7). The remote might have been dropped, tearing the small RF coil loose from the PC board. T1001 can be replaced easily because it is factory-tuned and locked. Check transistor Q1001 for open and/or leaky conditions (Fig. 11-8). Sometimes small children poke sharp

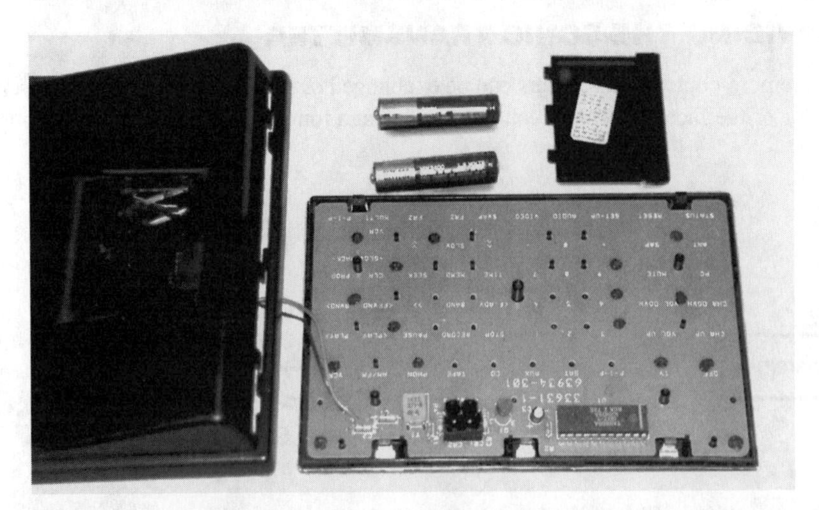

FIGURE 11-7 **Check the infrared diode, IC, transistors, and voltage measurements in the infrared remote.**

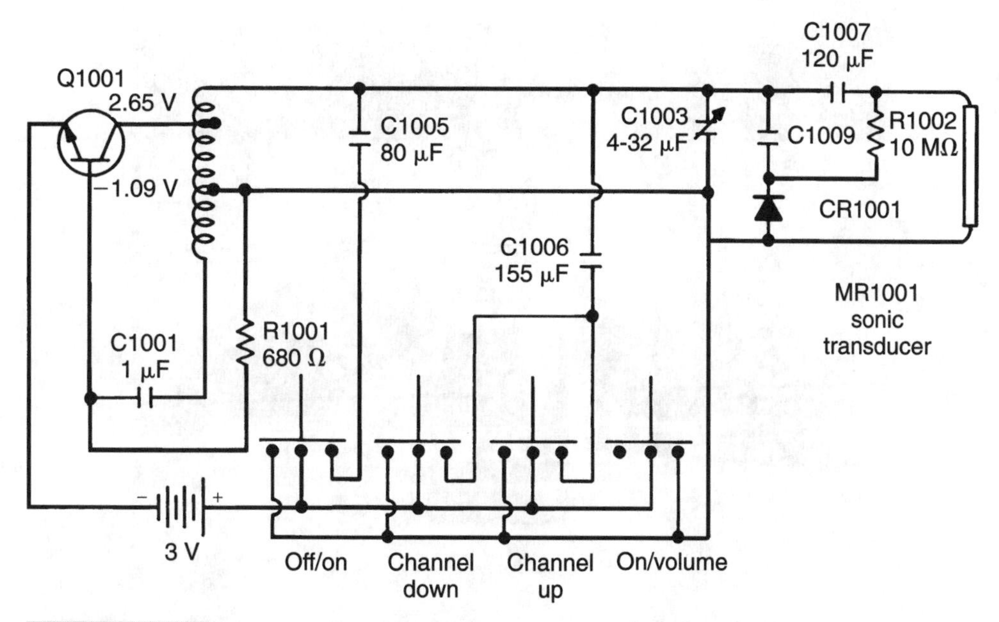

FIGURE 11-8 **Check the transistor and take voltage measurements with the tuning button pressed down.**

objects into the small sonic transducer elements, damaging the transducer unit. This component can be replaced by removing only two connecting wires. Small channel buttons can stick. A plastic button can bind against a plastic front piece and not engage the switch area. Spray the sticky buttons with a silicon cleaner/lubricant. Replace all remote-control transmitter parts with the original-type parts.

INFRARED REMOTE TRANSMITTER

The infrared remote transmitter comes in many sizes and shapes (Fig. 11-9). Some of today's remotes can change the remote-controlled VCR, as well as the color TV. At first, the digital control transmitter might appear quite complicated, but it is easy to operate. Besides operating the regular TV controls, the remote can operate the on-screen displays and the stereo audio system. The RCA Dimensia digital control operates the TV, VCR, AM-FM, phone, tape, and CD from one remote (Fig. 11-10).

The infrared remote transmitter must be pointed directly at the TV to make the remote perform the different operations. The remote receiver will not respond if someone stands between the operator and the TV. An infrared remote is generally very accurate and dependable; it will not interfere with any other TVs in the house.

Most infrared remotes are constructed like the regular sonic transmitter, except that LEDs serve as transducers. A regular keyboard assembly feeds into an encoder IC processor with crystal-controlled driver and amp transistors. One or more LEDs provide transmitting light power to the TV's remote receiver. The various operations of separate channel selection, channel-up and-down, on/off, volume, mute, and recall can be found within the infrared remote transmitter. Most infrared transmitters are operated by 3 to 9 V.

FIGURE 11-9 The infrared remote transmitter can operate a TV, VCR, receiver, and CD player.

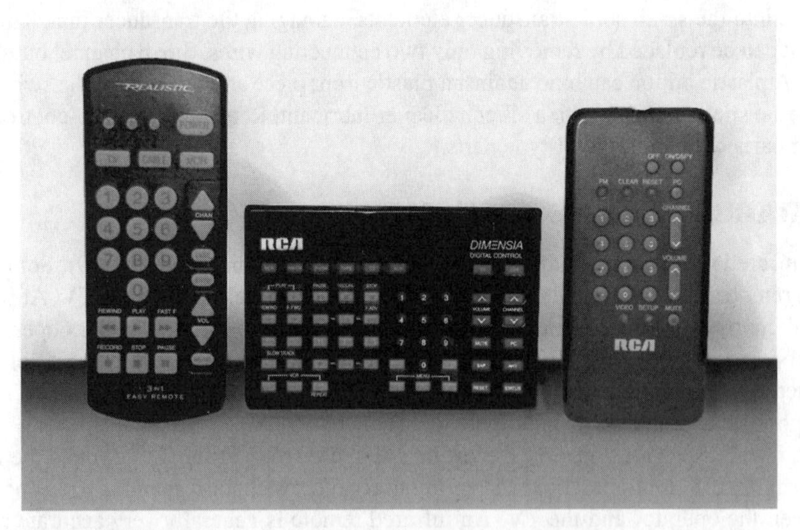

FIGURE 11-10 The digital remote can operate the TV and VCR.

TESTING THE INFRARED TRANSMITTER

Check the output of the infrared transmitter to determine if the remote or the TV is defective. The infrared transmitter can be checked on another TV using the same remote. If the remote changes all functions on the TV, suspect that the remote receiver in the TV is defective. Another method is to use an infrared indicator card in front of the remote transmitter. The infrared indicator card will change to a different color if the transmitter is working. The infrared indicator card is obtained through TV distributors. The RCA infrared indicator for checking the output of remote-control TV transmitters is stock number 153093. The infrared remote can also be checked with a portable radio. Each time a button is pressed, you will hear a gurgling sound in the portable radio (Fig. 11-11).

The infrared remote transmitter can be checked as any remote unit. Check and replace the 9-V battery. Determine if the remote transmitter is defective before removing the back cover. Take voltage measurements on the various transistors. Check each transistor with the diode-transistor tests of a DMM. The LEDs can be checked with the diode test function. The keyboard assembly should be replaced instead of attempting to repair it. Send the remote transmitter in for repair or exchange if simple repairs will not cure the malfunction.

INFRARED REMOTE-CONTROL TESTER

You can check remote-control transmitters in minutes with the infrared remote-control tester (Fig. 11-12). Place the remote a few inches from the tester and press any button on the remote. You should hear a chirping or audio tone when any button on the remote is pressed. Some infrared remotes produce a chirping noise, but others produce a loud tone from the piezo buzzer. This remote tester can check all infrared remotes for TVs, VCRs, and CD players. Move the remote transmitter back and forth in front of the tester until the loudest sound is heard. Check the remote for a weak signal by moving the remote away from the tester. With this checker, some remote transmitters can sound off at 2 feet. Keep the remote pointed toward the tester at all times.

Check each remote button for correct operation. If the button appears erratic or intermittent, suspect a dirty button. Suspect that the battery connection is broken or poor if the remote operates intermittently. If you have to press hard on the button or if it takes a little time for the remote to act, suspect that the batteries are weak. Check for poor connections or loose batteries when the remote control must be tapped or shocked into operation. No sound indicates that a control unit is defective or that the batteries are dead.

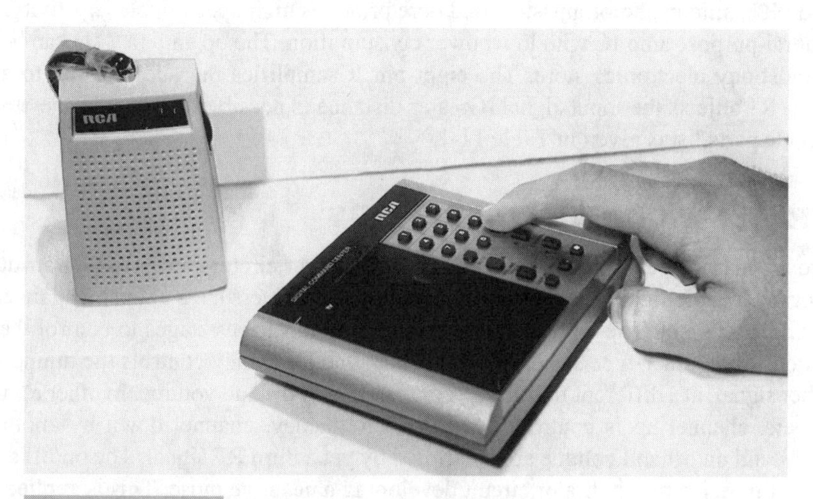

FIGURE 11-11 Check the remote beside a radio.

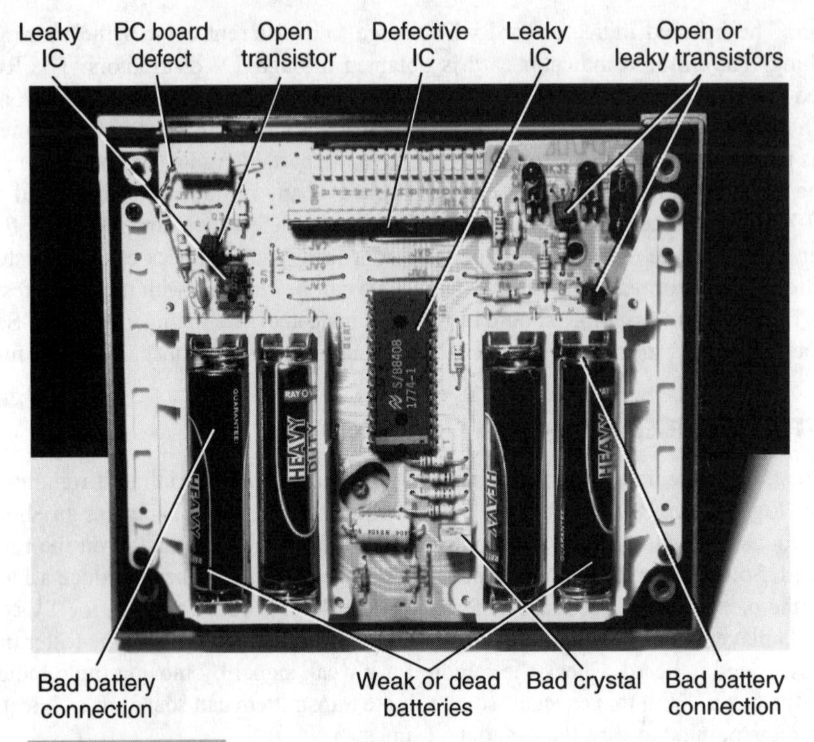

FIGURE 11-12 **Check the following parts in the remote transmitter for defects.**

The circuit The small remote transmitter tester circuit is built around an NPN infrared phototransistor. These phototransistors can be picked up at almost any electronics supply store. Sometimes the transmitting LED and phototransistor come as an operating pair (Fig. 11-13). The infrared beam is picked up by Q1 with the input signal applied to IC1. The infrared NPN silicon phototransistor used here provides high-speed photosensitivity. IC1 is a general-purpose amp IC with low power consumption. The op amp (a 741) can be found at almost any electronics store. The eight-pin IC amplifies the audio signal to a piezo buzzer. R1 adjusts the input signal. Greater distance is possible with ICI in the circuit. A complete parts list is given in Table 11-1.

RF REMOTE-CONTROL RECEIVERS

There are many types of remote receivers. The receiver circuit consists of a transducer or sensor fed to a high-gain remote amp. Some remote receivers have a separate gain control, while others do not. The high-gain signal is fed to the various stages to control the many functions of an older-type system (Fig. 11-14). One frequency controls the tuning motor; another signal, at a different frequency, controls the on/off and volume. In other remote RF receivers, channel up is controlled with one frequency, channel down by another frequency, and on/off and volume are controlled by yet a third RF signal. The on/off signal is applied through a multivibrator circuit developing a negative pulse. This is applied at two separate transistors, so the set can be switched on and off. The on/off signal is applied to a transistor amp with a power relay. The relay applies on/off ac voltage to the TV chassis. Although each manufacturer has a different-looking remote receiver system, they all do the same thing: They make TV operation easier for the operator.

STANDBY POWER CIRCUITS

The standby power-supply circuits must operate all the time so that the remote transmitter can turn on the TV chassis. A standby voltage must be applied to the infrared (IR) receiver so that the remote can turn on the TV. The standby-voltage circuit in an RCA CTC157

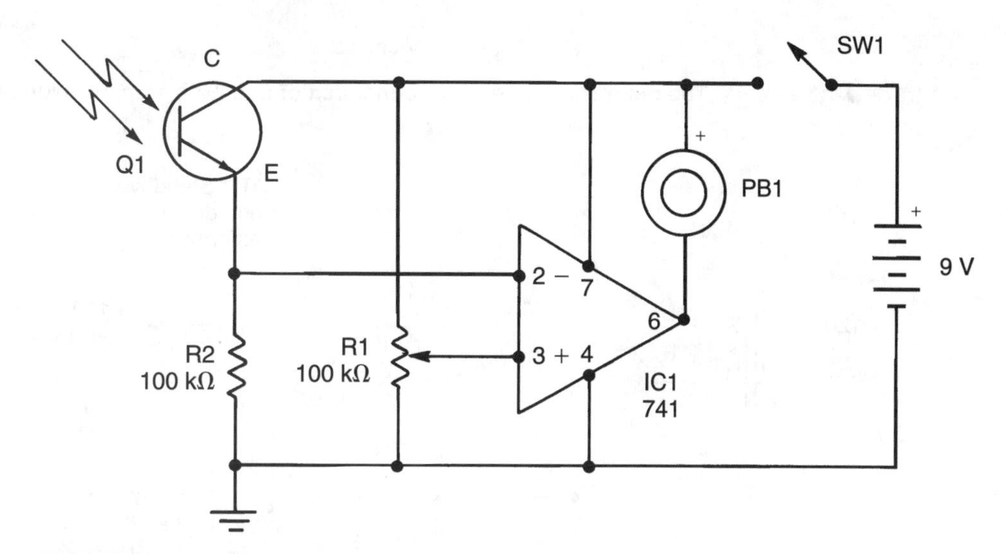

FIGURE 11-13 The schematic of the remote-control tester that you can build in one evening.

TABLE 11-1 A PARTS LIST FOR AN INFRARED REMOTE CONTROL TESTER

Q1	Infrared phototransistor npn type—276-142 or equivalent
IC1	741 general-purpose op amp IC 8-pin type—276-007 or equivalent
PB1	PC mount piezo buzzer—273-065 or equivalent
R1	100-kΩ, PC board PAT, thumbscrew or screwdriver adjust—271-220 or equivalent
R2	100-kΩ, 0.5-W resistor
SW1	Submini, SPST toggle switch—275-645 or equivalent
Battery	9 V
Case	3.25 × 2.125 × 1.125 project box—270-230 or equivalent
Miscellaneous	Piece of perfboard, 9-V battery, lead and socket, hookup wire, 8-pin DIP socket, 8-pin DIP socket, solder

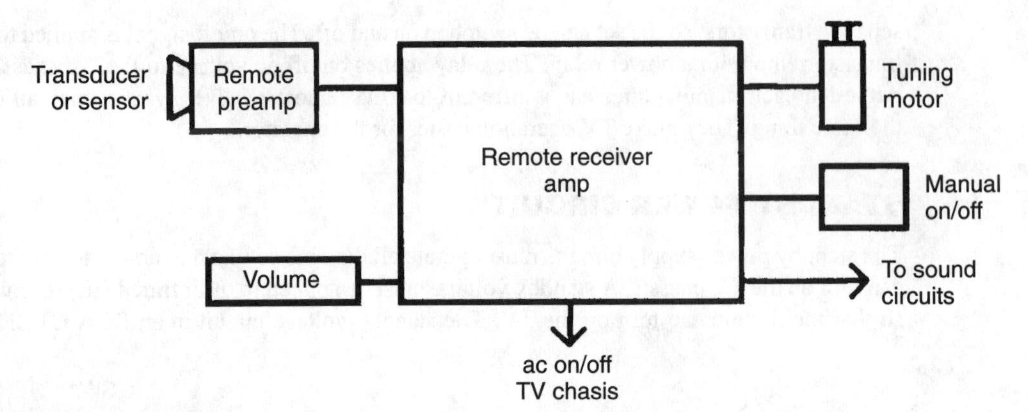

FIGURE 11-14 **The older remote receiver consisted of a remote amp, transducer, and motor.**

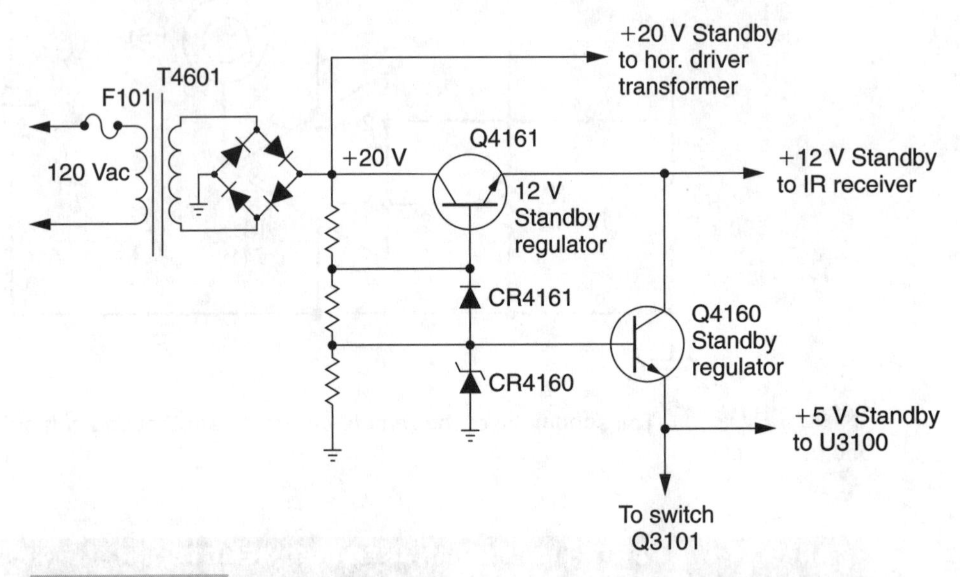

FIGURE 11-15 **A standby power supply for the remote receiver in the TV chassis.**

chassis supplies 12 volts to the IR receiver and 5 volts to system-control microprocessor U3100 (Fig. 11-15).

The power-line voltage is connected directly to the primary winding of T4601. The secondary voltage is applied to a bridge-rectifier circuit providing a +20-V standby voltage to the horizontal driver transformer. The +20 V is regulated by transistor Q4161 down to the infrared receiver (IR). Q4160 is a 5-V standby regulator that is furnished to the system-control microprocessor.

Low or improper 12- and 5-V supplies can be caused by a leaky or open voltage regulator (Q4161). Notice the two zener diode regulators in the base circuits. The absence of a 12-V source can result from an open voltage regulator. If the 5.6-V zener diode (CR4160) opens, both the 12- and 5-V standby supplies will go higher. The 12- and 5-V sources are very crucial for proper standby operation.

In-TV Remote-Control Circuits

INFRARED REMOTE RECEIVER

The infrared rays transmitted by the infrared transmitter are picked up by the light sensor of the remote amplifier that feeds the remote receiver circuits. The photodiode is called a *light sensor*, *pin diode*, or *remote sensor* (Fig. 11-16). The preamp is called a *photo amp board*, *remote amplifier*, or *preamp board*. The pulse-modulated infrared rays provide remote control for changing channels, volume-up and -down, power on/off, and many other functions. Most manufacturers request that the remote-control preamplifier unit be completely replaced instead of repaired. Some manufacturers do not supply parts for them.

From the high-gain preamplifier board, the FM signal is connected by cable to the remote-control receiver circuit. Many different kinds of remote-control receivers are used (Fig. 11-17). Some of these remote receivers are of the modular type. This kind of complete remote

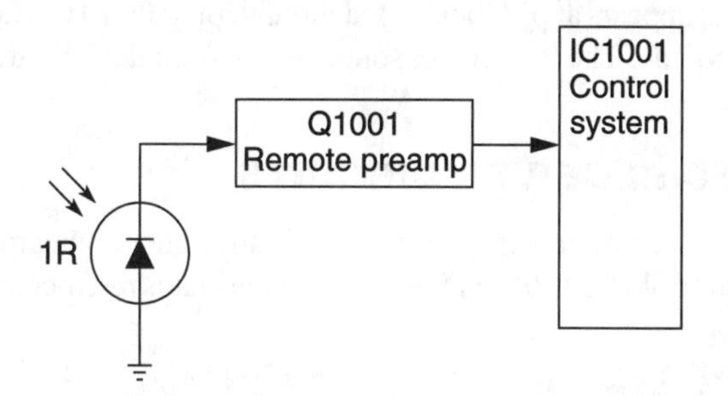

FIGURE 11-16 A block diagram of the infrared preamp board with the emitting diode.

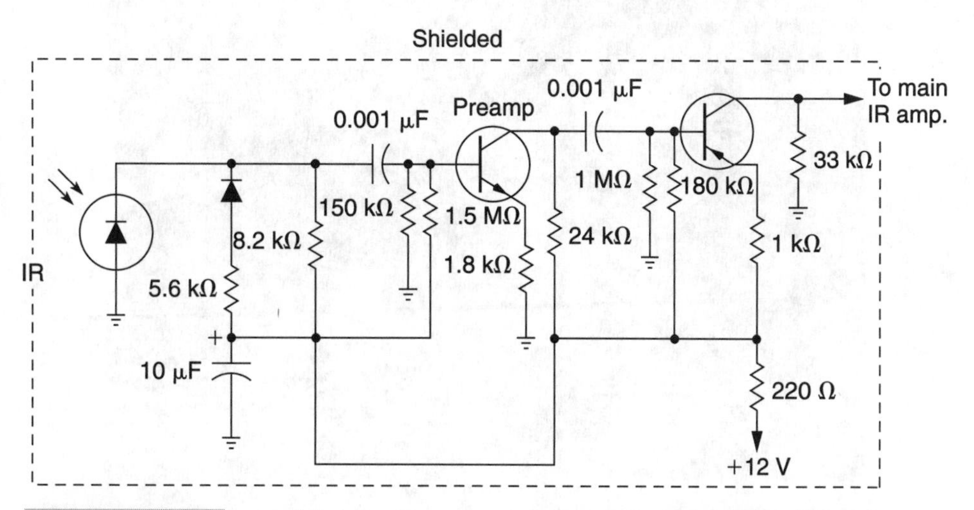

FIGURE 11-17 A typical transistorized remote preamp in the receiver of the TV chassis.

TROUBLESHOOTING AND REPAIRING SOLID-STATE TVS 2

receiver unit can be interchanged by unplugging a few cables, removing the original unit, and substituting a new one.

The new infrared remote receiver can have many functions. Besides channel selection, a scan-up and -down control might be included. Some units have a separate on/off switch. Besides controlling the volume, a remote might also mute the sound. A channel-recall button is provided in some remote-control systems.

The remote-control receiver can be exchanged at the manufacturer's distributor or sent in for repair at various tuner-repair depots. The remote receiver can be serviced by the TV technician if correct test instruments are available. Do not attempt to repair these circuits without the manufacturer's service literature and a schematic that provides the voltages.

SIMPLE REMOTE-CONTROL RECEIVER

You might find the remote receiver as a part of the memory or control module, or as a separate chassis in the color TV (Fig. 11-18). The remote receiver can be quite simple or rather complicated, depending on its functions. The simple remote receiver contains an IC amp, an infrared LED, and very few components (Fig. 11-19).

The infrared transmitter signal is picked up by LED1 and coupled directly to pin 7 of IC501. The amplified output signal appears at pin 1 and is fed through plug PG241 to the microcomputer within the control module. The 9-V power source is fed from the TV circuits.

RCA CTC157 INFRARED RECEIVER CIRCUITS

Two different inputs of customer control consist of a remote control and keyboard. The keyboard operates the system control (U3100), which in turn controls a microcomputer

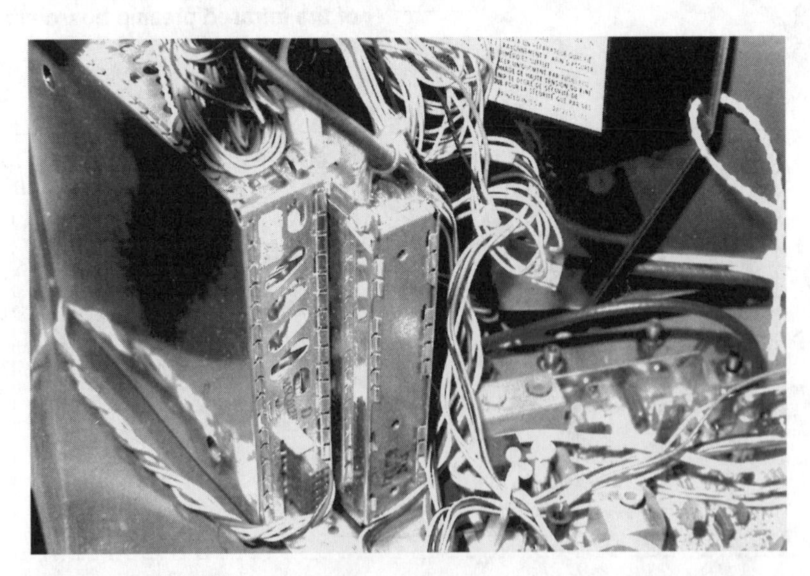

FIGURE 11-18 The remote receiver is located in the control module of the RCA chassis.

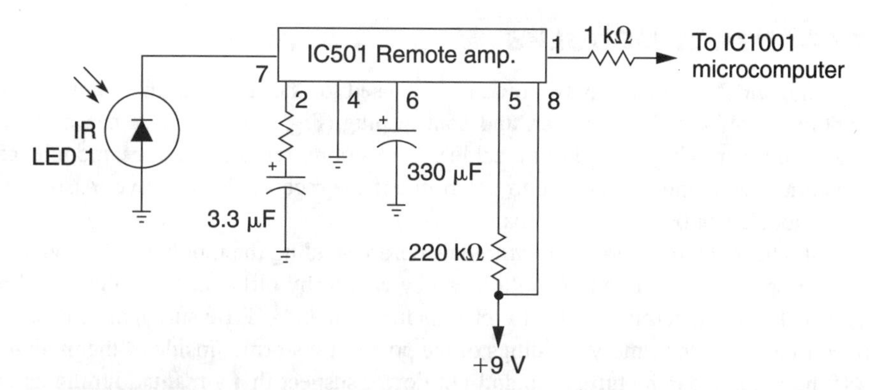

FIGURE 11-19 A typical remote receiver used ahead of the microcomputer in the latest TV chassis.

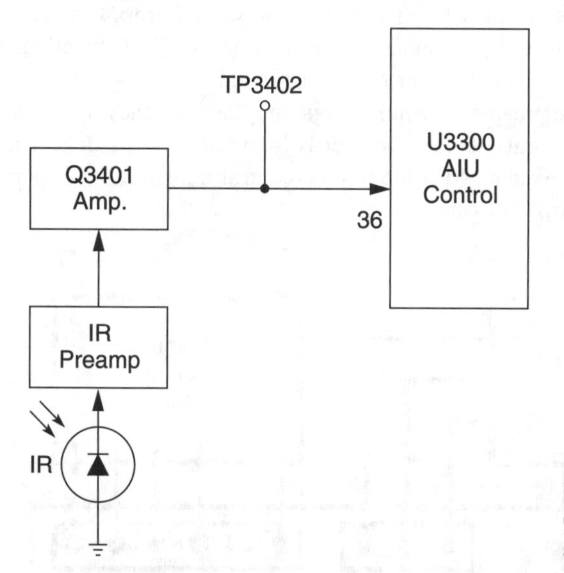

FIGURE 11-20 The RCA CTC157 infrared remote receiver schematic.

(U3300). The keyboard information is input at pins 21 to 24 and 25 to 28 of the system-control IC (U3111). All customer controls are generated electronically (Fig. 11-20).

The second input to the control interface (U3300) is an infrared receiver (IR). The IR detector picks up the remote control's signal and feeds it into a preamplifier circuit. Q3401 amplifies this signal and applies it to pin 36 of U3300. In turn, U3300 relays or controls each function selected by the remote-control transmitter.

If the remote-control transmitter will not control the TV, check the batteries in the remote. Does TV operate with keyboard controls? Try another remote. Notice if all functions are inoperative. Check for a remote input at TP3402 for scan pulses with the oscilloscope. If all customer controls do not operate, suspect a problem with the system-control IC (U3300). If the keyboard controls the TV chassis, but neither the remote functions nor the set can be turned on, check the standby voltage circuit.

TV CONTROL MODULES

Separate remote receiver-control modules are used for the remote amp, manual tuning assembly, on/off switch, ac power, and control plug (Fig. 11-21). These can be easily replaced by removing the plug-in cables. The entire remote receiver module can be substituted with another module to determine if the receiver is defective. Also, the complete module can be sent in for repair.

First, check the most common problems before sending the module in for repair. Intermittent operation of the on/off button can be caused by dirty contacts. Place a clip wire across J1104 to determine if the TV chassis turns on and off. Be sure that ac power is applied to the remote primary windings of the power transformer inside of the module.

If the volume can be turned up and not down, suspect that a manual tuning assembly, socket, or remote receiver is defective. Replace the remote module to isolate the manual tuning assembly (Fig. 11-22). Suspect a problem in the contacts within the volume-down area of the manual assembly if switching the receiver module does not help. Unplug the manual tuning assembly and short pins 1 and 4. The sound should drop. Likewise, if the channels will not scan up, but will go down, short pins 4 and 6. Each function can be checked in the same manner by using pin 4 as common ground.

In the latest TVs, the remote preamp assembly plugs into the frequency-synthesis tuner circuit-board assembly. No separate remote receiver is found in these models. The tuner circuit board module must be exchanged if the remote-control transmitter works normal, but the set has no remote-control functions.

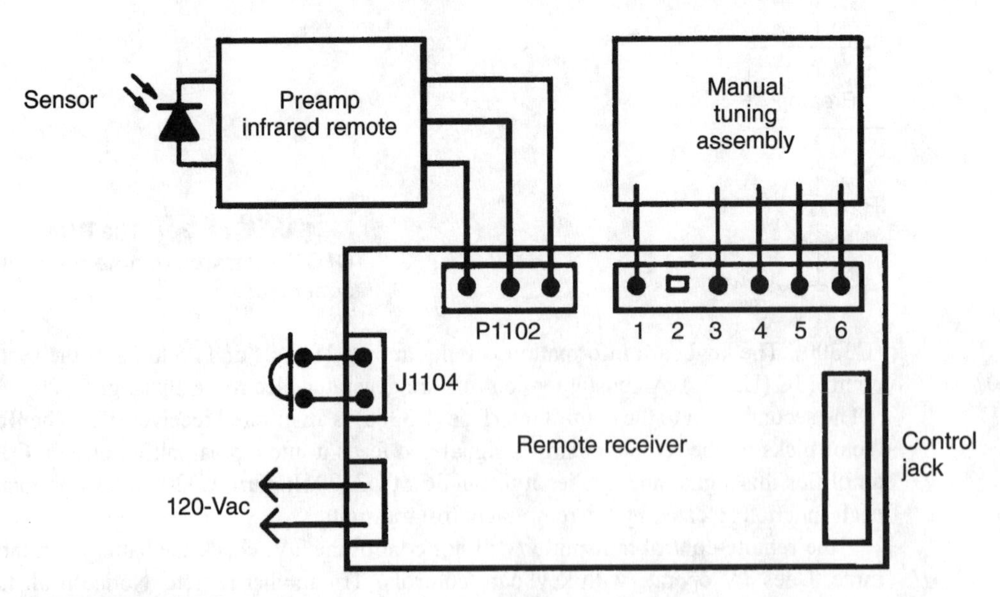

FIGURE 11-21 **The infrared sensor and preamp feeds to the remote receiver in a frequency-synthesis tuner.**

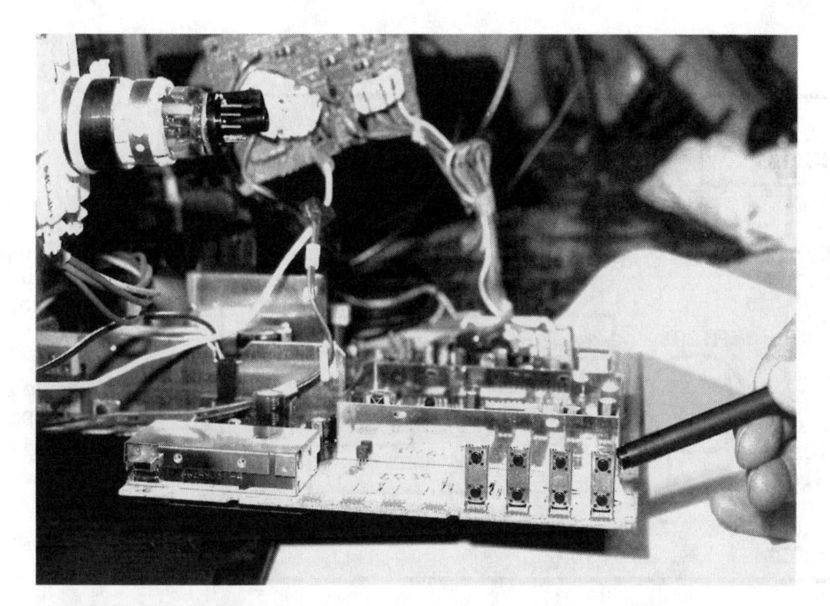

FIGURE 11-22 Small pushbutton microswitches feed into the microcomputer IC to operate each selected station.

ON/OFF CIRCUIT TESTS

One of the major functions of the remote is to turn the TV off and on. With a no-sound/no-picture symptom, the TV chassis or remote receiver might be defective. If the color receiver is entirely dead, check for 120 Vac at the input and output terminal of the remote-receiver relay-switch connections. An ac voltage at both pins (21 and 22) of a Samsung power circuit indicates that the remote-control relay is normal (Fig. 11-23).

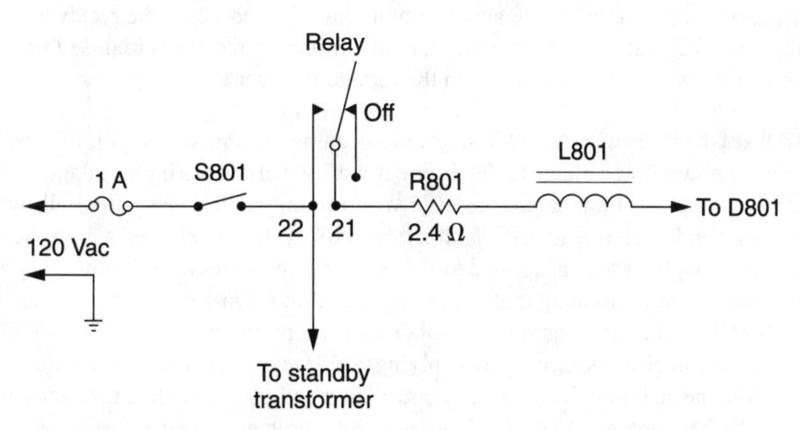

FIGURE 11-23 Shunt the relay contacts (21 and 22) to isolate the remote standby circuits.

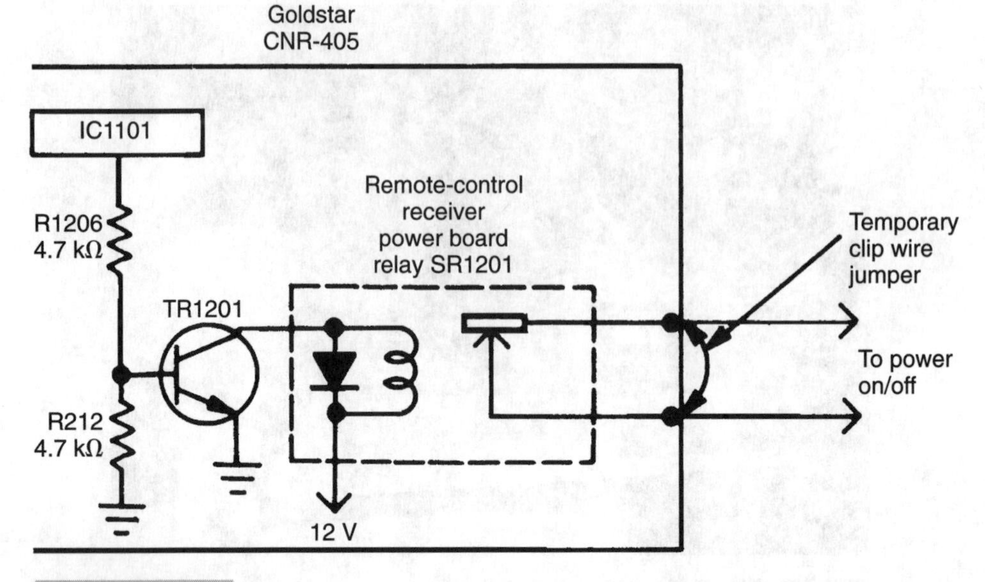

FIGURE 11-24 Poor relay contacts in a Goldstar CNR-405 remote produced a dead chassis.

If no voltage is found at the anode terminal of D801, the remote receiver unit might be defective. Listen for a click of the relay each time that the set is turned on. In case the relay clicks with no ac at D801, suspect burned or corroded relay contacts, or poor wiring connections at the remote circuit board. Check for ac voltage at pin 22, then at pin 21. A normal 120 Vac at pin 22 and no voltage at pin 21 indicates that the remote wiring is open. If normal ac voltage is found at pin 21, suspect problems within the receiver power circuits. Check components R801 and L801 for open conditions.

Temporarily clip a jumper wire across remote pins 21 and 22. If the receiver begins to operate, the defect is in the remote wiring, relay, or remote receiver circuits. Take voltage measurements and test each transistor in the remote receiver assembly.

100-W light-bulb test A 100-W light bulb can be used in series with the power line to reduce the power-line voltage to determine if the TV chassis has high-voltage shutdown. Of course, the 100-W bulb cannot be used with remote-control sets. The bulb will only flash for a second, then remain off. With a large bulb in series (clipped across the ac fuse holder), not enough voltage is applied to the remote receiver to make it function properly.

To control the ac line voltage, always use a variable or variac power-line transformer with a TV chassis that uses remote-control or power circuits (Fig. 11-24). This eliminates the remote-control circuits entirely by applying 120 Vac directly to the low-voltage power circuits. Now the ac line voltage can be controlled by adjusting the line transformer to determine if the TV chassis is shutting down with high voltage.

TV RELAY PROBLEMS

The defective relay can turn the TV chassis on without a remote or control function, might not turn the TV off or on, might cause a dead or intermittent TV chassis, or might cause a loud chattering noise. An open relay solenoid or shorted diode across the solenoid can cause no operation. Intermittent connections or dirty switch contacts can cause intermittent conditions. A weak spring or hung-up switch arrangement can allow the chassis to operate all of the time. The relay is used in the ac and dc power-supply circuits.

If the relay will not click in, suspect that the filter capacitor is defective. Relay chatter may also be caused by a dried-up filter capacitor (Fig. 11-25). The relay might come in by itself and let the receiver become hot unless someone turns it off. Relays that operate in transistor circuits could turn on by themselves, resulting in overheated components.

FIVE ACTUAL REMOTE CASE HISTORIES

The following are five real cases of remote-control-related problems in TVs.

No remote-check standby in a RCA CTC166 The remote would not turn the RCA CTC166 chassis on. The remote tested normal on another RCA TV. The 5-V standby voltage indicated that the standby voltage was very low. To isolate a possible leaky IR preamp IC, pin 9 was removed from the PC board with solder wick and iron. The 5-V standby voltage returned to normal.

In another CTC166 chassis, the standby voltage indicated only 0.17 V, but it should be around 5.3 V (Fig. 11-26). Measure the voltage on both sides of R3404 (47 Ω) to determine if CR3401 is leaky. Replace both U3401 and CR3401 if the zener diode is leaky.

No remote standby voltage in a Panasonic CTL1032R Very little standby voltage was found at the remote receiver IC1101 in a Panasonic CTL1032R. Manual operations were normal, but there was no remote action. Upon checking the remote voltage source in the power supply circuits, the standby supply voltage was missing. Regulator IC841 (7BL56) was leaky (Fig. 11-27). This voltage, supplied to the regulator, is on all the time, and it comes directly from the power line's 3-A fuse.

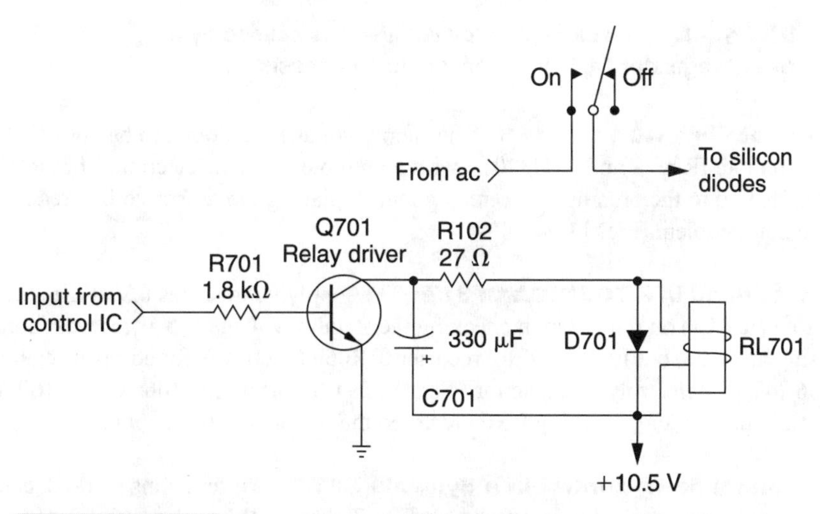

FIGURE 11-25 **The power relay can have pitted or dirty contacts, an open solenoid, or no voltage applied to the solenoid.**

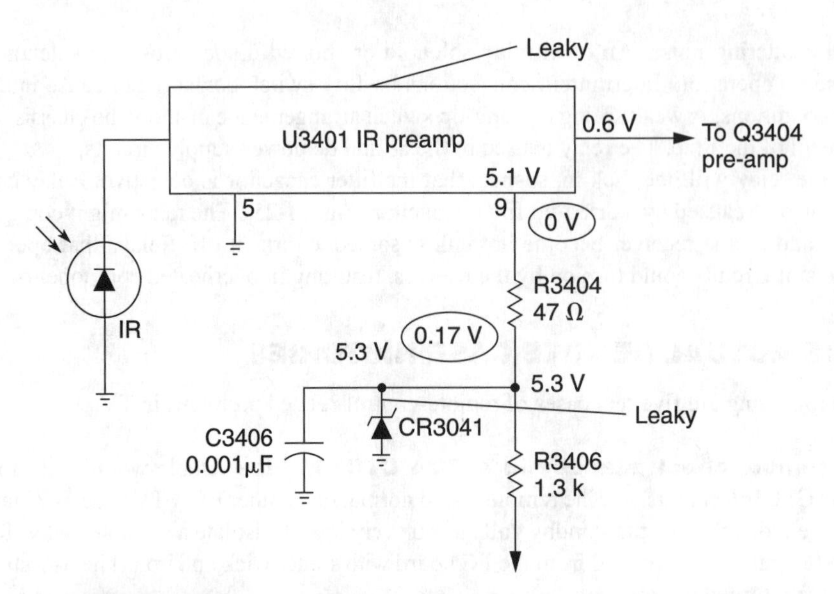

FIGURE 11-26 Low standby voltage might result in a leaky IR preamp IC or CR3041 in the RCA CTC166 chassis.

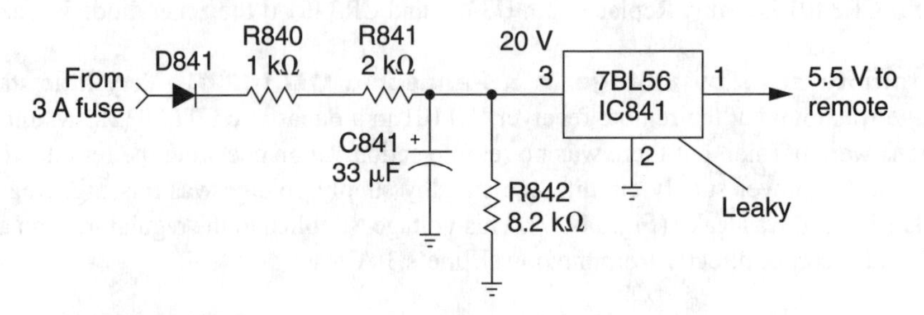

FIGURE 11-27 A lack of standby voltage was caused by a defective IC regulator C841 in a Panasonic TV chassis.

No remote in a RCA CTC166 The supply voltage was found to be normal at the receiver remote IR preamp IC (U3401), with no remote channel selection. The IR indicator and parts tied to the preamp IC seemed good. Replacing the defective IR preamp solved the remote problem (Fig. 11-28).

Remote dead in a TOSHIBA CF317 The supply voltage was 0 V at the remote signal amp (ICR01) on pin 8. Upon checking the standby voltage (+5 V), it was found to be missing at CR06, but high at DE02 rectifier. RR05 (47 Ω) was found open, with a leaky CR06 (47 μF) electrolytic capacitor (Fig. 11-29). Replacing CR06 with a 160-V electrolytic capacitor and replacing RR05 restored the remote-control operation.

No channel-down control in a Sylvania E24-7 All functions worked, except for the channel-down operation in a Sylvania E24-7 chassis. When the channel-down button was pushed, nothing happened. Also, the manual channel-down button on the chassis (SW1030) had no effect on toggling the TV channels down (Fig. 11-30).

After locating the manual low button and channel-down circuits on the remote receiver board, several voltage tests were made. The dc supply voltage was 23 V, with the same voltage at the base of Q34. Because the channel-down driver transistor tested off-value in the circuit, it was removed. Q34 tested good, but was replaced. The results were the same.

The power-line voltage (120 Vac) was found on the MT2 terminal of the channel-down triac (Q36), with no voltage at the gate terminal. The triac was defective or there

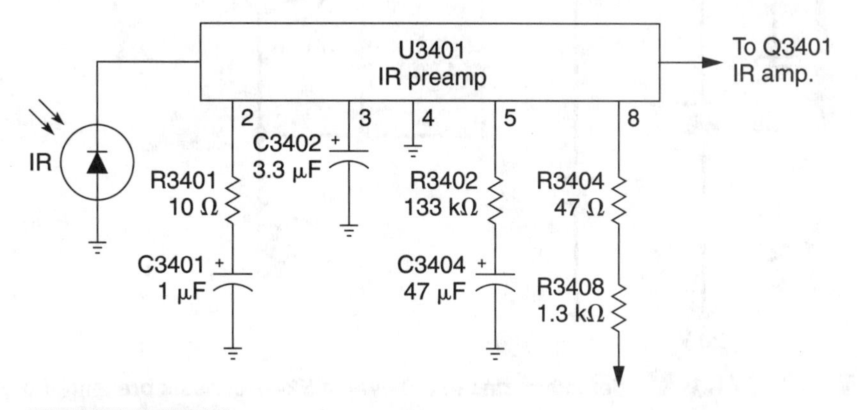

FIGURE 11-28 The dead remote in an RCA CTC166 chassis was caused by a defective IR preamp (U3401).

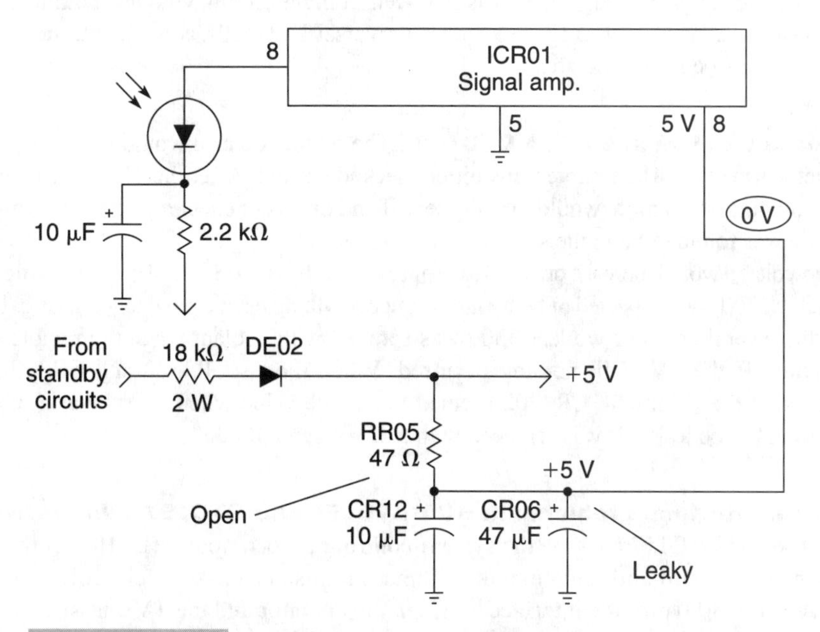

FIGURE 11-29 RR05 (47 ohms) was found open in the standby power supply of a Toshiba CF317 TV.

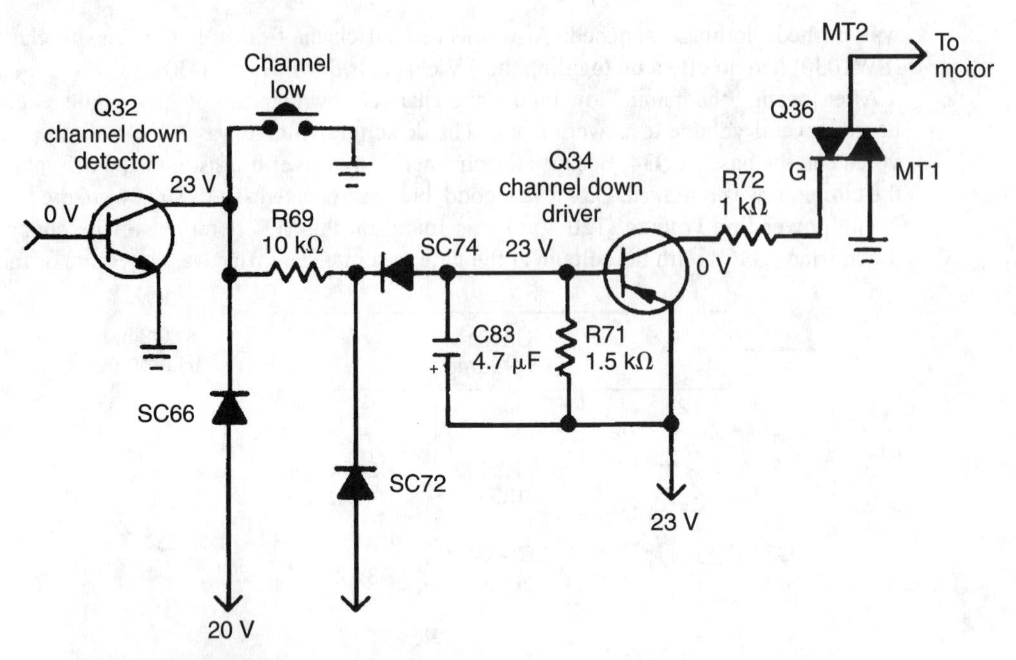

FIGURE 11-30 **A defective triac in a Sylvania E24-7 chassis prevented the remote from scanning the tuner downward.**

was insufficient gate voltage. Q36 was removed from the circuit, with no continuity across any terminals. After replacing Q36 with a universal ECG5604 component, the channel-down circuits began to operate.

Remote problem in an RCA CTC140 The remote control would not operate when the set warmed up. The remote transmitter checked normal. After the TV cooled down for several hours, the remote would turn the set off and on. After checking several circuits, the trouble was found to be in the system-control circuits.

The voltage would increase on the IR-in (pin 36) of AIU IC (U3300). Upon checking the IR circuit, U3300 was suspected of being faulty, but everything else seemed to work (Fig. 11-31).

After several hours of work, U3300 was sprayed with coolant when the remote would not shut off the TV. Still nothing occurred. When diodes CR3302 and CR3301 were checked in the IR circuit, CR3302 seemed to change value when it was warm. CR3302 was found to be leaky. It was replaced with a 4.9-V zener diode.

Control-interface problems in RCA CTC156 and CTC157 TVs In the RCA CTC156 and CTC157 chassis, the system-control microcomputer (U3100) controls the customer keyboard and remote-control input circuits. The system-control IC (U3100) and AIU (U3300) provide interface between the operator and the TV chassis. All of the interface functions are electronically controlled. U3300 controls the contrast, sharpness, color, tint, brightness, volume, balance, treble, bass, and on/off functions (Fig. 11-32).

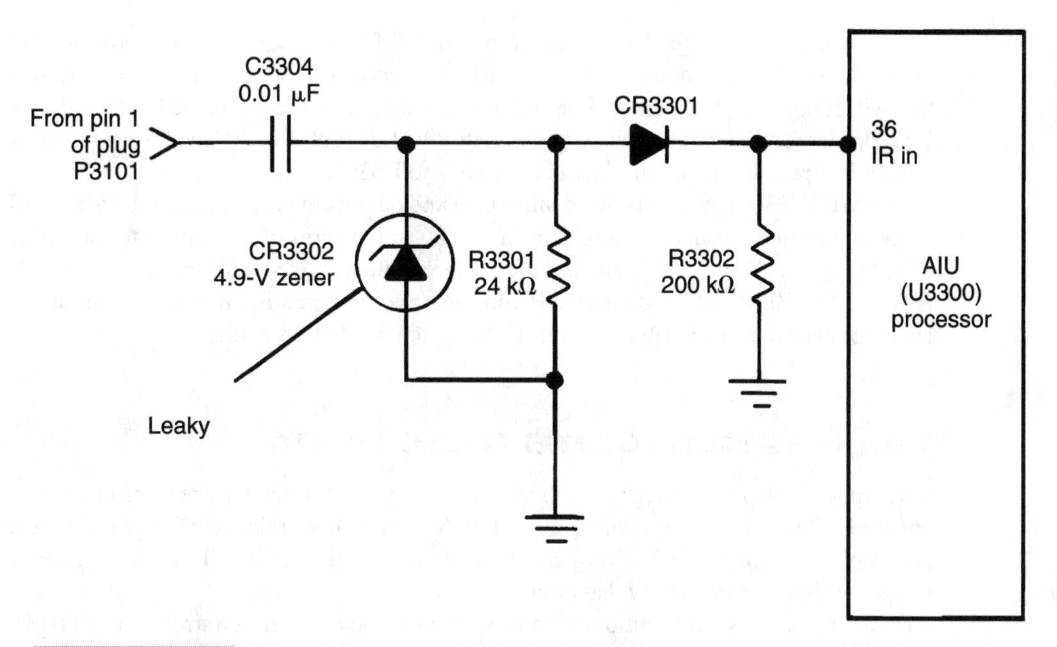

FIGURE 11-31 **A leaky CR3302 in an RCA CTC140 chassis caused the remote to not operate after three or four hours.**

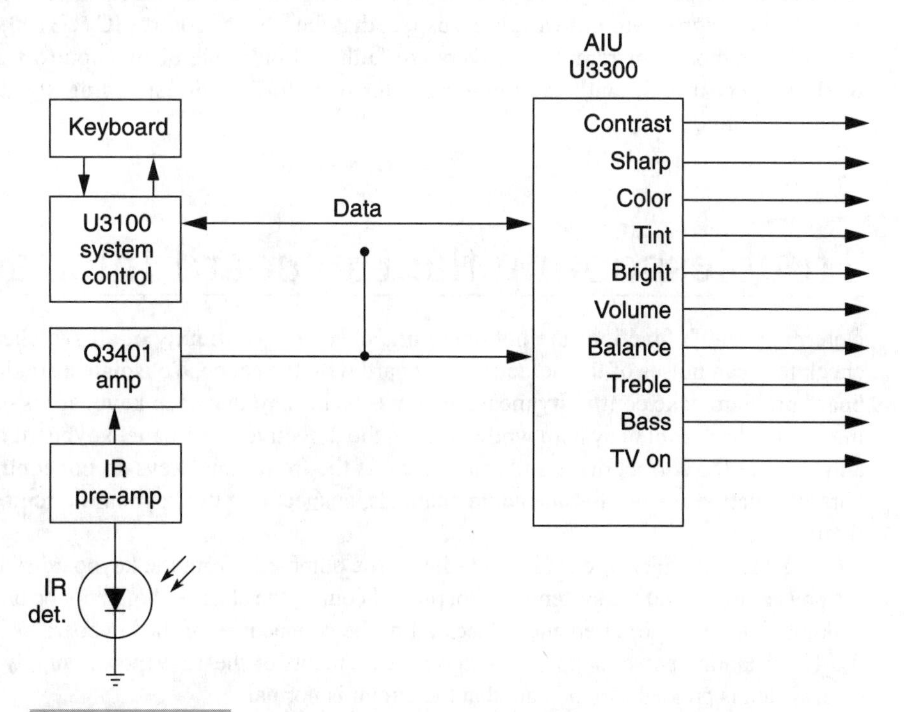

FIGURE 11-32 **A block diagram of the RCA CTC156 custom control interface circuit.**

The remote-control signal is picked up by the infrared detector LED and preamped to pin 36 of U3300. The manual keyboard signals are sent to U3100 and data is transferred to the AIU (U3300). The remote-control command signals are processed internally by U3300. They are then sent to the system-control IC (U3100), which decodes the information and outputs any necessary instructions back to U3300.

The AIU (U3300) interprets this control data and outputs the correct control signals. All control-operation signals are output from U3300 in the form of a binary-rate multiplier (BRM) signal. These BRM signals are different than pulse-width-modulated signals (PWM). The BRM duty cycle remains constant, but the repetition rate is changed to increase or decrease the desired dc control voltage to each low-pass filter.

TROUBLESHOOTING INTERFACE CIRCUITS

Determine if either the remote or keyboard controls operate. If the remote transmitter does not work, check the remote-control unit (1). If the remote is working, check the IR input at pin 36 of U3300 or TP3402. If only one function is operating, check the low-pass filter or check the signal processed by that circuit (4).

If the keyboard is not functioning, check keyboard and system-control IC U3100 (3). If the digital-communication signal between U3100 and the AIU does not change when a function control is pressed, suspect a problem with U3100. If the BRM outputs are not present, suspect U3300. Do not forget to check the voltage source at U3100 and U3300. If all signals are not working, suspect that the system-control IC (U3100), AIU (U3300), keyboard, or remote receivers are faulty. If only one or two controls do not work, suspect trouble with the low-pass filter or with the corresponding signal-processing circuits.

Troubleshooting the Keyboard Interface

Determine which functions are not operating, check the continuity of all switches, and check the scan pulses of the defective keyboard with the scope. To isolate a single keyboard problem, take continuity measurements. To be sure that each keyboard is operating, scope the scanning system while holding the defective front-panel keyboard button and monitor the correct drive and sense lines. If the front-panel keys do not control the correct functions or fail to tune cable channels, suspect that the keyboard circuit is defective.

Check the continuity of each keyboard with the ohmmeter from the keyboard switch to the power-source and the system-control pin. Of course, the chassis should not be on when making ohmmeter measurements. Check all of the connections of the keyboard switch to the 12-V standby power supply. Voltage measurements of the 12-V power supply when each switch is pressed can indicate that the circuit is normal.

Universal Remote Controls

Just about every manufacturer of TVs, as well as other electronics manufacturers, make a universal remote control that will operate almost all remote TVs, VCRs, audio and video components, and CD players. The General Electric RRC500 does the work of three remotes, and the RRC600 model does the work of four remotes. The RRC600 controls up to four infrared audio/video products, with over 200 key combinations, program sequencing, LCD display, and a low-battery indicator included (Fig. 11-33).

The Memorex CP8 Turbo universal remote control can operate home-entertainment systems without the user being home. The 16-event programming function allows the user to operate any audio or video component without being present at the scheduled operating time. For example, users can program the television to come on at 6:30 a.m. every day to serve as a wake-up alarm. The CD player can come on when they arrive home at 6:00 p.m. and go off at 7:00 p.m. when the TV is programmed to start. You do not have to even touch the audio/video system.

The CP8 Turbo unit has an advanced high-speed microchip that "learns" and "memorizes" the infrared codes of each of eight audio or video conventional universal remote-control units. The LCD display has a real-time clock and dual display light with large, easy-to-read characters, and it displays the commands in sequential steps. The remote even shows you when to change the batteries.

The universal remote operates on batteries like any other remote (Fig. 11-34). The remote can be checked like other remotes with the remote tester (covered earlier in this chapter). Most manufacturers request that the universal remote be sent back to the factory for major repairs.

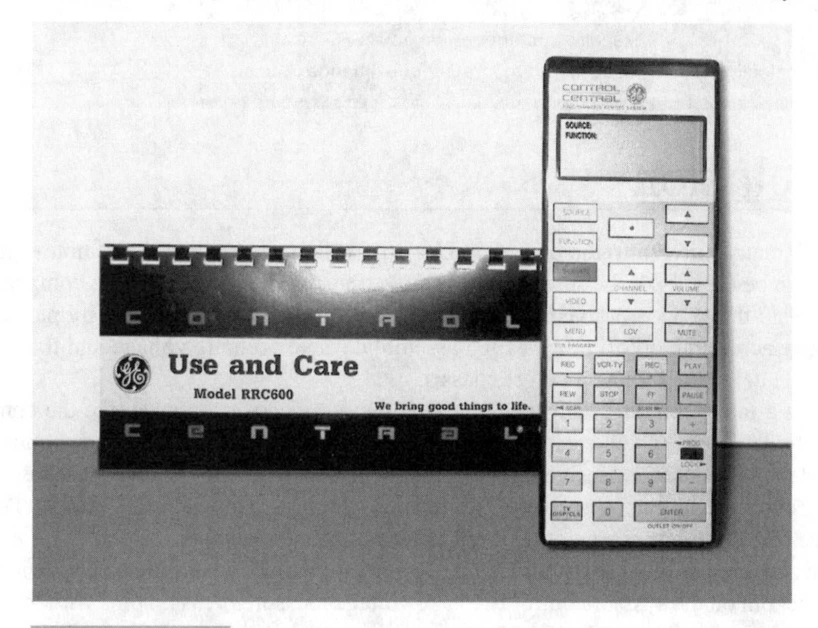

FIGURE 11-33 The General Electric RR600 remote controls up to four different machines with 200 key combinations.

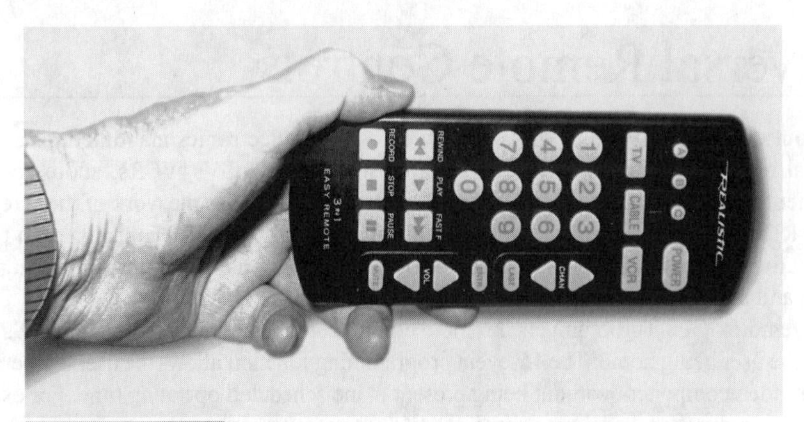

FIGURE 11-34 The new universal control can control the TV and VCR in most TV chassis.

TABLE 11-2 A TROUBLESHOOTING CHART FOR REMOTE CONTROL CIRCUITS

WHAT TO CHECK	HOW TO CHECK IT
No remote action.	Check batteries in remote.
Erratic operation.	Check battery terminals and dirty buttons.
Suspected remote.	Check remote with infrared tester.
Sub another remote.	Check remote receiver circuits. Check standby voltage. Repair remote receiver circuits.
Remote defective.	Send in for tuner repair or exchange remote.

Conclusion

Although many different remote-control TVs are available, servicing the remote-control system can be easy. First, be sure that the remote transmitter is working with comparison tests. Isolate the service problem to the TV chassis or the remote receiver by placing a shunt wire across the on/off relay switch assembly. Make accurate voltage and transistor tests on the defective remote receiver chassis.

Replace a faulty remote receiver with a new remote receiver, or substitute the control module to determine which component is defective. Remote-tuning components must be replaced in the event of extensive lightning damage. The remote transmitter, receiver, and modular components can be exchanged or repaired through manufacturer's outlets. Tuner repair centers service most remote-control transmitters and receivers.

When checking the keyboard circuits, measure the resistance across the button switch to the correct pin on the system-control IC. Take voltage measurements on pins with the button pressed. Scope the signal and sense waveforms at the system-control or IC pins with each keyboard button pressed to determine if the IC and keyboard is functioning. See the troubleshooting chart (Table 11-2) for remote controls.

SERVICING THE SOUND CIRCUITS

CONTENTS AT A GLANCE

Repairing the sound circuits is relatively easy. Besides applying test instruments, you can hear the sound symptoms from the speaker. The sound can be weak, garbled, distorted, intermittent, or just missing. The function of most sound stages is easy to understand. The solid-state sound circuit consists entirely of transistors, ICs and transistors, or one complete sound IC (Fig. 12-1).

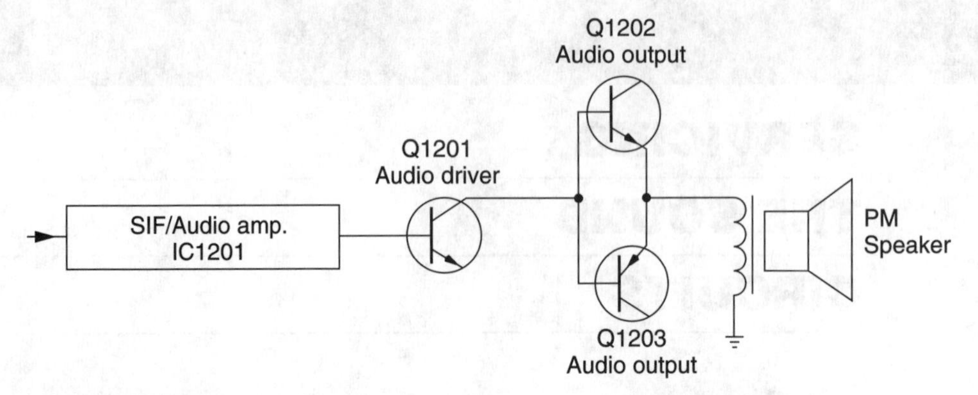

FIGURE 12-1 **The sound circuits might consist of a SIF/DRIVER IC with a transistor amp and output transistor.**

Sound problems can occur in or out of the sound circuits. Often, a combination of poor picture with garbled sound can occur because of a problem in the video circuits. An unstable picture with poor sound can develop in the AGC circuits. Running the sync with a buzz in the sound can occur in the sync or AGC stages. A normal picture and raster with no sound generally occurs in the sound circuits. Check out all other possible sources for combinations of picture-and-sound symptoms before tearing into the sound circuits.

In the tube chassis, the FM sound is taken from the plate of the video amplifier tube. In the solid-state chassis, the sound is tapped at the second or third IF video stage. A typical transistorized sound circuit consists of an IF sound detector, detector amp with audio driver, and push-pull audio output (Fig. 12-2). An IC can serve as an IF sound amp, detector, and audio amp to drive transistors in the audio-output stages. The latest TV chassis might have only two ICs in the whole chassis, with the complete audio stage combined with other receiver circuit functions (Fig. 12-3).

Required Servicing Instruments

The following test instruments can be used when servicing the sound circuits:

- VTVM, VOM, or DMM (Fig. 12-4)
- Sweep and signal generators
- Oscilloscope
- Capacitor tester
- External amp
- Audio analyzer
- Miscellaneous speakers, test leads, etc.

The VOM or DMM will solve most sound problems by helping you to take accurate voltage, resistance, and transistor tests. The external amp is handy for locating dead, weak, intermittent, and distorted stages. For crucial sound alignment, the scope and generators are a must-have item. Although sound alignment is only needed when crucial parts are replaced, a touch-up of the sound coil might be all that is needed to cure a TV with distorted, weak sound.

Transistorized Sound Stages

The sound stages were one of the first circuits to be transistorized in the hybrid TV chassis. Today, many different transistorized sound circuits are offered, but they basically all accomplish the same thing. The IC eliminates most of the transistors that were used in the early sound stages. A typical IC and transistor output circuit is shown in Fig. 12-5.

The audio signal is taken from the third IF video transistor and coupled by C412 and C415 through a sound take-off ceramic filter (CF401). A 4.5-MHz output signal from CF401 is fed to pins 1 and 2 of IC401. Internally, the limiter circuit is connected to the FM detector circuits. The sound coil (L401) is connected to the sound-detector circuit on pins 9 and 10.

The audio-amp circuit connects internally to the sound-detector circuit. R419 varies the volume of the sound circuits at pin 6 of IC401. The audio-output signal at pin 8 is connected to a driver transistor (Q403).

The NPN and PNP junction transistors operate in a push-pull fashion (Q401 and Q402). The audio signal is transformer coupled to the speaker through T401. Notice that two separate voltage sources from the power supply are fed to the IC401 circuits (18 V) and the audio-output transistors (120 V).

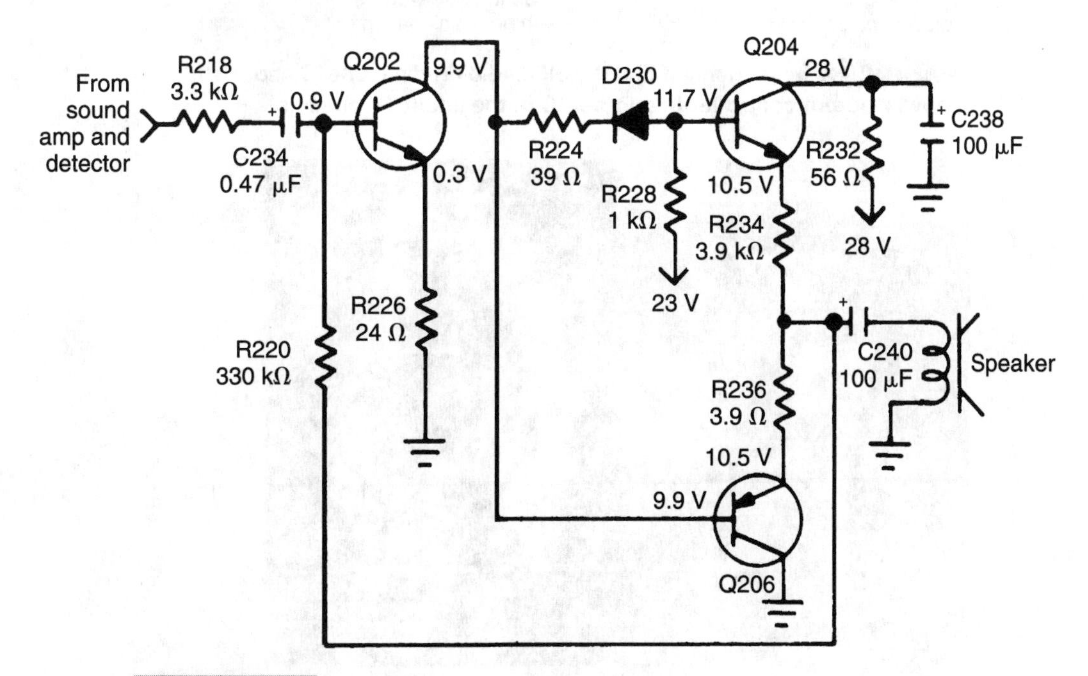

FIGURE 12-2 A typical solid-state sound circuit audio driver and push-pull output transistors.

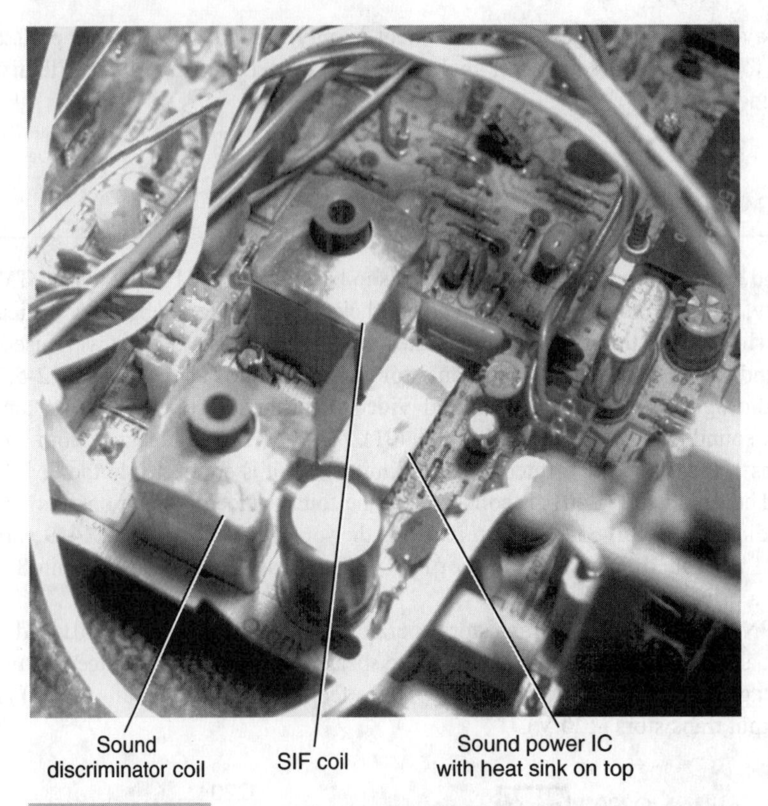

FIGURE 12-3 You might find the SIF/audio driver in one IC and a power-output controller in another IC in the latest TV chassis.

FIGURE 12-4 With a digital multimeter and a soldering iron, you can eliminate most sound problems.

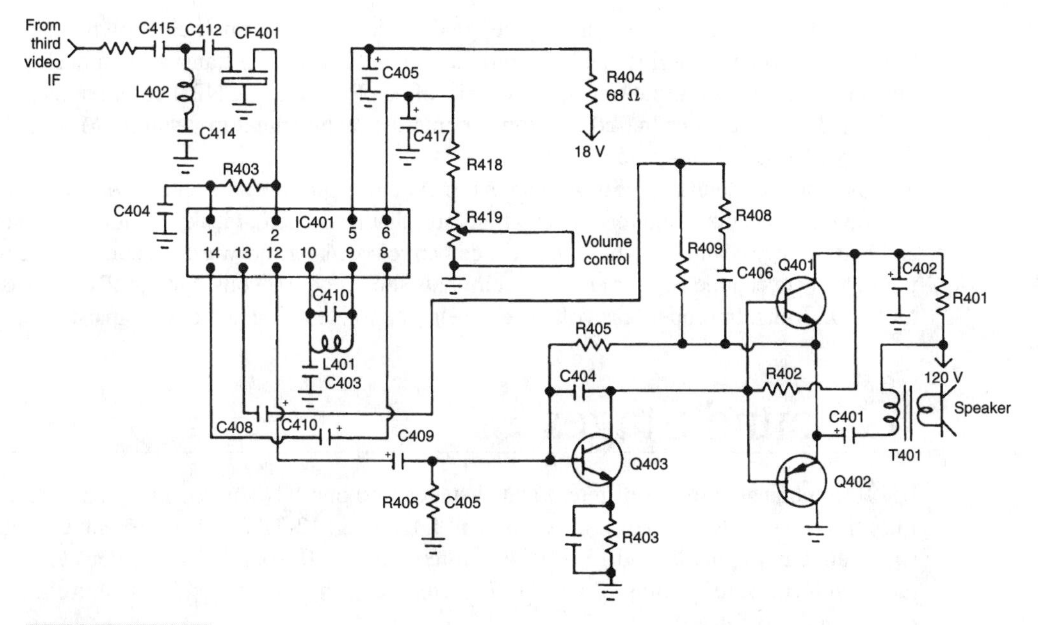

FIGURE 12-5 A typical IC and transistor sound-output circuit.

TRANSISTORIZED SOUND CIRCUITS ARE BACK

Often, sound circuits consist of either an IC IF and audio preamplifier, with an IC output or one large IC in the sound circuits. Recently, in the lower-priced portables and small-screen TVs, transistors are back. The IF/SIF sound systems, combined in one large IC with chroma/AFT/vertical and horizontal circuits, drive three audio transistors (Fig. 12-6).

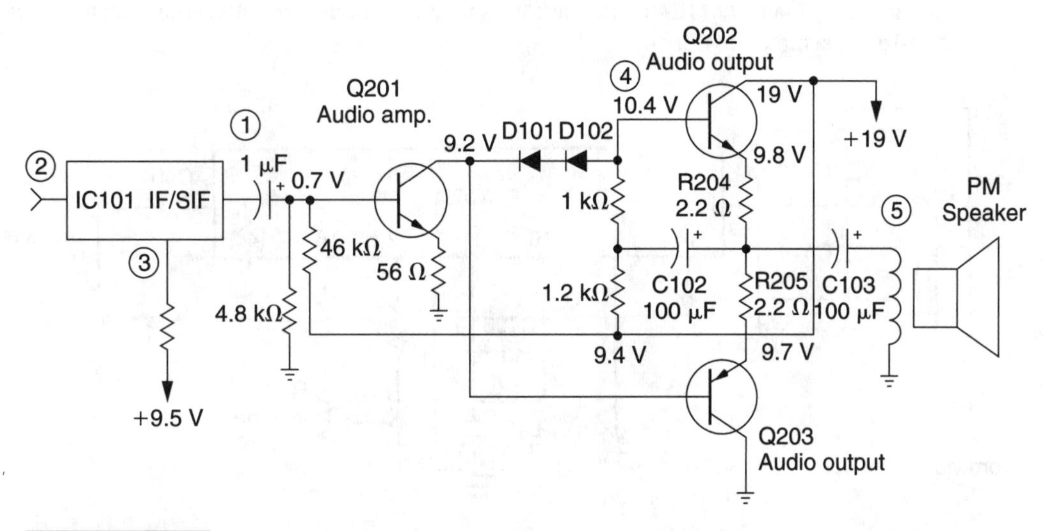

FIGURE 12-6 Check the audio-output circuit by the numbers with external amp and voltage measurements.

The first audio transistor is capacity-coupled to the IC with a small electrolytic capacitor. This audio amp provides audio amplification and drives two transistors in push-pull operation. The audio-output transistors consist of one PNP and one NPN output transistor. R204 and R205, emitter-biased resistors, couple the audio signal to a small FM speaker through a 100-µF electrolytic capacitor.

Signaltrace the sound circuits with an external audio amplifier or scope. Test each transistor within the circuit and remove any transistor that tests open or leaky. Crucial collector and bias voltages can help you to locate a defective transistor. Forward bias on the silicon transistor between the emitter and base terminals should be 0.6 volts, but the PNP transistor has 0.3 volts. Improper bias voltage can help you to identify a defective transistor.

IC Sound Stages

IC401 combines all of the different sound stages into one IC. The sound signal is taken from the IF video IC and fed to IF sound coil L402 (Fig. 12-7). L402 connects the sound signal directly to pins 14 and 15 of IC401. Internally, the IF sound signal is detected and tuned with coil L401 at pins 10 and 11. The volume of the sound circuits is controlled at pins 12 and 16 of IC401.

The AF amp and audio-output amps are contained inside the sound processor (IC401). C401 couples the amplifier and audio signal to the 16-Ω speaker. A 25-V source supplies power to the sound circuits of IC401.

Inject an audio signal at pin 9 or place the tip of a screwdriver blade (with your finger on the metal blade) on pin 9 and listen for a hum or tone in the speaker. If it is not emitting sound, clip a speaker across the suspected speaker. Keep the volume control wide open. Because the volume control is nothing more than a voltage-divider network, hum or tone tests cannot be made at the volume control, as in other sound circuits. Check the voltage at pins 1, 2, 4, 7, and 9 of IC401. Replace the sound IC if improper voltages are found with a no-sound symptom.

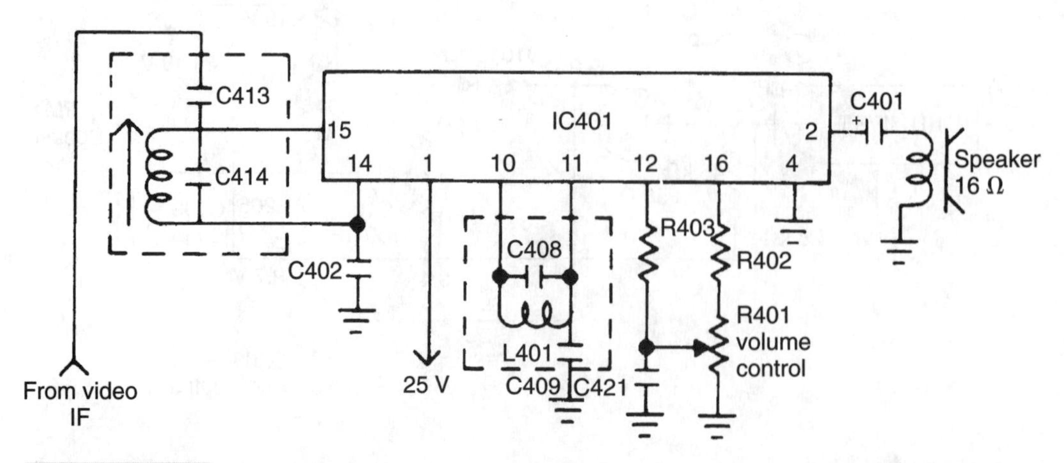

FIGURE 12-7 You might find only one IC as SIF/driver amp/power output.

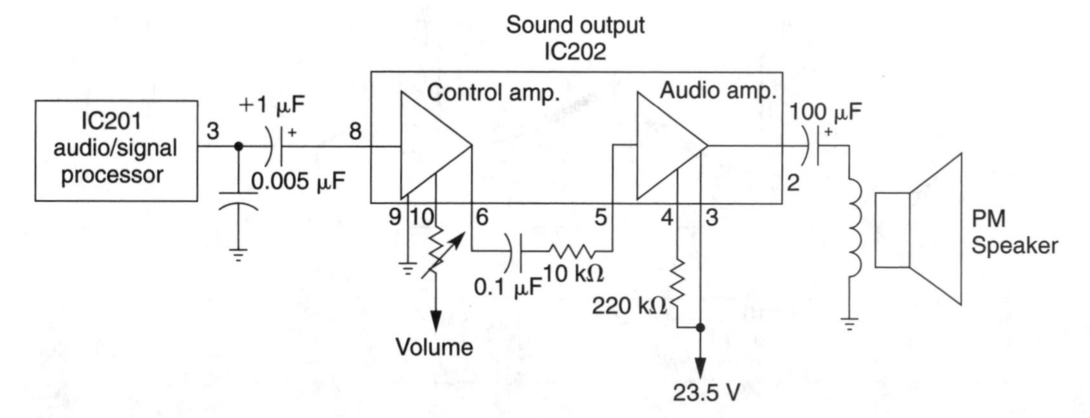

FIGURE 12-8 **The latest IC circuits contain an audio signal processor and power IC for greater volume.**

A quick touch-up of coil L401 can eliminate sound buzz or distorted sound. This sound adjustment should be made with a TV station tuned in; it will adjust L401 for maximum or clear sound. Often, only a slight turn is required to touch up the garbled sound. A defective IC401 can cause distorted, weak, intermittent, or "motorboating" sound problems. The sound circuits should be completely realigned when replacing an input IF sound transformer or sound coil. Always follow the manufacturer's sound-alignment procedures.

LATEST IC SOUND CIRCUITS

The sound-output circuits might consist of an audio and signal IC driving a larger-power output IC, stereo demodulation, and MPX stereo circuits. Stereo circuits are used in the more-expensive TVs. The sound-output IC is capacity-coupled to the SIF/audio signal IC with a small electrolytic capacitor and mono reception. A speaker from 16 to 35 Ω is coupled with an electrolytic capacitor to the audio IC. Usually, higher dc voltage is applied to the sound-output IC from the secondary voltages of the flyback (Fig. 12-8).

Troubleshoot the audio stages by checking the input audio signal at pin 8 of IC202. Trace the audio signal with an external amp or scope at pins 8, 6, 5, and 2. Take a crucial voltage measurement on pin 3 (connected to the supply-voltage source). Suspect that a coupling capacitor or speaker voice coil is open if audio signal is present at output pin 2. Do not overlook the possibility that IC202 is defective if audio can be heard at the input 8 terminal and no output is at pin 2, with normal supply voltage.

Signaltracing Sound Circuits

The sound circuits can be signaltraced with the audio signal generator and scope, audio signal generator and speaker, TV signal and external amp, or while doing sound alignment. Inject the signal at the input of the audio stages and use the scope to determine at which point in the audio stages to check for weak or dead circuits.

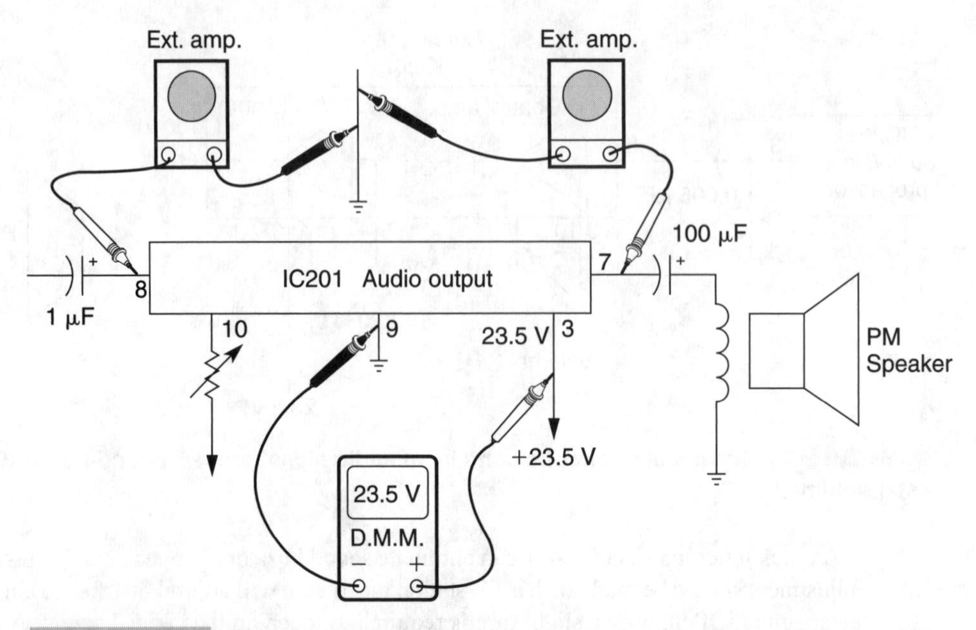

FIGURE 12-9 **Check the input and output terminals of the suspected sound IC with external amp and voltage tests.**

An external audio amp can be used to pick up the TV signal at various points in the sound circuits to locate distorted, weak, or dead stages. The audio signal can be checked up to the speaker terminals. Start at the speaker terminals and work back toward the input circuits of the audio stages. The audio signal can be checked on each side of a coupling capacitor, or from the base to the collector of each transistor stage. Check the input and output terminals of a suspected IC sound circuit with an external audio amp (Fig. 12-9).

STEREO SOUND

More than half of the new TVs have stereo sound. You might find a stereo/SAP decoder, expander, switching IC, IC sound processor, audio control, and audio output power ICs. In other stereo channels, you will find stereo broadcast, stereo demodulator, and video/audio in/out circuits. Often, the TV will indicate when it is receiving stereo sound by lighting an LED.

Servicing the Sound Circuits

Although the sound circuits are one of the easiest circuits to service, many different problems can occur in the sound circuits of the solid-state TV chassis. Here are the various sound symptoms, with quick practical methods to locate and repair the different sound stages.

NO SOUND

A no-sound symptom is a quick and easy repair, if the problem is in the sound circuits. The defective component can be quickly located with a few voltage and resistance measurements on the sound transistor and ICs. Another speaker clipped across the old one will help you to locate the defective speaker. Signaltracing the audio signal with an external audio amp can help you to locate where the sound stops. A small electrolytic capacitor and speaker with a clip wire can help you to signaltrace the audio output stages (Fig. 12-10).

Clip another speaker across the old one to see if the speaker is open. Take a voltage measurement at the collector terminal of audio-output transistor Q1802 (Fig. 12-11). This voltage measurement will indicate if the transistor is leaky (if the voltage is low), or if the transistor is open (when the collector voltage is high according to the schematic). Also, a voltage measurement here will indicate if T401 is open or if the voltage from the low-voltage power supply is insufficient.

Take a voltage test at the emitter terminal of Q1802. Zero voltage indicates the output transistor or 2.2 Ω is open. A voltage measurement between the base and collector terminals indicates the normal beta bias voltage. An NPN silicon transistor has a bias voltage of 0.6 V. To prove that the transistor is defective, take an in-circuit transistor test with the DMM.

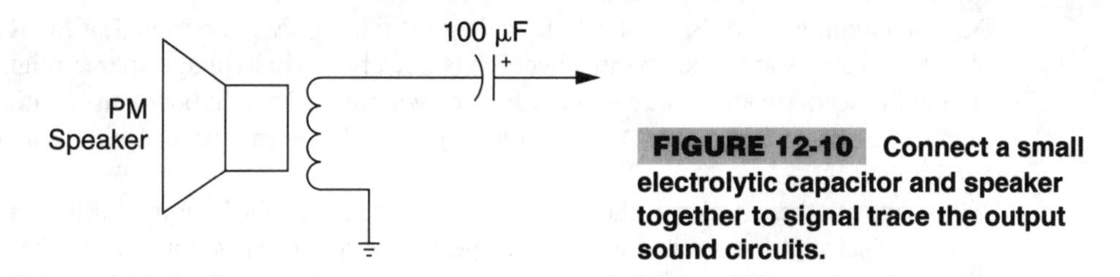

FIGURE 12-10 Connect a small electrolytic capacitor and speaker together to signal trace the output sound circuits.

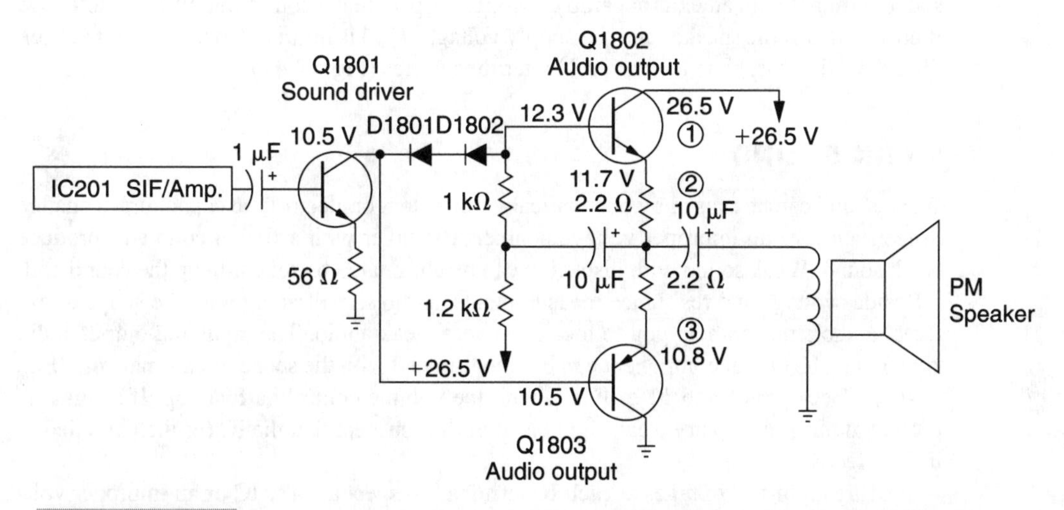

FIGURE 12-11 Crucial voltage measurements at the collector and emitter terminals of output transistors can indicate that the transistor is open or leaky.

The third voltage test at pin 14 of IC201 indicates proper supply voltage and normal IC operation. If the voltage is low, suspect that an IC is leaky. With the three quick voltage tests, you can quickly isolate most sound problems.

Another method to check the dead or weak stages is with the external audio amp. Check for audio at points 1, 2, and 3. Suspect that an earphone jack or speaker is defective if good sound is noted at the output transformer at point 2. Do not overlook the possibility that a shorting pin inside the headphone jack is bad. These jacks can act up after frequent earphone usage. A check at pin 10 (C) of the sound IC (IC201) will indicate if sound is normal at the driver or preamp stage of the sound IC.

Signaltrace the output stage with a small electrolytic capacitor and speaker. Clip one side of the speaker to common ground. Use one end of the capacitor lead as a probe and check for the signal at the output transformer and transistor. Normally, this signal will be loud. Now go to the output terminal (10) of IC201. Although this audio signal is weak, compared to the output terminal, you can still determine if IC201 and the input signal on the base terminal of Q1802 are normal. Check the audio signal through the earphone jack to the speaker in the very same manner.

NO SOUND IN IC CIRCUITS

Suspect that an IC, speaker, output coupling capacitor or power source is defective if the IC sound circuit is dead. First, check the voltage at the supply pin terminal of the IC. No voltage indicates that the low voltage source is defective. This voltage source might develop in the scan-secondary circuits of a flyback winding. Check for an open isolation resistor or regulator transistor. A very low supply voltage can be caused by a leaky sound-output IC.

Substitute another speaker if the voltage measurements are fairly normal. Shunt a 100-µF 50-V electrolytic capacitor across the speaker coupling capacitor. Check the input sound terminal with an external amp or scope. Suspect that a sound-output IC is defective if no sound is at the speaker and the supply voltage (V_{CC}) is normal. Do not forget to check for a defective earphone jack or broken terminal wires (Fig. 12-12).

WEAK SOUND

Weak sound can be caused by open or leaky transistors or electrolytic capacitors, a change in resistance, or an improper voltage source. A speaker with a frozen cone will produce weak sound. Weak sound with distortion can be eliminated by touching up the sound coil.

Besides voltage and resistance measurements, audio signaltracing with the scope or external audio amp can help you to locate the very weak audio. The input and output audio signal of an IC output component can be signal-traced with the scope or external amp (Fig. 12-13). Check pins 1 and 11 of IC501 with the volume control halfway up. If a signal is present at pin 1 and a very weak signal is at pin 11, suspect that the IC (or the parts tied to it) is defective.

Measure all of the voltages at each IC terminal. Suspect a leaky IC or an improper voltage source with lower voltage at pin 14. Determine if any other stages are fed from the same source. If these stages are normal, the voltage-supply source might be OK. Remove pin 14 from the circuit if the voltage is very low. Notice if the supply voltage returns to

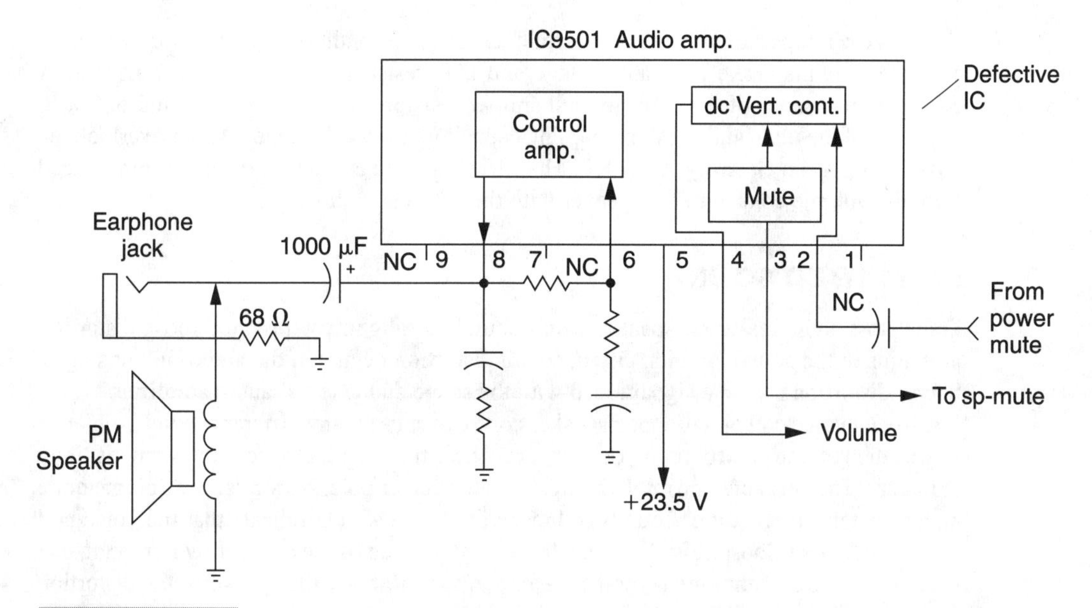

FIGURE 12-12 Check these sound components, which can cause a dead symptom.

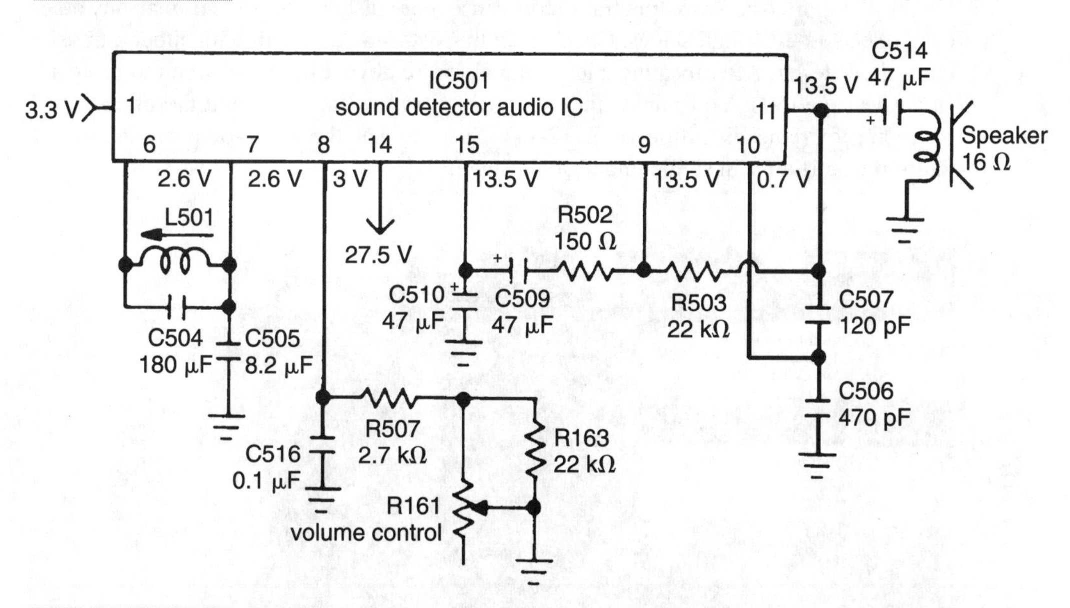

FIGURE 12-13 Use the scope or external amp to signal trace the sound-output circuits.

normal. Now measure the resistance from pin 14 to common ground. A low resistance (less than 500 Ω) indicates that the IC is leaky.

If the voltage at pin 14 is fairly normal, suspect either the IC or a component tied to one of the IC terminals if weak sound is at pin 11. Take a resistance reading from each pin to the common ground to determine if a capacitor or resistor is defective. Remove one end of the suspected component for accurate resistance measurements.

Check each transistor with the DMM for open or leaky conditions in transistorized audio-amplifier circuits. Take accurate voltage and bias resistance tests of each transistor. A change in resistance of the collector and emitter resistors can produce weak audio. Check each side of the input and speaker coupling capacitors for weak sound. Do not overlook the possibility that an electrolytic capacitor has dried up. The audio signal can be signaltraced from the volume control to the speaker with the external audio amp.

DISTORTED SOUND

Transistors, ICs, resistors, speakers, and sound misalignment produce most distortion problems in the sound circuits. Often, sound distortion occurs in the audio-output stages. Sound distortion can be a signaltraced with the scope and external audio amplifier.

A frozen voice coil or a dropped speaker cone can cause many different sound problems. Lower the volume control until you can barely hear the sound. Notice if the sound is tinny or mushy. The speaker cone might be frozen to the center pole. Now raise the volume quite high to listen for "blatting" sounds or loose vibrations, which indicate that the voice coil or speaker cone is loose (Fig. 12-14). Check the condition of the speaker by removing one voice-coil wire and clipping a good speaker across it. Replace the speaker if the distortion disappears. If not, troubleshoot the audio circuits.

Leaky or shorted audio-output transistors cause more distortion problems than any other components in the sound stages. Check each transistor in the circuit with either a transistor or diode tester. After locating a leaky transistor, remove it from the circuit to make accurate leakage tests. You might find one push-pull transistor open and the other leaky, producing extreme distortion (Fig. 12-15). Suspect that the audio-output transistor is faulty if distortion is in single-ended audio stages.

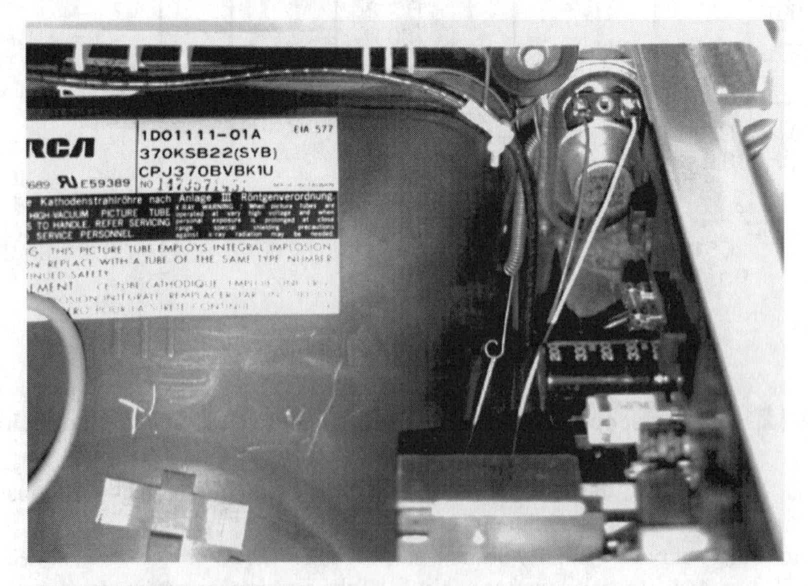

FIGURE 12-14 **Inspect the speaker for a torn or loose coil, dragging voice coil, and/or an open winding.**

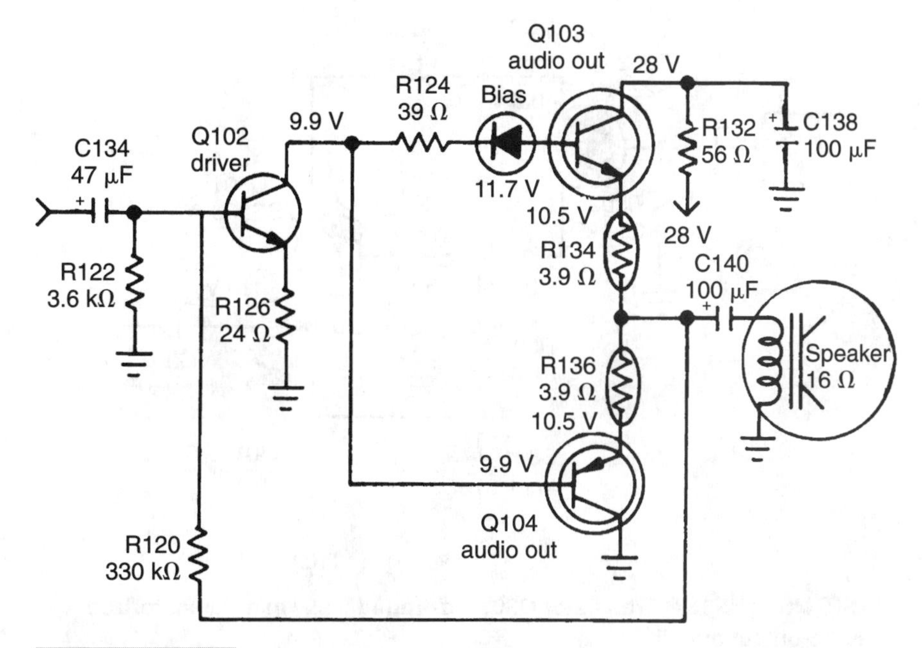

FIGURE 12-15 Take in-circuit transistor tests to obtain a leaky or open transistor. Check circled components for distortion.

Always check the bias and base resistors with the transistors out of the circuit. In many cases, the emitter bias resistors are burned because of shorted audio-output transistors. Remove one end of a bias diode when making leakage tests. These bias diodes can cause a low level of distortion after the audio-output transistors have been replaced.

In a K-Mart KMC1984A, extreme distortion was found in the speaker (Fig. 12-16). Q301 was found to be leaky via in-circuit transistor tests of the DMM. Voltage measurements on the base and emitter terminals indicated that a transistor was leaky. After Q301 was replaced with a universal ECG198 replacement, some distortion was still present. The base resistor (R302) (10 kΩ) was found to be open in the base circuit of the audio-output transistor.

Suspect that an IC or speaker is defective if the audio distortion is still there after other possibilities have been exhausted. Substitute a new speaker. Take accurate voltage and resistance measurements on the IC terminals to determine if the IC is leaky. Check the input and output signal of the IC for signs of distortion. If the input sound signal is normal and extreme distortion is noted at the output terminal, replace the defective IC.

EXTREME DISTORTION

Often, extreme distortion can be caused by a leaky output transistor or IC. Most extreme distortion is found in the audio-output circuits. Signaltrace the audio from the input of the output IC and on the output terminal. If the input audio is normal and distorted at the output terminal, suspect that an IC is leaky, electrolytic capacitors are open, or the supply voltage is improper. Leaky output transistor Q1902 and emitter resistor R1909 were

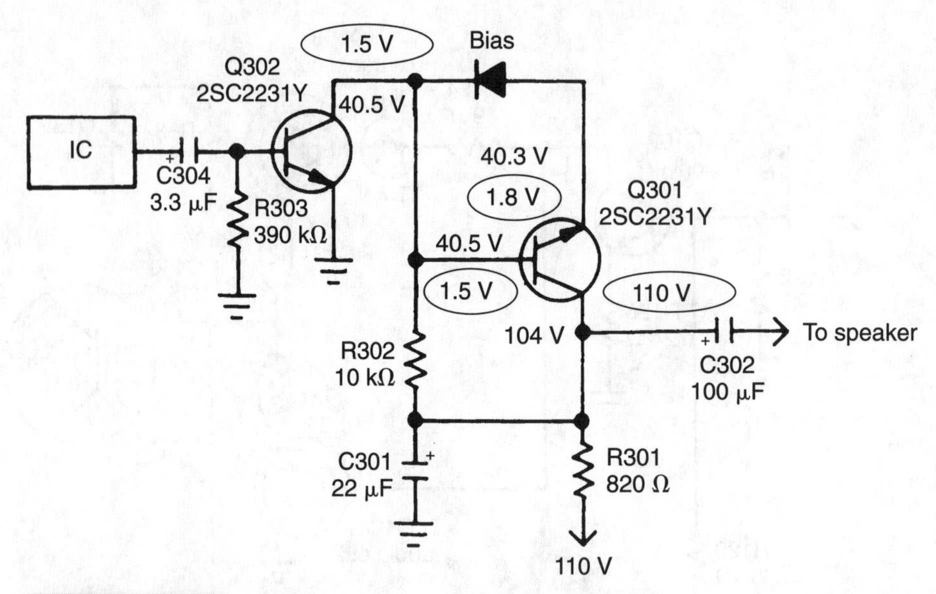

FIGURE 12-16 **Transistor Q301 was found leaky in a transistorized audio-output circuit.**

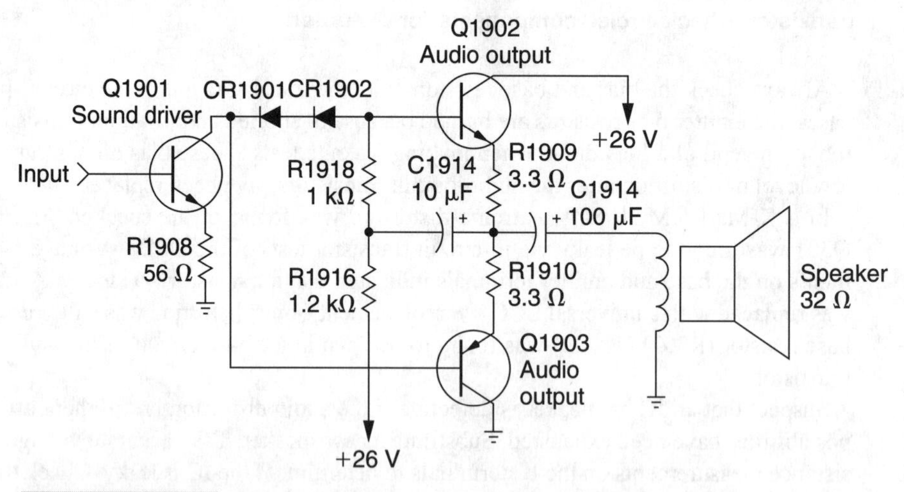

FIGURE 12-17 **Check for leaky audio-output transistors or an output IC and leaky bias resistors if the problem is extreme distortion.**

burned in an RCA CTC 157H chassis (Fig. 12-17). Check Q1903 for open or leaky conditions. Do not overlook the possibility of that a speaker is defective and causing distortion.

INTERMITTENT SOUND

Practically any component within the sound stages can produce intermittent sound conditions. The most obvious parts are transistors, power ICs, speakers, wiring, and component board connections (Fig. 12-18). Determine if both the picture and sound are intermittent. If so, the defective component is likely in the video or AGC circuits instead of the sound stages.

Try to isolate the sound stages by checking for intermittent sound at the volume control and output stages. Signaltrace the audio stages with the external audio amp. Notice if the sound is normal at the top side of the volume control in transistor circuits. Now go from the base to the collector terminal of each driver, AF amp, and output transistor until you notice that the audio signal is acting up.

Intermittent sound can be caused by poor board wiring or by bad component connections. Push up and down at the various sections of the sound board. Notice if the pressure points cause sound problems. Take an insulated plastic tool and push each component to determine if a poorly soldered connection exists. Pull or flex the speaker and the volume-control connecting wires for loose connections. Poor socket and lead terminals on remote-control sound modules can produce intermittent sound.

MOTORBOATING NOISES

A low or loud "putt-putt" sound is called *motorboating*. Often, motorboating noises occur in the input and output audio stages. A motorboating sound in the speaker can be caused by broken board wiring in the horizontal circuits. Suspect that other circuits are creating motorboating sounds if the picture blinks off and on with a motorboating sound in the speaker.

Check for defective AF or audio-output transistors creating motorboating noise in the speaker. Short the emitter and base terminals of the AF transistor to eliminate the transistor

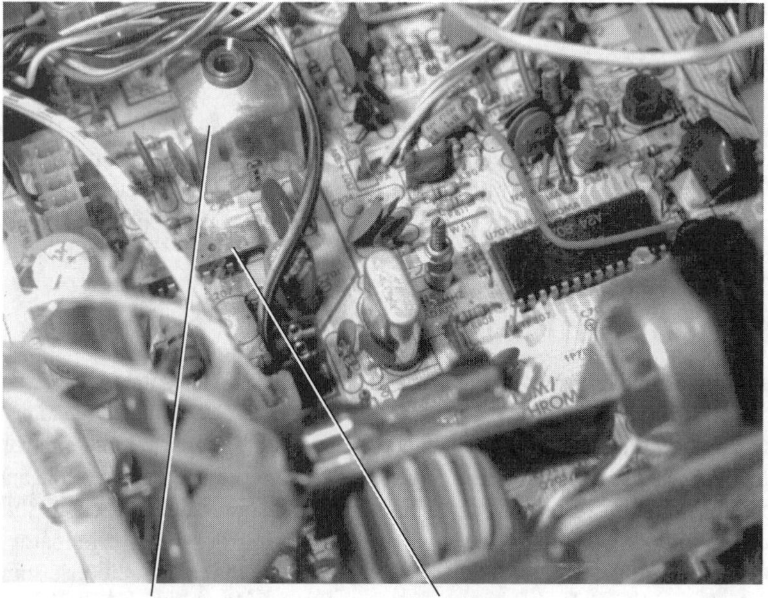

FIGURE 12-18 Check these components for intermittent sound.

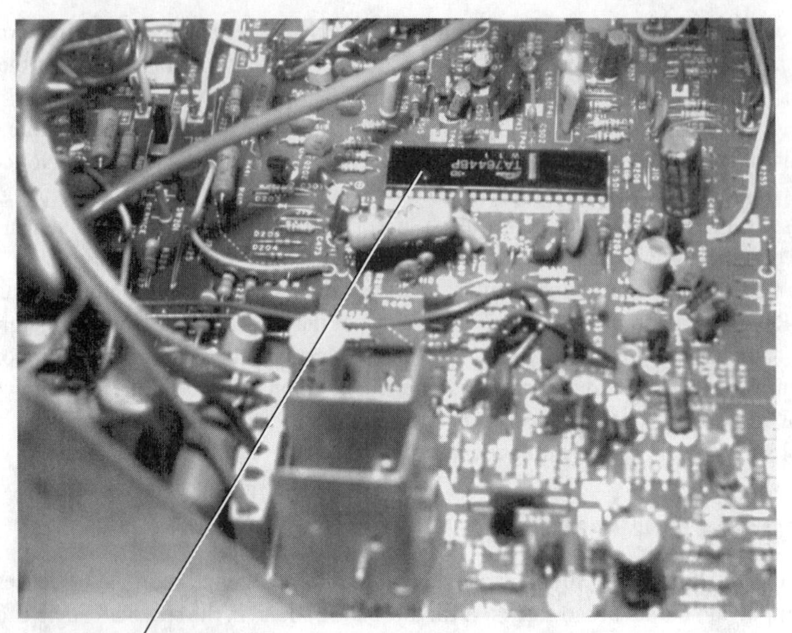

FIGURE 12-19 **Motorboating can be caused by bad transistors and audio ICs.**

that is thought to be causing the motorboating sound. Replace the suspected transistor if the sound disappears. Signaltrace the audio output transistor with the external amp to determine if the noise originates in these transistors. Replace the output transistor if you are in doubt.

The motorboating noise can be caused by a defective power IC (Fig. 12-19). Lower the volume control to determine if the noise is still present in the audio-output circuits. Momentarily shunt the IC input terminal to ground through a 100-μF electrolytic capacitor. Suspect a noisy component ahead of the power IC if the noise stops. Replace the motorboating IC if the noise is still present.

SOUND AND PICTURE DO NOT TRACK

After tuning in a good picture with the best sound possible and the sound has some distortion or poor tracking, suspect that a component is defective in the sound-detector stages or sound circuits are poorly aligned. Check the sound stages in the TV with remote-control or push-button tuning. Readjust the fine-tuning control in older-type tuners for best sound and picture.

If these adjustments fail to bring the best picture and sound together, check that the discriminator or sound coils are correctly aligned. Just adjusting the discriminator or sound coil can take care of the sound-tracking problem. Follow the manufacturer's sound-alignment procedures if the sound and picture cannot be brought together. Look for open coils and defective bypass capacitors in the sound-input stages if the sound cannot be correctly aligned.

SERVICING RCA CTC109 SOUND CIRCUITS

One IC contains the entire sound circuit of an RCA CTC109 chassis (Fig. 12-20). You might find this same sound circuit in many of the later RCA solid-state TVs. The input signal at pin 15 and the output signal at pin 2 can be signaltraced with a scope or external audio amp. A defective U201 can produce motorboating or weak, distorted, or absent sound.

Suspect U201 or C201 is defective or the voltage source (26 V) is incorrect with weak sound (Fig. 12-21). Check the 32-Ω speaker, U201, and the adjustment of L201 and C209 (7.5 pF) for distorted sound. A touch-up of L201 might cure the distorted-and-weak sound symptom. Replace C209 if the sound drifts. Intermittent sound can be caused by U201, the 32-Ω speaker, C201 (10 µF), or a dirty volume control (R4201).

For distorted-sound or no-sound symptoms, clip a new speaker across the original. Any 8-Ω or 10-Ω speaker will do as a test speaker. If the sound is still distorted, touch up the sound coil (L201). Check the input and output with the external speaker to determine if U201 is defective. Take voltage measurements on all IC terminals before removing IC201. Be sure to replace the copper heatsink on top of the audio IC.

FIVE ACTUAL SOUND CASE HISTORIES

Here are five different sound problems in different TV chassis, showing the methods used in locating the sound problem.

No sound/normal picture The color picture was normal, but there was no sound in an RCA CTC146B chassis. Audio was signaltraced and was normal at pin 28 of IF/SIF (U1001). No sound was noted on the collector pin of audio amp Q1201. No voltage was

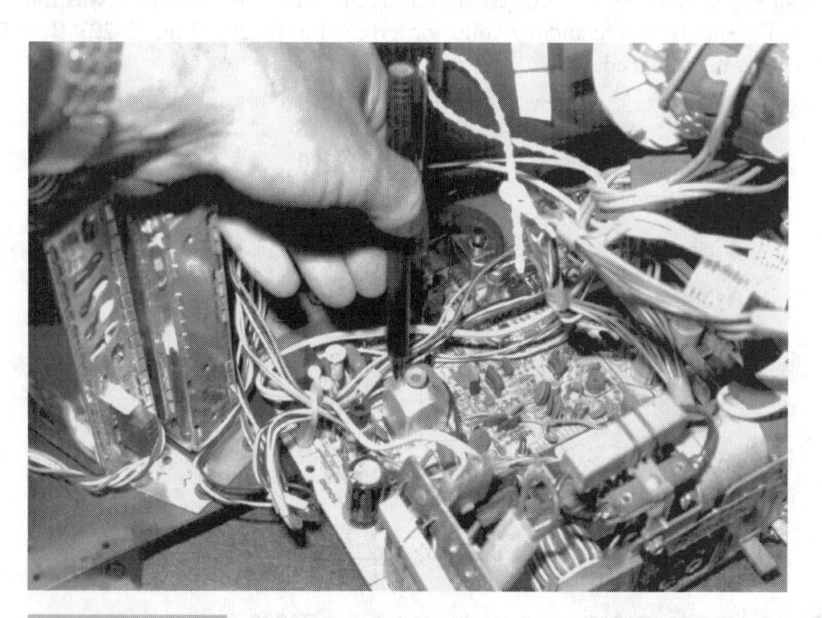

FIGURE 12-20 The complete IC sound section in a RCA CTC109 chassis.

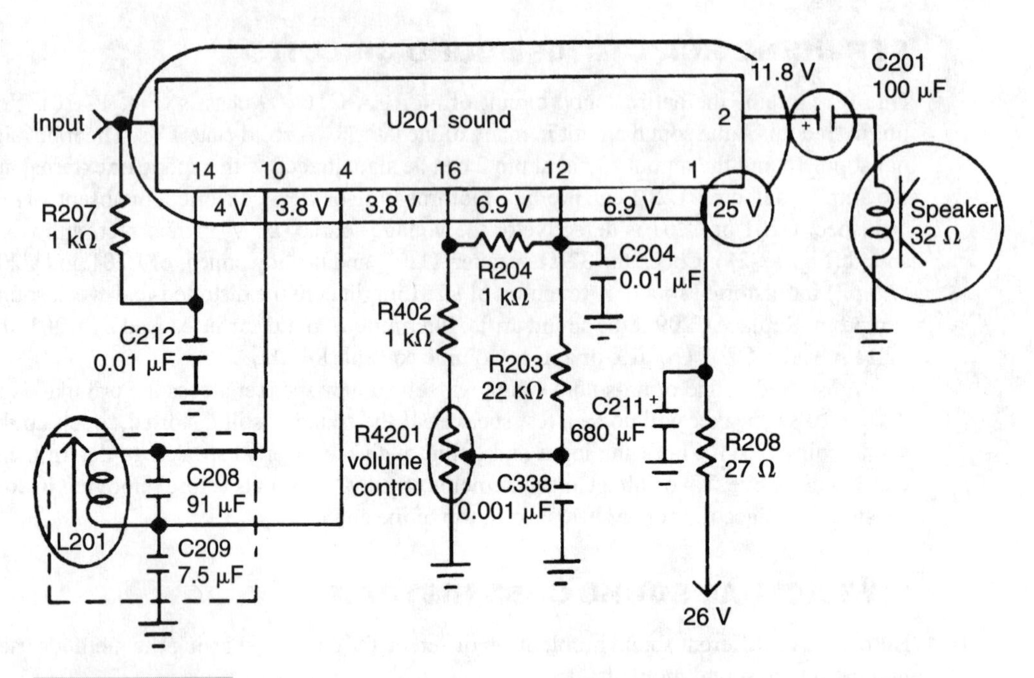

FIGURE 12-21 Check these components for weak or distorted sound in RCA CTC108 chassis.

noted on Q1201, Q1202, or Q1203. Checking revealed that the 18.5-V supply source in the power supply was higher in voltage than normal. R1211 (5.6 Ω) resistor was found open between the supply source and the collector terminal of Q1202 (Fig. 12-22). Both R1211 and Q1202 were replaced.

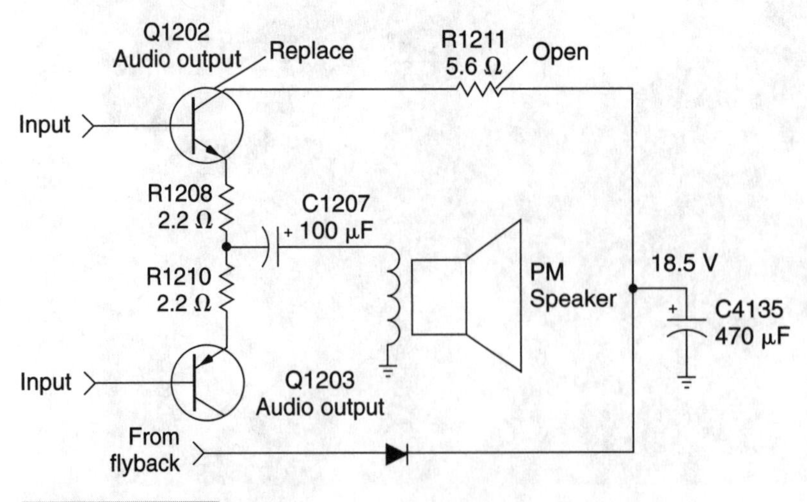

FIGURE 12-22 R1211 was found open; it and Q1202 were replaced in an RCA CTC146B chassis.

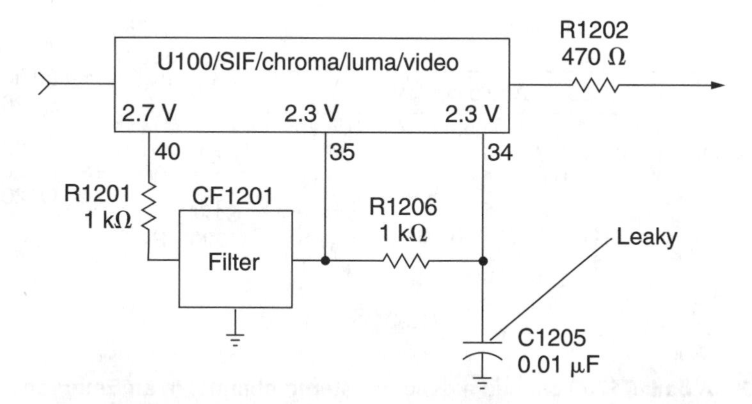

FIGURE 12-23 Replace leaky C1205 (0.01 μF) if the audio has a loud hum and makes a buzzing noise.

Audio hum and noisy sound in an RCA CTC159 An RCA CTC159 chassis contained hum and noisy audio. C1205 was checked for a leakage condition (Fig. 12-23). Measure the resistance from pin 34 to common ground of U1001. U1001 and C1205 were replaced with weak and noisy sound.

Low hum in the sound in a Panasonic CT1320V Sometimes the audio played normal, then would drop down with hum in the sound in a Panasonic CT1320V. IC201 was monitored at output pin 8. When the volume went down, it indicated that the output IC, varying supply voltage, or electrolytic capacitors were defective (Fig. 12-24). The supply voltage 7 was monitored with the DMM. It was normal throughout the change in sound

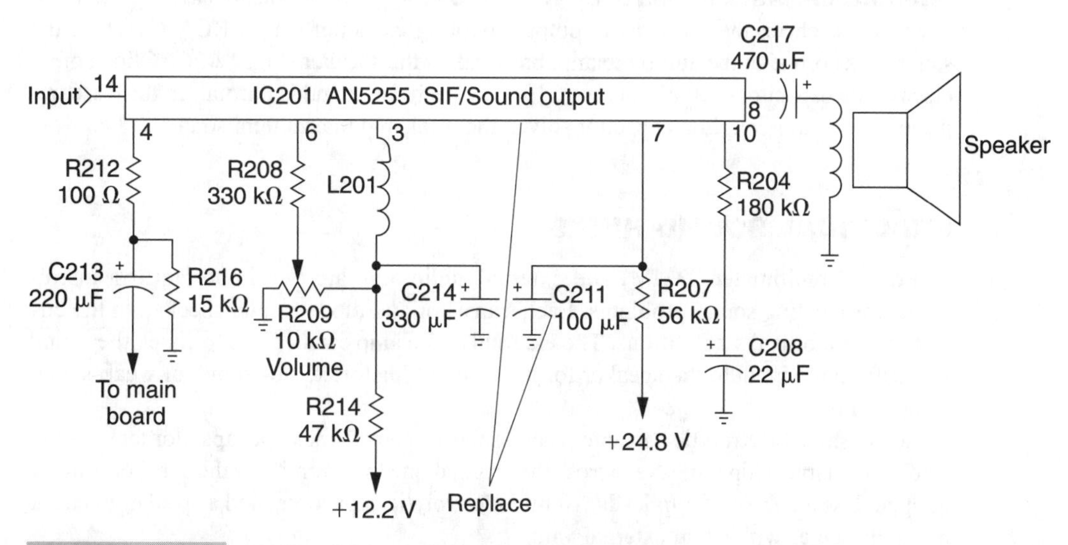

FIGURE 12-24 Audio with hum was caused by a faulty C211 and IC201 in a Panasonic CT1320 V portable.

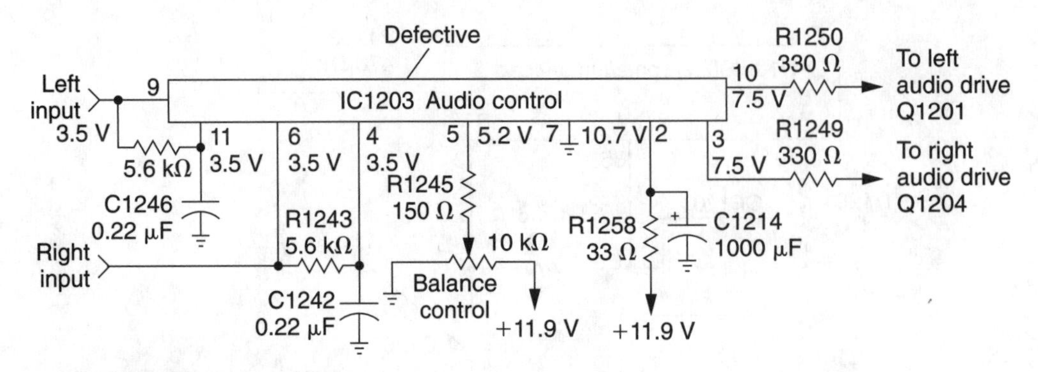

FIGURE 12-25 **A bad IC1203 caused a dead left stereo channel in an Emerson MS2250RA.**

level. C211 was shunted with another 1000-µF capacitor, which helped some in reducing the hum level. Defective sound IC AN5255 was replaced with a universal SK9324 IC.

Left channel dead, right normal in an Emerson MS2250RA This audio symptom indicated problems in the stereo audio-output circuits of an Emerson MS2250RA. The right and left channels were checked at the output IC, feeding each stereo channel of IC1203 (Fig. 12-25). No signal was found at the right channel (pin 3), but operation was normal in the left channel (pin 10). The balance control had no effect on the right channel. The supply voltage at pin 2 was good, with a 10.3-V measurement. The defective AN5836IC was replaced with an NTE1280 universal component.

Intermittent, weak sound in an RCA CTC157 If the sound appears to be weak or intermittent, check the electrolytic output-coupling capacitor. In an RCA CTC157, the sound was normal at the audio preamp, but weak at the speaker (Fig. 12-26). Clip another electrolytic capacitor (100 µF) across C1914 to see if the sound is normal. In this case, replacing the audio-coupling capacitor solved the weak and intermittent sound.

PRACTICAL SOUND HINTS

The digital multimeter (DMM) and external audio amp are two important test instruments for locating sound problems. Each transistor and diode can be checked in the circuit for open or leaky conditions. The external audio amp can be used to check the sound signal from the input to the speaker for intermittent, distorted, no-sound, or weak-sound components.

Tack in small electrolytic capacitors across the suspected ones if a capacitor tester is not readily available. Clip a speaker across the original one to determine if the speaker is noisy or open. Use a couple of clip leads, 100-µF electrolytic capacitors, and a speaker to check the audio stages without an external amp.

Check the input and output signal of a suspected IC for correct voltage and resistance measurements before replacing the IC (Fig. 12-27). Before removing the defective IC, be sure that you know where pin 1 is located on the PC board. Use a small soldering iron with very small solder to solder the IC terminals (Fig. 12-28). Clean out between each terminal with the back of a pocket knife to remove excess solder and flux. Universal transistors and ICs can be used successfully in the sound circuits as replacements.

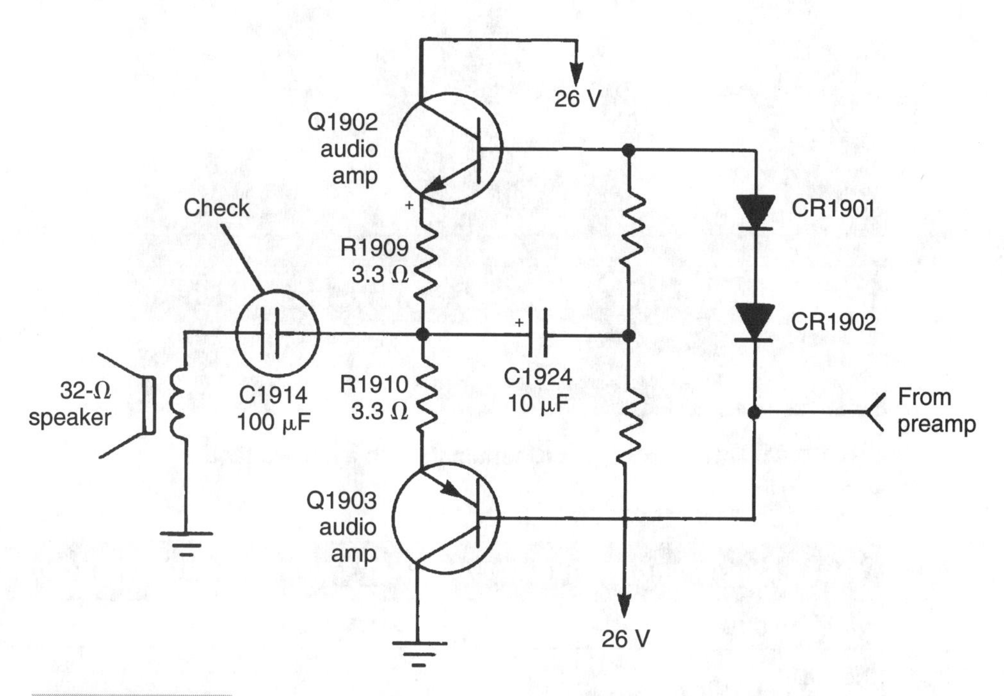

FIGURE 12-26 A bad capacitor C1914 caused intermittent weak sound in the RCA CTC157 TV chassis.

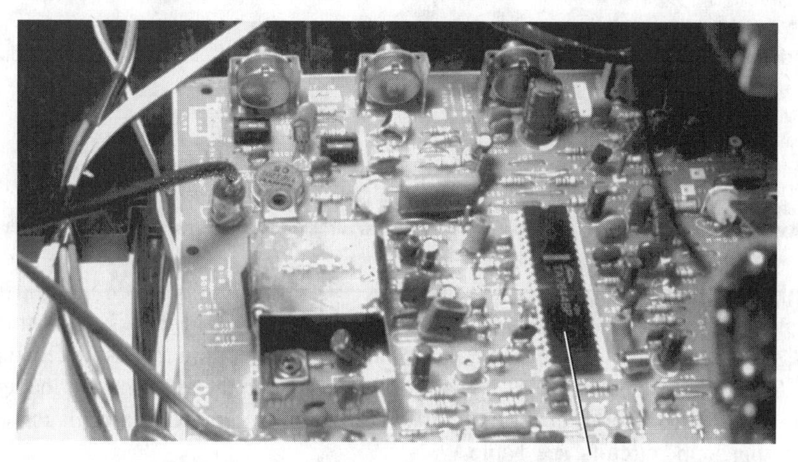

FIGURE 12-27 Check the input and output terminals of suspected output IC with a signal tracer.

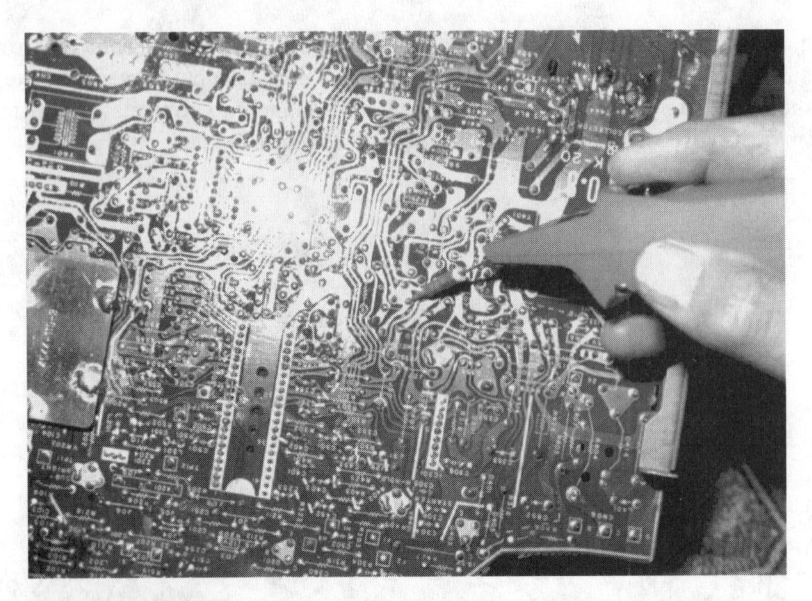

FIGURE 12-28 Solder the IC terminals with a low-wattage soldering iron.

TABLE 12-1 A TROUBLESHOOTING CHART FOR THE SOUND CIRCUITS

WHAT TO CHECK	HOW TO CHECK IT
No audio.	Sub another speaker.
Signal trace audio circuits with external amp.	Signal trace from volume control to speaker.
Suspect audio output transistor or IC.	Check for correct supply voltages. Test output transistors.
No sound at volume control.	Check sound input circuits.
Sound normal at volume control.	Replace transistors or output IC with sound input.
Signal trace sound input circuits with signal and scope.	Make sure picture is normal.
Distorted sound.	Audio output transistors and IC. Check voltage source.

Apply silicone grease on each side of the piece of insulation between the heatsink and transistor of the power-output transistors. Replace detachable heatsinks on all transistors and ICs. Look for burned or overheated base and bias resistors in transistorized audio circuits. Always keep the volume as low as possible and keep the audio circuit loaded down with the correct speaker after servicing the audio circuits. For more information on troubleshooting sound circuits, see Table 12-1.

13

TROUBLESHOOTING PICTURE-TUBE PROBLEMS

CONTENTS AT A GLANCE

Although the picture tube does cause some service problems, very few are replaced, compared to other components within the TV. Defective picture-tube circuits and circuits connected to the CRT can cause the picture to malfunction (Fig. 13-1). Check all connecting circuits before attempting to replace the picture tube.

Picture-Tube Components

The picture-tube components consist of a glass envelope, electron-gun assembly, focusing system, deflection circuits, and the phosphor screen. The black-and-white and color picture tube comes in many different sizes and shapes. Besides supplying the voltage applied to the anode terminal and electron gun assembly, a deflection yoke provides for the vertical and horizontal sweep of the raster. Static magnets and dynamic-convergence circuits with three color-gun assemblies are used to provide color pictures on the screen. The latest color picture tube might have a beam-bender magnet instead of dynamic-coil convergence circuits. Flat screens and traditional glass picture tubes are used.

The defective picture tube might have no picture, no raster, a weak or negative picture, low or no control of brightness, poor focus, an intermittent raster, and excessive arcing at the high-voltage anode terminal. The same symptoms can be caused by the picture tube circuits or outside circuits connected to the CRT. A defective picture tube can be detected by visual inspection, testing the tube, and taking voltage measurements.

Besides visual symptoms, low (or no) high voltage applied to the CRT will cause a no-raster symptom. Often high voltage can be detected at the picture tube by holding your forearm near the TV. The hair on your arm will stand up. Another method is to hold a small piece of paper

FIGURE 13-1 Video and CRT circuits with high-voltage measurements can indicate picture-tube problems.

FIGURE 13-2 The 27-inch RCA screen has a flat surface with square corners.

next to the screen and watch it pull inward. High voltage can be detected when you hear the deflection yoke expand when the receiver is turned on and collapse when the receiver is turned off. High voltage should be measured at the anode terminal with a high-voltage meter.

For successful picture tube and corresponding circuit tests, the following test instruments are required:

- A CRT tester. The picture tube tester checks the performance of each individual gun for proper emission. This tester can operate in conjunction with the tube charging tester, or it can operate separately. The CRT tester is a must-have instrument for the complete service shop.
- A VOM, DMM, and VTVM.
- A VTVM with a high-voltage probe.
- A separate high-voltage probe (42 kV).

LARGER PICTURE TUBES

Picture tubes are getting bigger each year. A few years ago, the 24-inch tube (with a 25-inch diagonal design around the corner) was the largest size. Today, 26-inch, 27-inch, and 31-inch screens are being produced (Fig. 13-2). The 27-inch screens are quite popular and are used in table models and consoles. The 27-inch and 31-inch picture tubes have a flat surface with square corners. Most picture tubes are measured diagonally from corner to corner (Fig. 13-3). Although the larger screens have the same number of scanning lines as the 10-inch screen, the 27-inch screen still produces an excellent picture. Very little distortion is found in the larger glass picture tubes. Of course, the larger the screen, the heavier the TV. It takes two people to lift, remove, and replace the larger picture tubes.

FIGURE 13-3 This 27-inch screen is square and has a flat surface.

Normal high voltage for the 27-inch picture tube is around 27.5 kV; it should not exceed 32 kV at maximum beam current. The 26-inch picture tube should have 26 kV applied and not exceed 29 kV. Check the schematic for the values of high voltage applied to larger direct-view screen picture tubes. Measure the HV with a VTVM (with high-voltage probe) or a high-voltage probe (meter).

CRT Bias and Driver-Board Circuits

You might find green, red, and blue bias transistor stages ahead of the color-output transistors. Each color bias transistor is directly coupled to the same color-output transistor. The G-Y, B-Y, and R-Y color difference signals from the IC processor are fed to the base circuits of each bias transistor. The color signal is applied to the base circuits, but the luminance or brightness signal is connected to each emitter terminal of the color bias transistor (Fig. 13-4).

The color bias signal is fed to each emitter terminal of color output transistors. Higher collector voltage is applied to each output transistor with a fixed lower voltage at the base terminal. The output of each color-output transistor is coupled through an isolation resistor to the cathode pin of the CRT. Each color-output transistor is coupled to the respective color-gun assembly within the picture tube.

Scope the color signal at the G-Y, B-Y, and R-Y terminals on the color IC processor. Next, check the luminance signal applied to the luma buffer transistor. Measure the voltage at the collector terminals of each bias transistor. Check the color signal at output of each output transistor. Measure the high voltage on each output collector terminal. A missing color can result from an improper input signal, leaky or open output transistors, low collector voltage, and a defective gun assembly inside of the picture tube.

Visual Symptoms

The following are common visual symptoms of problems in the picture tube and its associated circuitry.

NO RASTER

The front of the picture tube is entirely black with no scanning lines. An improper high or low voltage applied to the picture-tube elements can produce a dead raster. The electronic gun assembly will not produce a raster without the proper heater or filament voltage. Improper focus voltage can prevent the tube from lighting. A poor signal voltage at the cathode terminals can prevent a raster from operating—even with a normal picture tube and the correct applied voltages.

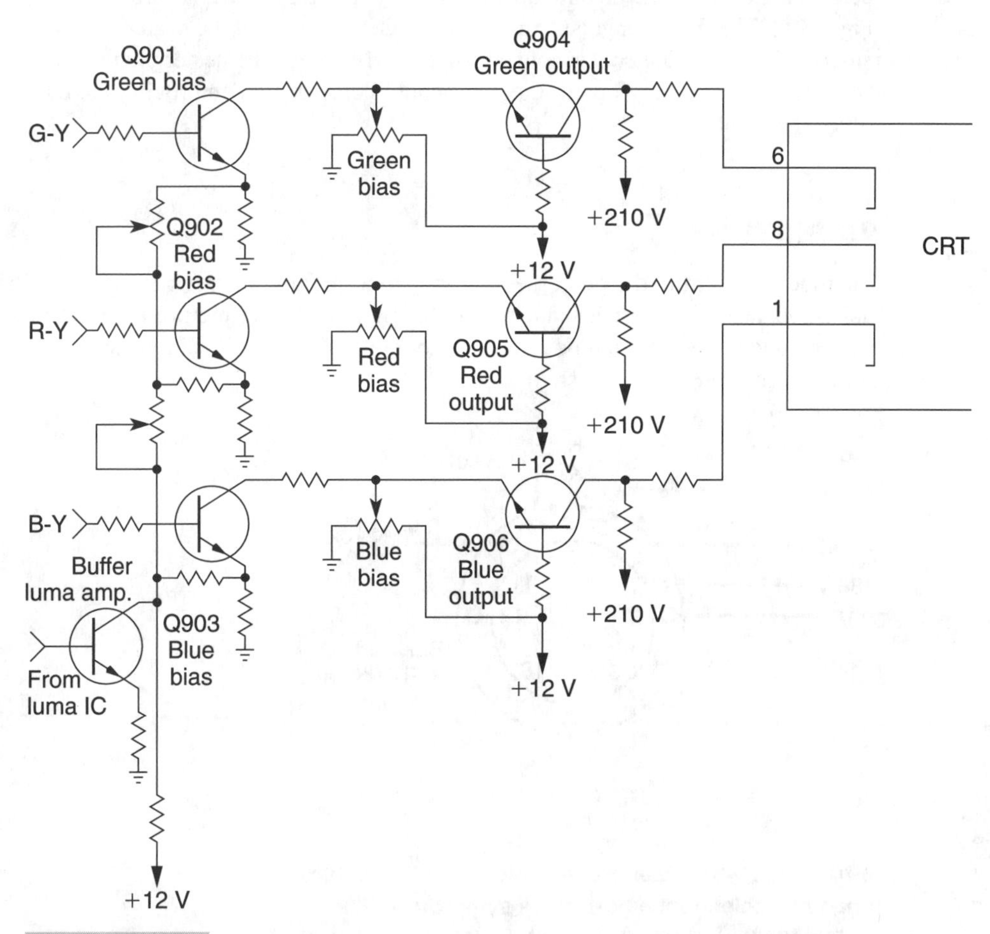

FIGURE 13-4 **The CRT green, red, and blue output with bias transistors are connected to cathodes in the CRT.**

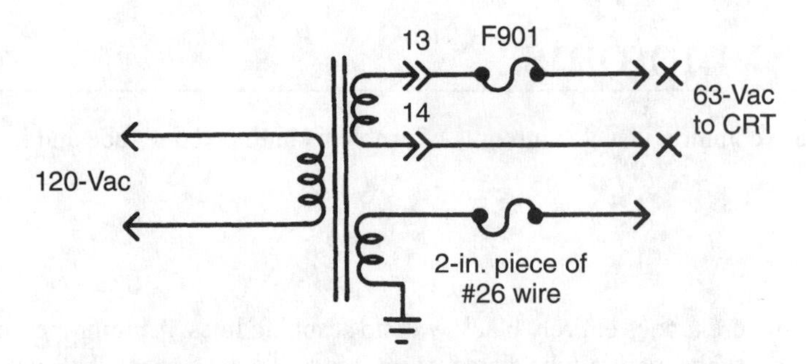

FIGURE 13-5 **A no-raster symptom in the Admiral 3M20 chassis was caused by poor heater plug.**

In an Admiral 3M20 chassis, the high voltage was normal, but no heater light could be seen from the end of the picture tube. In this model, the CRT heaters are fused with F901 (Fig. 13-5). The fuse was good. An ac voltage measurement at the heater transformer terminals indicated a poor socket connection at pin 14. When the heater plug was moved on the M900 power supply board, the tube would come to life. Improving the connection solved the problem.

NO PICTURE

The tube will light in the gun assembly with a no-picture/no-raster symptom. If a normal raster is found, suspect that the problems outside the picture-tube circuits. Improper voltages on the cathode and grid circuits of the CRT can cause a no-picture symptom (Fig. 13-6).

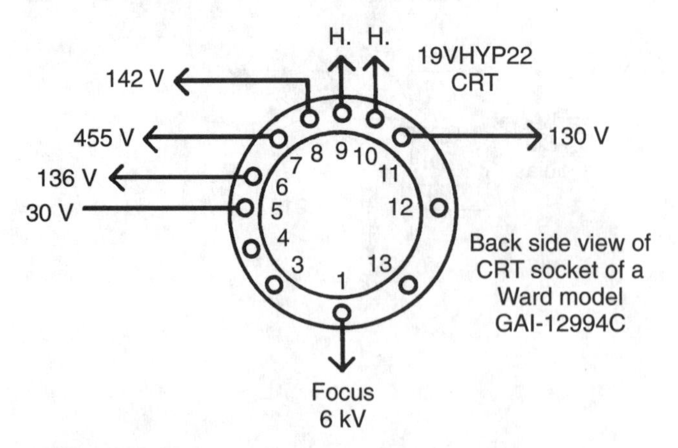

FIGURE 13-6 **Measure the voltage on the picture tube to troubleshoot a no-picture symptom with a normal raster.**

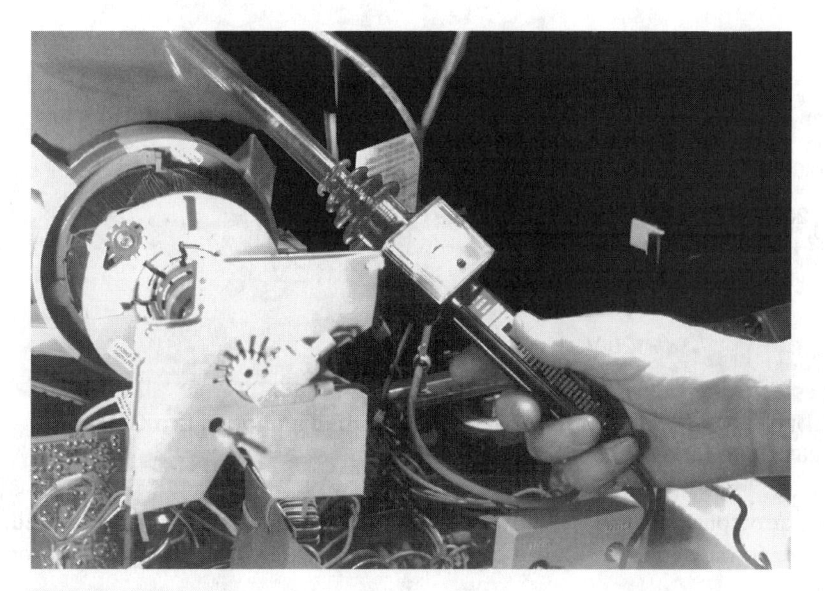

FIGURE 13-7 Check the high voltage at the anode button on the CRT to determine if high-voltage circuits are normal.

WEAK PICTURE

A poor or weak gun assembly can produce a weak picture. Suspect that a picture tube is defective if all three color guns test weak with correct element voltage and not enough brightness. Improper high voltage or voltage applied to the picture-tube elements can cause a weak picture. Usually with improper high voltage, the raster will pull in at the sides. Check the luminance and output circuits if the picture tube and corresponding circuits work properly.

NO BRIGHTNESS

High voltage might be present with a no-brightness symptom. Measure the high voltage with a high-voltage probe (Fig. 13-7). Of course, no high voltage will result in a no-brightness/no-raster symptom. An open heater element or no heater voltage produces a no-brightness/no-raster symptom. Improper voltages applied to the picture-tube circuits can also cause a no-brightness symptom.

NO CONTROL OF BRIGHTNESS

Improper brightness control can be caused by a shorted picture tube, improper element voltage, or improper control-signal voltage. With a tube tester and voltage measurements, determine if the picture tube or circuits are defective. Suspect that a luminance,

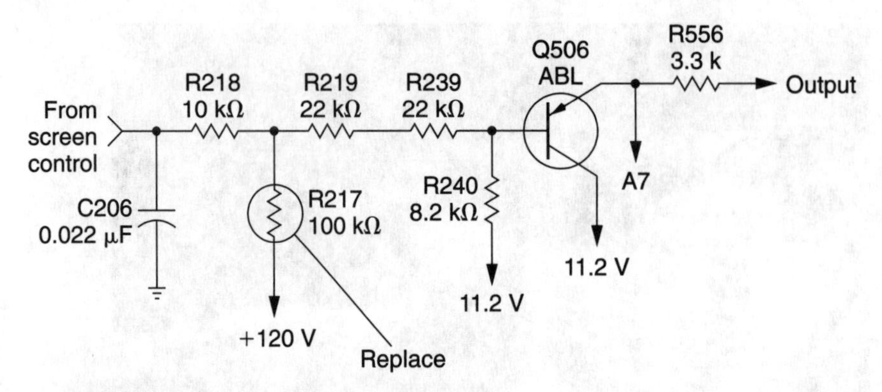

FIGURE 13-8 **Resistor R217 was found burned and open in a Goldstar CMT2612 with no control of brightness.**

video, or output circuit is defective if the picture tube and the voltages applied to the CRT are normal. High brightness can be caused by a defective service switch on some models.

The picture was very bright with no control of brightness in a Goldstar CMT2540. Replace resistor R217 (100 kΩ) to restore the brightness control (Fig. 13-8).

WASHED-OUT PICTURE

A weak picture tube or an improper focus voltage can appear as a washed-out picture. Suspect a defective video or luminance circuit with a normal raster. Often, a washed-out picture can be a combination of a weak gun assembly and an improper video signal.

NEGATIVE PICTURE

A negative picture can be caused by a shorted picture tube. Sometimes, just tapping the end or neck of the tube can cause a negative picture to come and go. An improper element and signal voltage can produce a negative picture.

BLOTCHY OR SHINY PICTURE

A weak picture tube can cause a shiny or blotchy picture. Notice if a close up of a person's face is extremely shiny when the brightness and contrast controls are turned up. Test the CRT for low emission. An improper screen or boost voltage can produce a blotchy picture. A defective degaussing coil assembly or poor purity can cause a blotchy-colored area on a section of the picture tube.

POOR FOCUS

A weak picture tube can cause poor focus. An improper focus voltage can produce a poor-focus symptom. Intermittent focus can be caused by a defective picture tube socket. Suspect that a focus pin has corroded at the CRT socket if the focus changes every few seconds. Poor focus can result from a cracked or broken focus control.

Check the focus voltage at the CRT socket with the high-voltage probe. The focus voltage should vary between 3.5 kV and 5.5 kV. You might notice only a slight change on the high-voltage probe because the high-voltage readings are quite close together. A VTVM with a high-voltage probe will give a greater voltage swing with correct focus adjustment.

Suspect that a defective focus control or leaky spark-gap assembly is at the picture-tube socket if a low focus voltage is measured at the socket. Only 1.5 kV was measured at pin 1 of the CRT in a Sharp 19C79A (Fig. 13-9). Remove the picture tube socket to measure the focus voltage if no focus voltage points are found outside of the socket area. Stick the end of a pigtail fuse inside of the focus socket. Be careful when taking focus voltage measurements. In this case, spark-gap SG855 had arced over several times, providing a low-resistance path with lower focus voltage applied to the second anode, resulting in poor raster focus (Fig. 13-10).

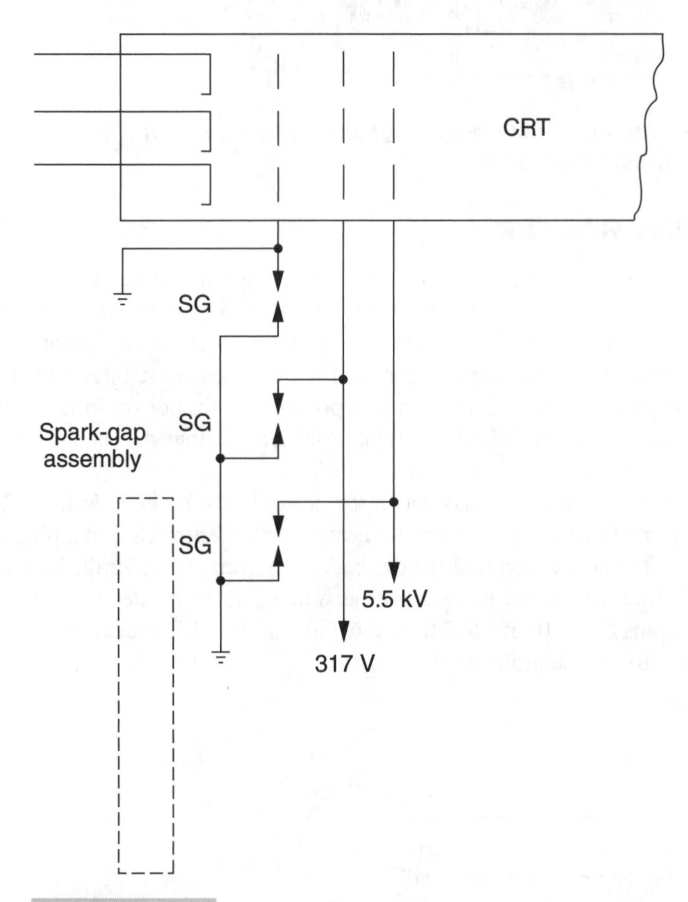

FIGURE 13-9 **A leaky spark gap (SG855) in a Sharp 19C79A reduced the focus voltage (1.5 kV).**

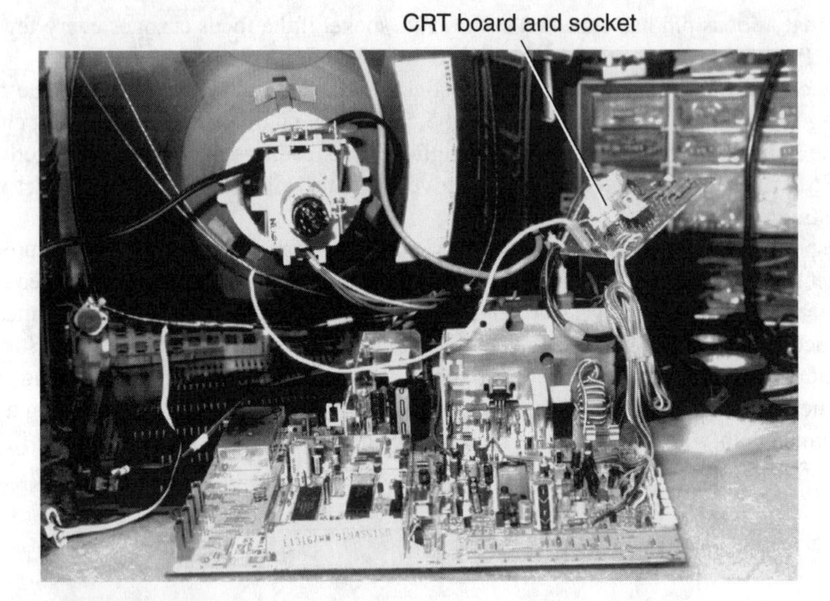

FIGURE 13-10 Arcing inside the spark gaps can cause a poor color missing or chassis shutdown.

INTERMITTENT RASTER

The raster can become intermittent with poor heater voltage, improper screen or anode voltage, or a defective picture tube. Sometimes the receiver will run for hours or days before going out. Careful voltage tests can help solve the intermittent raster symptom.

Improper heater voltage at the picture tube socket produces most intermittent raster problems. The heater voltage can develop from a power transformer or, in later models, from the high-voltage circuits or flyback winding. Poor plug fittings or a bad picture-tube socket can cause the intermittent heater voltage.

Replace the entire socket and harness or repair the heater leads. In the K-Mart SK1310A chassis, a twisted pair of heater wires go to the heater socket terminals and a plug on the chassis (Fig. 13-11). The socket commonly goes bad. Attempting to repair the heater plug-in on the chassis is fruitless. In one particular case, a new pair of twisted wires were soldered directly from pins 8 and 10 of T602 to pins 6 and 7 of the CRT socket, bypassing the defective socket and fixing the problem.

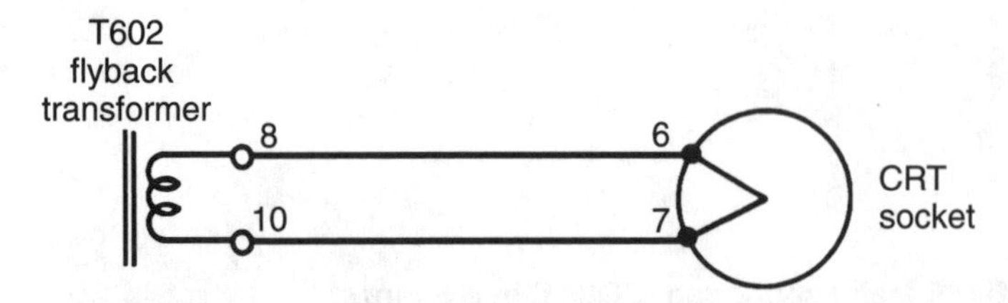

FIGURE 13-11 Replace the entire socket assembly if the heater pin terminals are in poor condition.

Besides the possibility of an intermittent voltage applied to the picture tube, suspect that a gun is shorted or that the internal heater assembly is intermittent. Take an ohmmeter continuity test of the two heater terminals after the light goes out. Remove the TV chassis and connect it to the bench tube to determine if the picture tube is defective on extreme and very difficult raster symptoms. Suspect problems with the picture tube if the chassis operates with the bench test tube for several hours. You might find that the CRT tester will not indicate an intermittent gun assembly.

RETRACE LINES

Retrace lines can be caused by a shorted picture tube or defective luminance and color-output circuits. Improper adjustment of screen and bias controls can produce retrace lines over the whole screen. A defective picture tube can have an extremely bright raster with retrace lines. High brightness and retrace lines with a defective CRT can cause high voltage and chassis shutdown in the latest TVs.

Improper boost- or screen-voltage sources can produce retrace lines in the picture (you might think it is an AGC or video problem). In a Goldstar CMT2612 portable (Fig. 13-12), scanning lines with no control in brightness were caused by very low boost voltage applied to the color-output transistors. D502 was burned and open in the 210-V boost source.

ALL-RED RASTER

Suspect a picture tube is shorted, color-output transistor is defective, or that the boost voltage or low voltage applied to the picture-tube circuits are incorrect if the raster is only one color. The video and sound might be normal, but the entire screen is of one bright color. Check all voltages on the picture tube. Test each gun of the picture tube with the CRT tester.

Improper adjustments of the screen and bias controls can cause one color to stand out. Try to complete the black-and-white adjustment. If one color will not adjust to a horizontal color line with the screen control when the service switch is in the service position, suspect

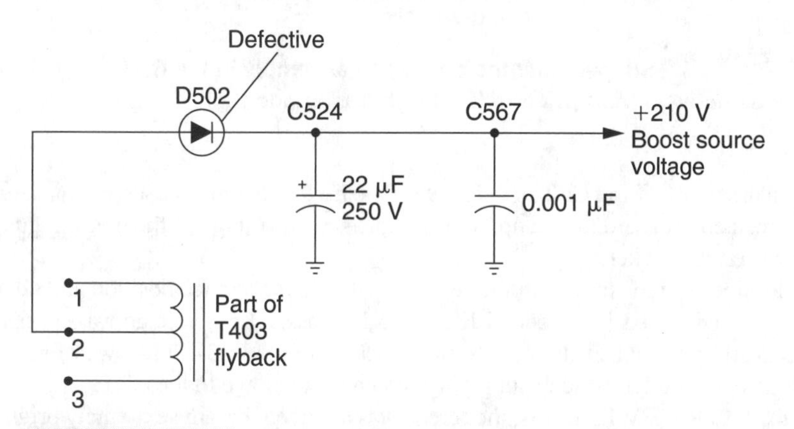

FIGURE 13-12 **Lack of brightness control and retrace lines in a Goldstar CMT-2612 portable were caused by defective diode D502.**

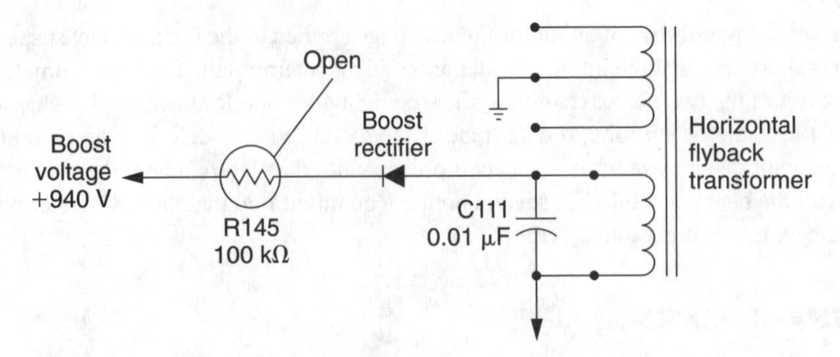

FIGURE 13-13 **Improper boost voltage in a General Electric TV was the result of open resistor R145 (100 kΩ).**

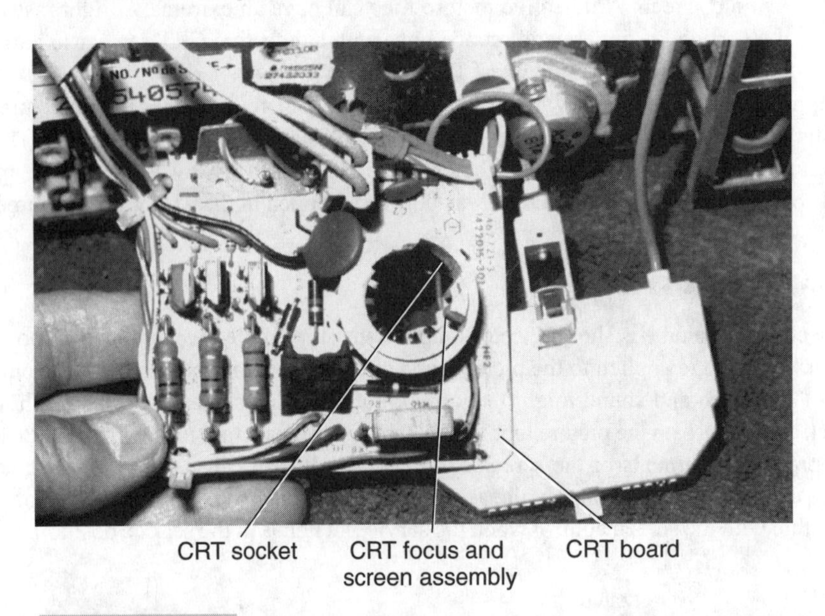

FIGURE 13-14 **Suspect that the spark-gap assembly in the CRT socket is defective if the symptom is a bright raster and then shutdown.**

that a gun assembly of the CRT is defective. Be sure that the master screen, bias, or individual screen controls are turned up. Measure the screen voltage of the missing gun color at the picture-tube socket.

No color lines were visible with the service switch in the service position and all screen controls were advanced in a General Electric KD chassis. Low voltage was measured at the screen-grid terminal of the picture-tube socket (Fig. 13-13). R145 was found to be open in the boost-voltage circuit supplying screen grid voltage to the CRT.

In an RCA FFR-498WK chassis, the screen was all green, became extremely bright, then shut down. With a very bright-colored raster, the spark-gap assembly must be checked for leakage (Fig. 13-14). Just 32 Ω were measured across the green-screen spark-gap assembly. In these latest RCA TVs, replace the whole spark-gap assembly with RCA part 148169.

ONE MISSING COLOR

Raise the screen control on all three colors to determine which color is missing. Suspect that a gun assembly or that the CRT circuits are defective if one color is missing with the service switch in the Service position. Test all three guns with the CRT tube tester.

A leaky color-output transistor can cause one color to be missing from the raster. Some of these color-output transistors are located on the picture-tube socket assembly (Fig. 13-15). A voltage test on the metal collector terminal can indicate that a transistor is leaky. Suspect that a transistor is leaky if the collector voltage is very low. Compare this voltage with the other two output transistors. Low voltage on all three output transistors can indicate that the supply voltage is improper.

The color green was missing from the raster in a Goldstar NF9X chassis (Fig. 13-16). After a quick test of the video color-output transistor, only 23 V were found at the collector terminal of Q353. The transistor was removed and a short was discovered between the collector and emitter terminals. R363 was replaced because it was running very warm. Replacing both Q353 and R363 solved the no-green problem.

INTERMITTENT COLOR LINE

Suspect an intermittent gun assembly if the raster operates normally for several hours or days, then a green line appears with a negative picture. Often, the brightness dims down with no picture. With very difficult intermittent picture-tube problems, the chassis should be connected to the bench test tube to determine if the picture tube or chassis is intermittent. Cover the chassis with a rug or blanket to keep the heat in because most intermittent

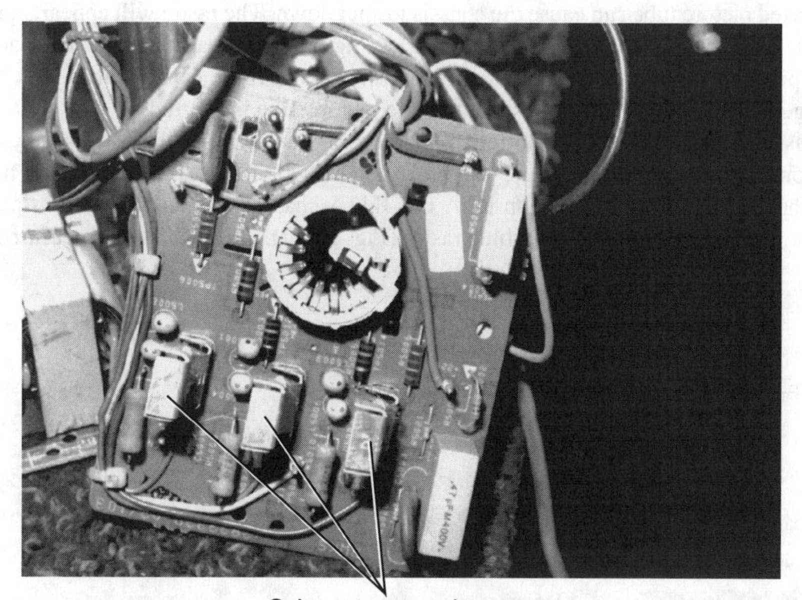

FIGURE 13-15 Inspect the color output transistors on the picture tube board.

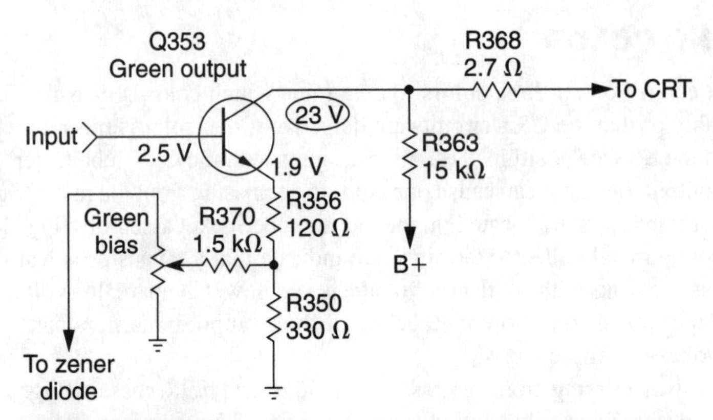

FIGURE 13-16 **Lack of green in the raster in a Goldstar NF9X TV chassis was caused by a faulty green output transistor, Q353.**

problems occur after several hours of operation. Replace the intermittent picture tube if the chassis operates normally for several days with the test tube connected.

Other High-Voltage Problems

CHASSIS SHUTDOWN

A shorted picture tube can cause the chassis to shut down. The raster will appear extremely bright, then shut down. Disconnect the anode high-voltage lead, insulate the lead away from picture tube or chassis, and plug in the power cord. If the sound comes up with high voltage at the anode terminal, suspect that the picture tube is shorted.

Chassis shutdown with only a bright red, green, or blue raster might be caused by a leaky or shorted spark gap. Measure the voltage at the cathode terminal of the picture tube before the chassis shuts down. In an RCA CTC111A chassis, the chassis would shut down after a few seconds with a bright blue raster (Fig. 13-17). Only 29 V was measured at the

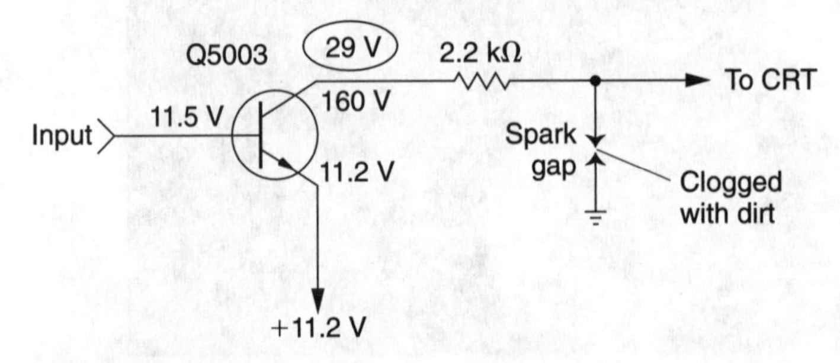

FIGURE 13-17 **A bright CRT, followed by chassis shutdown in the RCA CTC111A chassis, was caused by a leaky spark gap.**

FIGURE 13-18 Suspect a shorted yoke assembly if smoke curls up from it.

collector terminal of the blue driver transistor (Q5003). A resistance of 2.2 kΩ was found between the collector and emitter terminals. Q5003 was replaced, but this did not correct the chassis-shutdown problem. A short was found from pin 11 to ground of the CRT because of a leaky spark-gap assembly.

YOKE PROBLEMS

The horizontal and vertical deflection yoke is mounted near the bell and on the neck of the picture tube. The defective yoke can contain shorted turns, leakage between the horizontal and vertical windings, and excessive arcing between the windings. The shorted yoke can show a trapezoid pattern or picture on the screen. Arcing between the coil layers can show up as heavy horizontal firing lines in the raster. If excessive arcing is found, puffs of smoke might curl up from the yoke assembly (Fig. 13-18).

The defective yoke can be located with ohmmeter tests, with the flyback-yoke tester, and by feeling for warm spots inside of the yoke windings. In the black-and-white TV, the old-timer would disconnect the yoke after 30 minutes of operation and feel for warm areas inside the yoke (indicating shorted turns). A continuity check with the ohmmeter indicates if a winding is open or if the two windings are shorted together.

HIGH-VOLTAGE ARCING

Excessive high voltage can produce a loud crack or arcing sounds around the anode connection. Dust and dirt on the CRT anode terminal can cause some corona activity. Discharge the picture tube. Clean off the anode lead and tube area with cleaning fluid or detergent. Inspect the anode lead rubber for excessive firing areas. Replace the entire high-voltage cable and plug if evidence of arcing is found. Check the high-voltage anode for excessive high voltage. High-voltage arc-over at the CRT anode terminal can result from a defective hold-down circuit in the TV chassis.

High-voltage arcing can occur with a defective picture tube. With an open heater assembly, a blue area appears at the gun assembly when arcing begins. Check the glass neck for a cracked area. It might be inside of the yoke assembly. A shorted picture tube can cause high-voltage arcing around the ground strap and TV chassis.

FIRING LINES IN THE PICTURE

Sharp jagged or dotted lines across the picture can indicate outside interference or arcing in the chassis. Remove the outside antenna and notice if the interference lines disappear. Usually, man-made noise picked up from the antenna will appear on channels 2 through 6 (Fig. 13-19).

Thin firing lines can be caused by improper grounding of the dag (outside area) of the picture tube. Clip a lead from the chassis to the black, rounded area of the picture tube. Sometimes these small springs around the bell assembly become dirty and corroded, leaving a poor ground. Check to see if the ground strap from the chassis to the picture-tube

FIGURE 13-19 **A defective focus-screen assembly in the RCA CTC107 chassis can cause firing lines in the picture.**

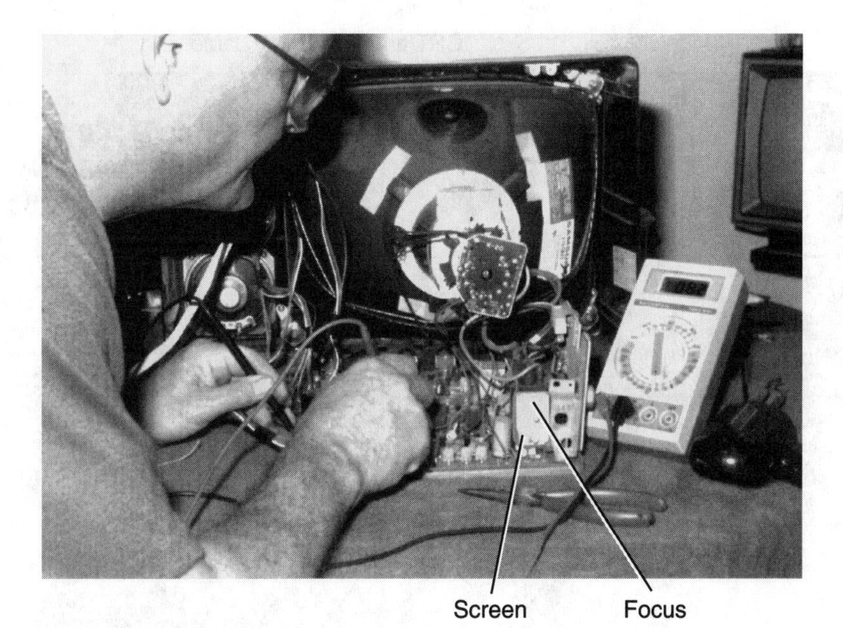

FIGURE 13-20 Rotate the focus control if firing lines are in the picture.

mounting bracket is in place. If broken, run a ground wire from the picture-tube mounting to the metal chassis.

Arcing lines can be caused by a defective flyback transformer or focus-control assembly. Check for corona arcs around the outside plastic body of the horizontal output transformer. Rotate the focus control to see if the lines disappear (Fig. 13-20). Replace the focus control if the focus is poor focus and firing lines are in the picture. These printed high-resistance focus controls have a tendency to break and arc over (Fig. 13-21).

DEFECTIVE CRT HARNESS

A defective picture tube socket can cause a blurry picture with firing lines. Poor focus can result from a corroded focus pin of the CRT socket (Fig. 13-22). Picture-tube spark gaps inside the CRT socket can cause arc-over. Sometimes you can hear the arcing by placing your ear near the tube socket. Besides arc-over, suspect that the heater terminals are faulty if an intermittent raster is found. If the heater terminals have changed color, forming a high-resistance connection, install a new CRT socket and harness.

A faint spitting noise was heard when the J.C. Penney CTC90JL chassis was first turned on. Intermittent firing lines would appear across the screen. (Under these conditions, suspect either a defective flyback transformer or focus control.) The arcing could be heard at the CRT socket. The arcing noise stopped when the picture-tube socket was removed. Installing a new CRT socket and harness solved the intermittent firing in the raster.

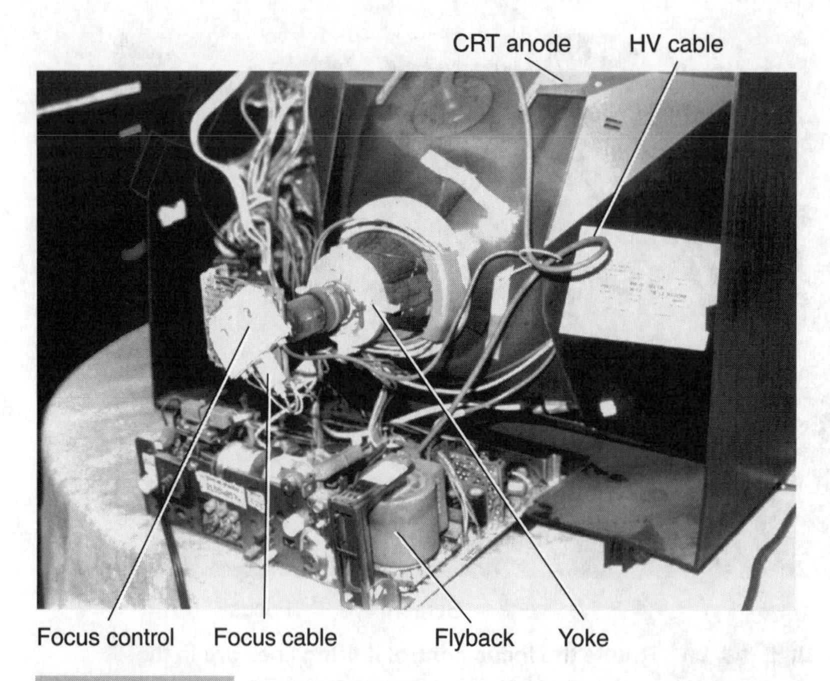

FIGURE 13-21 **Check the various components that could cause arcing and firing lines in the picture.**

FIGURE 13-22 **Check for a corroded focus pin in the CRT socket if the set exhibits poor focus.**

Testing the CRT

The defective picture tube can be located with a CRT tube tester and crucial voltage measurements. Check the high voltage at the anode terminal. Visually inspect the neck of the tube for light in the tube filament. Now test the picture tube with a good CRT tester.

The picture tube can be checked in the home, in the carton, or on the bench with the CRT tester. Intermittent conditions might not be located with the picture tube tester. Visual inspection and high- and low-voltage measurements on the tube elements can help you to locate the intermittent picture tube (Fig. 13-23). Sometimes tapping the end of the tube can turn up a shorted gun assembly. For very difficult picture-tube problems, connect the chassis to the bench tube for observation.

CRT REPAIRS

Operation of the weak picture tube can be extended by applying a tube brightener or by charging the tube with a picture-tube charging instrument. The tube brightener plugs on the end of the picture tube with the CRT socket plugged into the end of the brightness assembly. The brightener raises the heater voltage, increasing brightness. Although increasing the tube brightness is not a permanent repair, the life of the TV can be extended for several months. Charging the picture tube might prolong the life of a picture tube for several years. The test instrument renews each gun assembly by removing bombarded ions

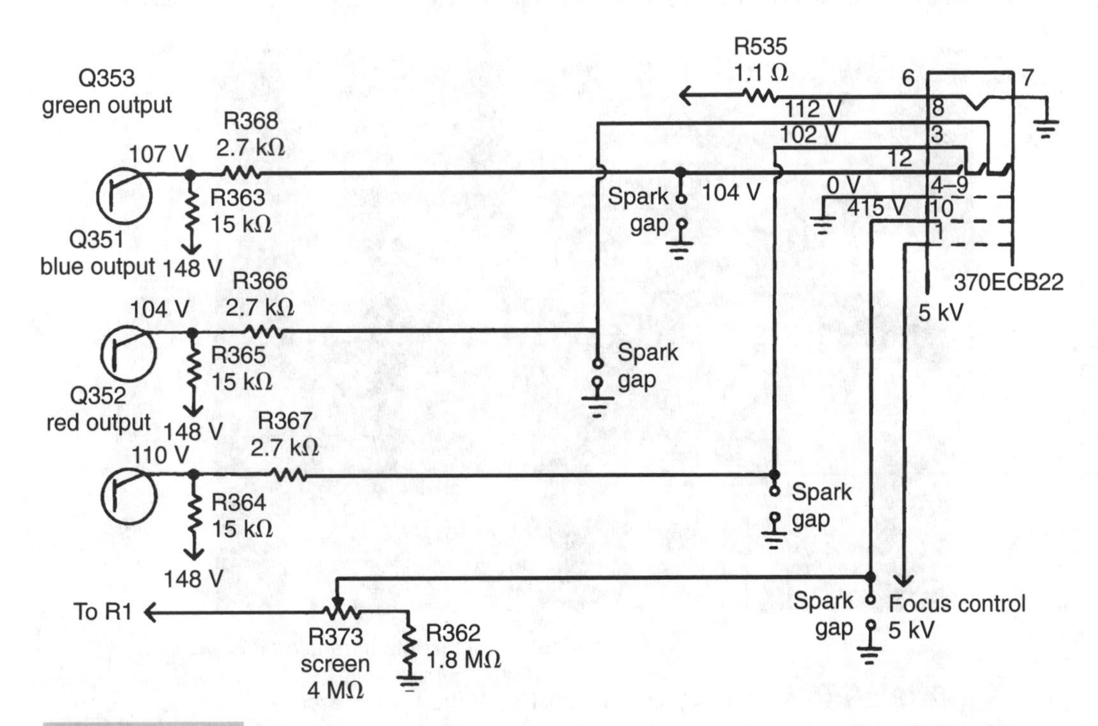

FIGURE 13-23 Monitor the voltages at the picture tube to determine if the CRT or the circuits are intermittent.

from the cathode area of the heater gun assembly. Charging the picture tube can help sell a used color TV. Use the instrument to sharpen and brighten the picture after TV repairs. A good picture-tube charging and testing instrument can cost more than $500.

IHVT-DERIVED VOLTAGES FOR THE CRT

Besides supplying low dc voltages for the various circuits of the TV, the integrated flyback transformer (IHVT) can produce high voltage, focus voltage, screen voltage, and filament voltage for the picture tube. The high-voltage winding of the IHVT horizontal output transformer contains high-voltage diodes molded inside of the transformer. The high-voltage lead goes directly to the anode button socket on the bell of the CRT. If any of the high-voltage diodes break down, the entire flyback transformer must be replaced (Fig. 13-24). The high voltage is measured with a high-voltage probe or VTVM on the high-voltage terminal of the CRT.

The filament voltage for the three guns inside of the picture tube is taken from a separate winding on the flyback transformer. This winding consists of one to three turns of large wire around the ceramic core and metal flange. The two filament cables plug into or are wired directly to the picture-tube socket. Poor socket connections, a defective CRT tube socket, or poorly soldered connections at the CRT board can cause intermittent or no raster because of improper filament voltage. Do not try to measure this ac voltage with any type of voltmeter.

The focus and screen voltages are usually taken from a focus-screen variable divider network of the high-voltage secondary winding. The focus and screen variable assembly can be found on the end of the CRT board or on a separate mount at the rear of the chassis, A

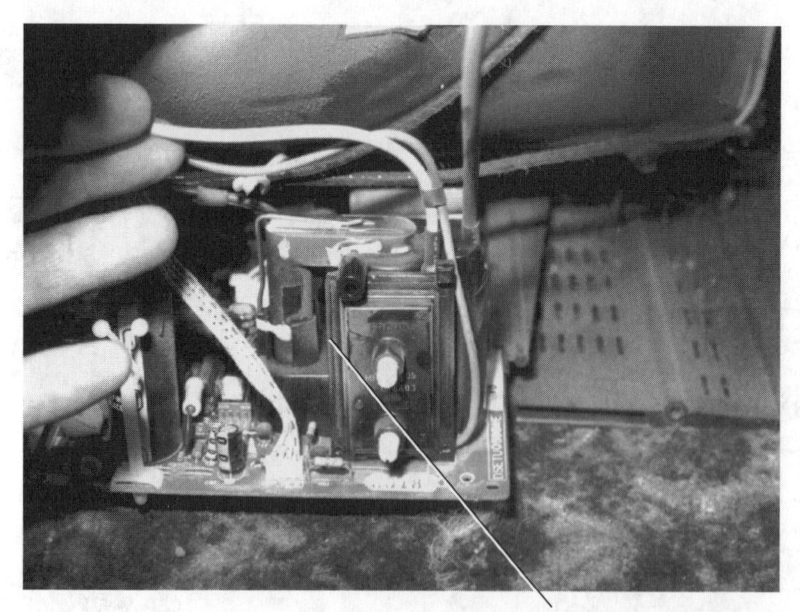

Flyback or horizontal output transformer

FIGURE 13-24 **Replace the flyback its HV diodes are arcing internally.**

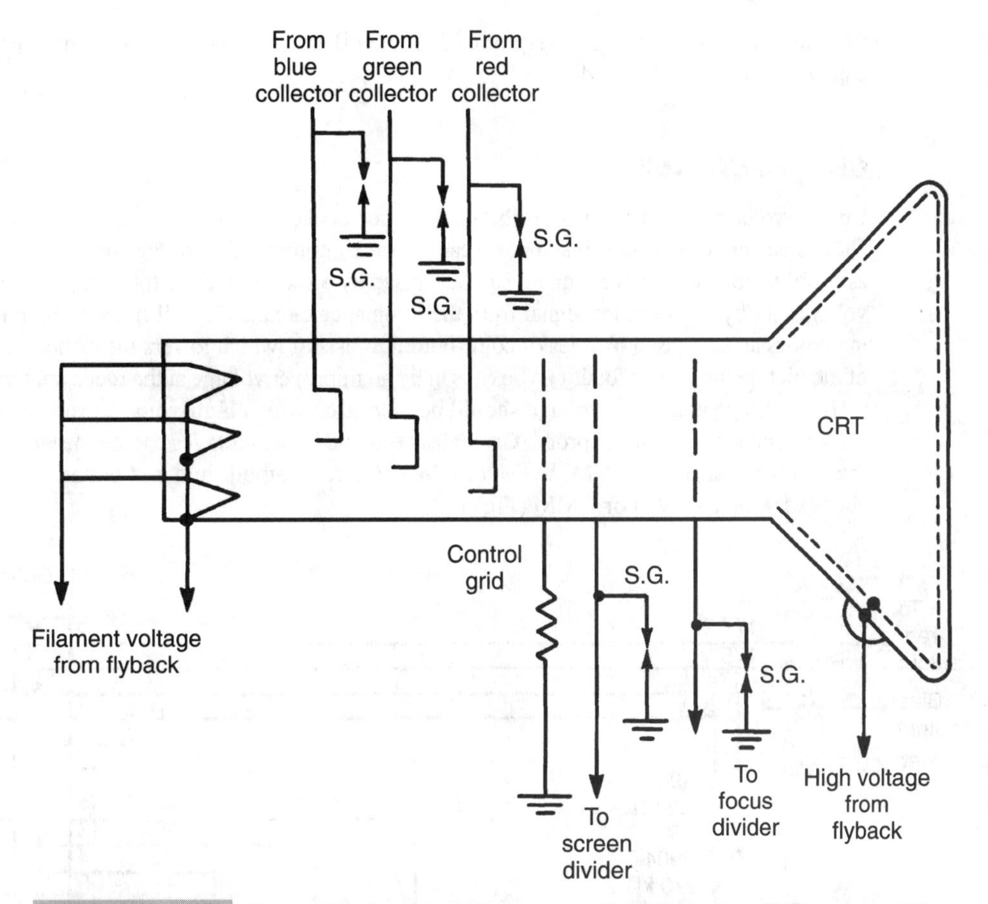

FIGURE 13-25 **Some TVs have spark gaps on all CRT pin terminals, except the heater pins.**

corroded or dirty focus control and CRT pin can produce an out-of-focus or blurry picture. In some TVs, the screen voltage can be taken from another secondary winding with a separate diode rectifier and filter network.

SPARK GAPS

You will find spark gaps at each grid, at the screen and focus grid terminals, and at all three cathode terminals of the picture tube. These gaps will arc over and protect the picture-tube gun assembly if extreme high voltage is present at any picture-tube element. If one of these gaps continually arcs over, it can cause an extremely blurry picture, a loss of color, a one-color raster, and/or chassis shutdown.

Sometimes when the picture-tube sockets become very dirty, excess dust collects in the gap area and causes the entire chassis to shut down (Fig. 13-25). If the raster comes on and goes into a blurry, distorted picture, check the spark gaps in the CRT socket. Usually, the spark gaps are molded into the CRT socket or are placed in separate components on the

CRT board. Replace the spark gap or CRT socket if excessive arc-over continues or if the gap areas cannot be cleaned.

CRT VOLTAGES

Crucial voltage measurements on the picture-tube socket can help you to determine if the CRT is defective. A no-raster symptom can indicate improper high voltage or a defective gun assembly. No control of brightness can be caused by a shorted picture tube, improper screen voltages, or by a poor video signal from the luminance circuits. The all-red screen or a missing color can be caused by a leaky color-output transistor, which lowers the cathode voltage of the picture tube. Poor focus can be caused by an improper voltage at the focus pin terminal.

High voltage and focus voltage should be measured with a high-voltage probe or with a VTVM and a high-voltage probe. Crucial screen and boost voltages can be measured with the 1500-V scale of a VTVM, VOM, or DMM. Correct cathode and grid voltage should be checked with a VTVM or DMM (Fig. 13-26).

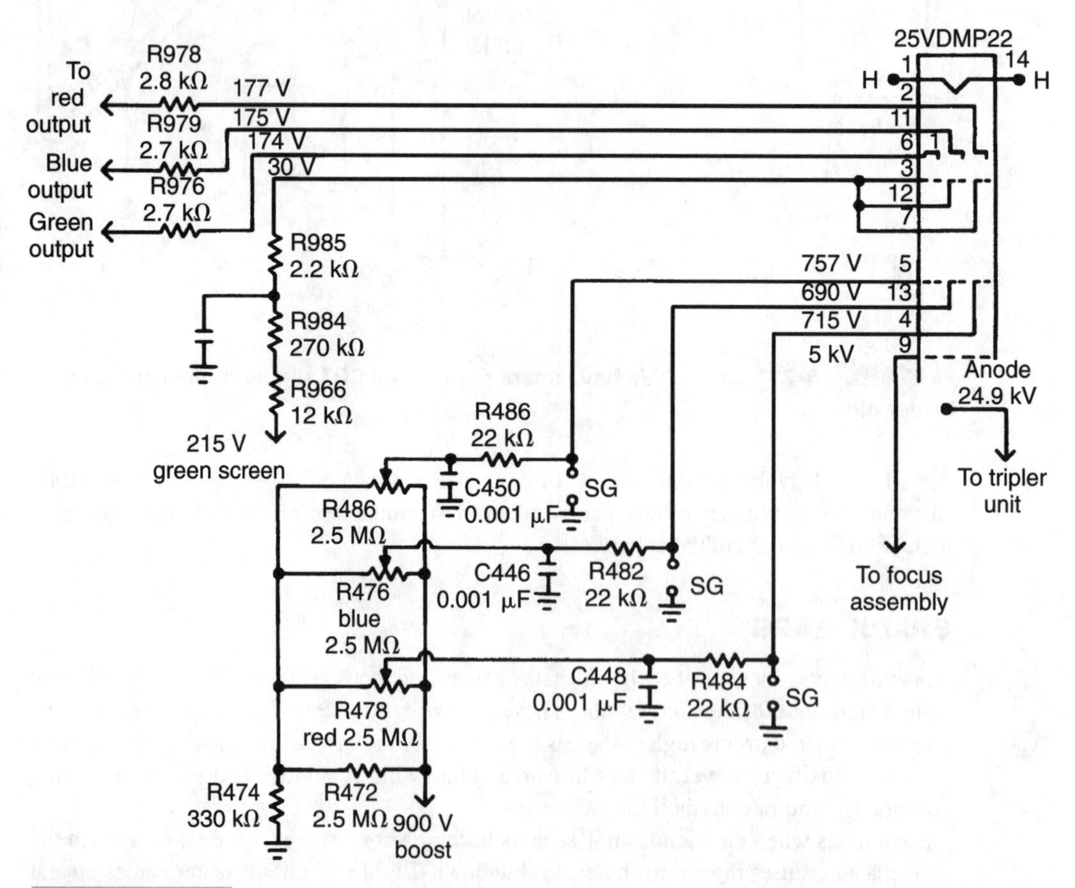

FIGURE 13-26 **Check for crucial cathode, grid, and high voltage to determine whether a picture tube is defective.**

PICTURE-TUBE PROTECTION

Remember, all picture tubes use an integral vacuum, so handle them with care. Do not remove or install the picture tube unless you are wearing goggles to protect your eyes from possible implosion and flying glass. Replace the picture tube with a correct tube. Do not handle the tube by the neck. Grab the picture tube with both hands on its heavy glass front. Do not force any components off the neck of the picture tube. Remember, some in-line picture tubes have deflection yokes that are permanently attached.

DISCHARGING THE PICTURE TUBE

Always discharge the high voltage at the picture tube when replacing or working around the high-voltage circuits. Remember, the picture-tube aqueduct glass around the high-voltage socket acts like a high-voltage capacitor and it can hold a charge for months. Stay alert when discharging or working around the high-voltage circuits to prevent shocks or serious injuries.

In the old black-and-white tube chassis, the picture tube was discharged from the anode connection to the chassis ground. Do not attempt to discharge the CRT in this manner. You can damage transistors and ICs in the solid-state chassis by discharging high voltage to the chassis. Always discharge the high voltage from the button area to the black aqueduct on the outside of the picture tube. Use a long-bladed, well-insulated screwdriver to get under the rubber socket connection, and ground a similar screwdriver blade to the outside ground of the CRT. Hold the metal shanks of the screwdrivers together to discharge the picture tube.

PICTURE TUBE REPLACEMENT

After determining if the picture tube is defective, take a visual inspection of those components surrounding the CRT (Fig. 13-27). Check the position of the purity and beam magnets on the neck of the tube. Measure the distance between the purity magnet and end of the tube for future reference when installing the new one (Fig. 13-28).

Carefully remove the back cover. Loosen the top screws and then the bottom. Hold the back so that it will not fall against the CRT and possibly break or crack the neck of the tube. Remove the chassis. Place the front of the tube or cabinet face down on a rubber mat or rug. Notice the mounting of each component from the neck of the CRT and around it. Remove the picture tube and lay it on its face on a rug or carpet so as not to scratch the screen area. Reverse the procedure to install a new tube. Degauss the picture tube before black-and-white setup.

PICTURE TUBE REMOVAL

Before attempting to remove the picture tube, remove the chassis and tuner from the cabinet. Place the cabinet face down on a blanket or soft surface to protect the front screen of the CRT. Remove the metal screw or bolt in each corner to free the bracket or shell that holds the degaussing coil. Sometimes the top or bottom degaussing coil is held in place with plastic ties and metal screws (Fig. 13-29).

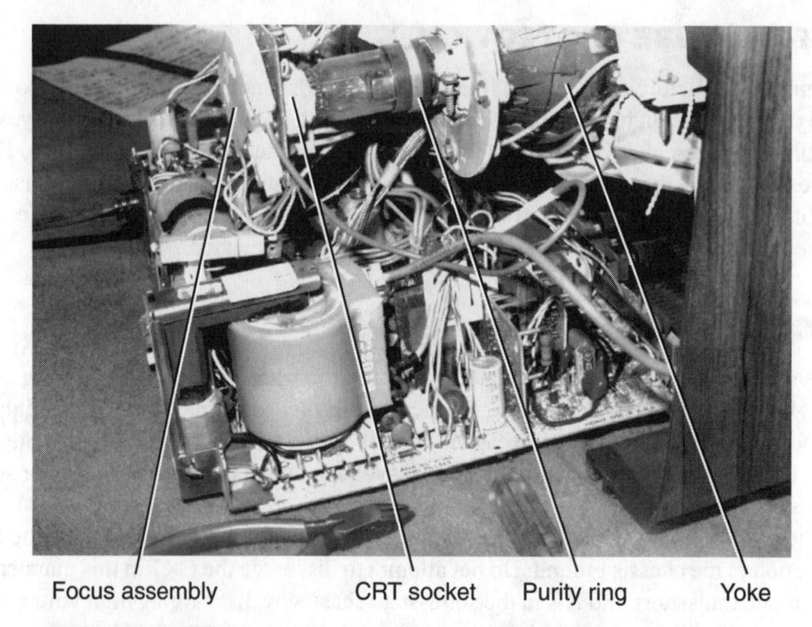

FIGURE 13-27 **Visually inspect all components mounted on a CRT before removing one from the chassis.**

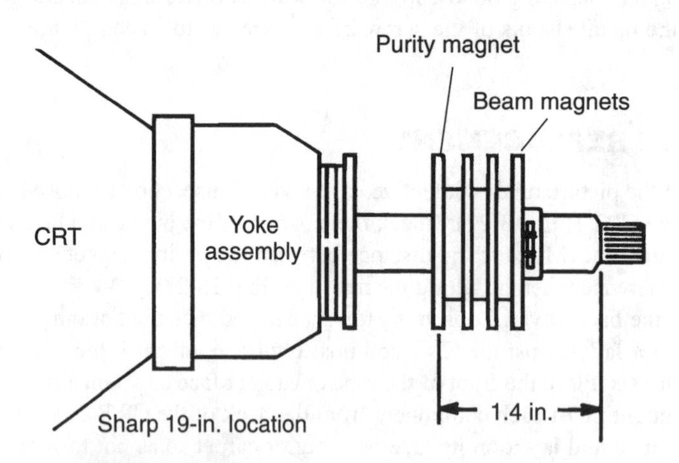

FIGURE 13-28 **Measure the distance of all components on the neck of the CRT for correct location at reassembly.**

Remove the screws or bolts on each corner of the picture tube so that the tube can be lifted out of the cabinet. It is best for two people to lift the CRT. Place it on a blanket or soft cloth to prevent scratches. Reverse the procedure when installing a new picture tube. Measure the distance from the end of the CRT to the mount tab magnets of the old tube to position the new replacement. Always wear safety glasses or goggles when removing and installing CRTs.

AN ON-SCREEN DISPLAY PROBLEM

The signal information for the on-screen display (OSD) is provided by the AIU (U3300). The system-control IC (U3100) determines when to output the OSD signals. During normal operation, the luminance information is applied to the base of Q2901. This output is mixed with emitter of Q5002, with chroma information applied to the base of Q5002. The collector of Q5002 drives the green gun assembly of the picture tube (Fig. 13-30).

When the OSD signal of U3300 is applied to the base of Q2903 and Q5004, both transistors are turned on, which causes CR5002 to become forward-biased. Now Q5002 is pulled low, resulting in a current path to Q5004 instead of Q5002.

Check the OSD circuits by scoping the input terminals of the OSD, luma, and OSD black signals. Scope the waveforms at the collector of Q2706 and Q5004. If Q2903 becomes shorted, the picture will be all green. A leaky CR2709 or Q2706 will eliminate black around characters; if it is open, the picture will be black all of the time. For a green OSD character, check the OSD output of U3300, Q2903, Q5004, and CR5002. If there is no black edge, check the output from U3300 and check Q2706 and CR2709. With no luminance, but okay chroma, check the output of U3300 and look for a leaky Q2709.

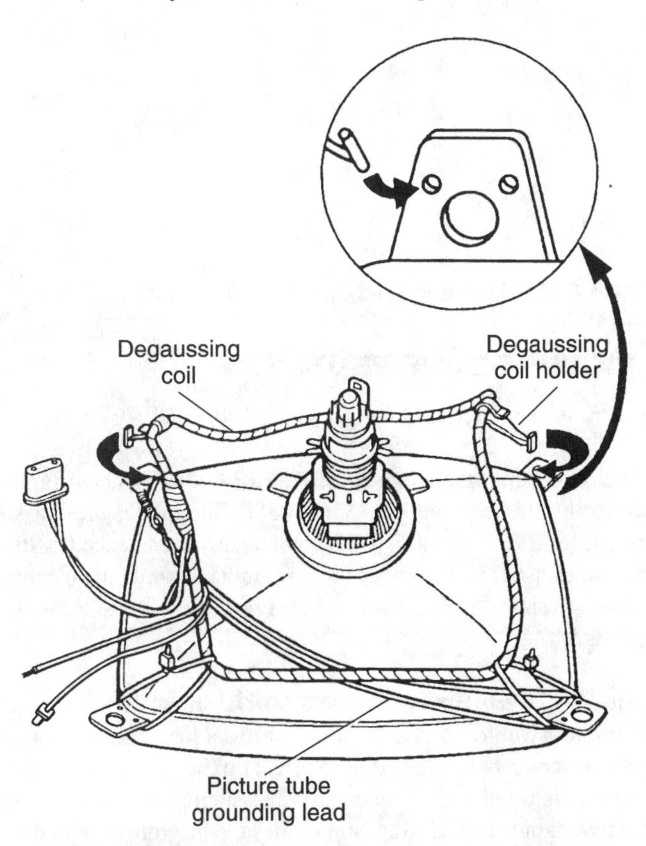

FIGURE 13-29 A picture-tube removal schematic for the Realistic 16-281 portable TV.

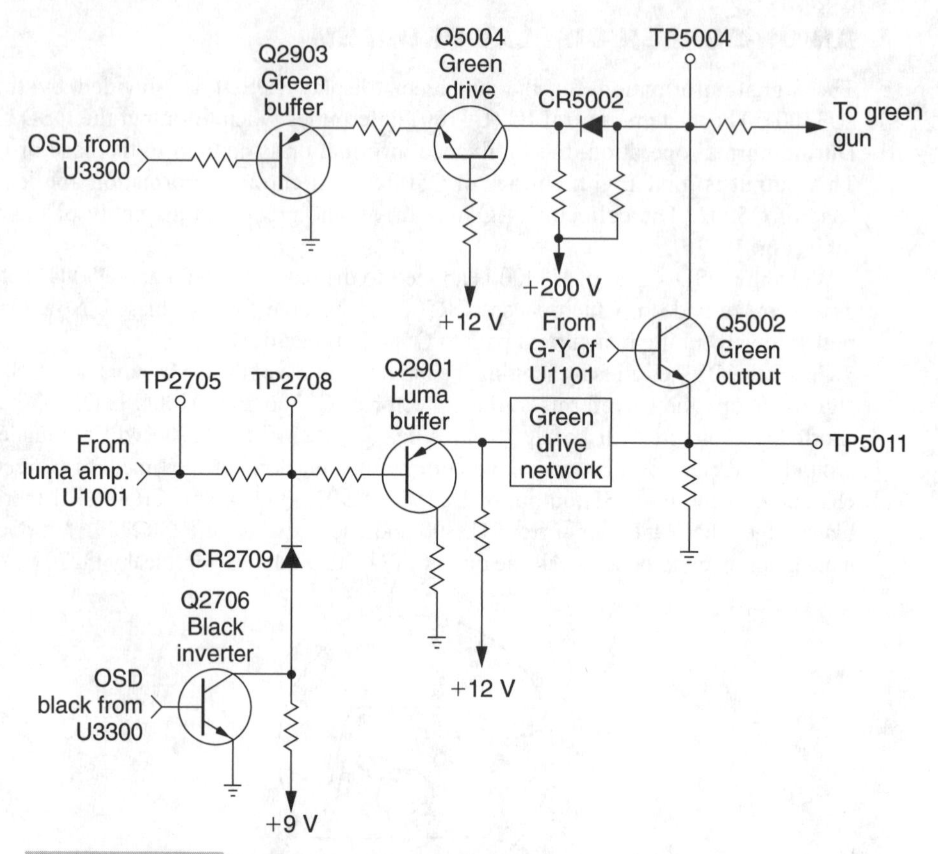

FIGURE 13-30 RCA CTC157 on screen-display (OSD) circuits.

FIVE ACTUAL PICTURE-TUBE PROBLEMS

The following five picture-tube-related problems actually happened.

High brightness and retrace lines in an RCA GJR2038P The brightness was extremely high and retrace lines were in an RCA GJR2038P. Sometimes the chassis would shut down after operating for a few hours. All CRT voltages were checked, with normal boost voltage to the output transistors. Improper voltages indicated that the brightness reference transistor (Q703) was shorted (Fig. 13-31). Several of these same transistors have been replaced in other RCA TVs.

Intermittent brightness in an Emerson MS250RA In an Emerson MS250RA portable, the screen brightness would operate for hours or just a few minutes, and come off and on. The brightness and screen controls had little effect on the picture. Both the 11.5-V and 18.8-V voltages were monitored in the video and contrast circuits, and were fairly normal when the brightness went out (Fig. 13-32). Sometimes the brightness would only dim. No scope waveform out of pin 42 or 22 occurred with the no-brightness problems. Replacing IC201 solved the problem.

Bright screen with shading in a Sanyo 91C510 The TV screen was very bright with some dark shading in a Sanyo 91C510. Transistors and voltages were checked in the video circuits and were normal. The boost voltage of 196 V was quite low at the picture tube cathode terminals. D311 and R342 were found to be good in the boost-voltage source (Fig. 13-33). C358 (4.7 μF, 250 V) was shunted, which solved the brightness problem.

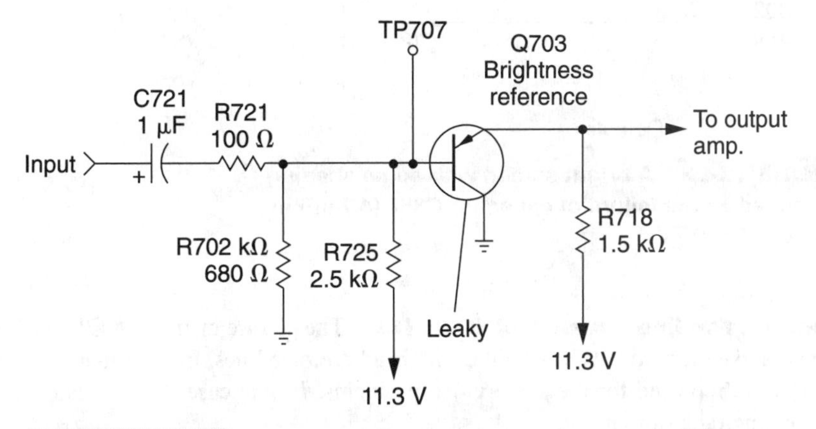

FIGURE 13-31 **High brightness with retrace lines resulted from a leaky transistor (Q703) in an RCA GJR2038P.**

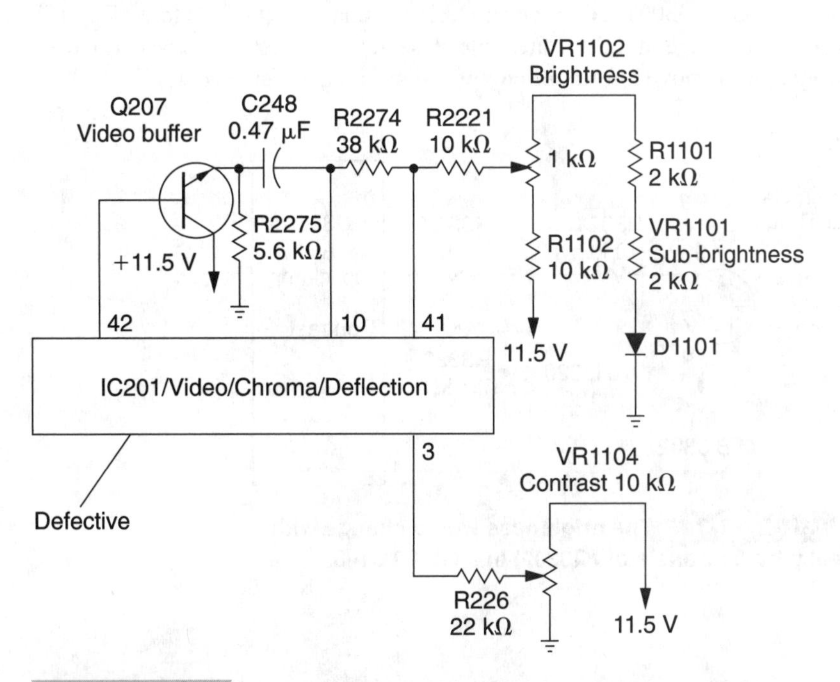

FIGURE 13-32 **An intermittent chip (IC201) caused intermittent brightness in this Emerson MS250RA portable.**

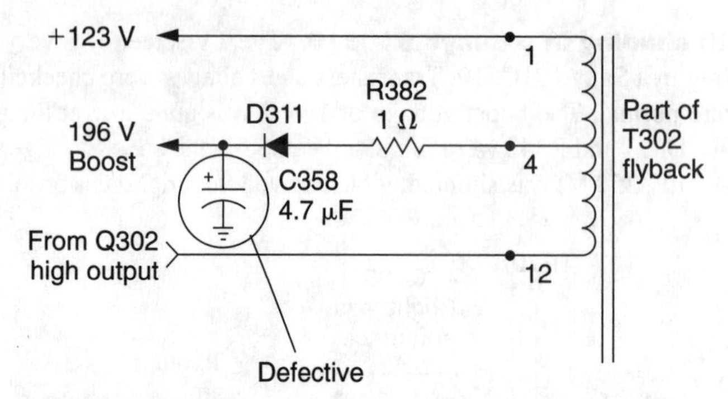

FIGURE 13-33 **A bright screen with some shading was caused by the failure of capacitor C358 (4.7 μF) in a Sanyo 91C510 TV.**

Bright retrace lines in an RCA CTC118A The picture in an RCA CTC118A chassis looked like there was AGC trouble, with bright retrace lines. In this model, several ICs (U701) were replaced for the same video problems. In this case, R114 was open in the color-output transistor circuit.

Brightness in and out in a GE CTC148 The brightness would change without adjusting the picture in a GE CTC148. Go directly to the input circuits of the PIX clamp and contrast of AIU U3300 IC. Check the LDR buffer transistor (Q3302) for leakage (Fig. 13-34). The normal voltages at Q3302 are quite close in value. Test the transistor in the circuit for leakage, then remove it from the circuit and test it again for leakage.

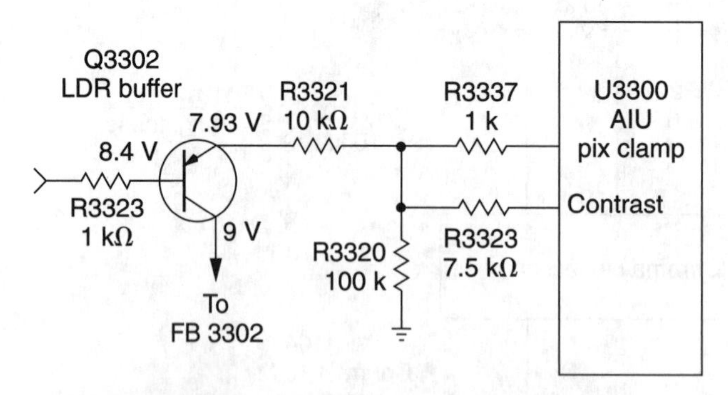

FIGURE 13-34 **The brightness would change with a leaky buffer transistor (Q3302) in a GE CTC148.**

TABLE 13-1 SERVICING THE PICTURE TUBE CIRCUITS

WHAT TO CHECK	HOW TO CHECK IT
No raster.	Check voltage at HV anode on CRT.
White raster no picture.	Check color output and video circuits.
Black raster with normal high voltage.	Check for no focus voltage.
Poor focus.	Check CRT and focus circuits. Check CRT socket.
Real bright raster, no control.	Check boost voltage. Check all voltages on CRT. Suspect brightness circuits.
Weak picture.	Test CRT; all three color gun assemblies.
Intermittent raster.	Check CRT socket and look for loose particles in grid elements of CRT.

Conclusion

Always keep the back cover on the TV after replacing the picture tube or repairing a chassis to prevent breakage. Carefully lay the defective tube face down on a rug or carpet to prevent scratching the screen area. Remember, you can't receive the turn-in value of the tube until the bad tube is returned and accepted.

Installing a rebuilt picture tube can place the TV back in working order for a lower price. Some picture-tube rebuilders will install a new gun assembly in those 19-inch in-line picture tubes with the yoke glued to the bell of the tube. This can be done at a much lower cost than purchasing a new tube. Charging the old picture tube can brighten the picture after repairs. Remember to clean the screen with window cleaner and polish up the cabinet before returning it to the customer. More information on servicing the picture-tube circuits is found in Table 13-1.

PART 3

TROUBLESHOOTING AND REPAIRING AUDIO AND VIDEO CASSETTE PLAYERS

14

BASIC CASSETTE PLAYER TESTS

CONTENTS AT A GLANCE

The cassette player is available in different general shapes and sizes with different features (Fig. 14-1). Thousands of cassette players are out there, broken down, collecting dust, in need of repair. Troubleshooting and repairing your own cassette player can be quite rewarding. Besides, it's lots of fun and you can save a few bucks in the process.

Servicing cassette players can be done by anyone who is handy with tools and can read a schematic diagram (Fig. 14-2). The electronic circuits are broken down so that everyone

FIGURE 14-1 The small portable and boom-box cassette players are the most popular units around the house.

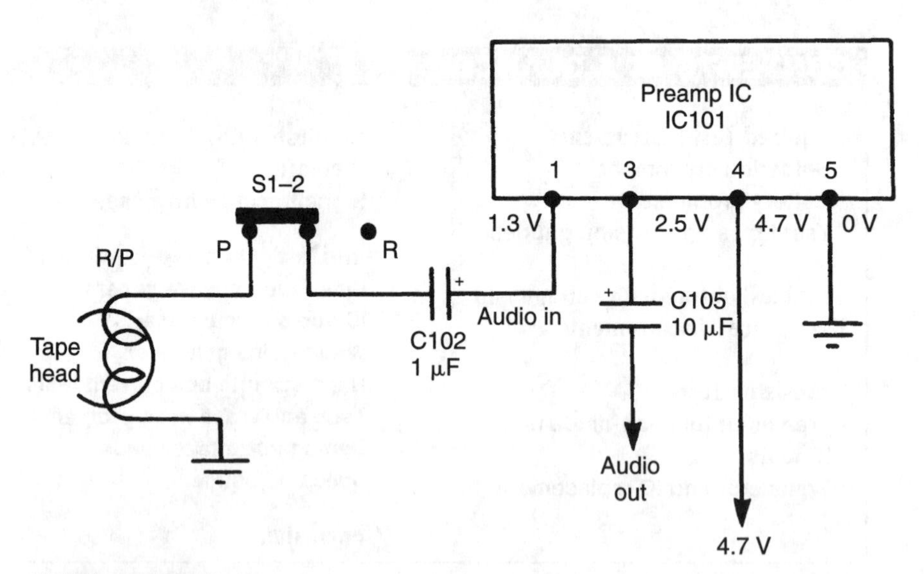

FIGURE 14-2 If you can read the simple schematic diagrams and handle a few tools, you're in business.

can understand them. If you have a little knowledge of electronics, if you are a novice, or if you are an electronics student, you can make most repairs found in this book. Very few tools and test instruments are needed. In fact, you can do the job with only one digital-multimeter (DMM).

Required Test Instruments

Several screwdrivers, a pair of long-nose pliers, and side cutters do the bulk of the work. A set of jeweler's screwdrivers take care of those tiny screws and bolts (Fig. 14-3). The small magnifying glass helps you to locate small parts and soldered terminals. If you do not have a VOM or a DMM, purchase one of the new digital multimeters that can check voltage, resistance, current, capacitance, and diodes with a frequency counter.

Today, several types of pocket DMMs can do all the tests needed to repair that broken cassette player. The tester might cost up to $150.00. The BK test-bench DMM (Fig. 14-4), model 388-HD, can check diodes, transistors, logic, frequency, current, capacitance, resistance, and voltage. Read carefully how each test is made from the manufacturer's literature included with the DMM.

This DMM can check transistors and diodes with the diode test. NPN and PNP transistors can be tested by plugging into a small transistor socket. The frequency counter has three different frequency ranges: 2 k (Hz), 20 k (Hz), and 200 k (Hz). Current from 200 μA to 20 A can be checked on five different ranges (Fig. 14-5). Small capacitors can be tested from 2 nF to 20 μF. Seven resistance ranges vary from 200 to 2000 MΩ. Dc voltages vary from 200 mV up to 1000 V, and the ac voltage range goes up to 750 V.

Besides taking crucial voltage measurements on the transistors and ICs, transistors and diodes can be tested with the diode-junction tests or with transistor-gain measurements.

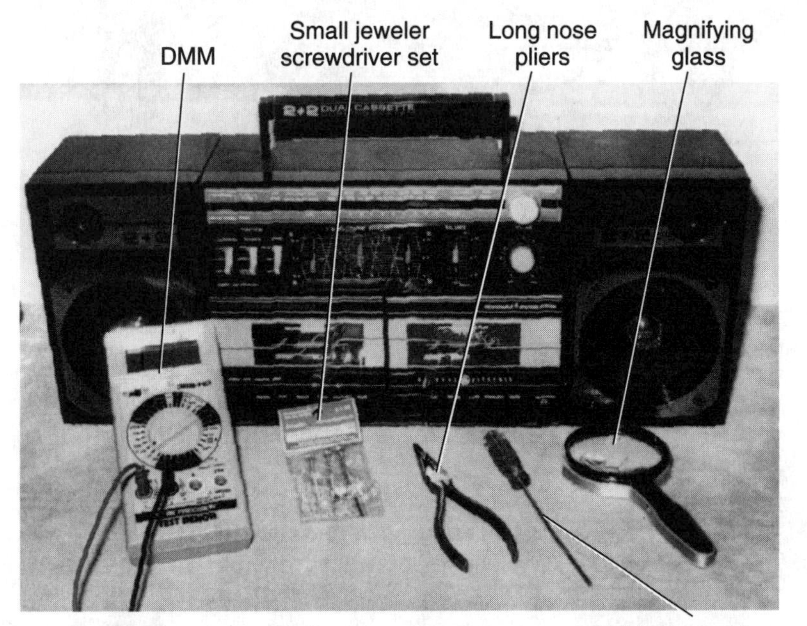

FIGURE 14-3 A few small tools might be required to put that cassette player back in tip-top-shape.

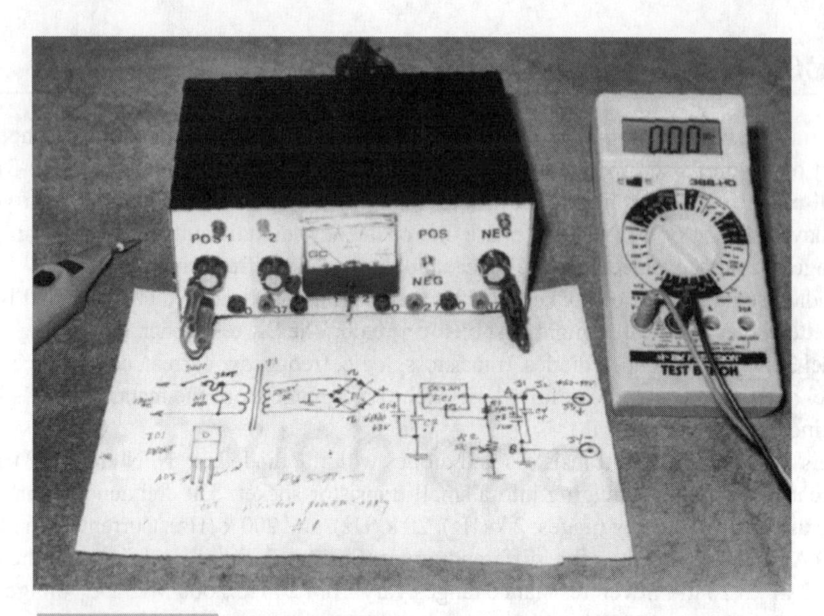

FIGURE 14-4 With a B&K test bench DMM, you can make all of the tests given in these pages.

FIGURE 14-5 Here, the DMM is used in series with batteries to check the total current (16.65 mA) of a pocket cassette player.

The low ac voltage range can be used in azimuth head alignment. Speed problems can be checked with the frequency-counter test. Total current drain of the small battery-operated cassette player can indicate that components are leaky. Defective or unknown small capacitors can be checked with this small DMM.

SOLDERING EQUIPMENT

The small 30-W soldering iron is needed to remove transistor and small board components (Fig. 14-6). Of course, the battery iron is ideal for surface-mounted and IC terminals, and the large 200-W soldering iron can remove larger components, mesh and shields, but neither iron is required. If you are going to service a large group of cassette players, the controlled-temperature iron is handy to have on the service bench (Fig. 14-7).

A small pencil iron can solder those tiny terminals or remove melted soldered connections with solder wick. Heating the solder-mesh material can take a few seconds longer with the small pencil iron, but it does a good job. Do not apply the iron tip too long to transistors or IC terminals; just do it long enough to melt the solder and make a good joint. Too much heat can damage the transistor or IC. Use the long-nose pliers to drain off excessive heat from the transistor leads.

After locating a defective transistor or IC, lift the solder from each terminal with solder wick. Do not apply too much heat if the transistor is to be tested out of the circuit. Sometimes transistors can test bad in the circuit and test good when removed. Be careful not to apply too much heat to the small PC wiring, or you could "pop" the wiring from the board. The defective IC can be quickly removed by applying heat to a row of terminals with solder wick. Remove excess solder from the board with solder mesh after the part has been removed.

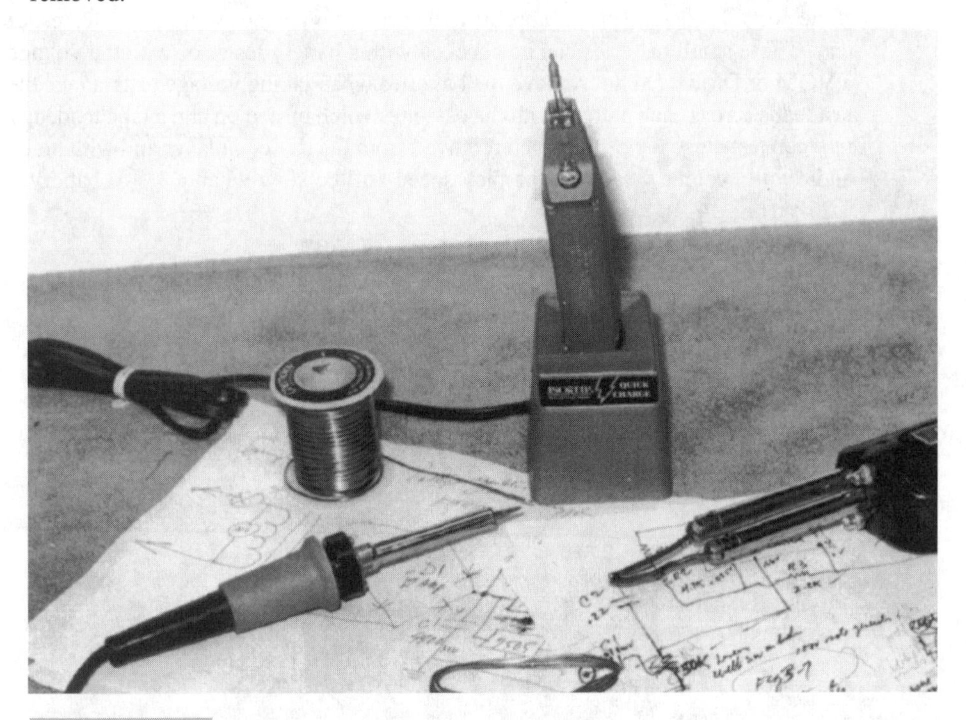

FIGURE 14-6 With a small, 30-watt soldering iron, you can make all soldered connections inside the cassette player.

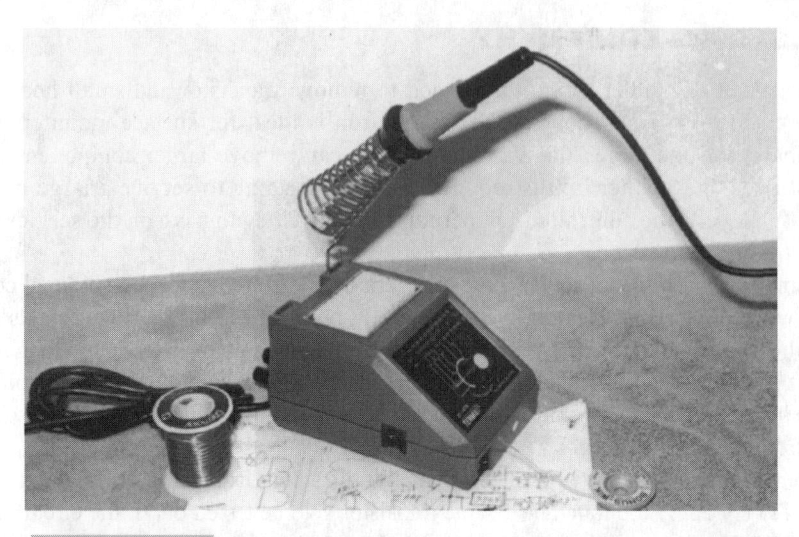

FIGURE 14-7 **The temperature-controlled soldering iron is nice if you have a lot of soldering to do when servicing cassette players.**

BATTERY PROBLEMS

If the portable cassette player appears to be dead or if the tape will not rotate, test each battery. These small batteries can be checked with a battery tester or with the voltage test of a VOM or DMM. Do not remove the batteries when taking voltage tests. Place the DMM test leads across each battery with the cassette switch turned on and a tape loaded. The battery can test close to normal when removed from the cassette player and with no load. The audio will become weak and the play speed will be slow when a 1.5-V battery drops to 1.25 V (Fig. 14-8).

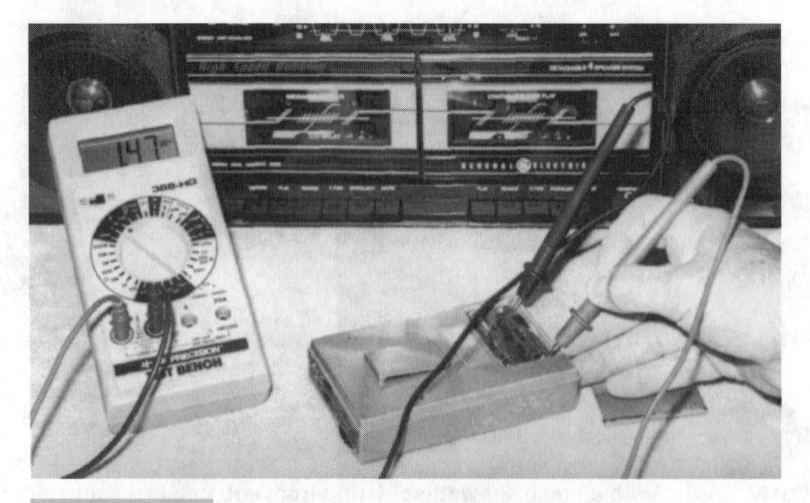

FIGURE 14-8 **Check the batteries in the cassette player while it is operating to determine if they are weak and used up.**

Replace the 9-V battery when the voltage drops below 7.5 V under load. You can locate only one dead or low-voltage battery with several other batteries. If all batteries were installed at the same time, remove and replace all of them if one drops below the required operating voltage.

Wipe the battery terminals on a cloth to clean off the contacts. Clean the battery contacts with a cloth and alcohol while the batteries are removed. Inspect the battery terminals for broken or corroded terminals. Do not leave batteries within the cassette player if the unit is not used in a three-month period.

Rechargeable batteries should be charged if the speed slows down or if weak audio is noticed. Like the battery shaver, charge them up before using. Some people say to discharge nicad batteries before charging. But, this is not necessary, just charge them. When these batteries will not hold a charge for only a few minutes, discard the rechargeable batteries. A defective rechargeable battery will not hold a charge very long within the cassette player. Overloaded circuits within the cassette player can cause the batteries to wear out in a short time.

TEST TAPES AND TENSION GAUGES

One or two test tapes are handy when checking audio stages, tape speeds, and head alignments. Tension gauges can be used to check the pressure roller, takeup torque, and tape tension (Table 14-1). Although these tension gauges are not necessary, they can speed up repairs. Today, some of these test tapes and tension gauges can be difficult to find. Try local electronic stores and cassette player manufacturer depots to locate them.

You can make your own 1- to 10-kHz test cassettes by applying the audio signal into the external microphone jack or by clipping the signal across the tape head terminals. Inject the signal from an audio signal generator or from a home-constructed sine-/square-wave signal oscillator given in home-built test equipment (Fig. 14-9). Check the exact frequency with the frequency counter of the DMM. Record at least one hour of audio signal on a new cassette for each test tape. Remove the input signal and play the recorded 1-kHz/10-kHz signal.

Always keep the volume low so that the audio input signal does not distort or clip the sine wave. Try another recording if too much volume, too little volume, or distortion is found on the cassette. These homemade test cassettes can be used for audio troubleshooting, azimuth head aligning, tape-speed checking, and for locating weak audio stages.

TABLE 14-1 TEST TAPES AND FREQUENCIES

FREQUENCY	TEST CASSETTE	FUNCTION
10 kHz	VTT–658 MTT–114 MTT–216	R/P head azimuth
6.3 kHz	Standard	Head azimuth and sensitivity
3 kHz	MTT–111	Tape speed adjust
1 kHz	MTT–118	Tape speed adjust
400 Hz	MTT–150	Playback-level sensitivity

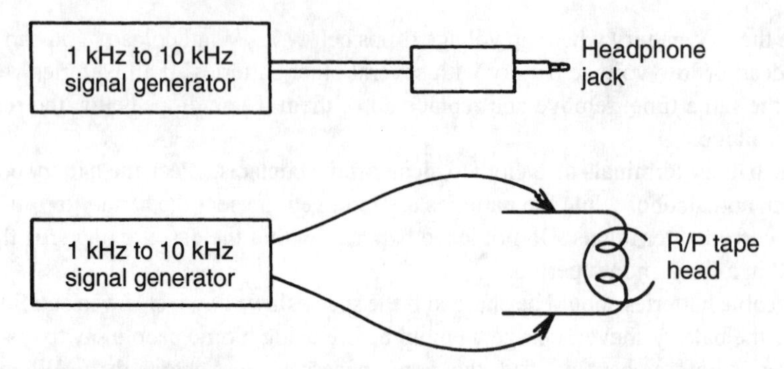

FIGURE 14-9 **How to inject a 1-kHz signal into the cassette player to record a test cassette.**

Troubleshooting with Voltage and Resistance Measurements

The defective transistor can be located with voltage and transistor measurements. Voltage measurements on each collector, base, and emitter terminal with common probe to ground can indicate that a transistor is defective. The open transistor can have a higher-than-normal collector voltage and no voltage on the emitter terminal (Fig. 14-10). An open emitter resistor or terminal can have zero measurement. Be careful not to short any two terminals together with the test probes.

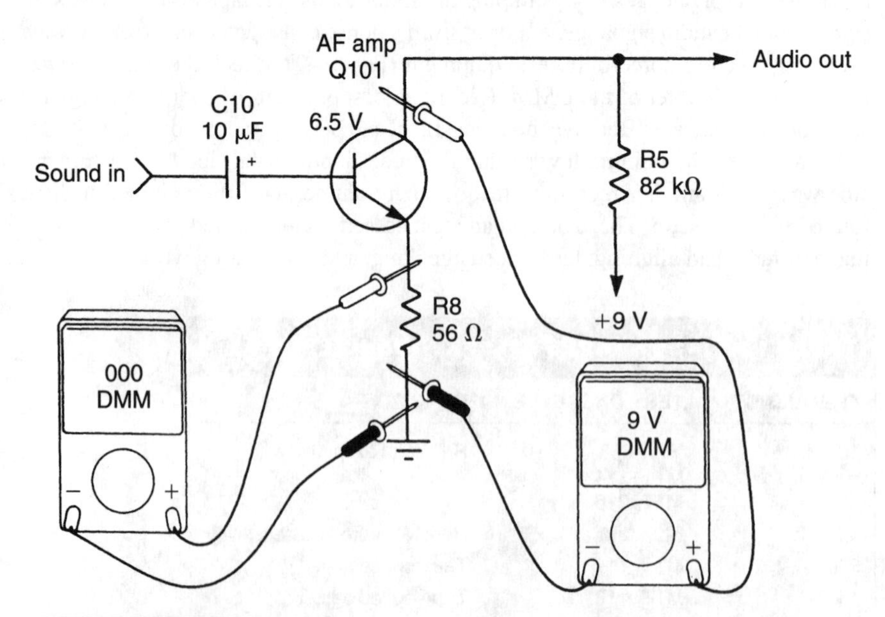

FIGURE 14-10 **Higher collector and zero emitter voltage indicates that a transistor is open.**

The leaky or shorted transistor can have close voltages on all terminals. Most transistors become leaky from emitter to collector (Fig. 14-11). Check the transistor with in-circuit transistor tests and then remove the components from the circuit and take another leakage test.

Another method to determine if the transistor is normal is to measure the bias voltage from the base to emitter terminals. Usually, the transistor is normal with either a 0.6- or 0.3-V measurement (Fig. 14-12). The NPN and PNP silicon transistor has a 0.6-V bias voltage and the PNP germanium transistor has a 0.3-V bias voltage. The difference in both voltage measurements from base to ground and emitter to ground should equal the bias voltage of a normal transistor.

The intermittent transistor can show different identical voltage measurements on collector and emitter terminals. The voltage can quickly change with the intermittent transistor. Sometimes, when the transistor is in the intermittent state, when touched with a test probe, the voltage will return to normal. The intermittent transistor might test open or leaky in the circuit, but test normal when removed from the PC board. If this is the case, replace the transistor.

Accurate resistance measurements from the transistor terminals to ground can help you to locate a defective transistor. Take crucial resistance measurements from each terminal to the other can indicate a leaky transistor (Fig. 14-13). Open emitter resistors or poor emitter terminal connection results in no or real high-resistance measurements. If low-resistance

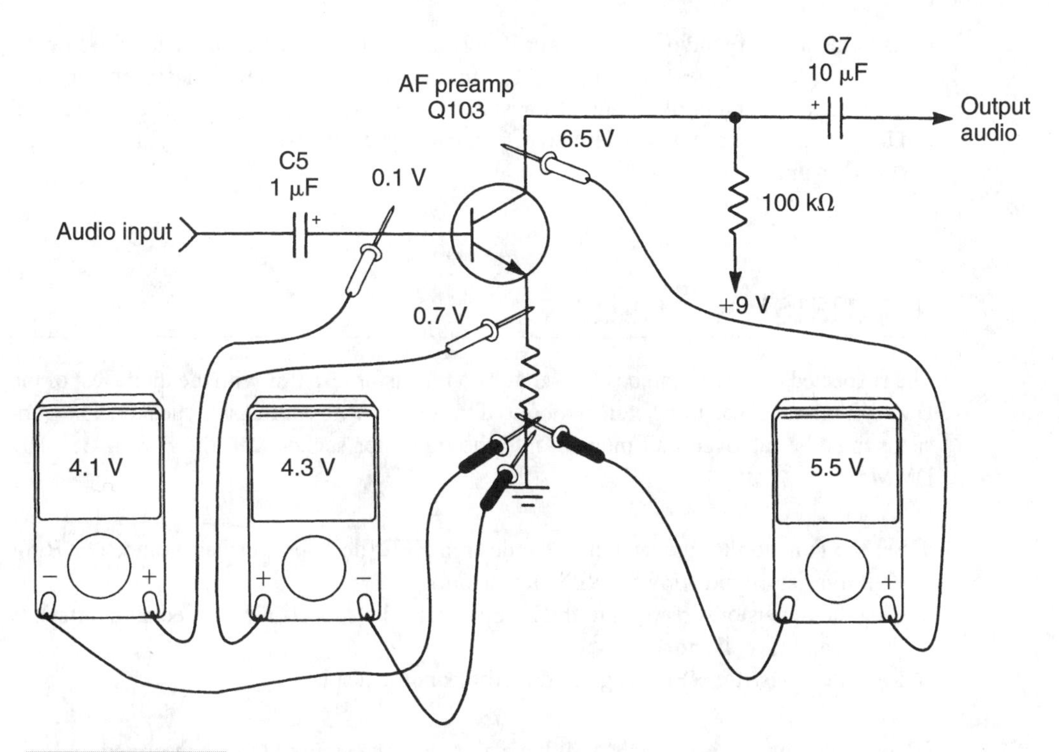

FIGURE 14-11 The crucial voltages indicate that a transistor is leaky if the voltages on all of the terminals are quite close.

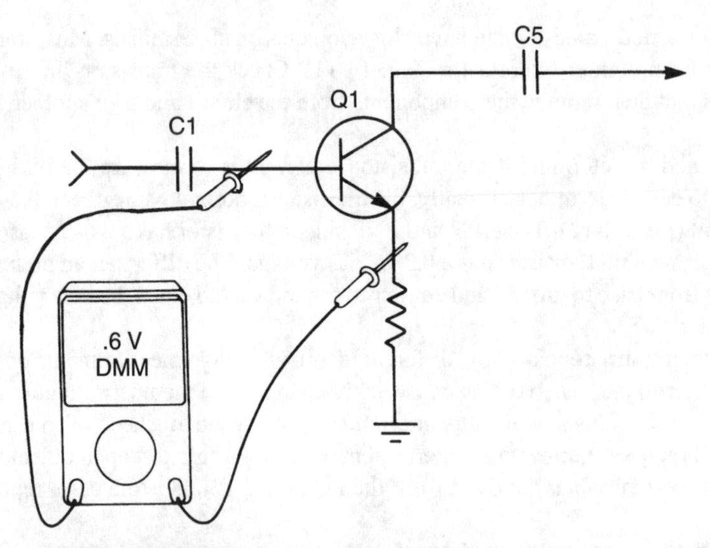

FIGURE 14-12 **The transistor is normal with 0.6 V between the base and emitter of a silicon transistor and 0.3 V with germanium transistor.**

measurements are found on transistor or IC terminals, check the schematic for real low-resistance resistors, diodes, and coils within that circuit. A resistance measurement from each element to ground of an output transistor within the directly coupled stereo channel can indicate that the channel is defective. Compare the same resistance measurements with the good channel.

Transistor Tests

The suspected transistor can be checked with a transistor tester or with the diode test of the DMM. You can check transistors in or out of the circuit with a transistor checker. The transistor must be removed and plugged into the transistor socket with the B & K 388-HD DMM.

1. Set the Function/Range switch to the desired HFE (dc transistor gain) range (*PNP* for PNP transistors and *NPN* for NPN transistors).
2. Plug the transistor directly into the HFE socket. The sockets are labeled E (emitter), B (base), and C (collector).
3. Read the transistor HFE (dc gain) directly from the display.

Notice that this B & K DMM will have an overrange symbol (1) when turned to diode test, logic, and resistance without probe connections in the circuit. All other ranges, such as resistance, frequency, current, capacitance, and voltage will have a (000) indication without probe connections. The overrange symbol will show up if the range is over that in which the selector knob is turned to. The low-resistance scale (200) can have a 00.1 resistance measurement on the display with the probes touching, which indicates the resistance of the test leads to the meter circuit.

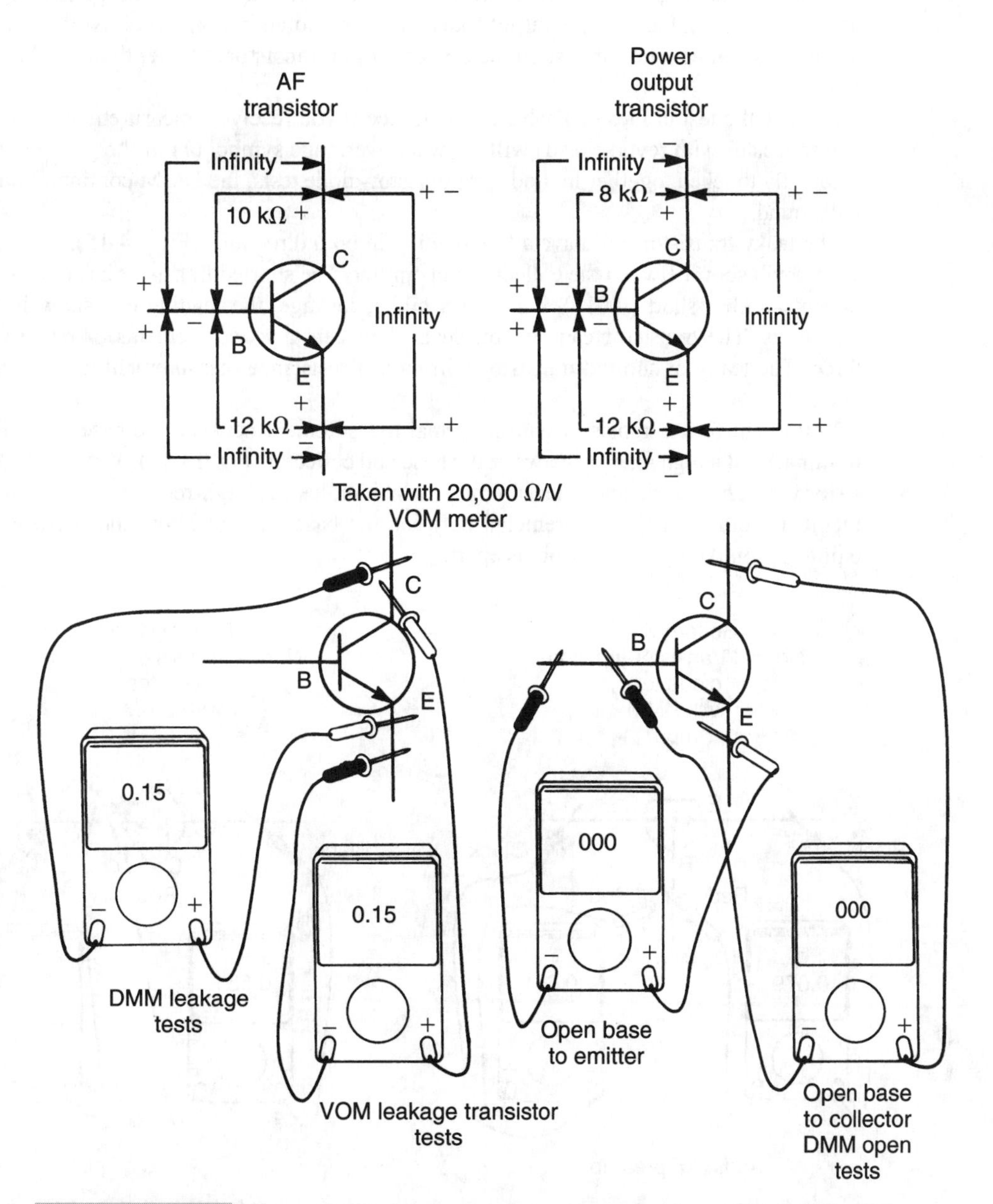

FIGURE 14-13 **The resistance and leakage test of the AF and power-output transistors found in the cassette player with a 20,000-V/ohm VOM.**

TRANSISTOR JUNCTION-DIODE TESTS

Set the function switch to the diode symbol tests. With NPN transistors, place the positive (red) probe to the base terminals. While testing PNP transistors, place the negative (black) probe to the base terminal. Leave the red probe on the base terminal and take resistance measurements between the collector and the emitter. The normal resistance measurements on the AF and output transistors are shown in Fig. 14-14. Notice that the normal resistance junction test on the power-output transistor is lower than the AF transistor.

Reverse the test probes on each test and notice if you receive a measurement. The normal transistor with reverse leads will show an overrange symbol (1). If the probes are accidentally touched together in diode or transistor-diode tests, the DMM continuity buzzer will sound.

The leaky transistor will have a low reading in both directions (Fig. 14-15). Most transistors will short between the collector and emitter. The shorted transistor has only a fraction of an ohm short (0.015), but with a higher leakage, the reading can show in both directions. The transistor can become leaky between any two elements or between all three. The leaky or shorted transistor will have a resistance measurement in both directions.

You might find a transistor with a normal measurement between the base and emitter terminals and a high reading between the base and collector (Fig. 14-16). Replace the transistor with a high-resistance measurement, which indicates a high-resistance junction. Remember, both normal measurements between the base and collector, and the base and emitter, should only be a few ohms apart.

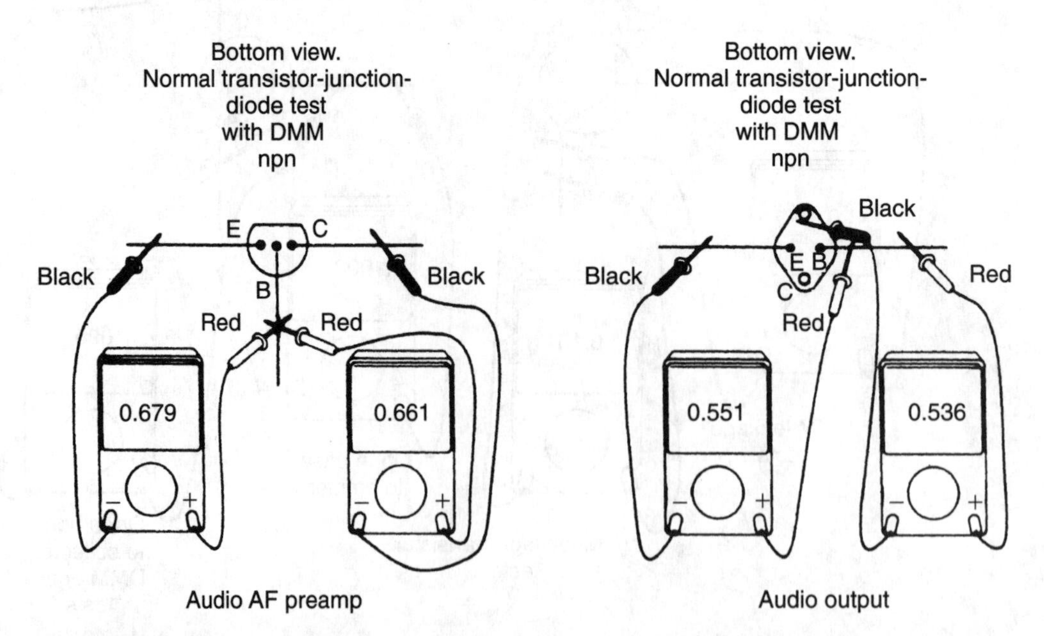

FIGURE 14-14 **Take normal transistor DMM junction-diode tests with the red probe at the base terminal of an NPN transistor.**

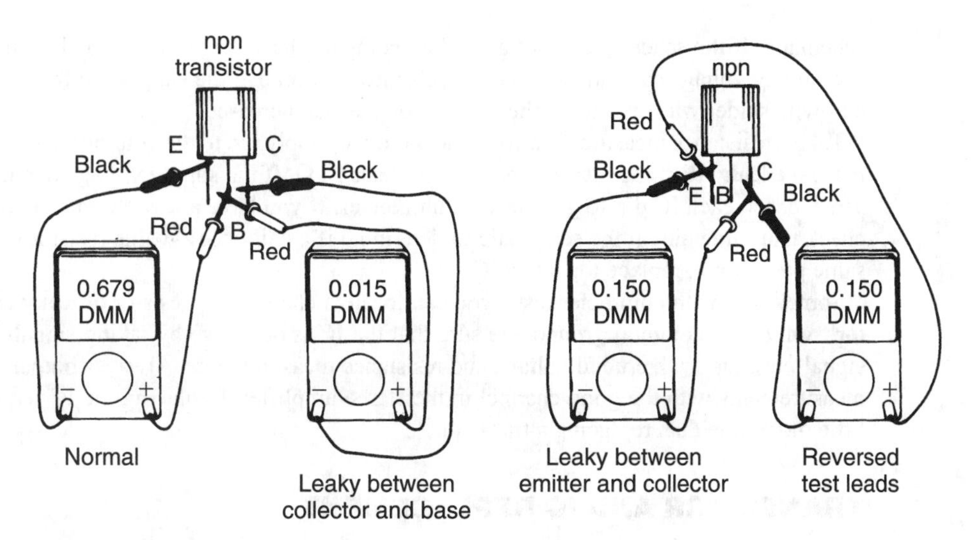

FIGURE 14-15 **The leaky or shorted transistor will have a very low resistance in both directions.**

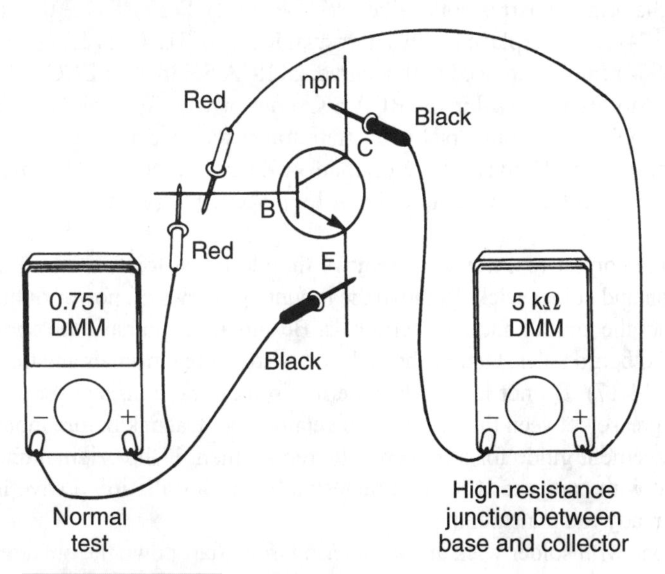

FIGURE 14-16 **Replace the transistor if you find a high-resistance junction measurement between the base and emitter or collector terminals.**

IC TESTS

The defective IC can be located with signal-in and signal-out measurements. If the audio signal is traced to the input terminal and not at the output terminal, suspect that an IC is defective. Crucial voltage measurements on the IC terminals can indicate that a component is defective. Measure the supply voltage and compare it to that which is listed on the schematic. If the schematic is not available, compare the voltage to the good channel. A low supply voltage can indicate that the IC is leaky. Remove the supply pin from the circuit with solder wick. Notice if the supply voltage has increased.

Take a resistance measurement from the removed supply pin to the common ground. Replace the leaky IC if the measurement is less than 1 kΩ. If the supply voltage remains the same, check each IC pin to ground with ohmmeter. If you have a low ohmmeter reading, check that same pin on the schematic to determine if a coil or low-value resistor is in the same path. If not, replace the leaky IC.

Sometimes, with ohmmeter tests, you can locate a change in a resistor or leaky capacitor from pin to common ground. Be sure that the IC is defective by taking signal-in and signal-out tests, with crucial voltage and resistance measurements. Always compare these measurements with the good channel in the stereo amplifier. Replacing the IC requires a little more time than replacing a transistor.

TRANSISTOR AND IC REPLACEMENT

After locating the defective transistor in the amplifier, the transistor must be removed and replaced. Most transistors found in audio circuits can be replaced with universal replacement transistors if the original part is not available (Table 14-2). For instance, the common AF transistor (2SC374) can be replaced with a universal RCA SK3124A or ECG289A. The AF transistor (2N3904) can be replaced with a universal RCA SK3854 or ECG123AP.

Look up the transistor number within the RCA SK series or with Sylvania's ECG series replacement guide books. Most universal solid-state transistors and ICs can be replaced with RCA, GE, Motorola, NTE, Sylvania, Workman, or Zenith replacements. Simply look up the part number and replace it with an universal replacement. Test the new transistor before installing it.

After obtaining the correct replacements, remove the old transistor from the PC board with a soldering iron and solder wick. Remove the mounting screws on power-output transistors, then unsolder the emitter and base terminals. Be sure that you have the correct terminals in the right PC board holes. Doublecheck the transistor wire terminals and the bottom base diagram (Fig. 14-17). Do not leave the soldering iron on the transistor terminals too long. Place silicon grease between the transistor, insulator, and heatsink before mounting.

Look in the replacement guide for a universal IC replacement if the original one is not available. Handle it with care; the IC cannot be tested before installation. Leave it in the envelope until it's ready to be installed.

Remove the old one with solder wick and a soldering iron. Start down the outside row of contacts and keep the iron on the mesh at all times. Move the mesh down as it picks up solder. Then, go back and be sure that each contact is unsoldered from each pin to PC board wiring. Flick the pin with a pocket knife blade or with a small screwdriver. Mark pin 1 on the PC board with a felt pen. Lift the defective IC out by prying underneath the component. Be careful not to damage other components nearby or to break the PC wiring.

Check for the terminal 1 dot, while the line or indexes determine how to correctly mount the audio IC (Fig. 14-18). Sometimes, only an index indentation is found on the IC. When looking down on top of the IC, pin 1 is to the left of the index. Pin 1 is indicated with a white dot; on other PC boards, the number one and last terminal numbers are marked (Fig. 14-19). Correct IC mounting is found in the solid-state replacement guide. Place silicon grease and a metal heatsink under large ICs.

TABLE 14-2 UNIVERSAL TRANSISTOR REPLACEMENT CHART

PREAMP NPN	AF AMP NPN
2SC458-SK3124A-ECG289	2SC372-SK3124-A-ECG289
2SCC536-SK3245	2SC458-SK3124A-ECG289
2SC693-SK3124A-ECG289	2SC536-SK3245-ECG199
2SC732-SK3245-ECG199	2SC644-SK3245-ECG199
2SC900-SK3899	2SC693-SK3124A-ECG289
2SC1000-SK3245-ECG199	2SC282-SK3931-ECG90
2SC1312-SK3899	2SC945-SK3124A-ECG289
2SC1740-SK3122	2SC1571-SK3245-ECG199
2SC1815-SK3124A-ECG289	2SC1740-ECG3122
2SC2320-SK3122	2SC2240-SK3122
POWER OUTPUT NPN	**BIAS OSCILLATOR NPN**
2SA537-SK3122	2SC537-SK3122
2SA634-SK3250-ECG315	2SC711-SK3899
2SC1030-SK3619	2SC1214-SK3124A-ECG289
2SC1096-SK3248	2SC1317-SK3124A-ECG289
2SC1383-SK3849	2SC1627-SK3449
2SC1568-SK9041	
PREAMP PNP	**AF AMP PNP**
2SB173-SK3004-ECG102A	2SB175-SK3004-ECG102A
2SB175-SK3004-ECG102A	2SB186-SK3003-ECG102A
2SB348-SK3004-EG102A	2SB346-SK3004-ECG102A
2SC732-SK3245-ECG102A	2SCD348-SK3004-ECG102A
POWER AMP PNP	**BIAS OSC. PNP**
2SB156-SK3007A-ECG102A	2SB75-SK3004-ECG102A
2SB178-SK3004	2SB172-SK3007A
2SB324-SK3007A	2SB186-SK3004-ECG102A
2SB376-SK3007A	2SB187-SK3004-ECG102A
2SB405-SK3004-ECG102A	2SB365-SK3004-ECG102A
2SB415-SK3004-ECG102A	
RCA SK series	ECG Sylvania series

Silicon and germanium diodes are used in many cassette circuits. These diodes can be checked with a VOM, DMM, or diode tester. To get a normal diode check, place the positive lead of the DMM to the negative (anode) terminal of the diode and the negative probe to the collector terminal. When checking it with the VOM for a normal reading, place the positive (red) probe to the collector and the negative (black) lead to the anode terminal of the diode (Fig. 14-20).

1 Set the function/range switch to the K1 diode position on a B&K 388-HD DMM.
2 Connect the red test lead to the V-Q-Hz jack and the black lead to the COM jack.

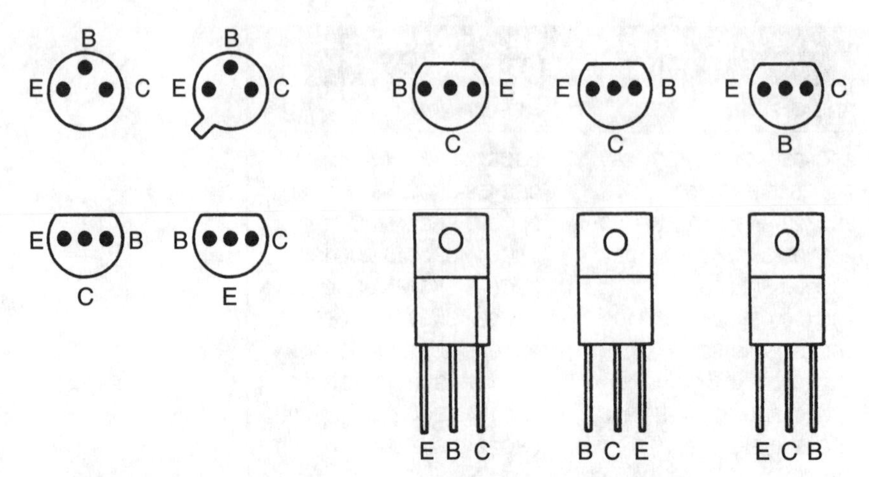

FIGURE 14-17 The many different transistor terminal connections with the AF, driver, and output transistors in the cassette player.

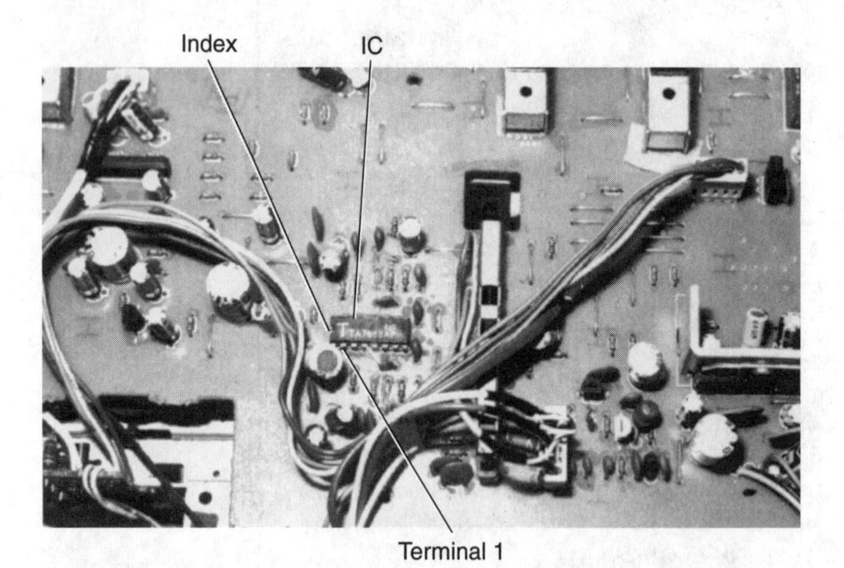

FIGURE 14-18 Check for number 1 dot, index, or a white line on top of the IC to mark and correctly insert the new replacement.

3 To check the forward voltage (FV), connect the red test lead to the anode and the black test lead to the cathode of the diode.

4 The display indicates the forward voltage if the diode is normal. Normal diode voltages are approximately 0.3 V for germanium diodes, 0.7 V for silicon diodes, and 1.6 V for LEDs. An overrange (1) indicates that a diode is open. A shorted diode reads O V.

5 To check reverse voltage, reverse the test-lead connections to the diode. The reading should be the same as with open tests leads (1) overrange. Lower readings indicate a leaky diode.

The normal diode will show a reading in one direction and a shorted or leaky diode will read with the test leads reversed in both directions. The open diode will not read in any direction. Most defective diodes are leaky or shorted.

Bridge rectifiers can be checked in the very same manner, connect the positive (red) lead to an ac terminal and negative (black) lead to a positive terminal of the bridge rectifier for a normal measurement (Fig. 14-21). The shorted or leaky bridge rectifier can show a low reading in both directions of two or more terminals. Replace the bridge rectifier if only one diode is shorted. A defective bridge diode can be replaced with four separate diodes if one is not available (Fig. 14-22).

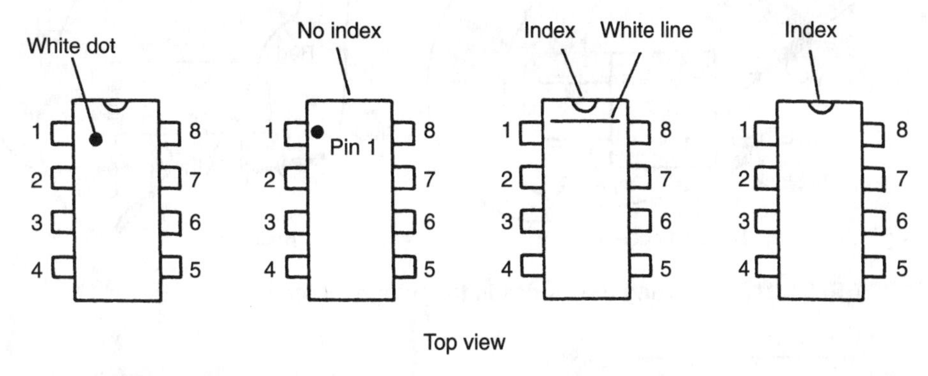

FIGURE 14-19 Locating pin 1 by looking down on top of the IC.

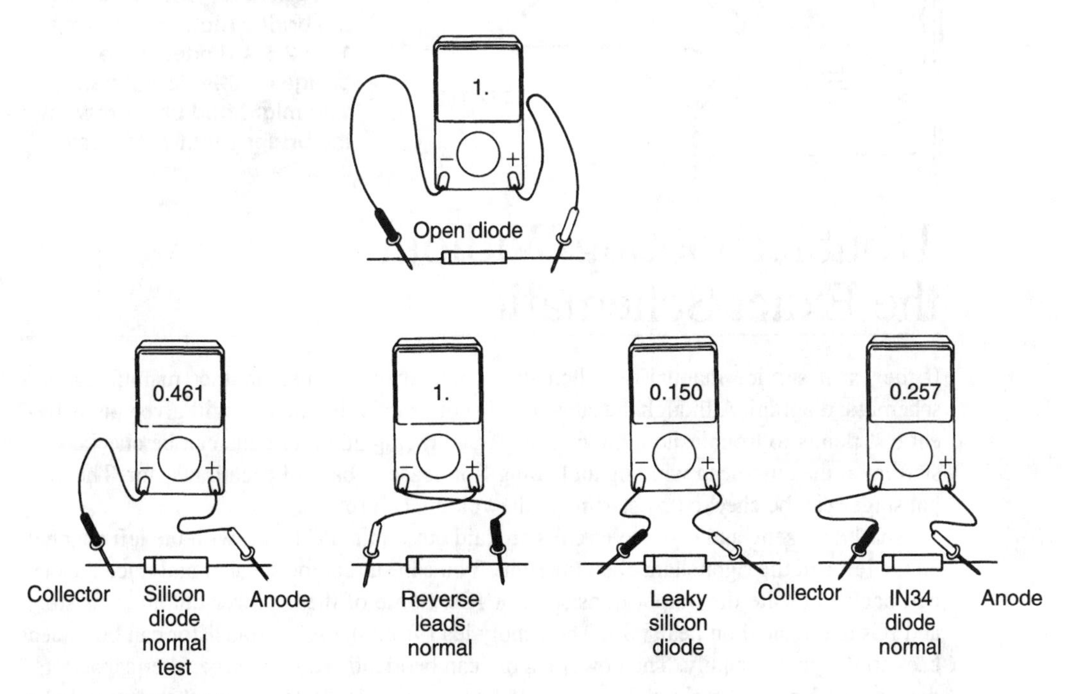

FIGURE 14-20 Checking the diodes with a DMM.

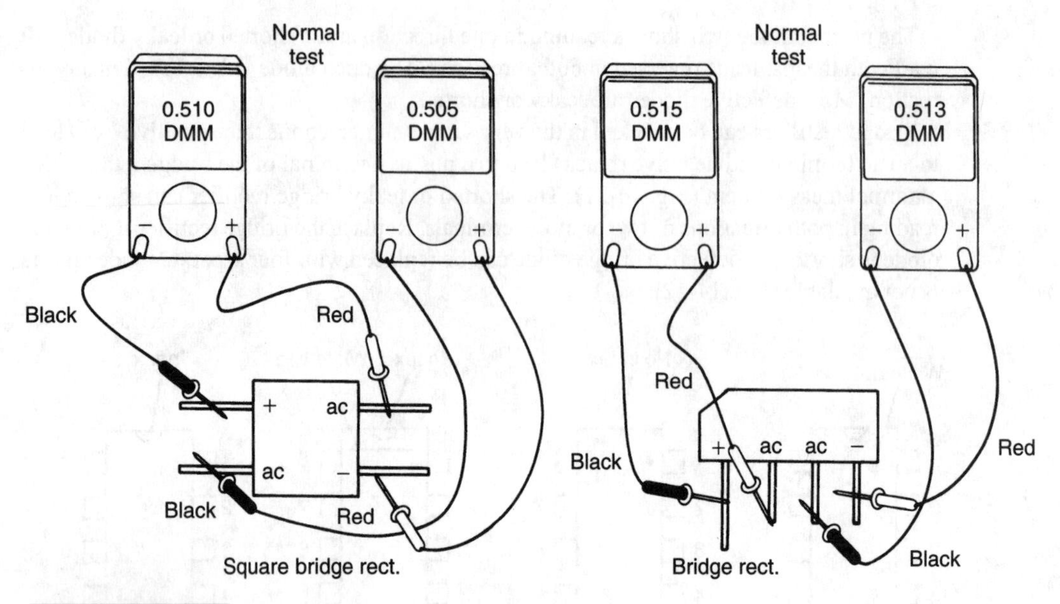

FIGURE 14-21 Checking the diodes in the bridge rectifier.

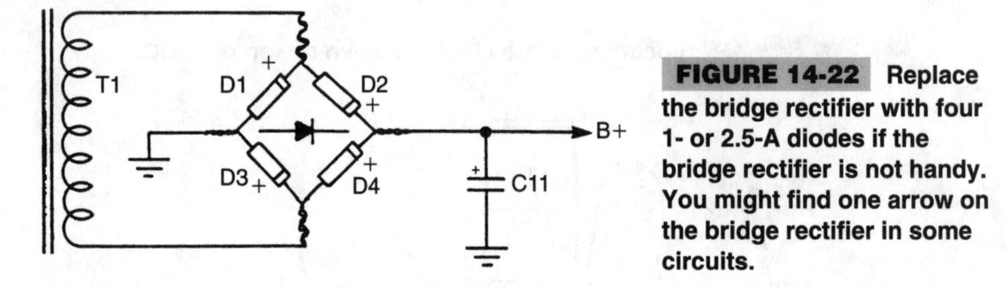

FIGURE 14-22 Replace the bridge rectifier with four 1- or 2.5-A diodes if the bridge rectifier is not handy. You might find one arrow on the bridge rectifier in some circuits.

Troubleshooting Without the Exact Schematic

If the exact service manual or schematic is not available, use another manufacturer's schematic diagram. Although the substitute is not exactly the same, it will give you different test points to troubleshoot the circuit. Signaltracing audio circuits can be checked by starting at the tape-head winding and going from base to base of each transistor. The output stages can be checked by starting at the volume control.

Usually, cassette audio stereo circuits are laid out on the PC board with the left channel on the left and the right channel on the right. You can start at the speaker and trace the circuit back to locate the output transistors or ICs. Some of these power-output transistors and ICs are located on heatsinks. The supply lead from the IC or transistor can be traced back to the power supply. The power supply can be identified with large filter capacitors, diodes, and bridge rectifiers. The suspected part can be checked against the identical one in the good stereo channel.

Transistor terminals can be identified by letters on the PC board, transistor markings, or with a transistor tester. You can locate the transistor leads with the diode junction test of the DMM (Fig. 14-23). Remember, the positive lead of DMM will have a common measurement if placed on the base terminal of an NPN transistor. The PNP transistor must have the black lead as the common base terminal. Locate the base terminal by getting two separate tests between the base and collector, and the base and emitter. The NPN collector terminal goes to a higher positive voltage, the emitter voltage is very low, and the emitter resistor goes to ground. The PNP collector lead is negative and the emitter has a higher positive voltage. Remember, a leaky transistor has a very low reading with reversed test leads. Also, check the numbers listed on transistor and compare them with the solid-state universal replacement guide.

SIGNALTRACING WITH A CASSETTE

A weak, dead, or lost signal within the cassette player can be signaltraced with a test cassette or one with music and the audio signaltracer or external amplifier. When the signal stops, the defective stage is nearby. Then, take crucial voltage and resistance measurements.

Insert a test tape and check the audio signal at the volume control. If the signal is normal here, go stage by stage from the volume control to the speaker. When the signal is weak or when no signal is at the volume control, start at the tape head and check from base to base of each transistor (Fig. 14-24). Check the signal in and out of IC terminals with amplifier or signal tracer.

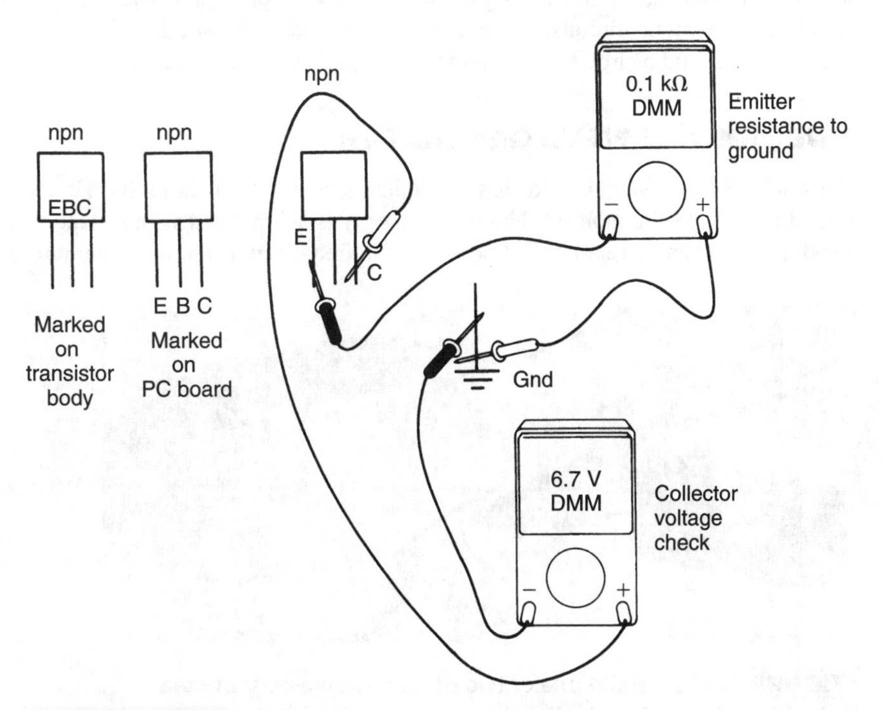

FIGURE 14-23 **Identify the correct terminals with the DMM if a schematic is not readily available.**

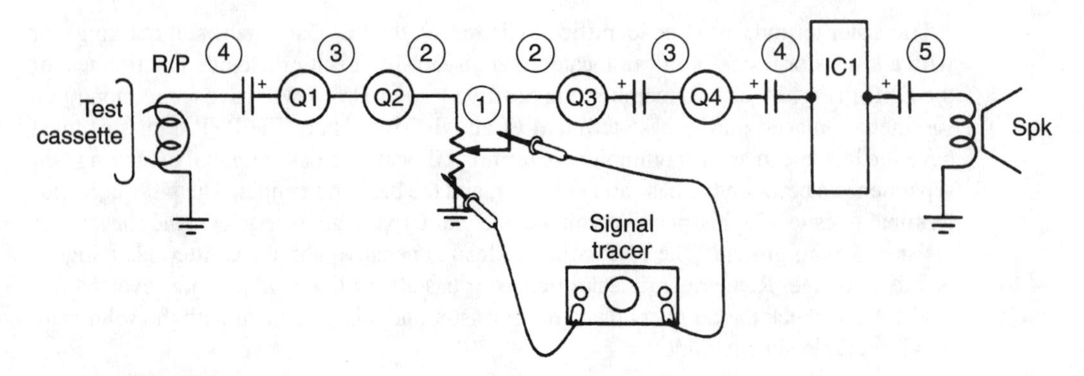

FIGURE 14-24 Start at the volume control with the signal tracer and proceed either way to check for signal loss.

Build Your Own Test Equipment

Besides the DMM or VOM, several homemade test instruments can speed up cassette player repairs. The sine/square-wave generator, 1-kHz audio oscillator, audio signal tracer, and speaker load and noise generator are small test instruments that you can build (Fig. 14-25). The noise generator and sine-/square-wave generator can help you to locate a dead, weak, or distorted audio stage. Use the signal tracer with the homemade 1-kHz test cassette to signal trace the audio circuits. Connect a speaker lead to the speaker connections while working on the audio circuits. Most parts can be picked up at Radio Shack (unless noted).

SINE-/SQUARE-WAVE GENERATOR

The sine-/square-wave generator has a frequency range of 20 Hz to 20 kHz and is built around an IC8038 function IC. The power supply is ac operated with a stepdown power transformer and an IC regulator. R1 controls the frequency. Build the generator on a perf

FIGURE 14-25 Build one or two of these home-constructed testers to help locate defective components within the cassette player.

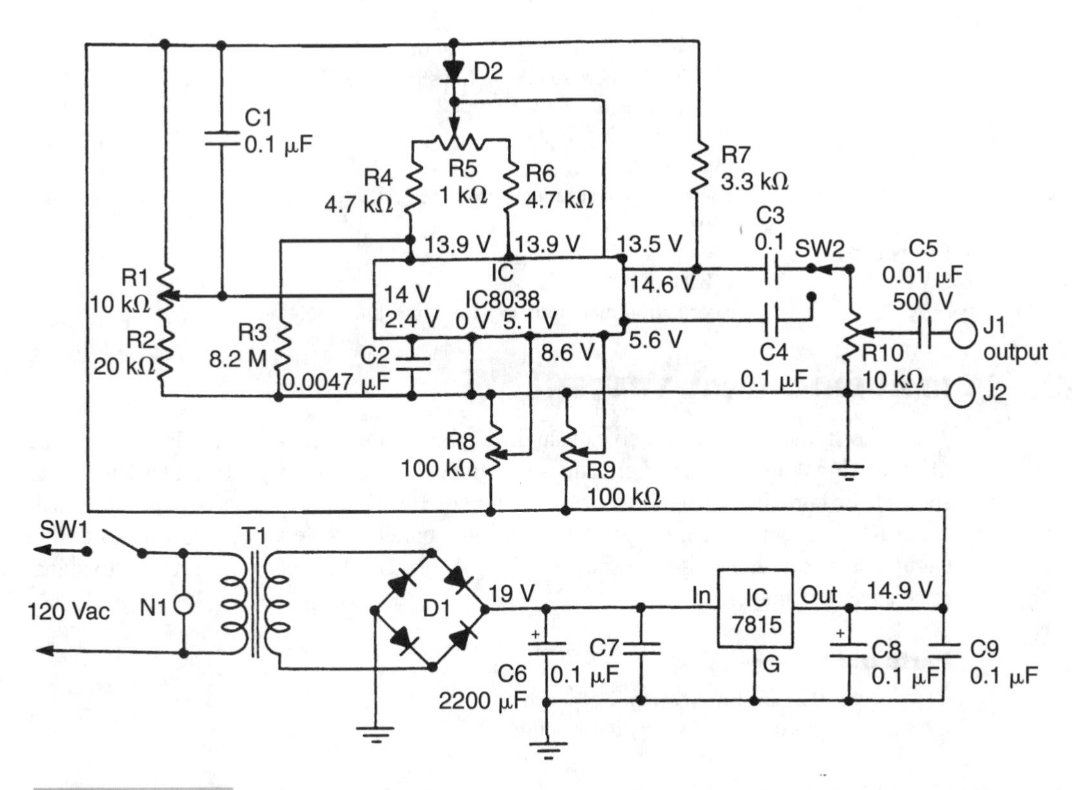

FIGURE 14-26 A diagram of a sine/square generator operated from the ac power line. This tester can be used for signal tracing, locating distorted stages, head aligning, and correcting speed problems.

board and enclose it in a 6-×-6½-×-2¼-inch plastic or metal cabinet (Fig. 14-26). This generator can be used to make many tests within the audio circuits.

Parts list

IC1	7815 15-V regulator.
IC2	8038 function generator, D.C. Electronics, P.O. Box 3203, Scottsdale, AZ 85271-3203.
C1, C3, C4, C7, C9	0.1-µF 50-V ceramic capacitor.
C2	0.0047-µF 50-V monolithic or high-ceramic disk capacitor.
C5	0.01-µF 500-V ceramic capacitor.
C6	2200-µF 35-V electrolytic capacitor.
C8	1-µF 35-V electrolytic capacitor.
R1, R10	10-kΩ linear control.
R2	20-kΩ 0.5-W resistor.
R3	8.2-kΩ 0.5-W resistor.
R4, R6	4.7-kΩ 0.5-W resistor.
R5	1-kΩ trimmer or thumb-variable resistor.
R7	3.3-kΩ 0.5-W resistor.
R8, R9	100-kΩ trimmer screwdriver or thumb-variable resistor.

N1	120-Vac neon indicator, 272-704 or equiv.
T1	12.6-Vac 450-mA stepdown transformer, 273-1365 or equiv.
D1	1-A bridge rectifier.
D2	1N914 switching diode.
J1, J2	Banana jacks.
Cabinet	MB-1C beige instrument enclosures, All Electronics, Box 567, Van Nuys, CA 91408.
Perf board	3" × 4.5"
SW1	On back of R1.
Misc.	ac cord, grommet, hookup wire, bolts, nuts, etc.

IC AUDIO SIGNAL TRACER

The audio signal tracer can receive the audio signal after the first AF or predriver stage and trace the signal up to the speaker. This audio tracer consists of only 1 IC and a 4-inch 8-Ω speaker. Be sure that C1 has a working voltage of 100 V to prevent damage to the signal-tracer IC. The signal tracer is operated from a 9-V battery. Place the components in a large enough cabinet to hold a 4-inch speaker (Fig. 14-27). Use the audio signal tracer to signal trace a cassette player.

Parts list

C1	0.22-μF 100-V ceramic capacitor
C2, C3	10-μF 35-V electrolytic capacitor.

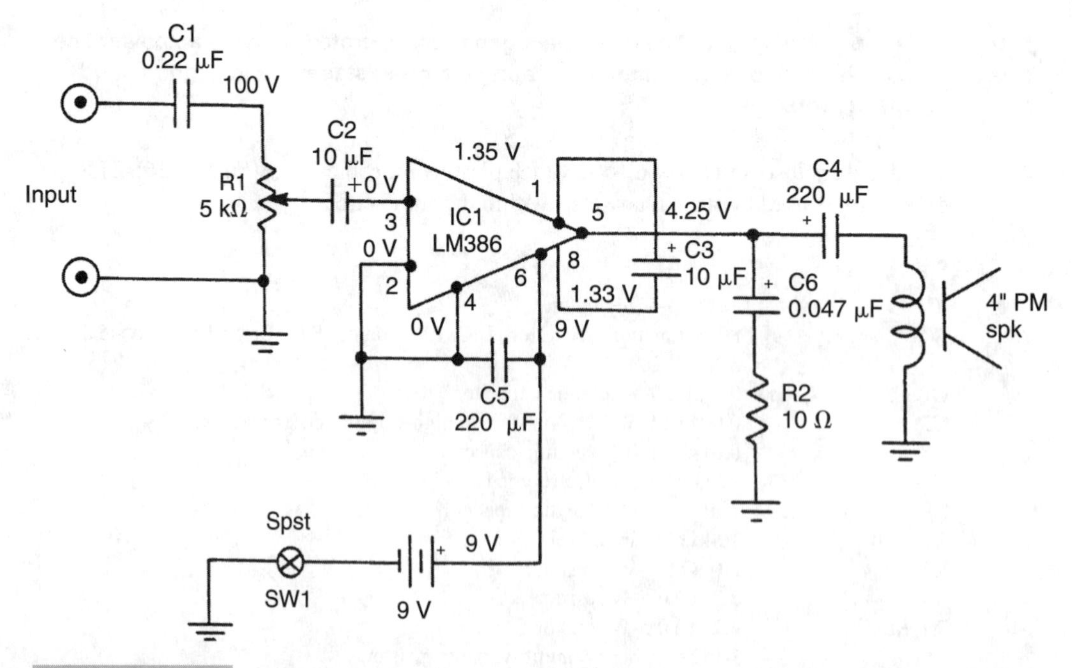

FIGURE 14-27 A simple audio signal tracer used in conjunction with a cassette to locate the defective stage.

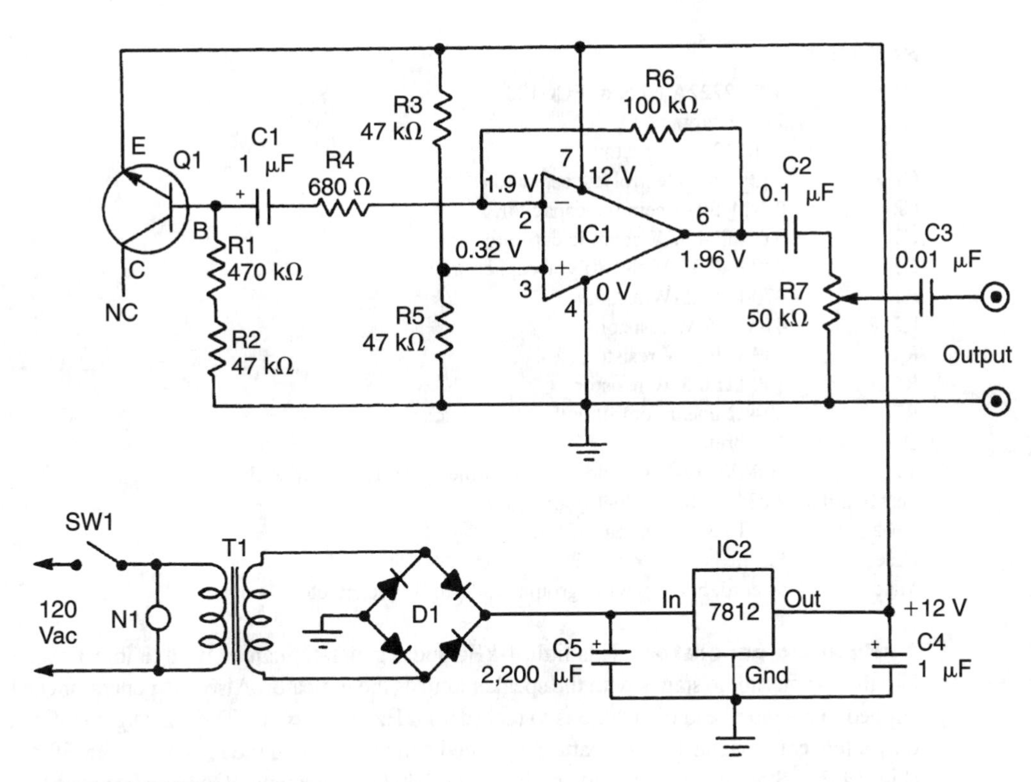

FIGURE 14-28 **The white-noise generator can be used to signal trace the AF, RF, and IF circuits of the cassette player.**

C4, C5	220-μF 35-V electrolytic capacitor.
C6	0.047-μF 50-V ceramic capacitor.
R1	15-kΩ volume control and switch.
R2	10-Ω 0.5-W resistor.
IC1	LM386 audio amp (276-1731 or equiv.).
Spk	4-inch 8-Ω speaker.
Cabinet	Large enough for 4-inch speakers.
Perf board	Multipurpose board (276-150 or equiv.).
Batt.	9-V battery.
Misc.	Battery cable, alligator clips, hook-up wire, 4-pin IC socket, nuts, and bolts.

WHITE-NOISE GENERATOR

The noise generator can be used in signal tracing audio, RF, and IF circuits. Inject the output noise generator to the various audio stages with the ground terminal to common ground of the amplifier. This white-noise generator uses a low-priced transistor and op-amp IC with a regulated power supply (Fig. 14-28). Start at the volume control or tape head and use the audio amp speaker as an indicator.

Parts list

Q1	MPS 2222A NPN or ECG123.
IC	741 op amp.
IC2	7812 12-V regulator.
C1, C4	1-µF 50-V electrolytic capacitor.
C2	0.1-µF 50-V ceramic capacitor.
C3	0.01-µF 450-V ceramic capacitor.
C5	2200-µF 35-V electrolytic capacitor.
R1	470-kΩ 0.5-W resistor.
R2, R3, R5	47-Ω 0.5-W resistor.
R4	680-Ω 0.5-W resistor.
R6	100-kΩ 0.5-W resistor.
R7	50-kΩ linear control with SPST switch.
D1	1-A bridge rectifier.
T1	300-mA 12-V secondary transformer (273-1358A or equiv.).
Perf board	2.83" × 1.85" (276-149 or equiv.).
SW1	SPST switch on rear of R7.
Case	Plastic box, 3" × 6" × 2".
Misc.	ac cord, hookup wire, grommet, 8-pin IC socket, etc.

1-kHz audio generator This little 1-kHz audio generator can be used to inject a signal into the various audio stages with the speaker amp as the indicator. Also, the generator can be clipped to the tape-head connections to record a 1-kHz test cassette. The signal generator has only a few components that are battery operated and built around the low-priced LM3909 IC (Fig. 14-29). Start at the tape head and inject the 1-kHz audio signal. Go from base to base of each audio transistor or to the input terminal of the preamp and power-amp IC.

Parts list

ICI	LM3909 LED flasher-oscillator IC.
C1	0.01-µF 500-V ceramic capacitor.

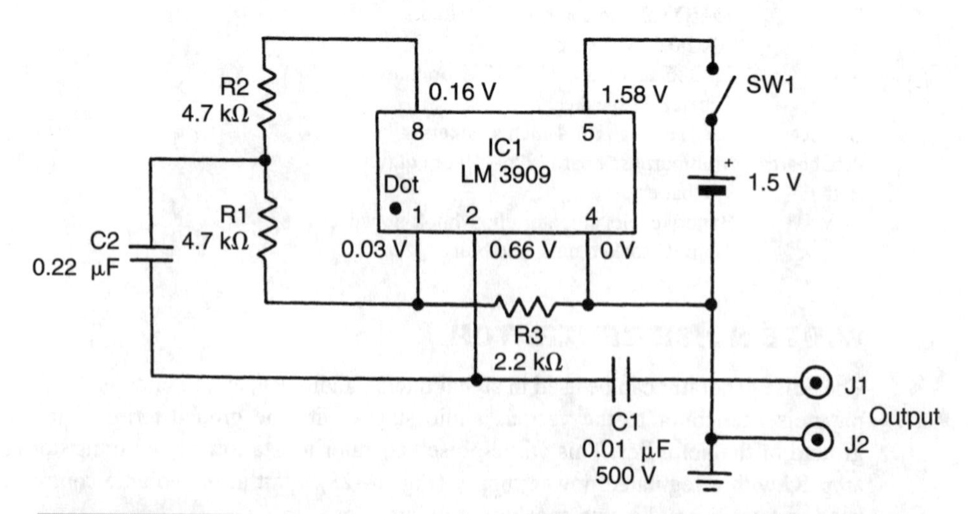

FIGURE 14-29 The 1-kHz audio oscillator can be injected for signal tracing or making that 1-kHz test tape.

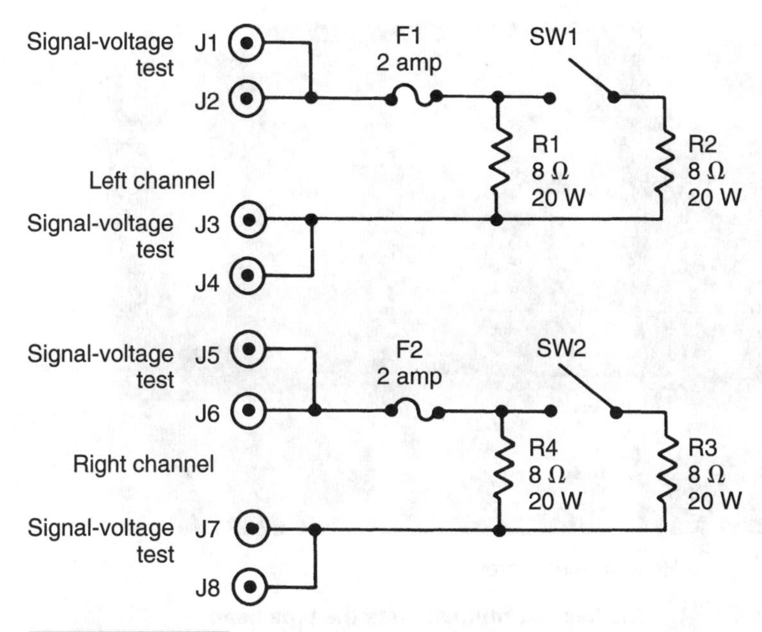

FIGURE 14-30 The dual-speaker load prevents speaker and amplifier damage while you service the audio stages of the cassette player.

C2	0.22-µF 50-V capacitor.
R1, R2	4.7-kΩ 0.5-W resistor.
R3	2.2-kΩ 0.5-W resistor.
Batt.	1.5-V alkaline AA battery.
SW1	Sub-mini slide switch (275-409 or equiv.).
Case	Plastic box, 4" × 2⅛" × 1⅝" (270-231 or equiv.).
Perf board	1 × 2 (cut from larger piece).
J1, J2	Banana jacks.
Misc.	8-pin IC socket, 4/40 bolts and nuts, battery holder, hookup wire, etc.

Speaker load The audio-output stage must be loaded down at all times to prevent damage to the output transistors or ICs. In directly coupled output circuits, a defective transistor can damage the speakers. The speaker load should be large enough to withstand the wattage of each stereo channel. Remove the speakers and connect the speaker loads to each channel while repairing the amplifier (Fig. 14-30). For amplifiers that output less than 20 W, use the 8-Ω load. For higher wattage amplifiers, switch in another 8-Ω 20-W resistor. Place a couple of extra jacks in the circuit for amplifier voltage adjustment.

Parts list

F1, F2	2-A fuses.
R1, R2, R3, R4	8-ohm 20-W resistor.
J1 through J8	Banana jacks.
SW1, SW2	Toggle switches.
Cabinet	Plastic economy box.
Misc.	Terminal strips, hookup wire, nuts, bolts, etc.

Head azimuth screw

FIGURE 14-31 The head azimuth adjusts the tape head horizontally with the tape. The azimuth screw is located alongside of the tape head, near the tension spring.

HEAD AZIMUTH AND CURRENT TESTS

The tape head must be properly aligned for optimum sound reproduction. Usually, one side of the tape head is fastened with a small screw and the other side with adjustable spring (Fig. 14-31). An improper head azimuth adjustment can cause distortion and loss of high frequencies. You can adjust the azimuth screw by playing a recorded cassette of violins or high-pitched music into the speakers.

Accurate azimuth adjustment can be made with a 3-, 6.3-, or 10-kHz test cassette with 8-Ω load instead of the speakers. Connect the low ac range of the DMM across the 8-Ω resistor or use the dummy-load speaker project (Fig. 14-32). Play the recorded cassette and adjust the azimuth screw for maximum readings on the DMM.

Current adjustment To be sure that bias voltage and the bias oscillator is operating, take a current test at the R/P tape head. Bias adjustment can be made at the same time. Insert a 100-Ω resistor between the ground (shielded) lead of the tape head and shield (Fig. 14-33). Place the cassette player in the record mode for this adjustment. Measure the voltage across the resistor with a VTVM or DMM. Most tape head current runs between 20 and 65 mV. Adjust the variable bias resistors at the tape head for correct voltages at each channel.

Tape head cleaning The tape head should be cleaned at least six times per year if the cassette player is in constant usage. Some manufacturers recommend cleaning after 40 hours of operation. Iron-oxide particles from the magnetic tape will build up on all components that come in direct contact with the tape. This excessive oxide can produce garbled or muffled sound during playback. One channel gap can be closed with oxide, which results in no sound or recording from that channel. Oxide deposits can cause an improper erase function and prevent automatic stopping. Keep magnetic metal away from the tape head at all times.

Clean the tape and erase heads with cleaning sticks and rubbing alcohol. This action can be done with the cassette door open or off (Fig. 14-34). Be sure that all packed oxide is removed from the tape head area. Wipe off the capstan drive shaft and pressure roller. Apply heavy pressure to clean the rubber pressure roller. Rotate the roller as it is cleaned so that you can see the rubber and it looks black.

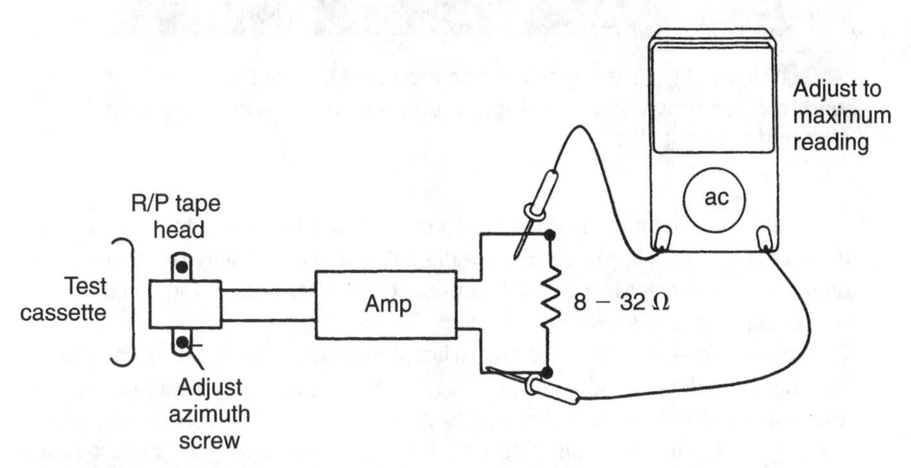

FIGURE 14-32 Connect the frequency counter of the DMM across the 8- to 32-ohm resistor at the speaker or headphone output jack.

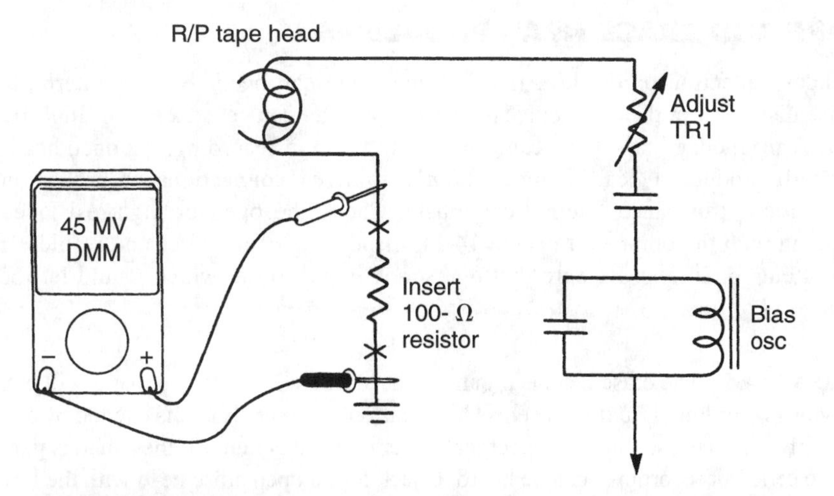

FIGURE 14-33 Adjust and check for tape-head current by inserting a 100-ohm resistor between the shield and ground of the tape head (20 to 65 mV).

FIGURE 14-34 The erase head is mounted before R/P tape head to erase previous recording. Clean off all oxide with a cleaning stick and rubbing alcohol.

Another convenient method is to use a cassette head-cleaning tape. Some use a dry formulation, but others apply a wet cleaning solution. Do not play the cleaning cassette too long. Follow the manufacturer's directions. Of course, it is best to clean everything with alcohol and cleaning sticks.

Always clean around the tape area when the cassette player has been repaired. Now is the time to get down inside the cassette rotation area for good clean up. Besides cleaning tape heads and the pinch roller, touch up the capstan, flywheel, idlers, and turntables. Keeping oxide dust from building on tape-path components can prevent future cassette player repairs.

TAPE AND ERASE HEAD PROBLEMS

Besides collecting oxide dust, the R/P tape head can become open, intermittent, or cause distortion in the speakers. The worn tape head can cause a loss of high frequencies. A magnetized tape head can cause extra noise in recording. An open head winding will produce a dead channel. Poorly soldered connections or broken internal connections can cause intermittent music. Check the open or high-resistance head winding with the ohmmeter (Table 14-3). Inspect the cable wires and resolder for intermittent conditions. Be sure that one screw is not loose, which would let the head swing out of line.

Erase head The erase head is mounted ahead of the P/R head so that it will erase any previous recording. Usually, the erase head has only two leads and is excited by a dc voltage or by the bias oscillator. Suspect that the erase head is faulty if the sound is garbled or if two different recordings can be heard. Check for an open erase head with the low ohmmeter range (Fig. 14-35). Be sure that the tape is pressed against erase head when operating. The erase-head resistance can vary from 200 to 1000 Ω.

DEMAGNETIZE TAPE HEADS

The magnetized tape head can produce background noise within the cassette recording. Hissing noise in the playback can result from a magnetized head. Keep tools that are magnetized away from the tape heads. Magnetized screwdrivers should not be placed near the cassette player. Do not place the cassette player near or on top of large speaker columns.

The tape head should be demagnetized at least twice per year. Insert an ordinary cassette demagnetizer, which looks like and loads like any cassette. The demagnetizer with probe can get down inside the cassette loading area to demagnetize the tape heads. Do not shut the demagnetizer off when it is next to the tape head. Pull the unit out, then shut off the demagnetizer. Head demagnetizers and head-cleaning kits can be found at most electronics stores.

SPEED ADJUSTMENTS

The speed of the cassette player can be checked with a test tape and a frequency counter (Fig. 14-36). Insert a 1- or a 3-kHz test cassette and place it in the Play mode. Connect a 10-Ω resistor at the earphone jack. Measure the frequency at the DMM frequency meter. If the reading is at 1 kHz, the speed is correct. A higher reading indicates faster speed and a lower measurement indicates a slower speed. Don't worry if the reading is around 1 kHz.

TABLE 14-3 TYPICAL TAPE HEAD RESISTANCE

MODEL	TAPE HEAD RESISTANCE
GE-3-54808KA	225 Ω Actual measurements
Panasonic RQ-L315	315 Ω Actual measurements
Sony M440V	348 Ω Actual measurements
Sony TCS-340	512 Ω Actual measurements
Typical R/P tape head resistance	200 to 830 Ω
Typical erase read resistance	200 to 1000 Ω

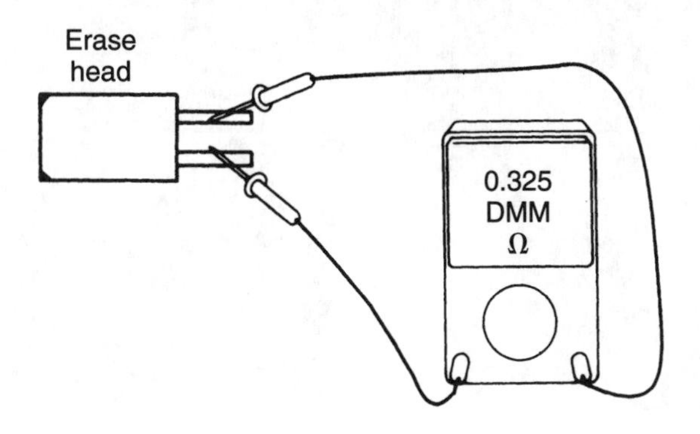

FIGURE 14-35 Check the resistance of the erase head with the DMM for an open winding.

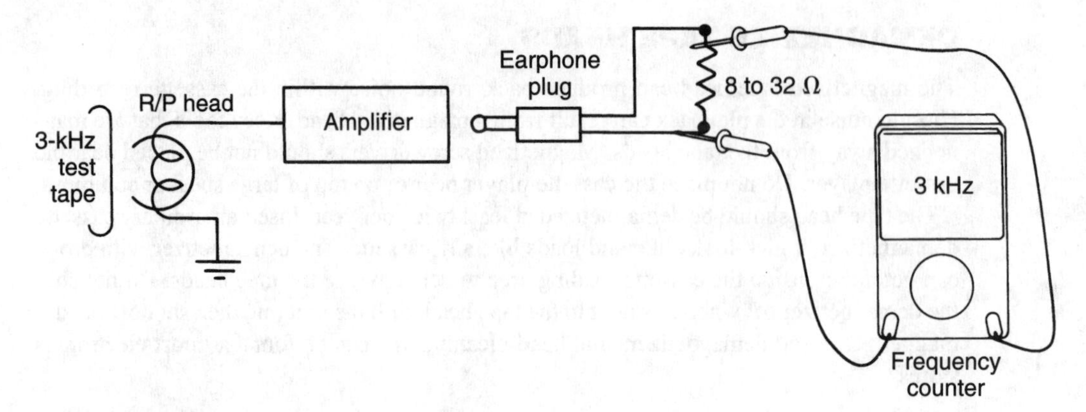

FIGURE 14-36 **Check the speed of the cassette player with a 1- or 3-kHz test cassette and the frequency counter of the DMM.**

Some larger cassette players have regulated speed adjustments or you can find a speed adjustment in the end bell of the dc motor. Slow speeds can be caused by dirty, oily, or loose belts. Oil deposits on the capstan/flywheel can produce slow speeds.

Conclusion

Yes, with a few hand tools, a DMM or a VOM, and one or two home-constructed test instruments, you can repair that broken cassette player. Remember, the sine-/square-wave project can be used for signal injection, distortion location, head azimuth adjustment, and speed tests. Just pick out the chapter that relates to the cassette player you are servicing and get started. Of course, if you read each chapter, many repairs are related to each cassette player.

15

REPAIRING TAG-ALONG AND PERSONAL CASSETTE PLAYERS

CONTENTS AT A GLANCE

Low-priced tag-along cassette players are used during physical activities or just for listening to music. Some of these units are only players and have no recording features, although some do record and play (Fig. 15-1). You can find some with only an AM radio or AM/FM radio combined with cassette player. The stereo circuits are the same as those in Chapter 16. These mini-cassette players or recorders fit into the shirt or coat pocket, snap on a belt, or have their own carrying strap.

The cassette player might operate only from headphones with no enclosed speaker (Fig. 15-2). Some units come with self-enclosed headphones while others have to be supplied. These headphones can be picked up anywhere with 8- to 40-Ω impedance. You can choose from those that clamp over the head to ones that are small enough to fit inside the ears (Fig. 15-3).

Both speakers and headphone operation are used in some of the tag-along mini-cassette players. Of course, the speaker is disconnected when earphones are inserted. The monaural speaker is usually located at the bottom of the cassette players.

Player Only

The lower-priced tag-along cassette player can be used for playing only, without any recording features. Here, only one small tape head has direct belt-drive features. Stop, fast forward, and play are the only pushbuttons, which makes this unit easier to operate (Fig. 15-4).

Only a motor belt, motor, and capstan/flywheel move. The tape motor rotates in only one direction and drives the flywheel in the same direction. The motor belt can be a square- or flat-type drive belt (Fig. 15-5). Very few speed problems are found with this type of drive system. Simply clean the motor pulley, drive belt, and flywheel for slow-speed problems.

CASSETTE TAPES

The mini or tag-along cassette player/recorder uses normal tape cassettes. Use C-90, not C-120, tapes in these machines. The tape is easily broken or stretched; if not used with ex-

FIGURE 15-1 The personal cassette player might only have play, auto reverse, auto shut-off, and an AM/FM radio in the same unit.

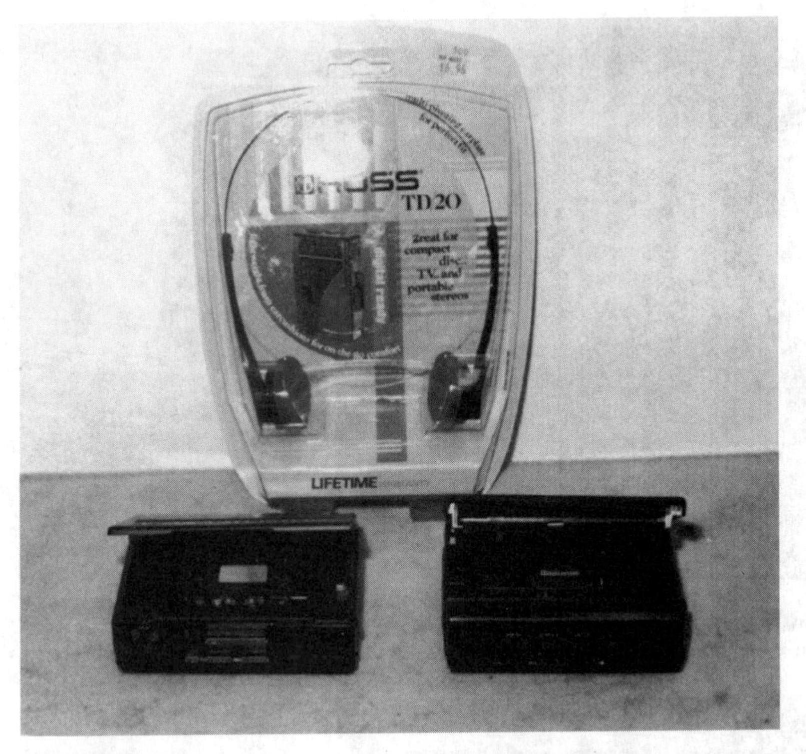

FIGURE 15-2 Some cassette players have no speakers, only headphone jacks.

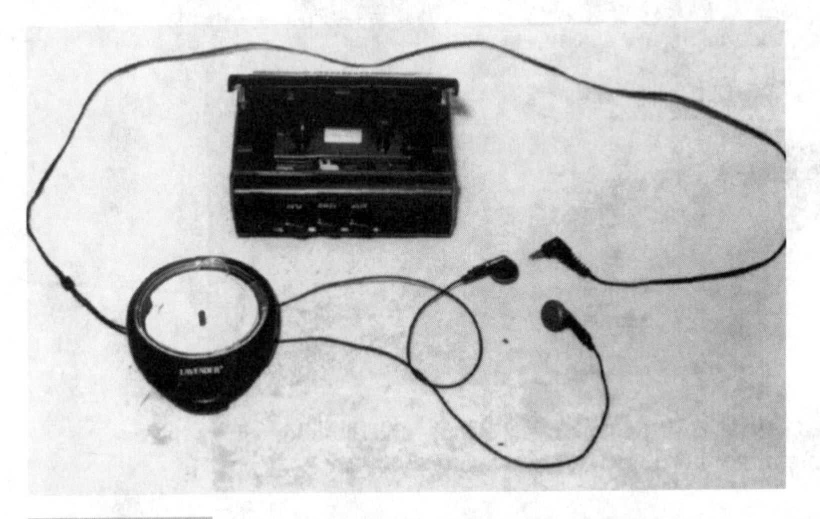

FIGURE 15-3 Many low-priced and expensive headphones are available. Some are placed over the ears and others fit inside the ear.

FIGURE 15-4 Only one R/P tape head and take-up operating reel are in the play-only cassette model.

FIGURE 15-5 The tape plays in only one direction with a flat or square-motor belt in play-only cassette models.

treme care, it can get tangled with the capstan or pressure roller. Cassette tapes, both recorded and unrecorded, should not be stored in locations with high temperatures, high humidity, or direct sunlight. Never place a recorded cassette near a magnetic source, such as a magnet, large speakers, or a TV.

FIGURE 15-6 The VAS recording switch and level control in the Panasonic RQ-L315 recorder.

Do not use a cracked plastic cassette in the tape player. A defective cassette can run slow, have sound dropouts, and crimped tape. Try another cassette if the player begins to drag or slow down and compare them. The defective cassette might have poor recording qualities with worn conditions and poor high-frequency response. Some cassettes can quickly spill out of the cassette and wrap around the pinch roller and capstan. Do not forget to check the tape head for magnetization.

VAS, VOX, OR VOR SYSTEMS

In the higher-priced mini-cassette recorders, you can find a voice-activated system. When recording using the VAS function, the sound is recorded automatically so that no tape is wasted. With the VAS/pause switch, the tape runs when sound is picked up by the built-in or external microphone. Then, when no sound is picked up, tape stops running automatically (about three or four seconds later).

The volume/VAS level control should be normally set to the (4 to 7) position. To record loud sounds only, rotate the control toward (1 to 3). To record low sounds, rotate toward (8 to 10). No sound is recorded at the 0 position. Remember to check the position of volume/VAS level control if the player will not record or start rotating when in VAS operation (Fig. 15-6). The recording level is automatically adjusted, regardless of the position of the volume/VAS control level control.

VAS recording with Panasonic's RQ-L315 cassette player has a VAS, Off, and a Pause switch. When switch is off, the player must be turned on by hitting the Play or Record switch. The VAS/pause switch can be used to stop the tape movement temporarily during recording or playback. Do not use the VAS/pause switch to stop the tape for a long period of time. Remember, the unit is not turned off when the VAS/pause switch is set to pause or on. Always use the Stop button to turn off the unit.

HEAD AND CABINET CLEANING

The sound quality of the cassette player might become weak and distorted with a dirty tape head. Remember that the tape head, capstan, and pressure rollers are in contact with the tape at all times. Make it a habit to clean these parts after 10 hours of playing (Fig. 15-7).

Clean the dirty tape head with alcohol and a cleaning stick or use dampened cloth with a little alcohol. Manufacturers rarely recommend using cleaning tapes because some contain

FIGURE 15-7 **Clean the tape head with a high grade of isopropyl alcohol and a cleaning stick.**

abrasives and can cause premature head wear. Just clean them with alcohol and a cleaning stick.

Do not clean the plastic cabinet with paint thinner or benzene. Clean the cabinet with a cloth dampened in a mild soap-and-water solution. Avoid excessive moisture. Wipe the cabinet dry with a soft cloth. Try to avoid spray-type cleaners when the cassette mechanism is in the cabinet because chemicals can discolor plastic body.

HEAD DEMAGNETIZATION

Use a cassette-tape demagnetizer tool. Several different demagnetizer tools can be found at most electronic supply stores. Carefully follow the instructions that are supplied with the device. Do not bring any magnetized metal objects or tools near the tape head.

THE CASSETTE MECHANISM

The topside of the cassette mechanism (Fig. 15-8) consists of supply and take-up reels, the capstan, R/P head, and erase head. The bottom view of the cassette mechanism can show the small drive motor, drive belt, capstan/flywheel, and various idler wheels. The motor rotates the capstan/flywheel with a rubber flat or square belt (Fig. 15-9). The idler wheels and arm are placed into the system by pushbuttons.

When the fast forward button is pressed, the take-up reel is pulled by an idler pulley, engaged against the flywheel. Often, the pinch roller is not engaged or the capstan is on the tape in Fast-Forward mode. When pressed, the Play/Record button applies voltage to the motor, rotates the belt, and the flywheel/capstan. The capstan with pinch roller pulls the tape from the supply reel and the slack is taken up by the take-up reel. You must push both the Play and Record buttons to set the unit into the Record mode (Fig. 15-10).

SLOW SPEED

Slow speed can be caused by a dirty motor pulley or oil spots on the pulley. A binding or dry capstan/flywheel bearing can slow down the cassette player. A worn, stretched, or oily belt can cause slower speeds (Fig. 15-11). Clean the tape head and the pinch roller. Slower speeds can result if excessive tape is wrapped around the pinch roller and in between the

FIGURE 15-8 The top view of the cassette mechanism in the Panasonic mini-cassette recorder.

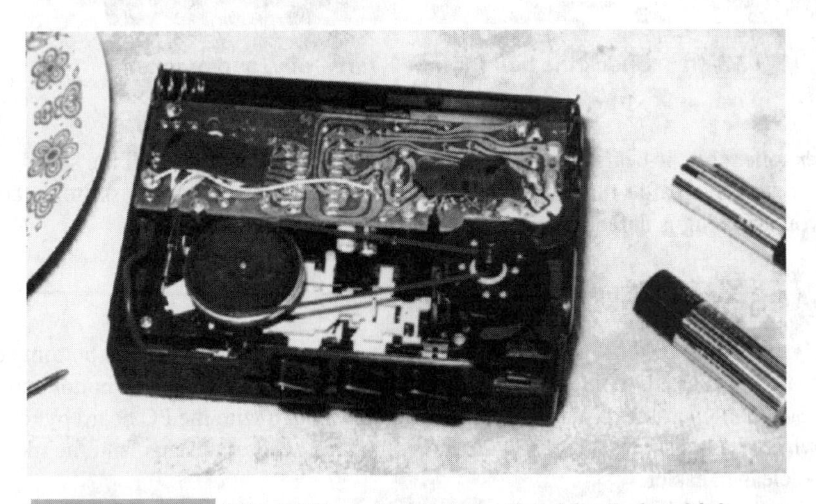

FIGURE 15-9 The motor pulley rotates the drive belt, which turns the capstan/flywheel for tape action.

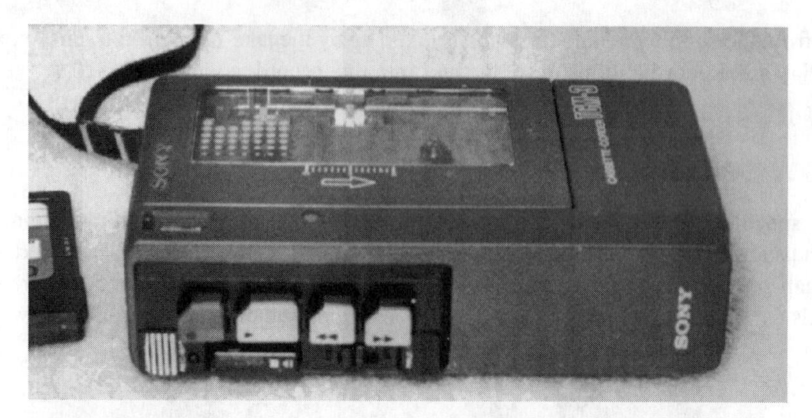

FIGURE 15-10 You must push both Play and Record buttons to record in most cassette players, including Sony's T6M-3 player.

FIGURE 15-11 Check the belt for loose, dirty, oily, and cracked areas.

rubber roller and the bearing plate. Suspect that the motor is defective if the speeds are erratic or slow even after the rotating parts are clean. Also, try another cassette, in case the one you are using is defective.

DISASSEMBLING THE COVERS

Simply removing the bottom screws lets the cover come off. Besides the bottom screws, one or two top screws inside the top lid must be removed to drop off the bottom covers in a GE 3-54808KA cassette player. Remove the mechanism with the PC board by removing two white chassis screws (Fig. 15-12). Now, the tape heads, capstans, and the motor belt can be cleaned easier.

Notice that within this cassette player only the take-up reel rotates. The supply reel is nothing more than a plastic post. The motor drive belt drives the plastic-metal flywheel. At the top of the flywheel, under the capstan, a pulley drives another small belt that rotates the take-up reel (Fig. 15-13). Here, the fast-forward speed is the same as the speed of the play take-up spindle, except that the capstan does not engage the tape. When replacing the PC board and mechanism assembly, be sure that the small wires are in the original position.

NO SOUND/NO TAPE MOVEMENT

Check for dead or weak batteries. Replace weak batteries. If the tape still does not rotate, inspect the leaf switch. Often, low-priced cassette players have a small leaf switch that makes contact when the Play, Record, Fast-Forward, and Rewind features are used. Clean the copper spring-type contacts with cleaning fluid and a stick (Fig. 15-14). Sometimes, pulling a piece of thin cardboard (match cover) through closed switched contacts cleans them. Suspect that the motor or motor circuit is defective if the batteries and the leaf switch work properly.

NO FAST FORWARD

In most surface-drive tape mechanisms, the idler wheel is shoved over to rotate the take-up reel. The idler wheel is rotated by a friction drive against a wheel that is attached to the capstan/flywheel shaft. If the player operates normally in the Play mode and slow in the Fast-Forward mode, suspect that the idler drive area is slipping (Fig. 15-15). Clean all drive surfaces. If the fast forward is belt driven, clean the belt and drive pulley. If both Play and Fast Forward are slow, clean the motor belt and flywheel surfaces.

FIGURE 15-12 **Remove six screws to remove bottom cover and PC board in the GE 3-54808KA cassette player.**

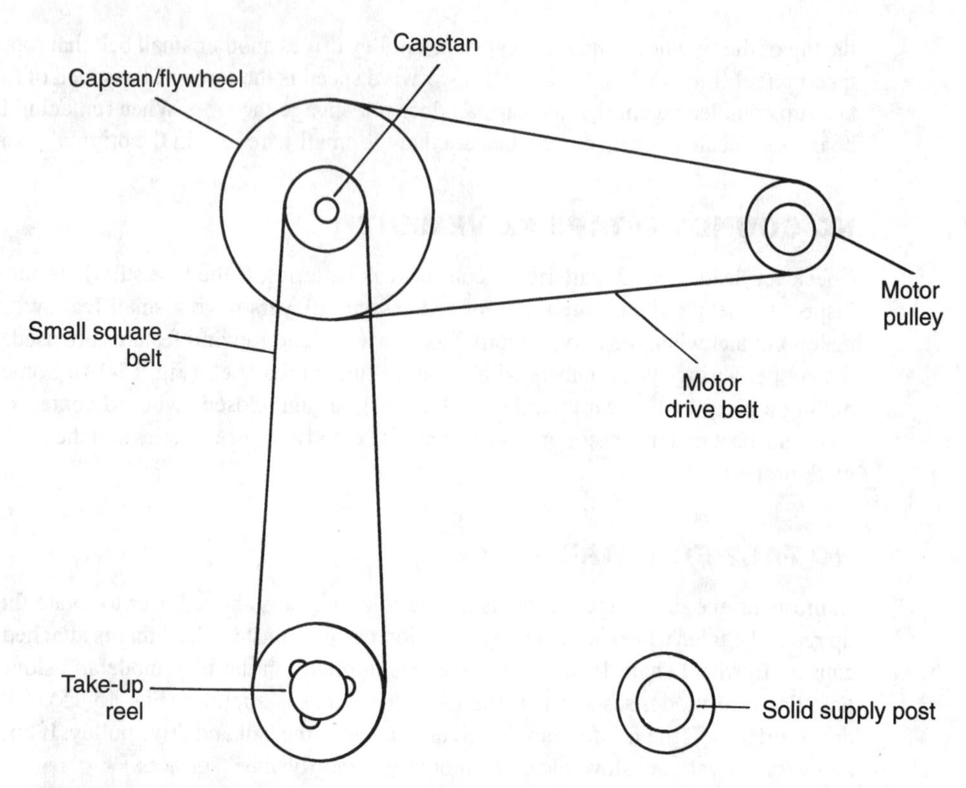

FIGURE 15-13 A motor belt drives the capstan/flywheel from the motor pulley. The take-up reel assembly is belt driven from a small pulley on the hub of the capstan.

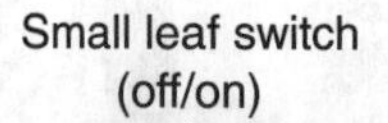

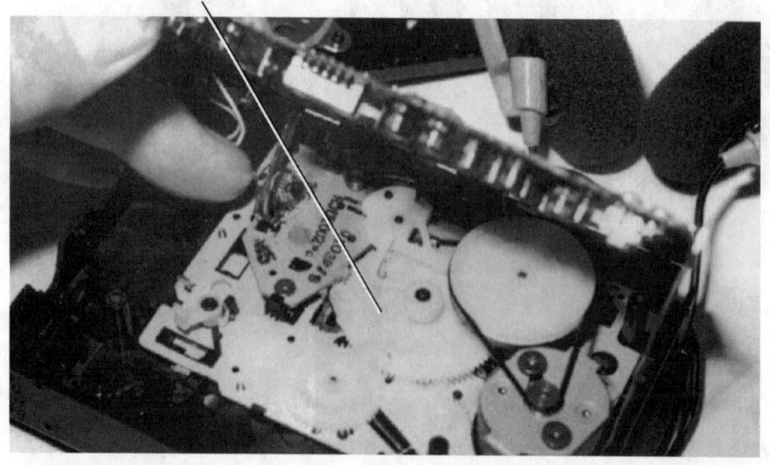

FIGURE 15-14 Small leaf-switch contacts are pressed together in the various modes to provide voltage for the motor and amplifier.

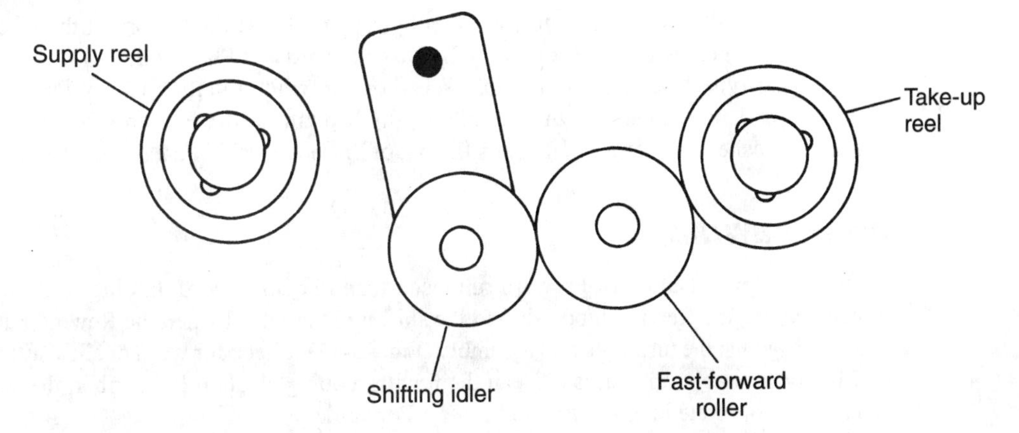

FIGURE 15-15 **The idler wheel is shifted toward the fast-forward roller. It then drives the take-up reel at a faster speed.**

Fast Forward and Rewind in the Panasonic RQ-L315 recorder are actually driven from small plastic gears. The small white plastic teeth mesh when switched to Fast Forward. The capstan gear rotates a large idler wheel and drives another shifting idler gear wheel and drives another shifting idler gear wheel (Fig. 15-16). The idler gear wheel is shifted toward

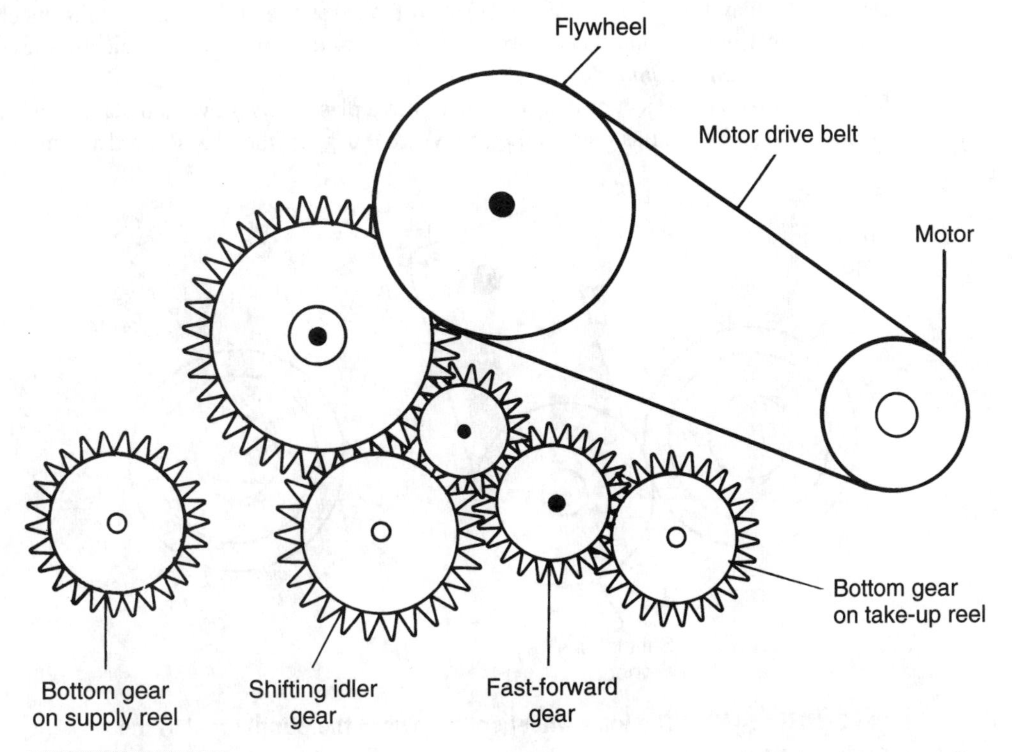

FIGURE 15-16 **Small plastic gears are shifted into position to operate functions in the Panasonic RQ-L315 recorder.**

the take-up spindle, which engages two small gear wheels. At the bottom of the take-up reel is a plastic gear wheel that rotates in the Fast-Forward and Play modes.

These gear-type assemblies rarely lose speed or slip while rotating. Check for broken gear teeth or jammed gears if the assemblies don't rotate in the Fast-forward mode. A missing C washer can let the small gears fall out of line and disable Fast Forward and Play.

POOR REWIND

Obviously, Rewind and Fast Forward run faster than Play or Record. In older and lower-priced players, the Rewind mode shifted the shifting idler wheel when the Rewind button is pushed against the turntable reel assembly (Fig. 15-17). Check for worn or slick surfaces on idler or turntable drive areas. Clean them with rubbing alcohol. Remember, the pinch roller does not rotate in either Rewind or Fast Forward.

With geared drive systems, the idler is shifted against the gear of the supply spindle. Usually, Rewind speed is lower than Fast Forward (Fig. 15-18). In Rewind, capstan gear rotates the large drive gear, which, in turn, rotates the shifting idler gear, and the idler drives the gear on the bottom of the supply spindle.

NO AUTOMATIC SHUT OFF

Excessive tension of the tape engages and triggers a small ejection lever that mechanically releases the Play/Record assembly and shuts off tape rotation. In the larger units, mechanical and electronic automatic shutoff systems are used. Sometimes the ejection lever is called a *detection* or *contact piece* (Fig. 15-19).

The automatic stop-eject or detection piece has a plastic cover over a metal angle mechanism, that can have adjustment at the end where it triggers the Play/Record assembly, the

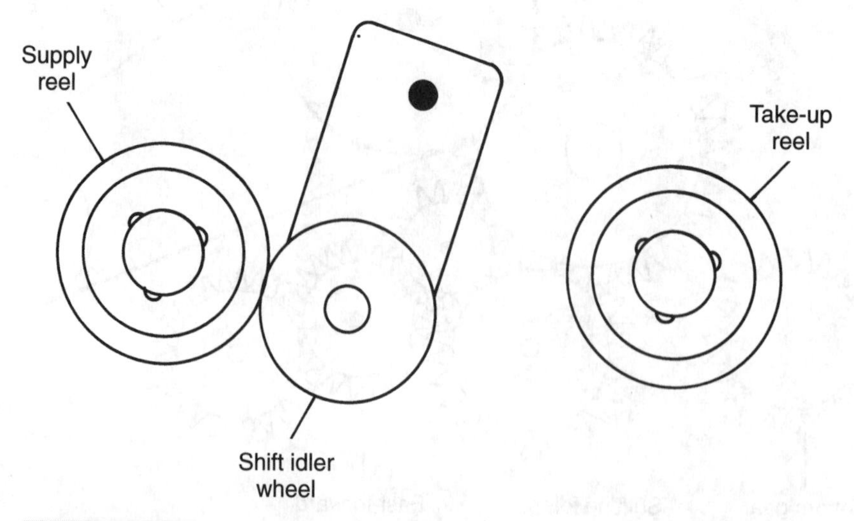

FIGURE 15-17 **The idler wheel shifts toward the supply reel in the rewind mode.**

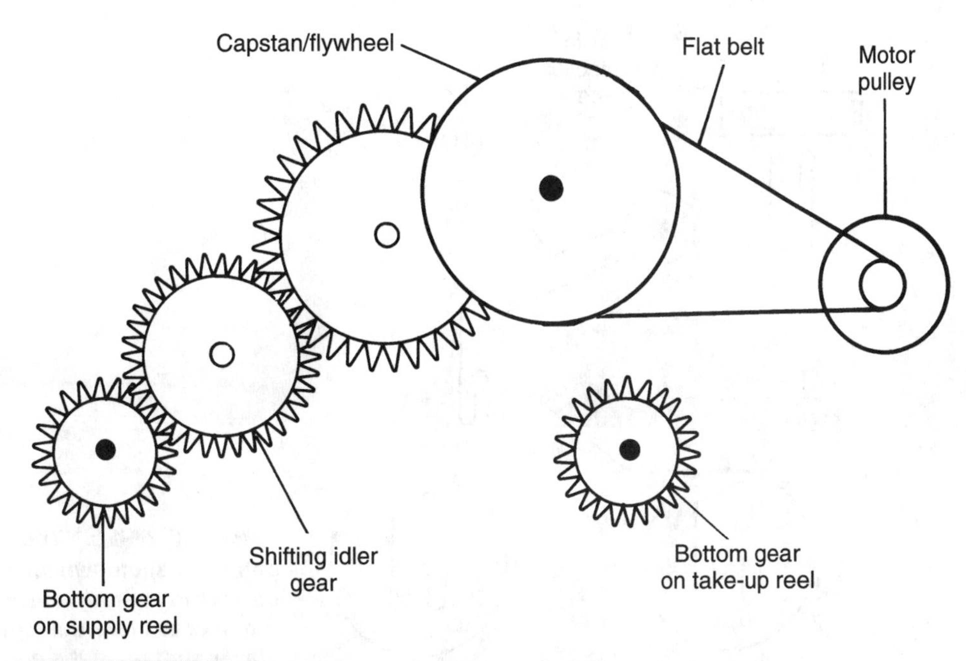

FIGURE 15-18 **The shifting gear presses against the gear, which is attached to the rewind spindle for rewind operations.**

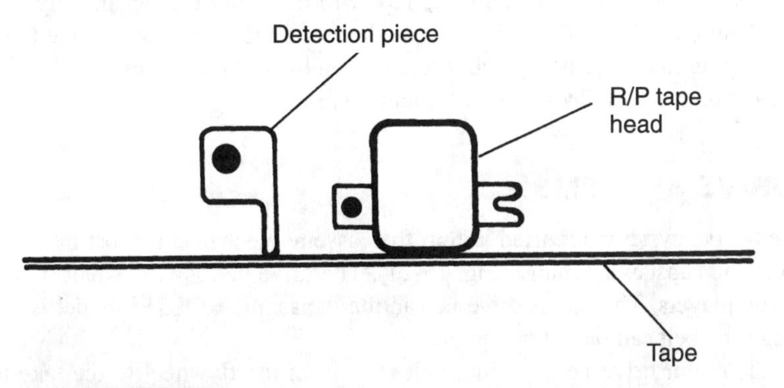

FIGURE 15-19 **The eject piece or detection piece is mounted close to the tape path to shutdown the player when tape end is reached, in automatic shutdown.**

automatic stop. The ejection piece is mounted alongside the tape head. When the end of the tape has been reached, the tape exerts pressure against the ejection piece and mechanically triggers the Play/Record mechanism.

The end of the eject lever contains a notched metal end that can be widened for understroke and pinched together with a pair of long-nose pliers for overstroke (Fig. 15-20). The adjustment of closing or opening notched lever should be made so the mechanism acts when the tape ends.

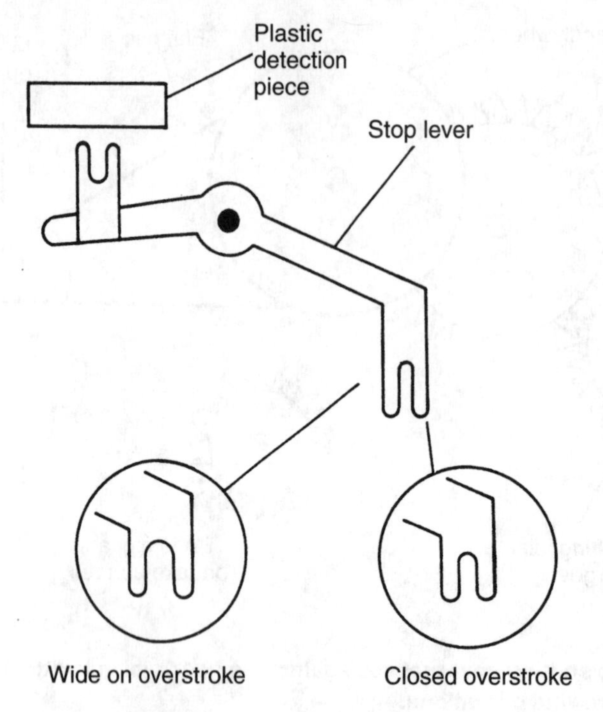

FIGURE 15-20 The detection shutdown lever can be spread apart or squeezed together to make the tape player shut off at the end of the tape.

Check the adjustment of the levers if the tape will not shut off mechanically. Notice if the lever is bent out of line. Does the eject or detection piece ride against the tape at the end? Straighten up the lever or replace it for auto shut-off. Place a drop of oil at the bearing if the ejection piece is binding or is difficult to move.

BELT-DRIVE SYSTEMS

Several belt-drive systems are used within the cassette mechanism. Most have a motor-drive belt to the capstan assembly (Fig. 15-21). The drive belt is very small in micro- or mini-cassette players. The motor drive belt in the Panasonic RQL315 model is only two inches long. The belt can be flat or square.

Besides the motor drive belt, another belt runs from the flywheel to the take-up reel. Some minicassette players have a Fast-Forward drive belt. Because these belts are very small and thin, they have a tendency to stretch and cause slow speeds. Clean each belt with speed problems with rubbing alcohol and a cloth. After clean up, if the speed is still abnormal, replace the motor drive belt.

CASSETTE SWITCHES

Many small switches are used in the personal cassette recorder. The sound-level equalizer (SLE) switch improves recording in locations away from the source—especially in class and conference rooms. The Pause and VAS are slide switches. Usually, the radio-tape switch is a slide switch. If these functions do not work or are erratic, spray cleaning fluid into the switch area. Use it sparingly so that you don't spill fluid over the belts and idler wheels.

The on/off switch that furnishes power to the motor and amp circuits might be a leaf switch. It is pressed on when Record, Play, Rewind, or Fast Forward are used. The small switch contacts might become dirty, but you can clean them with cleaning fluid and a stick. Suspect that a leaf switch is defective or dirty if the unit appears intermittent. This switch is squeezed shut with a metal lever (Fig. 15-22).

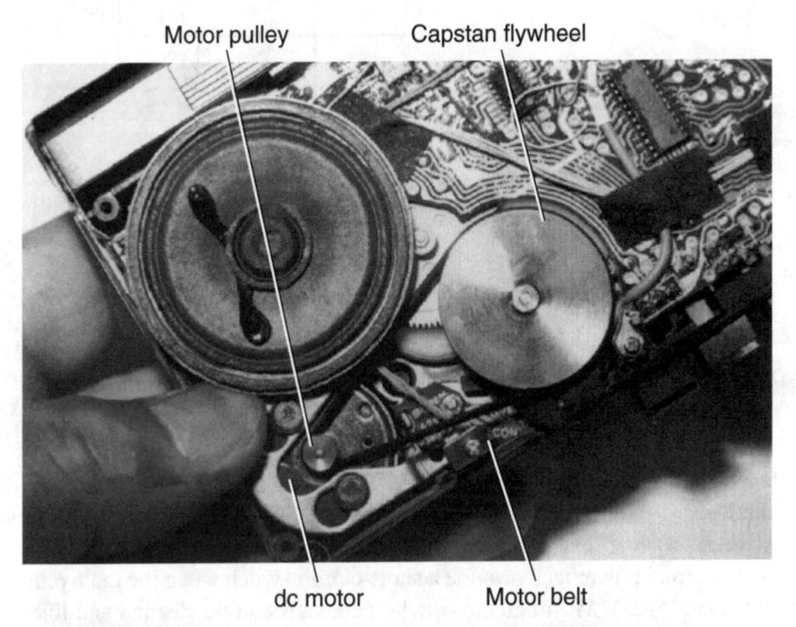

FIGURE 15-21 Only a two-inch, thin, square rubber belt drives the Panasonic cassette player.

FIGURE 15-22 The on/off leaf switch in the Panasonic RQ-L315 mini-cassette player.

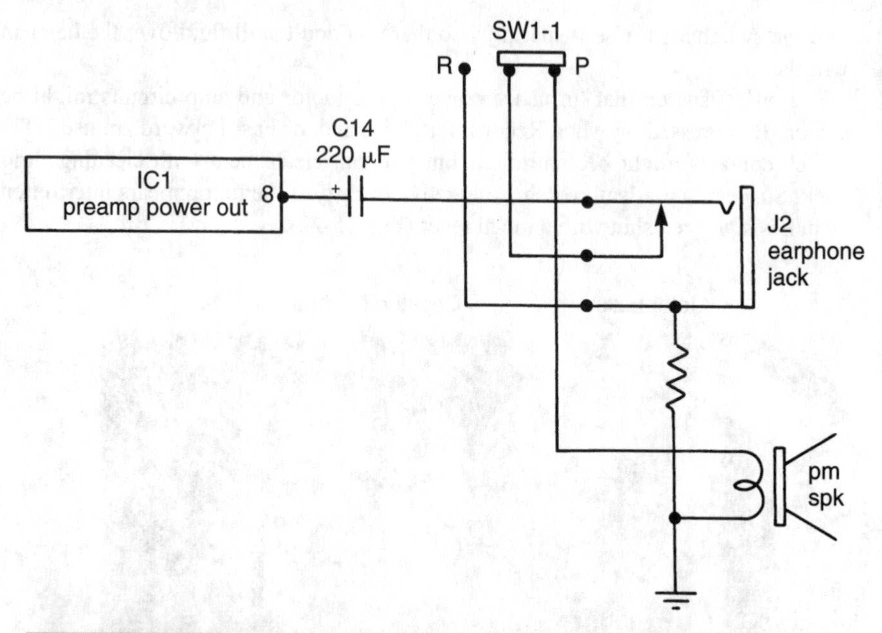

FIGURE 15-23 The headphone jack is switched on the speaker terminal when the plug is out of the circuits.

The tape-speed switch, VOR, and VOX switches can be slide types with several positions. The Record/Play switch in the Radio Shack VSC-2001 recorder has five separate contacts when switched in the Record or Play modes. Besides this switch, the earphone, remote and external power jack provide a short-circuit switch when the male plug is out of the circuit (Fig. 15-23). Most defective switch contacts can be cleaned and lined up. Replace them if they are broken.

EARLY AUDIO CIRCUITS

In the early cassette solid-state audio circuits, transistors were used throughout. The preamp, AF amplifier, and driver amplifier stages provided good audio to drive the transistorized push-pull output stages (Fig. 15-24). The two transistor outputs were transformer coupled to the speaker. A single oscillator transistor furnished bias to the Erase head and the R/P head.

The Record/Play head is switched into the first audio preamp stage in Play mode. The tape head is capacity coupled (C1) to the base of Q1. The base of Q2 is directly coupled to the collector terminal of Q1. The volume control (VR-1) is tied into the preamp circuit by capacitor C5 (Fig. 15-25).

VR1 controls the tape head and preamp volume that is applied to AF amplifier Q3. The base of Q4 is capacity coupled to the collector of Q3. Both Q5 and Q6 base circuits are transformer coupled to the collector terminal of Q4. The collector terminals of the push-pull output transistors are transformer coupled to the speaker and the earphone jack (Fig. 15-26). Notice that all of the transistors are early PNP types.

Next came transistor preamps and IC power-output circuits. The Record/Play head is capacity coupled to the transistor base (Q1) and to IC1 through a volume control. The amount of VR1 volume is applied to the input terminals of IC1. C14 capacity couples the audio signal to the speaker and to the earphone in the Play mode (Fig. 15-27). The audio circuit voltage is supplied from four 1.5-V batteries and a power-cord adaptor.

Today, most small audio circuits are located in one or two ICs (15-28). The preamp audio stages can have a separate IC from the power-output IC. The large IC found in the small cassette player contains all of the audio stages.

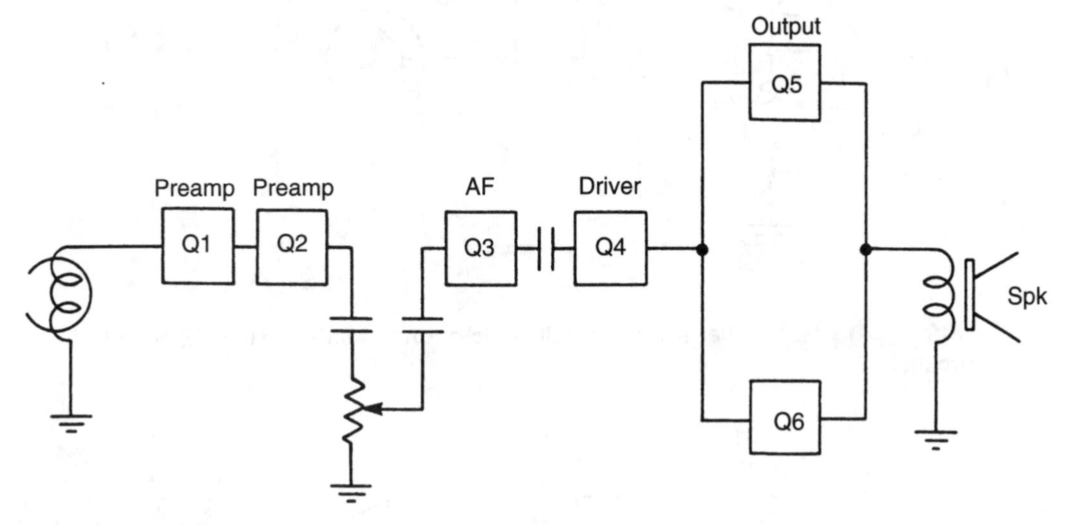

FIGURE 15-24 The block diagram of an early transistorized cassette player.

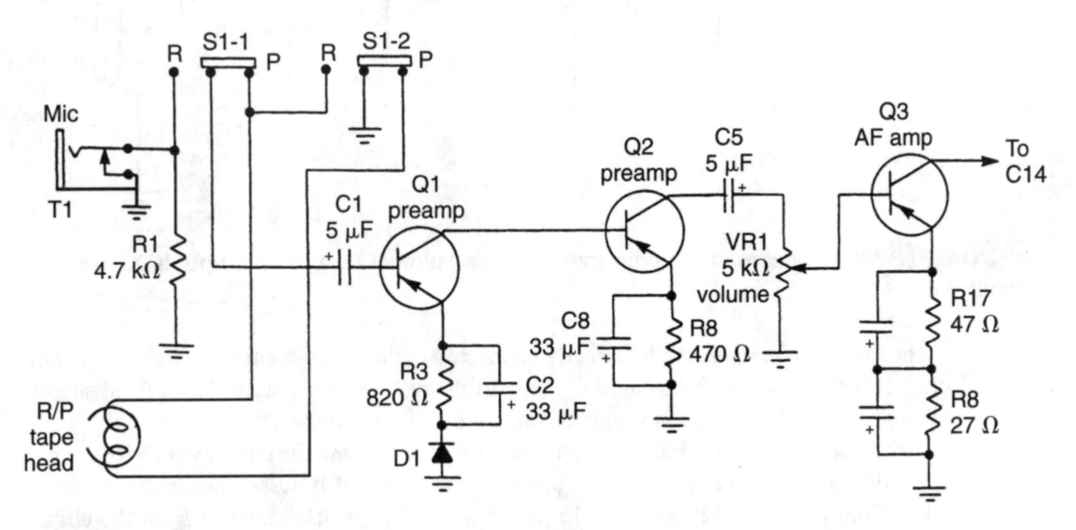

FIGURE 15-25 The audio signal path in the Play mode of early preamp transistor circuits.

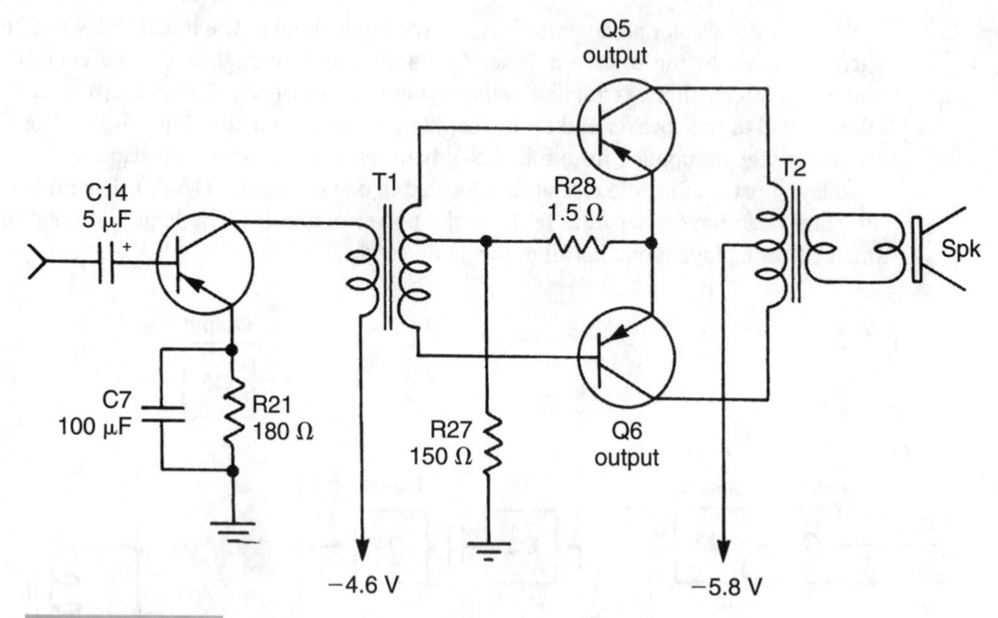

FIGURE 15-26 Driver and push-pull audio-output transistors in PNP output circuits.

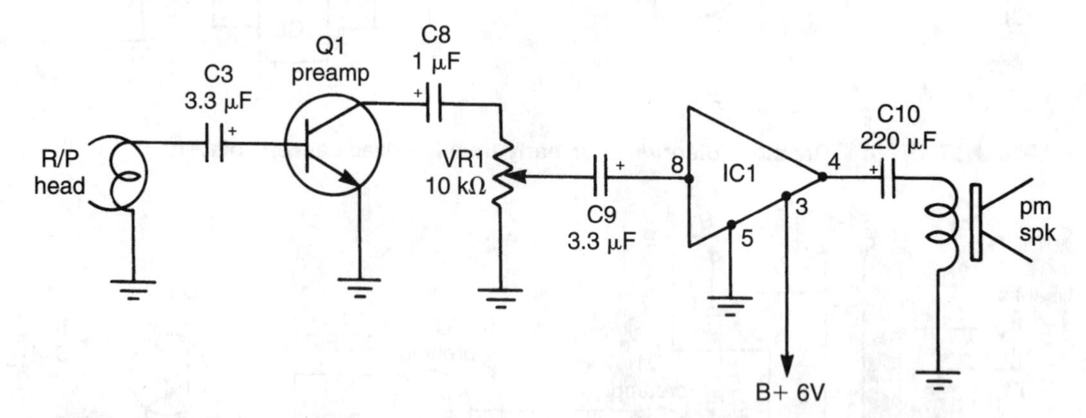

FIGURE 15-27 The simple preamp transistor and typical AF power-output IC audio circuit.

The microphone input can be directly connected to the input circuits through C105. The electret condenser mic that is used in many of today's cassette players has a fixed supply voltage. Often, the voltage is supplied from an RC-filtered network.

In the Play mode, the R/P tape head is switched directly into the circuit with S101, to pin 24 of IC1. The speaker is coupled through R113 to pin 8 of IC1. The headphone audio is taken from pin 10 and R114 (Fig. 15-29). The speaker is disconnected from the circuit when the headphone plug is inserted.

FIGURE 15-28 Only one audio IC is used in today's small cassette player.

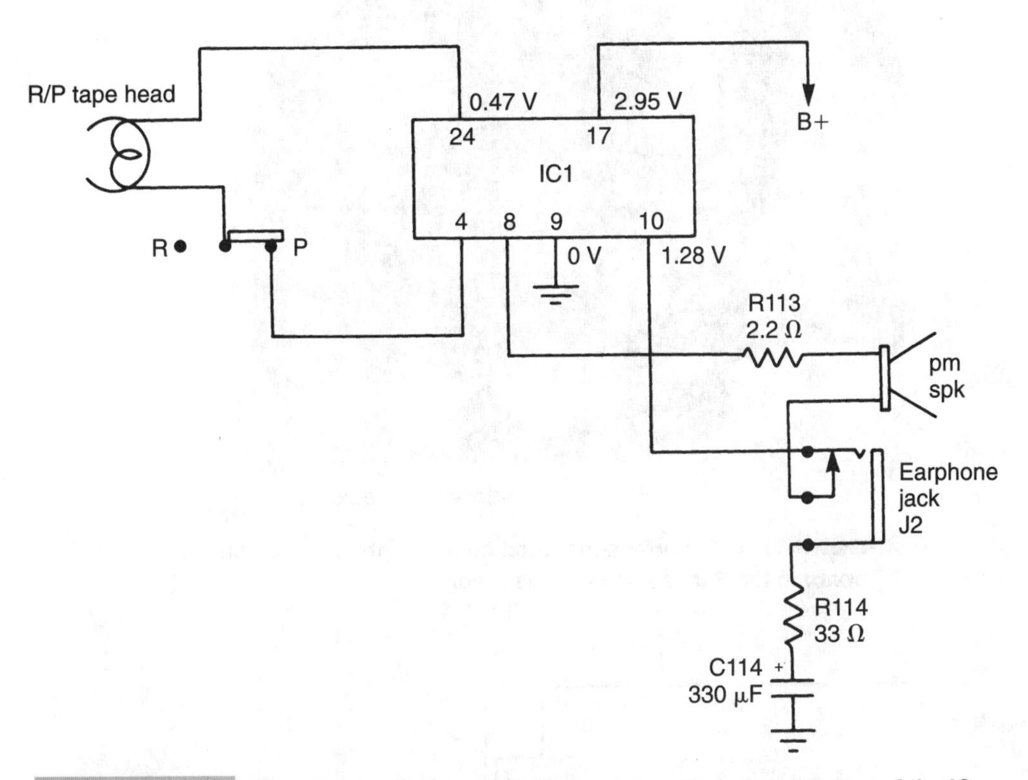

FIGURE 15-29 The signal path from the playback head input and output of the IC to the speaker circuits.

SURFACE-MOUNTED COMPONENTS

Surface-mounted components are used in the higher-priced personal cassette recorders. These surface-mounted parts are tiny, except for ICs and solid-state processors. They should not be touched, unless you have the patience and special equipment to remove and install components. The transistors and ICs must be replaced with original components. Some of the small resistors can be replaced with 1/8-scale parts. Surface-mounted components are located on the printed board side and regular capacitors and parts are mounted on the other side of the board (Fig. 15-30).

DEAD PLAYBACK

The tape is rotating with no sound from the speaker. Insert the headphones to see if the speaker is defective. If it is still silent, clip a small speaker with an electrolytic coupling capacitor and ground lead to the output circuit (Fig. 15-31). If it sounds, check for poor switching contacts in the headphone circuits. Check for broken speaker leads. Sometimes

FIGURE 15-30 **Surface-mounted components are used on the PC board in the Panasonic cassette recorder.**

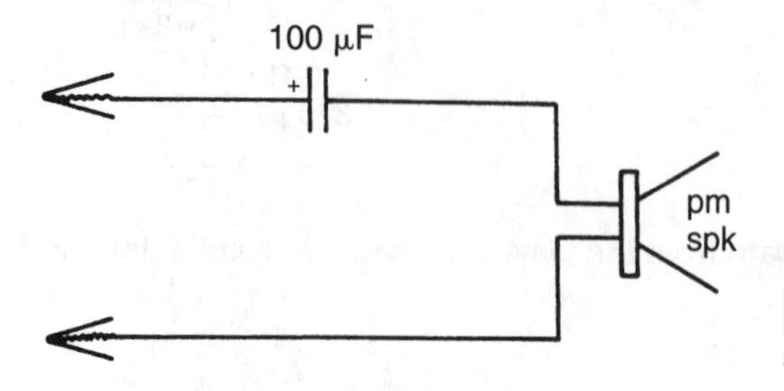

FIGURE 15-31 **A simple test speaker with only four small parts.**

FIGURE 15-32 **Just clip a small speaker across the suspected speaker with one lead removed.**

these small single-wire speaker leads get wedged between the PC board and parts, which breaks the leads.

If the audio-output stages are normal, insert a recorded cassette with music and signal trace the audio with an outside amp or an audio tester. Trace the audio from the playback head to the input terminal of the IC or base of the first preamp transistor. The signal is quite weak at this point. Signal trace the audio from the base to the collector of each transistor stage.

In IC preamps, check the audio at the input and the output terminals of the IC. You can break the audio circuits in half by checking the audio at the volume control. If you find adequate volume at the volume control, the output IC circuits are defective. If you find weak sound or no sound at the volume control, the defective stage is in the input circuits. Do not forget to check the supply voltage of each IC. No voltage or low voltage can indicate that the power supply is defective or that an IC is leaky.

DISTORTED SOUND

Muffled or distorted audio is often caused by defective audio-output stages. Leaky transistors and ICs can cause distorted audio. First, check the input audio that goes into the output stages for distorted sound. If the input is free of distortion, check the audio-output stages and the speaker.

Substitute another speaker for the enclosed one. Clip the speaker, preferably a larger one, across the speaker terminals (Fig. 15-32). Disconnect one lead to the built-in speaker. Replace small speaker if the audio is distorted.

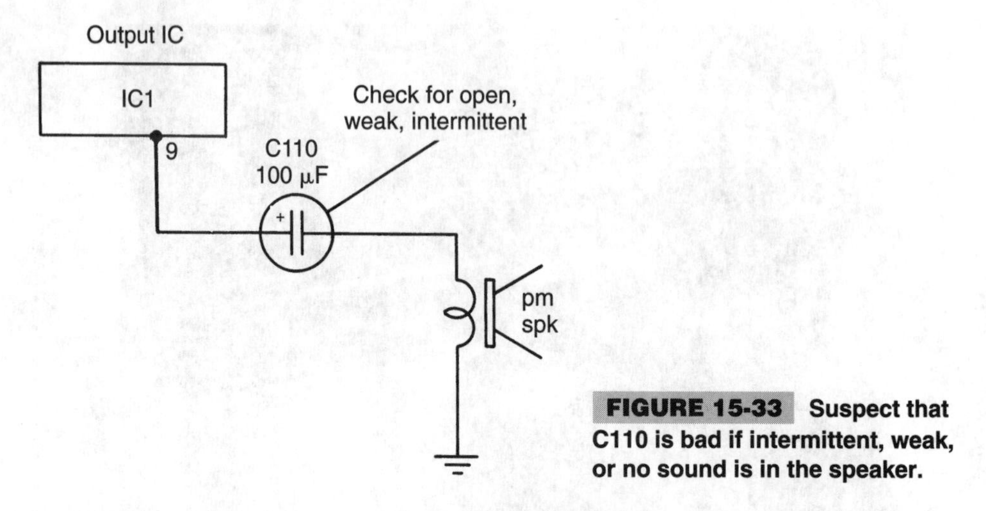

FIGURE 15-33 **Suspect that C110 is bad if intermittent, weak, or no sound is in the speaker.**

Check the supply voltage that is applied to the output IC. Measure the voltage at all IC terminals and see if any are different from those that are listed on the schematic. Replace the IC or transistor if the input sound is normal, all voltages are fairly good, and yet the speaker audio is distorted. Do not overlook the possibility that an electrolytic speaker capacitor is leaky or open if the audio is distorted, weak, or intermittent (Fig. 15-33).

REMOVING THE SMALL CASSETTE LID OR COVER

On personal cassette players, remove the lid by prying the small plastic hubs out of their sockets. These plastic lids have a tendency to snap off. The lid is quite easy to remove, but it is difficult to replace. After the cassette player is serviced, leave the main mechanism out of the unit until you replace the top lid. Sometimes the small spring lock on one end has to be loosened. Loosen these screws, but do not remove them (Fig. 15-34).

Now, place the plastic cover into top case, with leaf spring upward. Close the unit together and snap the other plastic end into the hub area. A miniature screwdriver can help you to pry the plastic end piece. Be sure that the top cover will open, stay apart, and snap closed after being replaced. Then, tighten the small screws on the metal mounting springs.

Replace the mechanism after the top cover is replaced. Be careful not to lose the miniature screws. Place them in a saucer or cap lid. You cannot buy these small screws over the counter. Of course, you can take some from a discarded or defective cassette player.

REPAIRING HEADPHONES

Besides cleaning the tape head and replacing small batteries, headphones cause the most problems. The cord breaks at the headphones or where it goes into the male plug and the male plug makes a poor connection at the jack. Any headphones can be used.

Monaural phones have only two connections at the male plug, but stereo earphones have three wire connections. Also, the stereo earphone jack has a common ground and two ungrounded wires. Use the ohmmeter to make tests at the male plug.

Clip the meter test leads to the common ground (long area at the back) and the outside tip of the pair of stereo earphones (Fig. 15-35). Then, clip the DMM test probe to the other male connection. The resistance should be from 7 to 50 Ω, depending on the headphone impedance. The higher the impedance, the greater the resistance. No measurement indicates that the cord wire or phone winding is open.

Flex the phone cord while it is clipped to the meter. If you find a poor connection, the meter hand will rapidly change. Clean the male plug for erratic or intermittent reception. Move the cord near the plug to determine if the breakage occurs at the male plug.

FIGURE 15-34 **Replace the plastic cover or lid by snapping the small ends into the hub and spring area.**

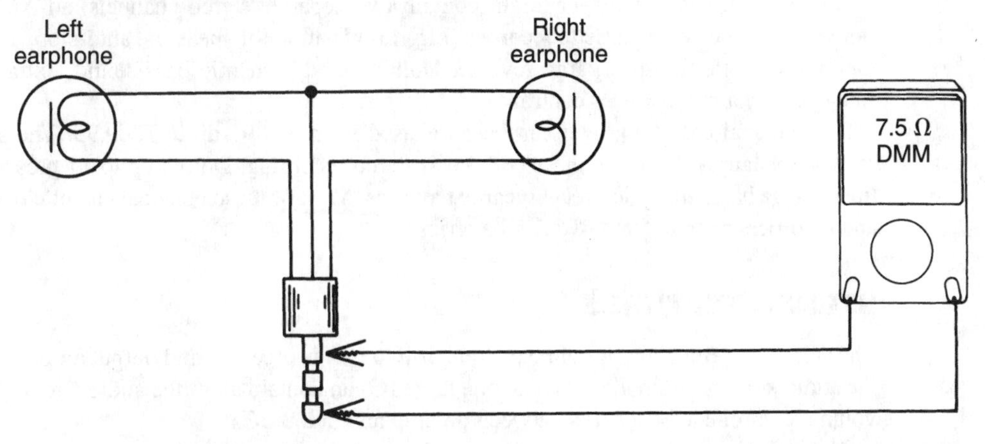

FIGURE 15-35 **Check the stereo headphones for continuity with the lowest range of the DMM.**

FIGURE 15-36 Some cassette players have no speakers, only headphone jacks.

Remove the rubber sponge cover and remove the speaker element (Fig. 15-36). Inspect the cord for breakage. Often, the small flexible wire is pulled out of the headphone-soldered connection. Resolder the poor connection for erratic audio. If the wire is broken, tin the flexible wires with solder and solder paste. Resolder the broken wire to the headphone.

Check the earphone for open internal wiring or an open winding at the earphone connections. Measure the total resistance of one of the headphones. Of course, this resistance is slightly lower than the impedance of the earphones. For instance, the 8-Ω impedance might have a resistance of 7 to 7.5 Ω.

If the cord is broken at the male plug or if the plug is damaged, replace it. Most electronic stores have standard male plug replacements. When soldering male plug connections, keep the bare wire ends short and make a clean soldered connection. Do not let the bare ends or wires touch the outside metal ground cover. Tin the back ⅛ inch of the cord and solder it to the small plug terminals. Place cellophane tape between the connection and the ground lug to keep it from shorting.

The personal cassette player might contain two separate stereo channels, an AM/FM stereo radio receiver, a cassette recorder (Fig. 15-37), monitor speakers, auto-stop, a four-function cassette transport, auto reverse, Dolby sound, dynamic bass sound, extra bass sound, or dynamic loudness control.

The personal cassette player can be purchased from $19.95 up to $269.95. The super models contain AM/FM synthesized tuners, stereo recording, Dolby, up to 14 preset stations, mega bass, and quick rechargeable batteries. Most of these personal cassette players and recorders operate from two AA batteries.

STEREO FEATURES

The stereo cassette player might have one or two ICs as preamps and output amplifiers in the audio system. In smaller players, one large IC can contain all of the audio circuits. The volume control can be located between preamp and output ICs.

Often, the input stereo circuits consist of two different head and external microphone jacks. If the external mic is in the circuit, the internal microphone is switched from the circuit. Both the microphone and R/P tape heads are switched in the right and left input circuits (Fig. 15-38). The left and right R/P tape heads and the erase head are excited from a transistorized oscillator circuit.

In Play mode, the tape head signal is switched to both left and right channels and amplified by the preamp stages. A dual volume control sets the amount of volume that is applied to the stereo IC power amps. The output audio signal is switched to the left and right speakers (Fig. 15-39).

When recording, the left and right microphones are switched into the circuit with S300 and applied to the preamp circuits. The audio signal is amplified and passed on to the audio output or to a separate IC record amp and switched back to the R/P tape heads. The audio signal picked up by the microphone is recorded on the tape by passing through the preamp and the audio amp stages (Fig. 15-40).

Also, when recording, the Erase head is excited by the bias oscillator. At the same time, the bias is applied to each stereo recording tape head for optimum recording. Voltage is only applied to the bias oscillator when recording. The bias oscillator circuit is dead in the Play mode (Fig. 15-41).

ERRATIC PLAY

Intermittent or erratic play can be caused by a dirty tape head, dirty R/P switch contacts, an intermittent preamp, an intermittent audio output amp, or by a bad speaker. Usually, the intermittent or erratic sound will only occur in one channel. If intermittent sound is heard in both speakers, suspect that a dual preamp IC, transistors, or a dual output IC are defective. An erratic voltage source applied to the amps can cause intermittent sound in both channels. If your cassette is defective, try another stereo cassette.

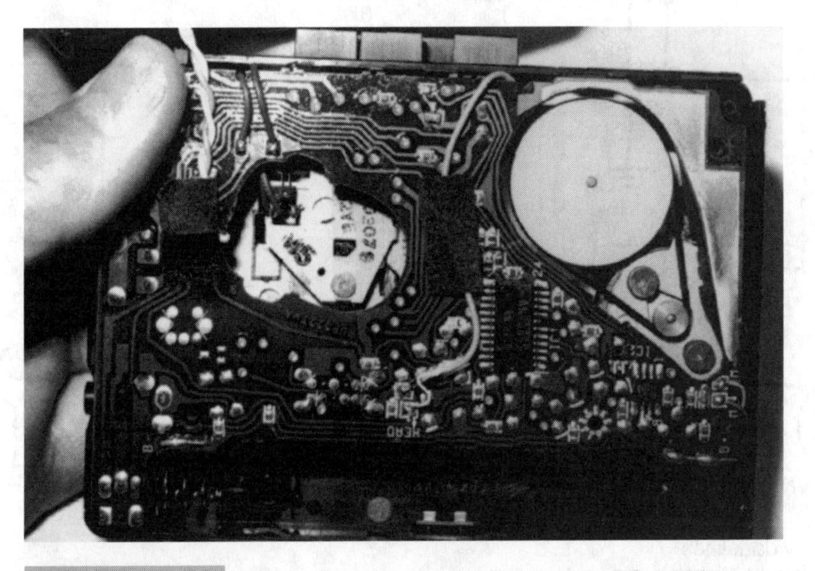

FIGURE 15-37 The inside view of the Panasonic RQ-L315 personal cassette recorder with some surface-mounted components.

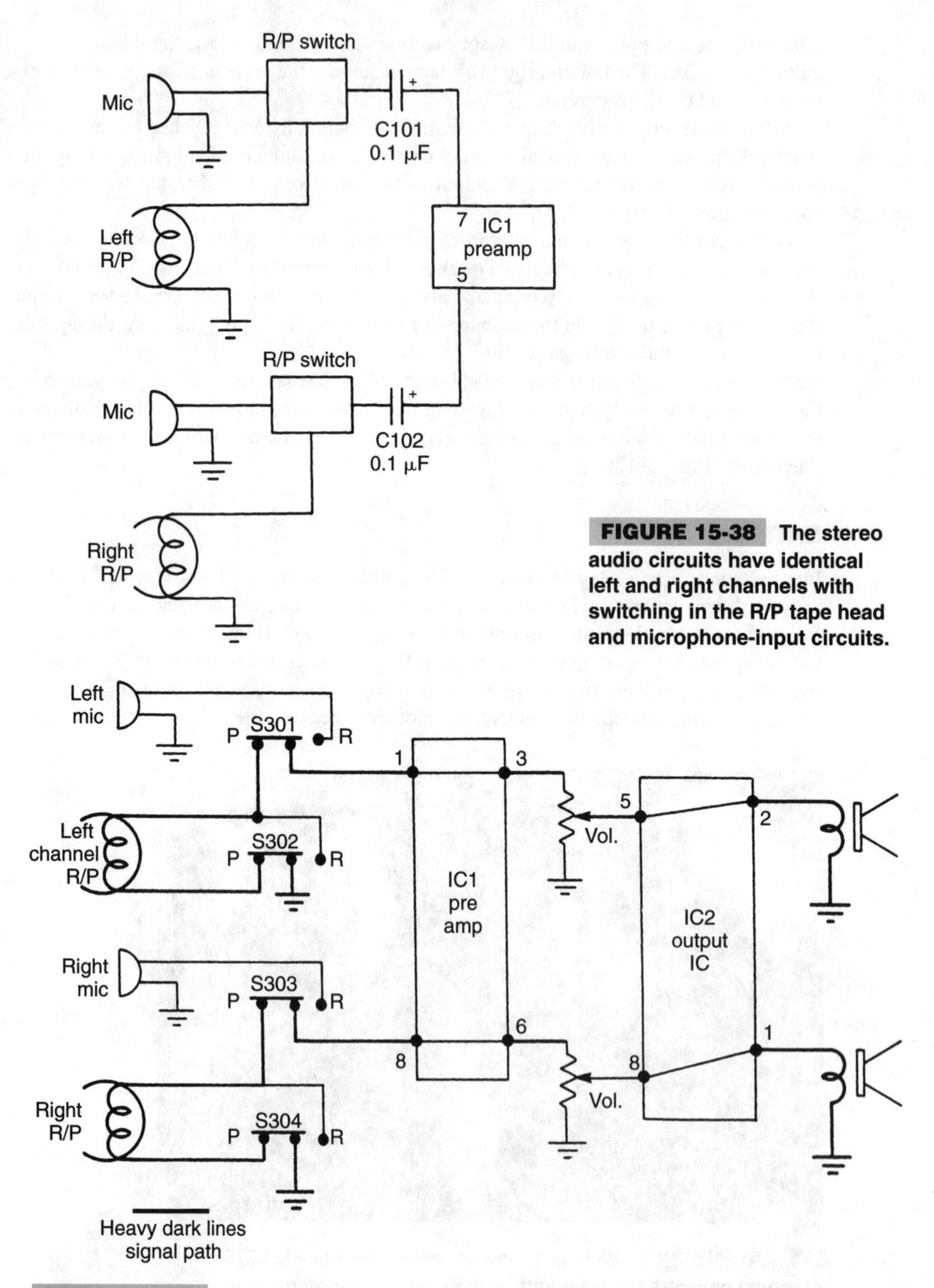

FIGURE 15-38 **The stereo audio circuits have identical left and right channels with switching in the R/P tape head and microphone-input circuits.**

FIGURE 15-39 **The dark lines show the audio signal from the tape heads to the speaker with audio ICs.**

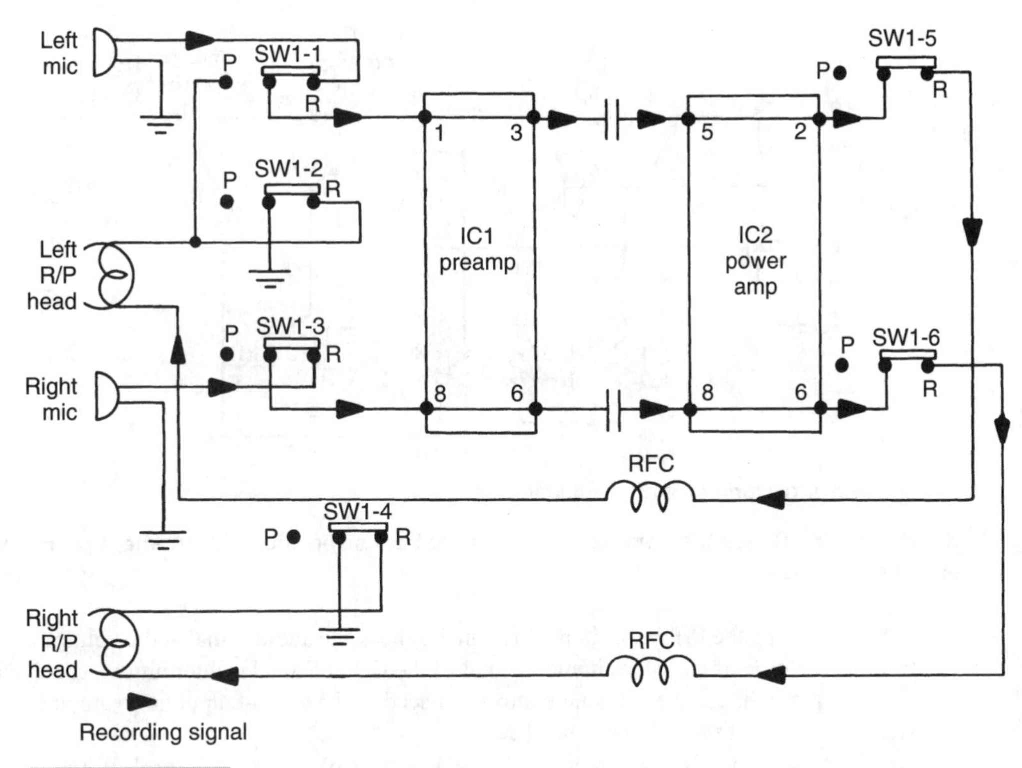

FIGURE 15-40 The dark arrows show the recording signal from the microphone to the R/P heads.

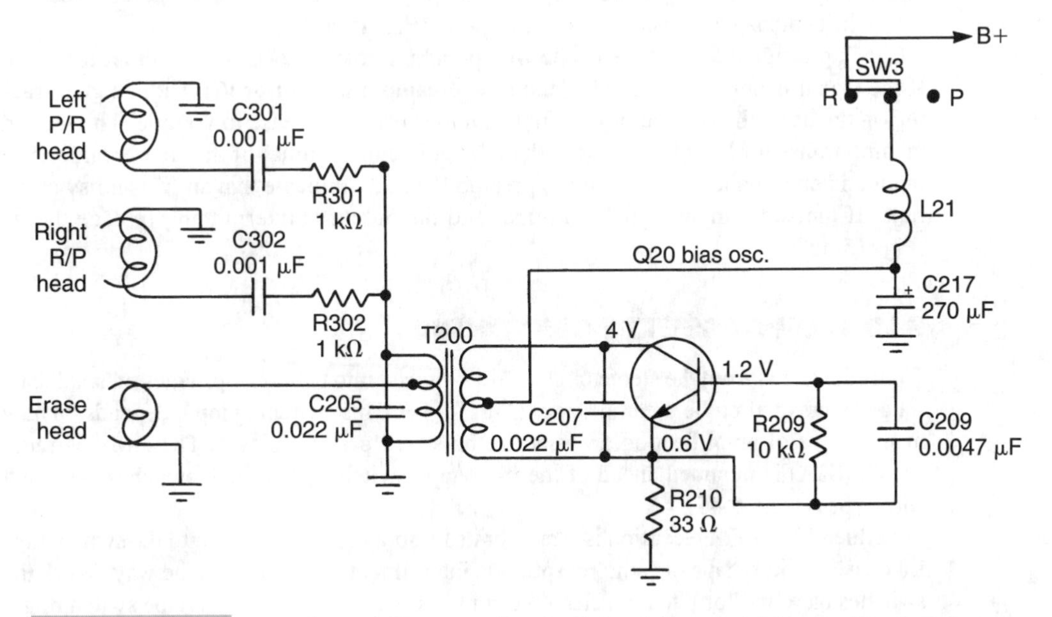

FIGURE 15-41 The bias oscillator excites both the left and right recording heads and the erase head.

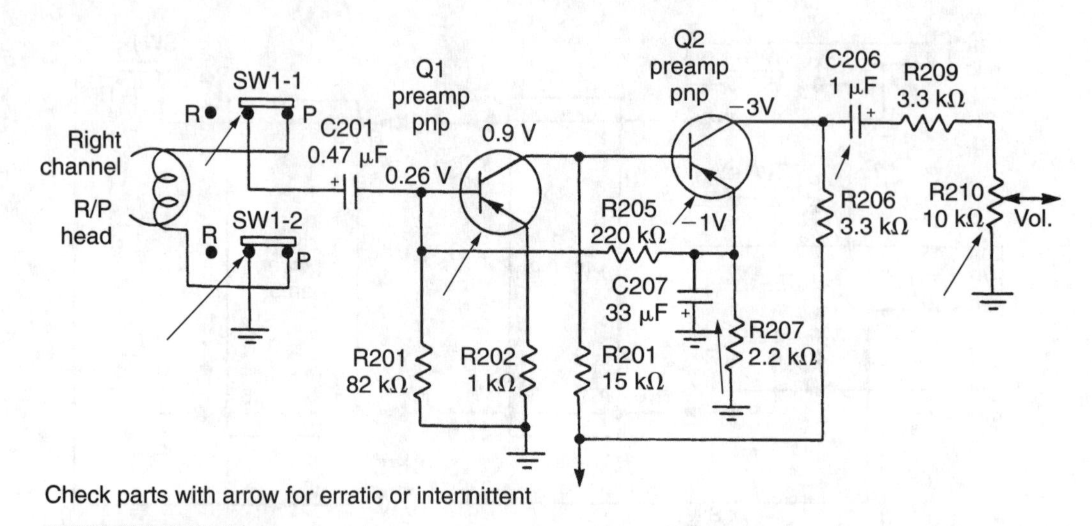

FIGURE 15-42 **Check the components that might cause erratic or intermittent problems in the audio channels.**

After cleaning the R/P heads in both channels, check the audio signal at the volume control for reception of the intermittent channel. If the right channel is intermittent, check the signal at the right-channel volume control. Proceed to the audio-output amp circuit if the signal is normal at the volume control.

If the audio signal is intermittent at the volume control, clean the control. With no improvement, signal trace the preamp circuits. Check for the intermittent audio signal at the tape head. You will have to crank up the volume on the outside amp or signal tracer because the audio signal is very weak at this point (Fig. 15-42).

Check the signal at the base of the first preamp transistor or at the IC input terminal. If the signal is normal, proceed to the next preamp transistor or IC. Audio signal tracing on the base, then collector terminal should show an increase in volume. The second preamp transistor has greater audio than the collector terminal of the first preamp. The input and output audio signals of the preamp IC might indicate that an IC is noisy or erratic. If the audio input signal is normal and the output is intermittent, replace the IC (Fig. 15-43).

RADIO OR CASSETTE SWITCHING

The audio signal from the stereo circuits are switched into the preamp stages on small cassette players or after the preamp stages (Fig. 15-44). This switching input depends on how many stages of amplification are used in the stereo cassette player. The AM/FM radio switch (S2A) is mounted ahead of the radio/tape switch. Often, both switches are small slide types (Fig. 15-45).

If either AM or FM reception is erratic or dead, spray cleaning fluid into the switch area. Likewise, if either tape or radio reception is intermittent, clean it the same way. Work the switches back and forth to help clean the contacts. The stereo FM radio/tape switch might be a DPDT slide switch (Fig. 15-46).

SINGLE STEREO IC

The personal stereo cassette player can have one complete IC for all audio circuits or have a separate power-output IC for both channels. Often, the dual volume control is located in the input circuit of the power IC. The audio amp signal is coupled to the volume control

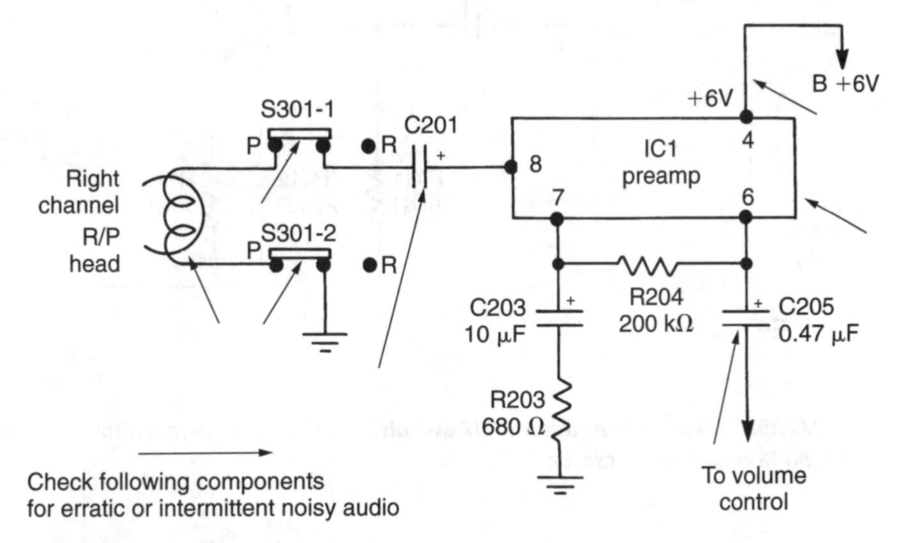

FIGURE 15-43 Check for erratic or intermittent noise in the audio channels.

FIGURE 15-44 Some personal cassette players have FM stereo, such as the Sony Walkman F1.

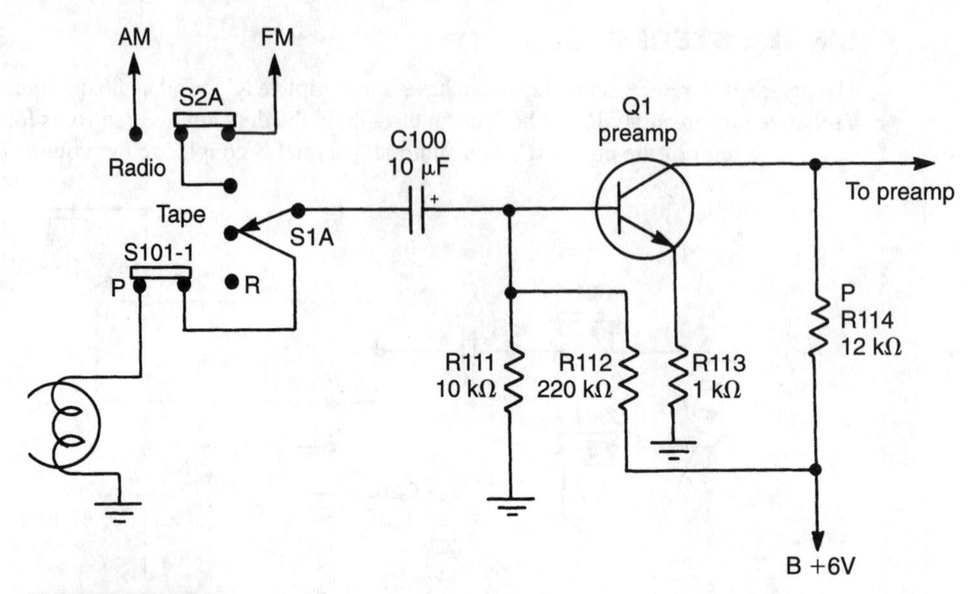

FIGURE 15-45 Clean the AM/FM switch (S2A) or radio/tape switch (S1A) if tape or AM/FM reception is erratic.

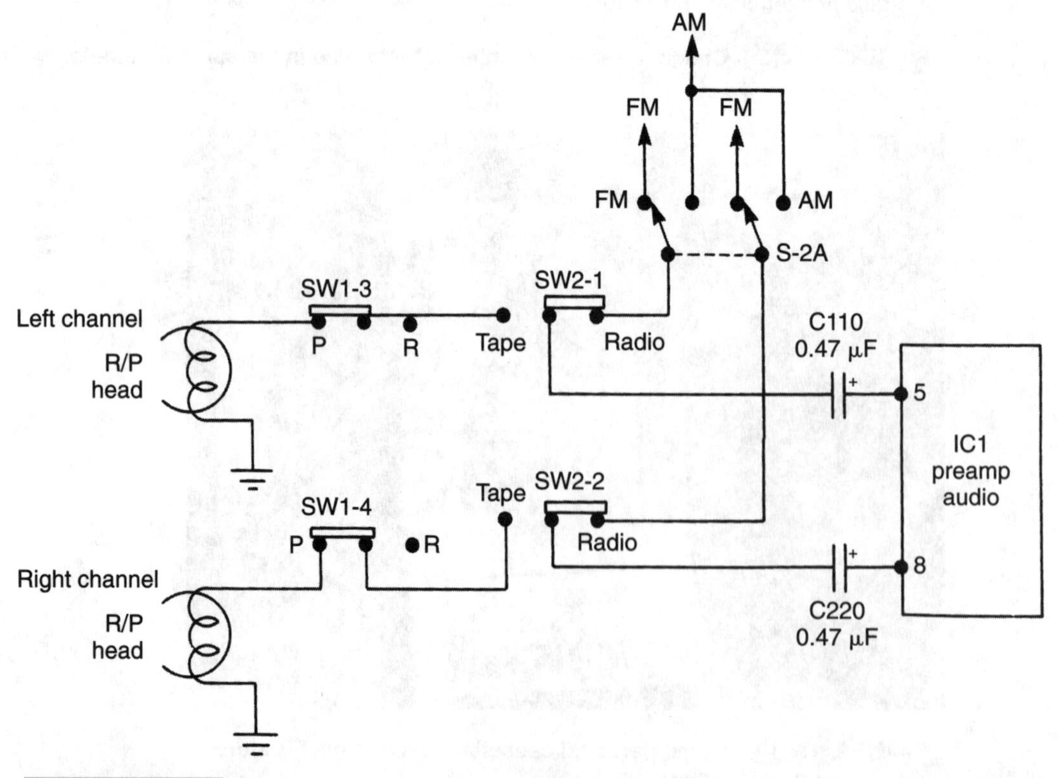

FIGURE 15-46 In FM stereo circuits, the FM stereo radio switch might be a DPDT switch.

FIGURE 15-47 The stereo portable might have one monitor speaker and a stereo headphone jack.

with a small electrolytic capacitor. Two separate capacitors couple the output audio signal to the separate speakers or headphones (Fig. 15-47).

In some players, each headphone switches in and out; the stereo speakers or one large stereo headphone jack does the same. Here, the loud audio signal is cut down with two separate 100-Ω resistors to the earphones (Fig. 15-48). Suspect that a switch contact is dirty if one audio channel is erratic or dead. Clean the headphone switch contact by squirting cleaning fluid into the plug hole. Work the headphone plug in and out to clean the points. The whole stereo jack must be replaced if one side is dead as a result of defective internal switch contacts.

POOR REWIND

The supply spindle or reel is engaged in the Rewind mode at a rapid speed. The tape is rotated backwards. Clean the surfaces of the spindle drive, idler wheel, and belt drive surfaces. Check for dry spindle and idler wheel bearings. While inside the unit, wipe off all moving surfaces with alcohol and a cloth.

The supply spindle can be fastened to the bearing shaft with a metal, plastic, or fiber "C" washer. Remove the small washer and slip off the spindle or supply reel. Clean the reel surface with rubbing alcohol and place a drop of light oil on the bearing. Wipe up the excess oil. Replace the supply spindle (Fig. 15-49) if it is broken. If the drive surface spindle is worn or cracked, replace it.

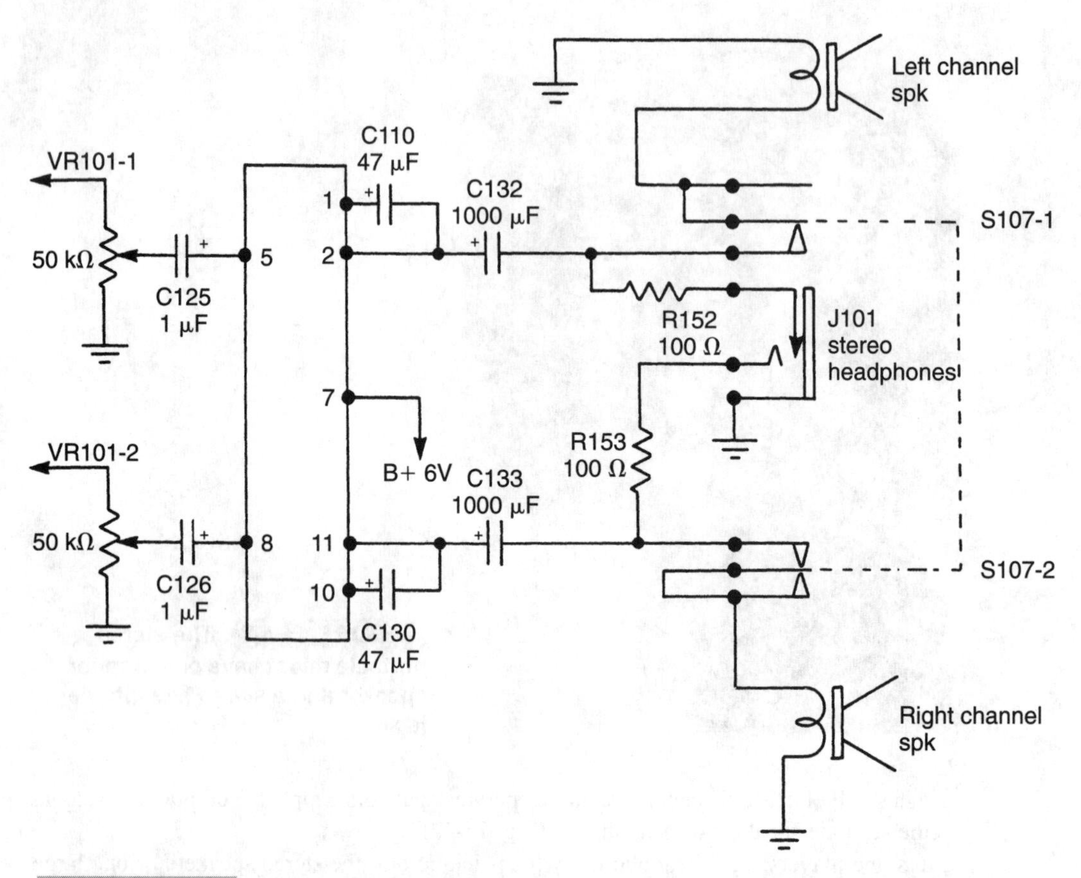

FIGURE 15-48 The dual-stereo headphone jack switches in two 100-Ω resistors for headphone listening when the plug is inserted.

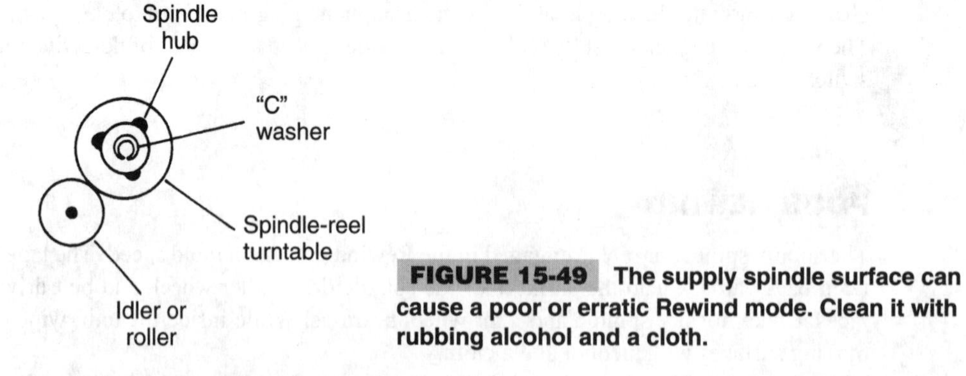

FIGURE 15-49 The supply spindle surface can cause a poor or erratic Rewind mode. Clean it with rubbing alcohol and a cloth.

DEAD CASSETTE PLAYER

When none of the cassette player's functions will operate, check and replace the batteries. Insert the ac adaptor to be sure that the batteries or terminals are defective. Doublecheck the polarity of the batteries. It's very easy to insert a battery in backwards. Inspect the battery terminal wires for breaks. Sometimes when the covers are replaced, a battery wire will be pressed against a sharp plastic edge and break into the battery wire.

Next, check the on/off leaf switch. Clean the switch contacts with rubbing alcohol and a cleaning stick. Test the switch contacts with the low RX1 range of the ohmmeter. When play is pressed, a direct short should be made across the switch (Fig. 15-50). Inspect the leaf switch for correct alignment.

With no tape action or sound, suspect a problem with the batteries or the on/off switch. Rotate the volume control to hear a noise in the speaker. If noise can be heard, the amplifier might be normal with no tape rotation. Check for dc voltage at the small motor terminals. No voltage might indicate that a wire is broken, a motor switch is dirty, or that the regulated motor system is faulty.

DEFECTIVE MOTOR

The small motor might be intermittent, erratic, slow, or dead. The intermittent motor might start one time and be dead the next. Then, when the motor pulley is rotated the motor will begin to run. Erratic motor rotation is often caused by a worn or dirty commutator. These motors are so small to take apart. Sometimes, with patience and care, the armature and wire tongs can be cleaned (Fig. 15-51). Other times, you can tap the outside metal belt of the motor and it will resume speed.

REPLACE DEFECTIVE MOTOR

The small end bearings can be dry and cause slow or erratic rotation. A squirt of light oil in each nylon plastic bearing can help. Be careful not to lose the small carbon brushes or end washers. Some motors can be repaired and others cannot be taken apart. Often, the defective motor is replaced instead of attempting to repair it. The small motor can be held on the main chassis with three end bolts (Fig. 15-52). Replace the motor if a normal dc voltage is applied at the motor terminals and it still doesn't rotate.

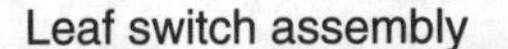

FIGURE 15-50 **Inspect the on/off leaf switch for dirty contacts or tongs or tines that are bent out of line. The leaf switch is under the PC board in Panasonic's RQ-L315.**

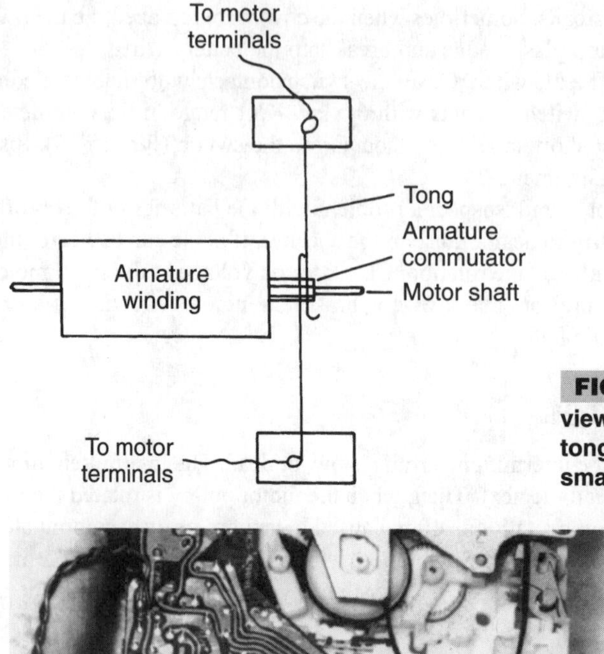

FIGURE 15-51 The inside view of the commutator and wire tongs or brushes that are on the small armature of the motor.

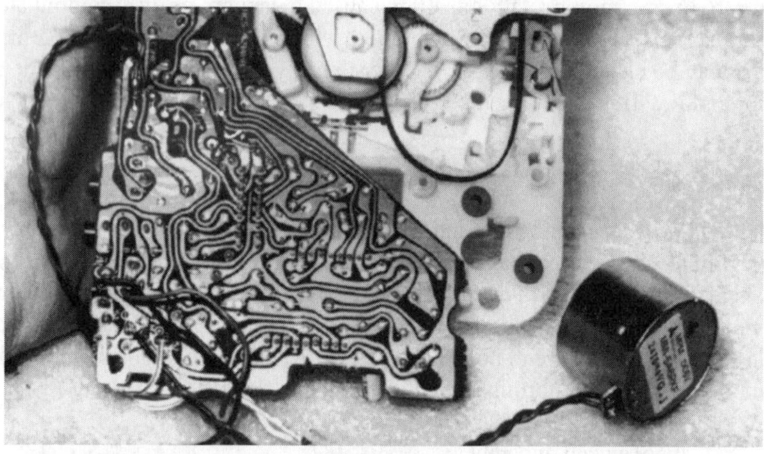

FIGURE 15-52 Three small bolts hold the motor to the metal chassis with the rubber grommet support.

TAPE SPILL OUT

Check for no movement or an erratic movement of the take-up reel if the tape spills out of the cassette as it plays. If the machine is not shut off within a few seconds, the tape can wind around the capstan and pinch roller. Usually, the cassette can be difficult to remove when the tape is wrapped around the capstan. Sometimes the tape must be loose to remove the cassette (Fig. 15-53).

Simply play the recorder without a cassette in the holder. Notice if the take-up reel stops or rotates unevenly. If so, clean the rubber drive surface on the bottom of the reel and also clean the drive idler assembly. If the take-up reel is driven with gears or a belt, clean them. Be sure that the take-up reel is running smoothly before trying another cassette.

If the take-up reel is rotating normally, suspect that a pinch roller is worn or uneven. Clean the pinch roller with rubbing alcohol and a cloth. Sometimes, a sticky substance or too much oxide on the pinch roller can cause tape to pull out. Remove all tape that is wound around the pinch roller bearing. Extra tape will make the pinch roller run slow with poor playback (Fig. 15-54).

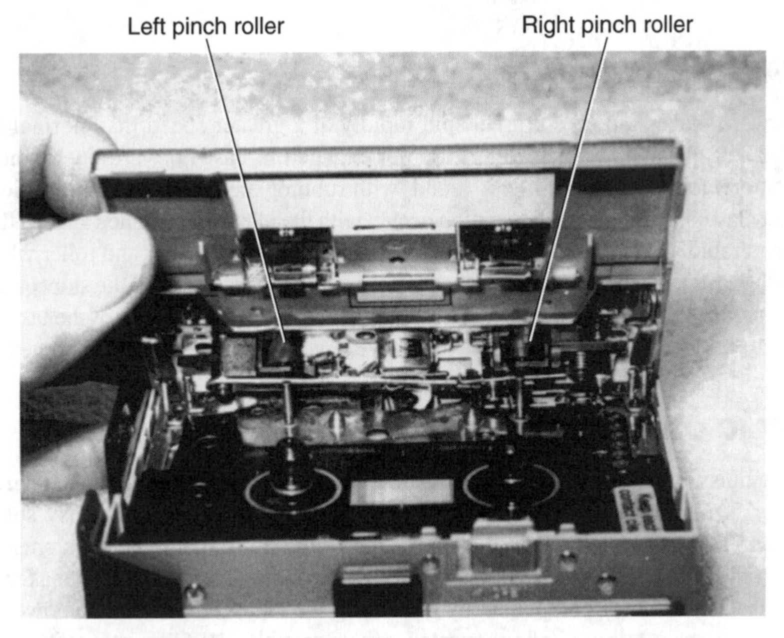

FIGURE 15-53 **A Toshiba KTA51 stereo cassette player has two top-side pressure rollers with auto-reverse action.**

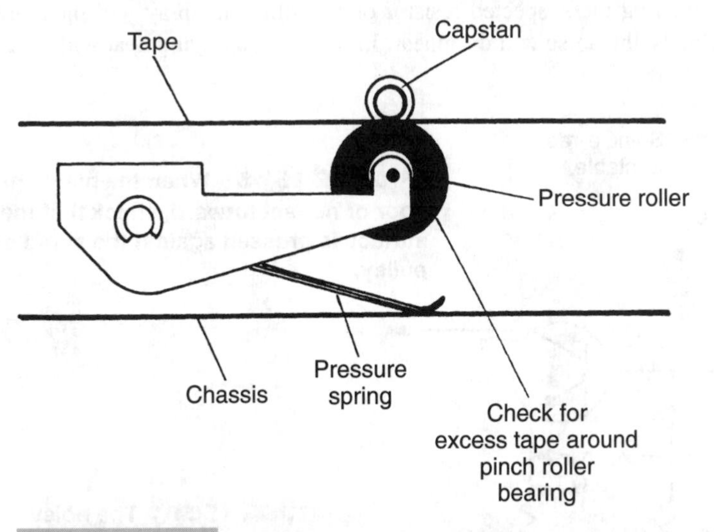

FIGURE 15-54 **When tape spills out of the cassette, check across the pressure roller bearings for excess tape. Remove it because the excess tape will slow down the tape action.**

Do not overlook the possibility that a cassette is defective if tape spills out of a certain cassette. Try another one and compare the results. The tape can be wound loosely or be crimped. A squeaky cassette can be caused by dry plastic hub bearings. Insert talcum powder into hub bearing with a toothpick.

NO FAST FORWARD

Fast Forward makes the take-up reel spin rapidly at a greater speed than normal Play or Record. The capstan and pinch rollers are not engaged in Fast-Forward operation. Clean the Fast-Forward drive belt, if one is found, with rubbing alcohol and cleaning stick.

In most players, Fast-Forward operation occurs with the idler roller pushed against the take-up reel at a rapid speed. Clean all idler and roller surfaces. Check idler and roller wheels for dry bearings. Slippage between the idler and reel surfaces causes most of the slow or no Fast-Forward problems. Notice if the idler wheel is engaging the bottom roller of the take-up reel assembly (Fig. 15-55). Clean or replace the Fast-Forward belt that is found in some players.

NOISY IC OR TRANSISTOR

Often the noisy transistor or IC can be located in the input and output sound stages. The hissing or frying noise that occurs with low recordings can indicate a noisy solid-state component. Lower the volume and listen for the frying noise. If the noise is present, you know that the defective component is between the volume control and the speaker.

Try to isolate the noisy component by grounding the input terminal of the power-output IC. Ground the base terminal of a suspected transistor with a 10-Ω resistor (Fig. 15-56). If the noise lowers or disappears, you know that the defective component is before this stage. If the noise is still present, replace the noisy IC.

Sometimes spraying the suspected resistor or IC with cold spray will make the noise louder. Other times, the noise will disappear. In this case, applying heat with a hair dryer

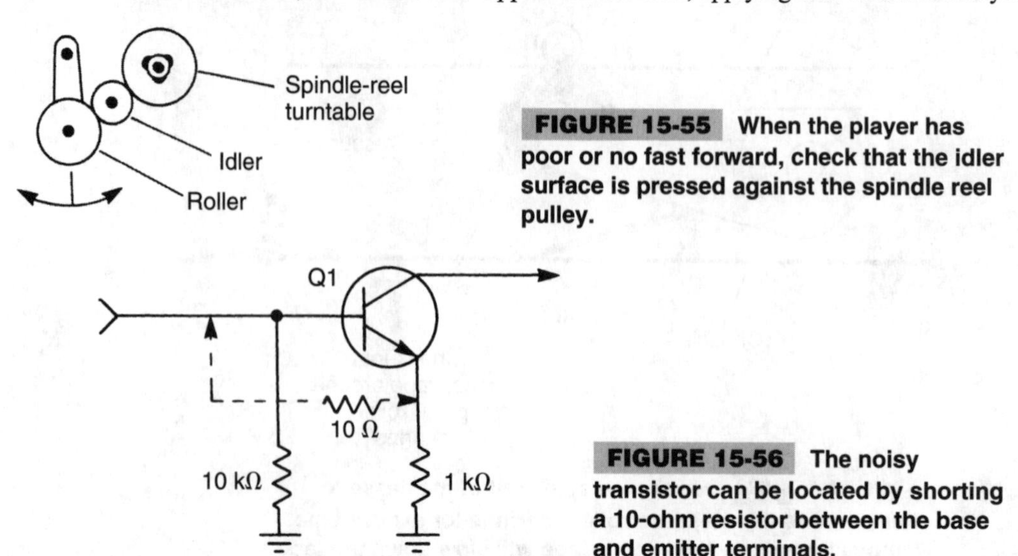

FIGURE 15-55 When the player has poor or no fast forward, check that the idler surface is pressed against the spindle reel pulley.

FIGURE 15-56 The noisy transistor can be located by shorting a 10-ohm resistor between the base and emitter terminals.

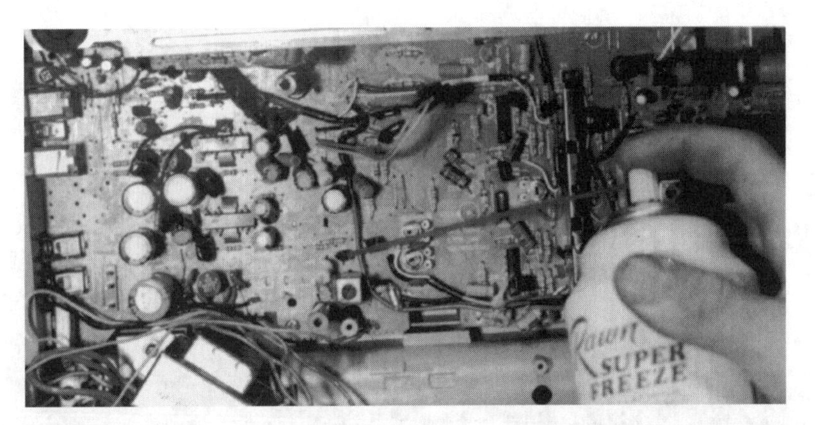

FIGURE 15-57 **Spray the noisy IC or transistor with a cold spray to make it act up or quiet down, to determine if it is defective.**

on the suspected transistor or IC will make the noise reappear after applying cold spray (Fig. 15-57). Do not overlook small ceramic bypass capacitors that can create noise with a B+ voltage on one side. Replace the noisy component with a good part.

If the noise disappears with the volume control turned down, the noisy component lies ahead of the volume control. Sometimes grounding the noisy preamp transistor base with a resistor or shorting the emitter resistor can cause the noisy transistor to quit making noise. Start at the preamp input transistor and proceed through the circuit. If the noise is present after grounding out the first preamp signal, the second preamp transistor must be noisy.

Usually, the noisy condition occurs in only one stereo channel. If both channels are noisy, suspect a problem with the stereo output power IC. The noise might disappear for several days, then reappear. Replace the power output IC if a loud frying or hissing noise is there at all times. A poor internal transistor or IC junction produces this type of noise.

The Personal Portable Cassette Player

Although only a few portable table-top cassette players are manufactured, thousands are still in the field (Fig. 15-58). The cassette player with large pushbuttons was the workhorse of yesterday. Besides regular tape-recorder troubles, the large pushbuttons caused many problems. Today's portable cassette recorder has most features, including one-touch record, auto-stop, cue and review, tape counter, and built-in condenser microphone. Many of these cassette recorders operated from four C batteries (Fig. 15-59).

PUSHBUTTON PROBLEMS

To engage the pushbuttons, extra pressure must be applied. If the switch became stuck or sluggish, the plastic button snapped off. Simply repair the cracked or broken plastic button by cementing it into position.

FIGURE 15-58 The table-top portable cassette player was yesterday's recording workhouse.

FIGURE 15-59 This Panasonic RQ-2103 cassette player operates from four C batteries.

A dry lever or shifting bar prevented the button from holding down. Clean the sliding bar and lever with rubbing alcohol and a cleaning stick. Apply light phone grease or oil to the sliding lever bar. Straighten the bent pushbutton lever with long-nose pliers. Sometimes, the button would pop out of the knob assembly when it was pressed (Fig. 15-60). Bond the plastic knob and metal together with speaker or model airplane cement.

DIRTY FUNCTION SWITCH

Some of these large portable cassette players had a horizontal sliding function switch. Like all sliding switches, the contacts would become dirty and tarnished, making intermittent or no switch contact. Clean them by spraying cleaning fluid in each end of the function switch. Work the switch back and forth by pushing down on the Play and Stop buttons. A lot of Record buttons were broken when the button was pressed without cassette in the holder. More pressure was applied and the button gave in.

Because these switches were fairly large, they had a tendency to cause intermittent or erratic electronic functions. It's best after clean up to solder each switch connection on the PC board. Inspect for cracked or broken wiring around the switch area.

NO PLAYBACK LOCK IN

Suspect that a lever assembly is bent or dry if the player will not go into any mode or seat properly. If the Play button will not seat, inspect the lever and catch of the sliding bar. The sliding bar is pushed to one side. When the button is seated, the spring pressure applies against the bar and locks the button into the hold position. Check for weak or loss of spring action on the lever bar. Sometimes, the small clip that locks the metal button tap is worn or bent out of line, which releases the button before it is locked into position.

Check for foreign objects in the loading mechanism if any one of the buttons will not function. Like all cassette loading assemblies, the tape heads are pushed forward and locked into position. Clean the loading mechanism and apply light grease on the sliding loading assembly. Excess dust and dirt can fall down into this area, which produces a dry or gummed up mechanism.

NO TAPE ACTION

If the spindles do not rotate and the motor does not sound like it is running, remove the bottom cover. Only four or five small screws hold the bottom plastic cover. Look for a small screw inside the battery compartment. Tape the batteries into the compartment, if needed,

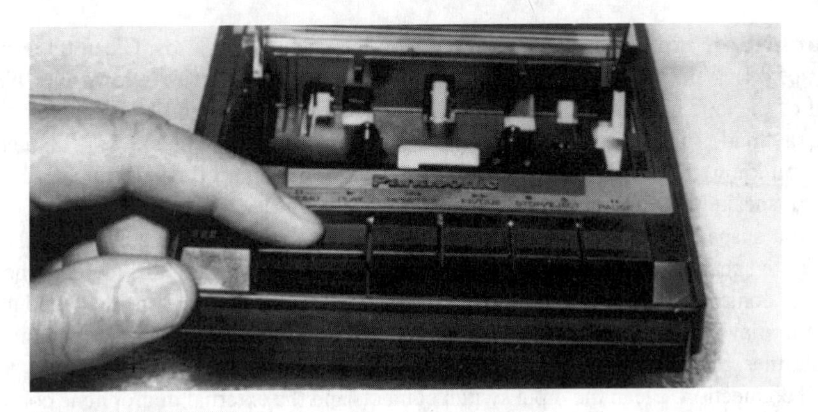

FIGURE 15-60 Cement broken plastic buttons with speaker or airplane glue.

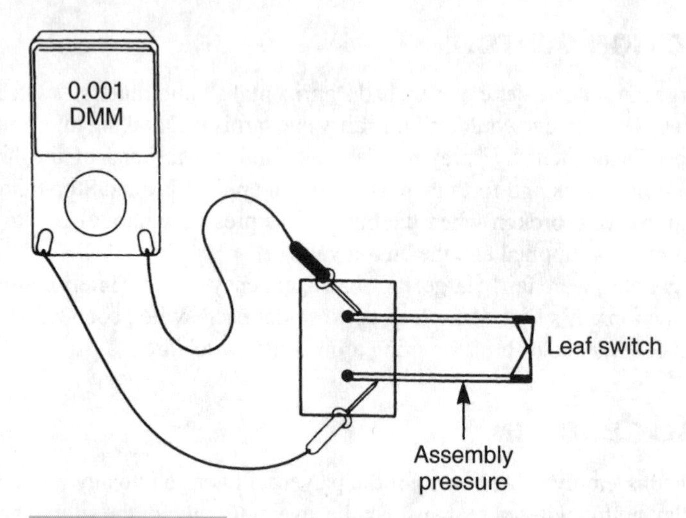

FIGURE 15-61 **Check the switch contacts for a direct short in the On position, with R × 1 range of the DMM.**

and press the Play button. Notice if the motor is rotating. Remove the motor belt and give the motor plug a quick spin.

Suspect a dirty or defective on/off leaf switch if the motor and amp are dead. Clean the leaf tines with cleaning solution. Check the switch contacts with the RX1 range on the ohmmeter (Fig. 15-61). Measure the voltage that is applied to the small motor at the PC board. A normal motor voltage indicates a defective motor. These small universal motors are sold at electronic and mail-order firms. Check to see if the cost of the motor is more than the recorder is worth before ordering.

Remove the motor terminals from the PC board. Check the motor continuity with the ohmmeter. The resistance should be less than 10 Ω. Suspect that brushes or commutator slip-ring connections are poor if the motor is open. Rotate the motor pulley and notice if the resistance changes. A dirty commutator can cause intermittent or erratic motor rotation. Often, three small screws hold the small motor into position (Fig. 15-62).

Intermittent audio channel First, clean the stereo tape heads. Clean the earphone and speaker jacks. If you still can't hear any sound, insert a small screwdriver blade in front of the tape head while it is in the Play mode. When the metal blade passes the tape head, a "thud" should be heard in the speaker with volume wide open. If the tape is rotating, you know that power is applied to the motor and amp sections. Eliminate the intermittent speaker by plugging in the headphones.

If you suspect the speaker, clip another speaker across the voice-coil terminals or remove the speaker leads and solder them to another good speaker (Fig. 15-63). Check the volume control for noisy or intermittent sound. Spray cleaning fluid inside the control.

If the player is still intermittent, check the tape head for loose or broken connections. Sometimes, by pressing the pencil eraser against each head terminal, it can show a poor internal connection. Clean the input switch contacts and the external microphone connection for intermittent recording or playback (Fig. 15-64).

Inspect the sine wave at the tape head and notice if the sound is erratic or intermittent. If so, proceed to pin 14 of the IC1. If an intermittent is at the input terminal, suspect that IC1 or speaker-coupling capacitor C15 is faulty. Monitor the B+ (6 V) applied to pin 9 and notice if the voltage changes or is removed from the circuit. Suspect that the contacts on the remote switch jack are dirty or poor if both motor and amp voltages are intermittent.

MUFFLED OR DISTORTED SOUND

Often, distorted sound is caused by a dirty tape head, a bad audio output component, or a bad speaker. In the early cassette players without recording features, the tape head was

FIGURE 15-62 This defective motor was removed from the chassis by removing three chassis bolts.

FIGURE 15-63 Disconnect the speaker terminals and clip a good speaker to the wires to determine if the speaker is intermittent.

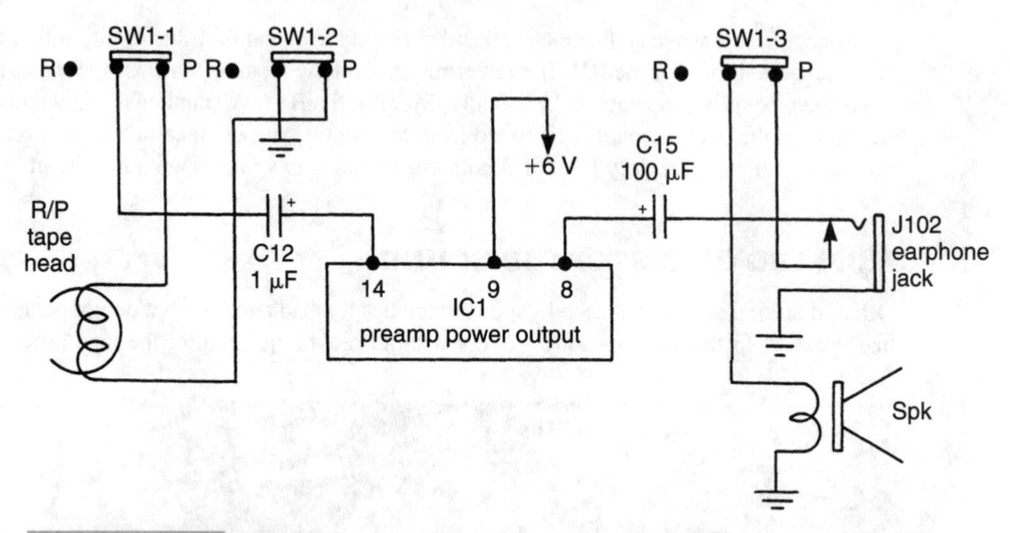

FIGURE 15-64 Clean input switch terminals SW1-1, SW1-2, and SW1-3 for dirty contacts and intermittent recording or playing.

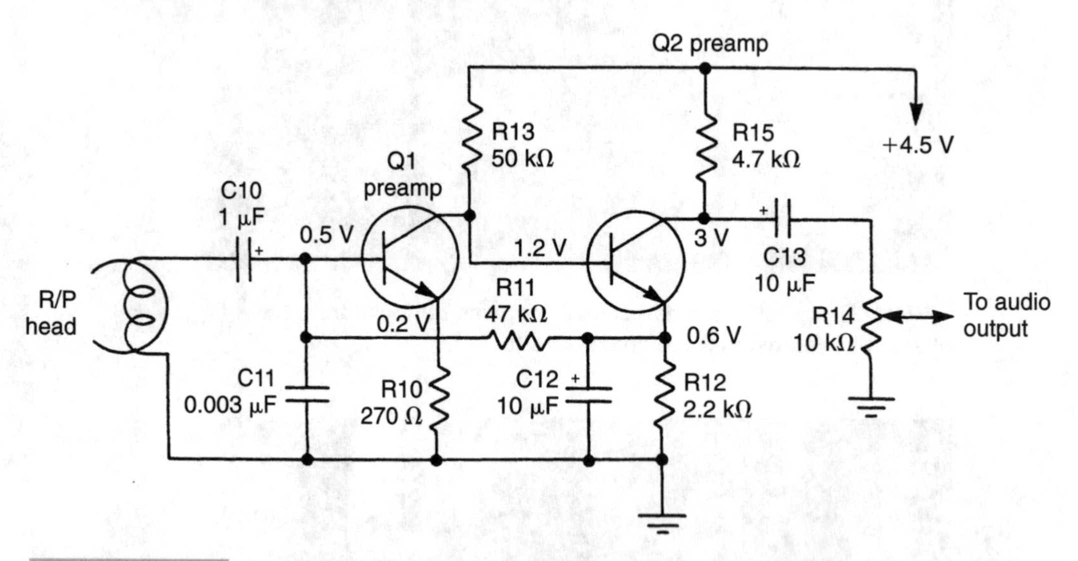

FIGURE 15-65 No switching or recording is in the early portable cassette players that have transistor preamp stages.

coupled directly to the audio preamp transistor (Fig. 15-65). No switching is found in the input circuits. The only switching in the push-pull output circuits are the headphone and speaker circuits. Clean the tape head that has playback distortion.

If the distortion is still present, clip another speaker in the output to determine if the speaker is defective. Next, check the voltages on the output transistor for poor bias or leakage (Fig. 15-66). Test each output transistor for leakage or shorts. Inspect the bias resistors on the base and emitter circuits with the transistors removed from the circuits. Do not overlook the possibility that coupling capacitors are shorted or leaky between the preamp and AF driver transistors.

Distortion or muffled sound within the IC1 circuits might be caused by a dirty tape head, a leaky IC, bad coupling capacitors, bad bias resistors, or a bad speaker. Take accurate voltage measurements on every IC terminal. Check the voltage on both sides of C2 and C10 (Fig. 15-67). Replace IC1 if the voltages are fairly normal and speaker has been tested for muffled or distorted sound. Compare the distorted sound within the stereo circuits by comparing the input and output circuits of both stereo channels.

ERRATIC COUNTER

Suspect jammed gears, a loose belt, or a gummed-up tape counter if the counter sometimes works. Often, the tape counter is belt driven from the supply reel. To check it, press the reset to rotate the reel numbers back to zero (Fig. 15-68). Rotate the supply spindle counterclockwise by hand. Notice if the indicator numbers start with number 1 and increase as the supply reel is rotated.

Suspect a broken or loose belt if no numbers appear on the counter. Remove the bottom cover and the PC board. Inspect the small counter belt. These belts are smaller than a regular rubber band. If the indicator gears are stuck, remove the indicator assembly and spray

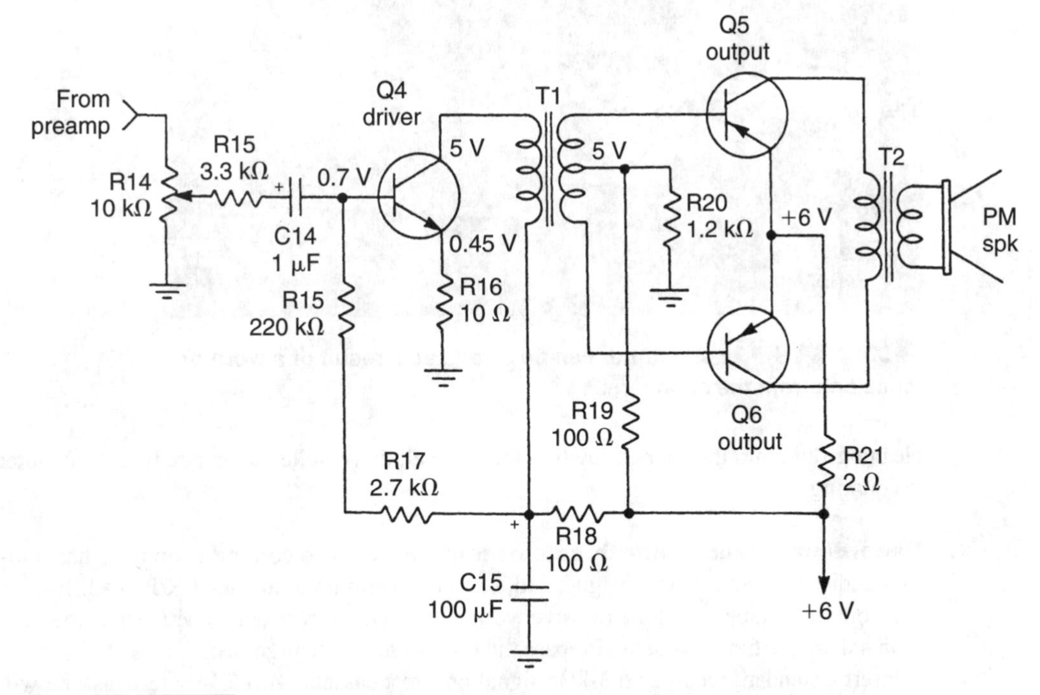

FIGURE 15-66 **Distortion is often found in the audio-output stages. Check the transistors and take accurate voltage measurements.**

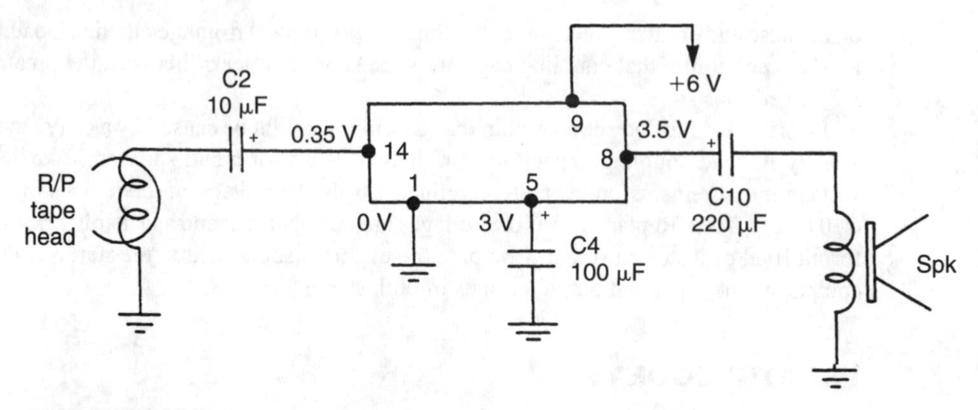

FIGURE 15-67 Check all voltages on the suspected IC and be sure that all components tied to the IC terminals are normal.

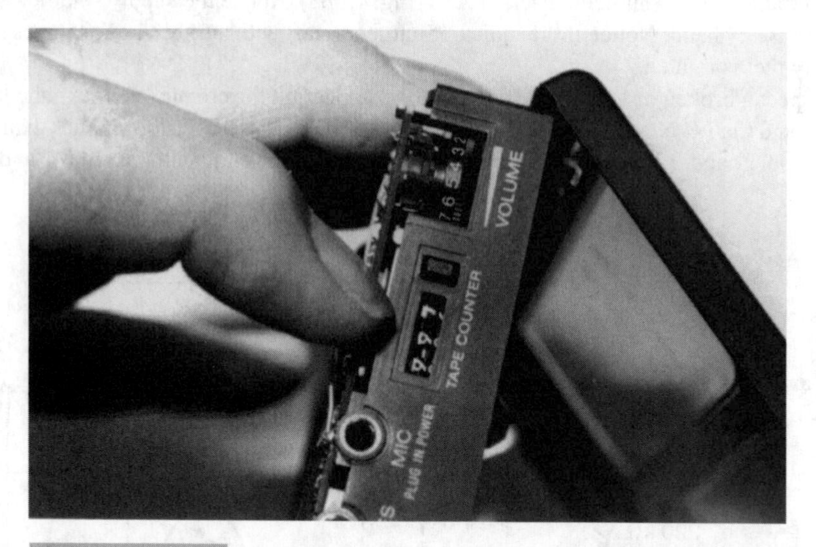

FIGURE 15-68 A counter can be erratic as a result of a worn or loose belt from the supply reel.

cleaning fluid into the gears. Slowly rotate the belt drive pulley to ensure that the counter is counting.

Record/play head azimuth adjustment When the recording is tinny or has muffled audio, suspect a worn, magnetized, or out-of-azimuth adjustment R/P head. Inspect the front of the tape head for excessive worn marks. Degauss or demagnetize the tape head with a demagnetizing cassette. Prepare the unit for azimuth alignment.

Insert a standard recorded 6.3-kHz signal on a test cassette. A 3.5-kHz test cassette will also do the job. Load the earphone jack with an 8-Ω 10-W resistor (Fig. 15-69). Clip the DMM test leads (ac voltage) across the resistor. Adjust the azimuth screw for a maximum reading on the meter. The correct azimuth screw is alongside of the tape head with a small spring underneath one side of tape head (Fig. 15-70). Sometimes this azimuth screw can be reached through a small hole in the plastic cabinet.

FLYWHEEL THRUST ADJUSTMENT

The motor pulley and belt drives a capstan flywheel, which, in turn, rotates the tape at the pinch-roller assembly. Sometimes the end play can become excessive and cause improper speeds. The flywheel can have a small adjustment screw in the metal brace that holds the capstan in place (Fig. 15-71). The metal brace can be bolted to the metal chassis.

Remove the two metal screws alongside the chassis-holding bracket. Check for another screw at the other end of the bracket. Remove the capstan thrust plate. Clean the capstan bearing and thrust plate with rubbing alcohol and a cloth. Remove the capstan/flywheel assembly and clean the capstan-bearing hole with rubbing alcohol and a cleaning stick.

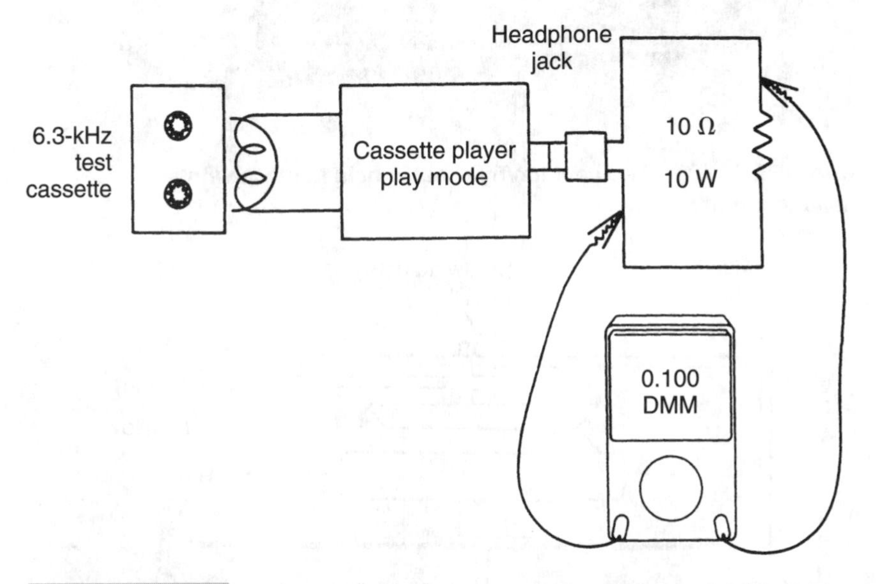

FIGURE 15-69 Adjust the azimuth screw while playing a 6.3-kHz test cassette for maximum reading on the ac DMM.

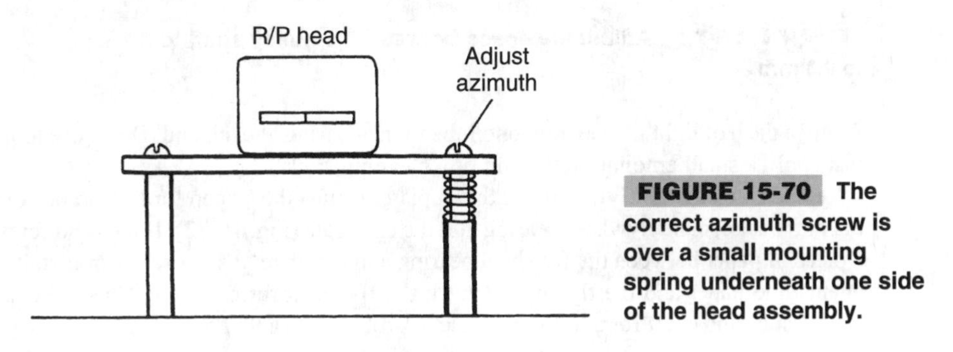

FIGURE 15-70 The correct azimuth screw is over a small mounting spring underneath one side of the head assembly.

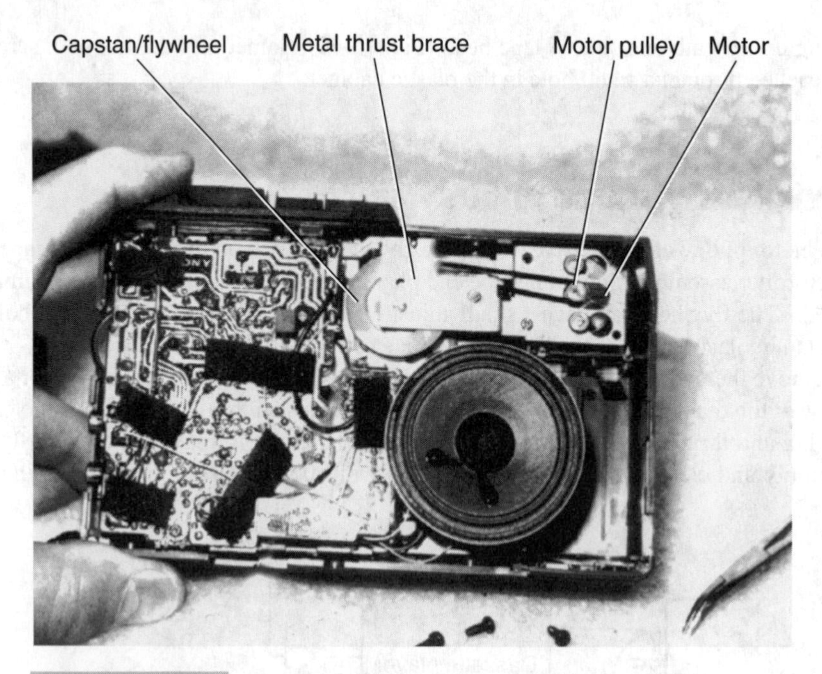

FIGURE 15-71 **The capstan/flywheel is held in place with a metal thrust bracket.**

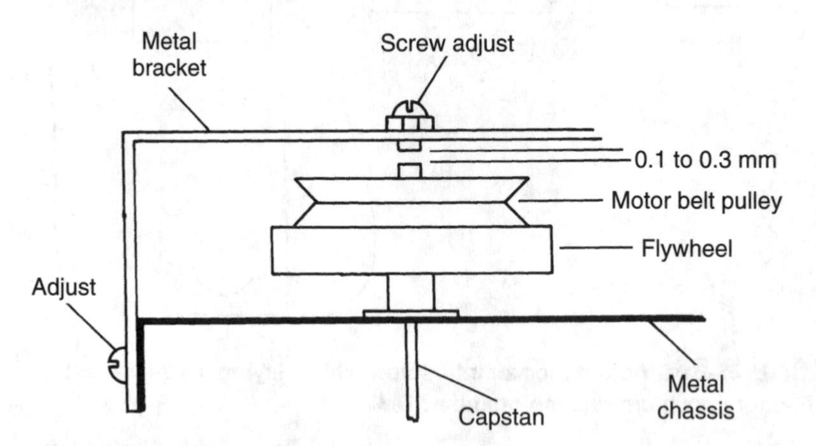

FIGURE 15-72 **Adjust the space between the pulley shaft to 0.1 to 0.3 mm.**

Apply a dash of light grease or phono lube to the bearing at each end. Don't use too much, just apply a small amount on the end of a cleaning stick.

Replace the capstan/flywheel and thrust plate. Adjust the screen for a clearance of 0.1 to 0.3 mm between the flywheel bearing and the end plate (Fig. 15-72). If it has no screw, slip a piece of paper between the flywheel bearing and the plate. Adjust screw or metal bracket against the paper and the flywheel. Check the flywheel for end play. Check the cassette troubleshooting chart for additional cassette problems (Table 15-1).

TABLE 15-1 CASSETTE TROUBLESHOOTING CHART

PROBLEM	CAUSE	HOW-TO-FIX
No sound, no functions	1. Ac power cord	Repair cord
	2. Dead batteries	Check voltage
	3. Bad on/off switch	Check leaf switch for dirty contacts. Clean up.
No ac operation. Batteries okay	1. Defective battery interlock switch	Check with ohmmeter
	2. Leaky silicon diodes	Check diodes with DMM diode test.
	3. Bad ac cord.	Replace or repair ac cord.
No playback	1. No voltage at amp	Check voltage at leaf switch.
	2. Defective leaf switch	Replace broken switch.
	3. Defective transistor	Check transistors for leakage or open.
	4. Defective IC	Check signal in and out of IC.
	5. Bad or open switch	Check and clean contacts.
	6. Defective speaker	Check with ohmmeter for open and replace.
	7. Broken R/P head wires	Replace wires, then solder.
	8. Dirty tape head	Clean with alcohol and cleaning sticks.
No or poor recording	1. Cassette tab broken out	Tape over the opening.
	2. Dirty R/P head	Clean with alcohol.
	3. Defective R/P switch	Clean with cleaning spray.
	4. Defective mic	Replace it.
	5. Defective amp	Check amp voltage and test transistors and IC parts.
	6. Dirty mic jack	Spray cleaning fluid.
	7. No bias voltage	Check bias oscillator circuits.
Poor or no erase	1. Dirty, packed erase head	Clean erase head with alcohol and stick.
	2. Wires off head	Inspect head wires and solder them.
	3. Dirty record/play selector switch	Clean with cleaning spray.
	4. No bias voltage	If dc operated, check the bias volt-age.
	5. Defective bias oscillator	Check oscillator transistor and circuits.
Weak sound	1. Dirty tape head	Clean R/P head.
	2. Amplifier failure	Check supply voltage and amp components.
	3. R/P head switching	Clean switch contacts.
	4. R/P head not resting against tape	Check for missing screw.
	5. Improper aziumth adjustment	Readjust azimuth for maxium output.
	6. Defective tape head	Replace R/P head.
	7. Defective speaker	Check with working speaker.
	8. Mic failure in recording	Substitute another mic.
	9. Excessive or insufficient bias current	Check bias current.
	10. Open speaker coupling capacitor	Shunt with good capacitor.

TABLE 15-1 CASSETTE TROUBLESHOOTING CHART (CONTINUED)

PROBLEM	CAUSE	HOW-TO-FIX
Poor tone quality	1. Defective transistor or IC	Check with DMM.
	2. Dropped speaker cone	Replace speaker.
	3. Lower dc voltage	Check power supply or battries.
	4. Poor high frequencies in recording	Make azimuth adjustments.
	5. Dirty tape head	Clean with alcohol.
	6. Poor recording insufficient bias	Check bias oscillator circuits.
	7. Defective mic	Substitute another mic.
	8. Defective cassette	Try another cassette.
	9. Jumbled recording	Clean erase head—check wiring on the head and check the bias current.
Excessive noise	1. Defective transistor	Test the circuit.
	2. Defective IC	Check input and output signals, also check voltages on the IC.
	3. Worn volume control	Clean or replace it.
	4. Poor wiring connections	Check the wiring and solder the joints.
	5. Motor rotating	Replace noisy motor.
	6. Hum noise	Replace bad filter capacitors in the power supply.
	7. R/P head magnetized	Demagnetize the tape head.

16

TROUBLESHOOTING BOOM-BOX CASSETTE PLAYERS

CONTENTS AT A GLANCE

The first large portable AM/FM cassette player contained a cassette player, an AM/FM radio, and a large speaker (Fig. 16-1). After a few years, the AM/FM radio contained a stereo receiver, left and right audio channels, and two stereo speakers. The boom box appeared with high-powered amps and separate speakers. Today, many portable stereo cassette players have detachable speakers and dual cassette players.

FIGURE 16-1 A large 5-inch pin-cushion speaker in the Panasonic RQ-542AS AM/FM radio cassette recorder.

The average boom-box portable cassette recorder has an AM/FM stereo tuner, built-in condenser microphone, automatic shutoff, cue and review controls, automatic recording level, high-speed dubbing, and headphone jacks. The boom box can be operated from line voltage or from six C batteries.

The deluxe portable has a digital synthesized tuner with 10 to 20 presets, auto reverse, clock timer, a four- or five-band graphic equalizer, megabass system, dual cassettes, auto recording level, automatic shutoff, a two-way four-speaker system, external microphone input jacks, Dolby B, and up to 10 W of output power. The deluxe portable can operate from six C or eight D batteries (Fig. 16-2).

Many of the latest boom-box players have two different cassette decks. Tape deck 1 is used for recording and playback. This deck can also record from cassette deck 2 or duplicate deck 2. Continuous playback of both decks can be provided with tape deck 2 first and deck 1 last. Each cassette deck can be controlled by its own pushbuttons (Fig. 16-3).

Often, dual tape decks are mounted side by side or one at the top and the other at the bottom. In Sharp's WQ-354, both tape decks are mounted in the cover housing. If the Stop/Eject button is pressed, the cover assembly moves outward, then down, showing tape deck 2 in front of deck 1. When continuous playing is requested, both decks are loaded with favorite cassettes and the door is closed. In this model, long spindles stick out and extend through both cassettes for playback.

Continuous Tape Playback

For extended tape playback of both cassettes, load both cassette holders. Place the tape you want to play first in deck 2. Then, place the second cassette in deck 1. Rotate the function switch to the tape position. Press Play on deck 2, then press Pause. Now, press Play for tape deck 1. Be sure that the dubbing switch is off. When deck 2 has completely played, the automatic Stop button in deck 2 engages. The Pause button will release and tape deck 1 will begin to play. When deck 1 has completed, the automatic stop shuts off the player. Always follow the manufacturers' instructions by the numbers with continuous-play operations.

Three-Band Graphic Equalizer

The three-slide switch lever control permits you to tailor the bass, mid-range, and treble frequencies to your listening pleasure. More controls are added in players that have the

FIGURE 16-2 The large boom box might operate from 12 Vdc or from eight D cells.

FIGURE 16-3 This GE boom box has dual cassette decks and an AM/FM-MPX stereo amplifier.

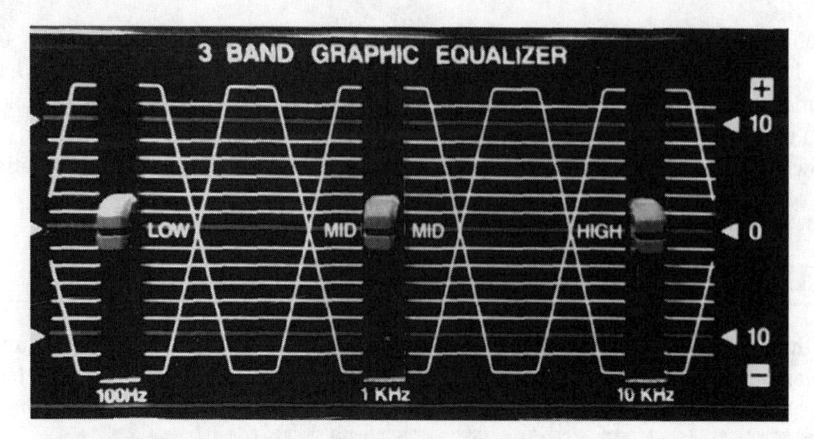

FIGURE 16-4 **This 3-band graphic equalizer is found in a GE 2+2 dual boom box.**

four- and five-band equalizers. Simply moving a control above zero emphasizes the frequency response; moving it below zero deemphasizes the frequency response. The low bass frequency range is 100 Hz, mid-range is 1 kHz, and the high treble frequency response is 10 kHz (Fig. 16-4). A five-band equalizer can cover the 100-Hz, 330-Hz, 1-kHz, 3.3-kHz, and 10-kHz frequencies.

Oscillator Switch

Some large cassette boom boxes or cassette decks have an oscillator switch that is located on the back to eliminate background interference or whistling when the radio is turned to the AM band for recording operation. When recording from the AM radio, slide the oscillator switch to either the A or B position to minimize the interference (Fig. 16-5).

Variable Monitor

A few cassette players, while recording from the radio program, can monitor the radio and have no effect on the recording via the AUX jack. The volume, balance, and graphic-equalizer controls can be adjusted to what level you prefer. In some players, the built-in ALC will set the record level for optimum recording level, regardless of volume, balance, or equalizer settings.

Dubbing

Some players have a dubbing feature, which enables you to dub (duplicate) a recorded tape cassette in tape deck 2 onto a blank cassette that is placed in tape deck 1. Do not copy or record copyrighted cassettes; that infringes upon copyright laws. Some cassette players have a high-speed, normal, and off switch. With the switch in the high-speed dubbing position, the recording rate is twice the normal tape speed.

Remember, tapes dubbed at high speed will be the same as those recorded at normal speed. Follow the manufacturers' instructions quite closely. Also, the erase head is pivoted out of the tape path in the dubbing mode.

DUBBING FROM TAPE DECK 2 TO TAPE DECK 1

Check for the erase head if you have no instructions or are in doubt of which tape deck does the recording. Place the blank cassette to be recorded in this deck or deck 1. This feature enables you to duplicate a recorded cassette placed in tape deck 2 onto a blank cassette placed in deck 1. Check your instruction book for correct decks because some are marked by numbers and others are marked with letters (Fig. 16-6).

Place the dubbing switch in the high or normal position. The high position records at twice the normal speed. In some cassette players, dubbing is only provided from deck 1 to deck 2. Place the blank cassette in deck 1. Now, press Pause. Then, press Record.

Drop the recorded cassette in deck 2 and press Play. When Play in deck 2 is engaged, Pause on deck 1 disengages and both decks start rotating at the same time. The volume control can be used to monitor the dubbing operation. Do not change the dubbing switch position.

When you have finished the dubbing operation, press deck 1 and deck 2 Stop/Eject buttons. If the whole cassette is to be duplicated, let the automatic stop of each deck shut off the player. After the dubbing operation has been completed, place dub switch in the Off position.

FIGURE 16-5 **The oscillator switch eliminates background interference or heterodynes when recording from the AM band.**

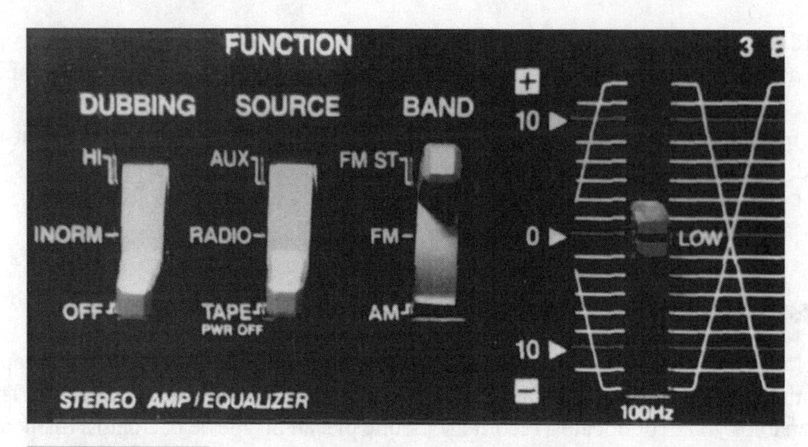

FIGURE 16-6 **High-speed dubbing from deck B to deck A with the dubbing switch in the High position.**

The Erase Head

The erase head removes any previous recording before the tape moves across the record/play tape head. Jumbled or excessively distorted music can be caused with packed dust oxide on openings of tape head, which prevents erasing the previous recording. Usually, only one erase head is contained in the cassette compartment. This head is smaller than the record/play head and is located to one side. The R/P head is mounted in the center.

In dual-cassette system, the recording deck contains the erase head. The erase head can be bolted next to the R/P head. Within Emerson's CTR961, the A and B dual-deck system/the erase head is found in the bottom B deck. Tape deck A is mounted at the top. The erase head in the bottom deck operates on a pivot and stays out of the way while dubbing and playing. When Record is pushed, the erase head is moved upward to engage the tape (Fig. 16-7).

Most erase heads are excited with a dc voltage or with an oscillator bias circuit. The dc voltage is switched in Record mode to the erase head (Fig. 16-8). This voltage comes from the dc power supply or batteries.

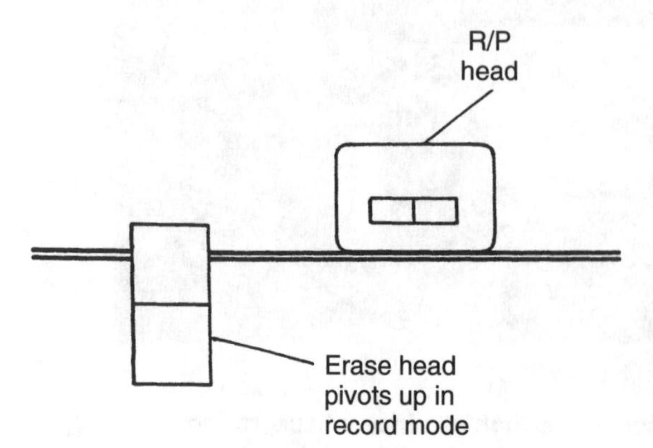

FIGURE 16-7 **The erase head is pivoted out of the tape area when in Dubbing or Play mode in some cassette models.**

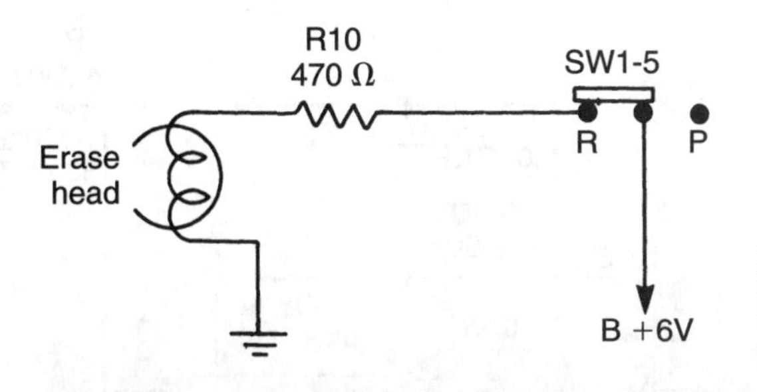

FIGURE 16-8 The dc voltage is switched to the erase head when recording.

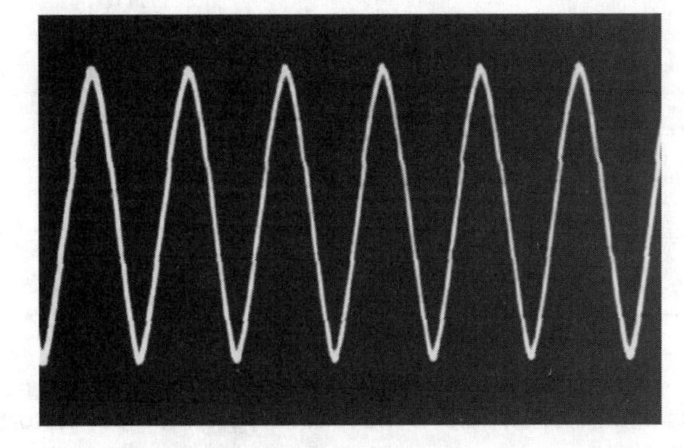

FIGURE 16-9 The bias waveform at the high side of the R/P tape head.

When Record is pressed in the deluxe boom-box player, B+ voltage is applied to the bias oscillator. This bias signal is applied at all times to both stereo R/P tape heads and can be seen with an oscilloscope (Fig. 16-9). Poor recording or no recording can be noticed when the bias signal is removed from the erase and stereo R/P tape heads. A defective bias oscillator circuit (Fig. 16-10) can jumble or distort music while recording.

The larger, expensive boom box can have an oscillator coil with a dual primary winding. The erase head is excited from the center top while both left and right stereo recording heads are excited from the top winding through a resistor/capacitor network. Q1 and Q2 are dual oscillators with B+ applied in the Record position.

Most bias oscillator problems are caused by a leaky or open transistor, dirty record/play switch, or poor transformer board connections. First, clean the erase head with rubbing alcohol and a cleaning stick. If the erase head does not remove the previous recording, check the voltage that is applied to Q1 and Q2. Test transistors Q1 and Q2 in the circuit. Check the continuity of each transformer winding. If one is open, solder the transformer board connection. Replace the transformer if a winding is open.

ERRATIC SWITCHING

If the boom box fails to record, suspect that the R/P switch contacts are dirty. Clean all of the switching contacts on the entire switch assembly with cleaning spray (Fig. 16-11). Place the nozzle inside by each contact and spray all switching wafer contacts. Work the switch back and forth to clean the contacts.

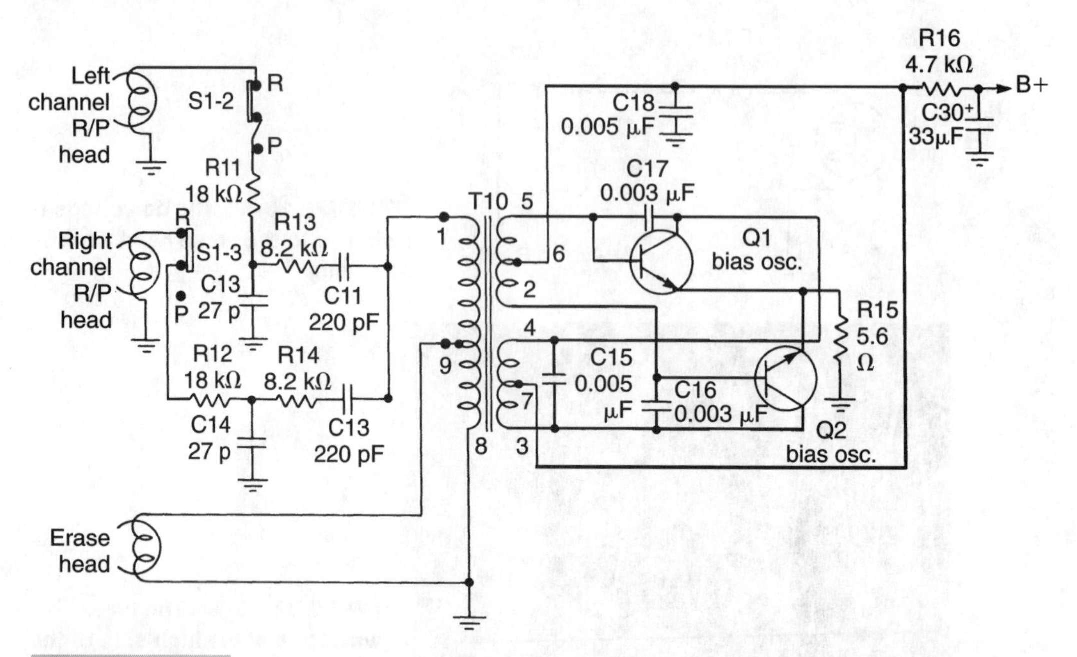

FIGURE 16-10 Check the bias oscillator circuits if the recording is muffled, distorted, or jumbled.

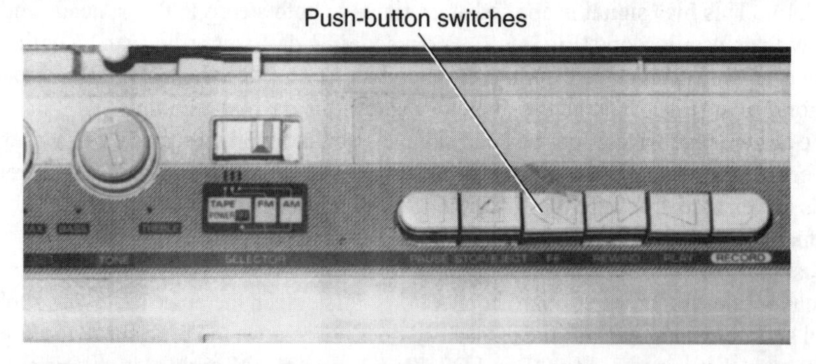

FIGURE 16-11 Clean the pushbutton switches to correct operation and erratic switching.

ACCIDENTAL ERASE

Break out the rear tab of the cassette when you want to keep a recording permanently (Fig. 16-12). When a cassette with the tab broken out is inserted into the player, accidental erase is prevented by a lever mechanism that keeps the Record button from being pressed down. If the symptom is that a certain cassette will not record, first check the tab at the back.

Most recorded musical cassettes have no tab at the rear of the cassette. This is to prevent erasing or recording over the purchased cassette. If you want to preserve a certain cassette, remove tab at the back. You can record over this cassette by placing tape over the opening. The small lever at the rear of cassette holder can be pushed and held in while pressing Record and Play for proper operation. Just insert another cassette with tab to be sure that the player will record.

Broken Soft Eject

In some models, the cassette lid will slowly open when Eject is pressed. The soft-eject system can operate from a gear type or plunger arrangement. In newer models, the cassette lid slowly opens with a lever-gear arrangement (Fig. 16-13).

Suspect that a gear is broken or suspect that the gear lever is off track if the door pops open with soft-eject system. The gear lever is attached to the cassette lid, which rotates a small plastic gear that has braking action. If the door is difficult to open, suspect that a gear assembly is gummed or damaged. Clean the gear assembly for smooth ejection with rubbing alcohol and a cleaning stick. Do not apply light oil to the braking gear assembly (Fig. 16-14). Remove any thread, twine, or foreign material from the gear assembly if it will not open.

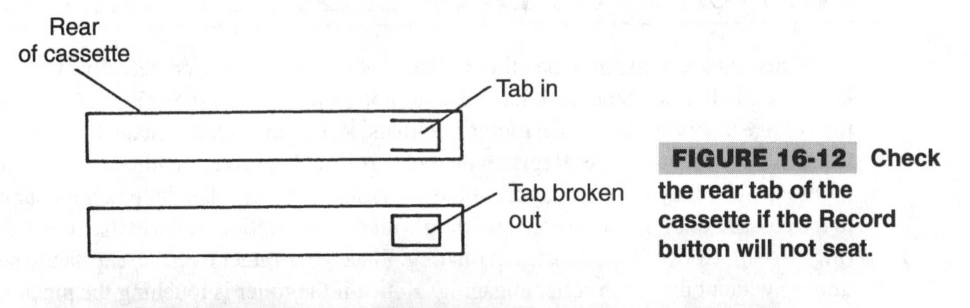

FIGURE 16-12 **Check the rear tab of the cassette if the Record button will not seat.**

FIGURE 16-13 **A soft-eject mechanism might consist of a breaking gear and plastic gear assembly fastened to the cassette door.**

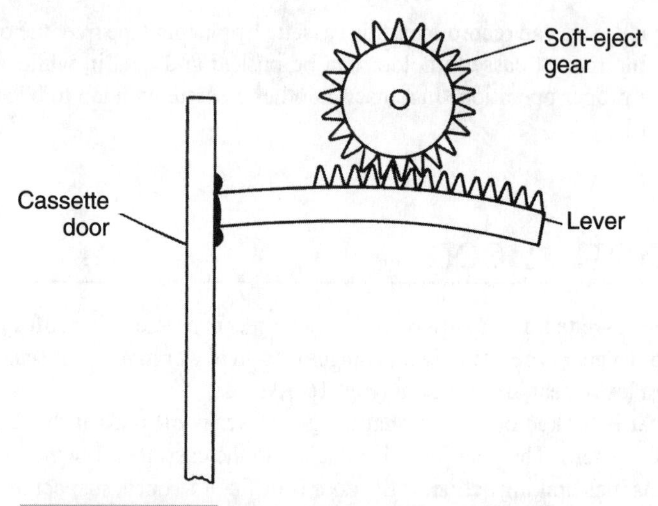

FIGURE 16-14 **When the door opens part way, suspect that a gear lever has jammed or that tape has spilled out of the cassette.**

No Action on Deck 2/Deck 1 Okay

If one deck is operating and the other will not rotate, tape suspected defective motor or broken motor belt. Each tape deck has its own motor and drive system (Fig. 16-15). Measure the voltage that is applied to the motor terminals. Rotate the belt if voltage is present and see if the motor starts to rotate. Replace the motor if it will not rotate with voltage applied.

Often, when Play is pressed, you can hear the small motor rotate by placing your ear next to the cassette door. If the motor is running with no tape motion, suspect that a belt is broken (Fig. 16-16). Notice if the capstan is rotating. Place your finger over the capstan to see if it's turning without the pinch roller engaging. Notice if the roller is touching the pinch tape.

Sometimes, the capstan/flywheel bearing will run dry and freeze. Simply move the flywheel by hand. If it starts off slow, suspect that the bearings are dry. Remove the capstan/flywheel assembly and mounting bracket. Clean the bearings and lubricate them with light grease. Check the motor belt for broken, loose, or stretched areas. Replace the motor belt if it shows signs of slipping.

Power-Supply Circuits

If there is no voltage or if you find an improper voltage at the cassette motor and amplifier, suspect that the power supply is defective. Does the unit play on batteries? If so, go directly to the ac power supply. Very seldom will you find any fuses in these power supplies. The early power supplies might have full-wave rectification with only two diodes (Fig. 16-17).

The low-priced power supply might consist of only two silicon diodes, a transformer, and an electrolytic capacitor. The power-supply output can be switched into the circuit with a remote switch and a leaf switch. Check the silicon diodes with the diode test of the DMM. Sometimes when one or more diodes are shorted, the primary winding of the power transformer (refer to Fig. 16-17) is taken out. Because the secondary windings are wound with heavier wire, this winding rarely opens. Check both windings with the RX1 ohmmeter range.

You might find the two small diodes inside one plastic component, such as the bridge rectifier. In other supplies, separate silicon diodes are used. Replace the entire part if one is shorted or open.

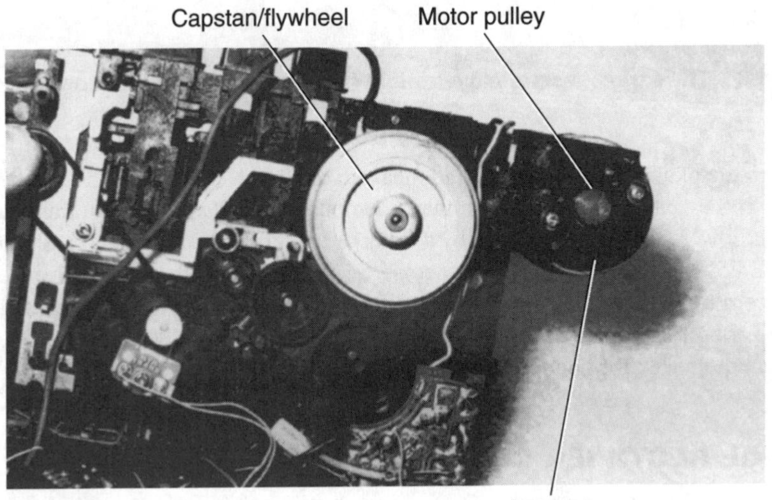

FIGURE 16-15 Check the motor, motor drive belt, and capstan/flywheel if one deck will not move the tape.

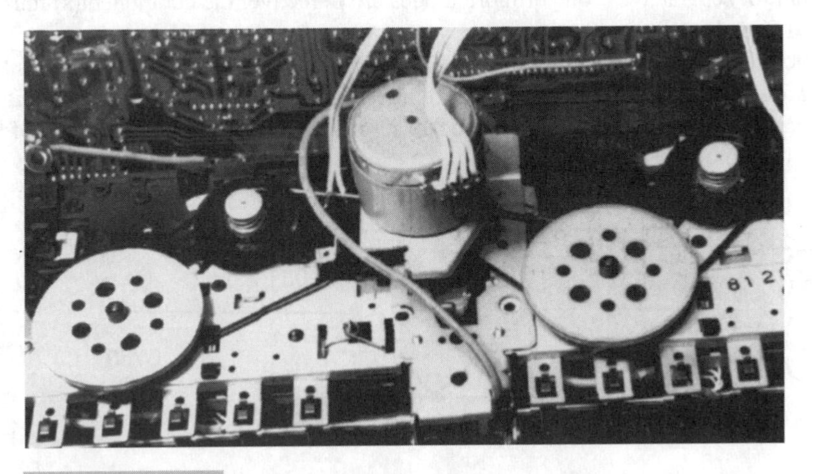

FIGURE 16-16 Each tape deck has its own motor, capstan/flywheel, and pinch roller.

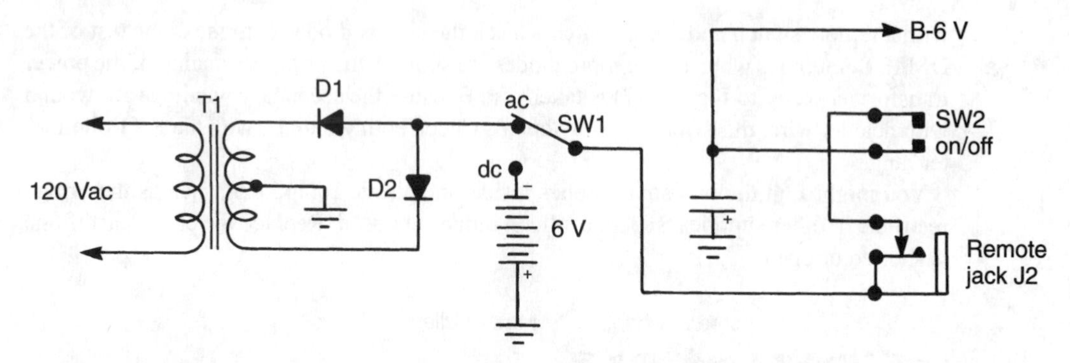

FIGURE 16-17 The early cassette player might have full-wave rectification with two silicon diodes.

Switch the player to ac operation and measure the dc voltage across the electrolytic capacitor. If the voltage is low, suspect that a component, such as an output transistor or IC, is leaky. Low voltage can be caused by a weak or open filter capacitor (C1). Shunt another 2000-μF 25-V electrolytic capacitor across C1. Clip it in with alligator leads with the power cord pulled from the outlet. Replace the electrolytic capacitor if the voltage is now restored. Observe correct polarity (+) of the capacitor when soldering it in. Check SW1 and the remote jack for poor contacts.

BRIDGE RECTIFIER CIRCUITS

In bridge rectifier circuits, four diodes provide full-wave rectification. Usually the transformer has only a primary and secondary winding without a center tap. The low ac voltage is applied across the bridge circuit and the dc voltage is taken from the two positive ends of diodes that have been tied together (Fig. 16-18). The bridge diodes can all be found in one unit or separately. If one or more diodes are defective, the component(s) must be replaced (Fig. 16-19).

When the ac cord is inserted into the outlet socket, the ac/dc switch switches the batteries out of the circuit. Now, an ac voltage is applied to the primary winding of the power transformer. An external dc socket is in series with the positive lead of battery (Fig. 16-20).

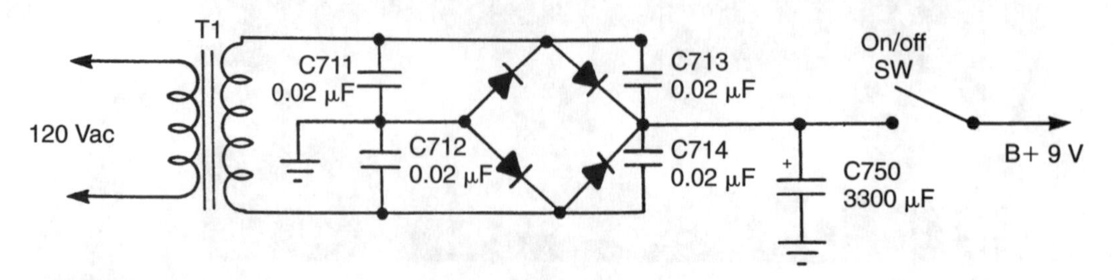

FIGURE 16-18 The bridge-rectifier circuit is used in the power supply of many cassette recorders.

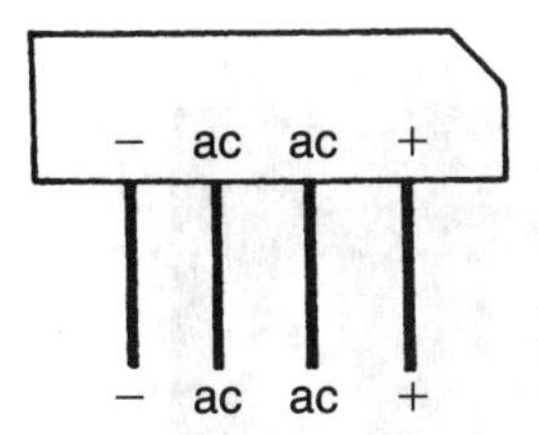

FIGURE 16-19 Replace the entire bridge rectifier when only one diode is leaky.

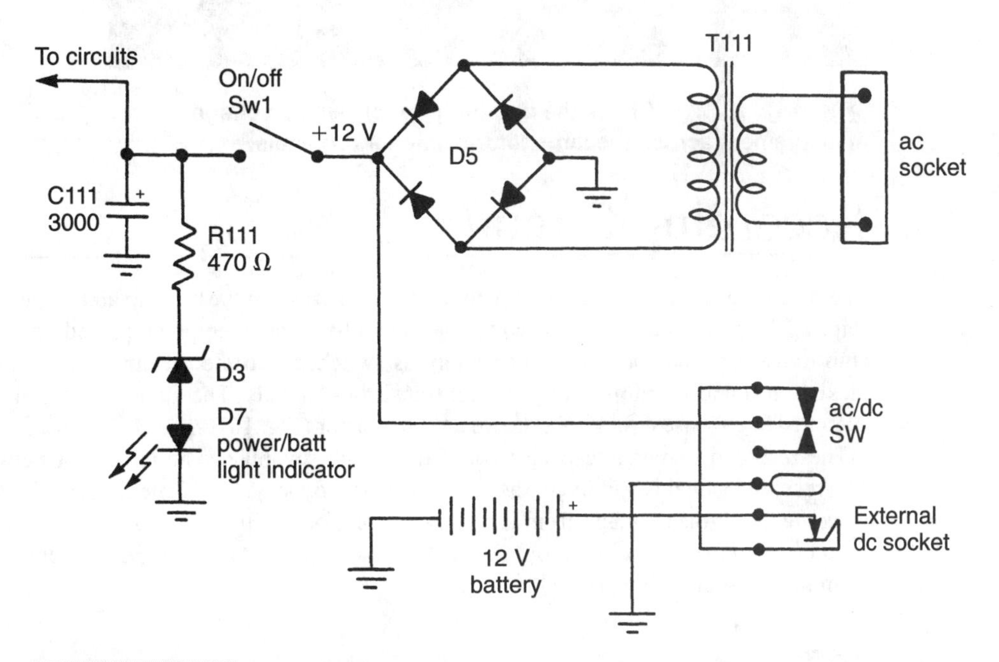

FIGURE 16-20 An external dc battery source or supply might be used with the external dc jack in the low-voltage circuits.

A poor contact in either the ac/dc switch or in the external dc socket can prevent battery operation. Sometimes when the unit is dropped on the ac cord outlet, the plastic socket splits, which can misalign a leaf switch. Sometimes, these switch assemblies can be removed, repaired, and placed back together with epoxy cement. The original part might be difficult to obtain.

If the cassette player has a power/battery light indicator, you can tell if the power supply or battery source is normal because of the lighted LED. If nothing occurs in battery operation, the light might be dim with one or two defective batteries. Likewise, when switched to ac operation, an unlit power light indicates problems in the low-voltage circuits.

Take a quick voltage test across the electrolytic capacitor (Fig. 16-21). Suspect that an off/on switch, bridge diode, transformer, or ac cord is defective if no voltage is available in ac operation. Always check the battery operation to determine if the ac power supply is defective. Suspect the external dc socket switch, ac/dc switch, or dirty or corroded battery terminals if the batteries test normal. Be sure that the ac cord is removed for battery operation on some models.

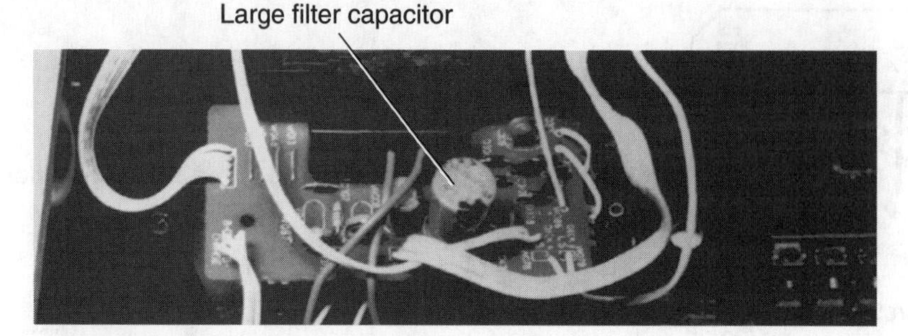

FIGURE 16-21 Check the dc power supply with a voltage measurement across the large filter-capacitor terminals.

Recording Circuits

The recording circuits in early boom boxes had transistors in the preamp and audio-output stages. The preamp transistor circuits consisted of two or three preamps and one record amp transistor. The condenser microphone is switched into the preamp circuits and the picked up sound is amplified by several transistor channels. The audio preamp signal is switched to the base circuit of a record amp transistor (Fig. 16-22).

The recorded output is capacity coupled through several resistor/capacitor networks. The recorded signal is applied to the R/P head and ground. At the same time, the R/P head is excited with bias voltage from a one-transistor bias oscillator. Also, the erase head is switched into the circuit in Record mode. The erase head can be bias operated or operated from the B+ source (Fig. 16-23).

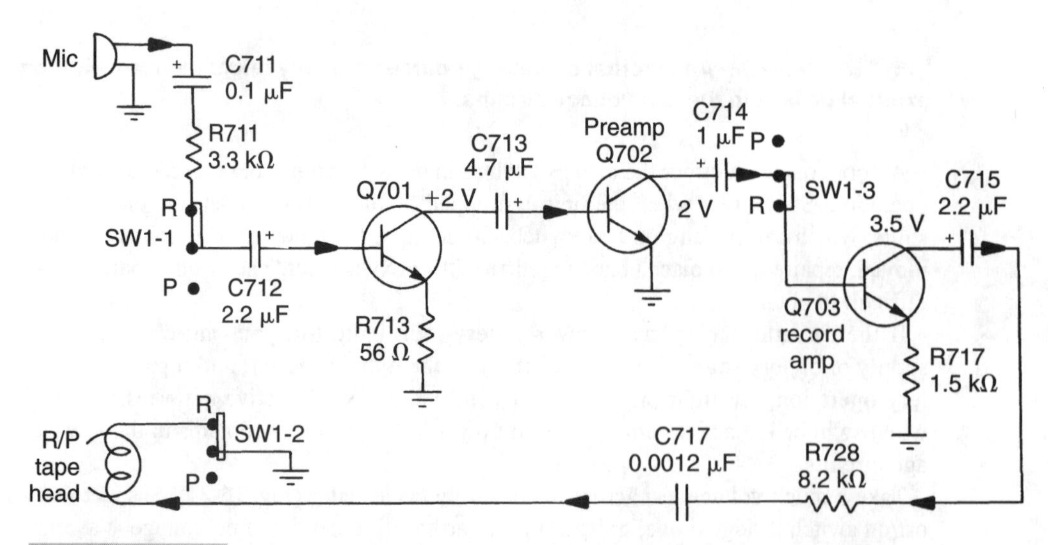

FIGURE 16-22 A block diagram of the recording circuits switched into the transistorized cassette circuits.

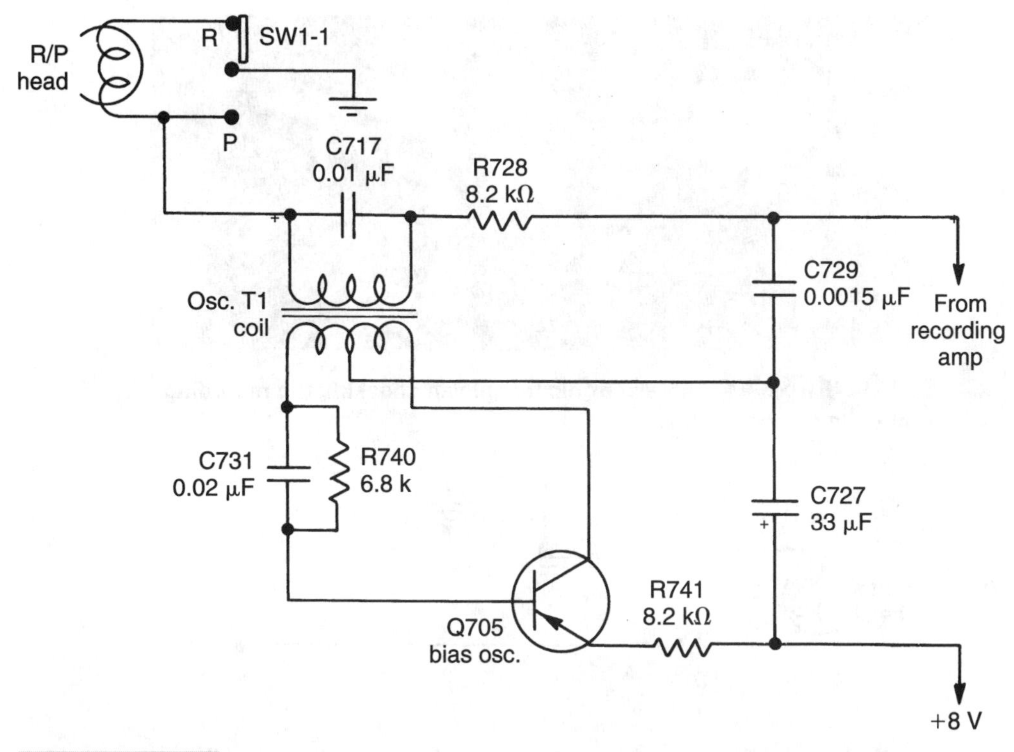

FIGURE 16-23 **A simple bias oscillator circuit that excites the R/P tape and erase heads.**

Clean the R/P head if the machine will not record. Clean the R/P switches (SW1-1, SW1-2, and SW1-3) with cleaning fluid. If the recorder operates normally when playing, suspect that the problem is with the microphone or Q703. Both Q701 and Q702 are in the circuit in Play mode. The only transistor not included in playback is the recording amp (Q703). Any malfunctioning part in the recording path from SW1 to SW3 to the R/P tape head can prevent recording (Fig. 16-24).

Check Q703 in the circuit with the transistor tester of the DMM. Take voltage measurements on all terminals of Q703. Shunt C715 and C717 with respective capacitors. Remember, the microphone pick-up audio can be signal traced through the preamp and record stages up to the R/P tape head with the external amplifier. Place a radio speaker near the mic and signal trace each stage with the audio amp and a pair of headphones. A defective bias oscillator stage can prevent a quality recording.

If the recording sounds jumbled with other recordings on the tape, suspect that an erase head is defective. The dc erase head is switched into the circuit in Record mode, with SW1-5 (Fig. 16-25). B+ 8 V is applied to the tape head through R787 (330 Ω of the erase head and ground). If the voltage is low across the erase head, suspect that a record-switch contact at SW1-5 is poor. Check the erase head for open windings or broken head wires. Notice if the erase head is riding against the passing tape. A loose mounting or a missing screw can prevent the head from touching the tape.

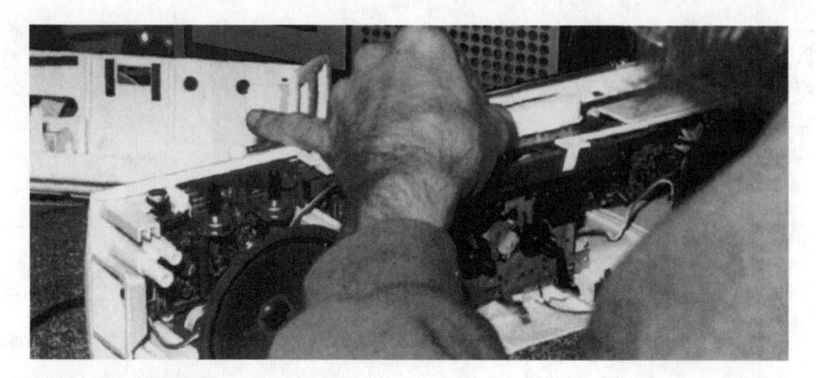

FIGURE 16-24 An electronic technician checking the recording circuits of a boom box.

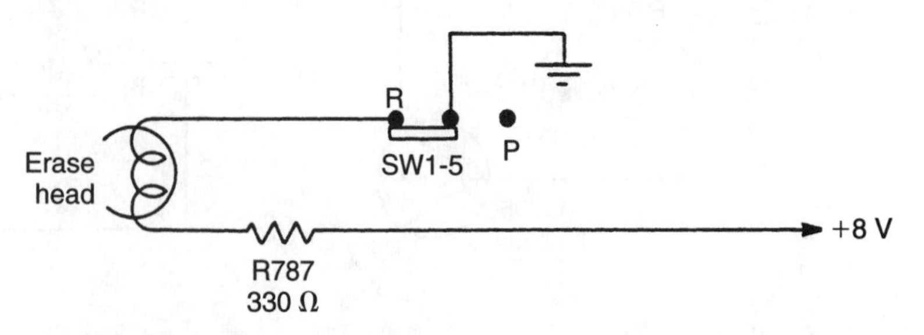

FIGURE 16-25 The dc-operated erase head is in the recording circuit when it is switched to record.

IC RECORDING CIRCUITS

You can find ICs in the preamp and in both left and right stereo channels. The amplified audio signal from the microphone and preamp IC can be fed back to the corresponding R/P head in Record mode after the preamp stage. Of course, the erase head and bias oscillator are switched into the circuit when recording (Fig. 16-26).

The two stereo microphones can be fed through an external mic jack and switched to the preamp, IC1101. The left-channel audio signal is switched into pins 9 and 10 of the preamp IC. The right-channel mic audio signal is switched into pins 6 and 5 of the preamp. Notice that the preamp stages serve both Record and Play modes.

The output signal from the left channel at pin 13 is coupled through a capacitor (C205) and switched to the left R/P head. The output signal of the right channel at pin 2 is fed to C305 and switched to the right R/P tape head. When recording, the dc-operated erase head is switched into the record circuit with SW1-7. Dc bias is applied to each left and right channel R/P head through R210 and R320 (Fig. 16-27).

Troubleshooting IC record circuits Before attempting to repair the recording IC circuits, determine if the playback circuits are normal. If the playback circuits are normal, clean the record and erase heads. Clean any switches that are related to the recording circuits. Check the oscillator bias circuits if you notice distortion or weak recording. Measure the dc voltage or bias voltage at the erase head if the music is jumbled (Fig. 16-28).

Check the schematic for another record IC. Larger boom boxes have a separate recording IC after the preamp circuits. Measure the supply voltage (V_{CC}) from the low-voltage power supply. If the voltage is very low, the recording IC might be leaky or the supply voltage might be improper. Remove the supply voltage from the IC by unsoldering the IC pin from the PC wiring. If the voltage rises above or is near to the required voltage, the recording IC is leaky. Replace it.

Check all voltages on each terminal of the IC and compare them to the schematic. If one or two pin voltages are way off, check the components that are tied to these pins for a possible

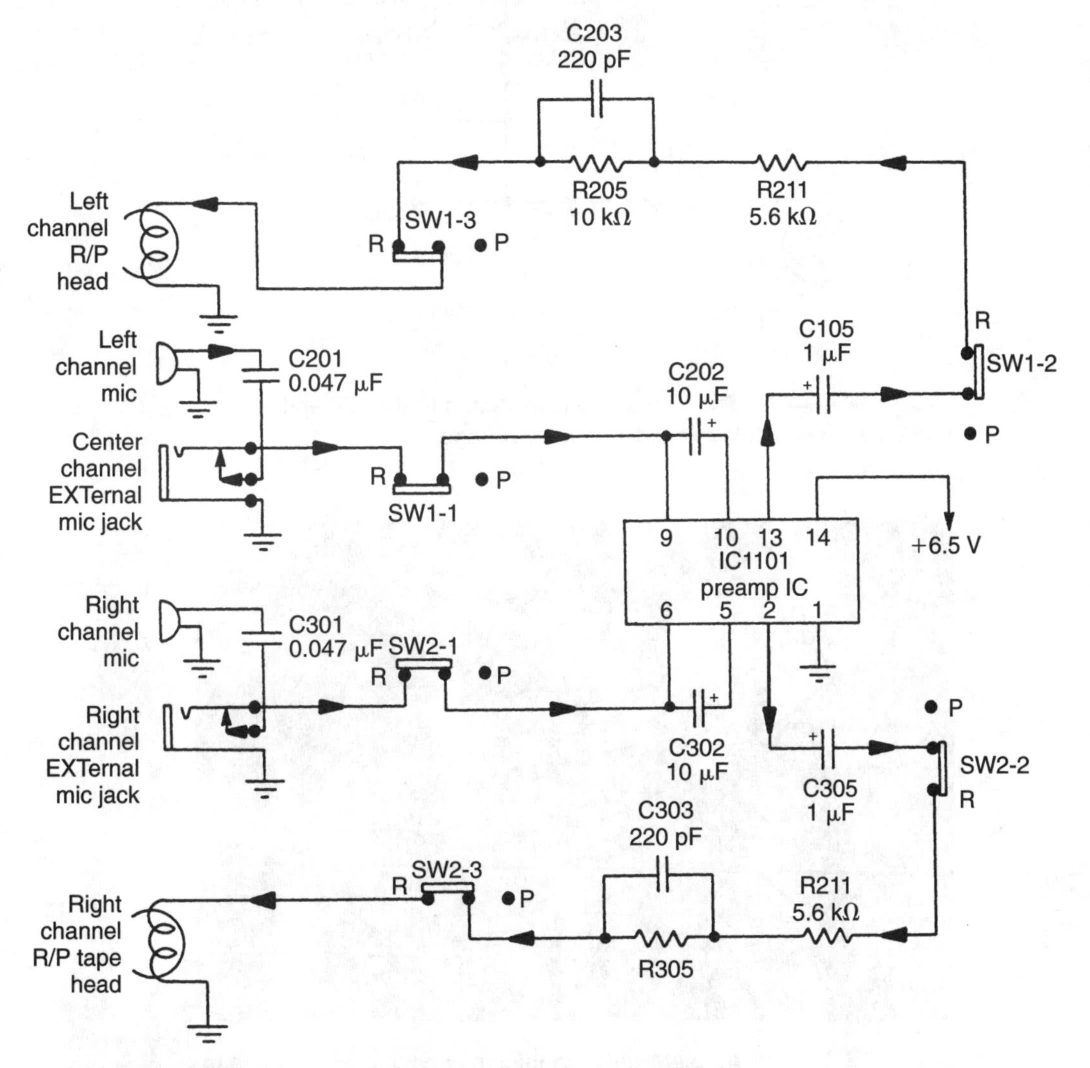

FIGURE 16-26 The IC recording signal path in the Record mode.

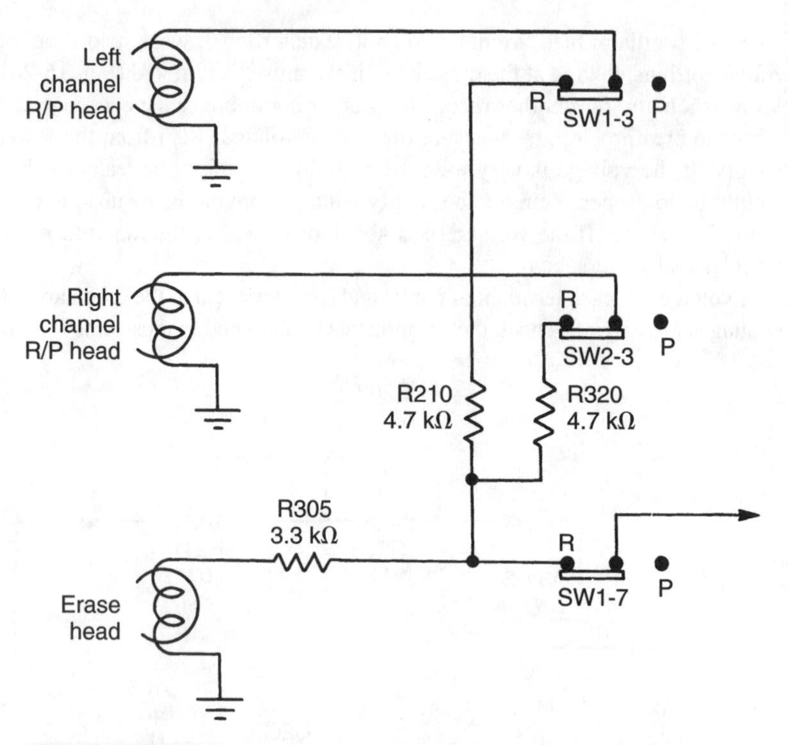

FIGURE 16-27 The dc voltage is switched to the R/P and erase head when it is excited with a dc voltage.

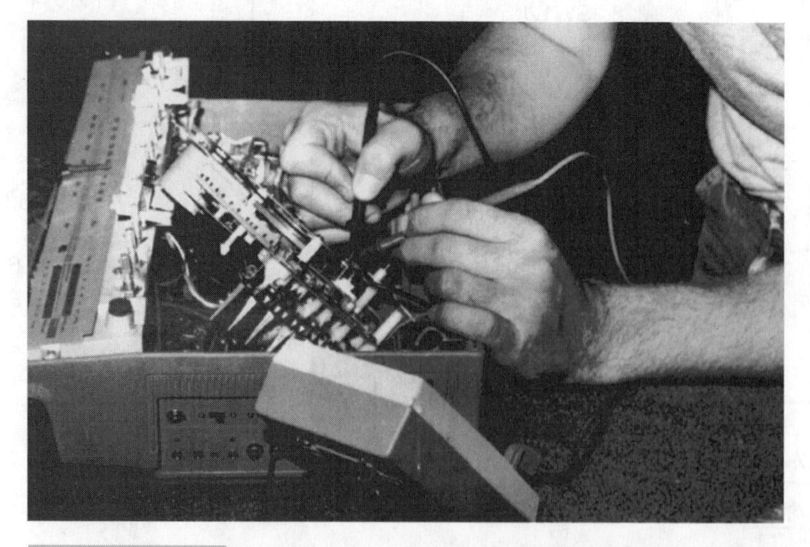

FIGURE 16-28 An electronic technician checking dc voltages in an R/P and erase head in a J.C. Penney 681-3950 boom box.

shorted capacitor or a change of resistance. If the recorded signal is coming into the IC and not at the output terminal, replace the leaky IC.

Signal trace the audio signal with an external amp or with a signal tracer from the microphone to the recording amp IC. Place a radio or another cassette player (with a tape playing) in front of the microphone. Check the signal at the input and at the output pin of the IC. Replace the IC if the voltage and input signal is normal, but no audio is being output from the IC.

DEAD RIGHT SPEAKER

The voice coil of a boom-box speaker can be damaged with excessive volume or by shorted audio-output circuits. The defective speaker might be dead, mushy, intermittent, or distorted. Clip another speaker across the terminals of the dead speaker. If another speaker is not handy, interchange the leads that go to the normal speaker for test only (Fig. 16-29).

Suspect that a voice coil is damaged if the sound is dead, intermittent, or noisy. The intermittent speaker can be located by applying equal pressure on the speaker cone with music playing. Pushing the speaker cone in and out makes the sound cut up and down. Check for an open voice coil with an ohmmeter at the RX1 range. Check speaker hinge connections on some speakers that hang alongside and use the hinges as speaker connectors.

Large speakers can be damaged or the voice coils can blow as a result of constant excessive volume. Push up and down on the center of the speaker and notice if the voice coil is dragging. The muffled sound can be caused by a frozen voice coil on the center magnet. The voice coil will not move if it is frozen to the magnet. When the dc voltage is applied to the voice coil in a dc-coupled speaker circuit, the voice coil can be burned open (Fig. 16-30).

Small holes punched into the speaker cone can be repaired with speaker cement. If the speaker is "blatting" on bass notes, the voice-coil diaphragm support or speaker cone can come unglued. Apply speaker quick-drying cement under the cone area. Replace any speaker that has large holes, a torn cone, or a damaged voice coil.

DEFECTIVE RECORDING METERS

In some boom boxes, a separate recording meter or LED circuit determines the range of the recording. The meter circuit can be switched in after the preamp stage or after the AF

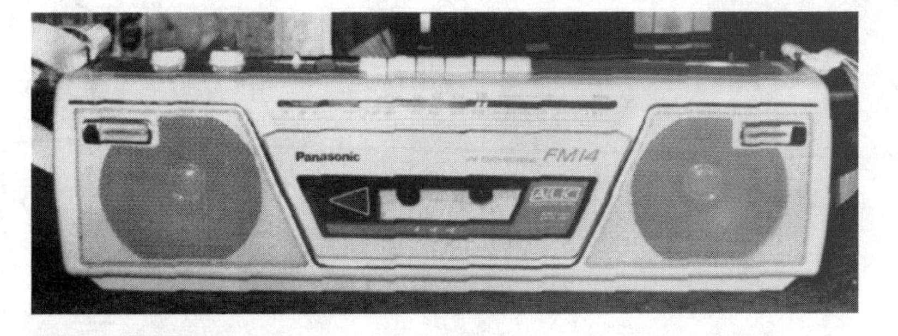

FIGURE 16-29 **Exchange speaker terminals leads to test the suspected dead stereo speaker.**

amp circuits. The audio signal can be checked right up to the diodes of the meter circuit. One or two separate transistors can be found in the meter circuits (Fig. 16-31).

When the recording meter is not moving during recording, check the audio signal through the AF stages to the small meter diodes. Trace the audio signal with separate external amp or with a signal tracer. Check the meter for an open winding if the audio is traced up to the meter circuit. Test each diode with the diode test of the DMM. The meter hand might be bent or rub against the dial area and not move with the recording.

FIGURE 16-30 A dc voltage or excessive volume applied to the voice coil and even damage high-wattage speakers.

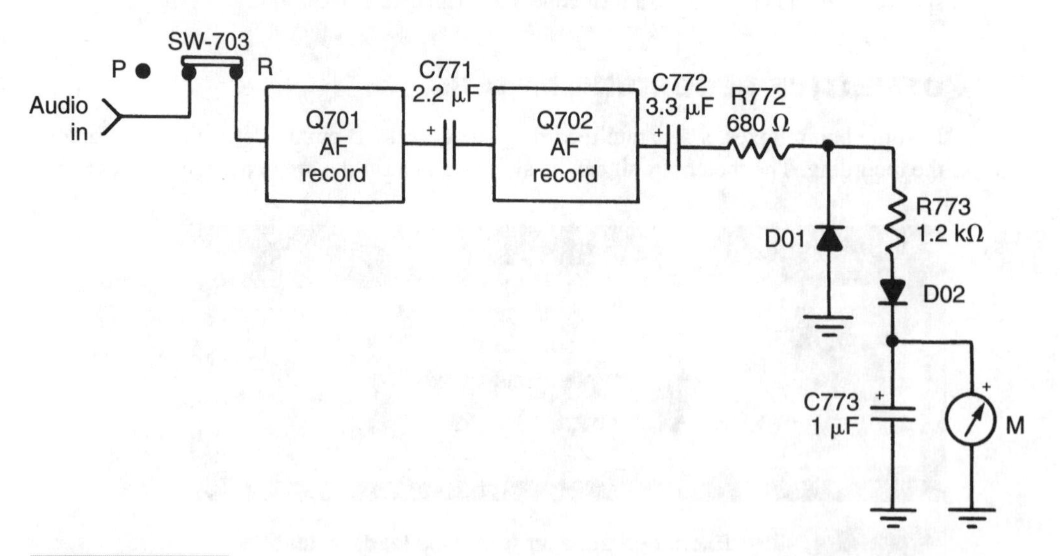

FIGURE 16-31 A block diagram of a recording meter in the output of a rectification circuit.

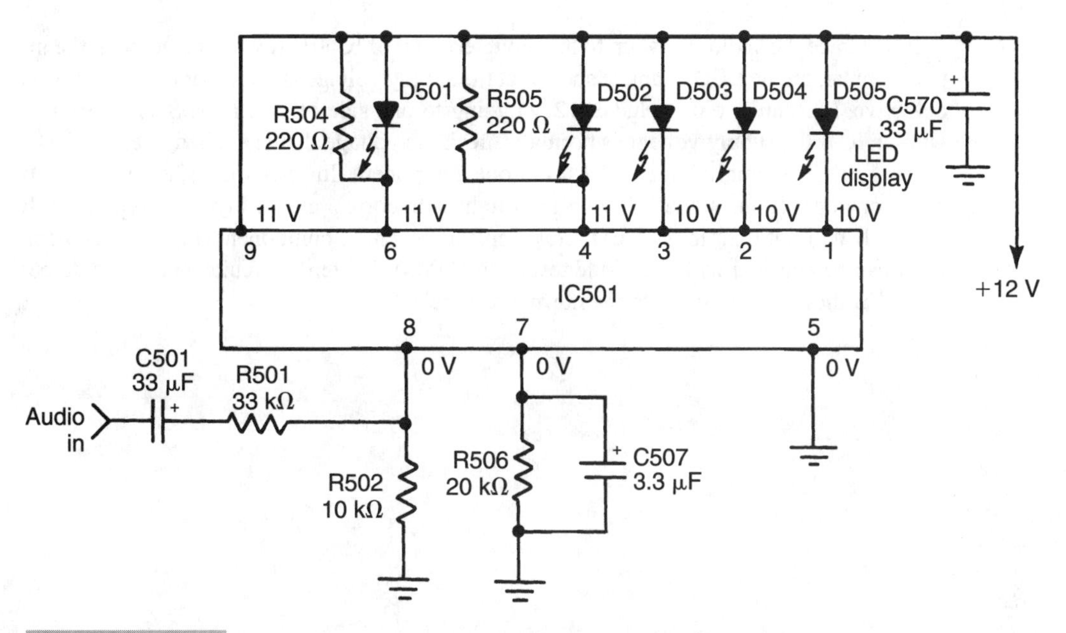

FIGURE 16-32 An LED meter-level display is operated from one large IC.

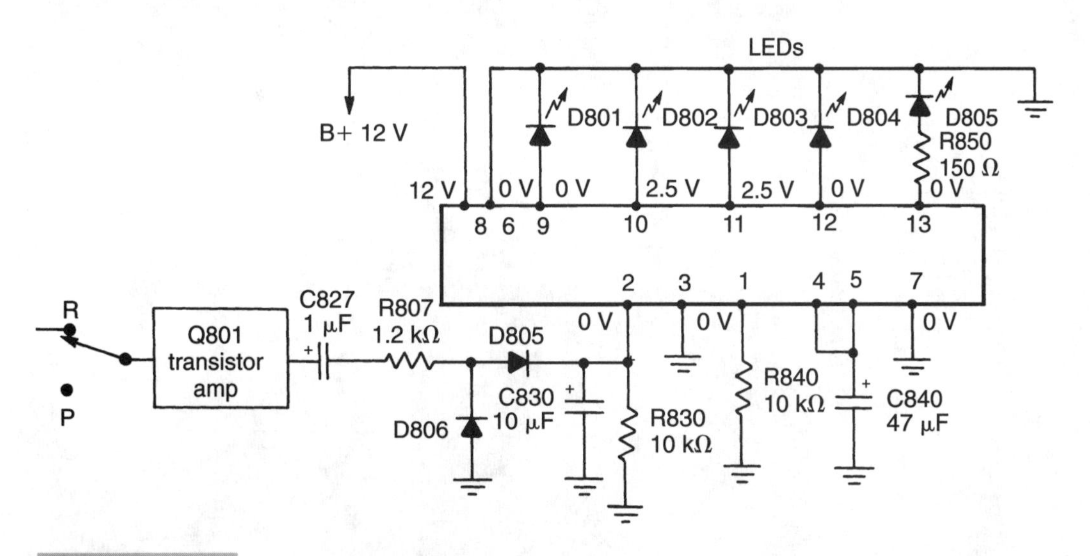

FIGURE 16-33 You might find single LEDs in a recording or sound-level meter circuit.

LED Meter Circuits

The peak-level LED display meter can be driven with an IC. The audio signal is applied to the input terminal with LEDs in the output audio-level circuits as indicators (Fig. 16-32). As the recorded or audio level increases, the LEDs will light up accordingly. You can find one IC for each set of display LEDs in stereo channels.

Signal trace the audio signal up to the input terminal of IC501. If you find audio at the input terminal, but no LEDs light, suspect that the supply voltage is improper or that the IC is defective. Measure the dc voltage (12 V) going to one side of all LEDs. Suspect trouble with an IC if the supply voltage is normal. Check the voltage that is applied to each LED.

The LED array might have all LEDs in one component. In earlier units, separate LEDs were mounted. If one signal LED will not light with correct applied voltage, replace it. If one LED will not light in the LED array, replace the entire component (Fig. 16-33). Each LED can be checked with the diode test of the DMM. In stereo circuits, check the defective LED indicator circuit with the normal channel.

17

TROUBLESHOOTING PORTABLE AM/FM CASSETTE/CD PLAYERS

CONTENTS AT A GLANCE

Besides the AM/FM stereo tuner, the cassette player can have a compact disc (CD) player. The CD player can be located at the top or front of the portable cassette system. Dual cassette decks are featured in some models. Auto reverse, remote control, Dolby B, and clock/timer operations are found in the deluxe units.

The programmable CD player can have up to 36 tracks. Detachable speakers and 10 watts of power provide quality sound. The auto-search music system restarts a CD selection or plays the next one at the touch of a button. Most of the portable cassette/CD players require 8 or 10 D batteries and 4 AA batteries.

The Power Supply

Besides operating from batteries, the player can be operated from a built-in power supply. Most players use a full-wave bridge rectifier system with a fairly large filter capacitor (Fig. 17-1). Each silicon diode is bridged with a 0.022-μF bypass capacitor to eliminate diode interference. D401 prevents accidental polarity that is played into the external dc jack (J402). C325 provides adequate dc filtering with on/off switch S305 in series with the power supply, amplifier, and motor.

The cassette motor operates directly from the 9-V battery or power-supply source. Isolating choke coil L301 and capacitor C326 provide extra filtering for the small cassette motor. C328 (0.1 μF) provides bypass arcing of the small commutator to ground (Fig. 17-2). Some expensive chassis have motor transistor regulation or voltage-regulator circuits.

If on/off switch S305 is dirty, the cassette player might be erratic or dead. Try all tape, CD, and radio functions before digging into the chassis. If the radio and CD player operates, but the cassette player doesn't, suspect that the function switch is dirty. If the unit operates on batteries and not ac power, suspect that the ac cord is defective, S306 switch contacts are dirty, or that the power-supply circuits are dirty.

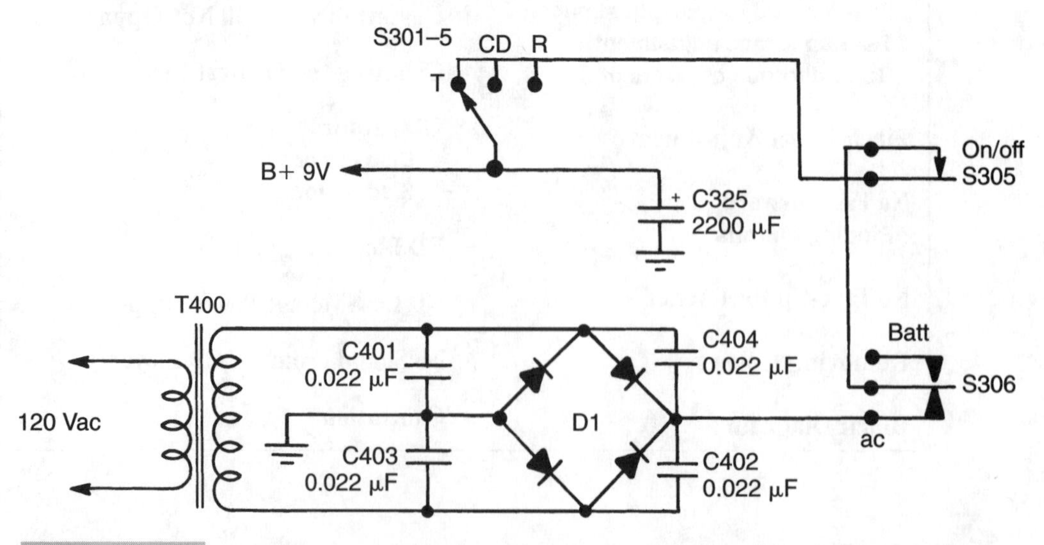

FIGURE 17-1 The full-wave bridge ac power supply with on/off S305, S306, and S301-5.

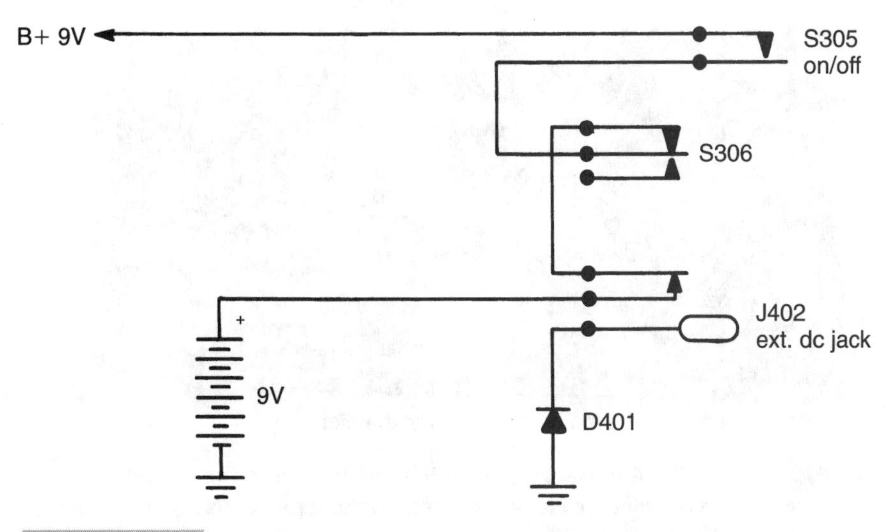

FIGURE 17-2 **The battery is switched out of the circuit when external jacks J402 and S306 are used.**

Check the dc voltage across the large filter capacitor C325 (2200 µF). Very low dc voltage might indicate overload leakage from the radio, tape, or CD circuits. Switch to all circuits to see what section is pulling down the voltage. Then, isolate the overloaded circuits. If the voltage is low on all sections of function switch, suspect that the power supply is defective.

Test each diode with the diode test of the DMM. If one or two diodes are leaky, replace the whole bridge component. With normal diodes and no dc output voltage, suspect an open primary winding of T400. Low dc output voltage can be caused by a leaky or open filter capacitor (C325).

Measure the resistance across the C325 terminals. Suspect that a capacitor is leaky if the measurement is less than 250 Ω with S301 switched to the radio function. Clip another 2200- or 3300-µF electrolytic capacitor across C325 with the power off to determine if the capacitor is open. A hum in the music when the volume is turned down indicates that a filter capacitor (C325) has dried up.

Head Cleaning

Open the cassette door and hit Play. This will push out the R/P and the erase head for easier cleaning from the door opening. In some models, two etcheon screws at the front of the cassette door allow the front plastic to be removed for easier tape head cleaning.

Clean the tape head with a cleaning stick and rubbing alcohol. Tape oxide that is packed on the front of the tape head can be removed with a wood dowel rod or a pencil eraser. Be sure that all small gaps on the front of the tape have a clean surface and are not clogged with tape residue.

At the same time, clean the pinch roller and capstan. Sometimes, if tape is wrapped around the capstan, the packed oxide dust can be difficult to remove. Be sure that all tape

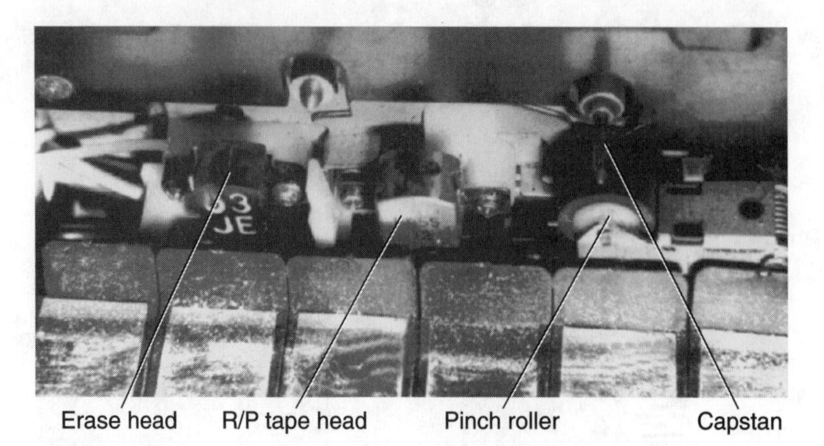

FIGURE 17-3 **A close-up view of the pinch roller in the cassette player. Clean the pinch roller after cleaning the tape heads.**

oxide is removed from the pinch roller. Hold the roller in one spot, clean it, then move to another section of the rubber roller. Do not apply so much rubbing alcohol on the cleaning stick that it runs down inside the capstan bearing (Fig. 17-3).

Erratic Play

Erratic Play or Fast-Forward modes can be caused by improper torque, a dirty or binding flywheel, or a dirty or loose drive belt or clutch assembly. Clean the flywheel, motor pulley, drive belt, and clutch assembly with rubbing alcohol and a soft cloth. The capstan and fast-forward/rewind belts should not be cleaned with rubbing alcohol and cloth. Replace belts that have been stained with grease or oil.

FAST-FORWARD TORQUE ADJUSTMENT

Insert a torque-adjustment cassette in the holder (Fig. 17-4). Measure the fast-forward torque. A torque of more than 50 g/cm is necessary for fast-forward operation. Check the manufacturer's service manual for correct torque adjustment. If the fast-forward measurement is less than 50 g/cm, wipe the flywheel, clutch assembly, and/or replace the drive belt.

TAKE-UP TORQUE ADJUSTMENT

Insert a cassette torque meter and measure the torque while it is playing. If the take-up is not adequate (30 to 60 g/cm), wipe the flywheel, belt, and take-up reel assembly. Replace drive belt if it is cracked or loose. Poor take-up torque might cause the tape to spill out of the cassette.

REWIND TORQUE ADJUSTMENT

Insert the cassette torque meter and if torque is below 50 g/cm, suspect dirty or worn parts. If the rewind torque is less than 50 g/cm, wipe the flywheel, clutch assembly, and/or replace the drive belt. Low rewind torque can cause slow or erratic rewind speeds.

Pinch-Roller Adjustment

After cleaning the pinch roller, it's wise to check the pinch-roller adjustment. In the Play mode, measure the pinch-roller contact with a spring gauge (0- to 500-g gauge). Hook the spring gauge to the pinch roller and pull it away from the capstan (Fig. 17-5). Measure the force at the moment when the pinch roller comes in contact with the capstan (when the pinch roller starts revolving). The gauge should read from 100 to 320 g. To adjust the

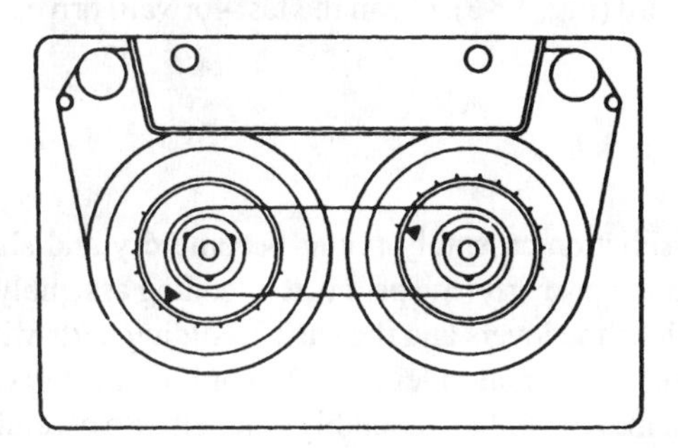

FIGURE 17-4 Check the fast-forward torque with a torque cassette. A torque of more than 50 g/CM is necessary for Fast-Forward operation.

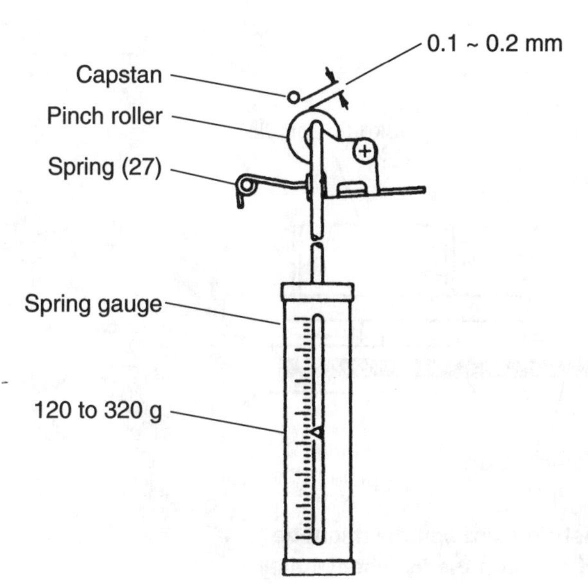

FIGURE 17-5 Be sure that the pinch-roller adjustment is between 100 and 320 g with a spring gauge.

contact pressures, bend the spring (27) and/or replace the spring. Remember, each manufacturer can have its own specifications, but most will read within these measurements.

No Fast Forward

Does the tape move normally in Play mode? If the tape is slow in both Fast Forward and Play, clean the capstan, motor belt, and motor pulley. Suspect that a motor or binding capstan is defective if the tape is still slow after clean these parts. Remove the motor belt and give the flywheel a spin with your fingers. A sluggish or dry capstan will not rotate a complete turn. Remove the capstan/flywheel, clean the bearings, and lubricate the bearings with light oil or grease.

Suspect an oily or slick fast-forward drive belt if the playback is normal and the Fast Forward is erratic. Some units have both a motor-drive belt that rotates the capstan and a separate belt that drives the Fast Forward (Fig. 17-6). Clean the fast-forward drive belt and all drive surfaces with alcohol.

BINDING BUTTONS

After several years of usage, the pushbutton assembly might become dry and sluggish, which results in some buttons not seating properly. Suspect that a loading assembly is dry if the Play button does not lock in. Clean the levers and the surface sliding parts with rubbing alcohol and a cleaning stick. Check each button lever for dry or binding parts.

Sometimes, spraying cleaning fluid into the button assembly can help. Be careful not to spray so much liquid into the assembly that it drips onto belts and driving surfaces. Often, if one button assembly sticks, the plastic button breaks off. Cement the button back in the lever assembly (Fig. 17-7).

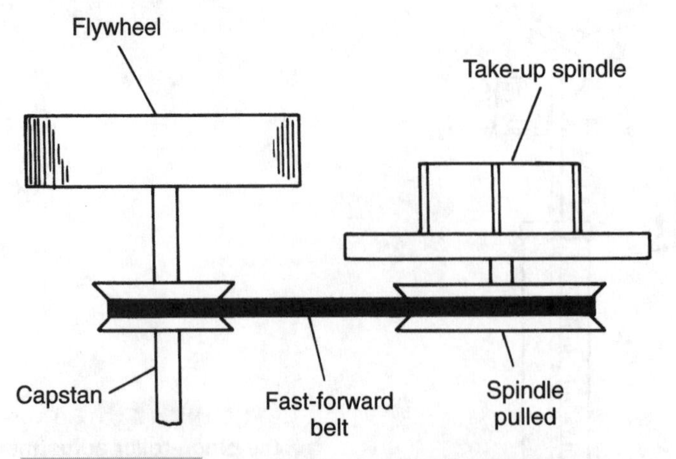

FIGURE 17-6 The Fast-Forward spindle might be operated with a drive belt between the fly-wheel pulley and the bottom of the take-up reel.

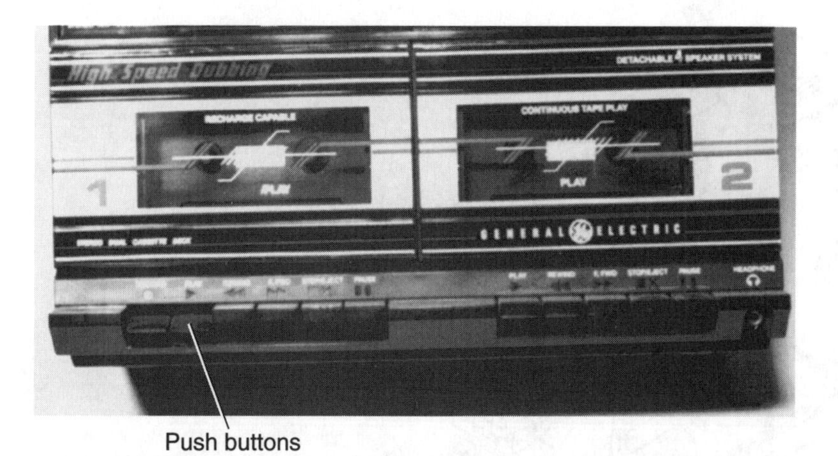

FIGURE 17-7 **Squirt cleaning fluid between the buttons that are binding against the plastic.**

No Take-Up Reel Action

The sluggish or erratic take-up reel results in pulling or spilling out tape in the capstan and pinch-roller area. The take-up reel must rotate smoothly to take off the tape from the capstan to load the take-up spindle. If the spindle or reel stops or operates in an erratic or intermittent motion, tape can spill out and destroy a good cassette.

Hit Play and watch the rotation of the take-up reel (on the right). Clean the idler, clutch assembly, drive belt, and capstan assembly. On units that have a gear-train drive spindle assembly, clean the gears and dry bearings. Place a drop of light oil on each gear bearing. Often, poor or no take-up reel action results in no fast forward in lower-priced cassette players.

Removing Covers

Although it might appear to be difficult to remove the back cover of the portable CD-cassette player, only a few screws hold the cover to the front cabinet. Always remove the battery cover and all batteries. Remove the several screws that hold the rear cabinet (Fig. 17-8). Remove the five screws that hold the power-supply board.

To disassemble the whole cabinet, remove the two screws that hold the PC board. Remove the three knobs (balance, tone, and volume) from the front. To remove the cassette player assembly from the cabinet, remove the screws that hold the large amplifier PC board. Now, remove the screws that hold the cassette mechanism (Fig. 17-9).

Remove the screws that hold the CD mechanism. Remove the three screws that hold the CD support and gear holder. Finally, remove the three screws that hold the display PC board. Lay out all the parts in a row so that they can be replaced properly. Sometimes listing the various components on a piece of paper can help you to reassemble the mechanism. Check (Fig. 17-10) on how to remove the speakers and the cassette door assembly.

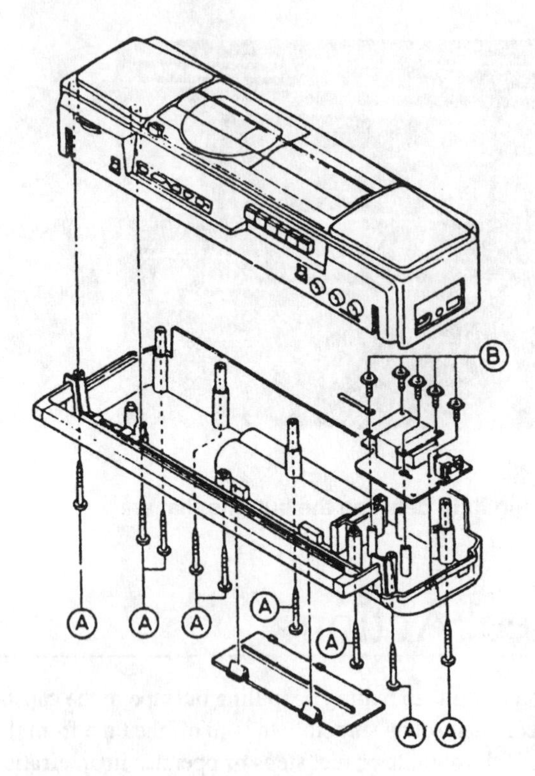

FIGURE 17-8 Remove screws (A) to remove the back cover on the CD/cassette player.

Block Diagram

Always check the block diagram before attempting to service the defective player. Try to isolate the defective component from the block diagram (Fig. 17-11). For instance, in this Radio Shack 14-527 compact disc and cassette player, a separate IC (IC301) is used in Record mode. If the cassette player played perfectly, but would not record, IC301 might be defective. In most cassette players, the preamp and AF amp stages are used for both Record and Play.

Notice that IC302 is a Dolby circuit IC. Here, the Record and Playback signals go through IC302. It's possible that the stereo Playback and Record signals enter IC302 at pins 1 and 14. If the cassette plays, but doesn't record, check the record signal out of pins 8 and 9. The Playback output signal appears at pins 6 and 11. IC302 could be defective only in one stereo record channel or in both and still not affect the playback (Fig. 17-12).

Both record and play use the preamp circuits (IC300). If the cassette player will not play or record, suspect a problem with the R/P tape heads (IC300 and IC302). Often, only one stereo channel will be weak, distorted, or dead. If both stereo channels are dead, weak, or distorted, check the R/P heads (IC302 and IC303). By trying to isolate each Record or Playback mode with the block diagram, you can quickly isolate the defective component (Fig. 17-13).

Stereo Amp Circuits

One advantage in servicing stereo circuits is that the other channel can be used as a gain signal reference in each stage. If the left channel is weak and the right channel normal, use the right channel to measure the gain of each transistor or IC. Check the signal at both volume controls and compare them. If the left channel is a lot weaker, check the preamp circuits. If both channels are fairly normal at both volume controls, suspect a weak AF transistor or IC output stage.

Transistors do not become weak; however, the circuit and components that are tied to the transistor or IC can create a weak stereo channel. A change in the base and emitter bias resistors can create weak audio. Leaky or open components that are tied to the IC amp can produce weak sound. Doublecheck each electrolytic coupling capacitor that is between stages. The signal should be the same on both sides of the capacitor. A low power-supply voltage can cause weak sound.

Just about any component within the stereo circuit can cause a dead channel—especially test transistors in and out of the circuit. Poor switch contacts can kill audio. Suspect that an

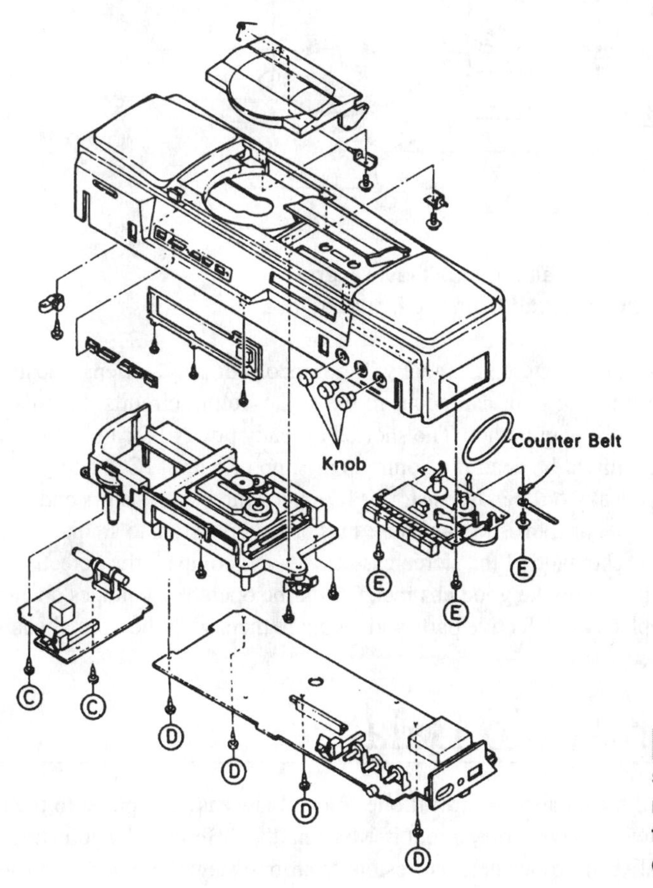

FIGURE 17-9 Remove screws (C) and (D) before removing three (E) screws to remove the cassette deck.

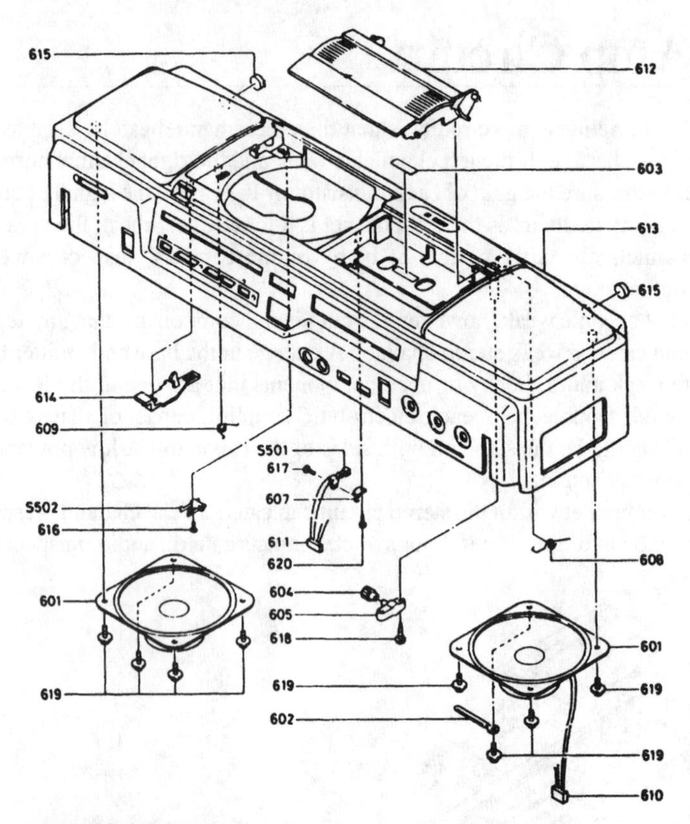

FIGURE 17-10 **The speaker, small levers, and cassette door can be removed from the front cabinet.**

R/P tape head is open if the stereo channel will not record or play. Often, a dead channel with excessive distortion or hum can occur in the audio-output circuits. Feel the power-output IC and notice if it is quite hot. The shorted or leaky power IC can become red hot. Both stereo channels might be dead if a common preamp or output IC is bad.

Before replacing a leaky or open transistor or IC, check the bias resistors and bypass capacitors. The shorted IC or transistor can cause bias or load resistors to smoke and run hot. If you do not have a schematic of the stereo cassette player, compare the defective channel resistors and capacitors with the good channel. Check the operating voltages on each component after you replace all defective parts and compare them with the normal channel.

Improper Audio Balance

Check the setting of the balance control if one channel is weak, compared to the other. If it's only slightly lower, do not worry about it. Make up the difference by adjusting the balance control. Usually, improper balance results from one stage having weak audio. Replace the erratic or worn balance control if it makes excessive noise in adjustment or when the audio cuts out in one channel.

The balance control is usually found ahead of the volume control in stereo cassettes (Fig. 17-14). If one channel will not balance, check for a loss of audio ahead of the balance control. Sometimes these balance controls can short internally to ground and cause one weak channel. Simply remove the grounded terminal of the balance control and notice if the control changes the audio. An internal broken ground terminal can cause improper balance. Replace the defective balance control.

Garbled Recording

Notice if the garbled music is coming out of one channel or both. Is the new recording garbled? Play a commercial recording to see if the recording or the playback is causing the

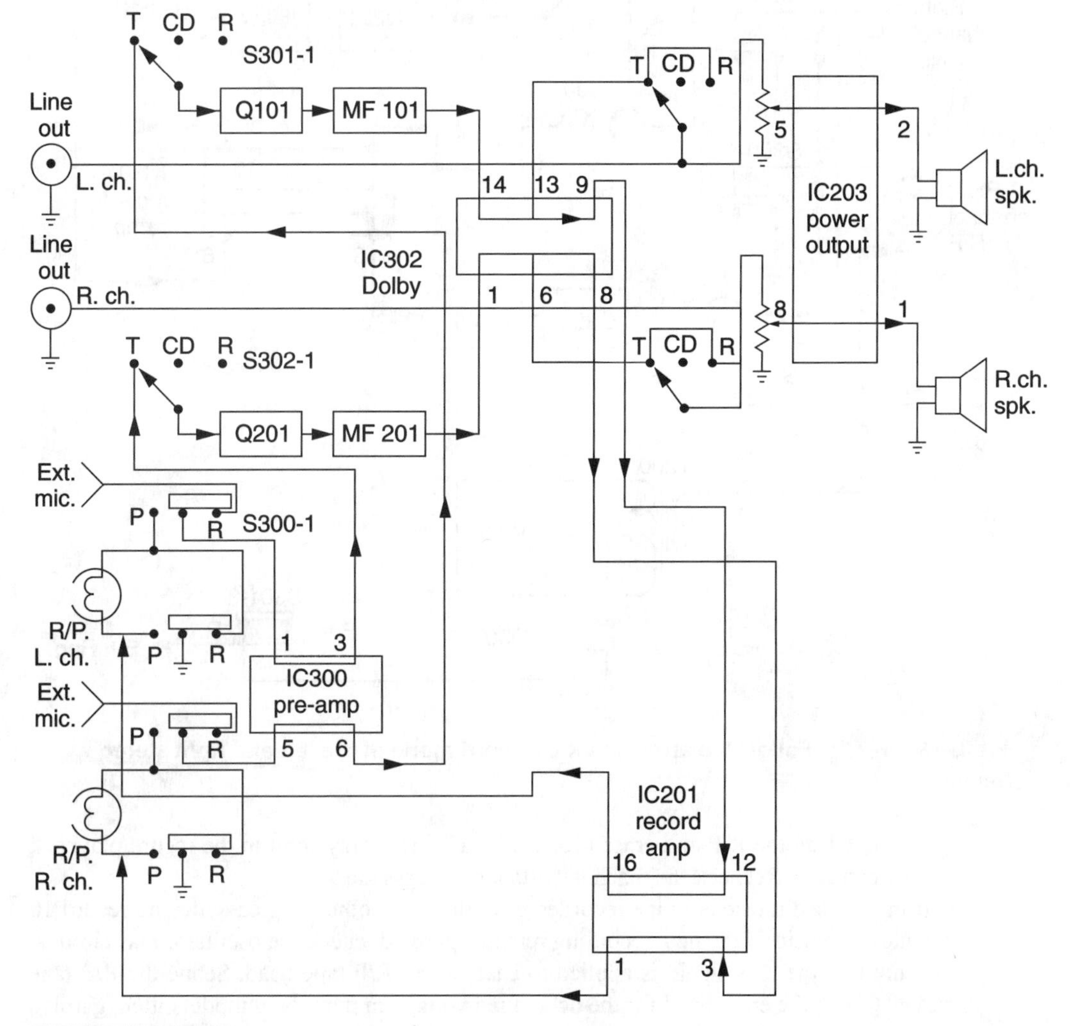

FIGURE 17-11 A block diagram of recording circuits in a cassette/CD player and radio.

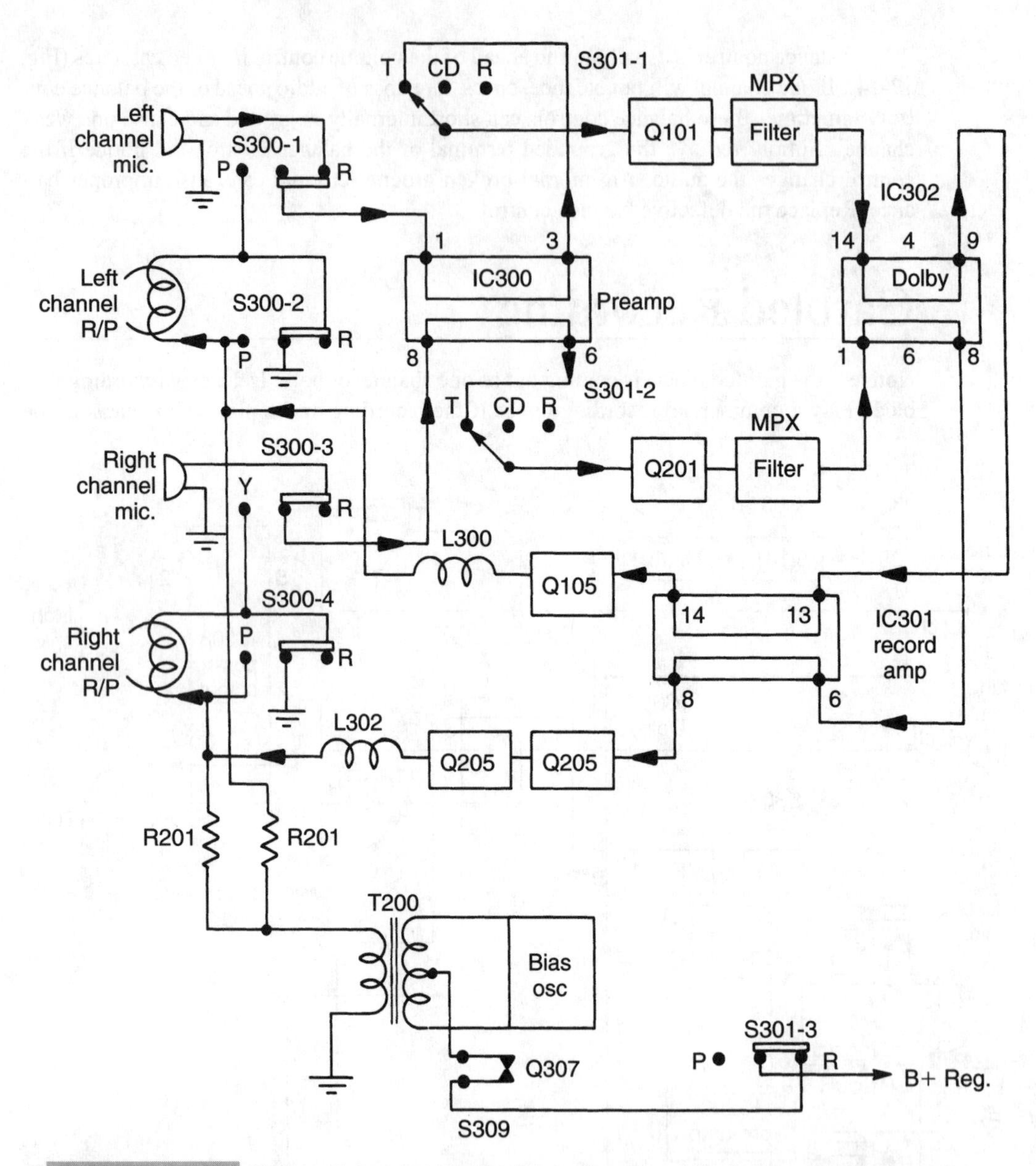

FIGURE 17-12 **Follow the arrows in the Record mode of the left and right stereo channels.**

problem. Clean the R/P and erase heads (Fig. 17-15). Spray fluid in the record/playback switch contacts. Recheck the machine with a good cassette.

If the garbled music is in the recorded cassette, try another new cassette and record 10 minutes of music. If the new recording remains garbled, check the oscillator bias circuits. Be sure that the bias signal is applied to each stereo R/P tape head. Scope the R/P tape heads. Check the erase head for the dc applied voltage in the record mode. Often, garbled music out of one channel indicates that a tape head or cassette is dirty. If the sound is garbled in both channels, check the erase and R/P heads, the bias oscillator circuit, and the defective amplifier channel.

Cassette Door Will Not Open

Suspect that tape has wrapped tightly around the capstan if the cassette door will not open (Fig. 17-16). Do not pry open the door. Remove the back cover and try to locate the capstan/flywheel. Rotate the capstan backwards by hand. This action will loosen the tightly wrapped tape so that the door can be opened.

Of course, the cassette tape might be damaged to the point where it cannot be used or saved. Sometimes the tape can be saved if it unwinds without any problems. You will have to cut the tape to remove all of it from capstan and pinch roller. Doublecheck the bearing of the pinch roller for excess tape wrapped between rubber roller and the outside bracket support. Clean the capstan, pinch roller, and spindles after the tape spills out.

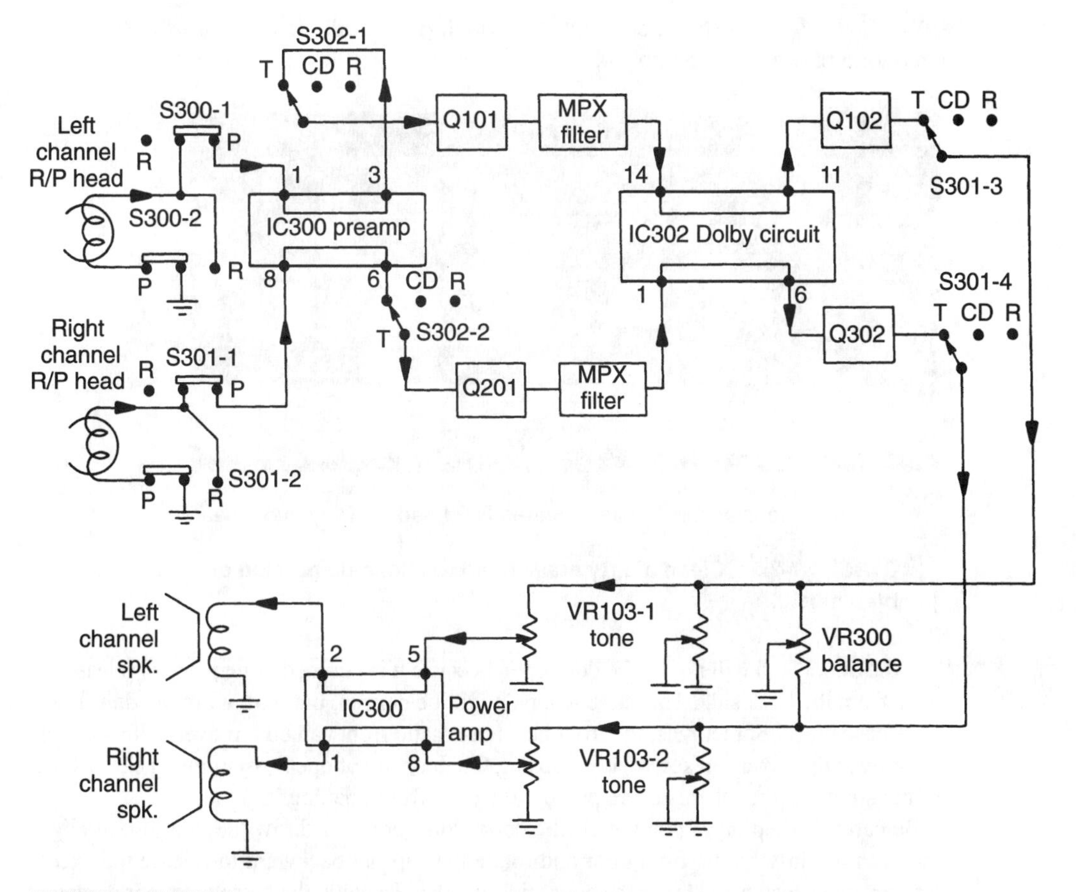

FIGURE 17-13 Follow the dark arrows in the stereo channels of the Play mode.

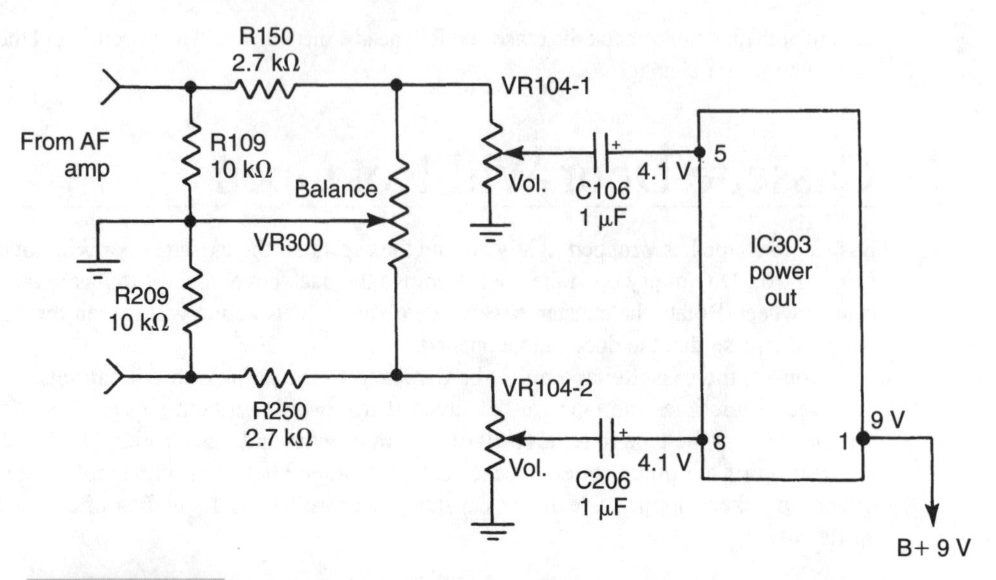

FIGURE 17-14 **The balance control, balancing the audio to ground with VR300, is often ahead of the volume control.**

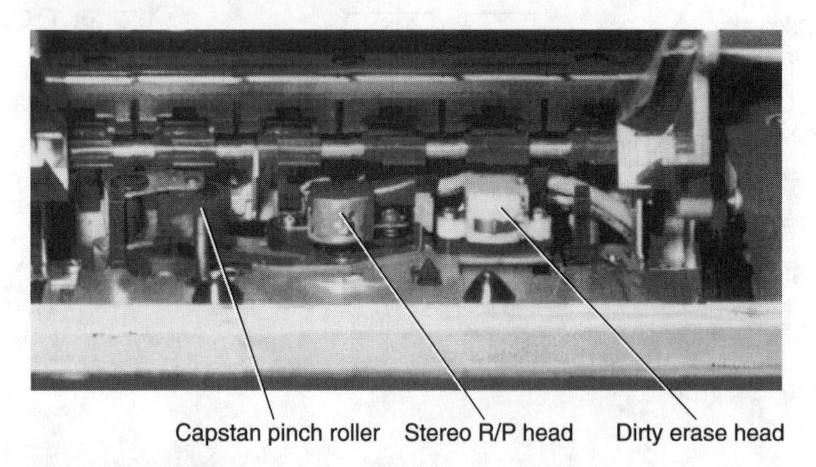

FIGURE 17-15 **Clean a dirty erase head to eliminate garbled or jumbled music.**

If the door catch will not let the door open because it is jammed or bent, try to release the door from the back side. The cassette tape might be normal, but have a broken door hinge or release piece. Sometimes, the front bezel or plastic front can be removed with a couple of screws; then you can see what is causing the door to not open. A jammed gear or lever of the slow-eject mechanism can prevent the door from opening.

Be careful when prying on the plastic door; don't break or throw the door out of alignment. Pry lightly on the front door and rotate the capstan backward to release the excess tape around capstan and leave the door slightly open. Usually, the cassette tape is damaged before the door is released and the cassette is removed. It's much easier to purchase another cassette than it is to locate another cassette door.

Cleaning the Optical Lens

Be very careful when working around the CD player to avoid exposure to laser-beam radiation. You can damage your eyes if you stare at the bare optical lens assembly while the player is operating, so keep your eyes away from that area. Obey the warning labels that are fastened to the laser optical assembly. Always keep a disc loaded while repairing the CD player (Fig. 17-17). Remember, the laser beam is not visible, like that of an LED or a pilot light. Do not defeat any interlock switches to check the laser operation.

A dirty optical lens can cause improper searching and playing of the laser assembly. Before you attempt to repair the optical sensor assembly, clean the lens area. Excess dust, cigarette or cigar smoke, stains, and tarnishes can occur on the optical lens. The laser beam

FIGURE 17-16 The jammed cassette with tape pulled out will not let the cassette door open.

FIGURE 17-17 Always have a CD on the turntable while servicing CD player circuits.

FIGURE 17-18 A closeup view of the laser lens assembly. Do not stare directly at the lens area.

can't reach the compact disc. Be careful not to apply too much pressure when cleaning the optical lens or you might damage the lens assembly (Fig. 17-18).

Clean the lens with lint-free cotton or with cleaning paper for camera lenses that is moistened with a mixture of rubbing alcohol and water. Cleaning solution for cameras is ideal. Wipe the lens gently to avoid bending the supporting spring. Blow excessive dust from the optical lens with a can of dust spray, which is available in camera departments.

CD Motors

In the standard CD player, a loading motor slides out to load the disc. In the portable CD/cassette player, push a button and the top lid raises up or comes out from the front so that the disc can be loaded. The lid is pushed down or in to load the CD manually (Fig. 17-19). The top lid (655) is hinged at the back so that the disc can be placed on the turntable of the disc motor assembly.

DISC MOTOR

The CD motor rotates the disc at approximately 500 rpm at the inside and slows down to about 200 rpm as the laser assembly moves toward the outer rim. The disc motor is also called a *spindle* or a *turntable motor* by other manufacturers. The disc-motor shaft is directly fastened to the disc table (Fig. 17-20).

The disc-motor voltage is supplied through a driver transistor or IC. The driving transistors are operated from a servo-controller IC (Fig. 17-21). The driver voltage is zero at the emitter terminals until a signal voltage is applied to the base terminals. This voltage can be less than 10 V in the disc-motor circuits.

Suspect a defective motor, drivers, and no servo signal if the disc motor will not rotate. Quickly measure the voltage across the spindle motor. If it has no motor voltage, check the voltage that is supplied to the driver transistors. Remove one lead of the disc motor and check the winding with the R × 1 ohmmeter range. Replace an open motor with the exact-type component.

SLED MOTOR

The slide, sled, carriage, or feed motors move the optical pickup assembly across the disc from the inside to the outside rim of the CD. The motor is gear-driven to a rotating gear (108 and 107 in Fig. 17-20). The gears glide the optical assembly down a sliding rod. Usually, the SLED motor is driven by a transistor or IC circuit (Fig. 17-22).

Check the slide motor-drive circuits with accurate voltage and resistance measurements. Be sure that both positive and negative voltages are present at the respective driver transistors or ICs. Take in-circuit transistor tests for each transistor with the diode transistor test of the DMM. Check the continuity of the motor winding with R × 1 ohmmeter scale. Remember, in intermittent operations, one transistor might be opening only under load conditions. It might be necessary to replace both transistors.

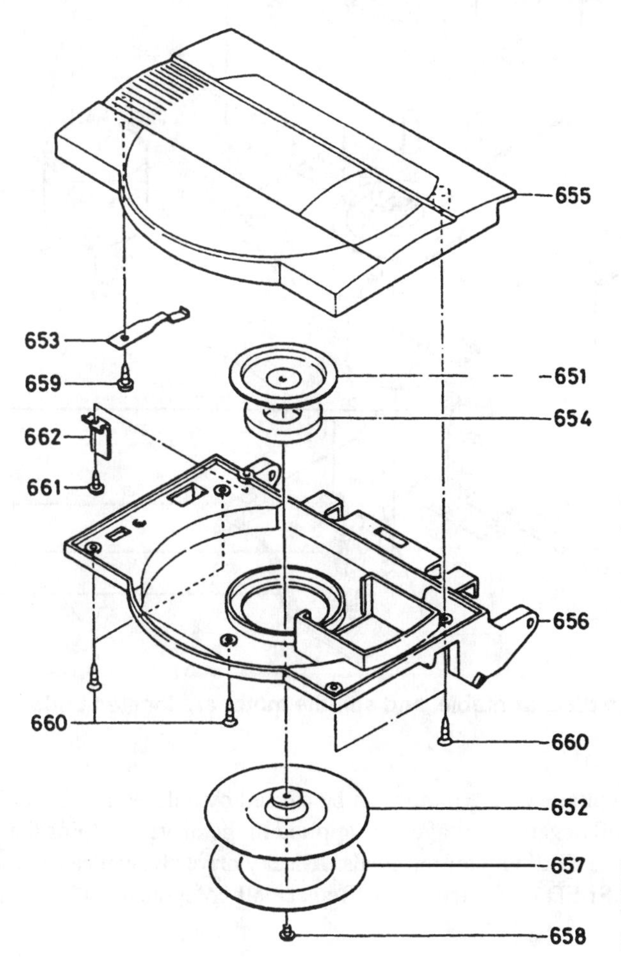

FIGURE 17-19 How the CD turntable and top lid are mounted in a CD case.

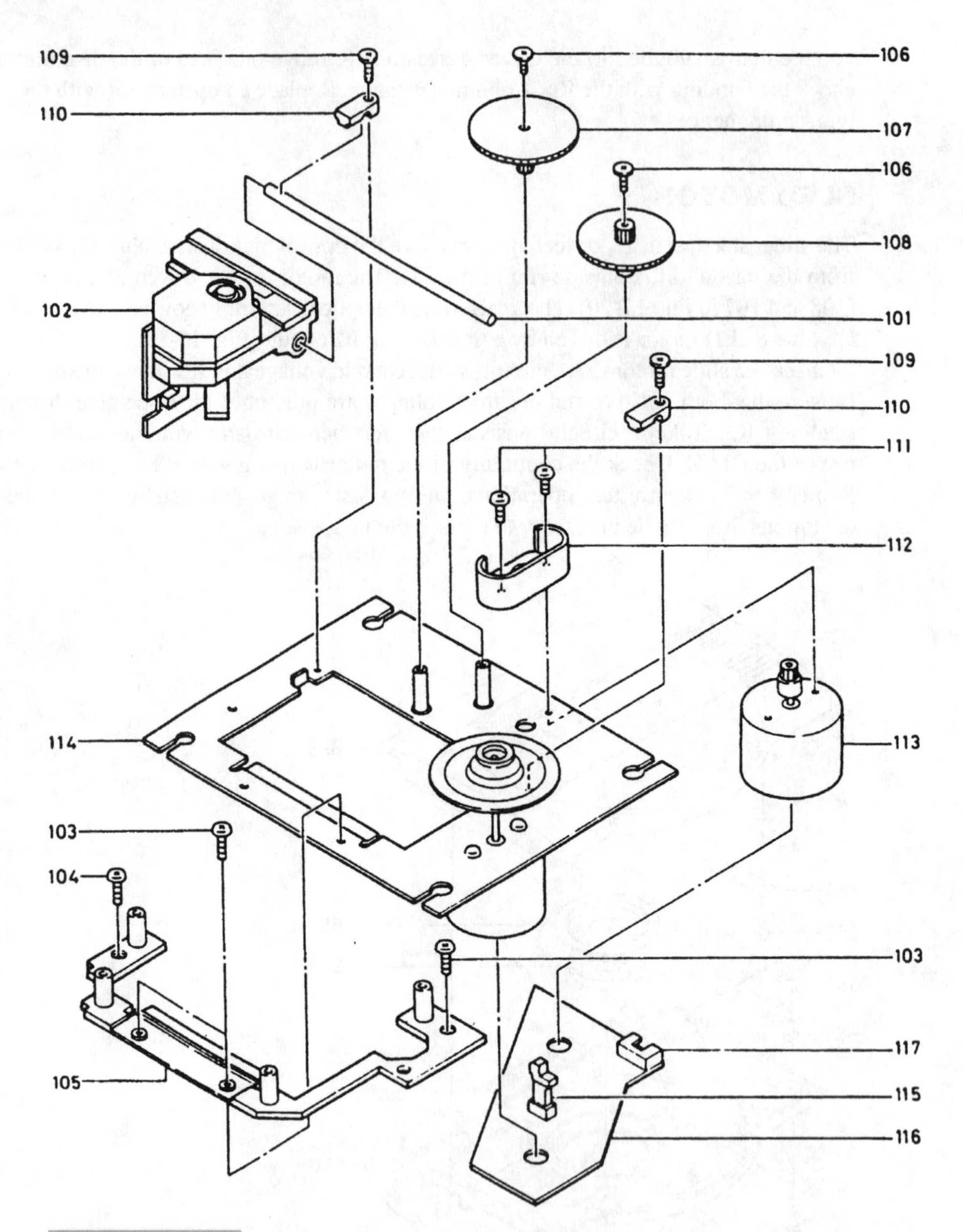

FIGURE 17-20 **The disc, turntable, and spindle motor are located under the CD.**

An improper output voltage on the motor can be caused by a defective driver IC. Be sure that the positive and negative voltages are applied to the driver IC (pins 6 and 10). Monitor the voltage at the SLED motor terminals. A leaky driver IC can cause improper voltages applied to the SLED motor terminals. Replace all components with exact-type parts.

CD Block Diagram

Although, the servo, Dac, DSP, and laser-pickup circuits are fairly complicated, many circuits can be checked in the CD player (Fig. 17-23). The disc and SLED motor circuits can easily be checked with voltage measurements on the DMM. The laser, door, and P-set

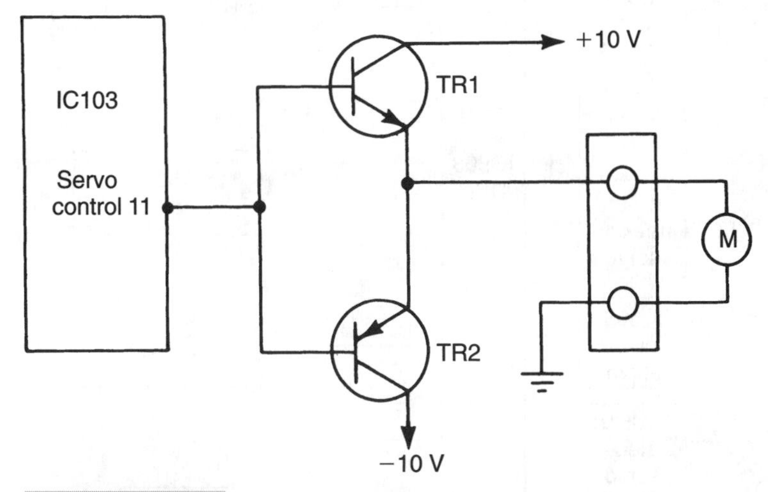

FIGURE 17-21 The disc motor voltage is controlled by driver transistors TR1 and TR2 with servo control IC103.

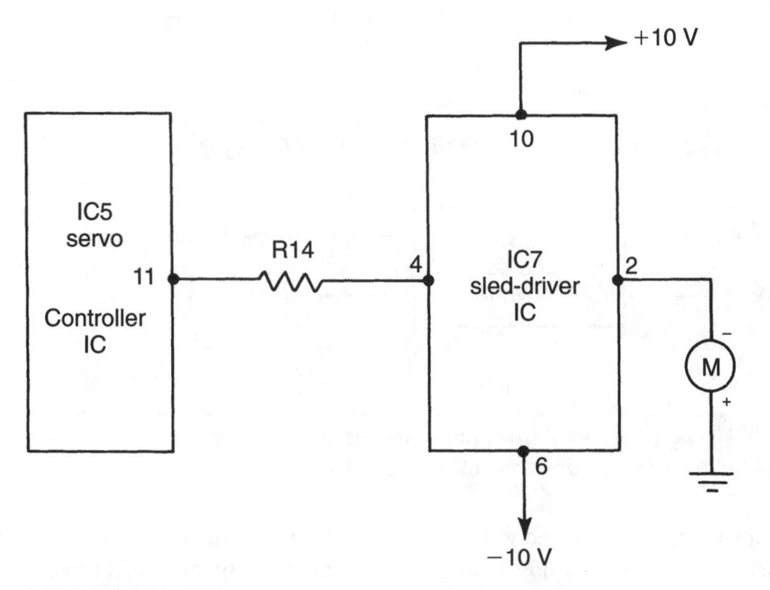

FIGURE 17-22 A block diagram of the SLED motor, driven by a voltage from the IC7 driver and a signal from servo controller IC5.

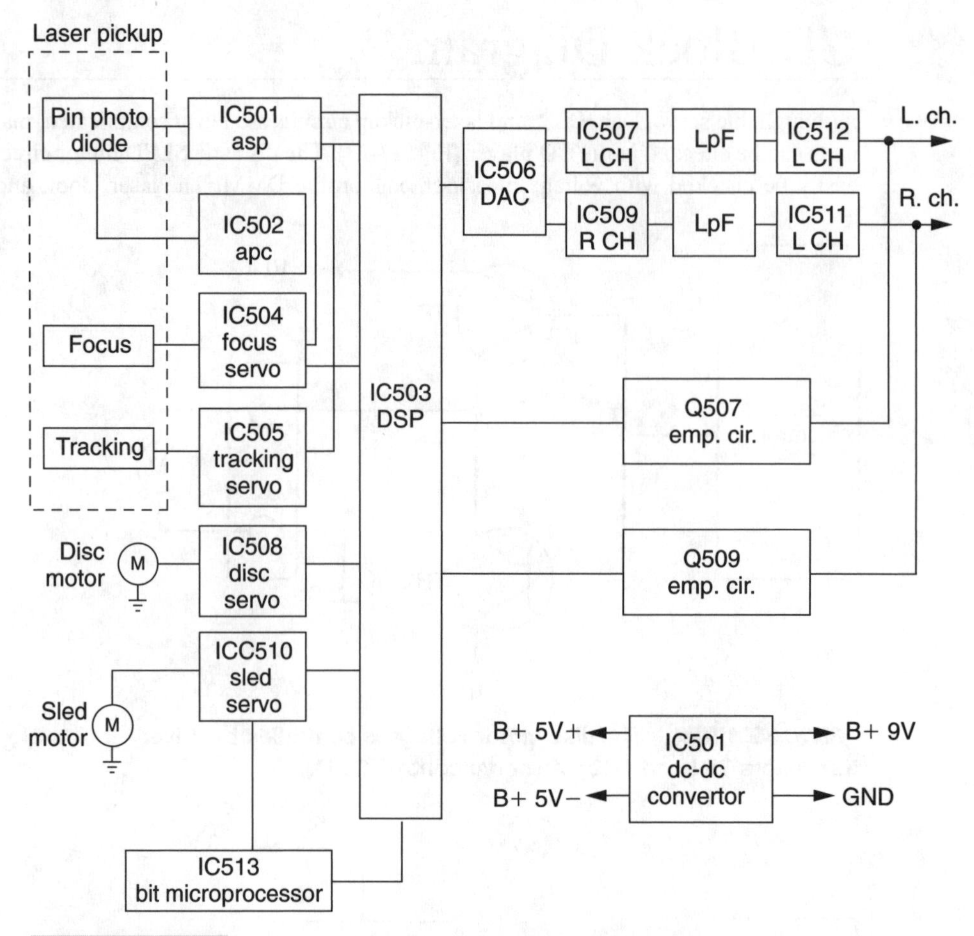

FIGURE 17-23 A block diagram of a portable CD player.

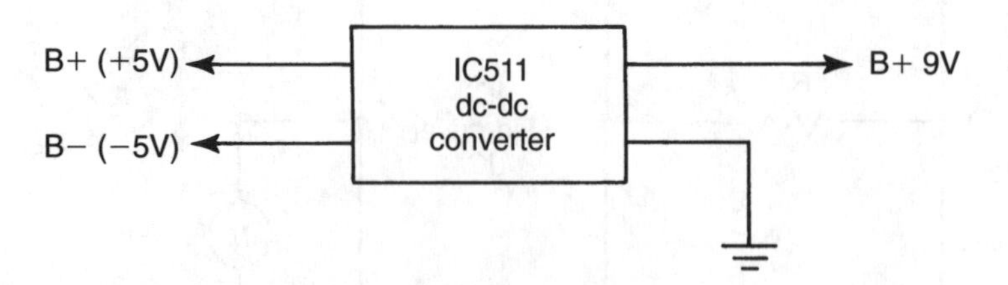

FIGURE 17-24 The +9 V from the batteries or ac power supply is fed to a dc/dc converter circuit IC511.

switch contacts can be checked with the ohmmeter. The dc/dc converter input and output voltages, and power-supply voltages cause many problems within the CD player (Fig. 17-24).

If either the positive or negative 5 V is missing, check the output voltage from the dc/dc converter IC. Check the B+ input voltage applied to the V_{CC} terminal of the converter IC

(Fig. 17-25). The supply voltage (+9 V) is applied to pin 9 of IC511. Check the 5-V source at TP22 and TP23. If the negative and positive 5 V are found to be normal, the power supply and dc/dc converter circuits are okay.

CD Low-Voltage Power Supply

The CD power-supply voltage (+9 V) is switched into the CD circuits with S301-5 and S301-6. Check S301-6 if the cassette player and radio circuits operate. Clean S301-5 and S301-6 if CD operation is erratic or intermittent. The CD player utilizes the same power-supply source as the cassette player. The dc voltage from the 9-V batteries or dc power supply is applied to S301-5. S301-6 switches the CD regulator circuits to the output voltage to the CD-player circuits (Fig. 17-26).

Check the B+ voltage that is fed to switch S301-5 from the ac power supply or batteries with the DMM (Fig. 17-27). If the voltage is low on both the batteries and the ac power supply, suspect that a circuit in the CD player is overloaded. Switch to the cassette player and notice if the +9 V returns. Check the CD player circuits if the voltage is low or normal on the cassette player or radio. Do not overlook the possibility that zener diode ZD315 might be leaky or R331 might be burned.

No Sound From the CD Player

The digital-analog converter (DAC) changes digital signals to audio signals at IC507. The analog audio-signal output of IC507 is applied to analog IC509-1 and IC509-2. The low-pass filters (L504 and L505) are located between outputs IC510-1 and IC510-2. The audio

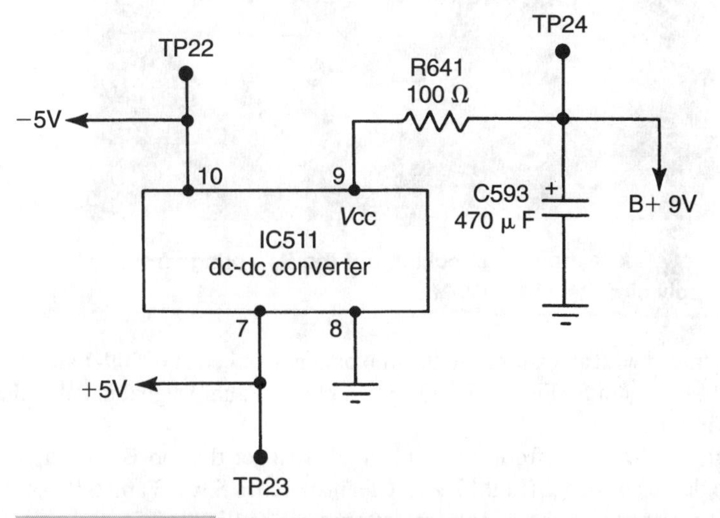

FIGURE 17-25 The +9V is fed into pin 9 of IC511 and the +5V at pin 7 and –5V at pin 10.

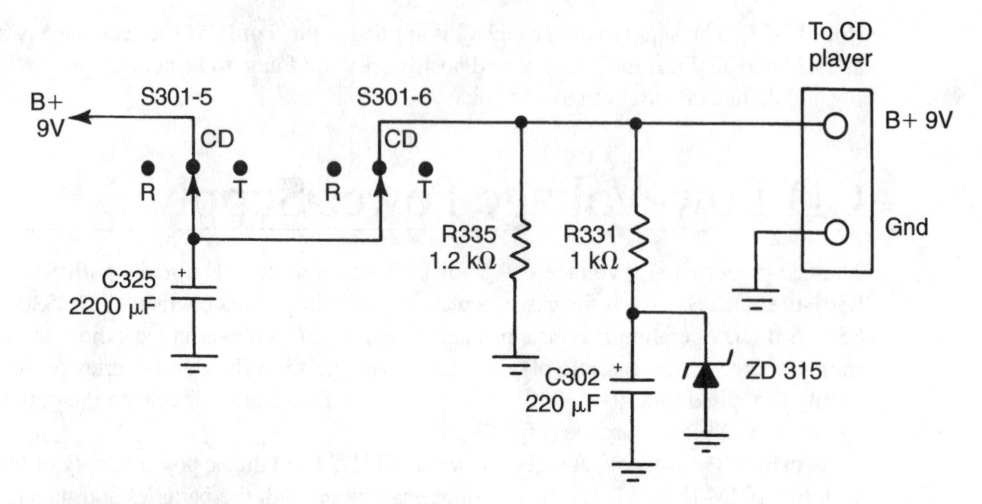

FIGURE 17-26 The +9 V must be fed through S301-5 and S301-6 before being regulated by ZD315.

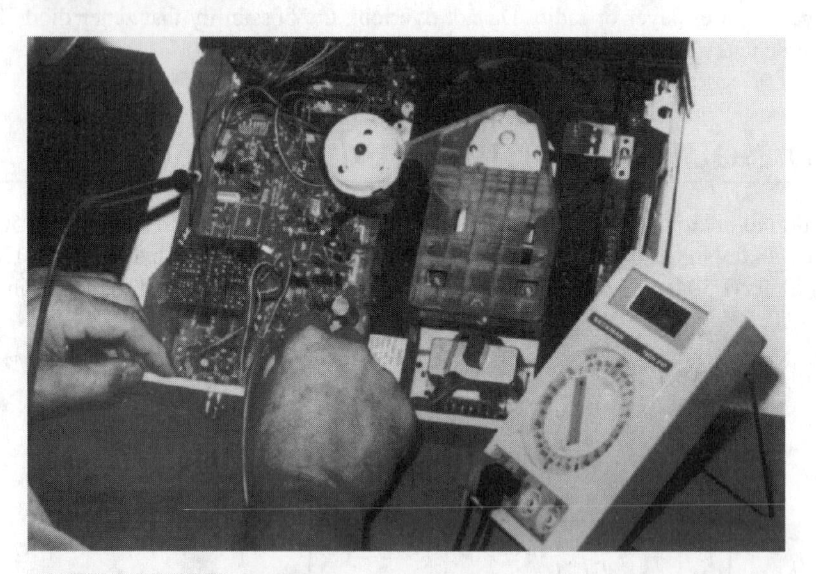

FIGURE 17-27 A technician checking out the B+ voltage from the power-supply circuits with a DMM.

signal can be traced with an external audio amp or signal traced at IC509-1 and IC509-2 to the RCH and LCH outputs (Fig. 17-28). The two-channel audio signal is fed to the audio-output amp circuits.

If you hear no CD audio from either channel, suspect that no B+ voltage is being passed to the dc/dc converter (Fig. 17-29). Clean switches S301-3 and S301-4 if the CD audio is erratic or nonexistent. If correct voltages are applied to IC509-1 and IC509-2, and IC510-1 and IC510-2, suspect that an IC is defective. If only one channel of audio is missing or weak, signal trace the audio at the input of IC509-1 and IC509-2 through to switch S301-3 and S301-4. If the audio signal stops or appears weak, check the voltage in the IC circuit.

Both of the stereo analog IC circuits are identical and can be compared to the normal channel (Fig. 17-30). Remember, the audio signal at pin 3 of IC509-1 and pin 5 of IC509-2 is rather weak; the audio signal will increase after each IC amplifier. If the audio signal is the same at pins 3 and 5 of IC509 and weak at pin 1 of IC510-1, suspect that IC509-1 (or a component in the circuit between pin 1 of IC509-1 and pin 3 of IC510-1) is defective. Also, C562 or L504 and IC509-1 could be open.

Likewise, if the audio signal is the same at output pins 1 of IC510-1 and pin 7 of IC510-2 and weak at CD player switch S301-4 of the right channel (Fig. 17-31), suspect IC510-2, R609 (1 kΩ), C575 (10 μF) and R614 (4.7 kΩ). Check the audio signal at TP17 and TP18 and see if the amount of volume is the same. If the signal is quite close, check S301-4 for dirty contacts. Always clean the function switch if either the radio, tape, or CD audio is missing.

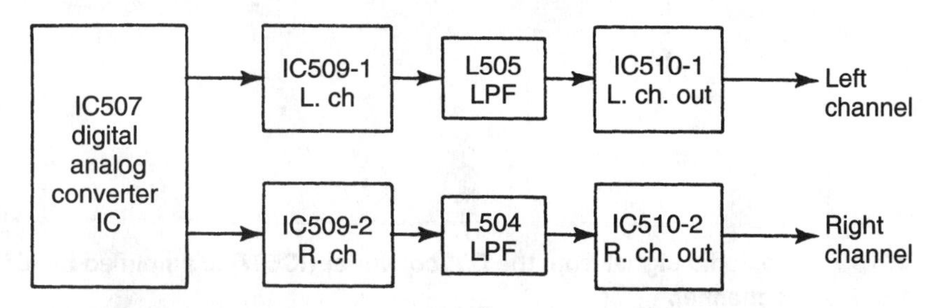

FIGURE 17-28 You can signal trace the audio signal from D/A converter IC507 with an outside amplifier or signal tracer.

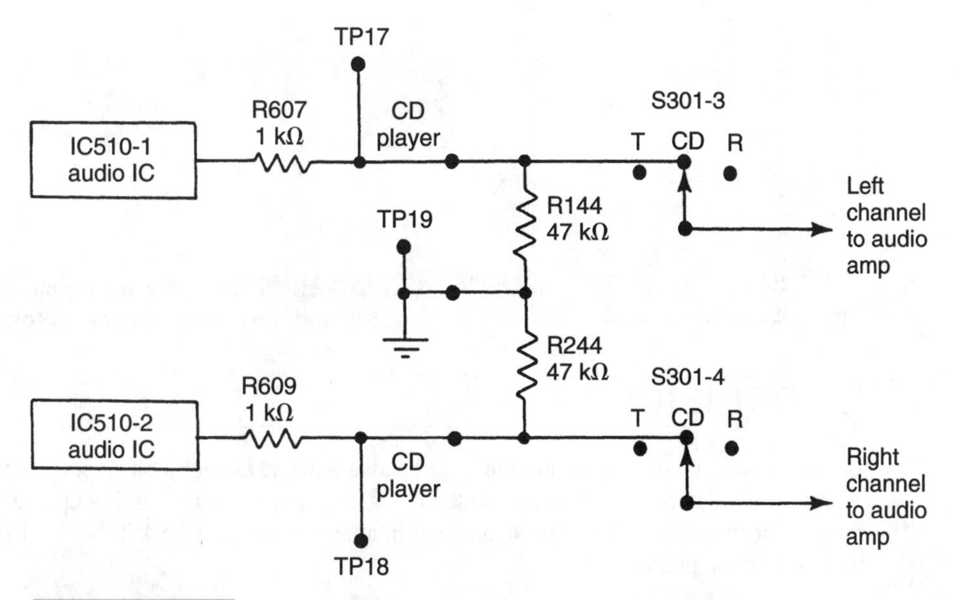

FIGURE 17-29 The CD stereo audio signal at TP17 and TP18 are fed to function switch S301-3 and S301-4 to the main stereo amplifier.

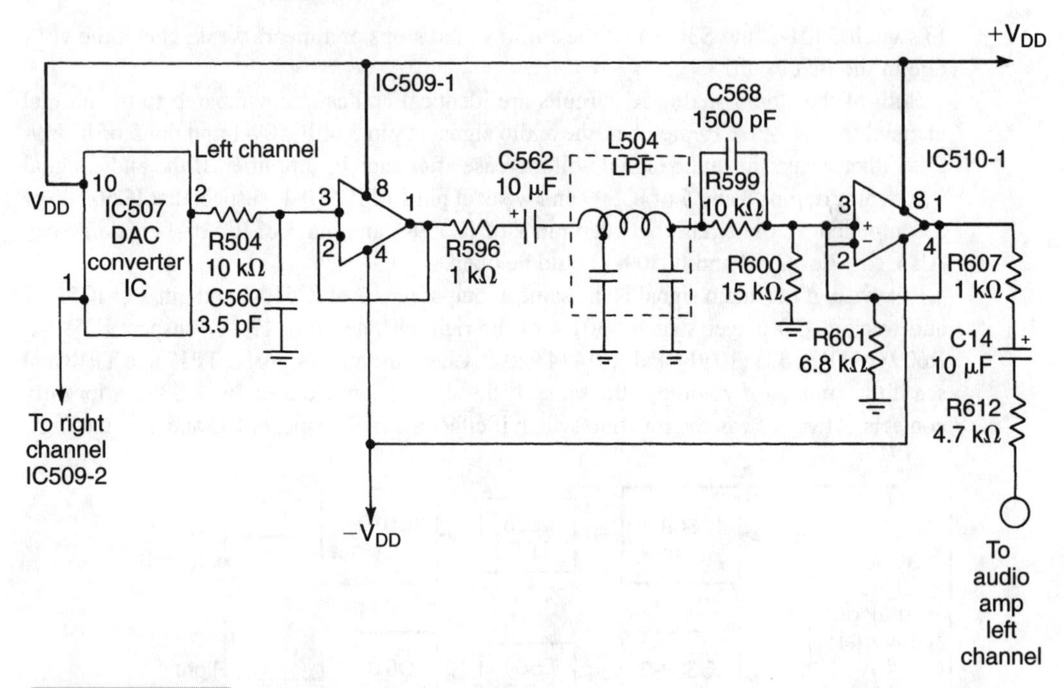

FIGURE 17-30 The audio signal from the D/A converter (IC507) is amplified by IC509-1 and IC510-1 of the left channel.

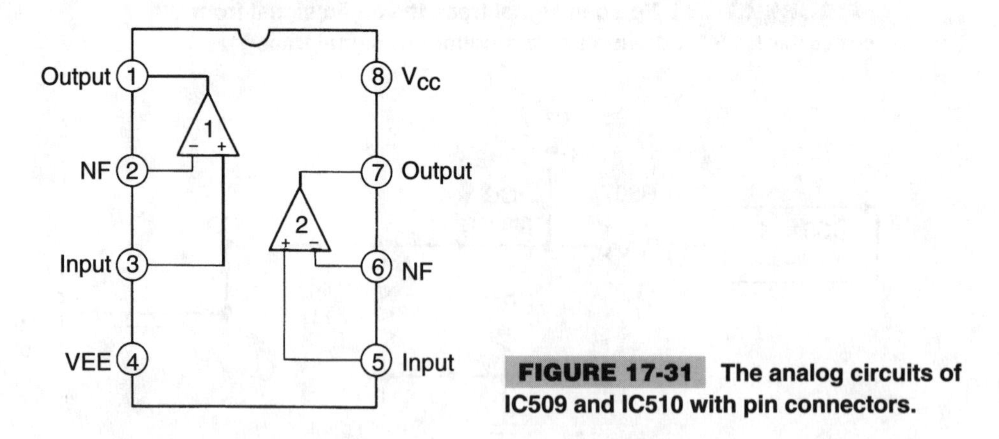

FIGURE 17-31 The analog circuits of IC509 and IC510 with pin connectors.

Conclusion

Be careful when working around the laser head assembly. Do not look directly at the assembly while operating it. You can watch the laser assembly search, going up and down, from a side view. Keep the CD disc on the turntable at all times. Check Table 17-1 for CD troubleshooting procedures.

TABLE 17-1 THE RADIO AND CASSETTE TROUBLESHOOTING CHART

SYMPTOM	CAUSE	REMEDYPUT
Output level too low	Power amplifier circuit	
Overall:	1. Faulty IC303.	1. Check and replace
	2. Shorted C131 and C231.	2. Check and replace.
Tape	Pre-amplifier circuit Faulty IC300.	Check and replace.
FM:	FM front/end circuit	
	1. Faulty IC1.	1. Check and replace.
	2. Open or shorted C2.	2. Check and replace.
	3. Open or shorted VC.	3. Check and replace.
	FM decode circuit	
	1. Faulty IC3.	1. Check and replace.
	2. Shorted C32, C33, C34, C35, R26 and R27.	2. Check and replace.
	3. Faulty C36 and/or C37.	3. Replace the faulty capacitor.
AM/FM:	FM-AM IF amplifier circuit Faulty IC2.	Check and replace.
Poor tape high frequency response	1. Incorrect head azimuth.	1. Adjust head azimuth.
	2. Faulty REC/PB head (62).	2. Clean and replace REC/PB head.
	3. Shorted C100 and C200.	3. Check and replace.
No sound	Power supply circuit	
	1. Faulty S301 and S306 or poor contact.	1. Check and replace.
	2. Faulty J401 and J402 or poor contact.	2. Check and replace.
	Power amplifier circuit	
	1. Open or shorted IC303.	1. Check and replace.
	2. Open or shorted C131 and C231.	2. Check and replace.
	Output circuit	
	1. Open or shorted speaker voice coil.	1. Check and replace.
	2. Faulty J301 or poor contact.	2. Check and replace.
No sound	Pre-amplifier circuit	
	1. Open or shorted IC300.	1. Check and replace.
	2. Open or shorted REC/PB head (62).	2. Check and replace.
	3. Open REC/PB head leads.	3. Check REC/PB head leads.
	4. Open VR104.	4. Check and replace.
	FM tuner circuit	
	1. Faulty S1 or poor contact.	1. Check and replace.
	2. Faulty IC1.	2. Check and replace.

TABLE 17-1 THE RADIO AND CASSETTE TROUBLESHOOTING CHART (CONTINUED).

SYMPTOM	CAUSE	REMEDYPUT
	AM converter circuit	
	1. Open or shorted L4, L5 and/or T2.	1. Check and replace.
	2. Shorted PVC.	2. Check and replace.
	AM/FM IF amplifier	
	1. Faulty IC2.	1. Check and replace.
	2. Open or shorted L3.	2. Check and replace.
	3. Shorted CF1, CF2, and/or CF3.	3. Check and replace.
Tape inoperative	1. Motor (64) dead.	1. Check motor lead-wires and replaced motor.
	2. Capstan belt (55) slipping.	2. Wipe flywheel (43) and replace capstan belt.
	3. Leaf switch (S305) poor contact.	3. Adjust or replace leaf switch.
Won't take-up tape	Capstan belt (55) slipping.	Wipe flywheel (43) and/or replace capstan belt (44).
No fast-forward and rewind	Clutch assembly (40) slipping.	Wipe flywheel (43), clutch assembly (40), and/or replace FF/rewind belt (41).
Excessive wow	1. Motor (64) defective.	1. Replace.
	2. Pinch roller (32) dirty.	2. Clean or replace.
Uneven speed	1. Motor (64) defective.	1. Replace.
	2. Motor pulley (54) slipping.	2. Adjust or replace motor pulley.
	3. Capstan belt (55) slipping.	3. Wipe flywheel, motor pulley and replace capstant belt.
No playback	1. REC/PB head (62) defective or open.	1. Replace.
	2. REC/PB head dirty.	2. Wipe REC/PB head with a cloth moistened with alcohol.
	3. Open or shorted REC/PB head leadwires.	3. Replace wire.
	4. No power to amplifier.	4. Replace leaf switch (S305).
	5. Defective component(s) in amplifier.	5. Check and replace defective component(s).
Low playback or distorted playback output	1. Amplifier defective.	1. Check and replace defective component(s)
	2. REC/PB head dirty.	2. Wire REC/PB head with a cloth moistened with alcohol.
	3. REC/PB head badly worn.	3. Replace.

TABLE 17-1 THE RADIO AND CASSETTE TROUBLESHOOTING CHART (CONTINUED).

SYMPTOM	CAUSE	REMEDYPUT
No erase	1. Erase head (63) defective or open.	1. Replace.
	2. Open or shorted erase head leadwires.	2. Check and replace.
No record	1. REC/PB head (62) defective or open.	1. Replace.
	2. Component(s) in amplifier defective.	2. Check and replace defective component(s).
	3. REC/PB head dirty.	3. Wipe REC/PB head with a cloth moistened with alcohol.

18

SERVICING MICROCASSETTE AND PROFESSIONAL RECORDERS

CONTENTS AT A GLANCE

CONTENTS AT A GLANCE

Most microcassette recorders operate from batteries for portability, but the ac adapter can be used on some models. These units require two AA batteries. In some units, nicad batteries and ac chargers/adapters are used. These small recorders use a total of three volts (Fig. 18-1).

The compact and easy-to-operate microcassette recorder might have a VOR system. This voice-operated recording system is economical for tapes, batteries, and operation time. You can start recording directly from the Play mode in several units. This function is convenient when correcting a previously recorded portion.

FIGURE 18-1 The small microcassette recorder might operate from two small AA batteries.

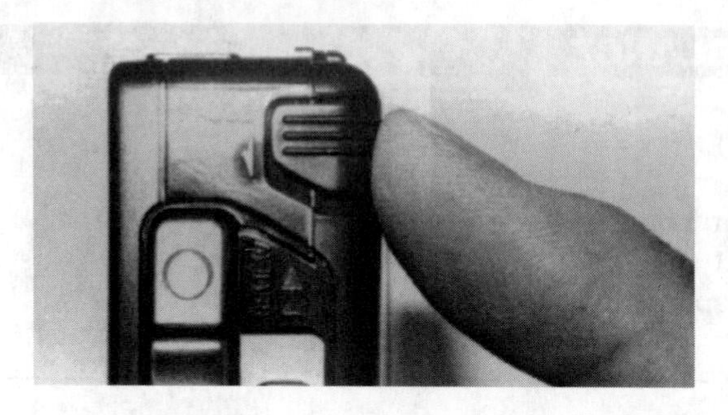

FIGURE 18-2 To pause momentarily, place the Pause switch in the direction of the arrow.

Some models have an automatic shutoff mechanism. When recording or playing, the tape stops at the end and the locked button will be released automatically (automatic shutoff). Besides a cue (fast forward) and review (rewind) control, some have a pause switch. The stop/eject button completely stops the motion and ejects or raises the microcassette. To stop the tape momentarily in the Play or Record mode, slide the pause switch in direction of the arrow (Fig. 18-2).

The deluxe microcassette recorder has two speeds (2.4 cm and 1.2 cm). The 2.4-cm speed is recommended for normal use. A 60-minute recording can be made using both sides of an MC-60 microcassette. A built-in tape counter and instant edit features are included in some models. A full-featured recorder might have auto reverse and a speaker.

Although most units have a built-in condenser microphone, the ultra-compact recorder might have a detachable external microphone system. This small microphone is usually located at the top end of the recorder.

Microcassettes

The microcassette measures only 1¼ × 2 inches and they only fit in the microcassette player/recorder. These cassettes can be purchased in 4-packs (Fig. 18-3). The standard microcassette has a small indentation on the top side of the plastic case. The nonstandard microcassette will not play in the standard recorders because their "L" dimension is different (Fig. 18-4).

The MC-60 microcassette will play for 60 minutes by using both sides at the 2.4-cm speed. Switch the tape speed to 1.2 cm for 120-minute operation. To prevent the cassette from accidentally being erased, break out the plastic tab at end of the top right-hand side. To reverse the process, cover the broken-out slot with tape. Remember, if a certain cassette will not record, check for broken tab at the end of the cassette.

TO RECORD WITH BUILT-IN MICROPHONE

Set the desired tape speed (2.4 cm) for normal use. Insert the cassette correctly. If the microcassette recorder has VOR recording, switch VOR to the L position (recommended for dictation). Switch the VOR off, if you don't want to use that operation. Push in the record switch and start recording (Fig. 18-5). Notice that when it is in VOR operation, the cassette will rotate only during periods of noise or talking.

For quick review, simply push the cue/review control toward review while recording. When the control is released, it begins to play back the recorded material. On some models, you can listen to the sound being recorded through the earphone jack. If you want to stop the tape momentarily when either playing or recording, slide the pause switch in the direction of the arrow.

FIGURE 18-3 **Microcassettes can be purchased in packs of four.**

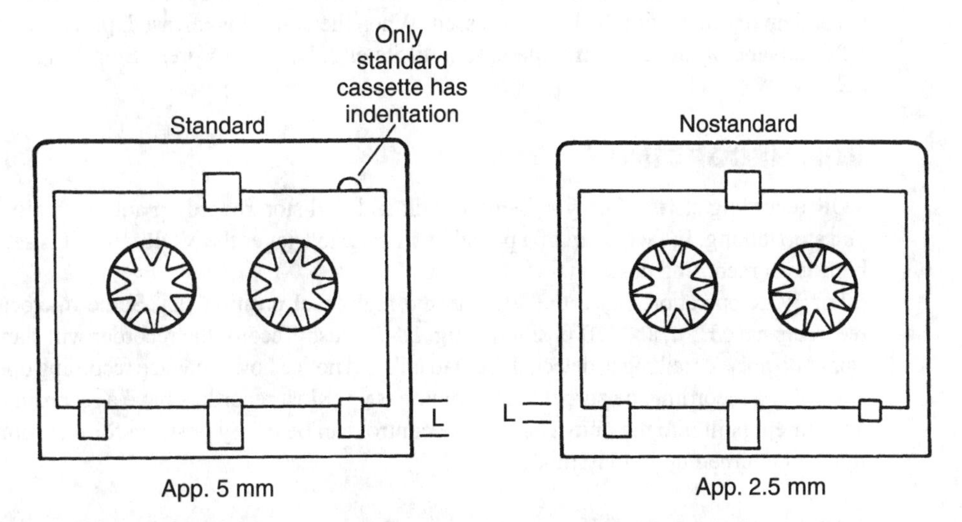

FIGURE 18-4 **The correct dimensions of the cassette indicates if it is a standard or non-standard cassette.**

FIGURE 18-5 The Record switch is located on the right side of this Sony M-440 V-VOR Pressman microcassette recorder.

FIGURE 18-6 To play, press the Play button.

PLAYBACK MODE

The same tape speed must be used for playing and recording. If not, the recording will sound too fast or too slow. To hear what you have recorded, press Rewind. To rapidly advance the tape, slide the cue/review control to the Cue position. To play the tape, simply press Play (Fig. 18-6).

To skip over or repeat a certain portion during playback, keep the cue/review control pushed up or down with the button pressed. When the control is released, playback begins. After advancing or rewinding the tape until its end, be sure to press Stop to release the cue/review switch.

VOR RECORDING

VOR recording starts when you begin to talk and will stop recording automatically when you stop talking. In fact, a tap of a pencil or finger can trigger the VOR when it is set in the high-level recording position.

For VOR operation, place the VOR in the high-level position (L). Some microcassette recorders have H, L, and Off positions (Fig. 18-7). Push Record; the recorder will start, then stop if no noise or talking is detected. Start to talk and notice how the voice recording operates.

To play a recording, be sure that the volume is up. Most recorders have a recording-level adjustment built into the unit. The volume control can be in any position for recording, but it must be turned up in Play mode.

CIRCUIT BLOCK DIAGRAM

The recording and playback circuits might use two or three ICs and several transistors. Only transistors were used in early microcassette recorders. Today, you can find ICs and transistors with surface-mounted components. Surface-mounted transistors, IC processors, capacitors, and resistors are ideal to fit into a small hand-held recorder.

The block diagram of the small microcassette recorder is rather simple with only a few components (Fig. 18-8). The solid black-and-white arrows indicate the signal paths when recording or playing. IC (U101) contains the play/record switch, automatic gain control, reference voltage, VD, audio preamp, audio power output, and alarm circuits. U101 can contain a 24- or 30-terminal flat surface-mounted component with wing-tip terminal connections.

FIGURE 18-7 Most standard VOR recording operation is done in the high-level position.

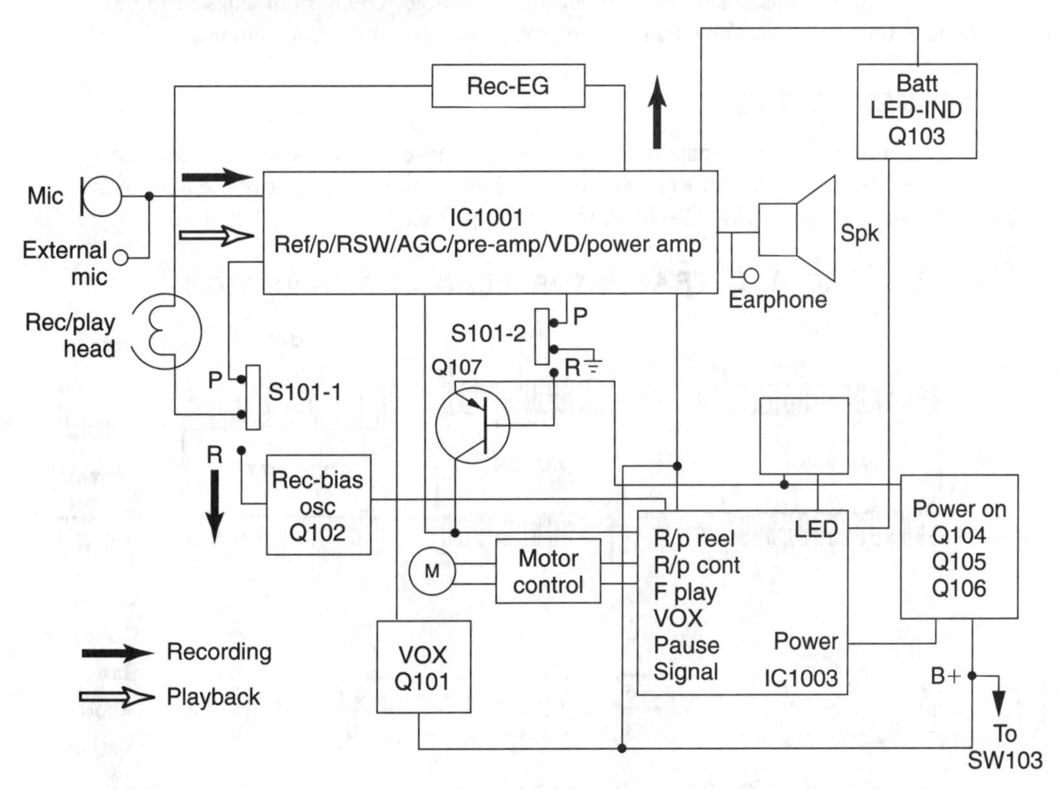

FIGURE 18-8 A typical block diagram of the microcassette recorder.

U104 is the system-control IC that controls record/play operations, motor operations, the VOX, the pause LED, and power and voltage regulators. U103 is an 8-pin motor-control surface-mounted IC. The recording bias, reel operation, battery LED circuits, power control, and the VOX are controlled with separate transistors.

First, determine where the trouble might be on the block diagram, then locate the same section on the schematic. Remember, different voltages can be measured when in different modes. Usually, this occurs in Play and Record.

SURFACE-MOUNTED COMPONENTS

Surface-mounted components are tiny in size compared to regular parts in the cassette player. Most of these parts are mounted flat against the PC board. Besides ICs, the transistors are three-legged components with flat leads to solder on the copper foil (Fig. 18-9). Notice that two diodes in one component look like a regular transistor.

Besides solid-state components, resistors, and capacitors are contained in small chunks or rectangular objects. Often, small capacitors are larger than resistors. The contact of each surface-mounted capacitor or resistor is located at each end. The tinned end piece is soldered directly to the PC board wiring. Usually, these small components are damaged when they are removed from the circuit.

Check and doublecheck each suspected component before deciding if it is defective. Be sure that the component is defective before removing it from the circuit. Some larger ICs are glued into position when they are manufactured. Often, both sides of the PC board have foil wiring to connect all components in the record/playback circuits.

DISASSEMBLY

To get at certain components for voltage, resistance, and inspection checks, the bottom cover must be removed. Remove the three bottom screws to remove the bottom cover in this microcassette (Fig. 18-10). Place the flat pan screws in a small dish, saucer, or lid so

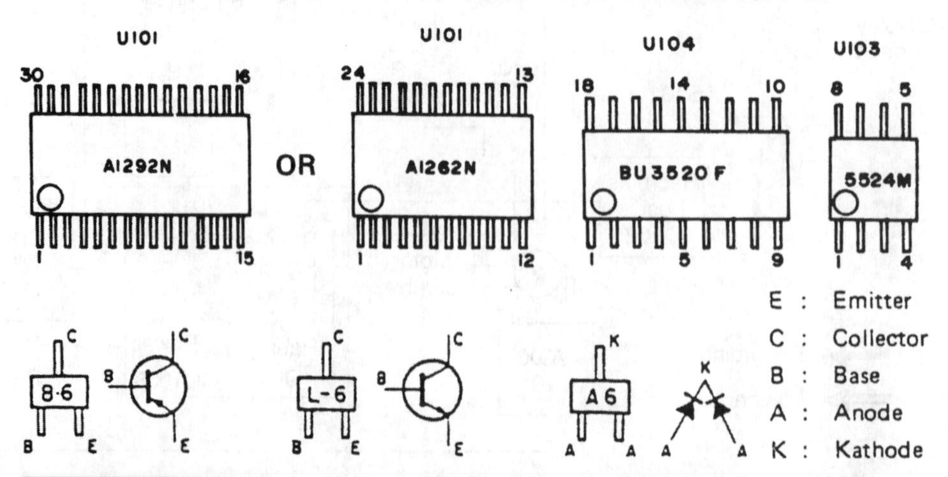

FIGURE 18-9 Surface-mounted IC and transistor lead identification in some microcassette recorders.

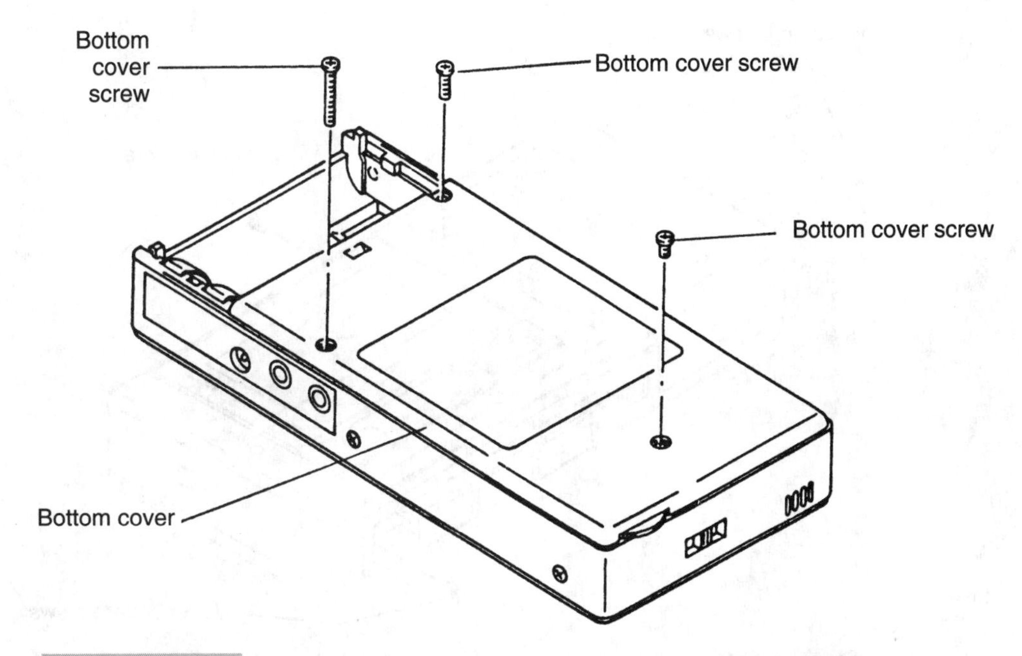

FIGURE 18-10 **Remove three screws to remove the bottom cover in the microcassette recorder.**

that they will not get misplaced. Remove three small Phillips screws to remove the bottom cover in the Sony Pressman model M-440V.

Remove the five upper cover screws in the cassette player body to remove the top cover (Fig. 18-11). This action will release the whole inside body of the recorder. In some models, after the bottom cover is removed, the main chassis can be removed by taking out several screws that hold the chassis to the top cover assembly.

REMOVING THE PC BOARD

Remove the bottom cover. Remove the two amplifier screws. Disconnect the following wire leads (Fig. 18-12):

1 Disconnect the two head leads (brown and black).
2 Remove three motor leads (red, black, and purple).
3 Remove two battery leads (red and black).
4 Disconnect the two speaker leads (green).
5 Remove the two mic leads (red and white).

REMOVING THE CASSETTE DECK

To remove the cassette deck from the cabinet and the PC board, remove the following (Fig. 18-13):

1 The bottom cover.
2 The upper cover.

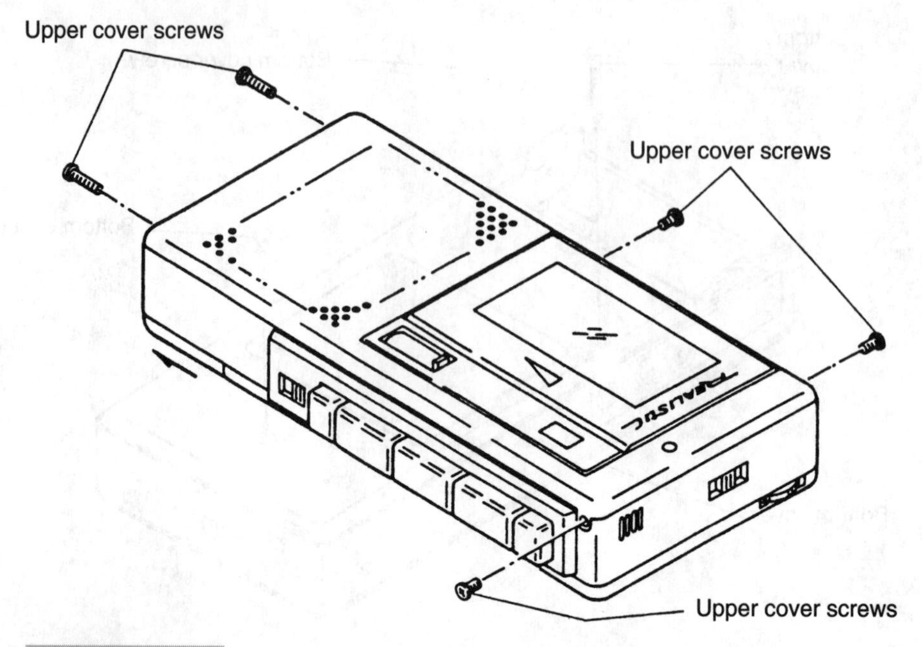

FIGURE 18-11 **Remove the five upper-cover screws to remove the top-side cover.**

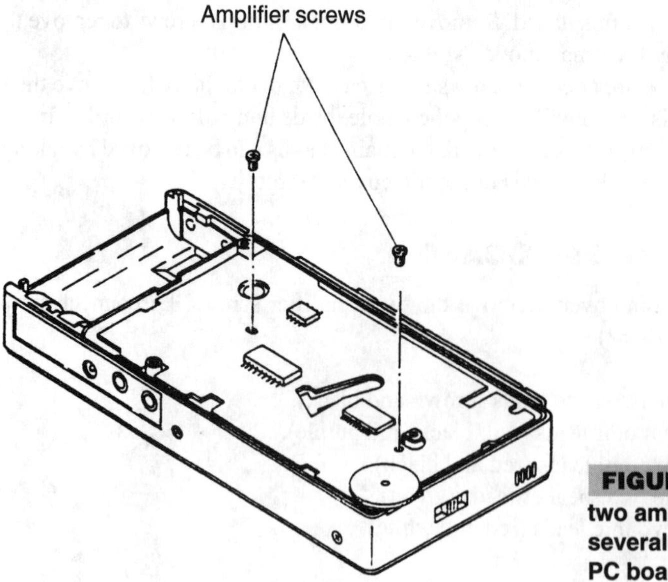

FIGURE 18-12 **Remove two amplifier screws and several wire leads before the PC board can be removed.**

3 The PC board.
4 The cassette screw.

If you do not have a service manual or instructions on how to remove the various assemblies, carefully peek at each assembly to see what screws are holding it in position. Do not remove more parts than necessary. Study the layout and proceed. It's best to lay out all removed parts in a line as they are removed so that you can easily reverse the procedure.

Only One Speed

If only one speed works in a 2-speed recorder, suspect that a speed switch is dirty. Sometimes, the speed is controlled by inserting a different size idler wheel. In other models, the speed is controlled with a speed regulator. Be sure that the speed problem is not mechanical in nature. Clean all idlers and pulleys with rubbing alcohol and a cleaning stick.

If only one speed is still working after cleaning, apply cleaning fluid into the contact areas of the speed switch. With no speed or if both speeds are incorrect, check the voltages within the motor regulator circuits (Fig. 18-14).

Here, the 2.4-cm and 1.2-cm speeds are changed with switch S102. Dirty switch contacts can cause no speed or erratic speed operation. Check the crucial voltages on Q102 and U103. Test Q102 with a transistor tester or with the diode-transistor test of the DMM. A leaky Q102 can produce fast speeds with S102 in any position.

If the supply voltage at Q102 and U103 is normal, suspect a change of resistance in the collector circuits or suspect that the motor is defective. Check the voltage across the motor terminals and notice the change when both speeds are switched into the circuit. If you find a voltage change, but no speed change, suspect that the motor is defective.

The tape speed of the microcassette recorder can be checked with a tape-speed cassette (W411). Adjust R503 for correct 2.4-cm/sec. speed with a 3000-Hz test tape. R502 is adjusted

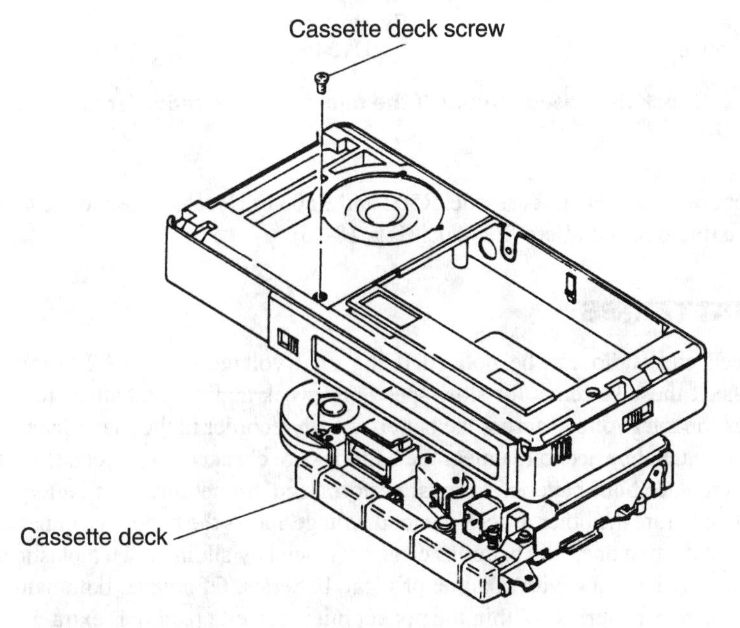

FIGURE 18-13 **Remove one cassette deck screw so that the cassette deck can be removed.**

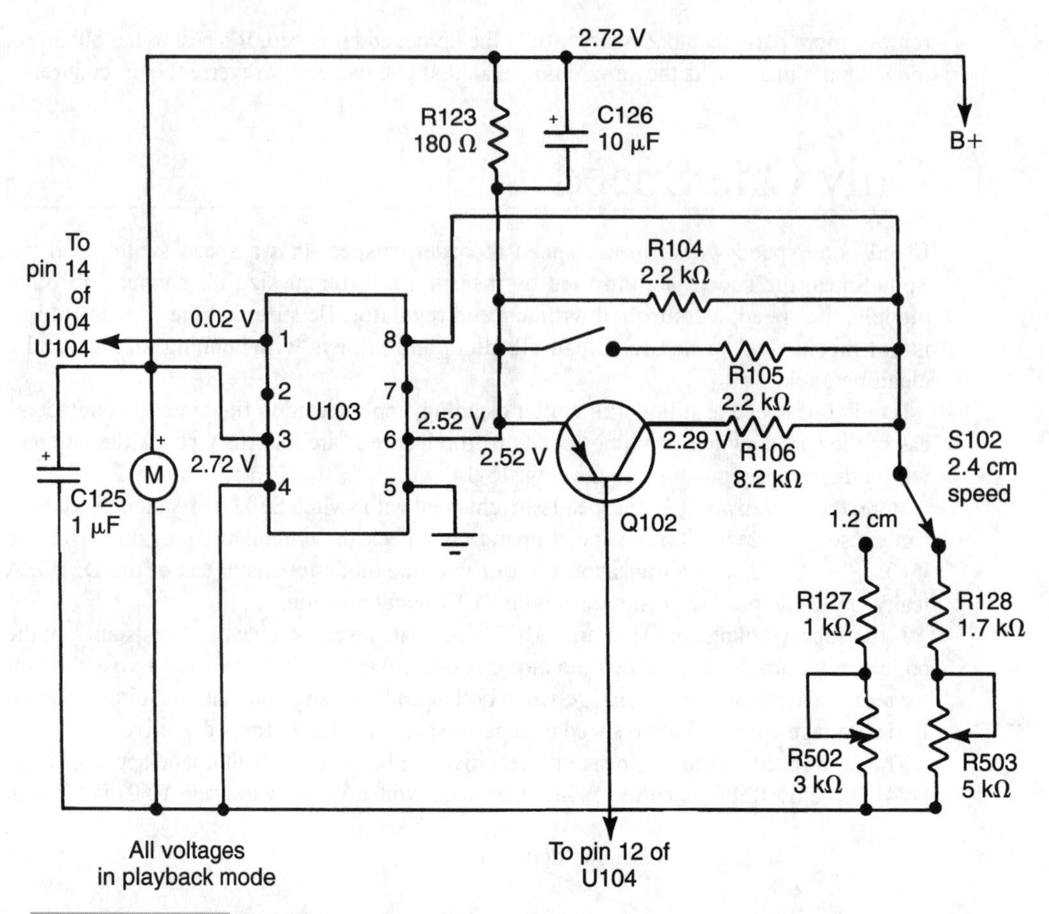

FIGURE 18-14 **Check the speed circuits if the microcassette recorder is running at an improper speed.**

for the 1.2-cm/sec. 1500-Hz test cassette. R502 and R503 are located close to the tape speed switch on the Radio Shack Micro-26 model (Fig. 18-15).

WEAK BATTERIES

Loss of speed and audio can be noticed if the total voltage drops to 2.5 volts in the recorder. Check those batteries for slow speed and weak audio. If a battery tester is not handy, check the total voltage across the batteries with recorder in the Play mode. Voltage measurements cannot be accurate unless the batteries are checked under load (Fig. 18-16).

It's best to replace both batteries for best performance. Sometimes, one battery can become defective before the other. Check the total voltage across the battery terminals of each battery to the defective one. The batteries can be replaced by sliding open a plastic cover.

Replace the AA batteries with alkaline or nicad batteries. Of course, both batteries are heavier than regular batteries. Within the pocket microcassette recorder, extra weight can become a problem, so why not install alkaline batteries?

Doublecheck the polarity of the batteries when removing the old ones. Look into the battery compartment for positive (+) or negative (–) polarity signs. Wipe the ends of the new battery on a paper towel or a cloth towel before installing. If the recorder does not operate, quickly shut off the unit and recheck for correct battery polarity.

RECHARGEABLE BATTERIES

Nicad batteries will last a lifetime if it is properly charged and cared for. Although the nicad batteries initially cost more, they can last the lifetime of the microcassette recorder.

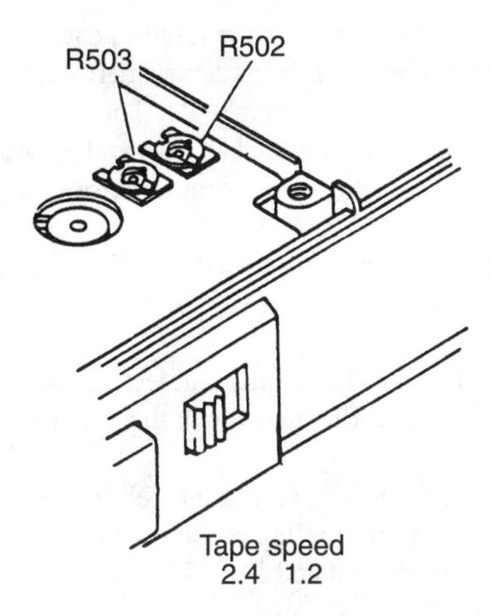

FIGURE 18-15 The two speed-control adjustments, R502 and R503, in a Radio Shack recorder.

FIGURE 18-16 Check the total voltages of both batteries with the cassette recorder operating with the DMM.

Some microcassette recorders are equipped with nicad batteries and an ac charger. Of course, the recorder should be operated on the ac adapter when on dictation or when ac power is available.

If the recorder is only battery operated, you can still use nicad batteries by purchasing a nicad charger. Nicad batteries should be charged when they are weak and will not operate the cassette player. Follow the instructions found on the charger. Discharge the nicad batteries if they start to operate for only a short time.

SLOW SPEED

Check the batteries for slow tape speed. If batteries are used quite rapidly, check the current drain of the recorder in operation. Just slip a piece of light cardboard between one battery and the terminal slip. Set the DMM to the 200-mA meter scale. Touch the end of battery to the battery terminal. Place the meter in series with the batteries and recorder circuits. Another method is to check the current by placing meter probes across the on/off switch with the switch in the Off position.

If the current is quite high, suspect a leaky amplifier or power circuit. The normal current of the Sony M-440V recorder is around 90 mA. If the current is higher than 125 mA, the unit might be defective.

Insert a new cassette and notice if the speed changes. Sometimes a defective cassette will drag down the play/record speed. Notice if the current fluctuates with the old or defective cassette.

Slow speeds can be caused by an overly large motor belt. Check the belt for oil or shiny spots. Replace the loose motor belt. Check for a dry or binding capstan wheel. If the capstan is sluggish, remove and clean it with rubbing alcohol and a rag. Place a drop of oil on the capstan flywheel bearing and replace the belt.

DAMAGED CASSETTES

Although the microcassette is small and fairly well constructed, a defective cassette can cause slow speeds, noisy operation, and sound dropouts. A cassette that is wound too tightly can cause slow speeds. Noise that is directly from the cassette can be caused by dry bearings (plastic against plastic). Dropouts can occur when the tape spills out of the cassette and becomes tangled around the capstan or pressure roller. The wrinkled tape is damaged and can't record or play as it was recorded.

Replace the defective cassette with a new one. Do not try to repair the broken tape. Excess tape spilled out of the center slot can be rewound by placing a pencil inside the take-up reel slot and rewinding (Fig. 18-17). If it is kinked excessively, replace the cassette. Check the plastic sides for a cracked case if the hubs will not rotate.

NO REVIEW OR CUE

The *Review* control rewinds and *Cue* fast forwards the tape. First, determine if the tape is normal in the Play mode. If it sounds normal, the motor belt and motor are good. If the fast forward is slow, suspect that the idler wheel or reel is slipping. In many units, a larger wheel or pulley is placed against the idler and reel for faster speeds (Fig. 18-18). Clean all

FIGURE 18-17 Check for wrinkled or excess tape at the center of the cassette. Take up the slack with a pencil in one of the reel hubs.

FIGURE 18-18 The idler wheel can be pressed against the hub or spindle for Fast-Forward or Rewind modes.

moving surfaces with rubbing alcohol and a cleaning stick. Check for oil on the running idlers and pulleys.

In some cassette players, Rewind operates at the same speed as Fast Forward. The idler wheel is just pressed against the reel in the opposite direction, which changes the direction. Often, with this type of operation, slippage is found in both the Review and Cue operations. Replace the idler or pulleys if they are worn or deformed, which produces slow speed.

NOISY HEADPHONE JACK

The headphone jack is usually wired in series with the small speaker voice coil. When the headphone is plugged into the jack, the speaker is cut out of the circuit (Fig. 18-19). In this

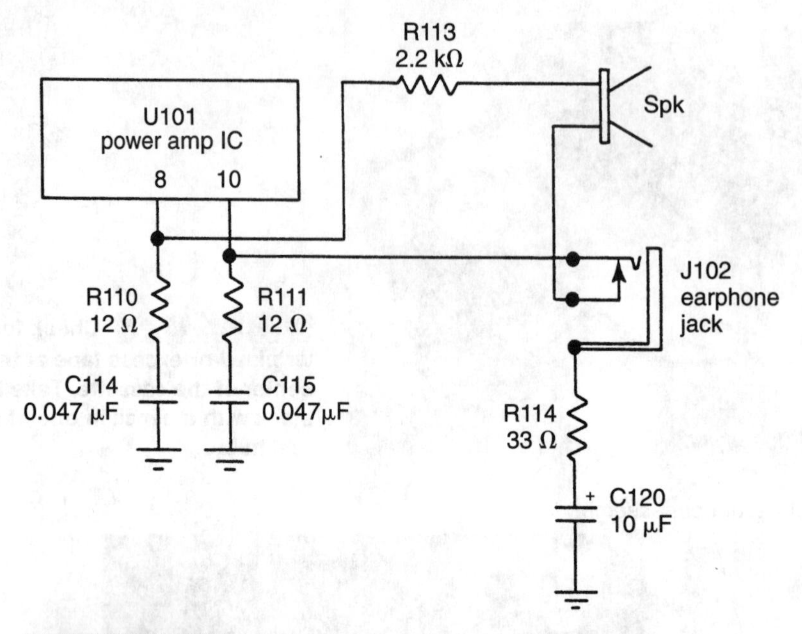

FIGURE 18-19 **The headphone jack is in series with R114 and C120 to common ground.**

circuit, the headphone sound-dropping resistor and isolation capacitor (R114 and C120) completes the circuit to common ground. In some recorders, you can listen to what is being recorded with the headphones.

The dirty earphone terminals can cause erratic sound or deaden the sound to the headphones. Within the stereo headphone jacks, one audio channel might be dead as a result of dirty or bent headphone terminals. Remove the cover and squirt cleaning fluid into the jack area. Insert the headphone jack a few times to help clean the contacts. If the headphone jack is molded, apply liquid into the jack opening. When these contacts are broken or extremely worn, the sound is intermittent or erratic. Replace the defective headphone jack.

ACCIDENTAL ERASE

Break out the small plastic tab at the right side end of the microcassette to prevent accidental content erasure. Break out both tabs if both sides are to be saved. If not, you can accidentally erase valuable information before you realize it (Fig. 18-20).

Remember that if a microcassette will not let the Record button press inward, the small plastic slot can be knocked out. Always play the cassette to be sure that the recording doesn't need to be saved. Apply a small piece of tape over the opening if you want to record over the cassette.

WORKS WITH BATTERIES, BUT NOT WITH AC

A lot of microcassette recorders or players will operate from the power line with an ac adapter. The dc voltage male plug is inserted into a small jack at the end of side of the recorder (Fig. 18-21). The internal batteries are disconnected when the adapter plug is inserted.

The male dc plug from the adapter breaks the circuit and switches the batteries from the circuit. Often, the dc voltage from the ac adapter is higher than normal when it is measured out of the circuit (Fig. 18-22). Inspect the switched terminal connections. Clean the contacts with cleaning fluid. Apply cleaning fluid to a soft cloth and wipe the male plug for a good contact. Replace the small dc plug, if it makes a broken or poor contact.

To determine if the plug is defective, plug the dc male plug into the dc jack and measure the voltage at the plug terminals. If no voltage registers, the contacts might be bent or worn. Check the voltage at the male plug to determine if the adapter or dc jack is defective. If in doubt, clip the male plug connections to the dc jack terminals with small clip leads. Try to repair the ac adapter with no voltage at the male plug. Doublecheck the dc voltage polarity at the dc jack.

INTERMITTENT VOR OPERATION

An intermittent voice-operated cassette player can be difficult to repair. Notice if the tape will operate in VOR position when the VOR switch is pushed or pressed to one side. This

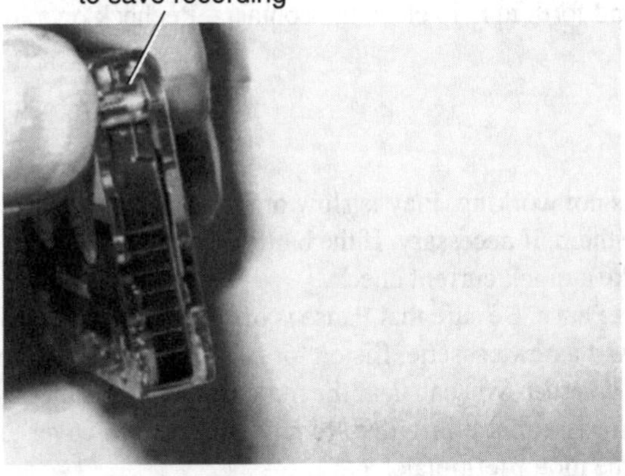

FIGURE 18-20 The small break-out tab helps protect the recording on the cassette.

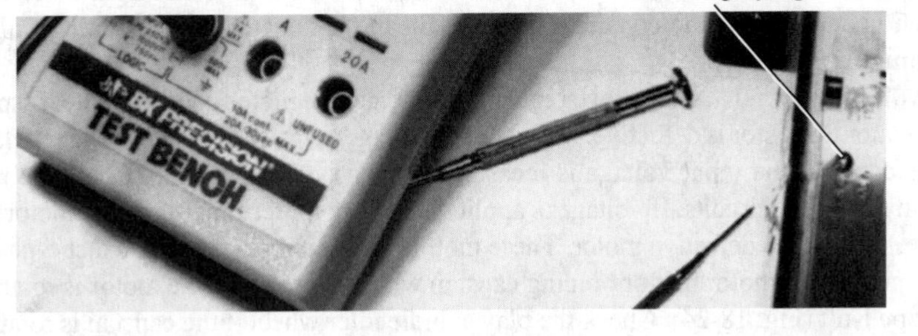

FIGURE 18-21 The dc voltage male plug from the ac adapter will plug into the side of this Sony microcassette recorder.

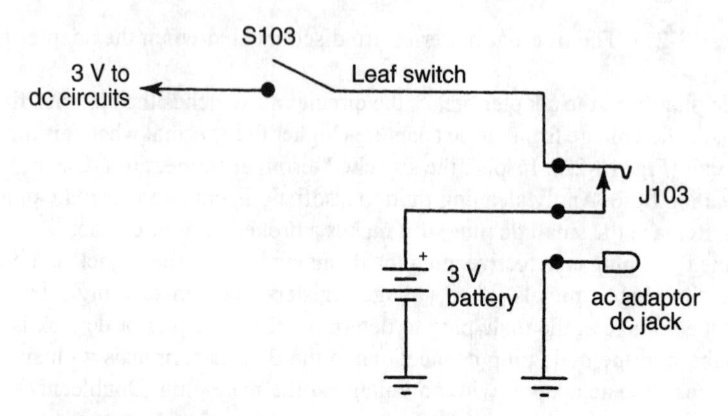

FIGURE 18-22 **When the ac adapter plug is inserted, the internal batteries are removed from the circuit.**

can be rather tricky because handling the switch or case can trigger the VOR circuits. Just keep talking while working with a suspected dirty switch.

Spray cleaning fluid in the switch area. Remove the back cover. Insert the small plastic tab into the switch area. Only a squirt or two is needed. Too much spray can drip on the drive belts or moving pulleys. Wipe off all excess with rubbing alcohol and a cleaning stick. Work the switch back and forth to help clean the contacts. Recheck operation of the VOR switch after clean up.

NO PLAYBACK

Usually, when Fast Forward is not working, Play is slow or the tape doesn't rotate. Check for weak batteries and replace them, if necessary. If the batteries do not last very long, suspect the overload circuits. Take a quick current check.

Check to see if recorder is in Pause. Be sure that Pause is off. Do the spindles rotate with a new cassette or without? Insert a new cassette. Listen for the tape or motor to rotate. Often, by holding recorder up to your ear, you can hear the motor rotate. If the motor does not rotate and new batteries and a cassette are installed, remove the bottom cover and check the voltage that is applied to the motor terminals.

If no voltage is at the motor, the leaf switch, regulator IC, or transistor might be bad. Check the voltage at the leaf switch. No voltage at the leaf switch might indicate that a switch is defective or the contacts are dirty. Clean the contacts with cleaning fluid and a cleaning stick (Fig. 18-23).

If voltage is found at the voltage regulator input and the motor doesn't rotate, suspect that a regulator transistor is defective. Measure the voltage at the IC regulator. The regulator can be defective if the input voltage is measured with no output voltage. Check all resistors within the motor circuits. If voltage is applied to the motor terminals and the motor doesn't rotate, replace the defective motor. These motors must be replaced with exact-type motors.

Suspect that a motor belt or binding capstan wheel is broken if the motor is rotating, but the tape isn't (Fig. 18-24). Check the play spindle idler wheel if the capstan is rotating. Do not overlook a locked brake lever that does not release. Often, when the brake or capstan wheel prevents tape motion, the motor pulley is rotating inside the motor belt. Sometimes, a burned rubber idler or spindle surface will have a slot burned in it, which prevents tape action.

NO RECORDING

First, check the microcassette for the removed plastic tab in the right end of cassette. The recording button cannot be pushed in when this tab is removed. Try another cassette, and clean the R/P tape head.

Does the recorder play other cassettes? If so, the head playback and recording circuits are normal in the path of audio signal. Check the bias signal with a scope at one side of the tape head in the recording mode. If a scope is not handy, check the voltages on the bias transistor. It might help to clean the record/play switch.

The bias exciting voltage from the bias oscillator is applied to the record/play tape head through S101 (Fig. 18-25). Practically no voltage (0.06 V) is found on the base and collector terminals of bias oscillator Q104. Check the bias voltage between the base and the emitter (0.32 V). If the collector voltage is low, suspect a leaky Q104 or an insufficient supply voltage.

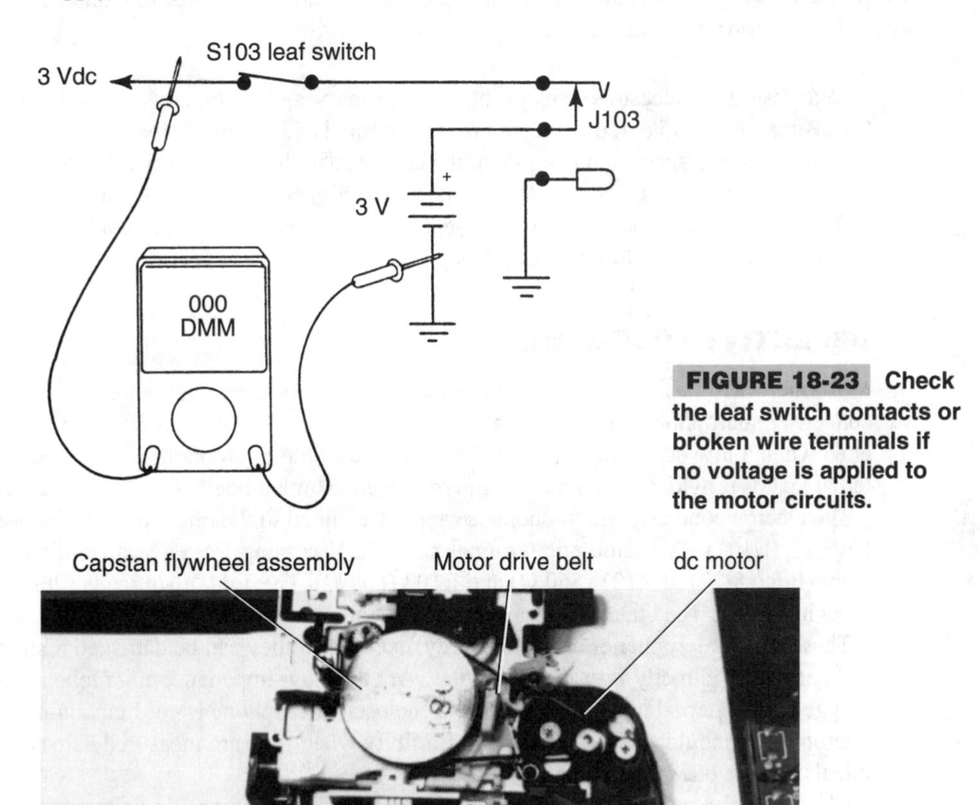

FIGURE 18-23 Check the leaf switch contacts or broken wire terminals if no voltage is applied to the motor circuits.

FIGURE 18-24 Check for a broken or loose drive belt if the motor is operating and the tape is not moving.

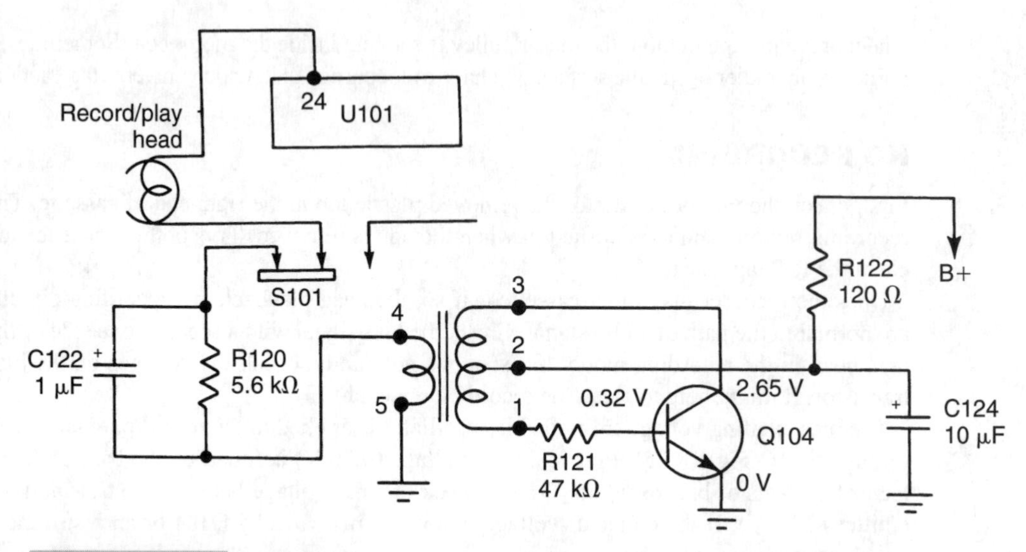

FIGURE 18-25 **The recording circuit consists of a bias oscillator to excite the record head during recording.**

Take resistance measurements on both the primary and secondary windings of T101. Check for poorly soldered contacts around terminals 1, 2, and 3, for erratic recording. Measure the resistance from one side of the tape head to the chassis ground (5.6 kΩ). If the resistance is lower, suspect that C122 is shorted. A higher resistance indicates that R120 is defective or that a winding of T101 is open. Also, a leaky C124 (10 μF) could cause a low voltage at the collector terminal of Q104.

DEFECTIVE MICROPHONE

Most microcassette recorders have condenser- or capacitor-type microphones. The electret condenser microphones are used in later models (Fig. 18-26). These miniature mics are excited when a low dc voltage is applied to one side of the microphone. Sometimes an external jack is provided for an outside microphone in some models.

The electret condenser microphone is capacity coupled to the mic input terminals of the large IC (U101). The same mic terminal has 1.32 V coming from a voltage-divider network (Fig. 18-27) of R101 (560 Ω) and R102 (600 Ω). The 1.44-V source is filtered with capacitor C101. The other mic terminal goes to common ground.

These small microphones are quite sturdy, except that they can be damaged if sharp objects are stuck directly into the mic holes. Another low-impedance microphone can be plugged into external mic jack to see if the enclosed microphone is not functioning. These microphones should show a resistance of infinity when they are measured across the terminals and are out of the circuit.

Check the bias voltage across the mic terminals when recording. No voltage might indicate that there is no supply voltage (1.32 V) or that R101 and R102 are open. Clean the contacts of the external mic jack. Remember, the external jack provides a shorting circuit for the internal microphone. Short C105 with a 0.1-μF electrolytic capacitor. Any small electrolytic capacitor will determine if C105 is open. Check the voltages on pins 1, 2, 3, 4, and 17 of IC U101.

NO SOUND

Check the setting of the volume control. Rotate it open and listen for sound. Insert the earphone and notice if the sound is okay. If the sound is normal during headphone operation, suspect that a speaker or a dirty headphone jack is cutting the speaker out of the circuit.

FIGURE 18-26 The electret condenser microphone in the Sony M-440V microcassette recorder.

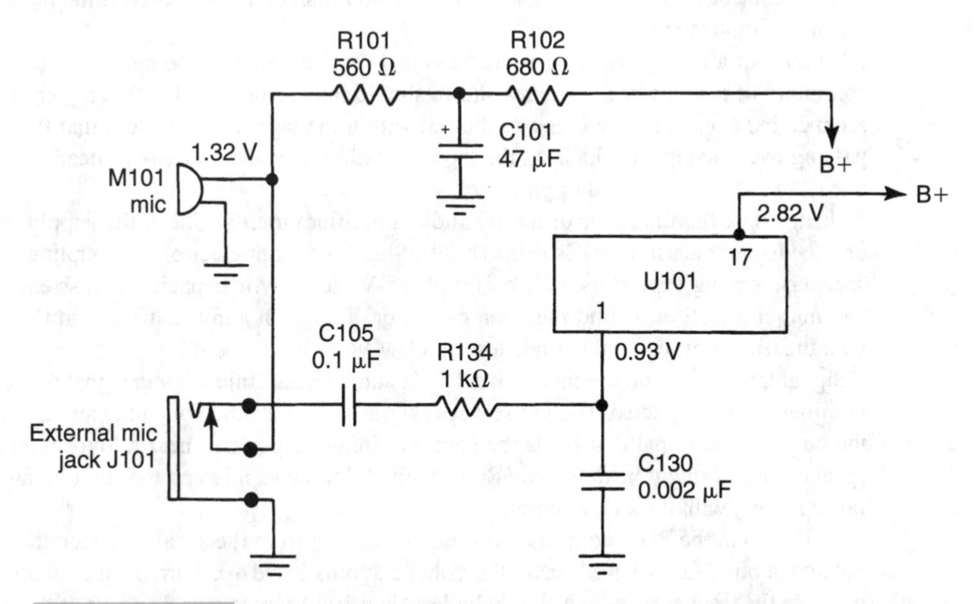

FIGURE 18-27 The small electret mic has +1.32 V to one side of the terminals from resistor network R101 and R102.

FIGURE 18-28 **Clip another speaker across the suspected terminals to determine if the speaker is open or distorted.**

Clip another speaker across the small speaker (Fig. 18-28). Replace the speaker if it is open. Be sure that the tape heads are clean. Often, a scratchy noise when the volume control is quickly rotated indicates that the amplifier is operating.

If you flick a screwdriver blade across the tape head with the volume wide open, you will hear a thumping noise if the amp is okay. Place the screwdriver blade next to the head ungrounded wire and listen for a hum. Check the tape head for broken lead wires. With the volume control wide open, you should hear a loud rushing noise—even if the head wires are off or if the tape head is open.

If you hear a noise, suspect that the head is defective with no tape motion. Measure the continuity of the tape head. The resistance should range from 200 to 400 Ω. Very low resistance indicates that a winding is shorted within the tape head. Be sure that the tape is passing over the tape head. Clean the tape head if one channel is weak or dead, or if only one channel will not record or play back.

Next, check the transistor or the IC audio amplifier circuits. Check the supply voltage that feeds the transistor or ICs (Fig. 18-29). Shunt the input electrolytic coupling and the speaker-coupling capacitors. Clip a 100-μF 15-V electrolytic capacitor across each input and output capacitor to find the open capacitor. Test each audio transistor in the circuit with the transistor test or the diode test of a DMM.

Signal trace the audio circuits with a 1-kHz audio signal. Inject audio signal at the input terminal of the suspected IC and use a speaker as the indicator. Touch the audio signal to the output IC terminal that feeds the speaker circuit. If you can hear a low sound in the speaker, the IC might be defective. Replace the IC if the supply and terminal voltages are fairly normal with no sound output.

If U101 is in the Play mode and no sound is coming from the speaker, check the supply voltage at pin 7 (2.75 V). Measure the voltage at pins 8 and 10. Turn off the recorder and measure the resistance, which should be less than 10 Ω across pins 8 and 10 of the power-output circuit. With no measurement, the speaker or external earphone jack is open. Check the resistance between 24 and 4. Measure the tape-head resistance. Suspect that the tape head or play/record switch contacts are open if the resistance measurements are very low or very high.

MUFFLED OR TINNY SOUND

A muffled or tinny sound can indicate that a voice coil is frozen against the magnet. The loose voice coil can cause a muffled sound. Substitute another speaker if the original is defective.

Check for weak batteries with distortion in the speaker. Doublecheck with the headphones to eliminate the speaker. If the headphones are distorted, suspect that an amplifier IC or transistor is leaky. Clean the tape heads before tearing into the amplifier circuits. Try another microcassette. Replace it if the cassette is causing the distortion.

Does the sound seem distorted when using the microphone for recording? If the recorder plays back normally, try another cassette. If the sound is distorted when a microphone is used in recording, check the microphone. Shunt small electrolytic capacitors within the mic circuits. Replace the microphone if it is defective.

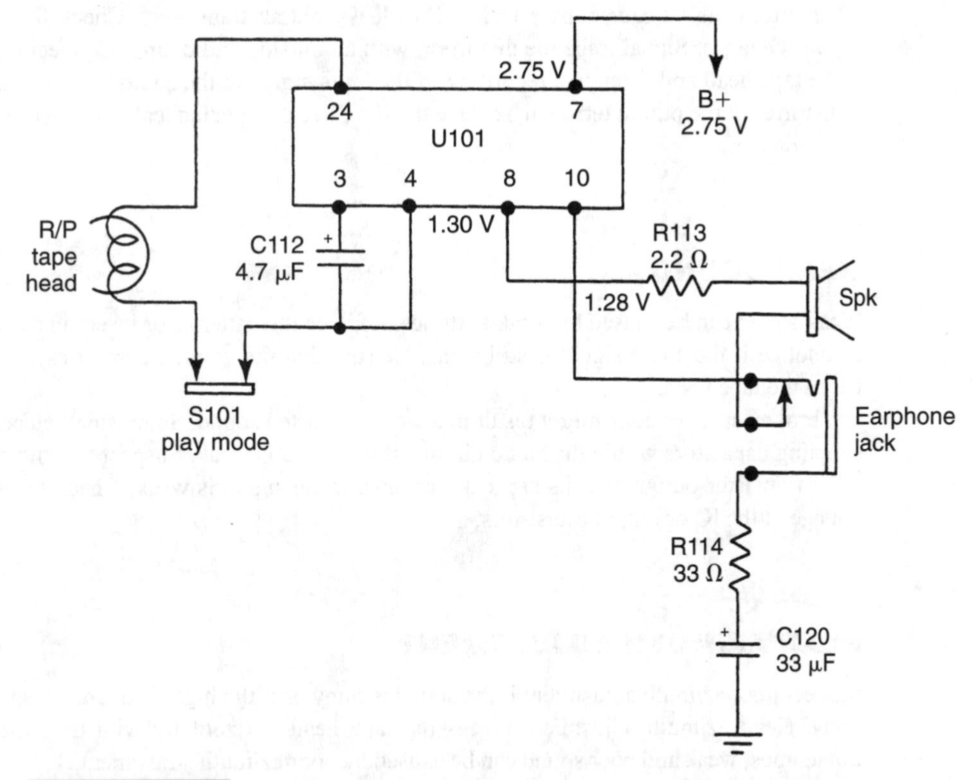

FIGURE 18-29 **Check the power supply voltage in pin 7 of U101 and check the continuity of the input and output components.**

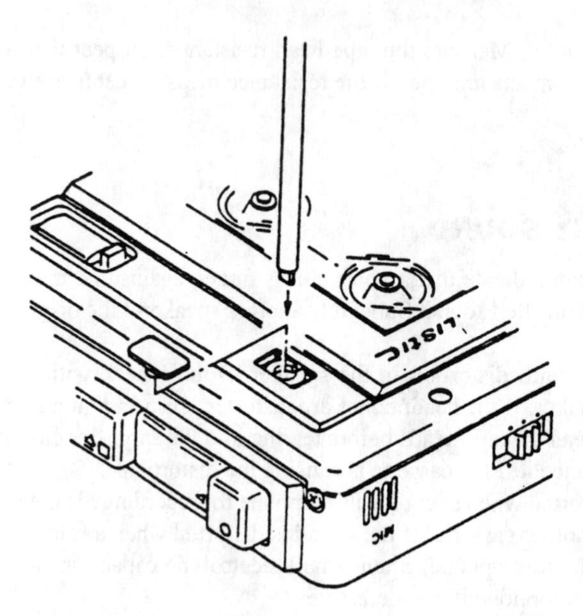

FIGURE 18-30 Adjust the head-azimuth screw for maximum output at the speaker with the azimuth A411 (3 kHz) test cassette.

Distortion can be caused by a leaky IC or leaky output transistors. Check for correct supply voltages. Signal trace the distortion with an outside audio amp. Connect the amp to the tape head and listen to a recording. If the input signal to the audio IC is normal, but is distorted at the output terminal, replace the IC. Check suspected leaky transistors in the same manner.

WEAK SOUND

Weak sound can be caused by weak batteries. Replace the batteries or insert an ac adapter and notice if the audio signal is still weak. Be sure that the volume control is turned up. Clean the tape head.

A broken speaker cone might result in a weak, distorted sound. Shunt small electrolytic coupling capacitors within the audio circuits if the audio is weak. Suspect that the power-IC or transistor-output circuits are leaky or open if the signal is weak. Check the supply voltage at the IC or output transistors.

HEAD AZIMUTH ADJUSTMENT

Suspect poor azimuth adjustment if the sound is tinny or if the high-frequency response is tinny. Head azimuth adjustment places the tape head horizontally with the tape path. Sometimes, weak and poor sound can be caused by poor azimuth adjustment. Try to adjust the sound with a piano or with the high notes of a recording. Insert a head azimuth test tape A411 (3 kHz). Turn the volume control up and adjust azimuth screw with an insulated tool for maximum output at the speaker (Fig. 18-30).

MICROCASSETTE CONCLUSION

Troubleshoot the microcassette recorder like any cassette player. Although small and surface-mounted components are crammed together to make servicing a little more difficult, you can still make many repairs. Check the most troublesome parts, such as jacks, speakers, transistors, capacitors, resistors, etc. Do not overlook the possibility that switches, headphone jacks, or microphone jacks are dirty.

A good tape head cleaning solves many problems with weak audio, distorted audio, no audio, lost stereo channels, and poor recording. Clean the tape head before tackling any other sound or recording problems. Change the batteries for weak audio, poor recording, and speed problems (Table 18-1).

TABLE 18-1 A TROUBLESHOOTING CHART OF THE VARIOUS SYMPTOMS FOUND IN THE MICROCASSETTE RECORDER

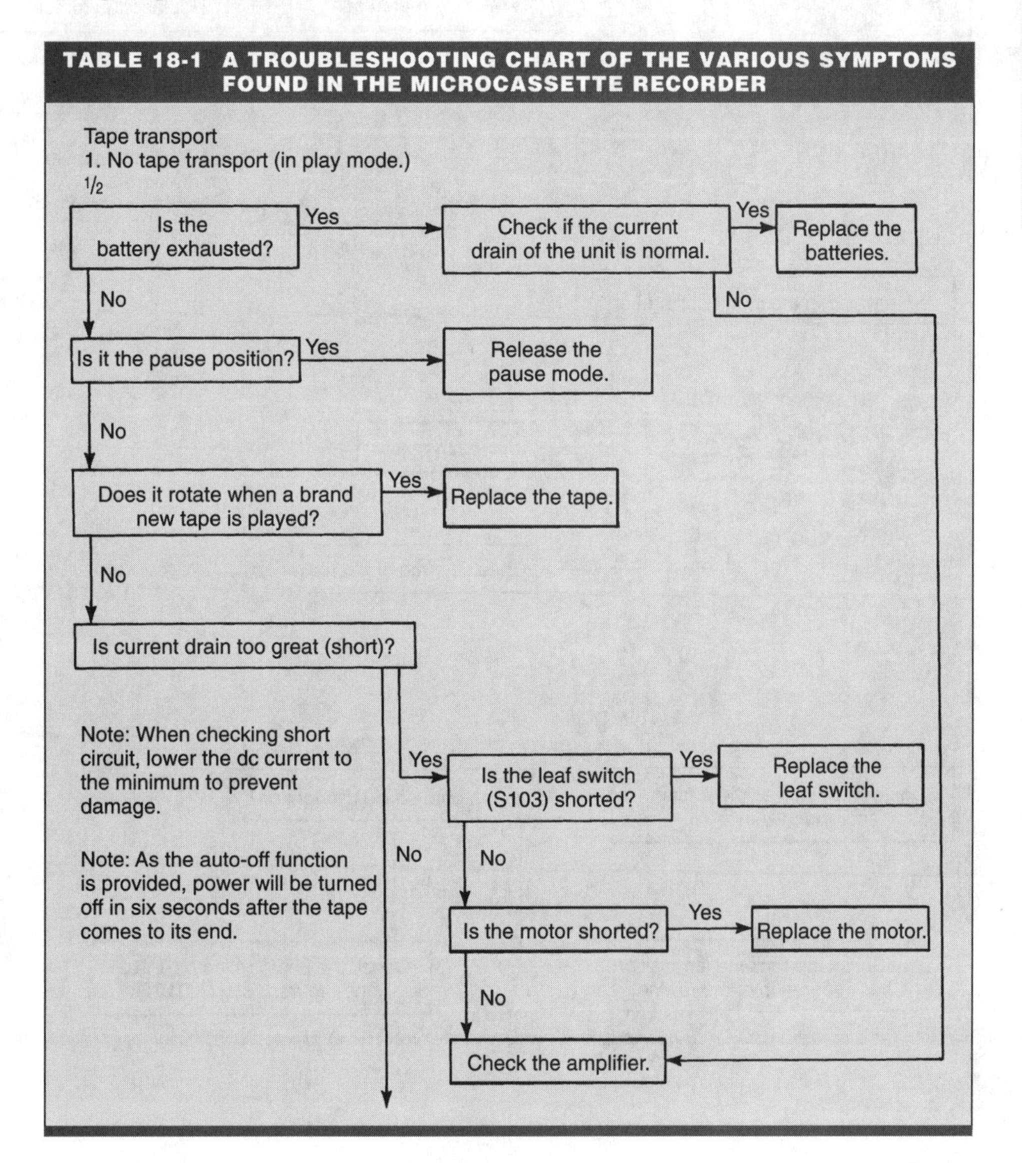

TABLE 18-1 A TROUBLESHOOTING CHART OF THE VARIOUS SYMPTOMS FOUND IN THE MICROCASSETTE RECORDER (CONTINUED)

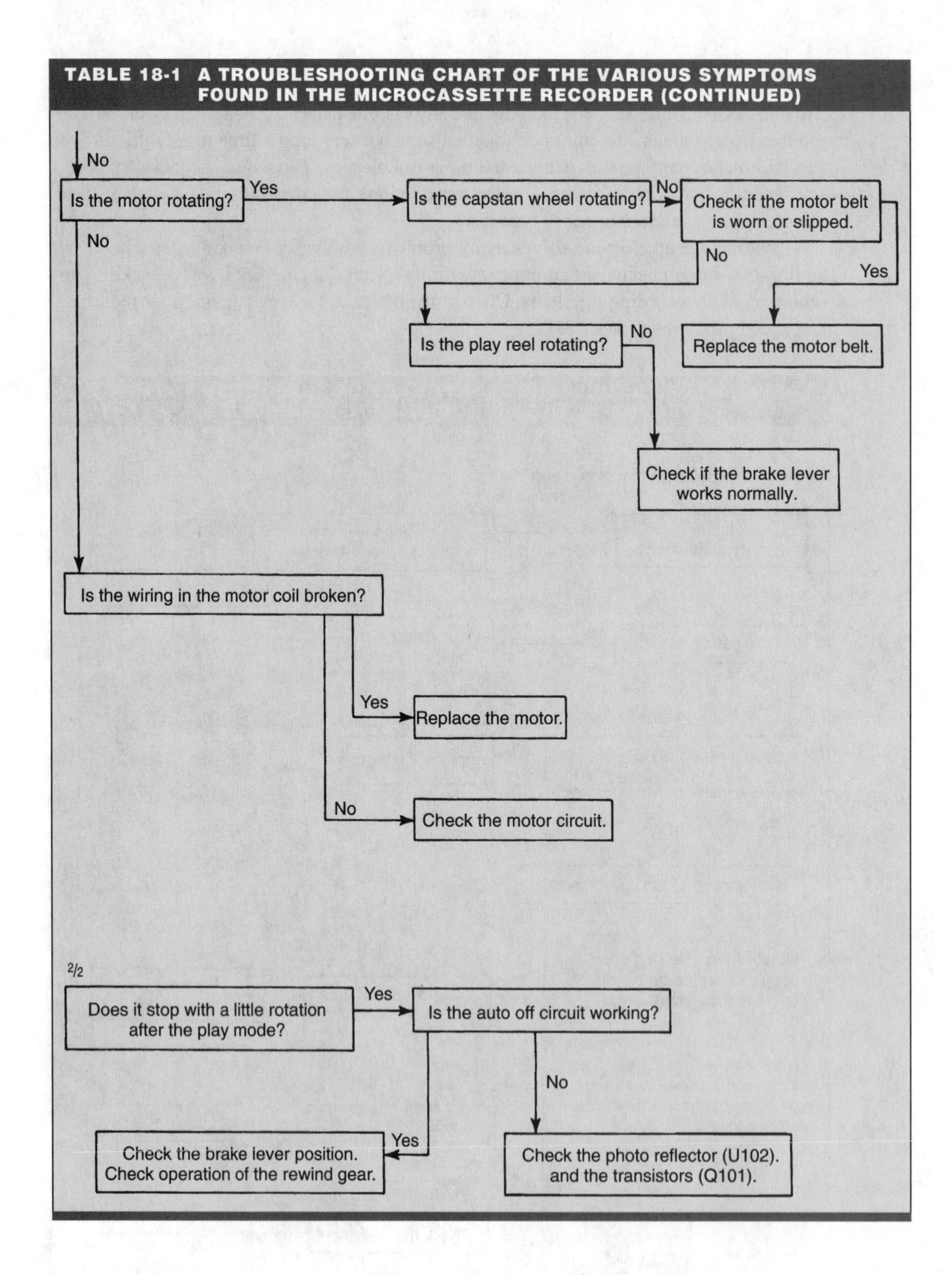

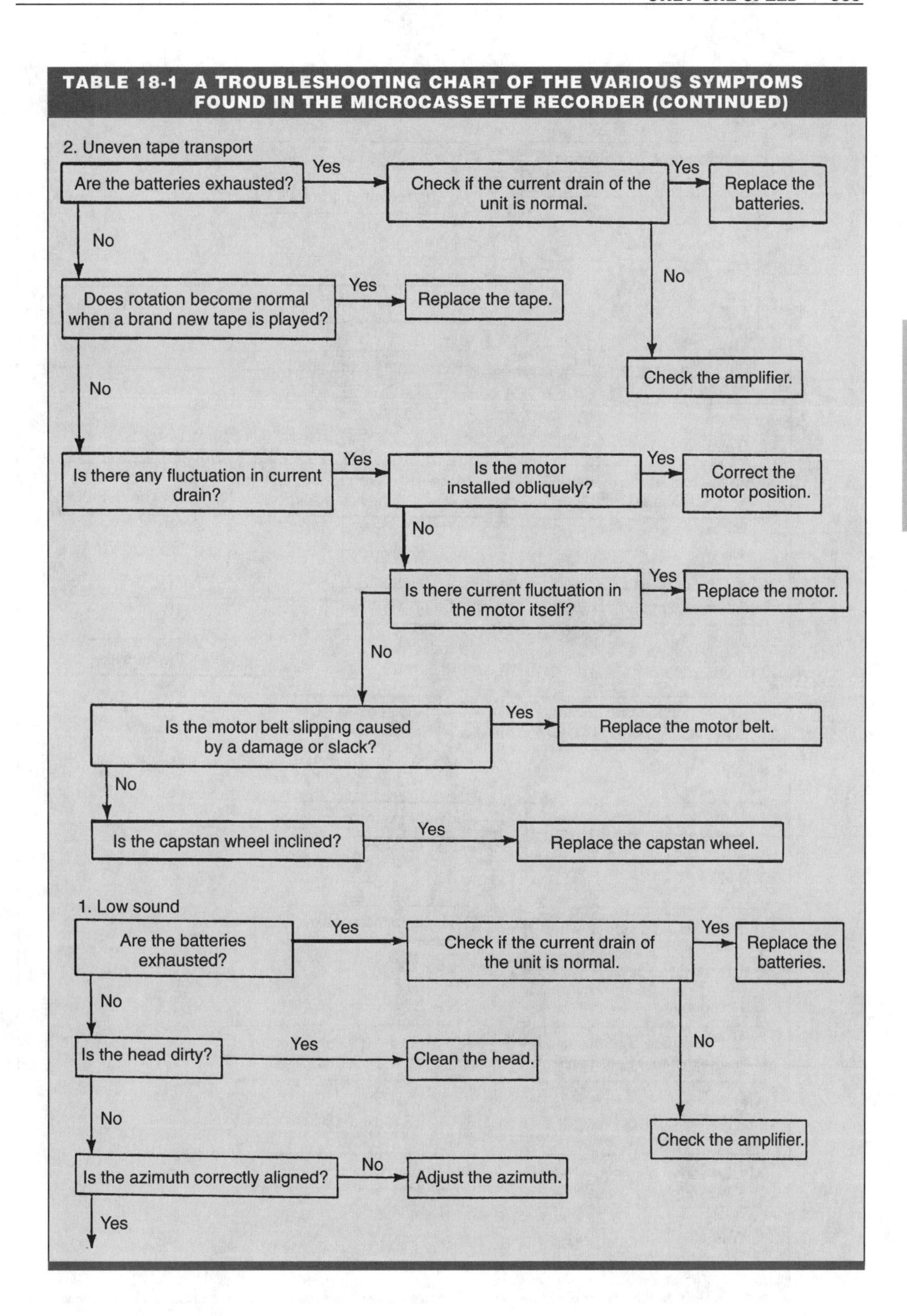
TABLE 18-1 A TROUBLESHOOTING CHART OF THE VARIOUS SYMPTOMS FOUND IN THE MICROCASSETTE RECORDER (CONTINUED)
2. Uneven tape transport
Are the batteries exhausted?
Yes
Check if the current drain of the unit is normal.
Yes
Replace the batteries.
No
No
Does rotation become normal when a brand new tape is played?
Yes
Replace the tape.
Check the amplifier.
No
Is there any fluctuation in current drain?
Yes
Is the motor installed obliquely?
Yes
Correct the motor position.
No
Is there current fluctuation in the motor itself?
Yes
Replace the motor.
No
Is the motor belt slipping caused by a damage or slack?
Yes
Replace the motor belt.
No
Is the capstan wheel inclined?
Yes
Replace the capstan wheel.
1. Low sound
Are the batteries exhausted?
Yes
Check if the current drain of the unit is normal.
Yes
Replace the batteries.
No
Is the head dirty?
Yes
Clean the head.
No
No
Check the amplifier.
Is the azimuth correctly aligned?
No
Adjust the azimuth.
Yes

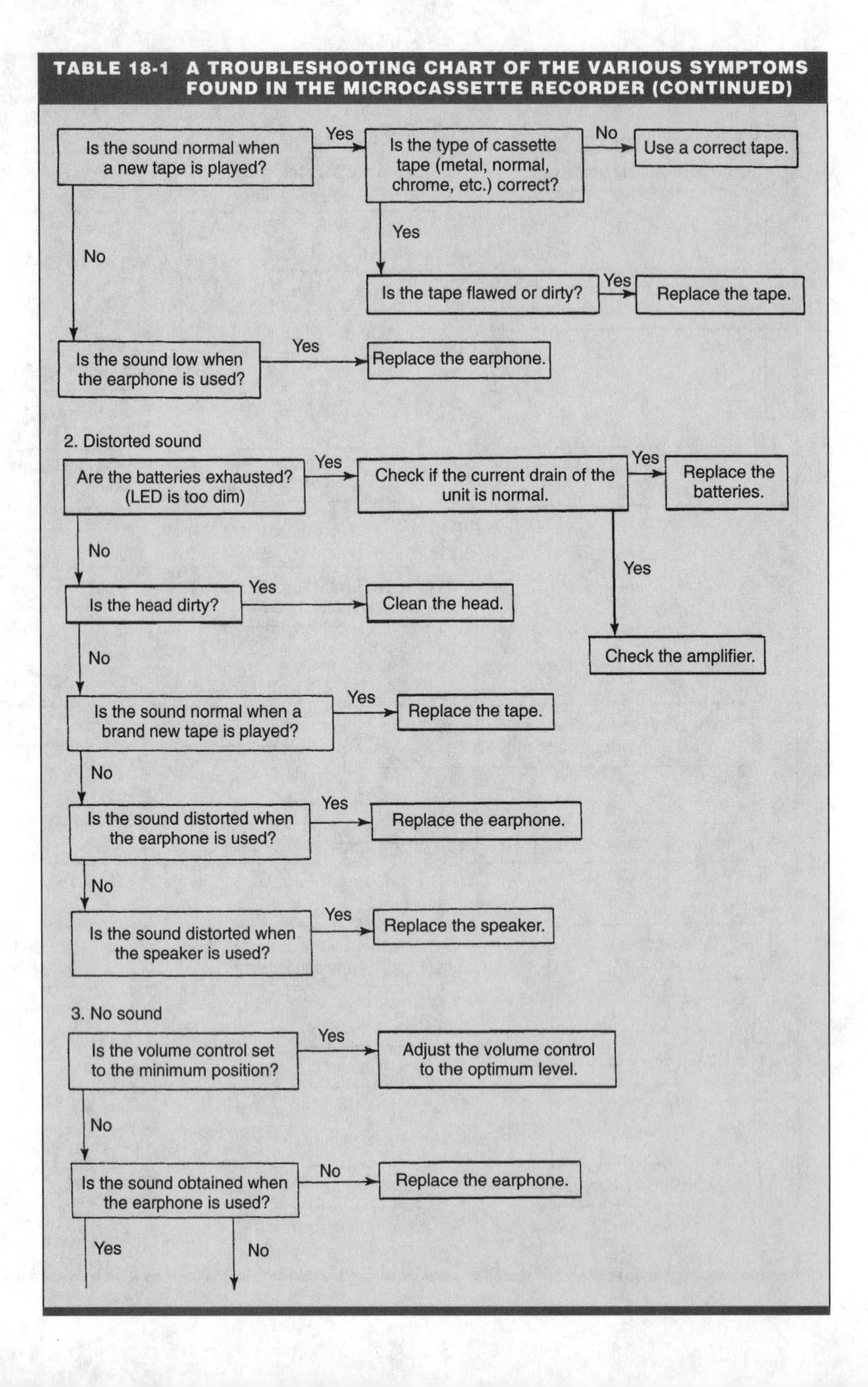

TABLE 18-1 A TROUBLESHOOTING CHART OF THE VARIOUS SYMPTOMS FOUND IN THE MICROCASSETTE RECORDER (CONTINUED)

TABLE 18-1 A TROUBLESHOOTING CHART OF THE VARIOUS SYMPTOMS FOUND IN THE MICROCASSETTE RECORDER (CONTINUED)

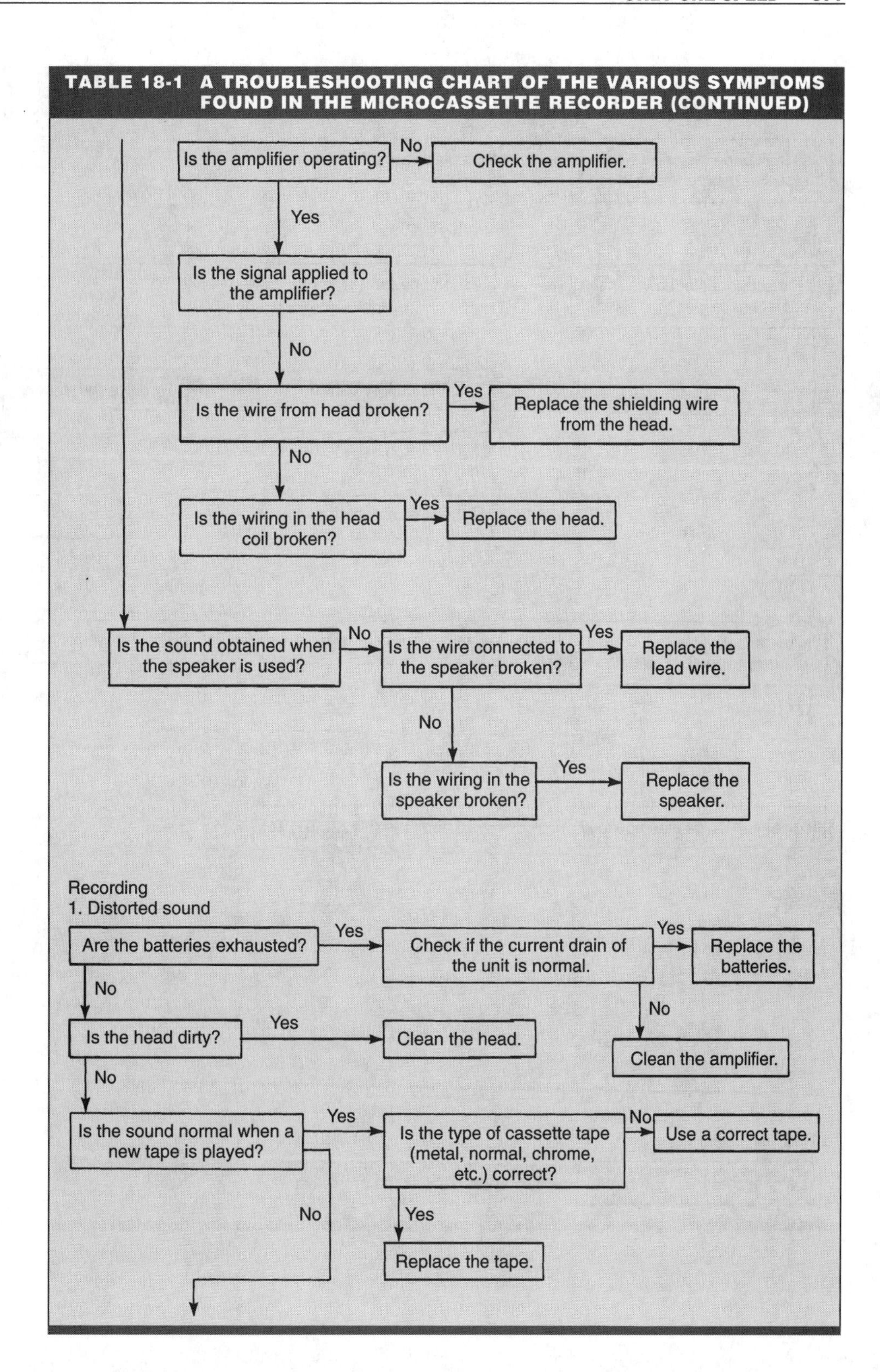

TABLE 18-1 A TROUBLESHOOTING CHART OF THE VARIOUS SYMPTOMS FOUND IN THE MICROCASSETTE RECORDER (CONTINUED)

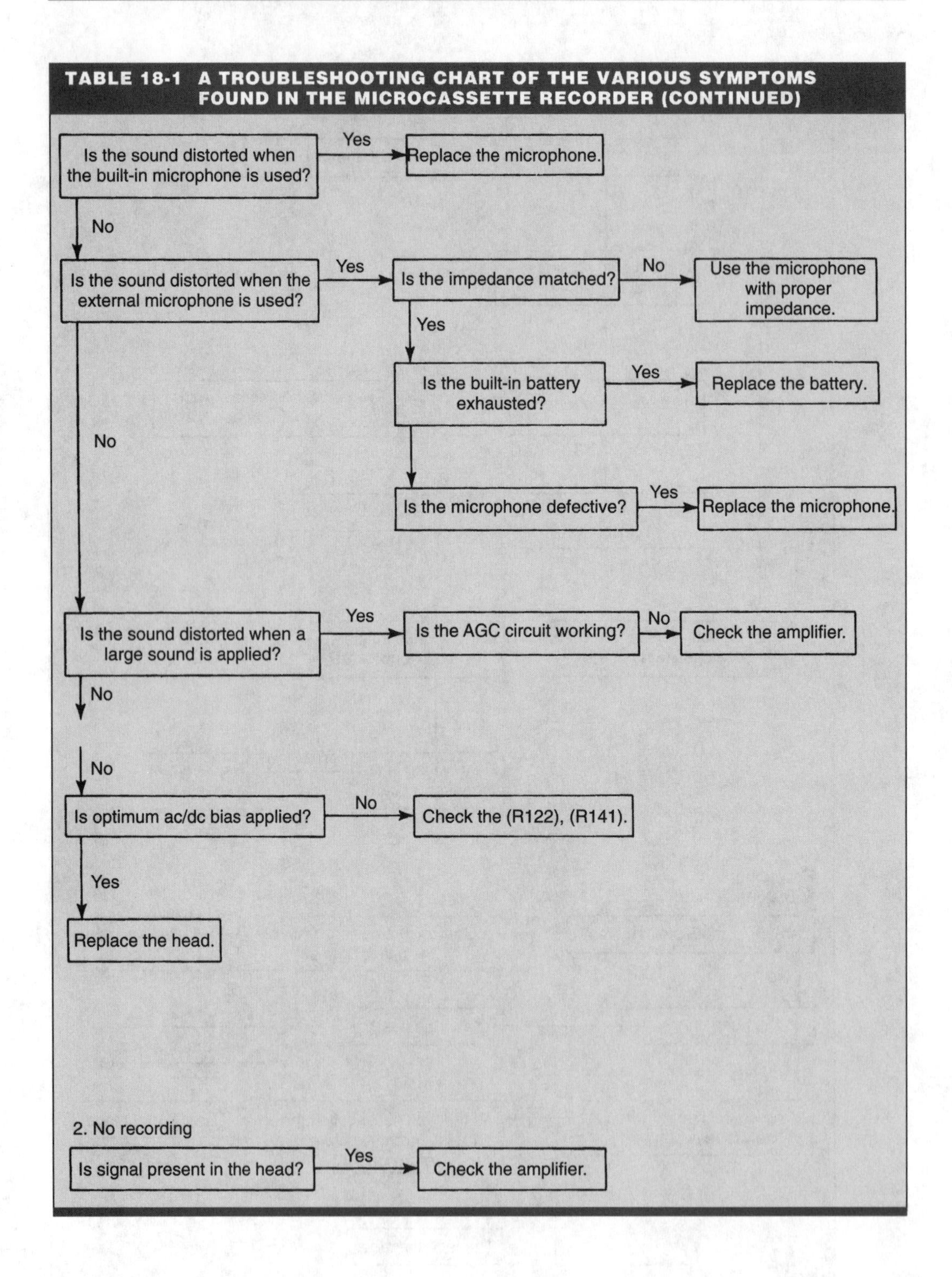

Professional Recorders

The professional portable cassette has many more features than other cassette recorders. The prices start at $99.95 and go up from there. Most professional recorders have stereo circuits with twin microphones and speakers. Some portable stereo recorders are equipped with stereo headphones, but a few have mono circuits (Fig. 18-31).

Some of the possible features on professional recorders are two-speed recording and playback with a variable speed control, up to three heads for true monitoring, Dolby, calibrated VU meters, pitch control, and variable speech control. Headphones, stereo output jacks, and one speaker can be used as a monitor. The professional cassette recorder is a rugged instrument that should stand up to years of rough treatment.

The smaller professional cassette recorders can operate from two AA batteries, and large units operate from two or three D cells, or four AA batteries (Fig. 18-32). The recorder can have direct Quartz lock drive or disc-drive features. Some have auto stop, pre-end alarms, and manual record lever controls.

The Sony TCS-430 professional cassette recorder is compact in size and features easy one-hand operation. Another feature is adjustable tape speed in the Play mode with an equalizer selector for optimum playback. The TCS-430 has a built-in one-point stereo microphone (Fig. 18-33).

10 PRECAUTIONS

1. When operating from ac, always use the required-voltage ac adapter.
2. When operating from the car battery, use the exact car battery cord that is recommended for the unit.

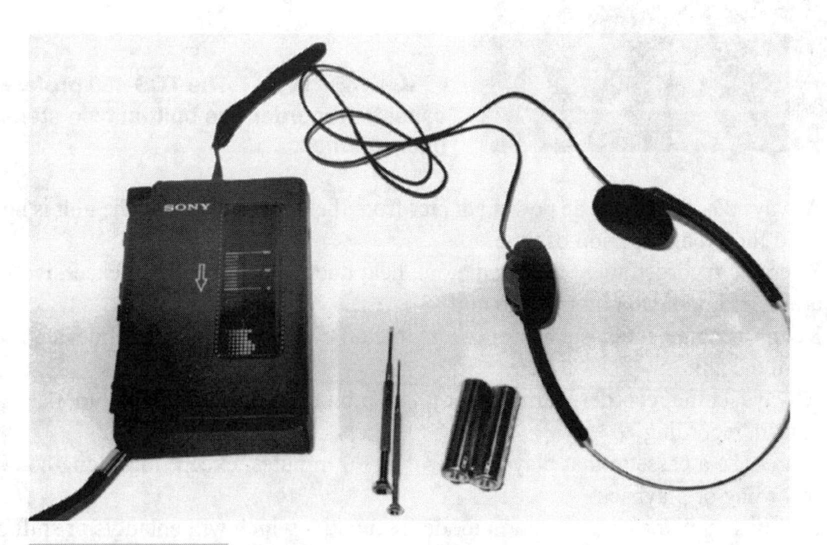

FIGURE 18-31 The Sony TCS-430 professional stereo cassette comes with a set of headphones and one speaker.

FIGURE 18-32 The small professional cassette player might operate from two AA batteries.

FIGURE 18-33 The TCS-430 professional cassette recorder has built-in twin stereo microphones.

3. Always disconnect the ac power adapter from the wall outlet when the unit is not to be used for a longer period of time.
4. Keep the recorder away from radiators, heat ducts, direct sunlight, excessively dusty areas, and rain. Do not drop the unit.
5. Keep strong magnets, such as speakers, metallic objects, and magnetic watches away from the unit.
6. Always let the recorder warm up for a few minutes before attempting to make an important recording.
7. Never use a cassette that plays longer than 90 minutes, except for a long continuous recording or playback.
8. Do not switch the tape operation mode frequently, which will entangle or spill out the tape.
9. Keep high volume from outside sounds low if the outside noise is loud and disturbing. Do not record at this time.
10. Protect your ears by lowering the volume in extended play.

CASSETTE FEATURES

All professional recorders use the standard cassette. To remove the cassette, press the Stop/Eject button in the Stop mode. To prevent accidental erasure of the cassette, break out the small tabs at the back. Break the tab out of side A directly to the right, when looking at the back side of the cassette. Tab side B is to the left. Tape can be applied over the removed tab to record over the cassette once again.

Some professional recorders use type II (CrO_2) tapes or type IV (metal) tapes. Always check with the manufacturer's literature to use the correct cassette tape. Only normal, type I tapes are used on most recorders.

BLOCK DIAGRAM

The manually operated deluxe recorder has only one signal source of sound, but stereo units have two separate channels. Sometimes the circuit diagram lists only one channel because both stereo channels are the same. The playback and recording signals in the recorder are duplicated in the stereo units. Of course, the speed circuit's motor control, system control, and clock signals are not considered to be stereo circuits.

The block diagram is quite handy to determine where the trouble lies before looking at the large, complex stereo circuits (Fig. 18-34). Remember, only one circuit can be shown

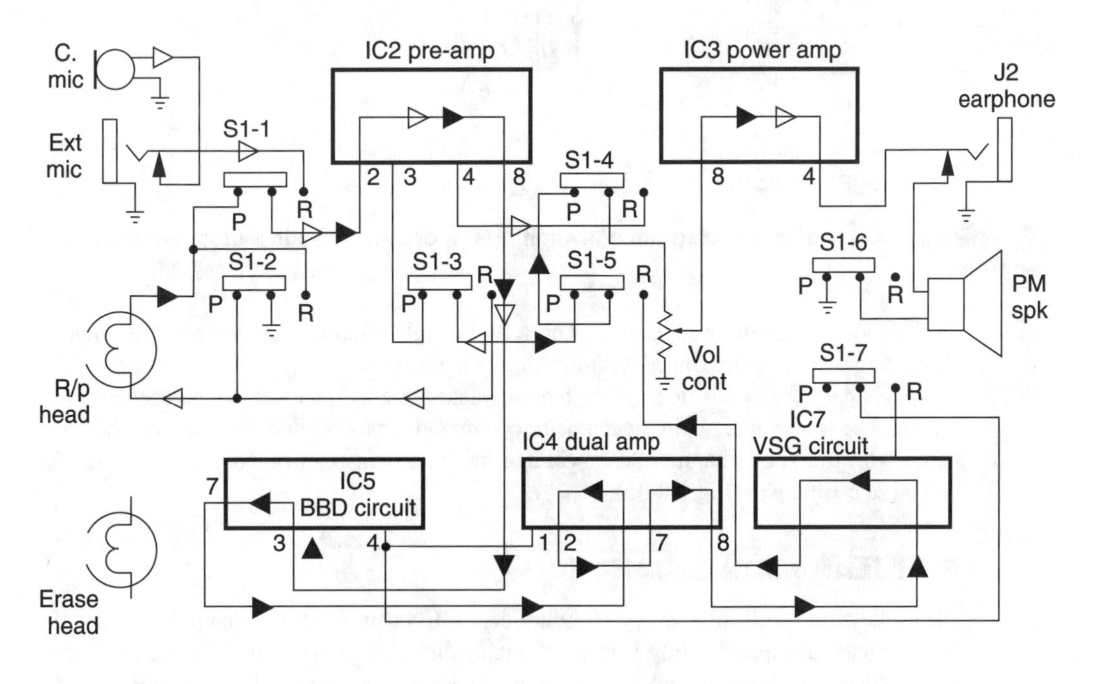

FIGURE 18-34 **The block diagram of a manual signal circuit in a professional cassette recorder.**

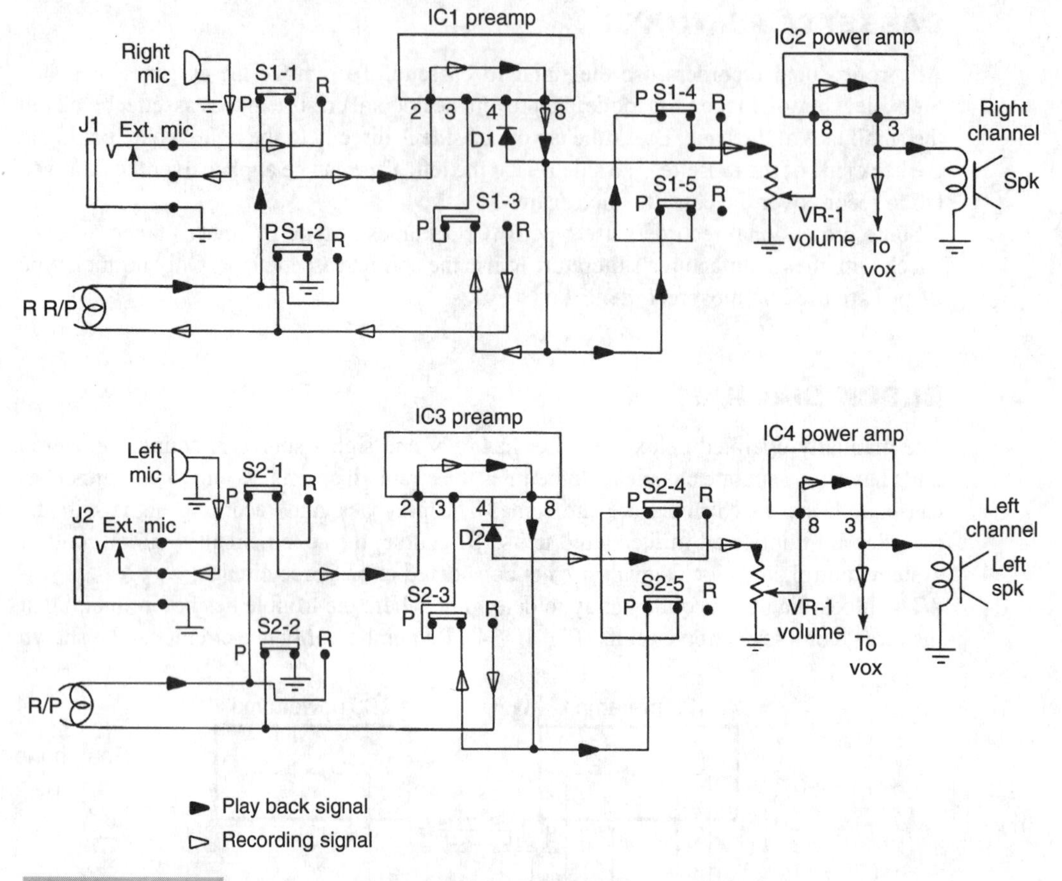

FIGURE 18-35 **The block diagram of switching Record and Play in a stereo cassette recorder.**

on the block diagram or schematic. The left and right channels are identical. Switching signal circuits are easily traced with the block diagram.

If one stereo channel is dead, distorted, or weak, the good channel can be used as a reference. The weak or dead circuit can be traced in both channels and the audio can be compared with the good channel. Also, voltage measurements can be compared to find the defective component (Fig. 18-35).

REGULAR MAINTENANCE

Clean the tape heads after every 10 hours of use to ensure optimum sound. Besides the heads, clean all tape-handling surfaces: pinch rollers, capstans, and the erase head with cotton swabs or cleaning sticks. You can easily clean the tape heads and pinch roller by pressing Play without a cassette in holder. This action brings the tape components out for easier excess. Remove the battery so that the recorder is not operating. Clean them with 90% denatured alcohol and a cleaning stick. After the heads and pinch rollers are cleaned, press Stop/Eject button to move the heads into position so that the cassette can be inserted.

Clean the casing with a soft cloth, moistened with a mild detergent solution. Do not use strong solvents such as rubbing alcohol, benzene, paint thinner, or fingernail polish remover on plastic parts. They can mar the finish or casing.

Keep all magnetized tools away from the tape heads because they might magnetize these parts. A magnetized tape head will result in increased noise, hiss, and a loss of high-frequency response. Use a standard tape head demagnetizer to demagnetize the record/play head.

For lubrication, use a specially formulated high-grade lubricant in the appropriate places. Lubrication is normally required only when parts bind, slow down, or make screeching noises after long periods of time. Use phono lube and lube gel for the areas to be greased. Use a drop of 3-in-1 oil or light oil on the motor and wheel bearings. The precision lubricator is handy to place grease or oil on the exact spot. Use all lubricants sparingly and avoid contact with other parts. Wipe the excess oil or grease with a cleaning stick.

DISASSEMBLY

Although the following instructions do not apply in every case, they do show how to remove the cabinet parts. Remove the battery cover and batteries. Remove the four screws that hold the rear cabinet. Remove the screw that holds the amplifier PC board. Remove the capstan and the counter belts. Remove the screw that holds the motor assembly. Remove the tape mechanism. Remove the two knobs (pitch and speed). Remove the screw that holds the control panel. Remove the spring and the cassette lid. Remove four small black screws to remove the bottom panel of the Sony TCS-430 stereo cassette recorder (Fig. 18-36).

MECHANICAL ADJUSTMENTS

Before attempting to repair or make mechanical adjustments of the recorder for slow or improper speeds, wipe all tape-contacting surfaces (R/P head, erase head, pinch roller,

FIGURE 18-36 Remove four screws to remove the bottom panel of the Sony TCS-430 stereo cassette-recorder.

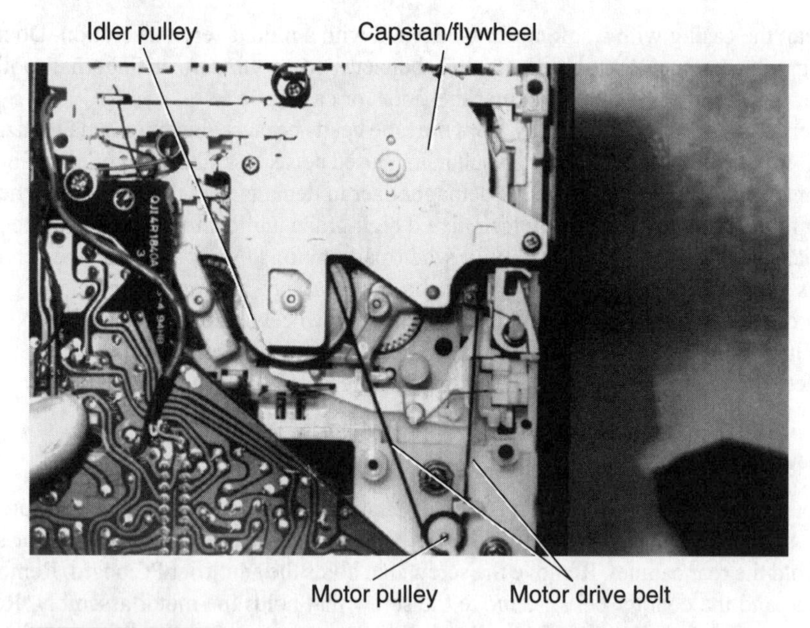

FIGURE 18-37 **Clean the contact surfaces of idler wheels, motor pulleys, belts, and flywheels with a cloth soaked in rubbing alcohol.**

tape guide, and capstan). After these components are cleaned with rubbing alcohol and a cleaning stick, clean the contact surfaces of the driving parts, such as the motor pulley, flywheel, fast-forward/rewind arm assembly, and all idler wheels with a piece of soft cloth soaked in rubbing alcohol (Fig. 18-37).

Of course, some drive belts, such as the capstan and the fast-forward/rewind belts can have a specially surface-treated surface and should not be cleaned with rubbing alcohol-soaked fabric. Replace any excessively greasy or oily drive belts. Clean all excess grease and oil with rubbing alcohol and a cleaning stick.

PINCH ROLLER ADJUSTMENT

Mechanical adjustments can be made with an apting gauge (0- to 500-g gauge) and a cassette torque meter. To adjust the pinch roller, clip the spring gauge to the pinch-roller assembly while playing the unit. Some of these pinch rollers require from 120 to 200 g of spring force. Simply hook the spring gauge to the pinch roller and pull it away from the capstan. Take a reading on the spring gauge (Fig. 18-38).

Now, measure the force at the moment when the pinch roller contacts the capstan (when the pinch roller starts rotating). If the pinch roller force is weak, adjust it by bending the spring (54) for greater tension. Replace the spring if it does not hold proper tension.

TORQUE ADJUSTMENT

Torque adjustments can indicate if the speed of the cassette player is very slow or uneven with a torque cassette (Fig. 18-39). Make both play and fast-forward torque adjustments.

The torque adjustment of a Radio Shack VCS-2001 professional cassette recorder in the Play mode is 35 to 65 g/cm. Fast-forward torque should be 50 to 120 g/cm while in the Rewind mode, the torque should be from 50 to 120 g/cm (Fig. 18-40).

If the play, fast forward, and rewind torque is very low and not in the rated area, clean the flywheel (23) and the fast-forward/rewind arm assembly (29), and check or replace the capstan belt (116) and the fast-forward/rewind belt (3). Doublecheck the seating of the torque cassette when making these tests. Do not overlook the possibility that the cassette drive motor (Table 18-2) is defective.

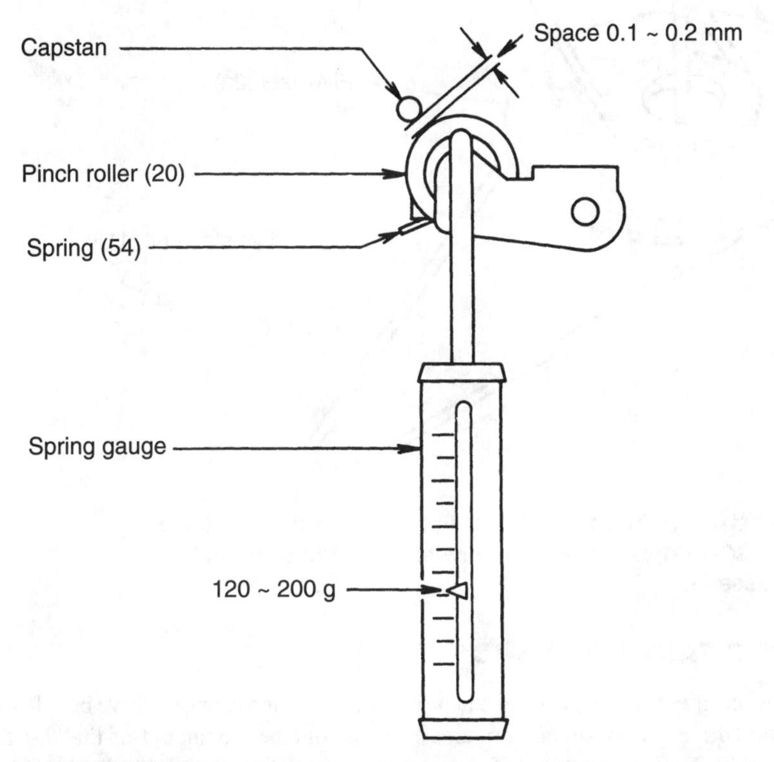

FIGURE 18-38 If Fast Forward, Rewind, and Play run slow, it might be caused by improper torque of the pinch rollers.

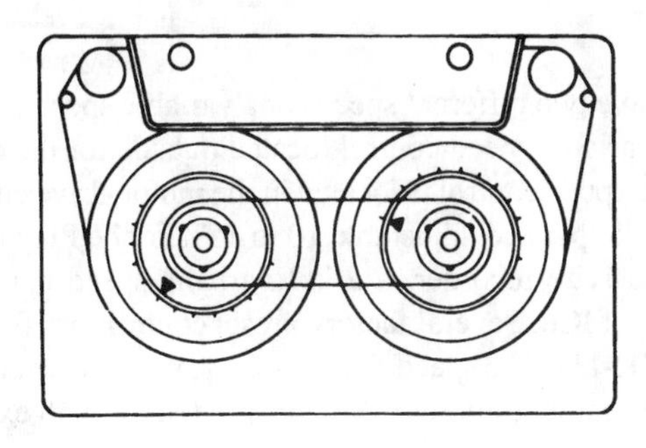

FIGURE 18-39 Check the Fast-Forward, Rewind, and Play torque with a torque cassette.

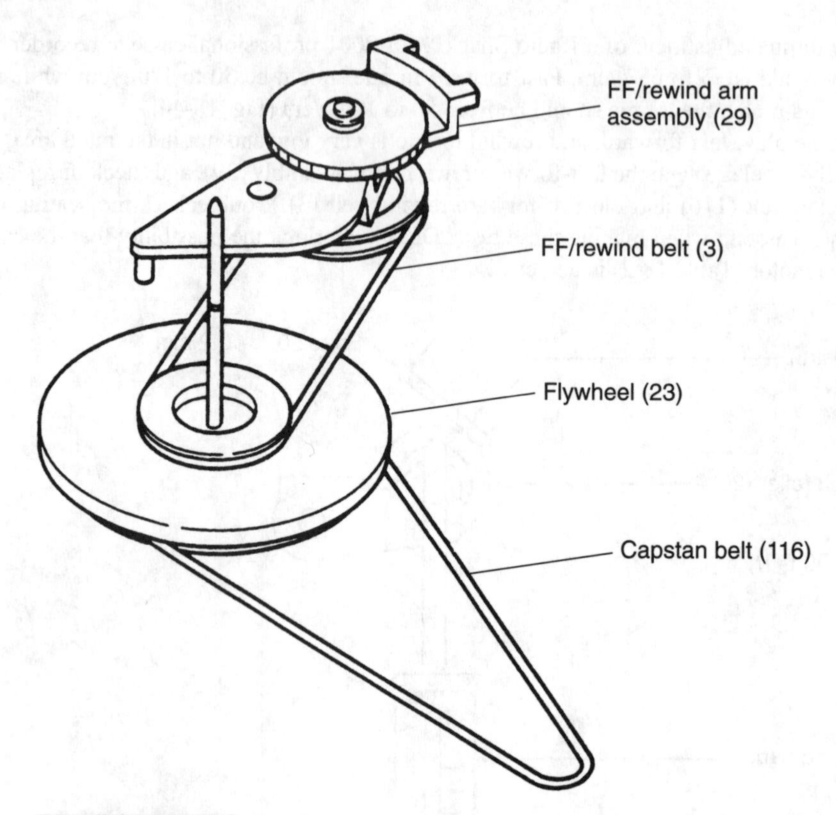

FIGURE 18-40 **The correct Fast-Forward torque for a Radio Shack's VSC-2001 cassette recorder was 50 to 120 g/cm on the torque cassette.**

ERRATIC FAST FORWARD

Often, uneven or erratic fast forward can be caused with a loose or oily drive belt. Inspect the fast-forward drive belt for oil or worn areas. Check for a bent or an out-of-line fast-forward arm assembly. Notice how the different idlers and pulleys engage. A dry bearing of an idler wheel or turntable reel can cause erratic fast forward (Fig. 18-41). Doublecheck the fast-forward speed with a torque cassette.

ONLY ONE SPEED

Some professional cassette recorders have two different speeds or a variable speed control. The Sony professional recorder has a variable speed control located outside for the operator to control (Fig. 18-42). The variable-speed control is located in the motor-drive circuits. The actual tape speed is 4.8 cm/sec (1.78 ips) and adjustable 10 to 15% in the Play mode.

Within Radio Shack's VSC-2001 cassette recorder, a 2-kΩ variable-speed control is found within the play or record circuits of IC6. Several factory preset controls are found in this speed circuit (VR-9, VR-10, and VR-11). The speed of the cassette motor is controlled by pins 3 and 4 of IC6 (Fig. 18-43). The cassette-motor 6-V supply is fed through external dc jack 4, leaf switch S4, and remote jack J3. Variable speed is only adjustable in the Play mode, not during recording.

Take crucial voltage measurements on terminals 3 and 4, and 6 and 8. Clean switches S2-B and S1-B with rubbing alcohol or cleaning fluid. Erratic speed can be caused by a worn or dirty speed control (VR3, 2 kΩ). Spray cleaning fluid into the controls. Suspect that IC6 is defective if a variable voltage is going into pins 6 and 8 and no change is at pins 3 and 4. Check all resistors in the speed-control circuits for increased values.

UNEVEN PRESSURE ROLLER

A worn or damaged pressure roller can cause the tape to spill out and feed unevenly. The pinch roller is quite small in the portable cassettes, although not as small as the roller in

TABLE 18-2 PRELIMINARY TROUBLESHOOTING CHART

SYMPTOM	CAUSE	REMEDY
Cassette cannot be inserted.	Cassette inserted improperly.	Check cassette. Check for foreign material inside. Notice if play or record button is depressed.
Record button cannot be depressed.	No cassette loaded. Cassette tab removed.	Reload cassette. Check for small tab removed at rear of cassette.
Playback button cannot be locked in.	Check for complete wound tape.	If tape is completely wound toward arrow direction, rewind tape with rewind button.
Tape does not move.	Incorrect battery polarity. Weak batteries.	Batteries in backwards. Test and replace if below 1.2 V.
	No ac.	Power adapter not connected.
No sound from speaker	Are headphones plugged in?	Remove headphone plug.
	Check location of volume control.	Volume turned down.
Tape speed excessively fast	Check setting of speed control.	Readjust speed control.
Weak or distorted sound	Weak batteries. Dirty heads.	Test or replace. Clean with alcohol and cleaning stick.
Poor recording	Weak batteries. Dirty stereo heads.	Test or replace. Clean with alcohol and cleaning stick.
Poor erasing	Improper connections.	Check all cord and wire connections.
	Dirty erase head.	Clean erase head.

FIGURE 18-41 Erratic Fast Forward might be caused by a loose or oily fast-forward belt.

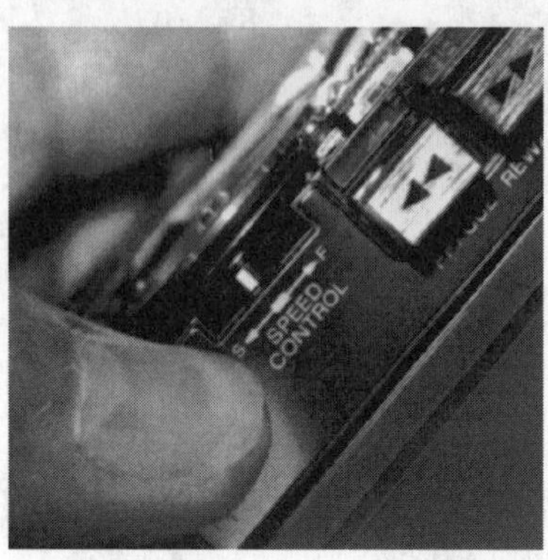

FIGURE 18-42 The variable speed control is located on the outside of this Sony professional portable recorder.

microcassette players. Check the pinch roller for worn spots. Determine if enough pressure is applied to the pinch-roller assembly with a small pressure gauge, if it is handy. Bend the tension spring for more pressure if the gauge measurement is low. Replace the entire pinch-roller assembly if it is defective (Fig. 18-44). Excessive wow can be caused by a dirty pinch roller.

NO TAPE MOTION

Check the batteries and replace them if they register less than 1.2 V. Clean the on/off leaf switch. The leaf switch is on in all functions (Fig. 18-45). This switch is in series with the external dc jack (J4), batteries, and remote jack (J3). Sometimes, these leaf-switch contacts become dirty and bent together.

If you find a voltage at the power-output IC and on both sides of the switch, suspect a defective motor circuit or motor when the motor will not rotate. Check for voltage across the motor terminals. If voltage is present and the motor doesn't rotate, replace the drive motor.

Sometimes you can hear the motor run, but nothing happens in the Play, Record, Fast-Forward, or Rewind modes. Remove the bottom cover and notice if the capstan/flywheel is rotating. Check for a missing or broken belt. Erratic operation can be caused by an oily or overly large motor drive belt. Wipe the flywheel surface and belt with rubbing alcohol and a cleaning cloth. Replace the capstan belt if it is loose or excessively greasy.

EXCESSIVE WOW

If wow is excessive, clean the pinch roller and capstan drive. Check the pinch-roller pressure with a pinch-roller gauge. Replace the pinch roller if it is worn or deformed. Check all drive belts and clean them. A defective drive motor might produce uneven speeds. Check the motor drive voltage from the IC or transistor voltage-regulator circuits. If the voltage is constant and the motor operation is erratic, replace the defective motor (Fig. 18-46).

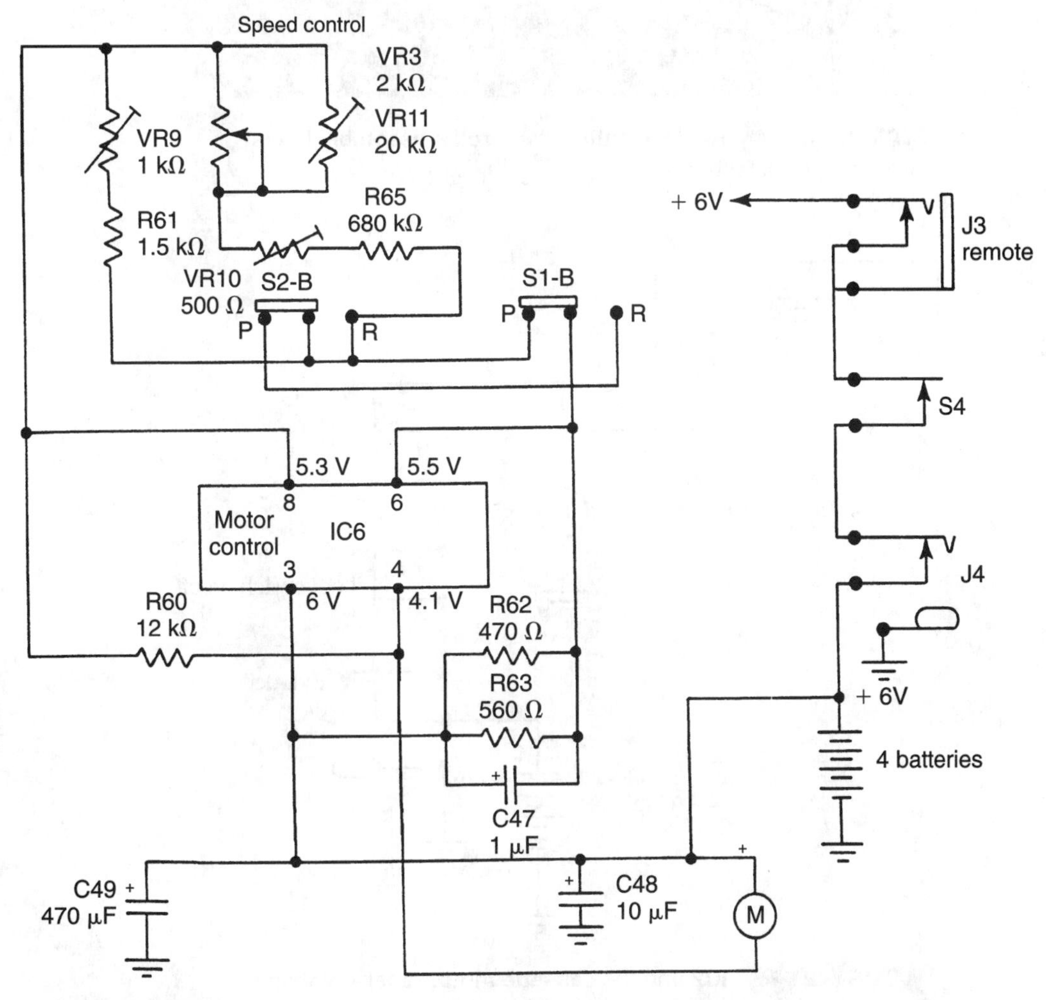

FIGURE 18-43 The variable speed control on the cassette motor-driver circuits.

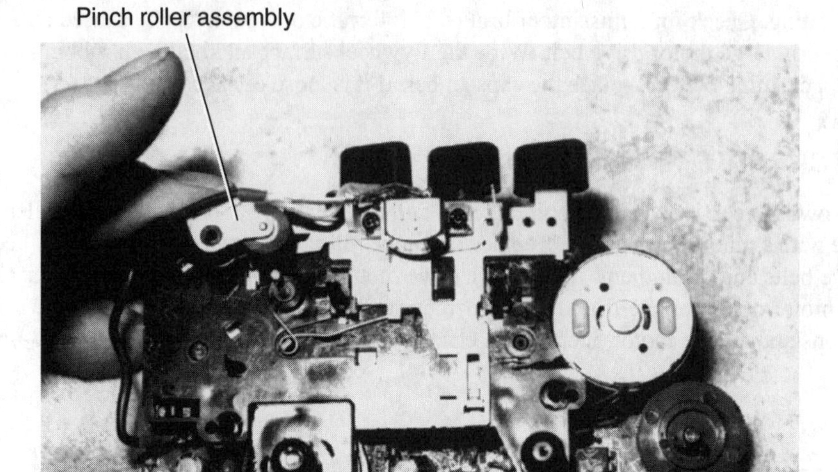

FIGURE 18-44 Replace entire pinch-roller assembly if it is worn, bent, or defective.

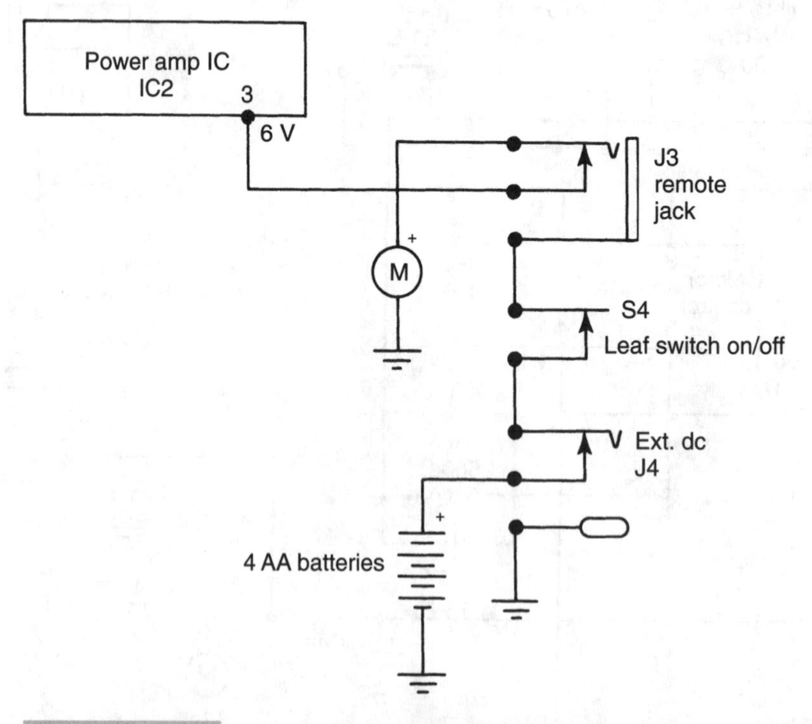

FIGURE 18-45 IC2 and the cassette motor receive voltage through the remote control, leaf switch (S4), and external dc jack (J4).

FIGURE 18-46 A close-up view of the small motor in Sony's TCS-430 stereo cassette recorder.

NO SPEED CONTROL

If the speed control doesn't work, check its setting. Excessive tape speed can be caused by an improper setting of the speed control. Clean slide switch S2 and replace it, if it is defective. Check the speed control for open or dirty contacts. The total resistance across the control is 2 kΩ. Measure the voltages on all terminals of motor-control IC6 (Fig. 18-47). Check the components around the IC. Replace IC6 if the voltages are fairly normal without the speed control.

NO LEFT MIC CHANNEL OPERATION

Stereo is used in most professional cassette recorders. Dual condenser microphones are contained in the left and right stereo channels of Sony's stereo cassette recorder (Fig. 18-48). Here, stereo headphones are used in playback and one speaker is used as a monitor. The sound recorded can be heard through the headphones. You can adjust the monitor volume with the volume control.

The condenser or electret microphone has a bias voltage applied to the ungrounded side of the mic terminal. With external mic jacks, the built-in microphones are placed in series with the external mic jack (J1). Suspect that a mic is defective if the right channel sounds normal and only the left channel is dead. Also, check the switch contacts on S1-1.

Try another external electret microphone in the external jack to determine if the internal mic and circuits are defective. If the external jack mic operates, take voltage measurements

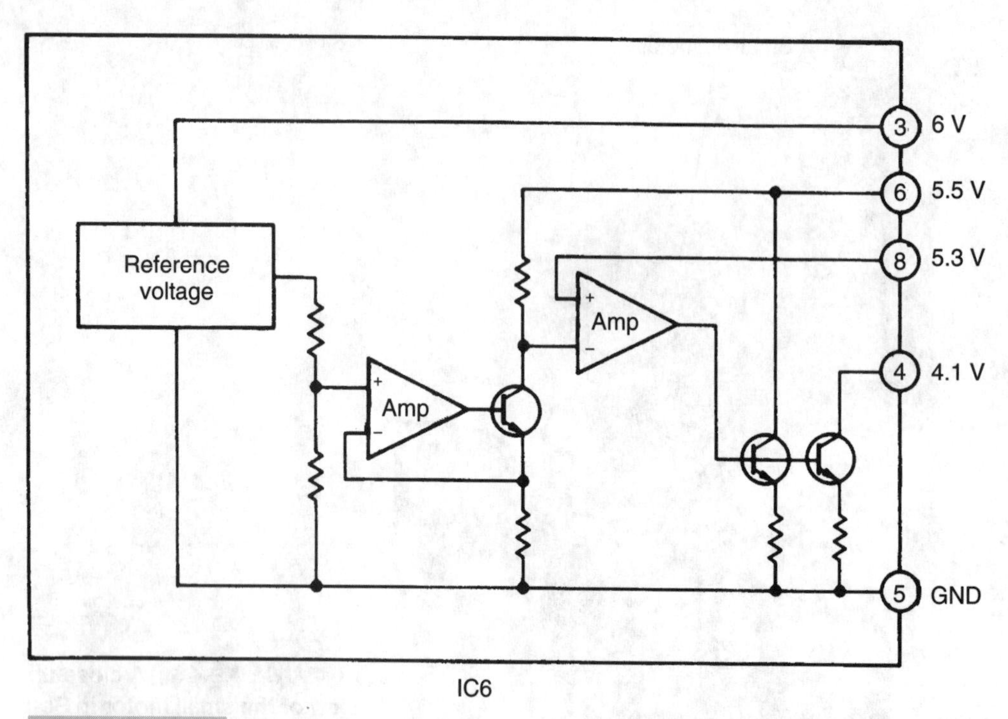

FIGURE 18-47 **The internal stages of motor-control IC6 with the operating pin-terminal voltages.**

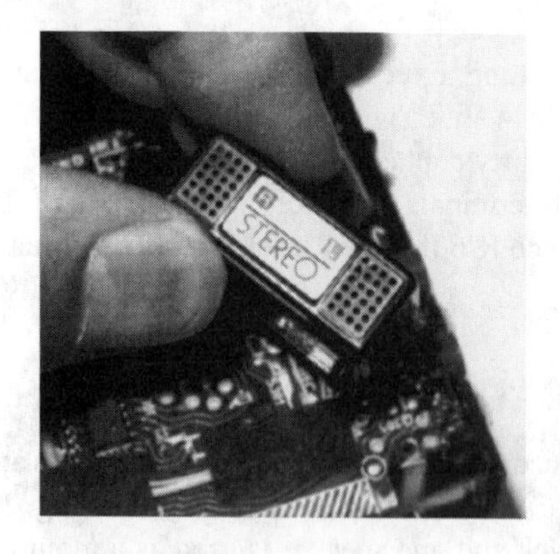

FIGURE 18-48 **Two electret mics in a stereo cassette player.**

across the microphone terminals (1 to 1.5 V). Suspect R4, R5, and C3 if you find no voltage at the mic terminals (Fig. 18-49).

If both the external and internal microphones are dead, check switch S1-1 and IC1. Does the left channel play a cassette normally? If so, the R/P switch or mic wires might be defective. If the left channel is completely dead on Play and Record, suspect that preamp IC1 is bad. Measure the supply voltages at pin 9 (5.1 V). Check all voltage terminals of preamp IC1 (Fig. 18-50). If the preamp stages are transistors, check the voltages in the microphone input and preamp circuits. Remember, two different identical preamp microphone circuits are used in the stereo recorder.

POOR LEFT-CHANNEL SOUND

If the left-channel sound is poor, clean the tape heads. Is the poor left-channel sound in the Playback or Record mode? Start at the preamp IC and work toward the monitor speaker or headphones of the stereo left channel if the poor sound is found in both the Recording and Playback modes. Try a new recording cassette to see if the left channel still has poor

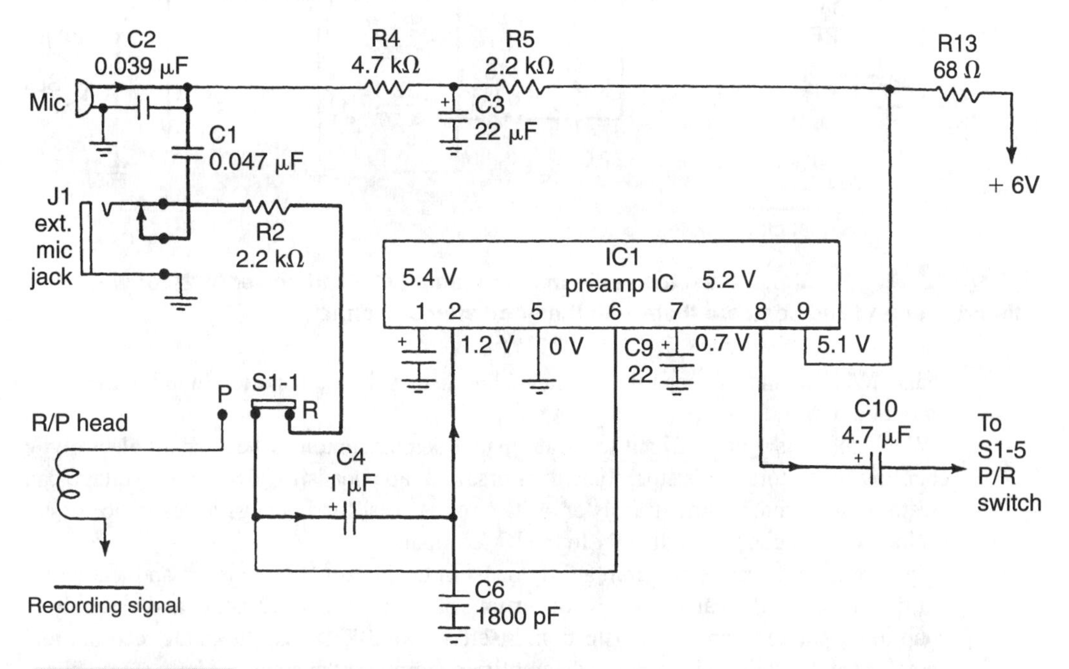

FIGURE 18-49 The condenser-mic recording-signal circuit of preamp IC1.

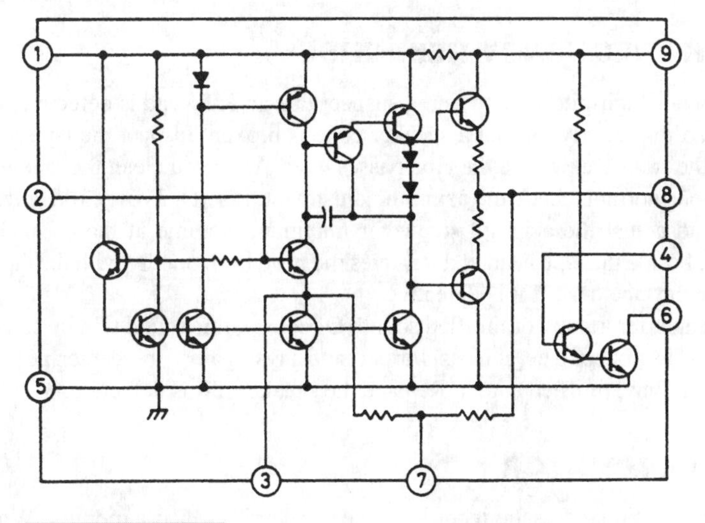

FIGURE 18-50 The internal preamp circuit of IC1.

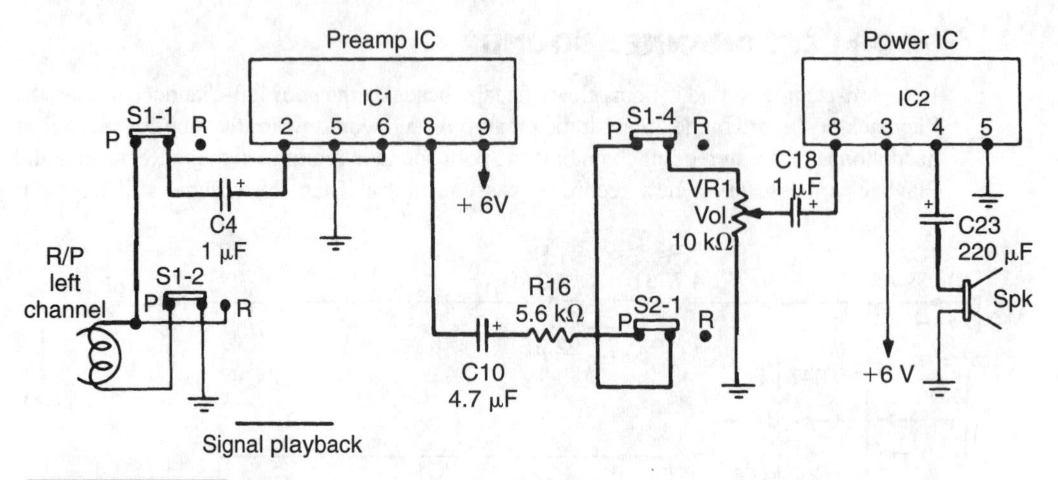

FIGURE 18-51 **Signal trace the audio in the preamp (IC1) and power (IC2) for weak or distorted sound and compare them with the good stereo channel.**

sound. Weak sound can occur in any stage, but distorted sound is often found in the audio output circuits.

Weak sound is more difficult to locate than a dead symptom. Check small electrolytic coupling capacitors, transistors, bias resistors, and tape heads. A low supply voltage can create weak sound in either transistor or IC circuits. Test for a change in resistance or capacitance when connected directly to the IC terminals.

Distorted sound can result from a dirty or worn tape head. Bias resistors and leaky output transistors or ICs can cause low distortion. Shorted or leaky resistors and electrolytic coupling capacitors can cause distortion. Insert a 1- or 3-kHz cassette in the recorder and signal trace the audio with an outside amplifier. Compare the signal of both stereo channels to locate the defective component (Fig. 18-51).

POOR HIGH-FREQUENCY RESPONSE

When music sounds high-pitched and tinny, suspect that an R/P head is defective, the azimuth is adjusted improperly, or that a speaker cone is broken. Inspect the tape head for worn spots on the front area where the tape passes over. After you clean the tape head, if the R/P head looks normal, check the azimuth adjustment. Insert a 3- or 6.3-kHz test tape (MTT-113N) and adjust the azimuth screw for maximum volume at the speaker. This screw is located beside the tape head and it moves the head horizontally with the tape (Fig. 18-52). Replace the tape head if it is defective.

Check the speaker for a tinny or muffled sound. Of course, this sound will be constant if the speaker cone is frozen. The cone is warped and lays against the center pole piece, which creates the tinny, muffled sound. Replace the speaker if it is defective.

NO SPEAKER/MONITOR

In many small professional cassette recorders, one speaker is used as a monitor. When the stereo headphones are inserted, the speaker is automatically removed from the circuit.

When recording, the sound can be heard through the headphones. Keep the headphone volume low or a howling sound (feedback) might occur. When you take off the headphones during a recording, lower the headphone volume control, then take them off. Otherwise, feedback might occur.

If the headphones are working, but the monitor speaker isn't, suspect that a jack contact is poor, speaker wires are broken, or a speaker is defective. Clip another small speaker across the built-in speaker terminals. If the outside speaker works, replace the defective speaker. If you hear no sound, check the speaker-wire terminals. Inspect the headphone shorting jack for possible damage or broken terminal wires (Fig. 18-53). Notice that R23 (33 Ω) is in the circuit to lower the headphone volume, compared to the monitor speaker.

Suspect that C23 (220 μF) is defective if the player has no speaker or headphone reception. These speaker-coupling capacitors can open up or become intermittent with poor internal wire connections. A poor internal terminal wire can cause weak, intermittent, or dead audio.

NO ERASE

If you find jumbled recordings on several different cassettes, suspect that the tape head is erasing poorly. Clean the erase head. Some erase heads are excited by a dc voltage or a magnet. The erase head is excited with the bias oscillator. The erase head might be open or shorted. Replace the erase head if it is open or defective. Check the erase-head continuity with the ohmmeter.

If you find both recording and erasing problems, suspect a defective bias-oscillator circuit. The waveform can be checked at the high side of the R/P head with a scope. If a scope is not handy, check the oscillator transistor voltage, T-300 windings, and resistance changes in the oscillator circuit (Fig. 18-54). Some recorders have a BBD clock and ramp generator in a VSC (variable sound circuit) chip, which feeds a BBD IC, which excites the erase and R/P tape heads.

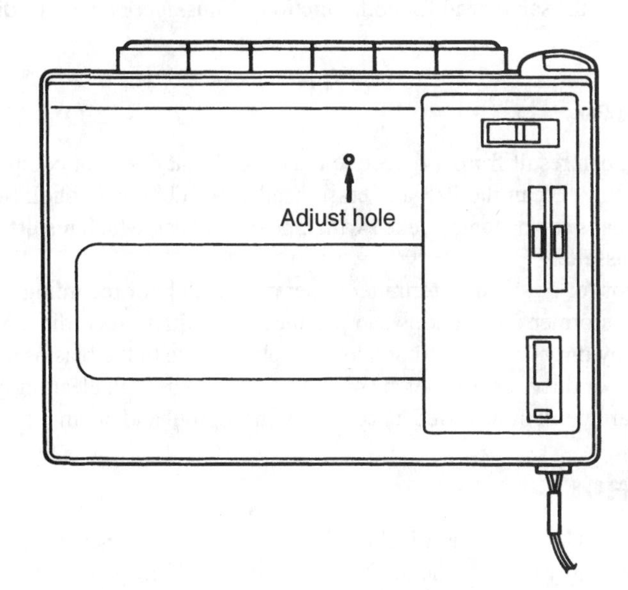

FIGURE 18-52 The location of the tape head azimuth screw in the Radio Shack VSC-2001 recorder.

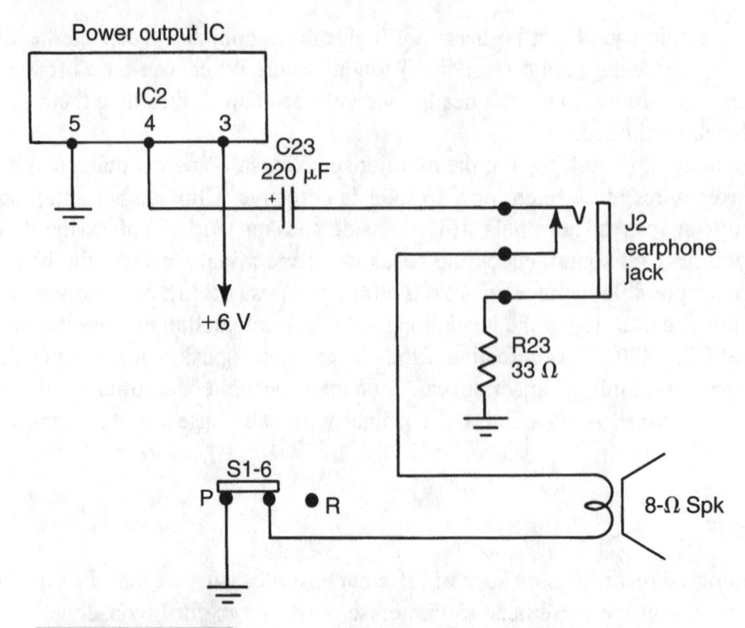

FIGURE 18-53 **The audio signal is taken from pin 4 of IC2 through C23, J2, and the speaker return wire through switch S1-6.**

In large professional cassette recorders, you can find a separate erase head. With the small units, the erase head can be located inside of the R/P head assembly. Peek at the front side of the R/P head to see if you locate another gap area. The stereo R/P and erase head has three gap areas. If the erase head is located inside the stereo R/P head, you will find six terminal hook-up connections or wires. Remember, in a stereo record/playback head, the Record and Play modes use the same head for both functions. Thus, it contains two different gap areas.

POOR RECORDING

Poor or erratic recording can result from a defective erase head that does not completely erase the previous recording. Clean the R/P and erase head with rubbing alcohol. Sometimes one of these small gaps can become packed with tape oxide dust, which results in no recording. Try another cassette.

Check the bias oscillator for erratic or intermittent operation with poor recording symptoms. Poor oscillator transformer connections can produce intermittent recording. Weak recording can be caused by dirty record head or a low supply voltage to the bias oscillator transistor. Check for poor or dirty record switch contacts. Clean them with cleaning fluid. If both Record and Play are weak or distorted, check the audio-output and preamp circuits.

NO PITCH CONTROL

The pitch slide switch (S2) is located in terminal pin 15 of IC5. S2-2 switches in a B+ 6-V supply voltage for IC5 or simply turns on the pitch-control circuits (Fig. 18-55). The pitch-

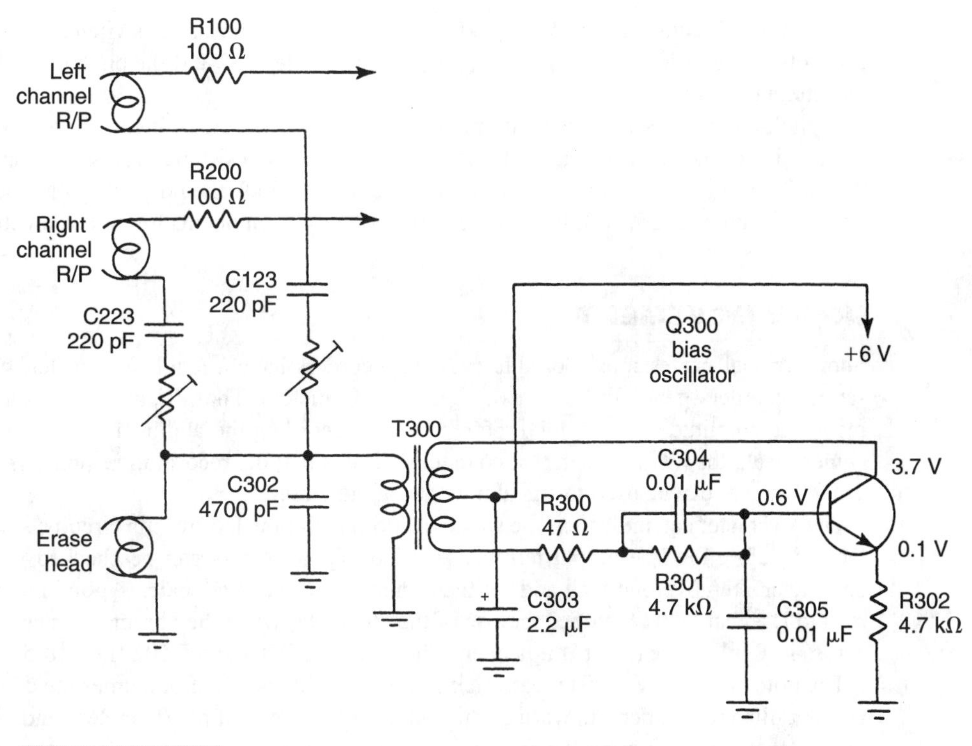

FIGURE 18-54 The bias oscillator might excite each stereo tape head and erase head in larger recorders.

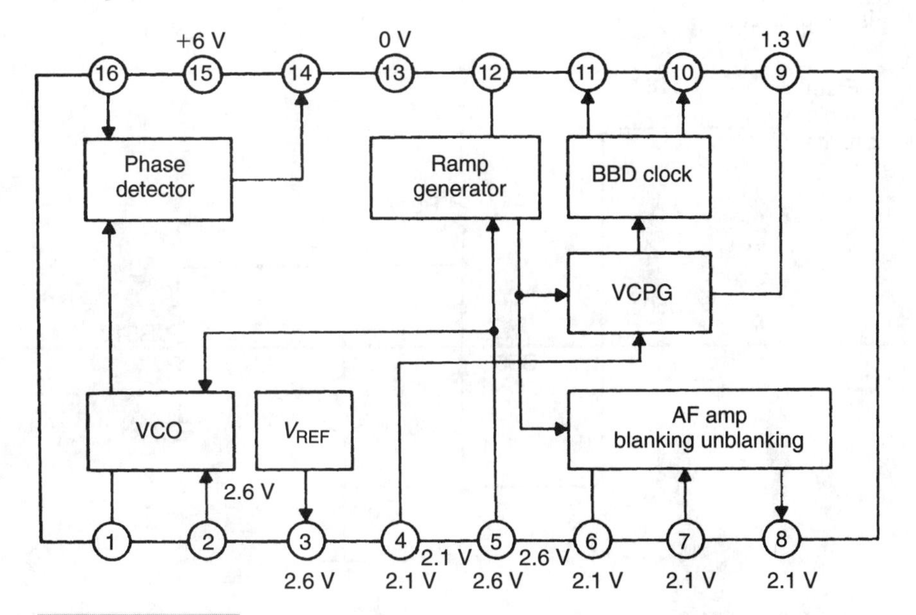

FIGURE 18-55 Check the supply voltage and the voltages on all pins of IC5 for no or improper pitch control.

control range is controlled by VR2 (5 kΩ) at pins 3 and 5. Check slide switch S2-2 if it has no pitch control. Measure the voltage (6 V) at pin 15. Clean S2-2 if the pitch control is intermittent or dead.

If pitch-control VR2 is open, the pitch control might be intermittent. Spray cleaning fluid inside of the control area if VR2 is dirty. Measure the resistance across the control to determine if it is open without pitch control. Do not overlook the possibility that the variable pitch control, IC5, IC3, IC4, or the corresponding components might be defective.

NO VU MOVEMENT

Some manual professional portable cassette recorders have a single VU meter, but the stereo recorders generally have two separate VU meters. The VU meter measures the recording amplitude or how loud the recorder is recording the audio. If one channel becomes weak, the VU meter in that channel indicates that the recording response is weak. The VU meter can be used to signal trace the audio circuits.

The VU meter can indicate if the sound is normal after the Dolby amp circuits; some are located before the line-output terminals. VU or signal meters can be checked with the ohmmeter. Remove one lead and set the ohmmeter to the R×1 scale. A good meter will read to less than 300 Ω. Notice the rise of the VU meter when the ohmmeter is connected across it. Replace the meter if it is open. Check diodes D501 and D502 (Fig. 18-56).

The pointer on the VU meter can stick or rub on the dial plate. Sometimes the dial cardboard will warp and bend upwards, which stops the movement of VU meter hand. Repair the meter by disconnecting the terminal wires and mounting screws. Remove the cabinet and the front meter glass. Reglue the dial plate and reassemble the meter. Turn the small set screw at the bottom to zero. Check for audio up to the VU meter if you suspect that it is the problem.

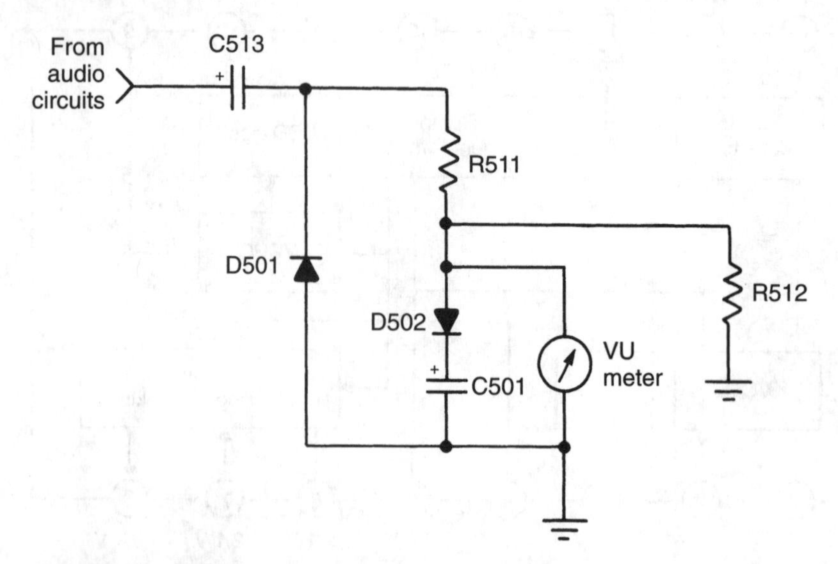

FIGURE 18-56 The VU-meter circuit attaches to the audio signal and is rectified by D501 and D502.

Troubleshooting

If possible, pick up a schematic diagram of the recorder for easy troubleshooting. A service manual of the exact cassette recorder has the PC board wiring layout and wiring diagrams to help you to locate defective components.

Besides a layout of electrical and electronic components, the exploded view of the tape mechanism helps you to see where the various parts tie together. Sometimes, the exploded view of the cabinet parts show how they are mounted and placed together. For additional troubleshooting symptoms, follow Table 18-2.

19

TROUBLESHOOTING AUTO STEREO CASSETTE AND CD PLAYERS

CONTENTS AT A GLANCE

CONTENTS AT A GLANCE

Today's average auto stereo cassette player has digital tuning with 12 FM and 6 AM presets, seek/scan tuning, and at least 6 W output (Fig. 19-1). The deluxe unit might have quartz electronic tuning, 18 or more presets, an LCD display, auto reverse, from 7 to 20 W output, and 4 to 6 speakers. High-power car players have 40- to 1000 watts stereo amplifiers (Fig. 19-2).

Surface-mounted components can be found in the electronic digital tuning, clock, system control, and AM/FM IF circuits (Fig. 19-3). These surface-mounted components can be mounted on the PC board side, and the other ICs and transistors appear on the other side. An intermittent surface-mounted component can be difficult to locate. High-powered ICs and transistors are used in the audio circuits.

Blows Fuses

Suspect that a power-output IC or transistor is leaky, connecting wires are pinched, or a filter capacitor has been shorted if the fuse keeps opening. Inspect the hookup harness for bare wires or improper connections (Fig. 19-4). Be sure that the American radio has a negative

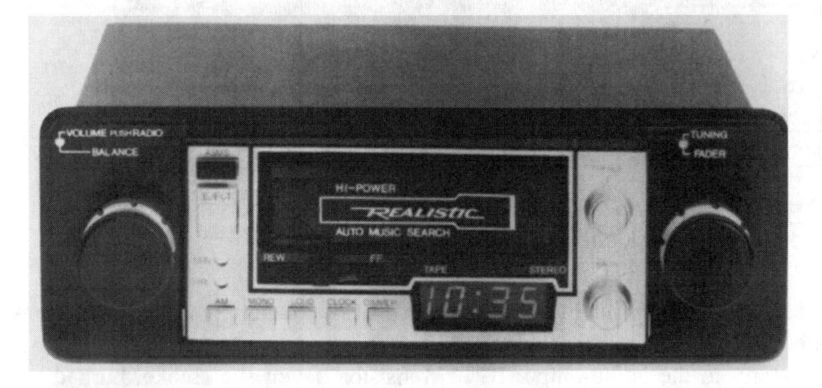

FIGURE 19-1 The high-powered auto cassette player might have audio music search and digital tuning.

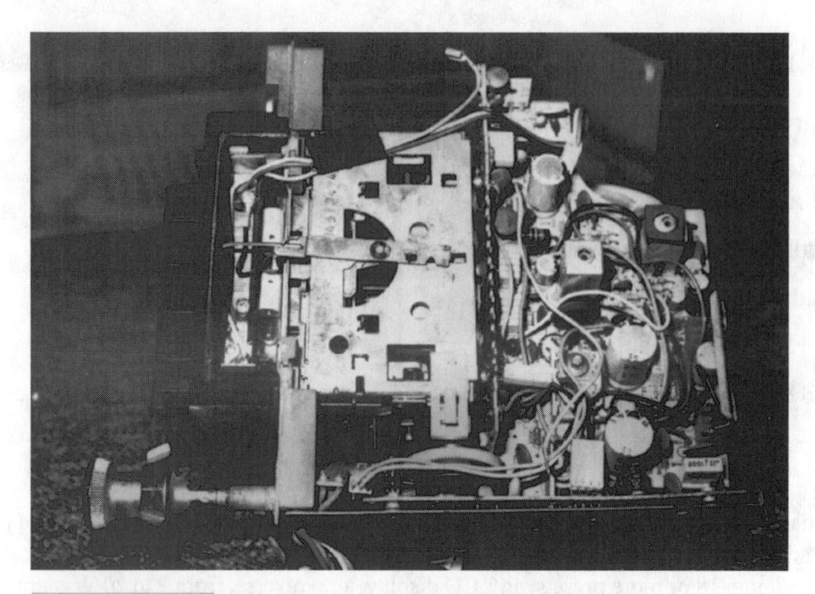

FIGURE 19-2 The auto radio-cassette might have a high-powered amplifier that could output up to 1000 watts.

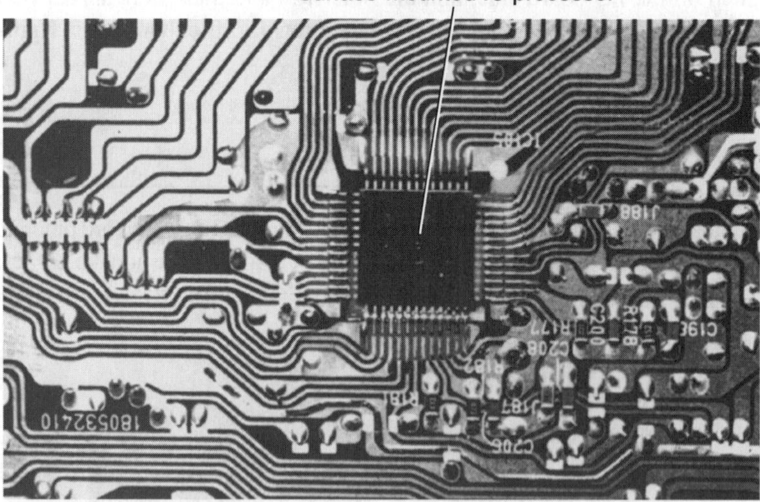

FIGURE 19-3 Surface-mounted components might consist of IC in the digital and clock circuits of the latest auto radios.

ground and a positive ground to foreign autos. The auto battery polarity might be switched or battery might be charged backwards.

Go directly to the audio-output IC or transistor if you see smoke, burned wires, or scorched PC board wiring. Often, the power IC uses the car radio chassis as a heatsink

(Fig. 19-5). Check for leakage from each terminal to ground. With burned PC board wiring, the defective output IC and transistor can be quickly located. Besides replacing all defective components, all wiring and PC board repairs must be made.

In the early auto radio chassis, power output transistors were used. Today, one large power IC or two different channel ICs provide audio to the speakers. Suspect a problem with the output dual-channel IC if both channels are dead, distorted, weak, or intermittent. Signal trace the audio up to the input terminals of the power IC.

The balance, tone, and volume controls are used in the input circuits of each power-output channel (Fig. 19-6). The left-channel input audio is at pin 13 and the right-channel IC pin is 2. The left-output signal is taken from pin 8 through coupling capacitor C51, and the right output from pin 7 through electrolytic capacitor C52. If C49, C48, C51, or C52 dries up, the output signal is weak. If either capacitor opens, the sound might die at the speaker.

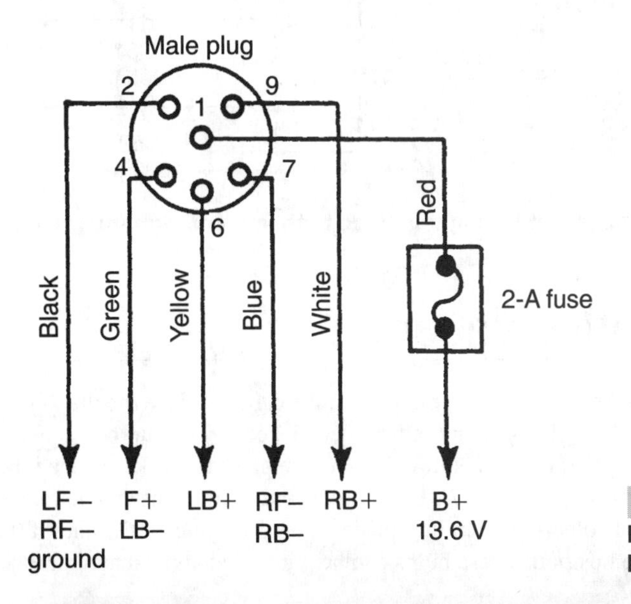

FIGURE 19-4 Inspect the hook-up harness for bare or burned wires.

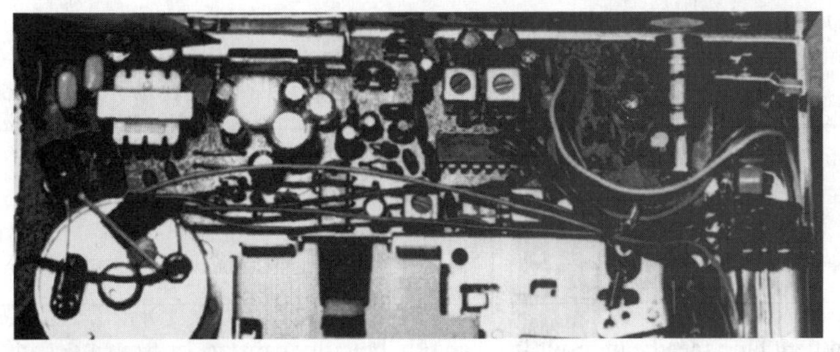

FIGURE 19-5 A leaky or shorted power-output IC might cause the fuse to blow or open.

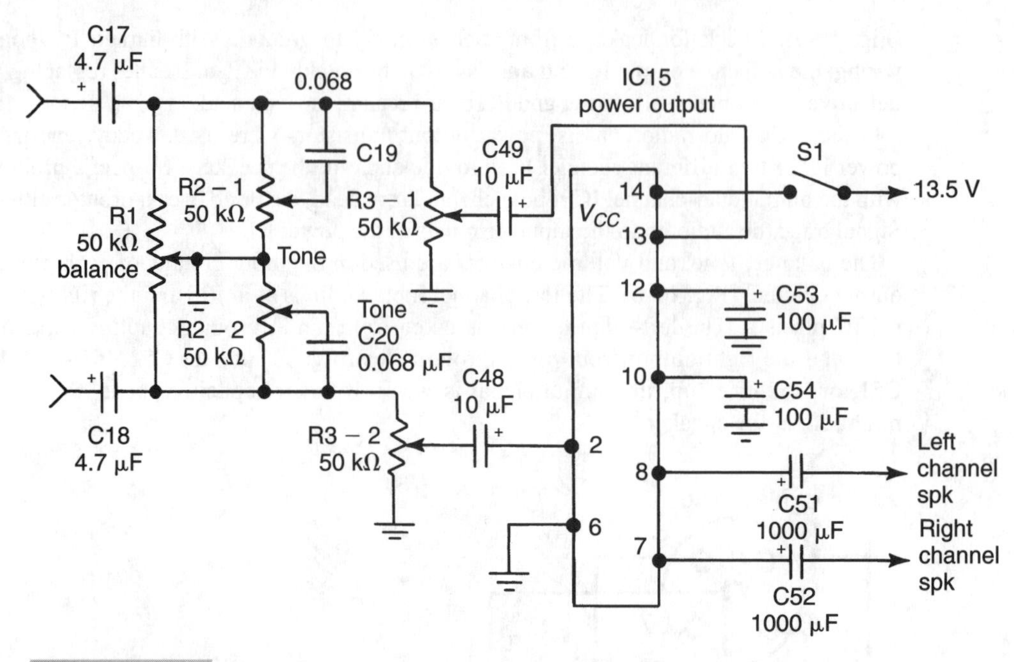

FIGURE 19-6 **One large power IC might include both left- and right-output circuits, with balance control.**

Pilot Lamp Replacement

In most cases, you must remove the auto radio from the car to replace the dial light bulbs. Regular 12- or 14-V light bulbs plug into a bayonet socket and should be checked when the covers are off the radio. If the bulb still lights, but the glass is black, replace it because they will die in a few months (Fig. 19-7). Light bulbs with long leads can be replaced by cutting off the leads and soldering them into place. A drop of glue on the end of the bulb helps hold the bulb into position. These bulbs can be purchased at most auto or electronic stores.

LEDs are often used as light indicators in the latest auto cassette players. LEDs are used as pilot lights, normal- and reverse-direction indicators, switching lights, and fast-forward and rewind lights (Fig. 19-8). In larger cassette players, Dolby, AMSS, and stereo LEDs are used. Usually, LEDs will last the life of the cassette player. Pilot lights are fairly easy to replace after the unit has been removed from the auto.

Erratic Speed

Intermittent tape speed can result from an oily belt, dry capstan, or from a defective drive idler, speed-control circuit, or motor. Check the voltage at the motor to isolate a defective speed circuit. Start with a new cassette. Clean all dry surfaces with cleaning stick and rubbing alcohol.

If you clean and replace the loose drive belts and the speed is still erratic, suspect that the motor is defective (Fig. 19-9). Monitor the motor voltage at all times. Sometimes tapping the end bell of the motor will make it change speeds. If the voltage is not constant at the motor terminals, suspect that a speed circuit is defective or that a radio/tape switch is dirty (Fig. 19-10). If the motor is working directly from the battery source without a speed-regulator circuit and is intermittent, install a new motor.

FIGURE 19-7 **Many different light bulbs and LEDs functions as indicators in this auto cassette player.**

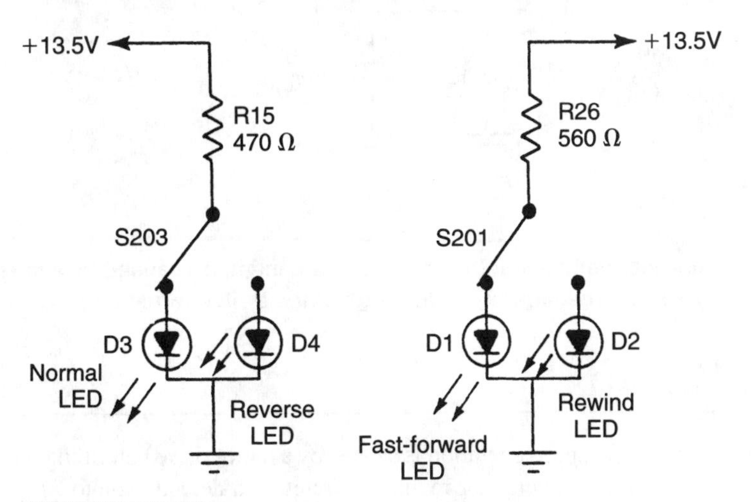

FIGURE 19-8 **LED indicators are used in the normal, Reverse (A), Fast-Forward, and Rewind circuits.**

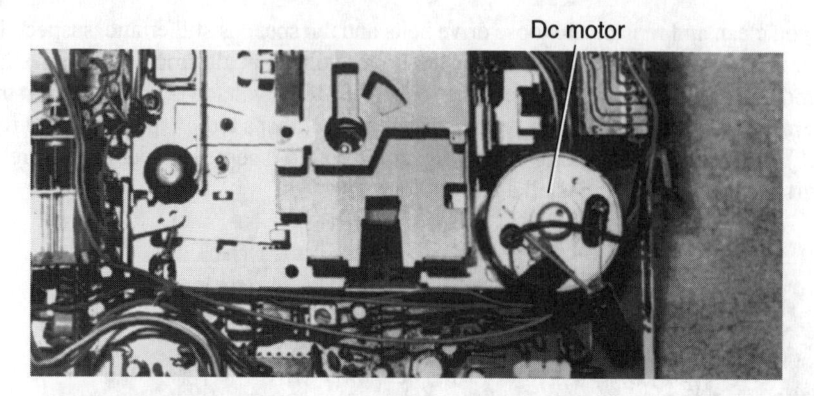

FIGURE 19-9 **A defective motor might cause erratic, fast, or slow speeds.**

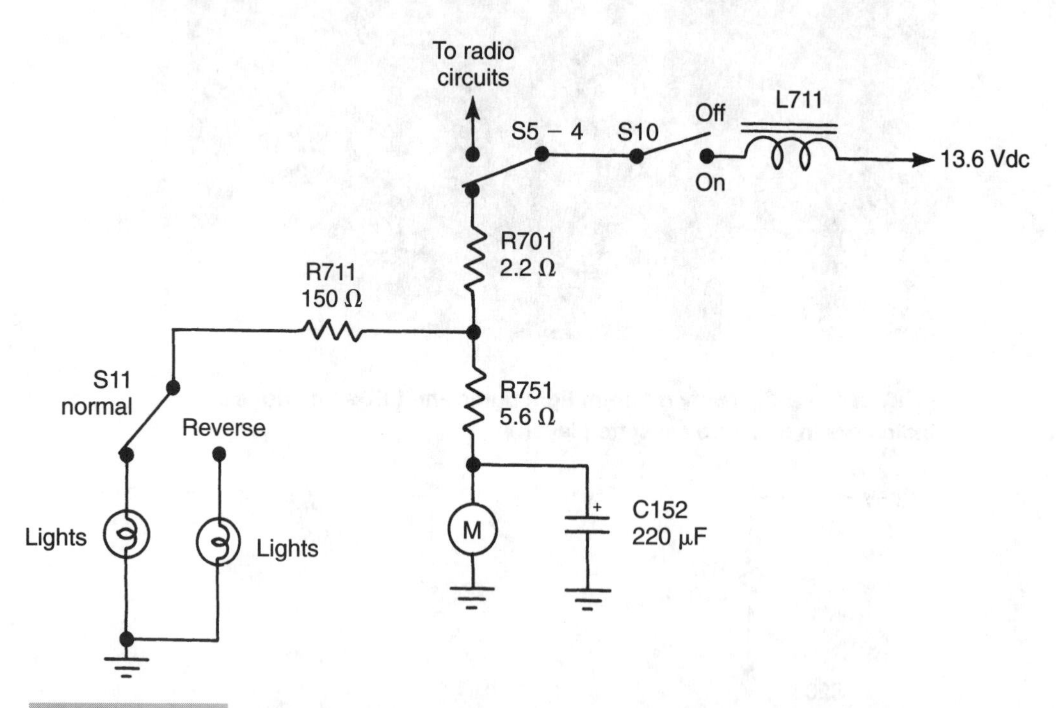

FIGURE 19-10 **Improper voltages at the motor circuits might be caused by a defective speed circuit, isolation diode, resistor, or by dirty radio/tape switch terminals.**

High Speeds

Most higher-than-normal speed operation is caused by a motor drive belt riding on the rim of the motor pulley, a defective motor regulator circuit, or a defective motor. A leaky IC or transistor motor regulator circuit can cause high speeds. Check and monitor the voltage at the motor circuits (Fig. 19-11). Readjust the speed control in the regulator circuits.

In some regulator circuits, two different adjustments are provided for normal and reverse rotation. Insert a 3-kHz test tape and connect a frequency counter at the speaker output. Adjust both the reverse and normal controls for a 3000-Hz reading on the frequency counter.

The motor-speed regulation circuit can be a transistor or an IC. The regulator circuit provides accurate voltage and keeps the motor operating at a constant speed. If the transistor or IC becomes open or leaky, the speed can increase or lower, according to the dc voltage that is applied to the motor (Fig. 19-12). Test each transistor within the motor-regulator circuit. Sometimes, these intermittent transistors will only operate under load. If in doubt, replace the transistor or IC. Check each diode and resistor within the speed-regulator circuit.

If the motor circuits and voltage applied are normal, replace the motor. The speed in a defective motor can run high or low. Often, the speed is low, sometimes caused by dry bearings. A squirt of light oil at each motor bearing might let the motor resume normal speed. A screeching or squealing motor indicates that the bearings are dry. Check the cost of a new motor before removing the defective motor.

Works On Radio/No Tape Action

Listen to the motor when operating in Fast-Forward or Play modes. If the spindles do not rotate or if you can't hear the motor, suspect that the voltage is not getting to the motor. Measure the voltage at the motor terminals. Suspect that an off/on or radio tape switch is dirty. Clean the switches. Check the dc voltage after each switch. If voltage is on both terminals, but none is at the motor, suspect that the isolation resistor is open (Fig. 19-13).

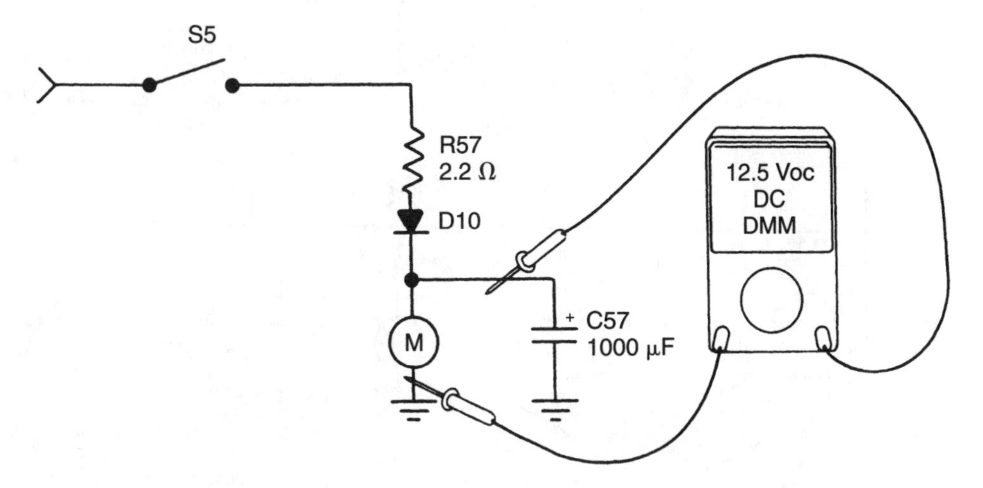

FIGURE 19-11 Monitor the voltage across the motor terminals if the motor is intermittent, dead, or rotating at a faster speed.

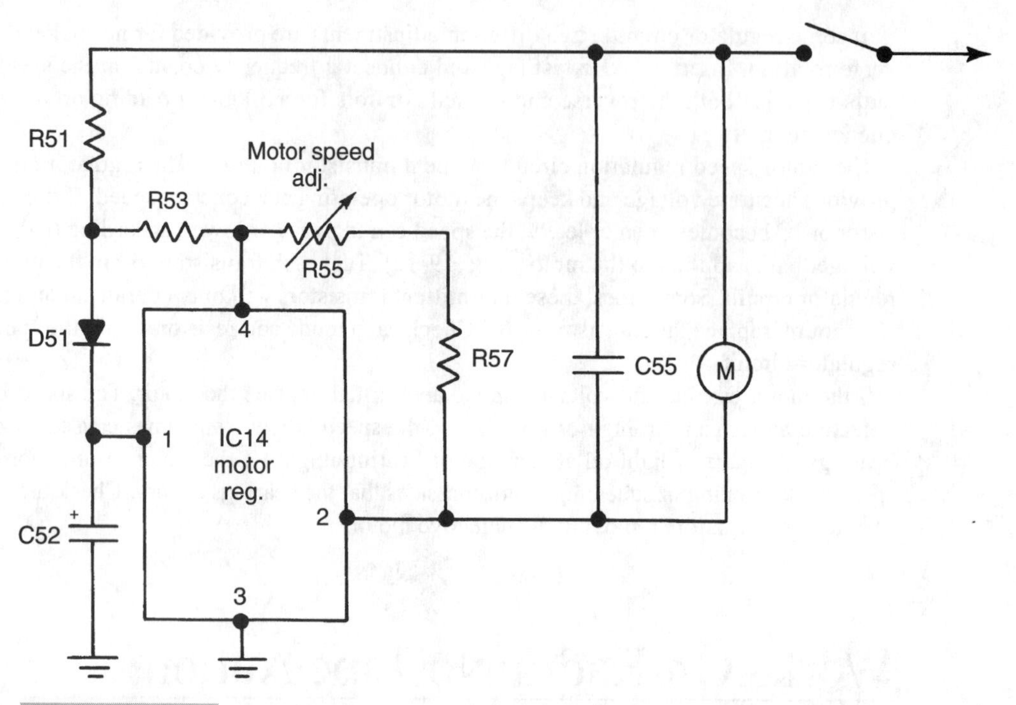

FIGURE 19-12 The defective motor-regulator circuit might cause the motor to rotate at high speeds.

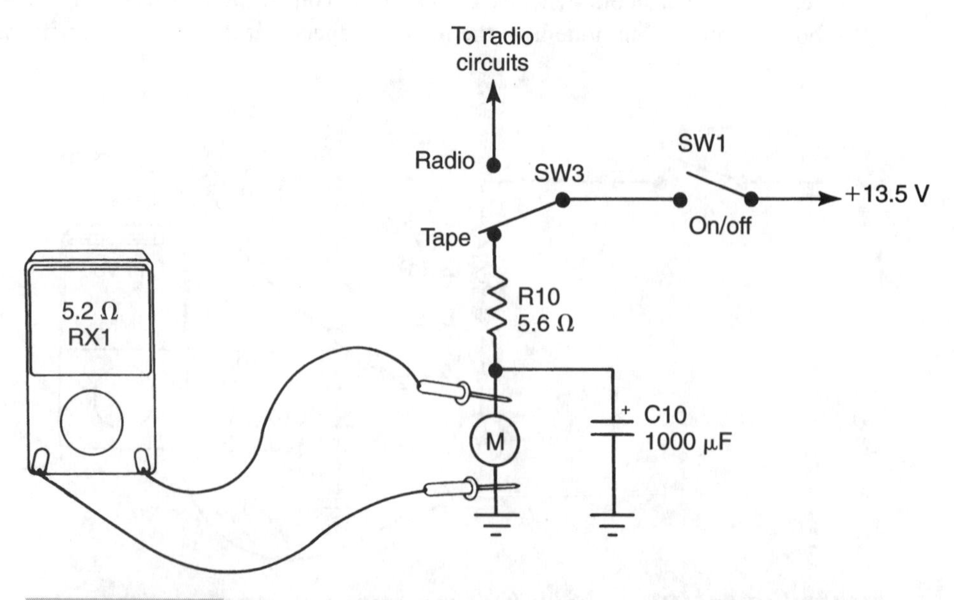

FIGURE 19-13 The motor might be dead with an open R10 (5.6 ohms).

FIGURE 19-14 **Suspect that the carriage or levers are bent if the cassette will not load in a Clarion 9100RT auto cassette player.**

Will Not Load

The cassette can load edgewise or full-on in the auto radio/cassette player. When the cassette is pushed in, a lever is engaged and the cassette platform snaps next to the capstan/pinch-roller assembly. A missing or broken tension spring will not let the holder load properly. A bent cassette or levers prevents loading (Fig. 19-14). Inspect all levers and bent areas. Apply light grease to the sliding areas.

Look for foreign objects inside of the cassette loading area. Sometimes gum wrappers or cigarette butts jam the loading area. If cassette player is near a cigarette tray, remove the player and clean out any cigarette ashes, which can cause binding or sliding metal parts to malfunction.

Jammed Tape

If the cassette will not unload, it might be caused by a malfunction in the loading platform or by a tape spill out. The tape might be wrapped tightly around the capstan and pinch-roller assembly. Sometimes, the capstan/flywheel can be reversed by hand to release the tape tension. Often, the tape must be cut loose because the cassette metal housing prevents removing excess tape and the cassette easily.

After removing the excess tape and the cassette, clean the capstan and the pinch roller. Clean them thoroughly with rubbing alcohol, a cleaning stick, and a cloth. Wipe off all excess tape oxide within the loading area. Clean the tape heads while the unit is open (Fig. 19-15). See if the cassette will load easily.

FIGURE 19-15 After removing a jammed cassette from the holder, clean the capstan, tape heads, and pinch roller with a cleaning stick and rubbing alcohol.

Erratic or Intermittent Audio

In auto cassette players with dual-capstan/flywheel and dual heads, switching the head contacts can cause erratic or intermittent operation. Spray cleaning fluid into the head-switching contacts. Check the small wire head leads for broken wires at the head terminals. When these heads are moved, the wires could break off (Fig. 19-16). Sometimes, the wire

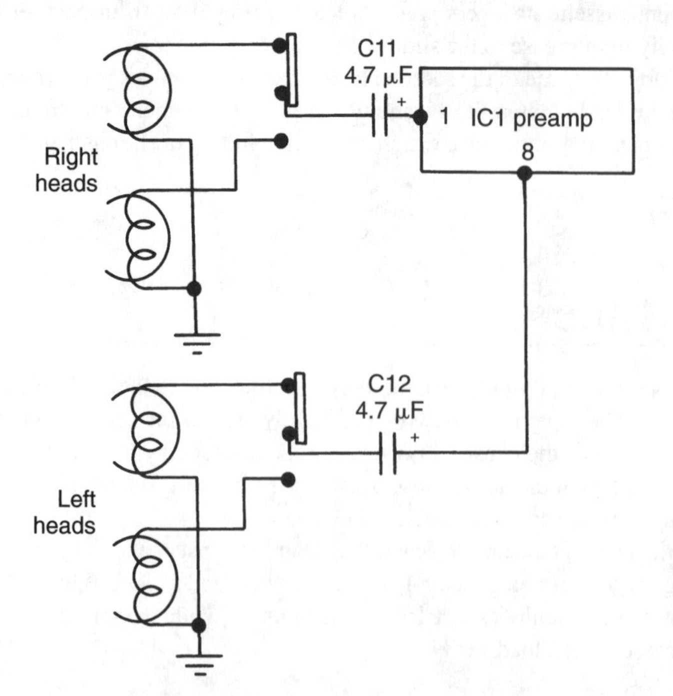

FIGURE 19-16 Check for broken tape-head terminal wires for erratic or no tape head switching.

breaks inside the insulation, but it still appears normal. Lightly pull on the small wires with a pair of long-nose pliers to uncover the broken connection.

Moving the head or switching assembly will eventually break the head wires. Pull back the insulation and tin the flexible wire before soldering it to the head terminal. Be careful not to leave the iron on the terminal so long that it dislodges solid terminal from the back of the head.

Auto Stereo Channels

In early solid-state car cassette players, transistors were used throughout the audio circuits. Now, very few transistors are used because ICs took over. The simplest IC audio channels consist of one dual IC preamp and a dual-output IC. The tape head is directly coupled to the preamp IC through a small electrolytic capacitor (Fig. 19-17).

If the preamp IC circuits appear to be defective, take crucial voltage measurements on IC1, check the continuity of the tape heads, and signal trace the audio output at pins 3 and 6. Do not overlook the voltage regulator circuit when a low or improper voltage is found at pin 4. Some auto cassette players have a separate voltage regulator that operates directly from the dc battery source (Fig. 19-18).

First, take voltage measurements on all terminals of Q9. Low or no voltage at the emitter terminal might indicate that a transistor is leaky or open. A low voltage output at the emitter might result from a leaky zener diode (D10). Check both the transistors and the diodes with the diode test of the DMM. If D10 becomes leaky, R40 might change in value or burn.

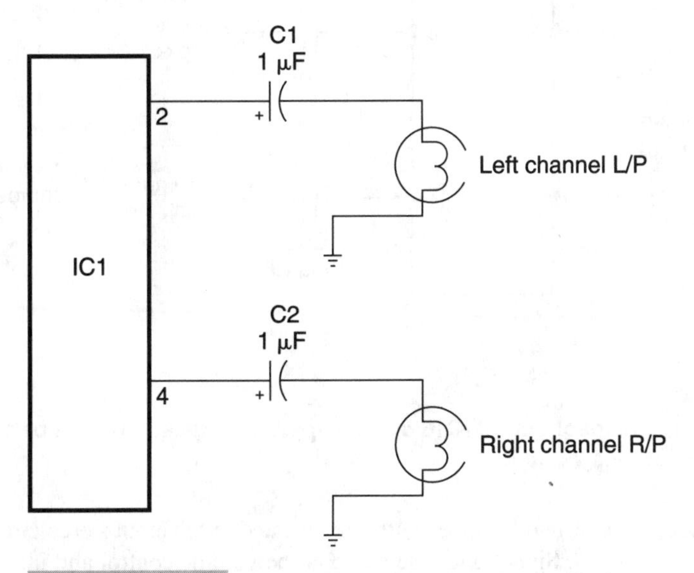

FIGURE 19-17 **The tape head is directly coupled through a small electrolytic coupling capacitor to the IC.**

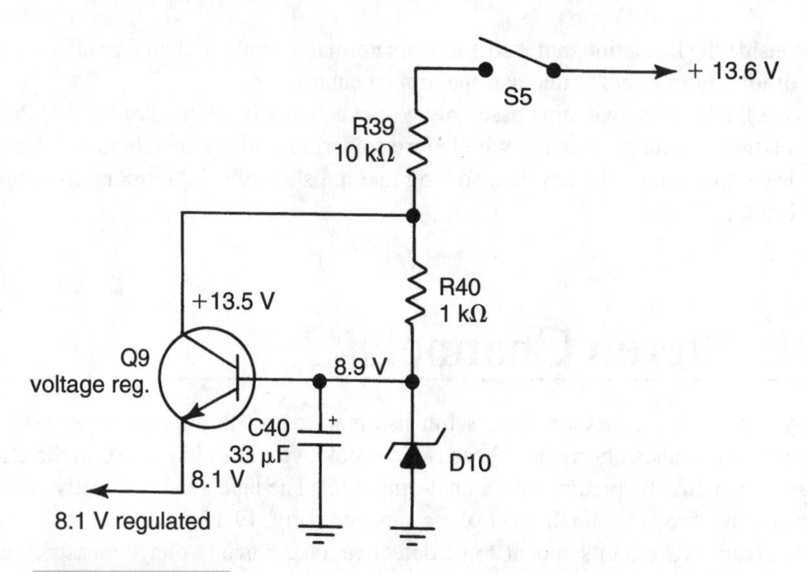

FIGURE 19-18 **Do not overlook the possibility that the preamp voltage transistor-regulator might be open or leaky if the voltage at the preamp circuits is improper.**

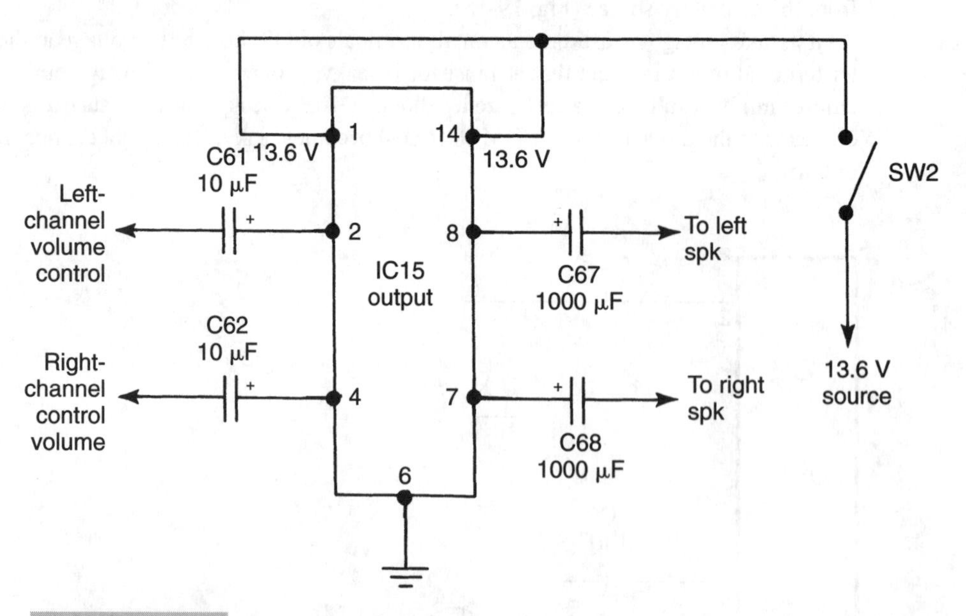

FIGURE 19-19 **Suspect output IC15 or the supply voltage at pin 14 is bad if no audio signals are at pins 7 and 8.**

Usually, the tone, volume, and balance control are located in the input stereo circuits of the audio-output IC circuits. Signal trace the audio at the volume control and into pins 4 and 2 of the IC output. Suspect that an IC or the circuit components are defective if the output signal is improper at pins 7 and 8 (Fig. 19-19).

Troubleshoot the output IC with crucial voltage measurements. Compare the good channel with the defective voltages. The supply voltage is at pins 1 and 14. If this voltage is low, check the dc source and the large filter capacitor. An open choke or fuse can prevent the voltage from arriving at pins 1 and 14 (Fig. 19-20).

Larger auto cassette players with higher wattages have separate power-output ICs or transistors. These power circuits are serviced in the same way as the dual units. However, in this case, each channel component is separate and when the dual IC becomes defective, the whole unit must be replaced—even if only one channel is dead or distorted. The left channel input audio is at pin 10 and the output is at pin 2 (Fig. 19-21).

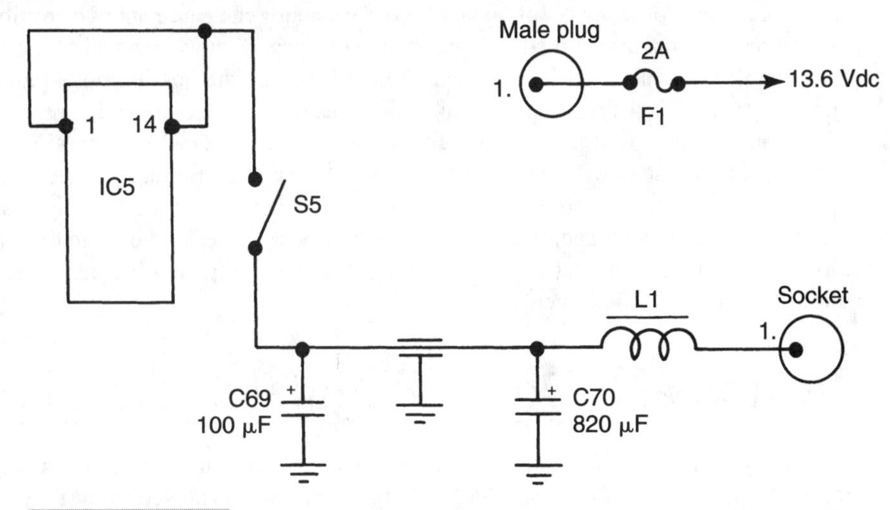

FIGURE 19-20 Open fuse (F1, 2 A) or choke coil L1 might prevent the dc voltage from being applied to switch S5.

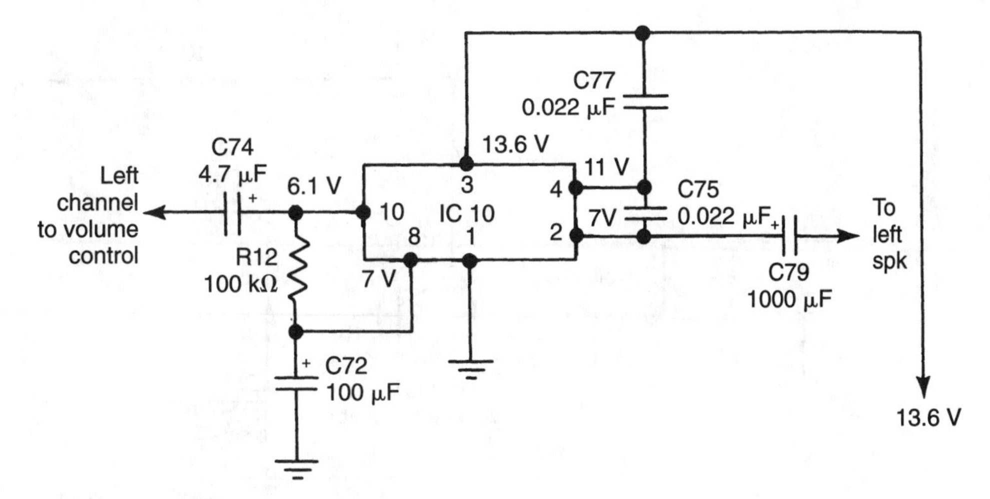

FIGURE 19-21 This separate power-output IC has the audio input at pin 10 and amplifies audio at pin 2 of IC10.

Distorted Right Channel

Often, audio-output circuits cause most of the audio distortion. Check both the right and left audio signals at the volume controls. If the right channel is distorted, proceed to the right input terminal of the output IC or transistor. Carefully feel the leaky IC or transistor because it will run quite warm. Measure all voltages on the IC or transistor terminals and compare them with the schematic. If the schematic is not available, compare the voltages with those in the good channel.

Check the bias and bypass capacitors that are connected to the IC terminals. A burned resistor or an open or leaky coupling and bypass capacitor can cause audio distortion. Replace all resistors that are burned or have changed values. Remove one end of the capacitor or diodes for an accurate leakage test. Do not overlook the possibility that an output coupling capacitor is leaky or open if the audio is distorted, weak, or dead.

Test leaky or open push-pull audio-output transistors with a DMM. Inspect and test all diodes and bias resistors with the transistor output of the circuit. Be sure that these low values of resistance are replaced with parts that have the exact resistance. Usually, a shorted output transistor has a burned or open bias resistor. Check for leaky diodes in the base circuits of some audio transistor circuits. Audio output transistors and ICs can be replaced with exact universal replacements.

Preamp and Dolby Regulator Circuits

The preamp and Dolby IC circuits' supply voltage can come from a transistor regulator circuit. Actually, the 12-V supply that feeds these circuits is isolated away from the B+ power circuits (Fig. 19-22). The power supply or dc battery voltage is fed from the switch through R74 to the collector terminal of Q51.

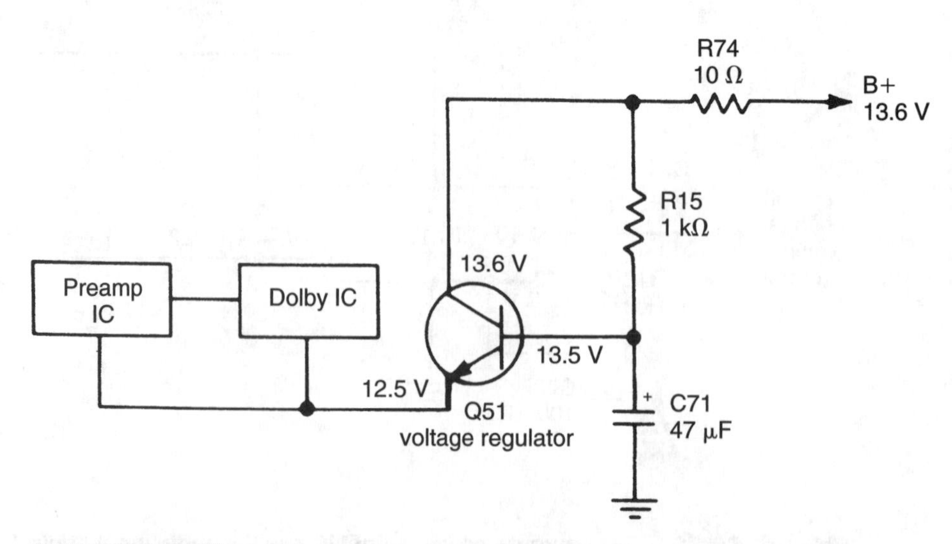

FIGURE 19-22 The preamp and Dolby IC circuits might have a voltage-regulator circuit to supply a 12.5-V source.

FIGURE 19-23 **Clean the noisy volume control by spraying shield or cleaning fluid inside the control area of each channel.**

The 12-Vdc supply voltage is taken from the emitter terminal. If Q51 opens, no voltage can be applied to the preamp or Dolby circuits. If Q51 leaks, a higher-than-normal voltage can be applied to these circuits.

Noisy Volume Control

After several years, the volume control will become noisy as the volume is turned up or down. Sometimes, the audio cuts in and out if the volume control is worn. This can occur in only one stereo channel. Temporarily spray shield or cleaning fluid inside the volume-control area. Don't spray fluid into the front end of the control shaft to clean the controls (Fig. 19-23). Remove the car player and the top cover to get at the control. Insert the plastic tip into each volume-control section and spray in the cleaning solution. Rapidly rotate the volume-control shaft to clean the noisy control. Replace the entire control if it does not work.

Hot IC

The leaky or shorted power IC could run red hot. Spot the discolored IC case and carefully feel the air above the overheated IC. Normal power ICs can run warm, but not red hot.

Often, when an IC shorts, resistors and PC wiring also burns. Sometimes the "A" lead terminal wire with fuse will be burned. The insulation might be charred and this wiring must be replaced. Replace all components that are burned. Repair burned or stripped PC wiring with pieces of hookup wire.

After all burned parts have been replaced, install a new IC. Carefully solder each terminal lead. Now, measure the terminal resistance to the chassis ground. Start with a 10-kΩ range, which can always be lowered. Compare the resistance measurement of each terminal with the normal IC. These comparable resistance measurements should be within a few ohms of each other. Be sure that the DMM resistance numbers stop before marking down the resistance. If the resistance of one measurement is way off, look for additional damaged parts in the hot IC circuits. Then, take accurate voltage measurements on each terminal and compare them (Fig. 19-24).

FIGURE 19-24 Take accurate voltage measurements on the dead channel IC before and after replacing it.

Dead Left Channel

If either channel is dead and the other normal, try to isolate the bad channel by signal tracing or by visually tracing the dead speaker wires back to the respective IC. Usually, the stereo IC or transistors are lined to the left and right when looking at the cassette player from the front end. Touch the left-channel volume-control center terminal with a meter probe with the volume wide open. You should hear a loud hum in the speaker.

If you hear no sound, go to the other center terminal and repeat the procedure. If both controls are dead, go directly to the dual-output IC and take voltage measurements. The dead left channel can be signal traced with an audio amp or an external signal tracer. Remember, a dead audio channel is much easier to locate than an intermittent one.

Proceed through the audio circuits until the signal is lost. Check the voltage within the transistor or IC terminals. Dead audio circuits are often caused by open or leaky transistors or ICs (Fig. 19-25). Do not overlook the possibility that an electrolytic coupling capacitor might be open—especially capacitors less than 10 μF. An open tape-head winding can also produce a dead left channel.

Keeps Reversing Direction

In some auto cassette players, when the tape reaches the end of rotation, the procedure is reversed mechanically or with electronic features. The automatic-reverse motion might be activated by a magnet that is attached to the bottom of the turntable or reel. Under the turntable is a stationary magnetic switch that operates in the auto-reverse circuits (Fig. 19-26).

The cassette player might quickly reverse directions with a defective switch, broken switch wire, or stationary turntable. When the turntable stops rotating, the switching circuit will automatically keep reversing directions. Check the turntable for a broken drive belt or a missing idler wheel. The turntable must keep rotating in either direction. When the turntable stops, the automatic-sensing circuit energizes a relay, which reverses the direction of the cassette motor.

Some auto cassette players have a commutator ring with spring-like tongs that keep the automatic-reverse circuits operating while rotating. A bent prong, poor prong contact, or dirty commutator ring can result in rapid reverse procedures (Fig. 19-27). Just clean the commutator and tongs with a cleaning stick and rubbing alcohol to solve the erratic changing of directions of cassette player.

No Auto Reverse

The auto-reverse procedure can operate mechanically, like the regular cassette player, or electronically. When AMSS is found in the car cassette player, it can operate by detecting

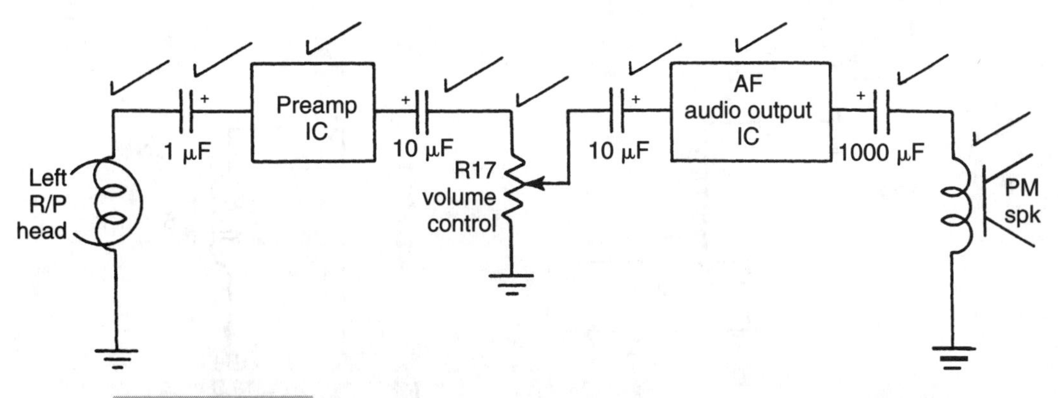

FIGURE 19-25 **Check these components for open or dead audio channel.**

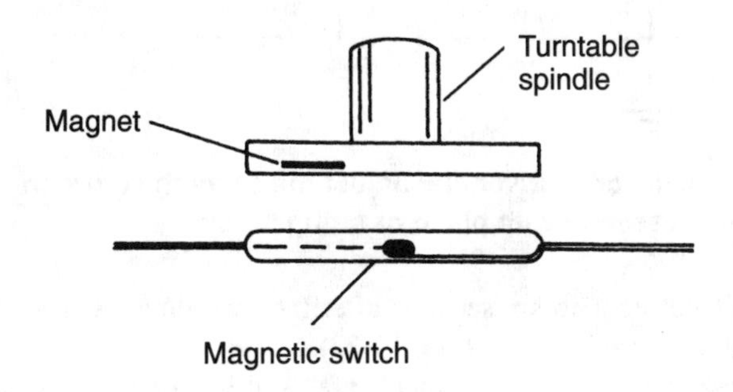

FIGURE 19-26 **The turntable might have a magnet and magnetic switch in the reversing circuit of the cassette player.**

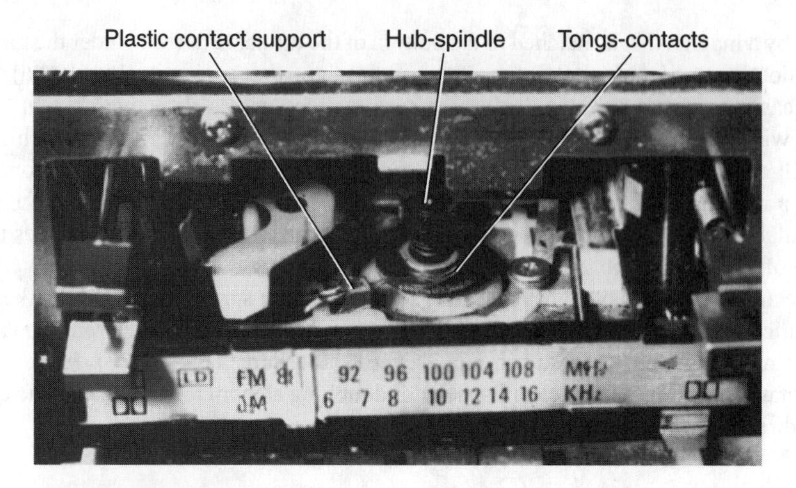

FIGURE 19-27 Clean dirty commutator rings if the cassette player changes directions.

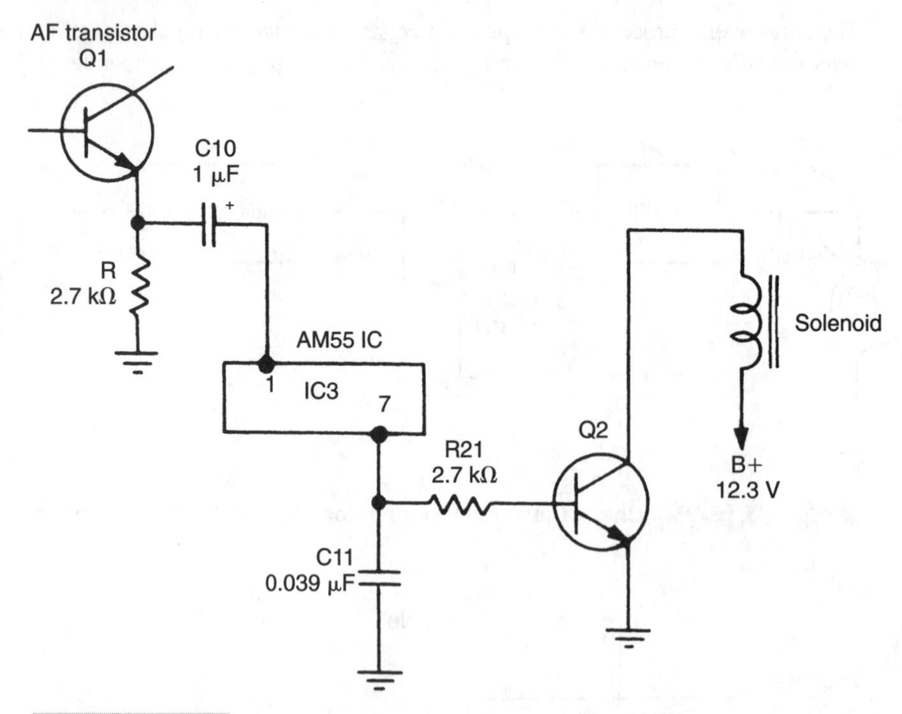

FIGURE 19-28 Take a small screwdriver and adjust the azimuth screw on the tape head while playing a cassette with piano or string music.

the no-signal segment of a tape. Four to six seconds after the recording has ended, the AMSS circuit reverses the direction of the tape (Fig. 19-28).

Often, dual tape heads, dual capstans, and flywheels are used in cassette players that have auto reverse. The solenoid plunger energizes and pulls in another set of tape heads and the capstan/flywheel assembly, while switching the direction of the dc motor. The AMSS LED indicator and reverse-normal LEDs can be switched in the circuit.

The AMSS circuits can fail to function if the tape heads are dirty or if the tape no-signal segment is shorter than four seconds. The circuits can fail if the tape recording is too low or if a high noise level is found on the tape. Clean the tape heads and adjust the azimuth head before attempting to adjust AMSS circuits. Closely follow the manufacturer's AMSS adjustment procedures.

If the AMSS electronic auto reverse still doesn't function, check all transistors and diodes in the AMSS circuits. Signal trace the audio signal to the input terminal of the IC. Check the solenoid winding for open or broken wires. Manually push the plunger in and out to see if it is stuck or frozen. Check if the solenoid winding cover is burned. Replace complete solenoid if the winding is charred. Accurate voltage and resistance measurements in the AMSS circuits will help you locate defective ICs, transistors, or diodes.

Auto Head Azimuth Adjustment

The head azimuth can be adjusted mechanically or electronically. Be sure that the tape head is clean. Either clean it with a cleaning stick and rubbing alcohol or with a cleaning cassette. Then, insert a cassette with music that contains piano or violin. Adjust the sound from the speakers for the greatest volume at the high frequencies.

The azimuth screw is located on one side of the tape head. When turning the screw, it will align the tape and the head horizontally. The screwdriver can be inserted through a hole in the cabinet (Fig. 19-29). Very slowly turn the screwdriver left or right, until it reproduces the highest frequencies.

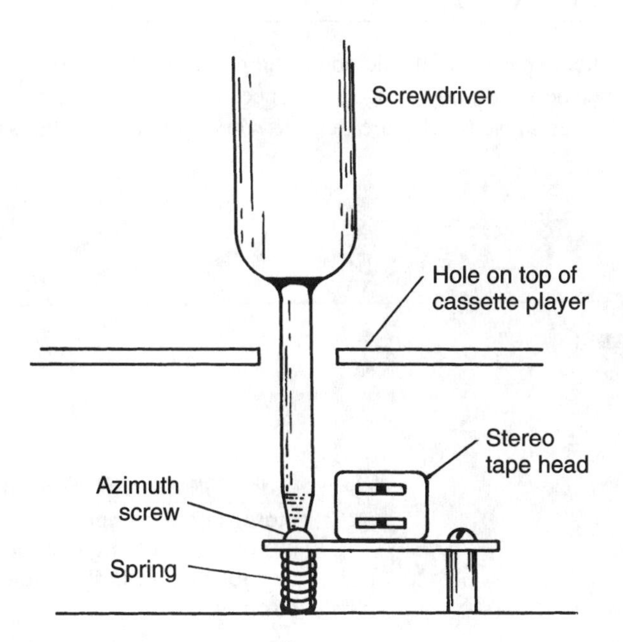

FIGURE 19-29 **A block diagram of AMS's audio-detection/auto-reverse circuit.**

Electronically adjust the azimuth screw for maximum meter reading with a 1- or 3-kHz test cassette playing. Connect the low-ac meter test probes to one of the channel speaker outputs. Leave the speaker connected and adjust the azimuth screw for maximum loudness and highest meter reading. Only a slight adjustment is needed, unless a new tape head has been installed. If the azimuth adjustment does not make much difference, suspect that either the head is worn, the audio circuit is defective, or that the tape head is magnetized.

Motor Problems

The 12-V cassette motor could operate intermittently, run slow or fast, and appear to be dead. A defective motor can cause excessive noise in the speakers. Monitor the voltage at the motor terminals. Often, the motor is blamed when the motor regulator circuit is actually defective. If the correct voltage is applied at the motor terminals and the motor does not rotate or if it runs slow, suspect that the motor is defective (Fig. 19-30).

A defective diode or isolation resistor in series with one motor terminal could be open. Check the voltage on both sides of the resistor and the diode. Do not overlook the possibility that off/on switch that applies voltage to the motor is defective or dirty (Fig. 19-31).

The intermittent motor could operate one hour and not the next. Sometimes, when tapped with the screwdriver handle, it will start to run or change speeds. The defective motor can run slow or speed up. Rotating the suspected motor pulley by hand will start the motor running. The dead motor might have an open winding or worn brushes.

Low Speaker Hum

Suspect an open or dry filter capacitor in the dc power supply if you can hear hum when the volume control is turned down. Simply bridge the suspected electrolytic capacitor with another of the same or higher capacity. Be sure that the working voltage is the same or higher (Fig. 19-32).

FIGURE 19-30 The defective motor might be open, intermittent, change speeds, or cause noise in the audio circuits.

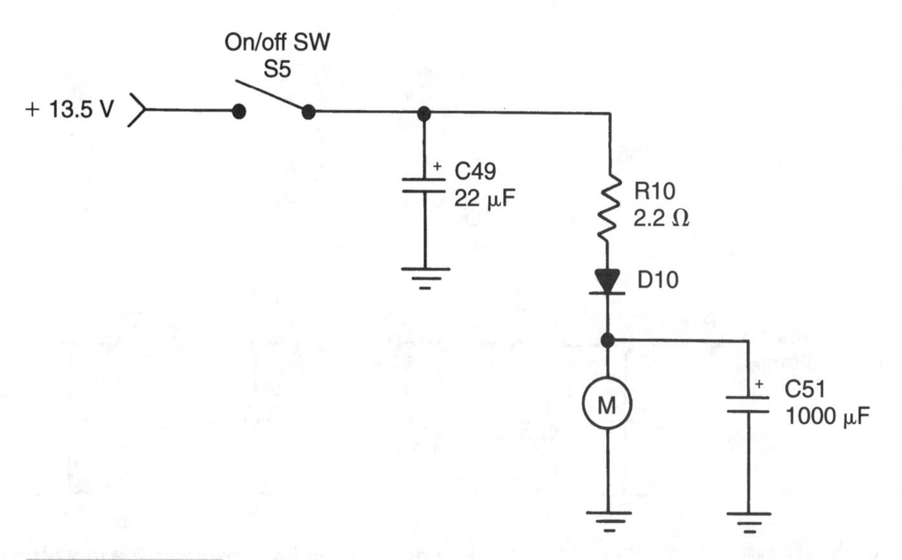

FIGURE 19-31 **Check on/off switch S5, R10, and D10 if improper voltage is at the motor terminals.**

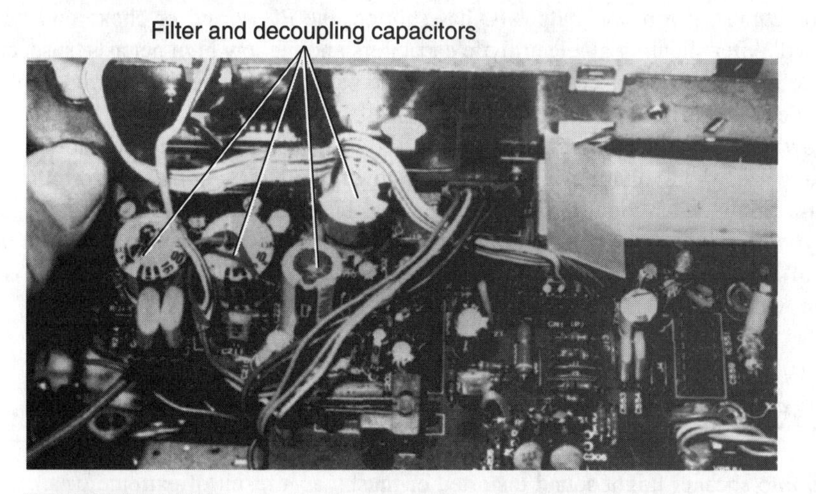

FIGURE 19-32 **Several filter and decoupling electrolytic capacitors are in the power supply of a Toyota 68600-00160 power board.**

Lightly tack the filter capacitor in with solder or clip into the circuit with test clips. Never shunt the capacitor while the player is operating or you can damage crucial solid-state components. Always check the correct polarity. The positive terminal goes to the positive battery ("A" lead) side.

Shunt all electrolytic capacitors in the power and decoupling circuits (Fig. 19-33). Sometimes real low hum that can be heard by certain people and not by others can be caused with a dry decoupling capacitor. Do not overlook the possibility that the hum symptom is caused by a leaky voltage regulator transistor.

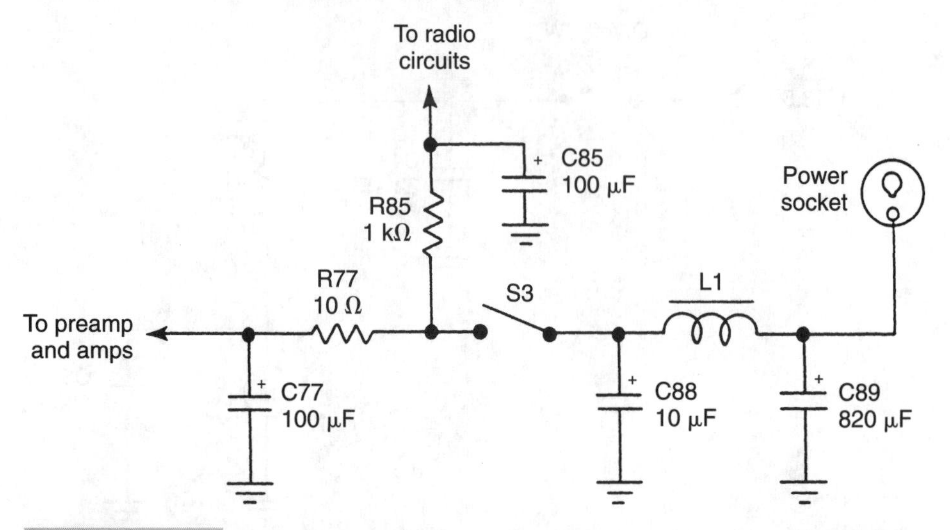

FIGURE 19-33 **Shunt the decoupling electrolytic capacitors if a low hum is in the speakers.**

Sometimes, if a power-output IC has shorted, the PC wiring or choke coil wiring is burned. After shunting all electrolytic capacitors and the low hum persists, suspect that a choke coil is burned or charred. If the outside paper is scorched or charred, replace the small choke transformer.

Low hum in the audio circuits can be caused by a defective IC or transistor. If the hum is canceled when volume control is turned down, the hum originates before the volume control. Shunt each input terminal of the IC or base of the transistor with a 10-μF electrolytic capacitor to the chassis ground and notice if the hum disappears. Start at the output of the tape head and proceed through preamp and AF stages until the hum is isolated. Hum with weak volume can be caused by a dry electrolytic coupling capacitor.

Speaker Problems

The auto speaker might sound distorted or mushy as a result of extreme weather conditions. The cone begins to warp and the voice coil rides against the center magnet. Excess dirt and dust can fall into the speaker cone and cause distortion or noisy reproduction. Speakers that are mounted upward in the front dash or rear ledge have a tendency to warp and mush up (Fig. 19-34).

An intermittent speaker might be caused by too much applied power, which damages the voice coil. Sometimes the small flexible wire that is soldered to the voice coil will break and cause an intermittent speaker. Simply remove the speaker from the mounting, connect the speaker wires and lightly press down on the cone of the speaker. Move the cone up and down as the music cuts in and out. Replace the speaker if it is defective. Check the speaker terminals and wires for intermittent audio.

Because at least four speakers are in the new auto arrangements, you might have four times more speaker damage (Fig. 19-35). The voice coil could be blown or torn loose from the cone of speaker as a result of excessive volume. Most speakers damaged by excessive volume are large woofers. Check the suspected dead speaker with R×1 ohmmeter range (Fig. 19-36). Replace the defective speaker with universal types that have the same wattage, size, and mounting holes.

Auto Cassette/CD Players

The auto cassette/CD player might consist of the auto reverse with CD changer controls. The cassette/tuner might have a CD changer in the trunk. The CD changer might operate directly from a cassette/receiver. The under-dash CD system can be connected to the auto cassette player high-power amplifier.

The Pioneer KEH-M3000 is one of the lower-priced cassette receivers with CD changer controls. You can place the CD changer in the trunk and select any disc or track and program up to 32 tracks for playback. The auto receiver has a quartz Supertuner III with 18 FM and 6 AM presets. The cassette player includes auto-reverse, key-off pinch-roller release with separate bass, treble, and speaker fader controls. The tape frequency response is a 40 to 14,000 Hz with 10 W output per channel.

The Sony XR-7500 cassette/tuner system with dual function controls operates Sony's CDX-A15/CDX-A100 CD changers. The CD changers can be mounted and operated in the auto trunk. This unit can be operated with a wireless remote. The cassette player contains

FIGURE 19-34 Press lightly on the speaker cone to locate an intermittent speaker.

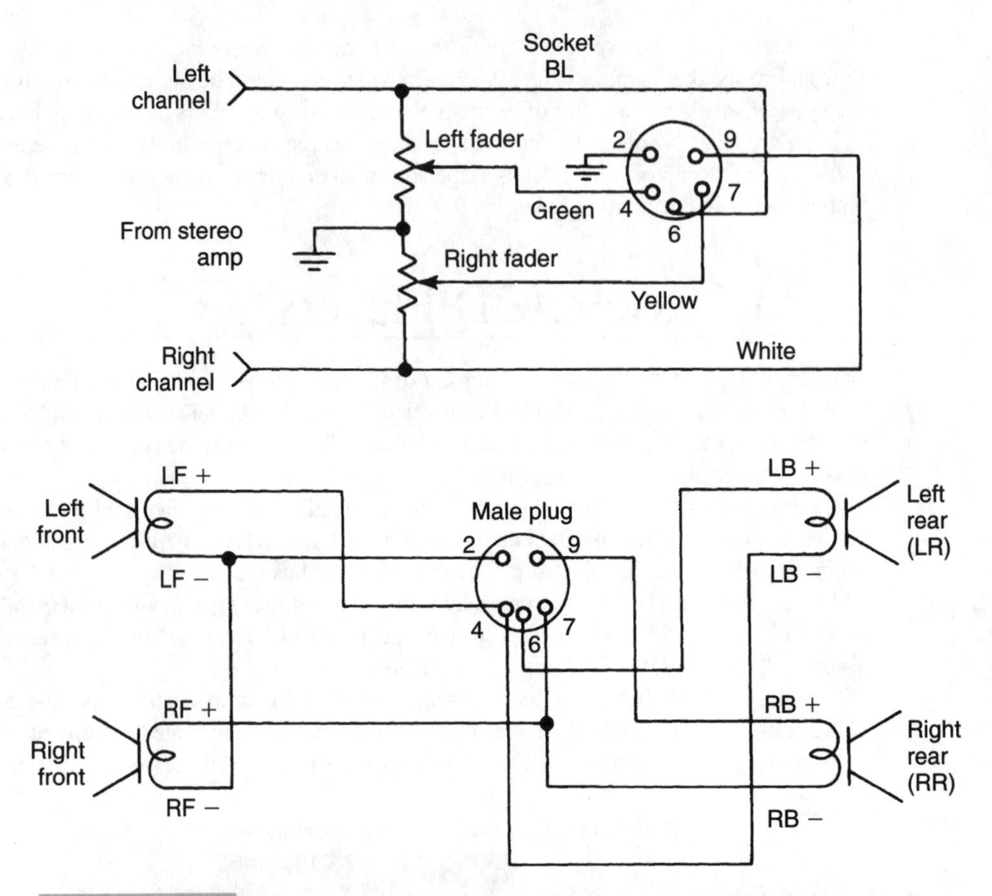

FIGURE 19-35 The fader, speaker harness, and speaker hookup to the front and rear speakers in typical stereo-cassette output circuits.

an amorphous tape head, 30- to 20,000-Hz audio frequency, and Dolby B and C to eliminate tape hiss. The preamp cassette/tuner is removable, but it must operate through a separate amplifier.

The Kenwood KRC-930 cassette/receiver contains the finest cassette receiver with the complete operating controls of the 10-disc KDC-C300/KDC-C400 changer. The cassette section has a flat response up to 21,000 Hz, Dolby B and C, full-logic auto-reverse, and programmable tape search. Other features are index, scan, blank skip, and key-off pinch roller release with separate bass and treble controls. Two sets of RCA preamp outputs are provided to an external amplifier. The cassette/tuner can be removed easily and operated with wireless remote.

The Radio Shack 12-1941 under-dash CD player can be connected to a cassette/tuner unit with relay control. The frequency response is from 40 to 20,000 Hz with separate line outputs to an externally connected high-power amplifier or self-contained 36 W per channel with a 4-Ω load. This high-power CD player can be operated independently, plugged into your existing car stereo, or speakers can be added. The player also has automatic play, auto search, two-way audible search, Pause, and LCD display.

Surface-Mounted Components

The latest cassette/CD players have both conventional and surface-mounted components. The surface-mounted components can be mounted on one side of the board with regular parts on the other. Several different boards are found within the cassette/tuner (Fig. 19-37). Very fine PC wiring connects the surface-mounted processor to the circuit.

FIGURE 19-36 Check the voice-coil winding of a dead speaker with the R×1 range of an ohmmeter.

FIGURE 19-37 Surface-mounted parts can be found on one side of PC board in the auto cassette or CD player.

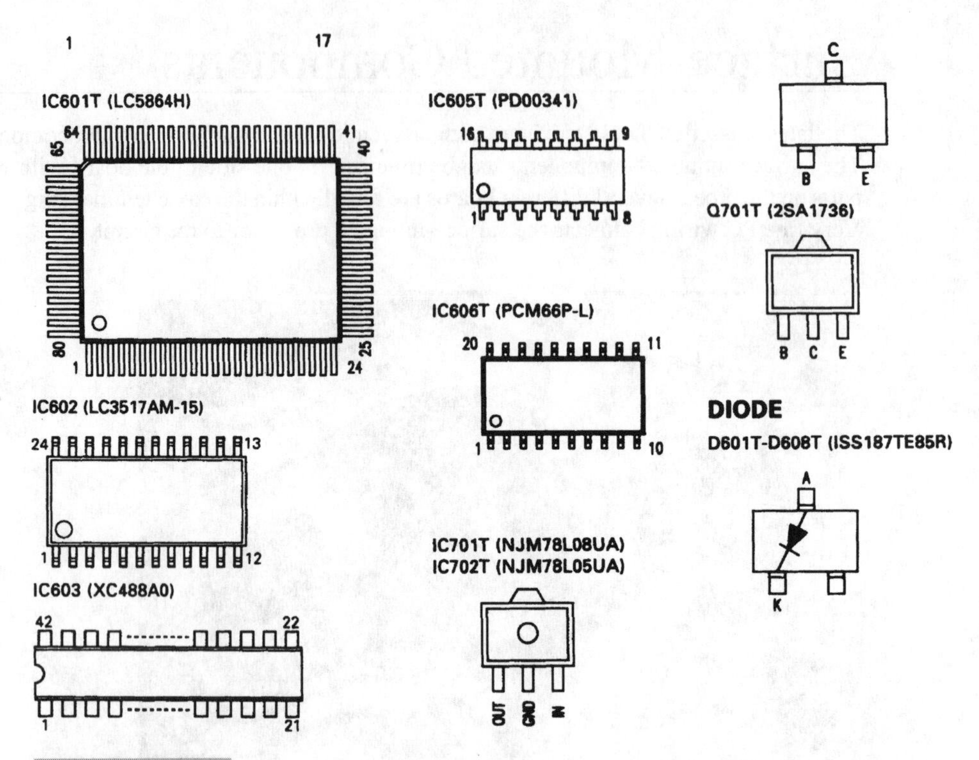

FIGURE 19-38 The surface-mounted terminals are a little different because they are mounted on a flat surface.

Surface-mounted transistors and ICs are used throughout the circuits. Besides surface-mounted ICs and transistors, diodes, capacitors, and resistors are used in many circuits (Fig. 19-38). Notice that diodes can have two active tabs, and the three-legged transistors can have different lead identifications. Power-output ICs should be bolted to heavy heatsinks.

Digital transistors are used in the audio preamp, and the mechanism and system control. The digital transistor can have an internal resistor in the base circuit or another bias resistor between the base and emitter terminal (Fig. 19-39). If these digital resistors are checked with a transistor tester or the diode test of the DMM, the resistance is higher in the base terminal. Likewise, the measurement from base to emitter with the internal base-to-emitter resistor is different. Compare another similar digital transistor with the low base-to-resistor test before discarding the transistor that you suspect is leaky.

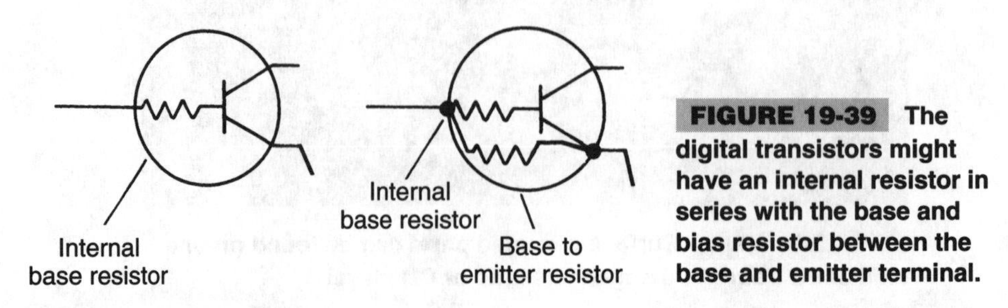

FIGURE 19-39 The digital transistors might have an internal resistor in series with the base and bias resistor between the base and emitter terminal.

Block Diagram

Before tearing into the CD/cassette player, always check for possible trouble with the block diagram. The block diagram helps to isolate the various stages for easier servicing. The CD mechanism consists of the disc, feed, and load motors with transistor or IC motor drives. Also, the focus and track coils have separate transistor or IC drivers. The servo controller and signal processor with power-control circuits are located in the CD mechanism (Fig. 19-40).

The dc-dc converter supplies +9 V and –9 V with +8-V and +5-V regulators, ICs, or transistors from the 14-V battery source. The +5-V source feeds the motor power on, MOS microprocessor (IC 601), and controller (IC901 and IC902). The +8- and -8-V source is

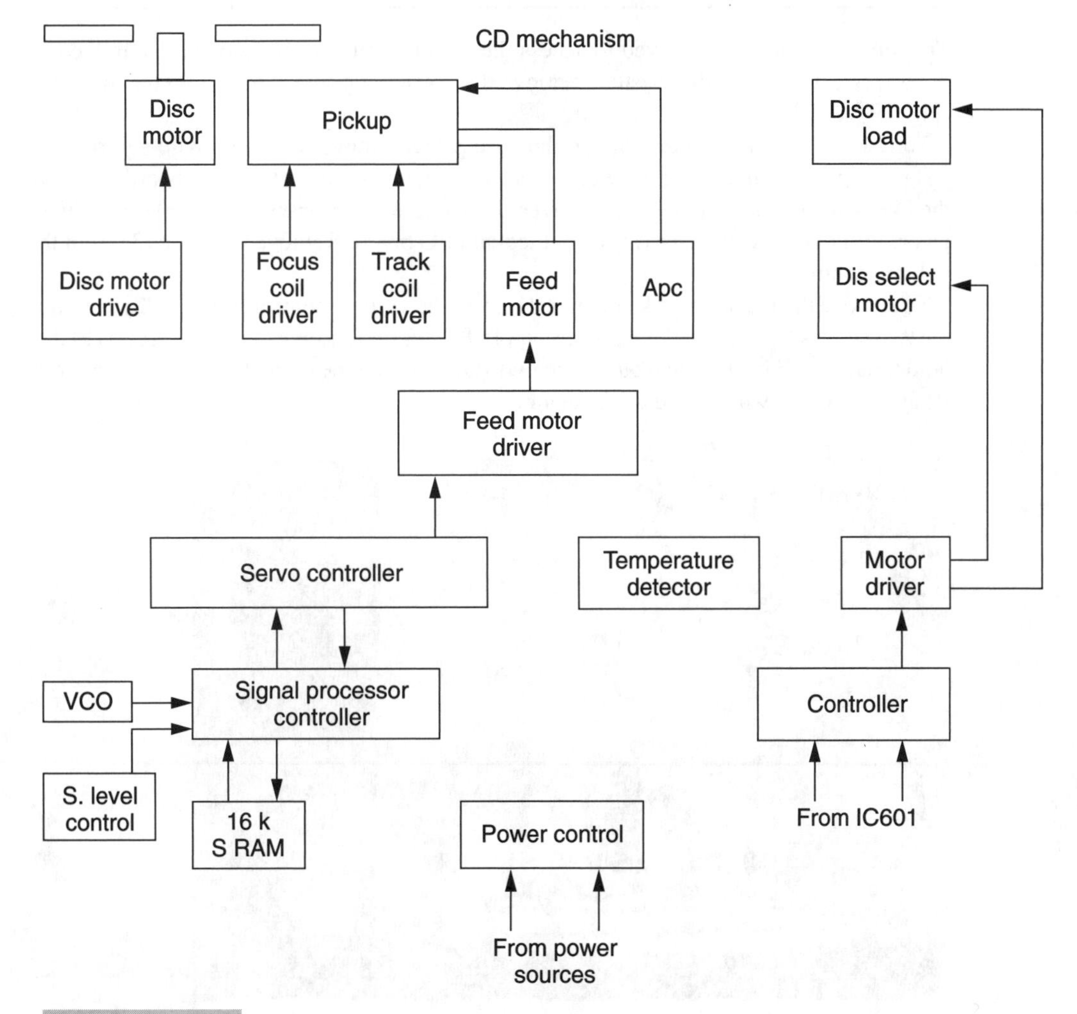

FIGURE 19-40 The block diagram might help to locate the defective part in a certain section of the CD player.

supplied by IC703 and IC704. Both +9 and -9-V sources feed the power-, signal-, and system-control circuits. Primarily, the 14-V source is directly applied to the line-out and power-output amplifier ICs.

The D/A converter (IC606) converts the digital signal to the audio and separates the audio into two separate channels. Separate buffer and low-power filter (LPF) ICs are found in each stereo channel. The stereo signal is separated after the mute and volume controls and dual-isolator and line-amp output jacks. The same stereo audio signal from the volume controls is fed to one dual power amp (IC504). The power amp provides 36 W to each speaker system.

Removing Covers

The top cover must be removed to access the various components and boards. In the Radio Shack car compact disc system, remove the six tapping screws that hold the top cover (Fig. 19-41). Then, slip the cover off.

To remove the CD mechanism and the nose piece, remove the clamp wire (a) from the CD mechanism. Pull out the three connectors (A, B, and C) from the main board. Remove the four screws (C) from the bottom cover. Take out the four screws (D) that hold the nose piece. Pull out the connector (D) from the main PC board. Remove the knob (3) from the volume control.

To remove the PC board, take out nut (H) from the volume control (Fig. 19-42). Remove the two screws (E) that hold the power-supply PC board. Remove the three screws (F) that hold the front PC board. Take out the screws (G) that hold the cord clasper. Remove screw (I) and screw (J), which hold the heatsink.

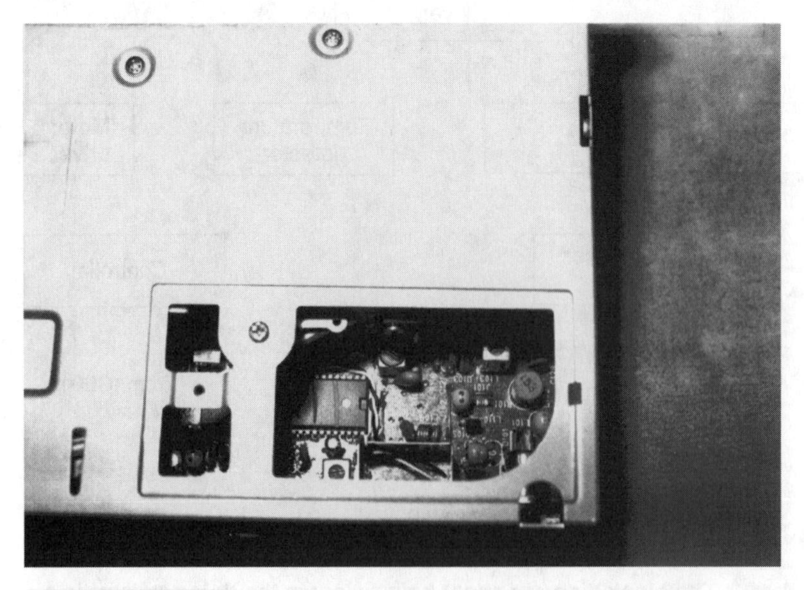

FIGURE 19-41 **The covers must be removed to get at the defective section.**

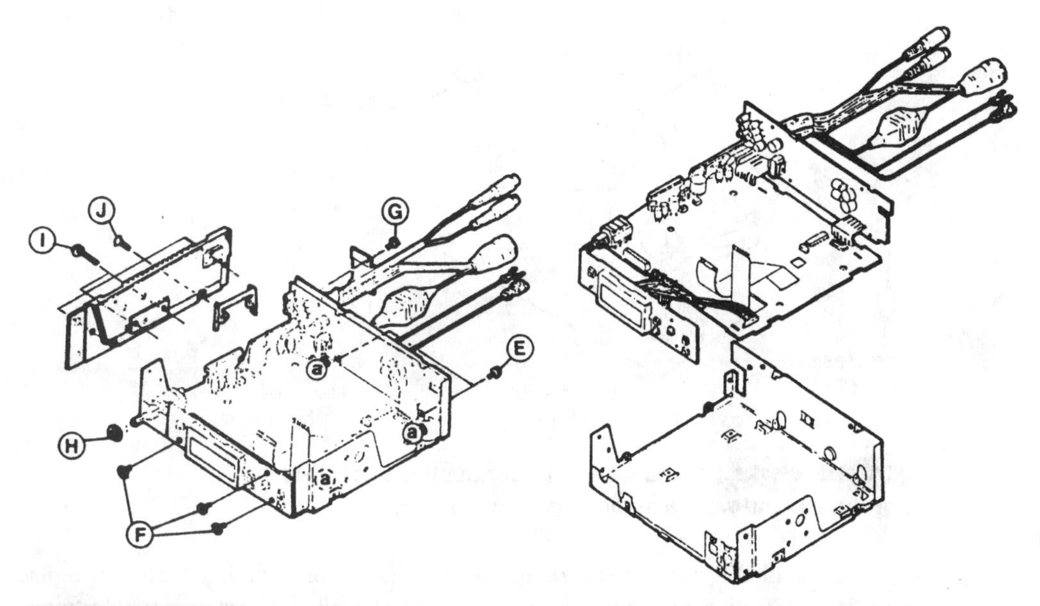

FIGURE 19-42 **Remove several components to remove the bottom cover in the underdash CD player.**

Be careful to prevent damage when removing the PC board and the separate components. Place all screws in a saucer or container. Make a list of parts as you remove them. Lay the components in line so that they can be easily replaced after you make the repairs. Carefully unsolder the mesh ground straps from the bottom chassis.

Safety Precautions

When servicing compact disc players or handling static-prone ICs, give extra care for ICs, processors, and laser-diode pickups, which are sensitive to, and easily affected by, static electricity. If static electricity is nearby, components can be damaged if you do not exercise the proper precautions.

The laser-diode pickup is composed of many optical parts and high-precision components. Extra care must be taken to avoid repair or storage where the temperature or humidity is high, where strong magnetism is present, or where dust is excessive.

Before attempting to repair or replace any component in the CD player, all equipment, instruments, and tools must be grounded. The CD player should be placed on a work bench that has a grounded conductive sheet (Fig. 19-43). The metal part of the soldering iron should be grounded. Any repair worker or technician should wear a grounded arm band.

When working on a CD player, never look directly at the laser beam, and don't let it contact fingers or any other exposed skin. You cannot see this beam with the naked eye. Try to keep a disc on the turntable or keep the laser beam turned away from your eyes.

The laser pickup has strong magnets and should never be brought close to magnetic materials. Keep metal screwdrivers and tools away from the pickup assembly. The pickup

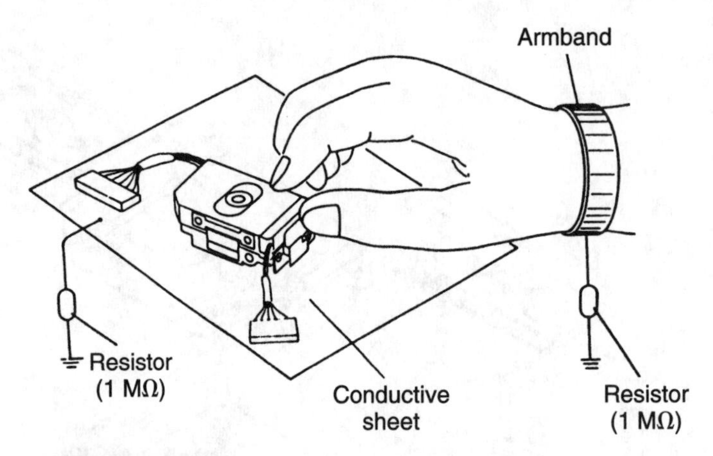

FIGURE 19-43 **Try to work on a conductive sheet that is grounded and wear a static wrist ground strap.**

should be handled correctly and carefully when removing or installing. Keep new replacements in a conductive bag until they are ready to be installed. Laser assembly testing and replacement should be performed by an electronic technician.

Laser Head Cleaning

Erratic operation and complete CD player shutdown can be caused by a dirty laser lens assembly. Wipe the dust off with soft cloth. Dust can be blown off with an air brush, such as those that are used to clean camera lenses. In fact, the camera soft cloth and brush are ideal for cleaning the laser lens assembly. Be very careful not to apply too much pressure because the lens is held by a delicate spring.

Tape Player and Tuner Operates/No CD

Check for an open fuse. Inspect the cable harness for poor wiring. Check the "A" battery lead connection. Measure for 14 V at the switched power lead. If it has no voltage, suspect that the power switch is defective (Fig. 19-44). If the CD player is connected to the cassette tuners, does the output relay operate? Listen for the plunger of the output relay. Check for poor choke (CH501) terminals or burned PC wiring.

NO SOUND/FAULTY OUTPUT CIRCUITS

Determine if both channel speakers are dead. If not, clip another speaker to the dead channel. If both channels are dead, suspect the dual IC (IC504), the power-supply circuit, or the power switch. Check the dead speaker for an open voice coil or faulty speaker connections. Measure the supply voltage source at the power-output IC.

Often, when the power-output IC becomes shorted, the fuse will blow and the PC wiring might burn. If the CD player has separate stereo line-output jacks that have normal sound, the output IC circuits are dead from the volume control to input terminals 7 (left) and 2 (right) of IC504 (Fig. 19-45). Both line-output preamp transistors and the input preamp power IC output circuits separate after the volume control.

Check the supply voltage at pins 9, 10, and 17 of IC504 if both output channels are dead. Touch one side of C529 and C530 with a screwdriver blade and listen for hum in the

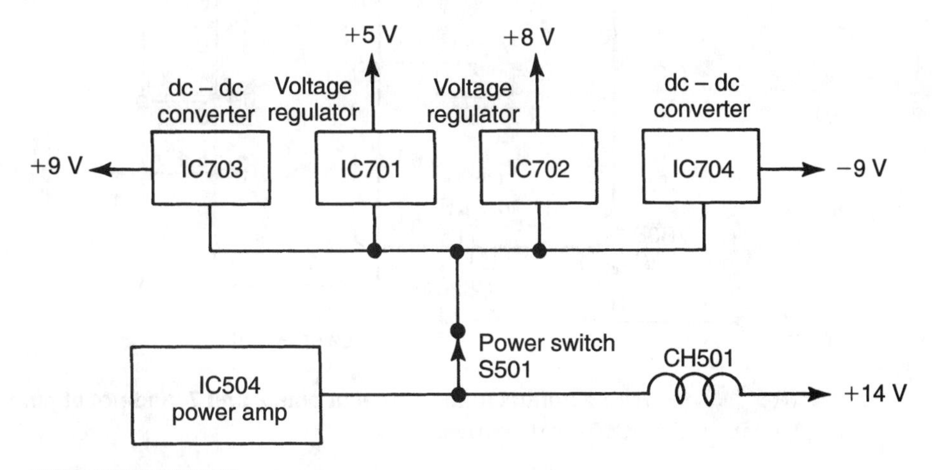

FIGURE 19-44 **Check for 14 Vdc across power switch S501.**

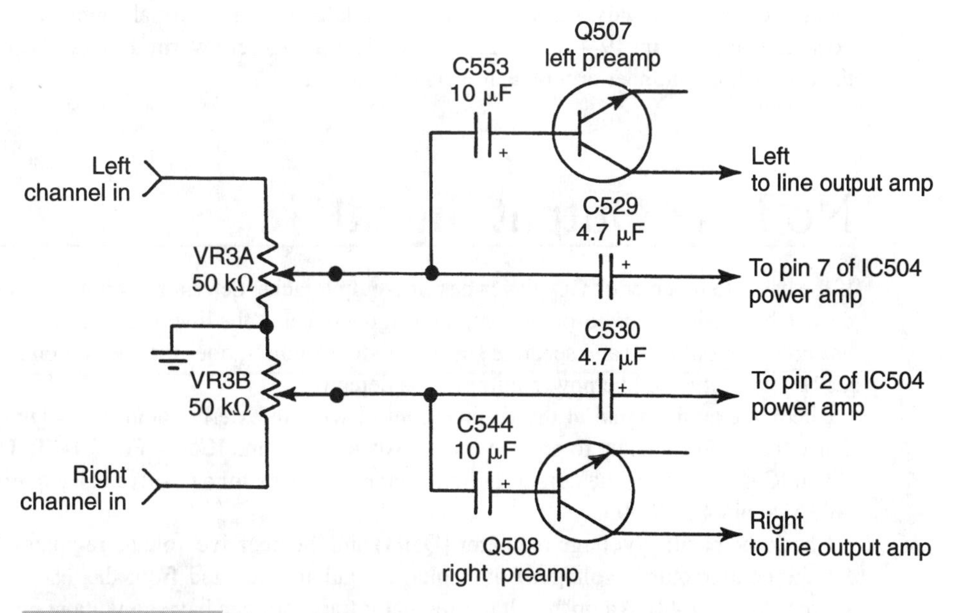

FIGURE 19-45 **In CD players with stereo line output and high-powered amplifiers, the audio stages are common to the volume controls.**

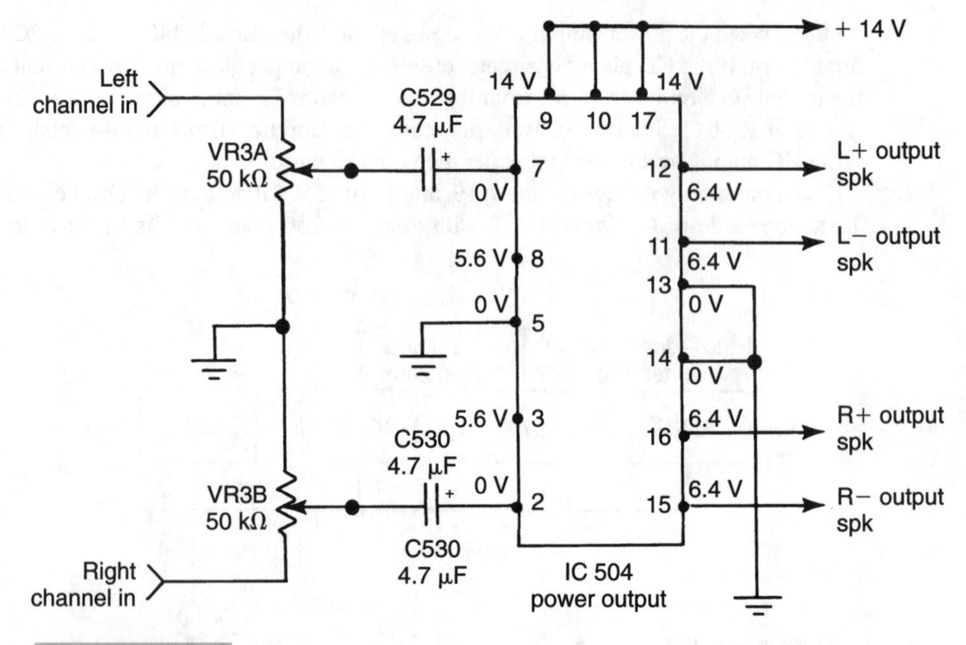

FIGURE 19-46 **With a normal audio signal at pins 2 and 7, and not at pins 11 and 12, suspect that IC504 is defective.**

speakers. Signal trace the input signal up to pin 2 and 7 of IC504. If the audio signal is normal up to these terminals, suspect that IC504 is defective with normal supply voltage and no sound output (Fig. 19-46). Notice if IC504 is running very warm. Replace IC504 with the original part number or with a universal replacement.

No Line-Output Signal

If the cassette/receiver or CD player has line-output jacks, determine if both channels are dead. Check with an audio-output amp or a signal tracer at the line output jacks. If the unit has both line-out jacks and separate speaker audio channels, does the speaker circuit operate? Often, only one line power-output IC is defective.

Check the audio signal at the volume control with an external amp or a signal tracer. Trace the audio signal up to preamp line transistor Q507 and IC505 (Fig. 19-47). Usually, a dual IC is the power-line output for both channels. Check for a positive 6.5 V at pin 8 and –6.5 V at pin 4 of IC505.

Check the positive voltage regulator (Q503) and the negative voltage regulator (Q504) for the emitter output voltage. Both voltage regulators are fed from the +8- and –8-V sources (Fig. 19-48). An open voltage-regulator transistor can have no voltage output, but the leaky regulator might have an increase in voltage.

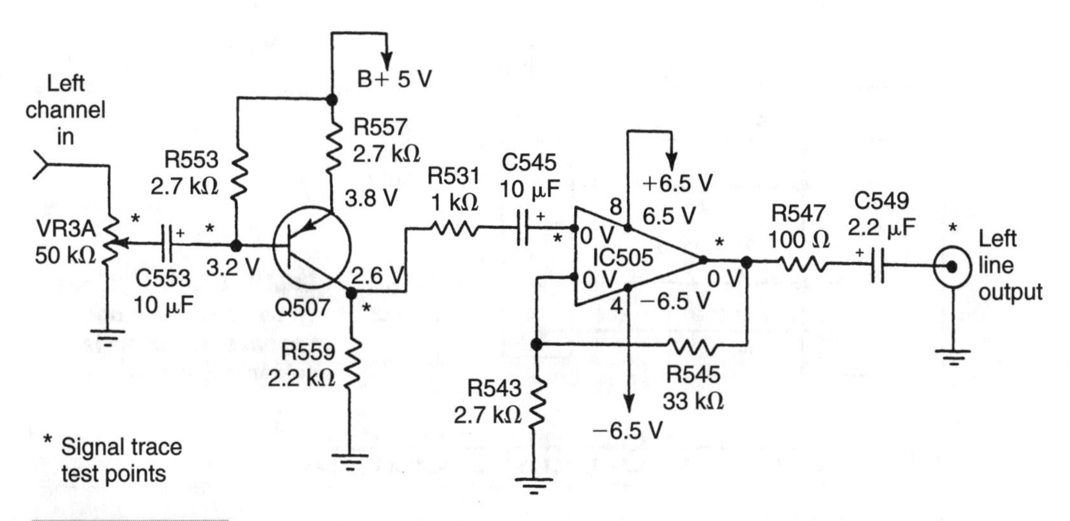

FIGURE 19-47 Signal trace the audio from the volume control (VR3A) to the line-output pins with the audio external amp or signal tracer.

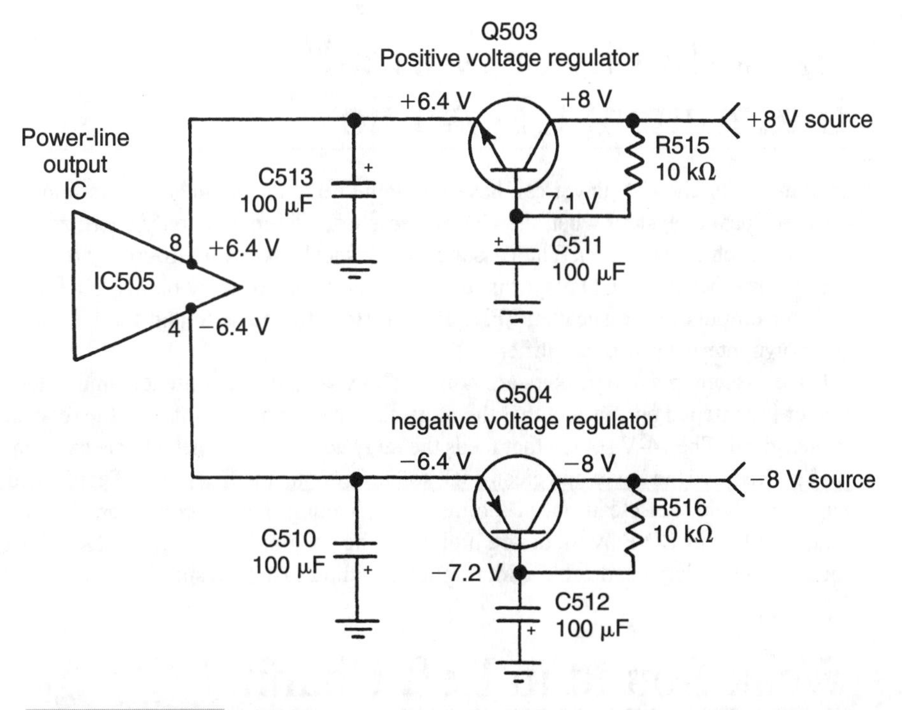

FIGURE 19-48 The positive and negative voltage (6.4 V) from the voltage regulator is applied to pins 4 and 8 of line-output IC505.

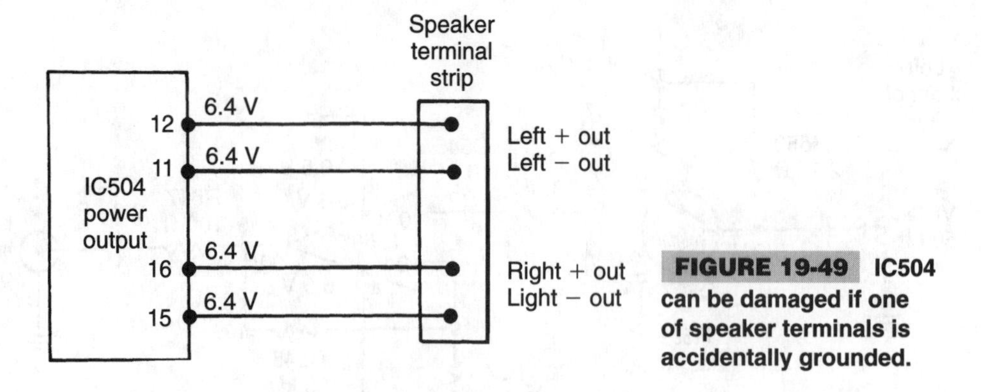

FIGURE 19-49 IC504 can be damaged if one of speaker terminals is accidentally grounded.

Ungrounded Speaker Outputs

When installing or repairing the CD-player speaker connections, notice if any of the speakers are grounded. With regular auto radios and cassette players, one side of the speakers are grounded. Here, both channels are above ground and work directly out of the dual audio-output IC (Fig. 19-49). If you ground one side of the speakers, you can damage the speaker and IC504. Remember, a dc voltage is at these speaker terminals.

Cassette Player Normal/ Intermittent CD Audio

In some units, the CD player can have an output relay that switches the radio/tape player into the speaker system when the CD player is off. The relay is only energized when the power switch is turned on. The cassette-player speaker-output terminals are switched to the speakers when the CD player is turned off. Four sets of relay points route the cassette speaker outputs to the speakers (Fig. 19-50). Both the cassette and the CD player have their own internal power amplifiers.

If the cassette player speakers are normal and you find intermittent sound when the CD player is switched on, suspect that the relay contacts are dirty. Notice if the relay solenoid is energized. The 14-V source that feeds the relay comes from a detachable cable in the CD player. Remove the cover and clean each solenoid contact with a piece of cardboard. Sandpaper can help, but clean out all points after cleaning them. Remember, the points that make contact in CD Play mode are high until the solenoid is energized. Do not overlook the possibility that a bad cable wire might be leading to a dead speaker.

Weak Sound in Left Channel

Signal trace the left channel at the volume control and compare it with right channel. If both channels are the same, proceed toward the speaker output terminals. Check the audio in both channels at the line-output terminals. If both line signals are normal, suspect that a power-output circuit is defective. Actually, the audio signal can be traced from the output D/A converter toward the speakers (Fig. 19-51).

Signal trace the audio from pin 8 of the D/A (IC605) toward the volume control with an external amp or an audio signal tracer. Check the input and output signal of each IC. Signal trace each side of an electrolytic coupling capacitor. Divide the audio circuit in half at the volume control. If the left channel is weak at the volume control, start at pin 8 of IC605 and proceed toward the volume control. If the signal is normal with both channels at the volume control, proceed toward the speaker output circuits.

Check the circuits where the signal becomes weak. Weak audio can occur as a result of bad coupling capacitors, ICs, or transistors. If the signal stops at one of the IC preamps, check the input and output terminals. Likewise, if the signal stops at a transistor circuit, test the transistor. Do not overlook muting transistors and ICs. A leaky muting transistor can partially close the audio signal to ground.

Dead Dc-Dc Converter

Usually, dc-dc converter circuits convert a high dc voltage down to a low voltage. Here, a positive and negative IC dc-dc converter provides a +9 V and –9 V to the various circuits. If one or both of these circuits become defective, the supply-voltage source cripples the

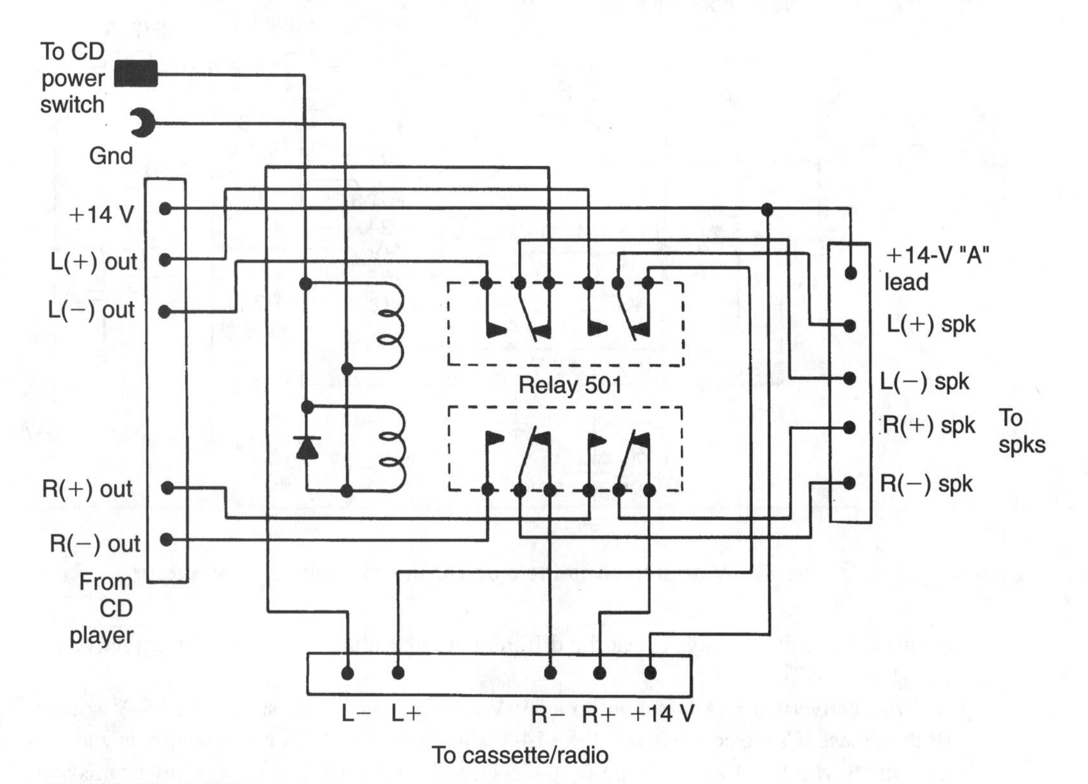

FIGURE 19-50 The CD player might have a relay that switches the speaker connecting from cassette/receiver and CD player to the connecting speakers.

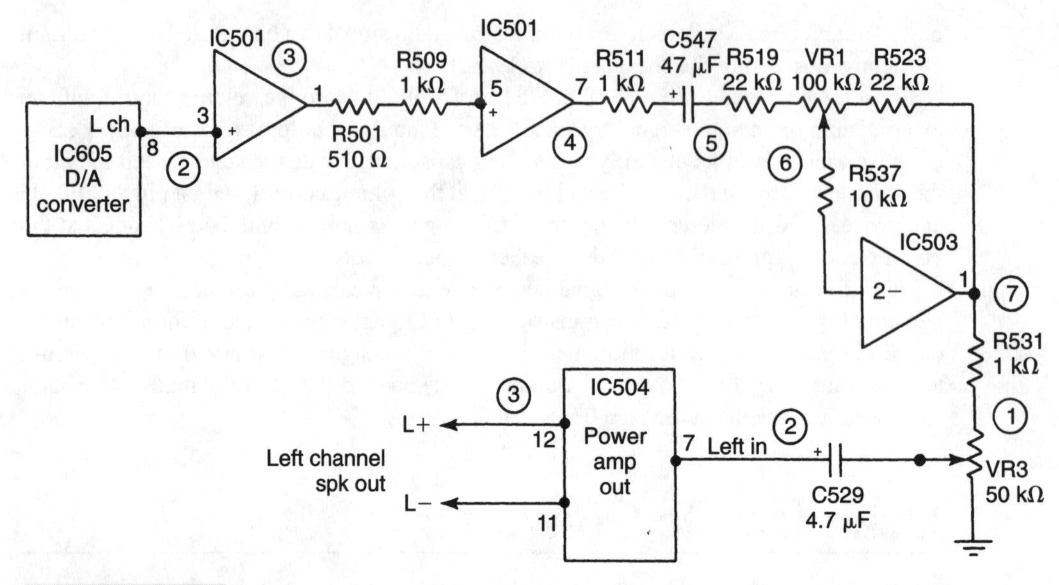

FIGURE 19-51 Signal trace by the numbers from the volume control in both directions.

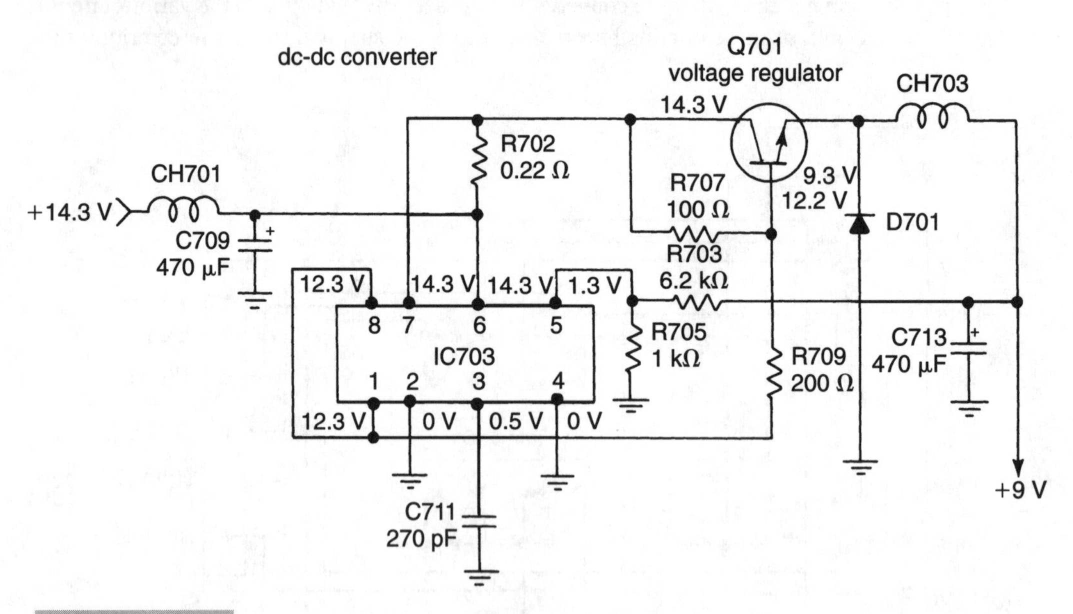

FIGURE 19-52 The 14.3-V source contains a dc CD inverter with a +9-V output.

connected circuits. Always check the different voltage sources to locate the defective circuit (Fig. 19-52).

IC703 converts a +14-V source to a +9-V source, and IC704 supplies a –9-V source. Both of these ICs are connected to the +14-V source. The +14-V battery supply is wired to terminal 6 with Q701 as a voltage-output regulator. If the +9-V source is low or missing, suspect IC703 or Q701. If Q701 opens, no voltage is at the +9-V source. If it is leaky with a short between the emitter and collector, the output voltage can be higher. However, an internal leak between the emitter and base will lower the output voltage to only a few volts. Remove Q701 if you suspect that it is leaking. Take crucial voltage measurements on IC703 if the +9-V source is low or improper.

The negative 9-V source contains only IC704 as the dc-dc converter. The positive 14 V from the battery feeds to pin 6 of IC704 and the negative 9 V is supplied from pins 2 and 4 of ground potential (Fig. 19-53). The leaky IC704 can lower the output voltage; an open IC will have very little negative voltage. Notice that D702 and C714 connections are reversed to common ground. Both IC704 and IC703 are identical ICs, but convert opposite voltages with the connection to ground.

No +5-V Source

Check the +5-V source if you find distorted sound or no sound in the CD output circuits. The 5-V source feeds D/A converter IC607 and also the system-control circuits. IC702 provides the +5-V source (Fig. 19-54). Pin 3 of IC702 receives +14 V from the battery source and the +5-V output at pin 1. Pin 2 is grounded. A leaky or open IC702 can cause no sound or low distortion. Take voltage measurements at all three pins. Replace IC702 if its output voltage is low or improper.

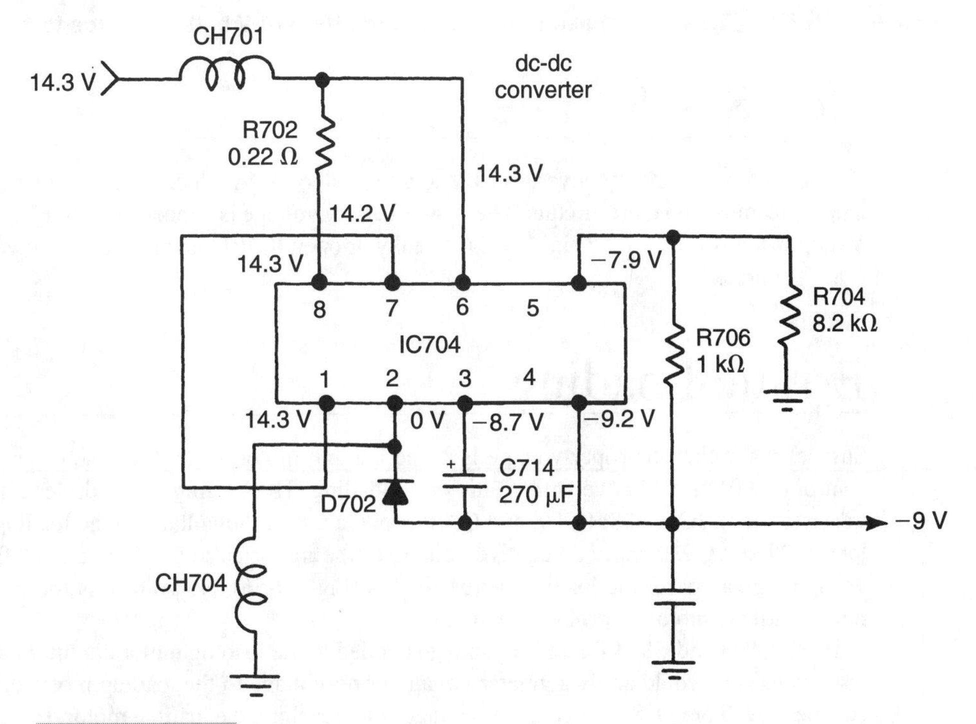

FIGURE 19-53 **The negative 9-V source is converted with dc/dc IC704 with –9-V source at pin 4.**

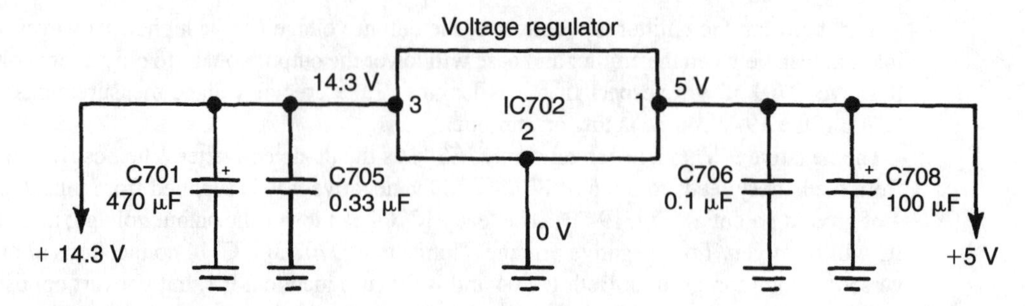

FIGURE 19-54 **IC702 provides a +5-V regulated source from the 14-V battery.**

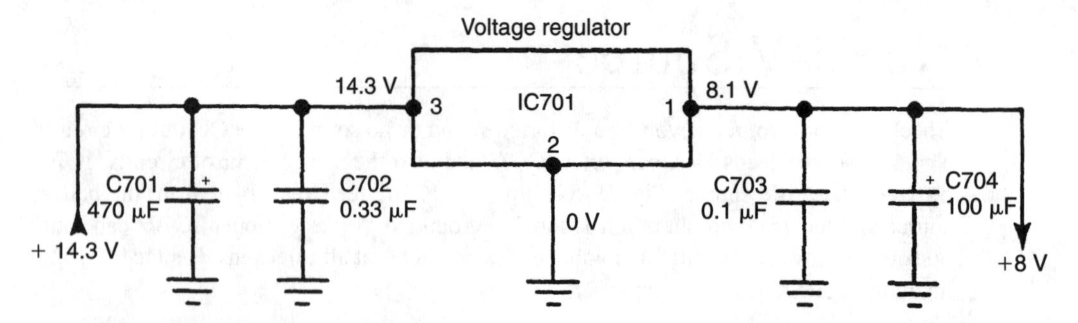

FIGURE 19-55 **The +8-V regulator (IC701) converts the +14-V battery source to 8 V.**

No +8-V Source

Voltage regulator IC701 converts +14 V to a regulated +8 V, which feeds the line-output amps, preamps, and mute circuits. The +14-V battery voltage is connected to pin 3 and +8-V output at pin 1 of IC701 (Fig. 19-55). A leaky or open IC701 can produce a low-voltage (+8 V) source.

Erratic Loading

Suspect that either improper voltage is at the loading motor, or that the motor or motor control (IC901) is defective with erratic or no loading. The loading motor driver can be a transistor or an IC. Q908, Q909, and Q910 provide a positive voltage to the loading motor. Q909 and Q910 apply a negative voltage to the motor circuits. Q903 and Q904 are voltage regulators for the loading motor circuits (Fig. 19-56). No voltage is found at the motor until the motor control is turned on.

Check Q903 and Q904 for correct voltage applied to the loading motor circuit. Leaky or open transistors could apply a greater voltage or no voltage to the loading motor drivers. An open Q909 or Q908 could prevent voltage from reaching the loading motor. If the voltage is present at the loading motor terminals and it isn't rotating, suspect that the motor is open. Check motor continuity with R×1 range of the DMM. Inspect the drive belts or gears for erratic speeds. Check if the loading mechanism is defective.

Disc-Select Motor Problems

The disc-select motor positive and negative regulated-voltage source feeds from Q901 (+) and Q902 (–). Q905 provides positive disc-select motor-drive voltage and A907 and Q906 apply the negative voltage to select the motor at the pin 11 terminal of C5903 (Fig. 19-57). If the motor will not go forward, suspect a problem at Q905 and Q901. For reverse motor operation, check Q906, Q907, and Q902. The negative and positive voltage that is applied to the disc-select motor is controlled by motor control IC901.

No Spindle or Disc Rotation

The spindle motor can be controlled by transistors or ICs. Here, the disc-motor voltage is fed from Q632 and Q631. Surface-mounted disc control IC603 provides voltage to driver

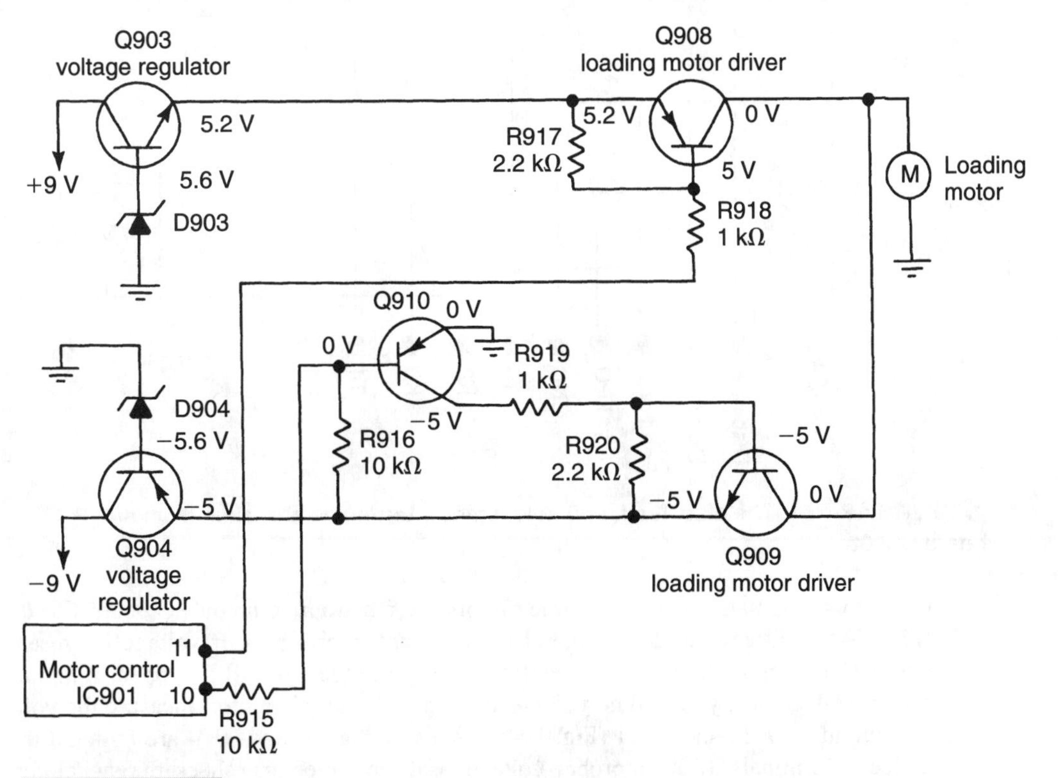

FIGURE 19-56 IC901 controls the loading motor from a positive and negative 9-V source with loading-motor transistor drivers Q908, Q909, and Q910.

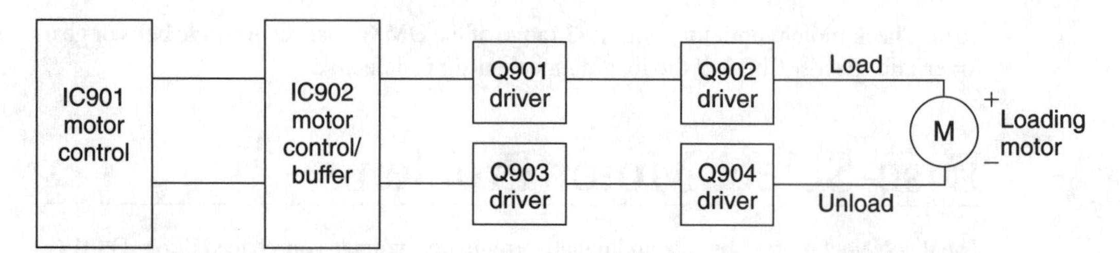

FIGURE 19-57 IC902 receives signal from IC901 to drive loading motor transistors and the loading motor.

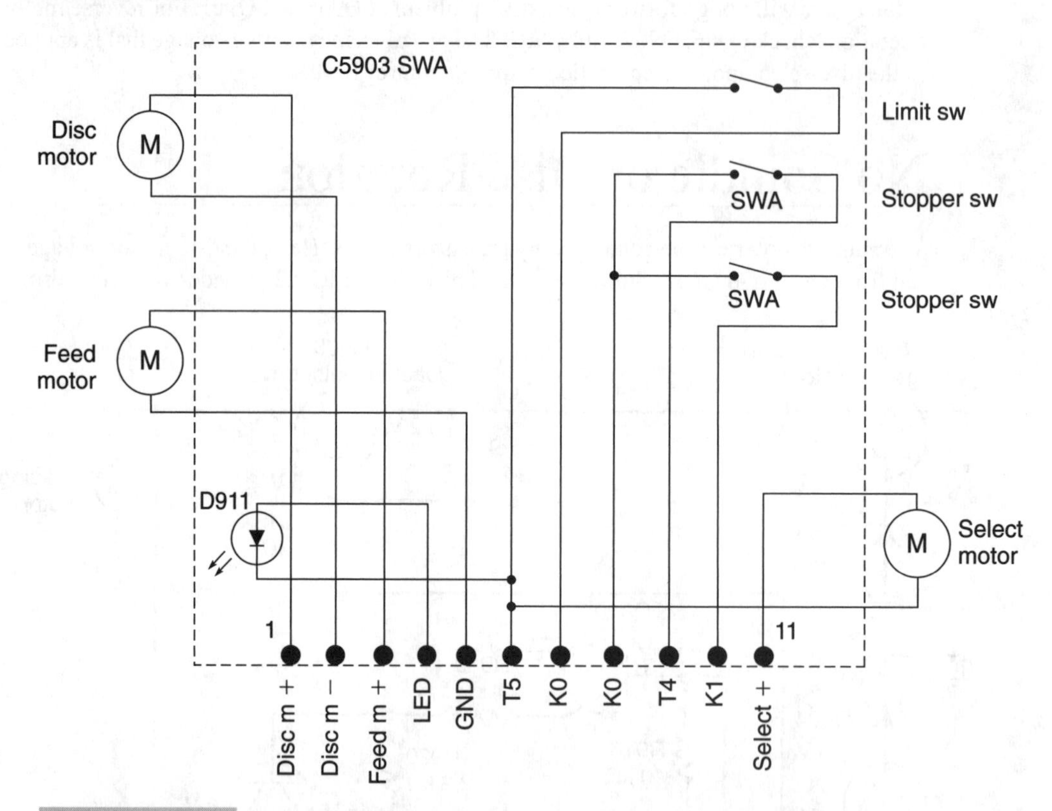

FIGURE 19-58 The disc, feed, and select motor terminals are taken from sub PC board CS903.

transistors Q632 and Q631. All three motors are fed from common PC board CS903 (Fig. 19-58). Check the dc voltage at the disc-motor terminals. If voltage is present without rotation, check the motor continuity. Replace the motor if it is open. If an improper voltage or if you find no voltage at the disc or spindle motor, measure the voltage output at Q632 and Q631 (Fig. 19-59). Be sure that +9 and –9 V are found at the collector terminals. If an improper voltage is at the collectors, check the switching-power voltage at IC607 and IC609. Do not overlook the possibility that the disc mechanism might be defective.

No Feed-Motor Rotation

The feed motor moves the laser assembly perpendicular to the disc assembly. The dc-voltage feeding motor comes from feed-motor driver transistors Q629 and Q630. The feed-motor drivers can be transistors or ICs. Check for both –9 and +9 voltages at the collector terminals of the driver transistors. Suspect that Q629 or Q630 are defective if no voltage is at the feed-motor terminals. Measure the continuity of the feed-motor terminals for an open motor with the R×1 range of the DMM.

Wiring Diagram

It is very difficult to troubleshoot the various CD player circuits without a circuit or wiring diagram. The various PC boards and units are wired together with a different wiring diagram (Fig. 19-60). The speaker cables, power leads, and line-output plugs are connected to the main PC board.

The feed-, disc-, and select-motor connections are on the SWA PC board with the disc-detect and disc-in sensors (Q916 and Q915) on the SWB PC board (Fig. 19-61). The loading motor is fed from the sub-PC board. If a schematic wiring diagram is not available, all leads must be traced to the respective boards, which wastes a lot of time.

Improper Search

With search problems, check the tracking-servo, kick-pulse, and feed-motor circuits. A defective tracking-servo circuit could be caused by Q625 and Q626. Measure the voltage on

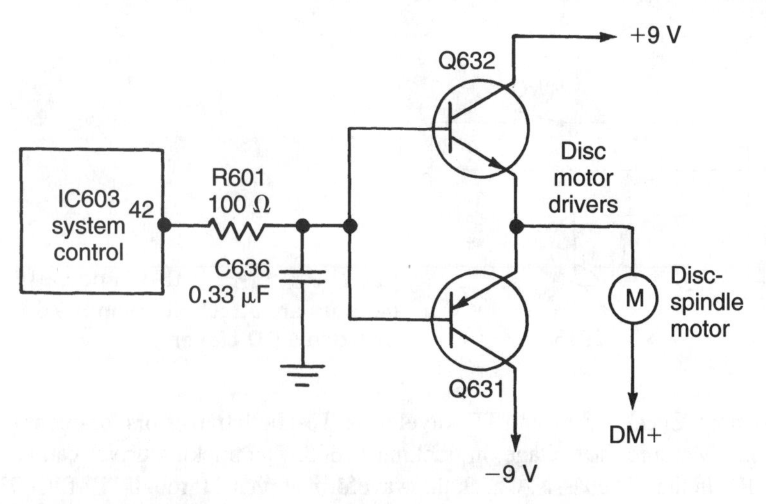

FIGURE 19-59 **Driver transistors Q631 and Q632 provide voltage to the spindle or disc motor.**

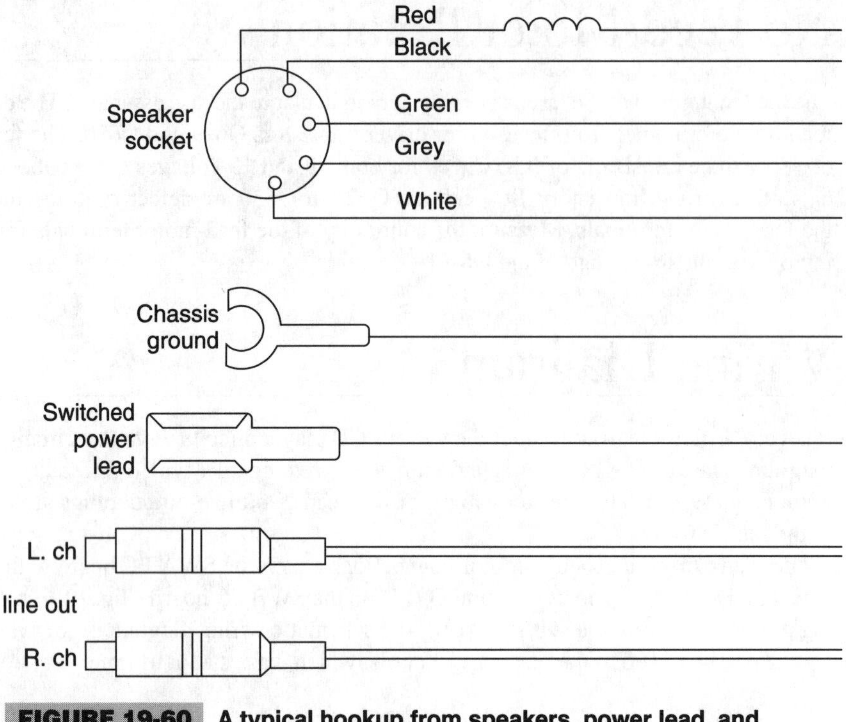

FIGURE 19-60 A typical hookup from speakers, power lead, and line-output cables.

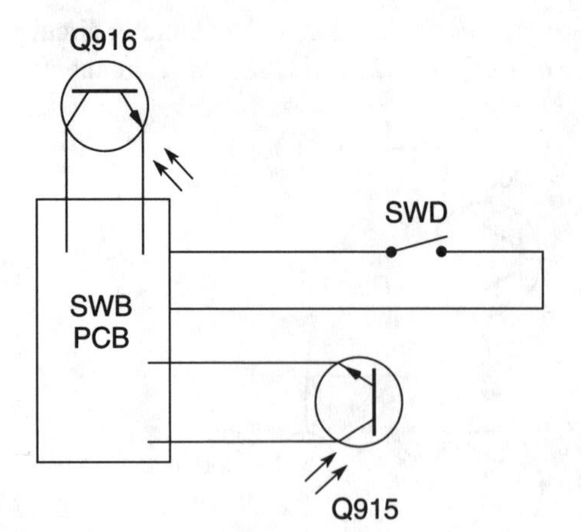

FIGURE 19-61 Q915 and Q916 sensors are attached on an SWB PC board in a CD player.

both transistors. Check the TP and TE waveforms. Test both transistors for opens or shorts (Fig. 19-62). Measure the voltage on tracking IC605. The tracking driver can have transistors or ICs in the CD chassis. Check the waveform at pins 8 through 11 of IC603 and at pins 22 and 33 of IC601. Check all voltages in the kick-pulse circuits. Measure the voltage at pins 0 through 8 of the servo processor (IC603) and at pins 22 and 33 of the signal-processor control (IC601).

Check the feed-motor circuits. Measure the voltage in Q629 and Q630 (Fig. 19-63). Test Q629 and Q630 for open or shorted conditions; remove them from the circuit, if in doubt. Check R715 for opens or burns.

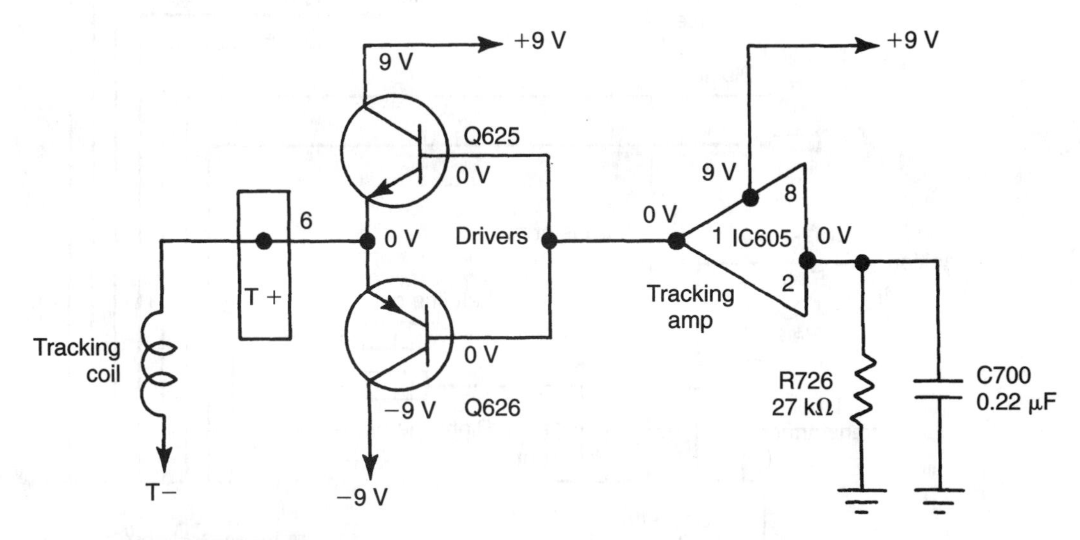

FIGURE 19-62 The tracking coil is driven with voltage and signal from driver Q625, Q626, and IC605.

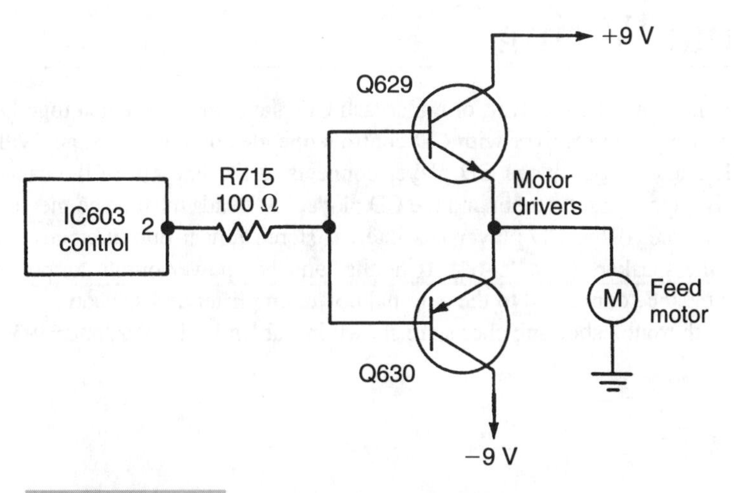

FIGURE 19-63 Check the voltage at the disc motor by transistors Q629 and Q630 from a 9-V source.

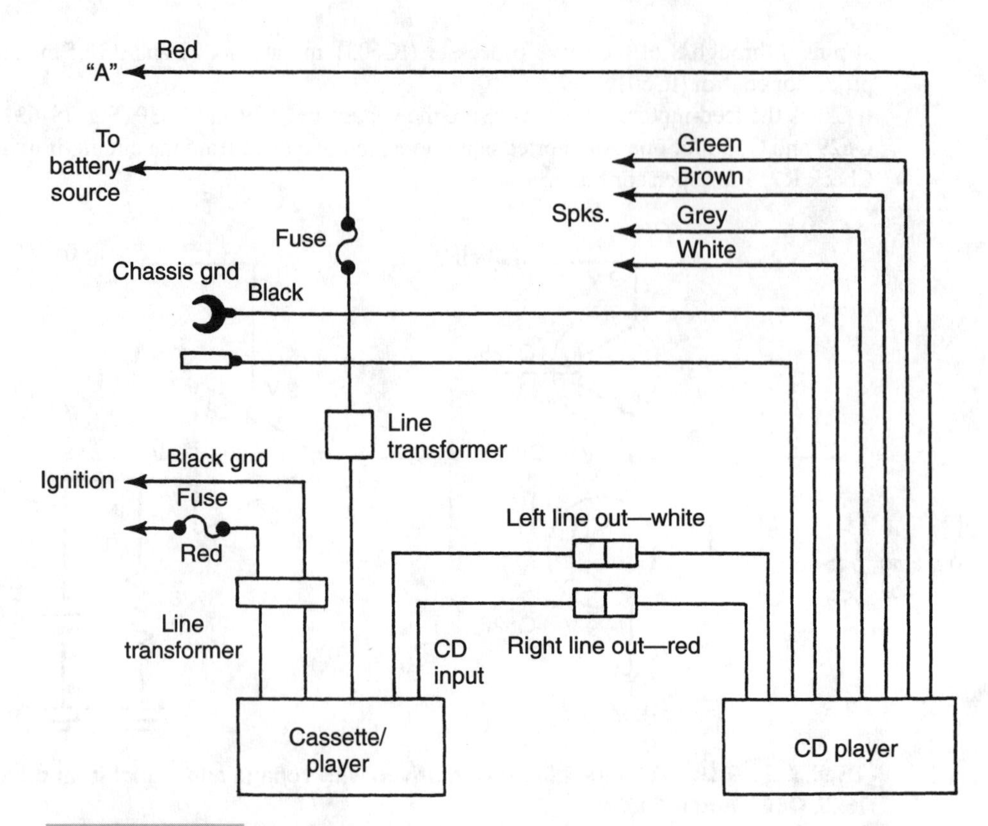

FIGURE 19-64 **The external wiring connections of a cassette/receiver and under-dash auto CD player.**

External Wiring

The cassette/tuner, cassette/receiver, or underdash CD player are connected together. The cassette tuner or cassette receiver with CD controls operates on a CD changer within the auto trunk. The line output of the CD player connects to the line out of the cassette receiver/tuner. Both the cassette/tuner and the CD player "A" leads must be connected to the 14-V battery source. If the CD player has internal stereo power-output stages, they are connected to the speakers (Fig. 19-64). If neither unit has power-output amps, the cassette/receiver must be connected to the external power amplifier and speakers.

CD under-dash troubleshooting charts are shown in Tables 19-1, 19-2, and 19-3.

TABLE 19-1 THE CD PLAYER TROUBLESHOOTING CHART

SYMPTOM	DEFECTIVE CIRCUIT	DEFECTIVE POINT AND CAUSE
No sound	Power supply circuit	• Fuse open. • Faulty connection between battery. • Power switch defective. • Check lead wire cold soldered.
	Output circuit	• Speaker voice coil open. Faulty connection between speaker and connection coil. • Power amplifier defective. Check each pin voltage of power amplifier IC504 and IC505.
	Control flat amplifier circuit	• Variable resistor VR1, VR2, VR3. • Line amplifier defective. Check voltage of IC505. • Voltage regulator defective. Check voltage of IC701, IC703, Q701, Q503, Q504. • BASS TREBLE circuit defective.
Distorted sound or insufficient sound	Output circuit	• Speaker wire grounded. • Power amplifier defective. • Check each pin voltage of power amplifier IC504, C529-C532, C535, C537-C540, R537-R540 defective. • BASS TREBLE circuit defective. • Line amplifier defective. Check voltage of IC505.
DISPLAY SECTION		
SYMPTOM	**DEFECTIVE CIRCUIT**	**DEFECTIVE POINT AND CAUSE**
No display	Power supply circuit	• Power supply circuit defective. Check voltage of IC702.
	Micro computer circuit	• Micro computer circuit defective. Check IC601. Check X601, X602, Q606, Q607, Q609.
Missing display segment	Micro computer circuit	• Micro computer circuit defective. Check IC601. • LCD defective.
DISC SECTION		
Disk loading inferiority	Power supply circuit	• Power supply circuit defective. Check Q901-Q904, D901-D904 and R901-R904 short or open.
	Sensor circuit	• Sensor circuit defective. Check Q915, Q916, D910 and D911.
	Motor driver circuit	• Motor driver circuit defective. Check IC901, IC902, Q905-Q910, R905-R910 and R915-R920.
	Mechanism	• Mechanism defective.

TABLE 19-1 THE CD PLAYER TROUBLESHOOTING CHART (CONTINUED)

SYMPTOM	DEFECTIVE CIRCUIT	DEFECTIVE POINT AND CAUSE
Disk turning inferiority	Power supply circuit	• Power supply circuit defective. Check IC607-IC609, Q613-Q616.
	Feed motor circuit	• Feed motor circuit defective. Check voltage of IC603, 2 pin and IC602, 25, 26 pin. Check Q629, Q630 short or open.
	Focus search circuit	• Focus search circuit defective. Check voltage of IC603, 35 pin and IC601, 13 pin. Check voltage of IC605, 7 pin. Check Q627, Q628 short or open.
Disk turning inferiority	Automatic power control circuit	• Automatic power control circuit defective. Check Q633 short or open. Check voltage of IC606, 1 pin.
	Disc motor circuit	• Disc motor circuit defective. Check voltage of IC601, 11 pin and IC603, 40, 42 pin. Check Q631, Q632 short or open.
	VCO circuit	• VCO circuit defective. Check frequency of TP VCO and readjust L601. Check voltage of IC604, 1-3 pin. Check waveform of IC601, 2-3 pin.
	Focus servo circuit	Check voltage of TP FE and readjust SVR605.
Search inferiority	Tracking servo circuit	• Tracking servo circuit defective. Check waveform TP, TE, and readjust SVR601, 602, 603. Check Q625, Q626 short or open. Check waveform of IC603, 9-11 pin and IC601, 19-21 pin.
	Kick pulse circuit	• Kick pulse circuit defective. Check voltage of IC603, 0-8 pin and IC601, 22, 33 pin.
	Feed motor circuit	• Feed motor circuit defective. Check Q629, Q630, R713 short or open.
No sound	Digital signal circuit	• Digital signal circuit defective. Check waveform of IC602, 1-24 pin. Check waveform of CN601 defective IC601.

TABLE 19-1 THE CD PLAYER TROUBLESHOOTING CHART (CONTINUED)

SYMPTOM	DEFECTIVE CIRCUIT	DEFECTIVE POINT AND CAUSE
Noise	VCO circuit	• VCO circuit defective. Check frequency of TP VCO and readjust L601. Check voltage of IC604, 1-3 pin. Check waveform of IC601, 2-3 pin.
	Slice level control circuit	• Slice level control circuit defective. Check voltage of IC604, 5-7 pin.
	RF circuit	• RF circuit defective. Readjust SVR604.
	Mechanism	• Mechanism defective. Check eccentricity of mechanism.
Disk eject inferiority	Sensor circuit	• Sensor circuit defective. Check Q915, Q916, D910 and D911.
	Motor driver circuit	• Motor driver circuit defective. Check IC901, IC902, Q905-Q910, R905-R910 and R915-R920.
	Power supply circuit	• Power supply circuit defective. Check Q901-Q904, D901-D904 and R901-R904 short or open.
	Mechanism	• Mechanism defective.
No sound	Mechanism	• Mechanism defective.
	RF circuit	• Digital filter, D/A converter, LPF circuit defective. Check IC605, IC606, IC501 and IC502.
	Display circuit (microcomputer)	• Micro-computer circuit defective. Check IC601.
Distorted sound or insufficient sound	RF circuit	• Digital filter, D/A converter, LPF circuit defective. Check IC603, IC604, IC605, IC606, IC501 and IC502.
	Mechanism	• Mechanism defective.

TABLE 19-2 MECHANICAL TROUBLESHOOTING CHART

SYMPTOM	TROUBLE	REPAIRS
No recording or playback	Motor not rotating	Open motor. Test with ohmmeter. Open series resistor or diode. Defective lead or on/off switch. Frozen motor shaft. Dead or loose battery. Defective power plug.
	Motor rotates	Capstan rotates with oil or dust on pressure roller. Worn pressure shaft.
	Capstan does not rotate	Dry or gummed up capstan shaft. Drive belt out of position. Loose screw on motor pulley.
	Take-up reel or turntable does not rotate	Insufficient pressure on take-up pulley. Slippage of idler wheel.
	Twisted tape	Pressure roller and capstan not parallel. Record/playback head out of line.
Defective fast forward	Take-up reel or turntable slow rotation.	Oil on drive pulley. Oil on drive belt. Insufficient belt tension. Insufficient pressure at drive pulley. Worn take-up reel shaft. Defective tape counter.
Defective rewinding	Supply hub or reel too slow	Oil on drive pulley. Oil on belt. Oil on rewind idler pulley. Insufficient pressure of rewind idler. Worn supply turntable or reel shaft. Defective tape counter.

TABLE 19-3 THE ELECTRICAL TROUBLESHOOTING CHART

Defective recoding on tape deck 1 or A	No recording bias	Check supply voltage on bias oscillator. Defective R/P switch. Defective leaf or on/off switch. Defective oscillator coil—check with ohhmeter. Defective R/P switch.
	Bias OK, but no recording	Level meter is operating—no record. Dirty R/P head. Defective R/P head. Dead no level meter—defective mic. Defective input jack. Defective R/P switch. Defective amp. Defective level meter. Defective meter diode or diodes.
	Defective erasing	Open or shorted oscillator coil. Defective bias transistor. Open or shorted erase head—check with ohmmeter. Defective capacitors in tank circuit of bias oscillator. Dirty erase head surface. Defective oscillator circuit. Improper supply voltage to oscillator circuits.
Defective playback	No sound, B+ supply	Open primary winding of power transformer. Defective leaf or on/off switch. Shorted B+ supply. Defective diodes in power supply.
	Voltages on all transistors and ICs are OK	Touch volume control for hum pickup—open R/P head or dirty switch. Touching volume control does not produce hum—defective speaker. Defective earphone jack. Bad speaker leads. Poor and dirty switch. Defective output transformer. Defective IC or audio transistors. Improper B+ voltage.
	No output, but noise and hum	Open R/P head.
	Defective tone	Shorted tone control leads. Defective tone control. Defective or dirty R/P head. Defective coupling capacitors.

TABLE 19-3 THE ELECTRICAL TROUBLESHOOTING CHART (CONTINUED)

	Low volume—abnormal voltage at transistors and ICs.	Check voltage on all transistors. Check voltage on all ICs. Defective coupling capacitors. Defective bias resistors. Defective small coupling capacitors. Defective large output coupling capacitor.
	Low volume output	Dirty R/P head. Defective IC output. Defective output transistors. Check voltages with normal channel.

20

REPAIRING STEREO CASSETTE DECKS

CONTENTS AT A GLANCE

Stereo cassette decks are available in many sizes and shapes (Fig. 20-1). The home audio cassette deck might have a frequency response from 20 to 20,000 Hz, a high signal-to-noise ratio, and low wow and flutter. The cassette deck might include a dual or double deck and dubbing operation (Fig. 20-2).

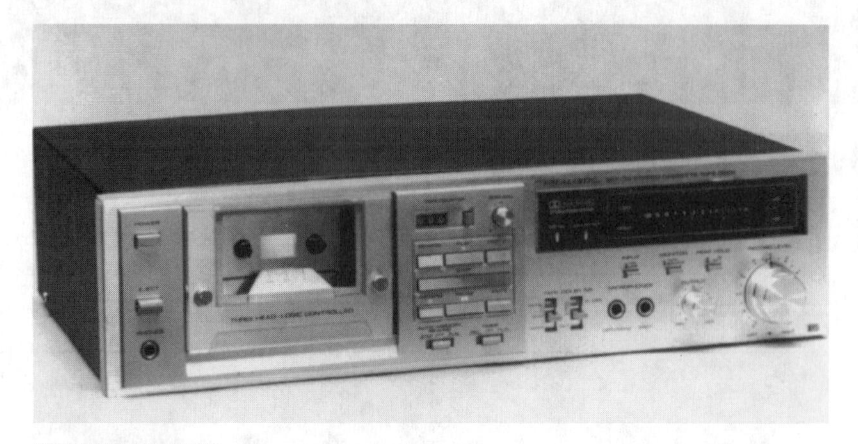

FIGURE 20-1 The cassette deck might include both Record and Play with stereo amplifiers or line-out connections.

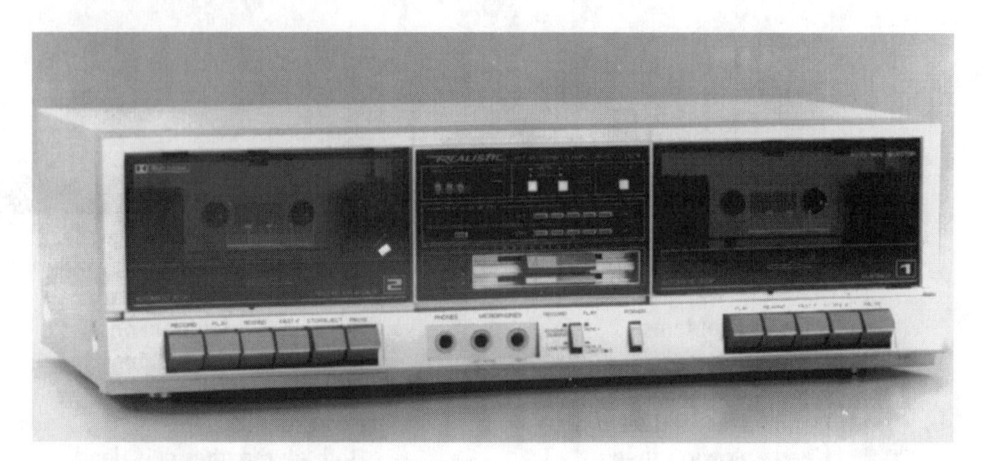

FIGURE 20-2 The cassette deck has dual decks; deck B records and plays and deck A only plays.

The deluxe cassette deck might have twin auto-reverse tape transports with Record and Play functions, a three-head transport mechanism with dual capstans for tape speed accuracy, a separate motor for fast forward and rewind, two auto-reverse record/play transports that let you record four sides nonstop or make two identical copies simultaneously, auto tape selector, blank skip, music search, reel tone counter, microphone inputs, recording level meters, and remote control.

The stereo cassette deck might also have an AM/FM-MPX stereo circuit. The deluxe radio is built to one side of the cassette tape deck (Fig. 20-3). You might find a phono turntable on top of some models. In older deluxe cassette decks, besides the AM/FM-MPX radio circuits, a stereo 8-track player is also included (Fig. 20-4).

Cassette Features

Many features require the best reproduction of tape possible, including wow and flutter below 0.08%, higher signal-to-noise ratio from 73 to 79 dB, and harmonic distortion as low as 0.6%. The deluxe cassette deck might have the following features:

- *Wow and flutter* The variation of speed with a lower reading indicating the best rotation. The wow and flutter measurement of 0.10% is excellent for most cassette players, but 0.6% is best.

FIGURE 20-3 The stereo cassette deck might include AM/FM/MPX receiver circuits.

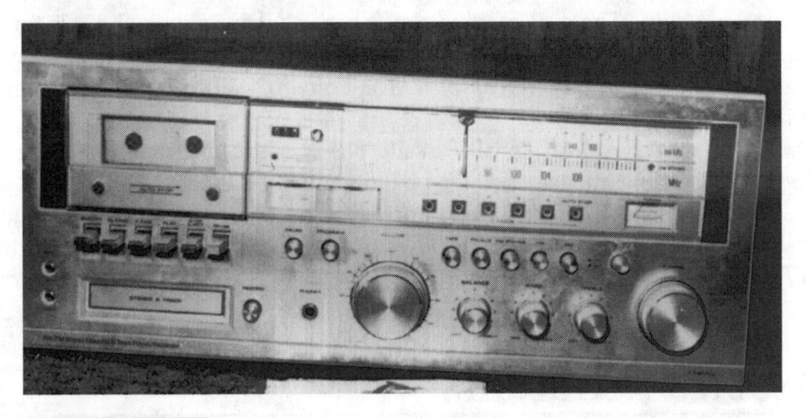

FIGURE 20-4 In older deluxe cassette decks, you might find the MPX-radio receiver, recording meters, tuning meters, and 8-track tape player.

FIGURE 20-5 **Most less-expensive decks use one motor for all functions, as in the Sounddesign deck.**

- *Auto reverse* The tape deck will play both sides of the tape without actually changing the cassette. The mechanism changes direction and often has two capstan flywheels and tape heads.
- *Music search* The player will automatically stop at the beginning of a song or music when operated in the Fast-Forward or Rewind modes. Only the deluxe cassette players have this feature.
- *Frequency response* The lower and higher the frequency, the more ideal the frequency response. The audible frequency range is 20 to 20,000 Hz. Often, women can hear above 15 kHz, but most men cannot hear above 10 kHz.
- *Noise reduction* The higher the signal-to-noise (SN) ratio, the better. Less hiss and better sound is ideal. Dolby C operates over a wider range, but Dolby B removes high-frequency hiss only.
- *Auto tape selector* Bias and equalization is set automatically when the cassette is inserted into the tape deck. A different bias is needed when metal tapes are used.
- *Number of heads* The dual-cassette player has one deck with dual heads for Record and Play. In this case, Play is contained on one head. The record, play, and erase heads might be called a *three-head transport*. The Sony TC-WR670 has high-density hard permalloy heads.
- *Number of motors* Most typical cassette players have one tape motor for normal speed. In the deluxe models, two tape motors might be used: one motor for accurate tape speed and another motor for Fast Forward and Rewind. The dual deck might have three or four different motors. In the lower-priced dual decks, one motor might rotate all functions (Fig. 20-5).

Cassette Problems

CAN'T OPEN DOOR

The front door might be warped or the door catch might not release when the Eject/Stop button is pressed. Do not try to force the door and break the plastic. If the front door cover can be removed, take it off to peer in at the door catch. Lift the catch mechanism with a small screwdriver. Replace the door if it is broken or warped and will not release. Repair the bent catch lever and check for loose or broken door hinges. Suspect that tape has spilled out if the door will not open and the cassette will not pop out (Fig. 20-6).

CASSETTE WILL NOT LOAD

Peer inside the tape holder for foreign material, such as cigarettes and gum wrappers. Be sure that the record safety lever will release. Inspect the plastic holder for cracked or broken areas. Try another cassette. The cassette might be cracked or broken.

Be sure that the door closes properly. Is the recorder in the Play mode? The mechanism might be high and not let the cassette load. Usually, small items inside the cassette holder prevent proper cassette loading.

DEAD CASSETTE DECK

Go directly to the power supply if the dial lights and nothing works. If the unit operates on batteries and not ac, suspect that a power supply is defective. Check the low voltage at the power-output transistors or ICs. With no voltage, check the primary winding of the power transformer.

Measure the B+ voltage across the large filter capacitor. If the voltage is normal at this point, suspect that a resistor or voltage regulator is open. If a schematic is handy, check the power-supply circuits for a voltage regulator (Fig. 20-7). The voltage regulator might furnish voltage to the motor and amp circuits. Q11, R10, and ZD1 might open and prevent voltage at the 17- and 15.5-V sources. The open or leaky voltage regulator can be replaced with universal transistors.

Simply trace the voltage source from the cassette motor or amplifiers back to the power supply if a schematic is not available. The voltage regulator should be mounted on the power-supply board. Sometimes it is mounted on metal chassis with leads to the PC board. Do not overlook the possibility that board connections are poorly soldered. By taking accurate voltage and resistance measurements, the voltage source can be traced back to the full-wave rectifiers and the main filter capacitor.

FIGURE 20-6 **Remove the front plastic piece if the door will not open. Look for pulled tape wrapped around the capstan and pinch roller.**

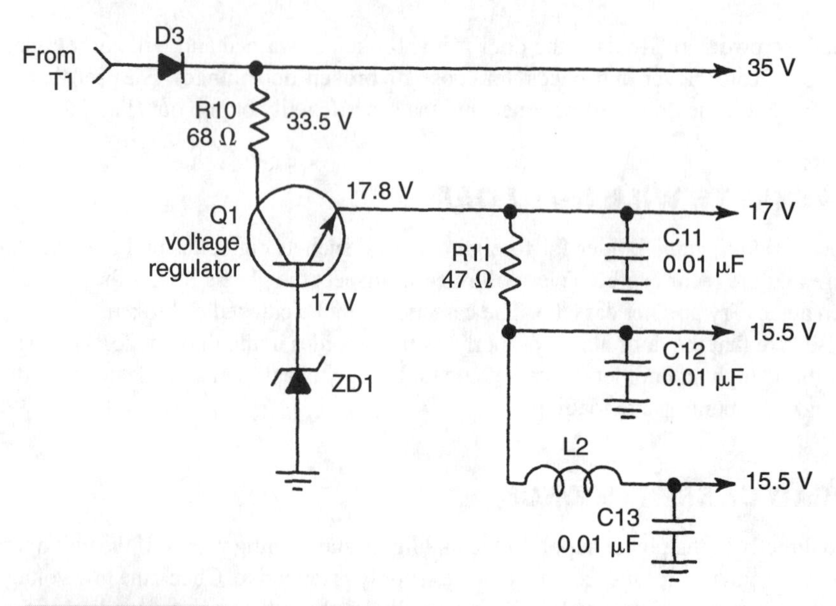

FIGURE 20-7 **You might find more than one voltage regulator circuit in the cassette-deck power supply.**

KEEPS BLOWING FUSES

Suspect that a fuse is blown in larger cassette players when nothing lights. Often, the fuse is blown by a shorted silicon diode, filter capacitor, or leaky output solid-state device. If the fuse keeps blowing, suspect that a power-output transistor or IC is leaky. Disconnect the output circuit from the power supply and cut a piece of foil or remove wires to suspected transistors.

Larger cassette players with higher power output might have four large transistors. Often, two are located in each output channel. Determine which channel is blowing the fuse by taking a low-resistance measurement between the collector terminal and ground. After locating the two output transistors, test each one in the circuit for leakage. Remove them and test them for leakage out of the circuit. You might find that one of the transistors is shorted and the other is open (Fig. 20-8).

While the transistors are out of the circuit, check for burned or open bias resistors. Usually, if a power-output transistor is shorted, the bias resistor opens. Also, while the two transistors are out of the circuit, check the driver transistor. Sometimes, the driver transistor becomes leaky and destroys the directly coupled power-output transistors. Most power-output transistors can be replaced with universal types. Leaky power-output ICs might also blow the fuse (Fig. 20-9).

STOPS AFTER A FEW SECONDS

If the tape deck keeps shutting off after only a few seconds, suspect that the automatic shut-off circuits are not working. In units with automatic shutoff, a magnet is fastened to the end of a pulley on the counter assembly. Some models have a magnetic switch behind

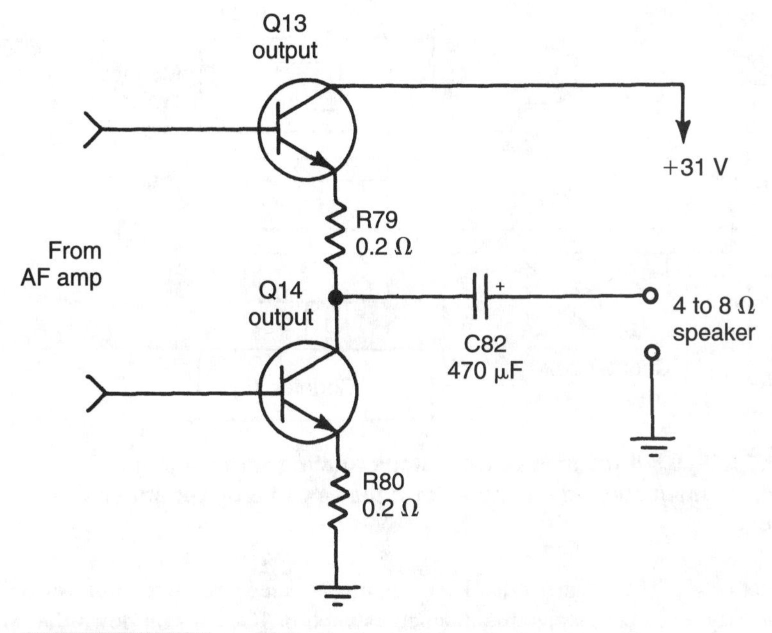

FIGURE 20-8 Check bias resistors if the power-output transistors are leaky or shorted. Test transistors out of the circuit.

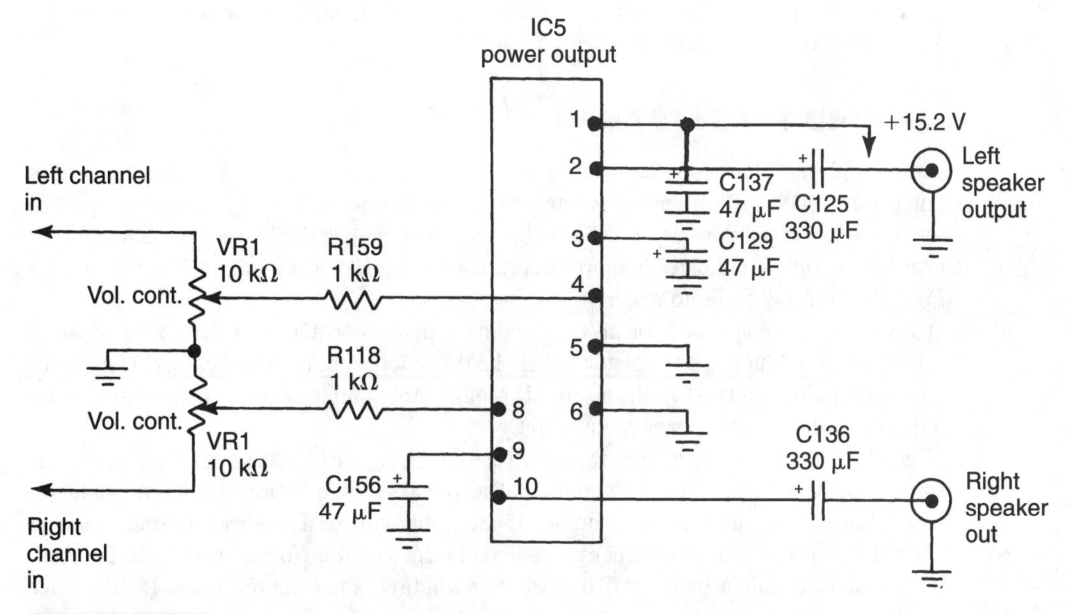

FIGURE 20-9 For leaky or shorted ICs, check the resistance at supply pin (1) to pin 6 at ground. If the measurements are under 100 ohms, the IC might be leaky.

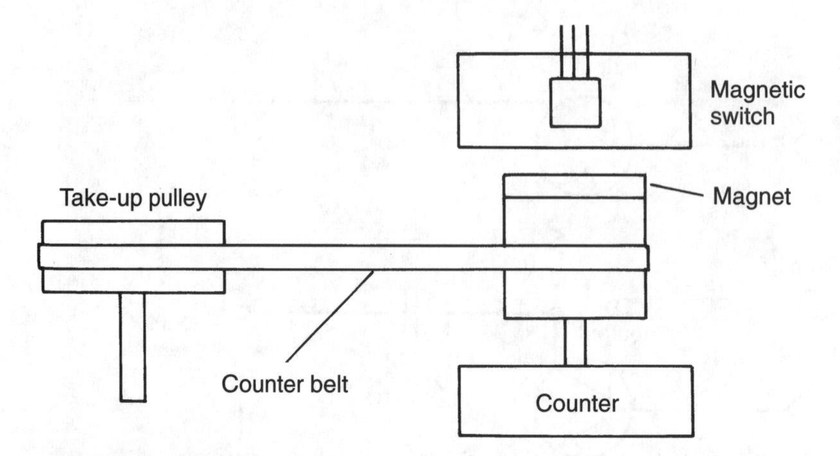

FIGURE 20-10 **If the tape counter stops rotating with magnet attached, the magnetic switch shuts down the cassette player after a few seconds.**

the magnet or IC. The magnet must keep rotating to keep the cassette player operating. When the magnet or tape stops, the magnetic switch or IC will shut down the operation (Fig. 20-10).

If the drive belt to the counter reel is broken, the cassette will start up and shut down at once. Look for a broken belt off of the counter pulley. Notice if the tape counter is rotating. If the belt rotates and the counter pulley and the unit shuts down, suspect that a switch or IC are defective. An IC is used in some Sharp models, but a magnetic switch is used in Sylvania tape players. The magnetic switch and IC are special components and must be obtained through the manufacturer or part depots.

SMOKING TRANSFORMER

Quickly pull the plug if the cassette player begins to smoke and groan when the unit is turned on. Check the primary winding of the transformer to see if it has now opened. Check the B+ supply voltage or check the resistance across the large filter capacitor. A level below 100 Ω indicates a short circuit. Check the B+ at the output IC or transistor (Fig. 20-11) if it is easier to reach.

Check each silicon diode for an overheating transformer. Often, a dead short or high leakage of an output IC or transistor will make the power supply draw heavy current. Arcing and dimming lights when the ac cord is plugged into outlet indicates a heavy overload. In this case, the transformer is running hot.

Remove the ac secondary leads from the rectifier circuits to determine if the transformer has been damaged (Fig. 20-12). If additional secondary windings are found, remove them. Now, plug the unit into the wall outlet. If the lights still dim, the transformer makes a noise; if it runs hot, the transformer must be replaced. The transformer is okay if it is cool. The transformer might be okay if the primary winding is not burned open. If the power transformer is defective, replace it with an exact-type part.

FIGURE 20-11 Suspect that silicon diodes, power-output transistors, or ICs are shorted if the power transformer runs too warm and smokes.

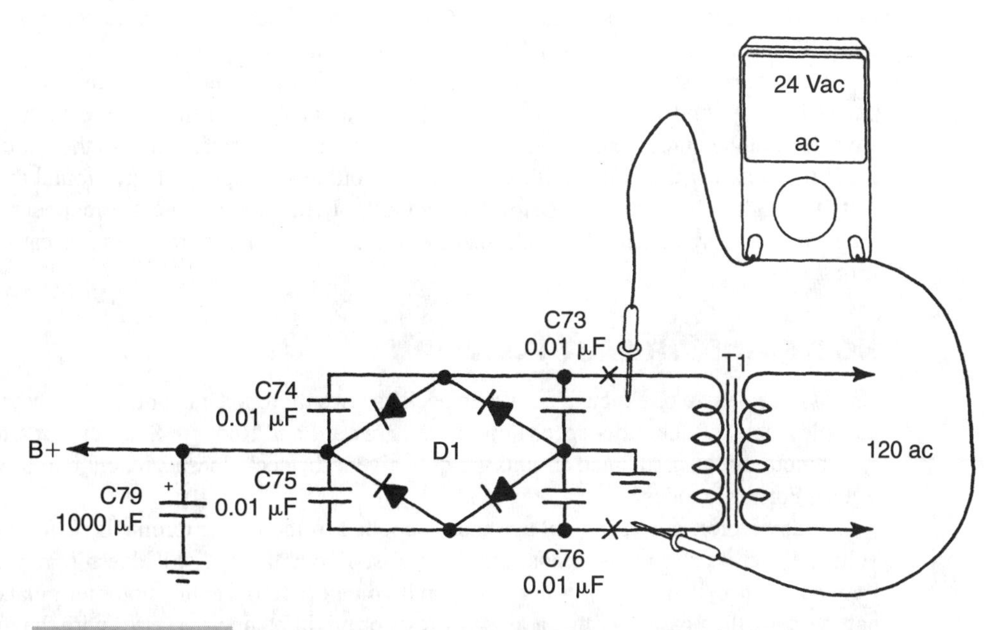

FIGURE 20-12 Remove the ac leads from the secondary of the power transformer and measure the ac voltage. Replace the transformer if it runs hot after the load is removed.

FIGURE 20-13 The noisy IC might quit making noise if it is sprayed with coolant. This action will indicate the defective IC.

NOISY OPERATION

Loud mechanical-type noises should be checked at once. If the noise is hissing or fuzzy audio, suspect that a transistor or IC is noisy. Check for noise in each speaker. If you find the frying noise in only one channel with the volume turned down, the sound is originating in the audio-output circuits.

Spray each transistor or IC with coolant and notice if the noise disappears or becomes louder. Spray several coats on each component before going to another. Sometimes when the coolant hits the transistor or IC, the noise stops immediately (Fig. 20-13).

If the noise persists after you apply the coolant, try shorting each base of transistor or input of IC to ground with a 10-μF electrolytic capacitor. Start at the volume control and work toward the output or speaker. Also, you can short the base terminal to the emitter to see if the noise ceases. If the noise is reduced in volume or stops, you have found the circuit. Testing the suspected transistor does not help. Replace the suspect component. You might find a noisy ceramic bypass capacitor in the AF or driver circuits that is causing a frying noise.

NO REWIND OR FAST FORWARD

Usually, the deluxe or expensive cassette deck has two motors. One motor is used for regular playback and the other speeds up to Fast Forward and Rewind. Suspect that a high-speed motor or the associated circuits are defective if the deck doesn't function in Rewind or Fast Forward modes.

Like any motor, check the voltage that is applied to the motor terminals. Connect the voltmeter across the motor leads and push Fast Forward. If it still doesn't move, try Rewind. Suspect that a motor is dead or open if voltage is found at the motor terminals. Do not overlook the possibility that a diode or resistor might be open in series with the motor leads if no voltage is measured at the terminals. Look for a broken belt if you can hear the motor rotate. In older models, do not confuse the two motors because one might operate an 8-track deck (Fig. 20-14).

ERRATIC TAPE SPEED/UNEVEN PRESSURE ROLLER

Erratic speed could be caused by a loose motor-drive belt, an oily belt, or a dry capstan. Check erratic speed with the torque cassette. If the torque is below 100 g, clean the motor belt, motor pulley, and capstan/flywheel. Then, again check the torque.

Uneven speed might be caused by a pinch or pressure roller that is not perfectly round or worn. Check along the rubber-bearing area for broken tape. Often, if the tape spills out and breaks, excess tape is wound around the pinch roller (Fig. 20-15). Remove the pressure roller, if necessary, and clean out the tape. Place a drop of light oil on the bearing. Wipe off any excess oil from the rubber area. Rotate the roller by hand to be sure that it is free.

Check the pressure-roller torque with a pressure gauge. Bend the roller pressure spring for more tension, if needed. The pressure roller helps pull the tape, with the capstan, across the tape heads and feeds it on the take-up turntable. The pinch roller should run smooth and even. Replace the pinch roller if it is not perfectly round or if it is worn on the edges.

JUMBLED OR BAD RECORDING

No doubt, the erase head is not working when more than one recording is heard in playback. The erase head erases any previous recording before the tape passes the R/P head (Fig. 20-16). Place the deck in the Record mode and do not record from microphones, radio, or the dual deck. Let the cassette record for 10 minutes. Time it with the tape counter. Now, rewind the same cassette to the beginning and hit Play. Notice if all previous recordings are removed.

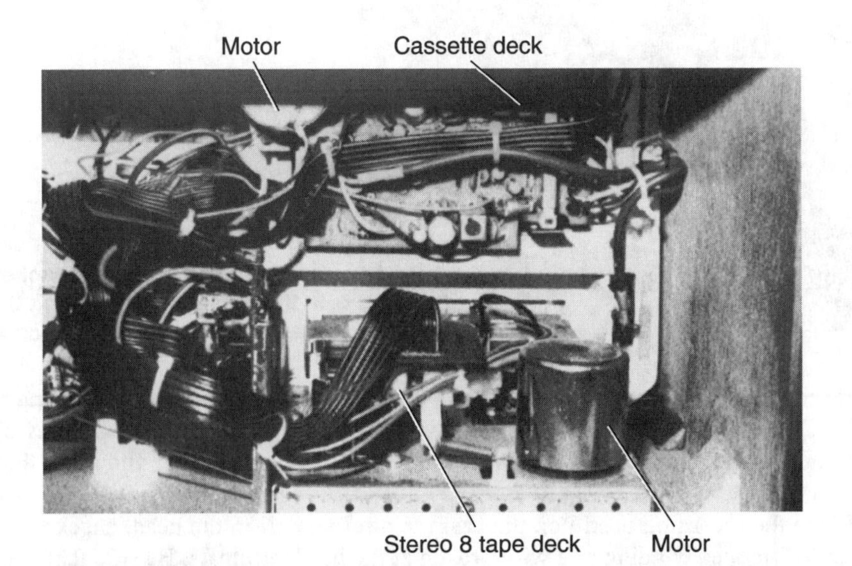

FIGURE 20-14 **In this J.C. Penney cassette deck, the cassette player is mounted at the top and the 8-track player is mounted at the bottom.**

FIGURE 20-15 A dry or worn roller, or one with tape wrapped around the bearing area, might cause uneven speeds. Check the pinch-roller pressure with a tension gauge.

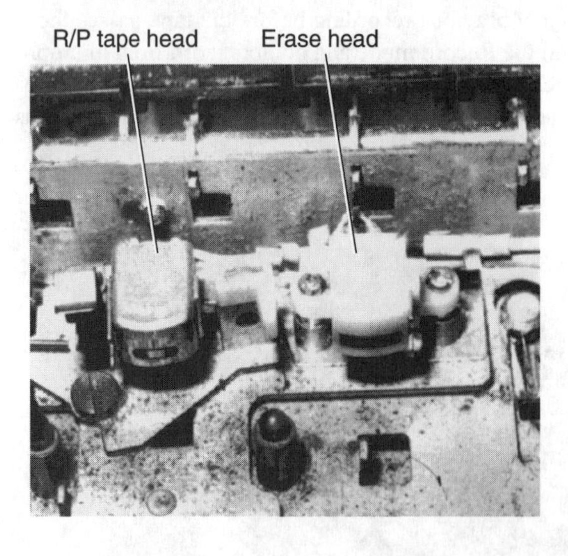

FIGURE 20-16 The erase head might have oxide packed in the gap area, preventing the erasure of the previous recording.

If the tape still has several recordings on it, the erase circuits are not working properly. Clean the erase head and all heads with rubbing alcohol and a stick. Spray cleaning fluid in the function and R/P switch assembly. Now, start over and erase the tape again.

When the recording is still on the cassette after you clean the head, check the erase head for an open winding or a wire broken at the head terminals. Be sure that one side of the erase head is grounded. Measure the dc voltage across the erase head if it is meant to be excited by a dc voltage. Check the bias-oscillator circuits with a bias waveform.

REVERSE SIDE OF THE TAPE DOES NOT PLAY

In some tape decks, both sides of tape might be recorded and played back with a press of a button or in auto reverse. This is done with two capstan/flywheel tape heads and switching. Clean all tape heads. Check the wiring of the dead tape head. Clean the tape-head switching circuits. Measure the continuity of the tape head. Hold a screwdriver blade on the ungrounded side of the tape head and listen for a hum. Often, the head or wires are defective because both heads use the same amplifier circuits (Fig. 20-17).

SINGLE-MOTOR FAST FORWARD

Fast Forward in a single-motor deck is performed with mechanical idlers or gears to increase the speed. In some of these decks, another winding runs at a faster rate of speed. When the player is switched to Fast Forward, the normal motor wire is out of the circuit and the fast-forward winding is switched into action (Fig. 20-18).

With normal speed, the B+ voltage is fed to the main winding and through R21 and C51. In Fast Forward, the switch places the fast-forward winding to the B+ voltage and takes

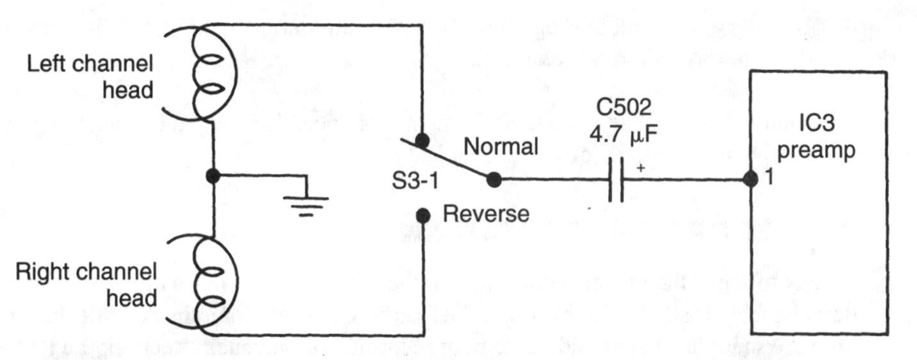

FIGURE 20-17 **Check the reverse tape head and the head switch (S3-1) if the cassette player will not play other side of the cassette.**

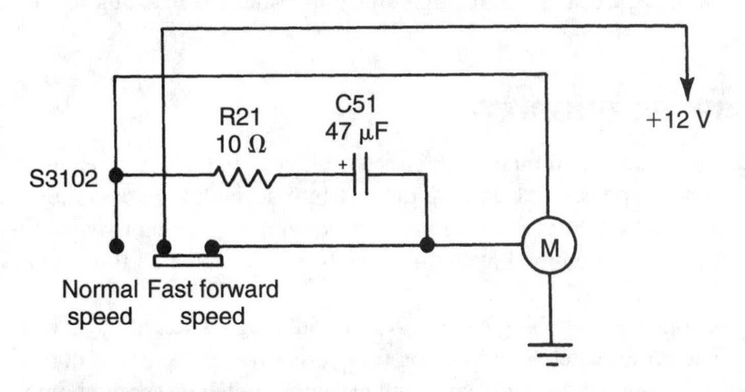

FIGURE 20-18 **Some cassette decks have a single motor with normal and Fast-Forward speed.**

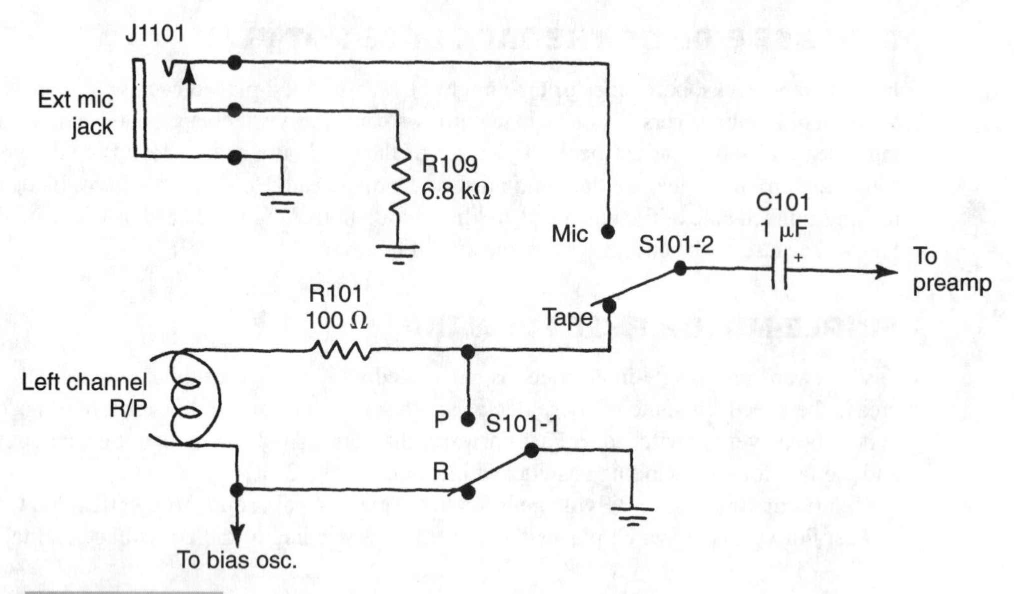

FIGURE 20-19 **Check and clean S101-1 after cleaning the heads if the cassette deck will not record in the left channel.**

R21 and C51 out of the circuit. Here, the motor speed is changed by switching the windings at the motor instead of via mechanical means.

NO RECORD ON LEFT CHANNEL

Check to see if the player will record on the right channel. Does the cassette play in the left channel? If so, the trouble might be a dirty R/P head. Sometimes oxide dust will pack into the small head gaps and cause poor recording or playback. Recheck the R/P head after you clean it to see if one of the gaps is packed shut. Next, clean the record/playback switch contacts (Fig. 20-19). Spray fluid into the switch area. Be careful not to let it drip on the moving surfaces. Be sure that the bias signal or voltage is reaching the left record head winding.

NO HIGH-SPEED DUBBING

High-speed dubbing is used in dual-deck cassette players. This feature enables you to duplicate a recorded tape cassette in tape deck B into a blank tape cassette placed in tape deck A. Some cassette players have high-speed dubbing, which is twice the speed of normal dubbing. Tapes dubbed at higher speeds are the same as those recorded at normal speed.

Check the position of the dubbing switch. A dirty dubbing switch might prevent high-speed dubbing. Clean the switch with cleaning fluid. Usually, this cannot be done from the front of the cabinet. Remove the back cover and clean the switch by spraying into it while flipping the switch back and forth.

LINE-OUTPUT DECKS

In many of the earlier cassette decks, the cassette player/recorder schematic consisted of tape heads, transport mechanism, preamps, Dolby, and AF amp circuits. The AF amp stereo output fed into a line jack, which was cable-connected to a larger amplifier (Fig. 20-20). These cassette decks have all the features of a regular cassette deck, except for the power-output audio stages and speaker terminals. The same troubles that exist in the line-output cassette players could occur in any cassette player/recorder.

Line-output power circuits

The power supplies might be somewhat different from ordinary low-voltage supplies. You might find two separate secondary transformer windings; one going to a bridge and the other to a full-wave diode rectifier. The full-wave transformer winding is center-tapped to ground and fed into two separate silicon diodes. The output is filtered with a large filter capacitor (C21) and fed into a voltage-transistor regulator circuit. The regulated output voltages are fed to the input preamp and Dolby stages (Fig. 20-21).

The full-wave bridge-rectifier circuit has a separate winding, which is fed to a large filter capacitor (C22). The high filtered voltage operates the motor-speed circuits and the collector terminals of the AF output line amplifiers. The bridge rectifier circuits might have pilot lamps in the transformer secondary and LED indicators within the B+ circuits.

NO AUTOMATIC STOP

The automatic stop feature might consist of a memory counter, reed switch, shut-off solenoid, and transistors. The reed switch is triggered with a magnet fastened on the back of

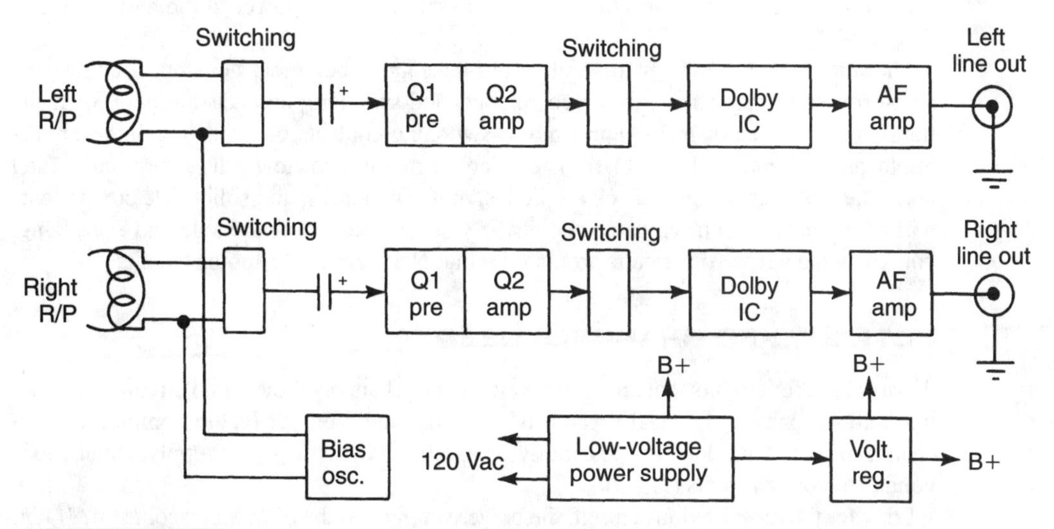

FIGURE 20-20 **The stereo cassette might have line-output connections cabled to a high-powered amplifier.**

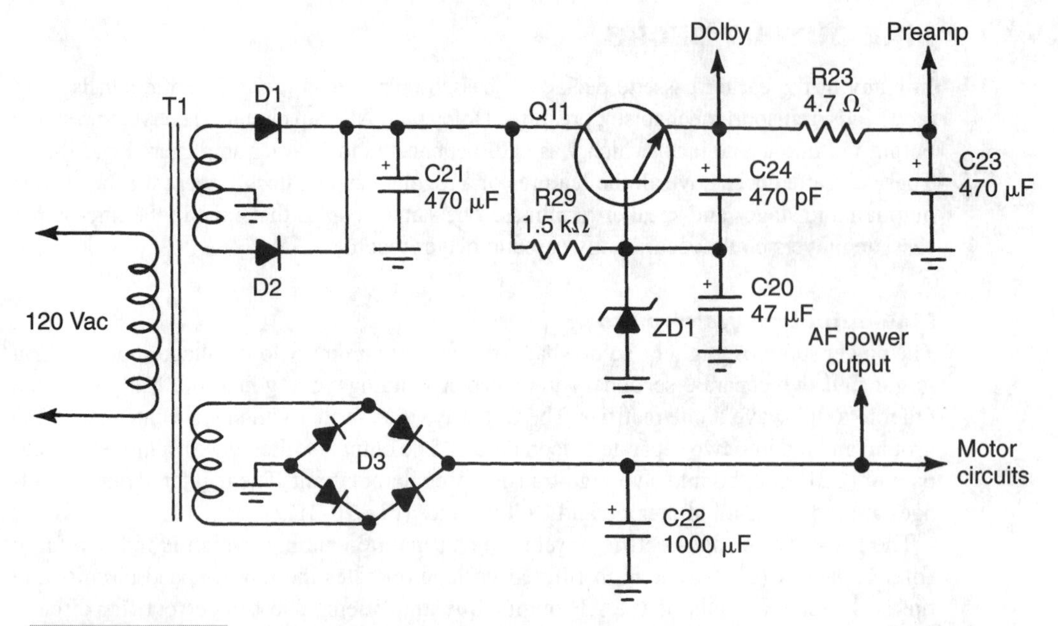

FIGURE 20-21 **Two different secondary-power transformer windings might have voltage regulators feeding Dolby and preamp circuits, although the regular bridge supply furnishes voltage to AF, power-output, and motor circuits.**

the tape counter. The reed switch practically remains closed at all times because the magnet is operated at a fast rate of speed. When the counter stops or when the tape has reached the end, the reed switch opens and breaks the bias on Q1, which turns on transistor Q2, and energizes the shut-off solenoid that turns off power to the motor circuits (Fig. 20-22).

The automatic shut-off system might fail if the counter belt breaks or stops rotating. The shut-off system might fail if the reed contacts and switching contacts are poor, or if the transistors are open or leaky. Dirty motor-switching contacts or motor-regulator circuits might prevent automatic shut off. The solenoid might shut down the tape rotation and leave the motor operating. Check all diodes, motor windings, and solid-state components within the motor-speed regulator circuits. If you can hear or see the solenoid energizing, you know the automatic circuits are functioning. Next, check the motor circuits.

BOTH STEREO CHANNELS DEAD

Always look for circuits that are common with both channels if the audio circuits are dead, intermittent, or noisy. A dual preamp IC and dual audio-output IC are common to both channels. Do not overlook the possibility of a problem with the power supply, which feeds voltage to both channels (Fig. 20-23).

Look for the most obvious circuit, the power supply. Do the pilot lights come on? If not, check the fuse and power supply. If the pilot lights are on, check the dual audio-output IC. Measure the dc voltage at the supply pin (V_{CC}). No voltage here might indicate that a B+ fuse is blown or that a power supply is defective.

Give the volume control the hum test on both channels or inject an audio signal. No sound indicates that a dual-output IC is defective. If the output stages show hum or life, proceed to the dual preamp and Dolby IC circuits. Insert a cassette and signal trace audio from the heads to the preamp, through the Dolby IC to the AF circuits and the volume control.

DEAD LEFT CHANNEL

If either audio channel is dead, start at the audio-output circuits. Most problems within the audio circuits are located in the audio-output and power-supply circuits. In transistor output stages, check the audio at the volume-control and output-transistor circuits. Poor speaker-terminal connections or an open speaker-coupling capacitor might cause a dead channel.

Quickly check the voltage at the collector terminals of the output transistors. Test Q12 and Q13 in the circuit (Fig. 20-24). If in doubt, remove one of them from the circuit for leakage tests. Remember, if one power-output transistor is leaky, the matching one might also be leaky or open. It's best to replace both.

Do not overlook the possibility that a driver transistor is open. If Q11 opens, the supply voltage (19 V) might be found on the base of both Q12 and Q13. Be sure that all resistors are normal—especially R72 and R77 (2.2 Ω). One or both of these resistors might be open. If either output transistor is open, the voltages will change on all terminals of the

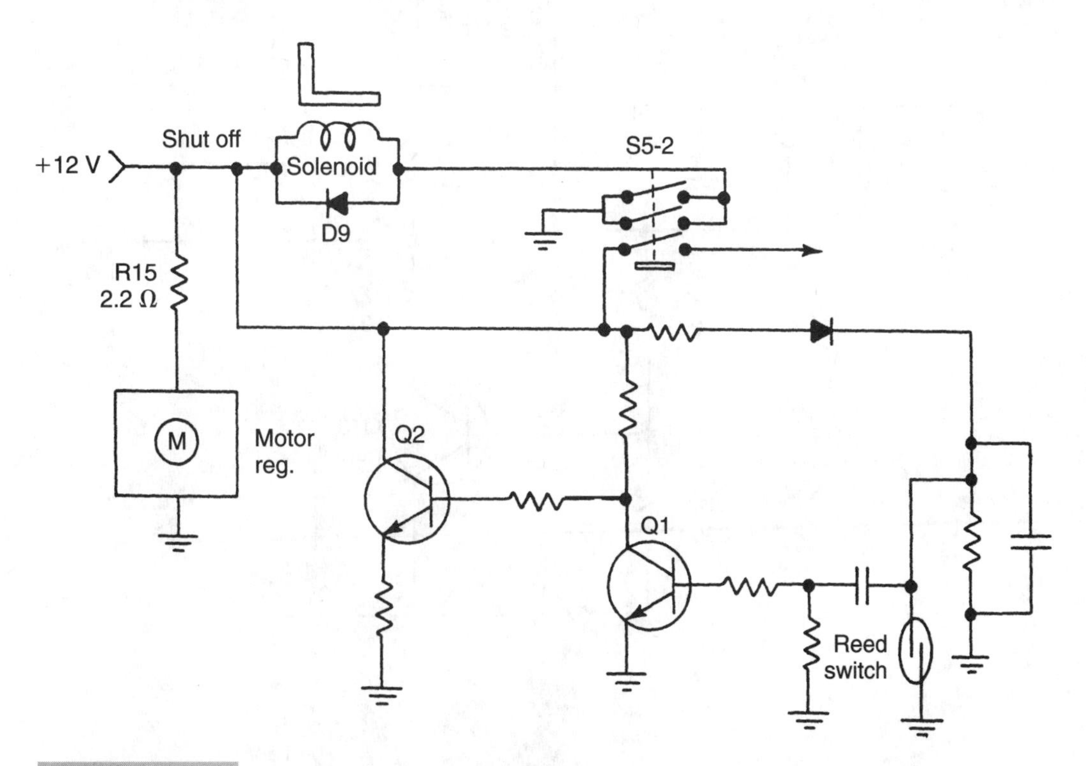

FIGURE 20-22 **The automatic-stop circuits have a reed switch triggered by a magnet on the tape counter, Q1, Q2, and a solenoid with switches.**

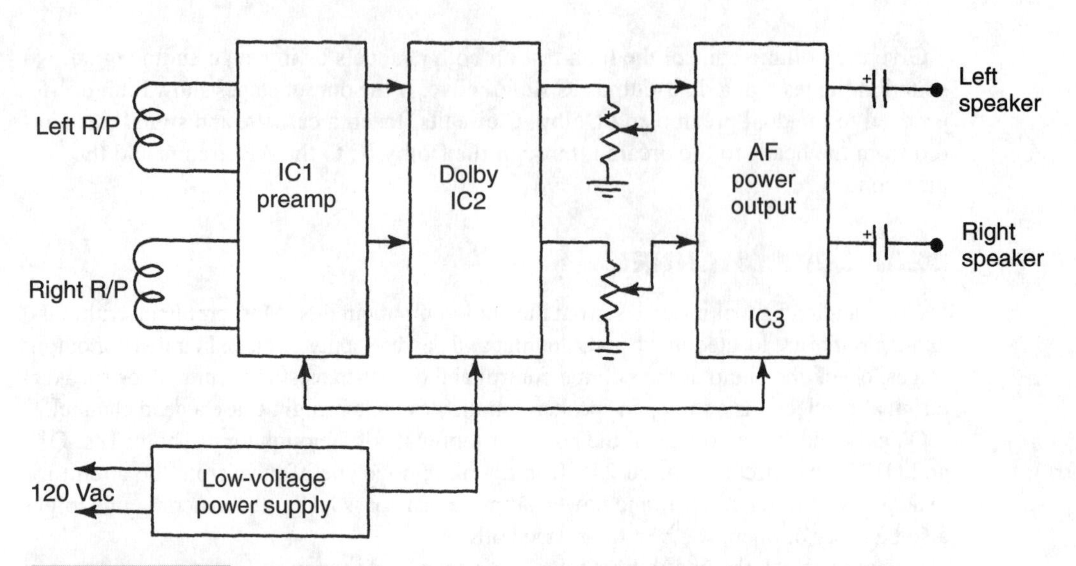

FIGURE 20-23 Check the power-supply voltage if both channels are dead.

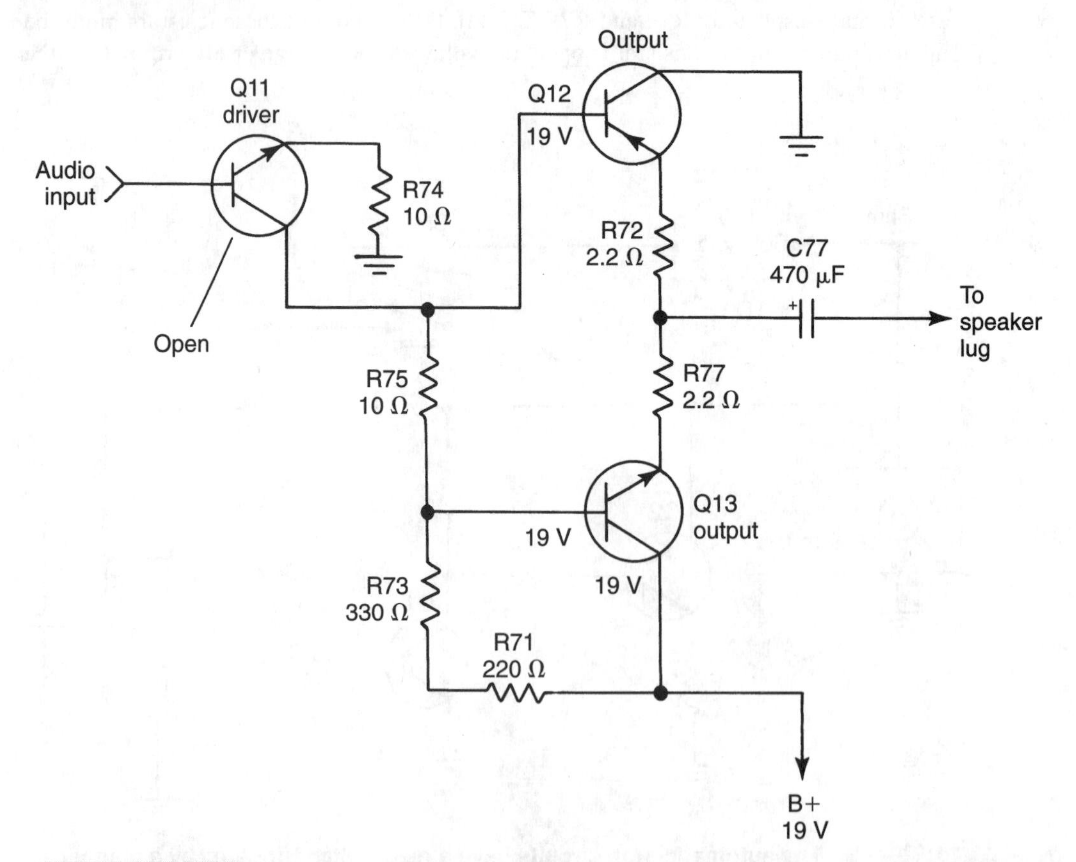

FIGURE 20-24 An open driver transistor might apply excess voltage on the push-pull power-output transistor with a dead audio symptom.

transistors, except those that are tied to ground. If a schematic is not handy, compare all voltage and resistance measurements with the normal audio channel.

After replacing the defective transistors, take comparable resistance readings on the collector, base, and emitter terminals to ground and compare them with the good channel. Be sure that all insulators are replaced under the power transistors. Take accurate voltage measurements on each terminal after the audio channel is working and compare the readings.

DEAD RIGHT CHANNEL

Proceed directly to the dead right-channel output circuits and take crucial voltage measurements on the transistors or ICs. Check the supply voltage terminal pin (V_{CC}) for the correct voltage. Measure the supply voltage at the metal body of the insulated transistors. Often, one output collector terminal is at ground and is not insulated away from heatsink. Be careful not to short the mounting screw to the B+ of the metal collector terminal (Fig. 20-25).

If the supply voltage is low at the power-output transistor pin, suspect that an output IC is leaky or a power supply is defective. Usually, a shorted IC will blow the fuse or cause the diodes to open or become leaky. Be sure that you are on the right channel when taking voltage and resistance tests. Remove supply pin 6 from the PC board to determine if IC1 is leaky or if the low supply voltage is in the power supply (Fig. 20-26). Most output transistors and ICs can be replaced with universal replacements.

INTERMITTENT LEFT CHANNEL

Intermittent audio is the most difficult problem to solve in the audio circuits. You might wait until the left channel quits or try to self-induce it to make the audio channels act up. If the left channel cuts out after a few minutes and stays out, try to isolate the intermittent sound ahead or behind the volume control. By isolating the front- and rear-end circuits, you can save a lot of service time. Take crucial voltage measurements on the transistors

FIGURE 20-25 Small audio-output transistors are bolted to a chassis heatsink in a J.C. Penney 683-1773 cassette deck.

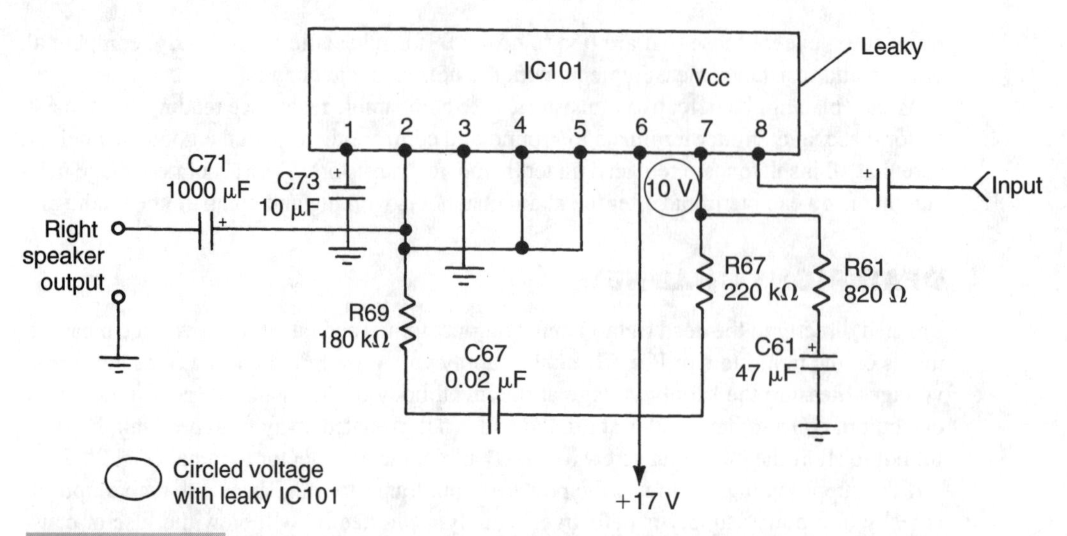

FIGURE 20-26 Often, the supply voltage at pin 6 (V_{CC}) will be low if IC1 is leaky.

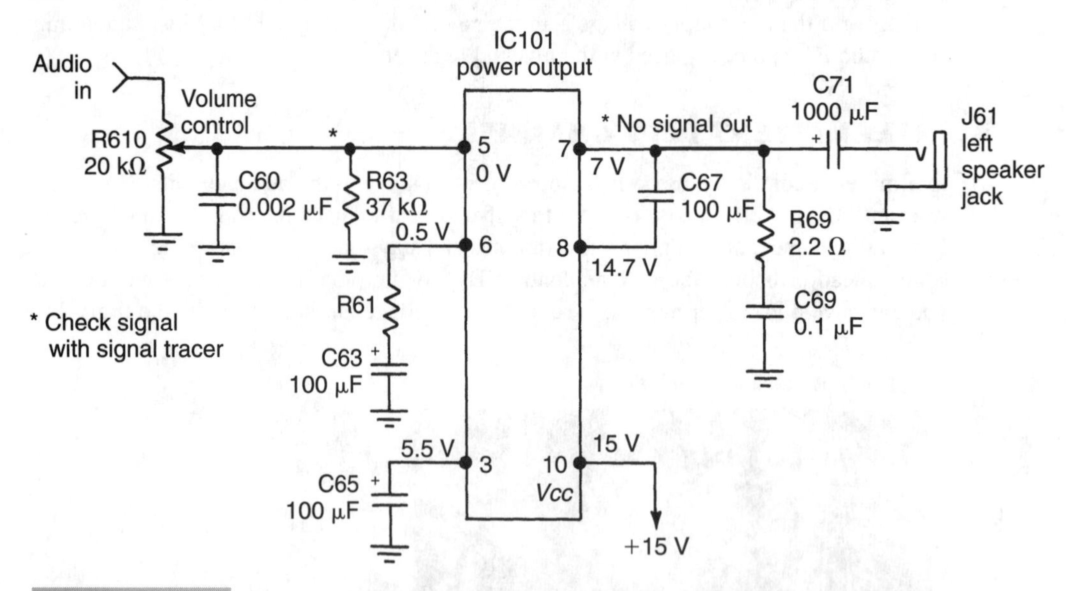

FIGURE 20-27 Check the signal at pin 5 and out at pin 7 of IC101 if a channel is intermittent.

and ICs in the dead channel. Don't be surprised if the dead channel comes to life when checking solid-state components.

Determine if the intermittent left channel has hum or frying noises if it is dead. Signal trace the signal up to the input terminal of the suspected IC with a signal tracer (Fig. 20-27). Sometimes, the voltage on all IC terminals might be quite normal if the IC is intermittent. Check all components that are tied to the IC for correct resistance and voltage. Replace the output IC if a frying noise is heard in the noisy channel with normal input audio.

WEAK RIGHT CHANNEL

Clean the R/P tape head. Isolate the right-channel circuits from the left. Determine if the weak channel is before the volume control. Signal trace the weak signal from the tape head to the preamp and AF circuits in the right channel. If the signal is very weak, stop, and take voltage and resistance measurements. If the signal is stronger at the emitter terminal of Q20 than at the base, and very little voltage is at the collector, suspect that a transistor is leaky (Fig. 20-28).

When testing Q20 and Q21 in this circuit, both showed leakage. The collector terminal voltage was high (17.7 V), compared to the normal schematic voltage of 10.5 V. The emitter voltage on Q20 was high at 2.58 instead of 1.9 V. The resistance measurements from the base terminal of Q20 to ground was low at 870 Ω. Both transistors were removed from the circuit and found leaky. Remember, with directly coupled transistor circuits, a voltage change on one transistor will make the voltage read different on another transistor.

The audio can be signal traced from the R/P head back to the preamp, A, and volume-control circuits. If the signal is fairly normal going into the left input terminal of IC5 and very weak at pin 3, suspect that an IC is defective. Measure the supply voltage at pin 9. All voltages can be compared to the normal right preamp IC (Fig. 20-29).

Check all components that are tied to the IC terminals in the right channel. If the voltages are normal, with input signal at pin 1, and very weak signal at pin 3, replace the IC. Do not

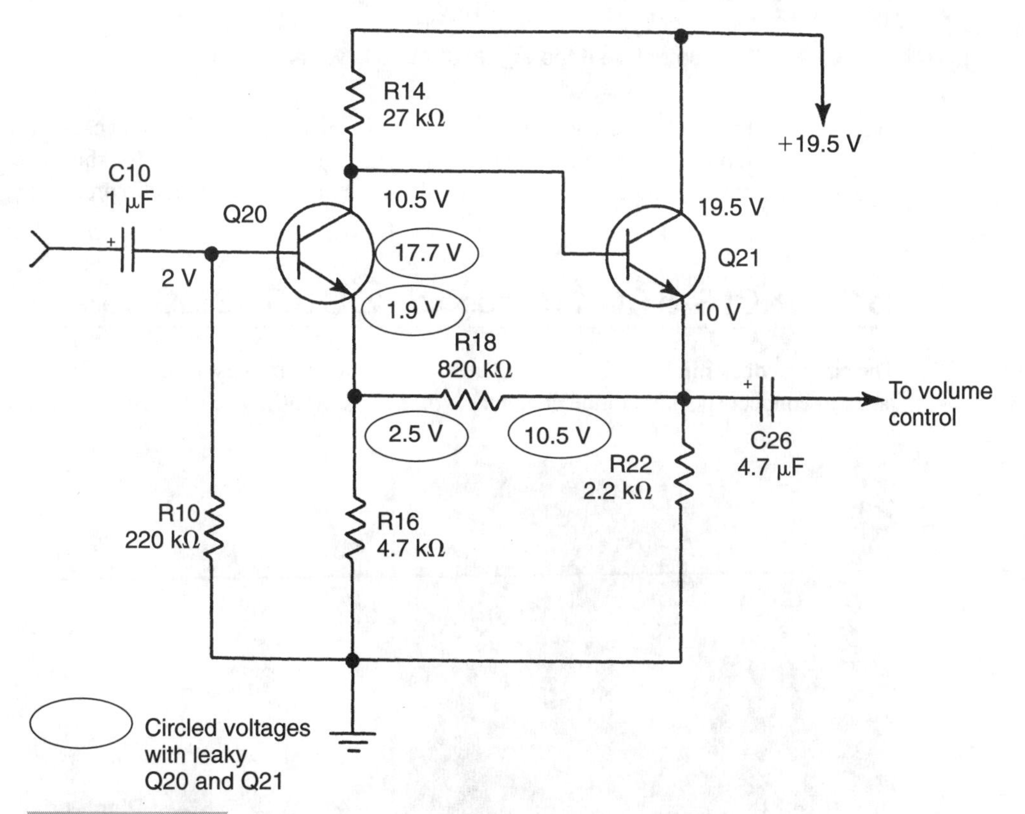

FIGURE 20-28 **Weak signals might be caused by leaky directly coupled transistors in the AF circuits.**

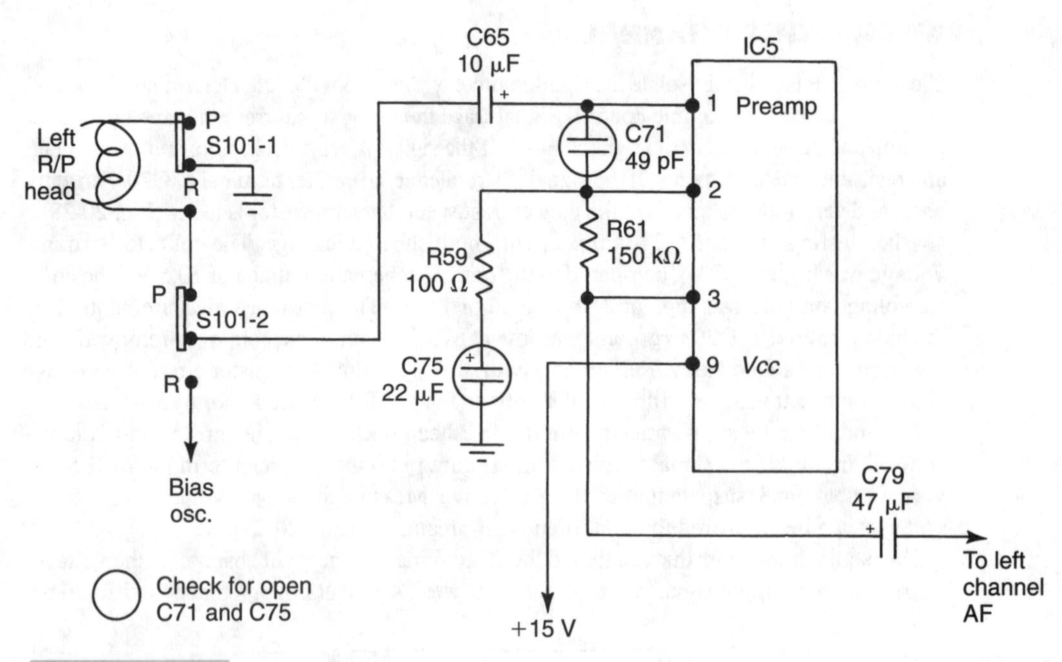

FIGURE 20-29 IC5 is defective if the signal at pin 3 is weak.

overlook small electrolytic capacitors tied to pin 2. An open C71 or C75 might cause a weak output signal at pin 3. Often, a resistance measurement is made to check for shorted components, but open capacitors and resistors might cause the same weak audio problem.

Speakers and Speaker Connections

The cassette deck might have speaker plates, pushbutton terminals, jacks, and RCA phono jacks to connect speakers to the channel outputs (Fig. 20-30). The early models had mul-

FIGURE 20-30 Pinch-type speaker connections are on the back of this Realistic 31-1995 model.

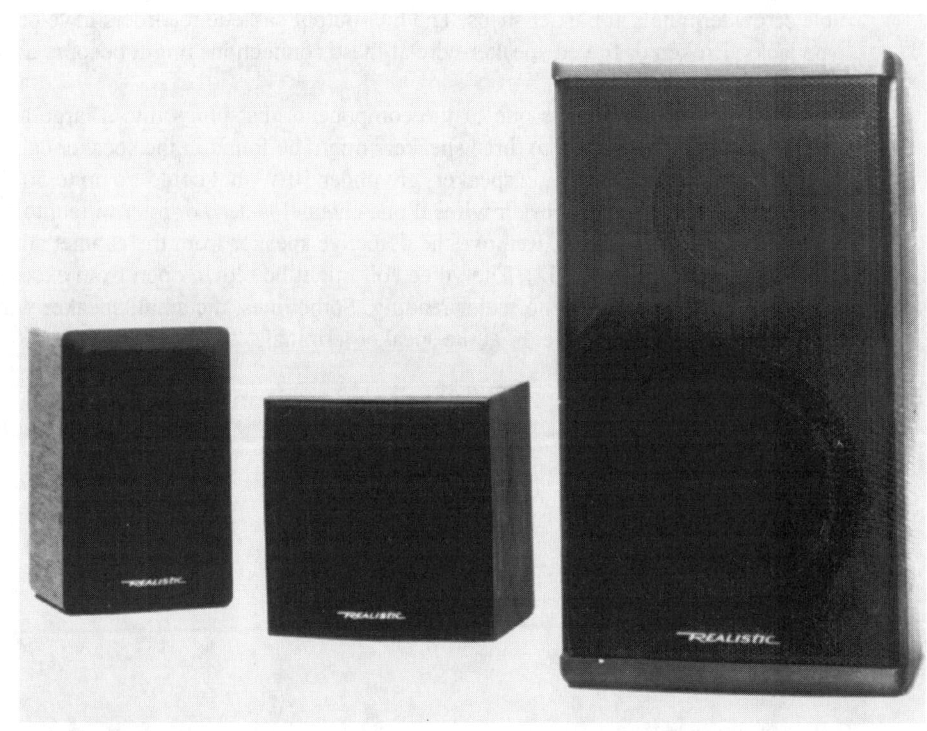

FIGURE 20-31 These outside shelf-type speakers, which are connected to the small cassette deck, can handle up to 50 W.

FIGURE 20-32 After isolating the defective speaker, check the voice-coil terminals with the R×1 position on an ohmmeter for the open winding.

tiple screw terminals or barrier strips. The line-output cassette recorders have headphone-type jacks. Broken or frayed speaker wire at these connections might become dead, intermittent, and noisy.

Unless the cassette deck is one of the components that plugs into a large amp, most speakers are shelf-types. Up to three speakers might be found in the speaker cabinet (Fig. 20-31). Most of these extension speakers are under 10 W, but could go up to 50 W.

Interchange the channel speaker wires if one channel is dead or intermittent to eliminate a possible defective speaker. Remove the defective speaker from the cabinet and check it with an ohmmeter (Fig. 20-32). The voice coil might be blown open from excessive volume, which would result in no meter reading. Sometimes, the small speaker wire might break if it enters at the cabinet or at the speaker terminal.

21

VCR REPAIRS YOU CAN MAKE

CONTENTS AT A GLANCE

Although you need VCR service technician to make crucial VCR repairs, you might save yourself some money by performing many of the common repairs. The minimum VCR service charge at most service centers is around $69.95. However, major service repairs, such as head replacement, might cost up to $250.00. The precision VCR is a very expensive machine, but you can keep those repair costs down by doing routine maintenance yourself (Fig. 21-1).

Simple repairs, such as head cleanings, demagnetizing, and checking for cracked or slipping belts can be made by the novice or beginner. A build-up of tape oxide might in time damage several components. Minor service problems can be made by almost anyone who can pick up a screwdriver or a pair of pliers. Preventive maintenance might eliminate costly repairs and damaged recordings. Knowing how your machine operates and how to check the VCR when it does not perform can save money.

The Video Cassette

Remember, the VHS and Beta cassettes will not interchange (Fig. 21-2). The Beta cassette is small in size, compared to the VHS. Actually, the whole Beta cabinet is smaller and lighter than the VHS machine. Today, even the VHS recorder is much smaller and lighter than the first VHS models that were introduced for home use.

The video cassette has a take-up and supply reel inside the cassette plastic container. These reels fit down over the respective spindle assemblies inside the VCR. Load the cassette with the label up. Insert the hinged door into the recorder. Be sure to rewind each cassette after playing or recording before you store it.

The cassette should be stored vertically in a cabinet. Place the cassette in a dust-proof box or container. Try to avoid excessive moisture. Do not leave the tapes close to a humidifier and expect to use them right away. Keep the cassettes away from large stereo speakers and

FIGURE 21-1 One of the first RCA videotape recorders to hit the market.

FIGURE 21-2 **The VHS cassette is a little larger than a Beta cassette.**

do not leave them on top of the machine. If the video recorder is mounted on top of the TV, remove the cassette each time and store it. A TV with strong stereo magnetic speakers might wipe out your favorite recording. Be careful when demagnetizing the VCR tape head and keep the demagnetizing tool away from the various cassette recordings.

When purchasing a new commercial recording, let it warm up to room temperature before attempting to play. Most VCR machines have a dew indicator, which might prevent playing or recording. In some models, if the machine does not operate for 10 minutes, the recording might be out of sync until the machine warms up. In the meantime, you have turned every control on both VCR and TV.

A defective tape might jam the VCR mechanism and result in a very expensive repair. Do not play a cassette that has a broken case or hinged door. Like the small cassette recorders, the cassette might have a dry plastic bearing that would slow down when recording or playing (Fig. 21-3). If a liquid is spilled inside the tape area, the cassette might slip, slow, and place liquid on tape heads and spindles; this would require a good

FIGURE 21-3 **Inspect the cassette for broken or cracked body parts.**

clean up before damage occurs to the recorder. Do not try to repair a broken tape with a piece of adhesive tape because this action might damage the video heads.

Pulled or unraveled tape could occur if the take-up reel stops or slows down, like any cassette recorder. The tangled tape might get inside the various spindles and gears, and require expensive repairs. The top cover must be removed from the front-loading VCR if the cassette will not pop up or cannot be pulled out after tape problems. Often, the top-loaded cassette can be removed by removing the top-loaded front panel. If the top-loading mechanism will not release, remove the top-panel cover of the VCR. Remove all excess tape and clean the heads and spindles with head cleaner and accessories. Always choose a good cassette for recording. Try another cassette before suspecting trouble inside the VCR.

Tools Needed

Only a few tools are needed for basic VCR repair. You might have them in your workshop or service bench. If not, choose small flathead and Phillips screwdrivers to remove the top and bottom covers of the VCR. Use long-nose and side-cutter pliers to remove and replace the soldered components. The low-priced 35-W pencil soldering iron is ideal to solder various electronic components. Pick a good liquid cleaning cartridge, head-cleaning kit, and cleaning spray. A pocket VOM or DMM is handy for voltage and resistance measurements.

How to Operate the Machine

The early VCRs have top cassette loading, but most of the present recorders have front-loading facilities (Fig. 21-4). Simply insert the cassette and push down on the top of the

FIGURE 21-4 Most early VCRs loaded from the top. Here, a cleaning cassette is inserted to clean the tape heads.

FIGURE 21-5 Today, most VCRs are loaded from the front.

loading platform in the top-loading units. The tape will automatically rotate when Play or Record buttons are pressed.

With front-loading machines, insert the cassette with the label face up. Push the cassette with your fingers until it is automatically retracted and loaded. In some recorders, a cassette indicator will light and indicate that the cassette is loaded properly. Check the cassette if it will not load properly.

To unload the cassette, push Stop. In top-loading machines, hit Eject and the platform will pop up. In front-loading machines, press Eject to open the loading slot. Remove the cassette (Fig. 21-5).

To Prevent Accidental Erasure

Every video cassette comes with an erasure-prevention tab (Fig. 21-6). When the tab is removed, the cassette will not record. To re-record such a tape, just place a piece of tape over the gap. Sometimes the VCR will not record and the machine will not operate. First, check the opening at the back of the cassette. If the tab is removed, the cassette will not record. In many cases, small cassette players and recorders have been brought into the shop for repair with only the tab removed from the back of the cassette.

Recording

Turn on the VCR power switch and let the machine warm up for 5 to 10 minutes. Select the channel that you wish to record with the VCR channel selector buttons. Rotate the

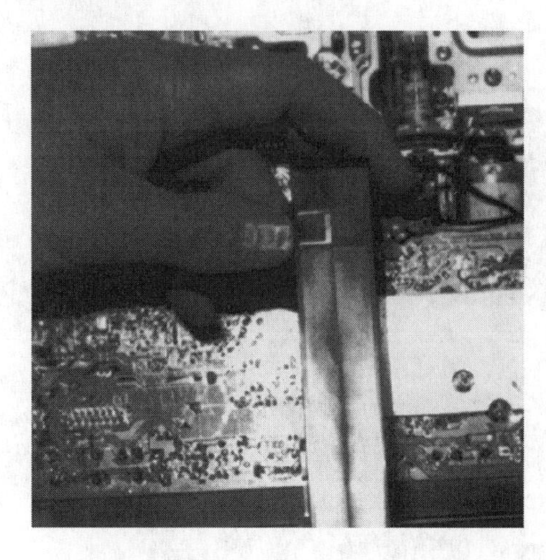

FIGURE 21-6 Remove the erase-prevention tab at the edge of the cassette to prevent erasure.

channel selector knob of the older models. Remember, the TV does not have to be turned on for station recording. However, you might want to see the channel you want to record. If so, set the video TV switch to the video position, turn on the TV. Select channel 3 or 4: whichever channel has the weakest TV station so that it won't interfere with the VCR playback. The video channel switch is located in the rear of the recorder (Fig. 21-7).

Now load the cassette. Set the tape speed switch to the desired tape speed, SP, LP, or EP. The SP setting is best for music programs, and lengthy movies and sport programs are best recorded with the LP or EP switch positions. The LP/EP recordings can only be played on VHS-SP recorders.

Press Record to begin recording. To keep commercials out of the recording, press Pause to stop the recording. Press Pause again to resume recording. Here, the remote control comes in handy. Of course, if the recording is being made while you are gone or watching another TV channel, the commercials will be included in the program recording.

FIGURE 21-7 Play the recording or cassette on channel 3 or 4 of the TV.

Press Stop to stop the recording. In some models with preset time recordings, the recorder will automatically stop. All machines automatically stop when the end of cassette is reached. In fact, some machines will automatically rewind at the end of the tape. Press Rewind search button to rewind the tape.

RECORDING ONE PROGRAM WHILE WATCHING ANOTHER

It always seems that the best TV programs are pitted against one another. Most VCRs will record your favorite program while you watch another. Select the VCR channel that you want to record and start recording. In manually tuned TVs, be sure that the program is tuned in properly. In electronically tuned receivers, the stations are tuned in automatically with the push of a button. Set the video/TV switch to the TV position and select the TV channel you want to watch, using the TV channel selector. If you want to check the picture during recording, set the video/TV switch to the video position. Then, set the TV channel selector to the video channel (3 or 4).

Playing

Turn the power switch on and let the player warm up for 5 or 10 minutes if the room is cold or damp. Sometimes, if the recorder is damp and cold, the recording might be jumbled until the recorder warms up. Load a prerecorded tape or play the cassette to check out the cassette you have just recorded. Set the TV channel to 3 or 4 (whichever channel that the video switch set in at the back of the recorder). Press Play. If noise or wavy lines appear, check the tracking-adjustment knob. This should be left in center position while recording and playing modes. Now press Stop. Press Rewind to rewind the tape. Always rewind the tape after it reaches the end. Store cassettes in cassette containers or in a cabinet for safekeeping.

Features and Functions

A good recording can be made when you know how and why each function switch works on the video recorder. Many times the VCR is brought in for repair when the operator did not know how to correctly operate the VCR. Go over the Record and Play modes several times. Read the VCR instruction book thoroughly before attempting to make a recording. These machines are not complicated; most children can operate them with ease. Here are the main features and functions of a Mitsubishi HS-328 UR model, to illustrate what each feature does:

- *Cassette loading slot* Cassette tape is inserted into this slot for loading (the recorder can be loaded manually or the cassette can be pulled in on some front-loaded machines).
- *Video/TV switch* When set to the TV position, the TV will receive off-the-air programs normally. When turned to video position, off-the-air programs selected by the VCR's built-in tuners are viewed on the TV. Set the video/TV switch to the TV position when recording one program while watching another.

- *Power switch* Press this switch to turn on power to the recorder.
- *Eject button* Press to remove the cassette.
- *Remote-control sensor* This is only found on recorders with remote-control operation. The sensor receives the infrared light from the remote-control transmitter.
- *Record button* Press to begin recording, both video and audio.
- *Start-time button* Press to select the start time for one-touch recording (OTR).
- *Record-time button* Press to record for 30-minute time intervals up to four hours.
- *Operation-mode display* Indicates operational mode.
- *Counter/timer display* This digital display covers operations in the counter/timer and program modes.
- *Channel indicator* The number corresponds to the selected channel.
- *Rewind/Review search button* Press to rapidly rewind the tape or to play the tape at high speed in the reverse direction.
- *Fast-forward search button* Press to rapidly wind the tape in playback or to play at high speed in the forward direction to search for a favorite recording.
- *Play button* Press to play a previously recorded tape.
- *Stop button* Press to stop all tape functions.
- *Pause button* When pressed during recording, it stops the movement of the tape. When pressed during play, a still picture will be seen, usually without sound. Press it again to continue tape movement.
- *Channel selector* Selects the correct TV channel for recording.
- *Counter memory button* When pressed, the tape will fast rewind until the counter display indicates "0000" and will automatically stop at this point.
- *Counter reset button* Press to reset the counter to zero.
- *Counter/timer button* Selects either the counter or present time mode of the counter/timer display.

How to Connect the VCR to the TV

You might find 300- to 75-Ω antenna connections on the back side of the video recorder. Often, the round cable (75 Ω) is found at the VHS in and out connections. If your system uses round cable (75 Ω), plug it directly into the VHS antenna (in) socket. The output lead (comes with the recorder) is 75 Ω. Only one round cable might be used for VHS and cable hookup (Fig. 21-8).

With a flat ribbon cable (300 Ω) from the antenna, a 75-Ω adapter or matching transformer must be placed on the antenna cable for a perfect match. A push-on VHS antenna adapter connects the 300-Ω flat cable to a 75-Ω (Fig. 21-9). You might find a 300-Ω UHF and VHS in terminal at the older VHS recorders. Also, a small push-on matching transformer might be supplied with the recorder.

When both UHF and VHF stations are located in your system, they can be connected to the VCR with a matching transformer (Fig. 21-10). The signals from both VHF and UHF are fed from the booster system into a UHF/VHF splitter transformer. Here, only the VHF stations are recorded. The VHF signal is fed to the VCR through a matching push-on antenna terminal connection to the VHF in 75-Ω terminal. The 300-Ω UHF lead connects di-

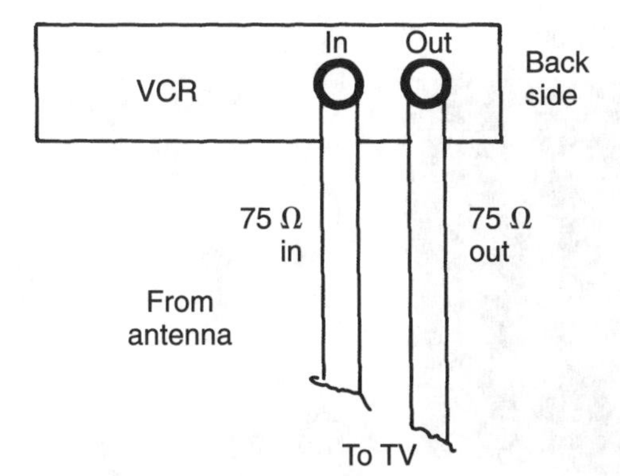

FIGURE 21-8 **Remove the round antenna lead from the TV and connect it to the 75-Ω VHF-input connections.**

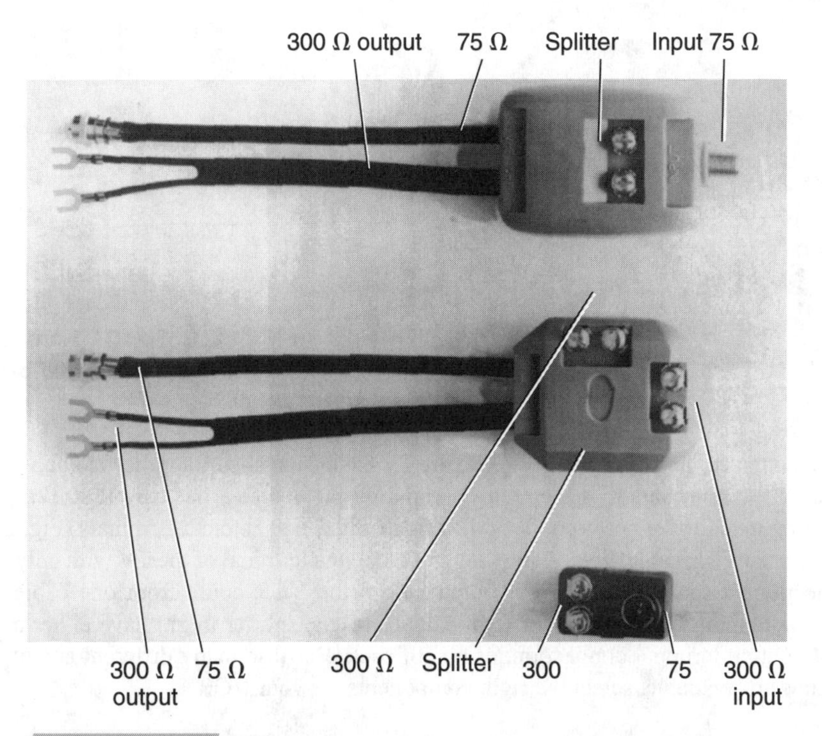

FIGURE 21-9 **Many different fittings can be connected to the TV and VCR. Use VHF/UHF splitter A for a single 75-Ω input cable.**

rectly to the UHF terminal of the TV. If you want to record both UHF and VHF, the UHF signal from the splitter is connected to the in connection of the VCR, which is 300 Ω. Then, connect the 300-Ω out terminal to the UHF terminal of the TV.

The UHF and VHF antenna might be connected to the recorder (Fig. 21-11). When only one antenna lead or cable hookup is used, connect it, as in drawing B. If only VHF is used in a given area, connect the antenna cable, as in drawing C. All antenna input connections can be made with the correct splitter, matching transformer, and lead-in wire.

FIGURE 21-10 Here, a VHF/UHF splitter is tied between the VCR and TV.

The output connections of the VCR to the TV can be matched for either 75- or 300-Ω receiver antenna terminals. If the receiver-input terminals use a flat wire (300 Ω), place a matching transformer between the VCR cable (75 Ω) and antenna terminals (Fig. 21-12). Connect the VCR cable directly into the 75-Ω antenna terminal of the TV with only a VHF connection. If the VHF and UHF outputs from the VCR come from one cable, use a VHF/UHF splitter at the receiver terminals. Here, the splitter might have either a 75- or 300-Ω output terminal connection, or one of each. Because many different antenna connections are available, select the right components for your VCR.

Head Cleaning

The dirty tape head might cause a loss of picture and sound. Excessive noise lines in the picture might be caused by a dirty tape head. If the tape head gap is closed with tape oxide, no picture or a partial loss of picture might be present. If the TV picture is clear, but during playback the picture is not clear, distorted, or snowy, expect that the tape head is dirty. The video heads might have a deposit of dirt or oxide built up, which would produce a poor picture.

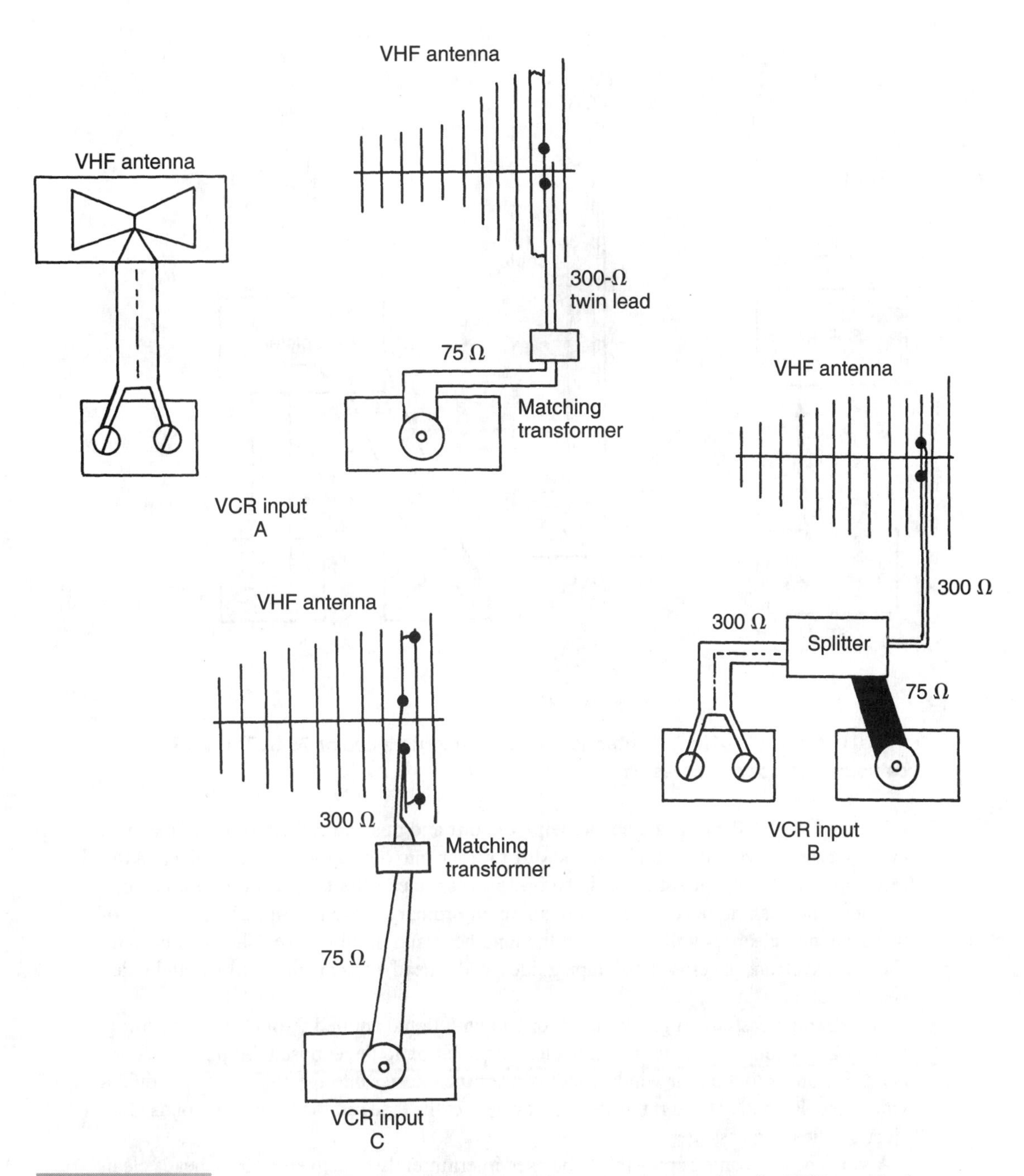

FIGURE 21-11 Both the UHF and VHF single cables are connected to the VCR terminals.

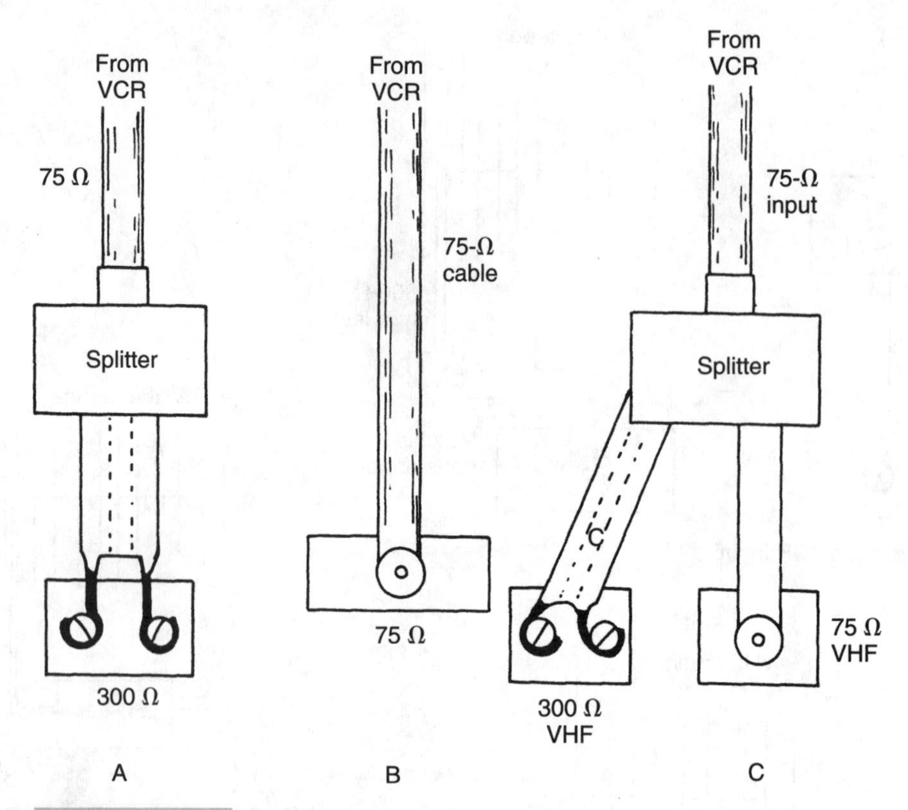

FIGURE 21-12 **The TV antenna terminals might be 300 or 75 Ω. The VCR cable can be connected as shown in A, B, or C.**

To keep the VCR in tip-top shape, wipe the dirt and dust from the top of the machine every week. It's very easy for dust and dirt to filter into the top-loading machines. After recording or playing cassettes back 10 or 15 times, clean the tape head with a cassette cleaner. The cassette head cleaner looks like an ordinary cassette (Fig. 21-13). Some of these cassette cleaners will only clean the tape head and nothing else. Choose a cassette cleaner that cleans the erase head, tape guides, audio head, capstan, pinch roller, and video head, all at once.

The cleaning cassette might work dry or wet with liquid applied through a slot opening. Simply apply four or five drops of the cleaning solution to the exposed felt pad. Also, apply a few drops to the four window slots. Insert the cleaner into the VCR as you would a video tape. Push Play to start the cleaning cycle. Let the tape run for 15 to 20 seconds, then eject the cleaning cassette.

A video head-cleaning spray might be used intermittently to help keep those heads clean (Fig. 21-14). Hold the spray can 4" to 6" from the head or other parts to clean. The force of the spray might clean dirt and dust from the tape heads or guide assembly. Use the extension tube for tight spots. In most cases, the top cover must be removed to get at the various heads and tape guides.

To clean the tape heads manually, the top cover must be removed from the VCR. The plastic front cover of the top-loading machine is held in place with several small bolts on

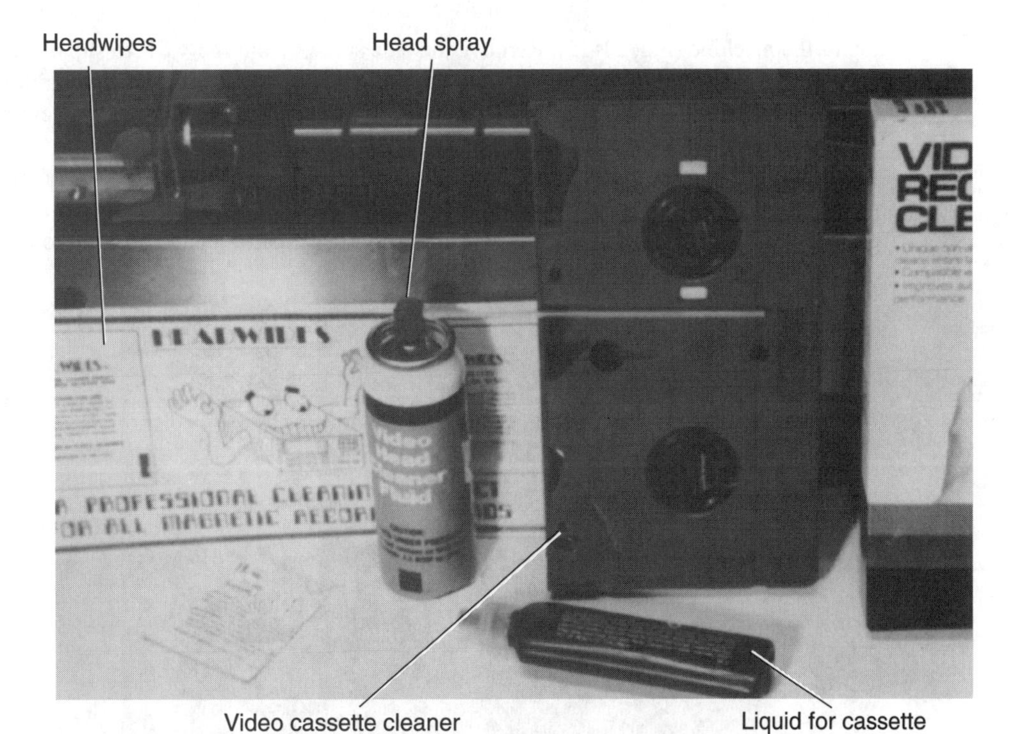

FIGURE 21-13 The video-cassette cleaner might look like an ordinary video cassette. Head-cleaning spray can be used to keep oxide off of the tape heads.

FIGURE 21-14 Head-cleaning spray can be used to clean dust from the various tape heads.

top of the machine (Fig. 21-15). Remove the side screws from the VCR to slip off the top cover of the front-loading machine. You might have to remove the top-loading platform or protective shield to get at the tape heads (Fig. 21-16). Usually, removing a few metal screws will release the metal shield.

Of course, the tape-head cleaning cassette should be followed up with a manual cleaning after 48 or 60 hours operation (at least once every three months for the average viewer). A manual cleaning consists of applying a cleaning pad and fluid to the var-

FIGURE 21-15 The top cover must be removed to access the tape heads in a top-loading recorder.

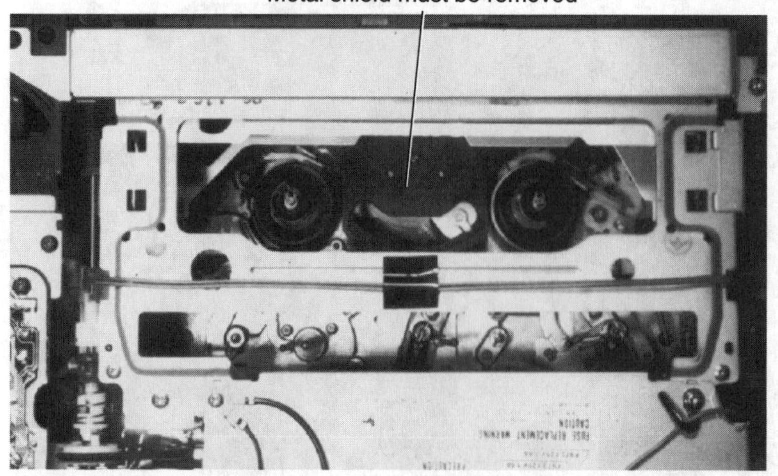

FIGURE 21-16 In this model, the metal shield bracket must be removed to manually clean the tape heads and guide assemblies.

FIGURE 21-17 **Keep your fingers off of the tape-head face area to prevent oil or moisture from being applied to the tape.**

ious tape heads. Do not grasp the video tape head on the front surfaces with your fingers (Fig. 21-17). Oil or residue on the fingers might be transmitted to the tape head and tape.

Lightly apply cleaning fluid around the large video head instead of using an up and down motion. You might damage the tape head gaps with too much pressure. Hold the head with your fingers on top of the video head and wipe the area with your other hand. Look for a packing of brown oxide at the tape head gaps. Wipe all the other heads and moving parts with a pad and cleaning fluid. A professional cleaning product for all magnetic recording heads, called Headwipes contains a saturated pad. Simply tear it open and pull out the wet pad, then clean the heads or tape-guide assemblies (Fig. 21-18).

Before Calling for Professional Service

If the VCR will not light, check the power cord, outlet, and on/off switch. On some models, the Play/Record switch must be switched off. A shaking picture or no color might be caused by an improper setting of the TV or of the recorder channel selector. Readjust the fine tuning. Be sure that the channel (3 or 4) is set correctly.

Suspect a defective cassette or that the tab is removed at the back of the cassette if the VCR will not record. Try another cassette. Be sure that the video/TV switch is set to the TV position. Check the VCR channel selector if the channel is active channel, but the TV isn't being recorded.

Notice if the Play/Record switch is on if Play does not operate. Switch it off. Check to see if the tape is at the end if the fast forward doesn't work. Likewise, a no-rewinding symptom could be caused if the tape is at the beginning. Readjust the tracking control if the picture is noisy or poor. Clean the VCR tape heads after the recorder has been used for

FIGURE 21-18 **Clean all tape heads and guide assemblies with commercial Headwipes or regular cleaning pads.**

a long time and the pictures are not clear after recording. Besides these minor adjustments, several visual and continuity measurements can be made before you cart the VCR off to professional service.

DEMAGNETIZING THE TAPE HEAD

The magnetized tape head might cause a loss of color, weak and erratic pictures, noisy sound, and flagging symptoms. Flagging occurs at the top or bottom of the picture, bending the picture. You might notice the picture bending when a straight object, such as a house outline or building, is shown. Flagging and glitches might result from a dirty or magnetized head, or from a defective tape.

Pick up a regular VCR head demagnetizer for correct tape head cleaning. Do not use the 8-track or cassette demagnetizer you have on hand. You might ruin the VCR tape heads. This same VCR demagnetizer can be used on all video, sound, and erase heads in the recorder (Fig. 21-19). Demagnetize the tape head each time you remove the cover for good head cleaning.

Do not apply power to the VCR when cleaning or demagnetizing the tape head. Turn on the demagnetizing tool and bring it directly to each tape head. Keep the end of tool from touching the head surface. Slowly rotate the video head drum. After demagnetizing all heads, slowly pull the demagnetizer tool away from VCR and shut off the power. If the tool is shut off close to the machine, you might magnetize several metal components on the VCR. This same method is used while demagnetizing a color TV screen.

SEVERAL CASSETTES WILL NOT RECORD

Check the knock-out tab at the rear of the cassettes if the VCR will not record. Place a layer of tape over the opening. Visually inspect the cassette to see if any of it is broken. Be sure that the cassette is loaded with the label up. A defective cassette might cause slow or erratic speeds. Always purchase a good brand of tape for the best recordings.

CHECK THE BELTS

Improper tape speed or no tape action might be caused by a broken or cracked belt. Inspect the belt for oil spots. A loose or stretched belt might produce slow speeds (Fig. 21-20). Clean all belts and pulleys with rubbing alcohol and a cloth. Slipping at the transport motor pulley might be cured with a coat of liquid rosin on the pulley area. Replace cracked, loose, or broken belts.

VISUAL INSPECTION

Check the pressure roller, reels, and turntables for easy rotation. A sticky or worn pulley might cause improper speeds. Inspect the plastic tape guides for frozen or dry bearings. These plastic spindles should spin freely. A drop of light oil at the top of the plastic bearing might help. Wipe up all excess oil from the tape-drive areas. Replace any broken or cracked components.

ERRATIC OPERATION

Check and clean the on/off switch when the machine erratically turns on or off. Notice if the pilot light is on. Inspect the plugs and contacts when a board or when portions of the PC

FIGURE 21-19 **Besides the large video head, demagnetize the erase, sound, and separate function heads.**

FIGURE 21-20 Be sure that the drive and capstan belts are not oily, loose, or broken.

board can be moved. Sometimes spraying contact cleaner inside the plug connection helps. Check for a broken wire or for a bad solder connection at the motor terminals if the motor movement is erratic. Burned or dirty relay contacts might cause erratic motor operation.

BURNED COMPONENTS

Overheated resistors, transistors, and ICs might indicate that a defective component is nearby. Take accurate voltage and resistance measurements on the IC and transistors with the DMM. A leaky IC or transistor might cause an overheated resistor to smoke. Check the schematic for correct resistor replacement after replacing the leaky component.

TAPE HEAD CONTINUITY

Check each tape head for low continuity or for an open head winding. A jumbled recording might result if the erase head isn't working properly. Check the continuity of the erase head. No sound might be caused by an open audio tape head. An open video tape head might cause the no-picture symptom when playing or recording.

NO TAPE ACTION

Inspect the belts and the motor if the tape doesn't rotate, but the tape indicator light still works. Listen for the motor rotation. Check the belt for rotation when the motor is rotating. If the motor does not rotate, check for faulty microswitches, relays, or a faulty motor. Measure the voltage applied to the transport motor (Fig. 21-21). Check for faulty key switches and Play and Record buttons.

KEEPS BLOWING FUSES

Suspect that diodes are shorted in the power supply if the fuse keeps opening after replacement. Look for a bridge rectifier or for four separate diodes in the power supply. Each one can be checked with the diode test of the DMM. Any low-resistance measurement in both directions indicates a shorted or leaky diode. Be sure that you replace the diode with another of the same amperage because some of these diodes vary from 3 to 10 A. Momentary transistor or IC flash-over leakage might cause the fuse to open. If the fuse holds after replacement, do not worry what caused it. Improper dc voltage applied to the various sections might result from a leaky or open voltage regulator transistor (Fig. 21-22).

FIGURE 21-21 A defective transport drive motor or pressure roller might prevent normal tape motion. Check the voltage at the transport motor if the motor does not rotate.

FIGURE 21-22 Suspect that a voltage-regulator power transistor is leaky if improper voltages are at the various circuits.

MECHANICAL PROBLEMS

Improper Play or Record functions might be caused with mechanical failure. Broken key and button functions might prevent manual loading or Play/Record functions.

Mechanical levers or parts might jam and prevent tape operation. If the bent levers operate switches, the Play or Record might not work. Most mechanical operations can be visually inspected to find the defective tape action.

SQUEAKY NOISES

Any dry component might cause a squeaky noise in the recorder. Remove the top and bottom covers and load the cassette. Set the recorder running and check each moving component for a squeaky noise. Dry motor bearings, turntables, plastic guides, and pulleys cause most rotating noises (Fig. 21-23). Often, a drop of oil on the bearing area solves these noise problems. Replace motors that have worn bearings.

SOUND PROBLEMS

Erratic-, distorted-, or no-sound symptoms might be caused by defective ICs in the sound circuits (Fig. 21-24). Locate the sound section and take crucial voltage and resistance measurements on the sound IC and transistors. A leaky IC might cause the problem. Check the audio tape head for excessive oxide, which could produce distortion. Take a continuity measurement of the sound tape head for a no-sound condition.

POOR PC BOARD CONNECTIONS

If the VCR records or plays intermittently, suspect poor board connections. Intermittent snowy pictures or sometimes no picture might be caused by cracked PC wiring. If the machine is intermittent, try to isolate the intermittent section with the block diagram.

FIGURE 21-23 Dry pulleys, motors, gears, or turntable bearings might produce a rotating squeaking noise. A drop of oil might cure the squeaky bearing.

FIGURE 21-24 Distorted, weak, or no sound might be caused by a defective output IC.

Many times, the wiring will expand and crack around heavy parts or metal standoffs. If the unit is dropped in shipment or accidentally knocked off the TV or table, these boards might break or crack around heavy components. Do not overlook areas around metal insulators and braces.

Check the wiring with a magnifying glass because the eye might not see a fine crack in the narrow wiring. Sometimes applying solder over the entire section might help. If the cracked area is finally located, solder bare hookup wire over each broken wire. It's better to repair the cracked board than to wait months for a possible replacement.

Take It to the Expert

After making the most obvious repairs and the recorder still does not function properly, take it to the dealer where the VCR was purchased, to a service depot, or to a local trained expert in VCR repair. Do not attempt to replace components in the crucial recording and playback sections. Leave all required adjustments to the professional who has correct tension gauges and test equipment for crucial adjustments. Do not turn the screwdriver bias or controls without correct test equipment. Whenever possible, replace all parts with the manufacturer's part number. Do not dig into the chassis unless you know what you are doing. You might cause more damage than the original repair. The VCR troubleshooting chart is found in Table 21-1.

TABLE 21-1 VCR TROUBLESHOOTING CHART

SYMPTOM	TROUBLE	REMEDY
Dead-nothing	Power ac	Check outlet. Check power plug. Check fuse and power supply.
Blown fuse	Solenoid (main)	Shorted winding. Shorted diode.

TABLE 21-1 VCR TROUBLESHOOTING CHART (CONTINUED)

SYMPTOM	TROUBLE	REMEDY
No video	Bad tape	Play another cassette.
	Poor connections	Check cables and wiring.
	TV adjustments	Check for correct settings.
	Dirty tape heads	Clean.
	Defective tape head	Look close at head. Take open tests with ohmmeter.
No audio	TV adjustments	Doublecheck TV and VCR settings.
Video fair	Poor tracking	Readjust tracking control. Is sound normal on TV?
	Leaky or open ICs	Take voltage and resistance measurements.
Poor audio recording	Tracking	Adjust tracking controls.
	Dirty tape heads	Clean audio ac head. Check bias oscillator circuits.
	Hum and buzz	Check audio cables and shields.
Poor color	Check tracking	Readjust tracking control.
	Check TV and VCR settings	Readjust.
	Dirty heads	Clean.
Erratic color	Color circuits	Press up and down on color board for loose components or connections.
	Poor socket connections	Clean and resolder connections.
Sound bars snowy pictures	Check tracking	Readjust tracking controls. Inspect video heads. Check video relay.
Noisy picture	Occurs at all speeds	Defective video head.
No video	Portion of picture clear	Check video head preamp, P/R switch, luminance circuits and rotary transformer. Check low-voltage power supply.
Wavy or picture wiggles	Tracking	Readjust tracking control.
	Dew circuits	Let VCR warm up 10 minutes.
	Dirty tape heads	Clean all heads.
	Defective cassette	Insert a good cassette.
	TV adjustments	Readjust manual TV tuner and horizontal setting.
Jittery picture	Erratic supply reel	Replace worn and dirty brake band.
Picture dropout	Cassette	Insert another cassette.
	Tracking	Adjust tracking control.
	Head	Clean all tape heads.
No recording	Input cables	Check and refasten cables and plugs.
	No record or play	Check power-supply circuits.
	Cassette	Inspect for tab out at back of cassette. Place scotch or vinyl tape over opening.
	Dirty heads	Clean all tape heads.

TABLE 21-1 VCR TROUBLESHOOTING CHART (CONTINUED)

SYMPTOM	TROUBLE	REMEDY
	Jumbled recording	Check dirty erase head. Check for open head winding with ohmmeter.
	Defective cassette	Replace with new cassette.
No playback	Cables	Check all output cables and connections.
	Channel switch	Make sure TV is on same channel (3 or 4)
	Dirty heads	Clean heads
Flickering color in playback	Video heads	Inspect video heads. Check automatic color-control circuits. Check video head preamps.
Erratic dropout	Cassette	Try another cassette.
	Dirty heads	Clean all heads.
	Tracking	Readjust tracking control.
	Tuner	Check for dirty tuner. Check tuner in TV set. Clean tuner contact.
	Tape head	Adjust tape head tension.
Tape will not load	Defective cassette	Try another cassette. Do not play broken or cracked cassette.
	Top loading	Check for binding levers or foreign material in holding area.
	Front loading	Check eject and front loading circuits. Check dc motor. Check foreign material like hair and gum wrappers in loading belt. Check for broken loading motor belt.
Cassette will not latch	Hold screw	Check for missing or loose screw.
No tape motion	Power supply	Check fuses. Check power-supply voltages.
	Power switch	Check and clean. Replace if bad or broken.
	Belts	Inspect and replace broken or cracked belts.
	Bad transport motor	Check motor continuity with DMM. Replace motor with exact part number.
	Pressure roller	Inspect roller for worn areas. Check pressure on roller and capstan. Check for broken or missing roller spring.
	Reel problems	Inspect takeup and supply reels. Are they rotating properly?
	Solenoids and relays	See if solenoids energize. Take winding continuity. Clean switching contacts.

TABLE 21-1 VCR TROUBLESHOOTING CHART (CONTINUED)

SYMPTOM	TROUBLE	REMEDY
No record or fast forward	Relays and solenoids	Check for poor contacts. Check for energizing of relay.
	Micro switches	Are they open or closed. Replace if erratic and dirty.
	Belts	Check to see if loose, cracked, or broken. Off?
	Poor tape tension	Readjust.
	Motor	Check voltage to motor. Check for motor continuity. Replace motor.
No pause	Pause switch	Check to see if pressure roller is against capstan. Check tape brake. Check linkage.
Squeaky noises	Belts	Replace old or dry belts.
	Tape tension	Readjust—too tight.
	Gears, pulleys, and turntables	Check for dry bearings.
	Guide spindles	Worn or dry-lubricate. Replace if excessively worn.
	Moving components	Notice if levers or wires are laying against moving parts.
HI FI sound	Problems on playback	Check audio heads on cylinder. Rotary transformer. Audio preamps. Audio switching relay.
Head wear	Excessive head wear	Check close up with microscope. Check with magnifying glass. Look for cracked or broken head.
Pulling or eating tape	Cassette	Replace with new one and try again.
	Mechanism	Check tape tension. Inspect turntable reels. Check reel brakes. Inspect defective pinch roller. Misaligned roller or tape guides.

PART 4

TROUBLESHOOTING AND REPAIRING COMPACT DISC PLAYERS

22

HANDLING AND CARE OF THE COMPACT DISC

CONTENTS AT A GLANCE

The compact disc looks somewhat like the 45 rpm record, except that it has a silvery surface and no grooves (Fig. 22-1). The CD has an outside diameter of 120 mm (less than 5 inches) with an inside hole of 15 mm. Just a little over one hour of music fits on a CD.

Rather than grooves and a pickup stylus, the little silver CD has a laser beam to pick up the recording. The CD starts out at an inside diameter speed of around 500 rpm and slows down to approximately 200 rpm, but the 45-rpm speed is constant. Instead of starting the recording at the outside edge, like the phonograph record, the CD disc is read from the center outward.

FIGURE 22-1 The compact disc is much smaller than the record.

Record Versus Disc

On the conventional phonograph record, the stylus (needle) rides through a groove that has been coded with varying amplitudes that correspond to the sound signal. The compact disc, however, has a track of microscopic indentations, called *pits*, rather than a groove (Fig. 22-2). These pits and the space between the pits are the encoded digital representation of the original analog audio information. The high-density information on these tracks is read by a laser pickup device that has no physical contact with the surface of the disc.

Digitizing the music signals eliminates both deterioration of the signals through the recording and playback process as well as the mechanical restrictions (physical wear). Also, with incorporating a high density and high fidelity that could not be achieved with conventional systems, it is now possible to reproduce a sound far superior to the limits attainable by analog systems.

The phonograph record contains two channels of information—one channel on each side of the groove. Because only one stylus must pick up both channels of information, a great amount of crosstalk exists between the channels. With the compact disc method of recording right and left channel information is in serial sequence. Channel separation is extremely good, which is important for accurate stereo reproduction.

The original audio signal is a smoothly changing (analog) waveform that is sampled at a 44.1-kHz rate (Fig. 22-3). No information is lost when the analog audio signal is converted to a digital pulse train—as long as the sampling rate is at least two times the highest frequency to be reproduced. The sampled audio is converted to 1s and 0s using 16 bits of resolution.

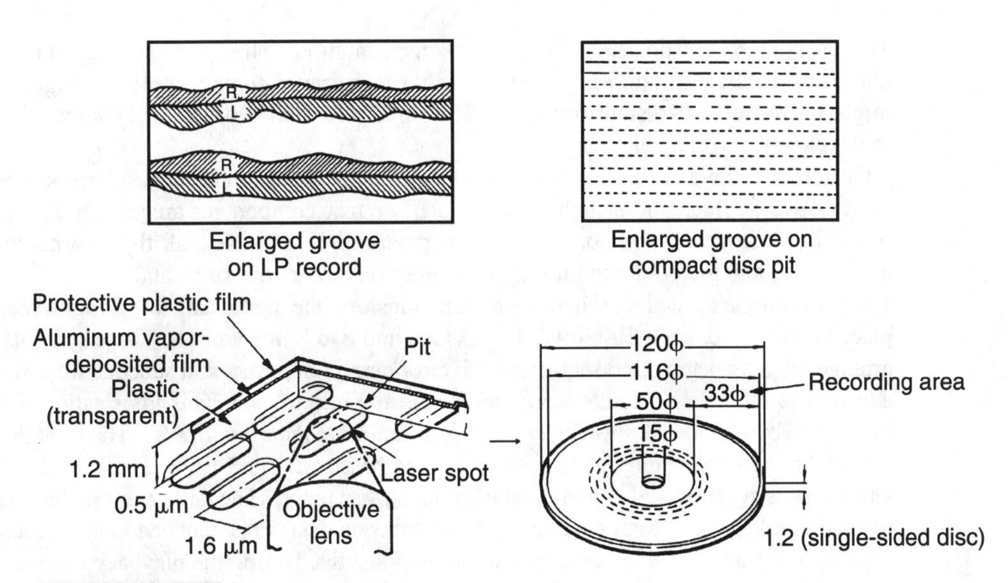

FIGURE 22-2 **The photograph record has large grooves and the compact disc has pits.**

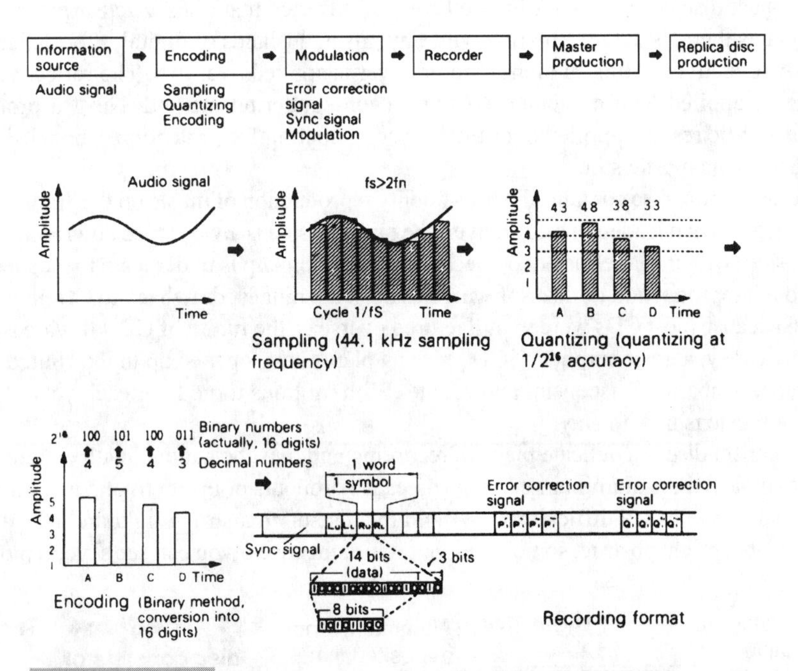

FIGURE 22-3 **The encoding and recording format of the compact disc.**

This provides 65,536 possible voltage level representations. The dynamic range of the system using 16-bit resolution is greater than 90 dB. Because some sampled voltage levels might fall between two steps of the 65,536 combinations, the voltage level is rounded off to the closest 16-bit level.

On records, the amplitude of the sound grooves where a loud sound is reproduced is large. Also, the fluctuations in the low-frequency music components are greater than those of the high-frequency components. With compact discs, however, all the information is pulse coded and incorporated into digital signals (combinations of 0s and 1s).

A very complex encoding scheme is used to transform the digital data to a form that can be placed on the disc. Each 16-bit word is divided into two 8-bit symbols. These symbols are arranged in a predetermined sequence with error correction, sync, and subcode information added. The subcode is used to store index and time information. This information is then modulated by a process known as *eight-to-fourteen-bit modulation (EFM)*. The 8-bit data is changed to 14-bit data through the use of a ROM-based IC. The EFM reduces the disc system's sensitivity to optical system tolerances in the disc player. The three merging bits added to each word (now 17 bits) contain sync and subcode data. This encoded data is recorded onto the discs as a series of small pits of varying lengths. During the playback process, the laser pickup reads the transition between the pit and the mirror—the island—not the pit itself.

Compact Disc Construction

The compact disc is composed of three layers of different materials. A clear plastic material contains the musical information with tiny pits and islands of digital information (Fig. 22-4). A reflective coating of aluminum or silver is applied over this. The reflective coating can be applied with a vacuum coating or ion-sputtering method. Next, a protective layer of acrylic resin is applied over the reflecting coating for protection. The music label is applied to the plastic side.

Hence, the shiny disc contains the true-fidelity reproduction of music on the "rainbow" reflection side. When loaded in a CD player, the rainbow side is always face down and the aluminum side (with the label) is on top because the laser pickup is underneath the loaded disc.

Although reproduction of disc software has been the biggest drawback to CD player production because most CDs were manufactured overseas, the future of CD software is much brighter. Today, a great number of CD pressing plants have opened up in the United States in Indiana, Alabama, Wisconsin, and Maine. With millions turned out each year, the true sound of music is here to stay.

The compact disc is a delicate piece of recording and must be handled with extreme care. Small pinholes in the aluminum coating can cause dropouts or errors in playing. Although these small pinholes are difficult to see, hold the disc up to a strong light and take a peak. The disc label might obstruct some holes, but return the disc if you can see several pinholes.

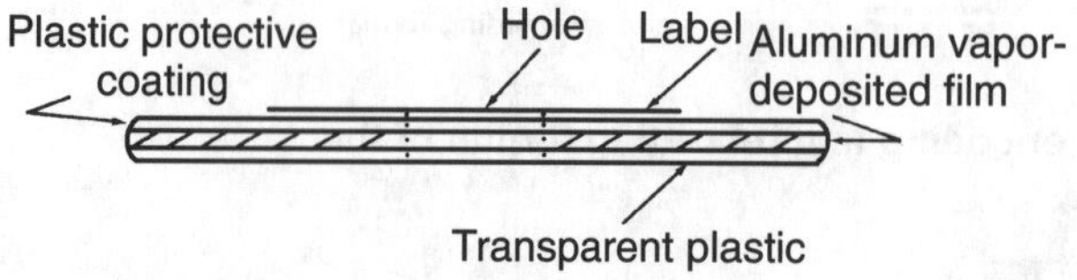

FIGURE 22-4 The CD disc consists of a transparent plastic layer, an aluminum layer, and a protective plastic coating.

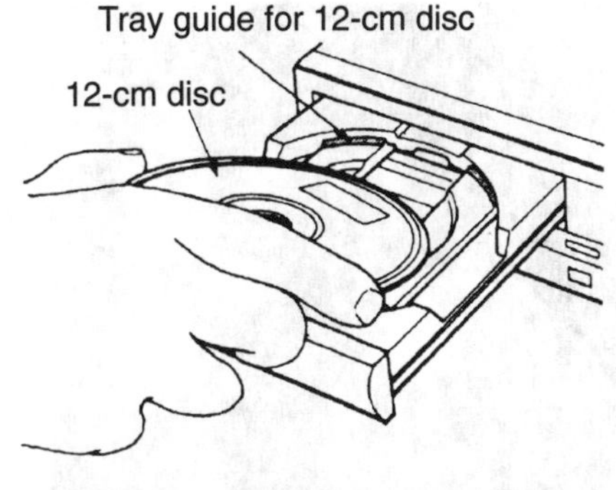

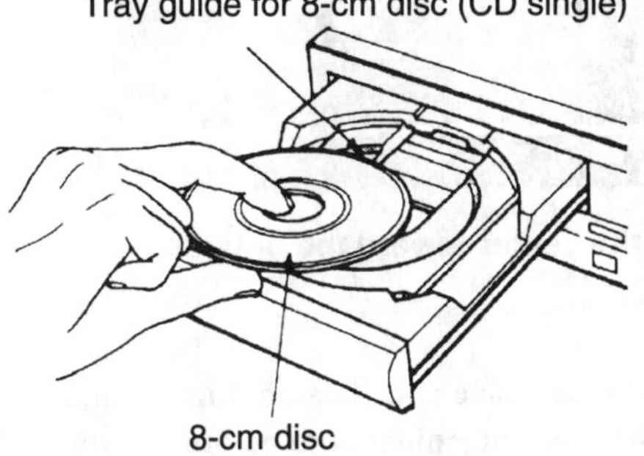

FIGURE 22-5 Load the CD only by its edges. Keep fingerprints and dust from the shiny disc area.

Handle With Care

Hold the compact disc by the edges (Fig. 22-5). Do not touch or scratch the rainbow side (opposite the label). Some players can play through a smear of fingerprints, but don't take any chances. Remember, the side of the disc with the rainbow reflection is the side that contains the audio information, so keep it clean. Do not stick tape to the label side, and don't write on it.

Keep the compact discs free of dirt and dust. To clean a record, wipe it with a circular motion. Do not try to clean the CD with this method. When fingerprints and dust adhere to the disc, wipe it with a soft cloth; start from the center out (Fig. 22-6). Excessive dust in the player can gum up the disc drive and the delicate mechanism. This prevents any scratches from covering a large area of corresponding data bits. If it is difficult to remove the smudges, wipe it with a moist cloth. Discs can become scratched the more they are wiped clean, just like a pair of eyeglasses. Excessive cleaning of dust can even help grind particles into the soft plastic. If it's not visibly dirty, leave it alone. Or you can blow off dust with a can of photo dust spray. Do not clean the disc with benzene, running alcohol, thinner, record cleaner, or antistatic agents. Like record cleaners, many different types of disc cleaners are on the market. Some might do more harm than good, so you have to be the judge and jury. Some of these units clean in a circular motion and should be avoided. Be sure that the commercial CD cleaner wipes outward on the disc surface. However, simply cleaning the disc with water and a soft cloth does a respectable job.

The CD disc must be handled and stored with care for long-lasting music reproduction. If the disc warps, the laser beam will not track properly. Avoid bending the discs. Scratches on the playing side can cause the disc to drop out and automatically shut off.

FIGURE 22-6 **Clean the CD from the center toward the outside edge (radial-type cleaning).**

Compact discs should not be exposed to excessive cold, heat, or high humidity. If a disc is left outside in the winter, let it warm up to room temperature before playing it. The disc can become brittle when it is subjected to cold. Excessive heat can warp a disc and cause a CD player to shut down. Keep discs in their plastic cases (Fig. 22-7).

If the disc is brought from a cold environment to a warm room, dew might form on the disc. Wipe off any dampness with a soft, dry cloth before inserting the disc into a player.

Store compact discs in a location protected from direct sunlight, humidity, and extreme temperatures. Many compact disc storage containers are on the market. Discs can be stored vertically or horizontally, provided they are kept in their cases. With proper care and cleaning, the compact disc should last for 10 years or longer.

WET AND DRY CLEANERS

To properly maintain compact discs and enjoy ultimate sound reproduction, use a CD-cleaning system regularly. Most makers of compact discs recommend that discs be cleaned with radial strokes—that is, from the center to the edge. A specially geared rotation cleaning mechanism of the cleaning system ensures even pressure of the moistened cleaning pad on the disc. The radial action of the cleaner removes any contaminates from the compact disc.

A dirty lens can cause the CD to skip and distort. To remedy this type of problem, a digitally encoded CD with a built-in ultra-fine brush will safely clean the optical lens of a CD player. It removes dirt, dust, and smoke residue from the lens in less than 10 seconds.

A CD cleaning kit can quickly remove fingerprints and smudges from CDs. These kits come with a fluid in a spray bottle, a lambskin wiper, a cleaning cloth and a hard brush. The automatic motorized chamois cleaning system provides true radial cleaning and it automatically stops when it is finished. The automatic cleaner can be used wet or dry with a kit of cleaning fluid. The automatic cleaner operates from four AA batteries.

A compact-disc polish system cleans and polishes compact discs, and it protects them from dust and fingerprints. These systems prevent CD skipping because they contain a special antistatic formula (see Fig. 22-28).

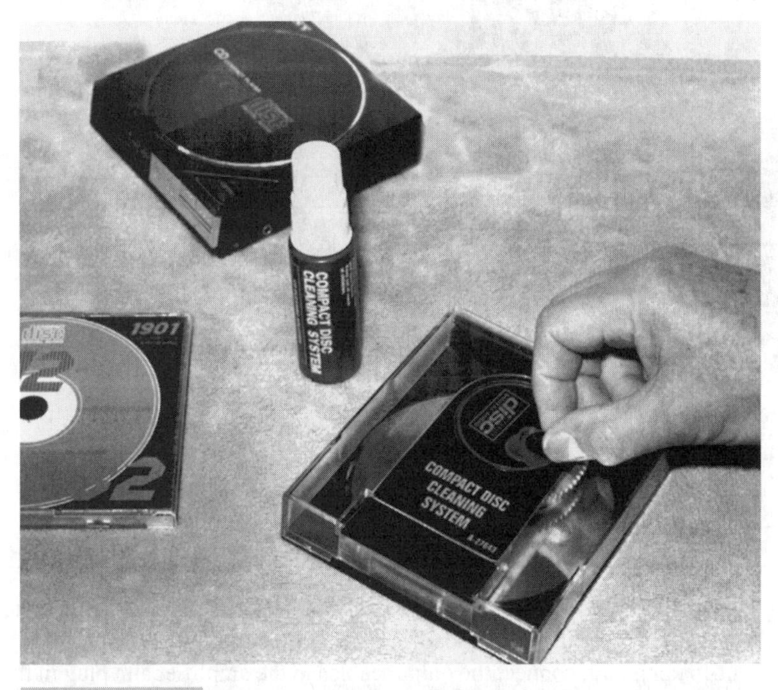

FIGURE 22-7 After cleaning and playing the disc, always place it back inside the plastic jewel box.

FIGURE 22-8 Many different CD cleaning kits are available.

FIGURE 22-9 **This keeper was found at the top loading area of a boom-box CD player.**

Loading the Disc

Although most CD players have front loading, a few models use a top-loading system. The following paragraphs explain how to load each type of player.

For a front-loading unit, connect the output cables to the amplifier and plug in the power cord. Press the Power button and be sure that the display is lighted. Push the Open/Close button and the loading drawer will slide out. Place the disc on the tray with the label up, and press the Open/Close button again to prepare the player to play.

With a top-loading unit, notice that a top-side keeper goes over the disc and rotates with disc in a portable or boom-box player (Fig. 22-9). Push down the top cover or lid until it locks into position.

The 5- or 6-CD changer can be loaded by pressing Open/Close button to push out the loading tray. Be sure that the door is completely open. Hold the disc by the edges and place it on the disc tray. Set the 8-cm discs in the center hole and 12-cm discs in the outer hole (Fig. 22-10). Load all holes in the 5- or 6-disc changer tray. Push the Open/Close button to play.

With a tray-mounted player, the disc tray must be ejected (open) to load the disc. Place the disc in the slotted area and push the Close button to take the disc inside a home- or auto-type CD player. All discs should be loaded in the auto changer CD player before closing or pulling the tray into the CD changer. After playing, press the Disc button once to select the next compact disc for playback. Press the button repeatedly to select a desired disc, as indicated on the display.

The Test Disc

Many different test discs are on the market for correcting alignment and making other adjustments (Fig. 22-11). Most manufacturers have their own test disc or they recommend one for alignment procedures (see Table 22-1). These adjustments can also be made with a regular musical disc if you are very familiar with that selection. The test disc is used to make EFM RF signal, grating, track offset, focus gain, and tracking gain adjustments. Some manufacturers use a regular disc to make the required test adjustments. Follow each manufacturer's special alignment and other adjustments in the manufacturer's service literature. Most test discs range in price from $9.00 to $50.

THE TEST DISC EFM SIGNAL

The test-disc EFM signal can be checked with the scope at the output terminal of the RF amplifier (Fig. 22-12). A very complex encoding scheme used to transform the digital data to a form that can be placed on a disc. This information is modulated by EFM. This EFM signal must be present or most CD players will automatically shut down if focus and tracing are missing.

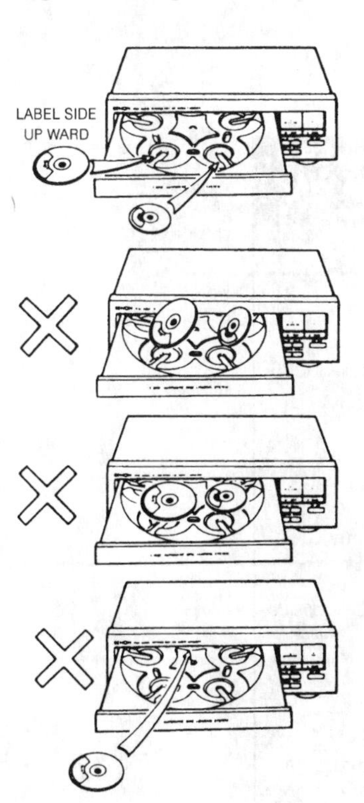

FIGURE 22-10 An 8-cm disc can be loaded in the center hole and the 12-cm disc can be loaded in the outer hole.

FIGURE 22-11 A test disc is used for alignment and adjustments. Keep the test disc within plastic container when it is not in use.

TABLE 22-1 A LIST OF VARIOUS CD PLAYERS WITH MANUFACTURERS RECOMMENDED TEST DISCS.

MAKE	MODEL	TEST DISC
Denon	DCH-500	CA-1094
Goldstar	GCD-616	YEDS-7 (Sony)
Magnavox	FD1040	NAP-4822-397-30085
Onkyo	DX-200	YEDS-18 (Sony)
Panasonic	SL-P3610	SZZP1014F
Pioneer	PD-7-10	YEDS-7 (Sony)
Quasar	CD8975W	SZZP1014F
Realistic	14-529	YEDS-43 (Sony)
Sanyo	CP-500	800104
		400088
		400067
Sony	CDX-5	YEDS-1
		YEDS-7
		YEDS-18
		YEDS-43
Sylvania	FDD104	NAP-4822-397-30085
		NAP-4822-397-30096

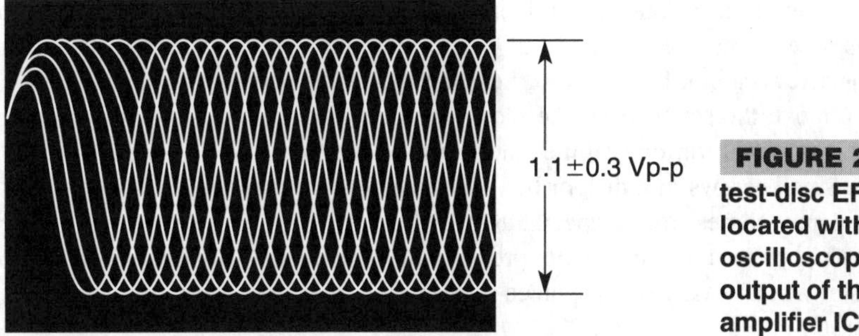

FIGURE 22-12 The test-disc EFM signal located with the oscilloscope at the output of the RF amplifier IC or transistor.

SONY YEDS-1 DEMO TEST DISC

The YEDS-1 stereo CD test disc contains 16 different music selections consisting of classical, jazz, and modern music. It contains a wide range of bass, percussion, and high notes throughout the various selections.

PHILIPS CC TEST SET

A two-set CD test disc is available at the NAP Consumer Electronics Corporation. It consists of a single audio-frequency test. One disc supplies audio signals for the CD technician to use to measure the performance of the CD player. The other contains a variety of simulated defects as fingerprints, dust, and scratches. Very fine lines represent fingerprints, dust is represented by large black dots, and scratches are simulated by interruptions in the reflective information layer. The set (17165500-40) costs $89.95.

THE ULTIMATE TEST DISC CD

A test disc for alignment and adjustment of CD players can be purchased from local or mail-order electronic firms. The *Ultimate Test Disc* gives a choice of volume levels, frequency responses, noise levels, resonance, phasing, and tracking tests on auto or home CD players.

This test disc contains 99 different tracks with sine waves of different frequencies, varying from 20 Hz to 20 kHz and sweeping left and right from 20 Hz to 20 kHz. The silicon track test counts from 1 to 99. The track test counting starts at 50 to 99. The test disc part is 80-505 and sells for $8.99 from:

MCM Electronics
650 Congress Park Dr.
Centerville, OH 45459-4072

Disassembly

After determining that the CD player is defective and the actual symptoms are located on the block diagram and schematic, check how the player covers are to be removed. Some

CD manufacturers have disassembly instructions on how to remove the outside covers and various assemblies. Be very careful when removing screws from the plastic areas. If a Phillips screw cannot be removed, heat the top of the screw with the soldering iron tip. Then, remove the screw from the plastic holder. Check the disassembly instructions on how to remove the various cabinet and boards in the Realistic CD-3302 boom-box player.

Place small screws in a dish or box container so that they will not be lost. Notice what size of screw comes from a special area. Different sizes and lengths of screws can delay the repaired product if they are not properly identified. Most TV benches are carpeted and small screws can easily be misplaced and possibly mar the plastic cabinet.

23

THE LASER DISC PICKUP ASEMBLY

CONTENTS AT A GLANCE

The laser pickup assembly is a crucial and delicate assembly that should be handled with care. The laser assembly can be dangerous to the eyes, so either keep a disc on the turntable or cover the laser optical lens with a piece of metal foil. Keep your fingers and

FIGURE 23-1 **The optical laser assembly on the side rails in a Sanyo CP-660.**

tools away from the optical lens assembly. Always follow the manufacture procedures when replacing a laser pickup assembly.

Basically two different types of laser pickup assemblies are available. Several of the compact disc players use the arc or swing-out arm mechanism, such as the Magnavox FD1040 and the Sylvania model FDD104. Others use the slide or sled mechanism that glides along metal rods straight out from the center of the disc (Fig. 23-1).

The optical laser pickup assembly consists of the objective lens, focus-tracking coils, collimating lens, beam splitter, semitransparent mirror, photodetectors, monitor, and laser diode. The basic optical pickup block diagram shown in Fig. 23-2 illustrates four photodetectors (A, B, C, D), two tracking photodetectors (E and F), the focus and tracking coils, monitor diode (MD), and laser diode (LD).

Photodetector Diodes

The photodetector diodes sense the EFM signal from the disc and pass it on to the RF amplifiers. Besides supplying the EFM signal, the photodetector diodes (A, B, C, and D) provide a tracking error signal. Because the EFM signal is very weak, a preamp is connected to the photodetector diodes. In some pickup assemblies, the photodetector diodes are called *HF sensors*.

The tracking-control photodetector diodes (E and F) provide a tracking error signal so that the error-control circuits help keep the beam on track. The tracking-error circuit generates an error signal if the laser beam spot moves away from the center pits. This error signal is used to ensure that the beam spot correctly tracks the line of pits. The tracking photodetectors are called *tracking sensors* in some players.

LASER PICK-UP PRECAUTIONS

Always handle the laser assembly with extreme care. Keep the pick-up out of dusty, high-temperature, or high-humidity areas. Be extremely careful not to accidentally drop the laser assembly. When exchanging the defective optical assembly, place it in the same box as the new replacement.

The laser beam can damage the human eye because the intensity of the focused spot might reach 7 × 103 w/cm^2—even if the intensity at the object lens is 400 mW maximum. The average laser output is from 0.25 to 0.9 mW. As the light beam spreads after being focused through the objective lens, it will not affect you at a distance greater than 30 cm. However, do not look directly at the laser beam—either through the objective lens directly nor another lens or a mirror (Fig. 23-3). Keep your eyes at least two feet from the optical assembly.

The LD chip contains poisonous arsenic, as GaAst GaA/As, although the poison is relatively weak in comparison to others (e.g., As2o3, AsC1s, etc.), and the amount is small. Avoid putting the chip in acid or an alkali solution, heating it over 200 degrees centigrade, or putting it into your mouth.

The LD could be damaged or deteriorated by its own strong light if a large current is applied to it—even if it is only a short pulse. Be sure that no surge current is in the LD driving circuit by switches. Be very careful to handle the pick-up because it can be damaged in a moment by human electrostatic discharge. The pins of the laser diode are short-circuited by solder for protection during shipment.

For safely handling of an LD, grounding the human body, measuring equipment, and a jig is strongly recommended. Wear a ground wrist strap. Use a grounded mat on the bench

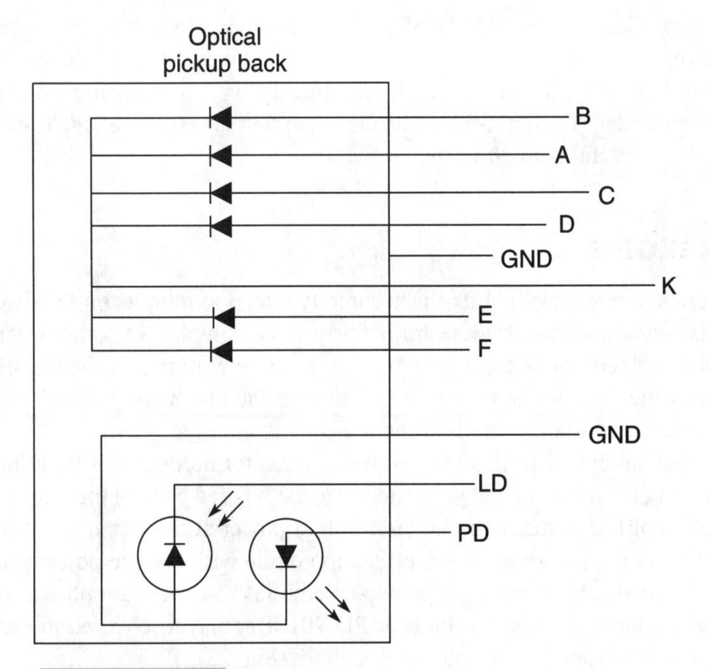

FIGURE 23-2 **The optical pickup block (KSS-168 AKP) in this auto CD player.**

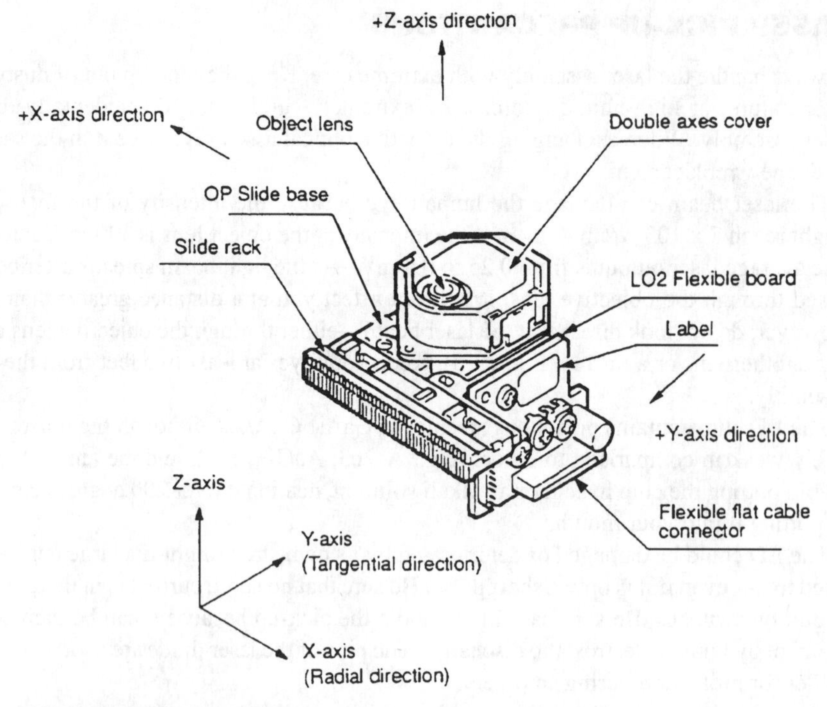

FIGURE 23-3 The components in a laser pickup.

and floor to handle the laser diode. Be sure that all test equipment is grounded to the mat under the CD player.

To open the short circuit, remove the soldering quickly with a soldering iron whose metal part is also grounded. Do not use a controlled iron that heats above 320°C or use a soldering iron more powerful than 30 watts.

THE LASER DIODE

The laser diode emits a beam of light that is accurately aimed at the pits on the disc surface. Although the semiconductor laser beam is fairly weak compared to other laser systems, you should never look directly at the optical lens while working on the CD player. Also remember that the laser diode is very susceptible to the effects of static electricity. Use extreme care when removing and installing a pickup assembly.

The laser diode optical output is also beamed onto a monitor diode that is built into the optical pickup. The light source current generated as a result is fed back to the minus input of the operational amplifier to maintain the laser output power at a constant level. In the Denon DCD-1800R pickup, there is a laser protection diode with reverse polarity across the laser diode (Fig. 23-4). Although most pickup assemblies have the auto power control (APC) circuits on the main chassis, the Pioneer PD-7010BK player, covered in the next chapter, has the APC IC located on the pickup assembly (Fig. 23-5).

Realistic CD-3370 laser pickup mechanism Although it appears to be smooth and shiny, the surface of the compact disc is covered with tiny pits that represent digital information stored on each disc. Each pit is approximately 0.5 microns wide; a disc can hold up to 2.5 billion pits. As the laser beam scans the pits on the surface of the disc, the rise and fall of the beam over a pit is detected as a binary "1," while no change in the beam on a smooth area is detected as a "0." The laser beam must focus precisely on the pits to accurately read the disc. Several control systems are used to maintain disc-reading accuracy.

The laser pickup uses four main photodiodes, A through D, to detect the laser light reflected from the disc and to check the focus of the lens (Fig. 23-6). Two additional diodes, E and F, are placed on either side of the main diodes to check for proper tracking. Finally, laser diode LD and diode PD monitor the laser-beam power.

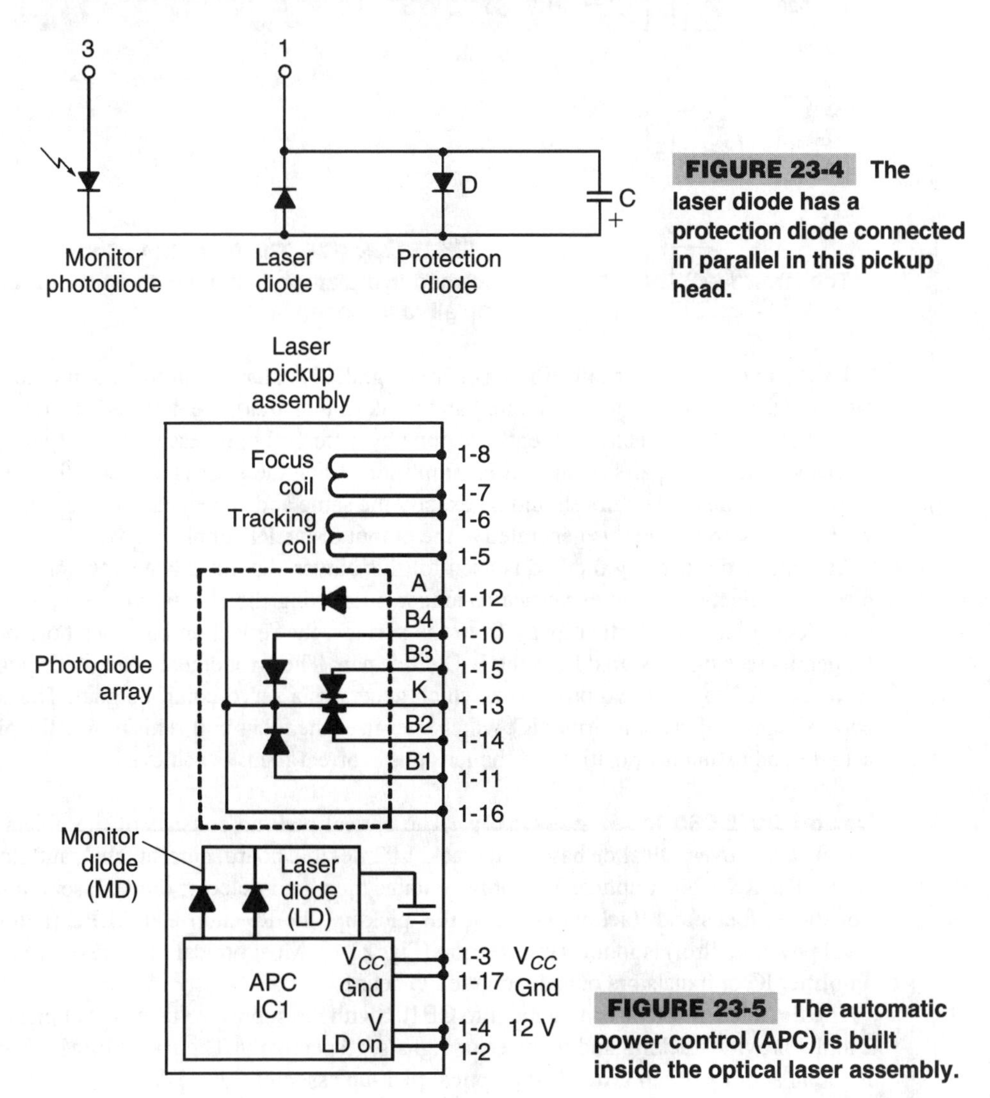

FIGURE 23-4 The laser diode has a protection diode connected in parallel in this pickup head.

FIGURE 23-5 The automatic power control (APC) is built inside the optical laser assembly.

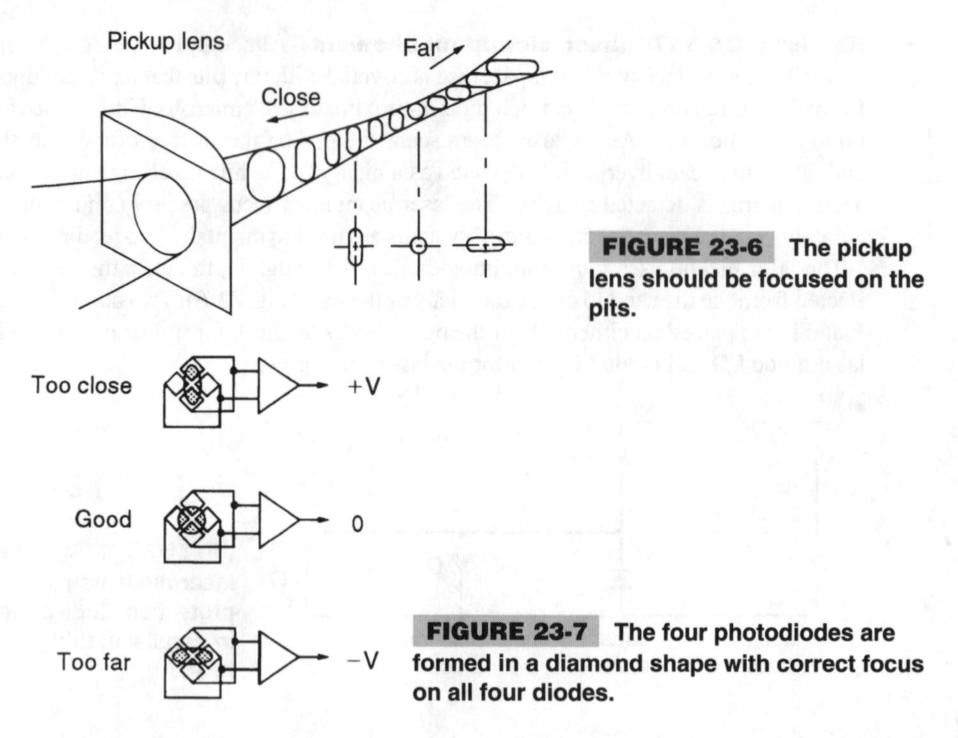

FIGURE 23-6 **The pickup lens should be focused on the pits.**

FIGURE 23-7 **The four photodiodes are formed in a diamond shape with correct focus on all four diodes.**

The main diodes are arranged in a square (Fig. 23-7). Changes in the magnitude of the laser reflection, strong (nonpitted area) and weak (pitted area), are detected and converted to digital signals. Focus adjustments are done by differently processing the outputs of the A-C and B-D diode pairs by the RF IC amplifier. When the laser is properly focused, the signals from the diode pairs should be exactly the same and cancel each other out. Otherwise, a focus error signal is generated at the output of the RF amplifier.

An astigmatic focusing method is used for this player. As the lens gets too close to the disc, the reflection becomes vertically longer, increasing the signal from the A-C diode pair. As the lens gets farther away from the surface, the reflection becomes horizontally longer, increasing the signal from the B-D diode pair. The error signal generated by the RF IC amp is fed to the servo processor, which generates a servo-control signal. The servo-control signal is fed to the driver IC, which energizes the focus coil. This moves the pickup actuator, adjusting the positions of the lens until correct focus is achieved.

Denon DCM-560 laser assembly The optical pick-up consists of the object lens, double axes cover, OP slide base, slide rack, L02 flexible board, current label, and flexible cable. The KSS-240A optical assembly contains the photo-detector diode, laser and monitor diode, focus and tracking coils. In this pick-up, the RF amp and ALPC (automatic level power control) is in the same section (Fig. 23-8). Most optical pick-ups have the RF amplifier IC or transistors outside, on the PC board.

The flexible connector is plugged into CB101 with the supply voltage (V_C) at pin 1. Pins 2 and 3 provide tracking and focus error signals, respectively. The RF or EFM waveform is found at pin 5. Pin 6 grounds the optical pick-up assembly. A +5-V supply connects to pin 7. The positive (+) tracking signal at pin 10 and negative (–) focus is at ground potential. The negative tracking signal at pin 11 is fed to tracking driver transistors TR183 and TR182. A positive (+) pin 9 is tied to focus driver transistors TR181 and TR180. The RF and EFM signal can be scoped at terminal 5 of CB101 or at pin 46 (RFO) of IC101.

The Optical Pickup

Optical pickups can have either one or three beams. Although there is very little difference in the sound output of the player, most incorporate the three-beam system, which is slightly more complicated in design. The semiconductor laser light source has a wavelength from 750 to 850 nm.

The path of the laser beam and arrangement of the optical elements in the Pioneer PD-5010BK optical system are shown in Fig. 23-9. The semiconductor laser emits a beam of light with a wavelength of 780 nm. It is barely within the range of visibility. The beam is produced from an extremely small point and has an elliptical distribution. It is dispersed in a conical shape.

To produce the beam used to detect tracking error, the beam is passed through a diffraction grating that splits the beam into three separate beams—the primary beam (zero order) and two side beams (one order). A small amount of higher-order elements are also produced, but these are lost and not used. Next, the beams are passed through a half prism, where 50 percent of the energy is lost.

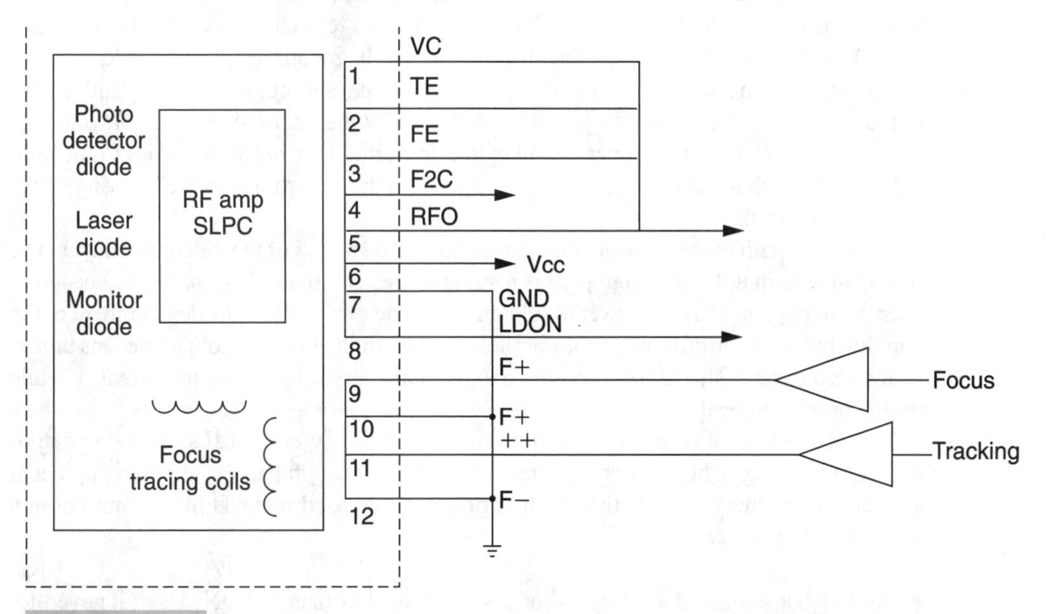

FIGURE 23-8 **The wiring diagram of the laser pickup assembly in a Denon DCM-560 CD player.**

FIGURE 23-9 **A lens assembly in a portable CD player.**

The collimator lens produces a completely collimated (parallel) beam. The diameter of the collimated beam is large enough to cover the movement of the objective lens. The beam is then condensed to a spot with an extremely small diameter by the objective lens before it is radiated to the disc. Part of the beam is then reflected back from the disc, diffracted, and routed back through the objective lens to be recollimated and condensed.

When this beam reaches the half prism, 50 percent passes through the grating and returns to the laser diode. The other 50 percent is reflected by the prism to the multiple lens that has the functions of both a concave and cylindrical lens. This beam then goes to the photodiode alley, where an electrical signal with a strength proportional to the intensity of the beam is produced.

The optical path of the compact disc can be compared to that of the videodisc player. The first feature is that the outgoing path is a straight line, so no auxiliary parts are needed to alter the light path. This way, overall tolerances can be minimized. The development of the double-shaft activator for use in the parallel-drive method allows the objective lens unit to be reduced in size. This makes it possible to maintain very satisfactory performance while using compact optical parts.

The second feature is the half prism. In the videodisc player optical system, the outgoing and incoming light paths are separated by a ¼-wavelength panel and polarizing beam splitter. The primary reasons that the half prism can be used in a CD player, but not in a video disc player, are:

- Although a semiconductor laser diode is much smaller than an HeNe laser, it nevertheless has a fairly high optical power output. Therefore, the energy loss caused by the half mirror is not a problem.

- Both video and compact discs tend to polarize light because they are made of a resin-based material that is not perfectly flat. In video discs, the amount of polarization is carefully checked against an established standard. In compact discs, the limitation is not very strict. Because of the lack of a strict standard, CD players normally use an extremely accurate ¼-wavelength plate. In actual use, however, this plate cannot function properly because of polarization of the laser beam caused by the disc. A half prism, on the other hand, is not at all affected by polarization of the laser beam. Consequently, a very stable optical path can be made.

Another feature of this optical system is the use of a parallel drive unit that allows optimum utilization of the objective lens at all times. As is shown in Fig. 23-10, the beam from the laser diode is converted into a completely collimated beam by the collimation lens. The parallel drive unit causes the objective lens to move parallel and perpendicular to the beam. Therefore, the optical path is usually not affected by movement of the objective lens within the collimated light cluster.

Still another feature of the optical section is the use of a multiple lens. This lens prevents the focusing point depth on the photodiode from becoming too shallow, a problem that has appeared as optical sections have become more compact. It is an effective way to permit lowering of the installation accuracy required by the photodiode. This multiple lens is cylindrical with the functions of a concave lens. Previously, there were two beams and both a concave and cylindrical lens. In this pickup, however, one lens performs both functions, thereby allowing a further reduction in size.

When designing the optical section of a CD player, the most important concern is accommodating any differences between various compact discs. To do this, it is desirable to

FIGURE 23-10 **The optical lens assembly in the roulette Magnavox CDC-745 changer.**

have a very short wavelength. The wavelength used is 780 nm because this is the shortest wavelength possible today with mass-produced pickups.

Laser Action

To reproduce signals encoded as a series of tiny pits, players use a laser-beam spot of approximately 1.6 micrometers in diameter. By rotating the disc and shining the laser beam on the series of pits, an optosensor (photodetector) detects the presence or absence of the pits within a fixed period of time. The changes in the reflected light correspond to the recorded signals.

The source of the laser beam is a laser diode with a 780-nm wavelength and a 3-nw optical output (Fig. 23-11). The beam from the laser diode is divided into three beams. The three beams pass through a half mirror, become a parallel beam through the collimator lens, get refracted by a prism, pass through the object lens, and focus on the disc. The light focused on the disc reads the disc data and is reflected. It then passes back through the objective and collimator lenses, through the half mirror and the flat concave cylindrical lens, then the beam strikes the photodetector. To be in line with the vertical fluctuations and the aberrations of the disc, the object lens moves up, down, left, and right so that the series of pits on the disc are always in focus.

DIFFERENCES IN ONE- AND THREE-BEAM LASERS

Both the one- and three-beam lasers use the objective lens, collimator, laser, and photodiodes. The one-beam system can have a semitransparent mirror and optical wedge besides

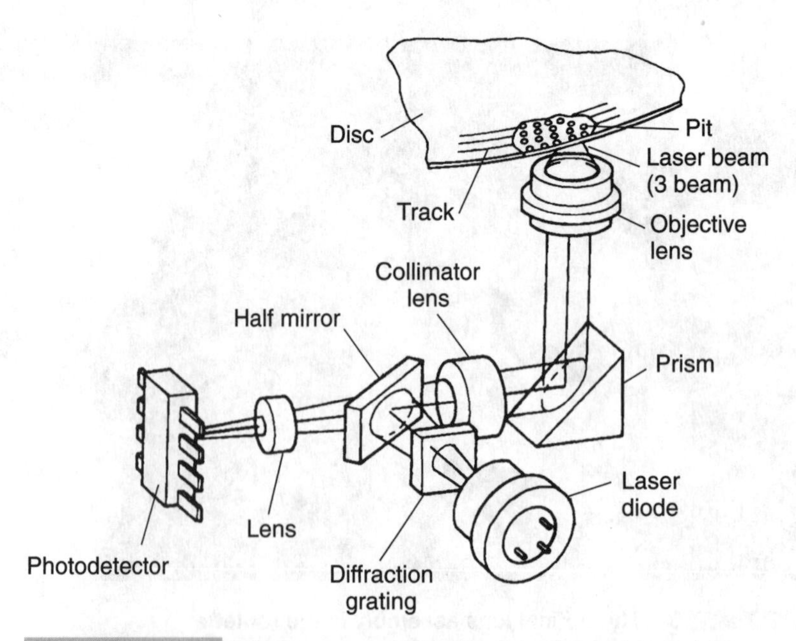

FIGURE 23-11 The laser parallel-beam in a three-beam pickup assembly.

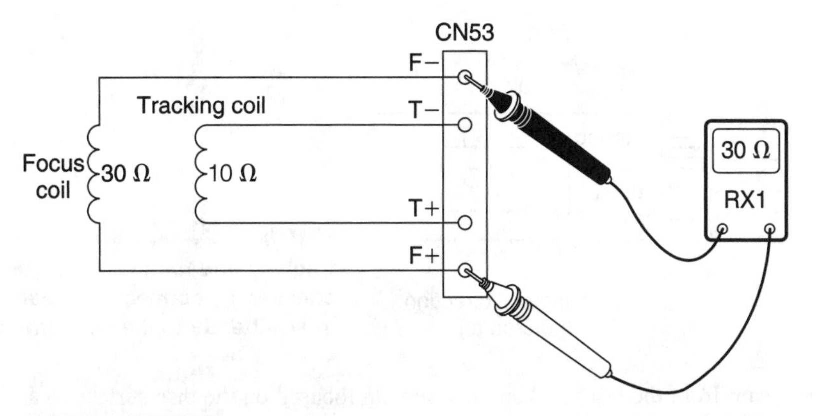

FIGURE 23-12 **Check the focus and tracking coil with the DMM's low-resistance scale.**

those components mentioned previously. The three-beam system can have a subbeam, quarter-wave plate, polarized beam splitter, diffraction grating, and cylindrical lens, besides those elements common to both systems. All of these components are located in the optical pickup section. Nevertheless, very little difference in sound is noticeable between the two systems.

Focus and Tracking Coils

The focus and tracking coils are located close to the optical lens. The beam from the laser must be focused on the disc surface at all times. The tracking must be precise as the disc rotates.

The focus and tracking coils within the pickup can be checked for continuity with an ohmmeter (Fig. 23-12). You might notice a shift in the optical lens when making resistance measurements. Both resistance measurements should be very low. In the RCA MCD-141 player, the focus coil measures 30 W, and the tracking coil 10 W. In a Radio Shack CD-1000 model, the focus coil has a total of 20 W, and the tracking coil is 4 W. A focus coil that reads less than 10 W and a tracking coil that reads less than 1 W could have shorted turns. Infinite ohms indicates the coils, lead wires, or socket connections are open. Both coils are controlled from the servo-control IC.

The focus and tracking coil can be checked with a C or D flashlight battery (1.5 V). Locate on the schematic diagram the focus and tracking coil terminal connections on the laser assembly. Remove the terminal leads for the test, writing down which was which. Place the battery across the tracking coil, and if normal, the optical assembly will move in a horizontal tracking direction (Fig. 23-13). Reverse the leads of the battery and it should move in the other horizontal direction if the tracking coil is normal. Now, place the battery across the focus coil terminals and the optical or activator assembly should go up or down. Reversing the battery polarity and the optical assembly should make it move in the opposite direction. If any one of the coils is open or shorted, the activator assembly will not move when tested with the 1.5-V battery.

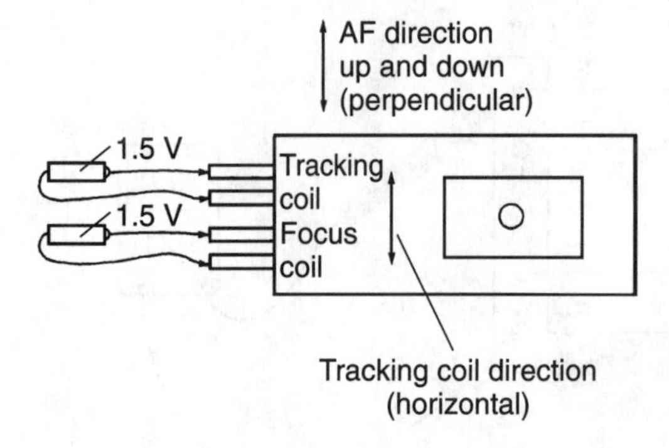

FIGURE 23-13 **Movement of the tracking and focus coil can be checked by connecting a battery intermittently to the coil terminals.**

The beam from the laser pickup must remain focused on the disc surface to accurately read the information. When the focus on the pits is no longer precise, the focus servo moves the object lens up or down to correct the focus. Under this system, when a beam is irradiated through the cylindrical and convex lenses, the beam is elongated and then becomes a perfect circle. When the laser beam is reflected from the disc, it is directed to the cylindrical lens by the prism and then to the optosensor (photodetector), where it is split into four and forms a perfect circle.

Latest Optical Systems

Besides all of the laser and photo diodes, focus and tracking coils, the pickup optical block assembly might have the spindle drive IC and SLED and spindle motor assembly (Fig. 23-14). Notice that the focus, tracking, and SLED components are driven from IC14 in a Denon DCH-500 trunk changer.

In the CDX-5 auto radio and CD player, PD1 and PD2 (photo- diodes), are connected together and apply to pins 5 and 6 of the RF amp and signal processor IC652 (Fig. 23-15). The laser diode has a laser drive transistor (Q651), with one half applied to the automatic power control (APC). The focus error signal is at pin 16 and tracking error at pin 17 of RF amp IC652. To determine if the laser optical assembly and RF amp circuits are normal, take an EFM waveform at pin 2 and 3 of IC652. No EFM waveform might indicate that the optical assembly, laser diode, and RF amplifier IC are defective.

Pickup Transport Systems

Pickup transport systems take the optical pickup assembly from the center to the outer edge of the compact disc. One system uses the arm or arc movement, where the optical pickup system travels in a semicircle to the outside of the disc. This action is controlled by a pickup motor. The optical pickup focus and tracking coil are mounted on the radial tracking arm (Fig. 23-16).

The other type is the sled (or slide) motor and gear assembly that moves the pickup across the disc on a rail or rod assembly. The sled assembly is now used in more CD players than the swing arm assembly.

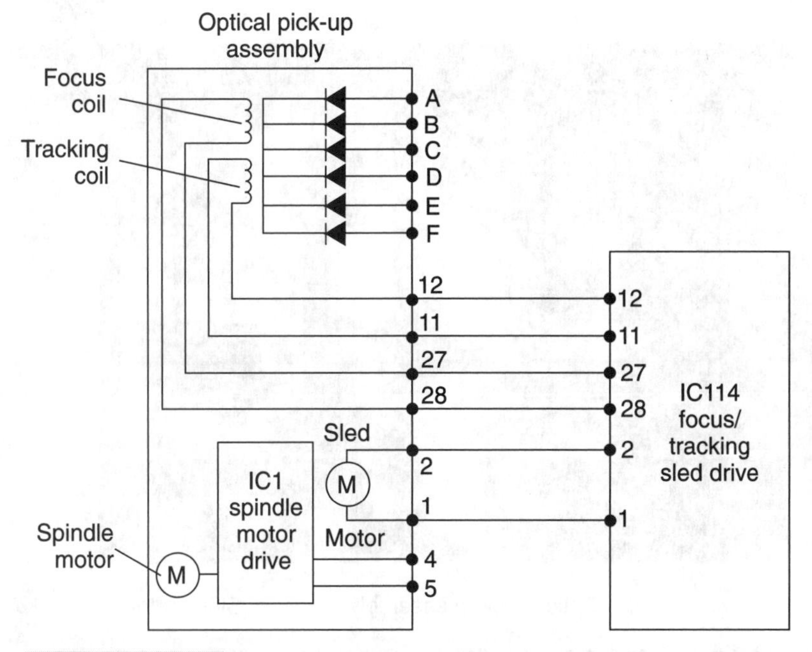

FIGURE 23-14 The spindle and SLED motor are mounted within the optical lens assembly in the Denon DCH-500.

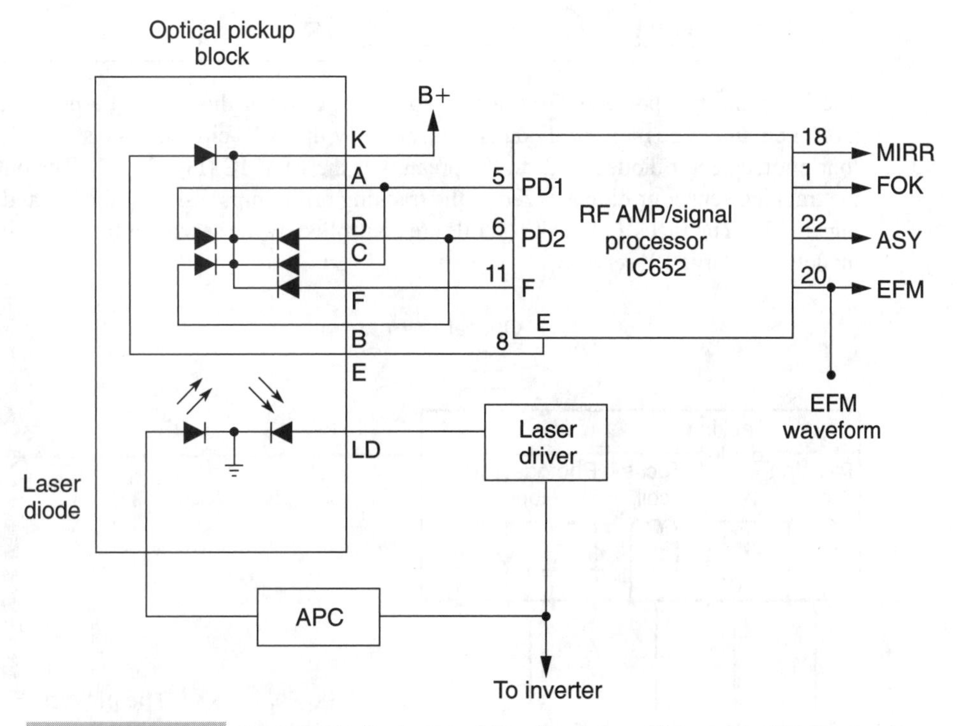

FIGURE 23-15 The photodiode 1 and 2 output signal is connected to pins 5 and 6 of RF amp IC652.

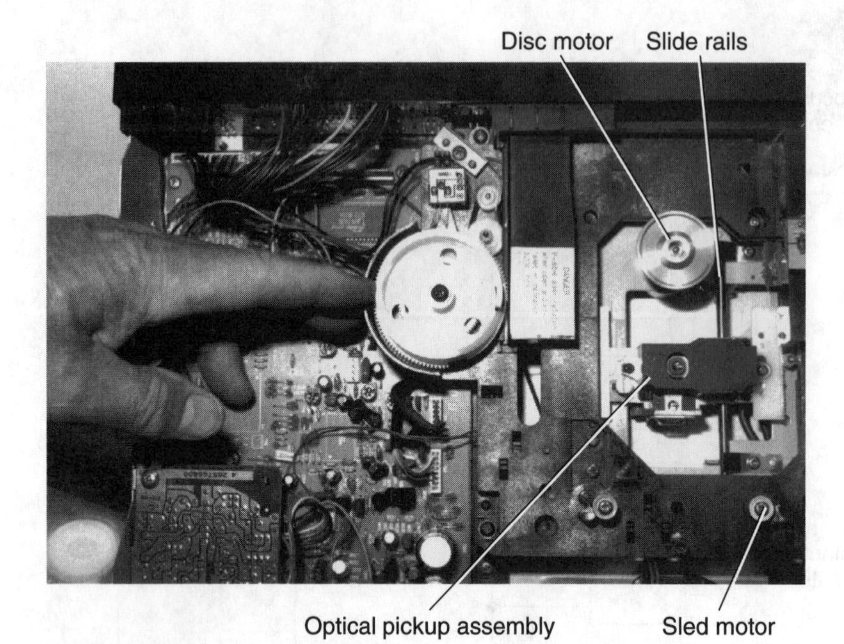

FIGURE 23-16 The CD optical pickup assembly on the side rails with disc and sled motors.

Laser Head Connections

The laser diode is powered from a 5-V source or through the automatic power control (APC) circuits. The monitor diode is controlled by the APC circuits. The signal from the four photodetector diodes feeds to the preamp of the EFM IC (Fig. 23-17). The output of the tracking sensor or diodes is fed to the tracking error amps inside of the RF and tracking amp IC. The focus and tracking coils are controlled by the servo-control IC, or in some models, one large RF servo IC.

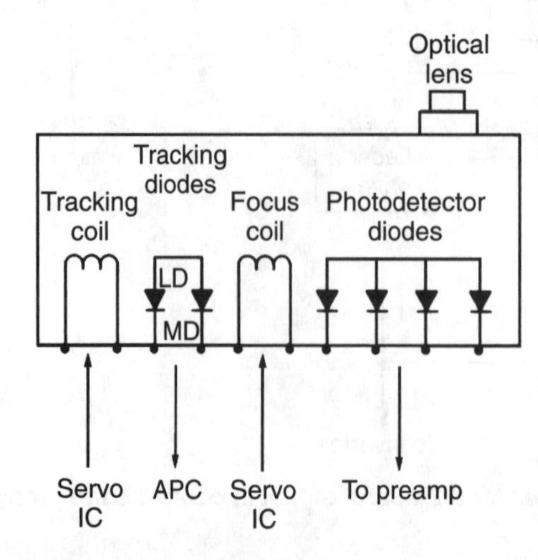

FIGURE 23-17 The pickup assembly components connect to the APC, servo IC, and preamp IC.

Testing the Laser Assembly

Weak or improper tracking of the optical pickup assembly could be caused by components outside the assembly, improper voltage sources, or a defective laser assembly. The laser diode light assembly can be checked with voltage and current measurements, light meters, or a homemade indicator. The commercial optical power meter or simulator should be used for crucial laser diode adjustments. Voltage and current measurements can be taken across a fixed resistor in the laser diode circuits, but not directly on the laser diode. The homemade indicator simply indicates whether the laser diode is emitting a beam to the disc. Never look directly at the optical assembly when making these tests. Do not take diode or resistance measurements on the diode terminals.

The commercial light meter or simulator slips underneath the flapper or clapper assembly over the optical lens assembly, instead of the compact disc. If the laser-beam assembly is not emitting, look for a laser-beam interlock assembly. This laser interlock is usually an LED/phototransistor combination located above the compact disc. When the disc is not in place, the LED provides light to the phototransistor and shuts down the laser-beam circuits. Place a piece of cardboard or paper between the LED and hole of the phototransistor (Fig. 23-18). Check the interlock switch when the clapper or flapper assembly is removed and the turntable or laser assembly are not working. The optical power meter or simulator can be purchased from manufacturer distributors. The service life of the laser diode could have expired if the meter indication was less than 0.1 mW. Also, when the EFM output is extremely low, the service life of the laser diode has probably expired.

Accurate voltage and current measurements within the laser circuits are the best way to know if the laser diode has become weak. Actually, the current of the laser diode is taken by a voltage measurement across a fixed resistor in the laser-drive emitter transistor circuit (Fig. 23-19). In this Realistic CD-1000 compact disc player, the laser unit is normally driven with a current of 40 to 70 mA. Check this laser-driving current by measuring the

FIGURE 23-18 This photodiode shuts off the laser beam when no disc is mounted.

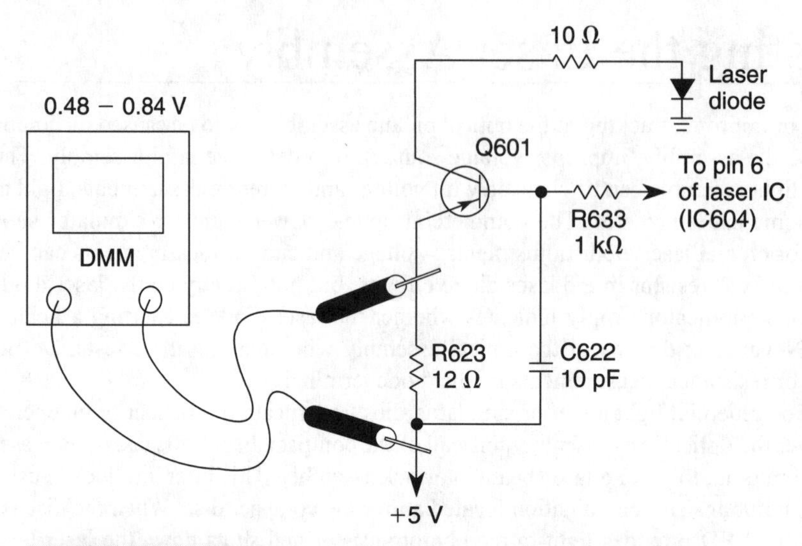

FIGURE 23-19 **Checking the voltage-representing current across R623 in this Realistic CD player.**

voltage across R623 (12 W). If it is over 100 mA, the laser unit could be defective. Remember, the voltage across the fixed resistor is less than 1 V. Here, the current value (0.48 to 0.84 V) across R623 is normal. The voltage value of 0.48 to 0.84 V equals 48 to 84 mA of current (a safe level).

The laser output should not be adjusted unless the laser pickup assembly or circuit is replaced. In the RCA MCD-141, the laser current can be checked by measuring the voltage across R209 (Fig. 23-20). The laser-diode current ranges from 40 to 80 mA (0.48 to 0.96 V) across R209. If the current is more than 120 mA (1.44 V across R209), the laser diode could be defective. Look for the current label on the pickup head assembly for the correct current value.

The electric current of the laser is usually indicated on the label on the pickup. For the Akai Model CD-M88T, connect a voltmeter between TP2 and TP3 on the servo PC board and measure the voltage. The electric current of the laser equals the voltage measured divided by 10. If the electric current of the laser exceeds 10 percent of the recommended value, replace the pickup assembly.

INFRARED DETECTOR/INDICATOR

You can make a laser indicator with only a few components obtainable from a local parts distributor or Radio Shack (see Fig. 23-21).

Q1 is an infrared photodetector (276-142) that picks up the laser-beam light and activates the buzzer and LED when infrared is detected. Wire terminal 2 (collector) directly to the piezo buzzer and wire terminal 1 (emitter) to the on/off switch. Terminal 3 is not used. The infrared emitter diode and detector comes as a pair, but only the photodetector is used in this small indicator.

The piezo buzzer (273-065A) was chosen because of its size and pleasant tone. The buzzer operates on 3 to 20 Vdc and has an operating frequency of 2.8 kHz. The positive

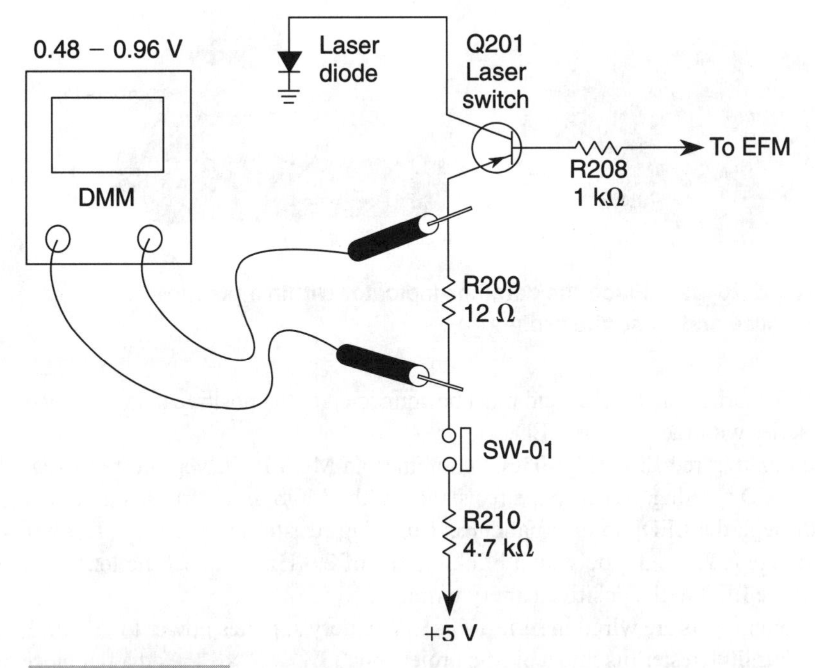

FIGURE 23-20 Checking the current and voltage across R209 to determine if the laser diode is defective.

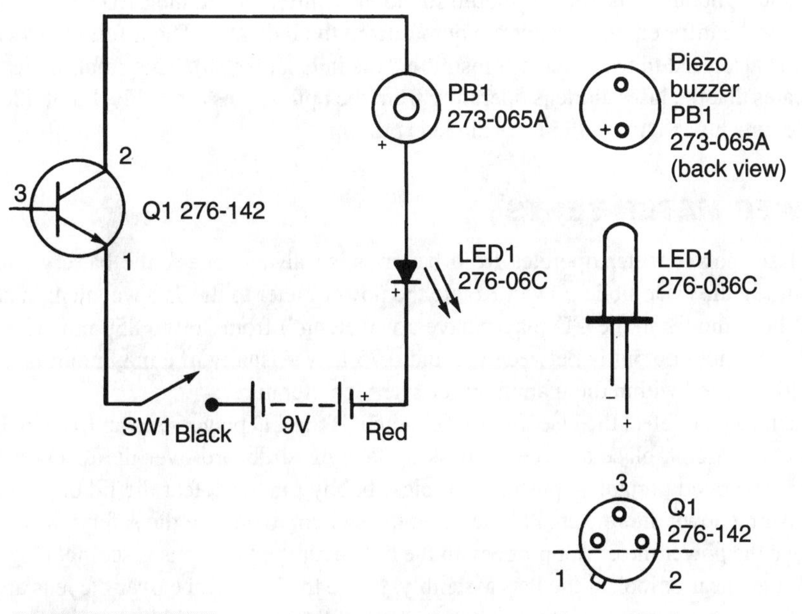

FIGURE 23-21 Build the infrared detector/indicator with only a few parts.

FIGURE 23-22 **Place the detector-indicator within a plastic project case and Masonite probe.**

pin (+) is marked on the case and must be connected to the positive side of the 9-V battery or in series with the flashing LED.

The blinking red LED (276-036C) combines an MOS IC driver and red LED within a plastic LED housing. Because the resistance of the MOS driver transistors limits the current through the LED, no external current-limiting resistor is necessary. The typical supply voltage is 2.5 to 3 Vdc with a blinking rate of 2.0 Hz. Connect the longest (positive) lead of the LED to the positive battery terminal.

All components are wired in series. The 9-V battery supplies power to Q1, PB1, and the LED. This little tester fits into a plastic project box (3¼" × 2" × 1". Add a flat piece of plastic to the end of the box, and mount Q1 to the end of it (Fig. 23-22). Cement the piezo buzzer to one end of the box and the LED to the other. The photodetector should fit under the flapper or clamper assembly and directly over the optical lens assembly.

The infrared indicator can be tested by holding it directly under the sun's rays or under a regular light bulb. The buzzer should sound intermittently and the LED should flash. The stronger the infrared, the greater the output from the indicator. The infrared detector/indicator is also used to test remote transmitters (as in Chapter 10). This small indicator only indicates that the laser diode is operating from the optical lens assembly, but it does not indicate how much current the laser diode is pulling.

POWER METER TESTS

The laser power meter operates from batteries, so always check the battery test before measuring the laser diode power. Rotate the power meter to the 750 wavelength at 1 mW. Most laser diodes in the CD player have a wavelength from 760 to 850 nm. The average laser maximum output is between 0.4 and 0.75 mW. Usually, the maximum laser output should be listed within the manufacturer's service literature.

Disengage or defeat the laser interlock switch. If the CD player uses an LED in the laser-protection circuit, place a piece of masking tape or cardboard over the LED opening. In battery-operated portables, push a toothpick, bobby pin, or defeat the lid or cover switch by pushing down into the small hole or on the switch, to engage the safety switch.

Place the power meter flat probe with the hole over the laser lens assembly (Fig. 23-23). Do not get near or look at the lens assembly. Move the laser probe over the lens area to get maximum laser beam or highest meter reading. If the meter hand hits the end scale, rotate selector to the next-highest measurement. A very low measurement indicates that the laser is defective and optical assembly should be replaced. Check this measurement with the manufacturer's specification.

LASER DIODE OPTICAL LIGHT METER MEASUREMENTS

Laser diodes usually have a wavelength of 765 to 850 nm with a continuous wave. Measure the laser power output over the optical lens assembly. Although the wavelength and power of each laser diode varies with each manufacturer, the range should be between 765 and 850 nm. The maximum or minimum laser output should be listed within the manufacturer's service literature. The laser-diode properties and output of several CD players are listed in Table 23-1.

FIGURE 23-23 **The laser power-meter probe is placed over the lens assembly to measure the power beam of the laser diode.**

TABLE 23-1 THE LASER DIODE MATERIAL, LASER OUTPUT, AND WAVELENGTH OF SEVERAL DIFFERENT CD PLAYERS

CD PLAYER	DIODE MATERIAL	WAVELENGTH	LASER OUTPUT (MAXIMUM)
Onkyo DX-200	GaAs/GaA1As	780 nm	0.4 mW
Mitsubishi DP-107	Ga-Al-As	765-795 nm	0.4 mW
Sanyo CP-660		775-830 nm	0.7 mW
Yamaha	Ga-Al-As	760-800 nm	0.5 mW
Pioneer PD-7010		780 nm	0.26 mW

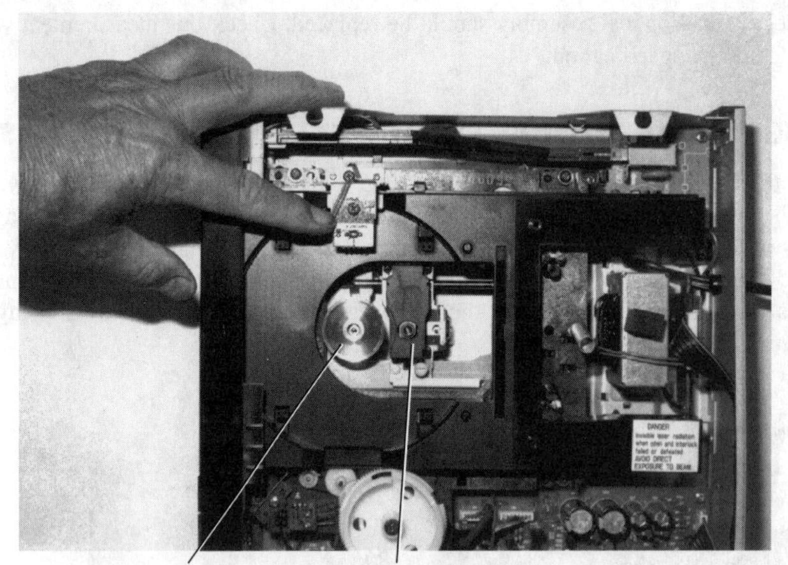

FIGURE 23-24 **The laser optical-lens assembly below the clasper CD assembly.**

Cleaning the Optical Lens

Before troubleshooting the optical activator or sensor assembly, clean the lens area. The optical lens could have become stained or tarnished with cigarette or cigar smoke. Excessive dust can cloud up the optical lens to the point where the laser beam can't reach the compact disc. Be careful not to apply too much pressure when cleaning the optical lens or you could damage the lens assembly (Fig. 23-24).

Clean the lens with lint-free cotton or cleaning paper for camera lenses moistened with a mixture of rubbing alcohol and ether. Cleaning solution for cameras is ideal. Wipe the lens gently to avoid bending the supporting spring. Blow excessive dust from the optical lens with a can of dust spray available in camera departments.

Protection of Laser Diodes

Laser diodes are so sensitive to such pulsive electrical noises as static voltage or surge current that their reliability can be decreased or destroyed. Take precautions against any kind of static voltage potential, in addition to electrically grounding the workbench and test equipment. Do not attempt to check the laser diode by applying multimeter or scope probes directly to the laser terminals. Do not apply voltage with a poorly made voltage source or with temporary contact pins or clips.

Some laser-diode assemblies have a shorting metal bar to neutralize any static electricity. Although a shorting bar is used to establish contact with the laser diode leads during shipment, it can lose its conductivity because of vibrations during transportation or oxidation. Take sufficient precautions—even when the shorting bar is on. Be sure that the optical laser unit is not handled with the shorting metal removed nor left near appliances that emit high-frequency surge voltages. For storing the unit, be sure to short the laser-diode leads with the shorting metal (or by soldering the leads together) and place in a conductive container.

When removing the laser-diode assembly, turn down the optical output (work current) and turn off the power. Short the laser diode leads with the shorting metal (or by soldering the leads), and remove the laser assembly connections. To install the laser assembly, make the right connections, remove the short circuit or bar, turn on the power, and then adjust the optical output. Always replace the whole optical laser assembly—not just the laser diode inside the optical assembly. Follow the manufacturer's procedure for removing and installing a new optical laser pickup or block assembly.

Handling the Pickup Assembly

Use extreme care when removing and replacing the pickup assembly. Be careful not to touch the terminals of the semiconductor laser or those that are attached to the board by hand or with a tool. New laser pickup assemblies arrive in a conductive bag.

Keep strong magnets (such as speakers or TV focus magnets) away from the pickup assembly because some pickups have a magnet within the assembly. Leave all adjustment set screws alone on the pickup. These are adjusted at the factory. Readjusting a pickup set screw could damage both the pickup and assembly. Do not adjust any semi-fixed variable resistor on the laser assembly; they are also factory-adjusted.

Handling Denon DCM-560 pickup assembly The DCM-560 laser pickup assembly is assembled and precisely adjusted using a sophisticated manufacturing process. Do not disassemble or attempt to adjust the optical assembly. Likewise, handle the laser pickup with kid gloves, so to speak.

Always handle the laser pickup with holding the side base (rosin molded part). When either a part of human body or tools happen to touch directly with the circuit part of PW board, it might cause deterioration. Be extremely careful handling this base section (Fig. 23-25).

DEFECTIVE LASER ASSEMBLY-SHUTDOWN

When the RF or EFM signal is not found at the digital or servo control IC, the turntable motor might start up, focus and tracking coils will begin to search, then the whole chassis will shut down. Sometimes the chassis shutdown occurs so fast, it's impossible to trace the EFM signal at the output of RF amplifier IC. You can't tell if laser assembly or the RF IC is defective.

To eliminate shutdown, determine if a separate RF amp IC is used or only one IC is found to control the focus and tracking coils, spindle and SLED motors, and the optical assembly. If a separate RF amp IC and servo IC is used, remove the low-voltage source (V_{CC}) pin from the servo IC and PC wiring. Simply use solder wick or mesh and a soldering iron to remove excess solder around the pin terminal (Fig. 23-26). Now, the servo loop is disconnected and you can check the RF or EFM waveform from the optical pickup assembly or RF IC, without chassis shutdown.

FIGURE 23-25 **The flapper or clamper assembly is located over the optical lens assembly.**

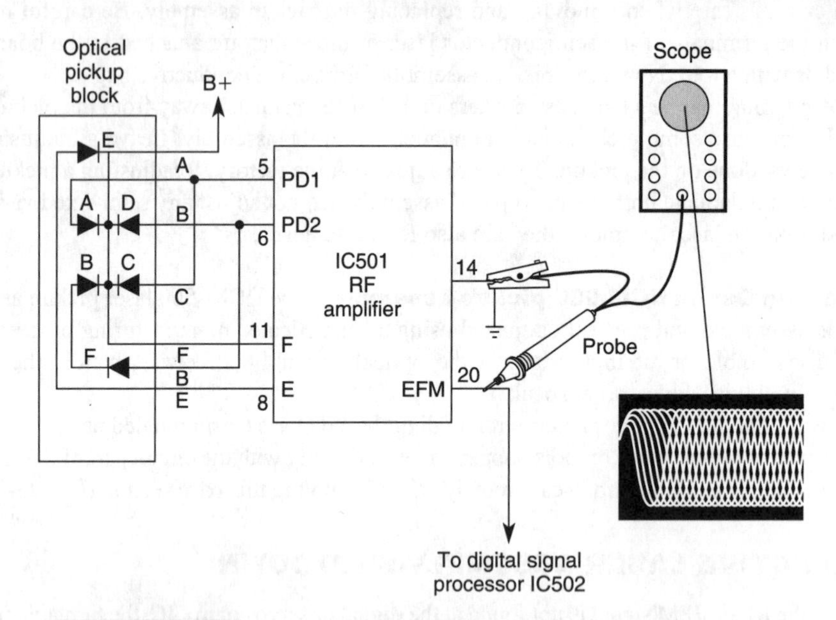

FIGURE 23-26 **The optical pickup waveform can be taken at the RF or HF output terminal with the oscilloscope to determine if the assembly is defective.**

REPLACING THE PICKUP LASER ASSEMBLY

Pull out the ac power cord. Be sure that the power is off before attempting to remove the different components necessary to get at the laser pickup assembly. The pickup is usually located under the flapper or clamper assembly (Fig. 23-27). In some players, the disc tray must be removed. Remove the bottom cover of the disc player. Some disc player chassis set right down into the cabinet and do not have a removable bottom cover. If you do not have the manufacturer's removal procedure at hand, write down the removal sequence for each component for easy replacement. After replacing the pickup assembly, make the laser current adjustments, if provided by the manufacturer.

Laser Warnings

Always leave the new optical assembly in the protective envelope until the old unit has been removed. Be careful when disconnecting the old cable from the pickup. Remove the soldered-shorting pin from the new pickup and reconnect the pickup cable to the new assembly within seconds to prevent damage to the laser pickup. Be sure that the human body, mat, and test instruments are grounded. Likewise, protect your eyes from the optical lens assembly.

The activator might be affected if magnetic material is located nearby because the activator has a strong magnetic circuit. Keep large speakers and magnets away from the CD player or optical assembly.

Check and clean the lens assembly for smoke, dust, and cigarette ashes. Clean the lens with a cleaning paper dampened with a little water, not by strongly pressing the lens with the cleaning paper. You can clean the lens with a camera cleaning kit. Clean the lens assembly before installing the optical assembly.

In CD players that comprise a built-in RF amp and APC circuit, remember that this assembly resists stronger against external electrostatic damages than with a pickup that does not include the RF amp. However, there is a possibility of pickup distortion in low HF level, or with great numbers of jitters. Check the tracking offset (EF balance) by making this adjustment according to the manufacturer.

Last, but not least, apply FROIL 946P lubricant when replacing the pickup. Place FROIL upon bushing and rails. Check the manufacturer's lubricating chart. Check the troubleshooting chart (Table 23-2).

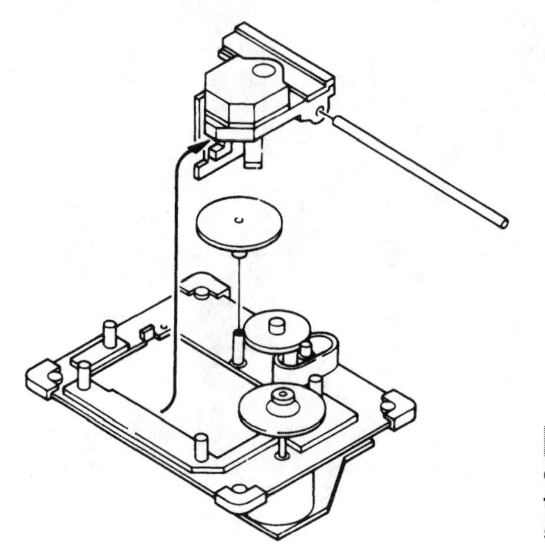

FIGURE 23-27 **Notice how the optical assembly is mounted inside the mechanical assembly with a sliding rod.**

TABLE 23-2 A LASER TROUBLESHOOTING CHART

TESTS	HOW TO
Power laser diode check	If a power meter is not available, use an infrared card or an infrared detector/indicator (figure 3-23) to see if the diode is operating.
Laser diode current check	Measure the voltage across the emitter resistor (figure 3-21) and determine the voltage and current. Check the current measurement against the numbers on the optical assembly.
Power meter tests	Measure the laser diode output with an infrared power meter. Start at the 1-mW scale and compare it with manufacturers laser output.
RF or EFM waveforms	Scope the RF terminal of the RF amplifier IC or transistors and determine if the EFM waveform is present and correct.

24

LOW-VOLTAGE POWER SUPPLIES

CONTENTS AT A GLANCE

The low-voltage power supply is one of the most-important circuits in the CD player; without adequate voltage, very few circuits can operate. Most electronic technicians first verify that the dc voltages in the power supply are correct before proceeding to the section that they suspect is faulty (Fig. 24-1). Extremely low voltages can indicate a defective power supply, or a leaky component in the voltage source could be overloading.

In addition to improper voltages, a malfunctioning power supply can produce excessive hum if the filter or regulator feeding the voltage source of the audio circuits is defective. Erratic relay operation could result from improper capacitor filtering. Finally, improper filtering of the dc sources can lower the dc voltages applied to the various sections of the CD player.

A digital multimeter (DMM), oscilloscope, and digital capacitance meter can help you locate most defective components in the low-voltage power supply. Of course, the VOM or VTVM can serve to take voltage measurements as the digital multimeter. But besides crucial voltage measurements, a DMM with a diode/transistor test helps when testing leaky diodes and regulator transistors. Overloaded circuits can be tested with the milliampere current range of the DMM. Use the oscilloscope waveform of the different voltage sources to indicate poor filtering. Finally, the digital capacitance meter is used to check for an open filter capacitor.

The Block Diagram

The low-voltage power supply provides the various dc output sources (Fig. 24-2). Besides voltages, the diagram can indicate zener diode and transistor regulation. A typical block

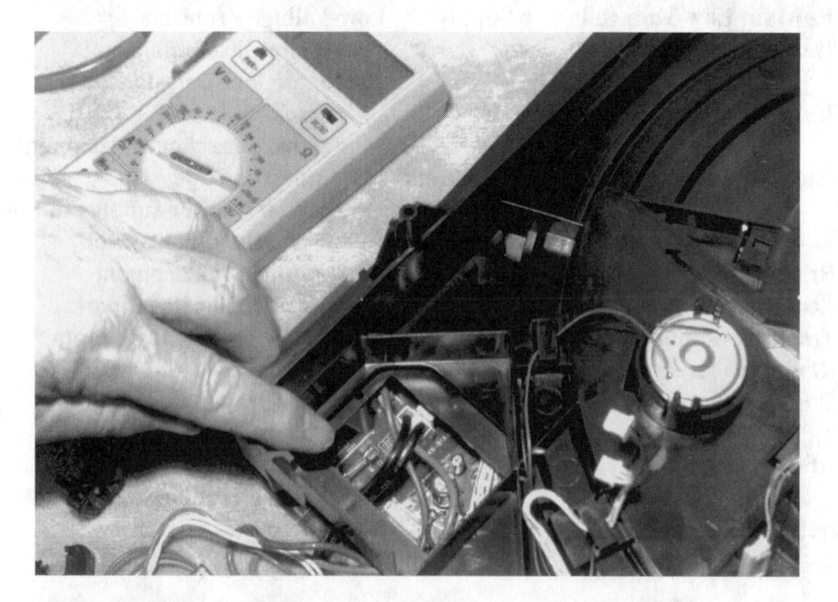

FIGURE 24-1 Checking the voltage in the low-voltage power supply of a CD changer.

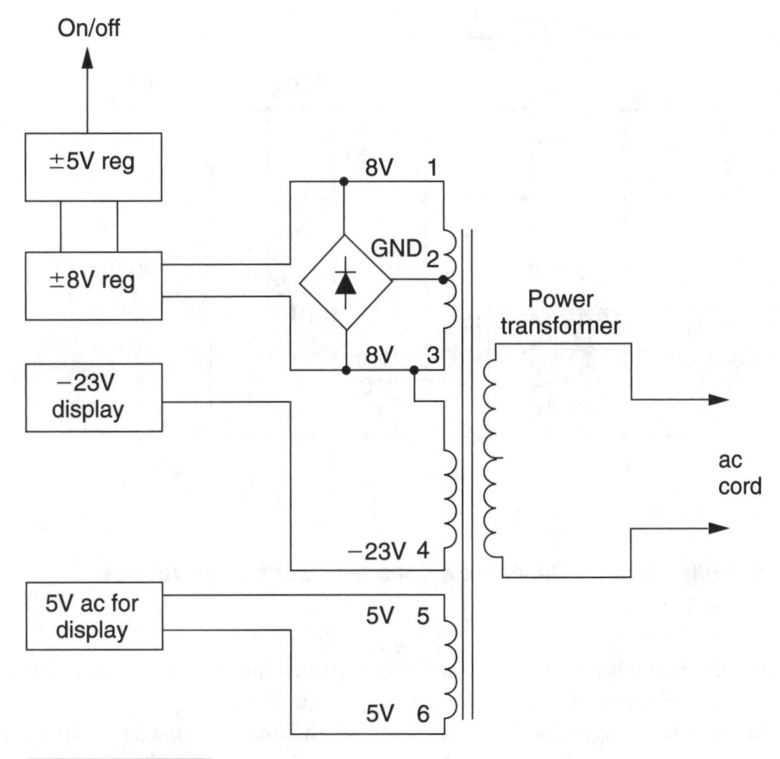

FIGURE 24-2 A block diagram of the Denon DCM-560 low-voltage power supply.

diagram shows bridge, half-wave, or full-wave rectifiers. Notice that the bridge rectifier has one diode symbol in the center of the diamond with the cathode terminal pointing toward the positive voltage source. The various transformer terminals are tied to the bridge half-wave rectifiers and fluorescent display tube. With the block diagram, you can quickly locate the possible improper voltage source.

The Main Low-Voltage Power-Supply Circuits

After locating the suspected low-voltage source with the correct symptom and block diagram, proceed to the main schematic diagram. The low-voltage circuits are tied directly to the other circuits in the main schematic or there might be a separate power-supply circuit. Here are shown the various voltage sources and where they are supplied to the various sections of the CD player.

Many of the manufacturers print the various dc and ac voltage sources in red ink. Notice the location of the actual ac voltages from the transformers, as applied to the bridge and diode rectifiers in Fig. 24-3. The input and output voltages are located on the regulator

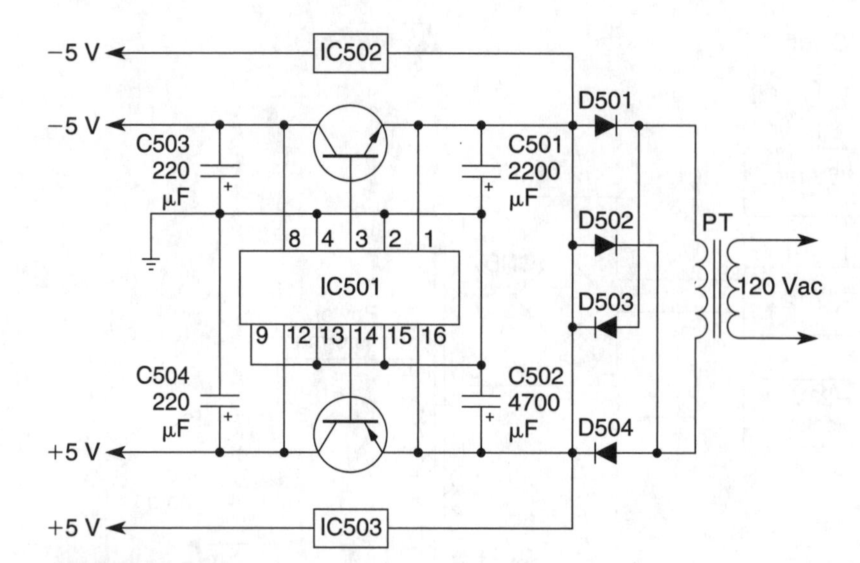

FIGURE 24-3 Denon DCM-460 low-voltage power-supply voltage sources.

components. Other manufacturers do not list voltages on the main schematic, but they are located on a voltage chart of ICs and transistors (Table 24-1).

Sometimes the dc voltages are drawn with red ink and boxes are used to help signal trace the various voltage sources (Fig. 24-4). A boxed red voltage might indicate the actual voltage measurement with no signal input. A solid red line indicates the positive B+ power supply, but the dotted red line indicates the negative B- power supply. The signal path might be traced with a halftone path (Fig. 24-5).

TABLE 24-1 A VOLTAGE CHART OF REGULATOR TRANSISTORS

REGULAR TRANSISTOR	Q603			Q604			Q602		
Terminal number	D	S	G	E	C	B	E	C	B
Voltage at play mode	−5.3 V	−9.2 V	−9.2 V	5 V	8.8 V	5.1 V	−5 V	−9 V	−5.6 V
REGULAR TRANSISTOR	**Q606**			**Q607**					
Terminal number	D	S	G	E	C	B			
Voltage at play mode	13.5 V	9.5 V	9.5 V	−16.7 V	−26 V	−17.1 V			

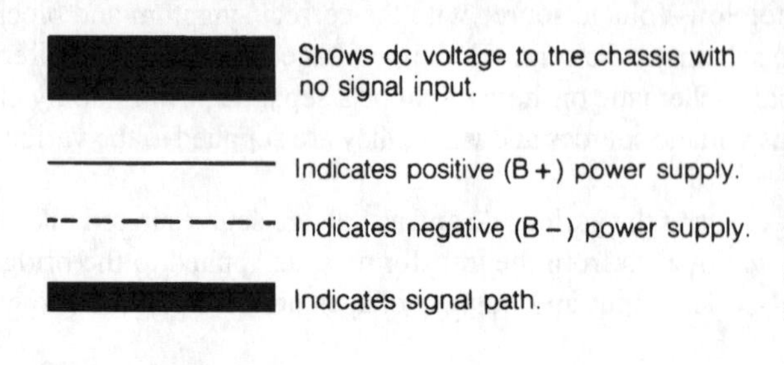

FIGURE 24-4 The low-voltage sources might be in red; solid lines denotes B+ voltages and dotted lines denotes B− voltages.

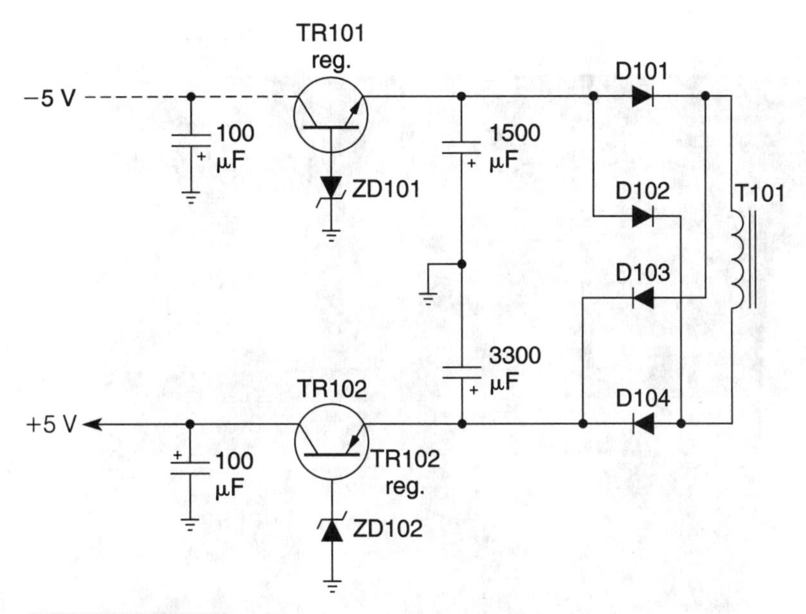

FIGURE 24-5 The low-voltage source might contain a solid or dotted line with a positive and negative voltage.

The Power Transformer

The ac power transformer in the CD players is quite small compared to a TV transformer (Fig. 24-6). Usually, it is the only transformer on the chassis. The power transformer is mounted on the chassis, and the other power-supply components are mounted on the main PC board.

Because CD players are made for many different countries, the primary winding must be adjustable to accommodate a range of power-line frequencies, currents, and voltages. A multi-voltage model would have an ac hookup of 120/220/240 V at 50 to 60 Hz. Various ac voltage taps are switched on the ac transformer primary winding (see Fig. 24-7). Notice when different input voltages are used, the protection fuse must change in amperage. In the United States, most systems use 120 V at 60 Hz.

The primary winding has a fuse at the power transformer in the Optimus CD-7105 changer (Fig. 24-8). Bridge rectifiers D11, D12, D13, and D14 provide a negative and positive voltage source to the regulator (IC20) at pins 16 and 1. The positive 8.3-V source is filtered with a 3300-μF capacitor and the negative supply with a 3300-μF capacitor with a working voltage of 16 V.

A +5-V regulated source is found at pin 12 and filtered by capacitor C7 (470 μF). The negative ($-V_{CC}$) source is found at pin 3 of IC20 and filtered by C28 (100 μF). The unregulated +8- and -8-V sources are taken directly from C25 and C26.

Notice a separate secondary winding from the power transformer to the display tube. The ac input transformer connection of an Onkyo C606 CD player for the countries of Switzerland, Sweden, the United Kingdom, and Australia are shown in Fig. 24-9.

The transformer secondary can have three or more separate windings providing ac voltage to bridge, full-wave, and half-wave rectifiers. The center tap of each winding might be

FIGURE 24-6 The power transformer and low-voltage sources in the Magnavox CDC-745 CD changer.

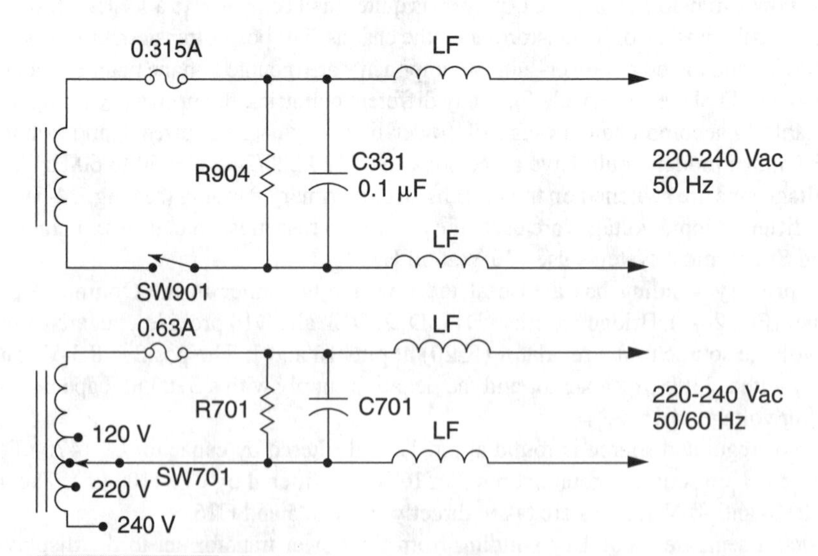

FIGURE 24-7 Because CD players are used in many different countries, the correct input voltage and frequency must be selected.

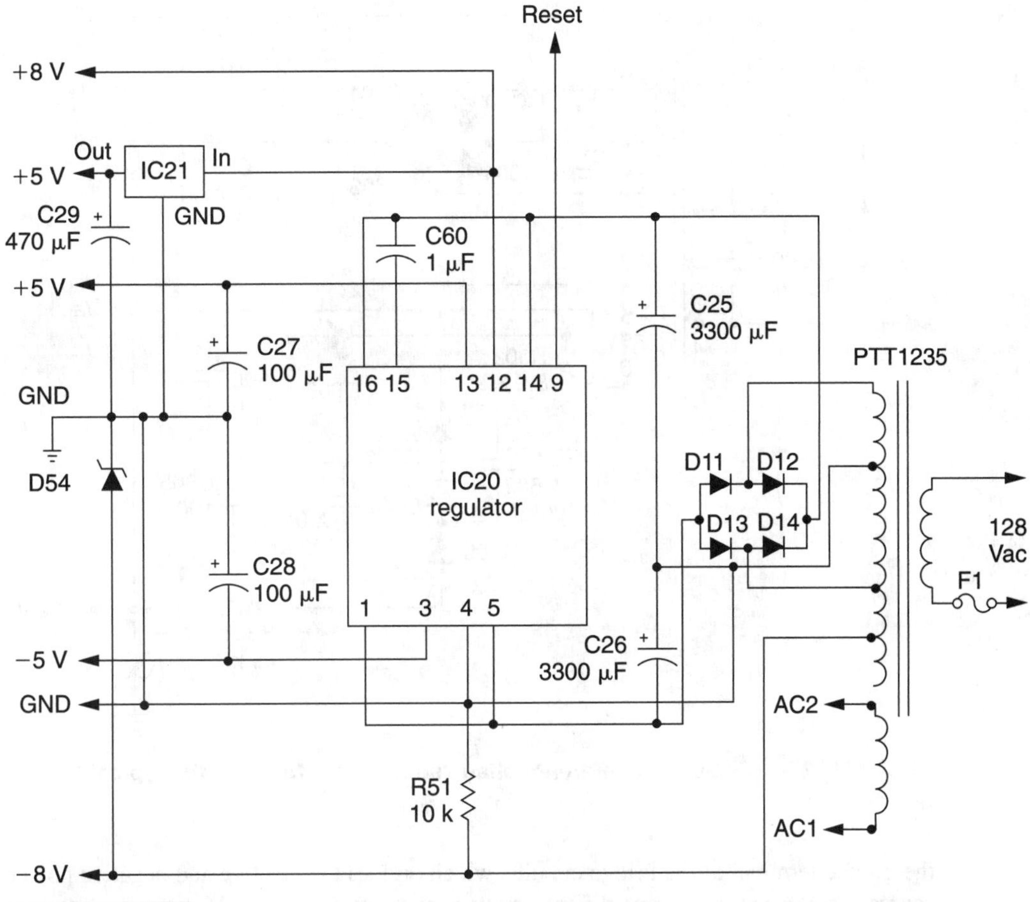

FIGURE 24-8 The low-voltage power sources in the Optimus CD-7105 changer.

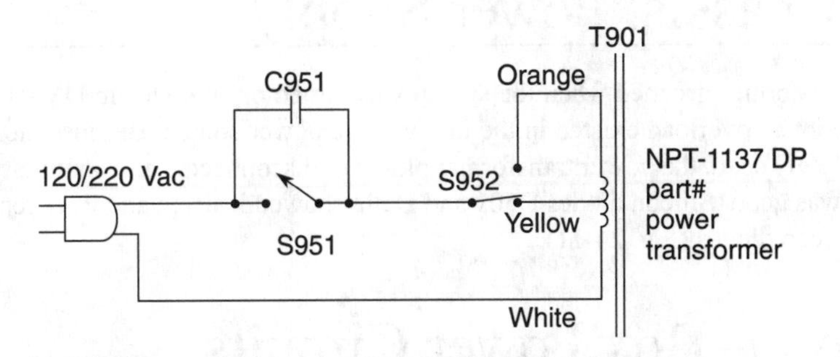

FIGURE 24-9 The primary winding of T901 must be switched to 120 V for North America products.

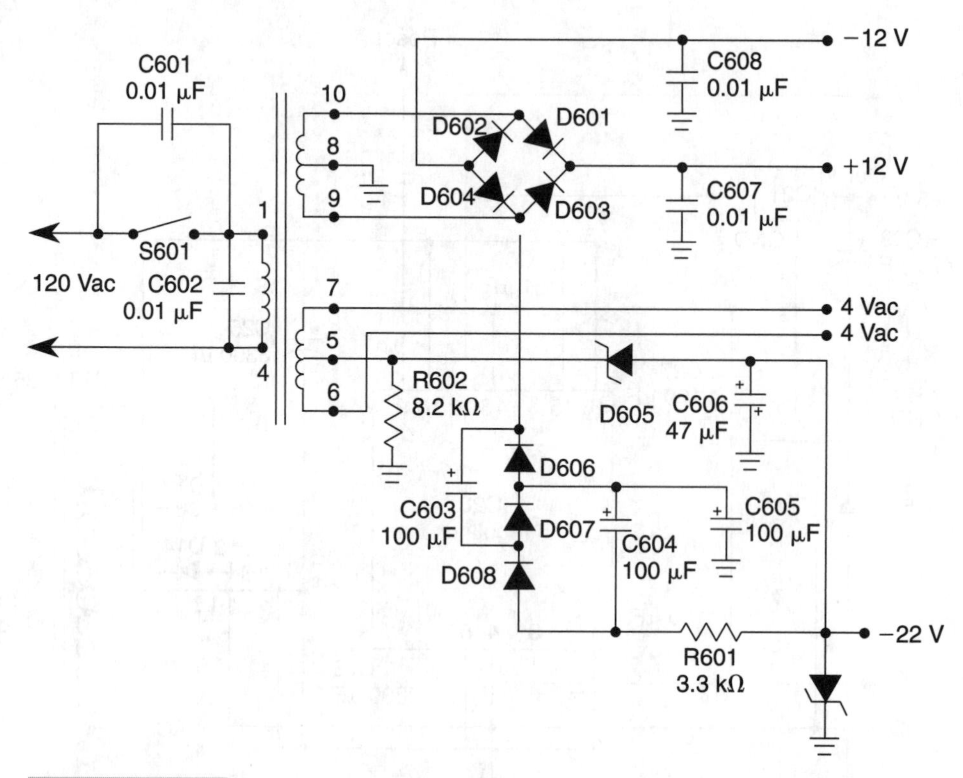

FIGURE 24-10 Several different voltage sources are found in the typical CD player.

the ground terminal of the bridge circuits, which divides the positive and negative power sources. In some models, three different voltage sources are produced from one secondary transformer winding (Fig. 24-10).

Dead Chassis: Power Supply

The power transformer groaned when the switch was turned on in a Denon DM-560 CD player. No doubt an overload existed in the low-voltage power source. Because the chassis was completely dead, the power transformer plug was disconnected and continuity was checked and was good. Silicon diodes D501 and D503 showed leakage and were replaced with 2.5-A silicon diodes (Fig. 24-11).

CD Boom-Box Power Circuits

The low-voltage power supply within the compact disc player might have a conventional bridge rectifier with large filter capacitor input. The cassette tape motor might have transistor and zener diode regulation. Separate transistor and zener diode regulation is used in radio and preamp audio cassette head circuits (Fig. 24-12).

Besides transistor and diode regulation in the regular low-voltage circuits, you might find a separate transistor and zener-diode voltage regulation within the CD circuits. The regulated +10-V source is fed directly to the driver ICs of spindle and SLED motors, to driver ICs of tracking and focus coils.

The voltage-regulated +5-V source feeds the rest of the circuits in the CD section. You might find a separate transistor regulator feeding +1.5 V to the laser diode of optical assembly (Fig. 24-13). The failure of the zener diode or transistor regulators might prevent CD operation in the various CD circuits. Low or no voltage can occur with an open or leaky transistor or a zener diode.

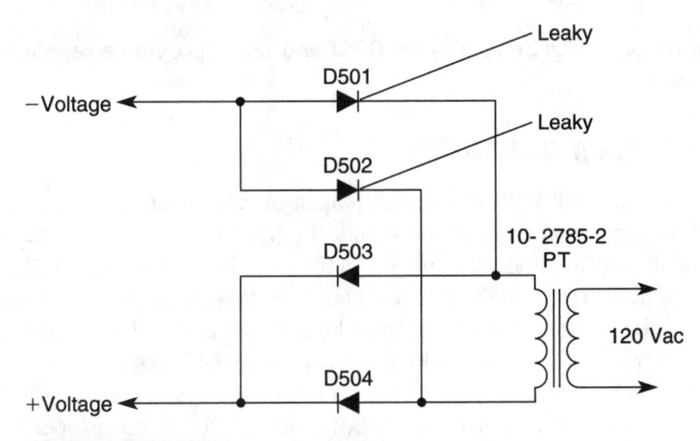

FIGURE 24-11 A leaky D501 and D503 produced a dead chassis in the Denon DCM-560 CD player.

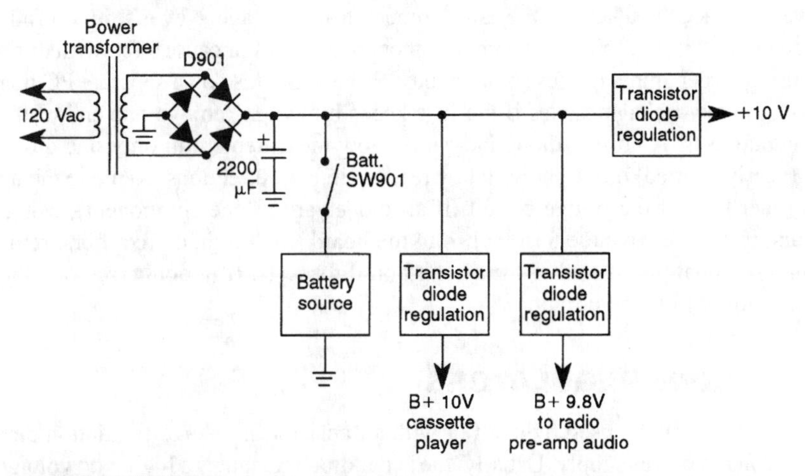

FIGURE 24-12 A block diagram of the various voltage sources in a boom-box CD-cassette player.

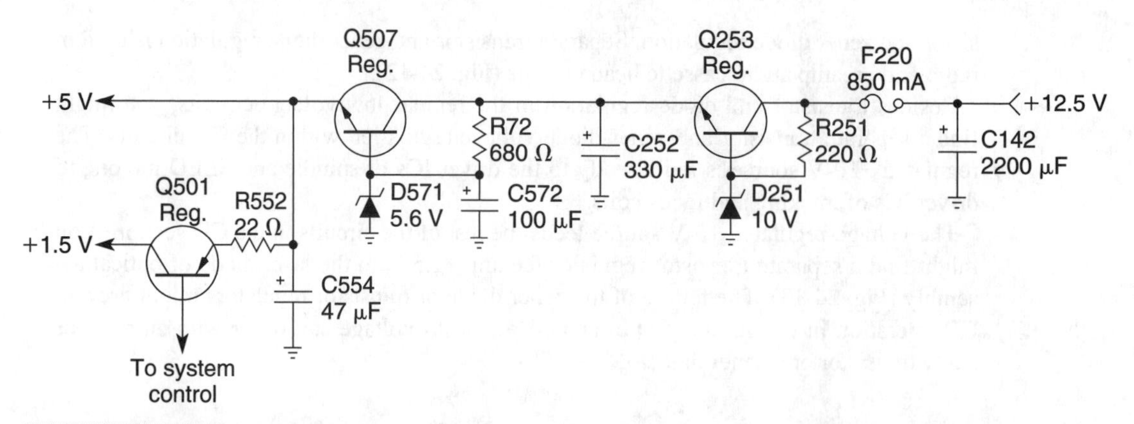

FIGURE 24-13 **Q253 and Q507, with Zener diodes D251 and D571, provide a regulated voltage source in this boom-box player.**

BRIDGE RECTIFIER CIRCUITS

Bridge rectifiers can be in the form of one component or four separate diodes, and they number two or three per low-voltage power supply. In Fig. 24-14, six duo-diode components form the bridge rectifiers and are mounted on the main PC board (Fig. 24-15). If one diode is leaky, it could open the line fuse. Two diodes are frequently faulty in a defective power supply. The whole bridge component must be replaced if one diode is leaky, but if four separate diodes compose the bridge circuit, replace only the leaky diodes.

Look for a shorted or leaky diode, voltage regulator transistor, or zener diode if the line fuse opens. Sometimes the line fuse opens because of flashover from a transistor or IC; however, when the fuse is replaced, the player operates normally. A change in power-line voltage can cause the line fuse to open. Suspect that a component in one of the different voltage sources is defective if the fuse keeps blowing with a normal low-voltage power supply.

Always replace the fuse with the same amperage and voltage. Never wrap tin foil around the fuse to just "get by." You will damage other components or cause a fire with this method or by using larger amperage fuses. Some models have the fuse on a separate PC board.

Check the low-voltage diodes if the fuse keeps blowing. Remove one end of the diode for accurate tests. A normal diode indicates a low measurement in only one direction. A leaky (nearly shorted) diode shows a low reading in both directions. Remove the anode or transformer lead of the bridge circuit (if all diodes are in one component), using solder wick and iron to remove the terminal from the board. In duo-diode rectifiers, remove the common terminal that is tied to the circuit board for tests. If in doubt, remove the entire diode assembly from the circuit.

ZENER DIODE REGULATORS

The zener diode can be used by itself or with a transistor in voltage-regulation circuits of the low-voltage power supply. Usually, the zener diode regulates a higher dc voltage to the display tube (Fig. 24-16). The ac voltage from the power transformer is rectified by D713 in the Denon table-top CD player. Notice that a negative voltage is taken from the anode terminal of D713.

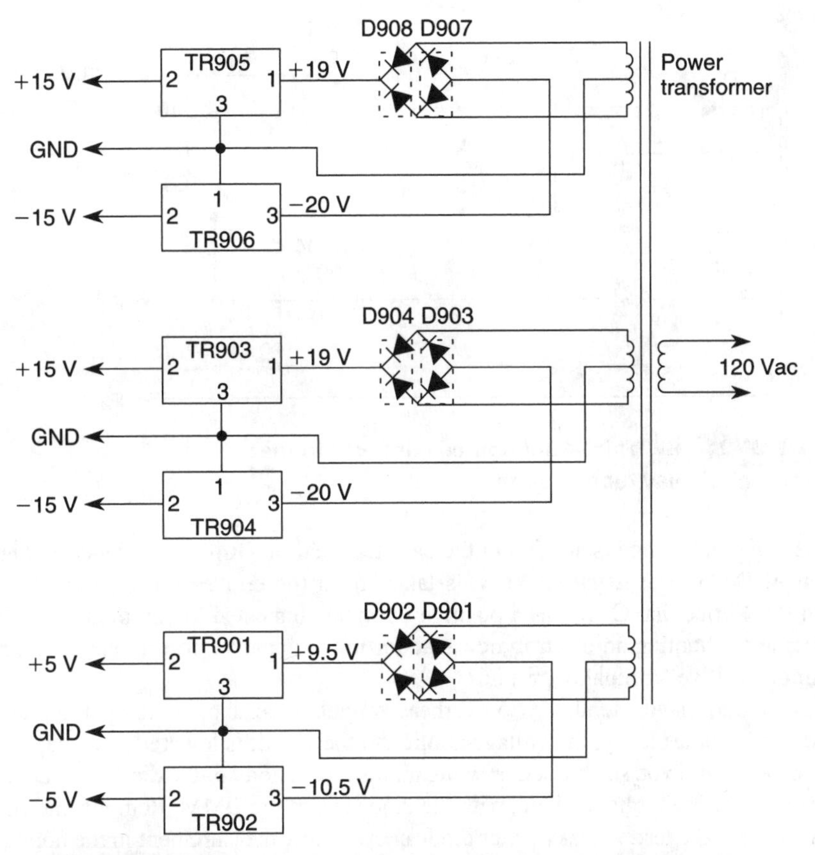

FIGURE 24-14 Three different bridge rectifiers with two separate sections provide three different positive and negative voltage sources.

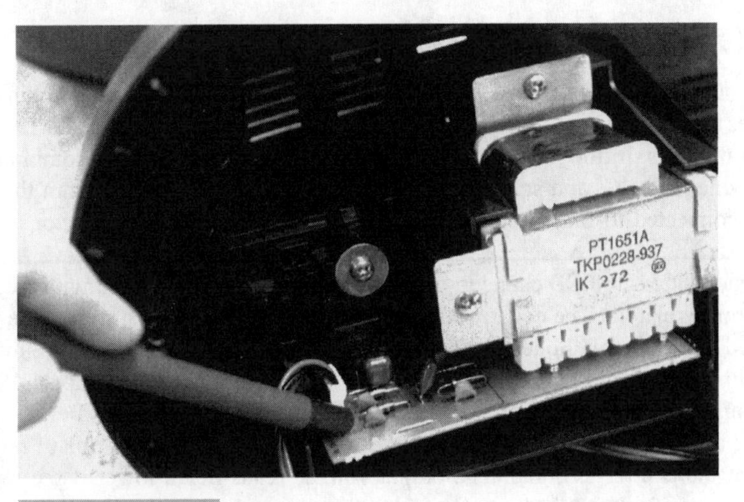

FIGURE 24-15 Silicon rectifiers are often mounted close to the low-voltage transformer.

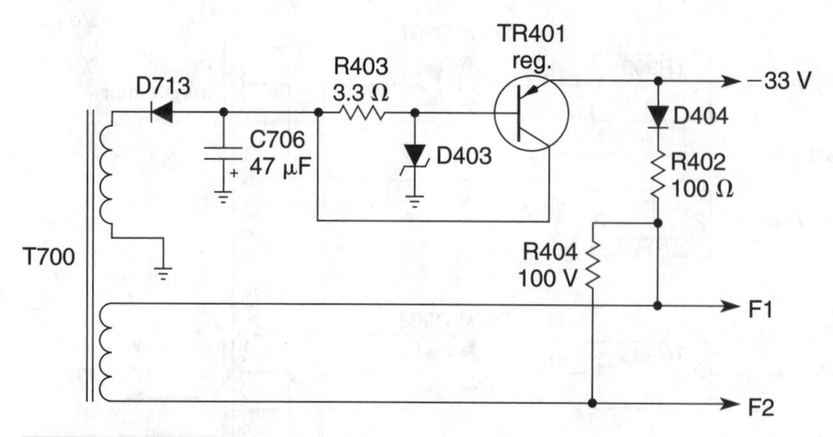

FIGURE 24-16 **This –33-V source connects to the fluorescent-display tube circuits.**

The negative voltage is applied to the base terminal of TR401 with zener voltage regulation of D403. The negative 33 V is taken from the emitter pin of a PNP transistor (TR401). Notice that C706 has a positive terminal connected to common ground. When replacing or shunting an electrolytic capacitor in the negative power supply, connect the common positive terminal to ground.

Zener diodes have a tendency to overheat, producing leakage or an open circuit. When the diode becomes leaky, the voltage applied to the circuit is lowered. Overheated diodes can be detected if you see burned or white marks on the body of the diode. Check the zener diode like any low-voltage diode, with the diode test of the DMM. Remove one end of the diode for leakage test. A leaky zener diode shows a low measurement in the normal direction and a higher resistance with reverse test leads. Replace the diode if a measurable resistance is found in both directions.

TRANSISTOR REGULATORS

The power supply might have power transistor regulators or be used in a combination of zener diodes and transistors. Large voltage-regulator circuits can have from 2 to 13 voltage-regulator transistors. In the latest low-voltage sources, you might find a combination of transistors, one large IC, and small IC regulators. Often, the dc voltage from the bridge diodes are connected to the emitter and others are connected to the collector terminals (Fig. 24-17).

In the Denon DCM-560 CD changer, one large IC has voltage fed into it with the help of two transistor regulators. The negative dc regulator has an NPN transistor with the emitter terminal tied to the silicon diodes. A positive regulated dc source is tied to the emitter terminal of a PNP transistor with regulated output at the collector terminal (Fig. 24-18). Separate –5 V and +8 V are taken from separate TR501 and TR502.

Regulator transistors have a tendency to open or leak. If the regulator opens, no output voltage is at the emitter terminal. The output voltage reads much lower if the regulator transistor becomes leaky. A leaky regulator transistor can damage the zener diode tied to the base circuit. Check all diodes and transistors in a defective voltage source.

Regulator transistors have a tendency to open or leak. If the regulator opens, no output voltage is at the emitter terminal. The output voltage reads much lower if the regulator transistor becomes leaky. A leaky regulator transistor can damage the zener diode tied to the base circuit. Check all diodes and transistors in a defective voltage source.

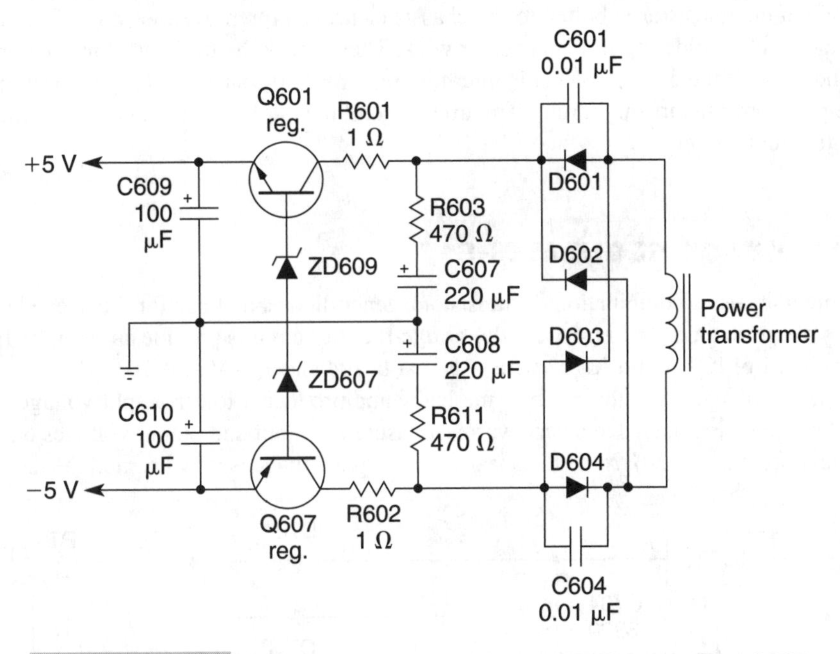

FIGURE 24-17 **Q601 provides positive regulated source and Q607 provides a negative voltage source.**

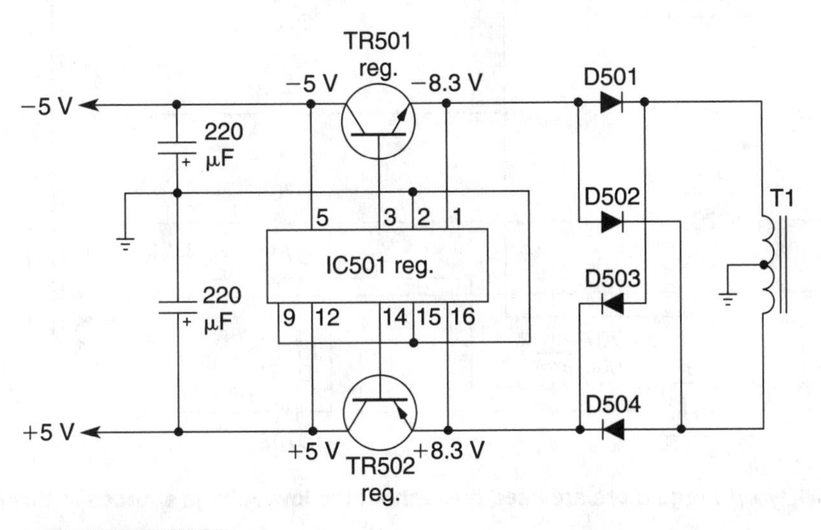

FIGURE 24-18 **A combination of regulator transistors and one large IC provides a –5-and +5-V source.**

Suspect that the voltage regulator transistor is leaky if an improper voltage is read at a given voltage source. Quickly check the transistor in the circuit with a transistor tester or with the transistor test of the DMM. Remove the base terminal for correct leakage tests. If in doubt, remove the regulator transistor from the board.

When the transistor is bolted to the chassis or heatsink, remove the solder from all terminals with a soldering iron and solder wick. Then, check the transistor for open or leaky conditions. If the output voltage is intermittent, replace the suspected regulator transistor. Regulator transistors might run warm after several hours of operation, but they should not be too hot to touch.

IC VOLTAGE REGULATORS

Some units use a combination of transistors, zener diodes, and ICs for the needed regulator voltage shapers. In Fig. 24-19, the bridge-rectifier circuits provide an input of 10 V at terminal 1 of IC701 and IC702 with –15 V at IC704 and +15 V at IC703.

An IC voltage regulator can become leaky and produce a lower output voltage. If it is leaky, the IC regulator feels very warm. Measure the input and output voltages on the IC regulators (Fig. 24-20). If the IC is leaky, the output voltage is very low and the input volt-

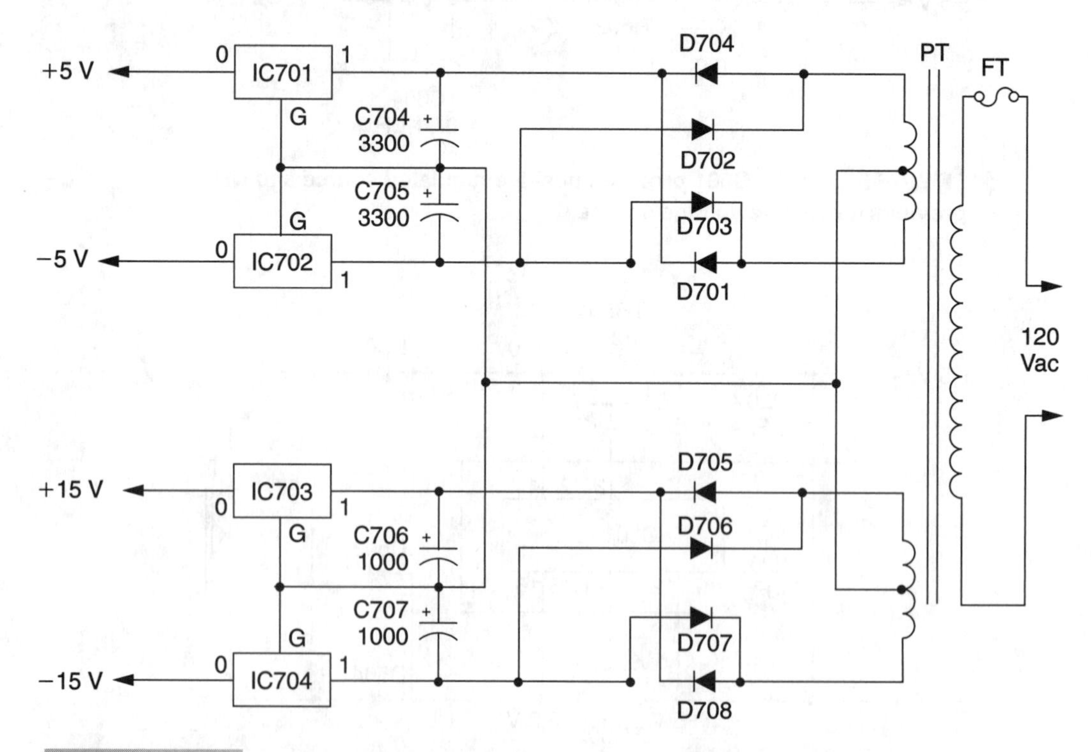

FIGURE 24-19 IC regulators are used throughout the low-voltage sources in this CD player.

In 1 IC101 Out
+25.2 V +10 V
GND

In 1 IC103 Out
−15 V −5 V
GND

FIGURE 24-20 **Check the output and input voltage sources on the IC or transistor regulators.**

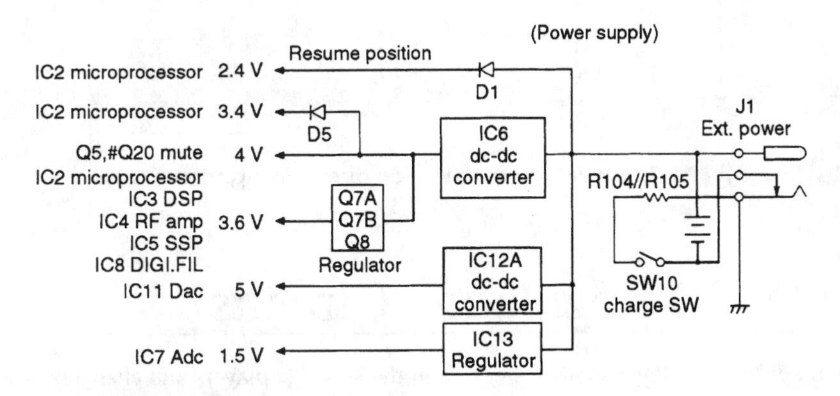

FIGURE 24-21 **A block diagram of the Realistic portable CD-3370 low-voltage circuits.**

age (pin 1) is somewhat lower. If open, no voltage is measured at the output terminal and a higher-than-normal input voltage is at pin 1. If the output voltage source is low after replacing the IC or transistor regulator, suspect that a component tied to the voltage source is leaky.

DC-DC CONVERTER

You might find a dc-dc converter in the low battery-operated CD portables or auto receivers. In the Realistic CD-3370 portable CD player, the player can be operated from an external ac power source or from two small batteries. The rechargeable nickel-cadmium batteries can be charged by the ac adapter by turning on charge switch (SWIG).

The dc-dc converter output terminals, pins 4, 5, and 29 (IC6), supplies the power circuitry with +4 V and the dc-dc converter (IC12A) supplies the player circuitry with +5 V, converted from the dc supply (2.0 to 4.0 V) (Fig. 24-21). Notice that IC6 converts the real low-battery voltage to a +4 V, and IC12A to +5 V. Before supplying the audio circuitry, the dc-dc converter output is filtered by C7 and C8 to eliminate ripple.

To detect a low-battery condition, IC9D compares the battery voltage with a 2.1-V reference. When the battery voltage falls below the reference voltage, IC9D outputs a low signal to the microprocessor, which then displays a low-battery indicator on the LCD (Fig. 24-22).

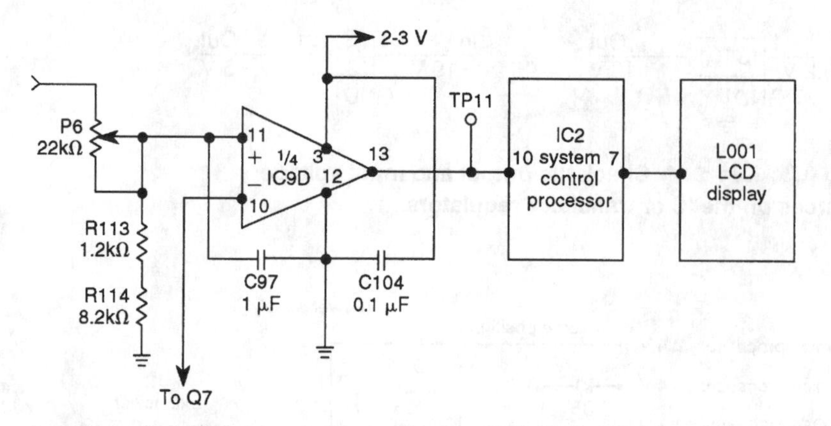

FIGURE 24-22 **The low-battery detection circuit in a portable CD player.**

Auto CD Regulation Circuits

Several different voltage sources are used in the auto CD players and changers. The battery 12.8-V source can be fed to several transistor, IC, and zener-diode voltage-regulator components. In the Denon DCH-500 changer, low-voltage circuits contain several 4.8-, 4.9-, 5-, 7.2-, and 9-V sources, which are derived from the 12.8-V car battery (Fig. 24-23).

The 4.8-V source is fed to a crystal-control signal processor (IC13), through the D902 diode to a 12.8-V battery source. Also, the 9-V source feeding IC13 and the motor drive IC (IC652) is found at pin 3 of IC902. The 5-V source at pin 3 of IC901 regulator with Q901 and Q902 as transistor regulators, and CD on terminal 2 of the system-control IC

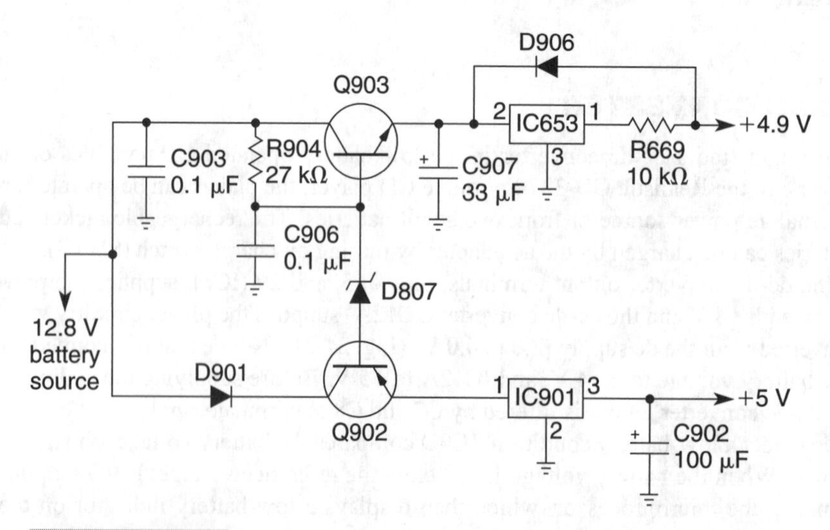

FIGURE 24-23 **Zener diode, IC, and transistor regulators are in a Denon CD changer.**

(IC601). The 4.9-V source feeds to system-control IC601 from IC653 and the Q903 voltage regulator from the 12.8-V line.

Overloaded Power-Supply Circuits

Unplug the suspected component in the lower-than-normal voltage source if any other components in the power supply seem normal. If the regulated voltage increases, a leaky IC, transistor, diode, or capacitor is pulling down the voltage source within the connecting circuits. Suspect a regulator IC, transistor, or zener diode is leaky if the voltage remains low after disconnecting the power source.

Locate what sections of the CD player are powered with the low-voltage dc source on the schematic diagram (Fig. 24-24). Take a low-resistance measurement. If the measurement is under 150 Ω, you can assume that a leaky component is lowering the voltage unless a dc motor is in the same circuit. Pull out each plug-in cable component with the ohmmeter attached as a monitoring device. If the resistance goes up at once, you have located the section with the leaky component.

Check each component in the overloaded circuit until the defective component is located. Take the ohmmeter probe and a resistance measurement of major components, such as transistors and IC terminals tied to the same low-voltage source. As you approach the leaky component, the resistance should be lower than at the common low-voltage source terminal.

Several Low-Voltage Sources

You will find many dc voltage sources within the CD player. Often, a pair of negative and positive voltage sources feed to the various circuits. You might have a +15-V and –15-V, +10-V and –10-V, and two separate +5 and –5-V sources in the larger CD players (Fig. 24-25). Check for a high negative source that connects to the display fluorescent-tube indicator circuits.

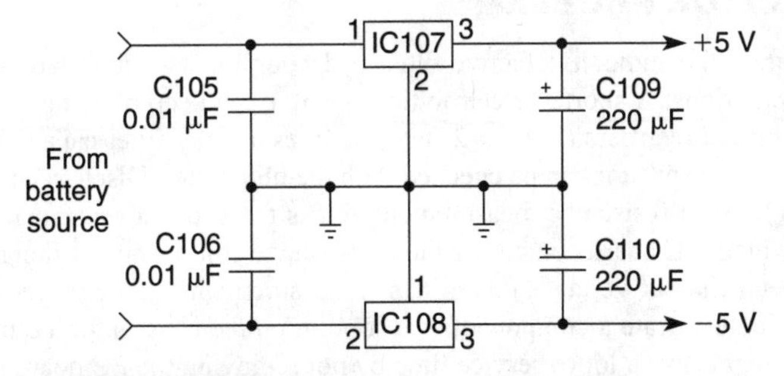

FIGURE 24-24 A typical low-voltage + and – source with ICs as regulators.

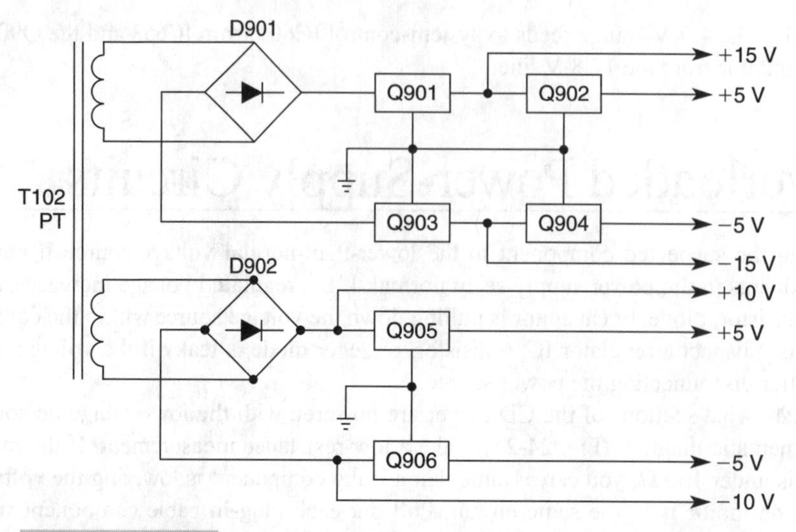

FIGURE 24-25 **Many low-voltage sources are regulated with ICs.**

Low-Voltage Problems

If the power-supply source is entirely dead, check for an open fuse, leaky silicon diodes, and open transformer primary winding. If some voltage sources are working and some circuits are not, suspect that a transistor or IC regulator is leaky. The open regulator transistor can produce a dead voltage source. Likewise, open or leaky IC regulators can cause a dead chassis (Fig. 24-26).

Go directly to the main filter capacitor and measure the dc voltage on the positive (+) terminal. If a fairly normal voltage is found here, check the output terminal of a regulator transistor and IC for normal voltage. Sometimes a small resistor in series with the regulator transistor will open with a shorted or leaky transistor regulator. Do not overlook the possibility that a component tied to the low-voltage source is leaky or shorted, loading down and producing very low voltage.

FILTER CAPACITOR PROBLEMS

Besides causing hum in the audio, a defective filter or decoupling capacitor can produce erratic or shorted conditions. A shorted electrolytic capacitor can keep blowing the main fuse or destroy the bridge rectifiers (Fig. 24-27). Sometimes a leaky filter capacitor runs warm. A suspected leaky capacitor can be checked with the ohmmeter. Discharge the suspected capacitor and take a resistance measurement across the capacitor terminals (Fig. 24-28). Because the main PC board containing the low-voltage components is mounted at the bottom of the metal chassis, voltage and resistance measurements on the suspected capacitor are very difficult. Locate a component or test point on the chassis for a quick capacitor check. You might save a lot of service time by not removing the PC board to find the electrolytic capacitor. If the test indicates that the capacitor is shorted, remove the PC board from the bottom of the metal chassis and remove the capacitor.

The defective filter capacitor might not be leaky, but it could have dried up or lost capacitance. Check the capacitor in the circuit with a digital capacitance tester. These low-priced capacitance meters are very handy in locating defective capacitors. A leaky or shorted capacitor shows a low resistance or no measurement at all. Here, a normal decoupling capacitor (33 μF) measures 34.9 μF in the circuit (Fig. 24-29). Replace the dried-up capacitor when it measures 5 μF less than normal. Remove the capacitor, if in doubt, for the accurate capacity measurement.

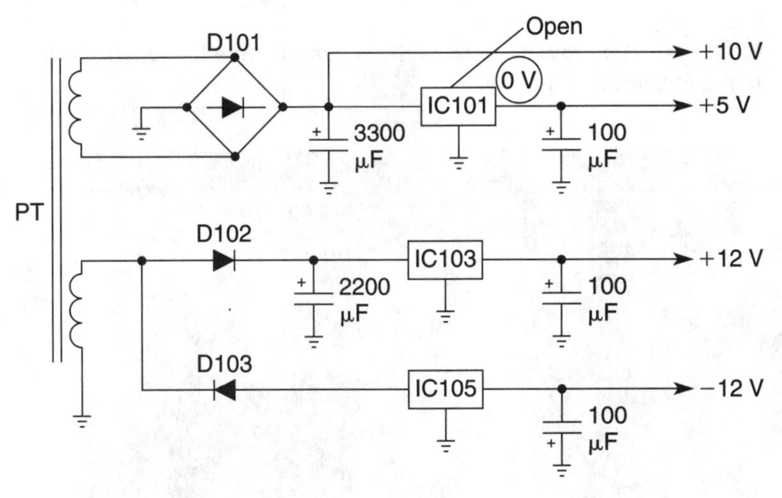

FIGURE 24-26 An open IC101 caused no +5-V source.

FIGURE 24-27 Power-supply diodes in the Magnavox CD changer.

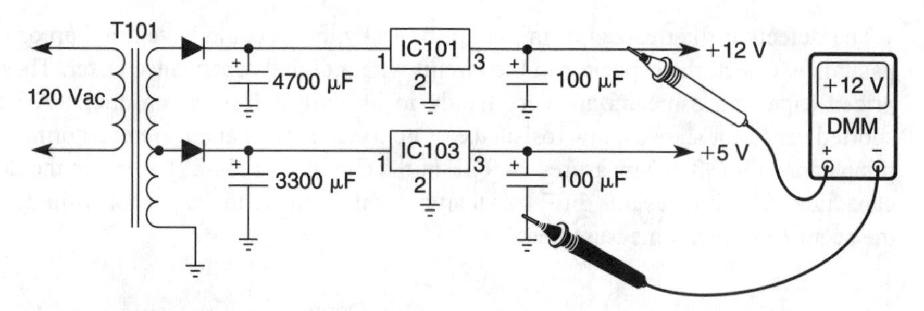

FIGURE 24-28 Defective filter capacitors can become leaky, shorted, or open in the low-voltage supply.

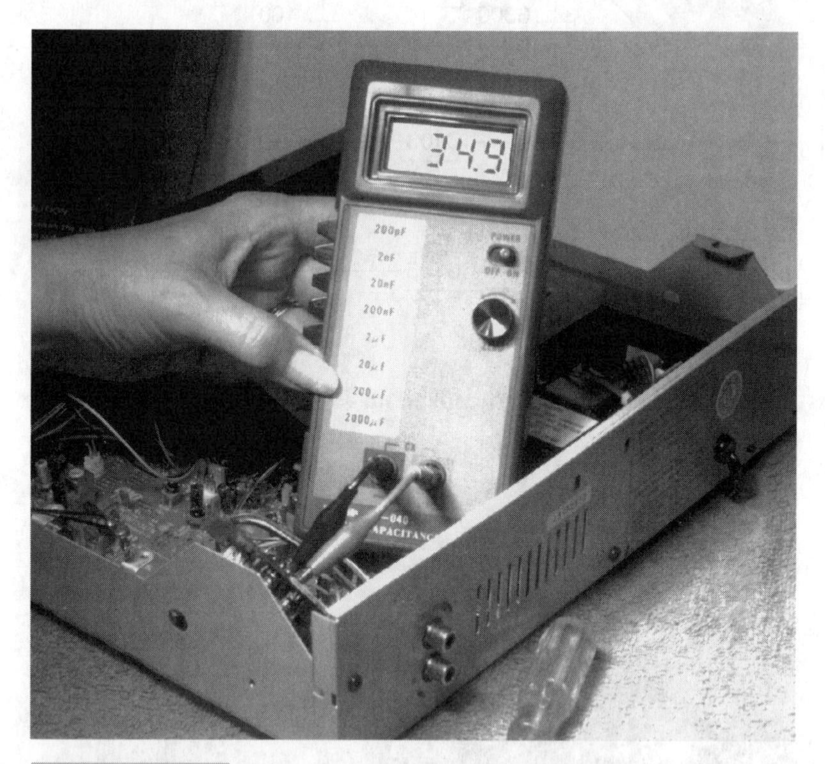

FIGURE 24-29 Check electrolytic capacitors with a digital capacitance meter.

Adequate filtering of each power voltage source can be checked with the scope. High ripple content can indicate an open or dried-up filter capacitor. Compare the waveform with the other positive or negative voltage source. For instance, if the +5-V source has higher ripple than the –5-V source, suspect that a capacitor is defective within the +5-V power source.

The main filter capacitors in the low-voltage source are 1000 μF (or greater). Turn off the power, discharge the voltage source, and clip another electrolytic across the one you suspect is defective. Never shunt another capacitor across the suspected one with the power on, or you could damage ICs and transistors. Replace the defective capacitor with one that has the same capacitance and voltage, or replace it with one that has a higher capacity.

Never replace a filter capacitor with one that has a lower voltage. The capacitor with lower voltage will overheat and could blow up. A 1000-µF, 25-V electrolytic capacitor could be replaced with a 2200-µF, 25-V type without any problems, provided the mounting space is adequate. Likewise, a 33-µF, 16-V decoupling capacitor can be replaced with a 47-µF, 16-V without any complications in the low-voltage power supply.

No Operation: Defective Filter Capacitor

Go directly to the low-voltage main filter capacitor on the PC board. This filter capacitor will be the largest electrolytic capacitor on the board. Check for the highest capacity. Measure the dc voltage across the filter capacitor (C507) in the positive power source. No voltage here might indicate that a capacitor, silicon bridge diodes, or secondary fuses are leaky (Fig. 24-30). You might not find a protection fuse within a low-priced CD player.

A leaky filter capacitor might have shorted out one or more silicon diodes in the bridge or full-wave circuits. If the CD player is left on too long, the primary winding of the power transformer (T103) might open. Low voltage is noted across the capacitor terminals if the electrolytic dries up or if an internal contact terminal is broken.

UNIVERSAL CAPACITOR REPLACEMENT

Most electrolytic filter capacitors can be replaced with universal types. Of course, the physical size of a larger replacement takes up a lot more room if the capacitor has a higher capacity and voltage rating. Always replace the capacitor with one that has a higher voltage and capacity rating if an exact-type capacitor cannot be located. These radial-type leads can be mounted higher up with reinforced leads or insulating "spaghetti" placed over each terminal for self support and insulation. Be sure that the capacitor terminals are tight so that the capacitor will not flop around (Fig. 24-31).

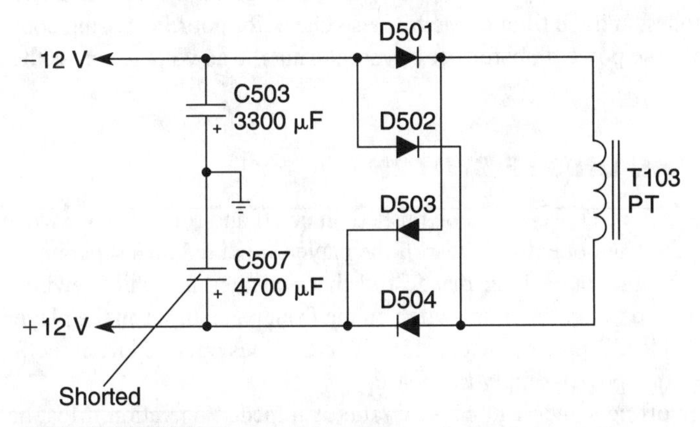

FIGURE 24-30 A shorted C507 (4700 µF) took out D504 and D503 in this low-voltage power supply.

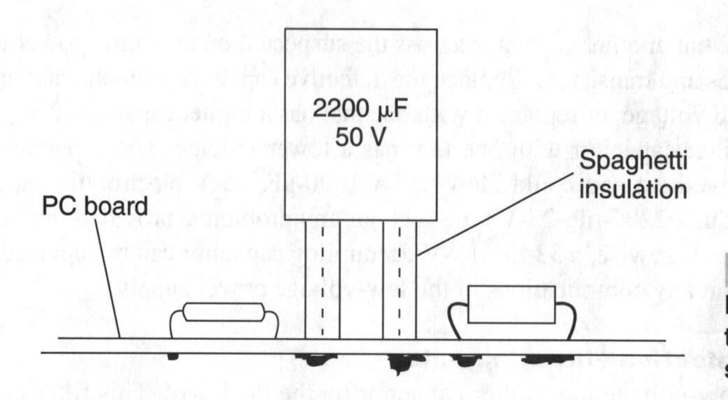

FIGURE 24-31 Mount large electrolytics above the chassis with insulated supports.

LIGHTNING DAMAGE

The power transformers and bridge rectifiers are at the highest risk when lightning strikes the outside power line. Power-line outage can cause the fuse to open. If the power lines are whipped together, 220 Vac can enter the CD chassis and damage the power transformer, fuse, on/off switch, or power-supply components. A direct lightning strike could damage the CD player beyond repair. So, be sure to make a complete estimate of all damaged parts before ordering replacements. Sometimes, when the ac power switch is off and lightning strikes, the charge jumps the switch and knocks out the power transformer. If very little visual damage is noticeable, most of the damage is probably contained within the power-supply circuits.

INTERMITTENT POWER-SUPPLY SOURCES

Intermittent power source within the CD player can result from a poor internal junction within the transistor or IC regulator. Poor component grounds can cause intermittent voltages. Do not overlook the possibility that a zener regulator diode is burned. Unsoldered or poorly soldered joints inside the ac cord tied to the primary winding can produce intermittent operation.

Monitor the dc voltage at the main filter capacitor and at the transistor regulator output terminal. If both meters become intermittent, suspect problems within the power transformer, silicon diodes, and the filter capacitor; also check for poor PC wiring connections (Fig. 24-32). Check the power transformer by monitoring the ac voltage across the ac secondary winding.

CHECKING THE ON/OFF SWITCH

The on/off switch in the CD player can be pushed on or off and rotated. The switch might be located toward the front or extreme rear of the player, mounted on a separate PC board (Fig. 24-33). You might find a long plastic rod that activates the on/off switch. In the boom-box CD player, a rotary function switch on the front panel turns on the player. In the Onkyo DX-C606 table CD player, the on/off switch (S951) is operated from the front and connected to a separate power-supply PC board.

The defective on/off ac switch can cause erratic or impeded operation. Most ac power switches are located in the primary-winding circuit of the power transformer. A few CD players have the primary winding of the power transformer connected directly to the power cord with the turn-on switch in the dc circuits. A dirty on/off switch can prevent normal contact operation. Spray cleaning fluid into the switch area and work the switch back and forth to clean the contacts. Replace an on/off switch that shows erratic or no operation.

Locate the defective on/off switch by clipping across the switch terminals with a short clip lead. The player should come on because it is now plugged into the power line. The on/off switch can also be checked by taking an ac voltage measurement across the switch terminals. Replace the switch if 120 Vac is measured across the switch terminals with the switch in the On position. These switches must be replaced with originals because they solder directly to a separate PC board.

TRANSFORMER REPLACEMENT

Power transformers pop, crack, or overheat if they have shorted windings or overloaded diodes. Remove the secondary leads from the circuit, plug in the power cord, and check the transformer for excess heat. If the transformer still runs very hot, replace it. In this case, the transformer has shorted windings. Suspect that diodes are shorted or leaky if the transformer operates cool and the fuse does not blow when the transformer is disconnected.

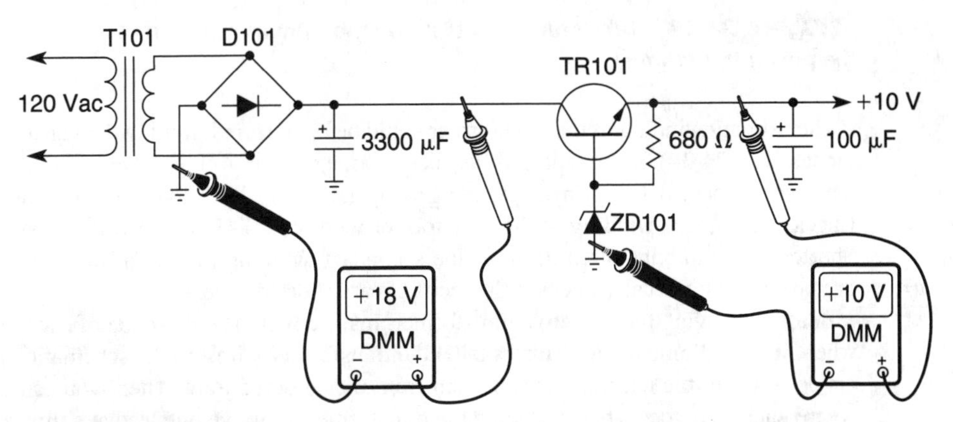

FIGURE 24-32 **Monitor the low-voltage power supply at two different sources to locate defective component.**

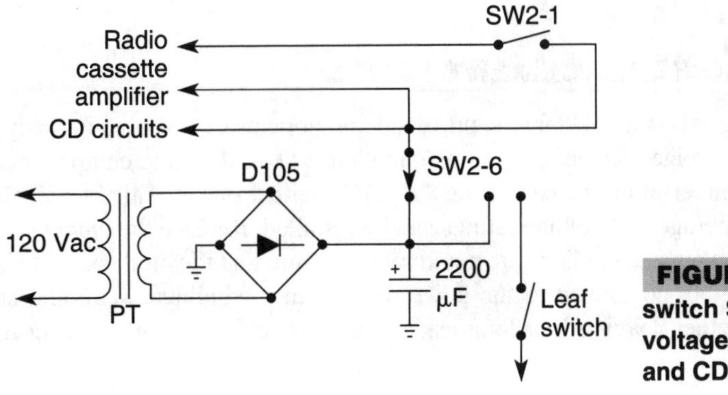

FIGURE 24-33 **Sliding switch SW2-6 switches voltage to the cassette, radio, and CD sections.**

FIGURE 24-34 Take continuity tests on the primary winding of the power transformer.

The primary winding of the transformer could be open from lightning damage or shorted diodes (Fig. 24-34). If the replaced fuse is too large or if tin foil has been wrapped around the fuse, suspect that a primary winding is open with leaky bridge rectifiers or diodes. Check the primary winding of the transformer with the 200-Ω range with the switch on. Shorted coil windings often occur in the secondary winding instead of the primary. The primary winding usually opens if the rectifiers are shorted or leaky.

Before removing the defective transformer, mark down the color code of each lead and where it goes. Some manufacturers label the transformer wires on the schematic, but others do not. Cut the terminal leads ½ inch from the soldered joint. Then, you can identify where each lead goes when replacing the transformer. Remove one lead at a time and solder on the new transformer wire. Whenever possible, replace the defective transformer with one that has the original part number. Transformer replacement in the CD player is not as costly as in the TV chassis.

UNIVERSAL REPLACEMENT PARTS

To prevent long delays in CD repairs, universal components can be used in many areas of the player. The bridge rectifiers, with one complete or two different components, can be replaced with universal bridge parts (Fig. 24-35). If neither one is available, single diodes of the same amperage and voltage ratings can be installed. Replace the diodes with single 3-A diodes when current readings are not known. Be sure that the diode polarity is correct with the ac symbol connected to the power-transformer windings. Wire and solder all diode leads together, leaving four long leads to go through the bridge-rectifier mounting holes.

Regulator transistors and zener diodes can be replaced with universal replacements. Cross reference the transistor and diodes for universal replacement. If the part number is not available, look up the operating voltage of the zener diode. The part number often provides the zener-diode voltage. For instance, an MTZ13C zener diode operates at 13 V. If in doubt about the wattage, replace with a 5-W zener diode. It's best to have one with higher wattage than a small one that could overheat.

Replace the IC regulator, the power transformer, and the on/off switch with the exact part number.

All filter and bypass capacitors can be replaced with universal parts. Replace with the correct voltage and capacity. The capacitor can have a higher capacity, but not a lower working voltage. Be sure that there is adequate room for mounting. Sometimes filter capacitors are mounted with the transformer on a separate PC board that is bracketed to the side of the metal cabinet.

QUICK LOW-VOLTAGE TEST POINTS

Inspect the line fuse before taking any voltage measurements. Check the voltage at each voltage source. Follow the schematic and locate the transistor, IC, or zener output regulator (Fig. 24-36). Each output voltage can be checked from the emitter terminal of the regulator transistor (1). Measure the collector voltage to locate a defective regulator transistor. Now check the input and output voltage at the IC regulator (2).

Go directly to the positive terminal of the bridge rectifier when locating a low-voltage or a no-voltage source (3). Check the voltage across the main filter capacitors, if accessible. Measure the voltage on each single diode in the bridge circuit. A higher positive voltage indicates that the B+ voltage is feeding the regulator circuits. Check the ac voltage applied to the bridge circuit with no dc output voltage (4).

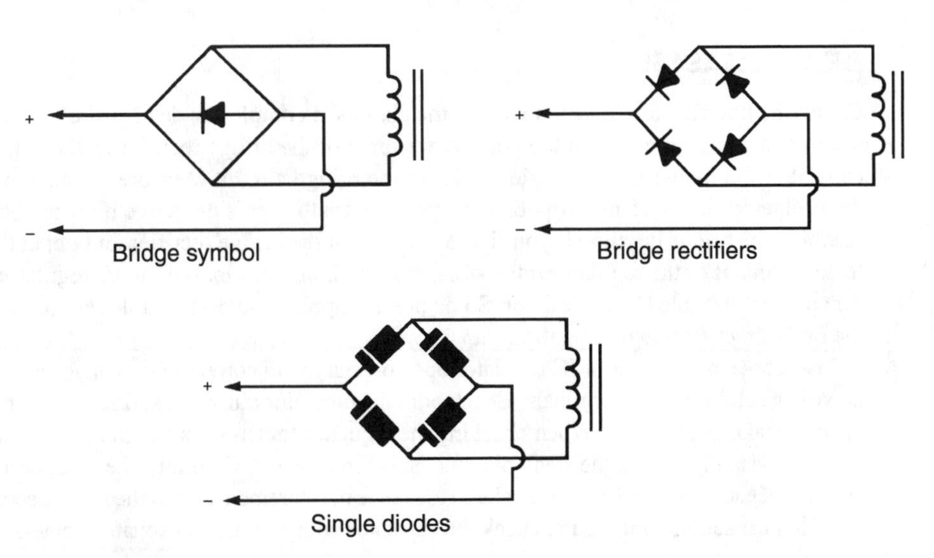

FIGURE 24-35 **The bridge component can be replaced with separate silicon diodes.**

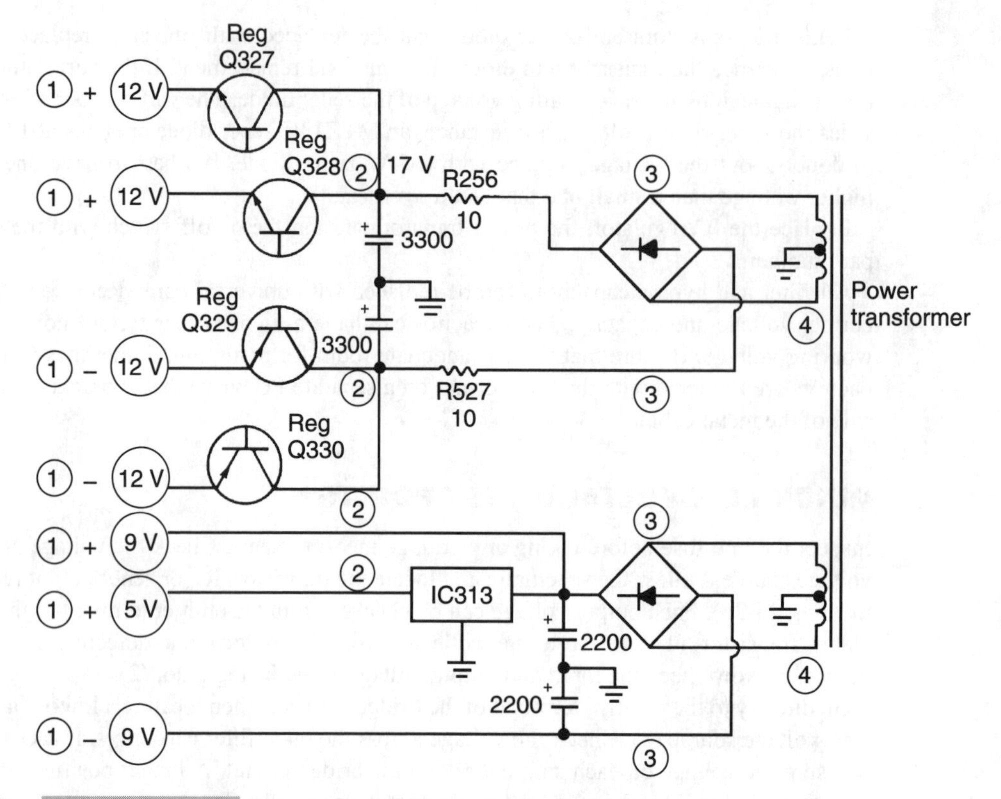

FIGURE 24-36 **Troubleshooting the low-voltage power supply using by-the-number tests.**

DEAD CHASSIS

Check the line fuse and on/off switch if the chassis is completely dead. Take a quick positive voltage measurement at the positive terminal of the bridge rectifier or the main filter capacitor. Do likewise with single diodes in the bridge circuit. Measure the ac applied to the bridge-rectifier circuit. Suspect that a power transformer is defective if no ac voltage is reaching the bridge circuits. If you find dc voltage at the bridge rectifier, but not at the collector terminal of the regulator transistors or at the input terminal of the IC regulator, look for an open fuseable (10 Ω) resistor. Some power supplies have a fuseable resister between the bridge rectifiers and regulation transistors.

Suspect that a transistor or IC regulator open or leaky with correct input voltage and low or no voltage at the emitter terminals. Check both the regulator transistor and zener diode for an open or leakage condition. When checking the regulator transistor out of the circuit for leakage tests, visually inspect the small bias and isolation resistors. Carefully check for correct resistance of each resistor. Likewise, take a resistance measurement across the small decoupling and filter capacitors. If in doubt, check the capacitor with a digital capacitance meter.

Dead Magnavox portable CD player The batteries were removed and tested in a Magnavox CD9510BK01 portable CD player. Very low voltage was found at pin 1 of the

K001 power-supply unit. Zero voltage was found at pins 6 and 7. Q1, Q3, and Q7 tested normal. IC TL145IC was found defective in the K001 power supply (Fig. 24-37).

IMMOBILE TRAY (NO LOADING)

If the disc tray will not move, suspect that a loading circuit or motor is defective, or that the dc voltage is improper. Locate the loading motor plug-in cable or wires. Measure the voltage at the motor terminals when the load button is pressed. Now check the power-supply voltage feeding the loading motor-drive circuits. Often, the voltage is a positive and negative 12, 10, or 9 V. Both positive and negative voltages must be applied to the output driver transistors or ICs of the motor-drive circuits.

Inspect the schematic and trace the voltage sources to the low-voltage power supply. Measure the voltage at the dc source. Low or no voltage indicates problems within the low-voltage source. If the regulated low voltage is normal, suspect that loading problems are within the loading drive motor circuits.

Erratic Magnavox CDC552 loading The five-disc changer tray seemed to operate erratically in a Magnavox CDC552 CD changer. Usually, the erratic condition seemed to occur before the tray was completely extended. The loading tray motor voltage was normal throughout loading and unloading of changer tray assembly. No doubt this was a mechanical problem.

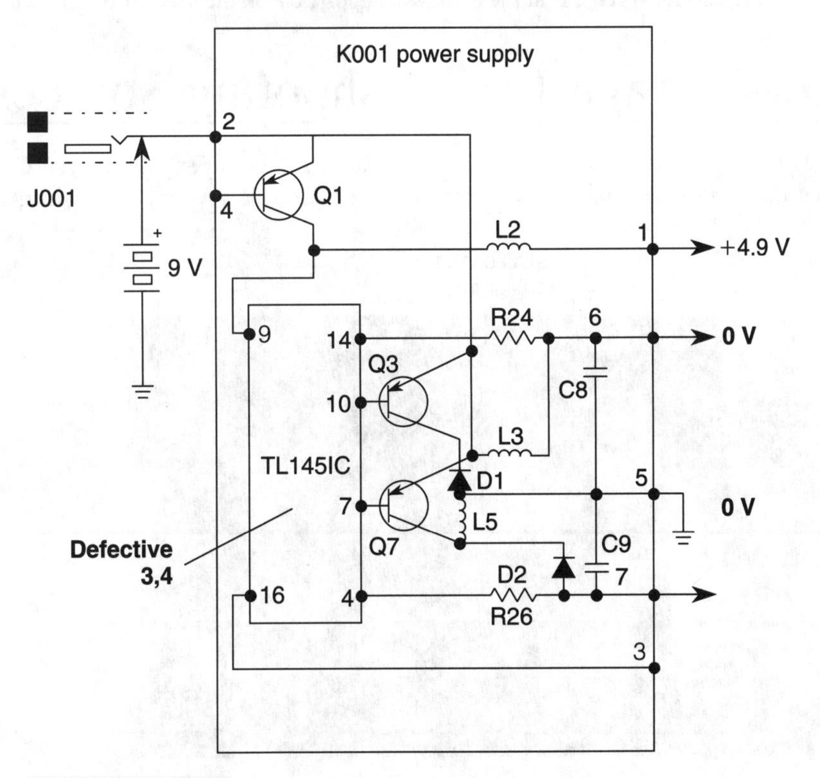

FIGURE 24-37 **A defective IC in the KOO1 power supply caused the dead Magnavox portable CD player.**

The tray was released and inspected. A careful inspection of the plastic-tray guide system seemed normal. Finally, after a closer inspection when the tray hung up, a couple of long wires were clinging to the bottom of the tray, just before it reached the outward position. These wires were disturbed and not properly replaced when the main chassis was serviced (Fig. 24-38).

Intermittent roulette rotation in a Magnavox CDC-745 Sometimes the turntable would rotate to change disc and other times not in a Magnavox CDC-745 changer. When the roulette turntable was removed, the motor seemed to operate normally. Like most motors in CD players, this one operates from two driver transistors or power IC. Monitor the voltage at the motor terminals (Fig. 24-39). Check the transistor or IC if the voltage at the roulette motor is intermittent. Replace the motor if it doesn't work, yet the voltage is normal at the motor terminals.

NO DISC ROTATION

If an improper voltage is applied to the spindle motor circuits, it could prevent the disc from rotating. Measure the voltage at the spindle motor with the disc loaded. Follow the schematic and check what voltage source is feeding the spindle-motor servo circuits. A 12- or 10-V negative and positive voltage frequently is used to feed the spindle servo-output drive transistors. If the correct voltage is at the dc power supply, check the servo drive circuits.

Low-Voltage Troubleshooting Sources

Check Table 24-2, a typical chart of the low-voltage sources of any CD player. Cross off each section of the schematic as you eliminate the various low-voltage components.

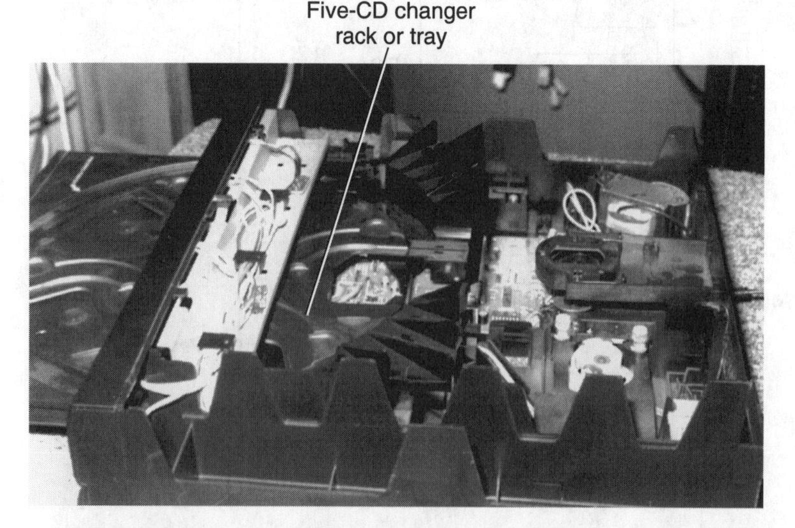

FIGURE 24-38 **Displaced connecting wires caused the loading tray to hang up in this CDC-522 changer.**

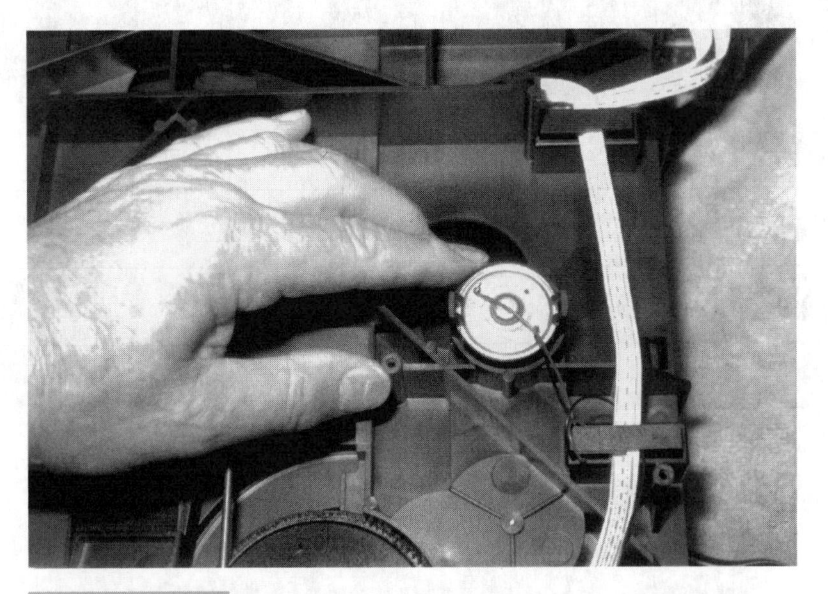

FIGURE 24-39 Intermittent turntable rotation in a Magnavox CDC-745 changer was caused by a defective roulette motor.

TABLE 24-2 A LOW-VOLTAGE POWER SUPPLY TROUBLESHOOTING CHART

SYMPTOM	CHECK
Dead chassis	Inspect the fuse if one is in the chassis. Check the dc voltage at the main filter capacitor. Test the silicon diode bridge circuit. Test the transistor or IC regulators. Check for an open primary winding of the transformer. Check other circuits for overloading.
Lights up, no action	Check the voltage at the main filter capacitor. Isolate stages not functioning. Check the transistor or IC voltage regulators. Check the outside circuits for overloading.
Intermittent chassis	Monitor the various voltage sources. Suspect a defective transistor or IC regulator. Monitor the ac voltage from the power transformer.
Hum in the sound	Shunt the filter capacitor with another. Check for a leaky IC or transistor regulator Do not overlook leaky zener diodes.
Loud groan, nothing	Suspect transformer overload. Check for shorted or leaky silicon diodes. Check for a leaky or shorted filter capacitor. If the transformer is hot, remove the secondary wires. Check the secondary ac voltage of the transformer.

25

THE RF SIGNAL PATHS

CONTENTS AT A GLANCE

The signal circuit consists of a preamp, signal-processing LSI (large-scale integration) chip, phase-locked loop (PLL), random-access memory (RAM), and digital-to-analog (D/A) converter. In some models, one LSI might handle most signal-processing circuits, except the D/A converter stage (Fig. 25-1). In others, the preamp, tracking, automatic focus, and signal-processing circuits are located in one large IC. Although two LSIs might do the work of the signal circuits in lower-priced compact disc players, several different LSIs and ICs are used in the most-expensive models.

The signal from the photodetector diodes is very weak and must be amplified before it can be used by the signal processor. In early CD circuits, the RF amplifier consisted of transistors; recent RF circuits can be included in one large IC that also contains the FOK, MIRR, FE, and TE circuits.

FIGURE 25-1 **The RF signal starts at the optical lens assembly and RF amp circuits.**

In the Optimus (CD-3380) analog RF amplifier, IC1 has a total of 30 pin terminals (Fig. 25-2). The laser output signal is applied to PD1 (7) and PD2 terminal 8. The amplifier EFM signal at pin 27 can be checked with the oscilloscope eye-pattern waveform. A MIRROR comparator signal is found at pin 22. The tracking error (TE) offset signal is found at pin 20 and focus error (FE) offset signal is at pin 19. Both the TE and FE signals are sent to analog servo control IC2.

Signal Path Block Diagram

The block diagram is useful in determining how the different stages and circuits are tied together. Use the block diagram to locate the problem area. Then, with the correct waveform, voltage, and resistance measurements, you can find the defective component. The signal from the signal circuits might also tie into the servo or system-control circuits.

Figure 25-3 shows the Realistic CD-3770 portable CD player block diagram. The RF signal from the four photodetector diodes and tracking diodes is fed into an RF amp (IC4). Besides tracking and auto-focus offset circuits, a signal is fed from RF amp IC4 to servo

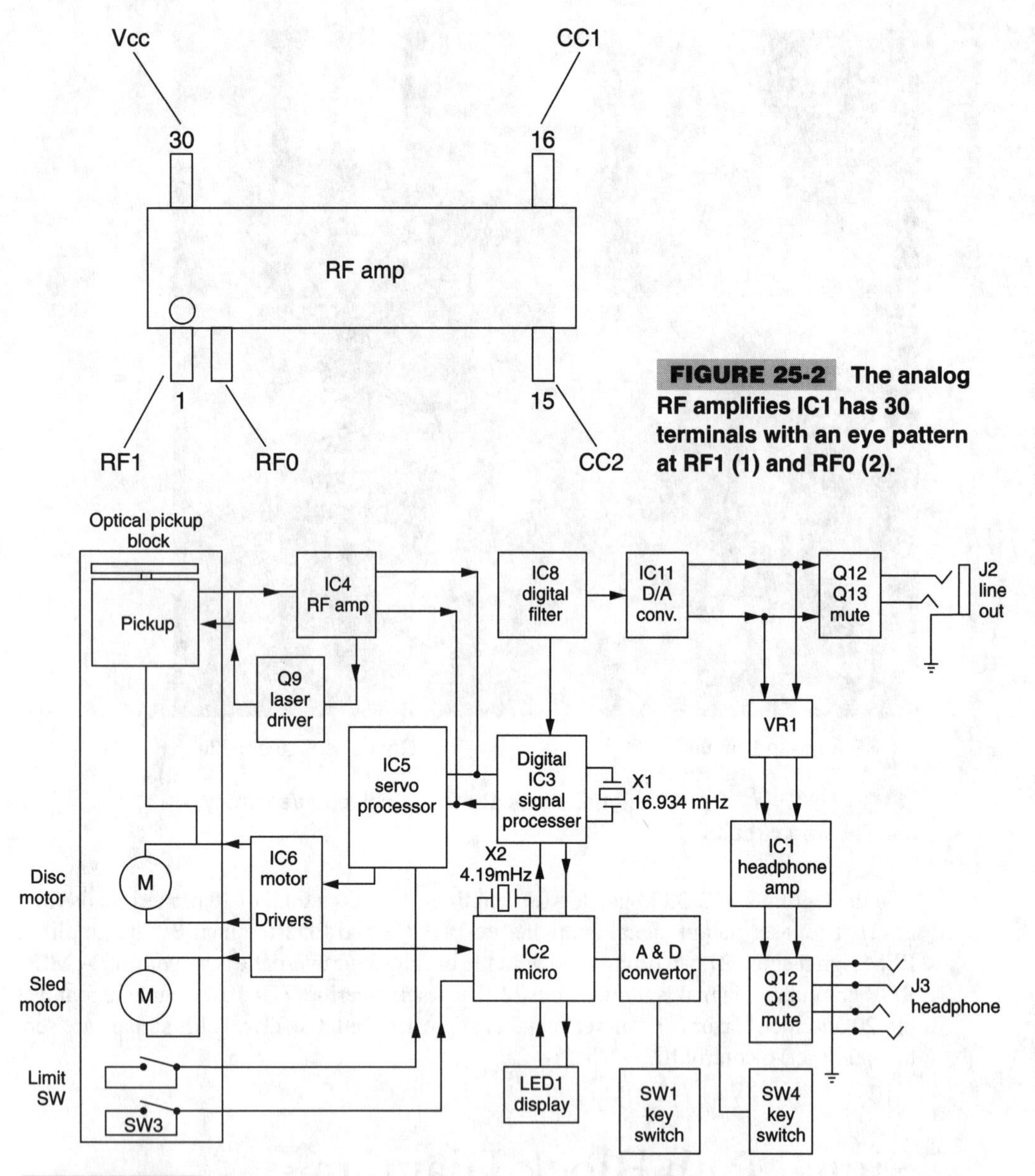

FIGURE 25-2 **The analog RF amplifies IC1 has 30 terminals with an eye pattern at RF1 (1) and RF0 (2).**

FIGURE 25-3 **The typical block diagram of a portable CD player.**

control IC5. The amplified RF circuit is fed from the RF IC to digital signal processor IC5. Q9 provides the laser driver signal for the laser diodes.

The EFM signal is fed from RF IC4 from pin 8 to pin 5 of the digital signal processor (IC3). The EFM signal of RF amp IC4 can be checked at pin 8 or the RF signal can be checked at test point TP1. If the EFM or RF signal is not present here, suspect that an RF amp (IC4) or optical pickup assembly are defective (Fig. 25-4). If the signal is missing from the RF amp to the servo signal processor or servo IC, the chassis will automatically shut down.

The EFM signal is fed to pin 5 of IC3 and the output signal of IC3 is fed to the digital filter (IC8). The D/A converter (IC11) separates the digital signal to the analog left and right stereo channels. Q12 and Q13 provide audio mute circuits of the line-output jack (J2).

Figure 25-5 shows a typical block diagram of the signal circuits from the optical pickup head to the audio output. Optical pickup diodes A, B, C, and D are combined and input at terminals 19 (PD1) and 20 (PD2) of RF amp IC11. The EFM eye-pattern signal taken at pin 8 of IC11 will determine if the optical assembly and RF amp are operating.

Pin 24 of the signal processor is connected from pin 8 and the output is found at pin 71. The RF signal is coupled to sample/hold IC14 and through D/A converter IC15. Here, the digital signal is changed into analog audio with two stereo channels. The line-output jacks are connected to IC15 and to the headphone IC, if it is contained in the audio circuits.

Replacing LSI or IC Signal Processors

After determining that the signal processor is defective with scope tests and voltage measurements, remove and replace it. The regular IC component exchanges like any other IC (Fig. 25-6).

The larger chips might be a little more difficult to remove. Be sure that the soldering iron has a flat, angled tip and is grounded. Check to see how the IC is mounted and mark the

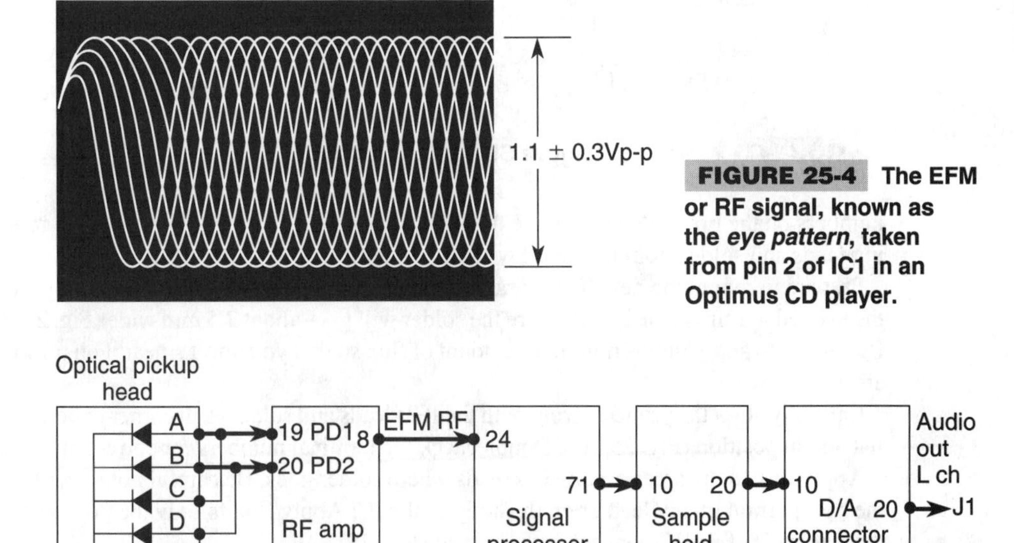

FIGURE 25-4 The EFM or RF signal, known as the *eye pattern*, taken from pin 2 of IC1 in an Optimus CD player.

FIGURE 25-5 The RF signal path from optical pickup to line-output jacks J1 and J2.

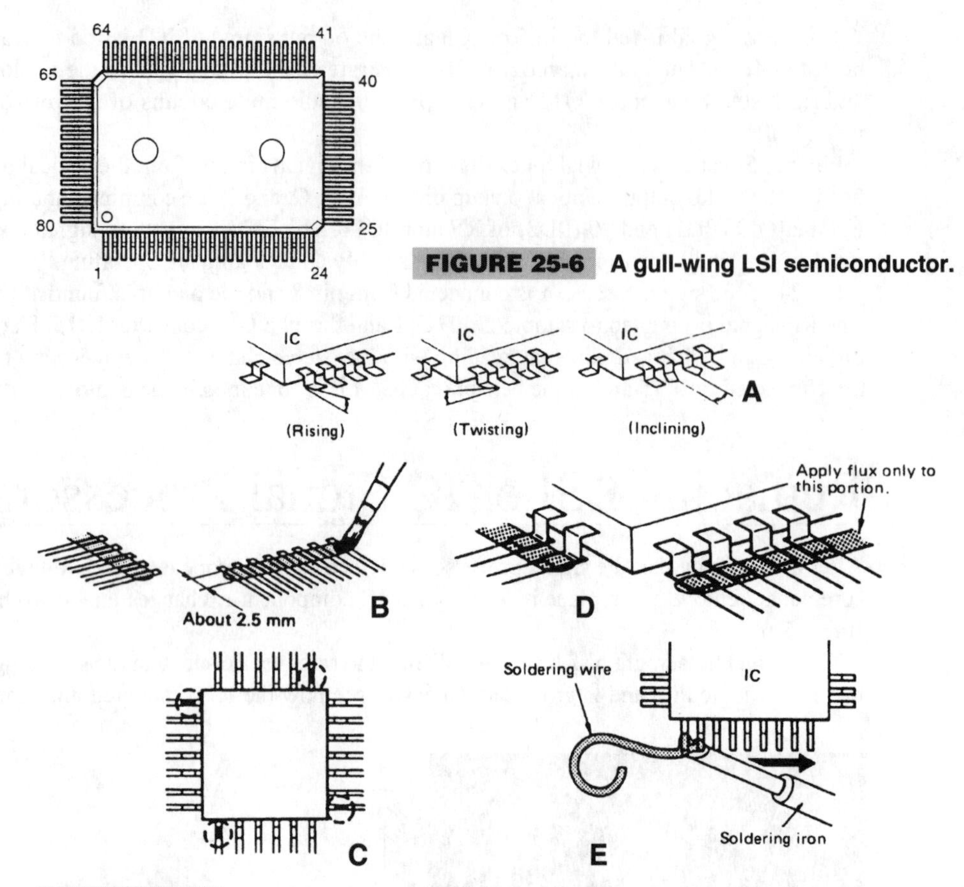

FIGURE 25-6 A gull-wing LSI semiconductor.

FIGURE 25-7 How to remove SMDs, LSIs, and ICs.

numbers on the board. Now, remove the IC by cutting each terminal. Remove the remaining leads and solder from the board with the solder iron and solder wick.

Prepare to mount the new IC by straightening each terminal (Fig. 25-7A). Apply flux to the printed wiring on the board where the solder will go—about 2.5 mm wide (Fig. 25-7B). Be careful to apply only a minimum amount of flux so that you don't smear it on unwanted areas.

Carefully align the printed wiring with the IC's leads and solder each corner to hold the IC flat and in position (Fig. 25-7C). Doublecheck the terminal numbers for correct mounting.

Apply flux to the areas on the IC's leads where solder goes. Be careful not to get flux on the root portion of any lead or on the body of the IC. Apply flux to only the portion where the flat leads of the IC connect to the wiring (Fig. 25-7D).

Place the flat soldering-iron tip on the junction, and feeding very thin solder to the joint, move the iron slowly in the direction of the arrow (Fig. 25-7E). Move the iron at the rate of approximately 1 cm in 5 seconds. Be sure that a clean fillet of solder forms on each lead, as soon as the flux melts. Moving the iron too quickly can result in loose solder or poor connections. Be especially careful when soldering the first lead (where loose soldering most likely forms). When the soldering is finished, check each terminal with a magnifying glass.

Surface-Mounted RF Components

Surface-mounted devices (SMDs) are used in many CD portables in the RF and signal-path components. These small transistors, ICs, processors, diodes, resistors, and capacitors are very difficult to locate—even with a parts layout diagram. They are located on the PC wiring side (Fig. 25-8). The values of resistors and capacitors can be numbered on top of the miniature devices. You might need a magnifying glass to identify the resistance or capacitance.

THE RF OR HF SENSOR PREAMP

The radio-frequency (RF) or high-frequency (HF) circuits receive the weak signals from the HF sensor or photodetector and amplify the signal to a level that can be used by the signal processor. The preamplifier might work in conjunction with the servo IC. The preamp might also be used for detecting and correcting errors for focus and tracking.

The RF preamp can operate from a separate IC, transistor, or a combined IC processor component (Fig. 25-9). The RF preamp is located between the pickup head and the LSI or IC signal processor. Here, the HF signal from photodetector diodes A/C and B/D are capacitor-coupled through C102 and C103 to preamp transistors TR101 and TR102. The output signal

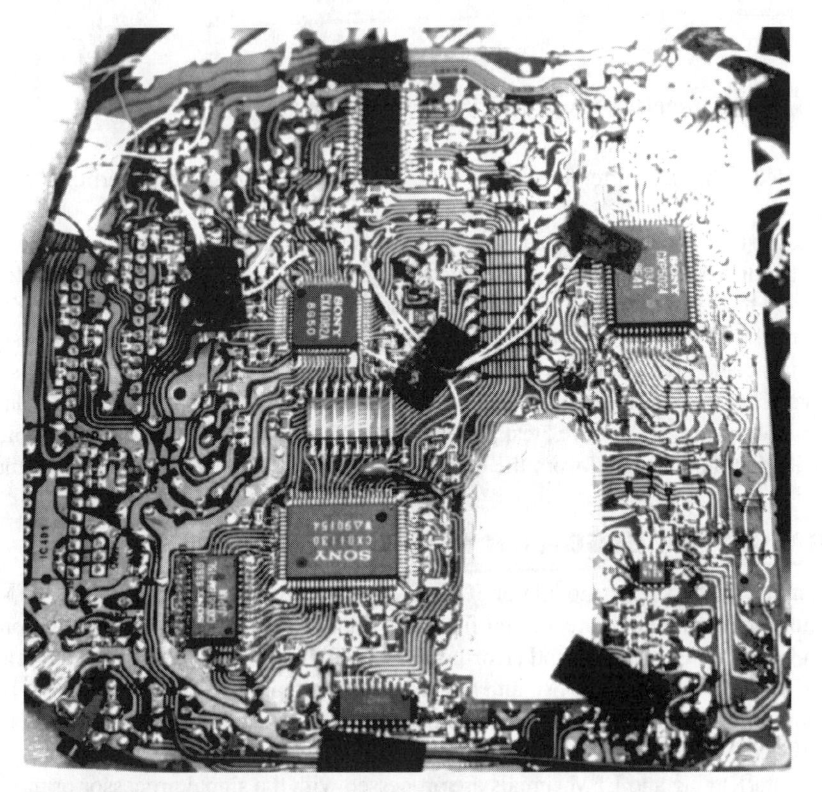

FIGURE 25-8 **Surface-mounted SMD ICs and LSIs are on the bottom board of the portable CD player.**

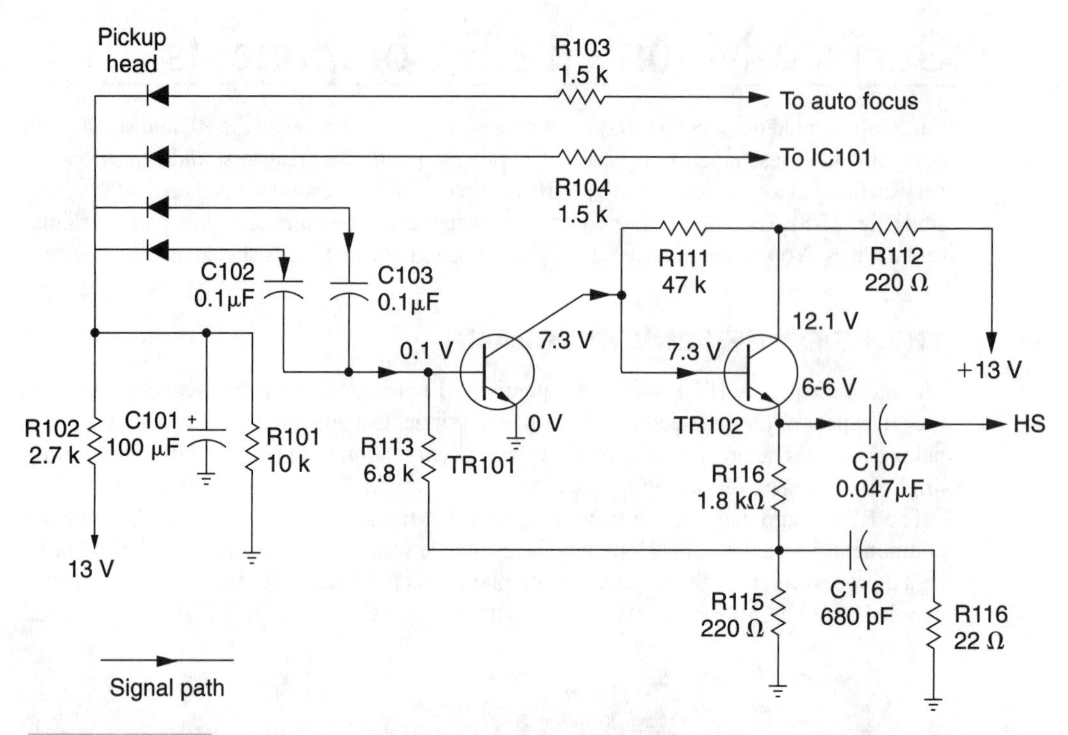

FIGURE 25-9 RF transistor amplifiers in older CD players.

is coupled from the emitter terminal of TR102 through C107 to the HS terminal of servo 1 board.

Because the RF preamp IC has many different circuits, it is wise to know what terminals go to the different circuits. The RF input from the photodetector diodes connect to terminal 2 and the EFM output at terminal 20, which ties to the EFM input terminal of the signal processor. The positive power source connects to terminal 24. An improper signal at terminals 2 and 20 with low or no voltage at terminals 24 shows you what circuits are defective. The various pin numbers of the RF amplifier in an Onkyo DX-200 CD player are given in Table 25-1. An "eye" test pattern at pin 20 or the designated "eye test point" can indicate that the RF signal from the head pickup and out of the RF preamp is normal.

SIGNAL PROCESSOR OR MODULATOR

The signal processor or modulator IC usually contains the clock generator, EFM, data luch, data concealment mute, digital filter, timing control, osc, subcode modulation, CLV servo, servo system control, and error-correction circuits. Besides the internal circuits, the VCO, disc motor control, RAM, and system control circuits are tied to it (Fig. 25-10). The RF or HF signal is fed into the signal-processor IC with the timing control and is sent to the D/A converter and servo-control circuits.

The interleaving and EFM signals are processed with the signal processor or modulator. This interleaving method is standard, determined when the discs are manufactured. The disc during playback returns the signals to the original state, in accordance with this method. The interleaving data is memorized inside the RAM IC and the data is recalled in exactly the same sequence of the original signals.

The signal processor might be a separate IC, LSI processor, or it could be combined with other circuits. The PLL IC circuit tied to the signal processor basically consists of an 8.6436-MHz VCO (voltage-controlled oscillator). Often, the RAM and PLL or VCO circuits are in separate IC components that are tied to the signal-processor IC.

The Realistic CD-1000 signal-processor circuit, in combination with both the data-strobe circuit and the section that detects and corrects the errors as soon as the data signals are demodulated, compensates when the sync signal is missing, and it identifies 1s and 0s. It is composed of circuits that control the whole signal-processor circuit and a RAM-control circuit, entering the data once into the random-access memory and rearranging it.

The rearrangement of the signals, as performed by this circuit, is particularly effective with a high signal loss. The continuous stream of audio signals is thrown into disarray, adroitly rearranged (interleaved) and entered into the disc. In the reproduction system, these signals are restored by a method that is the reverse of that used to throw them into disarray. In order to restore the signals, the data is memorized inside the RAM and then recalled in

TABLE 25-1 THE VARIOUS FUNCTIONS OF PIN NUMBERS ON THE RF IC

1. Focus OK output
2. RF input
3. RF summing amp output
4. RF summing amp inversion input
5. RF 1-V amp (1) inversion input
6. RF 1-V amp (2) inversion input
7. Ground of small signal analog system
8. EI-V inversion input
9. EIV amp output
10. FIV amp output
11. FI-V amp inversion input
12. Ground
13. Negative power supply
14. Focus error amp input
15. Focus error amp inversion input
16. Focus error amp output
17. Tracking error amp output
18. Mirror output comparator output (Active at H)
19. Mirror hold capacitor connection terminal
20. EFM output comparator output
21. Terminal for reference input level setting of auto assimilate control amp
22. Auto assimilate control input
23. EFM comparator system power supply
24. Positive power supply

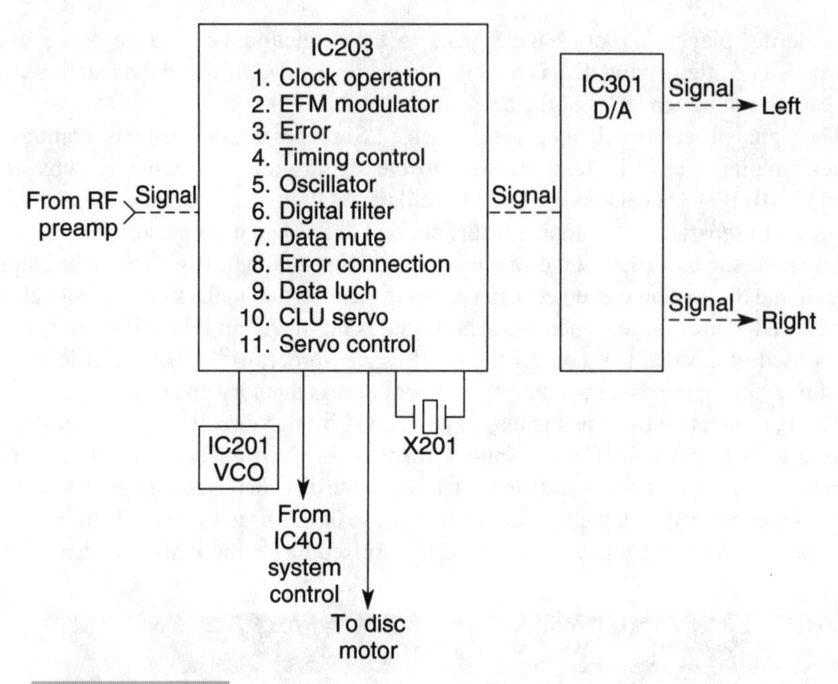

FIGURE 25-10 A signal processor (IC203) contains circuits in addition to the oscillator (VCO) and RAM circuits.

exactly the same sequence of the original continuous stream of signals. Even if there is a high signal loss, the sequence of the signals is dispersed at random, so the signals before and after those that were lost are still present when the sequence is restored. Hence, original signals can be compensated for. This interleaving method is determined when the discs are manufactured and the disc-reproduction equipment returns the signal to its original state.

Realistic CD-3304 EFM comparator The EFM comparator changes RF signal into a binary value. As the asymmetry generated because of variations in disc manufacturing cannot be eliminated by the ac coupling alone, the reference voltage of EFM comparator is controlled utilizing the fact that the generation probability of 1,0 is 50% each in the binary EFM signals (Fig. 25-11).

Because this comparator is a circuit SW type, each of the H (high) and L (low) levels does not equal the power supply voltage, requiring feedback through a CMOS buffer. R532, R528, and C546 form a low-power filter (LPF) to obtain $(V_{CC} + D_{GND})/2$ V.

AF (Auto Focus) and Focus-Error Circuits

The beam from the laser pickup must remain focused on the disc surface to accurately read the information. If the focus on the disc strays, the focus servo moves the object lens up or down to correct the focus. Under this system, when a beam is irradiated through a combination of the cylindrical and convex lenses, the beam is elongated and then becomes a perfect circle. If the laser beam is reflected from the disc, it is directed to the cylindrical lens by the prism and then to the optosensor (photodetector), where it is split into four, forming a perfect circle. These outputs from the four optosensing elements are supplied to the error-signal amplifier, and a zero output is produced. The focus error circuit is designed to detect changes in the distance to the disc, thereby ensuring that the laser-beam spot is kept in proper focus on the reflecting surface of the disc (Fig. 25-12).

In the Onkyo DX-200 focus error circuit, the changes in the laser-beam focus are achieved by what is called the *astigmatic method*. This method uses the fact that the shape of the laser spot reflected from the disc into a six-part photodiode in the pickup varies according to the distance from the disc. These changes result from the action of a cylindrical lens.

The A, B, C, and D photodiode light-source currents undergo a diagonal subtraction in three operational amplifiers in Q101, resulting in the generation of a focus-error (FE) signal (Fig. 25-13). Figure 25-14 summarizes the FE signal changes at different distances from the disc reflecting surface.

Realistic CD-3304 auto focus The focus OK circuit (FOK) generates a timing window to look on the focus servo from a focus search status. Pin 1 will receive the high-pass filter (HPF) output from an RF signal of pin 2, the low-pass filter (LPF) output (opposite phase) for the focus OK amplifier output (Fig. 25-15).

The focus-OK circuit output is inverted when $V_{RFI} - V_{RFO} = -0.37$ V. C501 is for determining the time constants of high-power filter (HPF) in the EFM comparator and mirror circuits, as well as the low-power filter (LPF) in the focus OK amplifier.

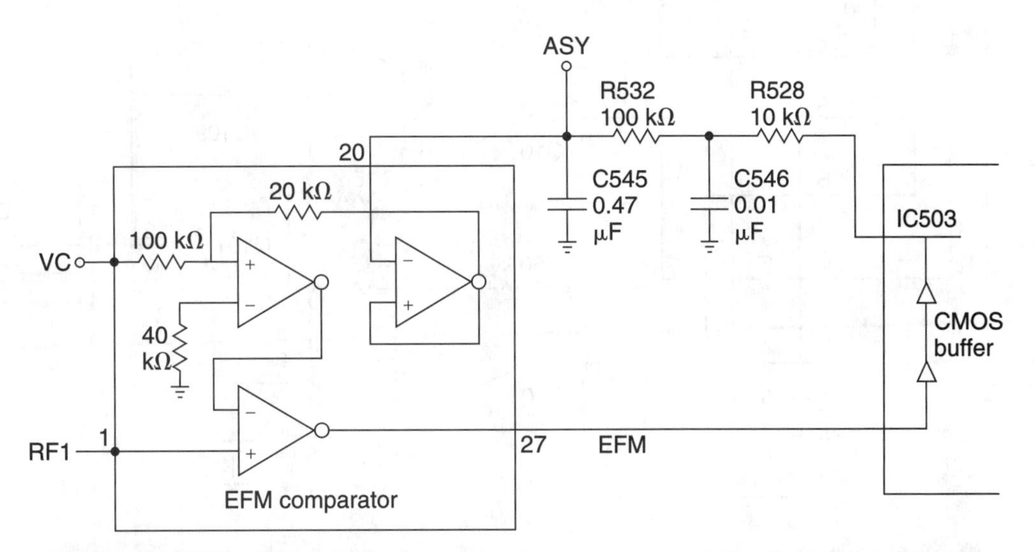

FIGURE 25-11 The EFM comparator changes the RF signal to a binary value in this CD player.

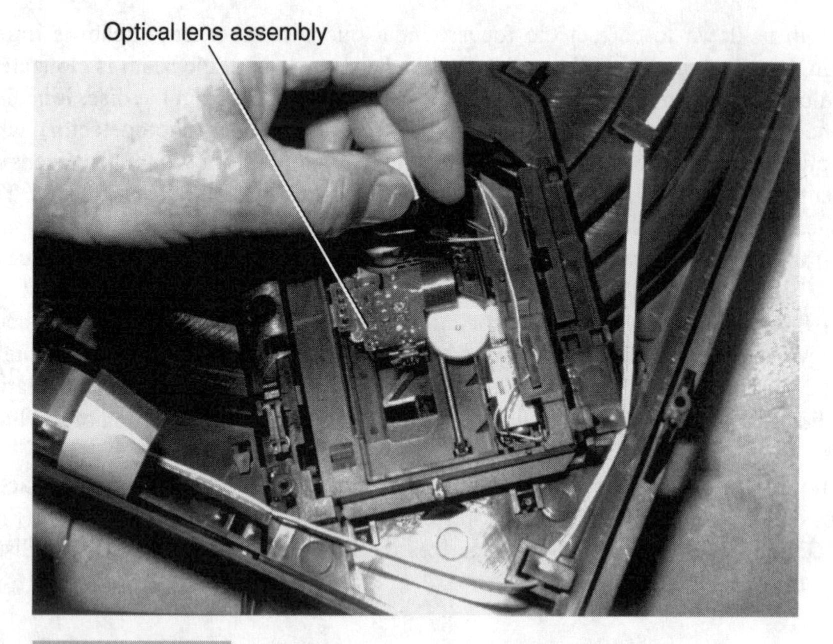

FIGURE 25-12 The auto focus and error circuits located in optical assembly of CD changer.

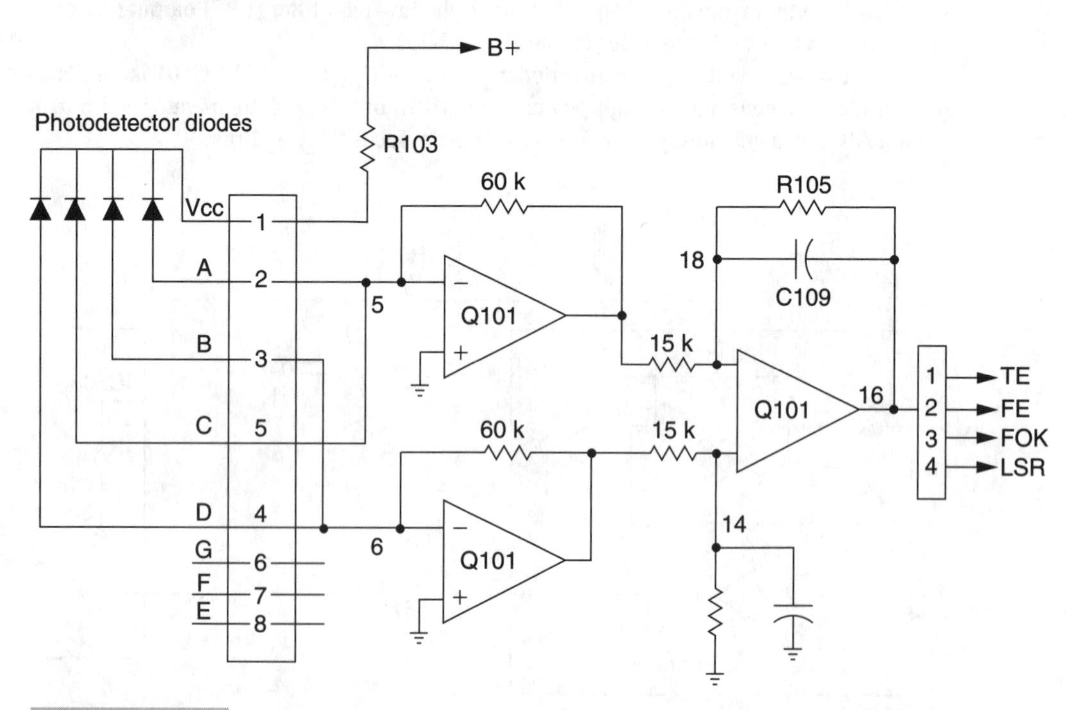

FIGURE 25-13 The focus-error output signal is taken from pin 2 of Q101 in an Onkyo CD player.

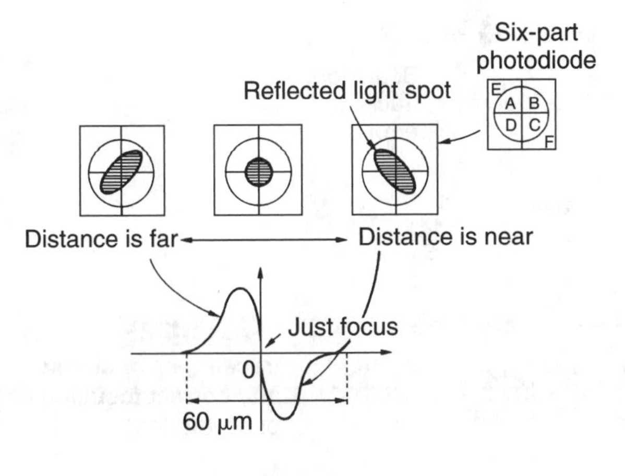

FIGURE 25-14 Correct focus is shown with a perfect circle in the center of these photodiodes.

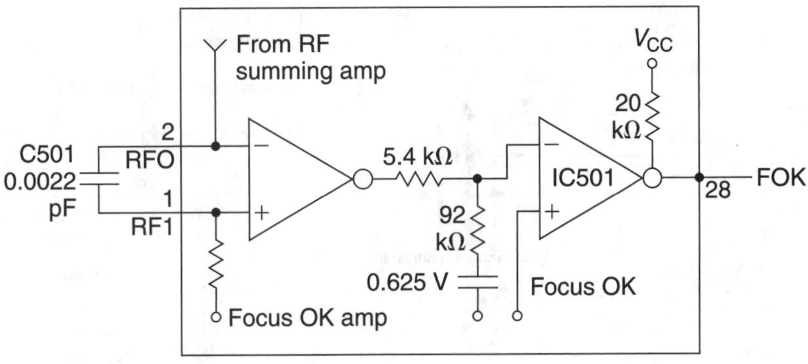

FIGURE 25-15 The focus OK circuit (FOX) in a typical portable CD player.

TRACKING-ERROR CIRCUITS

The tracking-error circuit generates an error signal if the laser-beam spot moves away from the center of the pits. This error signal is used to ensure that the beam spot correctly tracks the line of pits. The tracking-error circuits are inside the RF preamp or signal-processor IC.

For the Onkyo DX-200 tracking-error circuit, the mechanism divides the laser beam into three separate beams—a main beam and two auxiliary beams on either side of the main beam. These three beams are arranged at a slight angle (0.88˜) to the line of pits in what is known as the *three-beam method*.

If the beam tends to move away from the pits, as indicated, the degree of reflection in the auxiliary beams changes, depending on the direction of the shift. The reflected auxiliary beams are converted into electric signals by the E and F detectors at both ends of the six-part diode, and the mutual differences are obtained as a tracking-error (TE) signal (Fig. 25-16). The circuit includes three operational amplifiers. The E and F detectors are balanced by variable resistor R114, and the tracking offset is canceled.

Figure 25-17 shows the Realistic 8-MM tracking-error detection system. The laser spot must track the series of pits accurately at all times, regardless of the disc's eccentricity.

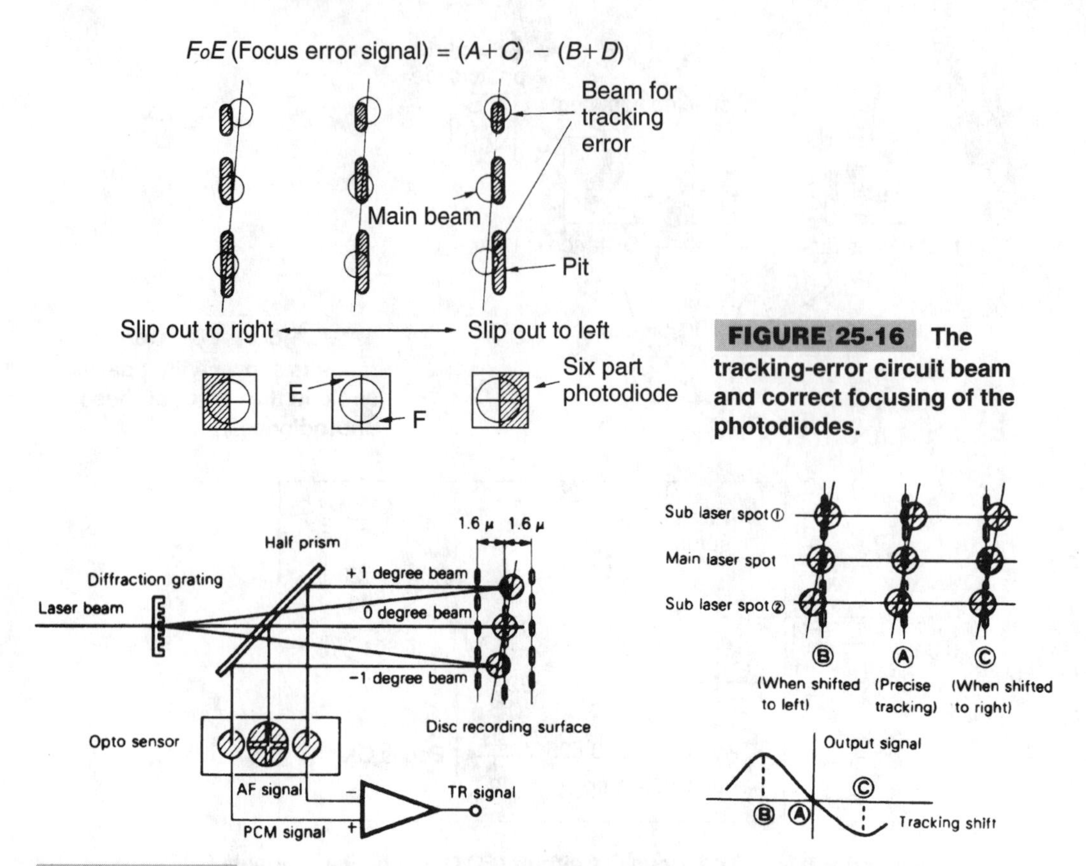

FIGURE 25-16 The tracking-error circuit beam and correct focusing of the photodiodes.

FIGURE 25-17 The tracking-error detection and output signal.

This particular unit uses an extremely accurate tracking mechanism known as a *three-spot system*. As shown in the figure, this system makes use of two subspots, obtained by passing the main laser beam through a glass diffraction grating, which are positioned before and after the main laser spot. They are aligned by shifting them slightly to the left and right. After the laser spot has been reflected by the disc surface, it is passed to the optosensor. However, the reflected light of the two subspots is detected first, converted into electricity, then enters the error signal amplifier. Any shift, no matter how slight, is translated into a difference in the input to the amplifier. This error-output difference enters the servo circuit to move the objective lens to correct the position of the main laser spot. Thanks to this system, the tracking servo is extremely precise.

DEFECT CIRCUITS

In the Realistic CD-3304 CD player, the RFI signal bottom, after being inverted, is held with two time constants, one long and one short. The short time constant bottom hold is done for a disc mirror defect more than 0.1 ms. The long time constant bottom hold is done with the mirror level prior to the defect. By differentiating this with a capacitor coupling and shifting the level, both signals are compared to generate the mirror-detect detection signal (Fig. 25-18). The defect signal at pin 21 of RF amp (IC501) is fed to pin 2 of the analog signal processor (IC502).

Realistic CD-3304 mirror circuits This circuit, after amplifying the RFI signal, holds the bottom and peak. The peak hold is done with a time constant able to track down a 30-kHz traverse and the bottom hold. This is done with a time constant able to track down envelope fluctuations in the revolving cycle (Fig. 25-19).

With the differential amplification of this peak and bottom hold signals, H and I, the envelope signal J (demodulated to dc) is obtained. Two-thirds of the peak value of the signal J is held with a large time constant for the signal K. When K is compared with J, a mirror output is obtained.

That is, the mirror output gives "L" on the disc track, "H" between tracks (mirror section) and also "H" in the defect section. The time constant for the mirror hold must be sufficiently larger than that of the traverse signal.

Sanyo CP500 chassis shutdown In a Sanyo CP-500 CD player, the laser diode lights, the focus and tracking coils began to search, then the chassis shuts down. The scope probe was connected to pin 20, to determine if the RF preamp IC101 was providing an EFM signal (Fig. 25-20). The RFO eye pattern can be checked at test point TP1 (RF) or terminal 3. Often, the servo IC will shut down the chassis if the EFM signal is not present at the RF IC amp.

Again, the chassis was fired up, with no sign of EFM waveform at IC101. The supply voltage was normal at pins 23 and 24 (+5 V). The +5-V source was good to the PD1 and

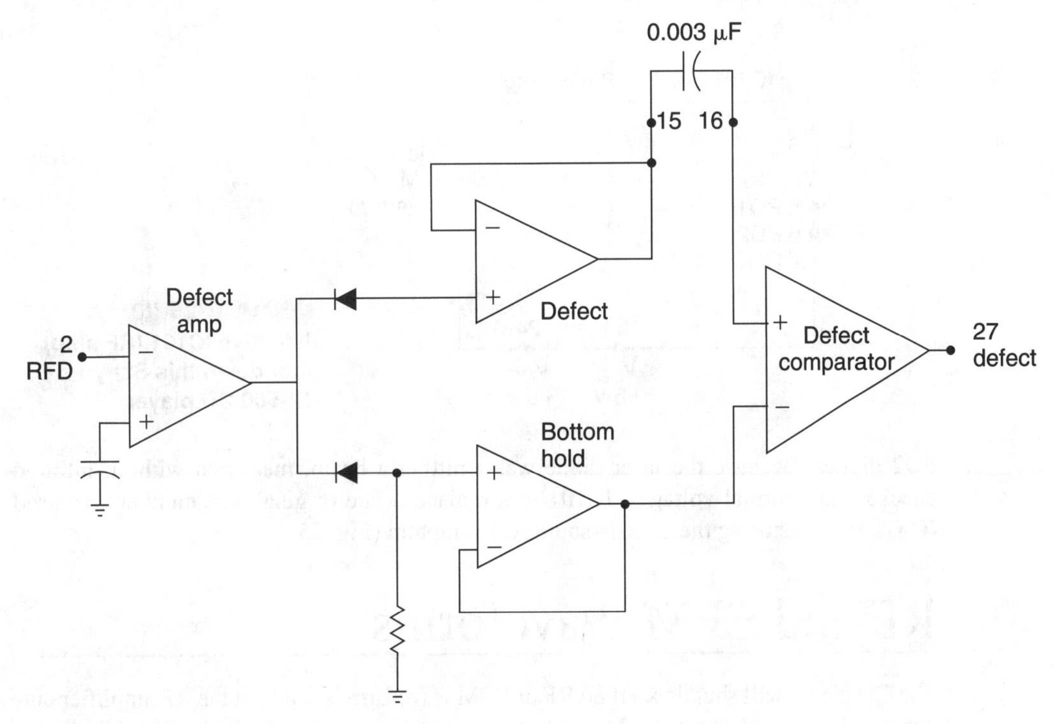

FIGURE 25-18 Operation of the detect circuit in the Radio Shack portable CD player.

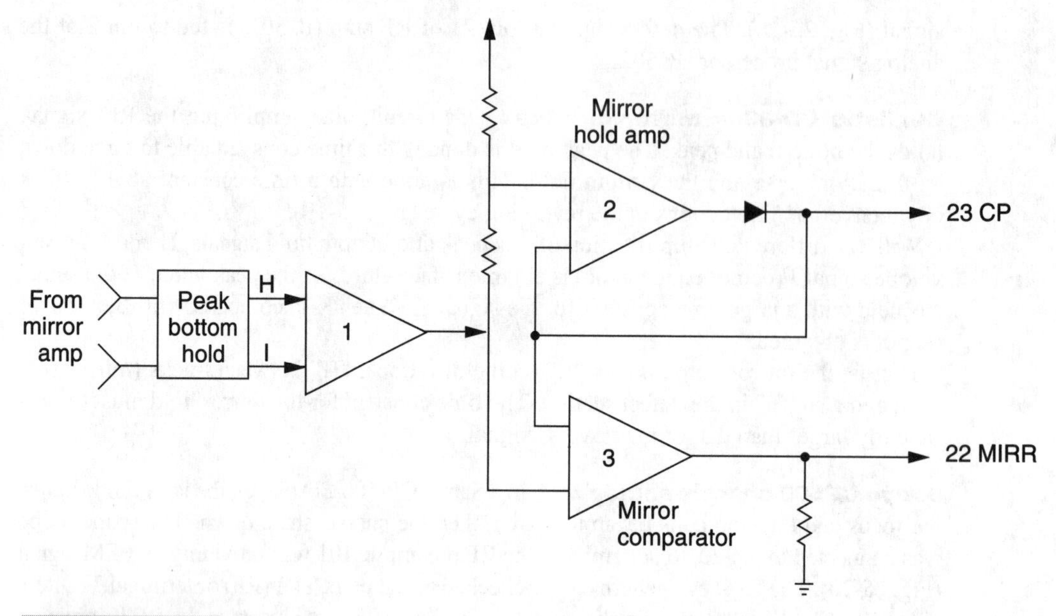

FIGURE 25-19 Operations of the mirror circuits in a Realistic portable CD player.

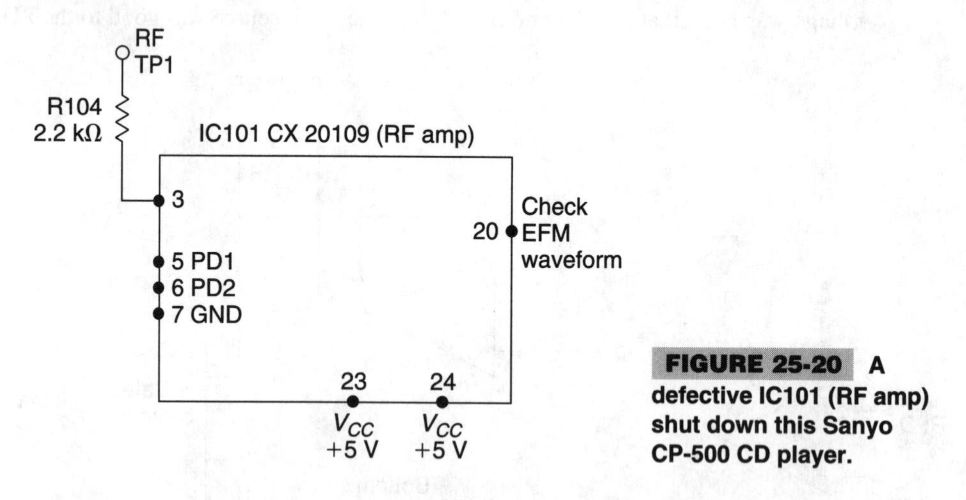

FIGURE 25-20 A defective IC101 (RF amp) shut down this Sanyo CP-500 CD player.

PD2 diodes. Because the laser diode was emitting a beam, measured with an infrared checker, and normal voltages, IC101 was replaced. The original part number was used (CX20109), restoring the chassis-shutdown symptom (Fig. 25-21).

RF and EFM Waveforms

The CD player will shut down if no RF or EFM waveform is found at the RF amplifier output circuits. Some CD circuits have a test point to take waveforms of the EFM eye pattern. Load a test CD disc and place the player into the Play mode. In the Optimus CD3380 player, TP1 is found off of pin 1 and 2 of RF amplifier IC1. Locate the RF or EFM signal pin of RF amplifier and take the eye-pattern waveform (Fig. 25-22).

No waveform might indicate that the optical assembly or RF amplifier is defective. Check the supply voltage (V_{CC}) on RF amplifier to determine if IC1 is defective. Check the troubleshooting chart in Table 25-2 for other symptoms.

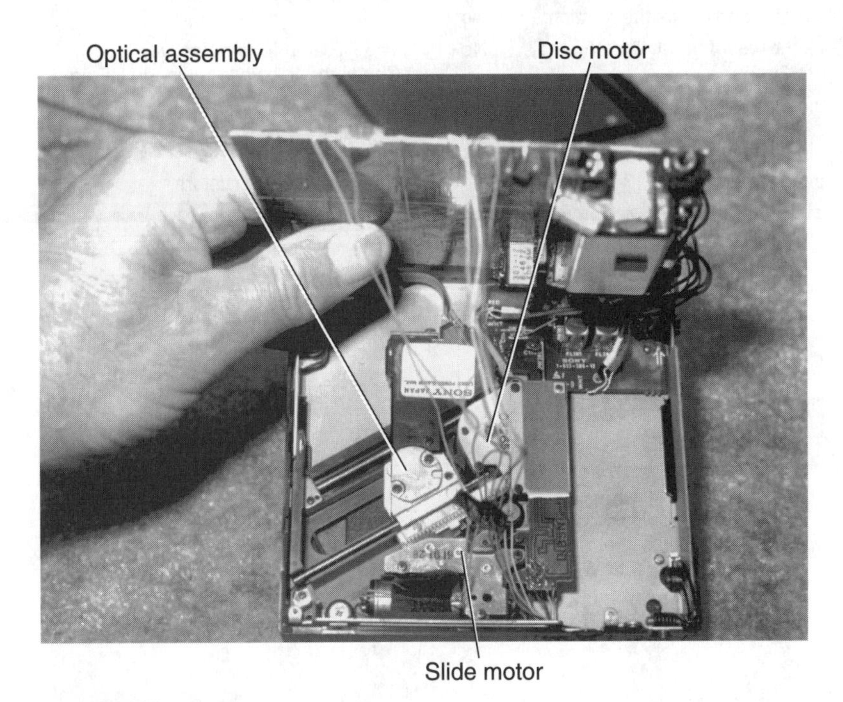

FIGURE 25-21 **The optical lens assembly and slide motor in a Sony CD player.**

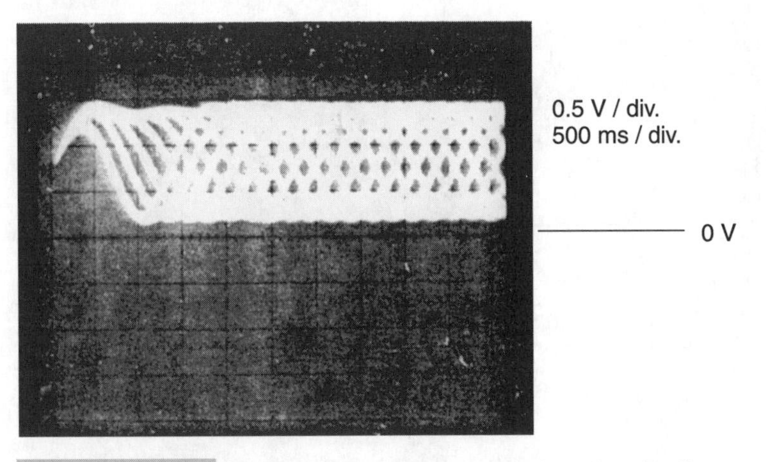

FIGURE 25-22 **The RF or eye-pattern waveform in a Radio Shack portable CD player.**

TABLE 25-2 THE RF LASER TROUBLESHOOTING CHART

Is the laser light on?	No	Check it with a power meter. Check with an infrared indicator.
Does the lens move up and down?	No	Look at the coils and see if they are moving.
Load a disc and check the rotation.	No	
Check the waveform at the RF amp (Figure 25-22).	No	The RF amp IC is defective. Check the V_{CC} voltage on the RF amp. Check the soldering pins of the RF amp. Check the wire and cable connector.
Does the CD player operate for a few seconds and shut off?	Yes	No EFM signal is at the RF amp IC or transistors.

26

THE SERVO AND MOTOR CIRCUITS

CONTENTS AT A GLANCE

The servo section provides signal and power to the focus and tracking coil drive, slide or sled motor drive, spindle servo, tray motor drive, track jump, and search circuits. In some models, the loading or tray motor is controlled by the system-control circuits. The servo circuits can be included in one or two large LSIs, or a combination of LSI, IC, and transistor circuits (Fig. 26-1). Some of the servo-control circuits might be combined within the signal-processor LSI (Fig. 26-2).

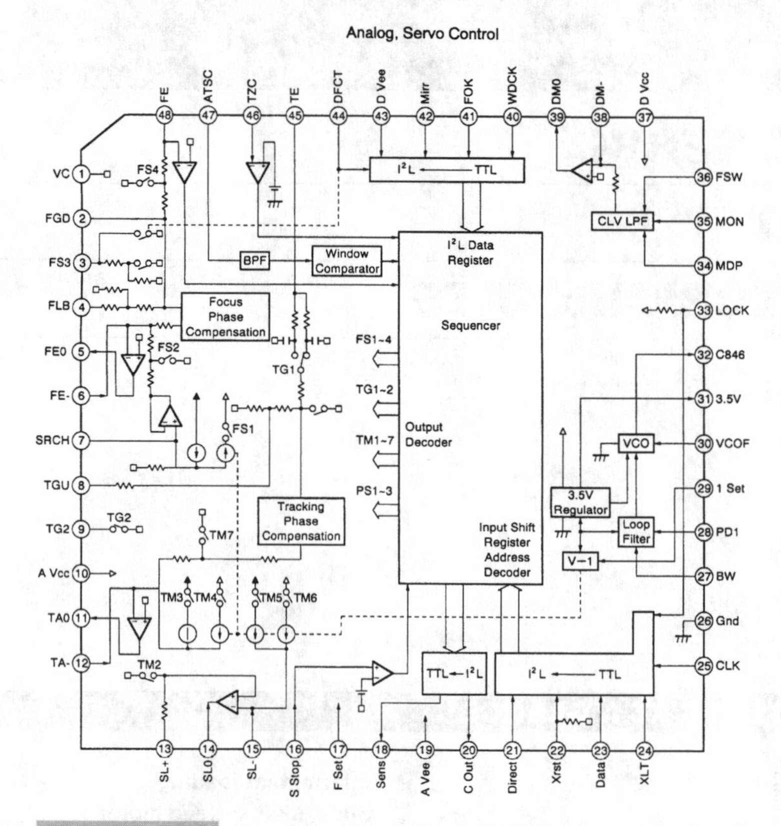

FIGURE 26-1 The servo IC can be an SMD or a conventional IC mounted on a PC board.

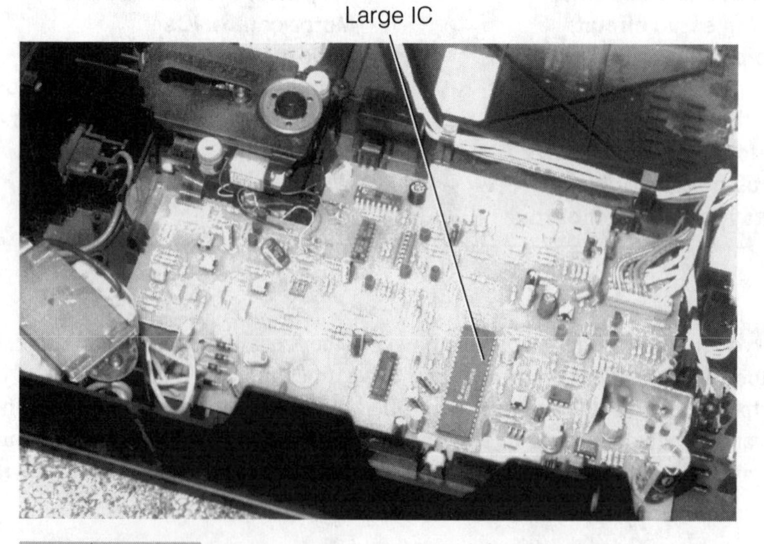

FIGURE 26-2 The servo-control IC is usually the largest on the PC board in the CD player.

If the TE, FE, DEF, and mirror circuits are not completed from the RF amplifier, the servo-control IC might shut down the whole operation. If the CD player comes on and then shuts down at once, be sure that the EFM or RF waveform is found at the RF amp. Be sure that the tracking and focus coils begin to search before the chassis shuts down. Besides these signals, the servo-control IC cannot function without proper voltage source (V_{CC}) from the low-voltage power supply.

The focus gain, focus offset, focus balance, tracking gain, tracking offset, and tracking-balance adjustments are located in the servo system. These adjustment procedures are given in detail in Chapter 28.

Block Diagram

The block diagram of any compact disc player indicates what circuits are controlled by the servo IC (Fig. 26-3). Here, in a Denon DCM560 player, the servo pickup system is fed signals to and from the RF amp IC101, and then operates the focus, tracking, slide, and disc motor assemblies. In the Sanyo CP-500 CD player, IC201 provides focus, tracking, and SLED servo control directly from IC201.

The focus, tracking, and sled servo IC201 accepts the TE and FE signals from the RF IC101 (Fig. 26-4). The tracking output signal (TAO) drives transistors Q204 and Q205 to the tracking-coil winding. FEO from pin 21 of IC210 drives three transistors Q201, Q202, and Q203 to the focus coil winding. The slide-motor output signal (SLO) drives transistors Q206 and Q207 to the SLED motor winding.

Servo IC

The servo IC is one large component with as many as 60 terminals, depending on how many circuits the IC controls (Fig. 26-5). The different pin numbers identify what circuit ties into each pin and the correct voltage measurement for each. Terminals 21 and 22 control the focus drive coil, and pins 27 and 28 control the tracking drive coil assembly. Terminals 23, 24, and 25 control operation of the slide motor assembly circuits.

Use extreme care when working around these large LSI or IC servo components. Be careful not to short together pins when taking voltage measurements or testing waveforms. Always use a small soldering iron when replacing a servo IC to avoid overheating and damaging the IC.

Denon DCC-9770 auto CD servo circuits The tracking-error (TE) signal taken from pin 1 of RF amp IC501 through the tracking gains control to pin 45 on servo IC502. The tracking-coil output (TAO) is fed from pin 11 of IC502 to coil drive IC402, Q414, Q415, Q416, and Q417 to the tracking coil (Fig. 26-6). Notice that two sets of transistors drive the negative and positive leads of the tracking coil.

The focus error (FE) signal is taken from pin 32 of RF amp IC501 through gain control and to pin 48 of servo control IC502. The focus-gain output signal (FEO) from pin 5 feeds the focus coil driver (IC402, Q410, Q411, Q412, and Q413). Again, each positive and

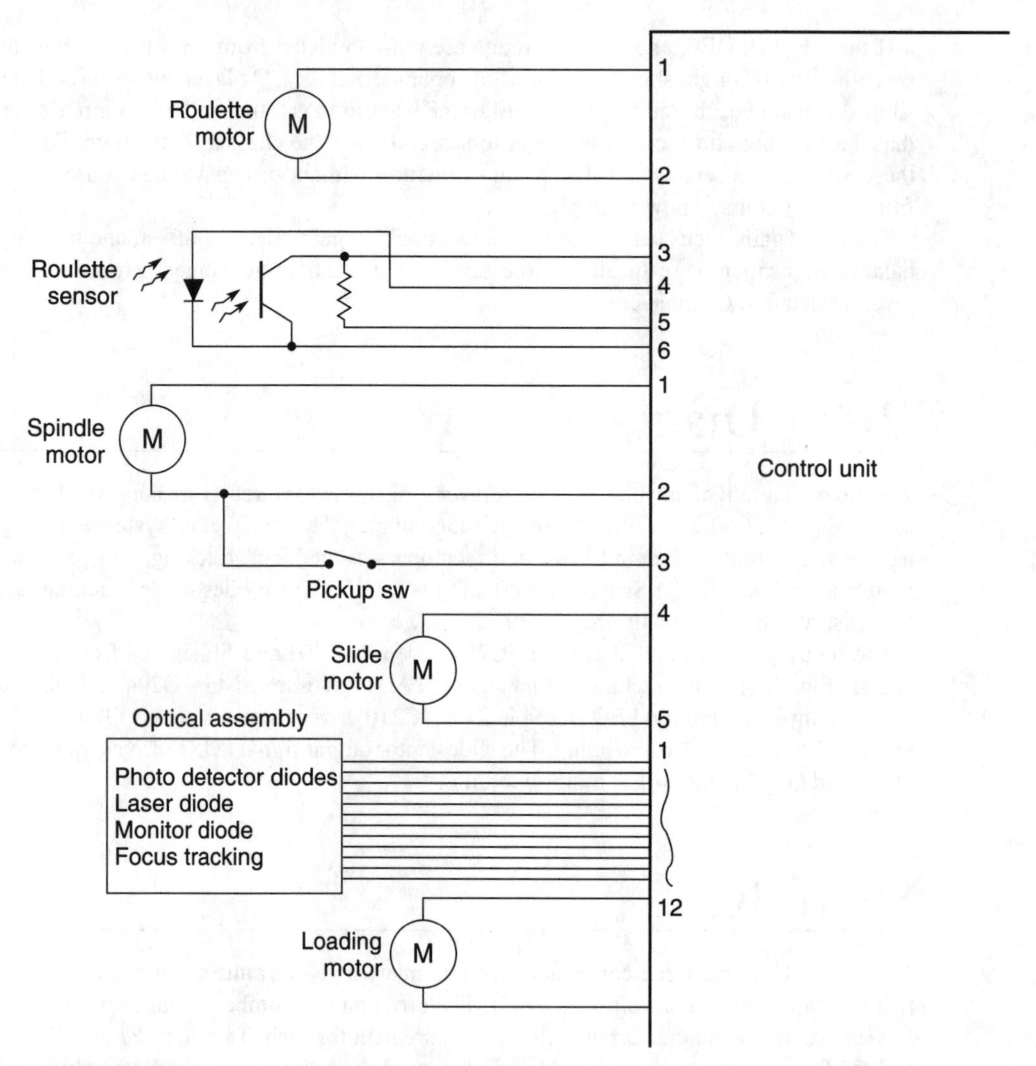

FIGURE 26-3 The block diagram of a typical control section of a CD changer.

negative focus coil terminals are driven with separate IC and transistors in this automatic changer.

The spindle and SLED motors are controlled by the servo-control IC (IC502) through the various ICs and transistors motor drivers. The loading motor is driven by driver Q418 through Q422 from loading and eject signal of the mechanism-control IC.

Denon DCM 560 servo circuits The RF amplifier and ALC control circuits are inside the optical laser-pickup assembly. You will find that in the most-recent CD players, the optical assembly (which includes photo-detector diodes, the laser diode and monitor diode, the focus and tracking coil, plus the RF amp IC). The tracking error (TE) signal is fed from pin 2 of optical pickup to pin 1 of servo IC101. Likewise, the focus error (FE) is at pin 3 and fed to terminal 5 of servo IC101 (Fig. 26-7).

IC101 controls the slide or SLED motor circuits to IC106, and transistor drivers TR185 and TR184 at pin 20. Terminal pin 11 feeds the FEO signal to the focus IC106, and driver transistor TR181 and TR180. The tracking-coil voltage is controlled by two transistor drivers, TR183 and TR182, with IC105 receiving control signals from pin 17 of IC101.

Optimus CD-3380 portable servo circuits The servo IC2 in an Optimus CD-3380 player, controls the focus and tracking coils, spindle, and disc motors. The tracking error signal is fed from pin 20 of RF amp IC1 to pin 45 of servo IC2. A focus error signal

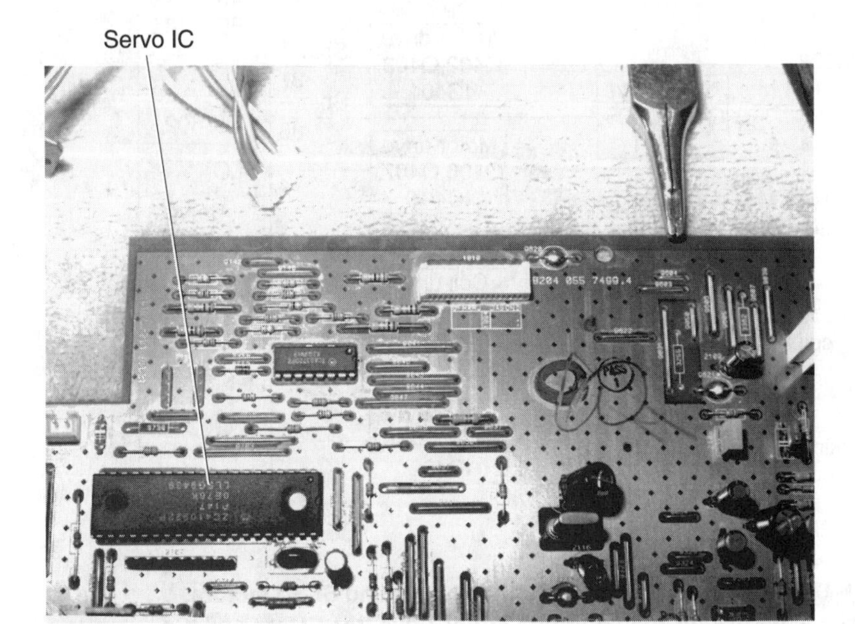

FIGURE 26-4 The servo section within the Magnavox CDC 745 automatic changer.

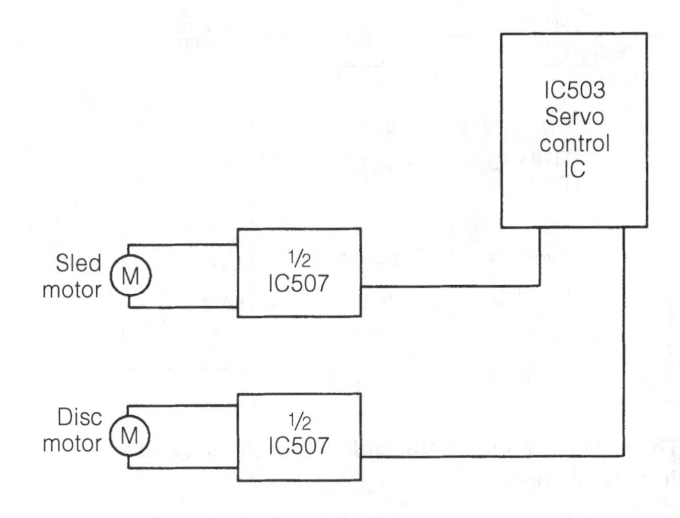

FIGURE 26-5 Servo control IC503 controls the SLED and disc motors through the respective IC drivers.

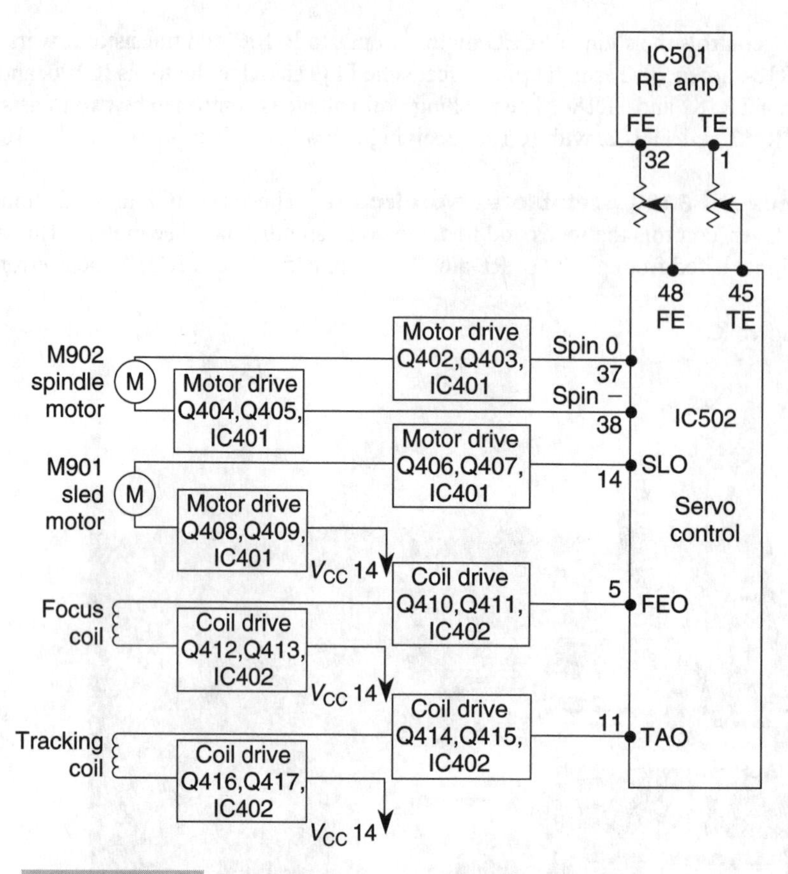

FIGURE 26-6 IC502 controls the spindle and SLED motors, focus and tracking coils, through motor-drive transistors and ICs.

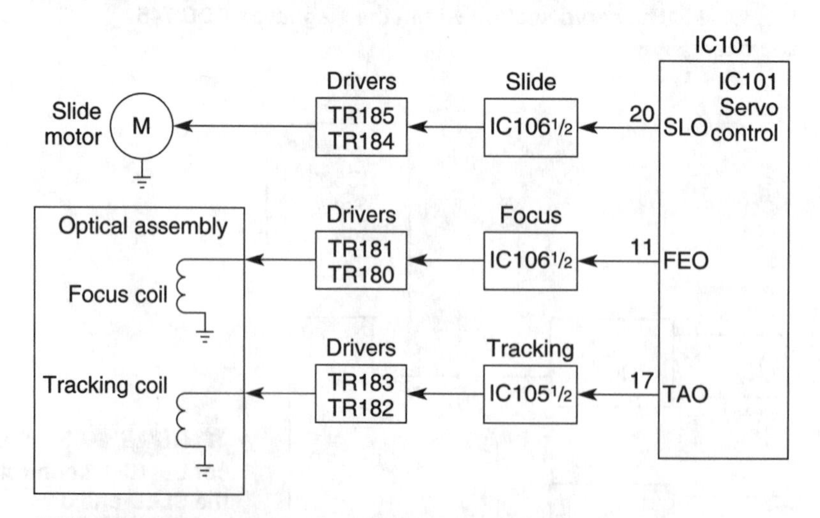

FIGURE 26-7 The servo IC controls the slide motor, focus, and tracking coils within a CD changer.

is fed from pin 19 of IC1 to pin 48 of servo IC2. The focus and tracking error signals are fed from the optical pickup assembly to input terminals 10 and 11 of the RF amp IC1.

The focus error offset (FEO) signal is fed from pin 5 of input pin 2 of IC5. IC5 is a separate IC that provides both focus offset and tracking-offset signal to the coil assemblies within the optical assembly. The tracking-offset signal is fed from pin 11 of servo IC2 to pin 7 of motor driver IC5. The amplifier focus voltage is applied to pins 15 and 18 and tracking signal to pin 16 and 17 of the optical pickup from driver IC5.

The spindle motor signal from servo IC2 (pin 39) is fed to input 9 of motor driver IC6. The motor winding is fed from pin 13 and 14 (D2+ and D1–) of driver IC6. A disc or sled signal is fed from pin 14 of servo IC2 to pin 2 of IC1. Both disc motor windings are connected to pin 17 (D1–) and 18 (D1+) of motor driver IC6.

Realistic CD-3304 boom-box servo circuits In the Realistic boom-box CD servo circuits, the focus error (FE) is taken from pin 19 of the RF amp IC501 through a focus gain control to pin 6 of the servo LSI IC502 (Fig. 26-8). The tracking error (TE) signal is taken from pin 20 of RF amp IC501 through tracking gain control to pin 3 of servo LSI IC502. The focus error output (FEO) is fed from pin 11 to IC506 driver and to the focus coil winding. Likewise, the tracking error output (TEO) is fed from pin 17 to driver IC506 and to the tracking coil winding.

Notice that the two separate driver op amps of IC506 are fed to the focus coil and two separate drivers to the tracking coil. Driver IC506 provides four op amps and 1/4 of each is applied to the different terminals of focus and tracking coils.

The spindle and slide motors are controlled by the servo-control LSI (IC502) with the spindle signal at pins 21 and the slide motor output at pin 20 of IC502. IC507 provides four op amps that drive the negative and positive terminals of each motor.

THE FOCUS SERVO CIRCUITS

The purpose of the focus servo circuit is to keep the laser-beam spot correctly focused on the pits of the disc surface. The focus zero cross (FZC) circuit detects the focus error signal

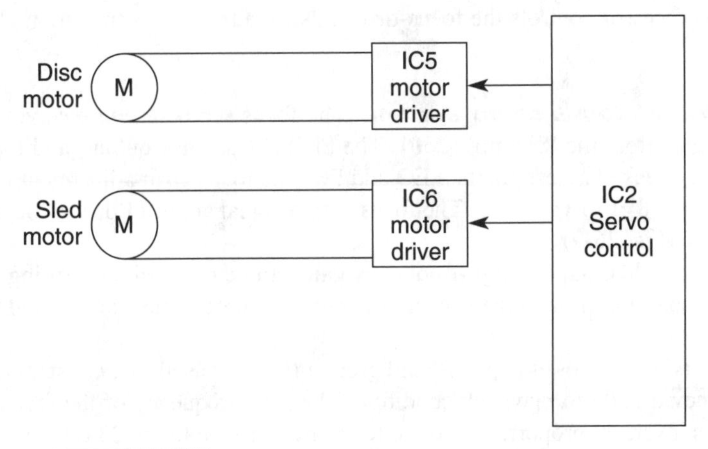

FIGURE 26-8 Servo IC2 controls motor drivers IC5 and IC6 of the disc and sled motors.

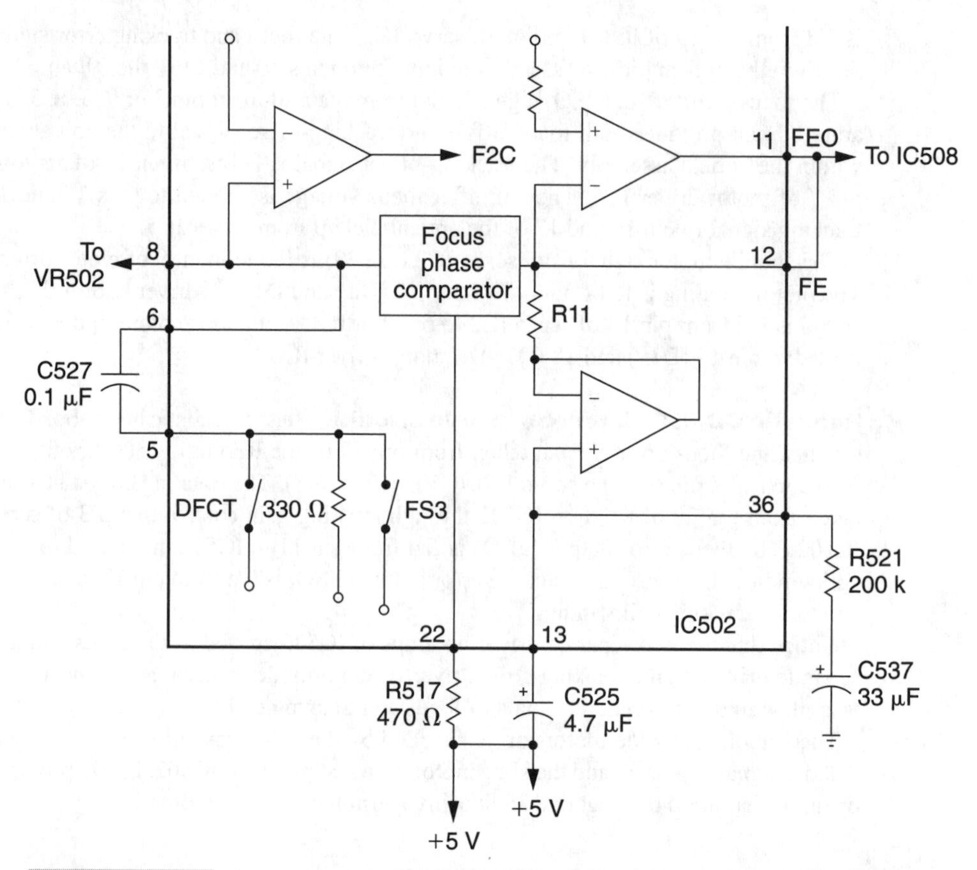

FIGURE 26-9 **The focus-control circuits in a Realistic boom-box CD player.**

and is used with the FOK circuit to determine the focus-adjustment timing. The focus-search circuit shifts the object lens up and down to find the correct focus point. The signal from the servo processor controls the focus-drive IC or the transistors that are tied to the focus coil.

Realistic CD-3304 focus servo system The focus servo system receives an FE (focus error) signal from the RF amp, IC501. The FEO (focus error output) and focus error signal can be checked at test points, FEO and FE. The focus error adjustment is made with VR502, on pin 8 of servo IC502. The focus output signal at pin 11 drives focus actuator driver IC506 (Fig. 26-9).

When FS3 is switched on, the high-frequency gain can be reduced by forming a low-frequency time constant through the capacitor connected across pins 2 and 3, and the internal resistor.

The capacitor (C526) across the pin 10 and ground (GND) has a time constant to raise the low frequency usually to playback condition. The peak frequency of the focus phase compensation is inversely proportional to the resistor connected to pin 23 (about 1.2 kHz when the resistor is 510 kΩ).

The focus search peak becomes ±1.1 Vp-p with this constant. The peak is inversely proportional to the resistor connected across the pins 35 and 37. However, when this resistor is varied, the peak's track jump and SLED kick also vary.

FOCUS COIL DRIVE CIRCUITS

Today's focus drive circuits can be a combination of ICs and transistors. Some portable CD players have a servo IC that provides signals to an IC driver up the focus coil. In other larger units, the focus is driven with transistors, the op-amp IC, and the servo IC. The input RF signal from the photodiodes connects to the servo IC. Then, the focus error signal goes to an amplifier IC that drives either a transistor or an IC driver. The focus coil in a Realistic CD-3380 portable CD player is operated by IC5 as driver and via a signal from servo IC2 (Fig. 26-10).

TRACKING SERVO CIRCUITS

The purpose of the tracking servo system is to control the laser-beam spot directly in the center of the pit track laterally or horizontally. The tracking-coil assembly movement is horizontal, where the focus-coil assembly moves closer or further. The tracking error signal from the RF amp or preamp IC goes to the servo-control IC, which drives a tracking-coil drive IC or transistor (Fig. 26-11). The tracking driver IC or transistors provide voltage to the tracking-coil assembly, located in the optical pickup assembly (Fig. 26-12).

The Onkyo DX-200 tracking servo circuit is shown in Fig. 26-13. The tracking error (TE) input from pin 1 of P201 is posed to the R226 variable resistor for gain adjustment before being applied to pin 13 of Q202. The signal passes via a phase-compensating operational amplifier (OP4) to pin 3, then to pin 2 to be passed through another phase-compensating operational amplifier (OP2). The output from pin 27 then passes to the optical-pickup tracking coil by driver Q204 to drive the object lens.

The purpose of R262, R263, C228, and C229 (located between pins 2 and 3 of Q202) is to switch the gain for the high frequency by pin 1 switch TG2. The frequency response can be switched by TG1 in combination with C227 connected to pin 1. These switching circuits are activated as a result of track kicking and track accessing in order to stabilize the tracking servo. TM1 is the servo on/off switch.

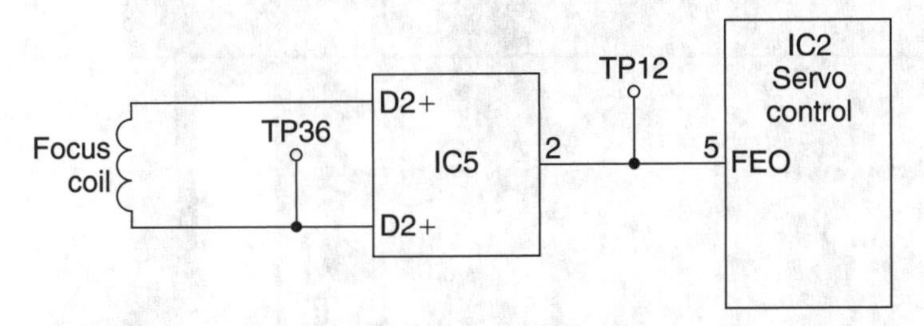

FIGURE 26-10 **IC2 servo control, controls focus driver IC5 and focus coil in a portable CD player.**

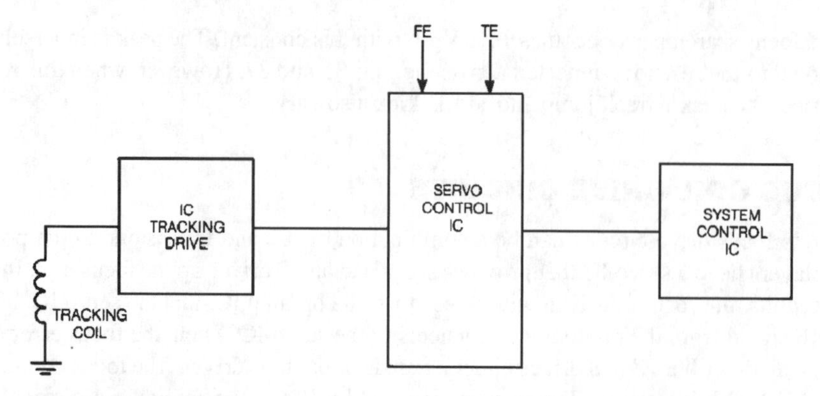

FIGURE 26-11 A typical tracking servo in the compact disc player.

FIGURE 26-12 The tracking coil can be driven with two transistors in the Sanyo CP-660 model.

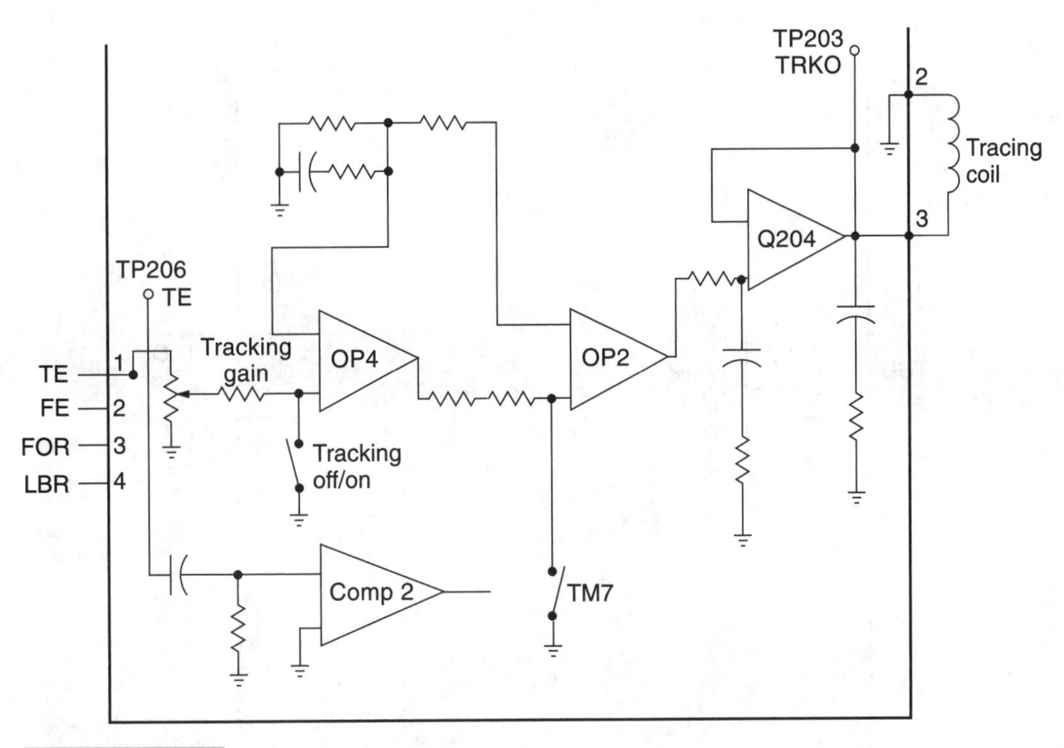

FIGURE 26-13 The tracking servo circuits in Onkyo CD player.

The same figure shows the Onkyo DX-200 tracking zero cross (TZC) circuit. The TZC circuit generates the timing for switching the tracking servo on and off, following a track-kicking action or when the number of tracks is counted together with the MIR signal during track accessing. The low-frequency components in the TE signal are removed by C215 and R225 before the signal is applied to the COMP2 comparator from pin 12 of Q202. Then, in response to an instruction from the microcomputer, the output passes from pin 5 (sense).

Realistic CD-3304 tracking servo system The tracking error (TE) signal is fed from the RF amp (IC501) to pin 3, through a reference adjustment of VR503. This TE signal can be checked at test point TEO. The output signal at pin 17 goes to the tracking actuator IC506 and then to the tracking coil (Fig. 26-14).

The capacitor across pins 14 and 15 has a time constant to lower the high frequency when TGZ is switched off. The tracking phase-compensation peak frequency is inversely proportional to the resistor connected to pin 23 (about 1.2 kHz when the resistor is 510 kΩ).

For a track jump in the FWD (fast forward) or REV (reverse) duration TM3 or TM4 are set on. At this time, the peak voltage fed to the tracking coil is determined by TM3 and TM4 current values and the feedback resistor from pin B. That is:

- *Track jump peak voltage = TM3 (TM4) current value × feedback resistor value*
- The FWD or REV sled kick is performed by setting TM5 and TM6 to On. At this time, the peak voltage added to the SLED motor is determined by the TM5 or TM6 current value and the feedback resistor from pin 21.

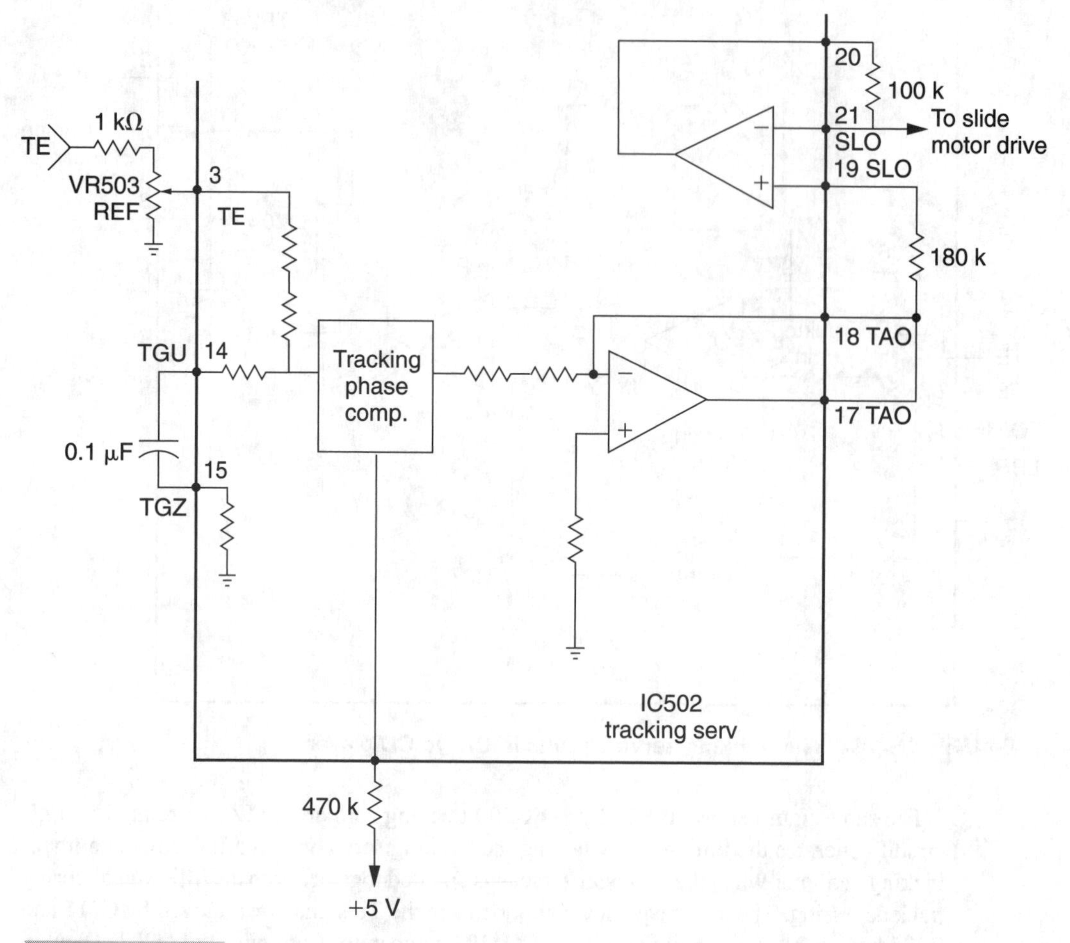

FIGURE 26-14 **The tracking servo system in a boom-box CD player.**

- *SLED jump peak voltage = TM5 (TM6) current value × feedback resistor value*
- Each SW current value is determined by the resistor connected to pin 35 and 37 when the resistor is set at about 120 kΩ.
- TM3 or TM4 is 1.1 mA and TM5 or TM6 is 22 mA. This current value is almost inversely proportional to the resistor variable within a range of about 5 to 40 mA for TM3. Stop is the on/off detection signal for the limit SW of the SLED motor innermost circumference.

Optimus CD-7100 changer tracking coil The CD-7100 CD changer consists of servo IC151, driver IC202, and the tracking coil. The tracking-coil signal is fed from pin 11 of servo control IC151 to pin 6 of op-amp IC202 (Fig. 26-15). IC202 is contained in a large IC that has three different op-amp circuits. Driver IC202 feeds the focus coil from pin 4. The tracking coil is located inside of the optical pickup assembly.

Optimus CD-7100 CD changer focus-drive circuits The focus-drive circuits in the Optimus CD-7100 changer is somewhat similar to the tracking-coil circuits. The servo control IC (IC151) feeds the FEO signal at pin 5 and connects directly to pin 1 of op-amp driver IC202 (Fig. 26-16). Output pin 3 of IC202 connects directly to the focus coil. The focus coil is located inside of the optical pickup assembly.

Realistic boom-box servo circuits The 200-Hz low-power filter (LPF) is formed with capacitor C542 (0.033 μF) and a 10-kΩ resistor connected to pin 42 and the secondary low-power filter (LPF) is formed with the built-in LPF (F_c up to 200 Hz, within 510 kΩ for pin 23), and the carrier component of the CLV servo error signals MDS and MDP is eliminated (Fig. 26-17).

In the constant linear velocity (CLV) –S mode, FSW becomes L (low) and the pin 42 low-power filter (LPF) F_C lowers, strengthening the filter V_{CC} (+5 V). F_C does not vary with power-supply voltage fluctuations.

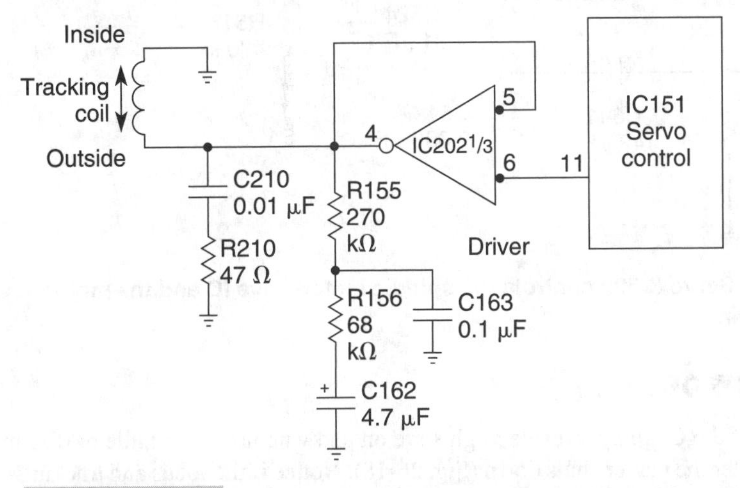

FIGURE 26-15 Servo control IC51 controls tracking driver IC202 and tracking coil in an Optimus CD player.

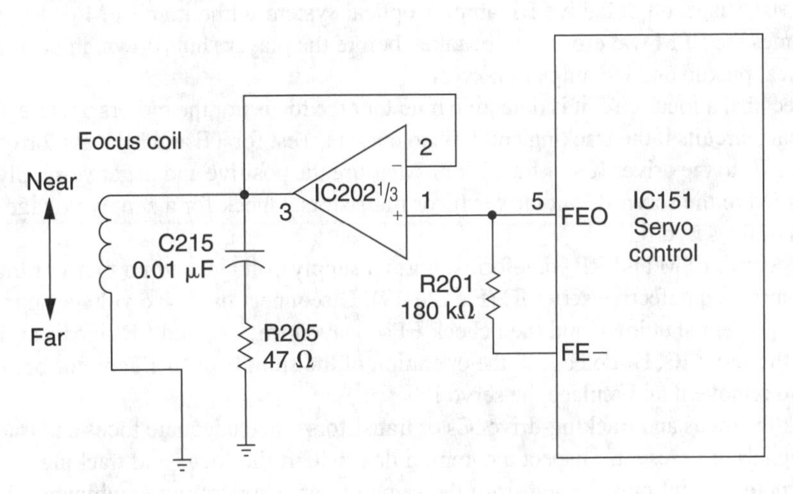

FIGURE 26-16 The focus driver circuit in a CD changer.

4 TROUBLESHOOTING AND REPAIRING COMPACT DISC PLAYERS

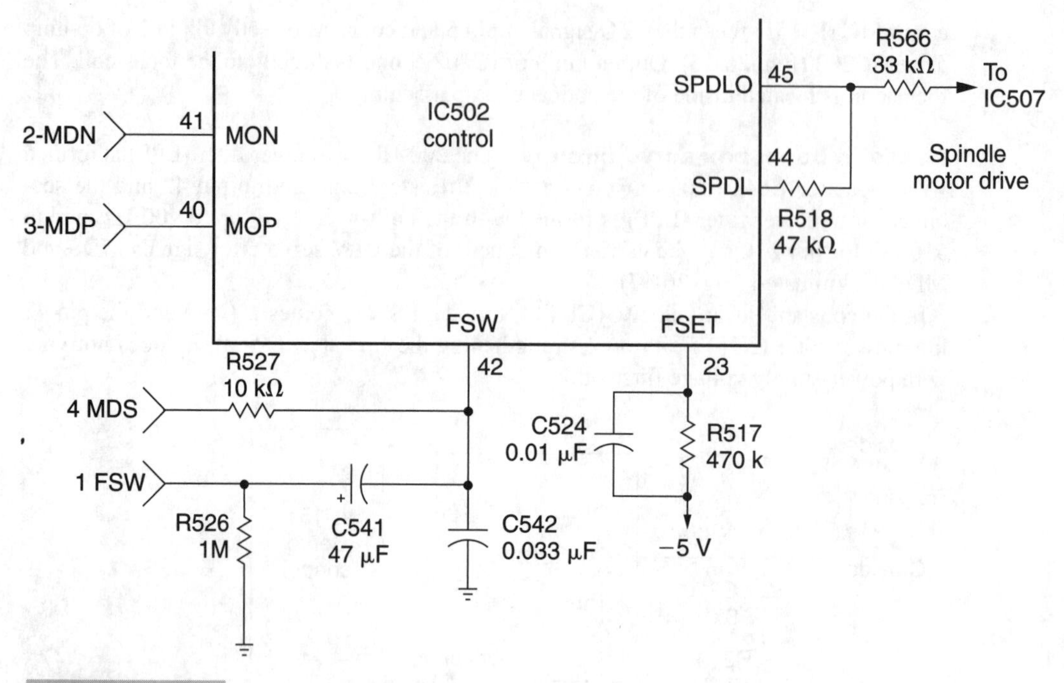

FIGURE 26-17 **Servo IC502 controls the spindle motor-drive IC and the motor in a boom-box CD player.**

SERVO PROBLEMS

Determine if the CD player display lights are on and whether the spindle or disc motor is operating after the player shuts down (Fig. 26-18). Notice if the focus and tracking-coil assemblies begin to search or move before shutdown. Check for FE, TE, and mirror signals from the RF amp IC. Does the RF amp have a constant EFM waveform before or after shutdown? Suspect a defective RF amp or optical system without an EFM or RF signal. Sometimes the EFM waveform can be taken before the player shuts down, indicating that the optical pickup and RF amp are normal.

Suspect that a focus circuit is defective if neither the focus nor the motors operate. Check the tracing circuits if the tracking coil fails to operate. Test for TE and FE waveforms from the servo IC to the driver ICs or transistors. Measure the positive and negative supply voltage applied to the servo IC and driver IC or transistors. Check for a supply voltage at the V_{CC} pin of the servo IC.

With correct EFM and RF waveforms, normal supply voltage, and no focus or tracking signal, suspect a defective servo IC (Fig. 26-19). Disconnect the servo voltage supply pin (V_{CC}) to prevent shutdown and then check EFM waveform, TE, and FE signals at the RF amp to the servo IC. Doublecheck the operation of the spindle or SLED motor before deciding to remove it and replace the servo IC.

Check the focus and tracking-drive ICs or transistors when adequate focus and tracking-drive signals are present. Suspect a common driver IC if the focus and tracking coil, and SLED motor do not operate, fed from the same driver IC or from a combination IC and transistors. Check each drive transistor with a transistor in-circuit tester.

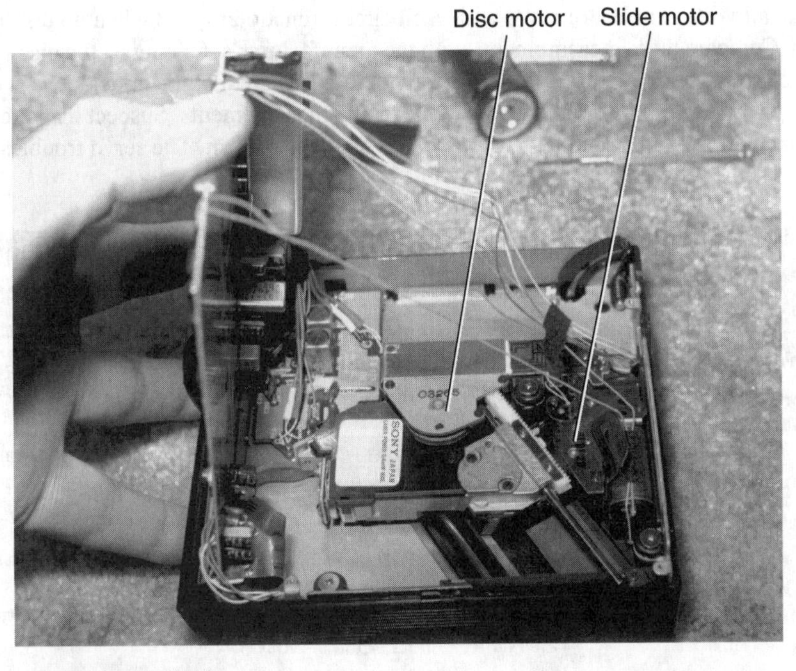

FIGURE 26-18 The spindle or disc motor in a portable CD player.

FIGURE 26-19 The large servo IC is located on the top side of the chassis in this table-top CD changer.

Take all voltage measurements with each circuit functioning and when no disc is in position. Compare these measurements with the manufacturer's. Of course, the disc interlock must be shunted, to operate without a disc moving. Be sure that the optical lens is covered or keep your eyes away from the lens while taking measurements. Suspect that a focus and tracking coil is open or defective if the drive voltage is present. The servo troubleshooting chart is shown in Table 26-1.

TABLE 26-1 TROUBLESHOOTING CHART OF SERVO CIRCUITS

SYMPTOM	CHECKS AND TESTS
CD player shuts down	Check the RF waveform at output of the RF amp.
	If there is no EFM waveform, check the RF amp and the optical assembly. Check the RF amp supply voltage (V_{CC}).
	Check the laser diode with the power meter at the lens assembly.
	Shunt the interlocks for power meter and servicing.
No focus	See if the focus and tracking coils move when first turned on.
	In some changer CD players, the optical assembly is buried and not visible.
	Perform a focus error adjustment test.
	Check the FE input at the servo IC.
	If there is no FEO output, suspect the supply voltage or the servo IC.
	Check the various test points from the servo IC to the focus coil.
	Perform a continuity test with the DMM at the focus coil.
No tracking	Test the input TE signal at the input of the servo IC.
	Check the TEO output from the servo IC.
	If there is no output waveform, suspect the servo IC or supply voltage (V_{CC}).
	Notice if the tracking coil moves when CD player is first turned on.
	Take a TE offset waveform across the tracking coil.
	Take TE offset alignment and adjustment.
	Suspect the driver IC or transistors.
	Notice if the servo output is normal when it is first turned on
	Perform continuity tests at the tracking coil.
No disc motor operation	Check the spindle or disc output at the servo IC. In some CD players, the disc motor is operated from the RF or signal processor IC.
	Check the voltage on the spindle motor and check motor continuity.
	Check the supply voltage to driver transistors or ICs. Test the transistors in the circuit.
No sled motion	Check the sled or slide motor voltage at the servo IC.
	Test the transistor driver or IC driver.
	Check supply voltage at the driver IC.
	Take waveform tests at test points from the servo IC to the motor terminals.
	Check the voltage on the motor terminals.
	Check the continuity at the motor winding.

TABLE 26-1 TROUBLESHOOTING CHART OF SERVO CIRCUITS (CONTINUED)

SYMPTOM	CHECKS AND TESTS
Intermittent action	Notice if the CD player shuts down or operates for several minutes or hours before the intermittent mode.
	Monitor the section that is intermittent.
	Check the voltages at all ICs and transistors.
	Monitor the various waveforms and test points in the intermittent circuit with voltage and resistance tests.

The Various Motor Circuits

Three basic motors are used in tabletop CD players. Usually, the CD boom-box player has top loading, which eliminates the loading motor. The auto CD changer might have four or more motors, and the tabletop changer might have up to five different motors (Fig. 26-20). You might find only two motors in the portable or combination CD and cassette player.

The tray or loading motor pushes out and pulls in the CD tray when the open/close switch is engaged. (A top-loading CD player has no loading motor.) A disc, spindle, or turntable motor rotates the CD disc at a variable speed, somewhat like the phonograph motor. Sometimes the disc or turntable motor is called the *spindle motor*. Although the phono motor operates at a constant speed, the disc motor travels faster at the beginning and slows down as the laser assembly moves toward the outside rim of the CD. The slide or sled motor

FIGURE 26-20 The motor-control board in a Magnavox CD changer.

FIGURE 26-21 **The slide and disc motors on the optical lens assembly board.**

moves the laser from the center to the outside of the CD on sliding rods (Fig. 26-21). Some players have a pickup motor that travels in a radial or semicircle motion.

TABLETOP CHANGER MOTORS

You might find a slide, disc, up/down, magazine, and loading motor within the tabletop changer (Fig. 26-22). The two new motors are the up/down and magazine motors. The up/down motor assists in loading and playing of disc, and the magazine motor rotates the turntable (carousel) or changes the different discs for playing. The magazine motor can be referred to as a *turntable motor*. The up/down and magazine motors can operate directly from a dc source or from a motor driver IC. In the Onkyo DX-C909 carousel player, the carousel, tray or loading, and chucking motors operate from microprocessor (system control) Q202 (Fig. 26-23).

Realistic CD-3370 portable CD motors The CD-3370 portable CD player uses two motors: a sled motor that moves the laser pickup over the disc and a spindle motor that rotates the disc (Fig. 26-24). A servo- control circuit is used to control each motor.

A tracking servo signal is used to move the pickup horizontally. If the position of the pickup is out of the tracking control range set by the tracking coil, a dc component appears on the tracking servo signal, pin 11 (TAO), of the servo signal processor IC5. The dc component is amplified and appears on pin 14 (SLO). It is then fed to pin 1 of IC6 to drive the sled motor, which brings the pickup to within the line-tracking control range (Fig. 26-25). The sled motor stops when the dc component becomes zero, and the line-tracking control regains control of the tracking.

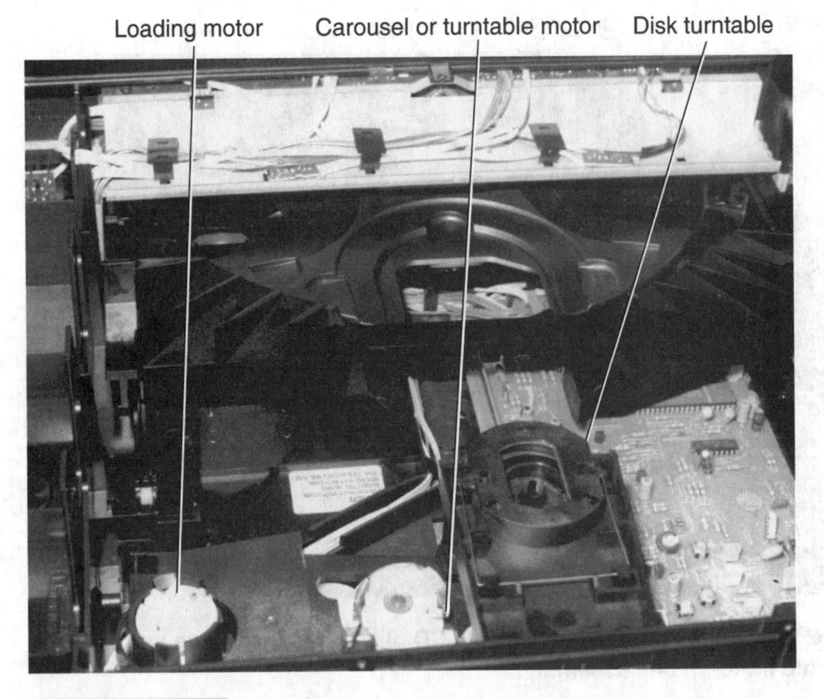

FIGURE 26-22 The table-top automatic disc changer can have slide, disc, up/down, clamping, and loading motors.

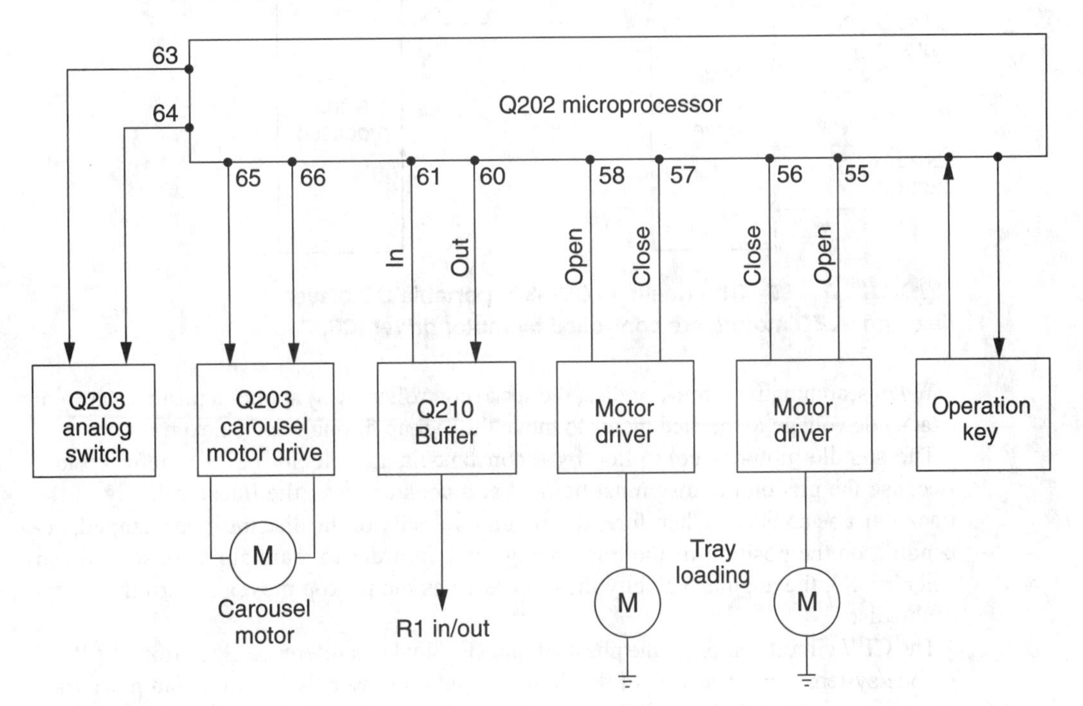

FIGURE 26-23 The tray, loading, and carousel motor operate from microprocessor Q202.

FIGURE 26-24 A typical portable CD player has a carriage and spindle motor in one assembly.

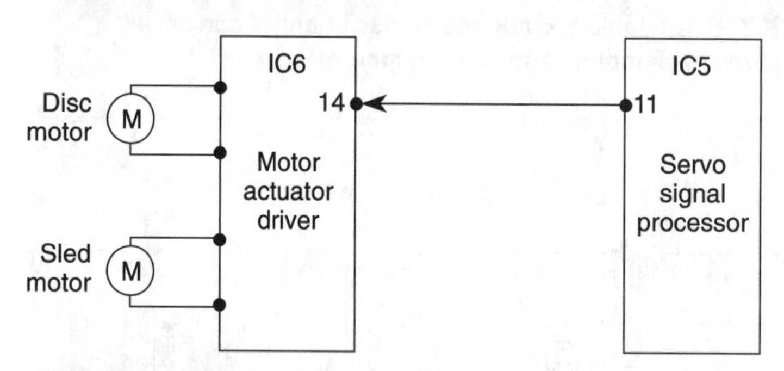

FIGURE 26-25 The Realistic CD-3370 portable CD player disc and SLED motors are controlled by motor driver IC6.

When searching for a track or disc, the servo controller (IC5) applies a positive (at +) or a (at –) dc voltage to the sled motor to move the pickup through to the next track.

The spindle motor is controlled by a constant-linear-velocity (CLV) motor circuit. Because the pits on the disc must be read at a constant rate, the linear velocity of the track must be constant. Therefore, the angular velocity of the disc must be changed, depending on the position of the track being read, in order to maintain a constant linear velocity. So, the angular velocity must increase as the pickup moves toward the center of the disc.

The CLV circuit compares the phase of the RF signal to a reference clock from the PCM decode/system control section of the digital signal processor (IC3). An output pulse train appears on pin 39 (SPDLO) of IC5. When the motor speed is correct, a 50% duty cycle signal appears; a shorter duty cycle indicates that the speed is too high. IC6 amplifies the signal, and uses it as a reference to control the spindle motor speed.

VARIOUS MOTOR TROUBLES

A defective motor could be dead, intermittent, rotate slowly, or it might be very noisy. An open or "dead" motor can be located with continuity and voltage measurements. After isolating the correct motor function and locating the correct motor, a continuity measurement across the motor winding will indicate if the winding is open (Fig. 26-26). Remember, the resistance measurement across these motors should be practically zero. Likewise, an improper voltage across the motor windings might indicate that a motor or motor circuit is defective.

Some of the motors are located under the chassis, so it might be difficult to get at the motor terminals. Trace the motor terminals up to the main circuit board. Usually, they are plugged into some sort of socket. Check the suspected socket if the motor is intermittent or if no voltage is present at the motor terminals (Fig. 26-27). This type of a socket connection can become loose after many hours of operation. Push the plug down tight. Sometimes the small clamps that dig into the connecting wire make a poor connection. This condition usually shows up after the player has been in use for some time.

Remember, dc motors can operate in any direction by simply reversing the polarity of the supply voltage. Most CD motors can be checked with a C or D battery. A suspected motor should operate with 1.5 V applied across the motor terminals (Fig. 26-28). Apply the battery voltage at the motor terminals or be sure that you have the correct pair of wires when injecting voltage into a socket. Always remove the socket from the main chassis before applying voltage, in case you have the wrong component. The slide or tray motor might not rotate if it is at the end of its operation. Reverse the motor battery terminals or remove the motor for a good test.

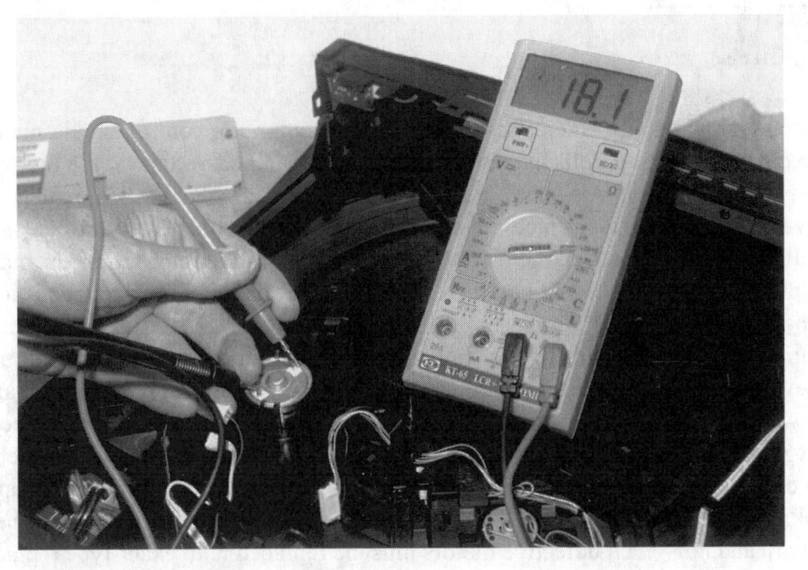

FIGURE 26-26 **Check the resistance of the suspected motor winding with the R×1 scale of the DMM.**

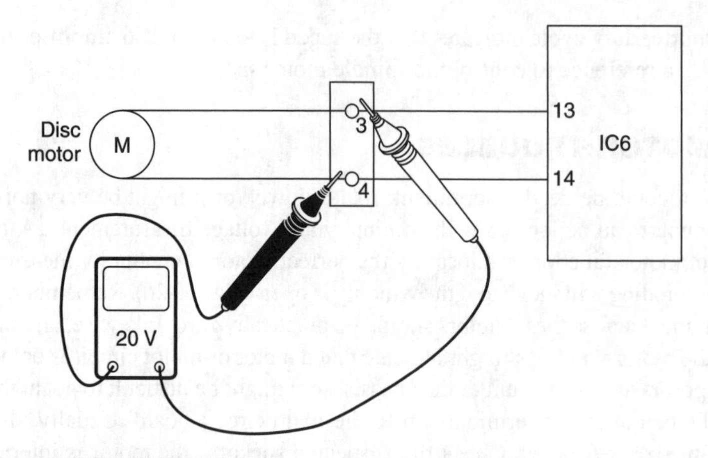

FIGURE 26-27 **Measure the dc voltage across the motor terminals to identify the defective motor.**

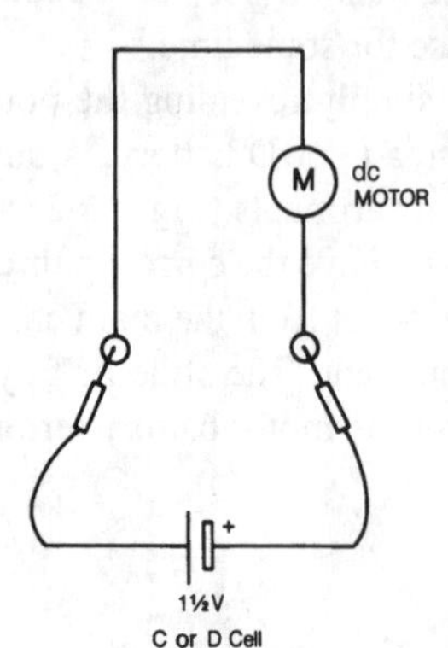

FIGURE 26-28 **Check the motor by connecting a 1.5-V battery to see if the motor rotates.**

An intermittent motor can result from an intermittent motor control circuit, poor motor cable connections, or a defective motor. Monitor the applied voltage at the motor terminals. Notice if the motor is intermittent, but has a constant voltage. If so, replace the defective motor. Suspect that a motor circuit is defective if the voltage varies when the motor speed acts up. Doublecheck the motor terminal wires at the motor terminals and where they plug into the main circuit board. Service the motor circuits if the voltage is intermittent or absent at the motor terminals.

Gummed-up motor bearings can slow the speed. This does not usually occur until after several years of operation. Clean the motor bearings with rubbing alcohol and cotton swab or other cleaning stick. The noisy motor might have worn or dry bearings. Most motors in the CD players are lubricated for the life of the player. Sometimes a drop of light oil on the motor bearing cures a noisy motor. Check for a worn bearing by checking end play and movement of the motor pulley. Replace the motor if the bearings are worn and noisy. All defective motors must be replaced with exact-type replacement motors.

THE TRAY OR LOADING MOTOR

The tray or loading motor, also called the carriage motor, moves the tray in and out for loading and unloading the disc. In most players, this process is activated with a push button (push to open or to close). Usually, the plastic tray is driven by a plastic gear box next to the tray assembly (Fig. 26-29). This same gear assembly might operate a large plastic gear, which raises and lowers the clamper assembly. When the tray is out for loading, the clamper (or flapper) assembly raises. As the tray is closing, the clamper provides spring-loaded pressure on the CD, holding it in position.

A dirty Open/Close switch can cause intermittent or erratic tray operation. Inspect the button terminals for poor contacts. Check the tray switch by shorting a test clip across the switch terminals. Check the interlock switches in the same manner. A test lead with two small alligator clips on each end can do the trick. If the drawer will not open, check all possible mechanical problems first. Be sure that the transit screw is removed (or loosened). Visually inspect the drawer gear assembly for foreign objects. Notice if the tray rails or gears are binding. Clean the area and apply a light coat of lubricant to the sliding areas (Fig. 26-30).

Be sure that the small tray motor is stopped when the tray is out. Often a small leaf switch is engaged by the large plastic drum that raises and lowers the flapper assembly. Dirty or poor contacts of this switch can cause erratic or no operation of the tray assembly. Clean the leaf contacts with contact cleaner. Place a piece of cardboard between the contacts, and, holding the leaf contacts tight, move the cardboard back and forth to shine the contacts.

The loading motor drives the loading tray and gear-box assembly with a small rubber belt. Oil and grease on the belt can cause erratic tray loading. Clean the belt with rubbing alcohol and cotton swab. A broken or stretched belt can prevent the tray from opening or closing (Fig. 26-31). Although phono compound or liquid rosin applied to the motor pulley can help temporarily, the drive belt should be replaced. The slipping compound can be used while you are waiting for the arrival of a new belt.

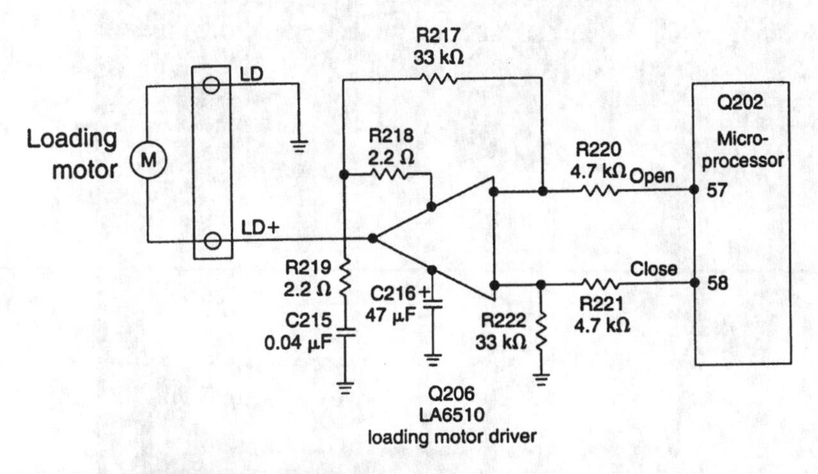

FIGURE 26-29 **The loading motor circuit might drive a loading tray from the IC (Q206) op amp and Q202.**

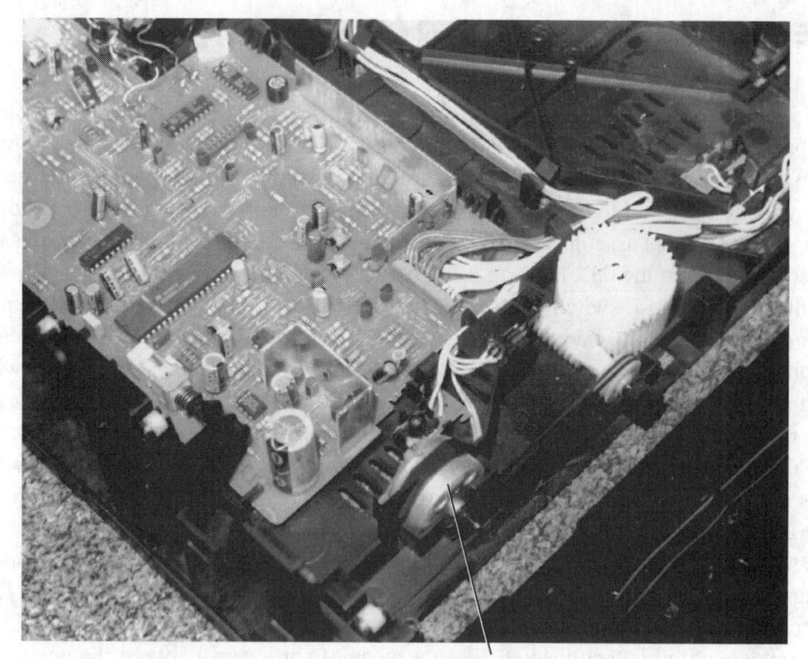

FIGURE 26-30 The loading motor drives a flat belt to move the turntable tray out and in the CD changer.

FIGURE 26-31 The loading motor belt loads and unloads the CD disc in a five- or six-tray automatic changer.

FIGURE 26-32 **The loading motor moves the tray in and out, and also raises cam and clapper assembly in a Sanyo CP-500.**

Stripped gears on the loading pulley or alongside the tray carriage can jam or prevent the tray from opening and closing (Fig. 26-32).

Carefully inspect both gear assemblies. These plastic gears might get stripped or broken if someone grabs the tray while loading or if you try to pry it open when it is stuck. Replacing the broken gear part is the only solution.

Pioneer PD-9010 carriage servo The return resistance voltage of the tracking activator current driver final stage is used as the input (Fig. 26-33). The carriage movement is performed by controlling the current supply in CX-2108 with the serial data so that the input is a dc voltage (Fig. 26-34). Because this type of carriage drive system is used, the final stage uses voltage drive. Because of the gain setting, the movement drive is limited at about ±11 V. Consequently, motor drive is a dc voltage if the unregulated voltage gets too high.

INTERMITTENT LOADING

Be sure that the tray assembly is not binding. Dress all cables and wires so that the turntable or roulette tray will rotate. Be sure that no wires or cables are holding up the loading tray. Likewise, check the belts and gears that are rotated by the different motors. A jammed gear will not let the motor rotate.

Monitor the various voltage and waveform signals of the motor control circuit. Clip a scope probe to the servo IC output that drives the motor driver IC or transistors. Monitor the voltage across the motor terminals. Replace the defective motor if it stops and if voltage is applied to the motor terminals. Remove the belt or gear train if you suspect that it is loading or binding the motor.

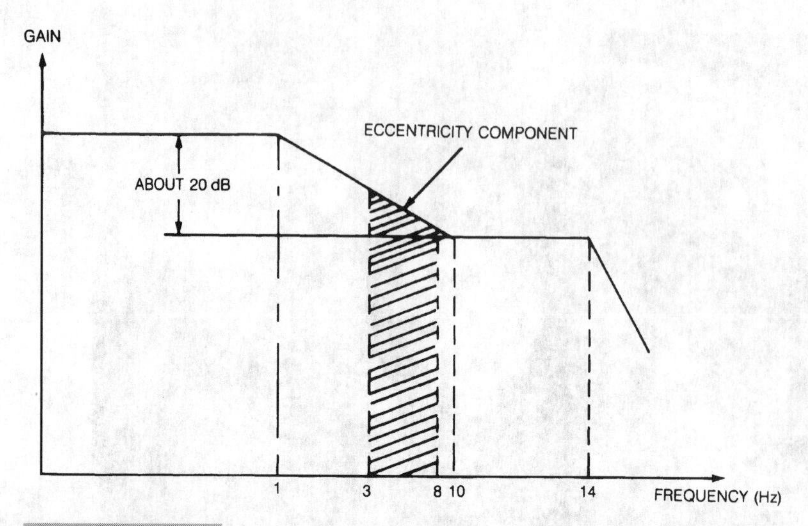

FIGURE 26-33 **A carriage servo chart of the carriage motor in a CD player.**

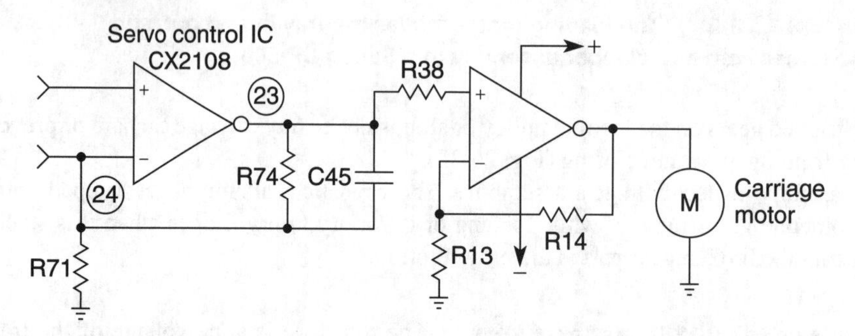

FIGURE 26-34 **By controlling the current supply in the servo-control IC, the output drive voltage goes to the carriage motor.**

Replacing the loading motor The loading motor is rather easy to remove in most CD players. The motor usually is bolted to the main chassis. Some units might have a separate loading assembly. Locate the loading or carriage motor near the tray and clamper assembly. The bottom cover of the CD player must be removed or the main chassis pulled up if there is no removable cover. Remove the small motor pulley belt. Remove the pulley if the motor won't fit through the chassis hole. Now remove the mounting screws that hold the motor to the chassis base.

Optimus CD-7105 loading motor The loading motor in the Optimus CD-7105 player is controlled from pins 39 (in) and 40 (out) of the system-control IC, IC351. This voltage is sent to pin 5 of loading driver IC202 to pull the tray in and to push the tray out at pin 6 of IC202 (Fig. 26-35). IC202 provides driving action of the spindle, disc select, and loading motors.

Driver IC202 has +8.3 V (pin 12) applied with –9.1 V at the negative terminal. The load voltage at pin 4 is applied directly to the loading motor terminals (CN203). Notice that the side of the loading motor winding is connected to common ground through pin 4.

Sanyo CP500 loading motor Remove the center screw that holds the plastic cam that raises and lowers the plastic disc pressure lever. Pull the plastic cam off and then remove the idler gear. Now, remove the loading drive belt. Remove the two small screws holding the loading motor to the plastic frame. Measure the distance from the top of the motor pull to the end belt of the defective motor so that the motor pulley can be installed in the right place on the loading motor. Reverse the procedure when installing a new motor.

SLIDE, SLED, OR FEED MOTOR

The slide, sled, or feed motor moves the optical pickup assembly across the disc from the inside to the outside rim of the CD, keeping the object lens constantly in line with the center of the optical axis (Fig. 26-36). The motor is gear-driven to a rotating gear that moves the laser beam down two sliding bars. In some players, the feed motor moves the laser pickup assembly in an arc or radial direction across the CD. The slide motor might have Fast-Forward and Rewind mode operation.

Erratic or intermittent operation of the slide motor might be caused by a gummed-up track or by poor meshing of the pulley and gears. Check the voltage on the slide motor terminals and note whether the voltage is intermittent. An erratic signal voltage can result from a defective transistor or IC motor circuit. Apply the battery voltage to the motor and note whether the motor and pickup assembly operate intermittently. Do not overlook the possibility that a motor is defective. Often, a voltage and continuity measurement across the motor terminals can help you to identify an open feed motor. Some slide motor circuits have a slide voltage test-point terminal.

In some models, the slide or carriage motor has a motor pulley that drives a worm pulley to slowly move the pickup assembly with a small motor belt. If the belt is loose or

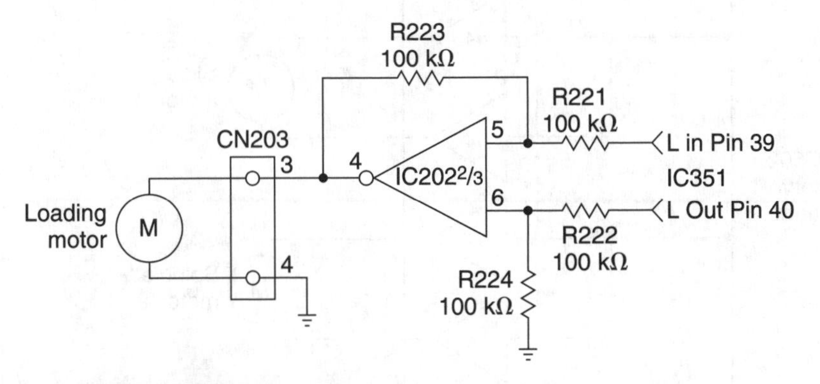

FIGURE 26-35 IC202 drives the loading motor in the Optimus CD-7105 changer.

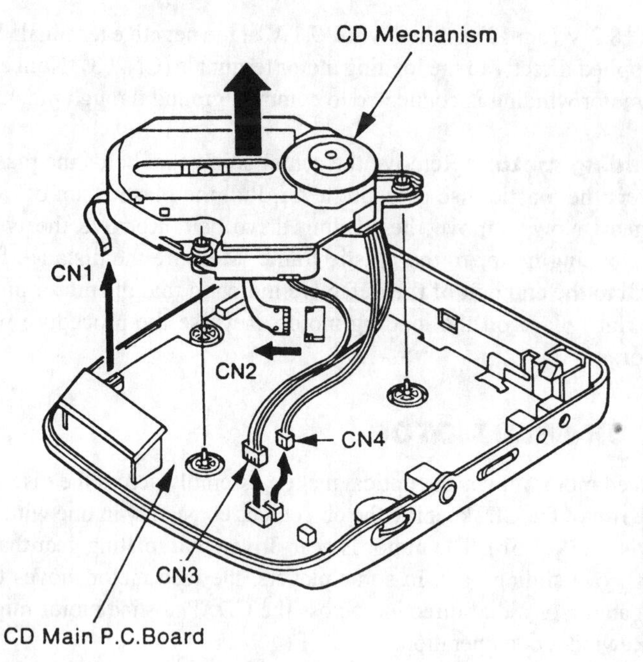

FIGURE 26-36 The slide and disc motors are on the mechanism in this Radio Shack portable CD player.

broken, the carriage or optical pickup will not move. Inspect the pulley belt for oil spots if the movement is erratic. Replace the belt if it shows any signs of slipping.

Realistic boom-box slide and disc motors The Realistic slide and disc motors are controlled by the servo control LSI (IC502). This signal is sent to 1/4 of IC507 driver IC, which controls both legs of the slide motor (Fig. 26-37). IC502 also controls 1/4 of the IC507 driver IC to operate the disc motor.

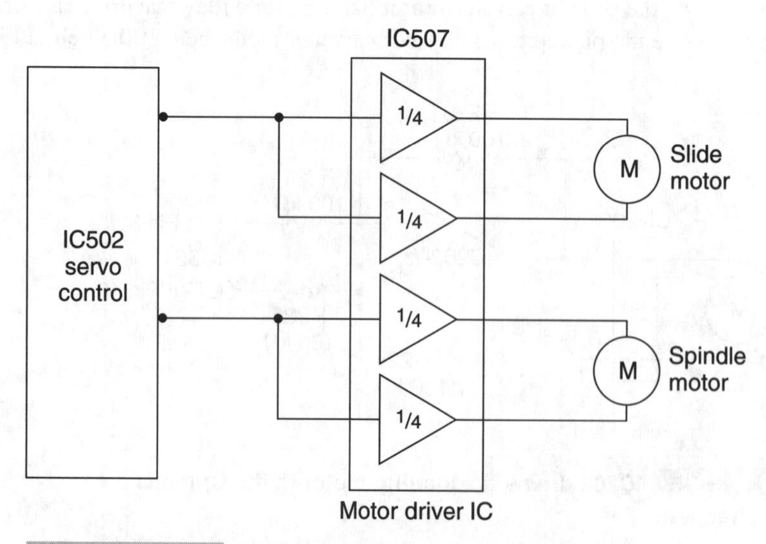

FIGURE 26-37 Driver IC507 controls the slide and disc motor with signal from LSI servo IC502.

REMOVING THE DEFECTIVE SLIDE MOTOR

Locate the slide or feed motor on the main chassis after you discover that it is defective. Usually, the motor is located close to the bar tracks (Fig. 26-38). Two small mounting bolts usually hold the slide motor to the bottom chassis. Remove the gear or pulley so that the motor shaft can be pulled through the chassis hole. Observe correct polarity when replacing the slide motor.

In some models, the plastic sled motor assembly must be removed before you can get at the slide motor. Four or more screws must be removed before the plastic assembly can be pulled upward. The two motor-mounting screws can now be removed, releasing the motor. Before replacing the new slide motor, also replace the gear assembly. Measure the distance between the outside gear and the top side of the motor end belt so that the gear can be correctly placed on the new slide motor (Fig. 26-39).

THE SPINDLE, DISC, OR TURNTABLE MOTOR

The spindle motor starts to rotate after the disc has been loaded. A small platform that is mounted on the turntable motor shaft spins the CD at a variable speed (Fig. 26-40). The spindle motor is located right under the clamper or flapper assembly. The disc starts out at approximately 500 rpm and slows down as the laser pickup assembly moves toward the outer rim of the CD (approximately 200 rpm).

Check the spindle disc motor with voltage and continuity measurements as you would the rest of the motors in the CD players. Check the spindle motor drive transistor or IC if the motor tests are normal.

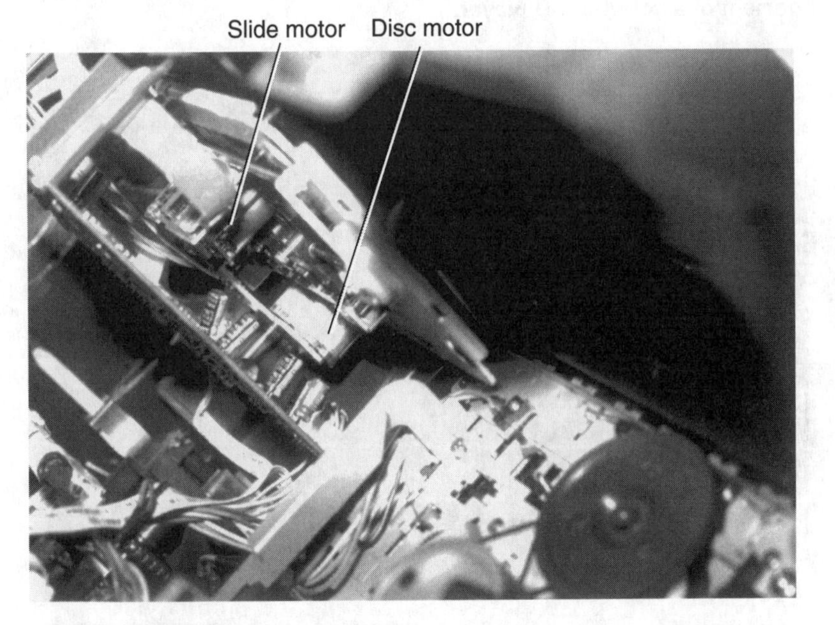

FIGURE 26-38 The slide and disc motors are mounted on a separate section of this Sharp boom-box player.

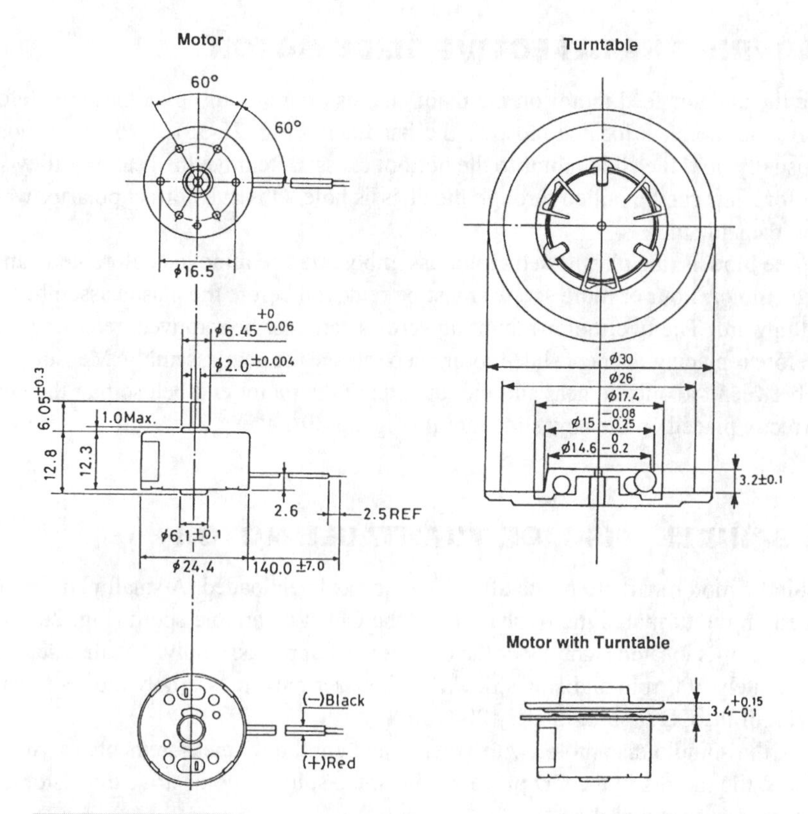

FIGURE 26-39 A typical motor dimension and mounting arrangement of a portable CD player.

FIGURE 26-40 The CD disc lays on a small platform of disc motor in a portable CD player.

TRAY MOTOR CONTROL CIRCUITS

The tray or loading motor is controlled by a loading driver IC or transistors, and a signal from the system-control processor IC (Fig. 26-41). Usually, a single positive voltage source goes to the balanced transistor circuit. After verifying that the motor itself is normal, take accurate voltage measurements on each transistor in the loading motor driver circuit. Test each transistor with the transistor or diode test of a DMM (Fig. 26-42). Suspect

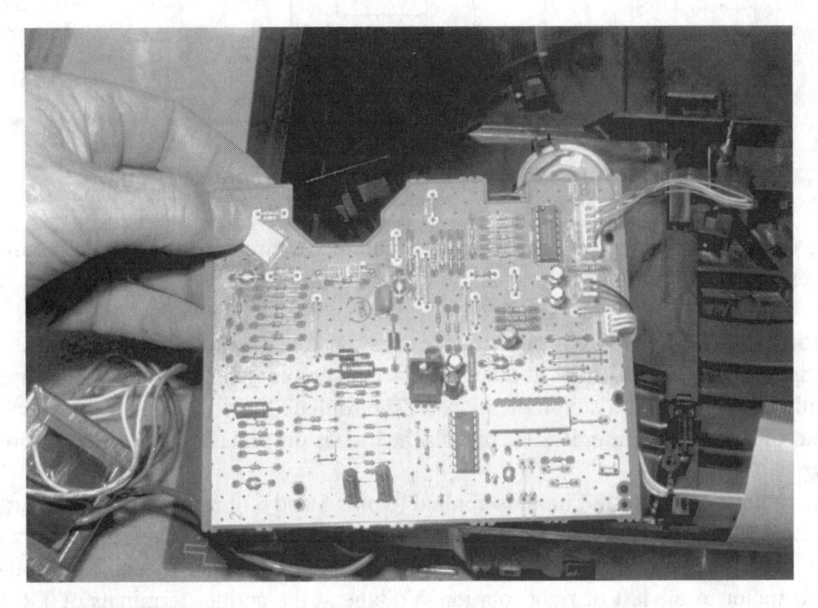

FIGURE 26-41 The motor-control board in a tabletop CD changer.

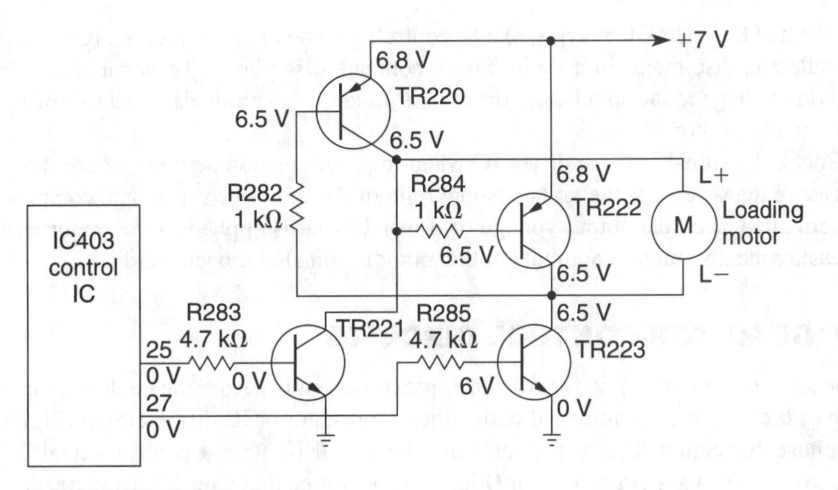

FIGURE 26-42 Take crucial voltage and transistor tests on the loading-motor circuit components.

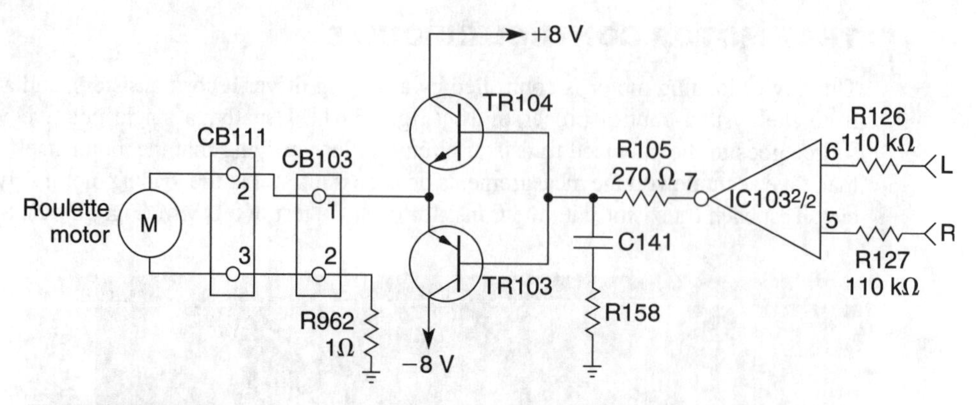

FIGURE 26-43 A typical transistor-motor roulette control circuit.

that a control IC is defective if the transistor and motor circuit test okay. Do not overlook the possibility of improper voltage at the voltage source.

Denon DCM-460 roulette motor circuits A roulette motor rotates the five- or six-CD disc tray around and stops at the right place on command. The roulette sensor circuits provide correct start and stop indications of the roulette motor. A signal from pin 21 (R-Speed) of system computer IC201 controls the transistors and ICs in the roulette motor circuits.

The roulette right and left input is applied to pins 5 and 6 of roulette motor driver IC103 (Fig. 26-43). IC103 provides a right and left voltage to the base of transistor drivers TR103 and TR104. The reverse voltage applied to the base of each transistor can control the rotation of motor to the left or right rotation. Voltage at the emitter terminals of TR103 and TR104 is applied directly to the roulette motor terminals through CB111 and CB103.

MOTOR-CONTROL ICS

In some of the latest CD players, one IC controls several different motors. IC6 controls the spindle and disc motor in a Radio Shack compact disc player. Terminals D1+ and D2– provide voltage to the spindle motors at pins 13 and 14. Terminals 17 and 18 of IC6 control the disc motor.

Check the signal into the driver IC when improper or no operation of the disc motor. Measure the voltage at the voltage-supply pin of the motor-driver IC and compare to the schematic. Check the output voltage of driver IC that is applied to the motor terminals. Measure the disc motor continuity at the motor terminals for open condition.

SLIDE MOTOR-CONTROL CIRCUITS

The slide or feed motor moves the laser pickup assembly from the center to the outside rim of the CD, and is controlled with either a transistor or IC driver circuit (Fig. 26-44). A phase-correction IC signal goes from the signal IC with a control signal from the control or servo system processor (Fig. 26-45). Notice that a positive and negative 11.3

V is fed to the collector terminals of Q104 and Q103 driver transistors. Also notice that Q104 is an NPN and Q103 is a PNP transistor. Only 1/4 of IC101 is used in the phase-correction IC. The slide motor-control circuit feeds from terminals 26 and 27 of the servo control, IC203.

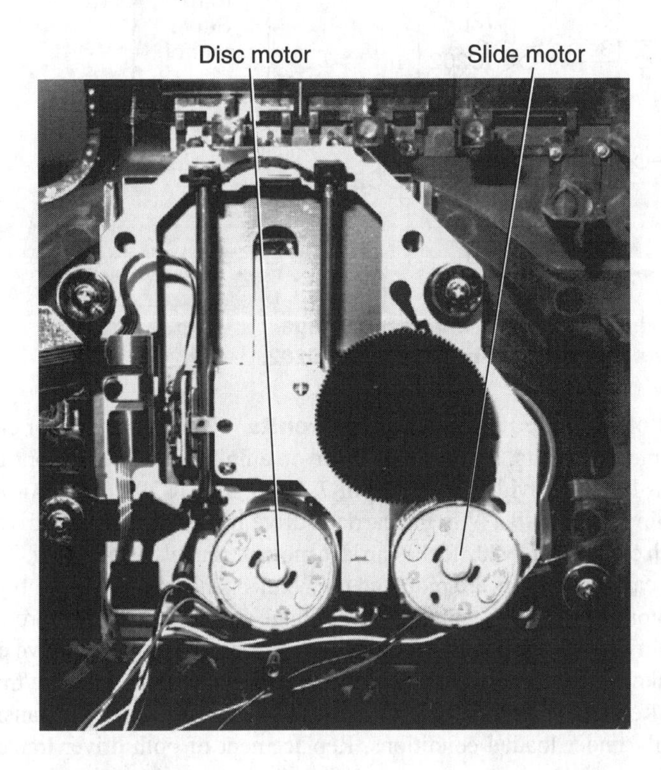

FIGURE 26-44 The slide and disc motors in a portable Sony CD player.

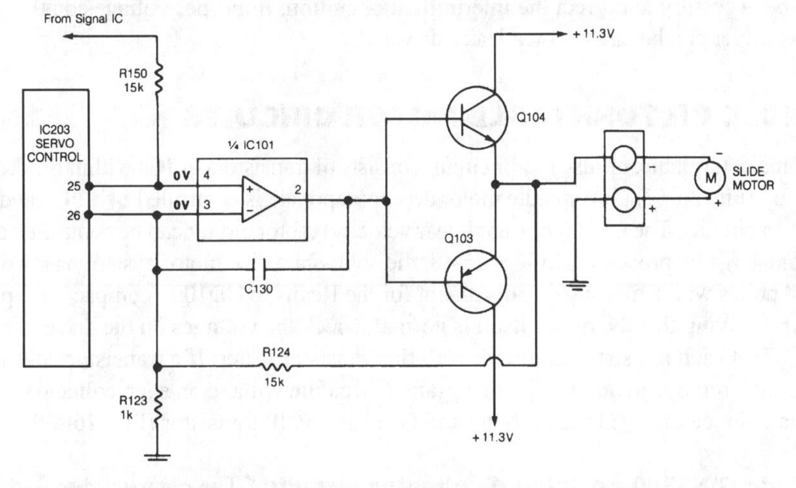

FIGURE 26-45 Servo IC203 drives slide motor driver IC101 and output drive transistors Q103 and Q104.

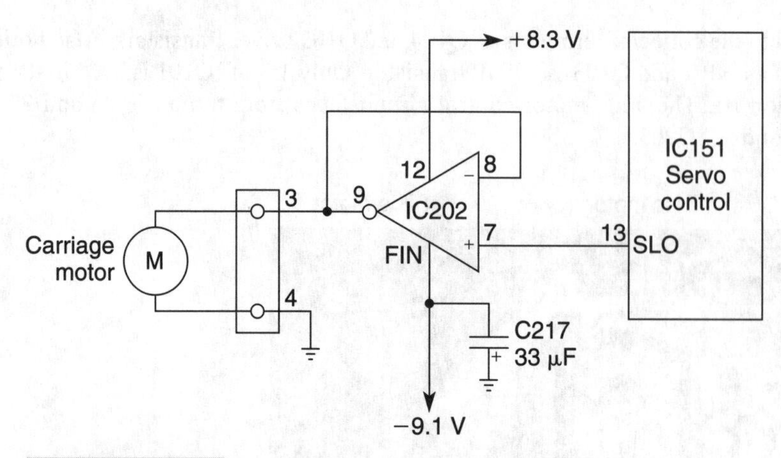

FIGURE 26-46 The Realistic CD changer carriage motor-control circuits are operated by IC151 and IC202.

Realistic CD7100 changer carriage motor circuits The carriage motor circuit, as in many different motor circuits, is controlled by an op amp from the servo-control IC. Terminal 13 of servo IC151 provides drive to pin 7 of op amp IC202 (3/3). An 8.3-V source is applied to pin 12 and a –9.1 V to the carriage drive IC. Pin 9 of IC202 drives the carriage motor through connector CN202 to the carriage motor terminals 3 and 4 (Fig. 26-46). Both the spindle and carriage motors are mounted on the mechanism board assembly.

Check the slide motor drive circuits with accurate voltage and resistance measurements. Be sure that both positive and negative voltage sources are present at the respective driver transistors or ICs. Take in-circuit transistor tests of each transistor with the diode/transistor test of the DMM. Remember, in erratic or intermittent operations, one of the transistors might be opening only under loaded conditions. Replacement of both driver transistors might be necessary to correct the intermittent operation. Improper voltages on the slide or feed IC driver can be caused by a leaky driver IC.

SPINDLE OR TURNTABLE MOTOR CIRCUITS

The spindle, turntable, or disc motor circuit consists of transistors or ICs within the drive motor circuit (Fig. 26-47). The spindle motor drive component is controlled by a PLL and servo-processor circuits. The CLV (constant linear velocity) motor circuit can be controlled directly from one large IC processor. In Fig. 26-48, the solid-state disc motor system has two different test points with a motor gain adjustment for the Realistic CD1000 compact disc player.

After verifying that the motor itself is normal, check the voltages on the drive transistors and IC. Test each transistor in-circuit with the transistor tester. If a transistor does not test as it should, remove it, and test it once again. Check the voltage on each collector and base terminal. Notice that Q315 is an NPN and Q316 is a PNP transistor (Fig. 26-49).

Realistic CD-3380 portable disc motor circuits The compact disc is recorded at a constant linear velocity. This means the disc rotation must be redirected as the laser pickup tracks toward the disc's outer edge. The revolutions vary from 500 to 200 rpm.

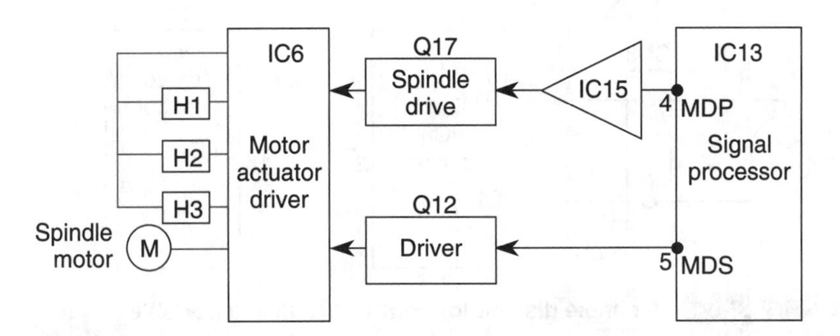

FIGURE 26-47 The spindle motor circuits in a Denon DCH-500 CD changer.

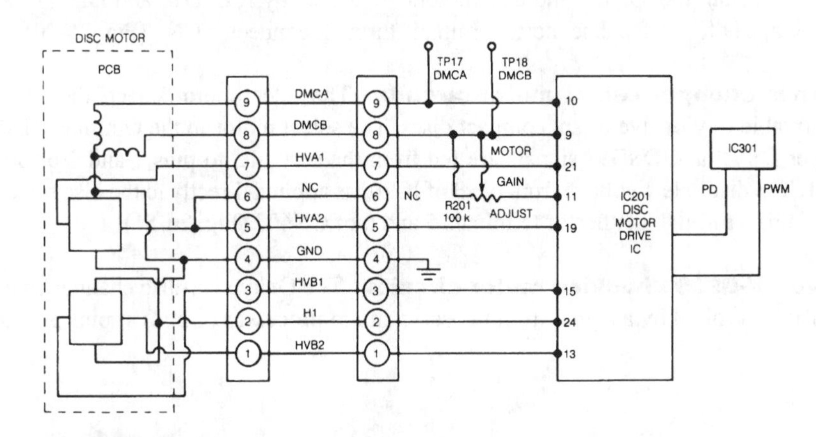

FIGURE 26-48 IC201 operates the CLV (constant linear velocity) circuit in this portable CD player.

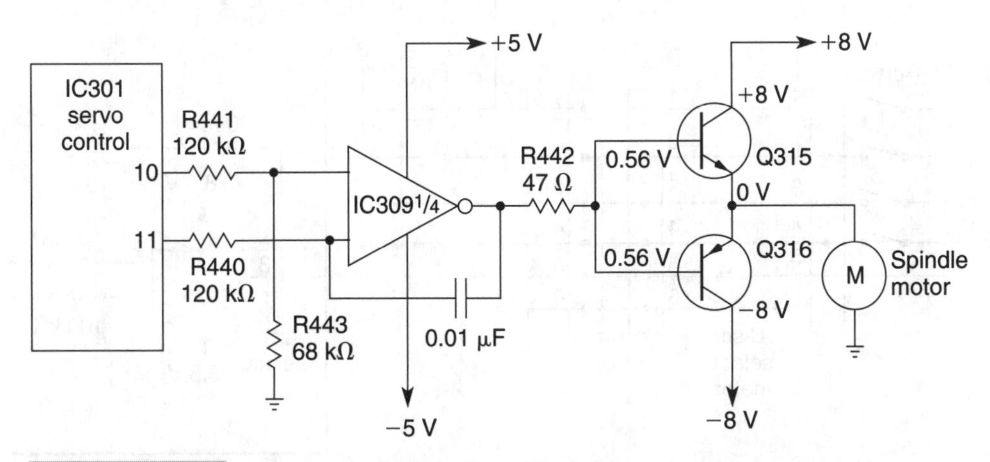

FIGURE 26-49 Servo controller IC301 controls driver IC309 and transistors Q315 and Q316 in this spindle motor circuit.

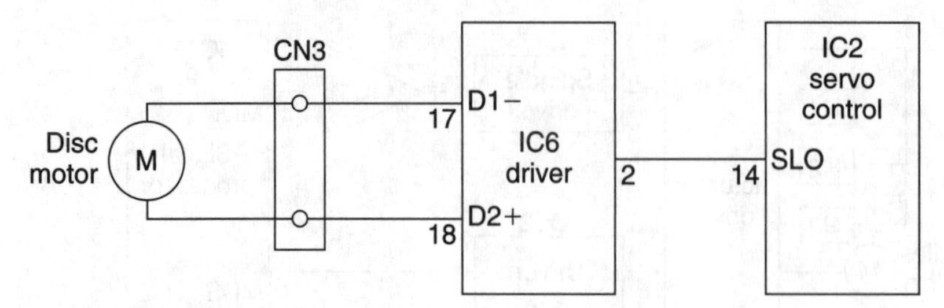

FIGURE 26-50 A simple disc motor-control circuit in a portable CD player.

The motor speed is controlled by signals from pin 14 (SLO) of servo IC2, fed to pin 2 of driver IC6. Both the spindle and disc motors are driven by IC6. D1– and D2+ (17 & 18) MV voltage is fed to the disc motor terminals through connector CN3 (Fig. 26-50).

Optimus changer select motor circuits The select motor selects the disc from the turntable tray of five or six compact discs. The select motor in the Optimus CD-7105 changer DSVP and DSDW signals are fed from the servo IC to pins 1 and 2 of op amp IC201. The disc select voltage from pin 3 of IC201 is applied directly to the disc select motor terminals through connector terminals 5 and 6 of CN602 (Fig. 26-51).

Onkyo DX-C909 chucking motor circuits The Onkyo six-disc changer chucking motor is controlled by a signal from the servo micro processor (Q202) at pins 56 and 57.

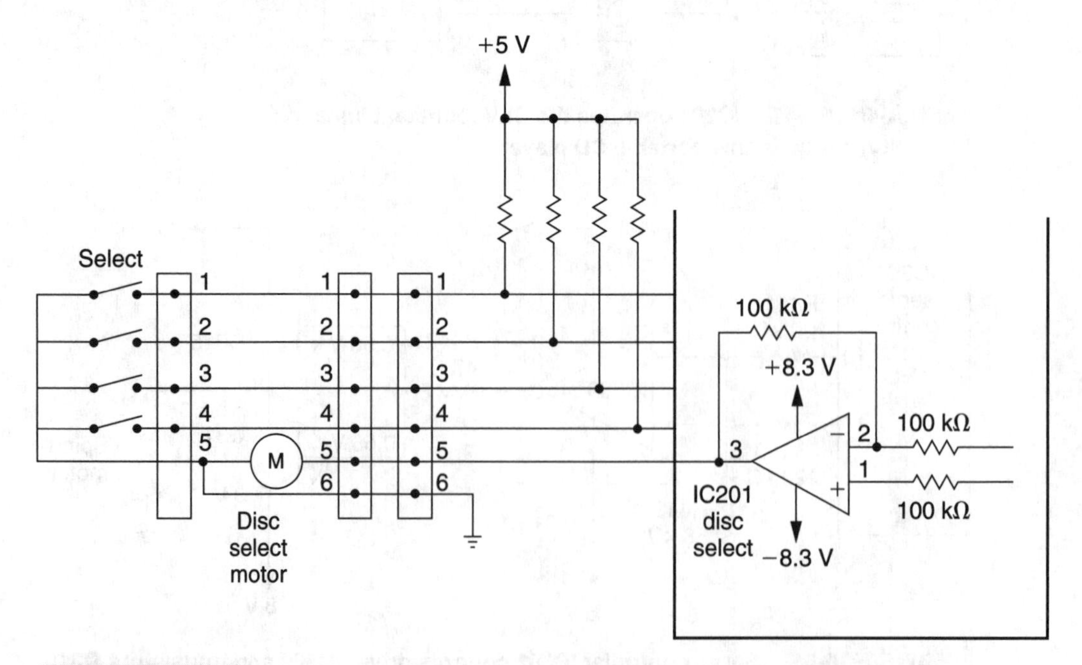

FIGURE 26-51 Optimus CD-7105 changer disc-select motor circuit.

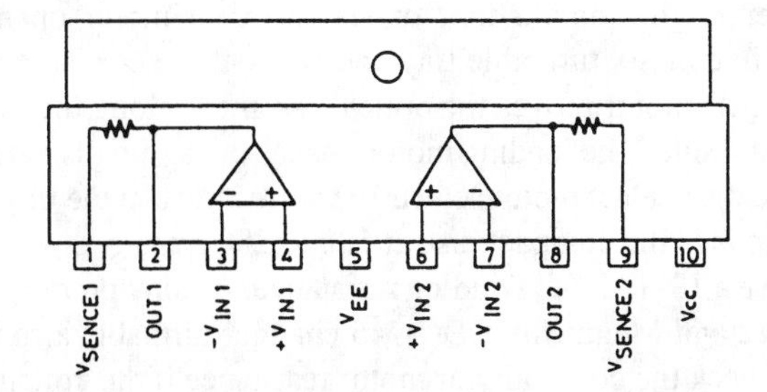

FIGURE 26-52 The motor-drive IC (LA6510) provides loading motor drive in this CD player.

This Open and Close signal is sent to the driver IC (Q206). The chucking motor terminals are tied to common ground and the positive terminal connects directly to driver IC pin 2. Because Q206 is a dual op amp, both loading and chucking motors are driven within the same IC (Fig. 26-52).

Denon DCH-500 elevator motor circuits In the Denon DCH-500 auto changer CD player, the elevator and disc motors are operated with a common IC driver (IC501). The elevator motor terminals are 2 and 10. The disc motor terminals are pins 3 and 10. Motor driver IC501 is controlled from pins 23, 24, and 25 of the system control IC601 (Fig. 26-53). Suspect that IC501 is faulty if both motors are dead.

Mitsubishi M-C4030 up/down and magazine motors The signal for controlling the up/down and magazine motors is taken from the micro computer control IC301. Both motors are controlled from a dual bidirectional motor driver, IC302. The input control signals from micro computer are fed to pins 4, 5, and 6 of IC302.

The magazine motor can be the same as a loading motor. The output voltage taken from pin 10 of IC302 driver, feeds directly to one side of the magazine motor terminals. The common motor terminals of magazine and up/down motors are pin 10 of driver IC302. The voltage on pin 3 of IC302 drives the magazine motor (Fig. 26-54). The up/down motor drive voltage is at output pin 2.

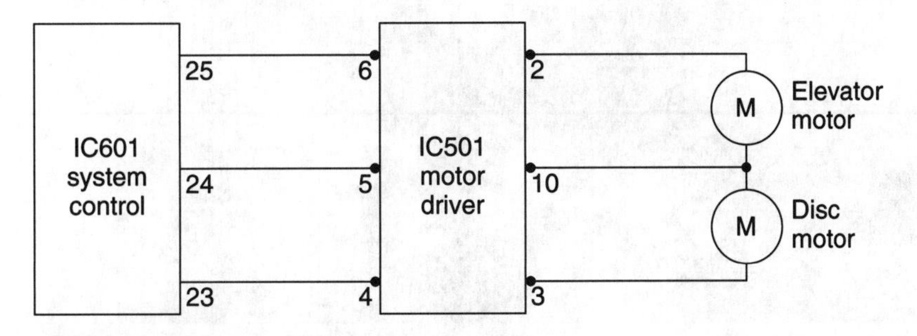

FIGURE 26-53 Driver IC501 provides drive motor voltage for both the elevator and disc motor in this CD changer.

THE DIFFERENT CD CHANGER MOTORS

The compact disc changer might contain four or six different motor operations. The roulette motor rotates the five or six turntable tray and stops at the correct disc selection (Fig. 26-55). A slide or SLED motor moves the optical assembly along the rails from inside of the CD to the outside edge. The loading motor pushes out and pulls in the turntable tray of five or six CDs. The disc-select motor selects the correct disc in the tray. Of course, the disc or spindle motor rotates the compact disc at 500 to 200 rpm.

Most of these motors have a 10- to 20-Ω winding resistance. A Sony portable SLED motor has a resistance of 12 Ω and Magnavox CDC-745 changer turntable a roulette motor has a resistance of 18 Ω. Check the continuity or motor resistance if the voltage is applied

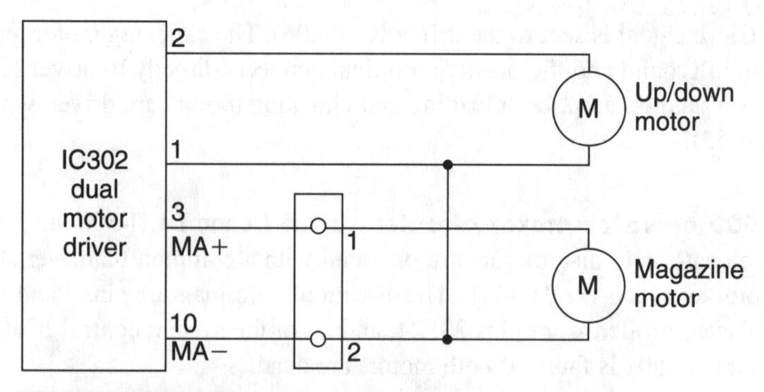

FIGURE 26-54 IC302 controls the up/down and magazine motors in this CD changer.

FIGURE 26-55 The turntable or roulette motor in a Magnavox CD changer.

TABLE 26-2 CD MOTOR TROUBLESHOOTING CHART

MOTOR	SYMPTOM	DEFECTIVE CIRCUIT
Loading motor	Dead	Inspect motor drive belt. Check motor continuity, test voltage applied to motor in operation. Check transistor or drive IC. Check signal applied to motor driver.
	Erratic operation	Inspect drive belt. Inspect area for foreign materials. Check for grease on motor belt. Check for binding track. Suspect erratic motor. Monitor voltage applied to motor terminals. Suspect erratic driver IC.
	Intermittent	Check for poor motor terminals and connections. Inspect motor plugs and jacks. Defective motor. Defective motor driver IC.
Sled—slide or feed motor	Dead	Check motor continuity. Check voltage at motor. Check supply voltage on motor driver IC. Check output voltage at driver IC. Check servo signal from signal processor or control IC.
	Intermittent	Monitor motor voltage. Suspect erratic motor. Inspect motor terminals and plugs. Check for gummed up or dry sliding surfaces on rails.
Disc or spindle motor	Dead	Check motor voltage. Check continuity of motor. Check motor voltage on motor driver. Measure VCC supply voltage on motor driver IC. Check CLV spindle motor IC. Check signal from servo control IC.
	Comes on shuts down	Notice if all functions are shut down. Check for RF or EFM waveform at RF amp. Check for defective motor circuits if RF or EFM waveform is found at RF amp. Suspect defective CLV spindle motor driver circuits.
Carousel or turntable motors	Dead	Measure motor continuity. Check voltage at motor. Check voltage at driver. Test for both + and – voltage at driver IC or transistors. Check signal at microprocessor IC.
	Intermittent	Monitor motor voltage. Suspect motor. Check motor driver IC.
	Erratic	Check turntable for improper mounting. Check for wires clinging to the bottom of turntable. Inspect start/stop switch. Monitor voltage at motor terminals. Suspect defective motor.

to the motor terminals, but the motor is not rotating. Check Table 26-2 for the CD motor troubleshooting chart and for motor problems.

Conclusion

Locate the defective motor on the chassis. Take crucial voltage and continuity measurements on the motor terminals. Check the driver transistors and ICs feeding the suspected motor. If the motor is normal, test each transistor with a transistor tester or with the

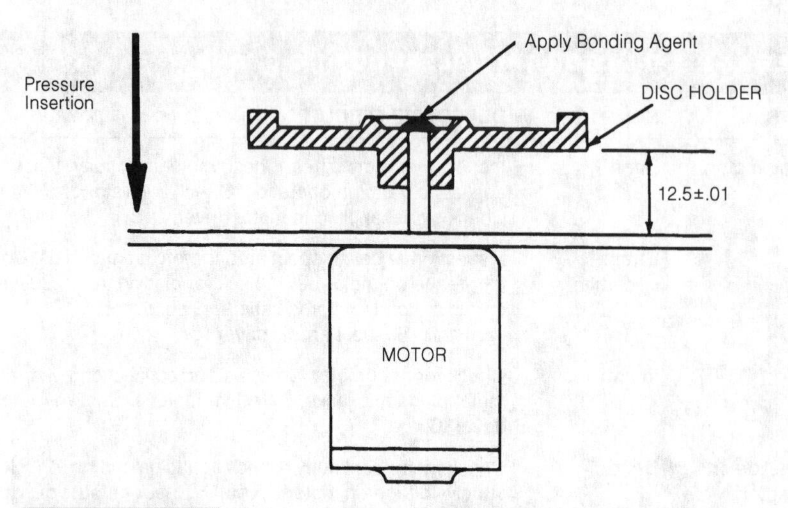

FIGURE 26-56 Be sure that the disc holder or table is at the right height after the motor has been replaced.

diode/transistor test of the DMM. If the motor is defective, take extreme care replace it. Do not apply too much downward pressure to misalign the motor assembly. Always measure the location of the motor pulley or gear so that it can be correctly replaced on the new motor (Fig. 26-56). Be sure that the motor is installed with the correct polarity in the circuit.

27

DIGITAL AUDIO (D/A) CIRCUITS

CONTENTS AT A GLANCE

The typical compact disc audio section consists of a D/A converter, a sample/hold, a low-pass filter network, audio IC amplifiers, and headphone circuits. The digital signal is fed into the D/A IC, then it is converted to an audio signal. This stereo signal is connected to separate sample-and-hold circuits that separate the left-channel sound for the left-channel

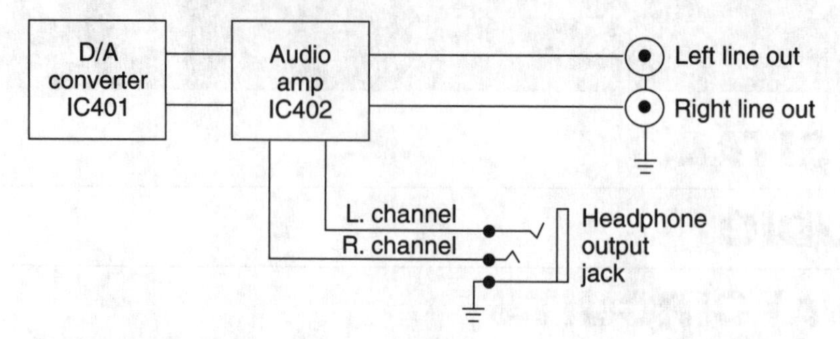

FIGURE 27-1 **The block diagram of the sound circuits from the D/A converter to the line output jacks.**

circuitry and the right-channel sound to the right-channel circuitry. Each audio signal is then filtered to remove the 44.1-kHz signal from the audio, which is passed on to the IC amplifiers. Separate IC audio circuits might be provided for headphone reception; however, some CD players do not have separate headphone circuits (Fig. 27-1).

Boom-Box CD/Cassette Player

The boom-box CD, radio, and cassette player uses the same audio amplifier circuits and speakers (Fig. 27-2). Instead of line-output jacks, this boom-box combination player has internal speakers and a headphone stereo jack. Some units have both headphone and line-output jacks. The output signal from the D/A converter might have one IC that amplifies

FIGURE 27-2 **The CD player in this Sharp boom-box player uses the same audio amp and speakers.**

both stereo channels before switching into the regular audio circuits. The CD audio signal is switched by a rotary function switch (Fig 27-3).

The rotary function switch selects audio signals from the radio, cassette, or compact disc player by simply rotating the function switch. The audio output from this CD player is then coupled by electrolytic capacitors into input AF transistors or into one large IC preamp (Fig. 27-4). Usually, one large IC is in the output circuits to drive a heavy-duty 4- or 6-inch speaker (Fig. 27-5).

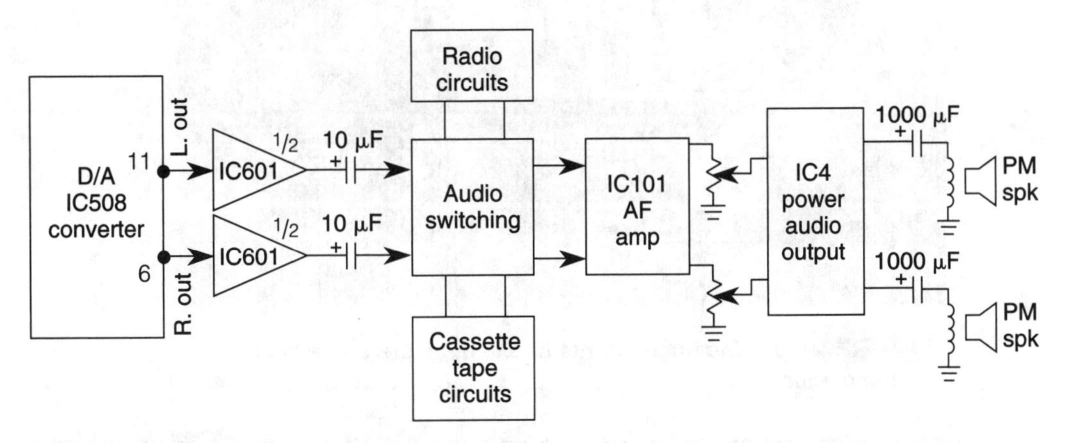

FIGURE 27-3 A block diagram of the Sharp QT-CD7 boom-box stereo audio circuits.

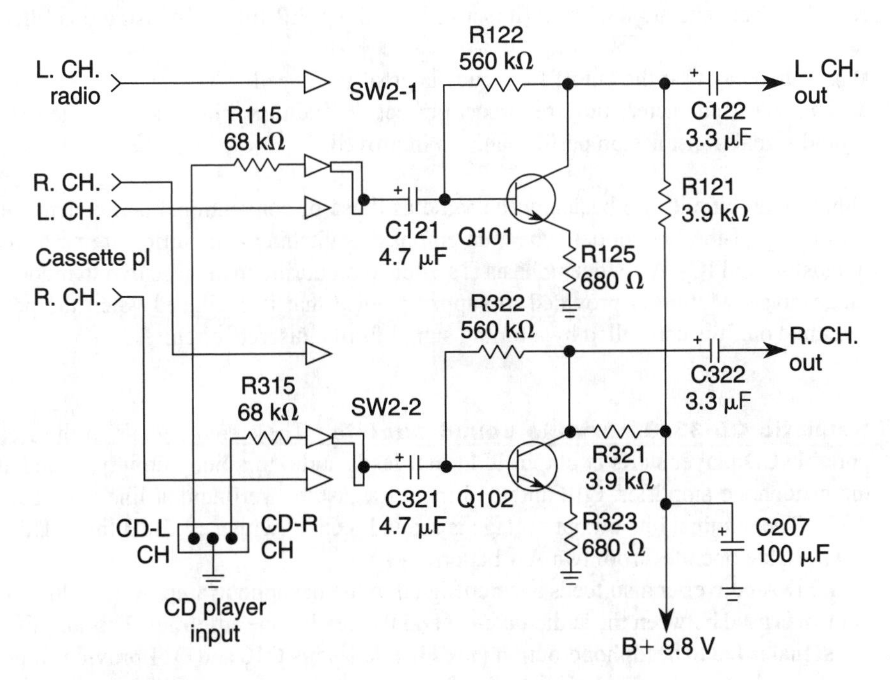

FIGURE 27-4 Both preamp Q101 and Q102 amplify the switched-in radio, cassette, and CD audio circuits.

FIGURE 27-5 The finger points to one large stereo vertical mounting sound IC.

Pioneer PD-9010X, PD-7010, and PD-5010 sound circuits In this high-end series, digital filters are used. This digital filter is IC CX23034 (Fig. 27-6). It doubles the sampling frequency to 88.2 kHz (it is a 16-bit, 96 tap FIR filter). By using this filter:

- Group delay near the cutoff frequency is greatly reduced.
- The cutoff characteristics are almost perfect. In addition, high-end distortion is lower and signal transmission performance is improved.

Other audio circuits are basically the same as those of conventional components, except for the top-of-the-line model, where de-emphasis switching and muting are performed by transistors and ICs. Also, deemphasis is switched on during manual search to reduce high-range noise. Muting is provided to suppress noise that is produced when the power is switched on. It is controlled by a timing signal from a discrete circuit.

Realistic CD-3370 portable sound circuits The stereo signals in the Realistic portable CD player starts at output IC11 and feeds audio to a line-output jack and also to the headphone amplifier. Q12 and Q13 mute transistors are found at line output J2 (Fig. 27-7). The nominal line output voltage is around 0.65 V, with a ±1.5-dB limit. This small CD portable operates from two AA batteries (3 V).

The D/A converter also feeds a stereo signal to the headphone amp, IC1. A dual volume control is used between the audio take-off and the headphone amplifier. This amplified audio signal is fed to headphone-output jack J3. Transistors Q10 and Q11 provide audio muting for each stereo channel.

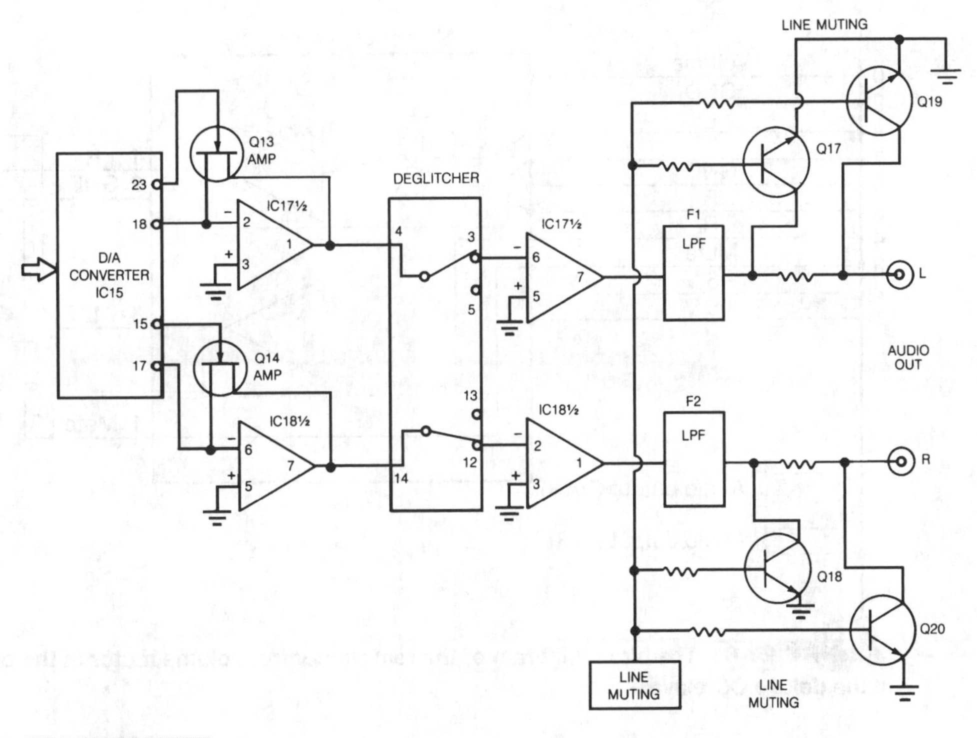

FIGURE 27-6 The block diagram of deglitcher and transistor muting used in the audio circuits of this CD player.

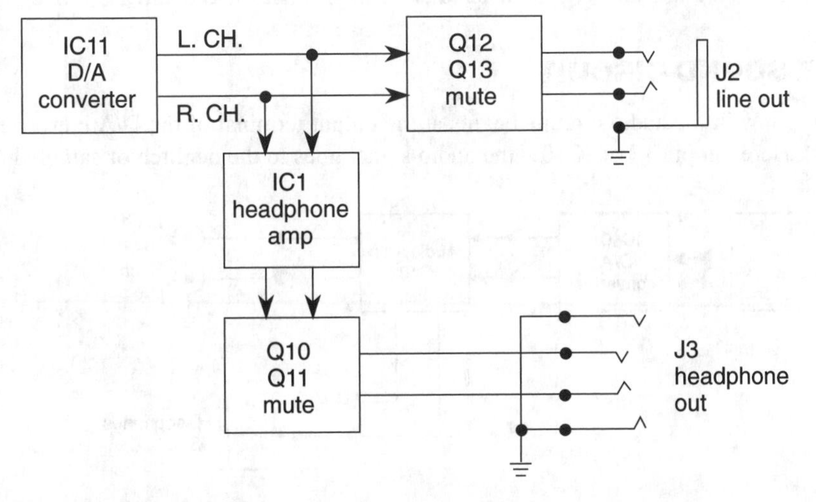

FIGURE 27-7 A block diagram of the sound circuits in the Realistic CD-3370 portable.

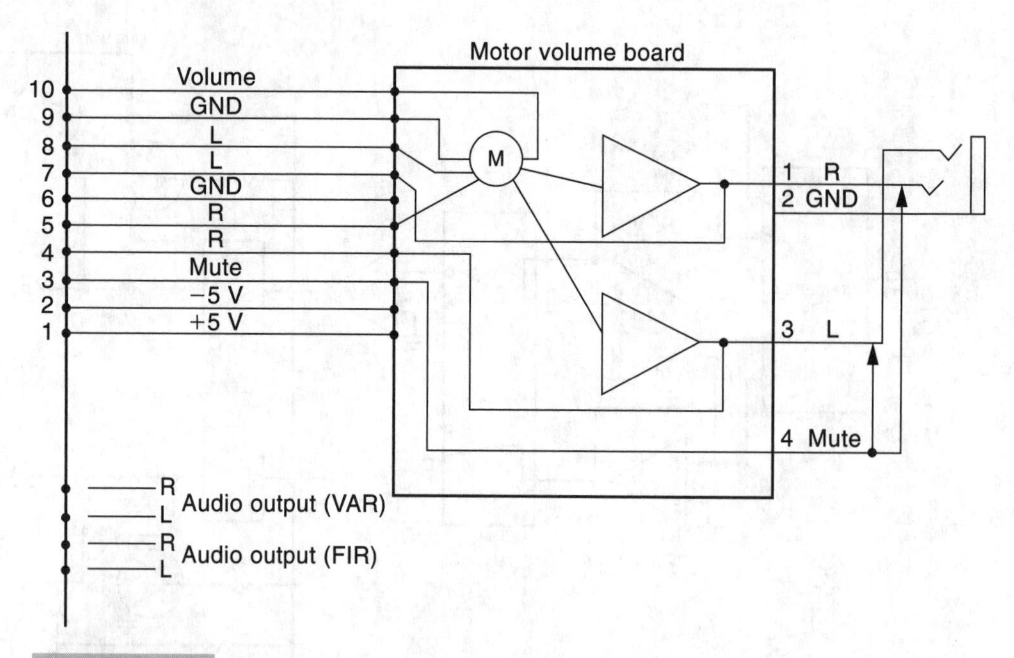

FIGURE 27-8 The block diagram of the remote-control volume motor in the output of the deluxe CD player.

Denon DCM-560 volume-control motor circuits The block diagram of the Denon DCM-560 audio circuits shows a motor that rotates the volume controls, and is controlled by the remote transmitter (Fig. 27-8). Both the headphone jack, variable line output, and fixed audio output jacks are controlled by the motor volume control. The motor volume circuits are in the IU-2788-2 and the headphone circuits are in the IU-2788-3 board assemblies.

THE SOUND CIRCUIT

Signal flow in the audio circuits begins at the output terminal of the D/A converter (Fig. 27-9). Here, on pin 17 of IC401, the audio signal goes to the deglitch or sample/hold IC.

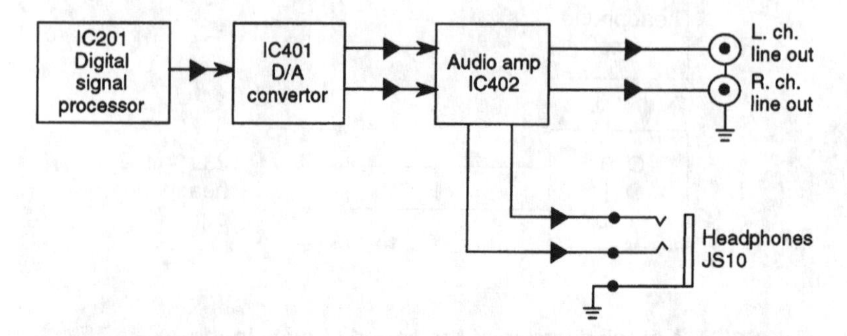

FIGURE 27-9 Follow the audio indicated by arrows from IC201 through to the line- and headphone-output jacks.

Next, the signal proceeds to the preamp and the low-pass filter (some models include the preamp in the low-pass filter network). From pin 6 of the low-pass filter network, the signal flows to the audio output IC402 (pin 3). Capacitor (C808) couples the amplified audio signal (pin 1) to the muting switch to the line-output jack. The audio signal can be traced as it flows from stage to stage with a scope or external audio amp. The audio-signal flow path might have arrows marking its route across the schematic.

D/A CONVERTER

The digital/analog converter IC actually changes the digital signal to voltage or audio. The input digital signal from the controller or signal processor is connected to pin 10 of IC401 (Fig. 27-10). In some audio circuits, a digital filter circuit might be between the digital signal and input terminal of the D/A converter for additional filtering. The stereo audio output signal is taken from pins 1 and 20. The left audio signal (pin 1) is fed to C419 and R431 (pin 2 of IC402) and the right audio signal (pin 20) is connected to C420 and R432.

SAMPLE/HOLD (S/H) CIRCUITS

The sample/hold or deglitch IC is usually located between the D/A converter and low-pass filter network (Fig. 27-11). Some circuits have a channel-output level control. The S/H circuit separates the right-channel data for the right-channel circuitry and the left-channel

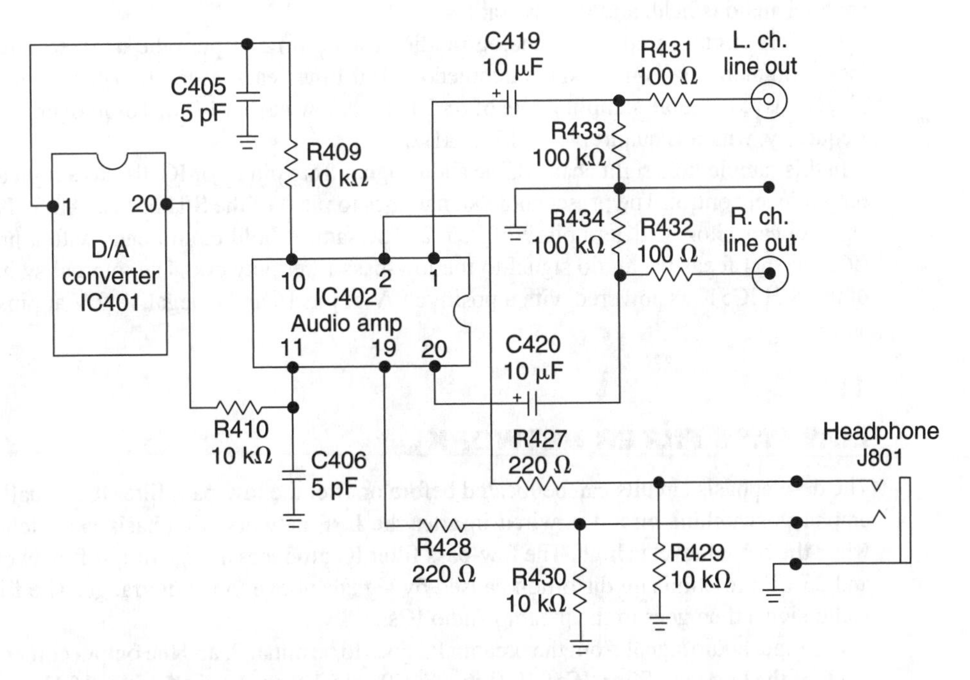

FIGURE 27-10 **Both the stereo line-output and headphone jacks are connected to audio amp IC402.**

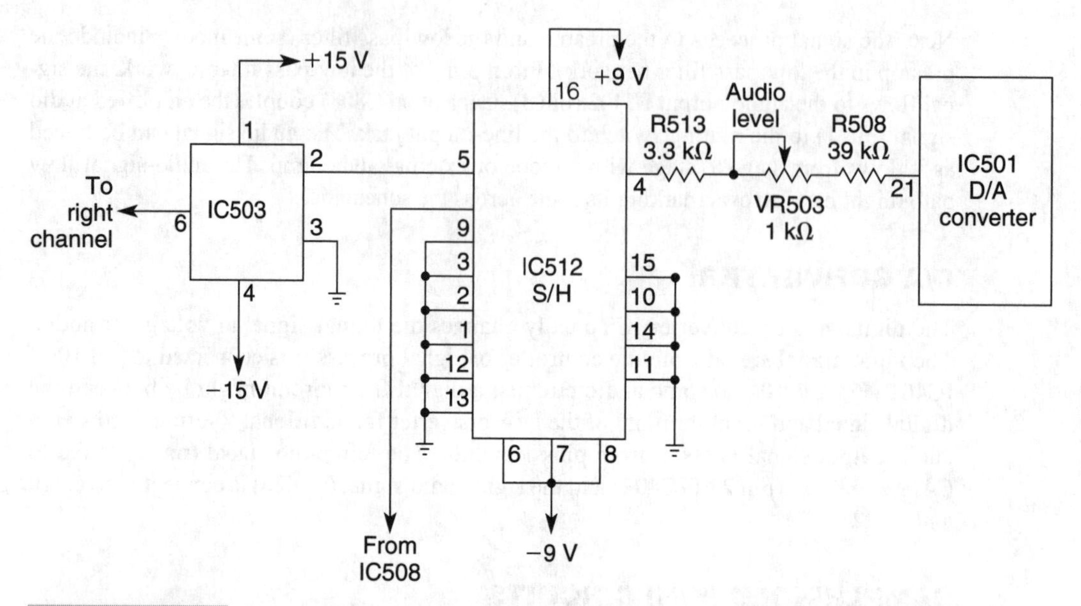

FIGURE 27-11 **A sample/hold IC is located between IC501 and line output IC503 in this CD player.**

data for the left-channel circuitry. When the left-channel sample audio passes, the right-channel audio is held, and vice-versa.

The S/H circuit samples the analog (audio) waveform at a periodic or fixed rate. The most common rate is 44.1 kHz (sampled 44,100 times each second). Today, some CD players have a double sampling rate of 88.2 kHz. A few have a 14- or 16-bit oversampling frequency, which is quadruple, at 176.4 kHz.

In this sample/hold right channel, the audio signal from pin 21 or IC501 has a right audio-output level control. The preset audio signal goes to pin 4 of the S/H IC512. An S/H right-channel capacitor IC ties to pin 9 of IC512. The sample/hold circuit ends with a preamp (IC503) that feeds the audio signal to the low-pass filter network. The internal switching of the S/H IC512 is powered with a positive 9 V at pin 16 and a negative 9 V at pins 6, 7, and 8.

LOW-PASS FILTER NETWORK

The de-emphasis circuits can be located before or after the low-pass filter IC. Usually, the emphasis switching circuit is wired in after the LPF network. Emphasis is switched on when the output level is high. The low-pass filter IC produces a large drop-off between 20 and 25 kHz to eliminate distortion caused by signals above the audio range. The filtered audio signal then goes to the preamp audio ICs.

The input audio signal from the preamp IC goes to terminal 2, and the output comes from pin 6 of the low-pass filter, IC504L (Fig. 27-12). –15 V goes to pin 4 with +15 V on pin 7. Terminals 1, 3, and 5 are at chassis ground.

MUTING SYSTEMS

Sound muting is often provided to suppress noise that is produced when the power is turned on. In some players, muting is automatic when the disc stops, during accessing operations, and during Pause mode. The sound muting circuits use relays, transistors, or ICs (Fig. 27-13).

Some units have line muting at both channel-output terminals. The same system mutes the signal to the headphone circuits. Muting might be controlled by an automatic muting IC. The control signal operates transistors, ICs, and relay components in the

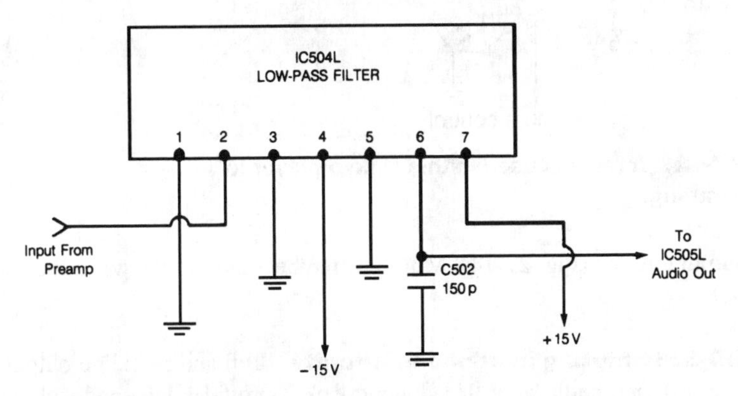

FIGURE 27-12 The low-pass filter eliminates distortion and signals above the radio range in this portable CD player.

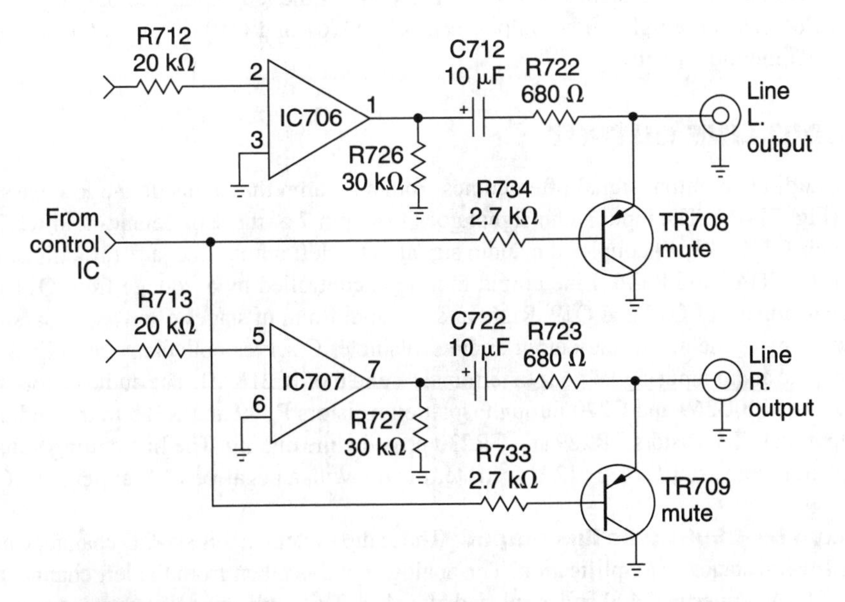

FIGURE 27-13 Typical transistor line muting in the line-output circuits.

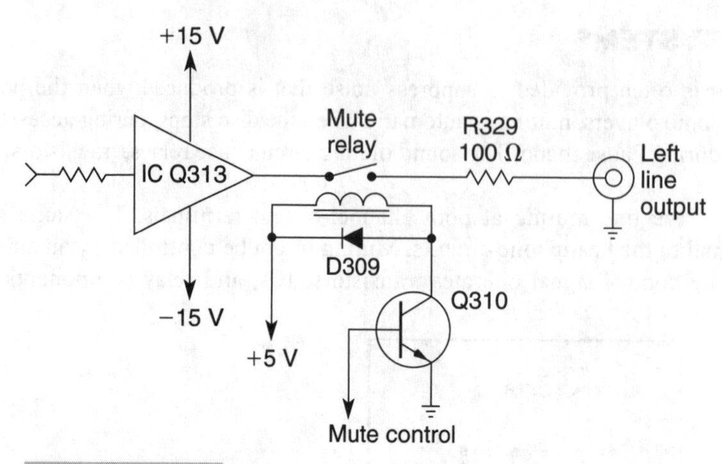

FIGURE 27-14 A relay is used in this Onkyo player to provide line muting.

line audio-output circuits (Fig. 27-14). Muting often becomes active when the output is at a high level.

Realistic CD-3380 muting headphone circuits Left and right line-output jack 2 connects to the external audio amp. Headphone jack 3 provides left- and right-channel headphone reception (Fig. 27-15). A volume control in each channel couples the audio signal from the line-output circuits. Capacitor C108 couples the left audio signal to IC9-1 and C208 couples the right signal to IC9-2. C112 couples the left channel to output jack 2 and C212 connects the right audio output to jack 2. Q101 and Q103 provide left- and right-channel muting circuits.

AUDIO LINE OUTPUT

The audio line output signal often comes from a preamp that is inside the low-pass filter IC (Fig. 27-16). The input audio signal goes from pin 7 of the sample-and-hold IC17. Capacitor C117 (47 μF) couples the audio signal to the left output line jack through isolation resistors R147 and R146. Line output muting is controlled by a voltage from Q21 to the base terminals of Q17 and Q19. Right and left audio-output stages often use separate ICs.

In many of the audio line-output circuits, a single IC serves both channels (Fig. 27-17). Here, the audio input signal goes to terminals 4 and 6 of IC214AB. The audio-output signal couples with C269 and C270 through isolation resistors R349 and R350 to the audio line-output jack. Transistors TR229 and TR230 provide line muting. The line muting voltage is applied to transistor Q243. +12 V feeds terminal 9 with a negative 12 V at pin 5 of IC214.

Onkyo DX-C606 audio line output The audio system in this six-CD changer carousel has several stages of amplification. The analog signal is taken from the left channel at pin 13 of D/A converter Q400 and amplified by Q401. This audio signal is coupled to pin 5 of Q405. C439 couples the signal through R441 and R447 before reaching the left line-output jack (Fig. 27-18).

The audio signal from the right channel starts at pin 16 of the D/A converter and is amplified by Q402 and on to Q406. Again, the right channel is amplified and is capacity coupled to Q406. C440 couples the amplified audio and ends up at the right channel line-output jack through R442 and R448.

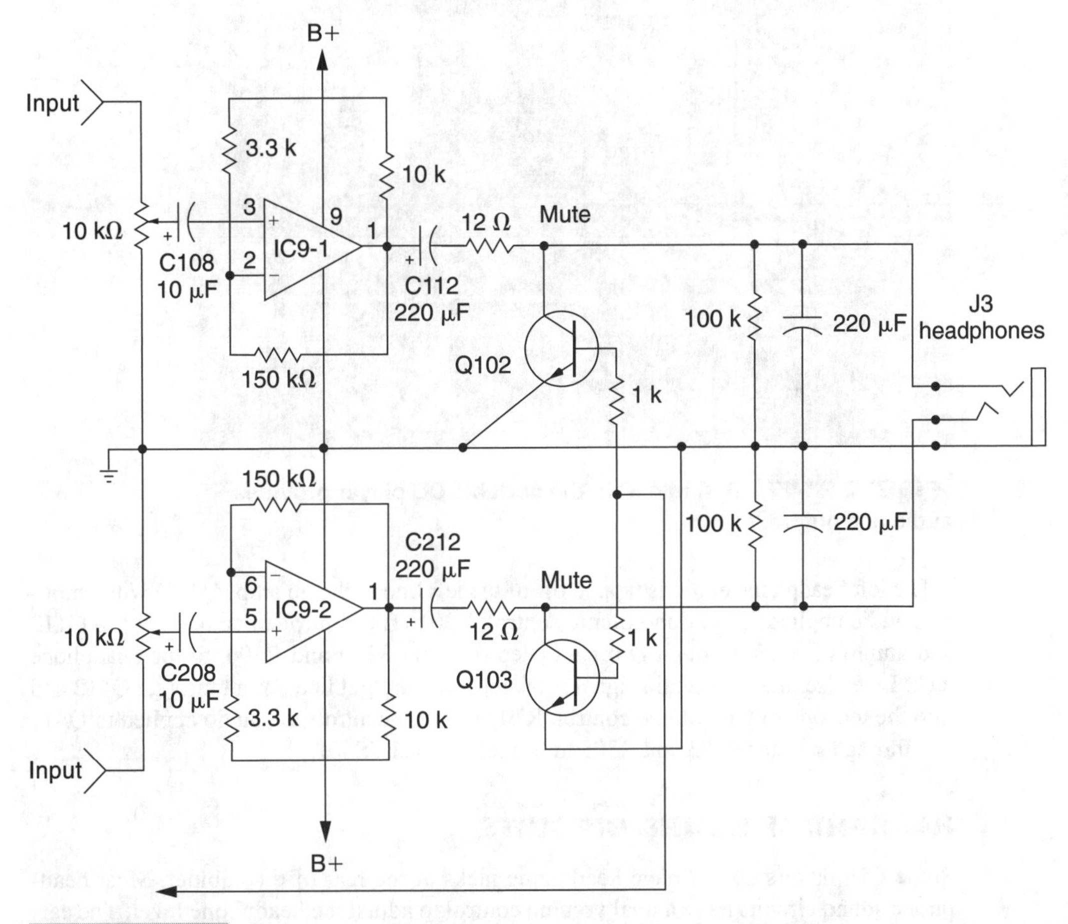

FIGURE 27-15 **Notice the headphone line-muting transistors in these headphone circuits.**

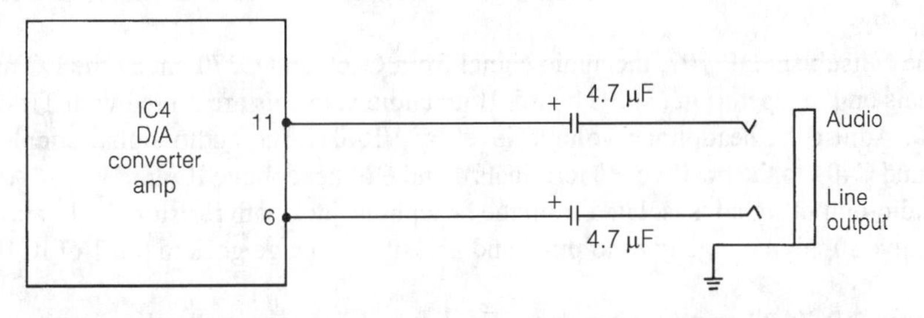

FIGURE 27-16 **IC4 has a stereo internal amp circuit within this D-A converter.**

FIGURE 27-17 A single IC in the portable CD player provides audio line output.

The left headphone audio is tapped off of the left line to the op amp (Q411) with amplified audio applied to the duo-volume control, R301. The controlled audio is fed to Q112 and amplified again through two paralleled resistors R304 and R306, to the headphone jack. Likewise, the right audio signal is taken from the right line output through Q412 and into the top side of the volume control. R301 (20 kΩ) controls the audio applied to Q411 and through resistor R303 and R305 to headphone jack JS401.

HEADPHONE SOUND CIRCUITS

Some CD players do not have headphone jacks at the rear of the cabinet. Most headphone sound circuits have a dual volume control to adjust the headphone level. The earphone circuits might consist of a dual-sound IC connecting directly to the preamp line-output ICs. The audio line entering the headphone circuits is muted in the latest models (Fig. 27-19).

In the Mitsubishi DP-107, the audio signal from C269 and C270 enters the LO and RO terminals on the Operation (3) PC board. Both audio channels are muted with TR405 and TR406. Adjust the headphone volume level by VR401. The audio signal couples with C403 and C404 to the positive (+) terminals 4 and 6 of headphone IC amp IC404 A and B. The audio-output signal is isolated from the headphone jack with resistors R417 and R418. A negative supply voltage goes to pin 5 and a positive voltage goes to pin 1 of IC404.

Optimus CD-7105 audio circuits The left and right channel audio circuits begin at the D/A converter (IC401). The left channel (LD) is found at pins 9 and 10 of D/A converter, while the right channel (RD) is found at pins 5 and 6 (Fig. 27-20). The left-channel audio is fed to pins 2 and 3 of op amp IC405 (1/2), through resistors R436 and R438.

IC405 is considered a low-pass filter IC. Pin 1 of IC405 in the left channel is coupled to the left line-out jack with C434 (22 µF). The right channel is coupled to right line-out jack with C433. Q403 mutes the right channel and Q404 mutes the left output channel. IC405 has a +5-V source at pin 6 and a –5 V at pin 4.

Realistic 42-5029 headphone circuits The left channel analog circuit from digital/analog converter IC11 starts at pin 11 and is capacity coupled by C61 and C48 to the

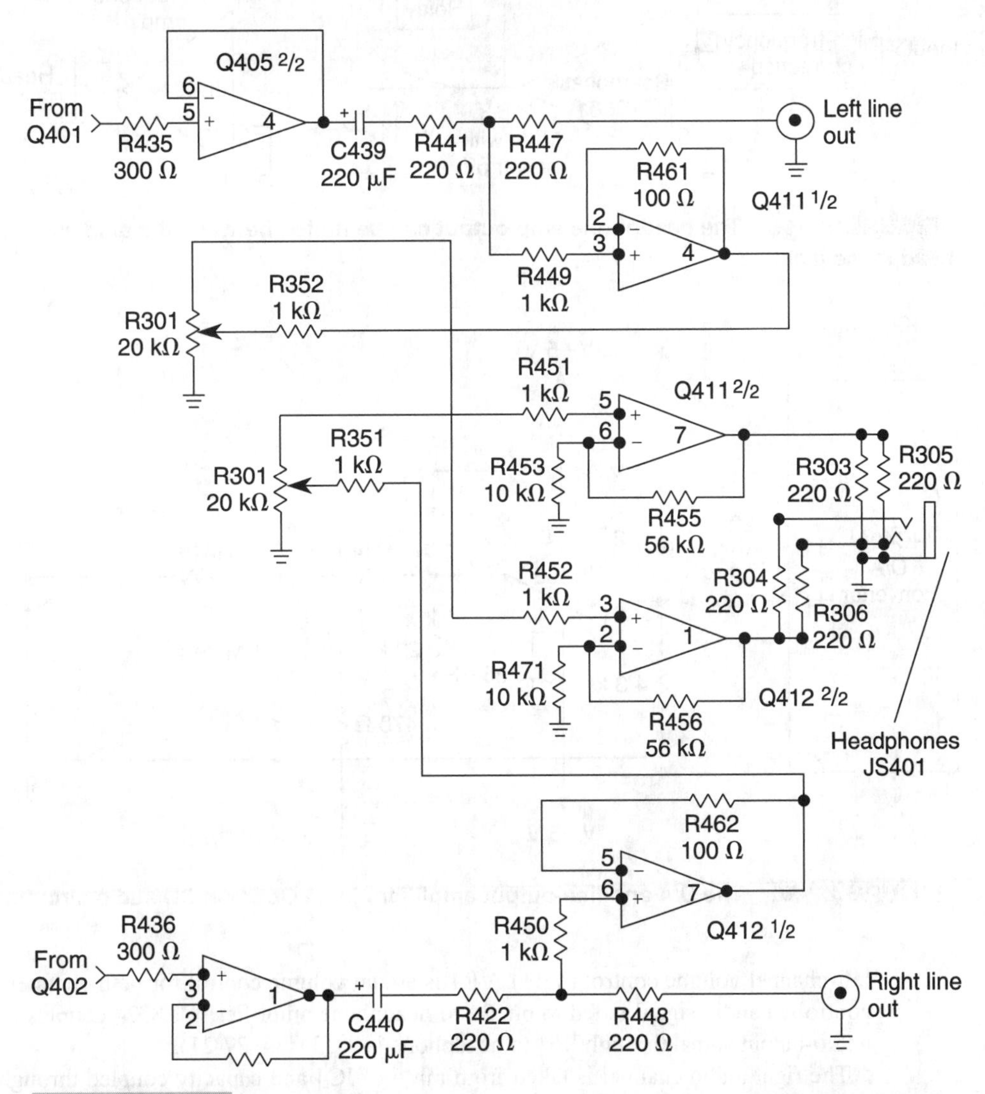

FIGURE 27-18 Q411 and Q412 provide headphone and stereo line-output audio to the corresponding jacks.

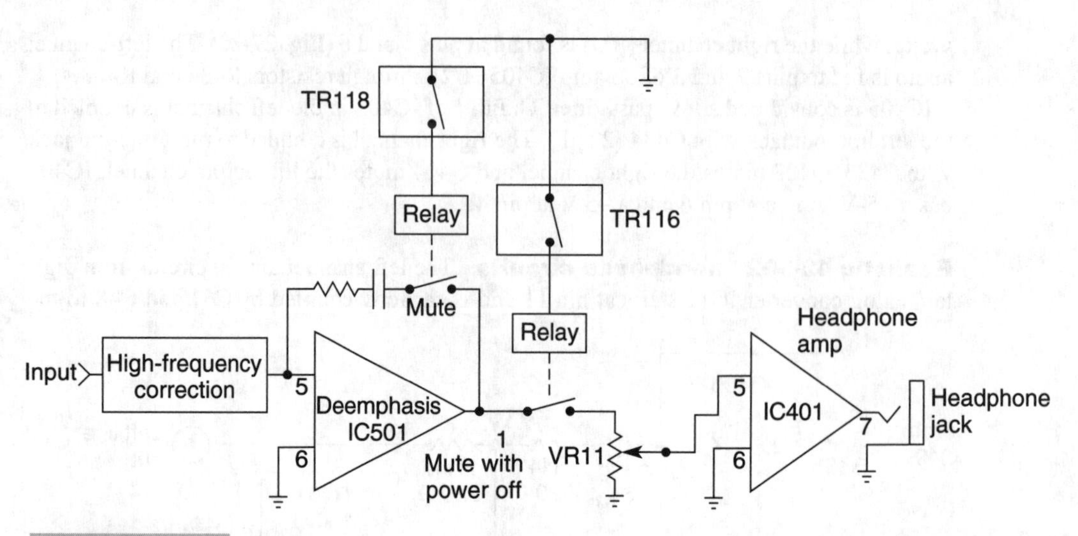

FIGURE 27-19 The headphone amp output can be muted between the amp and headphone jack.

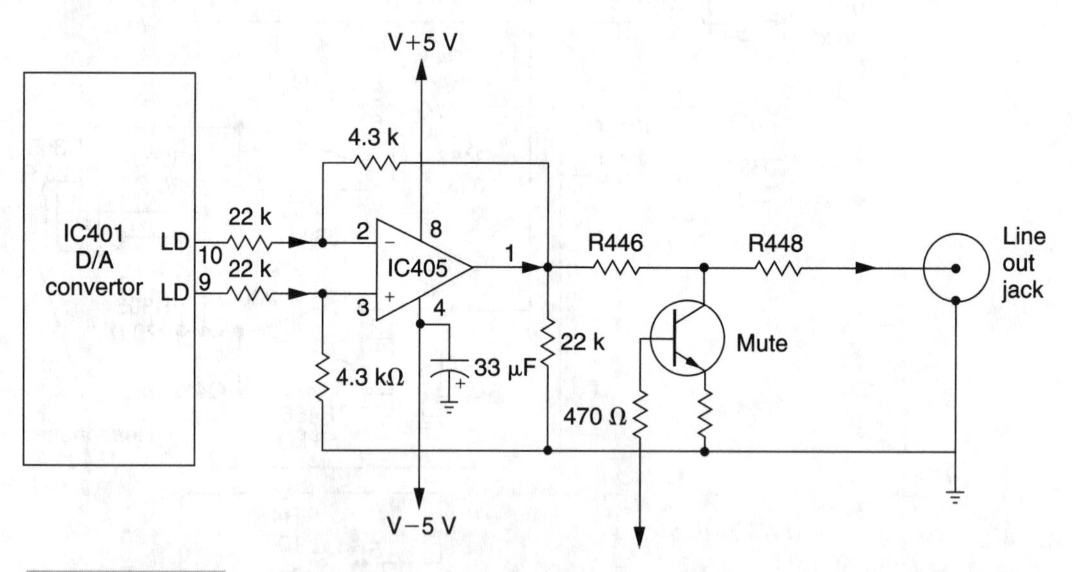

FIGURE 27-20 The D/A and line-output amplifiers in an Optimus CD audio circuit.

left-channel volume control (VR1). VR1 is a dual volume control for both channels. The controlled audio signal is fed to pin 24 of headphone amplifier IC1. C74 couples the left audio-output signal through R63 to headphone jack J3 (Fig. 27-21).

The right audio channel is taken from pin 6 of IC4 and capacity coupled through C88 and C41. The right line-output audio is tapped between these two capacitors. Again, VR1 controls the headphone right-channel audio and connects to pin 3 of IC1. The headphone

IC1 amplifies the audio and the output appears at pin 10. C114 couples the audio to the right stereo output jack, J3.

AUDIO OUTPUT VOLTAGE

Most CD players have an output voltage of 2 V. This is a fixed voltage if there is no audio level control. CD players with level or adjustable audio outputs should be adjusted to a level of 2 V. Check Chapter 28 for the audio-level adjustment. Some CD players have more than one set of output jacks. Frequently, one set of jacks has a fixed output and the other has an adjustable sound output. The audio-frequency response can vary from 2 Hz to 20 kHz, and the output impedance also varies (Table 27-1).

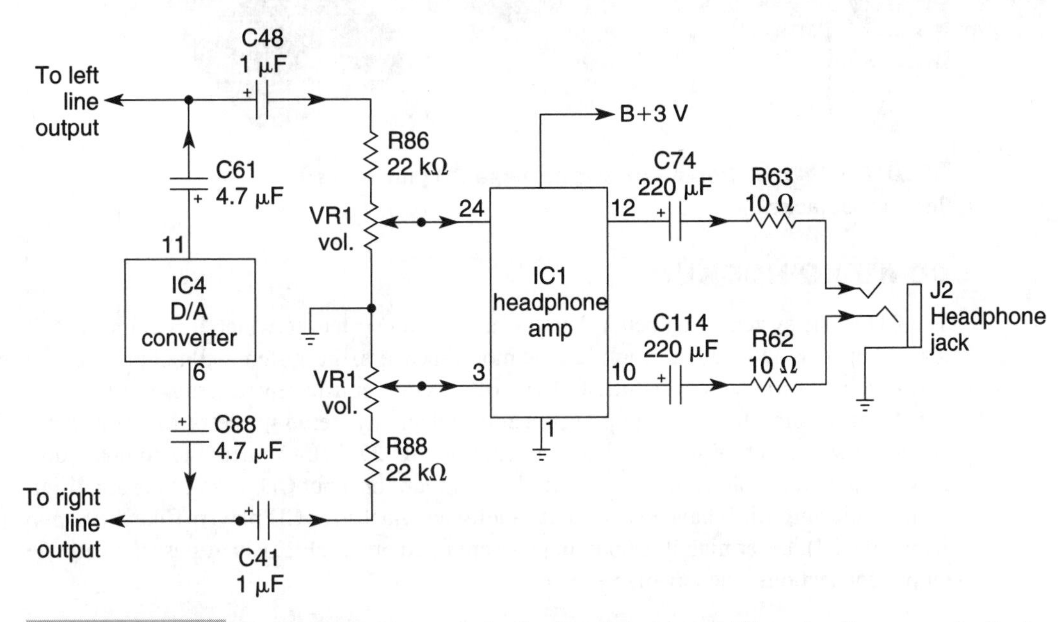

FIGURE 27-21 Both stereo channels in the headphone amplifier circuits are controlled at input terminals 3 and 24 of IC1.

TABLE 27-1 THE VOLTAGE AND OUTPUT IMPEDANCE OF SEVERAL CD PLAYERS

MODEL		OUTPUT VOLTAGE	OUTPUT IMPEDANCE
Akai	CD-M88T	2 V	1 kHz
Panasonic	SL-P3610	2 V	330 ohms
Pioneer	PD-7010	2 V	1 kHz
Quasar	CD8975 YW	2 V	330 ohms
Sanyo	CP500	2 V	470 ohms
Yamaha	CD-3	2 V	1 kHz

FIGURE 27-22 **Boom-box and portable CD players have line-output jacks.**

CD AUDIO HOOKUP

The CD music system is no better than the amplifying system to which it is connected. If hooked up to a low-wattage amplifier and mono speakers, the system will not produce the high-fidelity reproduction that the CD can offer. Great music reproduction results when the player is connected to a high-powered amp and quality stereo speakers. Most CD players come with a set of stereo connecting cables (Fig. 27-22). Connect the compact disc player output male plugs to the right and left auxiliary or input CD jacks of the amplifier. Some audio amplifiers have two separate audio input jacks for CD players (like the Sanyo JA540 model). Determine if the output jacks are fixed or variable. Always use the variable output connections when available.

Troubleshooting the Sound Circuits

A dead, weak, distorted, erratic, or intermittent symptom can exist in one or both channels. Locate the defective audio stage by performing component comparison tests, signal tracing, and individual component checks. By signal-tracing the audio circuits with a scope and an external amplifier, you can locate a defective stage or component. Accurate voltage and resistance measurements help locate defective transistors or ICs. You can inject an audio signal from a generator into the audio-output circuits to locate a dead or weak stage (Fig. 27-23). The speaker or scope can be used as an indicator. You can also use this method to locate a defective component in the headphone amplifiers.

TROUBLESHOOTING AUDIO DISTORTION

Check the audio-output circuits for a distorted channel. If both channels are distorted, suspect the common audio-output IC. Determine which channel is distorted. Very low distortion is difficult to locate, but in a stereo audio circuit, the good channel can be compared with the defective channel. Usually, distortion is caused by a leaky coupling capacitor, ICs, transistors, defective mute transistors, and broken or cracked resistors.

A sine- or square-wave generator can quickly compare the two channels with the duo-trace scope as monitor. Inject the square-wave signal at the volume control and compare each audio channel. Move the audio signal from one stage to the next with the scope at the headphone or line output jacks.

Inject the signal in one side and on the other of the suspected electrolytic coupling capacitor. Check the signal at the base and collector of AF or preamp transistors. If the square waveform is rounded at the top or is misformed, a distortion condition exists. Locate the defective audio IC with the input and output test of the generator and scope each power-output channel (Fig. 27-24).

CD PLAYER OR EXTERNAL AMPLIFIER?

Determine if the problem occurs in the CD player or sound amp. It's possible the CD player is fine, but that the amp has a defective channel. Interchange the connecting cables to see if the cable is at fault. The cable wires often break right where the wire enters the plugs (Fig. 27-25). A poor solder connection of the tip can cause a dead or intermittent channel.

Check the cable with an ohmmeter. Measure the resistance of the shielded cable from the shielding to the plug. There should be no resistance. Clip the meter leads to the center terminal of the male plug at both ends and flex the cable to determine if the continuity is broken or erratic. If the resistance changes at any time during these tests, suspect that an internal cable lead or plug is broken.

Substitute another amplifier to determine if your amplifier is dead or weak. Usually, both channels of a good amplifier don't fail at once. If a headphone jack is provided, determine

FIGURE 27-23 **Inject a 1000-Hz tone from the audio signal generator to signal trace the audio circuits.**

FIGURE 27-24 In this portable CD player, a separate IC is used for left and right output channels.

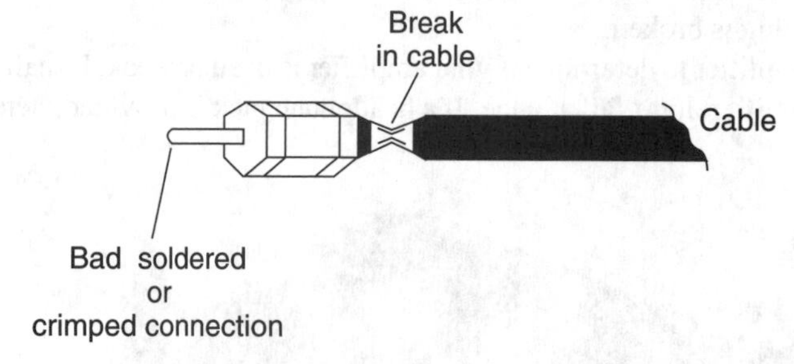

FIGURE 27-25 Inspect breakage where the cable connects to the line-input plug.

whether both channels are okay through those circuits. If so, the problem is in the final-amplifier output stages. Check the output cable with a scope or external audio test amp if the CD player is dead or weak.

HOW TO LOCATE A DEFECTIVE AUDIO CHANNEL

With a disc loaded and playing, check for audio at the audio line jack and the output terminal of the D/A converter with a scope to determine if the audio circuits are defective. If an audio signal is found at the output pin terminal of the D/A converter, but not at the line jack, suspect that the audio circuits are defective. An improper signal at the D/A converter might indicate that a D/A or signal processor is defective.

Next, check the low-voltage sources feeding the various audio circuits. Go directly to the power-supply voltage sources if all sound circuits are dead. Several different voltage sources might feed the audio stages (Fig. 27-26). Here, the D/A converter is powered with a +5 V, –15 V, and +15 V. The sample/hold IC receives a +9 V and –9 V. +15 and –15 V feeds the low-pass filter IC and preamp power audio line amp.

Signaltracing the Sound Circuits

Trace the audio circuits with a scope and an external audio amp. With the disc playing, check the signal at the output terminal of the D/A converter. If the signal is weak or missing, the trouble is in the D/A converter or before it. Check the signal into the D/A and take crucial voltage measurements.

Proceed to the input terminal of the sample-hold IC and output terminal. Check the preamp and low-pass filter in the same manner. If the signal is normal, proceed to the stereo preamp IC. Distortion often occurs in the preamp line IC. Measure for correct source voltages (both voltages must be present).

Check the input and output signal of each IC. When the signal becomes weak or distorted in the external amp, you have located the defective component. Then, conduct voltage measurements on the suspected component to verify whether the component is leaky or open.

SIGNALTRACING WITH AN EXTERNAL AMP

Just about any audio circuit can be checked with the external audio amplifier. If one stage is weak or distorted, check the signal output at the D/A converter with an audio amp. If it

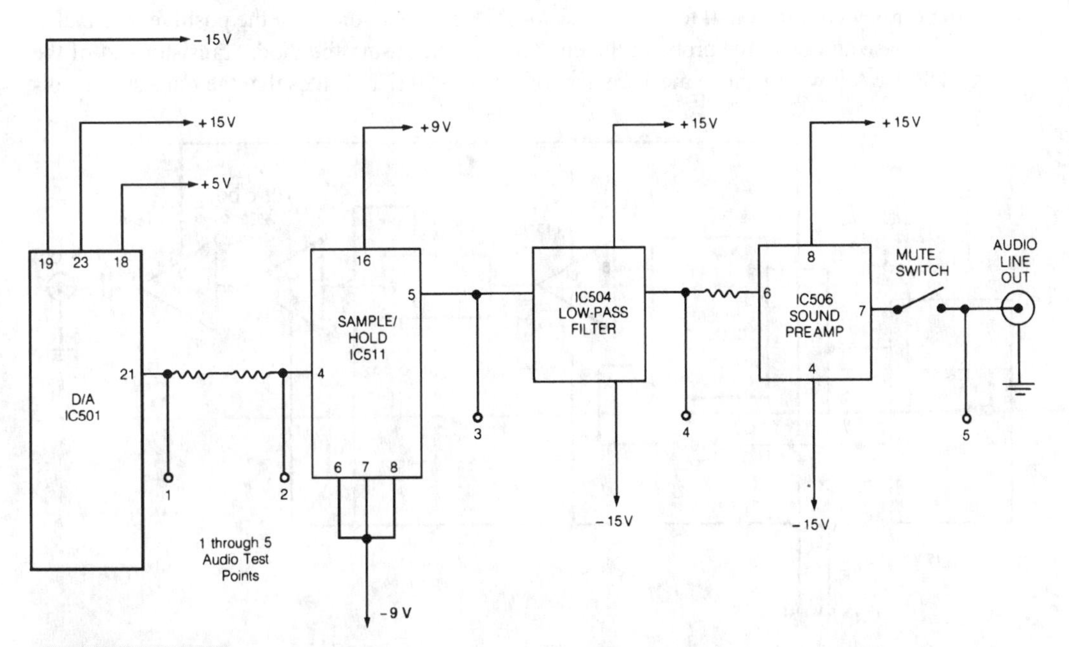

FIGURE 27-26 **Signal trace the audio circuits at test points 1 through 5 to determine which audio channel is defective.**

is normal, proceed to the sample-hold IC, through the low-pass filter network. Compare the left- and right- channel audio at this point. If the line-output stereo channels are normal and the headphone audio is distorted, signal trace the headphone circuits.

Sometimes low audio distortion is difficult to locate with the external amp. Weak audio and no audio conditions can easily be tested with the external audio amplifier. Of course, extreme distortion can be located with the external amp. Just compare each stage in the normal channel as you proceed through the audio circuits.

LOCATING DEFECTIVE TRANSISTORS OR ICS WITH THE DMM

A leaky or open transistor and IC can be located by doing accurate voltage and component tests with the digital multimeter (DMM). Take voltage measurements after locating the defective stage with signal tracing. If a scope or external amp is not available for signal tracing, you probably have to rely on accurate voltage measurements to locate an open or leaky IC or transistor. Identify a defective IC when a normal signal goes in and a poor signal comes out. Then, take accurate voltage measurements on the suspected IC. Then, accurate resistance measurements on pins with low voltage can indicate that an IC is leaky. The same procedure applies to transistors.

The desired voltage measurements are often indicated on the service schematic of most audio components (Fig. 27-27). Note the variety of supply voltages (5, 9, 12, and 15 V) on components in the audio circuits. Both negative and positive voltages go to the D/A converter, low-pass filter, preamp, and audio-output ICs.

Although most audio circuits use ICs, transistors might be used in the mute-switching and relay circuits. First, test the transistor for open junctions with the positive terminal at the base and negative probe at the emitter terminal, using the diode/transistor test of the DMM. A low-resistance measurement of 500 to 950 Ω indicates that the transistor is nor-

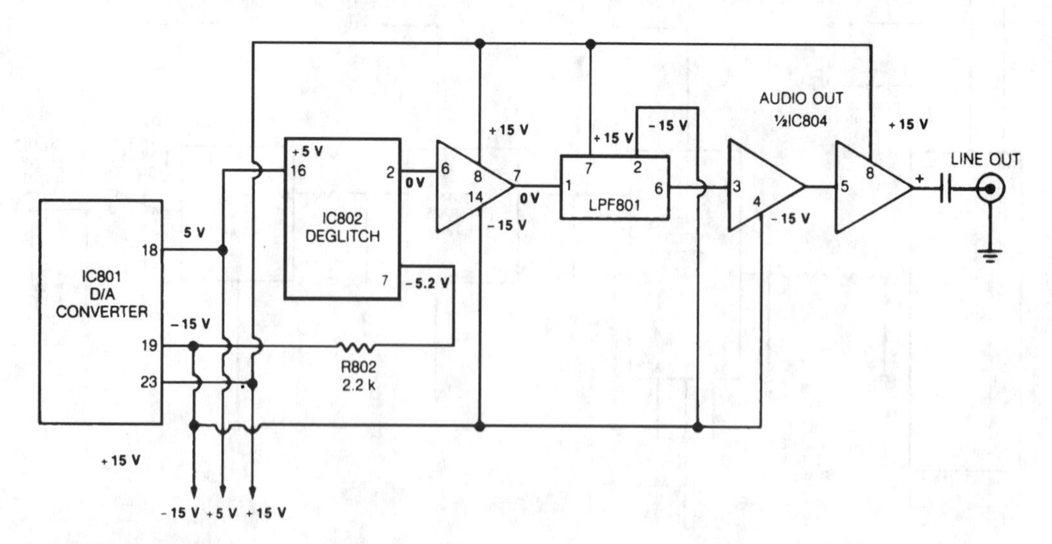

FIGURE 27-27 Take crucial voltage tests on each IC and transistor terminal if no sound is output.

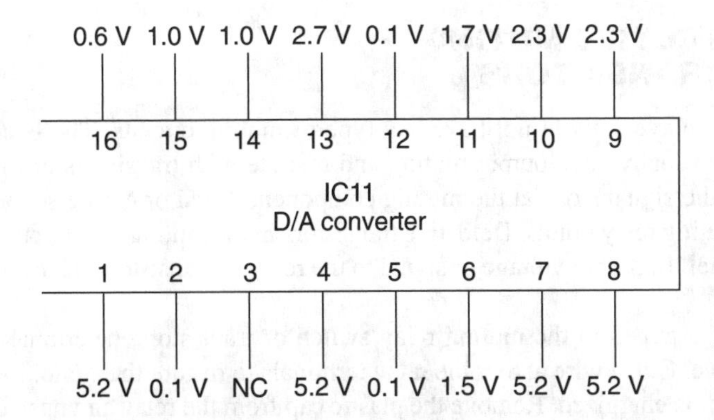

FIGURE 27-28 Take crucial voltage measurements on the IC terminals to determine if it is leaky or open.

mal. Now, reverse the test probes; an infinite or very high reading indicates that the transistor is normal. A similar measurement should result with the negative probe at the collector terminal.

A leaky or shorted transistor will have a shorted or low measurement, less than 1 kΩ in both directions. Doublecheck the emitter-to-collector terminals. Most transistors short between these two elements. Discard the suspected transistor if you get a low measurement in both directions between any two elements. Before removing the transistor when leakage tests are noted, doublecheck the schematic for diodes or other components (or even shorts) in the circuit that might be connecting the two elements being tested. If in doubt, remove the suspected transistor and test it out of the circuit.

IC RESISTANCE MEASUREMENTS

After locating an audio IC with a lower-than-normal voltage on any terminal/suspect that the IC is leaky or an improper low-voltage source. It's possible that both the negative and positive supply voltages are low when only one IC terminal shows leakage. Compare the various voltage measurements on the IC with the schematic. Always mark down the correct terminal voltage on the schematic after performing the correct repairs.

If a +15-V source at pin 8 is down to 7.5 V, this same voltage could be low at other ICs (Fig. 27-28). Apply power to the CD player and take another measurement. If the voltage remains the same, suspect that the power source or another component is pulling down the voltage. If the voltage returns to a normal +15 V, the IC is probably leaky. Take a resistance measurement between the low pin and common ground. Usually, a leaky IC pin reads less than 1 kΩ. A random resistance measurement between each terminal and ground can isolate a leaky IC, but check to be sure that there are no low-value resistors or diodes in the circuit (which would indicate a low reading before removing the suspected IC).

REPLACING THE TRANSISTOR OR ICS

It's always best to replace transistors and ICs with the original type of components, but when these components are not available, you must make substitutions. At this stage of development in CD players, the D/A converter and audio-output amps might have to be originals. But some of the sample-hold or analog switch/deglitch ICs and op amps are available with universal replacements.

LOCATING DEFECTIVE MUTING RELAYS AND TRANSISTORS

The lower-priced CD players might not have any type of muting circuits. The later and higher-priced players usually have output muting and operate with transistors or muting relays. Determine if the signal stops at the muting component. Dead or erratic sound can be caused by dirty muting relay points. Defective transistors in the muting-line circuits can produce a dead channel. Improper voltage or signal to the relay or transistors can cause the same problem.

If the audio signal is traced to the muting relay switch or transistors, determine what component is defective. Clip a wire across the relay terminals to restore the audio. Peek at the solenoid to see if it is energized. Remove the plastic cap from the relay and push down the flat metal piece that trips the contacts. Suspect that a contact is dirty if the sound does not appear. Clean the contacts or replace the relay assembly.

Suspect a defective transistor or IC circuit if the solenoid does not energize (Fig. 27-29). Measure the resistance of the solenoid and check it against the good one in the other channel (usually, both relays are not defective at the same time). Check the voltage across the solenoid coil and compare it to the good relay in the emphasis circuits. Suspect that a transistor or IC is defective if an improper voltage is at the relay winding. Do not overlook the possibility of an improper signal coming from the muting-control units.

Some of the line-output circuits have IC or transistor muting (Fig. 27-30). With accurate voltage and transistor in-circuit tests, you can often locate the defective muting component. When the power is turned on and the voltage is high, the muting transistors should be at cutoff; with the power off, the voltage should return to zero.

If no signal is at the line-output jack, simply remove both collector terminals of each transistor from the muting circuit. The audio should be normal. Measure the voltage at the base terminals of the transistors or the line muting transistor if you get a very low voltage or no voltage. Check each transistor for open or leaky conditions. Determine if the muting

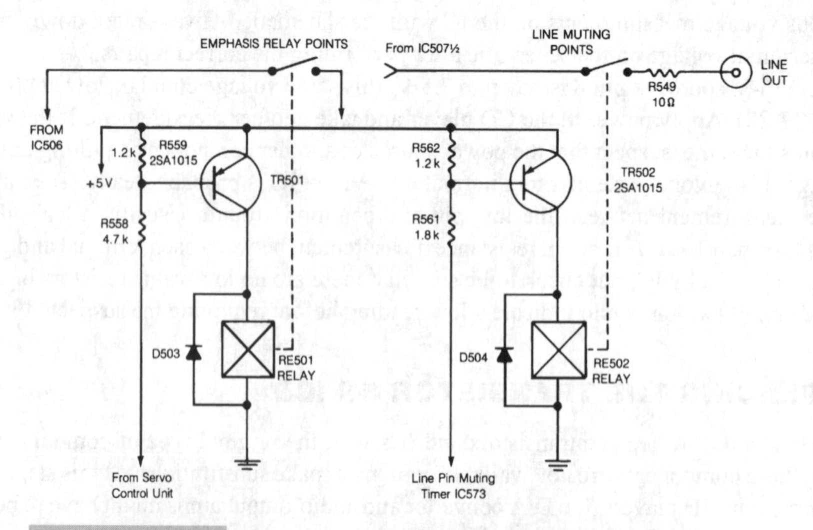

FIGURE 27-29 Suspect a defective transistor, relay, or an incorrect voltage if the muting circuits are defective.

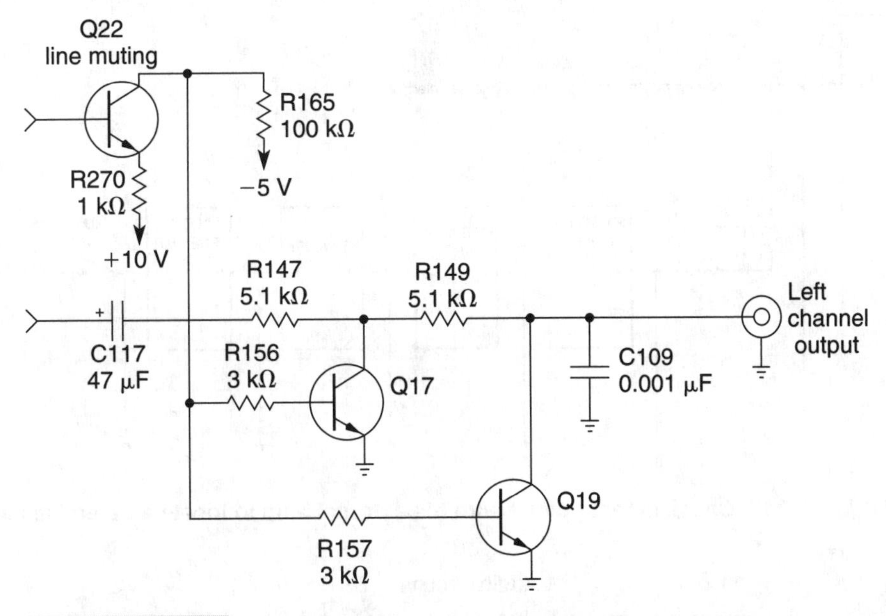

FIGURE 27-30 **The voltage is high on collector terminals of the muting transistor if the power is turned on.**

system is working if a voltage change can be detected at the collector terminal of the line-muting transistor (Q19). Take accurate voltage and resistance measurements on the line-muting IC.

DEAD RIGHT OR LEFT CHANNEL

Either channel can be dead and the other channel normal. For example, if the right channel has no output, check the audio signal at the right output line jack with the scope or external amp. Backtrack to the right sound output terminal of the preamp stage (Fig. 27-31). Because the stereo signal splits at the D/A converter output, check each component signal terminal up to the D/A IC or until the channel "comes alive." It's possible to have one dead and one normal channel coming out of a defective D/A converter IC. In this case, any component within the bad channel could be defective up to where the signal splits at the D/A IC.

ONE DEAD AND ONE WEAK CHANNEL

Any component common to both channels could cause one channel to be weak and the other dead. Because the power supply sources are common to both channels, check each voltage source first. Notice if either the negative voltage is lower than the positive voltage. It's possible that both negative and positive voltage sources are off because they usually come from the same common power supply.

In many of the audio ICs, one half of the same component is used in each channel. This is especially true of the preamp and audio-output ICs (Fig. 27-32). If a common audio IC leaks or opens, it can affect both audio channels. Take accurate voltage and resistance

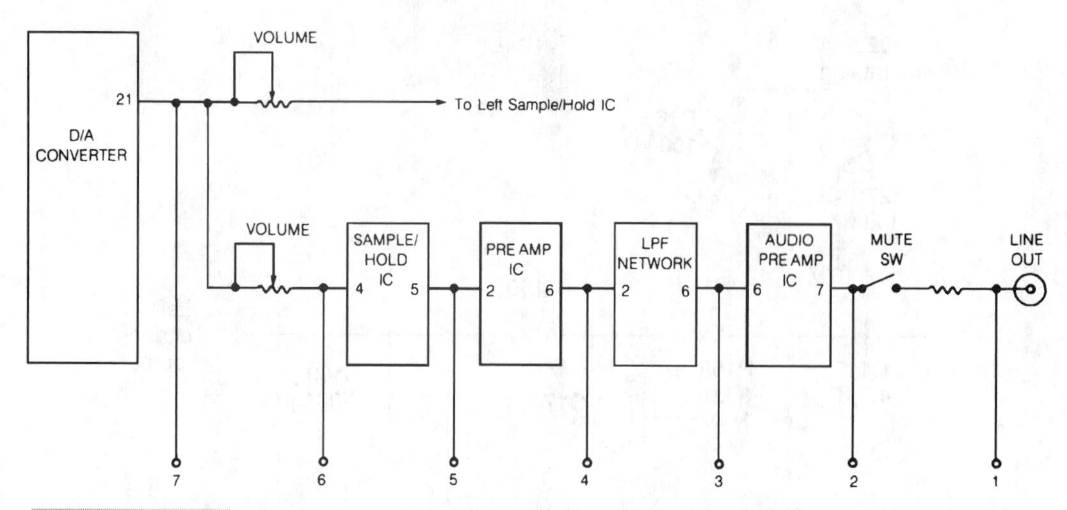

FIGURE 27-31 Check at test point 1 with the external amp to locate a defective circuit.

FIGURE 27-32 Locate the preamp or audio-output IC within the portable chassis.

measurements on the suspected IC. Signal tracing the audio up to the same IC in both channels can help you to locate the suspected common IC.

DISTORTED CHANNEL

Audio distortion in one particular channel can be caused by a leaky or open IC or transistor. Signal trace the audio signal to locate the stage causing the distortion. Take accurate voltage measurements after locating the suspected component. If only one channel is distorted, the problem must exist from where the signal separates after the D/A converter. Check the signal with the scope from the D/A through the LPF. Do not overlook the possibility of a distorted speaker system. Interchange the line output cables to determine if the amplifier and speakers are at fault. Check the power-supply sources if both channels are distorted.

INTERMITTENT SOUND

Check the most obvious intermittent plug by flexing the audio output cables. Be sure that all cables are plugged fully in at each plug. Inspect the plug and line-output jacks for corroded or noisy connections. Determine if the external amp is normal by checking headphone reception or by plugging into another amplifier. Notice what channel is intermittent.

Monitor the intermittent audio at the defective channel output jack with an external amplifier. Check the weak audio output at the D/A converter with an external test amp. Then, proceed to the input and output IC line amplifier to monitor the intermittent sound. Take crucial voltage and resistance measurement at the intermittent circuit. Disconnect the mute transistor collector terminal from the audio line output to determine if the mute circuits are malfunctioning.

TROUBLESHOOTING THE HEADPHONE CIRCUITS

Go directly to the headphone circuits if the audio line-output amp is normal, but the headphones sound dead or distorted. If no sound is audible at the line output and headphone circuits, suspect trouble in the amp or sample-hold stages. Notice if only one or both channels are dead. Suspect the output-audio IC if only one channel is defective in both systems because one half of the IC could be in the line output and the other in the headphone circuits (Fig. 27-33).

Substitute another pair of earphones to determine if the headphones are defective. Clip a small 8-Ω speaker to the headphone jack terminals to signal trace the audio output (Fig. 27-34). The auxiliary input with a male headphone plug and cable can serve

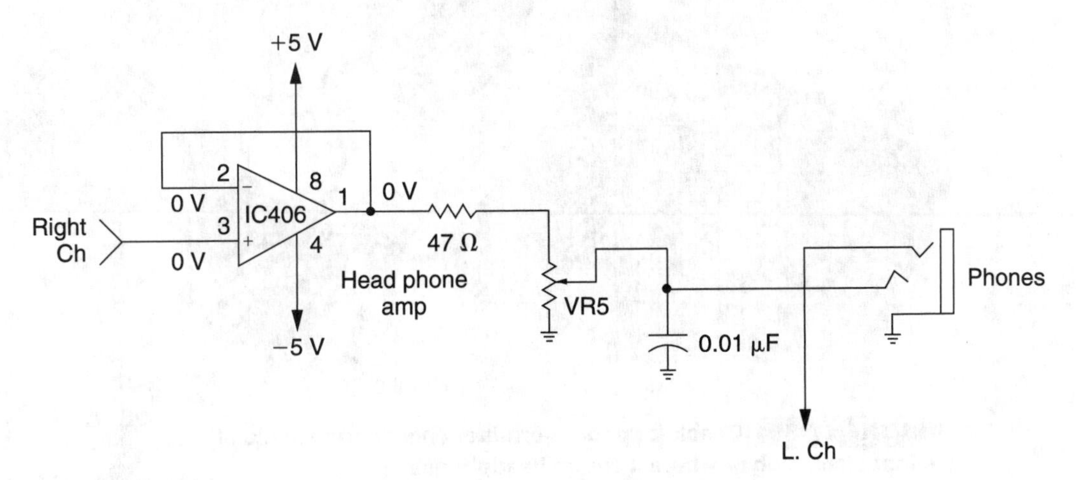

FIGURE 27-33 The stereo headphone circuits in a CD changer.

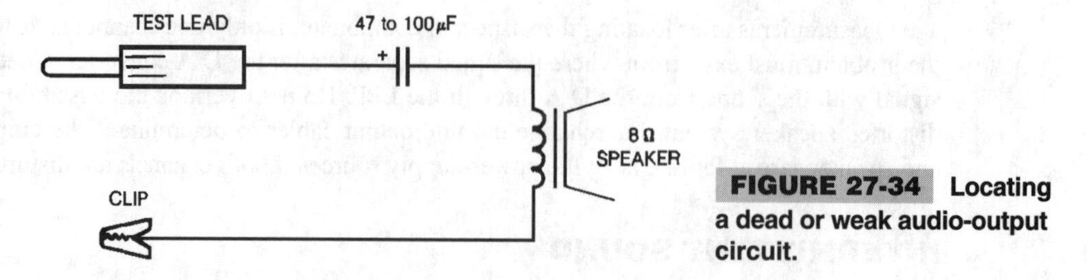

FIGURE 27-34 Locating a dead or weak audio-output circuit.

as a testing source. Signal trace the audio through the headphone circuits with the scope or external amp.

Cable problems Check the male plug on the headphones for poor cable connections. Headphone cables often break where the wire enters the male plug or right at the headphone case. Suspect that the connection is poor or broken if the sound is erratic. Check the resistance between common ground and the outside metal terminal of the headphone plug. Flex the cord and note if the meter jumps. If so, replace the plug or the whole cable. Erratic or intermittent sound can result from a dirty or worn female headphone plug.

Headphone problems After many hours of playing, jolting, and carrying around the portable CD player, the headphone cable often breaks right where it enters the jack or at the headphone unit (Fig. 27-35). First, clean the jack and plug area with cleaning fluid to

FIGURE 27-35 Check for poor intermittent headphone audio at the input jack plug or where it enters headphones.

eliminate erratic and noisy reception. A squirt of cleaning fluid into the phone jack area might help the noisy reception. Check inside of headphone jack for intermittent audio. Sometimes a wire can break at the jack terminals. Inspect the jack area for intermittent reception, resulting from poorly soldered connections at the headphone jack.

A broken headphone cable right at the plug or at each headphone can cause intermittent or dead audio. Usually, broken wires or cable can be repaired in a few minutes. Replace the headphone plug if it is broken inside of the molded male plug. These plugs can be located at electronic stores. Solder and tape all broken wires or cables. Check inside of the headphones for broken wires on the small speaker or phone.

Weak channel Notice if the defective channel is normal at the line-output amp. If it is weak, suspect that an IC or capacitor is defective, or that poor muting is in that channel. Remember, the headphone-output muting might be operating from the same line-output relays. Signal trace the audio into the headphone IC and out. Don't forget to turn up the headphone volume control.

Distorted channel Suspect the audio-output IC if excessive distortion is in the headphones. Signal trace the distortion with the scope and external amp. Look for the distortion in the early stages of the audio circuits if the distortion is also detected in one of the line-amplifier channels. Do not overlook the possibility of a defective set of headphones. Many compact disc manufacturers have their own audio troubleshooting methods.

28

CD PLAYER ADJUSTMENTS

CONTENTS AT A GLANCE

The electronic and mechanical adjustments should only be made after replacing a crucial component or a simple touch-up test. After installing a new pickup head assembly, the E-F balance adjustments should be made. When replacing the disc motor, the disc platform must be adjusted for maximum EFM signal at the preamp IC, if the disc platform is not found on the new motor. Focus or tracking-gain adjustments can be touched up after replacing the preamp or servo ICs. Crucial waveform adjustments might confirm if that particular circuit is performing (Fig. 28-1). If complete electronic adjustments are to be made, they should be checked in the correct order. Follow the manufacturer's order of adjustments, if they different from those described here.

Like the radio or TV receiver, adjustment screws do not get out of line by themselves. The same applies to the CD chassis. Do not touch them unless correct adjustment is needed. If the signal is missing on a certain component after you take scope or voltage tests, be sure that you don't make any electronic adjustments without the correct test

FIGURE 28-1 **Take crucial waveforms with the scope and a test disc.**

equipment. First, locate the defective component. The laser diode VR adjustments should not be touched because hazardous invisible laser radiation could result. Usually, this adjustment is made at the factory. Do not make any electronic adjustments without the correct test equipment or manufacturer's adjustment procedures.

Required Test Equipment

The oscilloscope, ac and dc digital voltmeters, audio AF oscillator, frequency counter, test discs, and various scope band-pass filter homemade test circuits are needed for correct CD electronic adjustments. Several manufacturers use special servo gain and conversion connecting cables. You might even want to use two separate scopes. The oscilloscope is used to take waveform adjustments with crucial voltage tests made with the DMM. In some tests, a two-channel scope is needed for anything over 30 MHz. The frequency counter is used in frequency adjustment of the VCO and PLL circuits.

The commercial laser power meter can help the electronic technician measure the laser output in compact disc players and infrared sources. This instrument can be used in servicing CD players, video disc players, VCRs, remote controls, and other infrared sources (Fig. 28-2). The laser power meter has a built-in load-simulation circuit, required to service many CD players. The range settings are 0.3 mW, 1 mW, and 3 mW with switchable wavelength settings of 633 nM and 750 to 820 nM.

The ordinary CD disc can be used to make most of the electronic adjustments. Some manufacturers use only one test disc for all of the electronic adjustments, but others use several different test discs. The eccentricity, surface oscillation, and scratch test discs are

FIGURE 28-2 Check the laser diode with a power meter.

TABLE 28-1 THE DIFFERENT TEST DISCS TO MAKE CD ADJUSTMENTS

TEST DISC	USE
800104	For demonstration
40088	
40067	
YEDS1	
YEDS7	For signal characteristics
YEDS18	
Eccentricity disc	Eccentricity width: 200 μM
Surface oscillation disc	Surface oscillation width: 400 to 500 μM
Scratch surface	Black scratch width: 300-μM black line
TCD-784	Type made in A-BEX

used to check the results of the electronic adjustments. The various test discs are given with each manufacturer's electronic adjustment procedures. Several manufacturers use special adjusting instruments, adapters, and connectors for their special adjustment procedures. See Table 28-1.

Typical list of instruments and tools

- Dual-trace oscilloscope (10:1 probe)
- Power laser meter
- Frequency counter
- Low-frequency oscillator
- Test discs (YEDS-7)

- Low-pass filter (check with manufacturer)
- Load resistors
- Standard tools on the test bench

Test Points

Most CD players have the various test points listed right on the PC board (Fig. 28-3). All test points should be located before trying to take scope waveforms. Besides making correct electronic adjustments, the various test points might help you to determine if that circuit is performing with a correct scope waveform or voltage test. Check for a chassis layout parts location drawing, which usually shows the various test points and ICs or LSIs (Fig. 28-4). Be careful not to touch other components or test points with the scope probe. Use a clip-on probe, which might cling tightly to the test point or IC terminal pins.

Often the different VR adjustment controls are located close together for easy adjustment (Fig. 28-5). Locate the correct control adjustment for the right adjustment procedure. Most manufacturers list these controls separately in the adjustment procedures or are clearly marked on the PC board. In some instances, a starting point might be shown of halfway rotation of the adjustment control. The slotted screwdriver adjustments are found in most models. The order of adjustments are:

1. Laser power meter
2. Focus offset verification
3. Tracking error balance verification

FIGURE 28-3 Check the various test points of the chassis for CD adjustments.

4 Pickup radial/tangential tilt adjustment
5 RF level verification
6 Focus servo loop adjustment
7 Tracking servo loop gain adjustment

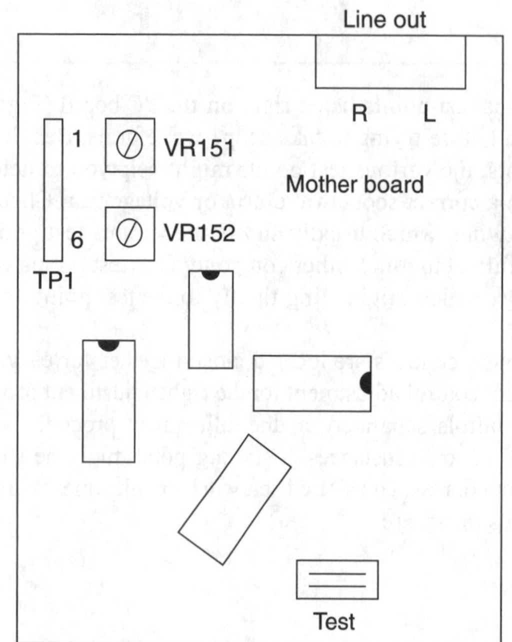

FIGURE 28-4 Locate the test points and variable controls on a CD chassis.

FIGURE 28-5 Look for very small adjustments on the CD chassis.

FIGURE 28-6 Check the laser diode with a portable laser power meter and check the manufacturer data.

LASER POWER ADJUSTMENTS

Laser power adjustments are made to ensure proper operation of the laser diode. You will know that the laser diode is operating correctly with the laser power adjustments. A low EFM output indicates a defective laser diode. In some optical pickup systems, the VR adjustment mounted on the assembly is adjusted at the factory and should not be touched.

Denon DCD-1800R laser power adjustment When you turn the power on, the laser beam is emitted. Be careful not to look at the laser beam directly. Place the laser power meter in contact with the pickup lens (Fig. 28-6). Adjust VR102 on the motor wiring board so that the laser output becomes 0.3 mW ± 0.01 mW.

Realistic CD-1000 laser adjustment Connect the oscilloscope to TP13 (TDET) and TP16, which is grounded on the adjustment board. Load a disc in the player and set the player to Play mode. Adjust R629 (laser gain adjust) so that the EFM signal level becomes 700 mV.

Optimus CD-7105 RF level adjustment The RF level verification adjustment should be made when the symptom is no Play or no Search, then shutdown. If the RF or EFM waveform is not present, most CD players will shut down after a few seconds. The objective to check the RF adjustment is to determine if the RF signal amplitude is correct or missing. Check the RF amplifier and optical laser assembly if the waveform is missing.

Connect the oscilloscope to TP1, pin 1 (RF) with the player in Test mode and Play. Use a YEDS-7 test disc. Move the pickup to midway across the disc with the track/manual search FWD or REV key. Then, press the PGM (program) key, the Play key, and then press the Pause key (in that order) to close the respective server and put the player into Play mode. Verify the RF signal amplitude at 1.2 Vp-p ± 0.2 V (Fig. 28-7).

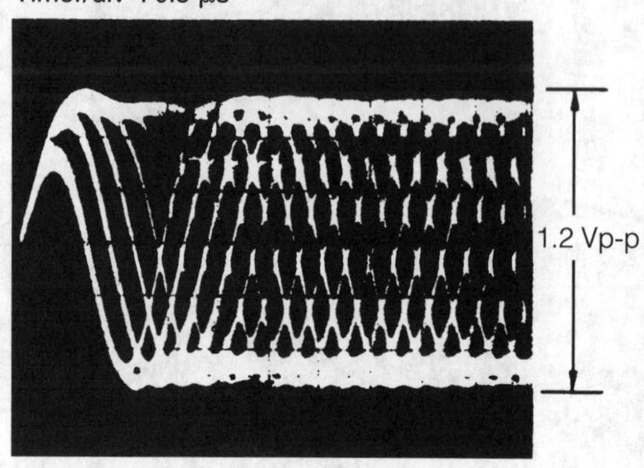

FIGURE 28-7 The RF eye-pattern waveform-level verification test.

PLL-VCO ADJUSTMENTS

The PLL circuit basically consists of an 8.6456-MHz VCO (voltage-controlled oscillator). The VCO oscillator output is divided (4.3218 MHz when locked) where the phase of the signal is compared with the edge of the EFM signal read from the disc. The PLL-VCO (phase-locked loop voltage-control oscillator) adjustment is made with the oscilloscope or frequency counter (Fig. 28-8).

JVC XL-V400B PLL free-run adjustment Use the frequency counter as the PLL adjustment instrument. Connect the frequency counter between ENN028 (TP11) and ground. Short the TP10 and ground. Adjust L301 to 4.32 ± 0.01 MHz, as indicated on the frequency counter with the regulating rod. Perform this checking adjustment no sooner than switching on the power.

Onkyo DX-200 CLV-PLL circuit adjustment Required test equipment: Frequency counter, adjusting rod, and shorting clip.

1 Turn on power to the set.
2 With the set in Stop position, use shorting clip to drop VC01 (TP-221) to ground.

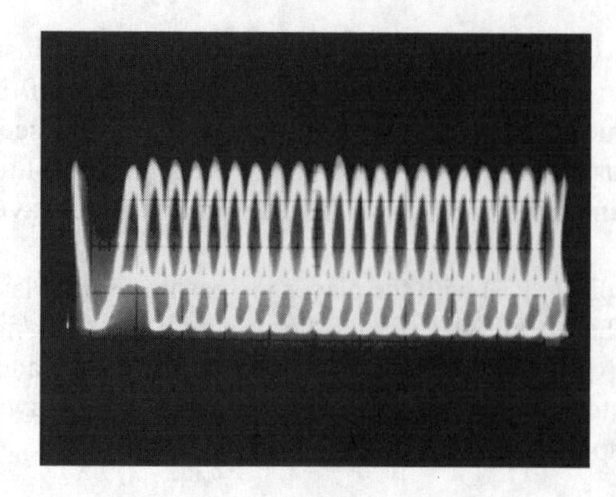

FIGURE 28-8 A typical PLL-VCO waveform of this voltage-controlled oscillator.

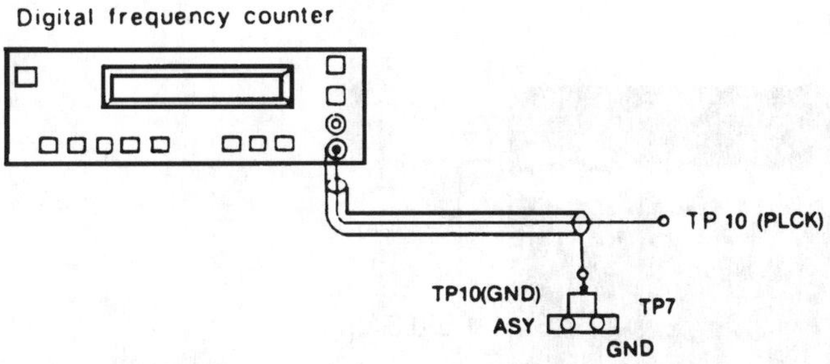

Digital frequency counter

TP 10 (PLCK)

TP10(GND)

ASY

TP7

GND

FIGURE 28-9 **The frequency-counter connections when adjusting a Realistic PLL-VCO.**

3 Connect frequency counter to WFCK (TP-283).
4 Turn the oscillator coil L206 to adjust frequency to 7.35 ± 0.01 kHz.
5 Remove the shorting clip.

Denon DCC-9770 PLL-free-run adjustment

1 Remove the soldering bridge (EFM) from the MD board.
2 Detach RS22 from main PC board. R552 is connected to pin 11 of IC503.
3 Connect the frequency counter to test point (PLCK).
4 Turn the power on and adjust RV505 so that the frequency-counter reading satisfies the specifications. (Adjustable limits: 4.3118 to 4.3318 MHz)

Realistic CD-3304 PLL (V_{CC}) adjustment

1 Put unit into Stop mode.
2 Short circuit test point (ASY) and test point (GND). Both terminal short (TP7).
3 Connect the digital frequency counter to test point (PLCK) and test point (GND) TP10 (Fig. 28-9).
4 Adjust VR504 so that the frequency-counter reading becomes 4.25 <SHADED> 0.01 MHz.

RF SIGNAL ADJUSTMENTS

In most CD players, the RF and skew adjustments are made for maximum RF (radio frequency) signal waveform at a test point. The radial-tangential screw (mounted under the flapper) is adjusted for the RF waveform on the oscilloscope. The correct adjustment is to make the diamond-shaped area as clear as possible without excessive jitter (Fig. 28-10). A special tool or a simple screwdriver slot are used for the radial screw adjustment.

Denon CDM-560 HF level adjustment Connect the oscilloscope through a 10-to-1 probe and clip to TP102 to check the HF level waveform (Fig. 28-11). Push the pause control that displays number (03) and play a test CD. Check the HF level of oscilloscope. Confirm that the waveform is in good shape. The diamond shape in the center of HF or EFM

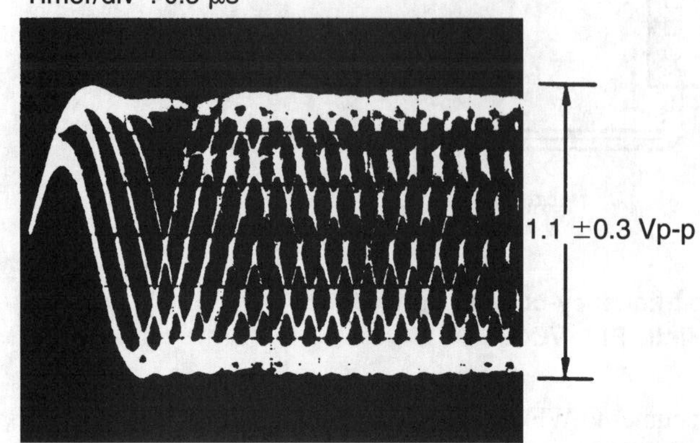

FIGURE 28-10 A typical RF eye pattern at TP1 in most CD players.

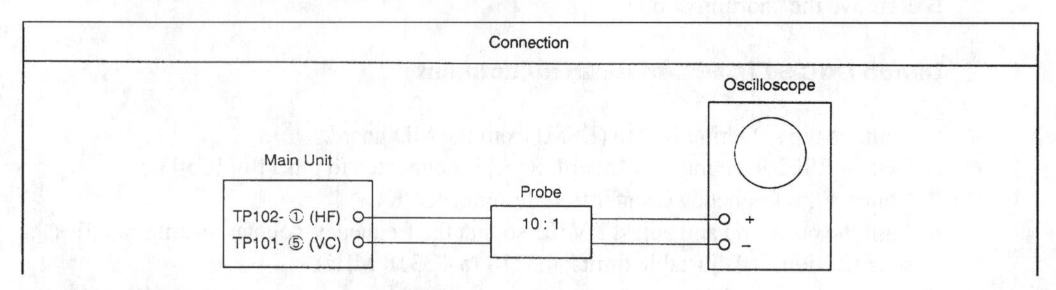

FIGURE 28-11 How to connect the scope for HF level adjustment in a Denon CD player.

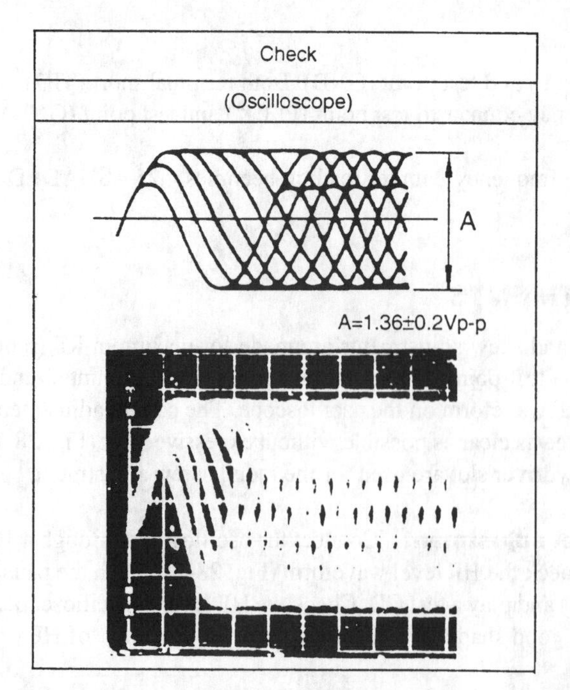

FIGURE 28-12 A typical RF eye-pattern in the Denon CD player.

waveform must be able to discriminate clearly. The amplitude should be around 1.36 Vp-p 0.2 V (Fig. 28-12).

FOCUS AND TRACKING OFFSET ADJUSTMENTS

The focus offset might contain the RF and FE adjustments, one of the same or separate adjustments, in different manufactured CD players. The focus offset adjustments can be called the *jitter* or *eye-pattern adjustments*. The focus offset adjustment is compared to the RF adjustment for less jitter and a clear diamond-shaped opening.

Onkyo DX-C909 RF adjustment It is not necessary to perform the adjustment of optical pickup. This configuration should be made when replacing the optical pickup.

1 Connect scope to test points RF and VC (Fig. 28-13).
2 Turn the power switch on.
3 Load the test disc YEDS-18 on the tray and press the Play button.
4 Confirm that the waveform on the scope is optimum eye pattern and optimum level as shown. *Optimum* means that the diamond shape can be clearly distinguished at the center of the waveform.

Denon DCC-9770 focus offset adjustment

1 Connect the oscilloscope to test point RF on the MD main board (Fig. 28-14).
2 Load a disc and set the mode to Play.
3 Adjust RV502 so that the waveform on the scope (eye pattern) is maximum and has a good shape. A well-shaped eye pattern has the mark diamond clearly distinguished at the center of the waveform.

Optimus CD-7100 focus offset verification When the player is out of adjustment, it will not focus in and the RF signal is dirty or not plain. Verify the dc offset for the focus error amp. Connect the scope to TP1, pin 6 with the player in the test mode and with the power switch turned on. No disc is needed for this verification test. Verify that the dc voltage at TP1 should be around 0 to 50 mV. If the specified values cannot be obtained or if no adjustment is possible by performing the verification or adjustments of focus offset verification, tracking-error balance verification, pickup radial/tanzential tilt adjustment, and RF level verification, the pickup block might be defective.

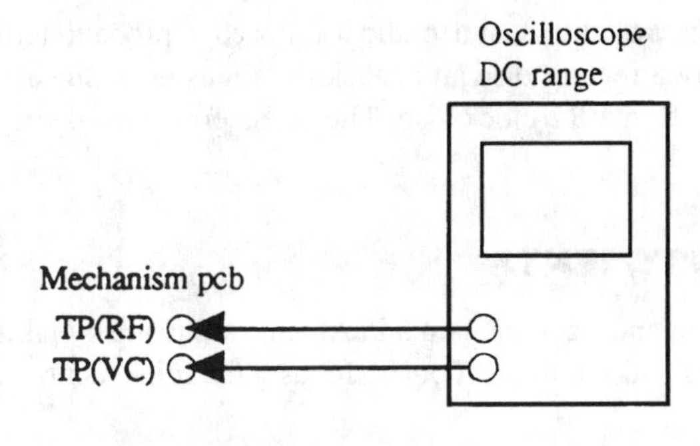

FIGURE 28-13 **Connect the scope to TP-RF and TP-VC for RF adjustment in a CD player.**

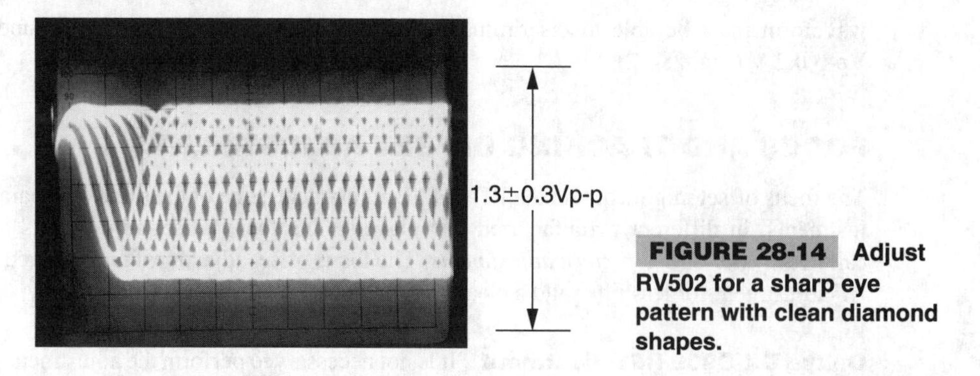

FIGURE 28-14 Adjust RV502 for a sharp eye pattern with clean diamond shapes.

Onkyo DX-200 focus offset adjustment The manufacturer's special jig is needed.

1 Turn on SW1 of the adjustment jig (1).
2 With the meter 2 in the range of 0 to 0.1 V, adjust R229 for minimum deflection of meter 1. (Note when the deflection of meter 1 is broad, adjust meter 2 to 0 V)

Denon DCH-500 CD changer focus offset adjustment Before starting adjustments, use a YEDS-18 test disc and be sure that the power supply voltage is 14.4 Vdc (not over 10 A).

1 Connect the oscilloscope to servo board test point RF (Fig. 28-15).
2 Place unit in Play mode with the loaded disc.
3 Adjust RV12 so that the scope waveform eye pattern is good. The diamond shape in the center should be clearly distinguished (Fig. 28-16).

Realistic CD-3304 focus offset adjustment

1 Insert the test disc and put unit into the Play mode.
2 Connect oscilloscope to TP1 (HF GND) (Fig. 28-17).
3 Adjust VR505 so that the eye pattern becomes clear and waveform (Vp-p) is maximum. When confirming the eye pattern, use the 10:1 scope probe (Fig. 28-18).

Realistic CD-1000 tracking offset adjustment Connect the oscilloscope leads to TP13 (TDET) and TP16 ground. Load a disc in the player and set the player to the Play mode. Adjust R116 so that the EFM signal amplitude becomes maximum (Fig. 28-19).

Optimus CD-3380 focus gain adjust Connect the oscilloscope probe to terminals TP30 and TP31 (Fig. 28-20). Play a regular disc and check the focus error signal waveform. To increase the focus gain, turn VR2 clockwise. The focus error signal should be about 400 mVp-p (Fig. 28-21).

TRACKING GAIN ADJUSTMENTS

In some CD players, the focus gain and tracking gain adjustments are made by taking oscilloscope voltage tests across the coil windings. For the focus gain adjustment, connect the oscilloscope across the focus coil. Play a test disc. Adjust the focus gain control between 500 to 600 mVp-p (Fig. 28-22).

For the tracking gain adjustment, the oscilloscope is connected across the tracking coil. Of course, one side of the tracking and focus coil is grounded, in which the ground lead of the scope is at ground potential. Now, play the test disc. Adjust the tracking gain control for a 1.8 to 2.2 Vp-p across the tracking coil. Always follow the manufacturer's adjustment procedures.

If the sound jumps when the machine is jolted or bumped, the tracking gain might be set too close or too small. If a test disc with a small scratch is played and the sound jumps, the tracking gain might be set too large.

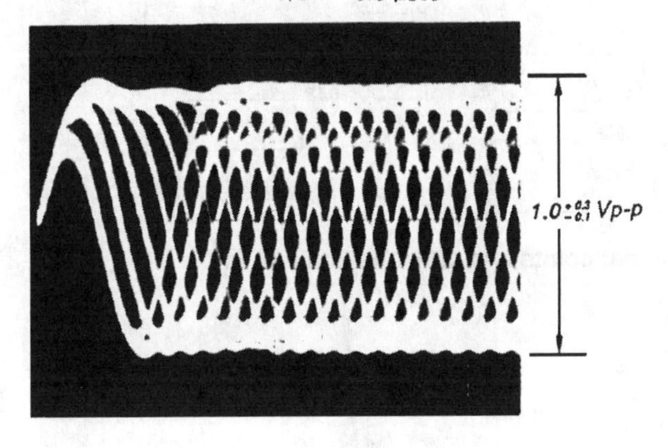

FIGURE 28-15 Notice the sharp diamond-shaped RF waveform in an auto CD changer.

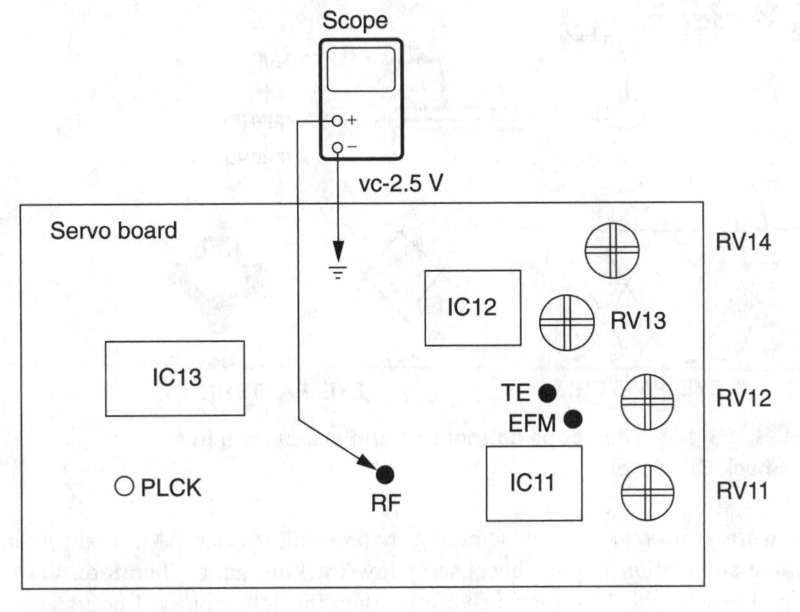

FIGURE 28-16 Connecting the oscilloscope to the RF connection and ground in an auto CD player.

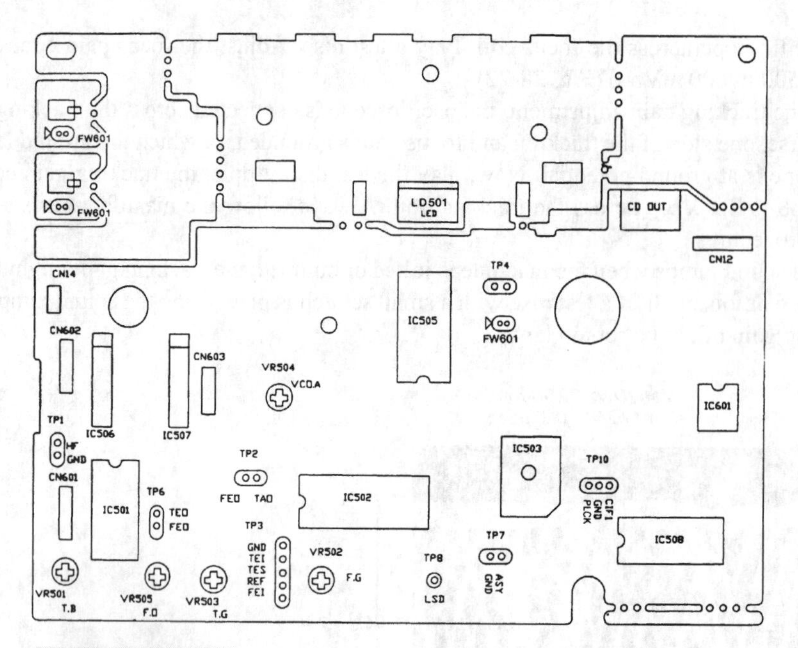

FIGURE 28-17 The test points and offset adjustment in the Realistic CD player.

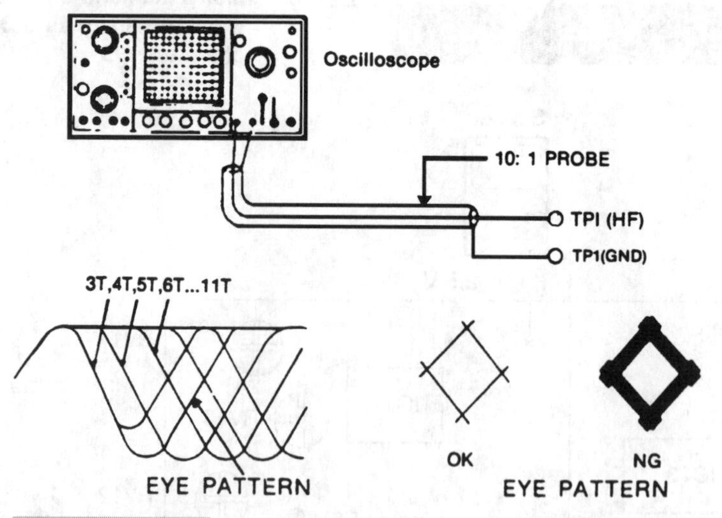

FIGURE 28-18 The scope connection and eye pattern in a Radio Shack CD player.

When gain is lower When selecting by pressing reverse AMS and forward button, brake application is poor because of low tracking gain. Therefore, the traverse waveform is after the 100 jump waveform, then the selection will be located slowly (Fig. 28-23).

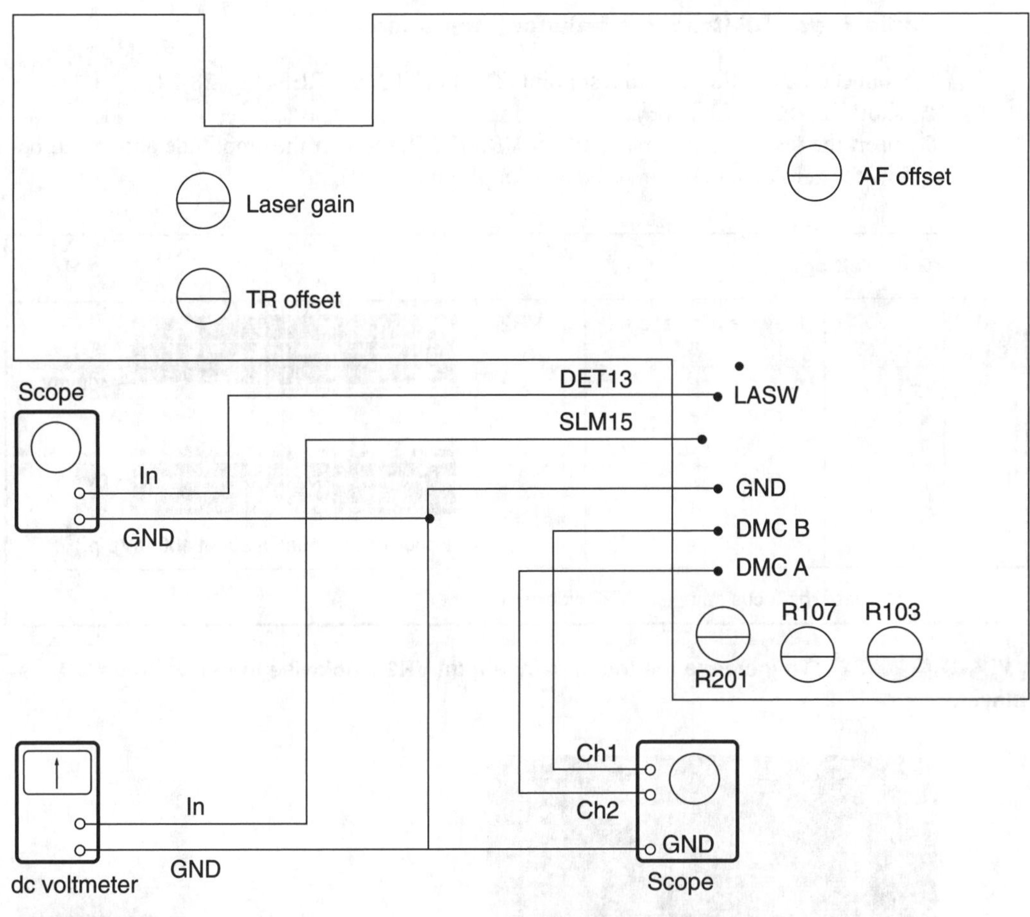

FIGURE 28-19 The various test points in the Realistic CD player.

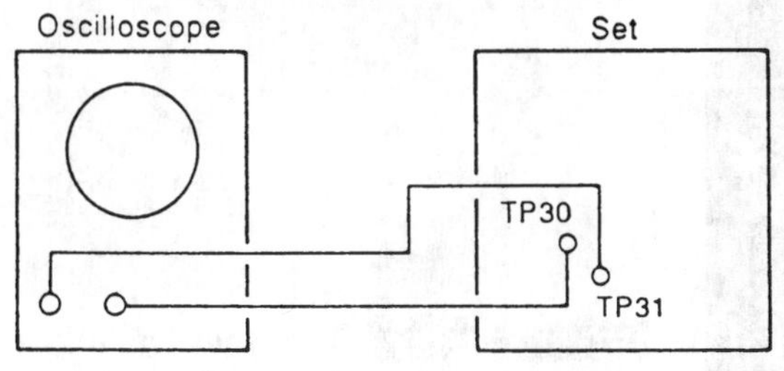

FIGURE 28-20 The focus servo gain scope connections in a portable CD player.

Radio Shack EF (tracking balance) adjustment

1 Connect the oscilloscope to test point (TE1) in TP3 and REF (Fig. 28-24).
2 Short TES and REF (TP3).
3 Insert the test disc, after play, adjust VR501 (TB) so that the amplitude above and below the zero dc line becomes equal (Amplitude A = B).

	Gain Setting			
1		Play the regular disc.	VR2	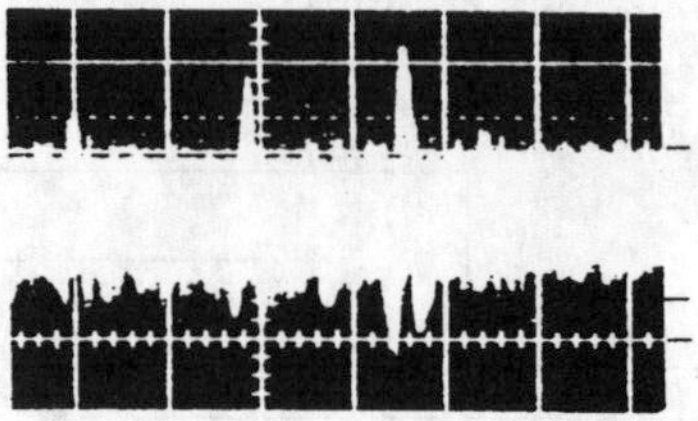 400 mV 100 mV 0V Focus error signal of about 400 mVp-p.
2	To increase the focus gain, turn VR2 clockwise.			

FIGURE 28-21 To increase the focus gain, adjust VR2 clockwise in the Optimus CD player.

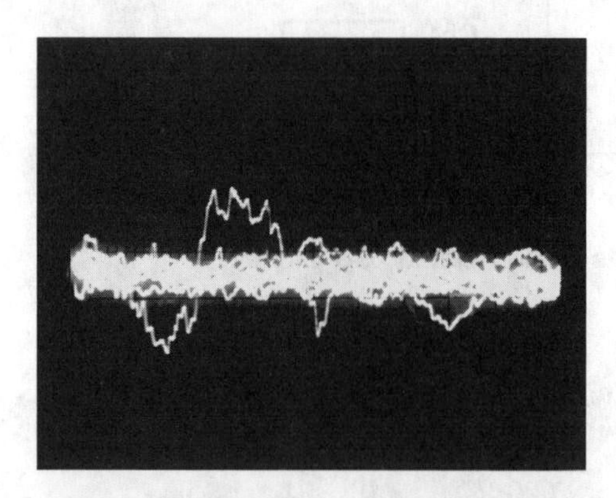

FIGURE 28-22 A typical scope pattern across the focus coil in the focus-gain adjustments.

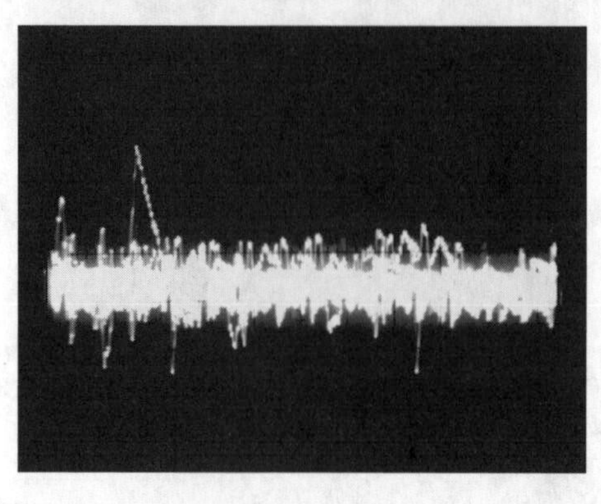

FIGURE 28-23 The typical gain adjustment across the tracking coil.

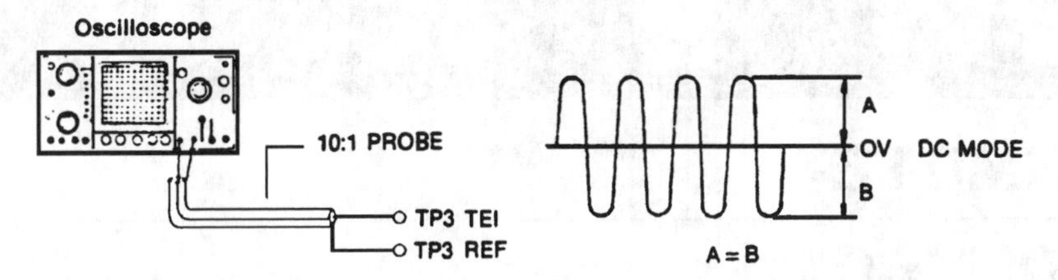

FIGURE 28-24 Oscilloscope connections for EF tracking balance in the Radio Shack CD portable.

AUTO RADIO ELECTRICAL ELECTRONIC ADJUSTMENTS

The adjustments within the car radio should be made with a 14.4-Vdc supply voltage (more than 2 A). Use a YEDS-1 or YEDS-7 test disc. Connect a two-channel oscilloscope with delayed sweep frequency counter and a light power meter to make the different tests. Practically all of electrical or electronic adjustments are like those found in the home CD player. Locate the test points and various controls on the RF amp PC board.

29

REMOTE-CONTROL FUNCTIONS

CONTENTS AT A GLANCE

Like the TV and VCR, the compact disc player can be operated via remote control. The system-control circuits can be operated with the remote or operated separately with buttons found on the front of the disc players. Usually, the remote-control receiver circuits operate within the control system.

The remote transmitter is constructed like the TV and VCR units. The remote transmitter can be a separate unit or it can be found in a digital command center (Fig. 29-1). Most of the buttons on the remote transmitter are like those found on the front of the CD player. The CD player can be operated from the remote, by itself, or by a combination of the two. The remote transmitter operates within the infrared spectrum.

The RCA Digital Command Center (MCD-141) controls practically all of the functions of the compact disc player. Some of these complex functions are track and index search,

FIGURE 29-1 **The hand-held remote transmitter for the Magnavox CDC-745 CD player.**

memory programming, and programmed playback. Also, when the compact disc player is turned on, the remote-control system automatically turns on the MSR-140 stereo receiver and chooses the compact disc player as the audio source.

Remote-Control Operations

A typical remote-control infrared transmitter has on/off, open/close, A-B, display, set/check, clear, play, pause/stop, positive +, negative –, fast-forward, and fast-reverse buttons. The more-complicated remote might have up to 24 different operations. One large IC with corresponding diodes might control all of these functions (Fig. 29-2).

AUTO REMOTE-CONTROL FUNCTIONS

Operate the remote-control unit in the auto while pointing it at the remote-control sensor on the in-dash auto player. Usually, the remote-control unit can be used up to 6 meters in a straight line from the in-dash player. This distance decreases when the remote-control unit is operated at an angle from the remote-control sensor.

Simply point the remote-control unit at the remote sensor when operating it. The in-dash player might not function if obstacles are between the remote-control sensor and the remote, so operate the control unit from directly in front of the in-dash CD player.

Do not press the operation button on the in-dash player and the remote control at the same time. This will cause a “miss” operation. The remote-control operation can be impaired if the remote sensor on the in-dash player is exposed to strong light (such as sunlight).

FIGURE 29-2 **A TV and VCR cassette recorder infrared remote transmitter.**

The Remote Transmitter

The remote transmitter is a hand-operated unit that controls at a distance most operating functions of the compact disc player. Most remote-control transmitter circuits are quite simple and revolve around one IC component. The unit is battery-operated from two or three AA batteries. The infrared-emitting LED (D7) sends out the transmitting signal. Some remote transmitters might have more than one emitting LED. Most transmitters are crystal-controlled (10X). Several diodes are used in the pushbutton switching network (Fig. 29-3).

REMOTE-CONTROL PROBLEMS

Often, the remote-control problems are quite simple. Defective batteries and broken components cause most remote-control problems. First, replace the batteries or test across each battery with the DMM or VOM (Fig. 29-4). Press the On button while making battery voltage tests. A weak battery might produce weak or no remote operation. Always replace the batteries with heavy-duty types. Be sure that the new batteries are installed correctly.

Check the battery terminals for poor connection. Inspect each battery terminal spring connector. Clean it with cleaning spray and a cotton swab. If the battery contacts are excessively corroded, clean them with a pocketknife and sandpaper. Like all contacts, a pencil eraser can clean dirty battery contacts. Inspect the battery terminals for torn or loose wires.

If the remote-control transmitter still does not operate on any given button, suspect that a switch button is dirty. Remove the bottom cover to get to the transmitter mechanism. The unit can be held together with screws or a plastic clip. Often, the top cover must be removed to get to the switch assembly. Remove the screws or clips and loosen the PC board from the pushbutton assembly. Clean the switch contacts with cleaning spray or TV tuner lube. Press the plastic tube down into each switch area. Some of these are difficult to access if they are sealed buttons (Fig. 29-5).

Rough handling of the remote unit could "kill" it or cause it to operate intermittently. If the remote is dropped or thrown, the PC board or a component might break. Inspect the PC board for broken connections or components. Resolder each connection to locate the intermittent problem. Check each switch and wiring connection with the meter. Inspect the transistor and LED terminals for broken or loose leads.

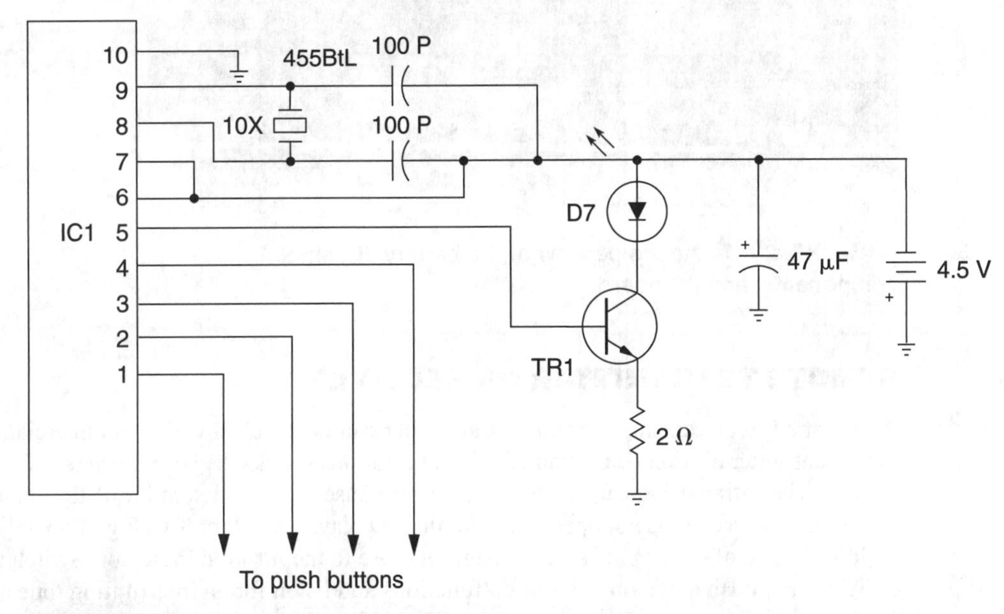

FIGURE 29-3 A typical remote control with buttons, an IC, a transistor, an infrared diode, and batteries.

FIGURE 29-4 Many different types of batteries are used in remote-control units.

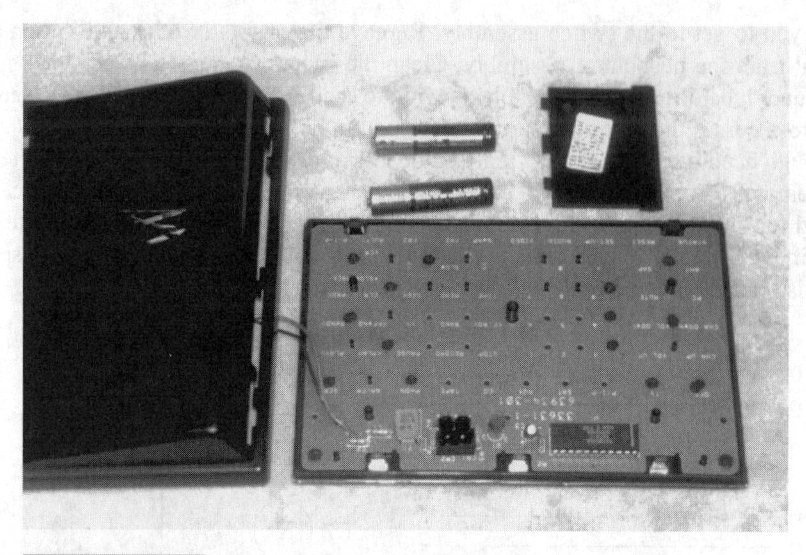

FIGURE 29-5 An inside view of the battery, IC, small components, and PC board.

REMOTE TRANSMITTER OR RECEIVER

After battery replacement, the remote transmitter can be checked with a commercial remote transmitter checker or the homemade indicator that checks the laser beam (see Chapter 23). The infrared indicator can be taken on house calls to determine if the remote transmitter or receiver is not operating with the CD player, VCR, or TV (Fig. 29-6).

Hold the end of the CD infrared transmitter close to the infrared indicator. Switch the indicator on. Push down on any button functions and listen for an interrupting tone and look at the LED on the infrared indicator. The infrared photodetector is located at the flat end. Move the remote control around until you get the loudest signal. Now check each function button; each one will produce a tone and a light when pressed. Suspect that a button contact is dirty or poor if the tone and light are intermittent. Check the receiver section within the control system of the CD player if the light works, but the receiver indicator doesn't.

FIGURE 29-6 Testing a universal remote with an infrared checker.

DEFECTIVE INFRARED TRANSMITTER

After replacing the batteries and checking the infrared transmitter on the indicator with the remote transmitter not operating, suspect that a transistor, diode, or IC is defective; components are broken; or that PC board wiring is cracked. Check for 3 V on terminals 8 and 9 of IC101 (Fig. 29-7). If you find no voltage, check the battery wire terminals and wiring connections between batteries and the IC pin numbers. Test each transistor and diode with the diode test of the DMM or transistor tester. The infrared-emitting diode can be checked with the fixed diode test. Do not be surprised if the diode resistance is greater than 1 kΩ. A leaky emitting diode will show leakage in both directions.

To check the IC, take voltage and resistance on the IC terminals to common ground (usually the negative side of the battery). Very low voltage at pins 8 and 9 might indicate that an IC is leaky. Very low resistance measurements from each pin to ground might help you to locate a leaky IC. Remove the IC pin with solder wick and take another resistance

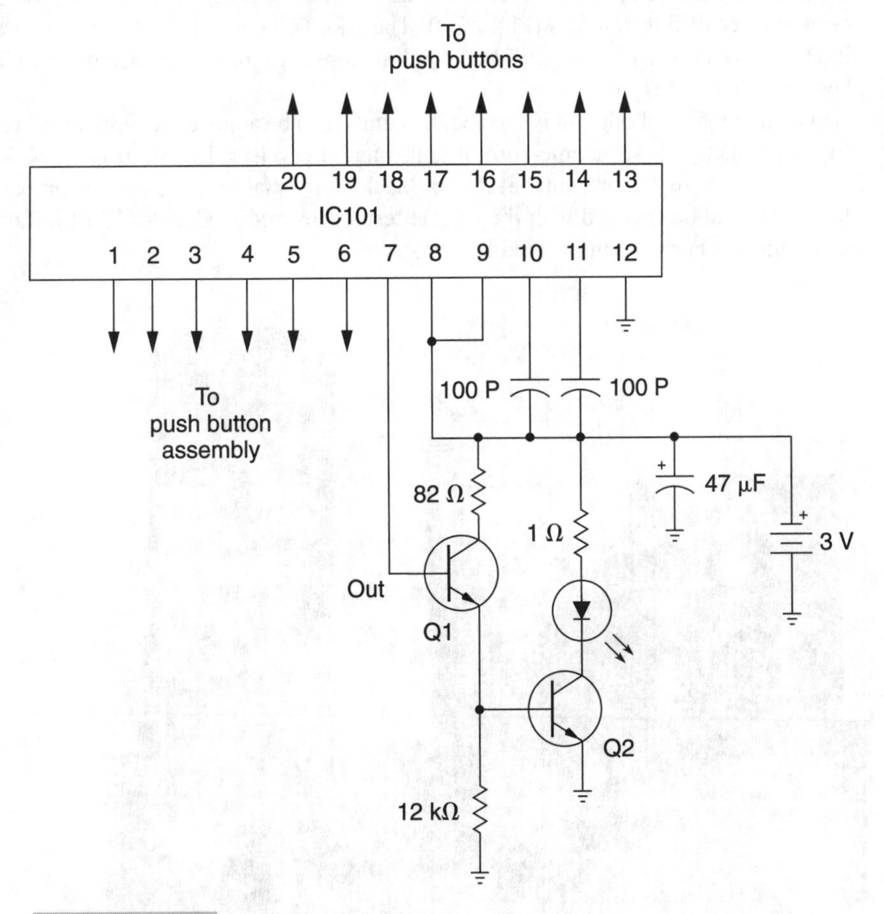

FIGURE 29-7 **Checking the infrared remote components with voltage, resistance, diode, and transistor tests.**

measurement on the 2-kΩ scale. Often, resistance measurements less than 1 kΩ with the pin connection isolated indicate that an IC is leaky. Check each IC pin number function and compare to the button that is not working for possible trouble. Replace each IC with one that has the original part number. Also, check each defective switch function and tie them to the respective switch on the schematic diagram.

Like the TV remote control, the CD player remote can be sent in for repair at various service depots. Many of the manufacturer CD player service centers will exchange or repair the remote transmitter. Check the TV module repair and exchange centers if they can repair the defective remote. Sometimes, if the remote is broken or is too costly to repair, a new remote transmitter is the only solution.

UNIVERSAL CENTRAL CONTROL REMOTE

The universal remote-control transmitter can operate most audio and video components. The General Electric RRC500 does the work of three remotes, but the RRC600 model does the work of four remotes (Fig. 29-8). The RRC600 controls up to four infrared audio/video products, with over 200 key combinations, program sequencing, an LCD display, and a low-battery indicator.

This remote, like all other universal control units, can be adjusted to work on all remote-control products. These remote-control units can be purchased at electronic TV dealers, hardware stores, mall stores, and at Radio Shack. Universal remotes operate on batteries. The remote can be checked with the remote tester (covered in Chapter 23 of infrared detector/indicator) or with an infrared laser tester.

FIGURE 29-8 A universal remote control that operates TVs, VCRs, receivers, and CD players.

FIGURE 29-9 **Checking remote-control transmitters with an infrared laser power meter.**

INFRARED POWER METER

The infrared power meter that you use to check the optical laser for light can be used to test remote-control transmitters. Although, the three scales on the power meter are designed for laser power ratings, start with the low scale of 0 to 0.3 mW. Check the best scale for infrared measurements of infrared transmitters (Fig. 29-9).

Place the remote three inches away from the pickup probe and register the measurement on the power meter. Check a new remote against the infrared power meter. Likewise, check a powerful remote-control unit. By making comparison tests with new remote, the defective remote measurement can be compared to the new infrared measurements. Move the probe back and forth to get the best reading.

INFRARED REMOTE RECEIVER

The remote-control receiver sensor is located on the front panel of the CD player (Fig. 29-10). The infrared sensor receives the transmission from the handheld remote transmitter. The weak signal is fed to a preamp stage, decoder, and system-control IC (Fig. 29-11). The infrared sensor is usually placed in a shielded container. The remote-control light sensor in the Denon DCD-1800R player feeds the remote signal to the remote-control decoder board (Fig. 29-12). IC001 receives both signals from the pushbutton keys and the remote receiver signal in the control system operation.

Onkyo DX-200 remote-control circuit After the remote-control input from the remote-control sensor (photosensitive diode D831) has been synthesized, detected, amplified,

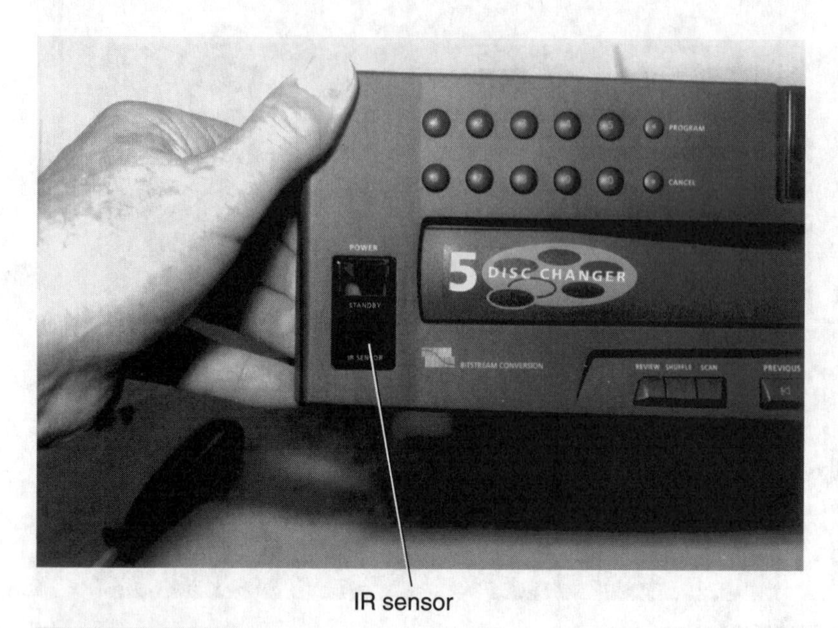

FIGURE 29-10 **Here, the IR sensor is located on the front panel of a five-disc changer.**

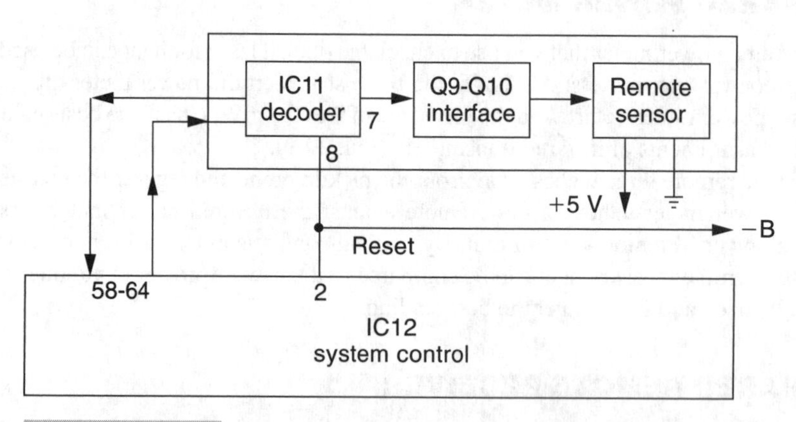

FIGURE 29-11 **The infrared-controlled transmitter signal activates the sensor in the IR receiver unit of the CD player.**

and rectified by the remote-control circuit, the signal is passed to the microcomputer. The remote-control circuit is incorporated in the IC BA6340 (Q501) with the synchronizing frequency (38 kHz), gain, and time constants determined by externally connected elements (Fig. 29-13).

The L501 core is adjusted to the point of maximum sensitivity, with respect to the remote-control input. The output (pin 1 of Q501) is switched to high level (+5 V) when there is no input signal, while data output is obtained at CMOS level when a remote-control input is applied. The shield around the remote sensor (D831) is very important. If the sensor is not properly shielded, noise might appear at pin 1 output when no input signal is applied, thereby preventing the remote-control mechanism from functioning correctly. The remote-control code is decoded by the main microcomputer (Q281).

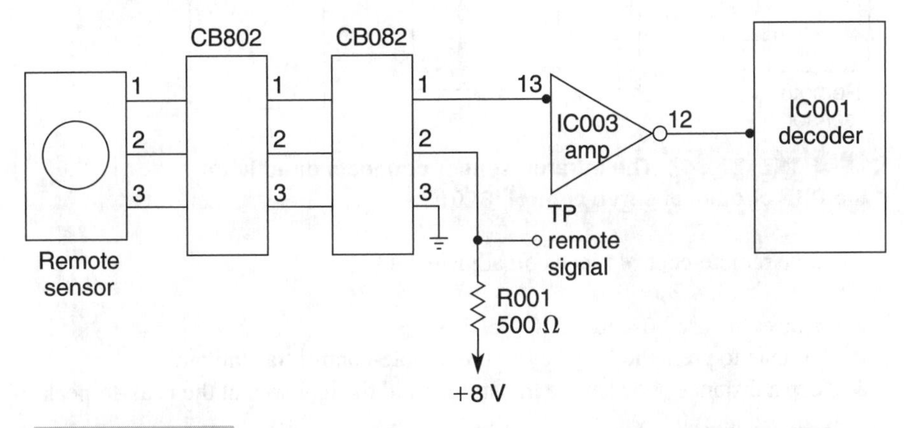

FIGURE 29-12 **The remote-control light sensor that connects to IC003 amp and decoder IC001.**

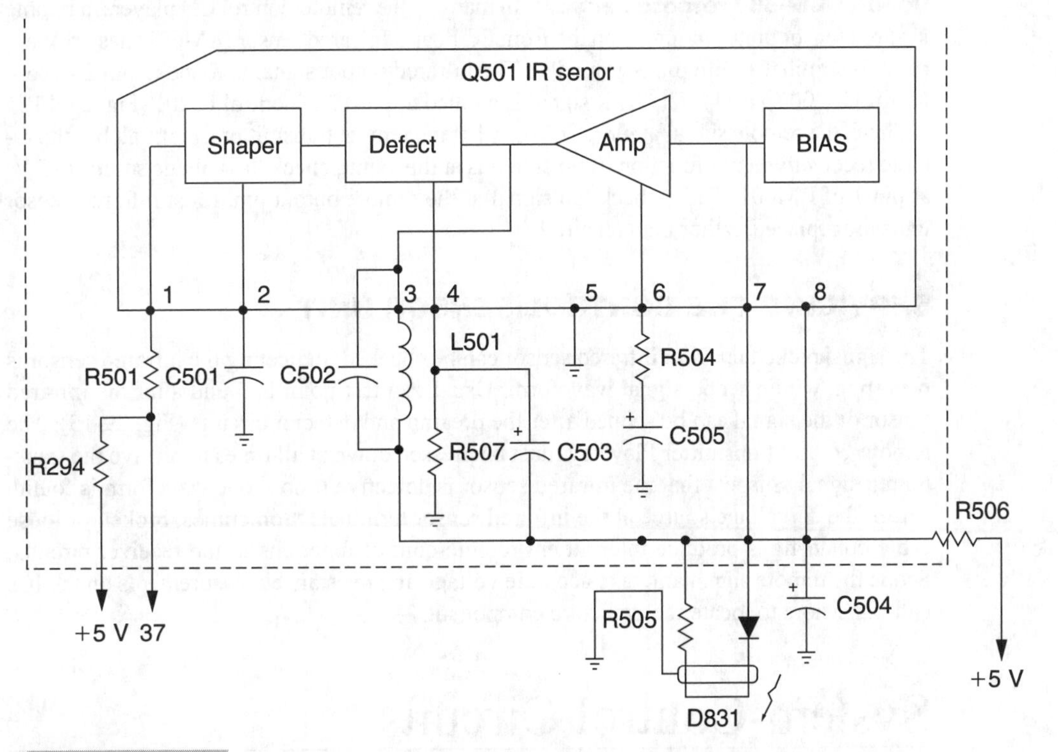

FIGURE 29-13 **The remote-control sensor circuit in this Onkyo CD player.**

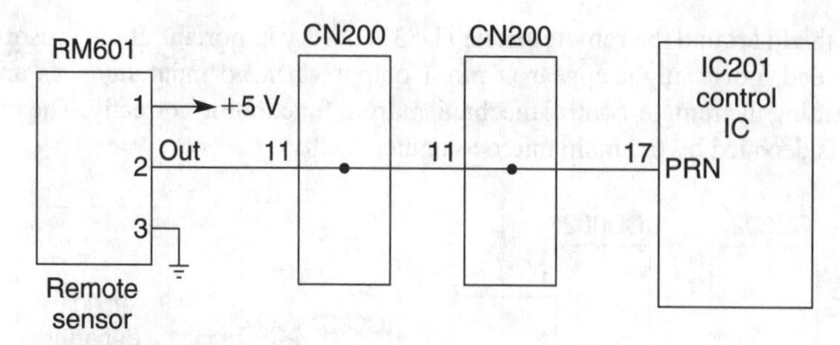

FIGURE 29-14 The infrared sensor connects directly to the PRN circuits of servo control IC201.

Onkyo remote-control tuned coil adjustment

1. Connect the oscilloscope to IC Q501 pin 3.
2. Continue to press the Play key of the remote-control transmitter.
3. Keep a distance between the transmitter and the unit so that the peak-to-peak of waveform becomes 0.5 V.
4. Adjust L501 so that the waveform attains maximum output.

Denon DCM-560 remote sensor In many of the remote-control CD players, a remote infrared sensor unit is mounted on the front PC board. Infrared sensor (RM601) has +5 V applied to terminal 1 with pin 3 grounded. The infrared output signal is found at pin 2 to connector CN200 (pin 11). The NPR signal is applied to pin 17 of control IC201 (Fig. 29-14).

Check the remote signal at pin 17 of IC201 if the remote transmitter is normal, but the remote receiver will not function. If no signal is at this point, check the voltage source (+5 V) at pin 1 of RM601. Then, check the signal at the remote output pin. Most infrared sensor units are replaced, rather than repaired.

SERVICING THE CONTROL SENSOR UNIT

The signal picked up from infrared sensor can be checked, indicating the infrared sensor is operating, with a scope signal waveform. Usually, a test point is found after the infrared sensor or the signal can be scoped after the preamp and detector circuits (Fig. 29-15). The remote-control transmitter Play key must be pressed down at all times to receive the transmitted signal. Suspect that the infrared sensor is defective if no scope waveform is found. Check the dc voltage source at the infrared sensor terminals. Sometimes, broken or loose cable connections produce intermittent remote-control functions in the receiver circuits. Scope the remote signal and take accurate voltage and resistance measurements on the ICs and transistors to locate the defective component.

System-Control Circuits

The simple system-control circuit consists of a decoder-control IC, display, and key input boards; in larger CD players, the system-control circuits consist of two microprocessors, display, and keyboards. In the RCA MCD-141 player, two microprocessors are in the system-control circuits (IC601 and IC901). IC601, the system-control microprocessor, is responsible for controlling the internal operation of the player (Fig. 29-16). The LMT, LID, and CHK switches inform the microprocessor of the location of the disc tray and the laser pickup assembly. IC601 also outputs time-shared scan/data information, which consists of keyboard scan signals and fluorescent display data to light the front panel.

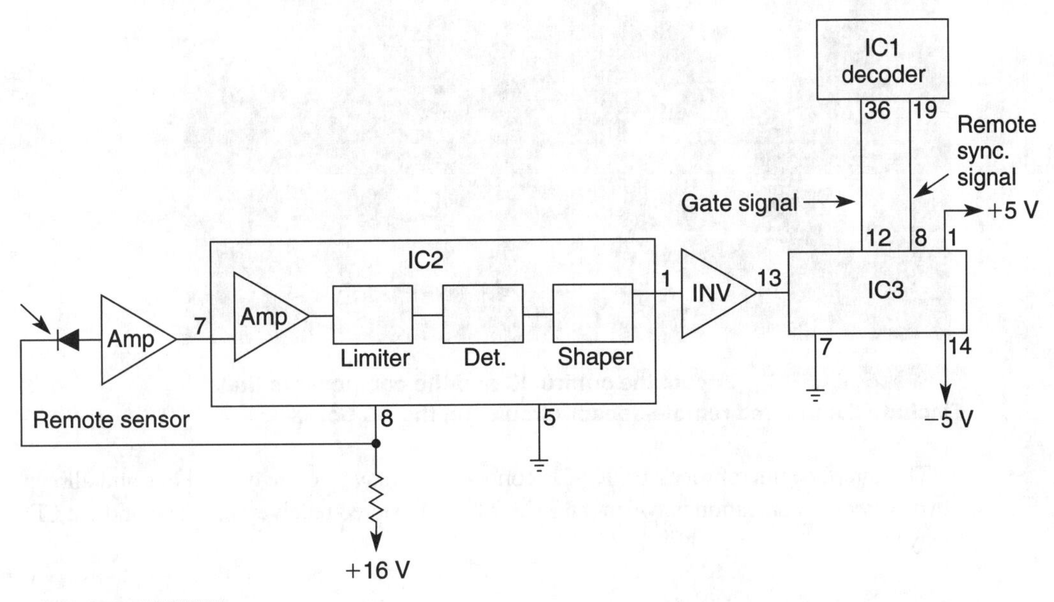

FIGURE 29-15 Check for remote signal at IC2 amp to determine if the infrared sensor is normal.

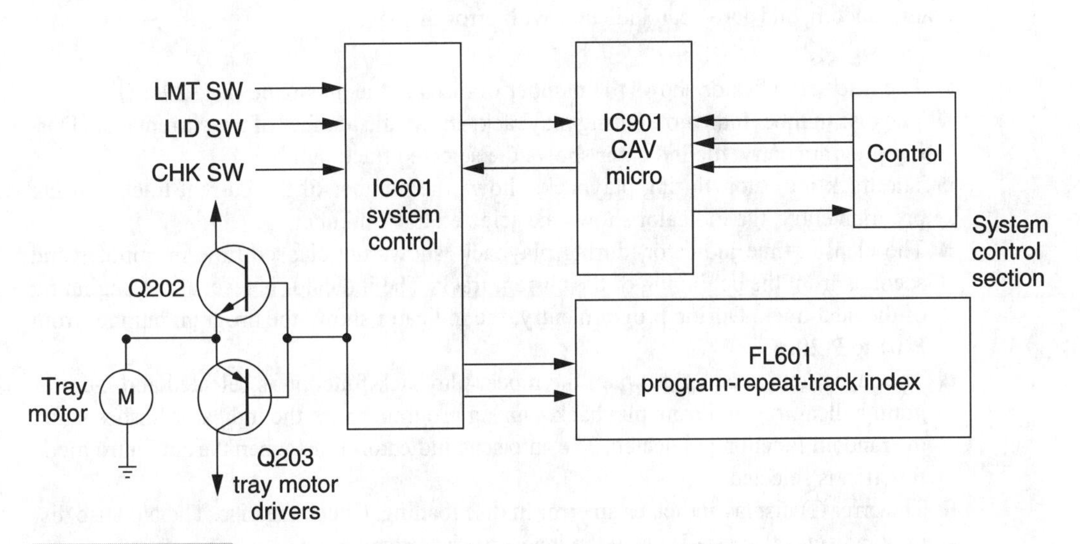

FIGURE 29-16 The keyboard, remote-control sensor, CAV microprocessor, and system-control IC601 control the display and remote circuits.

FIGURE 29-17 **Locate the control IC and the components that include the infrared remote-sensor circuits on the PC board.**

The interface microprocessor, IC901, connects the player to the control bus and allows two-way communication between the RCA MSR140 stereo receiver/amplifier and the CD player.

Mitsubishi M-C4030 display The Mitsubishi M-C4030 remote-operated tabletop automatic changer has a disc set, disc number, track number, elapsed time, reject, program, random, and auto-scan indicator with error display.

- The disc-set indicator shows the number of discs in the magazine (Fig. 29-17).
- The disc number indicator, during playback, shows the number of the current disc. During program entry, the indicator shows the selected track number.
- The track indicator, during playback, shows the number of the current track. During program entry, the indicator shows the selected track number.
- The elapsed time indicator, during playback, shows the elapsed time in minutes and seconds from the beginning of the current track. The indicator is reset at the beginning of the next track. During program entry, the indicator shows the program number from P-01 to P-20.
- The repeat indicator lights when the repeat playback function is selected and the program indicator is lit during playback. During program entry, the indicator flashes while the random function is selected. The auto scan indicator lights when the auto intro mode function is selected.
- The error (I) display indicates an error in disc loading. Check the disc. The No Disc display indicates that no disc is in the magazine or turntable.

TYPICAL CONTROL PROCESSOR

The typical microcomputer control processor can provide a gate signal and incoming receiver signal for the remote-control circuits. The display output data can be supplied to the fluorescent display board. Data is transferred to and from the servo and the signal-control section (Fig. 29-18). The control processor can provide loading in and out of the keyboard matrix and display data. The control processor can provide loading in/out and key matrix in from the keyboard. Data transfer in and out is on terminals 1, 33, and 34. The external clock oscillator terminals are 16 and 17. The control processor provides loading motor control data on terminals 9 and 10.

Crucial waveforms and voltages on the microcomputer control processor can determine if the IC is leaky or defective. Check for the correct supply voltage (+5 V) at terminals 41 and 42. Display the data waveform at pin 25, transfer the data waveform at pin 37, load the

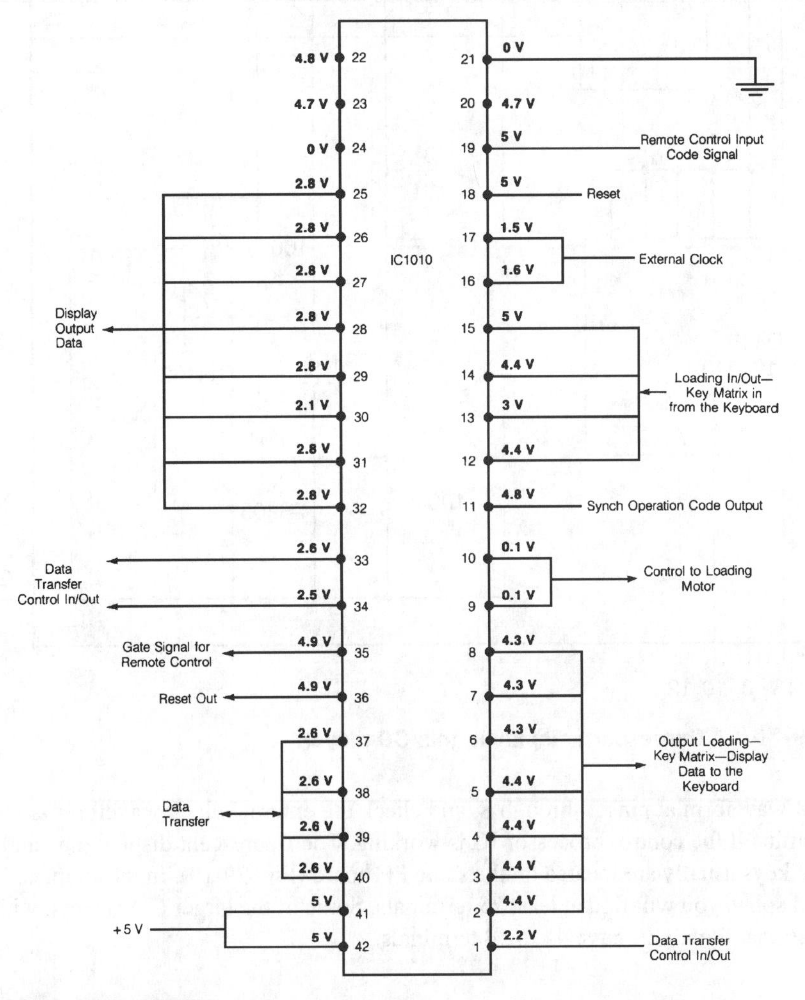

FIGURE 29-18 Check each pin for the correct voltage with the input signal at pin 18 of IC1010.

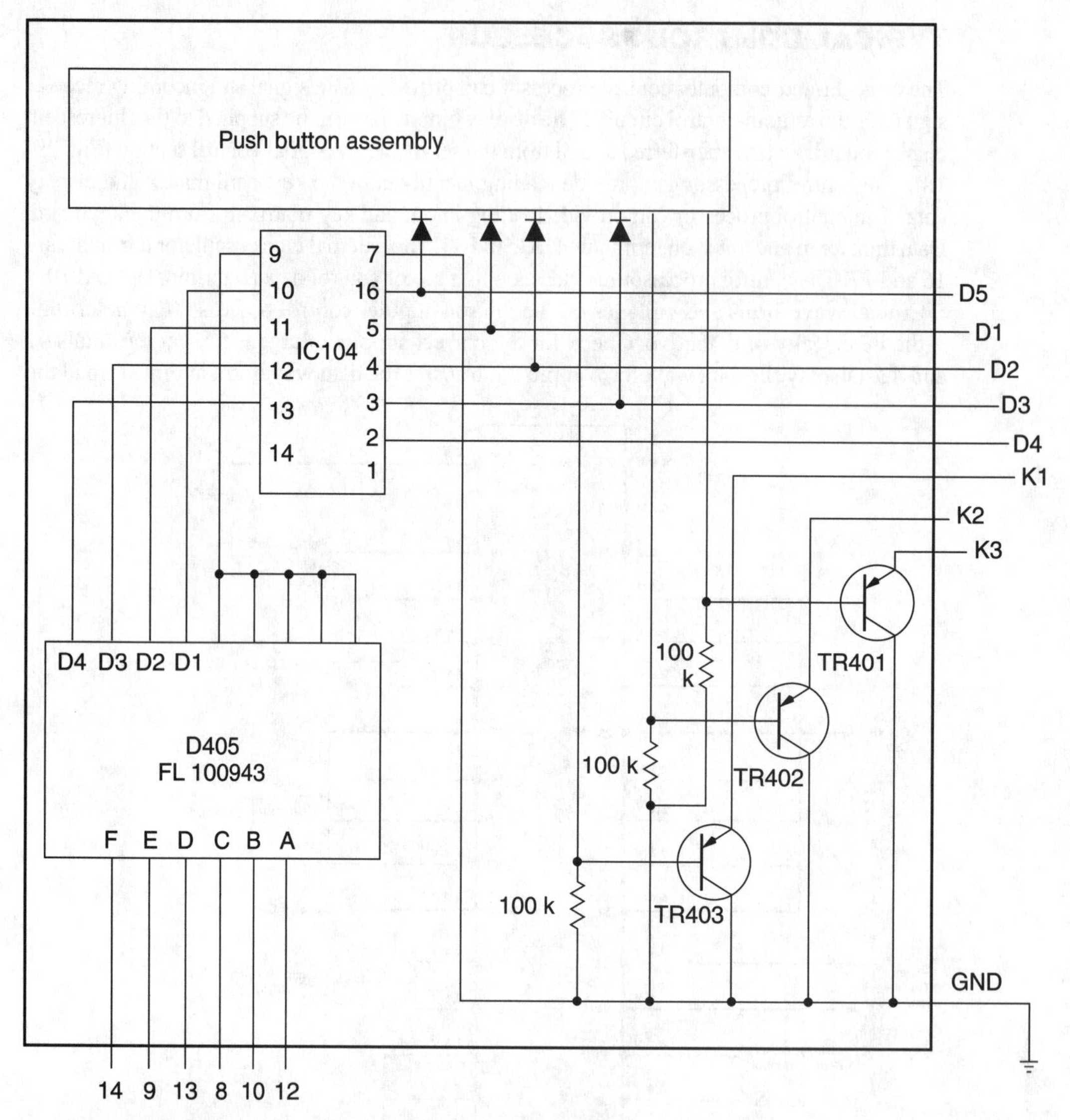

FIGURE 29-19 A fluorescent display in this CD player.

in/out waveform at pins 2 through 8, and check the external clock waveform at pin 17 to determine if the control processor IC is working. The fluorescent display tube and operational keys usually are located on the same PC board (Fig. 29-19). In the simplest fluorescent display, you will find at least 25 terminals. Some of the larger CD players, with many operational functions, have 44 to 54 terminals.

TROUBLESHOOTING DISPLAY FUNCTIONS

Suspect that a circuit or switch is defective if one function light does not turn on within the fluorescent display tube. Very seldom does the fluorescent display cause any service problems. Check for poor button contacts or jamming of the tact switch. Clean the switches and replace them with original-type parts. Inspect the flat or cable wires from the chassis to the display-key board (Fig. 29-20). Doublecheck the socket cable connections. Check for

FIGURE 29-20 The connecting cable wires to the display board in the Sanyo CP500.

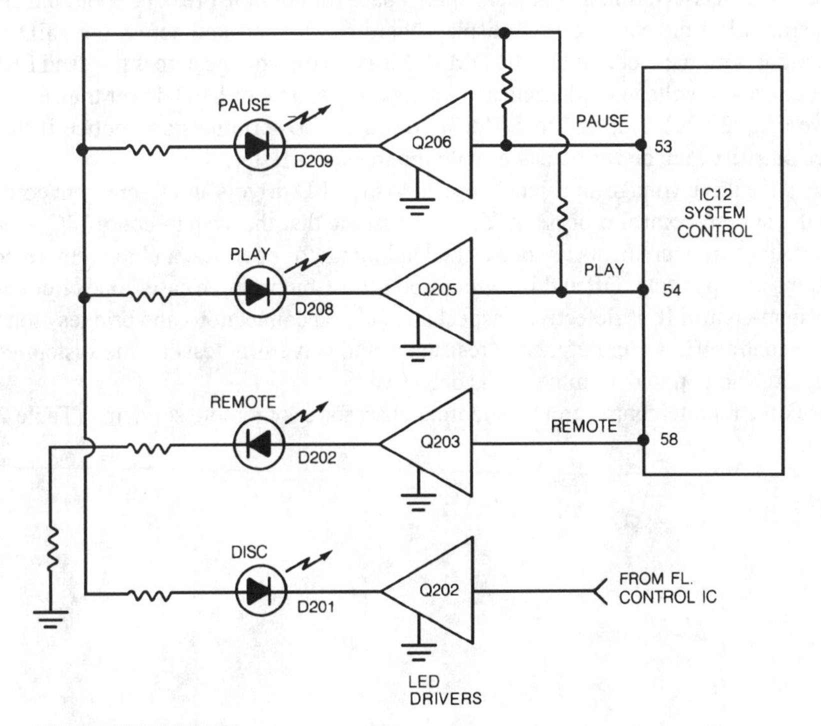

FIGURE 29-21 IC12 controls the Pause, Play, Remote, and Disc LEDs with IC drivers on the key/display board.

TABLE 29-1 THE REMOTE-CONTROL TROUBLESHOOTING CHART

SYMPTOM	SERVICE
Remote control dead	Check the batteries in the remote. Check the remote against the infrared indicator or the laser power meter. Substitute another remote transmitter.
Intermittent operation of remote	Suspect loose batteries. Remove the batteries and clean the terminals. Bend the terminals out if the remote has been dropped several times. Inspect for a broken PC board. Check for loose or broken parts.
Weak reception	Check the batteries. Replace all batteries if only one is bad.
Remote okay, no action	Suspect the infrared receiver or sensor. Take a waveform at the TP points in the receiver. Check the voltages on the IC. Test the transistors inside of the receiver. Replace the sensor unit.

poorly soldered or broken solder bridges around the cable area. Make a continuity check between the chassis terminal and the display board for possible break or poor connections.

Separate LED indicators on the display board can be checked across the LED as any fixed diode with the diode test of the DMM. Measure the voltage across the dead LED. No voltage or a low voltage might indicate that a power source or LED driver transistor is defective (Fig. 29-21). Apply 3 to 5 Vdc across the LED terminals and notice if the light comes on. Test each driver transistor with the transistor tester.

Check the drive voltage and signal applied to the LED drivers and fluorescent control IC from the system-control processor. You can assume that the system-control IC is normal if the voltage measurements are okay, the loading motor operates, and the remote-control functions are operating. Often this assumption is true, but in some cases, one circuit within the system-control IC is defective. Inspect all soldered connections and bridges around the system-control IC. Accurate voltage resistance and waveform tests on the system-control IC should help you to determine if it is defective.

Check the remote-control troubleshooting chart for easy remote servicing (Table 29-1).

30

SERVICING PORTABLE CD AND BOOM-BOX PLAYERS

CONTENTS AT A GLANCE

The portable CD player was designed for the ardent, on-the-go music lover. The portable player operates from batteries or a battery pack (Fig. 30-1). Some CD players operate from the ac power supply of a dc power jack. The CD player can be plugged into the power adapter with audio-output line jacks. You can listen through a pair of stereo headphones with either battery or ac operation.

Most portable CD players are top loaded; they don't have the loading motor that is in the larger tabletop models. Pushing the Open button makes the cover snap up. After the disc is loaded, push the top cover and it locks in place.

Many of the circuits used within the portable CD player are the same as in its big brother. The big difference is space. The parts are physically smaller and they are jammed together on several PC boards. Sometimes the components are hard to access. Extra patience and slower service methods must be used to prevent damaging and breaking small components. Be careful when trying to repair the portable compact disc player.

The Boom-Box CD Player

Today, the boom-box player can have a compact disc player, besides the cassette player and AM/FM/MPX radio circuits (Fig. 30-2). The boom-box CD player can have preamp circuits that switch into the regular stereo cassette and radio amplifier circuits. The outputs might include internal speakers and separate stereo line-output jacks (Fig. 30-3).

Because most CD players and boom-box players are loaded at the top of the plastic cabinet, no loading motor is used. Likewise, the CD mechanism can be hung from the top of the cabinet, on a separate PC board (Fig. 30-4). Because both units are mounted in such a small space, surface-mounted devices (SMD) are used on both sides of the PC chassis.

FIGURE 30-1 Portable and boom-box disc players are tag-along electronic products.

FIGURE 30-2 The boom box might include a cassette player, AM-FM stereo receiver, and CD player.

FIGURE 30-3 The portable boom-box CD player might have line output and headphone jacks with stereo speakers.

Test Equipment and Tools

Basically, the same test equipment and tools required for the larger CD players are used. Laser power meters, oscilloscopes, and test discs are required. The eccentric adjustment

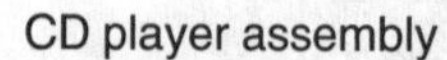

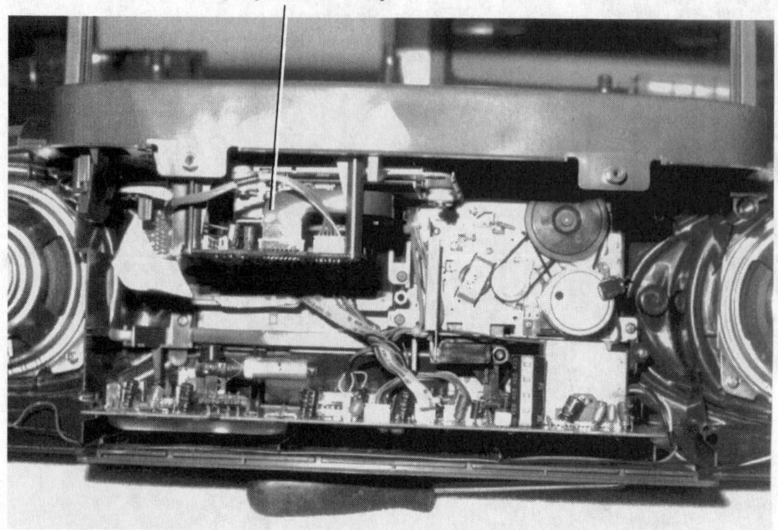

FIGURE 30-4 **The CD player is located at the top of the boom-box player.**

FIGURE 30-5 **Very small screwdrivers are needed to remove the outside covers.**

driver, small screwdriver set, and small-point soldering irons are needed for crucial adjustments and cover removal (Fig. 30-5).

You might find the PC wiring to be very thin—especially at the IC processor connections, where you will need a small point tip on your soldering iron.

Safety Requirements

The same safety requirements for working around the laser optical system and chassis apply in the larger table-top CD player. When defeating the top side door interlocks, do not look directly at the laser beam. Remember, the optical lens assembly is open and it points upward under the disc; the lens assembly is usually covered with the flapper assembly.

A wrist band should be used while servicing the portable CD player on a ground-conductive mat. Be sure that the soldering iron is also grounded. Less than 10 Ω should be measured between the test equipment, soldering iron, player, and common ground wire. Always replace marked components with original-type parts. The use of substitute replacement parts that do not have the same safety characteristics as specified in the parts list might create shock, fire, and other hazards (Fig. 30-6).

Several personal safety precautions should be made while operating the player or when the set is not to be used for an extended period of time. In the latter instance, remove the ac power adapter from the outlet. Do not leave the portable player near a hot radiator or hot air ducts, and don't place it directly in the hot sun. Keep the player away from excessive dust, rain, or any kind of moisture. Remove the player from the auto or outside temperatures of below 40°F or above 95°F. Keep foreign objects out of the safety slot (interlock) to prevent the laser beam from coming on when the lid is opened. Replace the top cover with an original-type cover so that the interlocks and laser beam will not cause damage to the eyes.

Top Loading

Most portable and boom-box CD players load the disc at the top of the machine. The top lid is lifted on the portable CD player to load the disc and, at the same time, cut off the laser beam with a prong on the top lid that engages a switch. In other players, a small plastic button will be engaged by the top lid, shutting down the CD player. This prevents the operator from eye damage because the laser lens assembly is pointed upward.

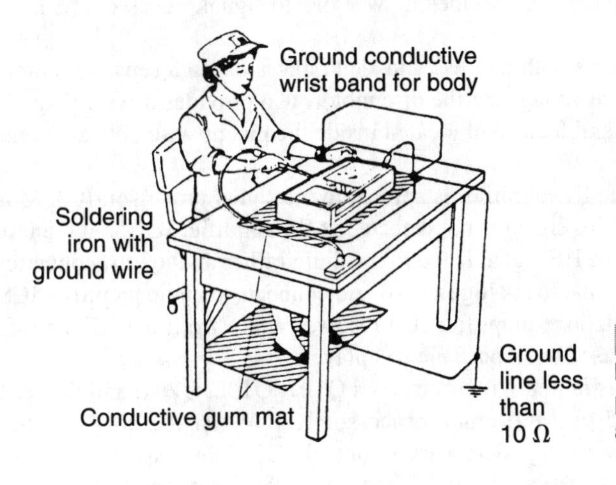

FIGURE 30-6 Ground the body, soldering iron, and CD player with an arm wrist band and a conductive mat.

FIGURE 30-7 **Top loading is also used in boom-box players.**

A boom-box plastic lid in loading is disengaged by pressing down on lid and it pops up (Fig. 30-7). The disc can now be loaded with a rotating hold down ring at the top of lid. The boom-box player will not play until the lid is closed.

The Block Diagram

The block diagram in Fig. 30-8 shows the actual electronic components in a portable CD player. A block diagram ties all of the components together with the major components listed with a minimum of connecting wires. RF amplifier IC1 receives the laser diode signal from the optical assembly and provides EFM signal to signal processor IC3 and servo control IC2.

Servo control IC2 controls both the disc and sled motors through separate motor drive circuits. IC5 provides drive voltage for the disc motor. IC6 provides drive voltage for the sled motor. The tracking and focus coil located inside the pickup assembly are controlled by servo IC2.

An EFM signal from the RF amplifier is applied to the signal processor, IC3. Most CD players will shut down if this signal is not detected at RF amplifier IC1 or at the laser optical assembly. The EFM or HE signal is fed to the digital filter and to D/A converter IC7. IC7 converts the digital signal to analog (audio) and connects to audio amplifier IC8. The line-output jack and headphone amplifier IC9 receives audio from the AF amplifier. A stereo headphone jack is used in all portable CD players.

The line stereo output is muted with transistors Q101, Q102, Q201, and Q202. These transistors are controlled by MOS microprocessor IC4. When the door switch S3 is open, IC4 does not provide signal to servo control IC2. The display and keyboard switches tie into microprocessor IC4. IC10, the power on reset, connects to the MOS microprocessor.

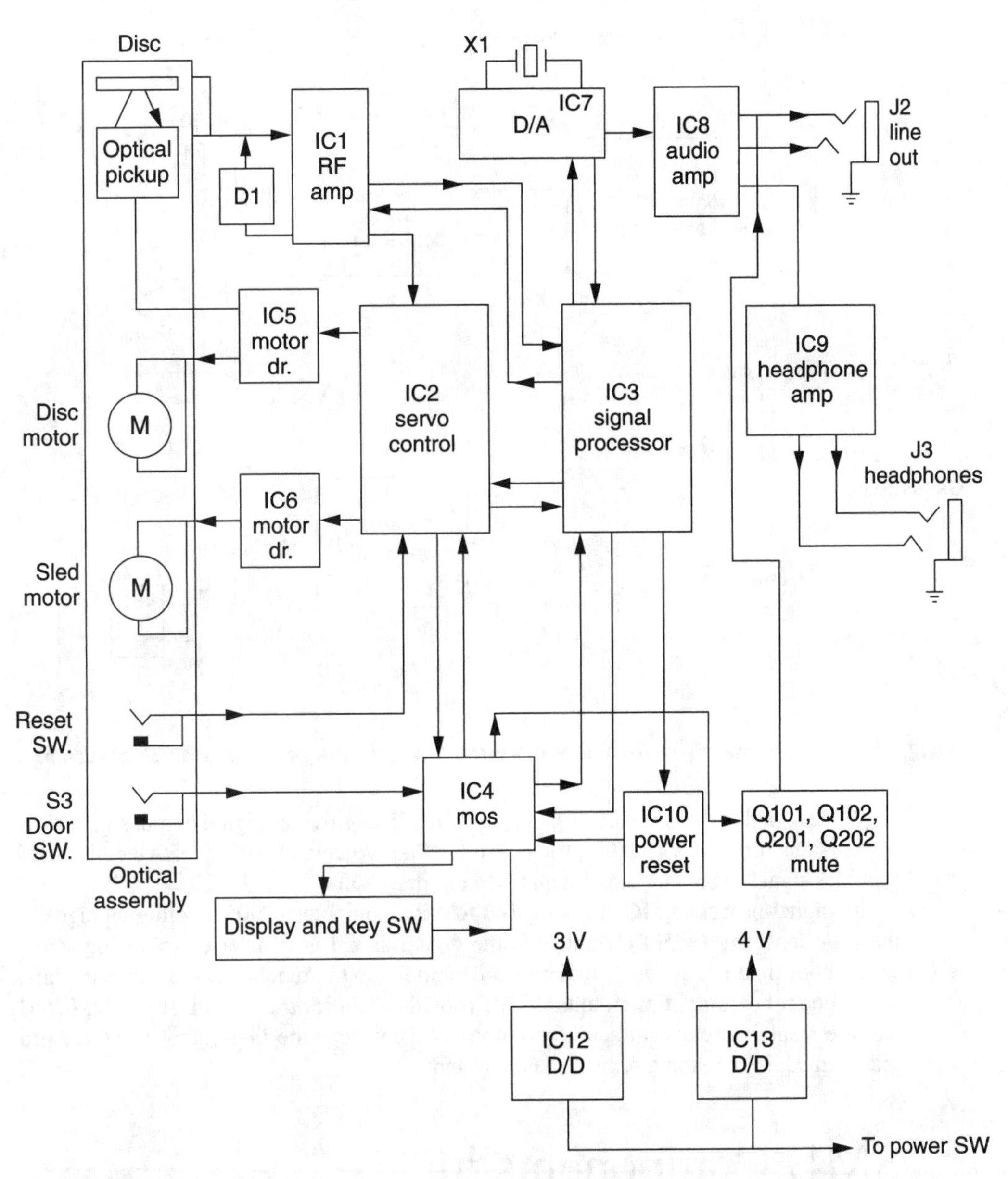

FIGURE 30-8 **A block diagram of a typical portable CD player.**

The power supply consists of two 1.5-V batteries, which supply +3 V to both D/D converters. IC12 is a 3-V D/D converter; IC13 supplies a 4-V output. In the schematic, TP16 indicates the 3-V source and TP17 indicates the 4-V supply.

BOOM-BOX CD PLAYER BLOCK DIAGRAM

The boom-box CD circuits consist of an RF amp (IC501) with an input signal from the optical laser assembly (Fig. 30-9). IC501 amplifies the weak laser signal and applies an RF

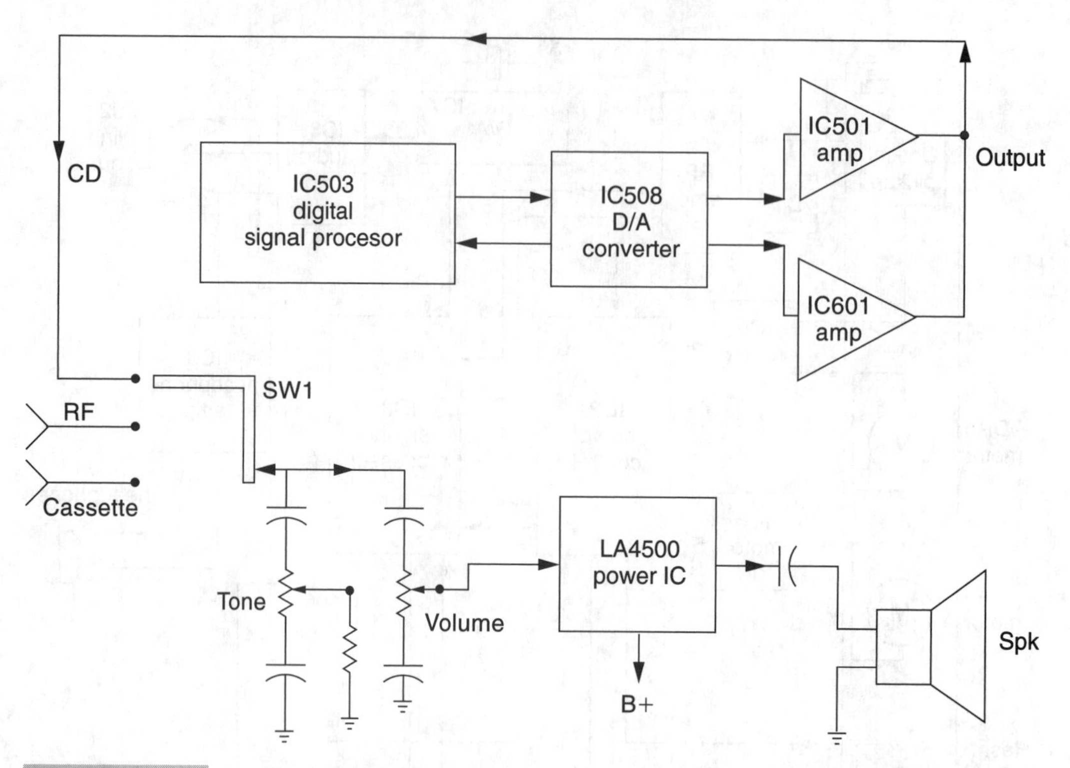

FIGURE 30-9 **Locate the defective audio circuits within the boom-box block diagram.**

and EFM signal to the servo LSI (IC502). Servo IC502 provides digital output signal to the digital signal processor (IC503), focus and tracking voltage to the respective coils, and it provides signal to operate the disc and sled motors.

The digital-processing IC provides data to microcomputer IC505, and digital signal to the D/A converter (IC508). At IC508, the digital signal is transferred to analog stereo, which both channels are amplified by IC601 and fed to the function switch or to the input of the boom-box amplifier circuits. IC505 provides data for the servo LSI IC502, IC503, and the display circuits. In some boom-box players, separate line-output jacks are provided, in addition to the internal speaker system.

SMD Components in Portable CD Players

Like the big-brother home CD player, the portable player has many SMD components on the bottom side of the PC board. Many IC and microprocessors with gull-type wings are soldered directly on the bottom side of the PC wiring (Fig. 30-10). Some portable CD players have a double-sided PC board with standard parts on the top side and SMDs on the bottom.

Besides SMD ICs, transistors, resistors, capacitors, and diodes are used on the PC boards. Check for poor connections or breaks where the part is soldered to the wiring. If the player has been accidentally dropped several times, cracked boards and SMD components can cause intermittents. Test, isolate, and install the new SMD part.

Battery Operation

The portable CD player can operate from house current, energy cells, or alkaline or rechargeable batteries. Most portable compact disc players have a separate battery pack or case to carry the needed batteries. Usually, four or six alkaline C batteries are used in the battery pack. The optical EBP-9LC battery case uses six alkaline C batteries to power the Sony D-14 compact disc player. The cells will power the CD player for 4.5 hours straight, and the alkaline batteries can operate for more than 9 hours with intermittent operation.

The nickel-cadmium rechargeable batteries can be inserted in the battery pack for a longer life span. The charging time and battery life of different nickel-cadmium rechargeable batteries might vary. The charging time of the KR-C-F rechargeable batteries in the Sony battery pack is approximately 15 hours. The fully charged batteries allow approximately 2.5 hours of continuous disc play in the Sony D-14 portable disc player. When the batteries become weak, a BATT indicator shows on the display window of the portable CD player. It's best to replace all batteries with new ones when the BATT indicator begins to flicker.

FIGURE 30-10 **SMD components are on the bottom side of the PC board.**

12-V CAR BATTERY

In some portable CD players, the unit can be powered by the car battery with a special cord. Sony's D-14 player is connected to the car battery with a DCC-120A battery cord that connects the dc 1N9V jack to the cigarette lighter socket of the vehicle. Before connecting, be sure to set the output voltage of the car battery cord to 9 V. Use only the headphones to listen to the CD player when it is powered by the car battery. Often, sound through the speakers will be noisy if the player is connected to a car stereo system.

Suspect poor battery connections when skipping or intermittent operation of the portable CD player is noted. First, clean the dc output plug and the stereo miniplug of the battery case with a dry cloth before connecting it to the player. Inspect the plug and jacks for possible breakage or poor wiring connections. Replace all batteries when the BATT indicator begins to flicker.

Check the battery terminals for corroded or dirty terminals. Always remove all batteries if the compact disc player is not used for a great length of time. Clean the battery terminals with a pocketknife, sandpaper, and cleaning fluid. Wipe the battery terminals and battery connections with cleaning fluid. Inspect the battery terminals for broken wire connections. A poorly soldered battery connection might cause no operation or intermittent operation.

The Power Supply

Most portable compact disc players have a power pack that operates directly from house current. The power pack is plugged directly into the back side of the portable player (Fig. 30-11). The ac power supply might contain a 9-Vdc output jack with a stereo line-output sound jack. Two separate line jacks at the rear of the power supply allow the portable compact disc player to be played through a regular stereo amplifier.

The typical 9-Vac power supply might consist of only a few components: transformer, bridge rectifier, filter capacitor, and a 9-V output jack (Fig. 30-12). Notice that the positive

FIGURE 30-11 Plug in the small power supply when the CD portable is operated indoors.

terminal of the power supply is on the inside. The male stereo jack connects directly to the left and right line-output jacks.

Within Radio Shack's CD-3000 compact disc player, the ac power supply furnishes 9- and 5-V sources (Fig. 30-13). The power switch can be operated at the power supply (S901) or inside the compact disc player with switch S961. Full-wave rectification is produced by diodes D901, D902, D903, and D904. Crucial 9- and 5-V regulation is accomplished with Q901, Q902, Q903, and Q904. The ac power supply board is held in the slanted adapter unit.

A voltage IC/transistor regulator circuit is found within Radio Shack's battery pack (Fig. 30-14). Notice that three different voltage sources are at J931 (9 V, 5 V, and 4 V). A protection thermal switch (S931) is between the battery terminal and the dc jack (J932).

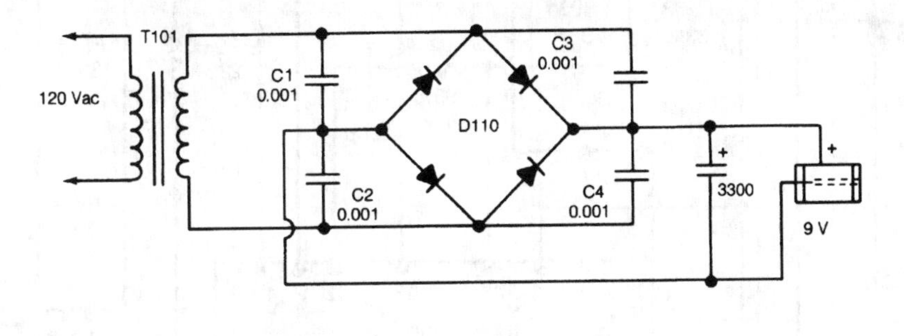

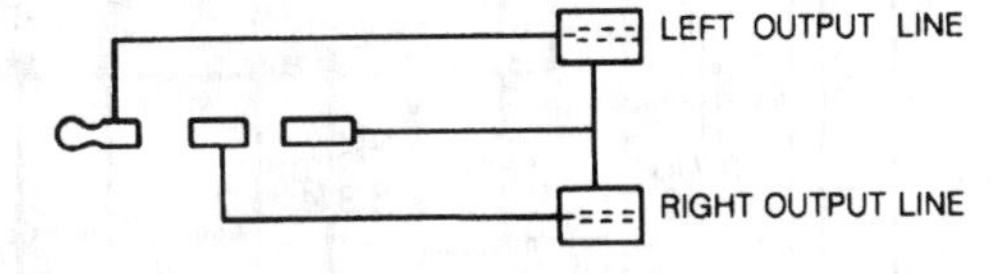

FIGURE 30-12 Only a few electronic components are in the ac power supply.

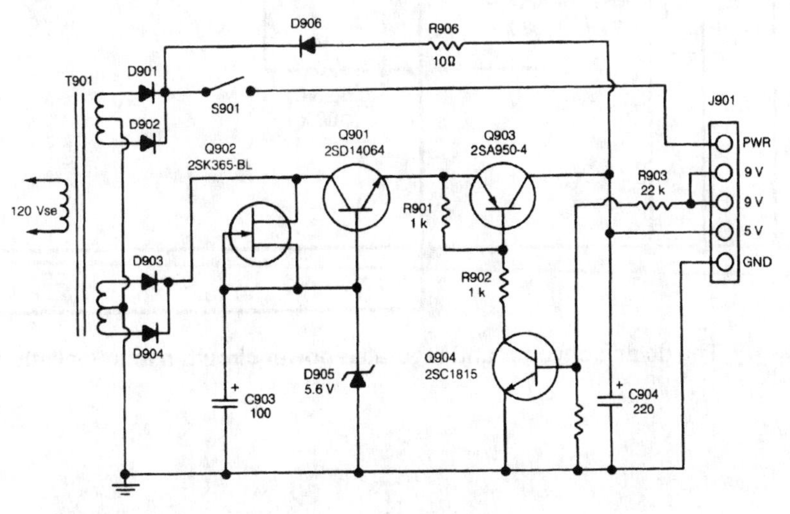

FIGURE 30-13 A dc/dc converter with regulator in the Realistic portable CD player.

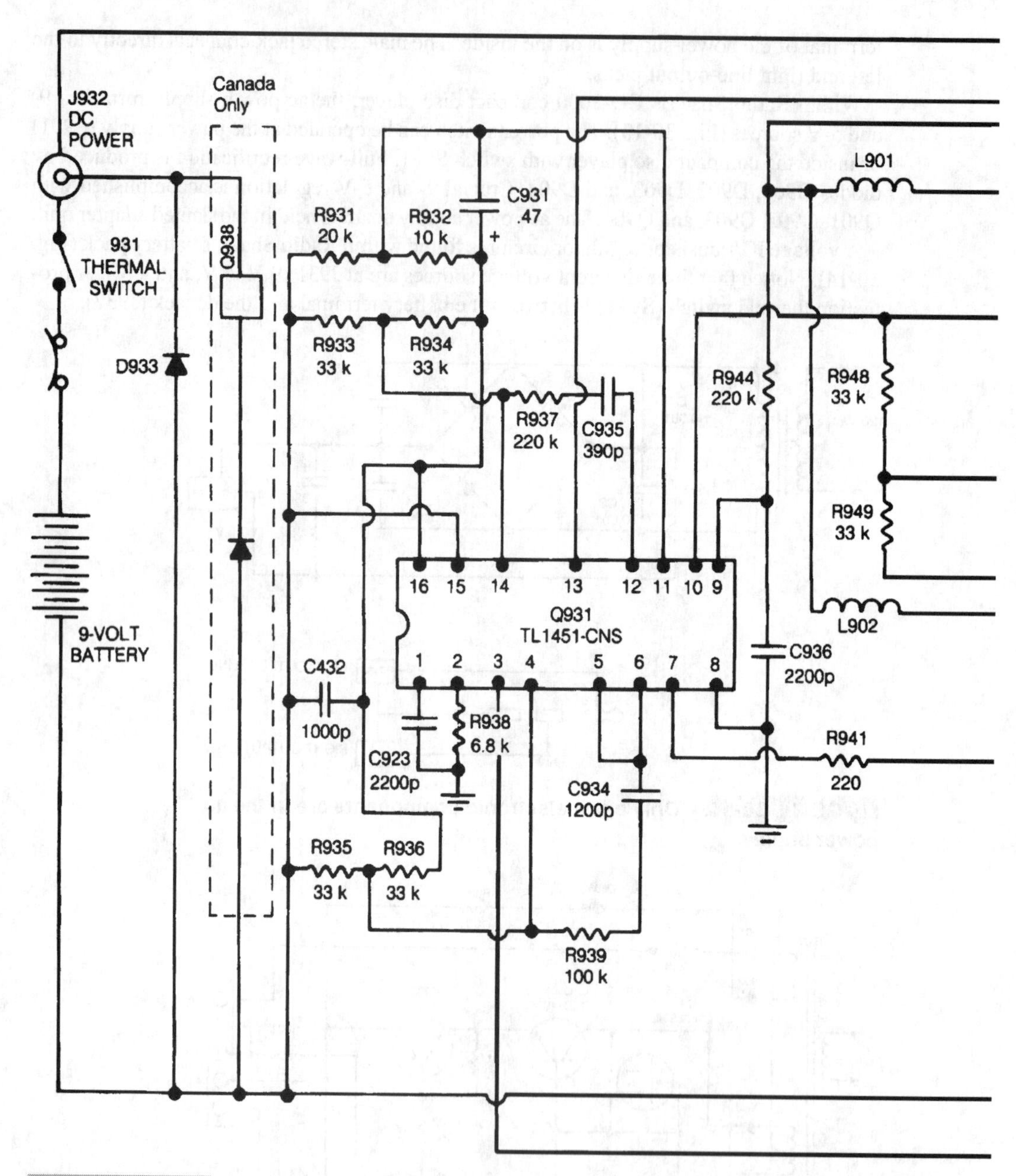

FIGURE 30-14 The dc/dc converter and regulated power circuit in this Realistic model.

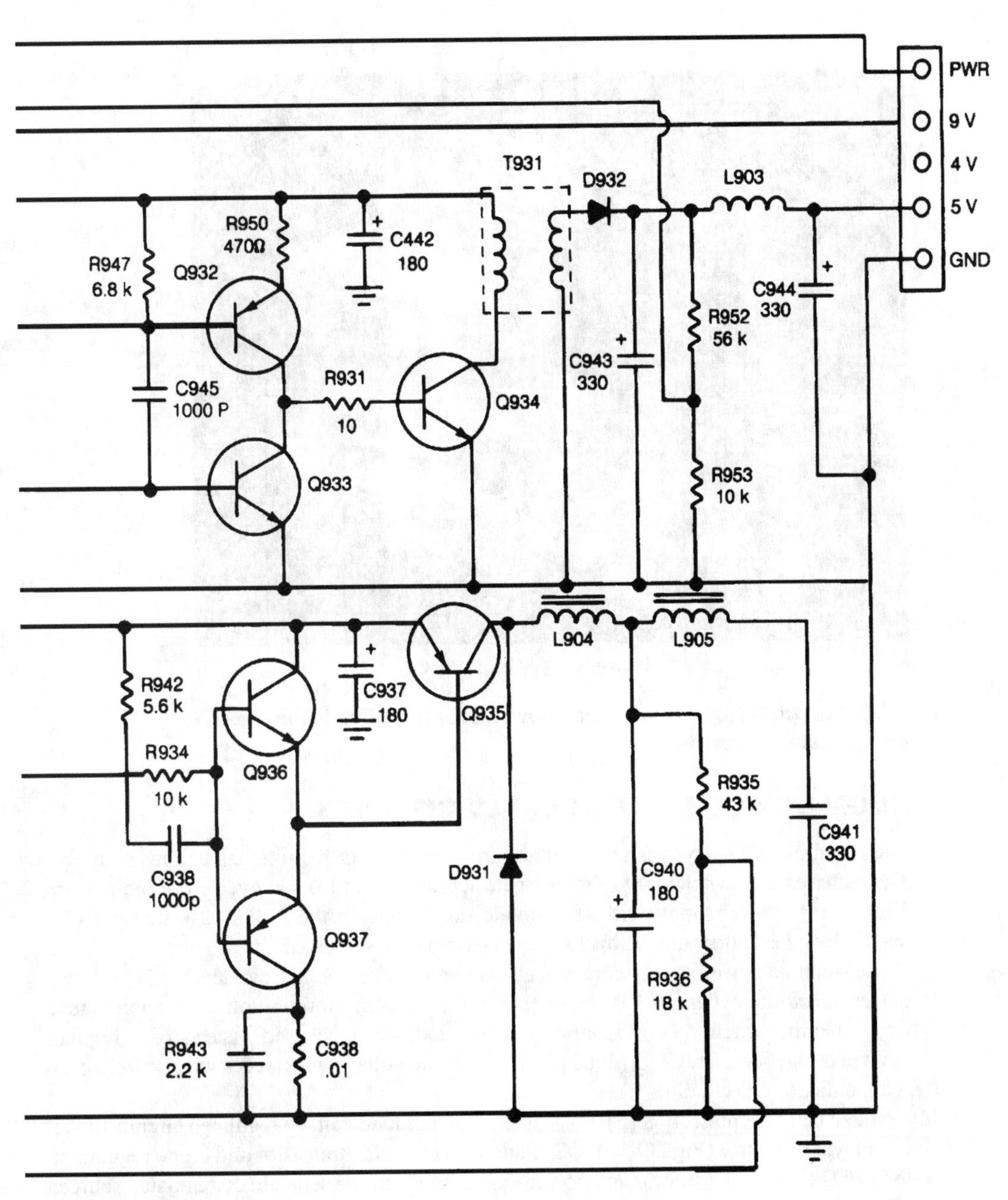

FIGURE 30-14 Continued.

FIGURE 30-15 **The boom-box power-supply source is mounted on separate PC boards.**

BOOM-BOX CD POWER-SUPPLY CIRCUITS

The boom-box CD player can be operated from batteries or the internal ac power supply. The batteries are switched into the circuit when the ac cord is removed from the player. The internal power-supply circuits provide dc voltage to the radio, cassette, and CD player. Besides battery and ac operation, an external dc source can be applied at J104.

The small ac power-supply components can be mounted on a separate chassis with the power transformer (Fig. 30-15). The step-down power-transformer voltage is applied to a bridge-rectifier circuit. Four separate silicon diodes can be used instead of a regular molded bridge rectifier. C142 (2000 μF) filters the dc voltage when it is switched from cassette, radio, or CD circuits.

Notice that fuse protection (F102) is used at the input dc voltage regulated circuits of the CD player dc source (Fig. 30-16). Q253 and D251 provide transistor and diode regulation to the CD circuits. The motor and power-amp circuits have a separate dc regulated source. Notice that diode D105 in the battery and external dc source prevents improper polarity of the voltage applied.

Phone and Line Output

The stereo headphone input circuits are taken directly from the low-power filter network (Fig. 30-17). The portable CD player usually is equipped with line-output jacks so that the player can be connected to a stereo amplifier inside the house. Excessive hum and noise might result when the player is connected to a car amplifier. These line-output jacks are located within the ac power-supply unit. The stereo headphone jack is located in the right side of the CD player.

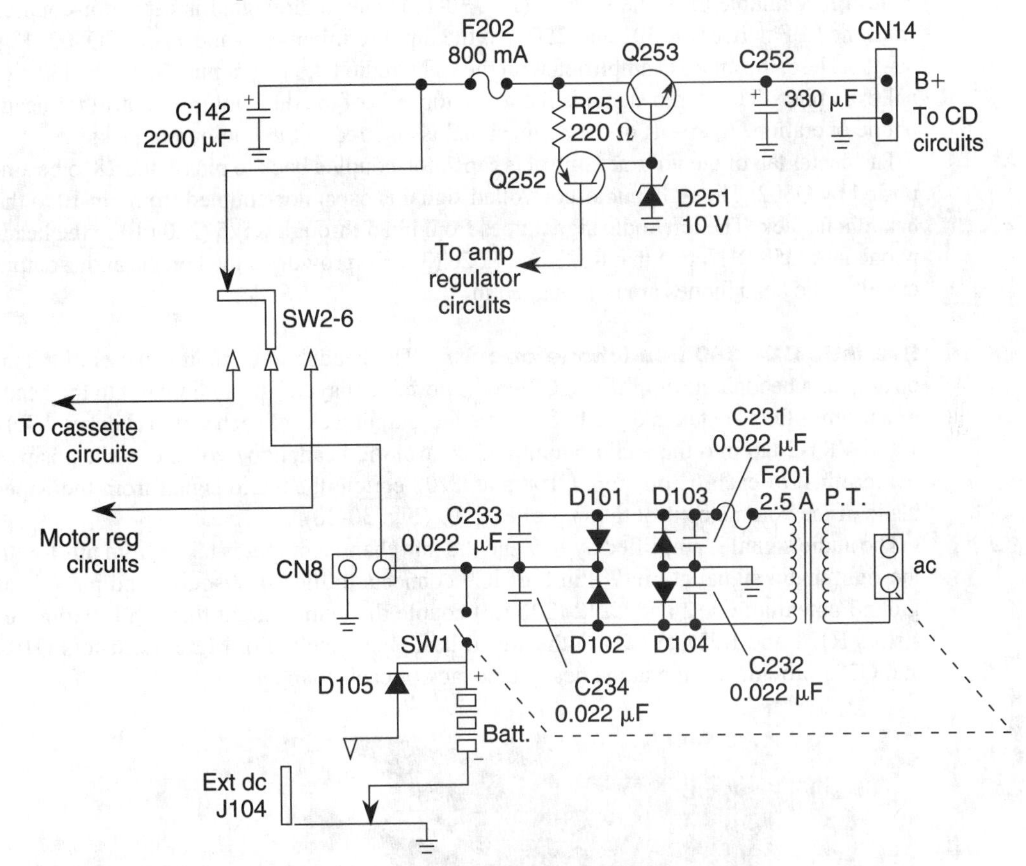

FIGURE 30-16 The boom-box power-supply circuit in a Radio Shack CD player.

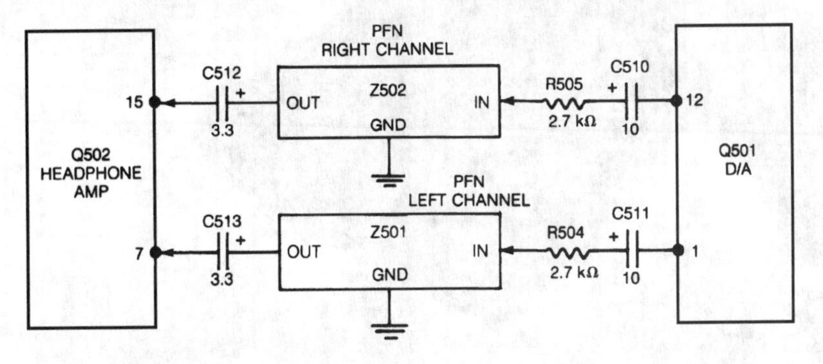

FIGURE 30-17 The audio output of D/A converter (Q501) is supplied through a power filter network (PFN) to the headphone amp (Q502).

Headphone Output Circuits

Often a dual stereo-output IC amplifies the audio for the headphones. A headphone level control is available on some players (Fig. 30-18). The audio signal is capacitor-coupled from the LPF network (2501 and 2502) to the input terminals (15 and 7) of ICQ502 (Fig. 30-19). Here, the audio is amplified with the right audio taken from pin 19 and the left signal from pin 5. The stereo signal is fed to the top side of the dual volume control for headphone operation. A tap-off of the audio signal is also fed to the line-output jacks.

The center tap of the volume control is capacitor-coupled back to pins 4 and 18 to be amplified by Q502. The right audio-controlled signal is capacitor-coupled from pin 10 to the headphone jack. The left audio signal goes from pin 5 through C525 (220 μF) to the headphone jack (J502). Notice that R527 and R526 (330 Ω) provides a load on the audio-output circuit if the headphones are not plugged in.

Realistic CD-3380 headphone circuits The headphone circuit consists of a dual op amp as a headphone amplifier (IC9) in the portable player. The audio input to the headphone amplifiers are taken directly from the line-output jacks of each stereo channel. VR1-1 and VR1-2 tap into the audio output and control the headphone volume to the positive (+) terminal of each IC op amp. C108 and C208 couple the audio signal from the wiper blade of the volume control to pin 3 of each IC (Fig. 30-20).

The audio signal is amplified by IC9 and the left-channel output is taken from pin 1 with the right audio signal at pin 7. Pin 8 of IC9 connects to the +4-V source and pin 4 is at ground potential. C112 and C212 (220 μF) couple the audio output through isolation resistors R121 and R221 (12 Ω) to the stereo headphone jack (J3). Mute transistors Q102 and Q202 provide muting at the headphone jack of each channel.

FIGURE 30-18 A separate headphone level control is in this RCA portable CD player.

Interlock Switch

All portable compact disc players have a top-lid interlock switch system that provides protection for the operation when the top lid is opened for loading the disc. Remember, the optical lens assembly in portable units shines upward toward your eyes, so the power to the laser diodes must be removed when the top lid is raised. In some models, the start limit and close limit switches are disengaged when the top lid is opened (Fig. 30-21).

The start limit switch consists of a metal pin, which shorts the limit switch contacts with the door closed. The start limit switch is connected to the decoder-operation key IC (Fig. 30-22). The close limit switch provides a dc voltage to the APC and laser-diode circuits. The square plastic ridge on the top lid pushes the close limit switch contacts together when the lid is closed (Fig. 30-23). To activate the close limit switch, place a piece of plastic inside the beveled area. A piece of solder or paper clip inside the start limit switch might defeat the switch (Fig. 30-24). Always keep a piece of metal foil taped over the optical laser assembly while servicing the unit.

Optimus CD-3380 interlock door switch When the top lid or door is closed, the microcomputer (IC4) allows voltage to be applied to the optical assembly; the IC6 driver

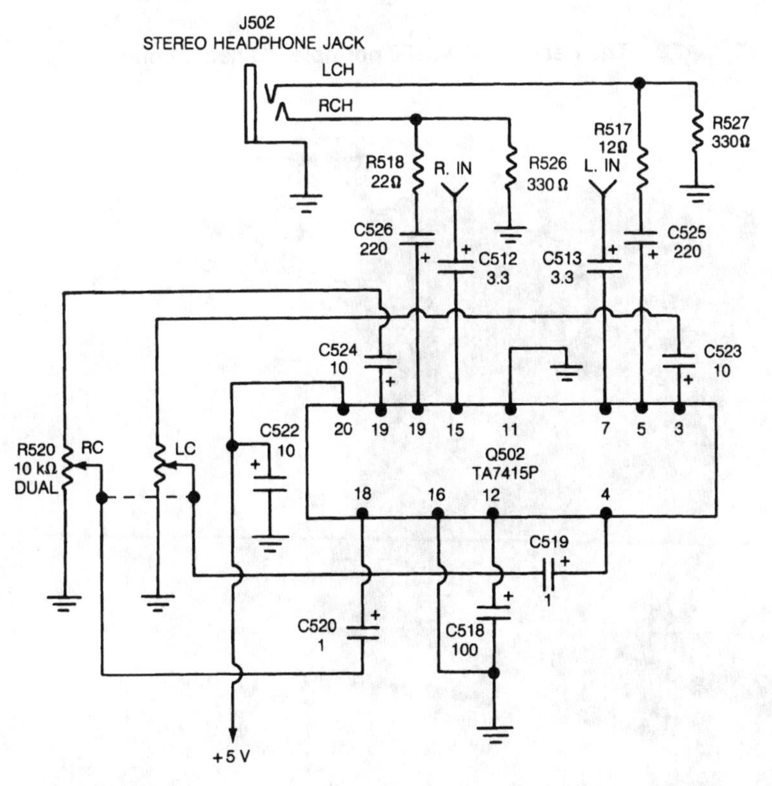

FIGURE 30-19 A typical headphone amp circuit in a portable CD player.

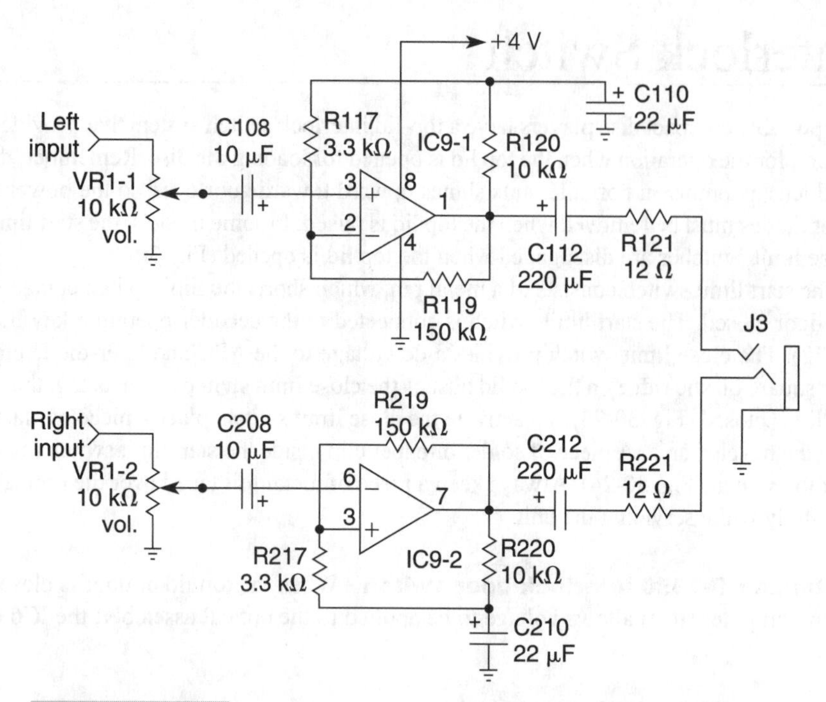

FIGURE 30-20 The Realistic CD-3380 portable CD headphone circuits.

FIGURE 30-21 The lid switch shuts down the player when the lid or door is open.

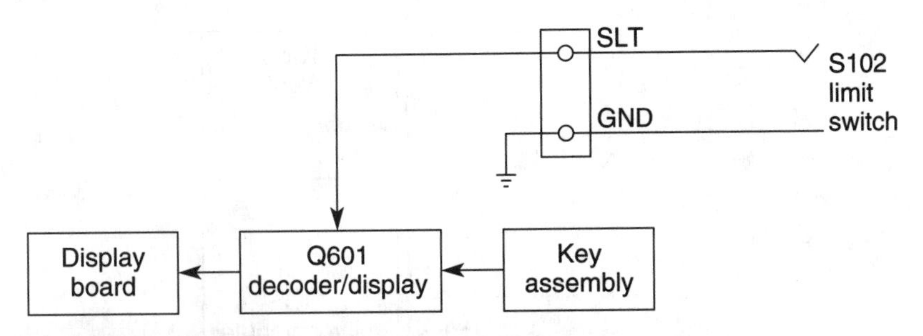

FIGURE 30-22 The start-limit switch is controlled by decoder/display Q601.

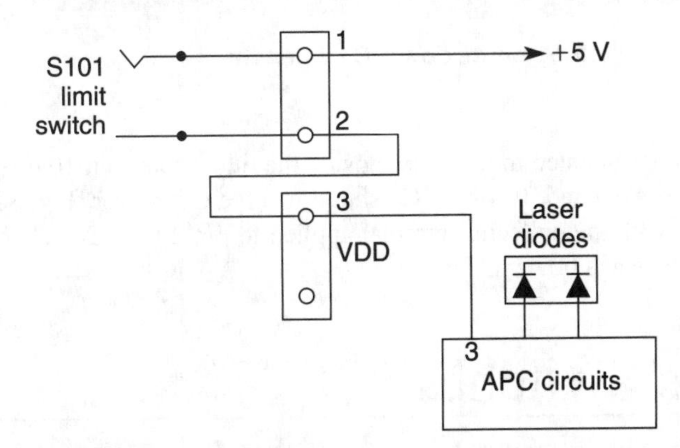

FIGURE 30-23 S101 interlock switch operates the laser-diode circuits when they are open or closed.

FIGURE 30-24 Shut the lid interlock switch when servicing and making laser power tests.

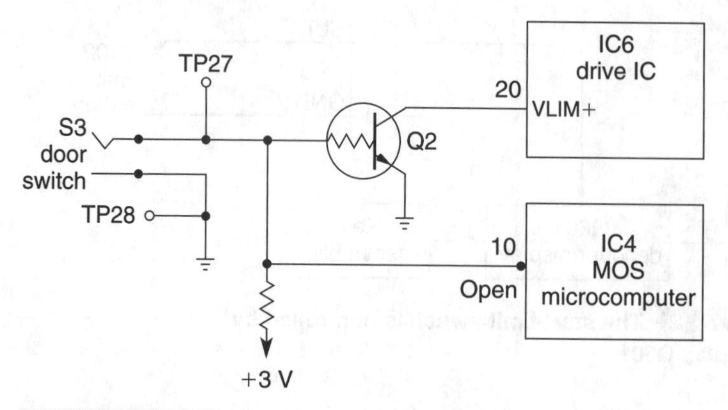

FIGURE 30-25 **Door switch S3 shuts down IC4 when the door is opened.**

provides rotation of the disc and sled motors. By closing the lid or door, pin 10 of IC4 is grounded. When the door is opened, IC4 and IC6 shut down the CD player (Fig. 30-25). Q2 is a digital transistor with the collector terminal applied to pin 20 of IC6. Check the switching action and voltage at TP27 and TP28.

The Various Switches

Many switches are used in the small portable compact disc player. Most of these switches are in the Off position, except a thermostat switch found in some models, which is always On. Here are several examples:

- *S101* The close limit switch is turned off when the top lid is open and on with the cover down (to play the disc). This switch provides a dc voltage to the laser and APC circuits. The close limit switch is located under the top panel.
- *S102* The start limit switch is off when the top lid is raised and it is activated with a metal shorting pin located in the top plastic cover. This switch grounds out the SLT terminal on the decoder/display IC. The switch is located along S101.
- *S601 through S605* The Play/Repeat, Pause/Stop/Memory, Track Down, Track Up, and Display switches are located on the front panel of the display PC board. These switches are connected directly to the decoder/display IC.
- *The Play/Repeat button* This starts the disc to play and repeat the disc-playing action (S601). The Pause/Stop switch (S602) stops the disc from playing for a moment or until the Play button is pushed again (Fig. 30-26). In some models, a Play/Pause key does the same functions. The Up (S604) and Down (S603) switches search for a particular section of the disc. A Mode or Search button, in some models, accomplishes the same thing. In other models, the automatic music sensor (AMS) switch is used to locate the beginning of the desired selection in either direction. Sometimes during pause, you can go back or advance faster than during playback.

- *S901* The adapter power switch is in the Off position and is located on the ac power-supply unit.
- *S931* The thermostat switch is always on and located on the battery-pack PC board of some models.
- *S961* The power switch controls the dc voltage applied to the portable CD player. The player will stand by and the disc playing will start simply when you press the Play key. The power switch should be turned off after use and when transporting the unit, so the player will not operate—even if any of the operation keys or buttons are pressed.

Removing the Boom-Box Case

The small side or bottom screws must be removed with a small Phillips or flat screwdriver before you can lift off the bottom plate cover. Some of these side screws are very small and short. Place them in a container so that you can see them. Be careful not to loosen these small screws when replacing them. Often, the mechanism assembly is fastened to the top assembly of the portable CD player.

Realistic CD-3000 mechanism assembly removal After removing the bottom cover (held with four tapping screws) remove the volume-control knob screw and knob.

1. Remove the headphone PC board held by three tapping screws. The volume-control knob must be removed first, before any of the PC board screws can be removed.
2. The main PC board assembly can be removed from the top section by removing four screws holding the assembly to the top section.
3. After removing the mechanism assembly, remove one screw from the display PC board, then unhook and remove the display board.
4. When removing the PC boards and mechanism, take care not to damage the connections of the flexible PC board.
5. To remove the mechanism assembly from the main PC board, unsolder the flexible cable to the flexible PC board.

Many of the assembly mechanism and display boards are removed in the very same manner. Carefully inspect the PC boards to be removed. Write down what screws and

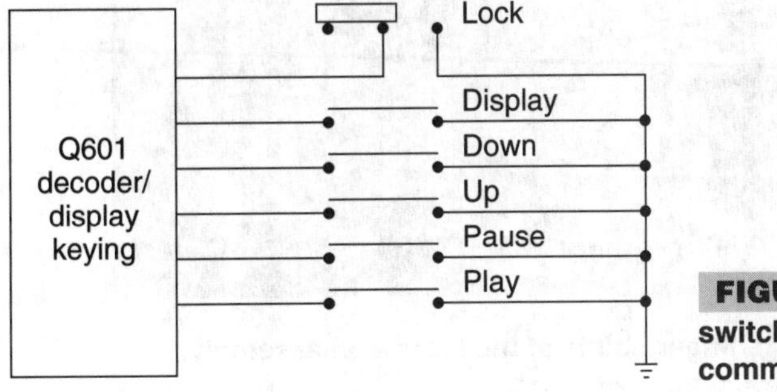

FIGURE 30-26 The keyboard switches are connected to common ground at Q601.

boards you removed so that you can replace them correctly. Place all small screws and components in a white container so that you can see them. Mark the various leads and cable connections on a separate piece of paper for easy replacement, as they are removed.

Realistic 14-529 boom-box disassembly

1. Open the battery compartment.
2. Remove the eight screws (A) that hold the front and rear case.
3. Remove the two screws (G) that hold the cassette-deck assembly.
4. Remove the four screws (D), (E) that hold the CD PC-board assembly.
5. Remove the two screws (C) that hold the audio PC-board assembly.
6. Remove the three screws (F) that hold the power PC-board assembly.
7. Remove the two screws (B) that hold the EQ PC-board assembly.

Laser Optical Assembly

Before checking the laser assembly, be sure that the lens on the disc table is always clean. Do not touch it. If you do, the lens might be damaged and the player will not operate properly (Fig. 30-27). Notice the slot in the plastic cover next to the disc turntable. The lens assembly is connected to a worm gear assembly, which is moved outward to the outside disc surface while playing with the feed or slide motor.

To remove the dust from the lens, blow or brush away the dust with a commercial lens blower. A small camera lens brush and blower will do the job. If anything sticky is on the

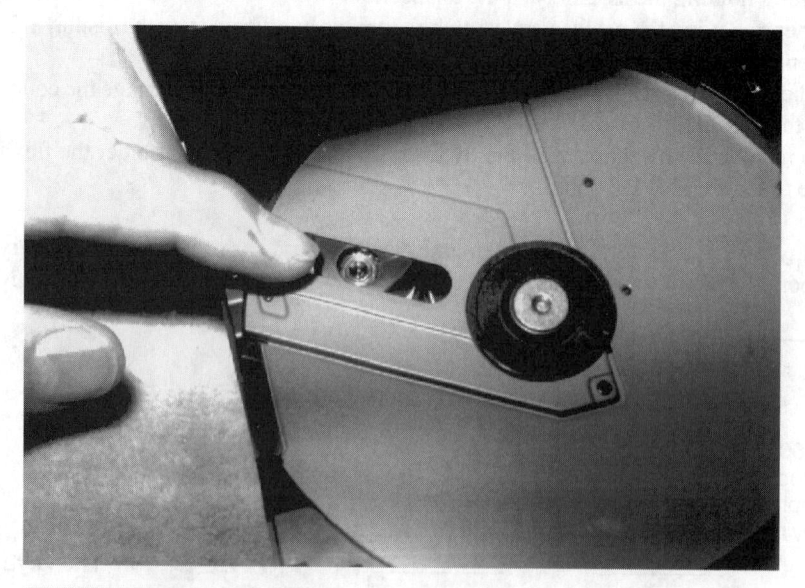

FIGURE 30-27 The finger points at the laser lens assembly, which reads the underside area of the disc.

lens assembly, clean it off with a soft cloth, moistened with a mild detergent solution. Be careful when using rubbing alcohol as a cleaning agent around plastic parts and cabinets. A dirty lens assembly might cause the player to skip or perform intermittently.

Required Test Equipment

The following components are necessary or helpful for troubleshooting.

- Laser power meter or infrared indicator
- Oscilloscope
- Test disc (YEDS-7 or equivalent)
- Eccentric adjustment driver and small-tipped screwdrivers

INFRARED INDICATOR

The infrared indicator described in Chapter 23 can be used to check the presence of the laser lens assembly. The top cover interlocks must be activated before the disc motor will rotate. Hold the photodetector over the lens assembly and push the Play button. Move the indicator around until you hear a loud tone from the piezo buzzer. Keep the indicator away from the disc turntable as it rotates. This sound indicates if the laser diode assembly is operating. Be very careful not to let the lens assembly appear outside of the photodetector board.

A laser diode mounted inside of the pickup is very susceptible to external static electricity. When replacing the laser assembly, use a conductive gum mat and a soldering iron with a ground wire to protect the laser diode from damage by static electricity.

Place the laser power meter over the optical lens assembly for weak or defective laser-diode tests. Make the correct interlock shorting procedures. Turn the unit on. Notice the power reading in mW. Move the laser meter around the lens assembly for a maximum reading of the power meter. Check the measurement with those required on the CD player. In some units, you will need to adjust the VR control for the correct reading required by the manufacturer.

LASER POWER METER MEASUREMENTS

The laser power light meter is much more accurate to test the laser output than an infrared indicator or card. Simply defeat the door or lid switch on the portable CD player and place the power meter probe over the laser line assembly (Fig. 30-28). Be very careful because the laser lens is activated when the door interlock is defeated. Do not look directly at the laser beam. Place the black sensor, with the cutout hole, over the lens assembly.

Move the probe back and forth to provide the greatest measurement. Keep a disc over the lens area by loading it when the probe is not being used. Mark down the measurement and compare to the manufacturer's measurement or compare with other portable CD player laser power movements. A very poor measurement indicates that the laser diode should be replaced by installing a whole new optical assembly.

FIGURE 30-28 Checking the laser diode with a laser power meter.

Realistic CD-3304 RF amp pickup In the Realistic boom-box player, the pickup assembly contains the laser diode, monitor diode, and four optical diodes that connect directly to the RF amp (IC501). The focus and tracking coils are located on the same assembly (Fig. 30-29). The spindle and sled motors connect to the same optical board connections. IC501 provides a digital signal to the servo and digital processing IC. The RF and EFM waveform should be checked at the RF amp if the player shuts down at once.

LASER OPTICAL ASSEMBLY REPLACEMENT

Always follow the manufacturer's laser head replacement procedures. Carefully view the laser head assembly for the best removal procedures. Usually, the laser pickup travels down a guide shaft, which is moved by the feed-motor gear assembly. The mounting screws at each end of the guide shaft must be removed to get to the pickup assembly. A flexible PC board cable must be unplugged or soldered before the pickup can be removed. Write down the numbers of all parts (on a separate piece of paper) that are moved if you do not have the instructions to replace the pickup head assembly.

Realistic CD-3000 pickup replacement Follow these steps for replacing the pickup assembly.

1. After removing four screws to remove the bottom plate, secure display PC board with two screws temporarily so that the buttons can be pushed during adjustments.
2. Remove the subgear and main gear assembly by removing the C washer (Fig. 30-30).
3. Remove the two screws holding the ends of the fixing guide shaft and unsolder the flexible PC board attached to the pickup. When replacing the pickup and flexible PC board mounted to the drive motor, remove the two screws that hold the pickup motor (Fig. 30-31).

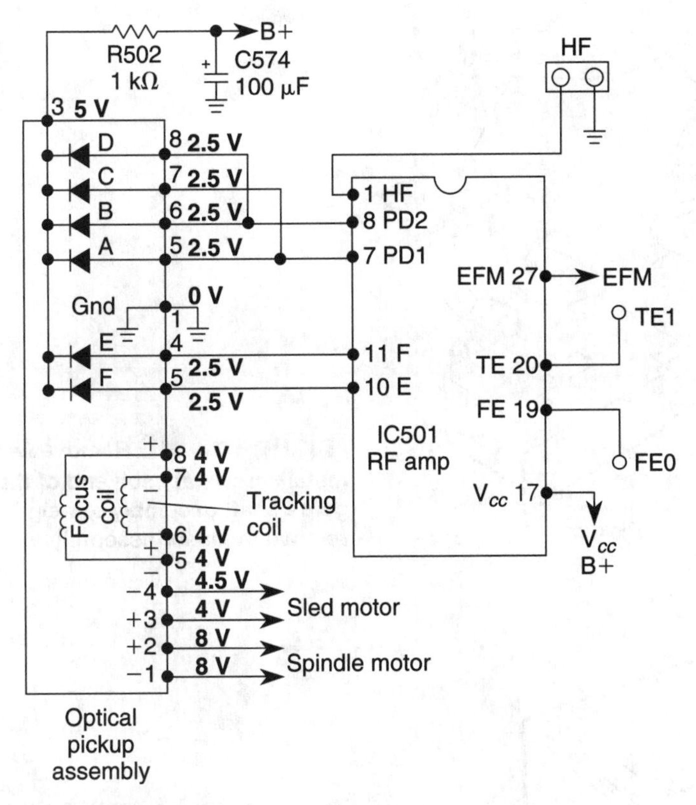

FIGURE 30-29 The optical pickup and RF amp voltages in the Realistic boom-box player.

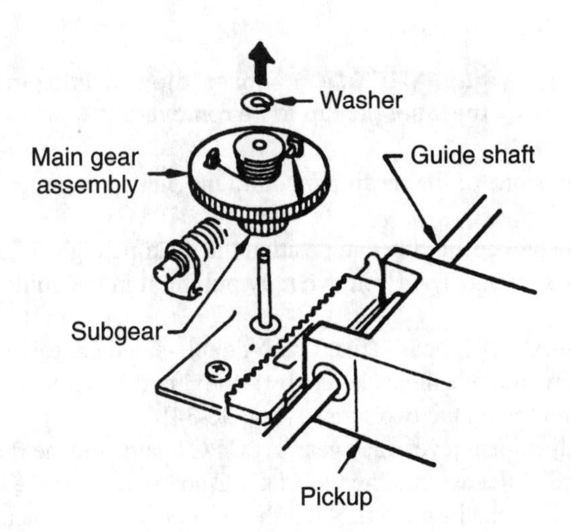

FIGURE 30-30 Removing the "c" washer of the main gear assembly so that the guide shaft can be removed.

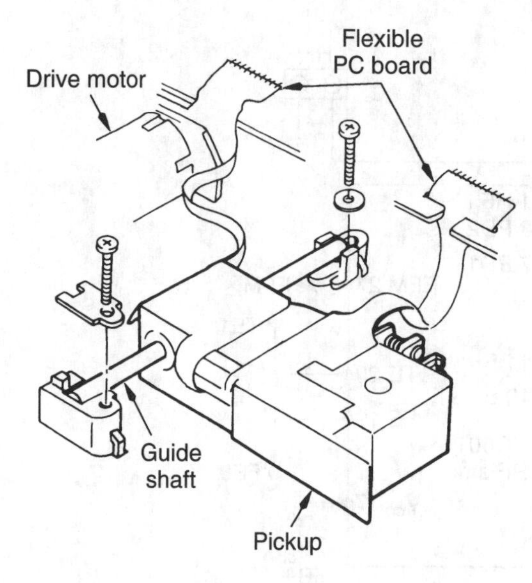

FIGURE 30-31 Remove two metal screws at each end of the guide shaft assembly to help remove the laser assembly.

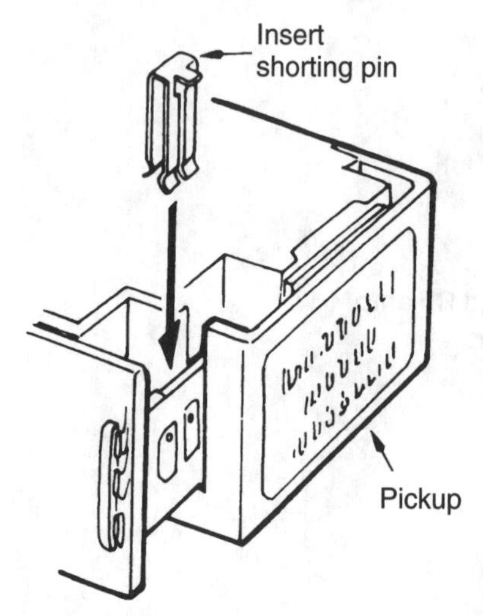

FIGURE 30-32 Insert the shorting pin into the laser pickup to be removed.

4 Insert the shorting pin, which is one of the repair parts, into the pickup to be removed (Fig. 30-32).

5 Lift the guide shaft side of the pickup slightly and position the pickup height adjusting lever to the mechanism chassis cutout. Then, remove it by pulling it in the guide shaft side (Fig. 30-33).

6 Unsolder the terminals of the APC PC board (fixed with double-sided adhesive tape) and remove the PC board. Now, remove the rack gear by removing the two screws.

7 Remove the pickup lever by removing the two screws (Fig. 30-34).

8 Assemble the pickup to the adjustment lever, rack gear, APC PC board, and the flexible PC board of the APC PC board. After assembling the APC PC board, remove the shorting pin. Now make the required laser adjustments. After the laser power adjustments are completed, install the pickup or mechanism.

Signal Processing and Servo Circuits

The signal-processing circuits are similar to those in any CD player. The photodetector diode signal is fed into terminals 1, 2, 4, and 7 of Q701 (Fig. 30-35). Besides sending the EFM output signal (pin 15) to Q702, the tracking servo amp output (pin 22) and focus servo amp output (23) are controlled by the signal-processing circuits.

Signal waveforms from pins 11, 15, 22, and 23 can indicate if the Q701 circuits are working. Check for correct operating supply voltage at pin 8 (+ 5 V). By taking correct voltage and signal waveforms on Q701, with a normal laser beam, the RF and EFM signal can be signal traced through the signal-processing circuits. Do not measure the voltage or touch pin terminals 1, 2, 4, and 7 of Q701 from the laser pickup assembly.

Realistic CD-3370 portable CD pickup circuits The optical pickup assembly is connected directly to RF amplifier IC4 in this Radio Shack CD player. Besides the RF amp, a laser transistor (Q9) drives the laser diodes within the optical pickup (Fig. 30-36).

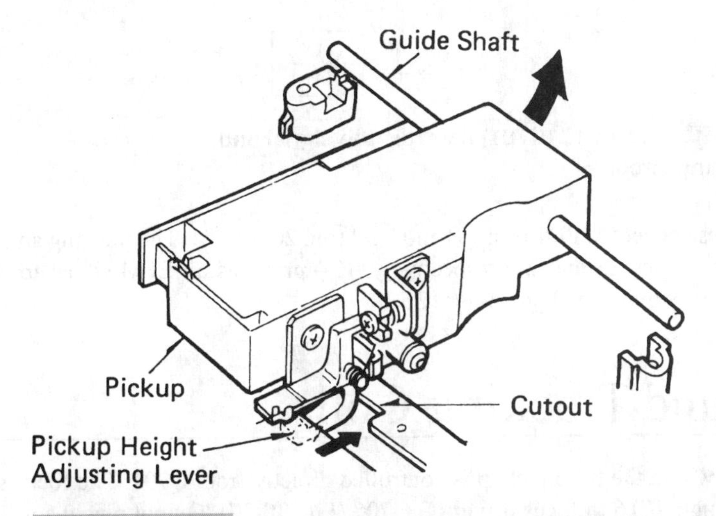

FIGURE 30-33 Pull the guide shaft to remove the laser pickup assembly.

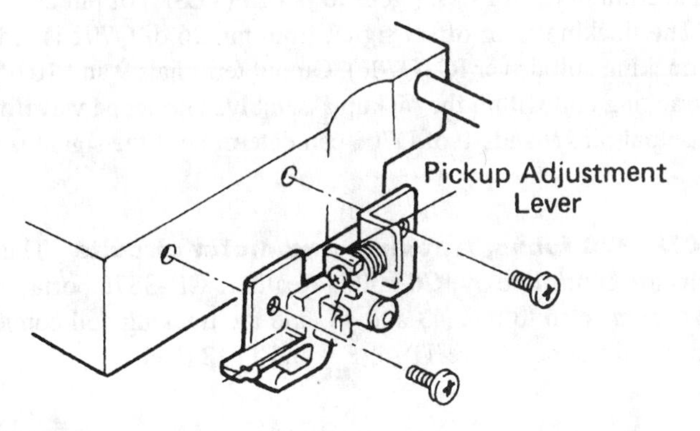

FIGURE 30-34 Remove two metal screws to remove the pickup-adjustment lever.

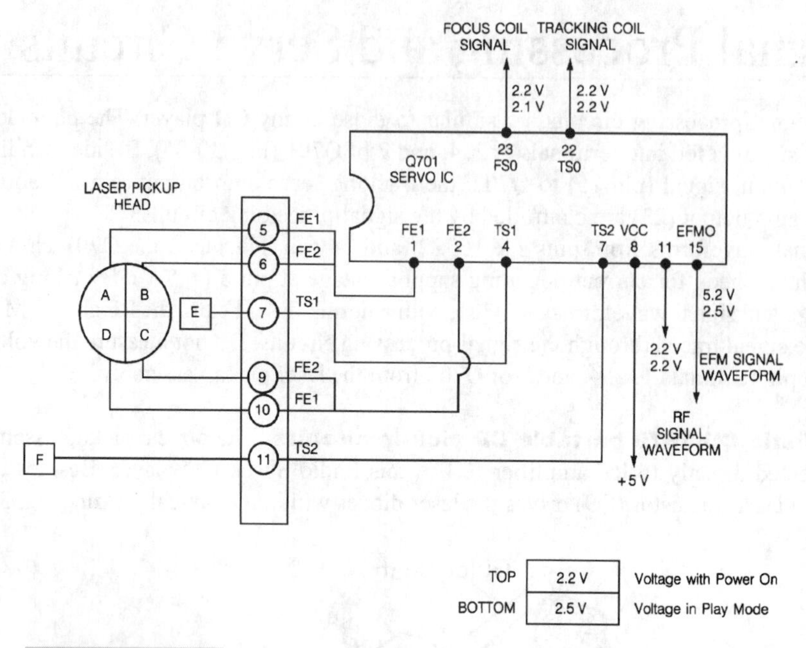

FIGURE 30-35 Servo IC (Q701) controls the signal and servo-processing circuits.

The photodiodes connect to PD1 (pin 19) and PD2 (pin 20) of IC4. The tracking and focus diodes connect to pins 23 and 22, respectively. IC4 provides an EFM signal to digital processor IC3 and servo signal processor IC5.

Focus and Tracking Coils

The focus coil within the head pickup is controlled directly from Q701. The focus signal is coupled through R716 to focus driver IC Q705 (Fig. 30-37). Output pins 9 and 16 are applied to the focus coil winding. The input signal at pins 2 and 7 and output waveforms at pins 9 and 16 of Q705 should determine if the focus coil circuits are functioning.

The tracking coil signal from pin 22 of Q701 goes to pin 25 (TESH) of pin 26 of servo IC Q702 (Fig. 30-38). The tracking error offset signal from pin 26 of Q702 is fed to op amp Q703, then to the tracking coil driver IC (Q706). Output terminals 9 and 10 of Q706 are tied directly to the tracking coil within the pickup assembly. The scope waveforms at input pins 7 and 2 and output pins 9 and 10 of Q706 can determine if the signal is reaching the tracking coil.

Realistic portable CD-3370 focus, tracking, and motor circuits The focus and tracking coil circuits are controlled by IC6 in this Realistic CD-3370 portable (Fig. 30-39). The focus coil is connected to pins 43 and 44, and the tracking coil connects to pins 41 and 42. The focus coil test points are TP10 (F–) and TP12 (F+).

IC6 also provides drive for the spindle and sled motors. The spindle motor connects to pins 36 and 37 of IC6. The sled motor terminals connect to pins 34 and 35. In this portable CD player, the coils and motors are controlled with one IC6 driver.

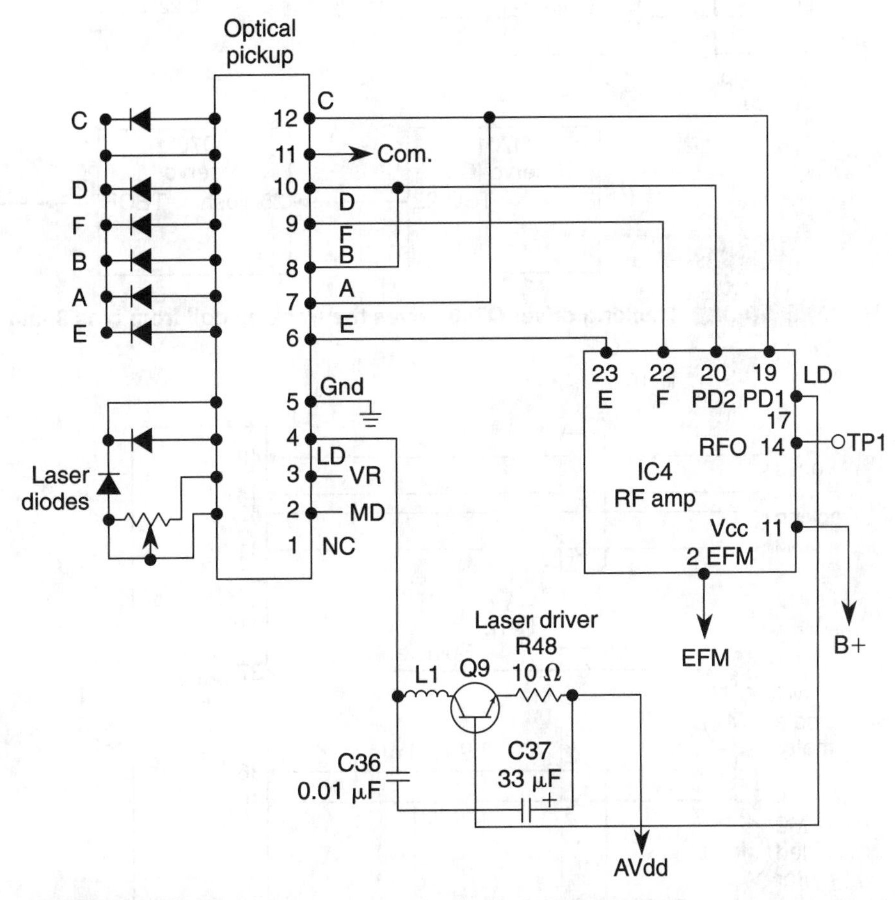

FIGURE 30-36 **The laser pickup circuits with RF amp IC4 in a typical portable CD player.**

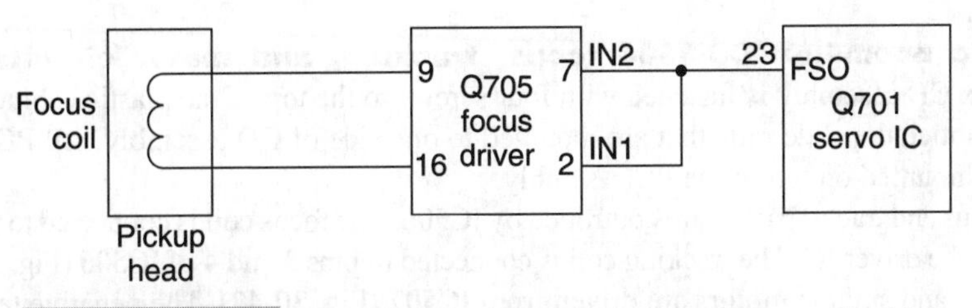

FIGURE 30-37 **Focus driver Q705 drives the focus coil within the pickup head.**

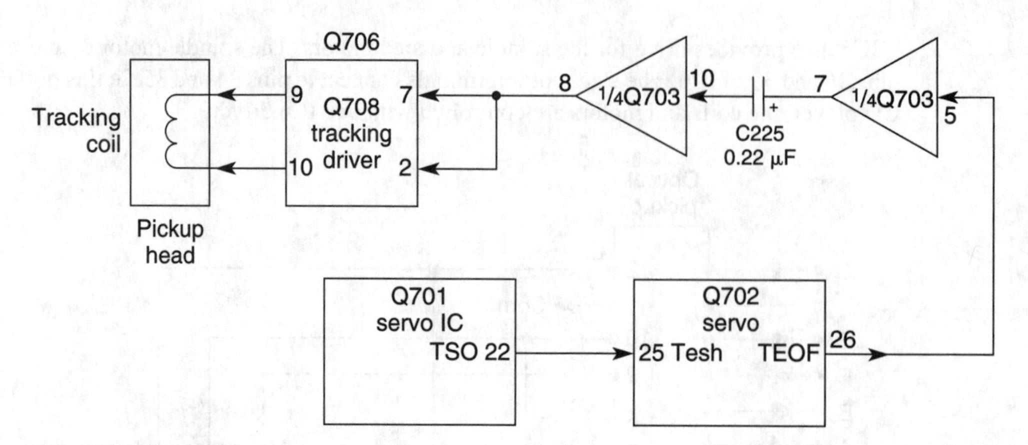

FIGURE 30-38 Tracking driver Q706 drives the tracking coil from pins 9 and 10.

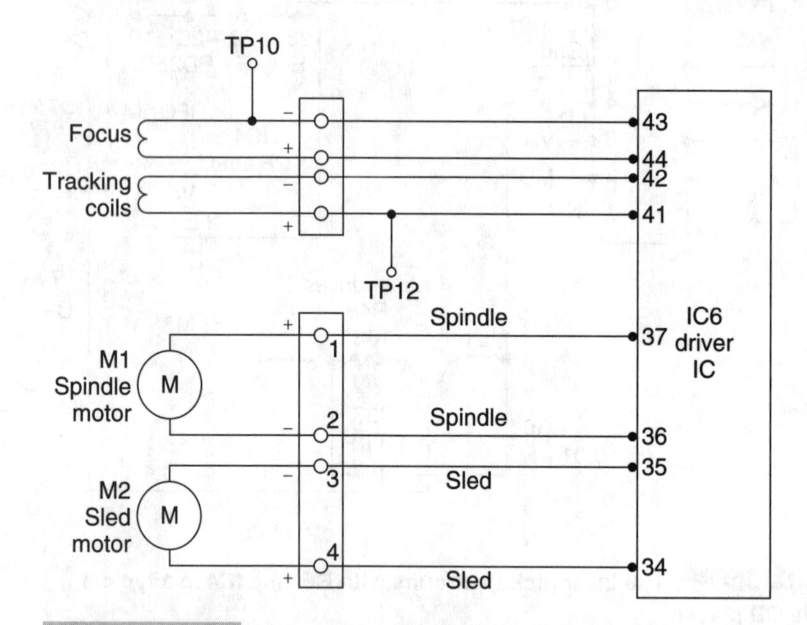

FIGURE 30-39 IC6 drives the focus and tracking coils, and spindle (M1) and sled (M2) motors.

Realistic boom-box CD-3304 focus, tracking, and motor circuits The boom-box CD assembly is fastened with four screws to the top of the plastic cabinet (Fig. 30-40). Notice the slide rails that are mounted to one side of CD assembly. All PC board parts are mounted on top of the PC assembly.

The focus and tracking coils are controlled by IC506. The focus coil is connected to pins 10 and 11 of the driver IC. The tracking coil is connected to pins 3 and 4 of IC506 (Fig. 30-41).

The sled and spindle motors are driven from IC507 (Fig. 30-42). The negative terminal of the sled motor (4.5 V) connects to pin 4 and the positive terminal to pin 3 of IC507. The positive spindle motor terminal connects to pin 11 and the negative terminal connects to pin 10 of motor driver IC507 (Fig. 30-43).

The following is a list of the main waveforms and abbreviations that are listed on each waveform:

VCOO	Voltage-controlled oscillator output
FOD	Focus drive
TRD	Tracking drive
FE	Focus error
TE	Tracking error
EFM	Eight-to-fourteen modulation
ASY	Asymmetry
CLK	Clock
XTAO	Crystal output
WDCK	Work click
SLD	Sled motor drive

FIGURE 30-40 The CD PC board is fastened to the top of the Sharp QT-CD7-3 boom box.

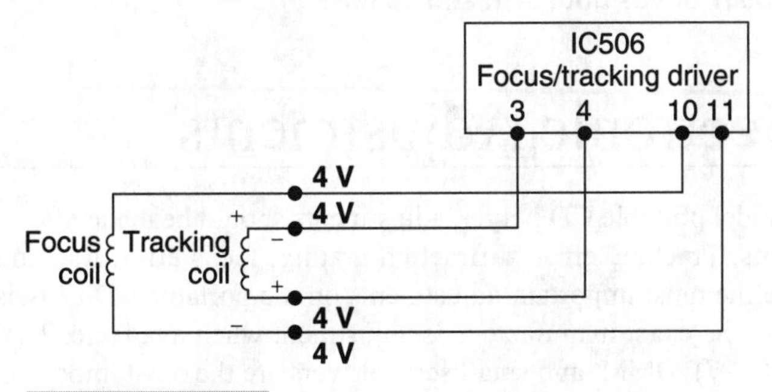

FIGURE 30-41 IC506 drives both focus and tracking coils in this Radio Shack boom-box player.

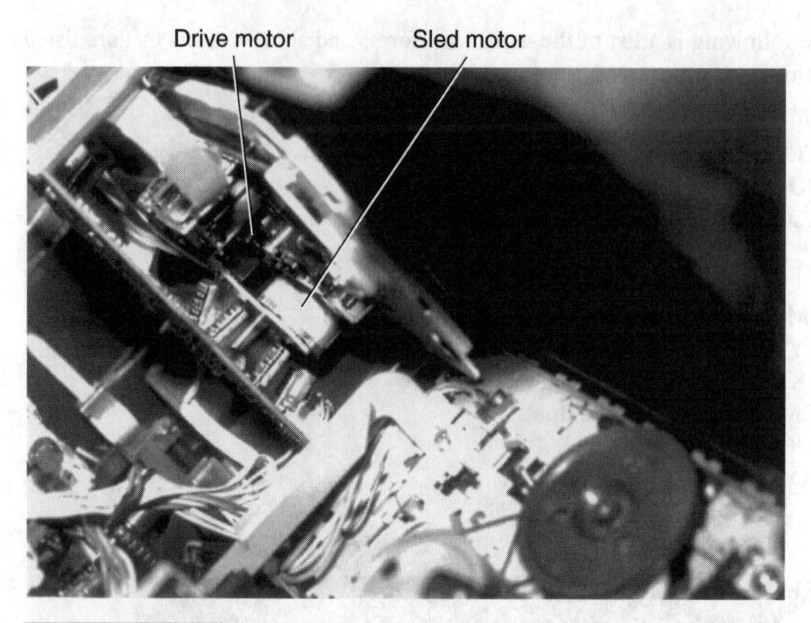

FIGURE 30-42 **The spindle and disc motor in this CD boom-box player.**

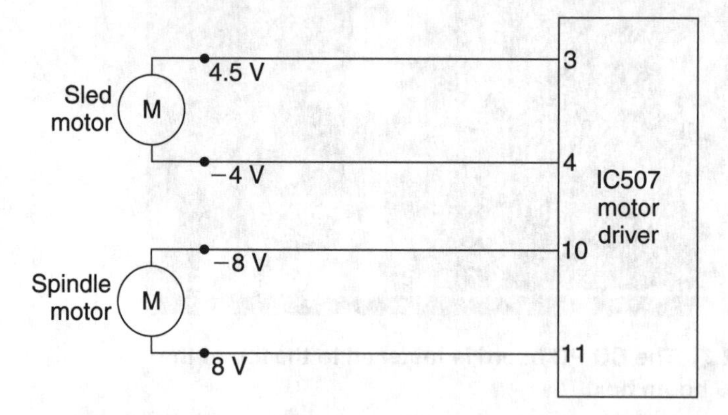

FIGURE 30-43 **IC507 drives both sled and spindle motors.**

Crucial Electronic Adjustments

Like with the table-model portable CD player, adjustments should be made after replacing crucial components. Tracking error, diffraction grating, focus error, tracking error balance, and APC are the most important adjustments on the portable CD chassis (Fig. 30-44). Always follow the exact manufacturer's adjustment when available. The oscilloscope test disc (YEDS-7), DMM, and small screwdrivers are the most important tools for electronic adjustments.

Realistic CD-3000 electronic adjustments Although it is impossible to list all of the portable compact disc player adjustments in this chapter, the most crucial ones were chosen from Realistic's CD-3000 adjustments. Locate the adjustment screws on the small PC board. Because the chassis might have to be pulled from the power supply to make the electronic adjustment, provision must be made to connect the voltage to the chassis. Proceed with batteries if the external battery can be used without tying the dc leads together.

Within the Realistic CD-3000, solder lead wires to pins 3, 11, 22, and 28 of Q701 temporarily for easy connection of the test equipment. Be careful not to short out the other IC pins. Do not keep the soldering iron too long on each pin. Now, connect the lead wires from the ac adapter to the main PC board. Be sure that the power switch will stay on or will short across the terminals with a soldered lead.

Tracking Error Balance Adjustment

Connect the oscilloscope to pins 3 and 22 of servo IC Q701 (Fig. 30-45). Load test disc YEDS-7 in the tray. Turn the unit on. Hold the pickup motor worm gear (feed motor) to stop its rotation while the unit is searching for music. Push the Up key and adjust R707 TE BAL VR so that the tracking error signal waveform becomes 10 mV. Once the Up key is pushed, operation continues for about 20 seconds. If the adjustment is not completed in 20 seconds, push the key again.

FIGURE 30-44 Locate the small adjustment pots on the bottom of the PC board in this portable CD player.

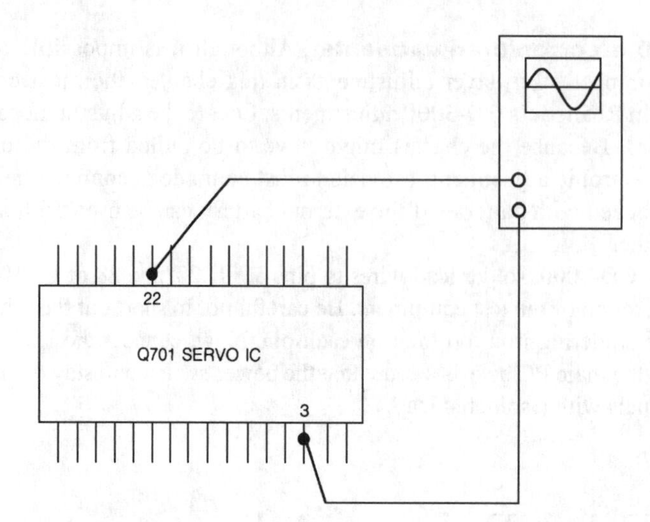

FIGURE 30-45 Connect the scope to terminals 3 and 22 for tracking-error adjustment.

FOCUS ERROR BALANCE ADJUSTMENT

Connect the oscilloscope to pins 3 and 11 of Q701. Play the test disc YEDS-7 and adjust FE BAL (R729) so that the RF eye pattern is clear.

Realistic CD-3304 PLL VCO adjustment

1 Put the unit into Stop mode (Fig. 30-46).
2 Short-circuit test points ASY and GND.
3 Connect a digital frequency counter to test points TP10 and GND.
4 Adjust VR504 so that the frequency counter becomes 4.25 ±0.01 MHz.

Optimus CD-3370 EF balance adjustment

1 Connect an oscilloscope to TP2 (TE) and TP8 (VC) through filter A. Notice that TP8 must be connected to ground on the scope (Fig. 30-47).
2 Switch on the power in the Test mode and push the Play button.
3 Next, switch on the P mode until LCD displays "ON TR SL"; then, switch it off and on again to display "OFF TR SL." During these displays, turn potentiometer P1 so that the waveform becomes vertically symmetrical with the VC level at the center.
4 Now, adjust potentiometer P1 so that TP2 (TE) voltage becomes 20 mV.

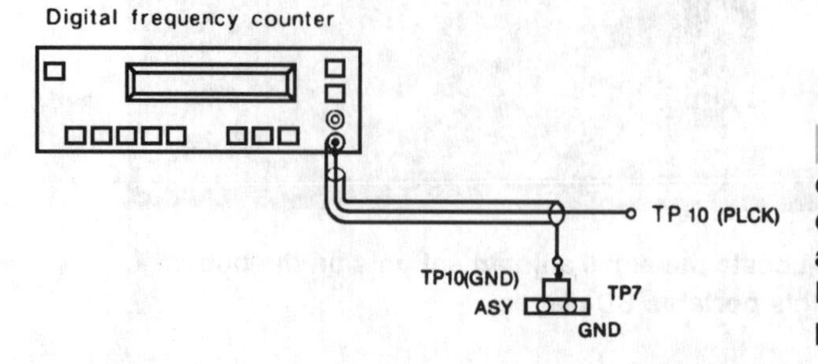

FIGURE 30-46 How to connect the frequency counter for PLL-VCO adjustment in the Realistic portable CD player.

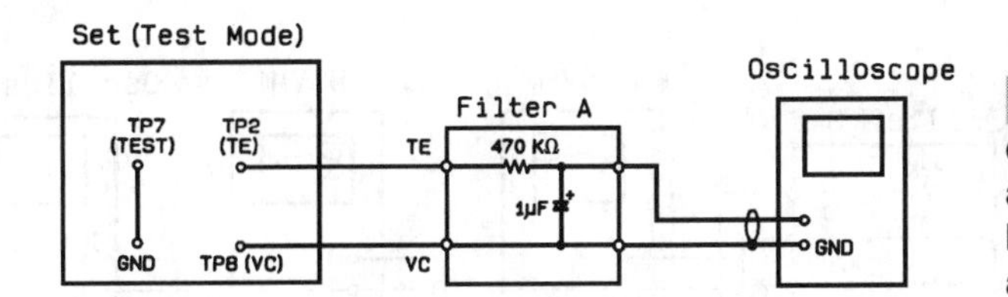

FIGURE 30-47 How to connect the scope through a filter network for EF balance adjustment at TP2 and TP8.

Realistic CD-3304 focus offset adjustment This focus offset adjustment is made without a jitter meter.

1. Insert the test disc and press Play.
2. Connect the oscilloscope to TP1 (HF, GND).
3. Adjust VR505 so that the eye pattern becomes clear and the waveform (VP-P) is maximum.

Note: When confirming eye pattern, use a 10-1 probe.

Optimus CD-3370 focus gain adjustment

1. Connect an oscilloscope to TP4 (FE), TP5, and TP8 (VC) through filter B. Notice that TP8 must be connected to ground on the scope (Fig. 30-48).
2. Cancel the Test mode, switch on the power, load the test disc, and push the FF button. Play the fifth track.
3. Input 1.3 kHz at 50 mV from the oscillator to TP5 and observe the signal output to TP4. Then, adjust potentiometer P3 to produce a level that is the same as the input lead.

Optimus CD-3370 tracking gain adjustment

1. Connect the scope to TP2 (FE), TP3, and TP8 (VC) through filter B. Note that TP8 must be connected to ground on the oscilloscope (Fig. 30-49).
2. Cancel the Test mode, switch on the power, load the disc, and push the FF button. Play the fifth track.
3. Input 1.7 kHz at 50 mV from the oscillator to TP3 and observe the signal level output to TP2; then adjust potentiometer P4 to produce a level that is the same as the input level.

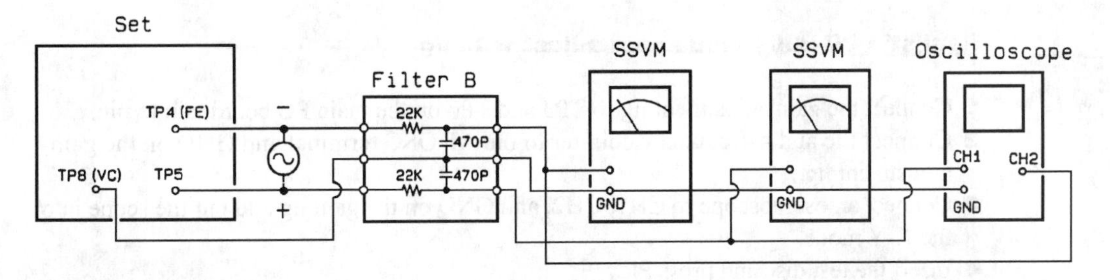

FIGURE 30-48 How to connect test instruments for focus-gain adjustment in the Optimus CD portable.

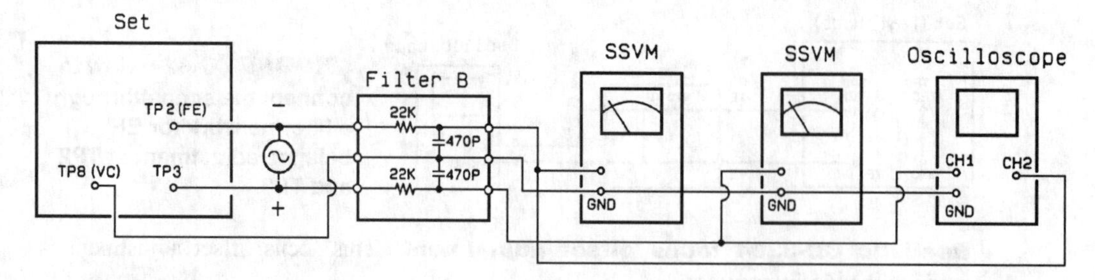

FIGURE 30-49 Connect the scope and test instruments through filter B at TP2 and TP3 for tracking-gain adjustment.

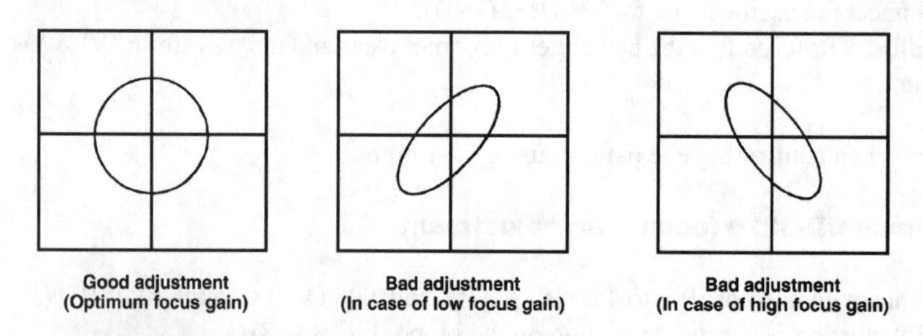

FIGURE 30-50 Adjust VR502 so that the waveform is a complete circuit in this CD player.

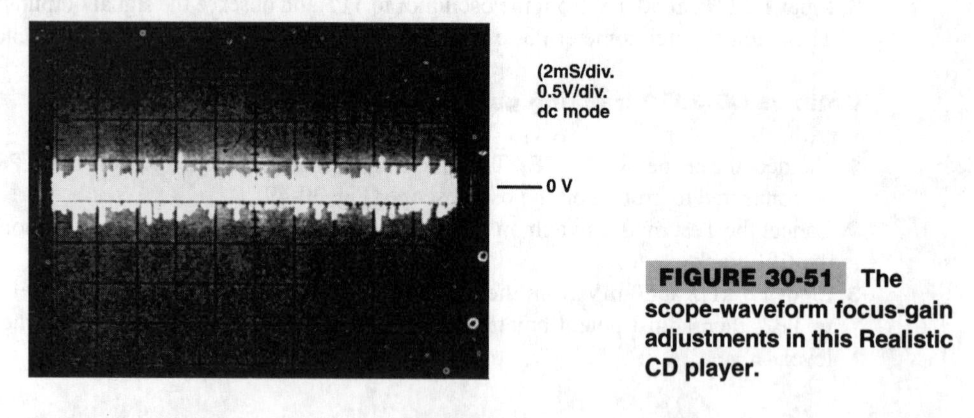

FIGURE 30-51 The scope-waveform focus-gain adjustments in this Realistic CD player.

Realistic CD-3304 focus gain adjust with jig

1. Connect the gain adjustment jig to TP3 and TP6 on the main PC board, pin to pin.
2. Connect the audio-frequency counter to the AF OSC terminal and GND on the gain-adjustment jig.
3. Connect an oscilloscope to CH1, CH2, and GND on the gain jig and put the scope into the X-Y mode.
4. Insert the test disc and press Play.
5. Set switch SW2 on the gain adjustment jig to the IF position.

6 Set switch SW1 on.
7 Adjust VR502 so that waveform on the scope becomes like Fig. 30-50.

Realistic CD-3304 focus gain adjustment without test jig

1 Insert the test disc and press Play (YEDS-43).
2 Connect the oscilloscope to IC506 pins 2 and 11.
3 Adjust VR502 so that the waveform on the scope becomes like Fig. 30-51.

31

REPAIRING THE AUTO CD PLAYER

CONTENTS AT A GLANCE

Most auto compact disc player circuits are quite similar to tabletop CD players. The same service precautions must be followed to protect your eyes from the laser beam during servicing procedures. The same test equipment, tools, and test discs are used to service both machines. Of course, the auto disc players are all front loaded. The big differences between the auto and the home CD player are the power supply and the physical size.

Naturally, the auto disc player operates from the dc car battery instead of the ac power line. The power supply in the auto player uses a dc-to-dc converter and isolated regulator circuits. The auto power supply is in a separate hideaway unit cabled to the main front-loaded player. In some players, five different discs can be automatically loaded with a CD changer in the car trunk area.

The big difference between the auto and home players is the physical size of each player. Because most components are placed in one container, several different PC boards are layered together. Like the auto radio, most components are jammed or fit tightly together so that all the parts will fit inside one metal cabinet. Because space is a premium, you might find chip resistors, capacitors, diode, transistors, and ICs. Although servicing the auto disc player might take a little more time to remove PC boards and locate the correct component, the reproduction of music is much greater than FM radio or cassette players.

Specifications for Three Auto CD Players

Here are three different auto CD players listed alphabetically with their specifications.

Denon DCC-9770 auto compact disc player Table 31-1 depicts a flowchart for removing each section of this unit and Fig. 31-1 shows the removed sections.

TABLE 31-1 FLOWCHART FOR REMOVING EACH SECTION

DENON	DCC-9770 AUTO COMPACT DISC PLAYER
Sampling frequency	44.1 kHz
Quantization	16-bit linear
Transfer bit rate	4.3218 Megabits/sec.
Frequency response	5 to 20 kHz ±1.0 dB
Dynamic range	96 dB
Signal-to-noise ratio	96 dB
Harmonic distortion	0.005%
Wow and flutter	Below a measurable level
Output voltage	1 V/10 kΩ
DENON	**DCH-500 AUTO CD CHANGER**
Laser diode	GaA1As 780-mW wavelength
Laser output power	Less than 40.0 mW
Frequency response	5 to 20,000 Hz ±1 dB
Dynamic range	90 dB
Distortion	0.005%
Wow and flutter	Below measurable limit
Outputs	Line output
Current drain	800 mA and 1.5 A during disc loading or eject
PIONEER MODEL	**CDX-1 CD PLAYER**
Disc diameter	120 nm; thickness 1.2 nm
Laser	Semiconductor; wavelength = 780 nm
Signal format	Sampling frequency = 44.1 kHz; 16-bit linear

TABLE 31-1 FLOWCHART FOR REMOVING EACH SECTION (CONTINUED)

Power requirements	14.4 Vdc (10.8 to 15.6 V possible); consumption = 18 W
Frequency characteristics	5 to 20,000 Hz (±1 dB)
Signal-to-noise ratio	90 dB (1 kHz)
Dynamic range	90 dB (1 kHz)
Wow and flutter	Below measurable range
Distortion factor	0.005% (1 kHz, 0 dB)
Number of channels	2
Output voltage	280 mV (when level switching high); –140 mV (when level switching low)

FIGURE 31-1 The auto CD player might have several sections to be removed before locating the defective component.

Auto CD Precautions

- Check the reset button with a ballpoint pen before starting the unit for the first time or after replacing the car battery.
- All auto CD players made for USA use a negative ground.
- Replace the blown fuse with one of the same amperage as shown on the fuse holder. If a fuse blows more than once, check the electrical connections and the CD player.
- In extremely hot weather, let the car's interior cool before turning on the CD player. Let the air circulate.
- Keep all foreign objects from entering the disc slot because the precision mechanism and disc could be damaged.
- Be sure to remove any discs from the CD player before removing the unit. Moving the unit with a disc in it could damage the disc and the unit.

- Some units will not play in extremely hot weather. Many of these units have a display that indicates that the temperature is too high and the unit will shut down.
- On cold days, condensation might form. Let the CD player set 30 minutes before playing.
- Sometimes a strong road shock will cause the laser head to skip.

The Block Diagram

The block diagram is useful in locating the defective section of the auto CD player. The auto RF and servo IC circuits are like those found in the tabletop CD player. The RF signal from the pickup photodiodes goes to a large RF AM signal-processor IC. The servo IC might include the RAM, clock regenerator, data compensator, digital filter, and VCD circuits, in addition to controlling the focus, tracking, and various motor circuits. A separate D/A converter IC might follow and feed the stereo signal to the sample/hold and de-emphasis IC. Separate tone and audio-output ICs are in the left and right output audio channels (Fig. 31-2).

THE OPTICAL PICKUP

A light laser beam generated by the optical pickup strikes the signal surface of the disc. The reflected beam is then sent to a photodiode, which reads the digital signals picked up from the disc. Because signals are read without contacting the surface of the medium, needle-generated friction (as in conventional phonograph systems) and the accompanying hiss are eliminated. Also, because the disc itself is coated with clear plastic, the signal surface is protected from the effects of dust, finger smudges, and dirt.

LASER OPTICAL PICKUP ASSEMBLY

The laser pickup assembly consists of the laser diode, collimator lens, beam splitter, crucial-angle prism, four-segment photodiode, reflection prism, quarter-wavelength plate, objective lens, and TE and FE diodes.

A diffused laser beam is emitted from the laser diode (LD), and is then made parallel by a collimator lens. This parallel beam is sent through a beam splitter, a reflection prism, and then a quarter-wavelength plate. The object lens then focuses the beam to the disc surface.

The beam is flashed onto a row of pits on the disc surface, then is reflected back through the objective lens. It passes through the quarter-wavelength plate, the reflection prism, and the beam splitter again. The LD emits a straight, polarized beam, which is then changed to a circular, polarized beam by passing through the quarter-wavelength plate. After the beam is reflected off the disc plane, it goes back through the quarter-wavelength plate and is changed back into a straight, polarized beam.

The polarity of the returning beam is 90 degrees out of phase with the outgoing beam. The beam is reflected off the disc plane, goes back through the quarter-wavelength plate, and is reflected off the beam splitters. It then goes through a crucial-angle prism and finally reaches the four-segment photodetector.

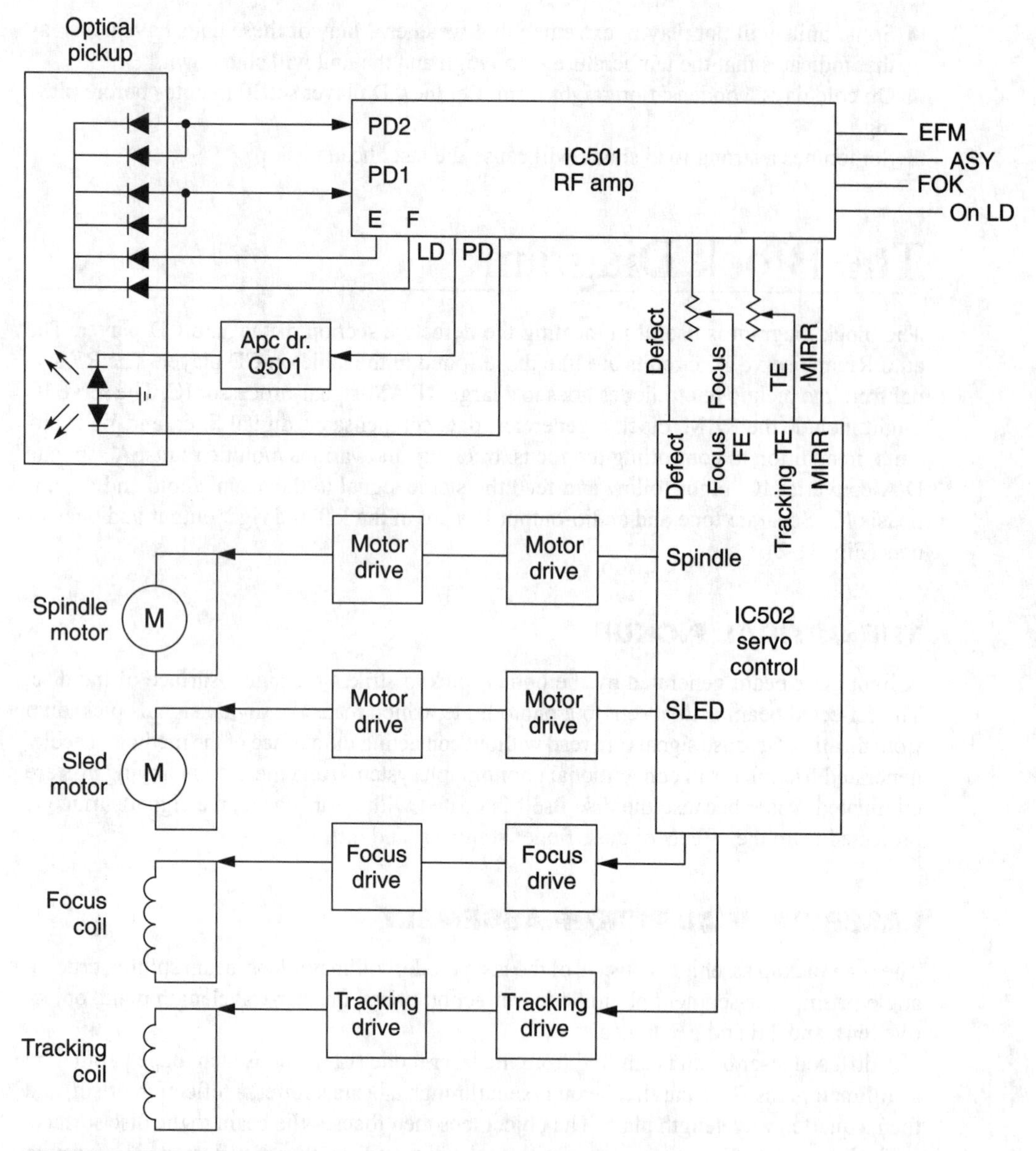

FIGURE 31-2 The block diagram of a typical auto CD player.

THE PREAMPLIFIER

The preamplifier amplifies the weak RF signal from the photodiodes within the laser pickup assembly. Usually, the preamp generates the RF, TE, and FE signals by amplifying and adding or subtracting the output from the pickup unit of the four photodiode detectors. The laser diode within the Sony CDX-5 auto CD player is furnished a voltage from a –5.2-V and +5.2-V power source with transistor regulation (Fig. 31-3). The tracking error (E) and focus error (F) diodes' output go to IC652, then to focus/tracking/sled servo IC504.

Denon DCH-500 CD RF circuits The optical PD1 and PD2 diodes are coupled to pins 19 and 20 of the RF amp IC11 (Fig. 31-4). The error and focus diodes are connected to pins 22 and 23 of the RF amp. Transistor Q11 provides laser drive for the laser diodes in the optical assembly. RF signal can be checked at the RF test point from pin 14 and the EFM waveform at pin 8 of IC11. The EFM waveform is fed from pin 8 to pin 24 of IC13.

Denon DCH-500 changer signal path The photodiodes of the optical assembly pick up the digital signal from the compact disc and apply it to pins PD1 (19) and PD2 (20) of RF amp IC11 (Fig. 31-5). IC11 amplifies the weak digital signal. The EFM signal is fed from pin 8 to pin 24 of IC13. The digital audio signal is sent to pin 10 of digital/analog IC301.

The right stereo signal is taken from pin 25 and fed to pin 3 of IC101. The left stereo audio is fed to the other half of IC101 on pin 5. The amplified audio signal is capacity-coupled to the right and left output, to be fed to the audio radio amplifier.

FOCUS/TRACKING/SLED SERVO

Often, one large servo IC controls the focus and tracking coils, in addition to operating the sled or carriage motor. Within Alpine's 5900 CD player, a MIRR signal is fed to pin 11, FE focus gain at pin 20, and tracking gain signal at pin 13 (Fig. 31-6). Both the focus error and tracking error signal have a gain control in each error leg. The focus error output from pin 21 controls focus driver Q505, Q506, and the focus coil. The tracking drive signal from

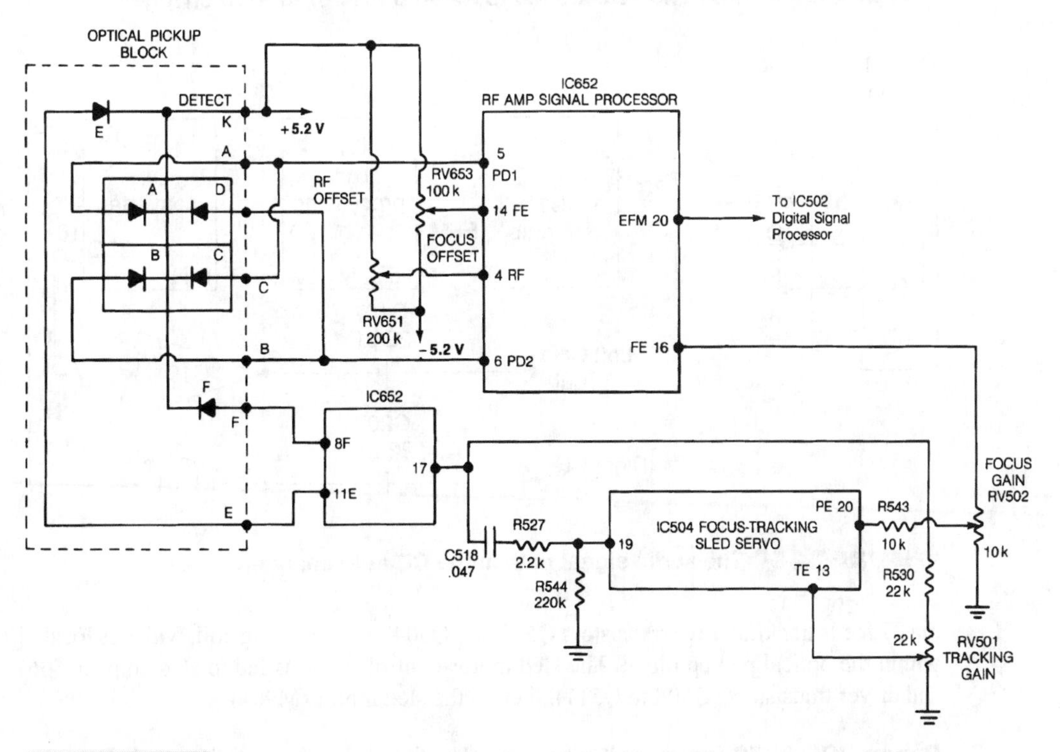

FIGURE 31-3 The photodiode signal is fed to the RF amp IC652 with the EFM output at pin 20.

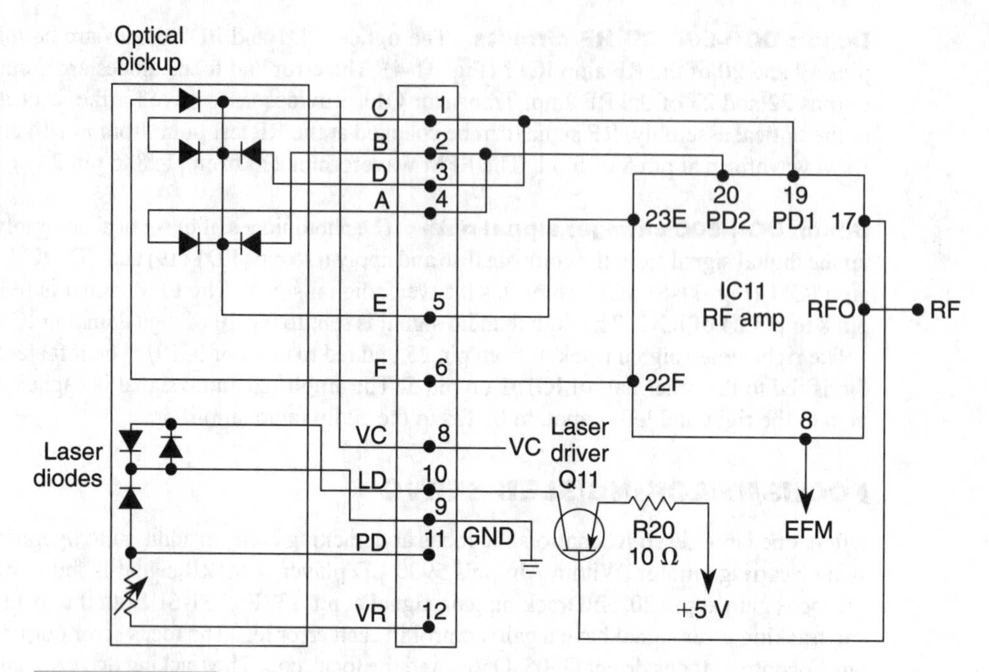

FIGURE 31-4 PD1 and PD2 are fed to RF amp IC11 or to a CD changer.

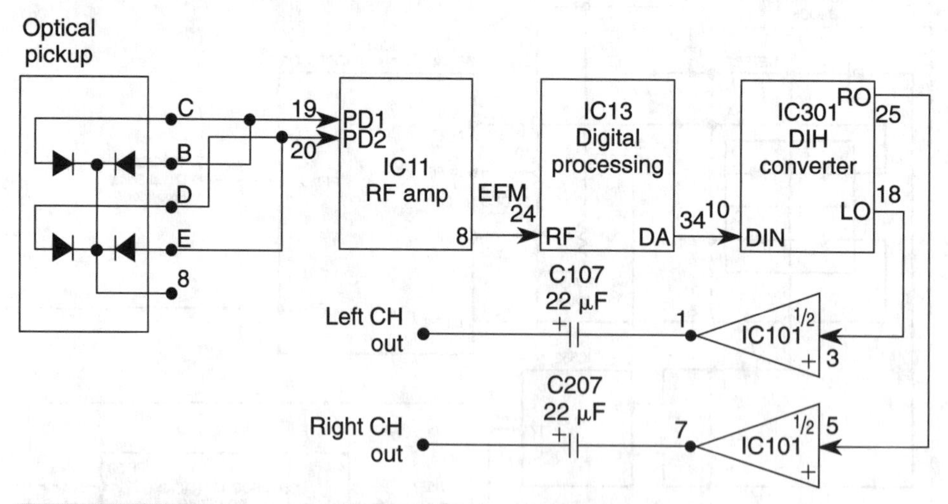

FIGURE 31-5 The audio signal path in the CD auto changer.

pin 27 feeds tracking drive transistors Q503 and Q504 to the tracking coil, which is located within the optical pickup block. The sled motor-control signal is fed to sled amp (IC506) and driver transistors (Q508 to Q511), then to the sled motor (M901).

Denon DCC-9770 focus coil circuits The focus coil circuits of Denon's DC-9770 auto CD player are taken from pin 5 (FEO) of the servo IC502. This signal is sent to amplifier and driver IC104 (Fig. 31-7). Transistors Q410 and Q411 are found between the two op amps of IC402. The output voltage signal is taken from transistors Q412 and Q413 to the positive terminal of the focus coil. The negative terminal focus coil lead ties in from output transistors Q410 and Q411. The collector terminals of Q412 and Q410 are taken from the B+ 14-V source. Both Q413 and Q411 collector terminals are grounded.

If the tracking activator starts to vibrate, activation could become impossible. To avoid this, the tracking servo is designed to close only when the relative crossover speed is low. It does so by detecting the disc movement with the RF and TE signals.

The tracking servo circuit is activated after the focus servo circuit is activated. The tracking servo circuit activation, however, is controlled to reduce the pickup speed relative to the movement of the disc surface. Also, a jump circuit accesses tracks randomly.

Denon DCC-9770 tracking-coil circuits The tracking coil (TAO) signal is taken from pin 11 and fed to IC402 (Fig. 31-8). IC402 contains four different op amps in one component. Q414 and Q415 receive the TAO signal and feed it from the emitter through R434 and IC402. The output signal on pin 7 couples directly to the base of Q416 and Q417. The positive tracking coil voltage at both emitters is fed to the tracking coil. The negative coil terminal ties in at the emitters of Q414 and Q415.

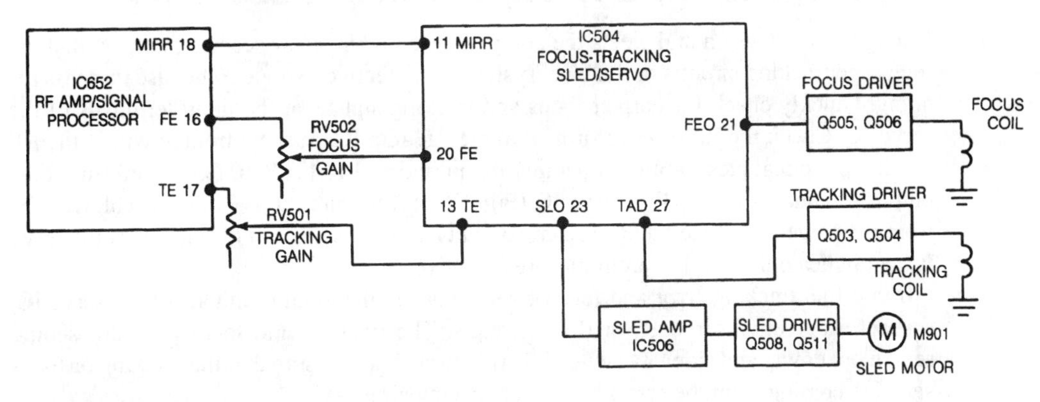

FIGURE 31-6 A block diagram of the focus/tracking in an auto CD player.

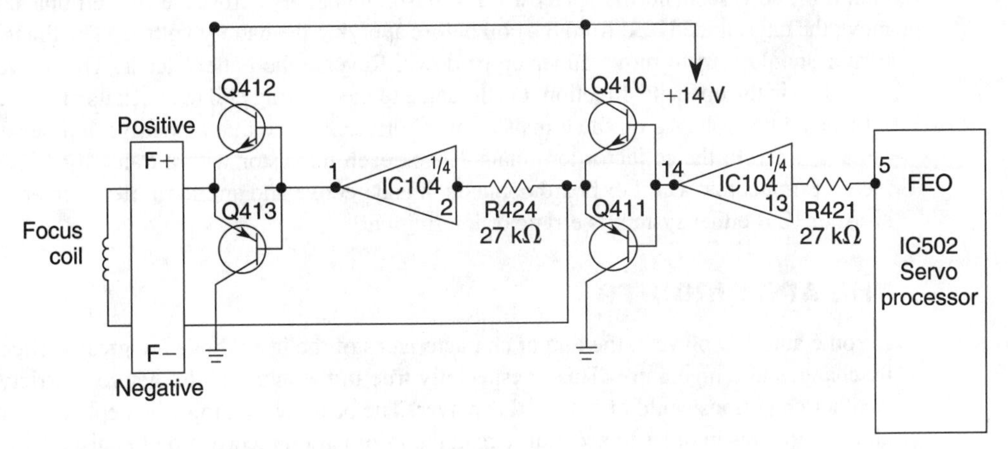

FIGURE 31-7 The focus and driver circuits of an auto CD player.

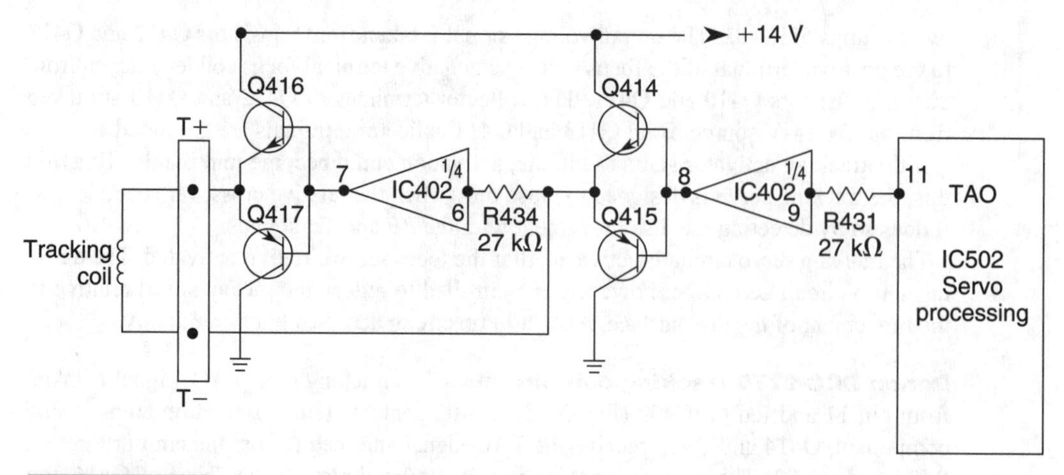

FIGURE 31-8 The tracking coils in a typical auto CD player.

TROUBLESHOOTING THE RF AND SERVO SECTIONS

If the player comes on and ejects the compact disc within a few seconds, suspect that the focus and tracking circuits or the eject system are defective. Notice if the disc motor is rotating. Quickly check for correct focus and tracking signals at the input terminals of the servo IC. Check the EFM waveform from the RF amp IC. Suspect trouble within the RF preamp or optical pickup block assembly if you find no EFM, TE, or FO waveforms at the output terminals of the RF preamp IC (Fig. 31-9). Measure the B+ and B– voltages fed from the power supply to the RF preamp circuits. Improper TE and FO signals at the servo IC can indicate that RF IC or circuits are defective.

If you find tracking error and focus error signals at the input terminals of the servo IC, suspect a defective servo IC or drive circuits. The tracking and focus activator should search up, down, and sideways when it is first turned on. Be sure that the tracking or focus signal is coming from the servo IC to each driving circuit.

The focus and tracking coil can be checked by taking a continuity measurement of the coil. If the coils seem normal, place a 1.5-V flashlight battery across the coil terminals. Remove the ungrounded lead from the coil before applying the battery voltage. The focus activator should start to move either up or down. Reverse the battery leads. The activator should go in the opposite direction. Do the same to the tracking coil to check its movement.

Measure the voltages on the transistor or IC driver. Notice that a positive and negative voltage is fed to the collector terminals. Check each transistor with a transistor tester or meter. Suspect that a coil or board connection is defective and an intermittent driver transistor or IC if either system is erratic or intermittent.

THE APC CIRCUITS

In some auto CD players, the output characteristics of the laser diode are greatly affected by changes in temperature. This is especially true in the automobile, where a variety of weather conditions could affect the CD player. The beam output must be kept constant at all temperatures in order to accurately read the information recorded on the disc. A monitor diode has been built into the pickup to monitor the quantity of beam output. The beam output from the laser diode is kept constant by applying negative feedback from the current (detected by the MD) to the LD drive circuit.

INTERLOCK CIRCUITS

Interlocks are supplied to protect you while operating the auto compact disc player. Interlocks can be added, so when a cover is removed, the CD player will not operate. Most interlocks are provided to prevent radiation from the laser beam of striking the eyes of the operator or electronic technician. Take extra safety precautions when the interlock is shunted around or defeated so that the player can be serviced. Interlocks should not be defeated and left in that position after the CD player has been repaired.

Yamaha YCD-1000 interlock operation The car compact disc player reads the disc signals by laser-beam operation. The human body must not be directly exposed to the laser beam. This unit is, therefore, equipped with an interlock to prevent the unwanted and unnecessary laser output. The laser outputs are controlled by the injection or cutoff of the constant voltage source to the laser diode at pin 30 (LS) of IC101 (Fig. 31-10) and also by an automatic laser power-control circuit. When pin 30 is in "H" (high) level, the laser emits the beam. When pin 30 is in the "L" (low) level, the laser does not emit the beam.

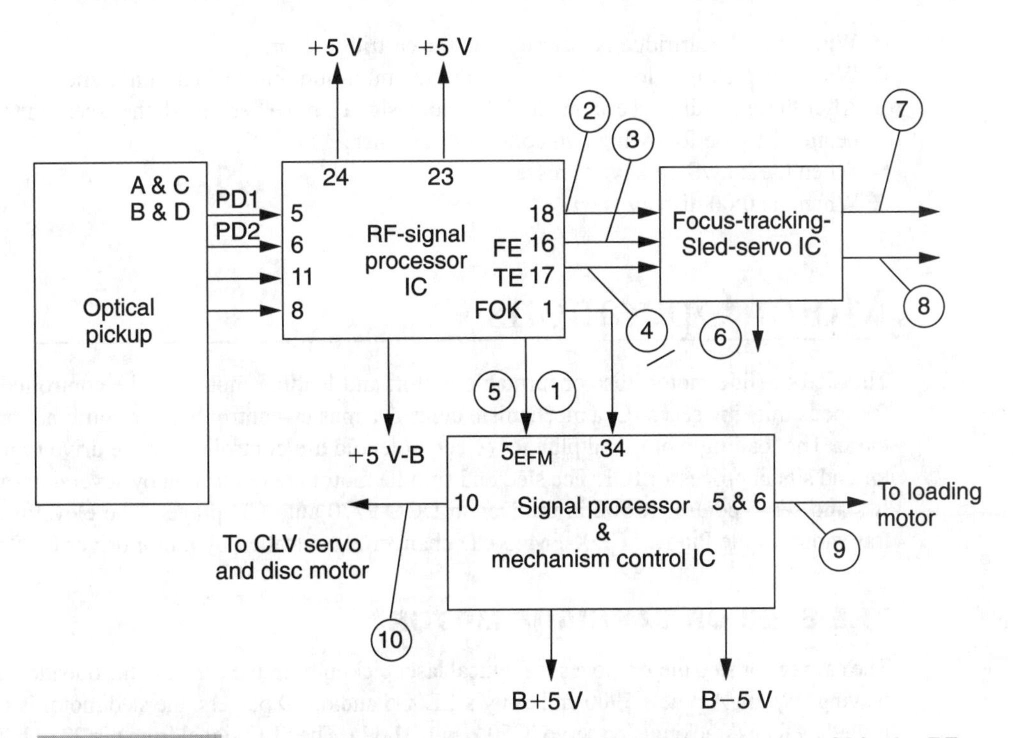

FIGURE 31-9 Take correct waveforms and voltage measurements within the RF and servo sections to locate the defective part.

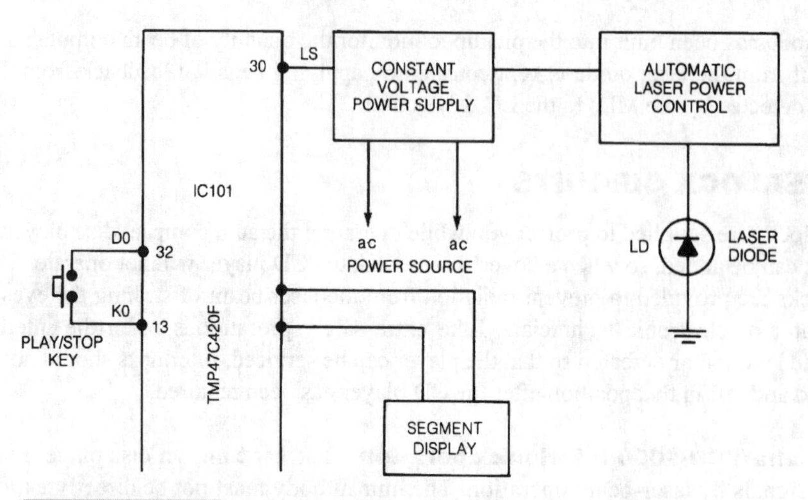

FIGURE 31-10 IC101 lowers the dc supply voltage to the laser diode when a disc is loaded.

Pin 30 is set in the "H" level when the unit is loaded with the disc and it reads the index signals or when the unit is set in the Play mode. When the unit reads the index signals and the following two conditions are met, the laser emits the beam.

- When the CD cartridge is inserted into the cartridge door.
- When the pickup is located at the area of the minimum internal circumference.
- After these conditions are met and the index signals have been read, the laser emits the beam when the following two conditions are met.
- When the Play/Stop key is pressed.
- When the 0:00 display is on.

Motor Operations

The sled or slide motor, disc or turntable motor, and loading motor can be controlled by the focus/tracking servo IC, a mechanism control, a master-control IC, or a combination of these. The loading motor and plunger (eject) solenoid are controlled by the drive transistor and signal processor IC1. The sled and spindle motor are controlled by several op-amp ICs and servo control IC502 in the Denon DCC-9770 auto CD player. The elevator and tray motor in the Pioneer CDX-FM45 CD changer is controlled by motor driver IC801.

THE SLED OR CARRIAGE MOTOR

The carriage or sled motor moves the optical laser pickup from the center to the outside while playing. Within Alpine's 5900 and Sony's CDX-5 audio CD players, the sled motor is controlled by focus/tracking/sled servo IC504 (Fig. 31-11). The SLO signal from pin 23 of IC504 is fed to the sled amp (IC506), then to sled driver transistors Q508, Q509, Q510, and Q511. The output voltage from the emitter terminals of Q508 and Q511 drive sled motor M901.

Denon DCH-500 changer sled motor circuits In this CD changer, the sled motor is controlled by tracking, focus, and motor drive IC14. IC14 is controlled by servo IC12 (Fig. 31-12). The negative motor terminal of the sled motor connects to pin 1 and the positive terminal to pin 2 of IC14. IC14 controls the drive voltage for the sled motor, tracking, and focus coils.

DISC OR SPINDLE MOTOR CIRCUITS

The disc or spindle motors rotate the disc, while the carriage or sled motor pulls the optical pickup head toward the outside edge of the disc. The disc motor in the Alpine 5900 and Sony CDX-5 auto players is controlled by digital signal processor/CLV servo IC502 (Fig. 31-13). The servo control signals from IC502 pins 1 through 4 are fed to a CLV servo control amp IC301 (1–2) and to the CLV disc motor.

In the Pioneer CDX-1 auto player, the spindle motor is controlled by IC9 (Fig. 3-14). A positive 16 V is applied to terminal 3. The spindle motor (CXM-405) is located approximately in the center of the main unit. Only two mounting screws hold the spindle motor assembly in place. When replacing the spindle motor, take care not to let the spindle motor

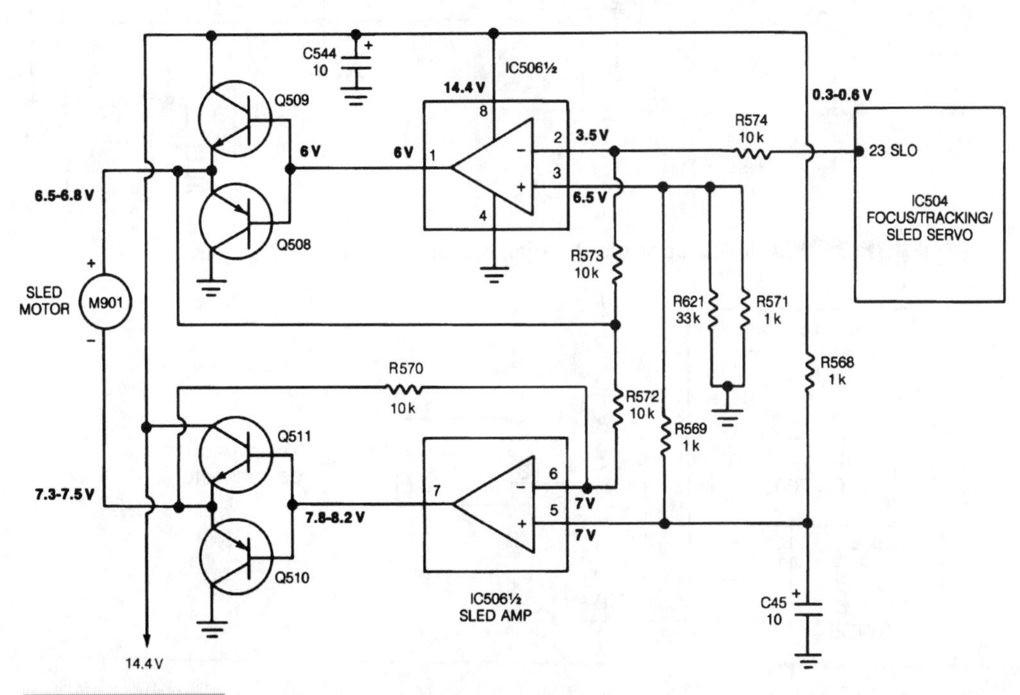

FIGURE 31-11 Q504 controls the focus, tracking, and sled motor in this auto CD player.

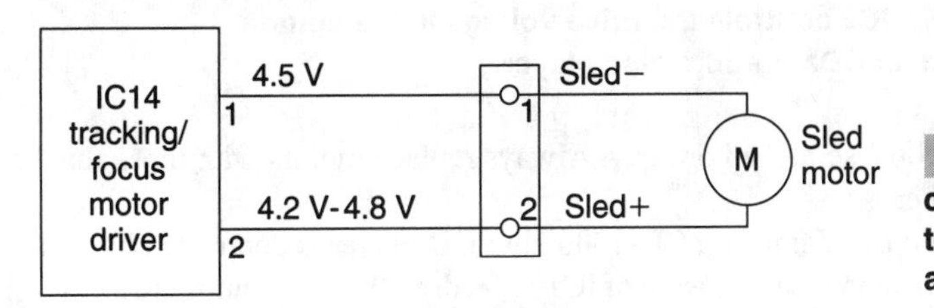

FIGURE 31-12 IC14 controls sled motor from terminals 1 and 2 of an auto CD changer.

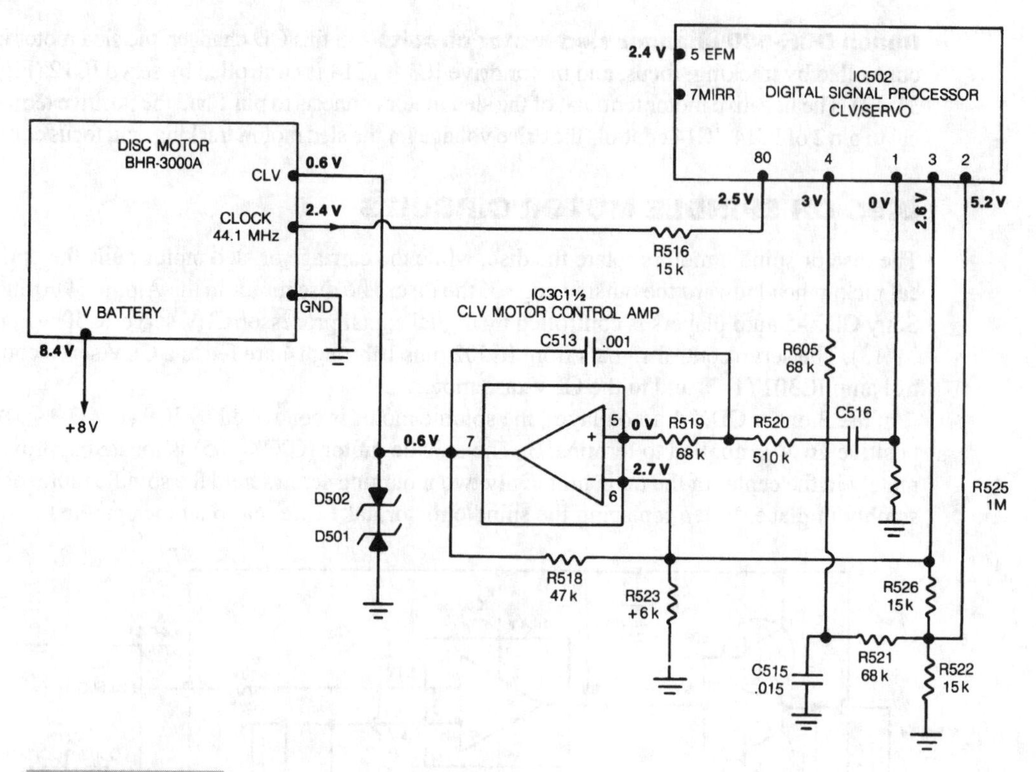

FIGURE 31-13 IC502 controls the disc motor in an auto CD player.

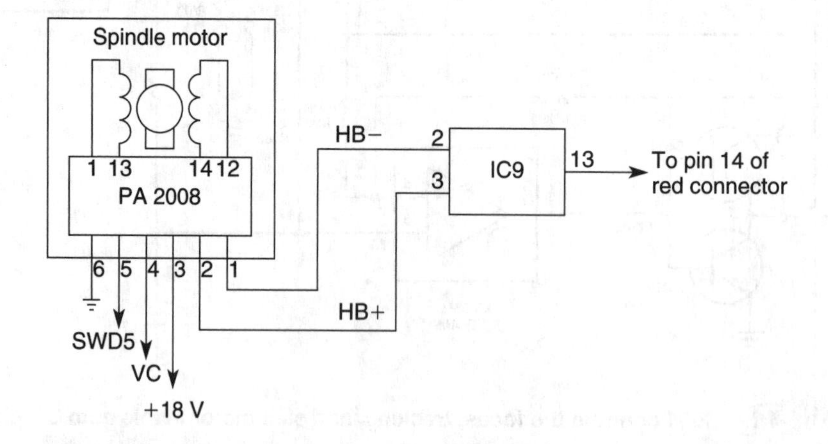

FIGURE 31-14 IC9 controls the drive voltage in the spindle motor of the Pioneer CDX-1 audio disc player.

flywheel shaft become scratched or dirty. Always replace motors with those that have the original part number.

The disc motor in the Yamaha YCD-1000 auto CD player is controlled by IC101 (Fig. 31-15). A disc drive amp located inside of IC101 feeds to the drive output IC214 (1–2). The positive terminal of the motor is fed from pin 10 of IC214, and the grounded side of the disc motor is fed through resistor R230 (2.2 Ω). If the disc motor will not rotate, check if R230 is open.

Denon DCH-500 spindle motor circuits The spindle motor drive signal is controlled by signal processing IC13 (Fig. 31-16). The signal is taken from pin 4 and is fed to pin 3 of spindle driver IC15. Pin 1 is directly coupled to the base terminal of Q17. The output emitter signal is fed to pin 8 of spindle motor IC1. IC1 and the spindle motor are mounted on the optical assembly.

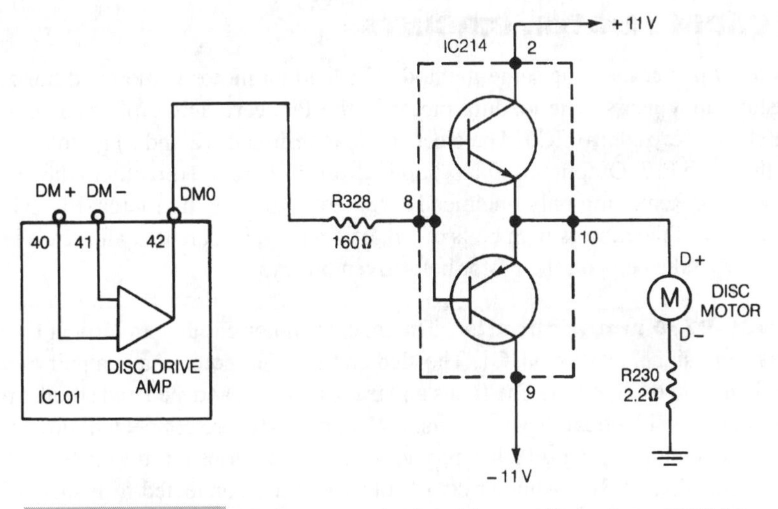

FIGURE 31-15 IC101 controls the driver voltage from IC214 to the disc motor.

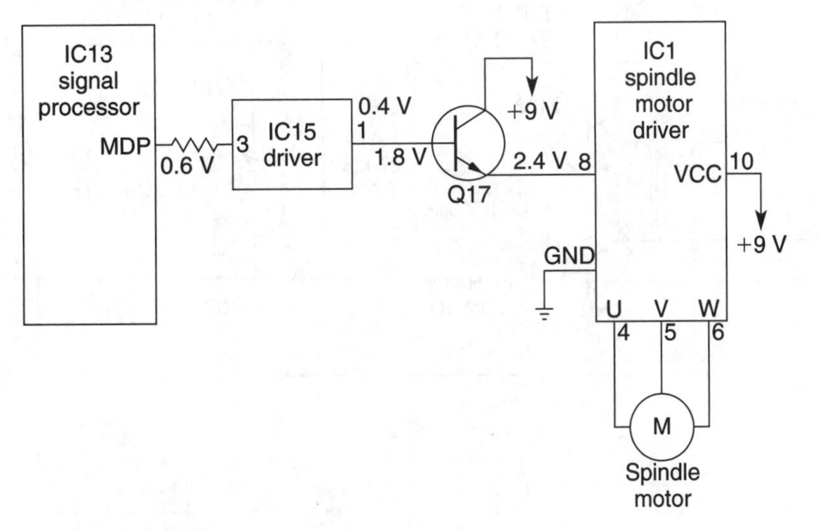

FIGURE 31-16 IC13 provides the MDP drive for IC15, Q17, and IC1 of the spindle motor.

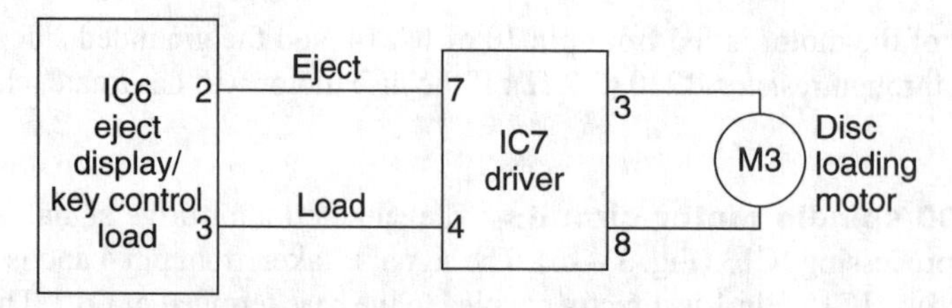

FIGURE 31-17 IC6 controls motor driver IC7 to the loading motor in this auto CD player.

THE LOADING MOTOR CIRCUITS

Alpine's 5900 player uses the same method. The loading motor is mounted horizontally with plastic pulley gears. The loading motor in the Pioneer CDX-1 player is controlled from the display key control IC6. The eject and load terminals (2 and 3) go to terminals 4 and 7 of the drive IC7. Output terminals 3 and 8 from IC7 are fed directly to the disc loading motor (M3). Removing only one metal screw frees the loading motor (Fig. 31-17). If you replace any of the motors with belts or a plastic pulley, be sure that all grease is wiped off. Do not twist the belts on the motor belt-driven pulleys.

Denon DCC-9770 motor circuits The loading motor circuits are driven by several transistors from signal processor IC1. The sled and spindle motors have a pair of transistors tied directly to each motor. The focus and tracking coils, and sled and spindle motors, each have the same identical drive circuits. Q408 and Q409 are connected directly to the spindle or disc motor. Op-amp IC401 provides drive in a series circuit for both sled and spindle motors (Fig. 31-18). Another set of transistors are connected to IC401, which is

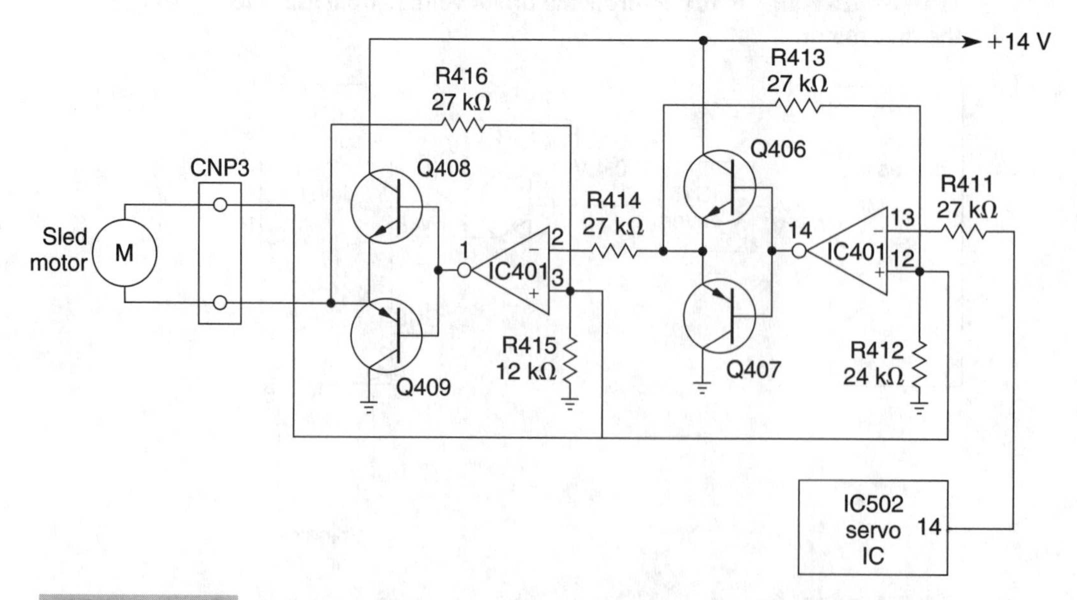

FIGURE 31-18 The sled, spindle, focus, and tracking coils are identical in this auto CD player motor and tracking circuits.

driven by a section of IC401. Four op amps are inside IC401. The SLD signal out of pin 14 of servo IC502 drives the SLED motor, and the signal from pin 39 of IC502 provides drive to the spindle motor circuits.

Troubleshooting the Motor Circuits

The suspected motor can quickly be checked by taking accurate voltage measurements across the motor terminals. This voltage measurement is rather low (1 to 10 V); for example, the Pioneer carriage motor operates at 6 Vdc. If the voltage is across the motor terminals, but the motor is not rotating, take a continuity measurement of the motor field coils. The resistance measurement can be less than 1 Ω across the motor terminals. If you are in doubt, clip a 1.5-V flashlight battery across the motor terminals. Always remove the motor plug or leads so that you do not damage other components. Do not apply the outside voltage across the input motor board terminal of a PLV solid-state motor assembly. If the motor appears to be normal, check the drive signal at the servo IC. Then, check the applied voltage to the power drive transistors or IC. Test each transistor with the transistor tester or a meter. Open or leaky transistors and ICs might produce an improper or no voltage to the motor terminals. Be sure that both positive and negative voltages are applied to the drive transistors.

POWER HOOKUP

Usually, at least two different fused "A" leads are in the auto CD player: one for the memory and the other for the main player unit. The audio line output connects to the CD input terminals of the cassette-receiver component. In some auto CD players with two separate units, the cables must be connected tightly between the main player and hideaway unit. The CD player can be controlled by the power switch of the cassette-tuner.

Denon DCC-9770 power hookup The Denon CD player connects power to the unit at the battery and ground cable (Fig. 31-19). This AM/FM CD player operates from the 14.4-V car battery. The orange and white wires connect to the light switch and the red wire (ACC) connects to the ignition power supply wire. A CD auto changer can be connected to the CD changer cable (such as DCH-500). Two separate stereo power amps connect to the rear and front external amplifiers.

THE POWER SUPPLY

Most auto CD players contain a dc-to-dc converter unit connected to the 14.4-V power "A" lead connection. The converter is controlled by an IC oscillator and two high-powered input transistors (Fig. 31-20). The output transformer (T1) produces several voltages on the secondary windings. Full-wave bridge and diode-rectification actions are produced in the secondary of T1.

Several IC and transistor regulators are used in the low-voltage power supplies. Transistors and zener diode regulators are used in some regulator circuits. Higher dc voltage (30 V) is fed to the display-tube drive/decoder in some early auto CD players. Some CD players use either a half-wave or full-wave power supply.

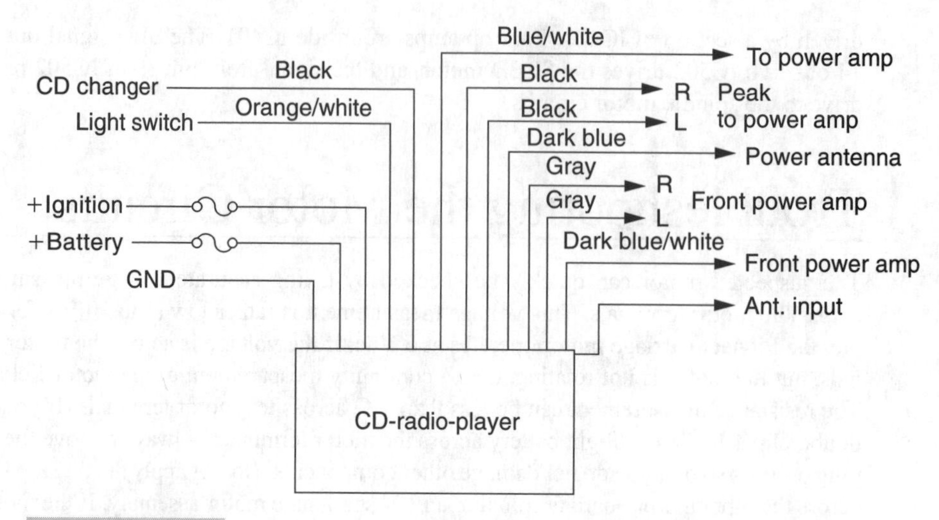

FIGURE 31-19 **The power hookup of a typical auto CD player.**

The Pioneer CDX-FM45 CD system has +5-, +8-, and +10-V regulation circuits. The +8-V regulator circuits provide working voltage to the D/A and audio-output circuits, and the +5-V regulated supply connects to the front-end signal processor, servo and optical circuits (Fig. 31-21). The +10-V regulator provides voltage to the motor driver circuits.

The dc-to-dc converter power supply in the Pioneer CDX-1 auto player has four different regulator ICs for the different power sources (Fig. 31-22). Two separate secondary windings provide +16-V, +12V, and +5V power sources. Notice that the +16V taps are taken before IC regulation. Check the regulator circuits like any voltage-regulated circuit.

Accurate voltage measurements and scope waveforms of the IC-to-transformer dc-to-dc converter solve most power-supply problems. Always check the output voltage from the IC or transistor regulator circuits that tie to the section that is not operating. Disconnect the voltage output terminal of the regulator of the power supply to see if the overload is in the power supply or connecting circuits with low voltages. Suspect components within the power supply if low voltage exists with the power-source lead disconnected. Troubleshoot the connecting circuits if the voltage is normal at the output terminal of the voltage regulator.

Denon DCH-500 focus offset adjustment

1. Connect the oscilloscope to the servo board test point RF (Fig. 31-23).
2. Put the set into Play mode by loading disc.
3. Adjust (servo) board RV12 so that the scope waveform eye pattern is good. A good eye pattern means that the diamond shape in the center can be clearly distinguished.

Denon DCH-500 focus gain adjustment The coarse focus gain adjustment is performed when replacing the optical assembly block or RV14.

1. Set RV14 (servo board) to the standard position (Fig. 31-24).
2. Check that there is not an abnormal amount of operation noise (white noise) from the 2-axis device. If there is, turn RV14 slightly clockwise. If the gain is higher, the set

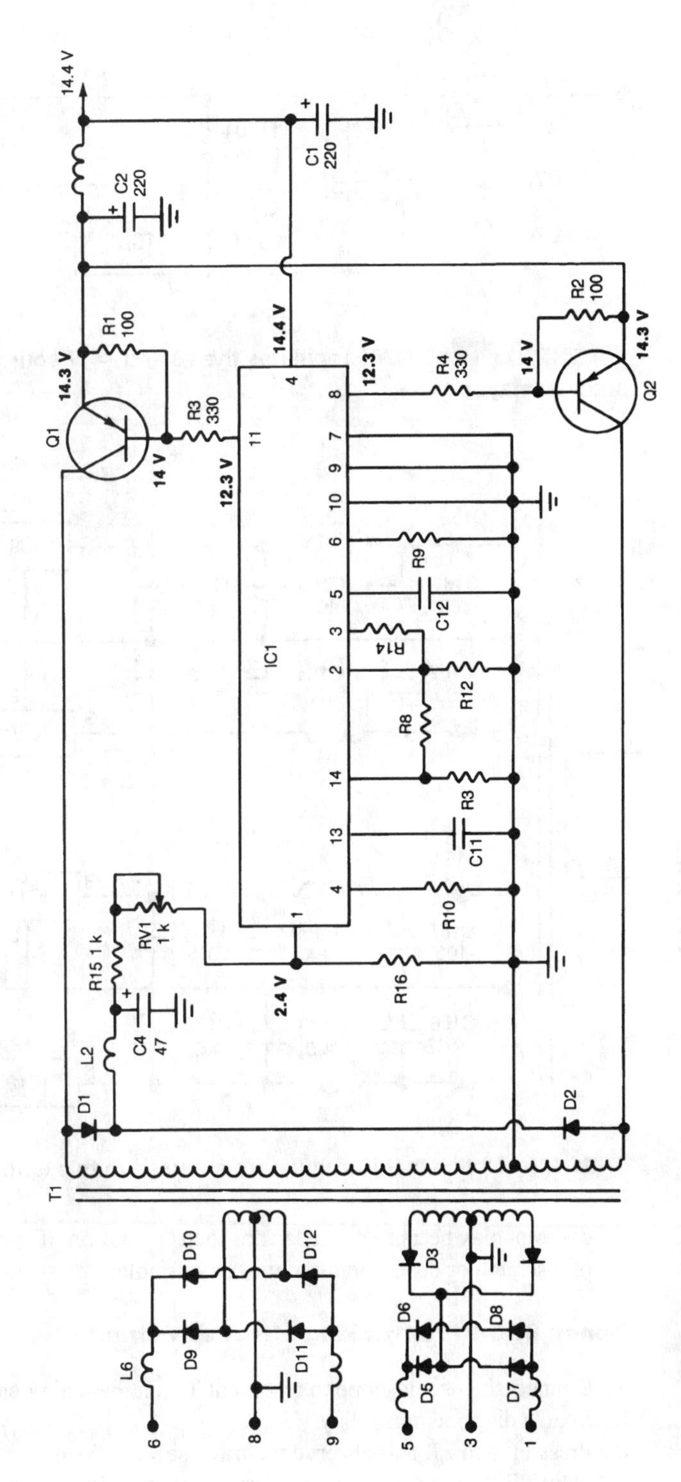

FIGURE 31-20 A typical IC dc/dc converter power supply.

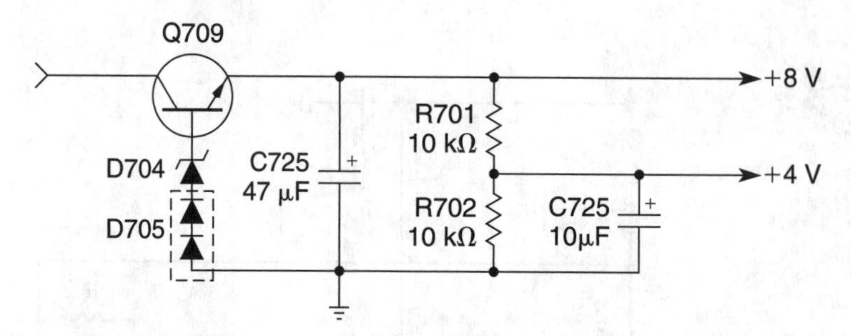

FIGURE 31-21 Q709 regulates the +8 and +4-V source in an auto CD player.

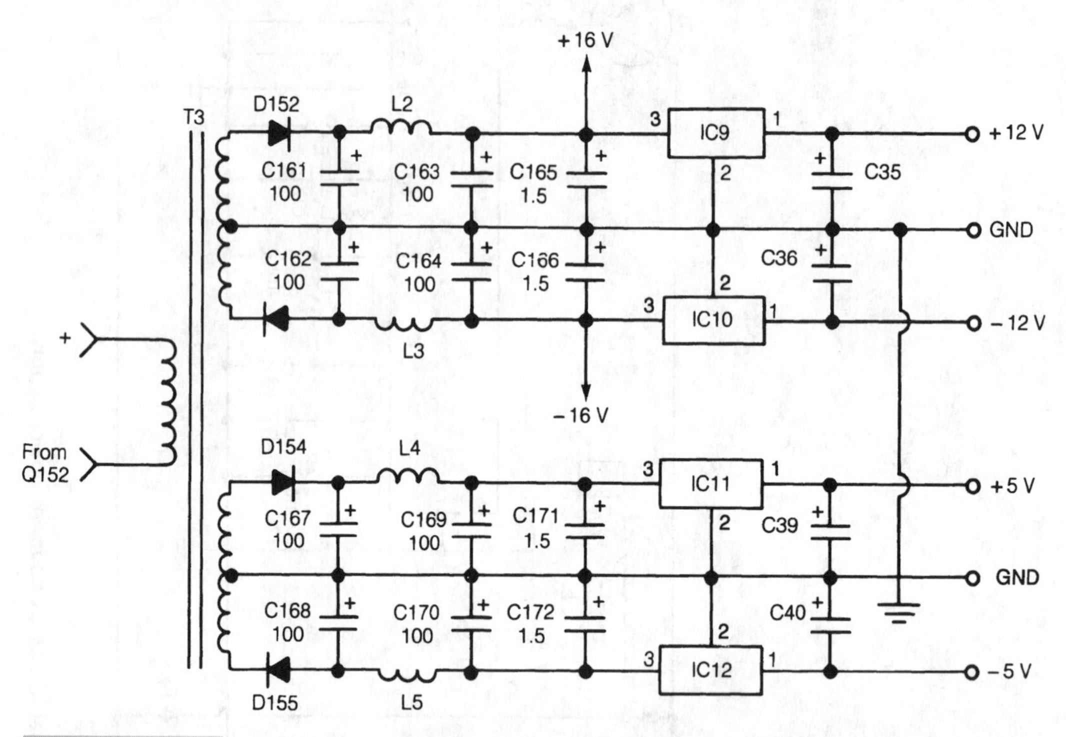

FIGURE 31-22 Notice that IC regulators are used in this auto CD power supply.

does not play because of the lack of focus operation; if operation noise is heard because of a scratch or dust, operation will be unstable.

Denon DCC-9770 tracking offset adjustment

1. Connect the oscilloscope to test point TE on the main board.
2. Load a disc and press Play.
3. Press FF and FR and observe the traverse waveform.
4. Adjust RV501 so that the waveform is symmetrical when centered at the level to be accessed (See Table 31-2).

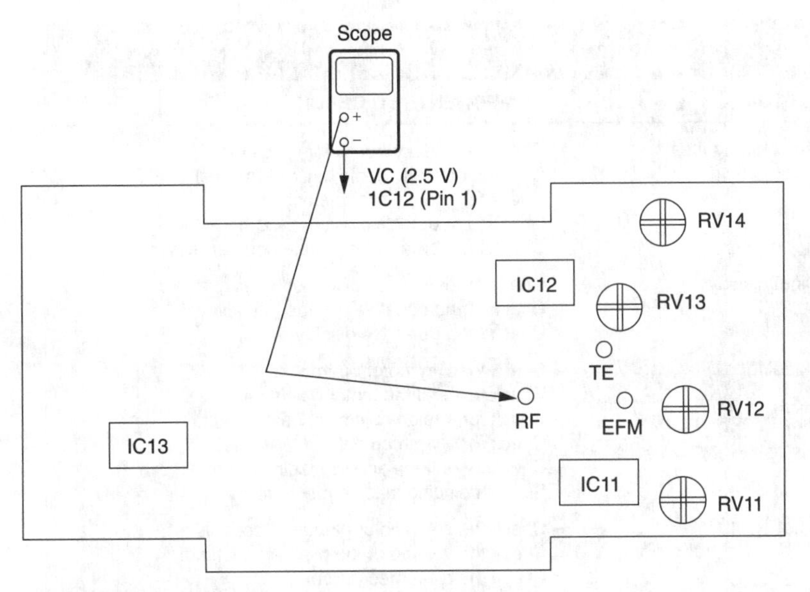

FIGURE 31-23 The focus offset adjustment for an auto CD changer.

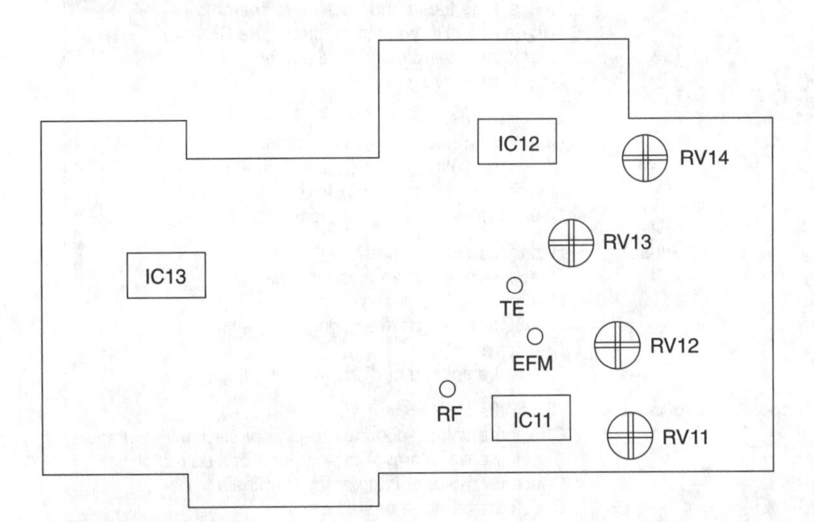

FIGURE 31-24 Adjust RV14 for focus gain adjustment in an auto CD changer.

TABLE 31-2 AN AUTO CD PLAYER TROUBLESHOOTING CHART

SYMPTOM	COMPONENTS TO CHECK
Dead	Check for the open fuse. Check the fuse holder. Check for an open transistor or IC regulator. Check the various voltage sources.

TABLE 31-2 AN AUTO CD PLAYER TROUBLESHOOTING CHART (CONTINUED)

SYMPTOM	COMPONENTS TO CHECK
Keeps blowing fuse	Check for a leaky transistor regulator. Check for a leaky audio output transistor or IC. Check for a leaky filter capacitor. Check for a defective voltage source. Check for a burned fuse cable and harness.
No light display	Check for defective powers sources. Check for no negative voltage to display. Check for a defective display circuit.
No loading	Check loading motor operation. Check to see if the motor belt is off. Check for foreign objects in gear operation. Check for a jammed belt. Check for voltage across loading motor. Suspect binding loading mechanism.
Will not eject	Check the solenoid or plunger mechanism. Check the voltage at the plunger solenoid. Check the eject mechanism.
Starts up and shuts down	Quickly scope for the RF or EFM waveform. Notice if the focus and tracking is searching. If there is no RF waveform, check the RF amp. Check the laser diode with a power meter. Check the servo loop.
No disc movement	Check the defective EFM. Check for a defective spindle circuit. Check the voltage on the spindle motor. Check for a defective transistor or IC driver. Check the servo or signal-processor circuits.
No disc rotation in playback	Has focus been locked? Check for a defective mechanism. Has the tracking closed? Check for a defective tracking mechanism. Check the EFM. Check the spindle or disc motor circuits.
Search operations abnormal	Check the EFM waveform. Check the voltqage at the carriage or sled drive voltage. Check the waveform at the carriage or sled motor. Check for motor voltage at the terminals. Check the carriage or sled motor. Suspect motor driver transistor or IC. Check the servo loop circuits.
No sound	Does the disc rotate? Has the focus been locked? Check for normal EFM. Check for sound at the output of D/A. Check the audio amp circuits. Check the headphone circuits.
Noisy sound	Check the sound circuits. Suspect a noisy output IC or transistor. Check the D/A converter circuits. Defective audio output circuits.

TROUBLESHOOTING CD PLAYER CIRCUITS

CONTENTS AT A GLANCE

Troubleshooting the compact disc player is not as difficult as it might seem. Simply isolate the various symptoms on the block diagram and apply them to the schematic to locate the defective section. A good schematic diagram is a must when servicing the compact disc player (Fig. 32-1). Of course, locating problems in the power supply and audio circuits

FIGURE 32-1 **Take crucial waveforms and voltage tests to locate the defective component.**

could be easily done without the schematic. Always have the schematic handy when servicing the RF path, focus, and servo circuits.

Crucial Waveforms

Besides taking correct voltage and resistance measurements, crucial waveforms are needed in most CD circuits. You will find actual waveforms and voltage charts in the service literature of each CD player. The service literature can be obtained from the local CD distributor, manufacturer's distributor, service depot, or from the manufacturer itself. If you are doing warranty work for certain brands of TVs and compact disc players, subscribe to the CD service literature.

You can follow the RF or EFM signal path with crucial scope waveforms (Fig. 32-2). Just by taking input and output waveforms on the crucial servo and focus ICs, you can determine if the component is defective. Often, each processor or IC is isolated with crucial waveforms. If you are working only on a couple of brands of CD players, take the crucial waveforms of a working model and mark them on the service diagram for future reference. Always mark on the outside area of the schematic the defective components and voltages found when locating a defective part.

A static wristband should be used while servicing the CD chassis to keep your body at the same potential as the compact disc player. Clip the ground clip to the CD chassis from the static wristband. Be extremely careful when attaching test equipment and taking voltage measurements and scope waveforms on the compact disc chassis. Otherwise, you could damage the fragile processors and ICs.

The Most Common Problems

Turn on/shut down is one of the most disturbing service problems with the CD player. This service symptom can be caused by a defective RF or EFM section (Fig. 32-3). When the CD player is first turned on, take an EFM waveform at the output of the RF amplifier. Al-

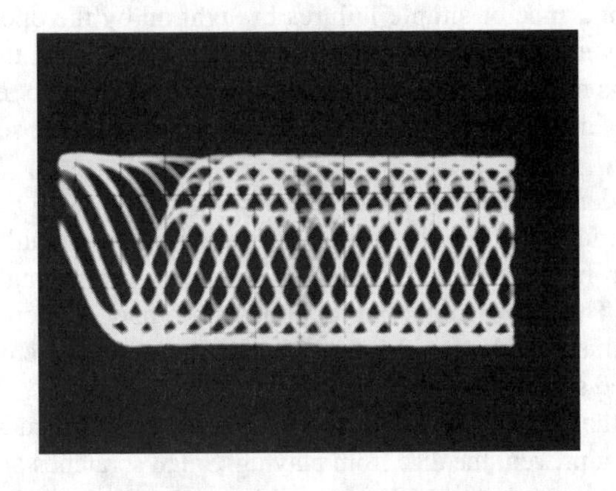

FIGURE 32-2 Observe the eye pattern (EFM) to see if the laser and preamp are normal.

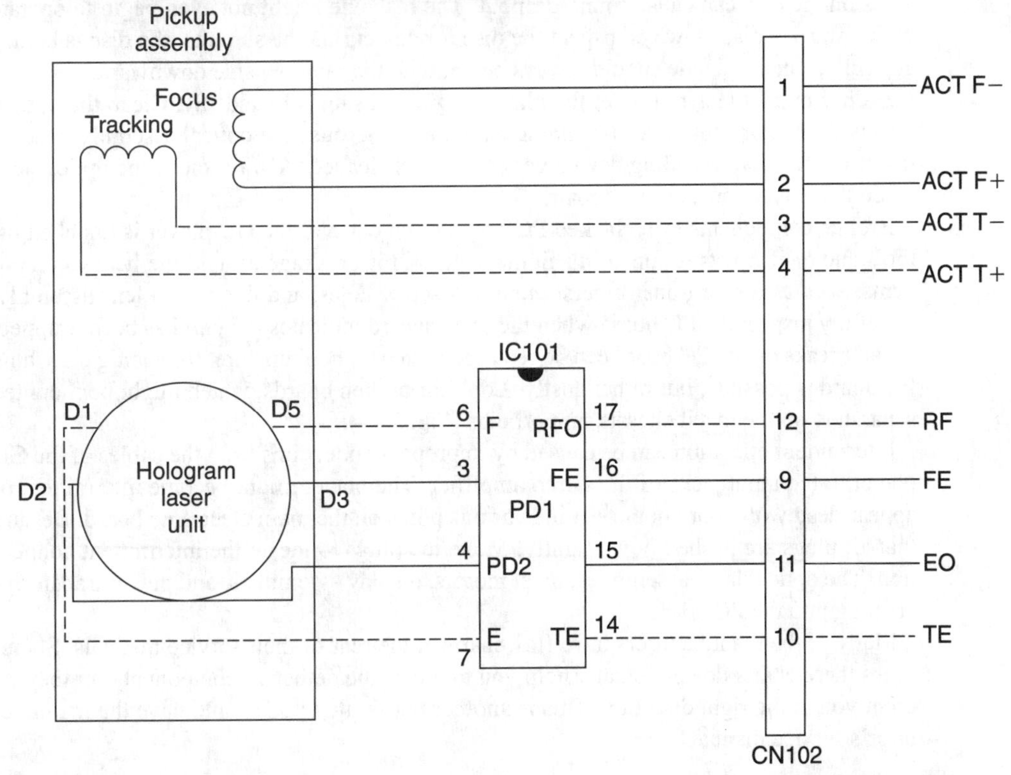

FIGURE 32-3 Defective laser assembly and preamplifier IC101 and cause shutdown in the CD chassis.

though the player shuts down in a few seconds, if the laser pickup and RF amplifier are operating correctly, you should have an eye-pattern waveform. If the waveform is missing, suspect an RF amp IC or laser pickup assembly is defective. Of course, the servo and signal processor will shut down after a few seconds with no EFM or RF waveform.

Many of the problems related to the compact disc failures are common to any player. These troubles can be listed as nuisance or simple failures brought on by the operator. Knowing how to operate the player can solve a lot of nuisance problems. Read the operating manual several times. This might also apply to the service technician who repairs many different brands of CD players. Knowing how they operate saves valuable service time.

Check the transit screws located at the bottom side of the player if the unit will not operate. These screws should be removed before the player is fired up. They should be in place when the player is moved from the shop to the home or vice versa, preventing damage to the laser optical assembly. The player might not turn on, start to play, or stop without removing the transit screws. In some players, the transit screws are removed entirely; in others, they can be loosened and still remain with the player.

Scratched or dirty discs make the player operate for a few minutes and then suddenly stop. Deep scratches in the disc can prevent the disc from playing. Small scratches or dirty areas on the disc can cause sound dropout. The machine might not even begin to operate with a warped disc. Always inspect the disc for defects and be sure that the disc is loaded correctly. The label side should always be up, with the rainbow side downward.

A dirty optical lens might let the player begin and shut off suddenly. Clean the optical lens with regular camera lens-cleaning equipment. Because the optical assembly is down under the flapper, a loading device very seldom is cleaned. Always clean the optical lens when the player comes in for repair.

Broken components or a cracked PC board often occur if the CD player is knocked off the stand or if it was dropped in shipment. Look for breakage around the heavy components, such as power transformers, mounting screw holes, and the optical lens assembly. Carefully inspect the PC board when the service card indicates the unit has been dropped. Small breaks on the PC board can be repaired with bare hookup wire. Replacing the whole PC board is possible, but rather costly. Look for broken boards, which might be mounted separately on the metal chassis support tabs (Fig. 32-4).

Intermittent operation can be caused by improper connections from the cables of the CD player to the input jacks of the stereo amplifier. The player might become intermittent or appear dead with poor contacts at the various plugs on the main electronic board. Be sure that all plugs are pushed down tightly. Move the plugs to locate the intermittent connection. The optical laser assemblies, most motors, display assemblies, and power transformers plug into the PC board.

Many of the manufacturers have flowcharts at the rear of their service manuals. Sometimes these charts do not actually help you to locate the defective component, but they do point you in the right direction. This is another reason that you should have the manufacturer's service manual.

Realistic CD-3370 portable flowchart Use the following flowchart to troubleshoot most portable CD players (Table 32-1). After completing a particular branch sec-

FIGURE 32-4 **PC boards mounted separately might be damaged if the boom-box player is accidentally dropped.**

tion, go through the main flowchart again to be sure that there are no other irregularities. By following this basic flowchart, you can apply it to any portable player. Use Fig. 32-5 to correspond with the chart and the various circuits.

Symptom: No power (defective power supply) Suspect that a fuse is blown or that the player is not plugged properly into the power outlet if the display will not light with the power switch pressed on. Do not forget to check the timer setting on some CD players. You might find that some CD player low-voltage circuits are not protected by a fuse. Inspect the power switch for breakage or poor contacts. If 120 Vac is measured across the power switch with the switch on, suspect that a switch is open. Dirty switch contacts can cause intermittent turn-on.

Turn the player off at once if you hear a loud groan from the transformer with the power switch on (Fig. 32-6). Remove the plug and check the low-voltage power supply for shorts. Often, if several diodes short out, the primary winding of the power transformer opens. Check the primary winding with the low-resistance scale of the ohmmeter. The bridge rectifier might be in one unit or there might be separate silicon diodes (Fig. 32-7).

Many electronic technicians go directly to the low-voltage power supply if the voltage is missing in a certain section of a TV or CD player. Because the power supply furnishes crucial voltage to the various circuits, test each voltage source for the correct voltage. You might find that the low-voltage power supply is at fault instead of suspected components in the CD circuits. It's wise to take Stop and Play mode voltages on each transistor or IC if you are servicing only one or two brands of CD players.

Leaky low-voltage and zener diodes, leaky or shorted voltage transistor regulators, and dried filter and decoupling electrolytic capacitors produce most problems in the low-voltage

TABLE 32-1 A PORTABLE CD PLAYER TROUBLESHOOTING FLOW CHART

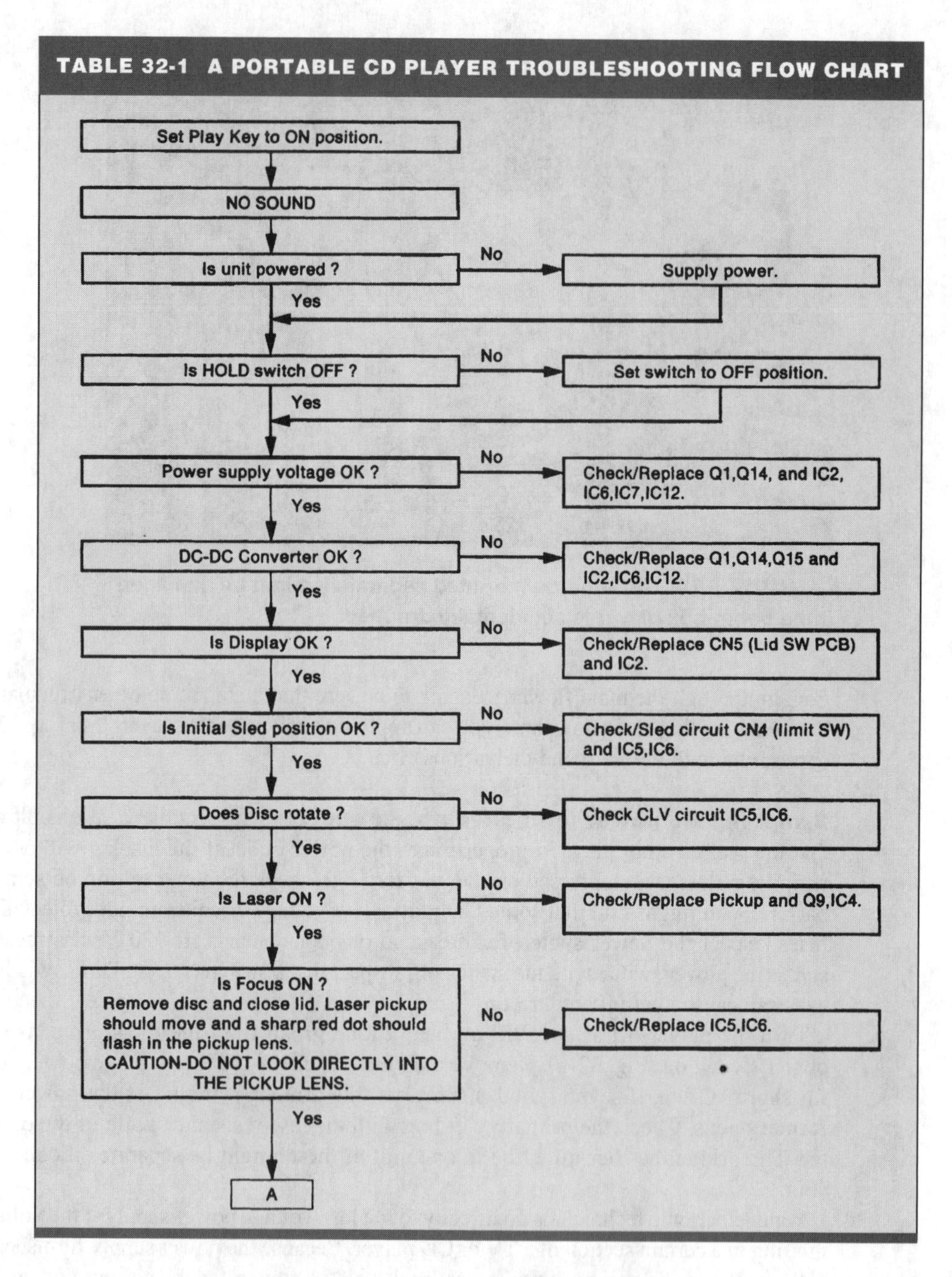

power supplies. Remove one end of each diode for correct leakage tests. It's best to remove the regulator transistors if you suspect that they are leaky. Test the transistor out of the circuit with the DMM or transistor tester. Check the suspected electrolytic capacitors in and out of the circuit with a digital capacitance meter (Fig. 32-8). For more on power-supply troubles, refer to Chapter 24.

INTERMITTENT POWER SOURCE

Intermittent components within the low-voltage power supply can cause intermittent operation. Monitor the various sections of power supply to determine where the intermittent component is located. Connect a voltmeter at the large electrolytic capacitor positive terminal and another at the output of the first IC or transistor regulator. If the intermittent is not in these circuits, proceed into the various voltage sources with a DMM.

Often, the intermittent component in the voltage source is a transistor or IC voltage regulator (Fig. 32-9). Sometimes, after the CD player operates a few hours and a regulator

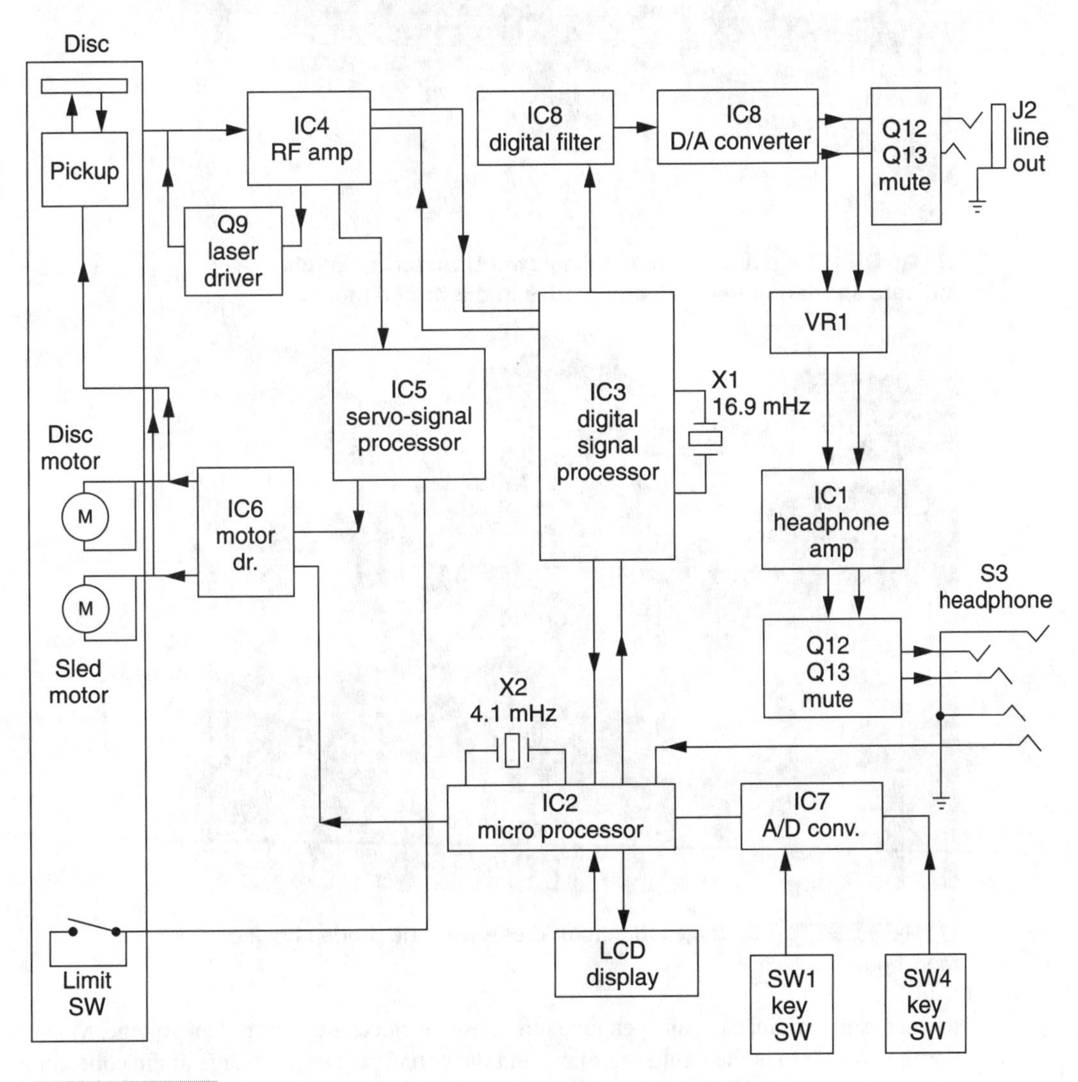

FIGURE 32-5 A block diagram of a typical portable CD player.

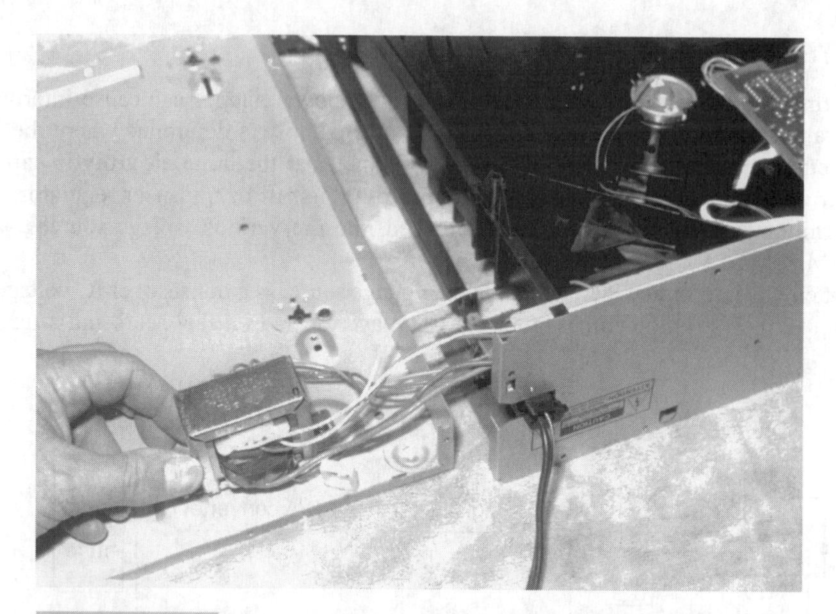

FIGURE 32-6 A loud hum in the power transformer might indicate shorted or leaky silicon diodes in the bridge rectifier.

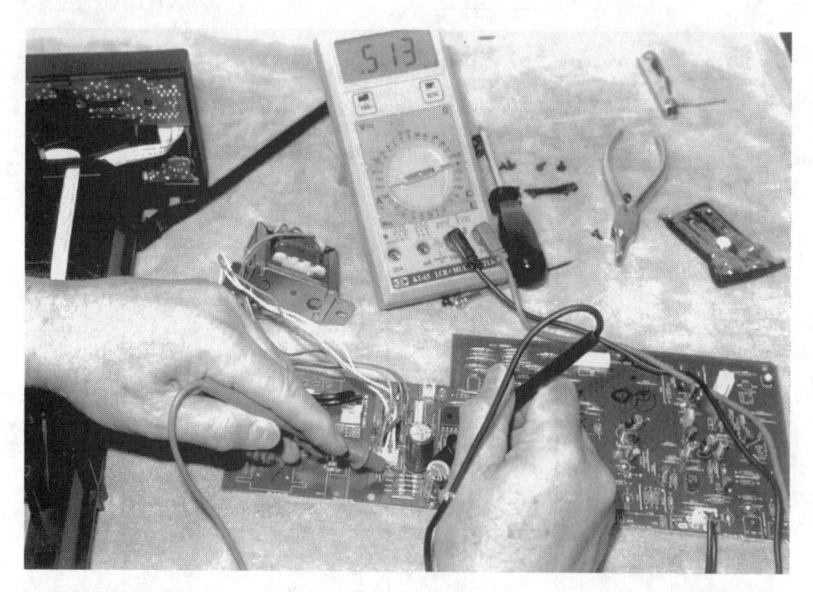

FIGURE 32-7 Check each silicon diode with the diode check of the DMM.

transistor gets warmed up, an open junction transistor can cause intermittent voltage. Monitor the voltage in at the emitter terminal and the output regulated voltage at the collector terminal. Regulator ICs can shut down intermittently after a few hours of operation. Inspect the transistor and IC regulator terminals for poorly soldered contacts.

Dead Sanyo CP500 No pilot light or display symptoms were found in this Sanyo CP500 compact disc player. A quick voltage measurement on the +14-V and +9-V bridge rectifiers indicated that there was no dc voltage source. In fact, diodes D601 and D602 showed leakage. The ac voltage measurement across the transformer secondary at pins 10 and 11 indicated no voltage.

The ac plug was pulled and a resistance measurement across the ac power plug indicated an open circuit. S901 tested normal. The primary winding across 1 and 5 were

FIGURE 32-8 Check suspected electolytic capacitor with a digital capacitor tester.

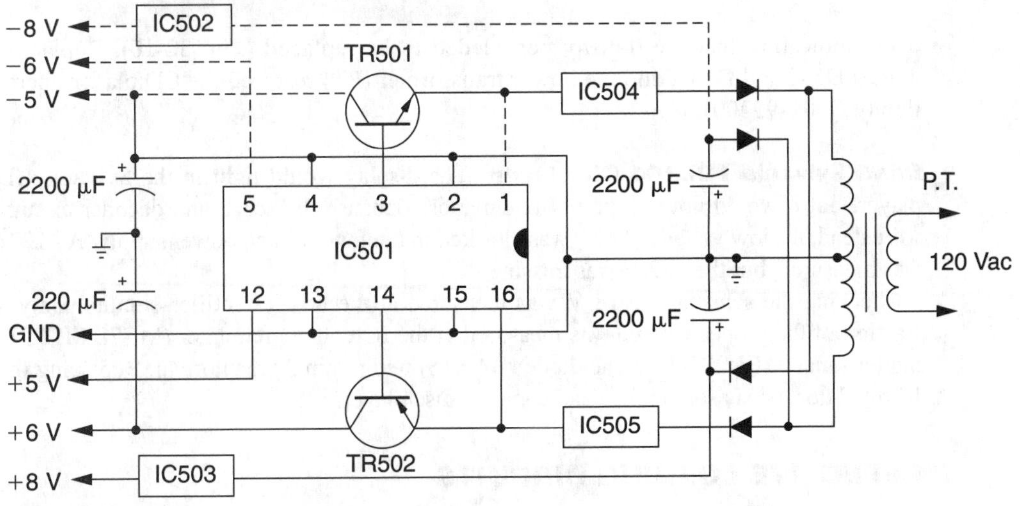

FIGURE 32-9 Check TR501, TR502, IC504, IC505, and IC503 for an intermittent voltage source.

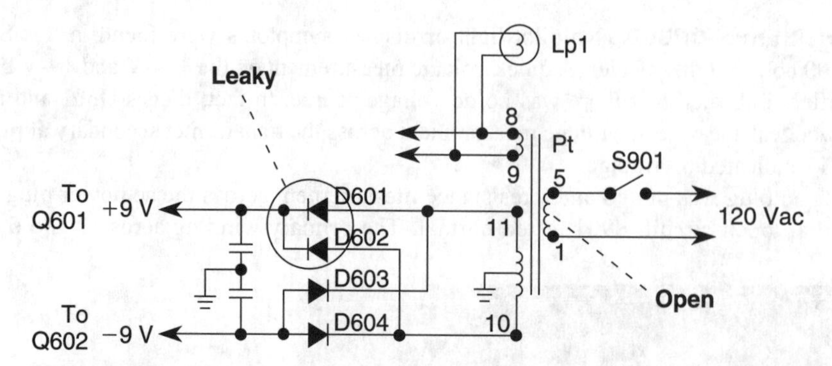

FIGURE 32-10 Check for leaky D601 and D602 diodes and for an open winding of the power transformer in this Sanyo CD player.

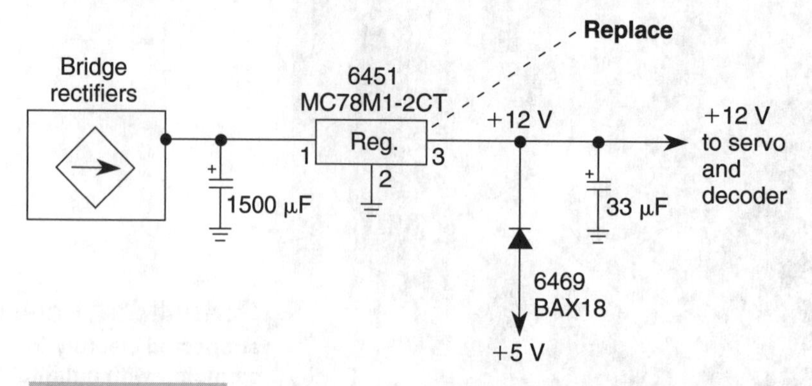

FIGURE 32-11 Replace the IC regulator (6451) in the 12-V source of a Sylvania FDD104 CD player.

open, indicating that the transformer needed to be replaced (Fig. 32-10). Replacing diodes D601 and D602, and the power transformer (PT) restored the CD player operation (4-300T-95200).

Dead Sylvania FDD104 CD player The display would light in the Sylvania CD player, but it would not accept any commands. Because the servo and decoder circuits were dead, the low voltage (12 V) was checked in the low-voltage power supply. A –12 V was measured, but the +12 V was missing.

Checking the schematic, +18 V was measured at the bridge-rectifier circuits. Only a fraction of the proper voltage was measured at the IC voltage regulator (MC78-M12CT) pin terminal 3 (Fig. 32-11). The diode (BAX18) tied to pin 3 was normal. Replacing the IC regulator (6451) solved the dead compact disc player.

DEFECTIVE LOADING CIRCUITS

When the power switch is turned on, the display panel lights come to life. In some models, the CPU Reset and Manual Search key must be on before the Open/Close button is pressed. When the tray is open, lower voltage is applied to the laser circuit, causing the laser not to emit light.

In some models, a leaf tray loading switch is engaged when the door is open, reversing polarity voltage to the loading motor. The disc is loaded into the tray. Push the Close switch and the loading motor pulls the tray so that the disc is loaded on the spindle or disc platform. The laser beam is lighted. Now the disc will start to rotate when the Play button is pushed. If no disc is loaded in the tray, the disc motor will not play music with an interlock governed by a phototransistor LED circuit.

In some models, ICs govern the opening and closing functions of loading the disc. When the Open button is pressed, the tray should start to move out because one of the pins on the IC control is low. The loading motor rotates. While the tray is open, one pin is low and the other is high. The disc is loaded. Press the close key or button and the low pin goes high and the loading motor rotates, pulling the door closed.

DRAWER OR TRAY DOES NOT MOVE

Listen to hear if the loading motor is trying to rotate. A damaged or jammed gear track can keep the motor from operating. If someone grabs the drawer during operation, it can strip or jam the plastic slide gear assembly. Go directly to the motor drive transistors and check the positive and negative voltage applied to the drive transistor collector terminals (Fig. 32-12).

The dc voltage applied to the collector terminal of Q325 is +10 V with –10 V on the collector terminal of Q326 (in a JVC XL-400B loading motor drive circuit). When the tray is open, the emitter terminals of Q325 and Q326 measure about –7 V; when the tray is closed, +7 V are measured there. Check the voltage at the motor terminals. No voltage might indicate a poor connection at the motor or plug-in connections (P305). Normal voltage at the motor terminals indicates a dead motor.

Remove one end of R503 from the base terminals of Q325 and Q326. Now, measure the voltage at pin 1 of the servo controller (IC301). You should measure about –7.5 V when the tray is open and +7.5 V when tray is closed. If not, suspect that IC301 or IC802 are defective.

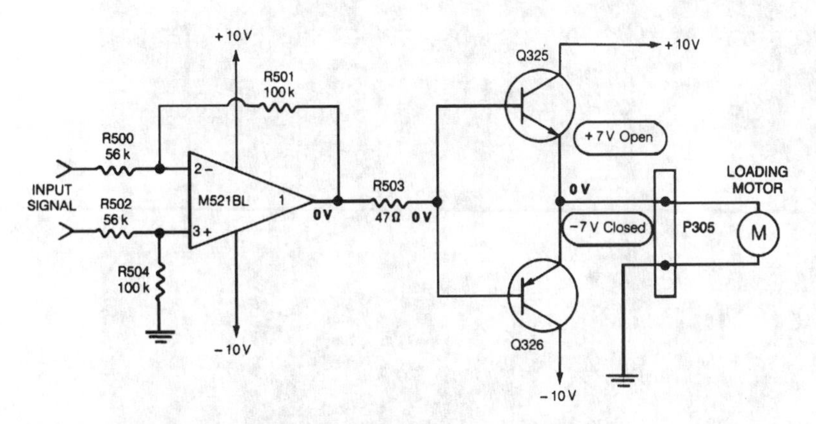

FIGURE 32-12 Locate a defective drive transistor or IC if the loading motor will not rotate.

No loading: Goldstar GCD-616 CD player The loading tray would not move out or in. When the tray or loading box was pressed, the sound of the loading motor operating or tray movement could not be heard. No voltage was measured on the loading motor terminals (M603). The motor leads were traced back to loading motor drive IC104. Voltage measurements on IC104 were taken with a 7-V supply on pin 6 (Fig. 32-13). Controlled voltage was found on pins 1 and 3, but not on pins 7 and 9. Replacing IC104 (BA61218) solved the no-loading problem.

Erratic loading: Magnavox CDC552 CD changer In this Magnavox CDC522 CD changer, the large carousel unit would sometimes catch and not move all the way out (Fig. 32-14). The drawer would not fully extend by about two inches. The large carousel track was inspected and seemed normal.

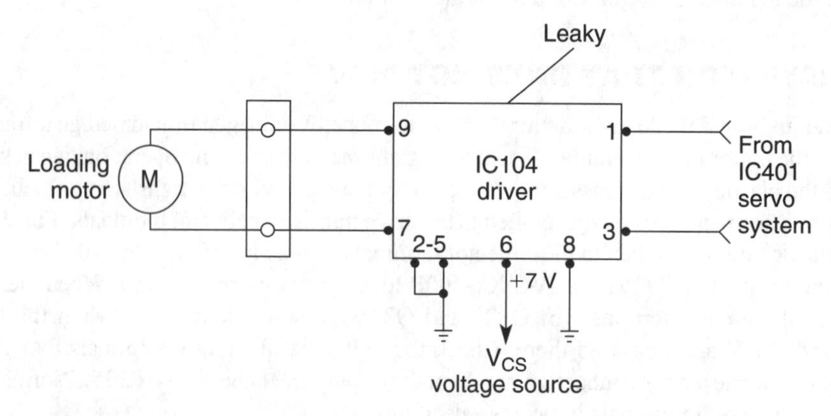

FIGURE 32-13 Replace IC104 in this Goldstar GCD-616 player, which had a no-loading problem.

FIGURE 32-14 Erratic loading in this Magnavox CDC552 changer was caused by loose wires.

FIGURE 32-15 Check for binding rails if the loading is intermittent.

The motor belt was tight, but showed some signs of slippage. After cleaning the motor pulley and belt, the unit still would not come all the way out consistently. Removing the large tray revealed that a long connecting wire was loose and sometimes would catch on the tray assembly and hold it back. The unit had been worked on before and the bunch of wires and cables laying in the bottom chassis area were not tied down.

INTERMITTENT LOADING

Check the loading tray for binding rails, foreign matter, a broken or loose loading motor drive belt, driver IC and transistors, and motor for intermittent loading problems. If the tray comes out and will not go in, check for poor or dirty switch controls and tray binding (Fig. 32-15). Monitor the loading motor voltage to determine if the problem is electrical, electronic, or mechanical.

Monitor the signal at the servo IC and motor drive IC or transistor. Determine if the voltage and drive circuits are normal. If the rotation of the motor stops, tap the end belt of motor and notice if it starts operating. Replace the intermittent motor. If the loading motor operates without any problems, check the motor drive belt for oil on the surface or a loose belt. Do not overlook the possibility of poorly soldered connections or plugs that go directly from the PC board to the motor terminals.

TRAY WILL NOT OPEN

Suspect that a motor or sliding gear is broken if the motor is rotating and the tray will not move outward. Inspect the sliding gear for broken spots or a dry surface. Do not overlook a damaged or broken tray open/close (loading switch). This switch is usually found close to the tray or plastic clamper gear (Fig. 32-16). A damaged or frozen tray or teeth missing from the plastic clamper gear might prevent the tray from moving (Fig. 32-17).

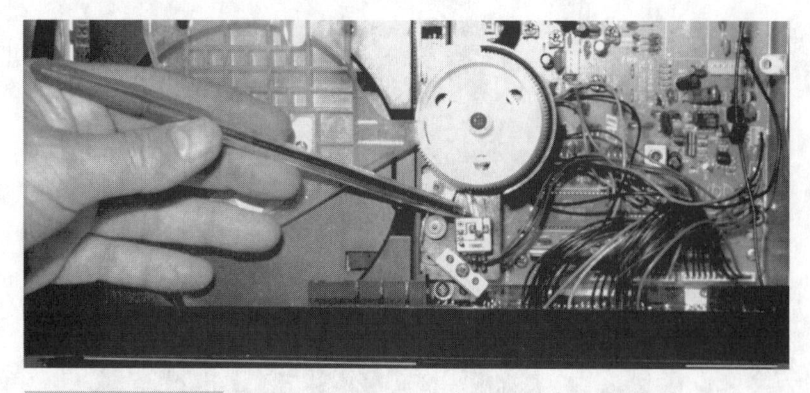

FIGURE 32-16 Check the Reverse Loading switch for dirty contacts when the tray will not open or close.

FIGURE 32-17 Check the teeth on the gear assembly if the loading tray is binding.

TRAY WILL NOT CLOSE

Inspect the rail-tooth gear slide assembly for broken teeth. Suspect dirty contacts on the close/open loading switch. Clean the contacts with a thin piece of cardboard. Hold the contacts together and push-pull the cardboard between the contacts to clean them. Be sure that the correct voltage is applied to the motor for the closing position. Also check for broken motor belts, missing pulleys, and broken plastic motor gears. (Check Chapter 26 on motor circuits for additional troubleshooting methods.)

POOR TURNTABLE ROTATION: MAGNAVOX CDC745

Intermittent or slow rotation of the roulette or turntable motor in the five-CD changer was caused by a defective turntable motor (Fig. 32-18). Sometimes, the turntable motor in a CD player is also called the *disc* or *spindle motor*. In this case, the turntable motor rotates the tray to select the correct disc to play or after the entire disc has been played and selected to go to the next disc.

Monitor the dc voltage at the motor terminals. When the turntable stops, notice if voltage is present at the motor. Tap the end belt of the motor and notice if the turntable motor starts or stops. Replace the motor if voltage is present, but the motor shaft does not rotate.

DEFECTIVE DISPLAY SECTION

Check the wiring and socket if the display section will not light when the player is operating. Inspect the display PC board for possible poor connections for intermittent or dead conditions. The wiring can be checked with the low-resistance range of the ohmmeter.

FIGURE 32-18 A defective turntable motor caused no rotation in this Magnavox CDC-745 changer.

Check the small pin, soldered or clamped section around each wire cable. Notice if the flat cable has been inserted correctly.

Suspect a defective LED or lead wire if only one operation light is out that should be on. The LEDs and diodes can be checked like any LED with the diode test of the DMM. Check for soldered bridges or connections going to a specific light. Measure the voltage across the suspected diode. Poor seating of a button or switch jamming can prevent operation. Sometimes a switch can be held up when plastic binds against a plastic switch. Often, a drop of cleaning spray solves the binding problem.

Measure the dc voltage applied to the LED driver transistors or ICs. Do not overlook the ac or dc voltage applied to the fluorescent tube or LCD display. In some models, ac is applied from a power transformer winding.

A defective reset circuit section can prevent the display from lighting. Check the dc voltage source from the power supply. Check for pulses at the pins on the control IC, which controls the display section. The fluorescent display should light immediately after it has been turned on. If not, check voltages on the IC that controls it. Inspect the FL PC board for poorly soldered connections. Check the external clock waveform found in some models of the display data IC, which might indicate that the IC is operating.

No display: Onkyo DX-200 This CD player seemed to operate without any display features. Often, the display circuits operate from the negative voltage source of the low-voltage power supply. A quick voltage measurement on the fluorescent tube indicated that the negative voltage was missing.

Negative voltage regulator Q907 is located with D906 wired in reverse of the positive voltage power supply. The collector terminal had a negative 40.7 V with no voltage found on the emitter terminal. Q907 was open (Fig. 32-19). Replacing Q907 enabled the display to light again.

DEFECTIVE EFM OR SIGNAL SECTION

Sometimes players start to operate, then shut down. Sometimes they make a second focus attempt, then stop. Improper or no EFM signal can prevent CD operation. Remember, the EFM signal must be present to send an FE and TE signal to the focus, tracking, and sled

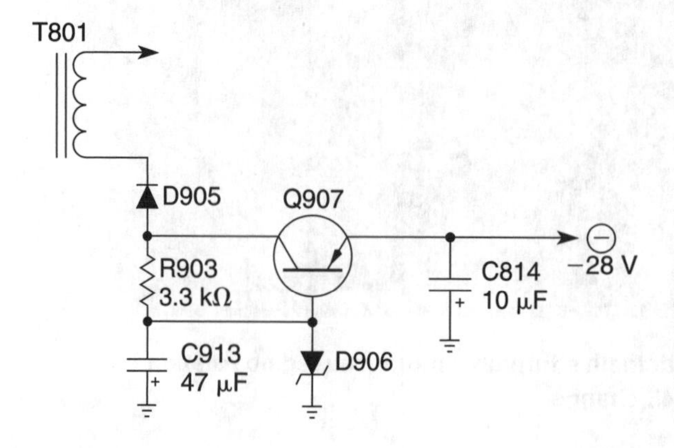

FIGURE 32-19 Open regulator transistor Q907 caused no negative voltage in an Onkyo DX-200 player.

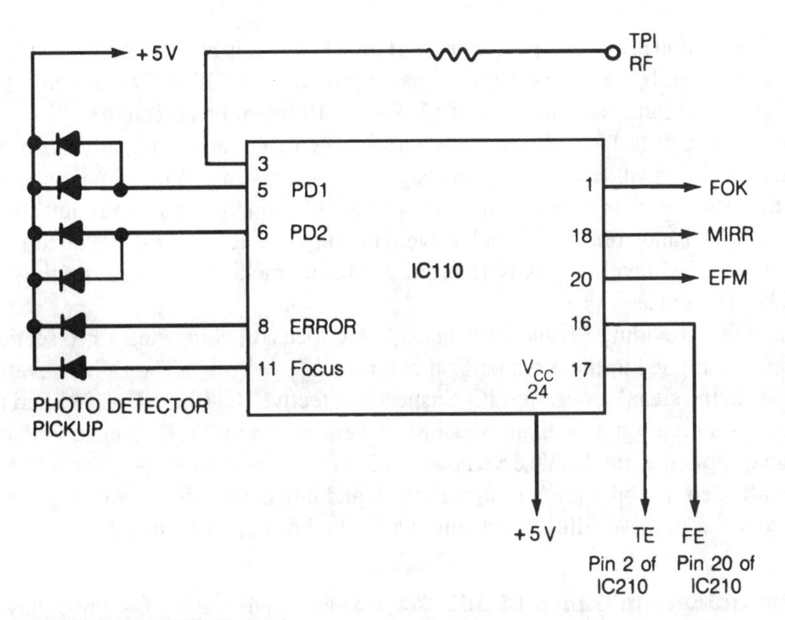

FIGURE 32-20 **The RF and EFM signal must be present for correct tracking, focus, servo operations, and shutdown.**

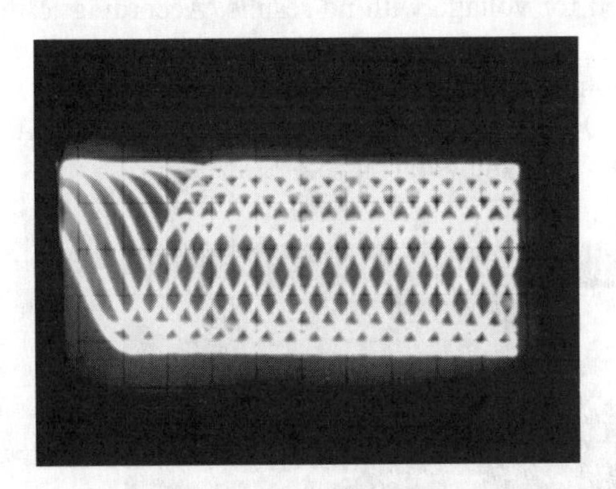

FIGURE 32-21 **Take the RF waveform at test point TP1.**

servo section. The EFM signal is sent to the digital control, CLV servo, and signal path to provide audio at the line output jacks (Fig. 32-20).

You might assume that the EFM signal is present from the preamp and RF IC when the disc continues to rotate. If not, take a quick RF waveform test at TP1 and pin 20 of IC110 (Fig. 32-21). Suspect a defective IC110 or no signal from the laser optical diodes. Improper laser voltage or a defective laser diode can result in low or no RF signal at the RF transistor IC.

LASER DIODE NOT LIT

The laser diode can be checked with the external power meter or indication tester. The crucial voltage of the laser test will indicate if the laser beam is present (check Chapter 23 for additional laser diode tests and procedures). If an RF signal is at TP1, the optical laser assembly is working. If there is no EFM signal at pin 20 of IC110, or if there are no focus and tracking error signals at pins 16 and 17, RF IC110 might be defective.

Before replacing IC110, check each terminal voltage and compare them to the schematic. Be sure that supply source voltage pin 24 is normal. You can assume the RF preamp transistor or IC is defective if there is an RF signal in and no signal out. By taking crucial waveform and voltages on each stage, you might be able to find the defective component. If the EFM level is greatly reduced when the tracking is closed, grating adjustments should be made.

In addition to providing FE and TE signals to the focus, tracking sled servo section, the EFM signal is applied to the digital signal control/CLV servo IC. The EFM waveform at pin 20 goes to the signal processing IC. Suspect a defective IC110 if RF at TP1 and the FE and TE signals are normal without an EFM waveform at pin 20. Of course, without the EFM signal applied to the CLV servo control IC, the disc will not rotate. The EFM signal might be affected if proper grating, focus offset, and tangential adjustments are not made. A quick touchup of these adjustments might help the level of EFM signal.

No laser indication: Sanyo CP660 CD player In a Sanyo CD CP660 player, the laser diode showed no signal of infrared power on the laser power meter (Fig. 32-22). At first, the laser interlock photodiode was suspected, but it turned out normal. The laser power IC121 terminals were checked for voltage with no results. According to the diagram, –9 V feeds the laser power IC.

Checking the wiring and voltage source indicated a –9 V was applied to R631, with no voltage at transistor regulator Q631. Q631 was tested and found open. Replacing the regulator transistor solved the no-laser power symptom (Fig. 32-23).

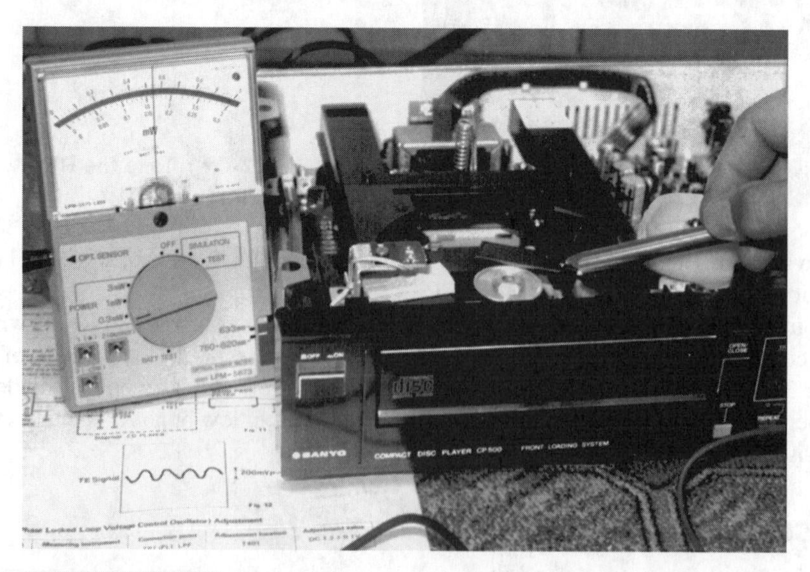

FIGURE 32-22 Check the laser diode with a power meter in a Sanyo CP660 CD player.

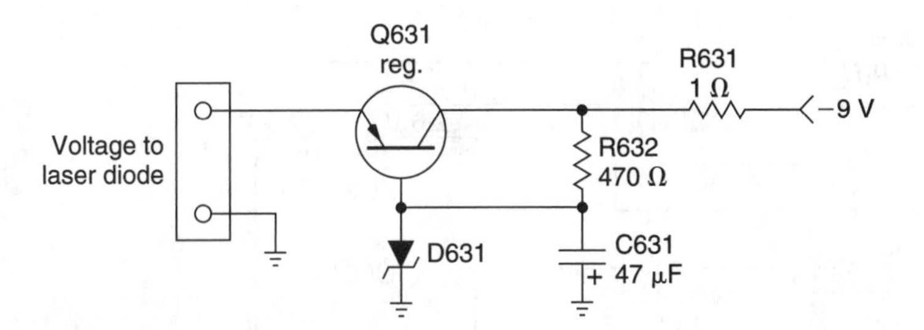

FIGURE 32-23 **An open Q631 provided no voltage to the laser diode.**

TROUBLESHOOTING OPTICAL SHUTDOWN CIRCUITS

The most difficult shutdown problems occur in the optical or digital signal-processing circuits. The disc starts to rotate; within a few seconds, the whole unit shuts down and the disc stops. Each time that the player is started, the unit begins to start, then stops. If the EFM signal is not sent to the signal-processing IC or data sent to the servo IC, the player will shut down. Also it's possible to have trouble within the servo circuits.

Notice if the tracking and focus coils begin to hunt when the unit is first turned on. Actually, there is not much time to get waveforms. Clip the scope to the circuits to be tested and start the player each time. You might have to start the player several times before all of the waveforms and voltage measurements are taken. A defective optical assembly, RF amp, and voltage sources can cause the chassis to shut down.

Place the laser diode cable over the laser lens to measure the infrared signal and if the laser diode is emitting power to the compact disc (Step 1). In some players, the interlock must have tape over the LED to take the place of a disc before unit will turn on. In battery-operated CD players, the small lid interlock has to be defeated by placing a paper clip or toothpick into interlock area. Keep your eyes away from the lens area while taking power measurements.

Turn on the laser meter and compact disc player. Notice if the disc or turntable is rotating. This usually occurs a few seconds before shutdown. If the laser meter indication is good, go to Step 2 (Fig. 32-24). If not, check the voltage applied to the laser diode from a laser driver or dc source. When the voltage is applied to the transistor or IC laser driver, check the voltage at the LD terminal. If there is no voltage, check the laser driver transistor or the voltage regulator circuits that feed the laser voltage.

Proceed to Step 2, when the laser diode power meter indicator is normal. Scope for EFM waveform at the RF amp (eye pattern). In some portable players, the RF amp and signal-processing circuits might be contained in one large IC. If the EFM signal is present for only a few seconds, you can assume that the circuits up to this point are normal.

If no EFM signal is present at the RF amp, suspect a defective RF IC, RF transistor, optical photodiodes, or low-voltage source (V_{CC}). Check the voltage at the supply pin (V_{CC}) 11 of IC501. If the voltage is missing or low, proceed to optical supply pin (K). Often, this voltage is supplied from a +5-V source.

Step 3. If the voltage is normal at the photodiodes and at PD1 and PD2, with no EFM signal at pin 8, replace IC501. Be very careful when replacing this RF amp. Do not apply too much heat to the IC. If after replacement, no EFM waveform or signal is present, suspect that an optical assembly is defective. Be sure that the IC was not damaged when it was

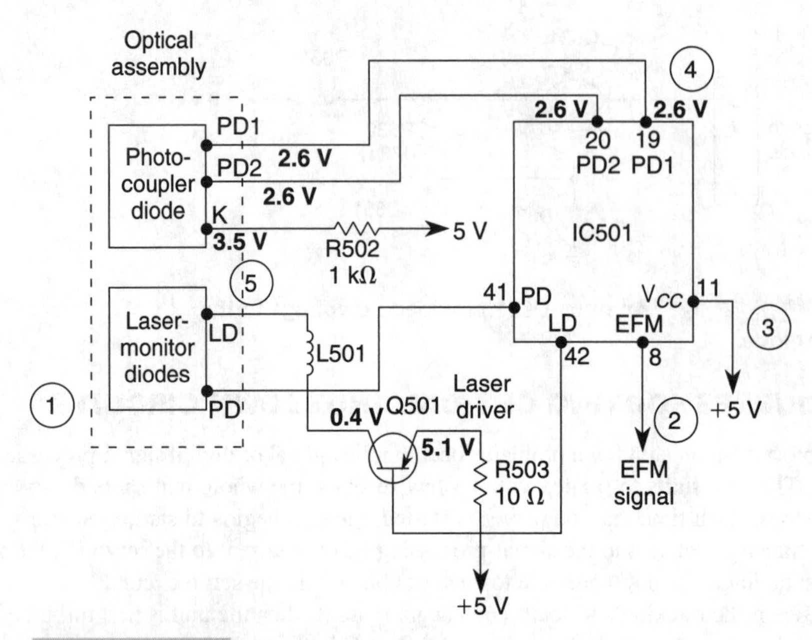

FIGURE 32-24 Check by the numbers to locate trouble in the laser pickup and RF amplifier circuits.

replaced; try installing another one. Also, check out the price of a new laser pickup assembly because they are quite costly.

Shutdown: JVC XL-V400B CD player In this CD player, the disc would start to rotate, then shut down. Often, when this happens, an improper RF or EFM signal is not being applied to the RF amp. The laser power meter indicates that the laser diode emits infrared light when the unit first came on. You cannot see the laser lens light because the infrared signal is invisible. Keep your eyes away from the lens area.

In these RF circuits, two RF transistors have an EFM signal from the laser photodiodes. No EFM signal or waveform was found at the signal processor or off Q201 and Q202. These measurements must be made quickly; several attempts must usually be made because the player shuts down. Q202 was leaky and was replaced.

DEFECTIVE FOCUS MECHANISM

Improper or no focus action can shut down the CD player. Does the lens actuator move up and down when the disc platter is closed by pressing the Open/Close button? The optical lens assembly might be covered with a flapper or clamped assembly, and it must be removed to check the actuator assembly. If the disc will not operate with the clamper assembly removed, check for an open interlock circuit. Sometimes the lens actuator mechanism can be seen with a reflective mirror when the disc is removed.

Focus lock is not achieved if the focus actuator does not move. Check for foreign or excessive dirt blocking the focus actuator. Clean the lens assembly at the same time. Quickly see if the EFM waveform is found at the RF signal-processing IC. Now check for a focus error waveform at the same IC. The focus error signal at IC110 should look somewhat like the waveform in Fig. 32-25. Often, an FE test point is found for this quick test (TP2). Doublecheck the FE signal where it enters the focus/tracking/sled servo IC (pin 20).

When an FE signal is applied to the focus servo IC, you should be able to check the waveform right up to the focus coil. The defective focus IC or component will indicate no waveform, only a white line at the focus driver stages and coil assembly (Fig. 32-26). If the focus circuits are performing, you should be able to take a waveform at pin 21 of IC211 and the emitter terminals of Q203 and Q202 (Fig. 32-27). The noisy-looking waveform is not present if the FE signal is missing.

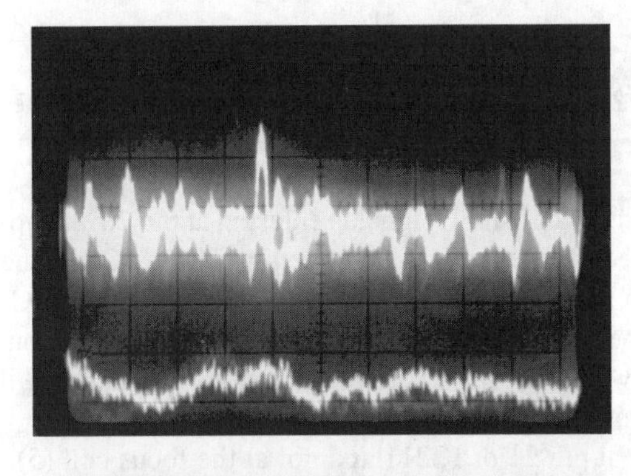

FIGURE 32-25 The focus error waveform taken from TP2 and IC110.

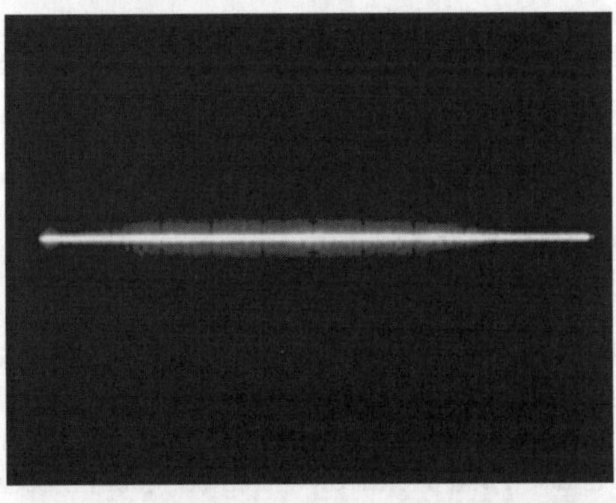

FIGURE 32-26 A horizontal white line on the scope at any test point means that no signal is present.

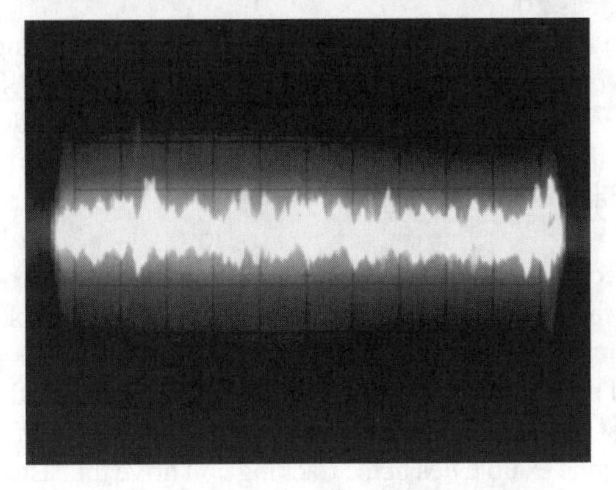

FIGURE 32-27 A focus waveform at pin 21 of IC211 and the emitter terminals of Q203 and Q202.

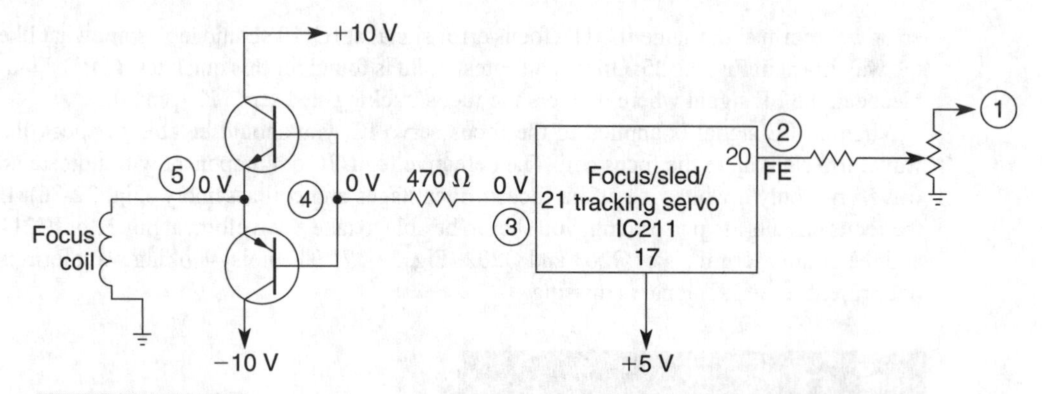

FIGURE 32-28 A waveform at TP2 and pin 20 of IC211 and no waveform at the focus coil indicates a defective IC211.

You can assume that IC211 is defective if an FE signal is entering at pin 20 and no waveform is at pin 21 (Fig. 32-28). Check the supply voltage feeding IC211. Often the same voltage (+5 V) feeding IC211 also supplies power to the RF signal processing IC110. If the IC110 EFM signal is normal, the same voltage should be applied to IC211. You might want to check all voltages at the low-voltage power supply before troubleshooting the CD circuits. Many electronic technicians check these low voltages first.

If you find a good FE waveform at pin 21 or IC211 and not at the focus coil (5), check collector voltages at Q202 and Q203. One of the voltages might be low with a leaky driver transistor. If one of the voltages is missing, go directly to the same source in the low-voltage power supply. Test each transistor in the circuit with a transistor tester. If in doubt, remove and test it out of the circuit.

DEFECTIVE TRACKING MECHANISM

The tracking and focus circuits can be checked in the reverse procedures if the player is shutting down without any actuator movement. Measure the resistance or take a continuity reading across the tracking coil. Some tracking-coil circuits have a test point at this connection. Take another resistance measurement from both emitter tracking transistors or the IC pin-to-ground to ensure there is not a bad socket or wiring connection between the drive component and the coil (Fig. 32-29).

Attach the scope test probe to the tracking-coil offset terminal feeding the tracking drive coil transistors. Now push Play. Notice if for a few seconds a noisy type of waveform is found at this terminal (Fig. 32-30). A noisy type of waveform might indicate that the tracking servo IC is performing. Go to the TE input terminal of the tracking IC (13) if only a white horizontal line is found at pin 27. No TE signal is present with a horizontal white line.

Remember, the tracking/focus/sled servo IC must have a signal from the EFM-RF IC before either will operate. Often, if a focus error signal is found at IC211, and the tracking error signal should be found at terminal 13 or test point TP4. The TE signal at TP3 is greater than that found at TP4 (Fig. 32-31). Of course, this waveform will be determined by the tracking-error offset or gain control. A quick waveform test at terminal (TE17) of IC210, TP3, TP4, and TAO-27 will indicate the trouble lies in the tracking coil drive transistor or IC.

You might find an IC or several transistors as tracking driver components. Some have two driven transistors within an IC (Fig. 32-32). Check each component in the same manner as all driving stages. Take a quick voltage measurement on each terminal and compare it to the schematic. Notice that the 10-V supply voltage is higher without a load. Zero voltage is at the emitter terminals until a tracking signal is applied. A typical tracking servo troubleshooting chart of a JVC XL-V400B player is shown with corresponding servo drive circuits in Fig. 32-33.

Do not overlook the possibility of breaks in wiring or board connections where the tracking and focus actuator takes off from the PC board. Check the flexible actuator leads for

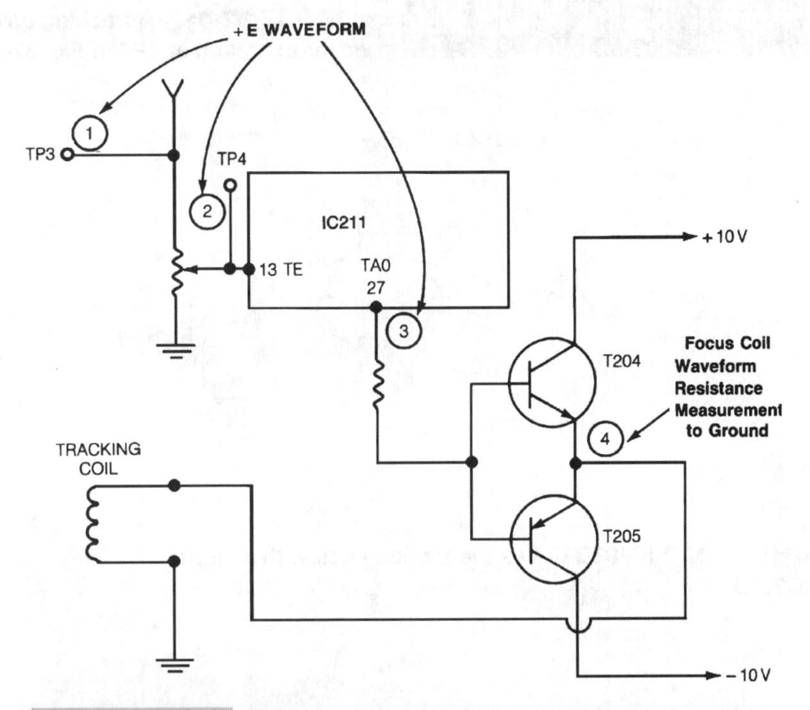

FIGURE 32-29 Take waveforms at TP3 and TP4 and pin 27 of IC211 to locate the defective tracking circuit.

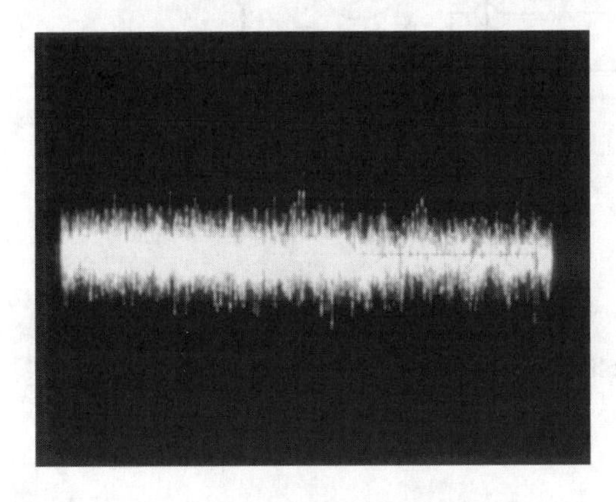

FIGURE 32-30 Take a tracking error waveform at pin 27 of IC211 in Fig. 32-29.

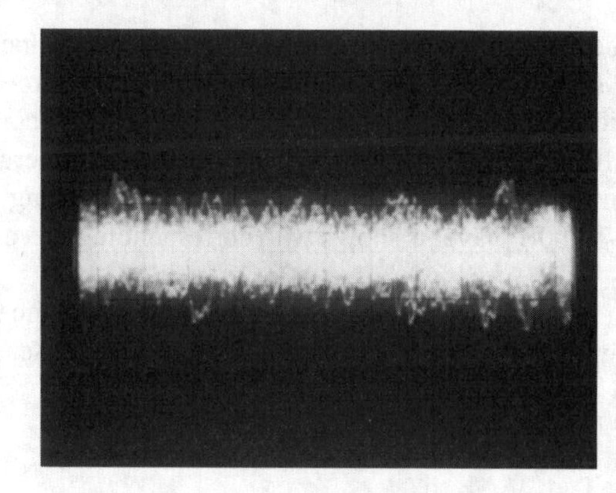

FIGURE 32-31 A tracking error waveform taken at TP4 in Fig. 32-29.

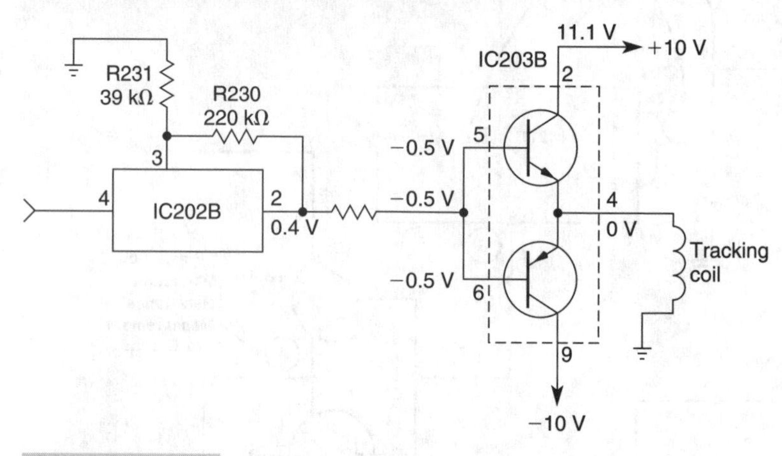

FIGURE 32-32 IC203B drives the tracking coil with a signal from IC202B.

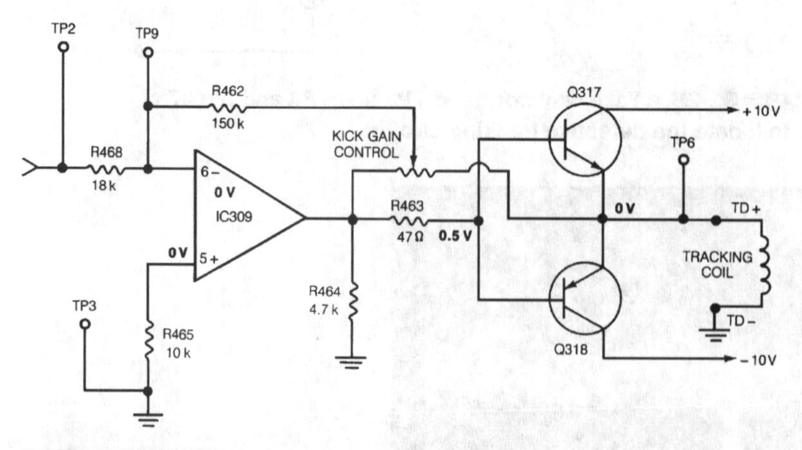

FIGURE 32-33 Check the voltages on IC309, Q317, and Q318 with no tracking-coil action.

breaks or intermittent wiring. Look for broken or cracked bridge wiring. Especially check the actuator lead sockets on the servo PC board. Doublecheck the wiring with resistance measurements.

DEFECTIVE CARRIAGE, SLIDE, OR SLED OPERATION

The carriage or sled motor operates from the same focus/tracking or servo control IC. The sled signal is taken from pin 23 of IC6 (Fig. 32-34). The SLO control signal might be fed to another IC amp and drive transistors, or it might be fed directly to the drive transistors and slide motor. Scope waveforms can be taken up to the slide motor terminals. The horizontal sled waveform will move up and down when operating (Fig. 32-35).

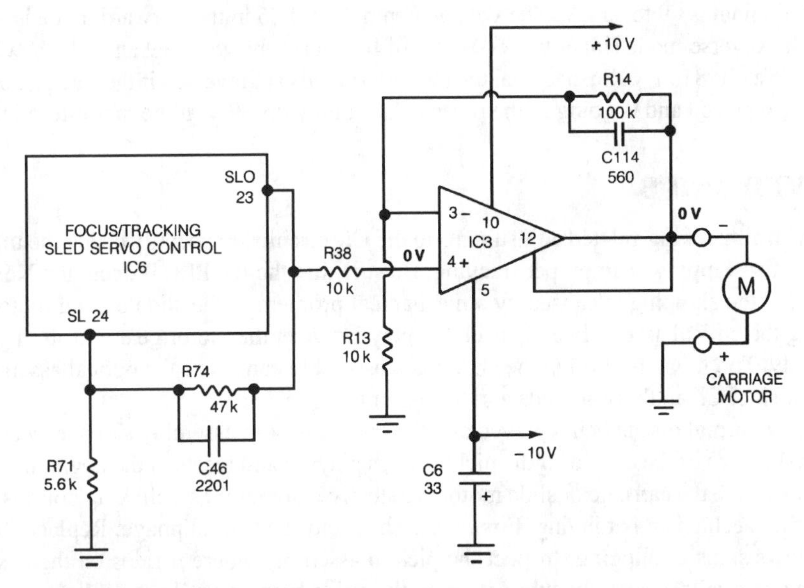

FIGURE 32-34 IC6 provides carriage motor signal to motor driver IC3.

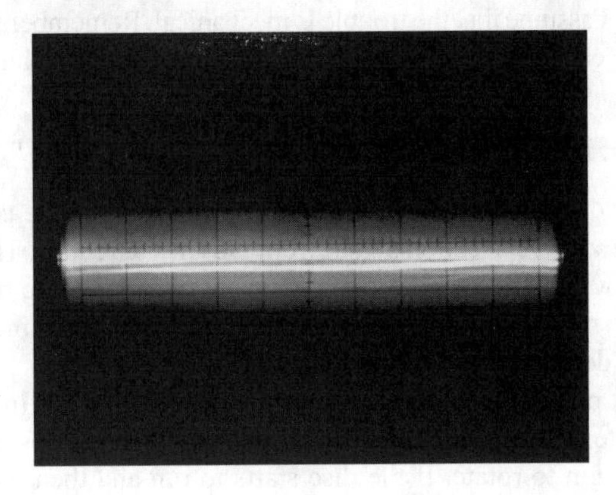

FIGURE 32-35 A slide waveform taken while motor is operating across the motor terminals.

Notice if the carriage motor is moving the optical assembly. If not, check for a broken or slipping belt or gear train at the motor pulley. Check the motor for an open winding. To check the motor, inject dc voltage from the battery motor box. Always remove the ungrounded lead from the motor terminals.

Check for normal voltage on IC211, Q104, and Q103. Suspect an improper voltage power source or a leaky driver transistor if one of the collector terminal voltages is missing or is low. Improper carriage motor polarity voltage can cause the motor not to move in the right direction. In many of the carriage motors, forward direction is with a negative voltage applied from the driver transistors. Check for a defective switch, mechanism, or soldered connections if the carriage does not continue after reaching the inside track.

In Stop mode, the voltages at pin 2 of IC3 and pin 23 of IC6 are "0." When the carriage mode (outer tracks) is in the Forward mode, the voltage on pin 2 of IC3 or at the carriage motor terminal is –9 to –11 V. The voltage at pin 23 of IC6 in the Forward mode is –0.8 to –1 V. In Reverse mode (inner tracks), pin 2 of IC3 varies between +9 and +11 V with pin 23 of IC6 at +0.8 to 1 V. Suspect that the motor terminals are reversed if the carriage or slide motor is replaced and is going in the reverse direction with +9 V at the motor terminals.

PLAYER SKIPS

Touch up all tracking-related adjustments to the CD circuits. Especially check the tracking balance for skipping or improper tracking. Be sure that the RF PLL is accurate. Next, determine if the skipping is caused by a mechanical problem at the slide or sled motor. Try holding the optical assembly and check the play between the motor gear and sliding track assembly. Too much friction in the slide track assembly can cause the optical assembly to jump or skip. Clean the slide rods and the assembly.

When a normal disc is being played and the optical pickup assembly skips, especially at the outside area of the disc, or if the pickup stops playing and resets to the beginning of the disc, suspect that a carriage or slide motor is defective. Sometimes a clicking noise is heard while the mechanism is moving. First check the motor belt for slippage. Replace the belt if it shows signs of slipping. Inspect the pickup assembly where it rides on the base and pickup guide rails. Check the lubrication on the guide base and rail assembly.

Monitor the tracking error signal at test point (TE) with the scope. Look for a change in signal when the skip occurs to determine if the trouble is electronically controlled. If there is no tracking error connection, you might assume that the trouble is mechanical. Remember, tracking gain can be set to close (small), causing the sound to jump when the machine is bumped.

DEFECTIVE SPINDLE OR DISC OPERATION

The defective turntable spindle or disc motor might be dead or not running at the correct speed. The disc motor is locked in with a correction signal from the CLV circuitry. The RF PLL and write frame clock (WFCK) must be accurate for CLV control. Touch up the RF PLL adjustment to ensure the disc motor will operate with the optical block assembly. If not, the dropout might occur with defects on the disc surfaces.

The EFM and focus signal must be present before the disc motor will operate. In some players, the disc will begin to rotate when the on button is pushed and then it stops. When Play is pressed, the disc should begin to rotate. If the disc starts to run and then the unit

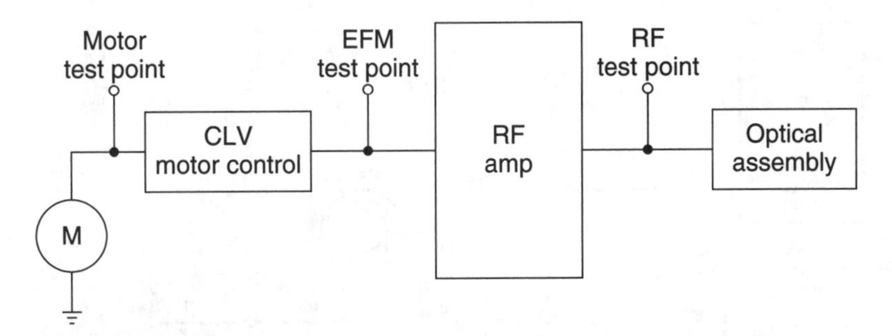

FIGURE 32-36 **The RF and EFM signal must be present for the CLV disc to operate.**

shuts down, suspect improper or missing EFM (eye pattern) or focus okay (FOK) signal. You must have an RF or EFM signal to get the disc to rotate.

To get the RF or eye pattern, the laser must be emitting and the focus servo circuitry working. Check the RF at the RF preamp and the EFM signal feeding the CLV motor-control IC with the scope (Fig. 32-36). If these signals are found at the CLV disc motor-control IC, suspect a defective motor-control or disc motor assembly.

The spindle motor can be controlled with a separate IC and transistor driver circuits. In some circuits, the disc or spindle motor is controlled by two transistor drivers fed directly from the large IC servo controller. The dead disc motor can be checked by starting at the motor and working toward the servo circuits. The quickest method is to check the voltage at the collector terminals of the disc drive transistors. If one of the positive or negative voltages is missing, check the low-voltage source at the power supply. A vibrating or unstable eye pattern can be caused by a poor turntable or disc motor.

SPINDLE MOTOR WON'T STOP

The collector voltage at Q315 is +10 V, and –10 V at Q316 (Fig. 32-37). If not, check the power supply. Zero voltage should be at the emitter and motor terminals in the Stop mode. If the spindle motor does not stop in Stop mode, suspect that the voltage at pin 1 of servo IC308 is higher. If not, replace or test both driver transistors (Q315 and Q316). A very high voltage at pin 1 of IC308 can indicate that IC308 or IC301 is leaky.

DISC DOES NOT START AFTER LOADING

Measure the emitter voltage at Q315 and Q316; +6 V should be found there. In some players, the disc motor voltage is 2 to 5 V at the beginning of the disc playing. A normal motor drive voltage at the motor terminals might indicate that the motor is defective. Check the motor winding with the low range of the ohmmeter. Inject disc motor voltage from the battery motor box and see if the motor rotates. Some motors will operate at 2.5 Vdc and between 1.5 and 3.7 V at 600 rpm. Inspect the motor harness and connections for poor wiring. Replace the motor if it will not rotate with a dc voltage applied to the terminals.

No voltage at the emitter terminals might indicate that servo controller IC301 is defective. If +2.5 V or more are found at pin 1 of IC308 and the motor does not rotate, suspect

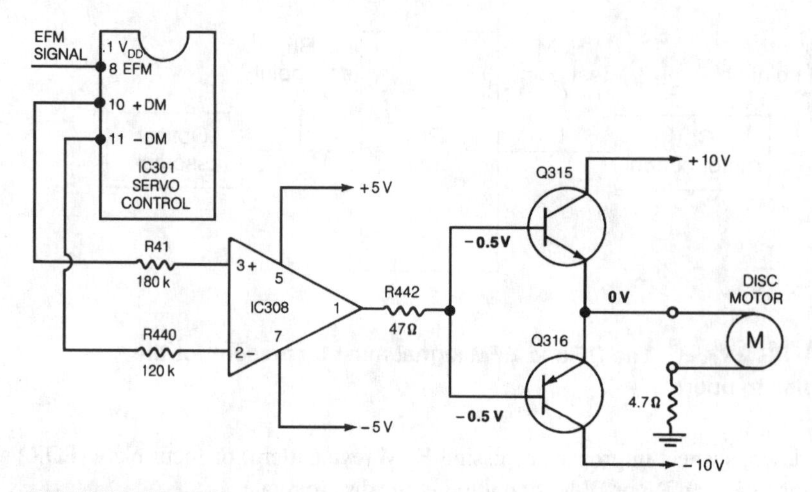

FIGURE 32-37 **A very high voltage at pin 1 of IC308 could indicate that IC308 or IC301 is leaky.**

that Q315 or Q316 are open. Replace IC308 if pin 10 is set to "H" without a motor-control voltage. Doublecheck IC301 if there is no control signal.

No rotation of turntable: Mitsubishi M-C4030 player The turntable or disc would not rotate in a Mitsubishi M-C4030 CD player. In the changer, the turntable motor sets up on the edge of the chassis and drives a worm gear assembly. At first, the motor assembly was suspected of a binding gear assembly. Two small screws were removed to drop the motor down to see if it was damaged. The DMM was clipped to motor terminals to determine if motor or circuits were normal. A voltage measurement of 10.7 V was found on the motor terminals, with turntable operation, except that the motor was not rotating. The motor was replaced.

DOES THE SPINDLE OR DISC MOTOR STOP AT ONCE?

The spindle motor should stop at once when the Stop mode is set, if all circuits are operating properly. If not, measure for about –6 V at the emitter terminals of Q315 and Q316 in the Stop mode. The spindle motor might be defective if it does not stop with negative voltage present. Check pin 11 of IC301 with a low negative voltage at IC308. If more than –3 V is present at pin 7 of IC308 in the Stop mode, check Q315 or Q316 for opens.

SPINDLE MOTOR RUNAWAY

Suspect poor pin connections around IC8 and IC9 if the spindle or disc motor starts in high speed (Fig. 32-38). Test pin 2 of IC8 for ground potential. Check for faulty bridges of wiring or pattern breaks. Rotate the spindle motor by hand and notice if it feels normal in rotation. A defective motor can cause high-speed problems. Are pins 5 and 6 of IC9 normal (EFM and ASY)? Check pin 7 (MIRR input) for a repeatedly "H" and "L" level waveform.

Remember, the spindle or disc motor must turn in a positive or clockwise motion. The laser beam must be lit and the FOK (focus) must be at high level. A normal eye pattern must be present with normal tracking and focus servo systems. Usually, the ungrounded side of the motor has a negative voltage for forward direction and a positive voltage for reverse direction. In some motors, 2 to 5 Vdc are found at the start; in others, it varies between 5 and 6 V. R15 and R11 might be open if the spindle motor is dead.

DEFECTIVE PLL CIRCUITS

The 4.3218-MHz frequency of the VCO-PLL circuits must be very accurate to ensure the correct recovery of disc dropout conditions (Fig. 32-39). Proper adjustment of the PLL frequency is needed so that the disc motor follows the optical lens assembly and responds to

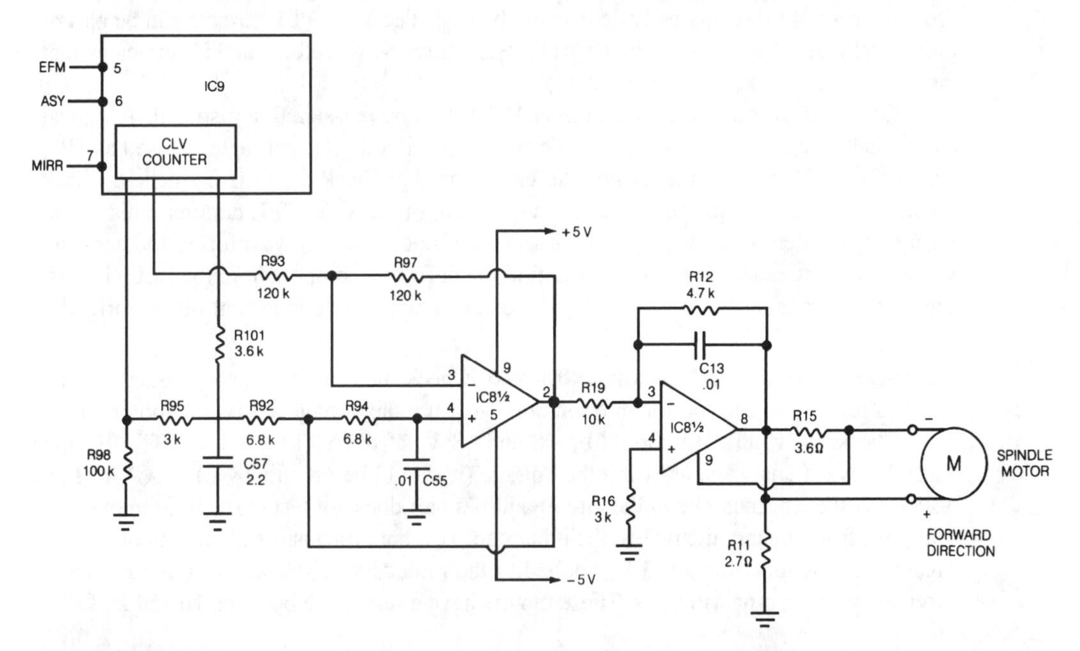

FIGURE 32-38 Suspect poor pin connections at IC8 and IC9 if the spindle motor runs at high speed.

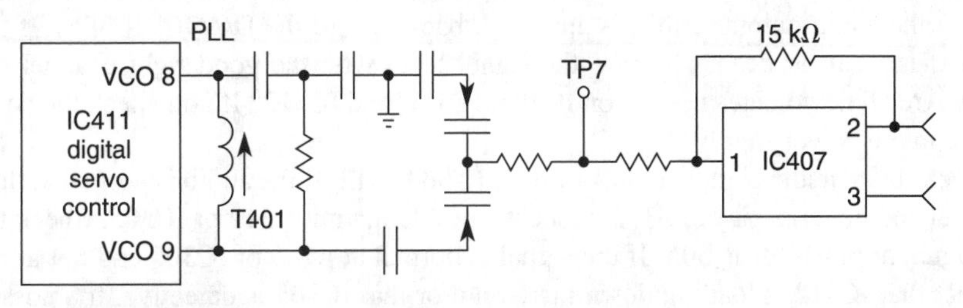

FIGURE 32-39 Adjustment coil T401 in the PLL-VCO circuit can correct disc drop-outs.

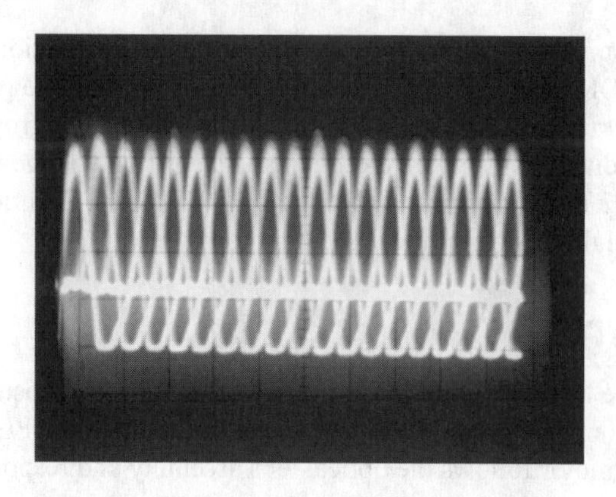

FIGURE 32-40 **A scope waveform of a PLL circuit in a Sanyo CP500 is taken at test point TP7.**

dropouts caused by scratches or defects on the disc. The VCO PLL circuits can be part of the digital control/CLV servo IC (IC411). Here, a scope waveform at TP7 indicates that the PLL circuit is operating (Fig. 32-40).

The PLL waveform is used for correct VCO-PLL output-signal adjustment. A typical adjustment of coil T401 is for a 1.2-Vdc waveform; if a digital voltmeter is used at TP7, adjust for 1.4 Vdc. A frequency counter can be used to check the PLL frequency. Check the manufacturer's literature for correct adjustment of the VCO-PLL circuits. This waveform can be taken with the player on or in Pause mode. Correct waveforms and accurate voltage measurements on IC411 should find the defective component in the PLL circuits. The eye pattern is fairly stable if PLL is in lock and turntable loop is controlling correctly.

No roulette rotation: Denon DCM-460 Check the roulette sensor, motor, transistors and IC drivers with no roulette rotation. Measure the voltage across the motor terminals. Check the voltage (it should be +8 and –8 V) at the TR103, TR104, TR101, and TR102 driver transistors. Measure the voltage (it should be +8 and –8 V) at IC106 (Fig. 32-41). Test each transistor in the circuit and if a pair does not test correctly, remove one transistor from the circuit and test the other one. Test both transistors in and out of circuit. Measure the supply voltage (5 V) applied to the roulette sensor. Take a continuity measurement at the motor windings. These motors have a resistance between 10 and 20 Ω.

DEFECTIVE AUDIO CIRCUITS

The audio signal path begins at the left and right output terminals of the D/A IC (Fig. 32-42). Start at the audio-output terminals and work back toward the D/A IC (IC305). If the left channel is weak or dead, compare the sound level with the good right channel or vice versa. Are all dc voltages normal on IC305, IC311, and IC312? If not, check the dc power source in the power supply.

Check for an audio signal at pin 1 of IC311 and IC312. If the audio is normal at the right channel and not at pin 1 of IC312, suspect a problem with IC312 or IC305. Check the audio signal at pin 13 of IC305. If the signal is normal at pin 1 of IC305 and not at pin 13, suspect that IC312 is loading down the circuit or that IC305 is defective. It's possible to have some circuits defective and some normal in the same IC.

Do not overlook the muting circuits; usually audio muting is found in the audio-output line circuits. If the audio is good up to the last audio IC or transistor, suspect improper muting at the line output. Disconnect the mute transistor emitter or collector terminal that is tied to the audio line and notice if the sound appears. If the sound returns to normal, check the mute transistor, voltage, and other components tied to the muting circuits. For further sound problems, refer to Chapter 27.

SOUND CHECK

The sound output of the CD player can be checked with a sound-noise distortion meter. Total harmonic distortion (THD) and intermodulation distortion also can be checked with

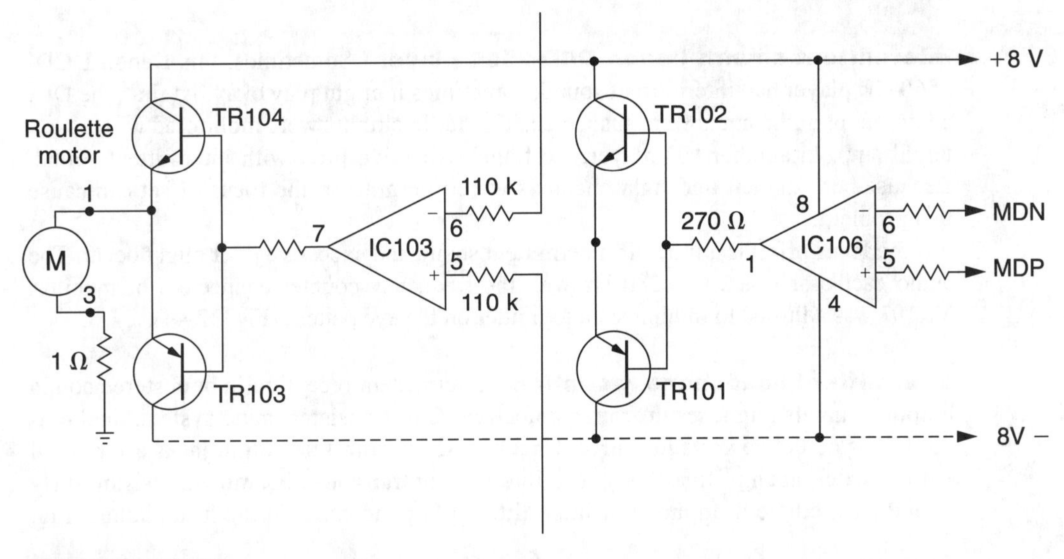

FIGURE 32-41 The roulette motor circuit consists of transistors TR104, TR103, TR102, and TR101, with IC driver IC103 in a CD changer.

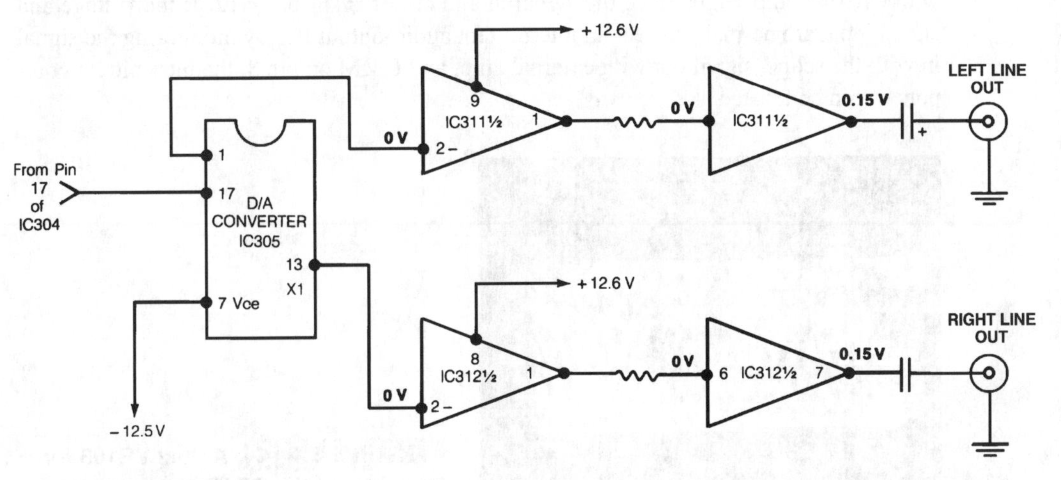

FIGURE 32-42 The analog (audio) signal is at pins 1 and 13 of the left and right stereo channels of IC305.

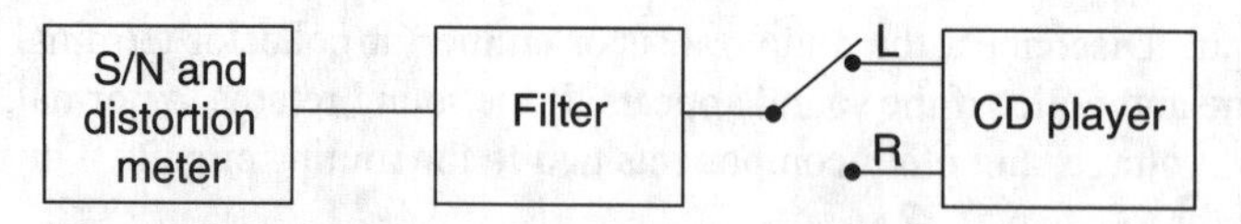

FIGURE 32-43 Connect a D/N distortion meter to the line-output jacks to check for distortion and signal-to-noise ratio.

the same hookup. Connect a filter between line output channel of CD player and distortion meter (Fig. 32-43).

Intermittent sound: Denon DCD-2560 player Sometimes, this Denon DCD-2560 CD player had intermittent sound; sometimes it might play okay. At first, the D/A converter or audio amps were suspected. The audio circuits were monitored with an external amp. In another CD player, we found excessive jitter with intermittent sound. Because both the left and right channels were intermittent, the focus offset can cause this problem.

If the focus offset is a little off, intermittent sound and excessive jitter might occur. The audio oscillator was set to 580 Hz with the frequency counter connected for monitor. VR103 was adjusted to minimize pattern jitter on the eye pattern (Fig. 32-44).

Intermittent headphone reception Intermittent reception in both stereo sound output channels might result from a defective IC or transistor, mute system, dual output IC, or defective voltage source. Check to see if both line output jacks are normal with an external amplifier. Suspect an intermittent transistor, IC, mute transistor, dirty phone jack, single headphone or intermittent plug and cable to the headphones (Fig. 32-45).

Check the connection on the stereo headphone jack. Clean the dirty contacts of the plug and jack. Substitute another pair of headphones. Check the output of intermittent channel with external amp. Signaltrace the signal in and out of Q411 or Q412. If the voltage and signal input are normal, suspect an intermittent audio-output IC. By monitoring the signal in with the scope, signal out with external amp, and DMM on pin 8, the intermittent component can be located.

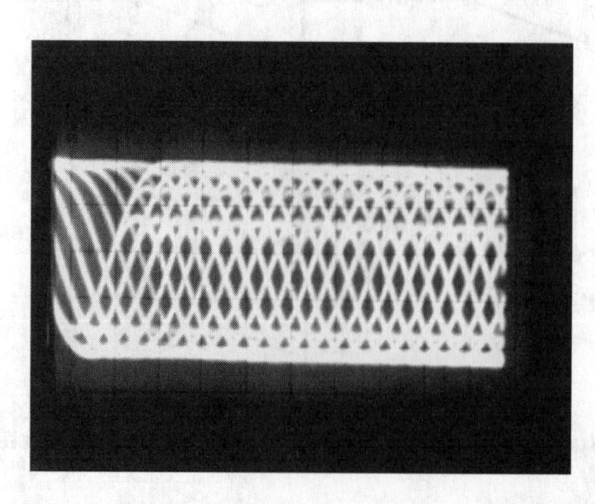

FIGURE 32-44 Adjust VR103 for less jitter at the RF IC amp and intermittent sound in a CD player.

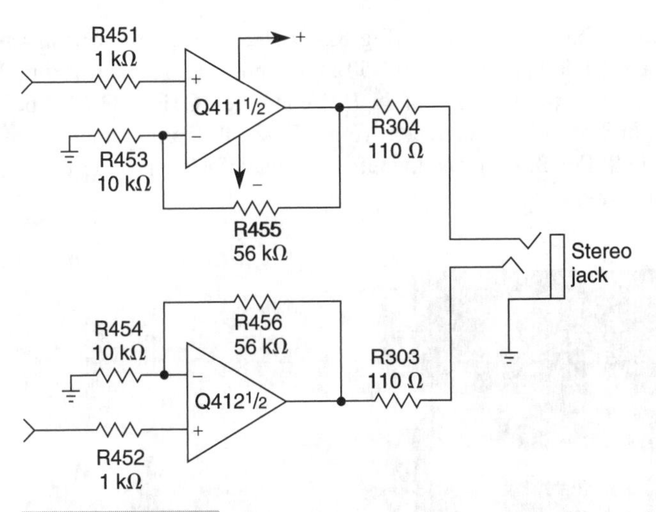

FIGURE 32-45 **Intermittent headphone operations can be caused by defective Q411 and Q412, a dirty headphone jack, broken headphone cable, and an improper supply voltage.**

Major Waveforms

It is wise to take waveforms on crucial ICs and mark them on the schematic diagram when servicing only a couple of different brands of CD players. These normal waveforms might, in time, point out a defective IC or transistor when all voltages and other waveforms on the IC are normal. You might find many different waveforms on some manufacturer's CD schematics; in others, there are only a very few. This also applies to actual voltage measurements.

The following are a number of major waveforms that are crucial in troubleshooting the CD player, along with a few that might not be found in the service manual. The RF or EFM signal found on the RF amp IC (Fig. 32-46); the focus coil waveform taken across focus coil winding (Fig. 32-47); the tracking coil waveform taken across tracking coil (Fig. 32-48);

FIGURE 32-46 **The important RF or EFM eye pattern waveform at the RF IC amp.**

the focus error waveform fed to the focus coil (Fig. 32-49); the 8.4672-MHz crystal waveform on pin 53 of the Sanyo CP500 player (Fig. 32-50); the input data waveform from the digital control/CLV servo IC at pin 10 (DIN) of the D/A converter IC (Fig. 32-51); the WLCK input waveform at pin 8 of the D/A converter (Fig. 32-52); the right output waveform of the D/A converter pin 8 (Fig. 32-53); and the left audio waveform on pin 13 of the switching audio IC504 (Fig. 32-54).

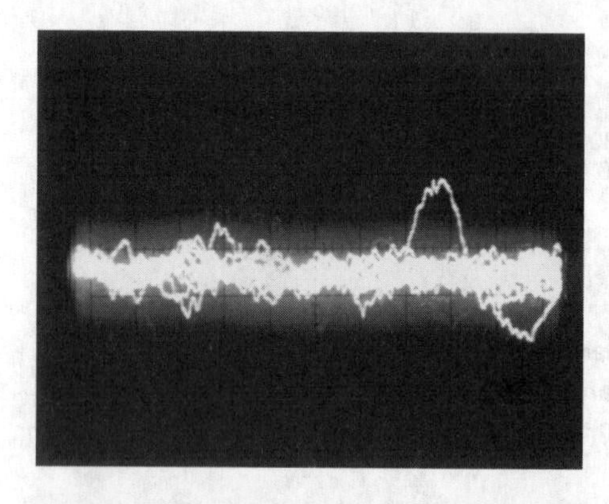

FIGURE 32-47 The focus coil waveform taken across the focus coil winding.

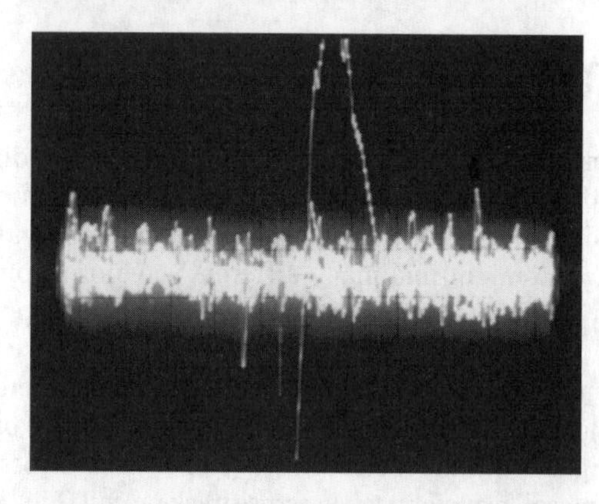

FIGURE 32-48 The tracking coil waveform (TE) fed to the tracking coil.

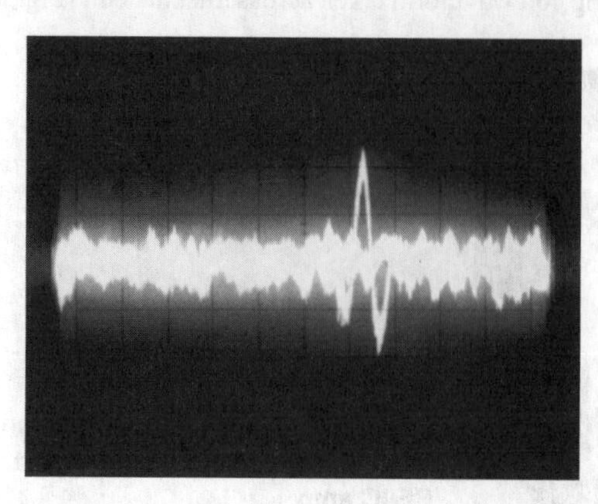

FIGURE 32-49 The focus error (FE) waveform.

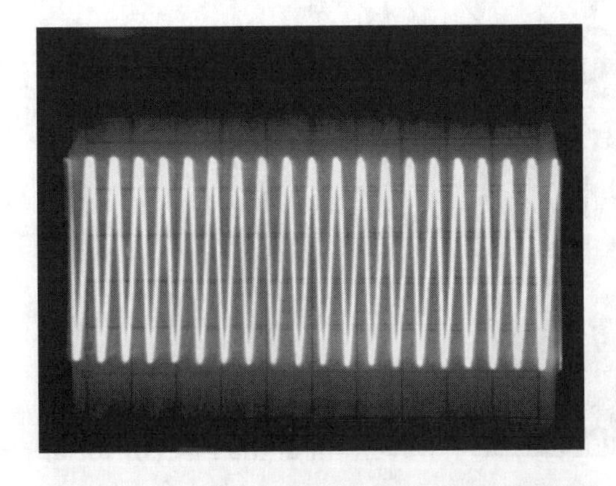

FIGURE 32-50 The 8.4672-MHz crystal waveform at pin 53 of the Sanyo CP500 player.

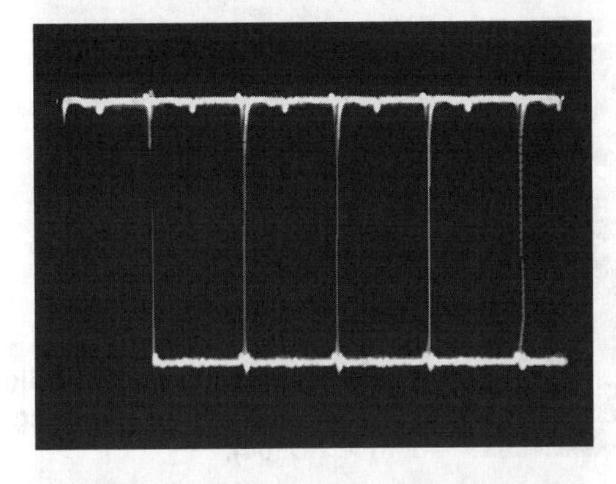

FIGURE 32-51 The input data waveform from the digital control/CLV servo IC at pin 10 (DIN) of the D/A converter IC.

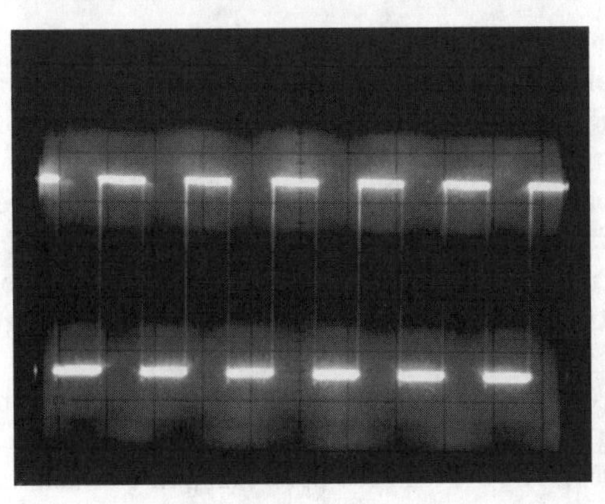

FIGURE 32-52 The WCLK clock input waveform at pin 8 of the D/A converter IC.

Service Notes

- Sound blasts might occur with disc scratches when the focus gain control is set too large. Most focus gain controls increase in a clockwise rotation.
- The focus pickup might fail if the focus gain control is set too small. Too small of an adjustment might offer low resistance to vibration.

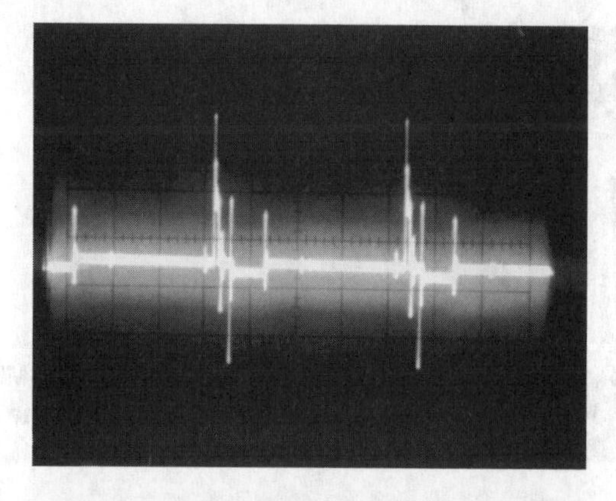

FIGURE 32-53 The right output waveform of the D/A converter pin 17.

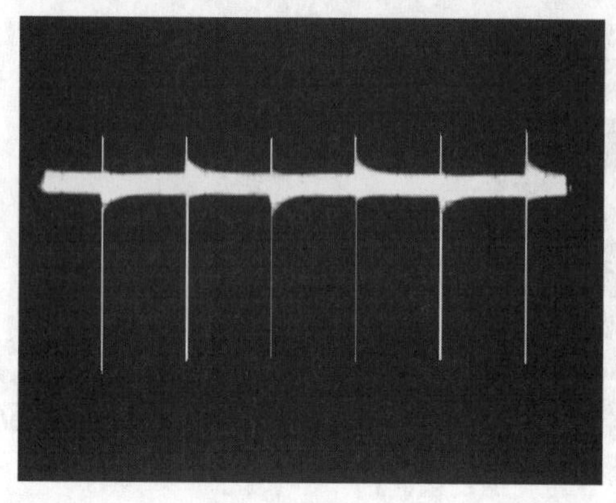

FIGURE 32-54 The left audio waveform on pin 13 of switching audio IC (IC504).

- If the focus offset adjustment is too far from the 0-V level focus pickup, failure will occur readily. Also, if the RF signal jitter is extremely large, the audio output is distorted.
- When tracking gain is too large, track jumping can occur easily because of a disc scratch. Also, during operation, the mechanical noise becomes large.
- When the tracking gain is too small, sound blast might occur when using a large eccentricity disc. Also, head take out is delayed and there is low resistance to vibration.
- If the tracking offset adjustment is too far from the 0-V level (particularly in the case of an eccentric disc), the possibility of continuous sound blast occurring during performance exists, and sound blast occurs easily because of disc scratches.
- Always put a soft cloth under the unit to prevent the plastic and case from bench scratches—especially in packing the unit for shipment.
- Be sure that the lock or transport screws are fastened before shipping or delivering the CD player for any long distance.
- Service bulletins and a change of part lists are often produced by the manufacturer. Subscribe to or ask for these bulletin changes.
- When replacing the pickup assembly, place paint over the lock screws so that they will not loosen.
- It is very important to select a ground point as close as possible to the test point when taking voltage measurements or waveforms.

- Irregular working of the display when the set is opened and playing might have been caused by incidental body effect in the region of the crystal oscillators. Switching off and on the main voltage might eliminate this effect.
- If the eye pattern is present, you might conclude that the laser is working, the laser is in focus, and that the turntable motor is running.

General checkpoints before servicing:

- Be sure that the disc is clean and not damaged.
- Check all clock frequency waveforms.
- Measure all low-voltage power-supply sources.
- Be sure that the mute circuit is inactive.
- Be sure that the CD player and all test equipment has warmed up before attempting to make adjustments.
- The laser beam can be disconnected when servicing circuits that do not require RF or EFM signal to protect the eyes.
- Keep a disc on the turntable or spindle at all times to prevent damage to your eyes or from touching the laser while servicing the CD player. Remember, you cannot see the laser beam. Your eyes are more important than fixing any CD player.
- If the CD player does not focus in and the RF signal is dirty, take a focus offset adjustment.
- If it does not play or if track search is impossible, make a tracking error balance test.
- If some discs can be played and others not, adjust the radial/tangential tilt screw.
- When there is no play or no search, check the RF level verification test.
- If the playback does not start or if the focus actuator is noisy, make a focus servo loop adjustment.
- If playback does not start during searches, the actuator is noisy, or if tracks are skipped, make the tracking servo loop gain adjustment.
- Make a list of all plugs and cables and how they are inserted before disconnecting them.
- Dress cables and wires so that they will not interfere with the loading trays or the roulette turntable.
- Small power transformers can be mounted in the bottom metal cover. Remove the transformer so as not to pull out the transformer leads.
- Use the correct size of screwdrivers, nut drivers, and star screwdrivers to remove outside screws without marring the screw or cabinet.

TROUBLESHOOTING AND REPAIRING CAMCORDERS

33

CAMCORDER CASSETTE FORMATS

CONTENTS AT A GLANCE

Camcorders are nothing more than a combination electronic camera and video recorder in one package (see Fig. 33-1). The tools commonly found on the electronic technician's bench are all that is required to service the VCR section. Many of the same tools needed to repair the VCR recorder are required. Besides those tools found on the service bench, a good vectorscope, color monitor, lighting equipment, reflection charts, and a light meter will be needed. A lot of camcorder maintenance can be performed by using just the reflection charts and dual-trace oscilloscope for electronic adjustments.

When video cameras were first used with color TVs, the picture was seen instantly on the TV screen or monitor. With camcorders, the scene can be recorded, played back at once, or viewed at a later date. Some camcorders have playback features, but the smaller units record only. Like the VCR, several different camcorder tape formats are available.

New Features

Besides more total pixels (270,000 to 570,000) added to the CCD unit, image stabilization, automatic focus, fuzzy logic, and greater zoom operation with 10-to-1 or 12-to-1 power are found in many new camcorders. Image stabilization helps the picture to remain steady—even if your hands are not. The area of white balance has improved greatly in the new camcorders. No longer are the video pictures tinged blue, brown, or green.

The word lux refers to light. And in the camcorder, it means the amount of light needed to get a normal picture. Today, the camcorder requires less light, supplies better resolution, and produces brighter colors. Most new camcorders have a flying erase head, with less glitches and noises picked up. Several new camcorder models have a color electronic viewfinder instead of black and white, and have video lights built right into the camcorder. Even with all these added features, camcorders have become smaller and weigh less.

FIGURE 33-1 The RCA VHS camcorder with pro-Edit and power zoom.

FIGURE 33-2 **The VHS-C, Beta, 8-mm, and standard VHS cassette.**

The smaller camcorder, VHS-C and 8-mm, have taken hold in the past five years. The 8-mm camcorder now accounts for 35 to 40 percent of camcorders sold. These camcorders are light in weight (around three lbs.) and can record up to three hours of tape.

A new camcorder developed by Hitachi will be one-third the size and weight of the present camcorders. The camcorder is so small that it can fit in the palm of your hand. This means there will be no video tape or a servo section to record and play back the tape. A method of compressing and flash memory of chips makes it possible to record up to 30 minutes of video. Of course, the small camcorder is a few years away, and will probably be quite expensive at first.

Beta

The Beta camcorder matches up with the Beta VCR machines. They cannot be plugged into a VHS recorder. The Sony Beta format was one of the first VCRs on the market, and it furnished the best-quality recordings. Beta camcorders can take pictures on a Beta cassette, but they have no playback features. The Beta camcorder has about 280 lines of resolution, compared to the 250 lines of VHS machines. The Beta cassette can play and record from 15 minutes up to 5 hours. The L-750 cassette is slightly smaller than the standard VHS cassette (see Fig. 33-2).

VHS

The video home system (VHS) camcorder employs the same cassettes as the VHS video tape recorder (VCR). They are the most stable and popular units at present. The VHS cassettes are low in price and can be played back in camcorders or VCRs. The VHS tape however, is not interchangeable with the Beta cassette. Many VHS camcorders can be played directly through the TV set or monitor. In standard play (SP), two hours of recording will fit on the standard T-120 cassette. Or, up to six hours of recording will fit on a VHS tape in the extended play (EP) mode.

Since the early VHS format, both the cassettes and VCRs have been improved. The HQ recording system improved the picture with sharper definition, truer colors, and less snow. Today, the super (S-VHS) camcorders provide more lines of resolution, which means more detailed pictures. The standard VHS camcorder usually has 240 to 250 lines, while the super VHS can have over 400 lines of resolution. The S-VHS camcorder can be played through the TV or monitor, but not the regular VCR unit. The VHS cassette can be viewed on an electronic viewfinder. Regular VHS cassettes can be played back through the S-VHS machines.

VHS-C

The compact VHS-C camcorder is small in size, light to carry, and uses the small VHS-C cassette. The VHS-C camcorder is much easier to take on vacations because of the physical size, compared to the VHS model. The VHS-C camcorder, however, must be supported for steady pictures. The VHS-C cassette can be played directly into the TV set or color monitor, but it must be placed in the regular VHS-size plastic holder before inserting it into the VHS VCR (see Fig. 33-3).

Although the tape itself and the recorded magnetic patterns are the same as a VHS cassette, the VHS-C cassette is only one-third the size. The tape movement is actuated by a plastic, geared reel inside of the regular VHS cassette. The take-up function is driven by a gear. The VHS-C cassette contains a supply reel disk, but there is no take-up reel disk.

When the VHS-C cassette is inserted into the VHS cassette, the adapter extracts the tape and positions it in the same manner as the full-size VHS cassette. Now the tape can be inserted into the VHS-format deck (see Fig. 33-4). When inserted, the supply reel disk of the VHS deck drives the supply reel of the VHS-C cassette. Remember, the take-up reel disk of the VHS recorder drives the take-up operation via the pulley and gear of the adapter (see Fig. 33-5). A small amount of noise is sometimes created by the VHS-C and VHS adapter.

Naturally, with smaller camcorders, all parts are reduced in size. Here, the regular size and weight of the 41-mm-diameter rotating head drum is about one-third the size of the full VHS format. To retain compatibility with the VHS format, the rotation speed is in-

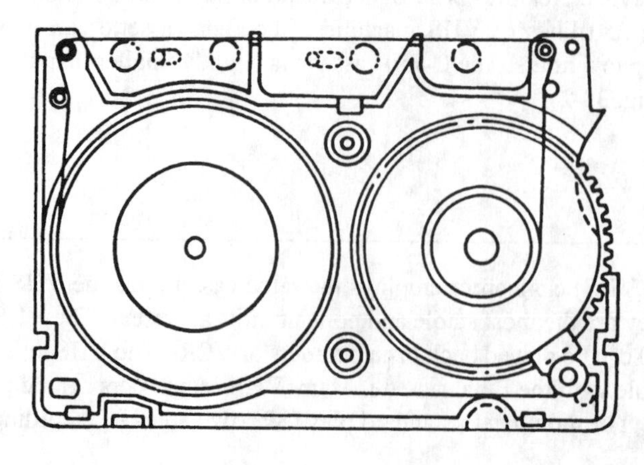

FIGURE 33-3 An inside view of the supply reel and gears of the VHS-C video cassette.

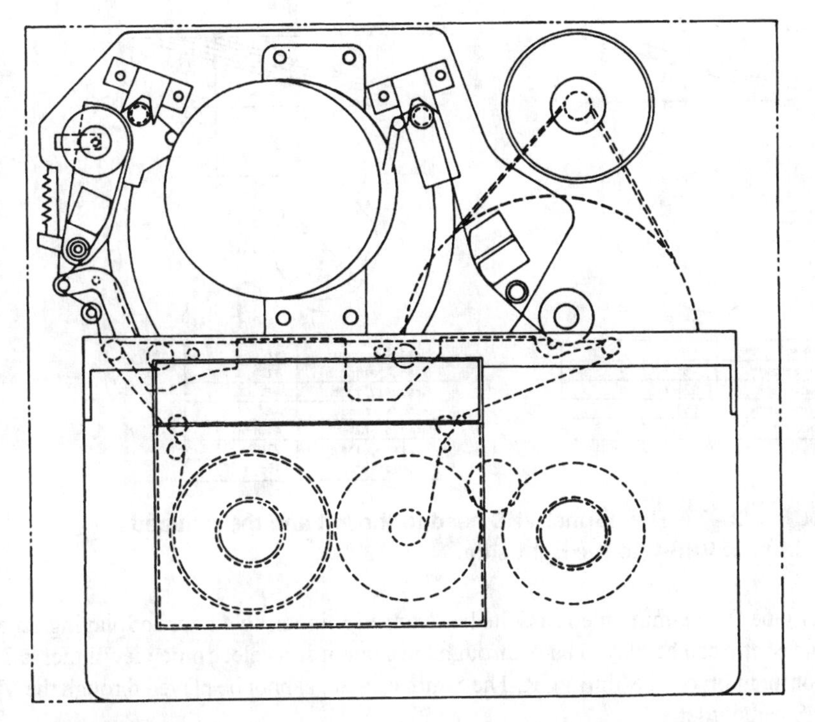

FIGURE 33-4 Detailed drawing of the VHS-C cassette mounted inside the VHS adapter for Play mode.

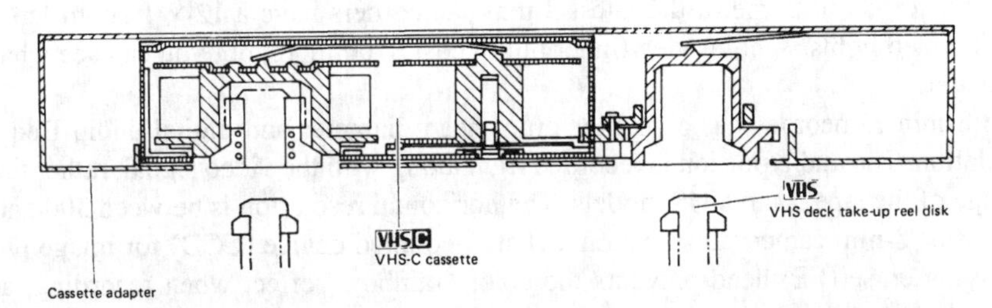

FIGURE 33-5 The end of a VHS-C cassette inside of a VHS cassette adapter.

creased to 45 revolutions per second, while the tape-wrapping angle is increased to 270 degrees (see Fig. 33-6). The conventional cylinder drum diameter is 62 mm and has a 180-degree tape wrap-around angle.

8 mm

The 8-mm design is the newest type of camcorder. It is made by both electronics and camera manufacturers. The 8-mm camcorder operates with a small lightweight format and

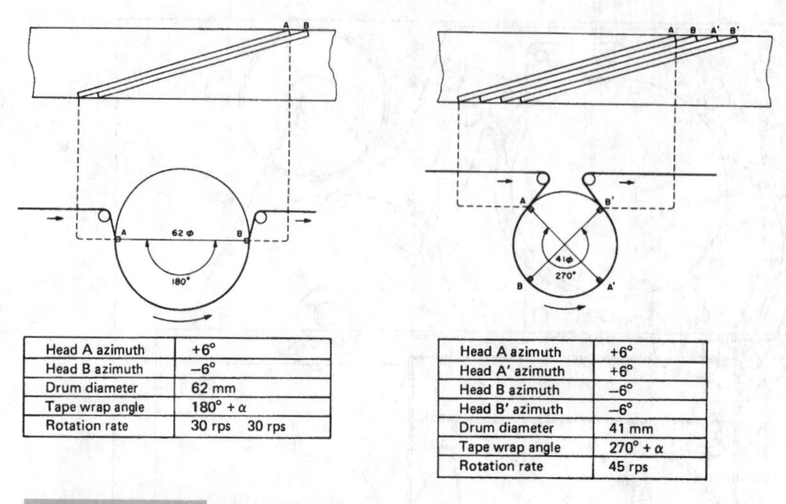

Head A azimuth	+6°
Head B azimuth	−6°
Drum diameter	62 mm
Tape wrap angle	180° + α
Rotation rate	30 rps 30 rps

Head A azimuth	+6°
Head A' azimuth	+6°
Head B azimuth	−6°
Head B' azimuth	−6°
Drum diameter	41 mm
Tape wrap angle	270° + α
Rotation rate	45 rps

FIGURE 33-6 **The normal VHS head to the left and the reduced head drum of VHS-C to the right side.**

thinner tape. The 8-mm video cassette has from 15 minutes to 4 hours of playing time. The 8-mm cassette can be played back through the camcorder's electronic viewfinder, a TV set or color monitor, or an 8-mm VCR. The 8-mm cassette cannot be played through the VHS-C or VHS camcorder.

The latest 8-mm camcorders have optical image stabilization and electronic stabilization. The optical stabilization is considered better with no reduction in picture detail when the stabilizer is on. Some of the latest 8-mm camcorders have a 12-×-1 zoom lens with built-in video lights. A color viewfinder might cost $100 more, but you can see what you take better.

The 8-mm camcorder has a smaller drum head diameter and digital audio frequency modulation. The audio portion is recorded right along with the video signal, rather than on the edge of the tape, as in VHS models. The horizontal resolution is between 300 and 330 lines. Most 8-mm camcorders contain a charge-coupled device (CCD) for image pickup. The flying erase (FE) head prevents the color "rainbow" effect when recording, and is mounted in the same drum or cylinder as the video heads (see Fig. 33-7).

Video Cassette Problems

The video cassette itself, like the audio cassette, can cause a lot of audio and video problems. Tape spilling out can be caused by improper alignment, or by a tape that is wound too loose or tight. To prevent jamming of the tape or cassette, inspect the cassette before inserting it into the camcorder for a broken or bulging case. Poor tape recordings in both video and audio sections can be caused by a defective tape head. Try the cassette in both the VCR and the camcorder to determine whether the cassette or camcorder is defective. If in doubt, try a different recorded cassette that is known to be good in the camcorder.

Camcorder Features

Knowing how the camcorder operates is important when servicing the unit. In fact, many reported "problems" are operational. Since the camcorder has become popular, several new features have been added that some users do not fully understand.

AUTO FOCUS

The focus is always automatically and precisely adjusted when the camcorder is set in auto focus. Manual focus adjustment is also possible in most cameras. You will probably find optical focusing in small and low-priced camcorders. Most camcorders use the infrared beam for auto focus, except for the NEC V50U (VHS) camcorder, which has a piezo auto focus circuit.

Auto focus works via infrared rays that are emitted from the camcorder to the object, then reflected back to a receiving lens (see Fig. 33-8). Here, the reflected rays strike two photodiodes. The focus lens moves until the two photodiodes receive an equal amount of light, correcting the camera focus.

FIGURE 33-7 The 8-mm RCA camcorder with an adjustable EVF.

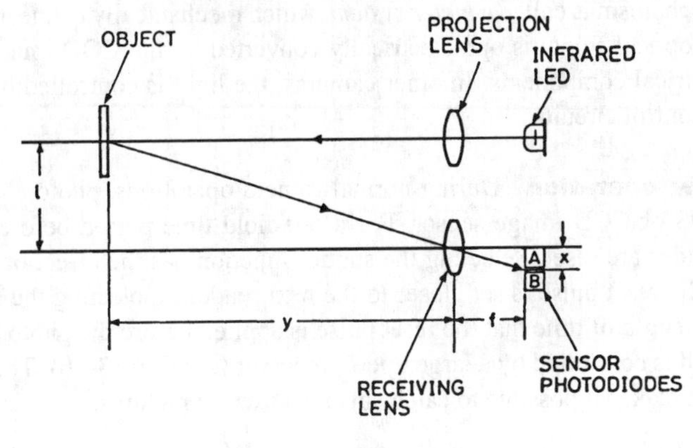

FIGURE 33-8 The measurement principle of an infrared autofocus system.

AUTO WHITE BALANCE

The white balance is fully automatically adjusted, and it continuously changes with fluctuations in illumination during shooting. The automatic white balance circuit controls the gain of the red and blue chrominance signals to maintain white balance (color temperature) under various lighting conditions in the RCA CPR300 VHS camcorder. *White balance* refers to the adjustment of the recording system to the color temperature of the light illuminating the subject. The auto white balance feature has white balance sensors at the front of the lens assembly.

CCD AND MOS SENSOR

The early TV cameras and camcorders have pickup tubes called the *vidicon*, *saticon*, and *newvicon* tubes. Today, most camera sections use either CCD or MOS sensors. A charge-coupled device (CCD) is a semiconductor that consists of orderly arranged MOS-cell capacitors. It consists of photoelectric conversion, charge accumulation, and charge transfer and operating time.

In the older camcorders, the CCD device could have up to 250,000 pixels. In the new camcorder, the average is around 270,000, with several going from 410,000 to 570,000. Charged-coupled devices (CCD) are used almost exclusively today. Most CCD camcorders have 1/3-inch CCDs. Minolta has come out with camcorders that have two CCD units, and Sony has developed three-chip devices. The super VHS and VHS-C camcorders have more than 400 lines of resolution compared to the old 250 for VHS camcorders. Of course, picture quality is excellent, but these units require a special tape, camcorder, and a television TV rated at high resolution for best results.

Camcorder models that use MOS image devices include Minolta, Pentax, Radio Shack, RCA, and Hitachi. (The others use CCDs.) The MOS image sensor operates in the same manner with picture elements (photodiodes with npn three-layer construction). The advantage of the CCD and MOS devices over the tube pickups are that the former have longer life, no image lag or burn, no figure distortion, strong chip, instant on, lower power consumption, and are small and lightweight (see Fig. 33-9).

AUTOMATIC IRIS

The automatic iris mechanism is called a *meter system*, which mechanically connects drive and brake coils. The optical signal is optoelectrically converted by the COD images and transformed into electrical components. In other cameras, the light is controlled by electronic shutter-speed control circuits.

High-speed shutter operation During normal camera operations, photo electrons are stored in the MOS or CCD image sensor. For a one-field time period before being scanned, the photodiodes are reset. Selecting the shutter function has the effect of delaying the reset pulse. The reset pulse is set closer to the next readout. Selecting the shutter speed determines the length of time that the reset pulse is applied before the photo tube is scanned. Shutter speed is controlled by a large microprocessor (see Fig. 33-10). The high-speed shutter function makes it possible to catch super-fast action pictures.

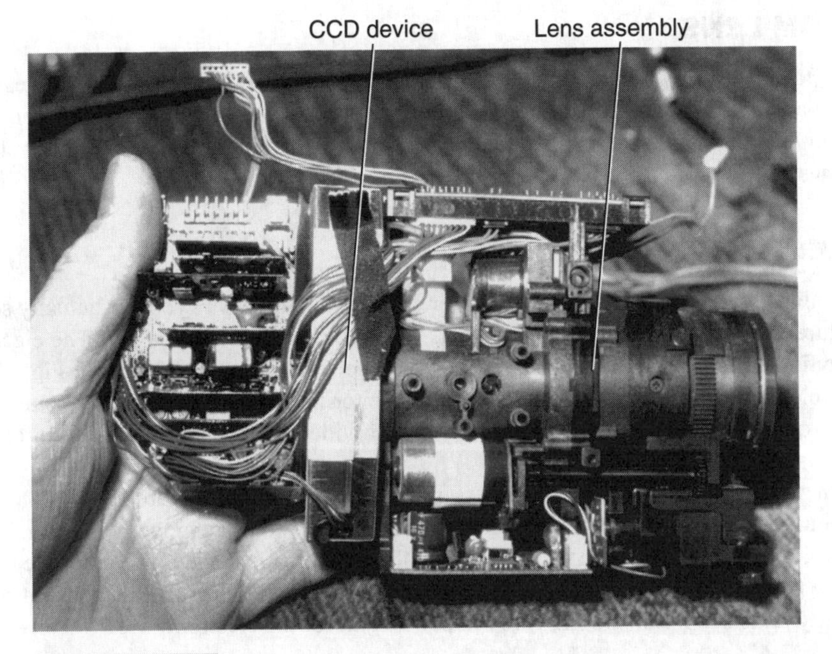

FIGURE 33-9 The CCD image sensor is located in the shielded box behind the lens assembly in a camcorder.

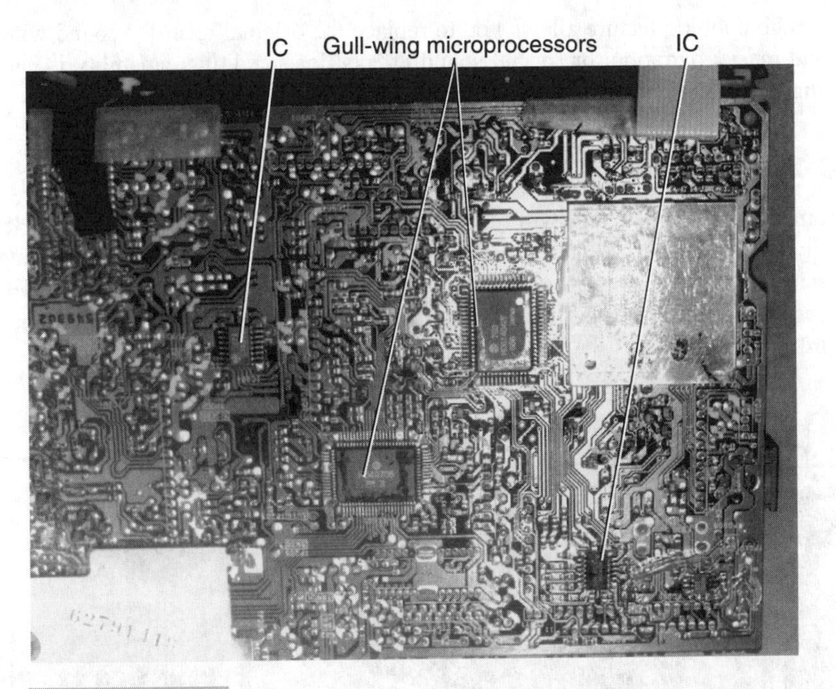

FIGURE 33-10 The large ICs in the camcorder all serve many different camcorder circuits.

ZOOM LENS

The power zoom switches move the lens assembly for close-ups or distant pictures or for wide-angle shots. The zoom buttons are easily controllable while operating the camera. A zoom lever manually zooms the picture in or out. A microbutton can provide for closeup scenes.

ELECTRONIC VIEWFINDER (EVF)

Not all camcorders have an electronic viewfinder. With the EVF, you can actually see the picture you are taking on a small screen. What you see is what you get. The electronic viewfinder image is in black and white, while the recording is in color (see Fig. 33-11). The electronic viewfinder can be used as a monitor in playback operations. Actually, the electronic viewfinder is a tiny TV receiver with video, high voltage, deflection circuits, and a small picture tube.

On about half of the latest camcorders, the electronic viewfinders are now in color. You can quickly play back the tape and see whether the color is adjusted properly (see Fig. 33-12). The viewfinders are rated in pixels, just like the CCD image device. The color viewfinder in the Canon ES1000 uses a 0.7-inch color LCD with 140,000 pixels for excellent picture quality. This can mean another color chip or circuit for the electronic viewfinder.

AUDIO DUBBING

The audio dubbing feature allows you to replace the original recorded sound with background music, narration, or special sound effects. Insert editing capability is handy for editing tapes by inserting new scenes into the already-recorded video cassette.

HQ TECHNOLOGY

In some late camcorders, you will find the HQ symbol mark feature of the new VHS high-quality picture system. HQ technology improves the picture quality while retaining VHS interchangeability. In some models, this means raising the white clip level and detail enhancer. Remember, the HQ camcorder system will operate in the conventional VHS recording system.

FIGURE 33-11 The electronic viewfinder is located at the rear of a Hitachi camcorder.

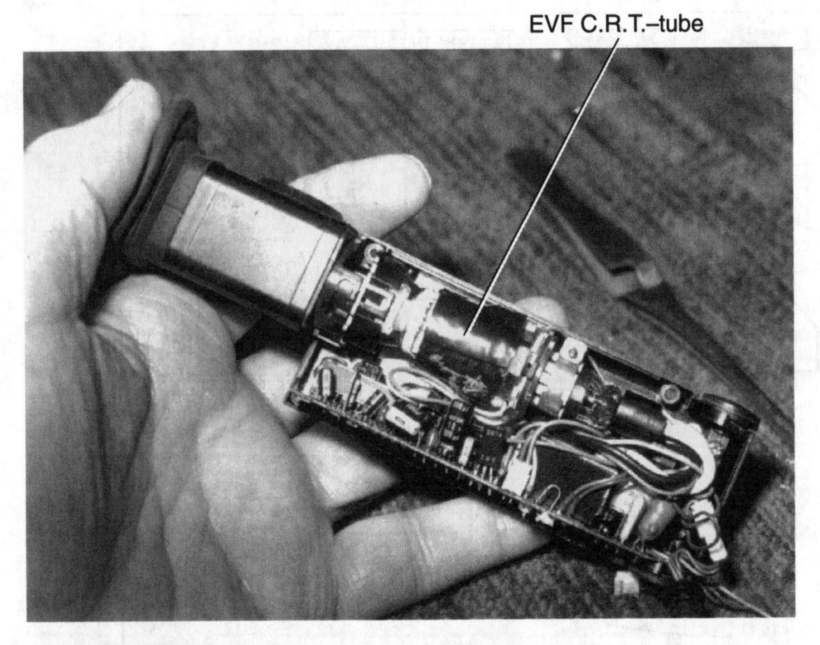

FIGURE 33-12 The black-and-white viewfinder of a small tube, packed tightly with video and sweep circuits.

POWER REQUIREMENTS

For greater portability, the camcorder operates off of a camcorder battery. These batteries slide in and out of the camcorder for recharging. Most output 6 V, 7.2 V, 9.6 V, 10 V, or 12 V at 1 Ah to 3.5 Ah. The JC Penney Model 5115 operates with a 1-V battery, Fisher model FVC73U at 6 V, Minolta 8200 camcorder at 7.2 V, and JVC model GR-S70V at 9.6 V. You can sometimes find a dc/dc converter with several different voltage sources.

A lithium 3-V battery is found in Canon's ES1000 camcorder as a back-up source for the quartz circuit. The lithium 3-V power is supplied when the battery pack is unloaded. While the battery pack is not mounted, the lithium battery outputs 3 V for backing up the quartz circuit. When the main battery is mounted, the lithium 3-V battery is charged with the regulated power from IC230.

HOOKUPS

Most camcorders can be connected directly to the TV set or color monitor with a VHF connecting cable and an audio/video cable. An RF adapter might be required between the camcorder and the TV antenna input terminals (see Figs. 33-13 and 33-14). If the TV or color monitor has video and sound input jacks, the audio/video cord is connected between the camcorder and the TV or monitor (see Fig. 33-15).

When recording from other equipment, the video cassette recorder is connected between the TV and the camcorder. In some models, the ac video adapter is used when operating the camcorder from the power line. Editing the recording hookup can be accomplished with the VCR connected between the camcorder and the TV or monitor. Dubbing introduces

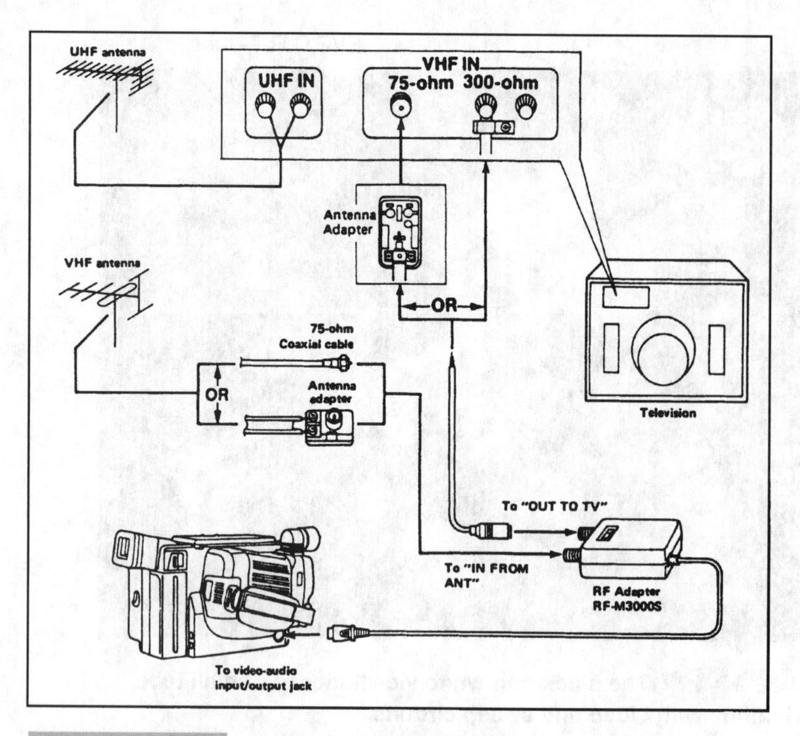

FIGURE 33-13 Use an RF adapter to connect the camcorder to the back of the VHF terminals of TV set.

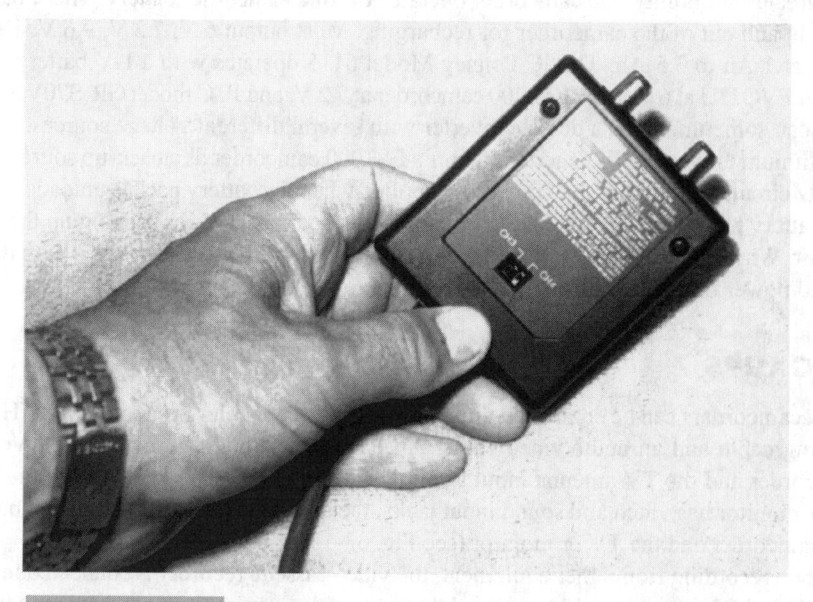

FIGURE 33-14 RF adapters usually come with the camcorder to connect to the TV antenna input.

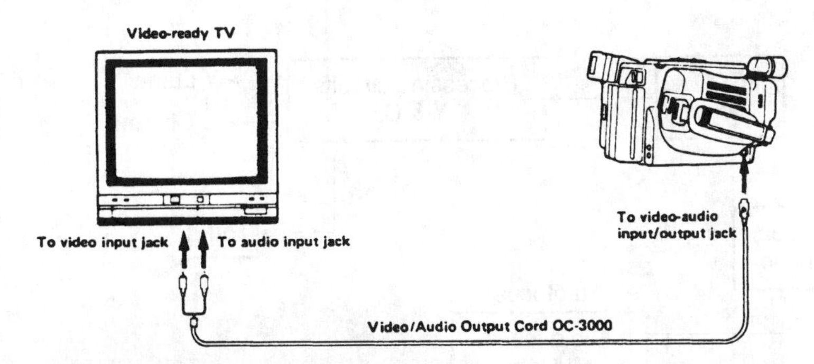

FIGURE 33-15 Connect the camcorder directly to the TV's video and audio jacks, if provided.

some degree of picture-quality deterioration. It is recommended that the video camcorder recorder be placed in SP mode.

NEW WEIGHTS

The older VHS camcorder can weigh 5 lbs. or more, while the VHS-C and 8-mm camcorders weight less than 3 lbs. with the battery. The Canon ES1000 weighs 2 lbs. without the battery, Hitachi VM-ES8A is 1.7 lbs., the Panasonic VHS PV-900 weighs 4.2 lbs., and the RCA CC710 VHS-C is 2 lbs. without the battery. Today's camcorders are very compact, easy to operate, and light to carry.

Block Diagram

Looking at the camcorder block diagram can help to determine where the trouble is. The manufacturer's service literature is a must-have item in troubleshooting the camcorder. Not only does it contain the block diagrams, but it breaks down each stage and shows how the components are tied together in each separate schematic. The schematic contains not only the circuits, but also the voltages and crucial waveforms.

VHS

The early VHS camcorder camera block diagram has a pickup tube or CCD stage, processor, AWB, auto focus, and EVF stages (see Fig. 33-16). The CCD stages consist of a CCD image sensor, CCD sync generator, CCD driver, and sample-and-hold circuits. The iris motor, AIC operation, chroma, and sync pulse generator can be found in the process circuits. The AWB circuit contains an AVT sensor, AWB gain, switcher, and decoder. The focus lens, zener and iris motors, motor-control circuits, demodulator, AGC, and clamping and digital converters are located in the auto-focus circuits.

The RCA CC547 VHS camcorder has a digital zoom, built-in video light, and color viewfinder. This camcorder uses a ½-inch CCD device with 270,000 pixels as image sensor,

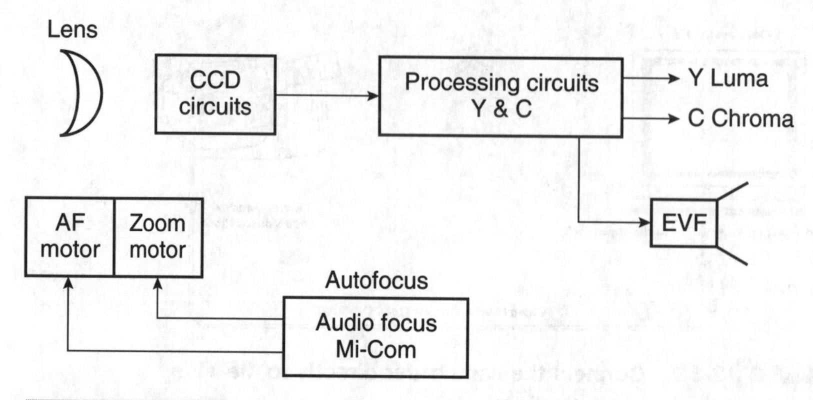

FIGURE 33-16 A typical camera section with various components and circuits.

with minimum light required is a low 1 lux. It has a 12× optical zoom lens that can be boosted to 24 times with zoom plus. The RCA CC547 has a color viewfinder of 120,000 pixels. Other features can include a built-in titles, flying erase head, A/V dubbing, and auto date/time.

The electronic viewfinder consists of horizontal and vertical sync, horizontal and vertical oscillator, deflection coils, flyback transformer, and CRT. The input of the EVF stages connects to the video output of camera circuits. The EVF circuits are often contained in one unit.

VHS-C

The VHS-C camera section consists of the AIC circuits, autofocus, automatic white balance, chroma processing, encoder/NTSC signal processing, luminance, MOS or CCD image sensor and circuits, matrix and filter, pulse generator, preamplifier, signal processing, sync generator, power-supply distribution, and electronic viewfinder circuits.

The VHS-C camcorder requires an adapter in order to play back through a regular VCR. The small C-type tape cartridge can be placed in a VHS adapter. The VHS adapter can then be inserted into the VCR for playback. These VHS-C camcorders can also be played back through the antenna connection of the TV.

8 MM

The 8-mm Samsung SCX854 camcorder camera section consists of the lens, zoom lens, CCD image sensor, V drive, D & A converter, Y signal process block, C signal process block, CDS/AGC/R, digital signal process, and an 8-bit A/D converter on the process board (see Fig. 33-17). The auto board holds the auto focus Micom/EVR, EEPROM, iris block, Hall detect block, auto-focus motor drive, and zoom motor-drive circuits.

The camera section of a Canon ES1000 camcorder contains VAP unit, zoom reset, zoom lens, zoom motor, IG meter, focus lens, focus reset, focus motor, and CCD unit upon the lens board. The CCD PC board contains a V driver (IC1052) and connecting circuits.

The VAP PC board contains a gyro sensor, gyro amp, D/A, gain cons, buffer amp, VAP Mi-Com, PSD sensor, PSD amp, D/A, gain cons, buffer amp, driver, pre-driver, PWM, and E2 PROM circuits. A camera key unit consists of several switches of date on, title, AFM/off, shutter, sensor on/off, C reset, fade, and BLC switch circuits.

The camera PC board contains the timing generator, S & H and AGC IC, camera/AF Mi-Con, focus motor drive IC, zoom motor drive IC, flip-flop IC, digital processing, D/A converter, and several camera, iris, and analog regulators.

The Camera Section

The camcorder features are broken down into the camera and the VCR sections for easy operation. A brief description of each stage in the camera section follows. Although some of these circuits are sometimes called by another name, the circuit functions are the same, as explained in Chapter 34 (see Fig. 33-18).

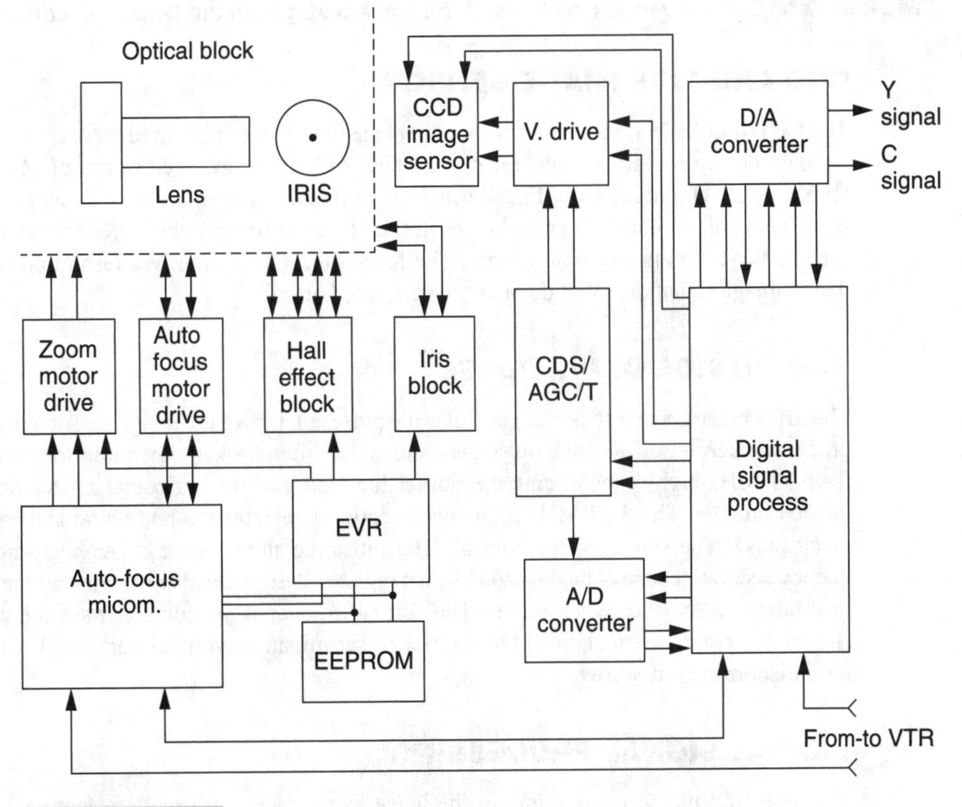

FIGURE 33-17 The block diagram of an 8-mm camcorder.

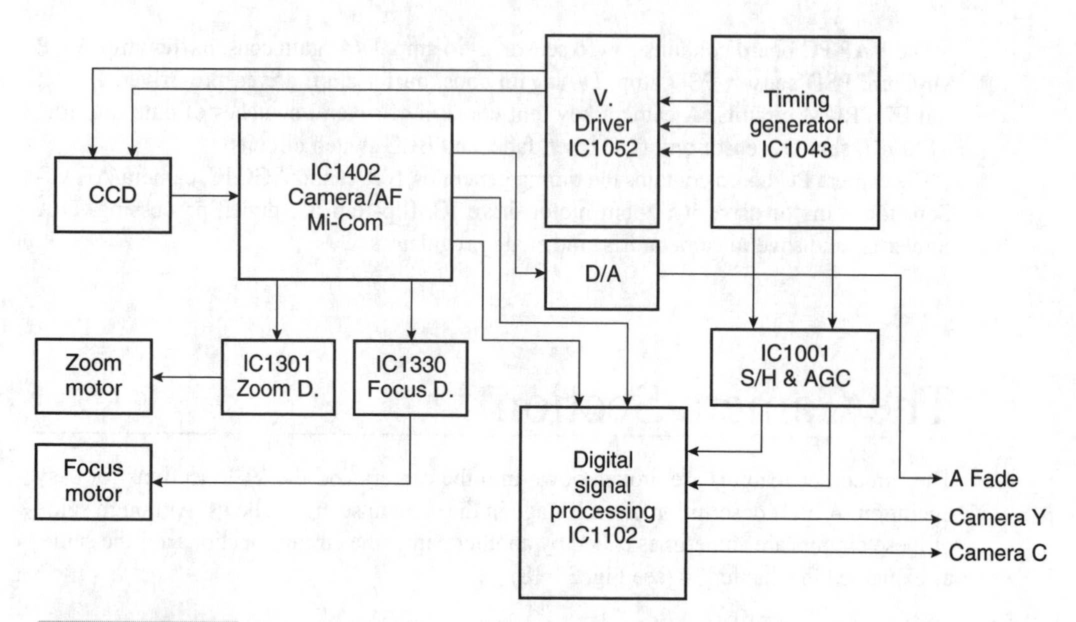

FIGURE 33-18 The various sections of the camera stages in the 8-mm camcorder.

CCD AND MOS IMAGE SENSOR

The CCD and MOS image sensors take the place of the older pickup tube. A CCD charge-coupled device is a semiconductor that consists of orderly arranged arrays of MOS cells (capacitors). The MOS color image sensor uses picture elements with a structure of npn three-layer photodiodes. The color-resolution filter arranges white, yellow, cyan, and green color filters in a mosaic pattern. The four-color output improves resolution with reduced image retention. Both devices are integrated circuits.

CCD OR MOS DRIVE PULSE

The drive pulse generator circuit generates the pulses that drive the image sensor and signal-processing circuits. The drive pulse generator IC is often broken down into four sections: the 5.37-MHz high-speed circuit, the shutter function, and the horizontal and vertical frequency circuits. The 5.37-MHz high-speed section generates the horizontal shift register clock pulse. The shutter speed control determines the shutter speed. The horizontal frequency section generates the horizontal shift register start, vertical shift register clock, vertical buffer, reset, and sweep pulses. The vertical frequency section generates the vertical shift register start, vertical optical block, field discrimination, vertical start, and FA and FB field discrimination pulses.

DIGITAL SIGNAL PROCESSING

You will find many new circuits in the latest camcorders with modern features. In the Canon 8-mm camcorder, the VC 2HI/VCS3A circuits consist of the CCD, CDS and AGC,

A/D, DSP, AF process, AF Mi-Com, camera Mi-Com, sub DSP, C. sup and D/A converter with a luminous (Y) and chroma (C) output connections (see Fig. 33-19).

Within the new camera section of the Canon 8-mm camcorder, the Sub DSP IC and AF IC are incorporated in the digital-processing IC that has been used for UC2HI. The new circuits consist of a CCD, CDS and AGC, DSP, and a camera Mi-Com, with Y and C output terminals. This new circuit arrangement reduces the physical size of the camera section.

SYNC GENERATOR

The sync generator provides signals that synchronize the operations in the color camera. It is usually one large IC. It supplies signals to the iris driver, auto white balance, date generator, process, luminance enhancer, chroma amp filter, and MOS image sensor drive pulse.

SIGNAL PROCESSING

In the signal-processing circuits, the cyan, white, green, and yellow are driven simultaneously by the 5.43-MHz sampling frequency. The matrix circuit produces luminance (Y) and chrominance (R, B, and G) signals.

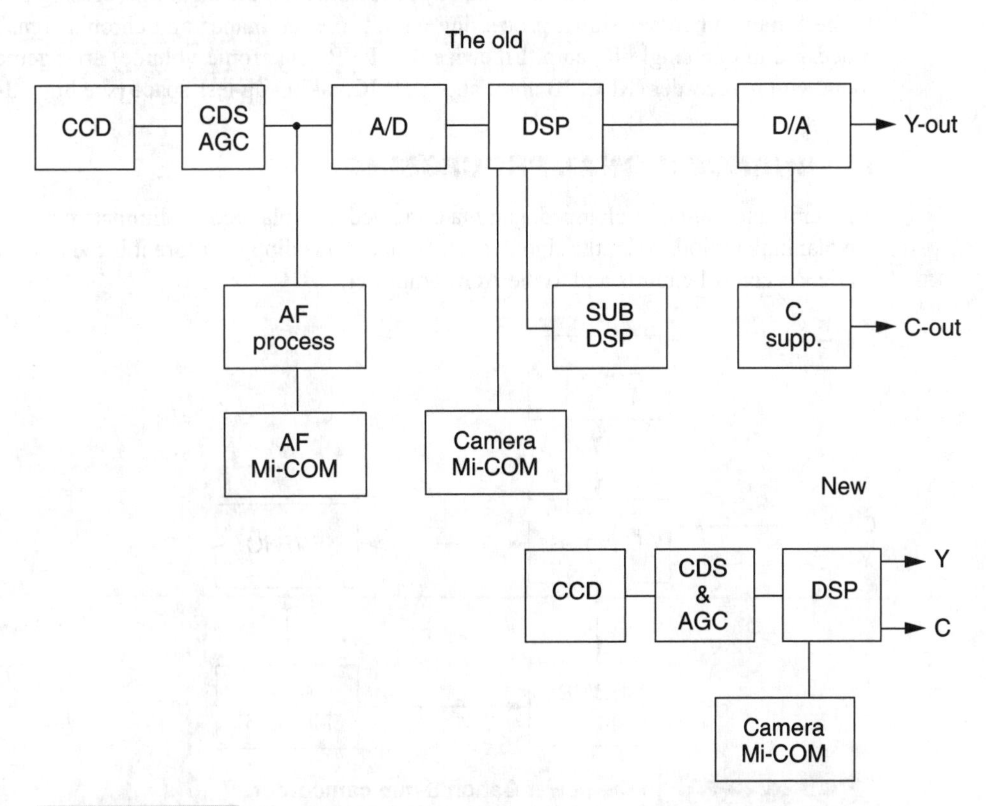

FIGURE 33-19 The digital-processing circuits of older and newer Canon 8-mm camcorders.

PREAMPLIFIER

The yellow, cyan, green, and white signals from the MOS or CCD image sensor are amplified by the preamplifier circuits. The input impedance is very low and controlled by a single IC.

MATRIX COLOR CIRCUITS

The amplified cyan, green, yellow, and white signals are fed into the matrix IC to produce the luminance, red, blue, and green signals.

RESAMPLING PROCESS

The white, cyan, green, and yellow signals are fed from the preamp circuits to the sampling IC. Improved high-frequency response and signal-to-noise ratio are obtained by averaging the value of every picture element (pixel). The luminance (YL and YH) signals are fed to the process board.

Canon Y/C signal processing In the conventional Canon models, two ICs were used for the luminance (Y) and chroma (C) process within the signal-processing circuits. In the 8-mm camcorder signal-processing circuit, the luminance and chroma signals are processed in one single IC chip. Likewise, the EVR (electronic volume) arrangement is found within recorder (REC/PB amp) and EVR IC, taking up less space (see Fig. 33-20).

LUMINANCE SIGNAL PROCESSING

The luminance signal is clamped, gamma corrected, and blanked to eliminate noise during the blanking period. Also, the signal is white- and dark-clipped before it is fed to the automatic iris control circuits and to the AGC amplifier.

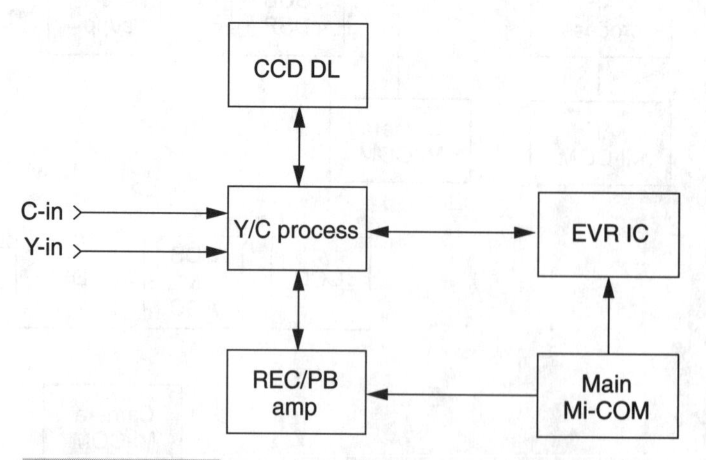

FIGURE 33-20 **In the newer Canon 8-mm camcorder, the luma (Y) and chroma (C) all included in a single Y/C process IC.**

CHROMA PROCESSING

The red, green, and blue signals are applied to the input of the color processing IC. Here, the red and blue signals are applied to the white balance circuits. These signals are mixed to produce the –R –YL and –B –YL color difference signals that are applied to the encoder processing circuits.

ENCODER

The luminance signal is applied to the NTSC signal-processing circuits. The luminance signal is clamped and white clipped to reduce the color output in the bright areas of the picture. The white-clipped luminance signal is blanked, applied to the YC mixing circuit, and added to the chrominance signal. The –R –YL and –B –YL signals are applied, clamped, and balance-modulated by the 3.58-MHz signal. These two modulated signals are added and applied through a 3.58-MHz bandpass filter. The reduced noise signal is applied to the Y/C mixing circuits and then added to the luminance signal.

AIC OR IRIS CONTROL

The AIC (automatic iris control) circuit controls the level of the video signal. This circuit controls the iris opening as detected from the processing circuit. The AIC circuit consists of the low light detection, AGC killer, and iris motor-drive circuits.

AUTOMATIC WHITE BALANCE

The automatic white-balance circuit maintains correct white balance under various lighting conditions. These circuits correct the white area of a picture so that it does not have a red or blue cast or tint.

AUTOMATIC FOCUS

The automatic focus circuit transmits an infrared signal that strikes the subject and bounces back to the infrared receiver located on the camera. This signal is detected by two photodiodes that produce electrical current, according to the infrared light received. The auto-focus circuit uses these two signals to move the lens in the proper direction. When the two signals become equal, the lens is focused and the auto-focus motor stops.

VTR OR VCR SECTION

The VTR or VCR section consists of the video head, head switching, system control, trouble-detection circuits, on-screen display, servo system, luminance record, chroma record process, luminance playback, chroma playback, and audio circuits. Although there are many other smaller circuits within the VTR, the most important ones are given here. In addition to these circuits, many mechanical movements are described in detail in Chapter 38.

Video heads The upper cylinder or drum in the VHS and VHS-C video recorders have a 41-mm diameter with four heads (see Fig. 33-21). The tape wrap for this system is 270 degrees, and it rotates at a speed of 2700 rpm. This system produces a more compact camcorder, providing compatibility with previous VHS recorders (see Fig. 33-22). The 8-mm drum is smaller in diameter and contains two channels and a flying erase (FE) circuit. The FM audio signal is recorded along with the video, rather than at the edge of the VHS tape.

Audio tracks on tape The VHS camcorders have a single (mono) linear audio track at the bottom edge of the tape. The linear track runs in a straight line along the edge of VHS tape. The sound quality of a single mono track is only fair, with a frequency response from 50 to 10 kHz (see Fig. 33-23). The stereo camcorder and VCR split the audio track into two different sound tracks in the stereo machines. A control track is found at the top edge of tape, while the middle area combines the video signal track. The 8-mm FM audio signal is recorded right in with the video signal.

System control The system-control microprocessor controls the reset, clock, power, battery detection, function switch, capstan, on-screen display, servo, luma/chroma, trouble detection, tape run, loading motor, mechanical control, and character generator in most camcorders. Usually, one large IC operates all of these functions.

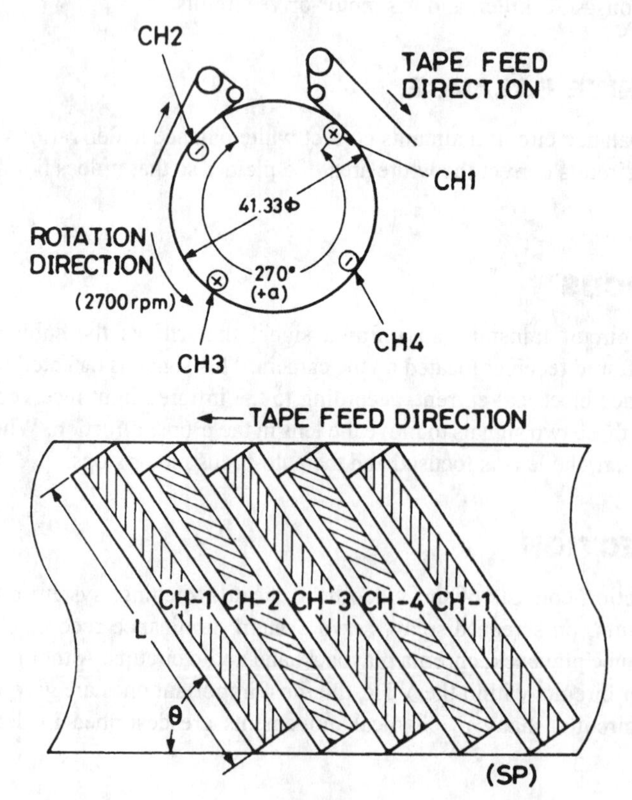

FIGURE 33-21 The VHS/VHS-C video head configuration.

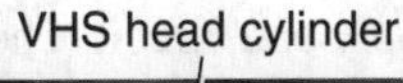

FIGURE 33-22 The VHS head assembly in the RCA CPR300 camcorder.

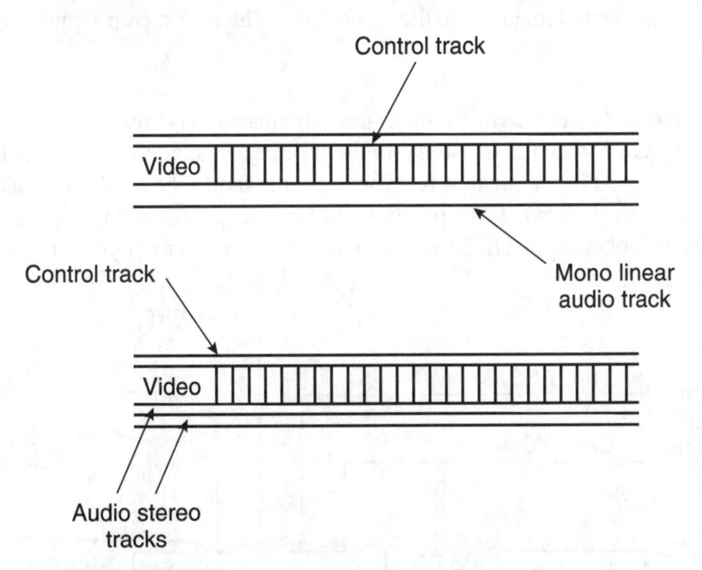

FIGURE 33-23 The mono and stereo audio sound tracks on a VHS or VHS-C cassette tape.

Head switching The head-switching circuits consists of a preamp and the switches of the four head channels in the VHS and VHS-C models. Head switching is done in the Record and Playback modes.

Trouble detection circuits The VHS and VHS-C trouble detection circuits consist of the end circuit, dew sensor, mechanism switch, take-up reel sensor, supply reel sensor, and supply end circuit. The battery-detection circuit is usually controlled by the system-control IC. The clog head-detection circuit warns that an accumulation of magnetized tape dust is in the head gap of the 8-mm head assembly.

On-screen display When the display switch is on, the system-control IC produces a low signal that is applied to the character generator in the VHS-C camcorder. This signal from the character generator displays the battery level, tape counter, shutter speed, and operation mode. This character signal is amplified and mixed with the video in a video amp IC. From here, the video signal is fed to the EVF and AV output.

Servo system In the VHS-C, the upper cylinder must rotate at 2700 rpm in Playback and Record mode in the VHS-C circuits. This means that the phase and speed of the capstan and cylinder motors must be controlled. The 1/2-V sync, REF 30-Hz, 30-Hz PG pulse, cylinder FG (CYL FG) pulse, capstan FG (CFF) pulse, and CTL pulse signals are in the servo system.

Luminance record process The luminance IC, in the VHS and VHS-C units, contains the record AGC, luminance signal extractor, detail enhancer, preemphasis, clipping, frequency modulation, and the E-E amplifier circuits (see Fig. 33-24). The signal is sent through a high-pass filter (HPF) to the record mixer and amp circuits. The luminance and chroma signals are mixed before going to the record amp. The record amp signal is applied to the video heads.

Luminance playback process In the 8-mm luminance playback circuits, two different ICs might be used. The pick-up signal of the tape heads is fed to a preamp, channel switch, and AGC circuits in the preamp IC. This signal is fed to the 5.8-MHz peak, trap, and the phase comparator between the preamp and luminance-processing IC. The luminance-processing IC contains the HP limit, demodulator, DO detector, dynamic deempha-

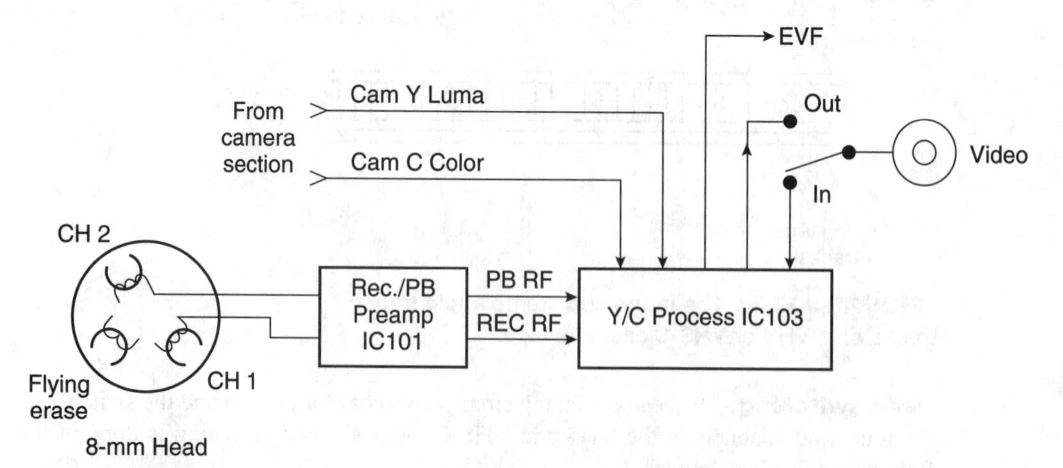

FIGURE 33-24 The luminance and color reording process in the 8-mm camcorder.

sis, noise cancel, line noise expand, Y/C mixer, and video amp. This video-playback signal is fed to the EVF and AV connector.

The luminance-playback signal is picked up by the four heads, amplified, and switched to the luminance playback IC in the VHS and VHS-C camcorders. This signal is demodulated and mixed with the chroma playback IC to produce a video signal. The video signal is fed to a character mixer IC, then out to the EVF and AV out connector.

Chroma playback process In a typical 8-mm camcorder chroma playback circuit, the signal from CH 1 and CH 2 is amplified by IC201. The signal goes through a 1.3-MHz LPF, AGC, burst deemphasis, PB balance modulator, chroma comb filter, chroma deemphasis, and a PB color killer and is mixed with the luminance signal in the Y/C mixer. The video signal is amplified to the EVF and AV connector circuits.

The chroma playback signal in the VHS and VHS-C chroma playback circuits are quite similar to the 8 mm. The signal from the four channels are switched and amplified in the preamp stage. The signal passes through a 1.3-MHz LPF, PB balance modulator 3.58-MHz BPF, AGC, burst deemphasis, and a comb filter and is mixed with the luminance playback signal in the Y/C mixer. This signal is fed to the character mixer to the EVF and AV out connector.

Service Notes

Like the video tape recorder, only a few manufacturers make camcorders. You might be working on one model that looks somewhat like another brand on the sales floor, but inside the camcorder body, the components are the same, except for a different part number. Most camcorders are made in Korea and Japan (see Table 33-1).

TABLE 33-1 THE DIFFERENT CAMCORDER MANUFACTURERS

MANUFACTURER	MAKES
Canon	Canon
Hitachi	Hitachi
	Kyocera
	Minolta
	M<itsuishi
	Pentax
	RCA
	Realistic
	Sears
JVC	JVC
	Samsung
	Toshiba
	Zenith

TABLE 33-1 THE DIFFERENT CAMCORDER MANUFACTURERS (CONTINUED)

MANUFACTURER	MAKES	
Matsushita	Chinon Curtis-Mathes Elmo General Electric Instant Replay JC Penney Kodak Magnavox NEC	Olympus Nikon Panasonic Philco Quasar Sylvania Technika
NEC	NEC	
Sanyo	Fisher Sanyo Vivitar	
Sharp	Sharp	
Sony	Aiwa Fuji Kyocera Pioneer	Ricoh Sony

The electronic technician should take a few precautions before attempting to repair the camcorder. The camcorder is small, with parts jammed together. To get at one defective component you might have to remove 10 others. You need steady fingers and small thumbs. Sometimes the little things save valuable service time and money.

WARM-UP TIME

The camcorder should warm up at least 3 minutes for operation and at least 5 minutes before attempting to make camcorder adjustments. This provides sufficient time for all parts to function properly and remove any moisture inside the camcorder. If moisture condensation is detected in the camcorder, the dew sensor prevents operation of camcorder, so as not to produce jamming of tape from possible sticking.

TAB LOCK

Be sure that the tab at the back of the cassette is in or "turned on" for recording. If the tab is out, the cassette cannot be used for recording. Often, when customers bring in the camcorder for repair because it will not record, it's because the tab is broken out of the cassette. Place a piece of tape across the opening where the tab was to record. Then remove it to prevent recording over a recording that you want to keep intact.

WRITE IT DOWN

Place the components in a regular order or write down where the parts go as you remove them. If you have to leave the bench or order a part, you might not remember where it belongs after you return or when you receive the part, which could be months.

SERVICE CLOTH

It's best to place a servicing cloth down on the bench—even if the bench is carpeted, to prevent damage to the camcorder's plastic body. Even small scratches might not polish out. Also, you don't want to lose a small screw or nut underneath of the camcorder.

SERVICE LITERATURE

This is a must-have item. You cannot troubleshoot the various stages, provide adequate electronic adjustments, replace components, take crucial voltage measurements, or find the correct part numbers without the service literature. Camcorder service literature is not cheap, but it will pay for itself in the first repair.

WRIST STRAP

Always wear a wrist strap when servicing and replacing delicate components, such as image sensors, microprocessors, and ICs.

KNOW WHEN NOT TO TOUCH

Do not attempt electrical adjustments when you don't have the correct test equipment to make these tests. Don't haphazardly tear into the camcorder when you do not have the correct service literature. If you cannot locate the defective component, check with the field service representative, wholesaler, service depot, or your fellow electronic technician. Ask for help—it's out there.

34

THE CAMERA CIRCUITS

CONTENTS AT A GLANCE

The basic camcorder circuit can be broken down into the camera and VCR sections. The camera section often consists of the lens, CCD or MOS, image sensor, V drive, CDS/AGC/r, B bit A/D converter, D/A converter, Y signal processing, C signal processing, auto focus, and electronic viewfinder (Fig. 34-1). The auto board in the 8-mm camcorder often consists of auto-focus Mi-Com, zoom motor drive, auto-focus motor drive, Hall detect block, and iris block.

FIGURE 34-1 **The camera section includes the lens assembly, CCD unit, and camera PC board circuits in the camcorder.**

VHS/VHS-C Camera Circuits

The VHS and VHS-C camera section can contain the lens assembly, automatic focus, iris or AIC control, and zoom motor assembly. The image sensor might have a pickup tube, as in the early VHS camcorders, and the later VHS and VHS-C models use a CCD or MOS pickup device. The drive pulse generator and preamplifier circuits are contained in the MOS or CCD circuits. Other VHS and VHS-C camera circuits are the sync generator, sample and hold, signal processing, matrix and filtering, luminance processing, chroma processing, automatic white balance, encoder, and power-distribution systems (Fig. 34-2).

CANON 8-MM CAMERA-SIGNAL CIRCUITS

The charge-coupled device (CCD) picks up the various images from the lens assembly, with the V output applied to terminal 36 of S/H IC1001. Timing generator IC1003 provides S/H pulses one and two and the playback timing signal to IC1001. The picked up signal passes through the S/H, AGC, and BLK stages with output at pin 10 of IC1001. The camera signal passes through transistor clamp circuit (Q1003) to pin 26 of digital signal processor IC1102 (Fig. 34-3).

The A/D input signal is processed within the digital IC1102, with camera Y and C outputs at pins 44 and 46, respectively. The luminous (Y) output signal passes through a buffer (Q1102), low-pass filter (FL1101), and buffer transistor Q1101. Camera color signal (C) goes through buffer transistor (Q1101), band-pass filter (C1125 and L1106), to buffer

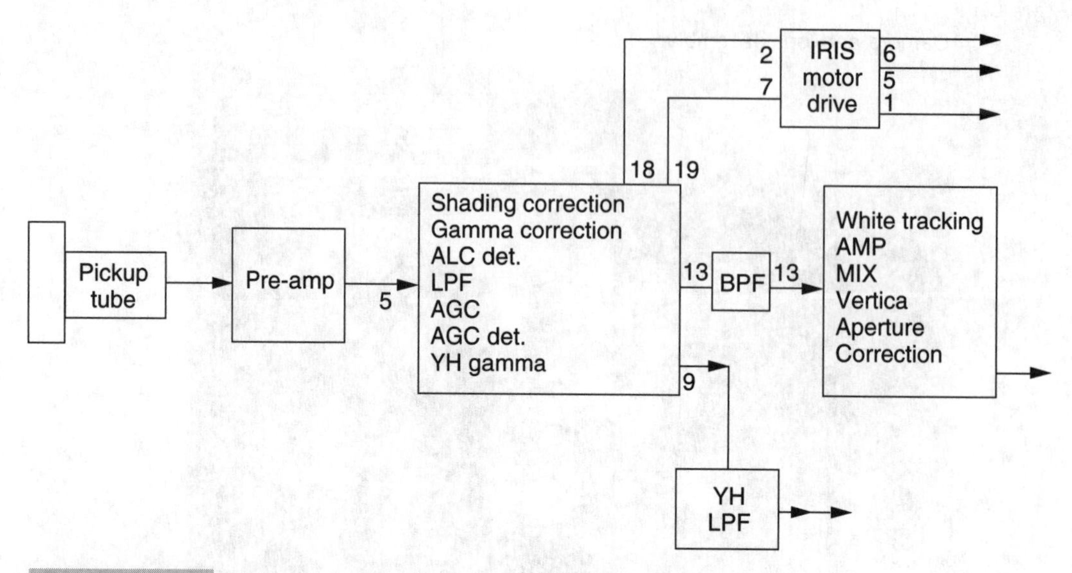

FIGURE 34-2 The block diagram of a VHS camcorder front end.

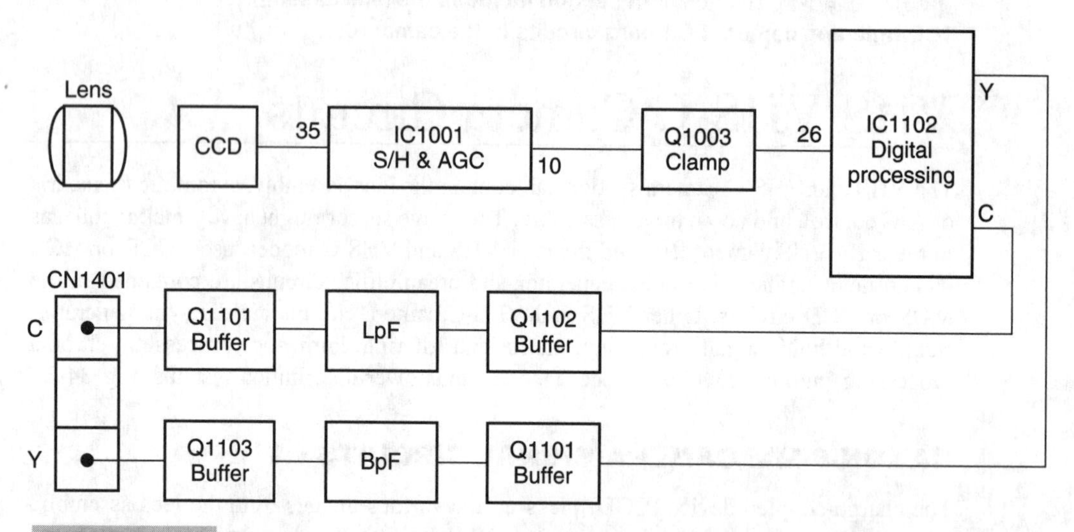

FIGURE 34-3 The Canon 8-mm camera-signal circuits to the video and color output.

Q1103. Both the chroma and luminance camera signals are fed to terminal connection CN1401, to the recorder PC board.

Pickups

In the early VHS, VHS-C, and Beta camcorder, a pickup tube was used to detect the image and convert it to an electrical signal. However, the tube's disadvantages have been overcome by CCD or MOS pickup devices.

CCD PICKUPS

The charge-coupled device (CCD) is a semiconductor that consists of orderly arranged arrays of cells (capacitors). When light shines on the CCD element, an electric charge is developed that varies with the intensity of light. The CCD device can store the charge in a well or a hole when certain voltages are applied. The depth of the well or hole varies with the voltage applied to the electrode. The charge collected in the adjacent cell or hole is moved to the cell where voltage is applied. With the many different cells or holes, the self-scanning process keeps repeating.

The surface area of the chip is divided into thousands of light-sensitive spots, called *pixels* (picture elements). For example, the Hitachi VME56A camcorder has a ⅓-inch CCD with 270,000 pixels as image sensor. A Fuji FUJIX-H18 H128SW camcorder has a ⅓-inch CCD with 410,000 pixels as image sensors and close to 400 lines in the camera. The Minolta Master C-570 camera section has a ⅓-inch CCD image sensor with 270,000 pixels with Digital Electronic Image Stabilization (EIS). The average camcorder uses a CCD image sensor with around 270,000 pixels.

MOS PICKUPS

The MOS image sensor is a semiconductor chip device. When light falls on the picture element, the electron/hole pair is generated inside the NT layer and PT layer, which forms the photodiode. As the electrons flow out, the NT layer in the PT layer, the hole remains. With positive and negative voltages applied to the metal electrodes, the electrons are moved through the silicon base, providing vertical and horizontal picture scanning. Low infrared and well-balanced sensitivity is obtained by using photodiodes as the picture element. The MOS (metal-oxide silicon) image pickup device is in the Hitachi, RCA, and Realistic camcorders. You can find the MOS device in some Kyocera, Pentax, and Sears models.

TYPICAL (VHS-C) MOS IMAGE SENSOR

The HE98245 MOS color image sensor has an image size of 8.8 (H) × 6.5 (V) mm, matching a ⅔-inch optical system. The chip size is 9.9 (H) × 8.0 (V) mm. The number of picture elements are 570 (H) × 485 (V) with a total of 276,450 pixels. With this type of MOS color image sensor, sensitivity in the infrared region is low, and well-balanced special sensitivity characteristics are obtained by use of picture elements (photodiodes) with an npn three-layer structure. Blooming is also suppressed with this structure.

The color-resolution filter is made by arranging complementary white (W), yellow (YE), cyan (CY), and green (G) color filters in a mosaic. The four color signals, W, YE, CY, and G, are read out via four output lines, so interlaced scanning with high resolution and without residual image can be performed.

When light falls on a picture element, an electronic hole pair is generated inside the n+ and p+ layer. In the n+ layer, the electron remains and the hole flows to the p+ layer. The photoelectrons are stored in the n+ layer as the photoelectronic conversion signal (Fig. 34-4).

A positive pulse is applied to the gate of the vertical switching transistor (TV). When reading the signal, the barrier of the gate lowers, the photoelectron stored in the n layer is drawn into the n layer as drain. When a positive pulse is applied to the gate of the horizontal switching transistor (TH), the barrier of the gate lowers, the photoelectron in the n

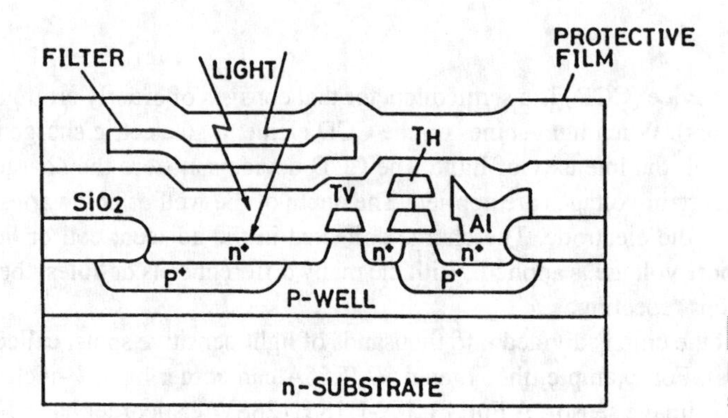

FIGURE 34-4 The MOS picture element of a VHS-C sensor.

layer as the source is drained into the n layer or a drain connected to the signal line and is derived as the output.

When the positive scanning pulse generated from the vertical scanning circuit opens the gates of the vertical switching transistors (TVs) and the horizontal scanning circuit opens the gates of the horizontal switching transistors (THs) beneath them, from left to right (in sequence), the photoelectrons of the photodiodes of two horizontal lines are output in sequence to perform horizontal scanning. When the gate opened by the vertical scanning circuit is changed in sequence and the same is repeated, the photoelectrons of all the photodiodes are output in sequence, and all the picture elements are scanned.

TYPICAL (VHS-C) SYNC GENERATOR CIRCUIT

Many operations within the color camera section are synchronized to provide correct operation. IC1107 provides the generated signals to synchronize the different operations in the color camera circuits (Fig. 34-5).

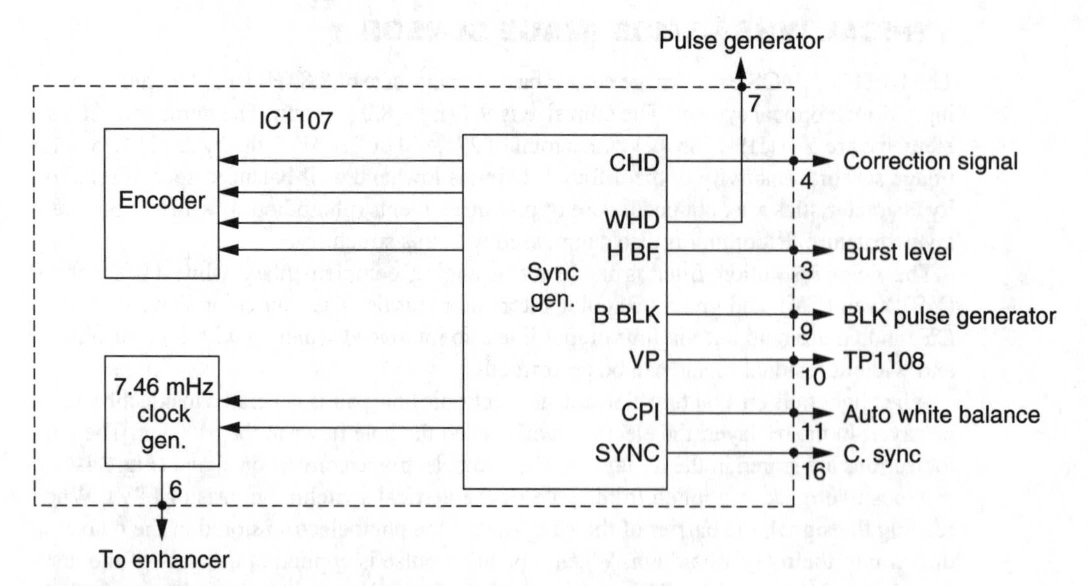

FIGURE 34-5 A typical VHS-C sync generator circuit.

The reference frequency signal inside IC1107 is controlled with a crystal oscillator (14.31818 MHz). All start and synchronization output signals are counted down from the reference frequency. IC1107 generates the following signals: Composite sync (C.Sync), pin number 16; Blanking pulse (BBLK), pin number 9; Clamp pulse (CPI), pin number 11; Wide horizontal drive pulse (CP2), pin 1, Composite sync (C. Sync), pin 12; Horizontal burst flag (H. BF), pin 3; Vertical pulse (VP), pin 10; Camera horizontal drive pulse (CHD), pin 4; and 7.16-MHz clock pulse (CLK GEN), pin number 6.

Signal-Processing Circuits

The VHS and VHS-C signal-processing circuits often consist of the CCD or MOS image sensor, preamplifier, resampling, chroma, and luminance circuits. The 8-mm signal-processing circuits can consist of CCD image sensor, signal-separation S/H, AGC, AGC detector, color-separation S/H, and white balance (Fig. 34-6). In the 8-mm signal-processing circuits, a CCD image sensor picks up the images and feeds them to the signal-separation S/H and AGC circuits. The output of the camera-processing circuit consists of luminance (Y) and chroma (C) signals. The camera has two video output signals that are applied to the EE video signal-processing circuits in the VCR or VTR. The other is the chroma signal, which is supplied to the chroma signal-processing circuit in the VFR.

TYPICAL (VHS-C) PREAMPLIFIER CIRCUITS

The weak signals from the image sensor are amplified by preamplifier circuits. The four complementary-color signals (W, YE, G, and CY) produced by the MOS color image sensor (IC1001) are applied to IC1002 and IC1003, which compose a preamplifier (Fig. 34-7). The MOS color image sensor (IC1001) is a current source with a low output at about 200 mA. Therefore, the camera's overall signal-to-noise ratio (S/N) depends on the preamplifier.

The preamplifier has the low input impedance of about 500 Ω because it must read signals with the short clock interlude of 186 ns (5.37 MHz). To reduce the input impedance, the dc output of the second-stage preamplifier (IC1006) is negatively fed back to the FET amplifier in the first stage (IC1002). The amplified color signals come out of pins 4, 5, 12, and 13 and enter the resampling circuit (IC1007) and matrix circuit (IC1100).

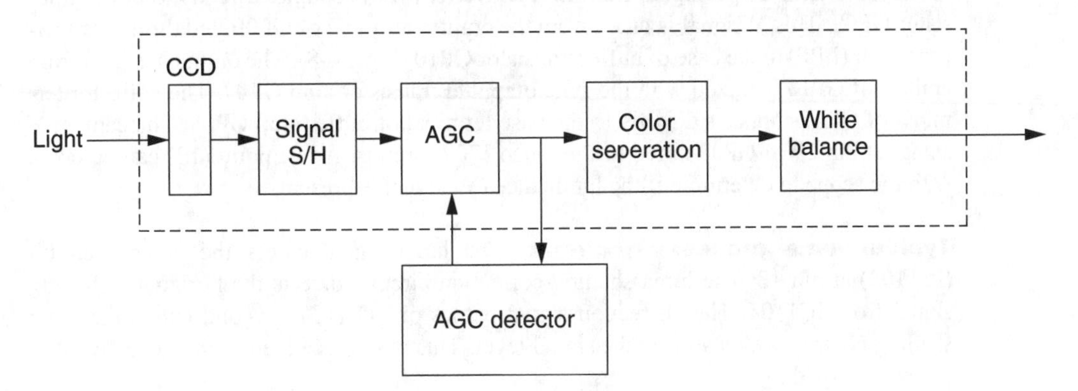

FIGURE 34-6 The Block diagram of an 8-mm signal-processing circuit.

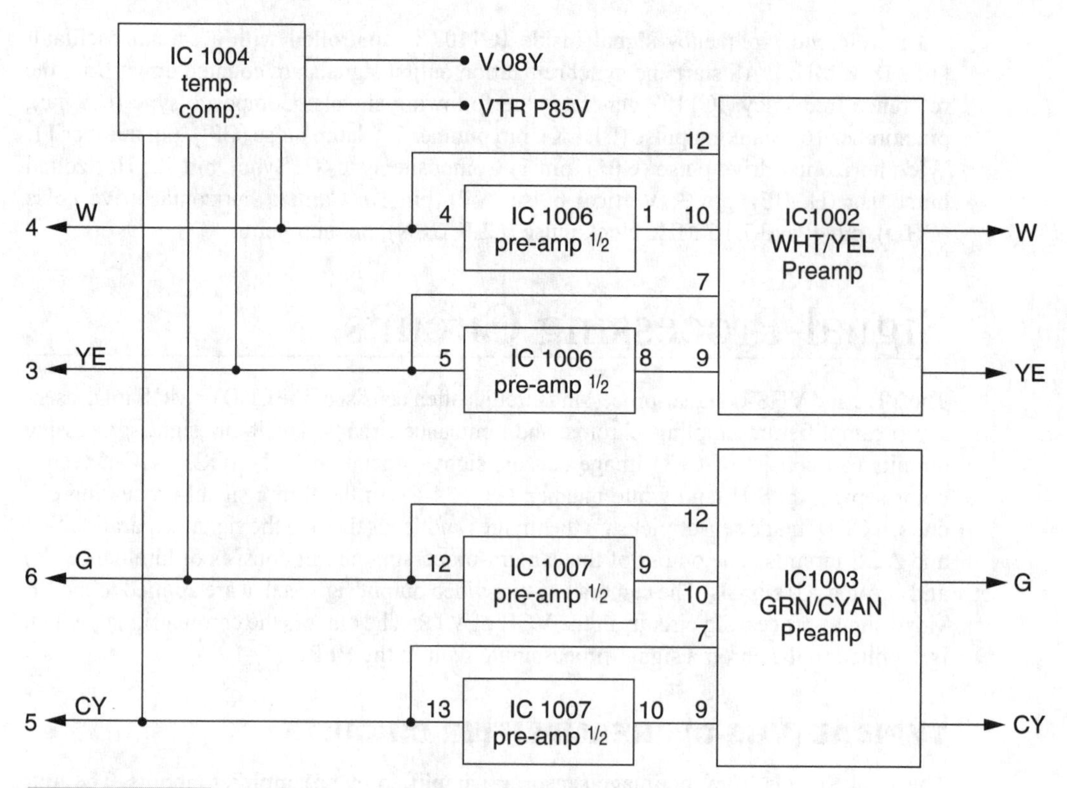

FIGURE 34-7 Typical VHS-C preamplifier circuits.

LUMINANCE SIGNAL PROCESSING

The purpose of the luminance circuits is to separate the luminance signals and form the composite video signal. The VHS and VHS-C luma circuits consist of a shading connection, setup, feedback clamp, gamma correction, blanking and linear chip, AGC, AGC detector, and gamma-control circuits. While in the Samsung SCX854 camcorder, the luma (Y) signal is fed from D/A converter to Y signal-process block.

The luminance output signal from D/A converter (pin 7) couples directly to buffer transistor GP09. This Y signal is taken from the emitter terminal of GP09, fed through a low-pass filter (LPF) to the base of buffer transistor GP10 (Fig. 34-8). The output Y signal from emitter of GP10 is mixed with the sync output at transistor amp GP07. The collector terminal of GP07 connects directly to the base terminal of buffer amp GP08. The camera Y signal at pin 29 of GP08 is fed to the video VTR circuits. A test point (10) can be taken with the scope to determine if the luminance (Y) signal is present.

Typical luma processing circuit The luma signal enters the process circuit (IC1103) at pin 18. The luma shading-correction circuit corrects the horizontal shading signal from IC1105. The correction signal passes pin 11 (IC1105) and enters the luma shading control, which adjusts the signal level. This output is added to the luma signal to correct the shading.

To fix the black level of the luma signal, control RM 1101-5 determines the set-up voltage, which is added to the luma signal. The black level of signal, which serves as the reference level, must be fixed to perform gamma correction and white clipping. The feedback clamp feeds back a sampled voltage to the input signal to fix the black level. The black level of the output of the gamma-correction circuit is sampled and held at the timing of the vertical optical block pulse from pin 15 (Fig. 34-9). The sampled voltage is compared to

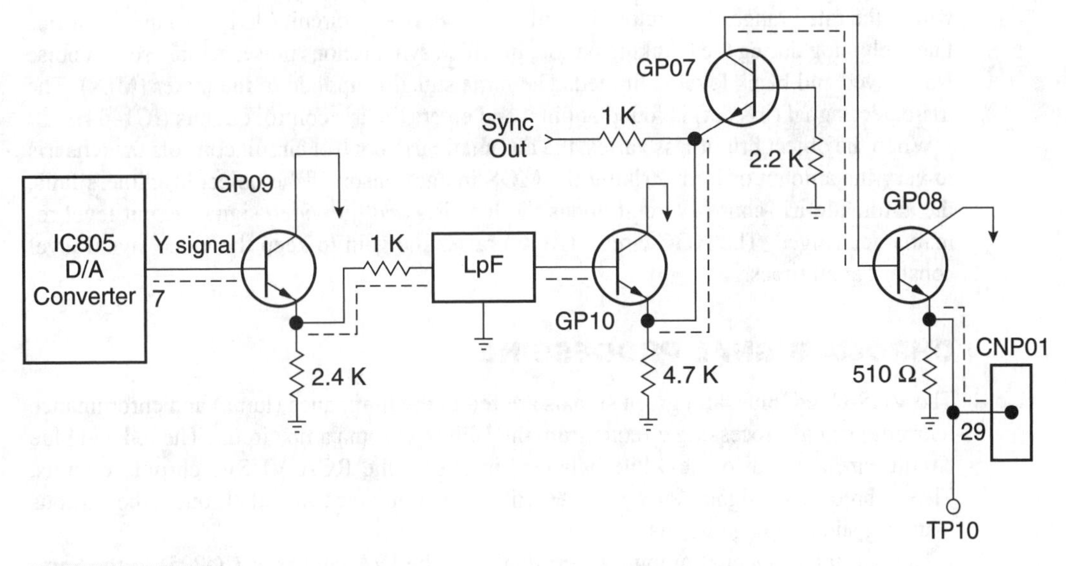

FIGURE 34-8 A typical VHS-C luma-processing circuit.

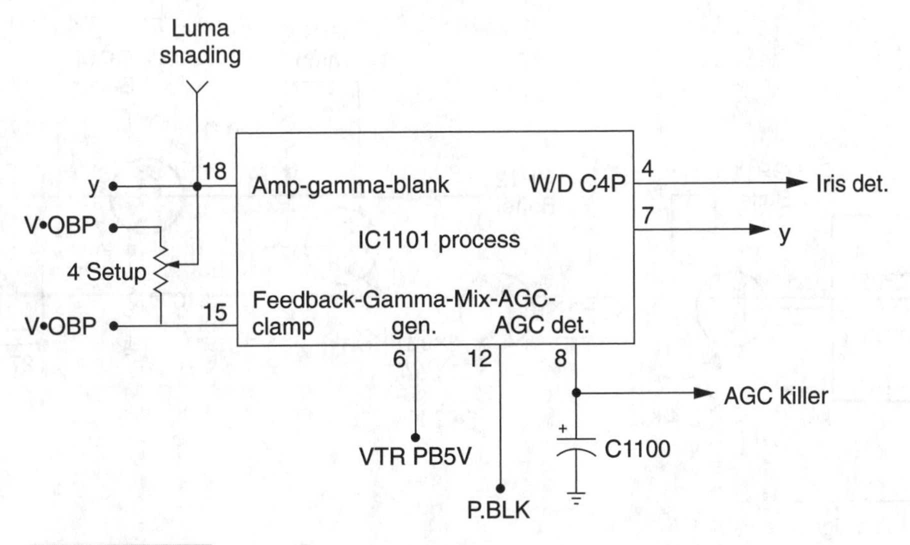

FIGURE 34-9 VHS-C luma-processing circuits.

the set-up voltage generator. This difference voltage is fed back to the input signal when the V.OBP pulse is applied. Now the black level of the luma signal is fixed.

The gamma-correction circuit amplifies the luma signal nonlinearly according to the gamma (1) and gamma (2) (dc voltages) supplied by the gamma-control circuit in order to set the overall gamma characteristic value.

The blanking (blank) and linear clip (white/dark clip) circuits remove the high-level noise produced by the MOS color image sensor and perform linear clipping to limit the white level within the rated range. The output is applied to a dark clip circuit (dark clip) that performs linear clipping during the blanking period, in which synchronous noises remain so that noise is removed and black level is limited. The luma signal is applied to the mixer (MIX). The iris-detect signal (iris det) is found at pin 4 and enters the iris control circuits (IC1404).

When the object brightness varies, the automatic iris-control circuit controls the lens iris to vary the amount of light striking the MOS image sensor. If the object brightness fails, the automatic iris-control circuit opens the lens iris and the video signal output level remains unchanged. The AGC circuit (AGC) raises the gain to keep the video signal level constant at all times.

CHROMA SIGNAL PROCESSING

The VHS-C red, blue, and green signals are fed to the luminance (luma) and chrominance (chroma) signal-processing circuits from the LPF or chroma amp circuit. The red and blue circuits are applied to the white-balance circuits in the RCA VHS-C chroma circuits. These three color signals are corrected like the luminance signal through the various blanking and clipping circuits.

The chroma (C) signal outputs at terminal 5 of the D/A converter (ICP05) in the Samsung chroma-processing circuits. This C signal is connected directly to buffer transistor GP11 and the emitter signal is capacity coupled to buffer transistor GP12 (Fig. 34-10). The

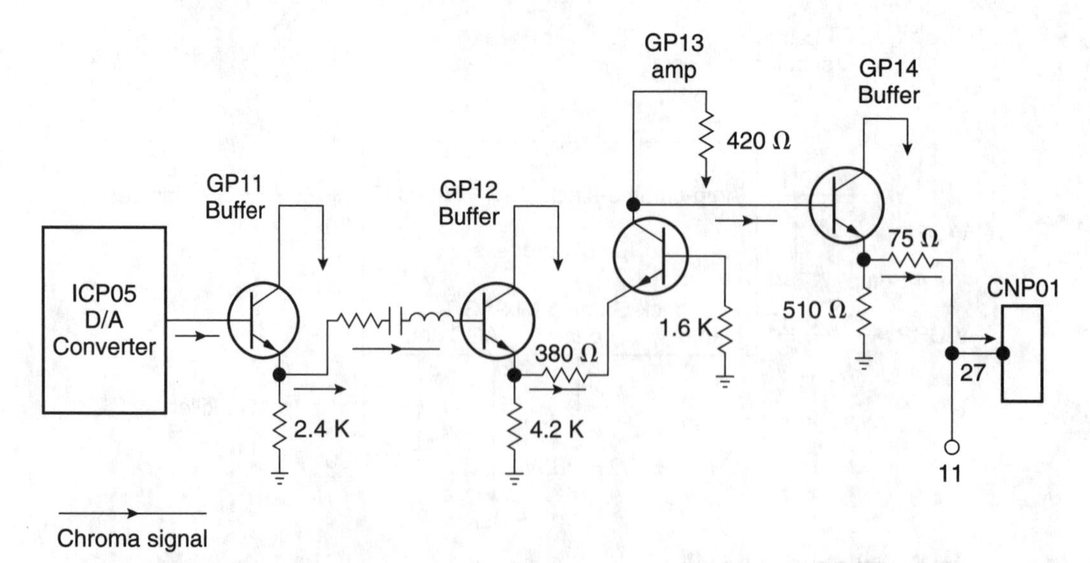

FIGURE 34-10 The 8-mm camera chroma-processing circuits.

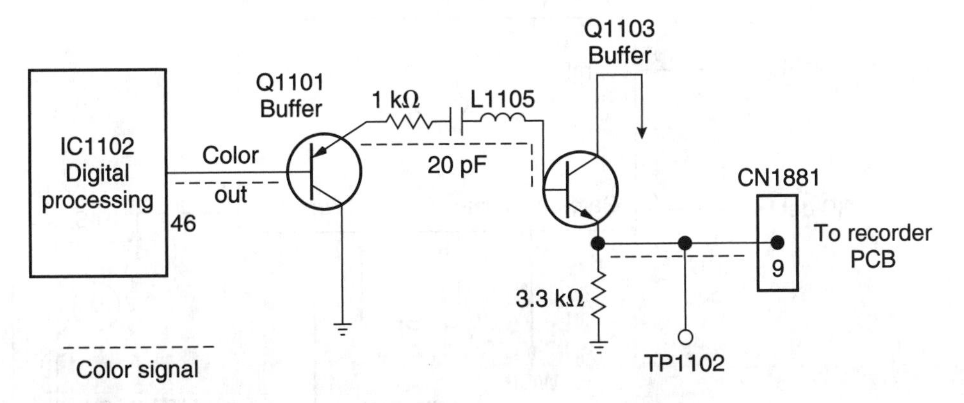

FIGURE 34-11 Typical 8-mm camera chroma-processing circuits.

C signal is taken from the emitter terminal of GP12, fed to the emitter terminal of chroma amp GP13, and coupled directly to buffer output transistor GP14. Scope the chroma signal (11) at the emitter output of GP14 to pin 27 of CNP01. This chroma output signal is then fed to the main VTR circuits.

Typical 8-mm chroma processing circuits The chroma (C) output camera signal is taken from pin 46 of the digital processing IC1102. This C signal is directly coupled to buffer transistor Q1101 (Fig. 34-11). The emitter terminal of pnp transistor is coupled through 1-kΩ resistor/20-pF capacitor, and coil L1105 is coupled to the base of buffer transistor Q1103. The chroma signal is taken from the emitter of Q1103 and directly fed to the connector terminal 9 of CN1881. A color (C) camera output signal waveform can be taken from TP1102 or pin 9 to determine if the camera chroma circuits are functioning.

AUTOMATIC IRIS CONTROL (AIC)

The AIC circuit controls the opening of the lens iris according to the object brightness. The AIC circuit can consist of controlling IC, iris motor, and manual iris control. The iris-detect signal is applied at pin 2 of the AIC control (IC1404). The gate cuts off the signal during low (Lo) periods and permits the signal to pass during high (Hi) periods (Fig. 34-12). The level detector provides an average iris-detected signal. If the luma signal level falls, the iris motor operates to close the iris. The AIC (IC1404) controls the iris motor at pin 9 (Fig. 34-13).

AUTOMATIC WHITE BALANCE

White balance refers to the adjustment of the recording system to the color temperature of the light that illuminates the subject to be recorded. When properly adjusted, white balance produces accurate recording of the colors in the scene. Keep the camcorder set at automatic white balance at all times.

Automatic white balance generates gain control signal of the R and B signals from the two color-difference signals and feeds them back to the gain-control circuit. Now a white object is white, regardless of the light source (color temperature). –(R-YL) and –(B-YL)

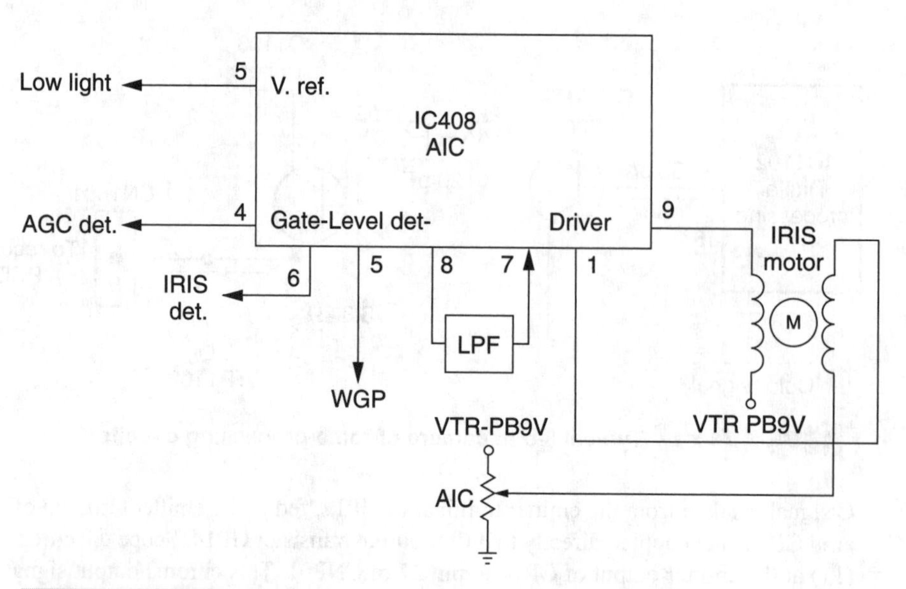

FIGURE 34-12 A VHC-C automatic iris control (AIC) circuit.

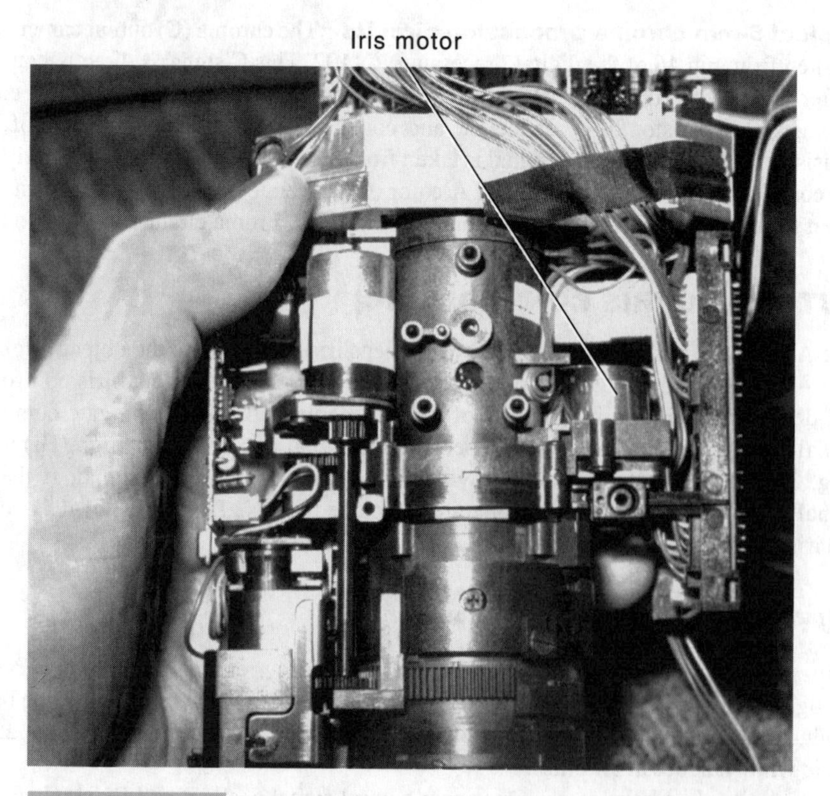

FIGURE 34-13 Some camcorders have an automatic iris motor located on the lens assembly.

signals from the process circuits are fed to the automatic white-balance control circuit (IC1103) through pins 2 and 3 (Fig. 34-14).

The dc level of the color temperature signal is fixed with a clamp pulse at pin 9. The gate permits the input signal to go to the following stage only when the preblanking pulse is at pin 5. The clip circuit clips the color-difference signals at the high and low levels. The color-temperature detector consists of a differential amplifier that subtracts the –(R-YL) from –(B-YL) to generate the color temperature R-B. The dc amp controls the white-balance control voltage to control the gain of the R and B signals.

AUTOMATIC FOCUS CONTROL

The auto focus system measures the distance from an object by emitting infrared rays toward the object and detecting the reflection, using the principle of triangulation. Infrared rays from an infrared LED pass through a protecting lens to the object (Fig. 34-15). The

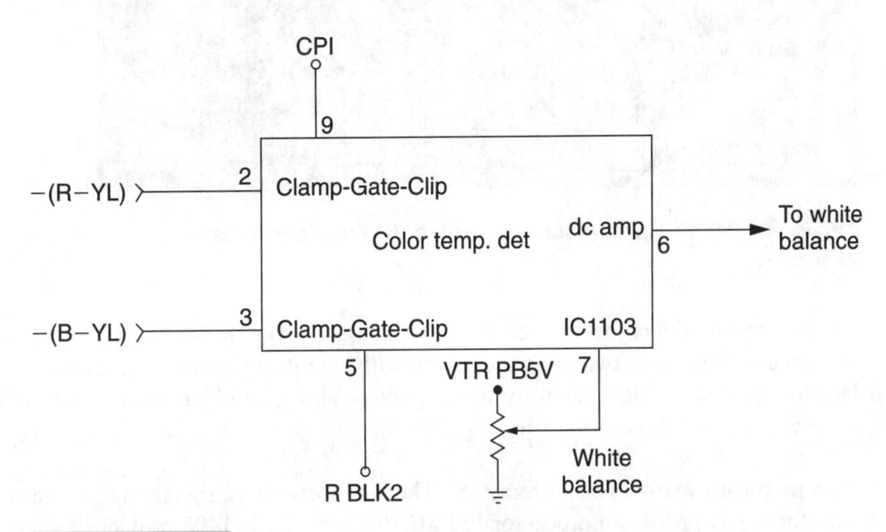

FIGURE 34-14 The VHS-C automatic white-balance control circuit.

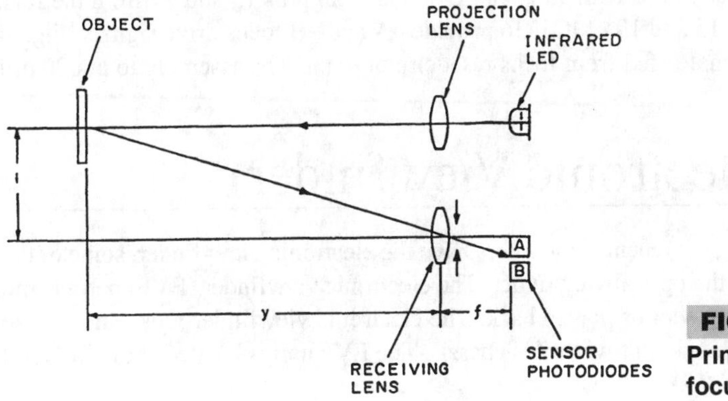

FIGURE 34-15 Principles of automatic focus.

FIGURE 34-16 **The macro lever on a VHS camcorder lens assembly.**

infrared rays hit the object and reflect back through a receiving lens and enter the sensor. The sensor consists of two photodiodes. Now the autofocus system moves the receiving lens to equalize the light intensity of the photodiodes. The object lens is moved by the same amount as the receiving lens (Fig. 34-16).

Canon 8-mm auto focus circuits The auto focus drive signals start at the camera/auto focus (AF)/Mi-Com processor IC1402 (Fig. 34-17). IC1402 provides a focus stop, focus power saver, focus CW and H/CCW, focus pulse, and focus reset signals to the focus motor drive IC1336. Inside IC1336 consists of a decoder and drive circuits. IC1336 drives the focus motor with an A and B signal from pins 13 and 17 from the focus motor drive IC. Pins 15 and 19 of IC1336 provide –A and –B focus drive signals (Fig. 34-18). A focus reset signal is fed from focus reset circuit in the lens assembly to pin 20 of IC1402.

The Electronic Viewfinder

Although the most recent camcorders have the electronic view-finder, some of the smaller cameras have the optical viewfinder. The electronic viewfinder (EVF) permits monitoring the image being shot or played back. The electronic viewfinder looks and acts somewhat like the small black-and-white TV chassis. The EVF unit is located at the front of the camcorder (Fig. 34-19).

The EVF circuits consist of a miniature picture tube with horizontal and vertical deflection circuits. The flyback transformer provides high voltage to the CRT. Vertical and horizontal sync circuits are generated and fed to the EVF deflection and VCR system-control circuits. A small amplifier and sync-separation circuit round up the EVF circuits (Fig. 34-20).

TYPICAL 8-MM EVF CIRCUITS

The electronic viewfinder shows a color picture on an LCD screen assembly. The LCD assembly connects to the EVF PC board. The black light assembly also connects to the EVF PC board. All of these assemblies are controlled from the recorder PC board. A regulated 5-V source is fed to the EVF board at pin 1 of CN2302.

FIGURE 34-17 **The focus control motor is located in the lens assembly of a VHS camcorder.**

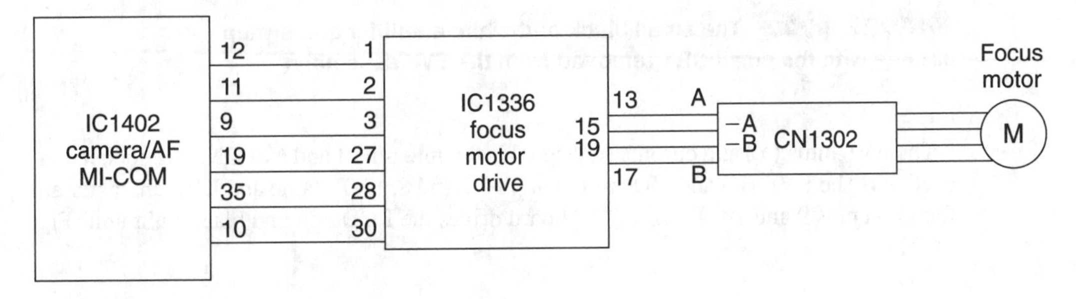

FIGURE 34-18 **The 8-mm autofocus circuits.**

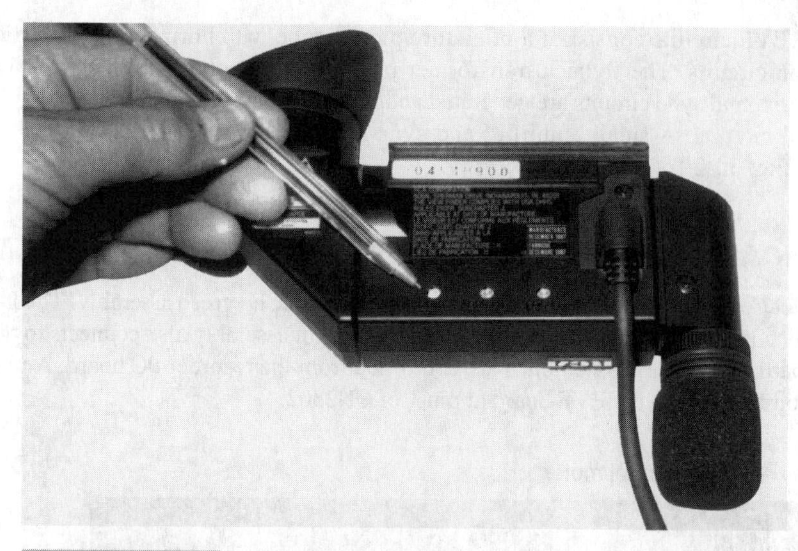

FIGURE 34-19 The separate EVF assembly from RCA CPR300 with various adjustments under assembly.

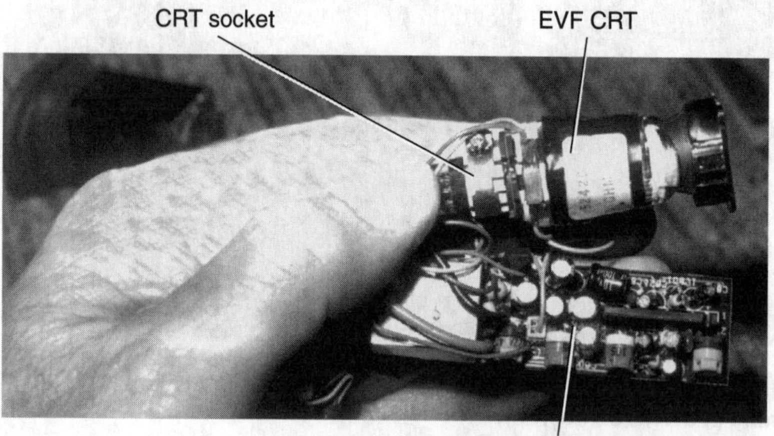

FIGURE 34-20 The small black-and-white amplifier and sweep circuits with the small CRT removed from the EVF assembly.

The EVF luma (Y) and chroma (C) signal is fed into pins 3 and 6 of CN2302 socket connections. The EVF HD and VD are fed at pins 7 and 8. EVF flame and EVF character are found on pins 9 and 10. The EVF PC board drives the LCD unit and black light unit (Fig. 34-21).

TROUBLESHOOTING EVF CIRCUITS

With a no raster symptom on the EVF tube, check the +5-V source that feeds the EVF circuits. Remember, the EVF heater circuit is fed, like the TV circuits, from a flyback winding. Improper HV or horizontal circuits can produce a no-raster symptom. Measure the HV to the CRT. With no high voltage, check the waveform at base of the horizontal driver transistor. If there is no waveform, check for a waveform at the horizontal output terminal of the horizontal and vertical sweep IC.

If the high voltage is normal and the CRT heaters are not lit (this can be difficult to see), check the continuity of the heater terminals. Remove the small CRT socket for this test. Then check the continuity of the heater winding of the flyback and tube. Note that a small series heater resistor is included in the heater lead (Fig. 34-22).

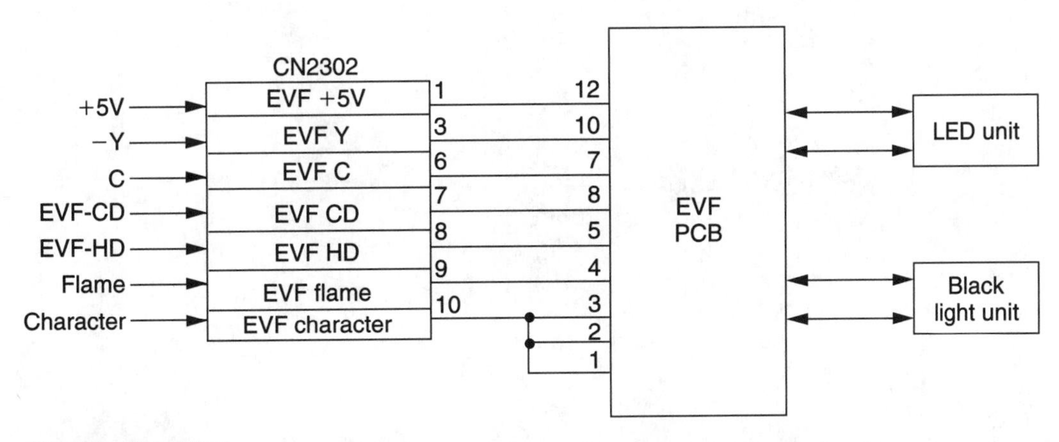

FIGURE 34-21 **The 8-mm EVF PC board, LCD, and black-light unit circuits.**

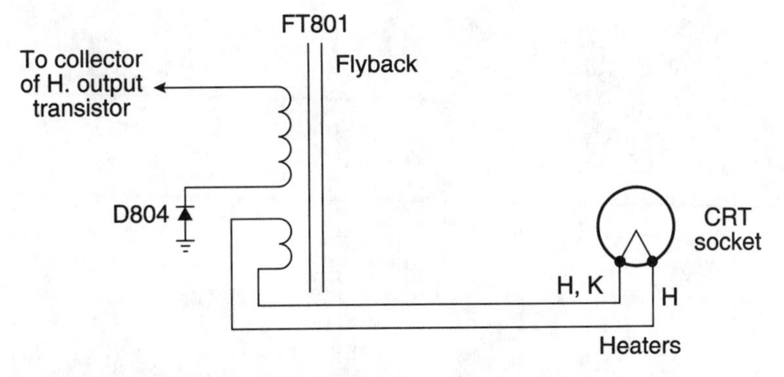

FIGURE 34-22 **The heater circuit of the EVF CRT socket.**

Check the vertical section for only a horizontal white line or insufficient vertical height. Readjust the vertical height control for insufficient vertical sweep. Check the vertical waveform at the vertical drive output pin on the sweep IC. Check for an open yoke winding or a bad socket connection if a normal vertical drive waveform is found. Suspect that the yoke-return resistor is open if the resistance is low (around 3.3 Ω).

For only a white raster and no picture, check the video input and output waveforms. Take a waveform fed to the sweep IC and output that feeds to video amplifier. Signal trace the video signal right up to grid G1 of the CRT. On color EVF circuits, check the color (C) input waveform and follow through the various color circuits. Usually, if a flyback transformer or small CRT is found defective, it is cheaper to replace the whole EVF unit.

35

VIDEO AND SYSTEM-CONTROL CIRCUITS

CONTENTS AT A GLANCE

The video circuits consist of the video in and out circuits, head switching, Record and Play circuits, luminance signal Record and Play circuits, and color signal recording and playback circuits. All of these circuits appear in the early camcorders, but the flying erase head is used in the 8-mm camcorders. Today, many of the latest VHS, VHS-C, and 8-mm camcorders have a flying erase head for better video pictures (Fig. 35-1).

FIGURE 35-1 The camera video section is located toward the front of the camcorder.

Video Signal Input/Output Circuits

The video signal input circuits are switched into the input—either from the camera or external devices connected to the AV connector. Within the Samsung 8-mm camcorder, the camera luminance (Y) and chroma (C) signal is fed to the Y/C processor board video section. The video jack serves as the input and output jack with switching the video in or out from Y/C process board (Fig. 35-2).

In the older VHS and VHS-C camcorders, the video signal had a separate input and output jack. In the Canon 8-mm camcorder, the video is switched in and out of the same jack on a jack unit. The video in from the jack unit, appears at pin 30 of IC2101. The Y/C process IC switches the video out at terminal pin 32. The video in and out waveform can

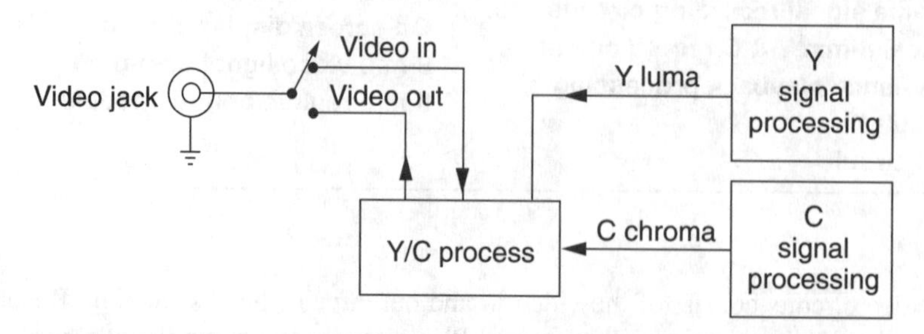

FIGURE 35-2 A typical 8-mm camera Y and C input/output circuits.

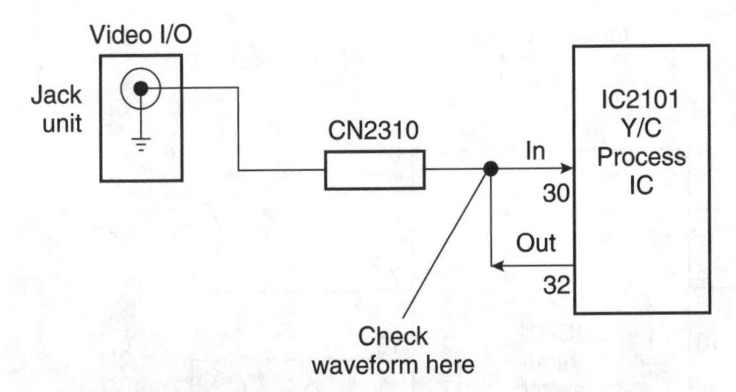

FIGURE 35-3 **The 8-mm video in and out from the Y/C processing IC2101 circuits.**

be checked right at the jack or checked on pins 30 and 32 to determine if video is present (Fig. 35-3).

Head-Switching Circuits

The head-switching circuits in the camcorder are often controlled by an IC. The head-switching circuits set the video heads in the Record or Play mode with several control signals.

TYPICAL 8-MM HEAD-SWITCHING CIRCUITS

The head-switching control signals in this model are: REC inhibit, SW 30 Hz, ASBL/PB, REC signal, squelch, and head SW signals (Fig. 35-4).

The REC inhibit signal in pin 18 of the head-switching circuit (IC201) is generated by the servo circuit during the Record mode. This signal, with the SW 30 Hz, controls the recording amplifier.

The SW 30-Hz signal with the input at pin 16 of IC201 is supplied from the servo circuit. This signal, along with the REC inhibit signal, controls the recording amp. During the Play mode, the head output will be selected to make the signal continuous.

The ASBL/PB signal with input at pin 26 of the head-switching IC (IC201) is supplied from the system-control microprocessor, selecting the mod\e (either Record or Play) of the video heads.

The Record (REC) signal input at pin 17 of IC201 is supplied from the system-control microprocessor. This signal controls the recording amplifier.

The squelch signal input at pin 15 of IC201 is supplied from the system-control microprocessor (IC901). It controls the recording amp and holds recording in check during record pause.

The head SW signal at pin 24 of IC201 is supplied from the system-control microprocessor. This switches the CH 2 head to the still head with the same azimuth as the CH 1 head during the slow and still modes.

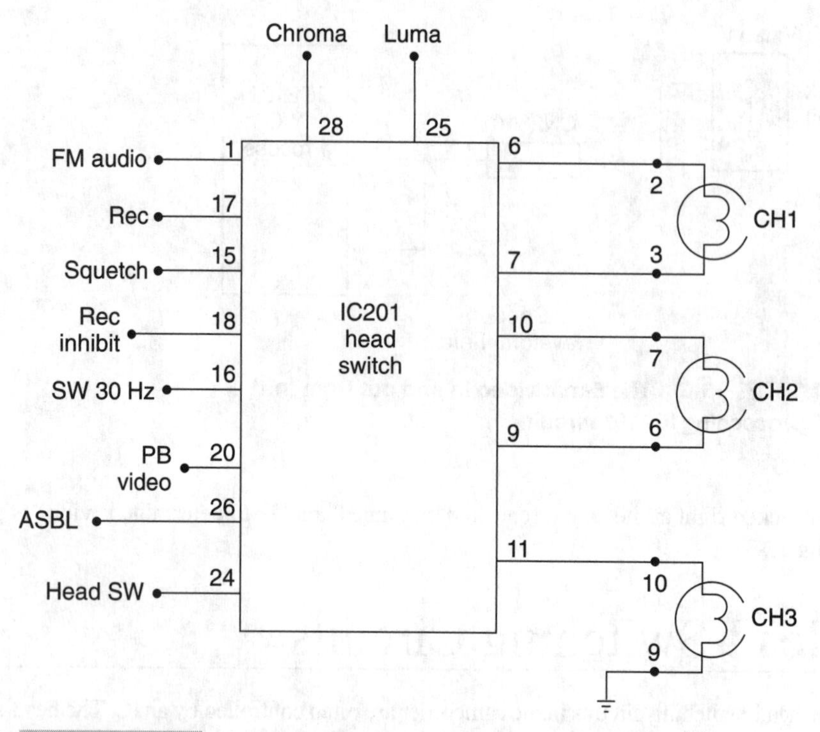

FIGURE 35-4 The typical 8-mm head-switching circuits.

Typical Video Record Circuits

The camera luma (Y) signal inputs at the syscon servo block of IC605 and IC604. The signal is fed from pin 7 of IC605 to pin 30 of the Y/C process IC201. Likewise, the chroma (C) is fed in at terminal 53 of Y/C process IC201. The video record signal appears at output pin 39 (Y) and 8 (C) of IC201. The luma signal is fed through a delay line and buffer stage to the preamp IC101 (Fig. 35-5). The chroma output of pin 8 of IC201 is fed through an audio trap and buffer transistor to pin 2 of preamp IC101.

The chroma and luma signal is mixed inside of IC101 and switched to the recording heads. Both CH 1 and CH 2 have a record amp circuit inside of IC101. The recorded video signal is coupled to CH 1 and CH 2 of the head cylinder. The head cylinder also includes an 8-mm FE (flying erase) head, which ties to an FE amp and a flying erase oscillator circuit.

THE 8-MM Y & C RECORD MODE

In Record mode, the Y/C process IC2101 provides recording output at pin 8 and feeds this signal to pin 4 of the head amplifier IC2001. The recording waveform can be taken at pin 4 or 8, whichever is easiest to get to. The head signal is amplified, switched, re-entered on pins 37 and 40, and output at terminals 27 and 34 of head amplifier (Fig. 35-6).

Channel one (CH 1) output at pin 34 of IC2001 is applied to pin 10 connection CN2001 and applied to CH 1 of drum head. Channel CH 2 output at pin 27 and appears at pin 8, to

CH 2 of drum head. The flying erase head signal ties to terminal pin 1 and onto the flying erase of drum head.

Operation in the Play Mode

In the typical VHS luminance and chrominance circuits in Play mode, the signal is picked up by tape heads in Play mode, the signal is picked up by tape heads R1, L1, R2, and L2. The Play heads feed the signal to the head amp IC3501. The luminance signal is fed from pin 20 and 21 to phase-compensator circuits Q3501, L3501, and C3526, while the chroma signal path comes out of pin 15 of IC3501 and feeds to Q3502 amp through a 630-kHz low-power filter (LPF) going to pin 34 of IC8001, the chrominance processing IC.

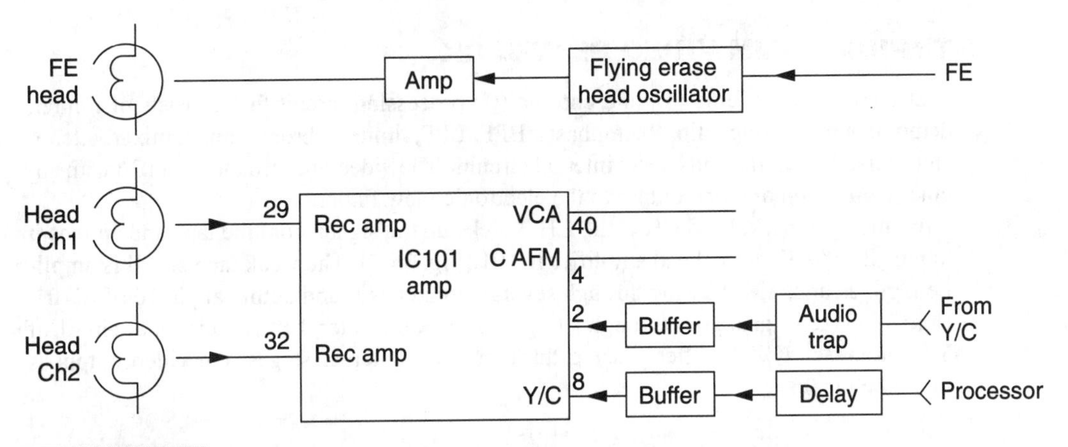

FIGURE 35-5 Typical 8-mm video recording circuits.

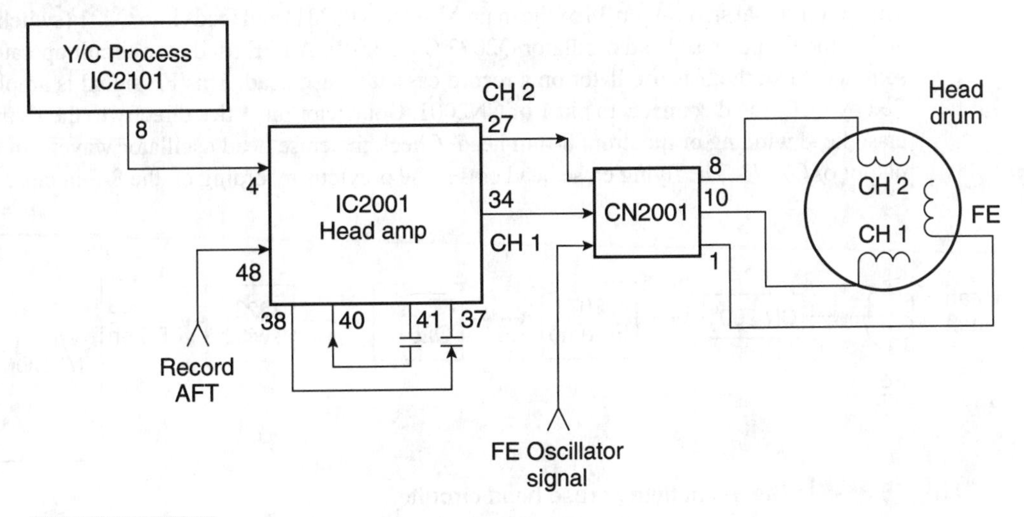

FIGURE 35-6 A typical 8-mm Y/C recording circuit.

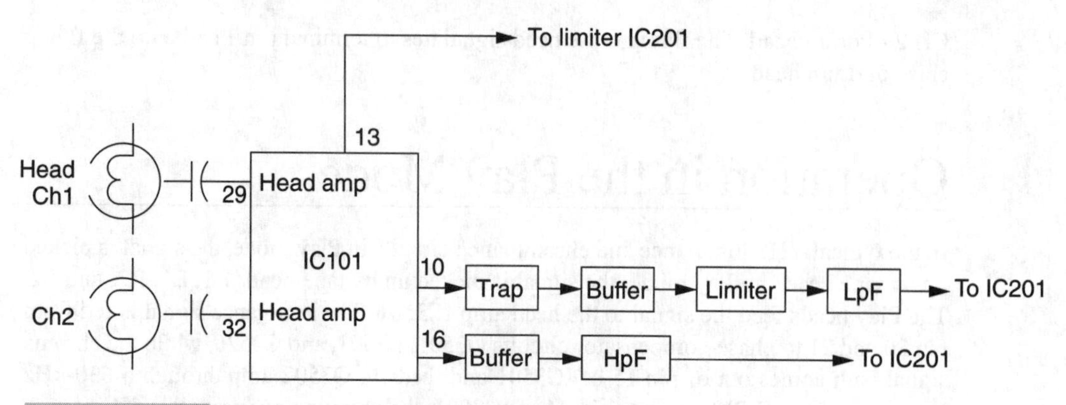

FIGURE 35-7 The 8-mm video playback head circuits.

TYPICAL 8-MM VIDEO PLAYBACK

IC201 contains the luma (Y) and chroma (C) processing circuit that consists of limiter, demodulator, peaking, clip, Reemphasis, BPF, LPF, limiter, chroma, luma, mixer, attenuator, burst, VCA, trap and cancel internal circuits. The video out is found at pin 36 with another connection of video output to the electronic view finder.

In video playback, head CH 1 and CH 2 picks up the signal from the tape and applies it to pins 29 and 32 of the head amplifier IC101 (Fig. 35-7). The weak tape signal is amplified, mixed and, after passing through several circuits, is found output at pin 10 of IC101. Here, the video signal goes through a trap, buffer, soft limiter, LPF, and input at pin 41 of Y/C processor IC201. After passing through many internal stages, the video output is found on pin 36.

TYPICAL 8-MM FLYING ERASE HEAD

The FE on signal starts at pin 38 of the main Mi-Com (IC231) and feeds into Q231 (switch), and to the flying erase head oscillator Q2073 (Fig. 35-8). An FE oscillator circuit operates somewhat like the bias oscillator on a record cassette erase head. This FE signal is amplified by Q2071 and connects to pin 1 of CN2001. Connector pin 1 ties directly to the flying erase head winding of the drum 8-mm head. Check the erase head oscillator waveform at output of Q2073. The flying erase head erases the previous recording on the 8-mm tape.

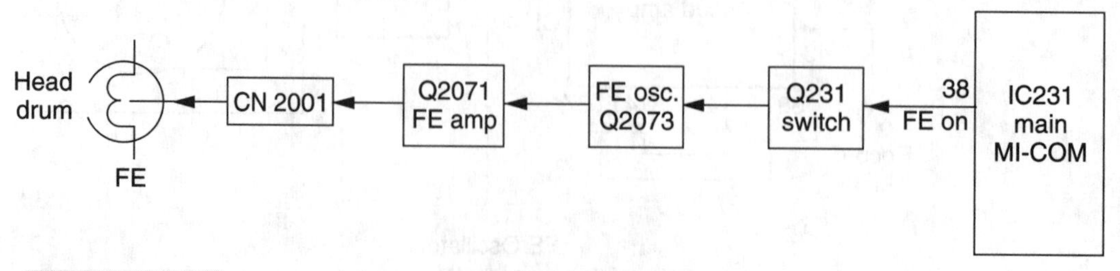

FIGURE 35-8 The 8-mm flying erase head circuits.

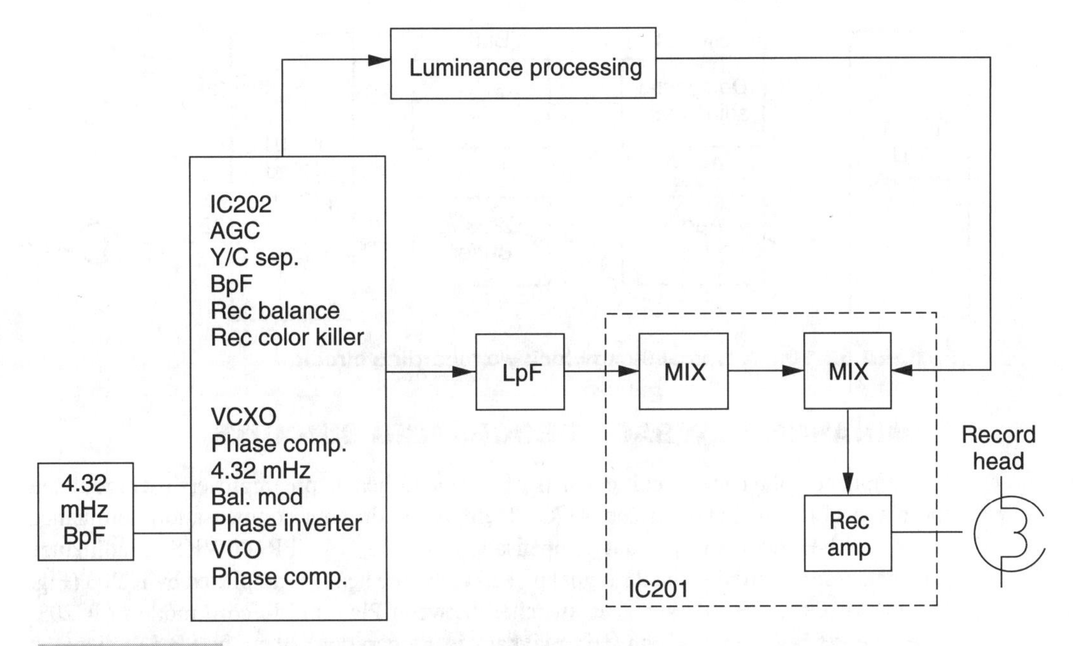

FIGURE 35-9 The 8-mm signal recording circuits.

CHROMA SIGNAL RECORDING CIRCUITS

During the camera mode, the 743-kHz down-converted chroma signal is generated from the 3.58-MHz chroma signal coming from the camera. In the VTR mode, the 3.5-MHz chroma signal is extracted from the input video signal and the 743-kHz down-converted chroma signal is generated. IC202 operates in the Record mode when the REC signal is high (Hi), and in Play mode, when it is low (Lo) (Fig. 35-9).

IC204 and IC205 contain the camera/external device video signal selector. IC202 contains the luminance signal record/processing (AGC and chroma signal separator), chroma signal record/processing (AGC, burst preemphasis, chroma preemphasis, frequency converter, 3.58-MHz generator, and 47 1/4-FH generator). IC201 contains the recording Y/C mixer and recording amplifier.

TYPICAL 8-MM Y & C RECORD CIRCUIT

The Y and C recording circuits in the Samsung SCX854 camcorder starts at Luma and chroma processing IC201. The Y signal is fed out of pin 39 and the C or chroma signal from pin 8 of IC201 (Fig. 35-10). The luma signal passes through a delay time adjustment circuit and to buffer amp. You can check the record Y signal at TP204 before entering preamp IC101.

The chroma (C) signal is found on output terminal 8 of Y & C IC201, feeds through an audio trap and buffer transistor Q228 to input pin 6 of IC101. At the emitter terminal of Q228 (pnp), you can check the color waveforms (C) at TP228 before it enters pin 6 of preamp IC101. The head preamp (IC101) amplifies the Y and C signals with CH 1 output at pin 29 and CH 2 at pin 32.

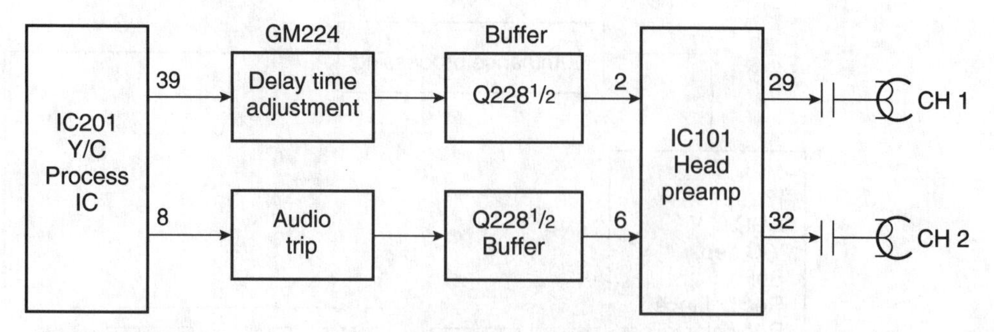

FIGURE 35-10 A typical 8-mm block y/c recording circuit.

LUMINANCE PLAYBACK PROCESSING CIRCUITS

The luminance playback circuit consists of the video heads, preamplifiers (demodulated and playback), signal synthesizer, AGC, Reemphasis, dropout compensation, luminance mixer, and 1-H delay for drop-out compensation. In the RCA CPR100 VHS-C luminance playback process circuits, the RF signal picked up by the heads is amplified by IC203 (Fig. 35-11). Also, the PB 5-V signal is switched between Play and Record mode in IC203. When the PB 5-V signal is high (Hi), playback is in operation; when the PB 5-V signal is low (Lo), recording operation is switched in. IC203 contains the luminance chroma (Y/C) mixer and preamplifier. The FM demodulator, reemphasis, and Y/C circuits are contained in IC201. IC204 amplifies the luminance playback signal and is fed to the electronic viewfinder (EVF) and AV out connector.

Luminance playback circuits The signal picked up by the tape heads is preamplifier and switched by IC203 (Fig. 35-12). It is then fed to a 4.6-MHz peak playback equal-

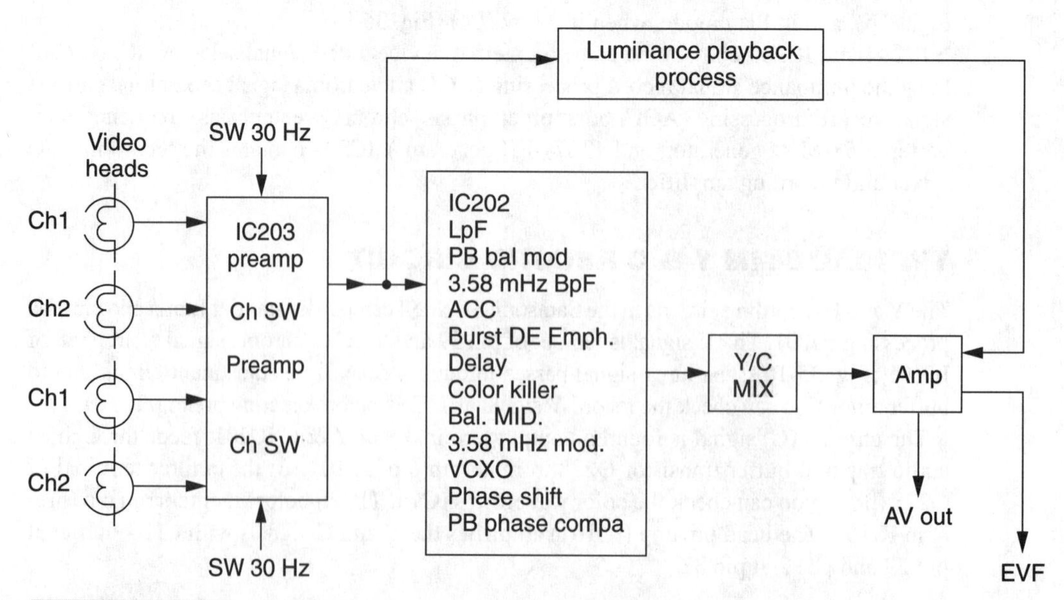

FIGURE 35-11 Luminance playback process circuits in a VHS-C camcorder.

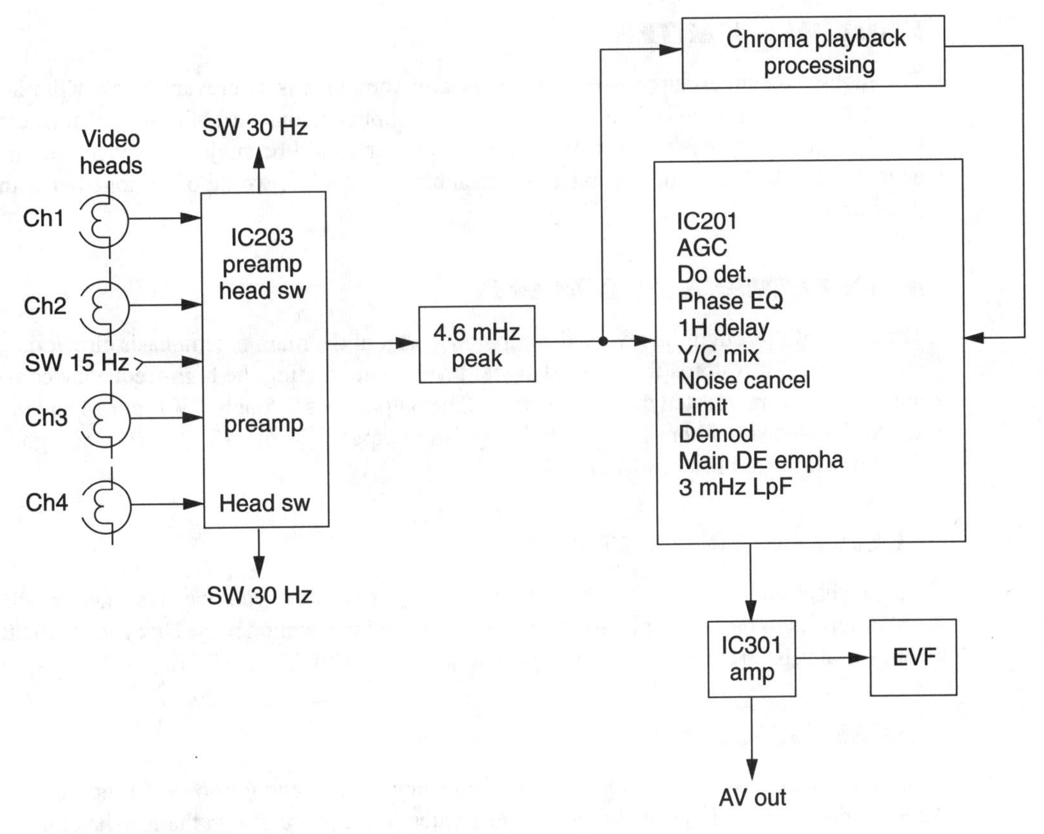

FIGURE 35-12 **A VHS-C chroma signal playback circuit.**

izer (CP201). Because the preamplifier has a flat frequency response, it corrects the head frequency-response playback output. It raises the response around the upper limit (4.6 MHz) of the FM carrier to flatten the overall frequency response.

AGC CIRCUIT

The AGC circuit corrects the deviation, including interchannel deviation, of the outputs of the video heads. The AGC detector (AGC DET) detects the input level and feeds the output back to the AGC circuit to control gain. The signal passes through the dropout (SW 8), SW 4, and pin 4 before coming to the phase equalizer Q201, DL201, and the dropout detector (DO DET), which are included in the IC.

PHASE EQUALIZER

Q201 is the phase equalizer stage. This circuit corrects phase distortion that occurs when the playback equalizer (CP201) equalizes the amplitude so that distortion of the playback waveform is reduced. The output passes through a buffer (Q214) and input at pin 2 before entering IC201 once again.

HIGH-PASS LIMITER

The high-pass limiter suppresses lower sideband components to prevent black/white inversion. If the recording is done with preemphasis applied to that part of luma signal where the black level changes to the white level, the carrier would be subject to dropout at the edge during playback, and inversion between black and white would occur together with deterioration of S/N.

MAIN REEMPHASIS CIRCUIT

The effect of this circuit is reversed compared to that of the main preemphasis circuit during recording. It restores the original signal level by alternating the high-frequency components that were boosted during recording. The output goes through SW 1, pin 14 and the 4.1-MHz bandpass filter (L213 and C252). The output is at pin 13 of IC201 and goes through a low-pass filter (3 MHz).

PLAYBACK EQUALIZER

The playback equalizer contains Q207, Q208, and Q213. This circuit operates when the PB 5 V is applied to the base of Q207. The output enters H-correlation noise canceller/dropout compensator IC204. The luma signal goes out of pin 6 to pin 23 of IC201.

NOISE CANCELER

The noise canceler circuit removes high-frequency noise (random noise) in the played-back luma signal to improve the S/N. The greatest effect occurs with the low-level luma signals. The output luma signal passes through SW 5, pins 24 and 28, before coming to the sync expander (SYNC expand) and passes through SW 2 before coming to the sync separator (SYNC SEPA).

SYNC SEPARATOR

In the sync separator (SYNC SEPA), the output video signal passes through SW 2 and comes to the sync separator (SYNC SEPA), which separates the sync signal, as in recording. The output appears at pin 20 and feeds to chroma sync (IC202).

CHROMA SIGNAL PLAYBACK CIRCUITS

After being processed in IC202, the chroma signal mixes with the luma signal and becomes the video signal in the Radio Shack 150 chroma playback circuits (Fig. 35-13). IC203 contains the preamp and continuous signal generator. The chroma playback processing, which contains chroma signal separator, AGC, burst reemphasis, playback balanced modulator, and crosstalk cancel, is contained in IC202. The luminance and color (luma/chroma) mixer is located in IC201 with the output video amplifier (IC301).

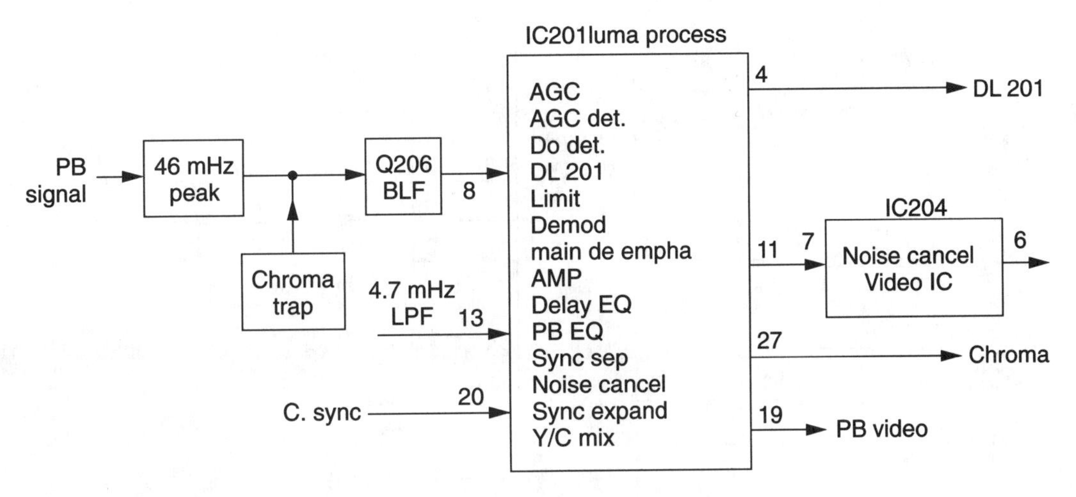

FIGURE 35-13 The VHS-C chroma signal playback circuit.

System Control

The most important component in the system-control circuits is a microprocessor, also called a *microcomputer* or *IC*. This system-control IC can have from 48 to 82 connecting terminal leads (Fig. 35-14). Usually, the microprocessor controls all functions of the camcorder, including the camera and VCR functions. In the VHS models, the system-control

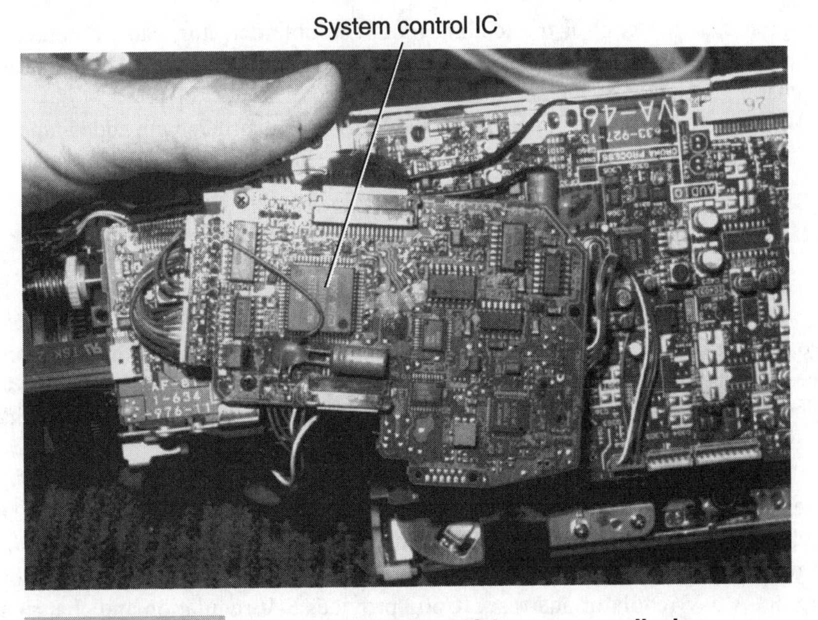

FIGURE 35-14 The system-control IC has many gull-wing terminals.

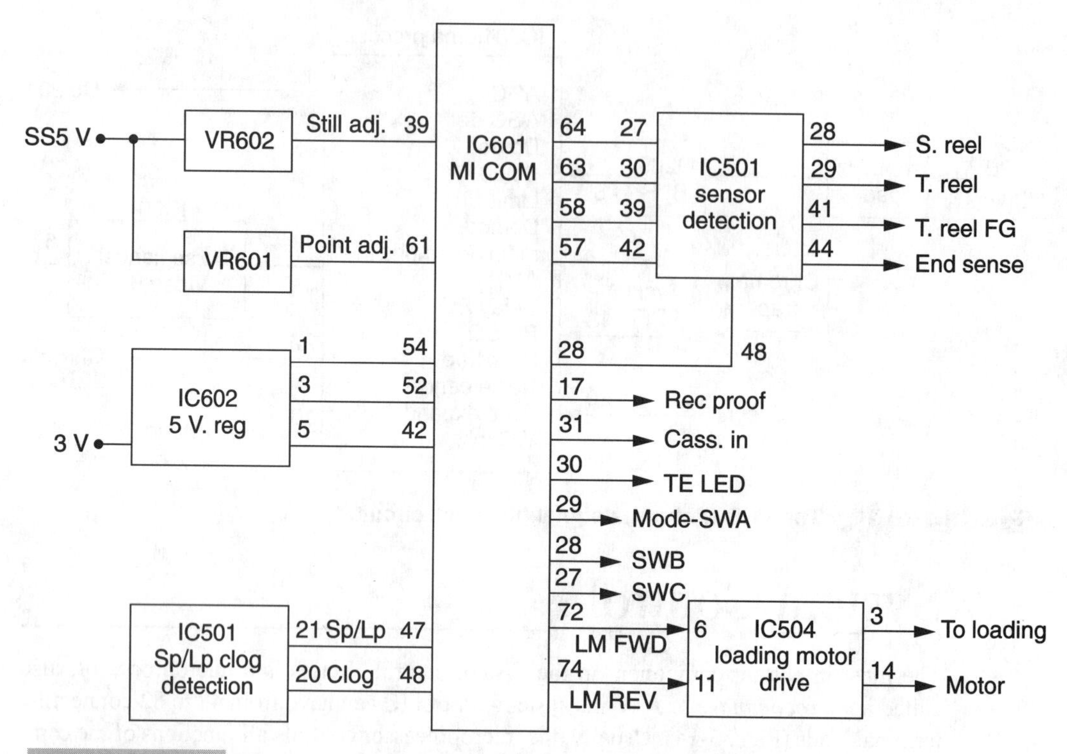

FIGURE 35-15 The block diagram of an 8-mm system-control circuit.

microprocessor can control the loading motor, trouble detector, camera control, power control, battery overdischarge detector, servo control, function control, luma/chroma control, character generator, and audio circuits (Fig. 35-15).

All of these functions are controlled in the VHS-C camcorders with added tape run control, mechanism control, tape speed detect, and screen display.

SYSTEM-CONTROL CIRCUITS

The system microcomputer in 8-mm camcorders provides signals to the warning detector, loading motor, sync generator, D-D converter, continuous recorder, video circuits, digital servo, capstan, LED indicator, and key and mode-switch circuits. The CPU or microprocessor IC is a very delicate component and it must be handled with care. Be sure that all possible tests are made before replacing this system-control component. Besides, it is very expensive to replace.

In the Samsung 8-mm system-control circuit, system-control IC601 (micom) controls the sensor reel; end sense; top sense; dew in; record proof; cassette in; T/E LED; mode switch A, B, and C; loading motor load; and unload loading motor. A lithium 3-V battery operates the system control while the battery pack (5 V) has been removed for charging. A 5-V regulator and reset IC602 provides 5-V regulation and charging of the lithium battery. SP/LP clog detector IC501 provides correct SP/LP operation and a clog-detection circuit.

Notice that the loading motor operates from loading drive IC504 and is controlled by IC601. IC601 provides loading motor forward and reverse drive voltage. The Mi-Com IC601 controls the various sensor circuits.

Typical 8-mm control circuits Main Mi-Com IC231 in the Canon ES1000 camcorder controls the loading motor circuits and circuits in the mechanism unit. Pin 1 of IC231 controls the unloading of load motor through loading motor drive IC101. The cassette in load is at pin 7 of main Mi-Com IC. A load voltage is applied from 2 and 3 of IC101 and unload voltage is at pins 18 and 19 (Fig. 35-16). Check the loading voltage at test points FP101 and FP102.

Mode SW 1, 2, and 3 are fed from pins 22, 23, and 24, respectively, to connecter CN102, which connects to the mechanism PC board. The dew sensor is H at pin 121 and fed to pin 14 of CN102. Recording proof is controlled at pin 21. Supply reel sensor + and – connects at pin 7 and 8 of CN102. Take-up reel + and – sensors are at pins 5 and 6 of CN102. All four reel sensors are controlled by reel sensor amp IC110. The supply and take-up reel sensor signal is at pins 120 and 115 of Mi-Com IC231.

Function switch key input circuits Some of the function switch input circuits are tied into the system-control microprocessor. Usually, the function input switch circuits consist of the key matrix, the input to the camera section, and the system-control microprocessor. The VCR section of the switch key input is connected to the Pause, Stop, Rewind, Play, Fast Forward, Reset, Display, and Receiver buttons (Fig. 35-17). The clock output pulses place diodes in series with the key input switches in some camcorders, supplied by two phase signals from the system-control IC. IC901 supplies two phase signals to the key matrixing circuits. When the function switch is pressed, IC901 detects the switch by monitoring the data input. IC901 now sends out the required signals to carry out the function of the key pushed. Often, the key input buttons are

FIGURE 35-16 The system-control microprocessor in this RCA VHS camcorder is on the main circuit board.

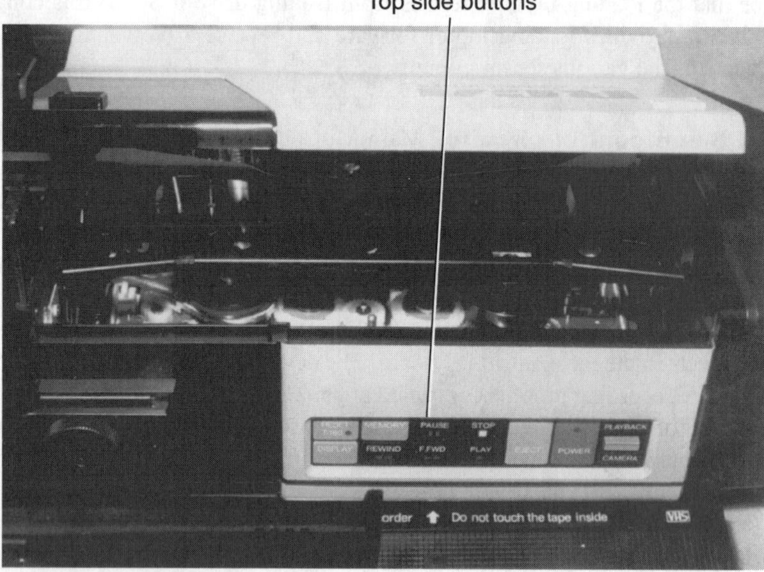

FIGURE 35-17 The top function switch assembly on the RCA CPR300 VHS camcorder.

FIGURE 35-18 Top keyboard buttons on an RCA VHS camcorder.

mounted on a separate PC board that slides into position with one of the side panels removed (Fig. 35-18).

In the more expensive camcorders, a shutter and shutter-speed circuits are tied into the function switch input circuits. Shutter switches S607 and S608 are only accepted in the

Record or Record/Pause modes in the RCA CPR300 VHS camcorder (Fig. 35-19). With S607 set in the Normal position, the shutter speed is 1/60 second. When S607 is in the high-speed position, switch S608 can be used. Each time that S608 is pressed, the shutter speed advances to the next sequence 1/120 to 1/250 to 1/1000, etc. Now this shutter high speed is registered in the electronic viewfinder. The shutter speed switches' (S608 and S607) key inputs are applied to the shutter IC701; these signals are applied to IC901 at pins 19, 20, and 21.

POWER-CONTROL CIRCUITS

The power supplies for the camcorder are also controlled by the system-control microprocessor. The correct voltage source is furnished by batteries or the ac power line via the ac adapter/charger. Often, these voltages are regulated with ICs or transistors. A relay can be found in some models to apply the regulated voltage to the function-control microprocessor. In turn, these output voltages are fed to the various camcorder circuitry. The power-control operation consists of turning the power off and on, VTR power, camera power, and eject operations.

The Mi-Com control IC601 in the Samsung SCX854 camcorder supplies signal to sensor detector IC501, which, in turn, controls the supply and take-up reel sensors, end and top sensor. The cassette in operation is found at pin 17, while the tape end sensor is controlled from pin 31, through LED drive transistor (Q602). The forward and reverse voltages correspond with the load and unload voltages to loading motor from drive IC504.

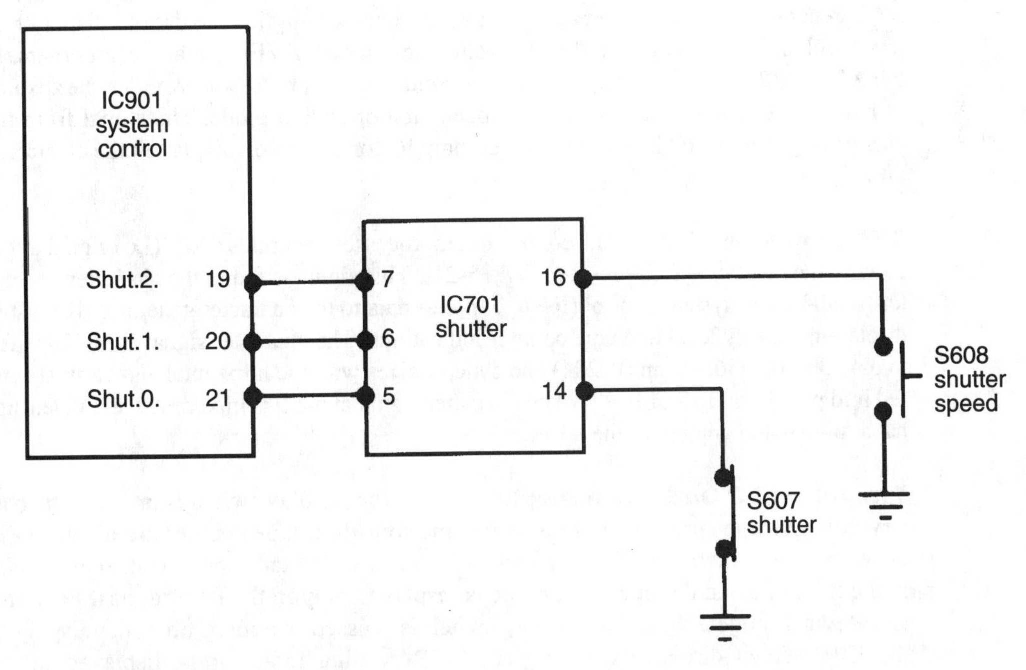

FIGURE 35-19 The function shutter-control circuits in the RCA VHS camcorder.

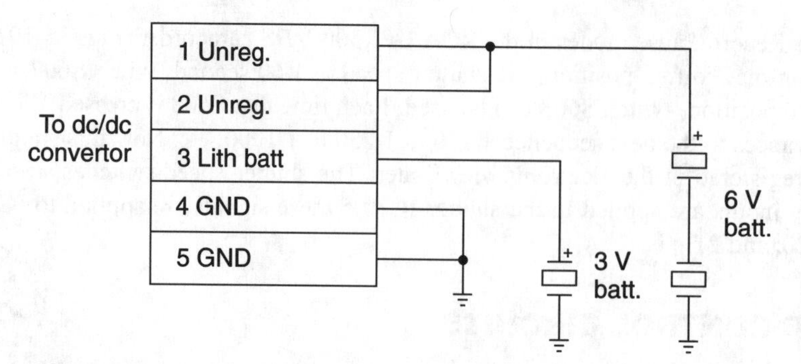

FIGURE 35-20 Samsung battery terminal connections.

Samsung battery terminal circuits The battery terminal connections consist of an unregulated 6-V camcorder and a lithium 3-V battery. The 6-V battery input connection is at pins 1 and 2 of CNB01, which connects to the dc/dc converter (CN901). The lithium (Li) battery connects to pin 3, with 4 and 5 terminals of CN801 at ground potential. These battery connections go to the dc/dc converter terminal connection, CN901 (Fig. 35-20).

ON-SCREEN DISPLAY CIRCUITS

When the Display button is pressed, the system-control IC generates a signal to a character generator that the system-control microprocessor applies the data to display battery level and tape counter in the electronic viewfinder (EVF) of many camcorders. In the RCA CPR300 camcorder, the system-control microprocessor provides the display of battery level, tape counter, shutter speed, and operation mode. The signal from the character-generator IC is fed to the video-amp IC for on-screen display in the electronic viewfinder.

VHS on screen The system-control microprocessor generates a low (Lo) signal at pin 25 with the display switch pressed (Fig. 35-21). This signal is fed to the character generator (IC904). The system control (IC901) applies data to the character generator (IC904) in displaying battery level and tape counter indications. The character signal at pin 10 is fed to pin 13 of the video amp (IC204) and synchronizes with the horizontal and vertical sync fed into pins 14 and 15 of IC904. The character signal at pin 9 is mixed with the video signal at pin 16 and applied to the EVF.

Typical VHS-C On Screen Display When the Display switch is pressed, the battery voltage, tape speed, operation mode, and four-digit tape counter are displayed in the electronic viewfinder (EVF) screen (Fig. 35-22). The tape speed is displayed only in the SP mode and the operation mode is displayed only in the Record, Fast Forward, and Rewind modes. When the display switch is pressed one more time, the display in the EVE screen disappears. However, the Recording indicator is displayed in the Record mode.

When the power is first supplied, the system control (IC901) and character generator (IC904) are reset. When the character generator (IC904) is activated after being reset, the system control (IC901) provides a high (Hi) signal at pin 28 of the character generator (IC904). When the display switch is not pressed, no display data and no output data is supplied. When the display switch is pressed, the character signal data showing the tape counter, operation mode, tape speed, and battery voltage are output, synchronized with the C.G. clock pulse.

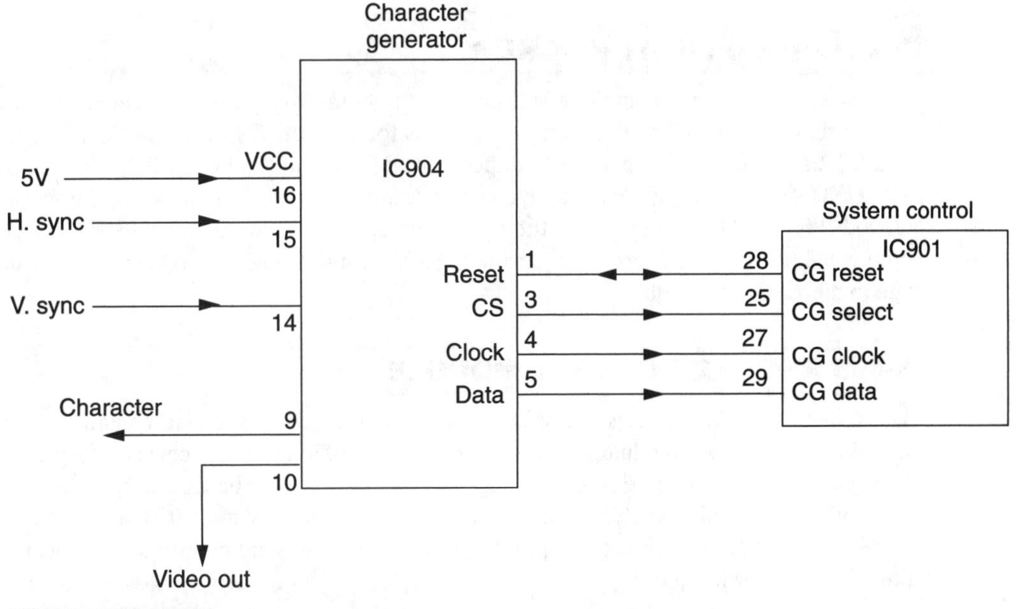

FIGURE 35-21 Typical VHS-C on-screen display circuits.

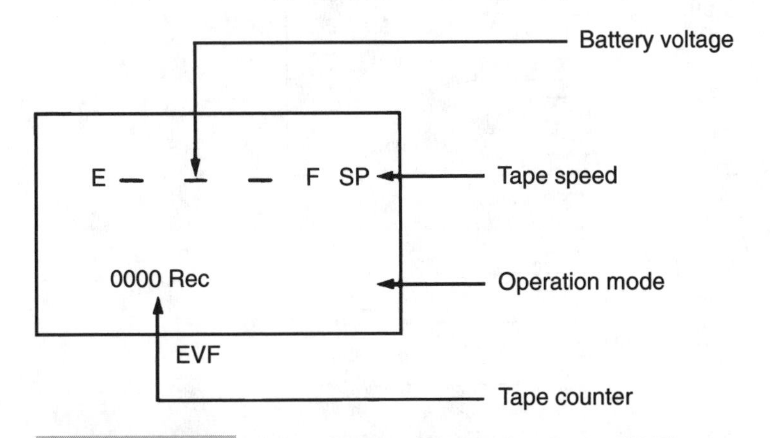

FIGURE 35-22 The VHS-C electronic viewfinder (EVF) screen display.

The character generator (IC904) outputs a character data signal, synchronized with the horizontal and vertical sync signals (H. and V. sync) input from pins 14 and 15. The character signal is supplied to pin 13 of IC301 and is mixed inside with the luminance or video signal at pin 16. When the display mode is changed, the system control (IC901) outputs low (Lo) at pin 28 to reset the character signal generated by the character signal generator (IC904). The character generator (IC904) makes pin 1 (reset) low (Lo) when no V.SYNC signal input is at pin 15 to inhibit the display data output of the system control (IC901).

8-MM VIDEO LIGHT HOOKUP

The video light is used when there is insufficient light in a given area for proper lux readings of the camcorder. The light turn on switch is found in the C-key unit with the camera and VTR switches. The light on (L) connects from C-key unit to pin 8 of CN109 and pin 8 of CN002. The on return is applied to switch transistor Q1901 and light drive transistor Q1902 (Fig. 35-23). The output of the light drive transistor goes to one side of the video bulb while the other lead ties into the battery +6-V terminal. RR973 provides correct voltage to the video light bulb.

8-MM REMOTE CONTROL MODULE

The remote module connects many different circuits together and also contains the infrared remote receiver module. The remote module RM050 terminals consist of a ground, IR signal out, and a positive supply voltage (V_{CC}) (Fig. 35-24). The IR out signal ties into pin 9 of CN050, while the SS 5-V supply is at pin 10, through resistor R050 (47 Ω) to pin 3 of RM050. The IR out connects to pin 9 of CN601 of the system control and connects to pin 85 of IC601 Mi-Com.

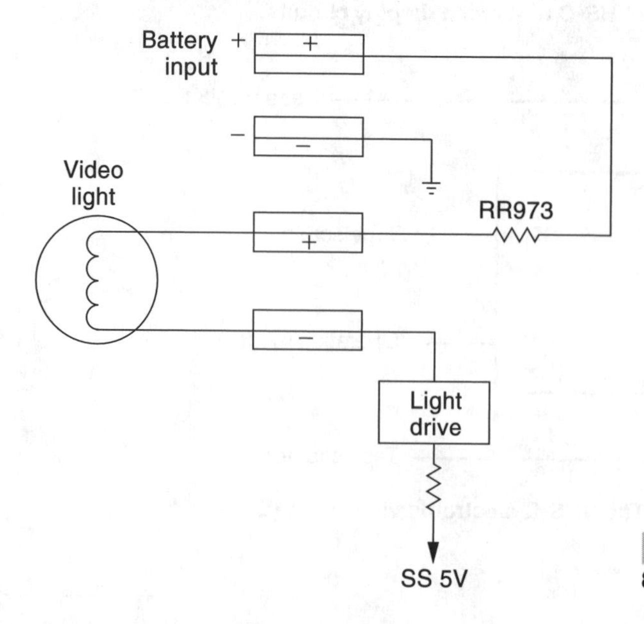

FIGURE 35-23 The 8-mm video light circuits.

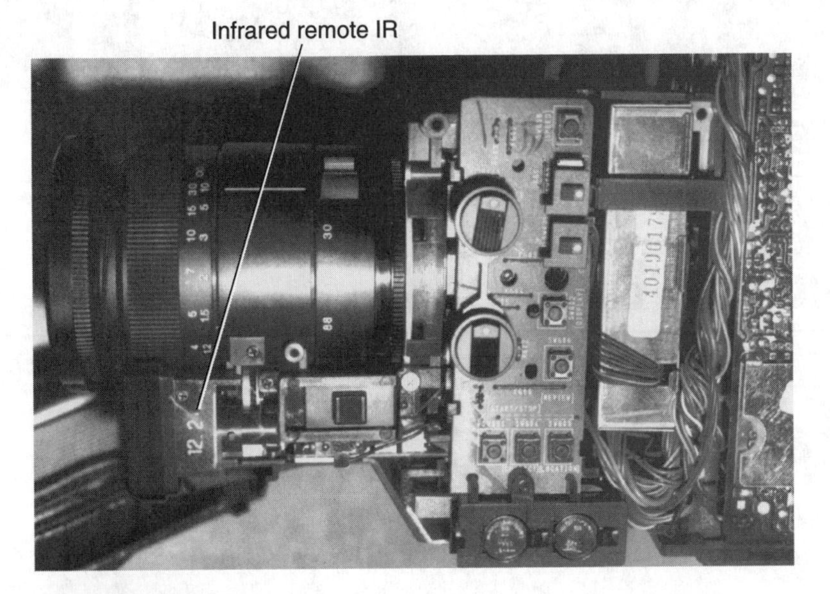

FIGURE 35-24 The remote-control infrared unit is located to the front of the lens assembly.

36

SERVO AND MOTOR CIRCUITS

CONTENTS AT A GLANCE

The trouble-detection circuits troubleshoot the tape-transport system, mechanism, tape type, and recording condition for tape-protection, mechanisms, and the actual recording process. Usually, the various sensor indicators are controlled by one large microprocessor. The system-control microprocessor monitors the supply-end sensor, take-up reel sensor, dew sensor, supply reel sensor, safety tab switch, cassette switch, and the mechanism state mode switch (Fig. 36-1). If the correct signals are not applied to the system-control microprocessor, the VTR will stop functioning.

Tape-End Sensor

When the tape reaches its end, the clear leader of the tape allows the end sensor light to shine through the tape to fall on the end photo sensor. The end photo sensor then applies a voltage to the system-control IC and the control microprocessor stops the tape movement. The operator then knows it is time to change the cassette.

In a Canon ES1000 8-mm camcorder, if the videotape is run beyond its end, the tape guide might be damaged or the head drum might be squeezed with the tape. To prevent such action, the end-of-tape check is conducted to detect the end of tape during operation. Upon detection of the end of tape, the tape is stopped immediately.

The end-of-tape detecting LED indicator is turned on/off with the tape-sense LED signal appearing at pin 63 of Main Mi-Com IC231. If the signal input across pins 22 and 26 of Main Mi-Com IC231 goes low twice in succession, it recognizes the end of tape.

FIGURE 36-1 A top view of the VTR RCA CPR300 (VHS) showing the different detection circuits.

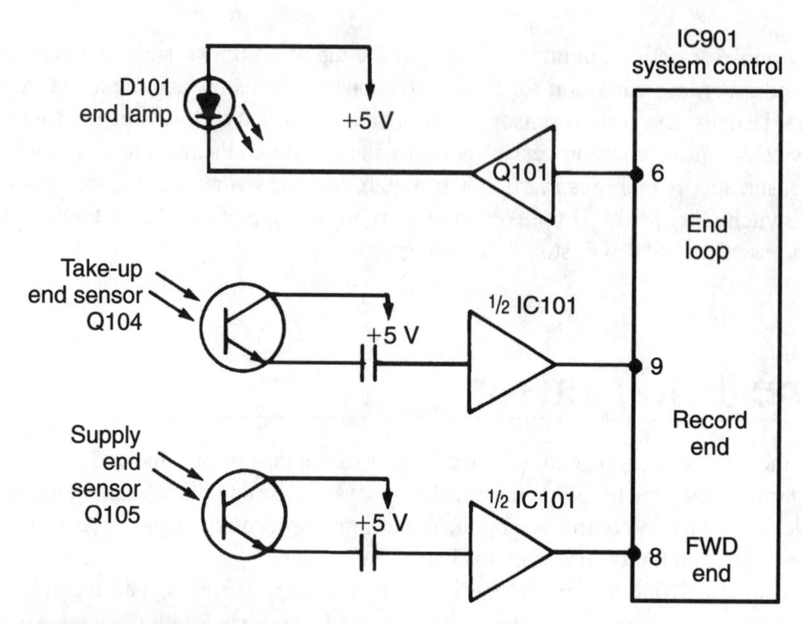

FIGURE 36-2 The 8-mm tape-end sensors.

TYPICAL 8-MM TAPE-END SENSOR

Sensors that detect failures in the tape transport system are the tape-end sensors (Q104 and Q105) on the take-up and supply sides, reel sensors (Q102 and Q103), and cylinder lock sensor. The condensation sensor and cassette-holder sensor (S103) detect the mechanism condition. Added to these are the tape-thickness sensor (S101) and the tab sensor (S102) (the latter detects whether a tab is on the cassette).

For tape-end sensors (Q104 and Q105) on the take-up side, phototransistor Q104 detects light from an LED (D101) through a transparent part at the end of the tape (Fig. 36-2). Likewise, phototransistor Q105 checks for the end of the tape at the supply side. D101 is driven by the pulse signal developed at pin 64 of IC9O1 to save power.

After being shaped by comparator IC101 to protect against optical noise, the outputs of Q104 and Q105 are applied to pins 9 and 8 of IC901. The IC901 checks the inputs at pins 9 and 8 synchronously with the LED drive pulse and sets the input to the Stop mode as soon as pin 9 or 8 turns high (Hi). Also, it detects how the tape is installed from both inputs.

VHS-C END-SENSOR CIRCUITS

When the tape reaches the end while moving in the forward direction, the light shows through the clear leader of the tape and allows light to shine on the surface of the supply-end photo sensor. The end-sensor lamp is turned on by control microprocessor IC901 (Fig. 36-3). Now the supply-end photo sensor applies 5 V to pin 5 of comparator IC902. A high signal from IC902 is applied to pin 18 of IC901. Then IC901 detects this high signal and shuts off the tape movement.

SAMSUNG 8-MM TAPE END-SENSOR

When the tape end reaches its final destination, a transistor signal is sent to pin 7 of the W501 board, where it enters pin 41 of sensor detector IC501. This signal is detected inside IC501 and output at pin 39 to end sense at pin 56 of Mi-Com IC601. IC601 stops the tape at once (Fig. 36-4).

VHS-C SUPPLY REEL AND TAKE-UP REEL SENSORS

With no take-up end sensor in the reverse direction, the tape is stopped at the end by sensing reel pulses from the supply reel sensor and take-up reel sensor. Input pins 20 and 21 of the system-control microprocessor (IC901) receive the applied pulses. The speed is reduced at a point near the end where the tape can be stopped without damaging it. The Rewind operation is stopped when the tape reaches the end of tape travel. Q126 and Q119 are the supply reel and take-up reel sensors, respectively (Fig. 36-5).

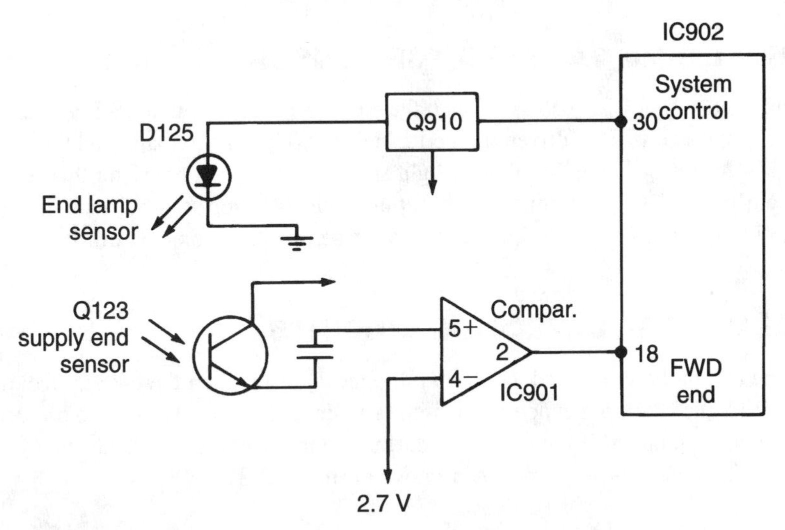

FIGURE 36-3 The VHS-C tape-end sensor circuits.

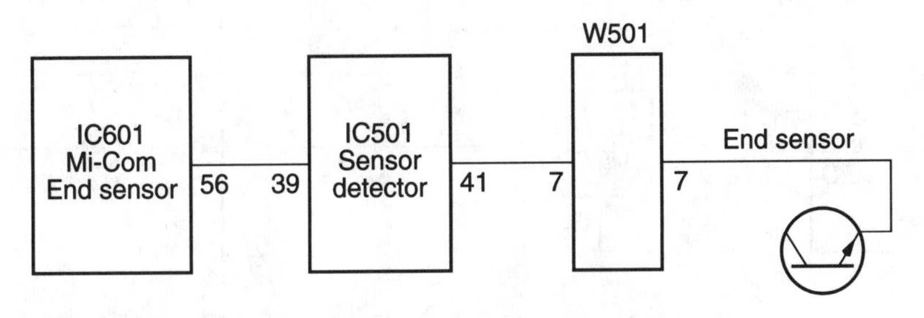

FIGURE 36-4 Samsung 8-mm tape-end sensor circuits.

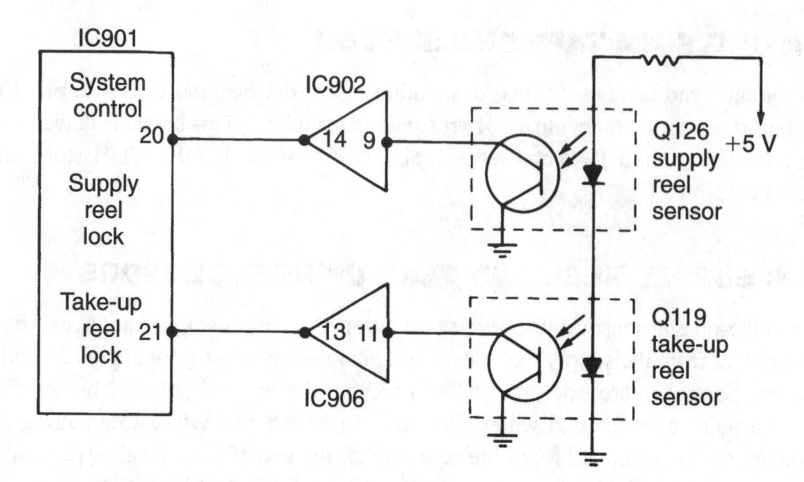

FIGURE 36-5 The typical VHS-C supply and take-up reel sensors.

TYPICAL 8-MM TAPE LED TOP SENSOR

The tape-end sensor LED shows through the clear end of tape with a SS 5-V source at one end. The tape-end LED is driven by the LED driver (Q602) from tape-end terminal 31 of Main Mi-Com IC601 (Fig. 36-6). This light shines upon a phototransistor that sends a signal to pin 10 of W501. The signal appears at input pin 44 of top sensor detector IC501. This signal output at pin 42 of IC501 and is fed to top sense pin terminal 57 of Mi-Com IC601.

TAKE-UP REEL DETECTION CIRCUITS

The tape can pull out, unwind, or clog up the mechanism when the take-up reel does not rotate. With the take-up reel detection circuit running, the tape is stopped to protect it. The tape might get jammed if the reel stops during the unloading period and cannot eject the cassette. Then the loose tape must be removed from the VTR.

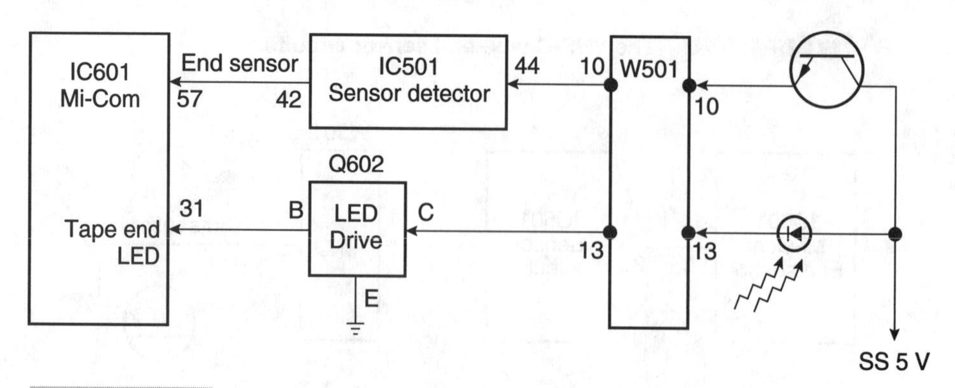

FIGURE 36-6 The block diagram of 8-mm tape-end sensor circuits.

Dew Sensor

The dew sensor detects moisture in the VCR or VTR section of the camcorder. With moisture increasing, the dew sensor resistance increases, sending a voltage to a comparator that shapes and applies pulses to the control microprocessor, not allowing tape operation. If tape operation began with moisture on the tape heads, the tape and mechanism can be damaged. The dew sensor is often located in the center of the VTR section (Fig. 36-7).

TYPICAL 8-MM DEW SENSOR CIRCUITS

Dew moisture in the camcorder can cause tape sticking and can possibly result in jamming the machine. When moisture occurs in the camcorder, as when the unit is brought into a hot room from the cold, the camcorder will not operate if moisture is found around the tape heads or cylinder. The moisture-detecting device resembles a resistor and it changes resistance when moisture is in the camcorder.

The dew sensor in the Samsung SCX854 camcorder contains dew + and dew – terminals. The dew – terminal is soldered to ground and the dew + connects to pin 47 of IC501, a sensor detector (Fig. 36-8). The output terminal of dew + output terminal 46 of IC501 is fed to terminal 58 of Mi-Com (IC601) with the Dew In signal. IC601 prevents camcorder operation with moisture in the camcorder.

When dew indicator keeps indicating moisture that is in the camcorder and the camcorder will not Play, Rewind, or Fast Forward, be sure that no moisture is in the head area. This can be an intermittent problem, caused by a poor pin connector and socket, in which

FIGURE 36-7 **The dew sensor in an RCA CPR300 camcorder.**

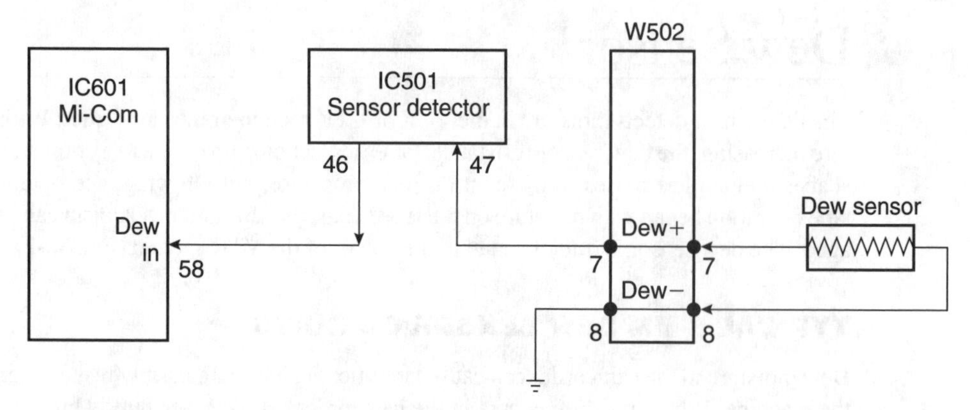

FIGURE 36-8 Typical 8-mm dew-sensor circuits.

the dew sensor connects. Even though the soldered contacts look normal, a high-resistance connection can cause the same problem as the dew sensor.

Solder all pin socket connections to the PC board. If trouble still exists, remove the wire from the dew sensor to the socket and solder both wire terminals together. This same condition can exist when one pin connector corrodes and places a high-resistance connection in the dew socket, resulting in camcorder shutdown. Sometimes this condition can occur when the covers are removed from the camcorder or when the unit is roughly handled.

CANON 8-MM DEW-CONDENSATION CIRCUIT

If moisture condensation is detected during operation, the power LED indicator blinks for warning. Also, the "DEW" mark blinks in the viewfinder screen to let the user know of the detection of moisture condensation. With this condition, the mechanism is put into the Stop state and the tape-loading sequence is not carried out—even if the tape cassette is inserted. Upon detection of moisture condensation, the internal one-hour timer of the Main Mi-Com IC231 is made active to hold dew mark indication unless the battery pack is removed.

The dew sensor is connected to ground terminal and pin 121 of the "Dew H" terminal of IC231 (Fig. 36-9). A positive voltage is also applied through a voltage-dropping resistor of pin 121 from the SS 5-V source.

Cylinder Lock Circuits

VHS-C CYLINDER LOCK CIRCUIT

The cylinder lock (IC601) detects pulse width of SW 30-Hz signal obtained by the servo circuit for detecting a drop in the cylinder motor speed. If the pulse width is less than the rated value, microprocessor IC901 judges the cylinder lock (SW 30-Hz cyl. lock).

Detection is made in the VTR mode and enters the Stop mode. In camera mode, it enters the Stop mode and will not go to any other mode, except Eject or Off.

TYPICAL 8-MM CASSETTE HOLDER SENSOR

The cassette holder sensor (S103) detects the cassette holder status. The output is applied to pin 7 of IC901. IC901 drives the loading motor to change the mechanism from the Eject mode to Stop mode with a high signal at pin 7 (Fig. 36-10).

TYPICAL 8-MM REEL SENSOR CIRCUITS

The supply and take-up reel sensor circuits consists of two optoisolators in the Samsung camcorder. An optoisolator is a coupling device in which the coupling device is a light beam. The optoisolator component consists of an LED and phototransistor located in one part. The supply reel optiosolator (pins 1 and 3) connect to the SS 5-volt source. Pin 4 of the opto part goes to another optoisolator that operates the take-up reel (T. reel), while pin 2 goes to pin 28 of sensor detector IC501 (Fig. 36-11).

Pin 3 of the take-up reel of the opto part has an SS 5-V source to the collector of the phototransistor with the corresponding LED connected to pin 9 of W501 through a resistor to ground. The phototransistor emitter terminal of optoisolator pin 2 connects to pin 29 of the sensor detector. The supply and take-up reel signals are amplified in the sensor detector (IC501 and IC601) and applied to terminal 64 (S. reel) and 63 (T. reel) terminals. IC601 controls the operation of the two reel sensors.

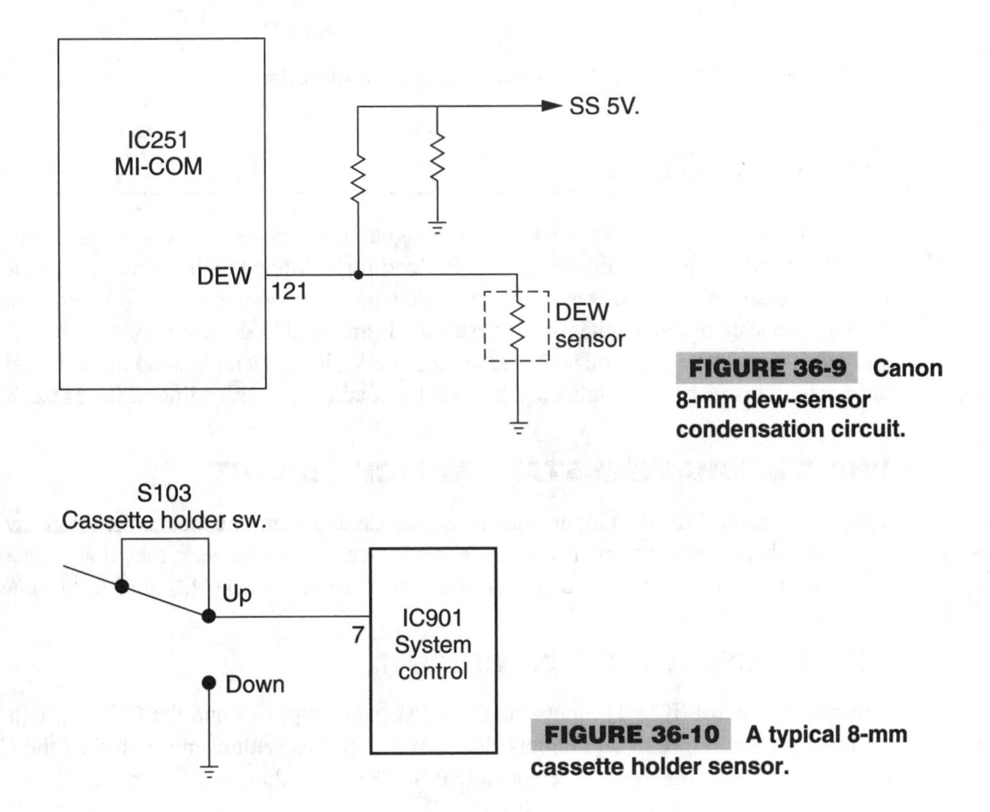

FIGURE 36-9 Canon 8-mm dew-sensor condensation circuit.

FIGURE 36-10 A typical 8-mm cassette holder sensor.

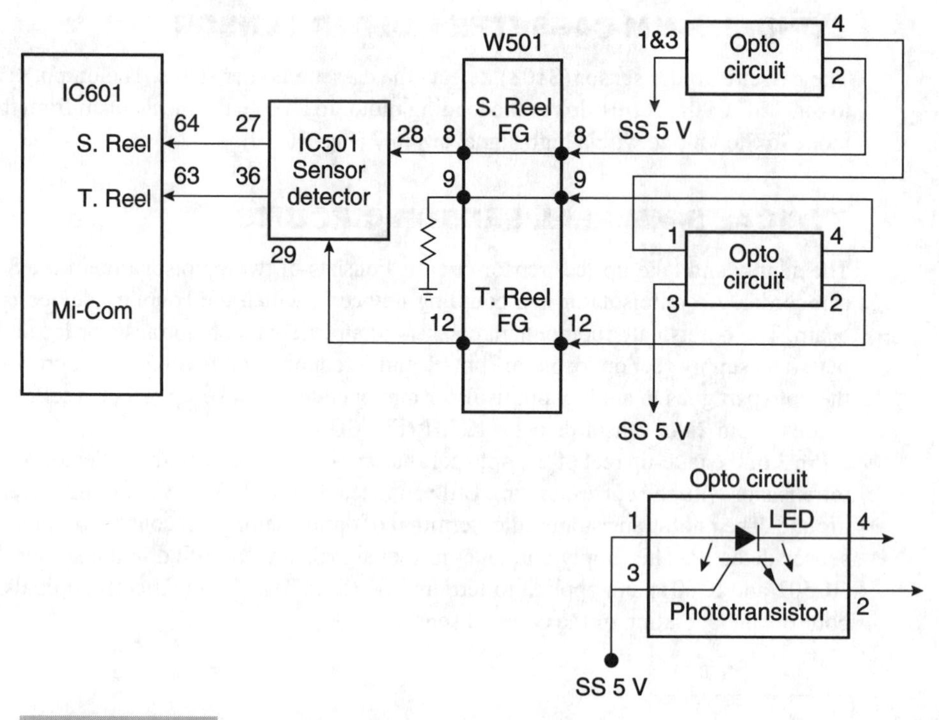

FIGURE 36-11 The typical 8-mm reel-sensor circuits.

Mode Switch

The mode or mechanism-state switch applies signal to the system-control IC microprocessor. At the end of the unloading operation, the loading is stopped with a signal applied to the microprocessor. Also, the mechanism state switch provides signals to the microprocessor to indicate the state of the mechanism. These signals are used to determine if the mechanism and mechanical-state function switches agree. The VTR is placed in Stop mode if they do not agree. The mechanism-state switch is used in loading and unloading of the cassette.

VHS-C MECHANISM-STATE SWITCH CIRCUIT

The system control (IC901) inputs mechanism mode data from the mechanism state switch to decide whether to select mode and mechanism mode are the same. If they do not become the same in 10 seconds, IC901 judges the mechanism to be locked (M. state 0 to M. state 3).

VHS-C TAPE-DETECTION CIRCUIT

The system control (IC901) inputs the FLW.FG pulse at pin 49 and the CTL pulse at pin 50 from the servo circuit and counts the FLW.FG pulses within one period of the CTL pulse to detect the playback tape speed (SP or EP), according to the formula. The tape-speed detection circuit is shown in Fig. 36-12.

When the Play/Pause mode continues for more than 5 minutes, the Stop mode is entered to protect the tape with the 5-minute timer control. When the Record/Pause mode continues for more than 5 minutes, the REC Lock mode is entered to protect the tape.

Loading Motor Drive Circuits

The loading and capstan motors are not considered to be in the servo circuits, but are controlled by the system-control microprocessor. In some camcorders, the capstan motor is driven with the cylinder motor circuits. The output terminals of the system-control IC control the loading motor drive circuits, which drives the loading motor that ejects and loads the cassette and loads and unloads the tape.

TYPICAL 8-MM LOADING MOTOR CIRCUITS

The main Mi-Com IC231 controls the loading motor to set the mechanism in the Operating mode. IC231 drives the into loads out of pin 120 to pin 13 of the loading motor drive IC (Fig. 36-13). Unloading terminal 1 of IC231 is connected to pin 14 of the loading motor drive IC. IC101 provides a loading voltage out of pin 2 to the loading motor and the unloading voltage of pin 18 to loading motor. These dc loading motors can be reversed by

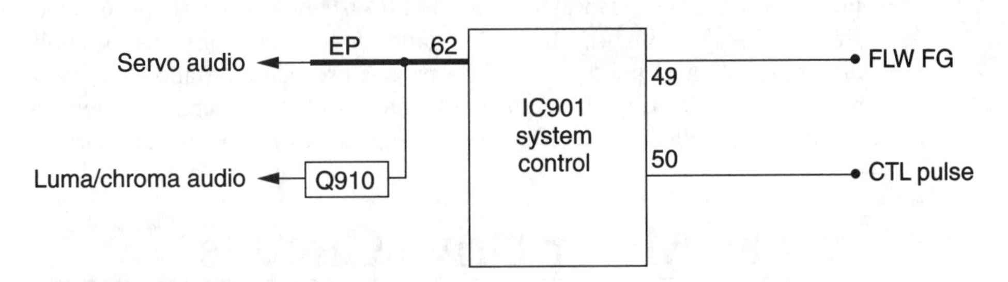

FIGURE 36-12 The tape-speed detection circuit of a VHS-C camcorder.

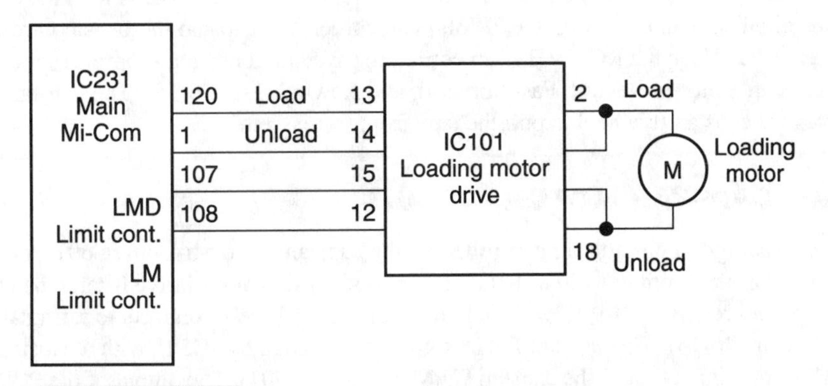

FIGURE 36-13 The loading motor is controlled through IC101 by main MI-COM IC231 in this 8-mm camcorder.

5 TROUBLESHOOTING AND REPAIRING CAMCORDERS

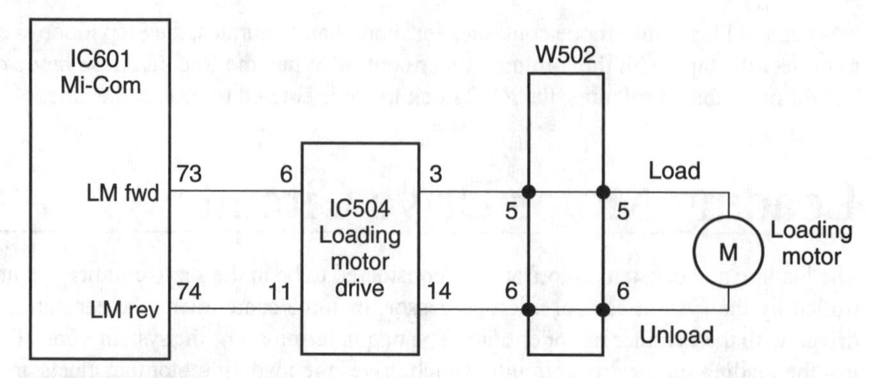

FIGURE 36-14 **The loading motor circuit in the 8-mm camcorder is controlled by IC504.**

changing the polarity of the voltage applied to the motor terminals. IC231 can prevent the loading motor action of the dew condensation in the camcorder.

TYPICAL LOADING MOTOR CIRCUITS

IC601 (Mi-Com) in the 8-mm camcorder controls the loading motor circuits. The loading motor signal (LM FWD) is located at pin 73 of IC601 and connects to pin 6 of loading motor drive IC504 (Fig. 36-14). The loading motor voltage is at pin 3 and is applied to the loading motor through pin 5 of W502. In reverse procedure or unloading, the LM REV signal at pin 74 of Mi-Com IC601 applies to pin 11 of IC504. Here, the unloading voltage out of pin 14 is applied with a different voltage polarity to the loading motor terminals.

Capstan Motor Drive Circuits

In many of the small VHS-C camcorders, the capstan motor is driven by the control microprocessor; in other VHS and 8-mm camcorders, the capstan motor is with the cylinder motor circuits. Within the RCA CPR 300 camcorder, the capstan motor is controlled by the servo IC601; in the RCA VHS camcorder, the system-control microprocessor controls the capstan motor in Record, Fast Forward, and Rewind (Fig. 36-15). The capstan motor rotates the capstan flywheel to pull the tape from the cassette.

8-MM CAPSTAN MOTOR CIRCUIT

The capstan motor circuits are controlled by the capstan servo Mi-Com IC601. A capstan On signal is sent from pin 100 of IC601 to pin 7 of capstan motor drive IC and the capstan forward and reverse (CAP F/R) signal out of pin 99 of IC601 connects to terminal 18 of IC503 (Fig. 36-16). The capstan PWM signal is amplified by IC501 with a capstan error (CAP error) signal sent to the capstan PWM circuits (IC901). The output of the PWM circuits connects to terminal 4 of IC503. A capstan FG (CAP FG) signal is inserted at pins 70 and 77 of IC601 from IC503.

The capstan motor-drive IC consists of a matrix, level shaft and amp circuit for capstan rotation. A wave-shaper circuit provides the FGA and FGB signals from the FG generator of capstan motor circuits. The three capstan motor windings (U, V, and W) connect to terminals 27, 28, and 3, respectively, to the capstan motor drive (IC503).

Servo Circuits

During recording, the servo circuits control the tape speed. The tape speed can be 1800 rpm with 8 mm and 2700 rpm in VHS-C camcorder servo circuits. During playback, it ensures accurate tape speed, aligning the video track with the scanning of the video heads. The speed and phase-control circuitry of the capstan and cylinder or drum motor are located in the servo circuits. This tape-speed control keeps the speed of the video head track constant; phase control is performed by the tracking control system.

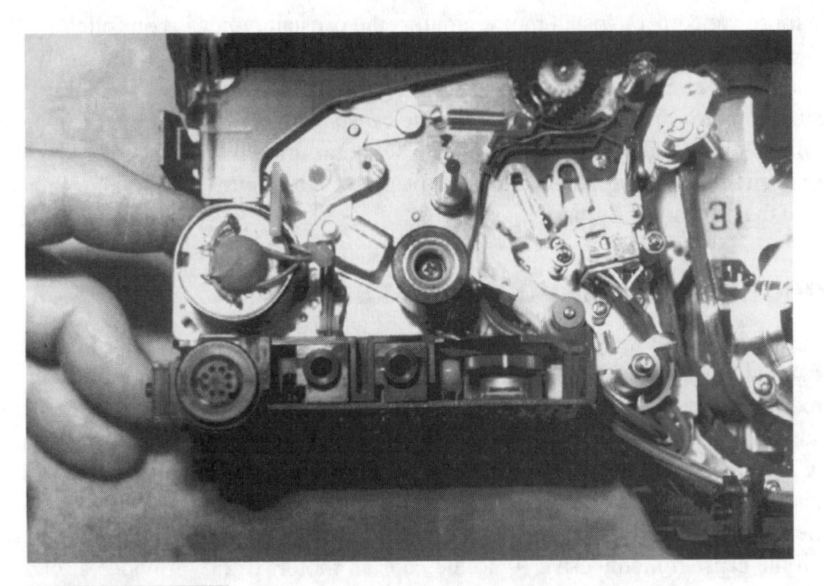

FIGURE 36-15 The finger points toward the location of the loading motor in a VHS camcorder.

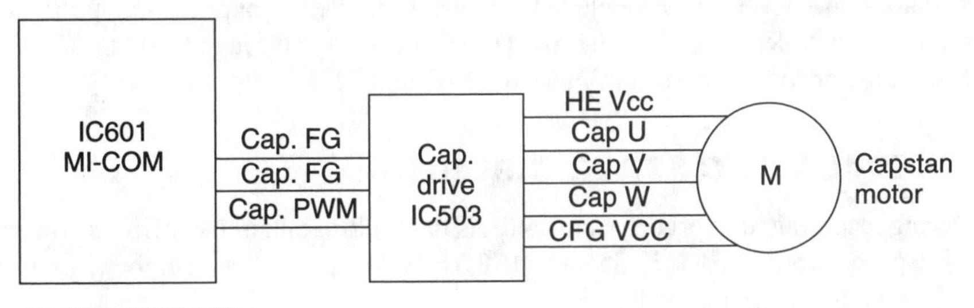

FIGURE 36-16 A block diagram of the 8-mm capstan motor circuits.

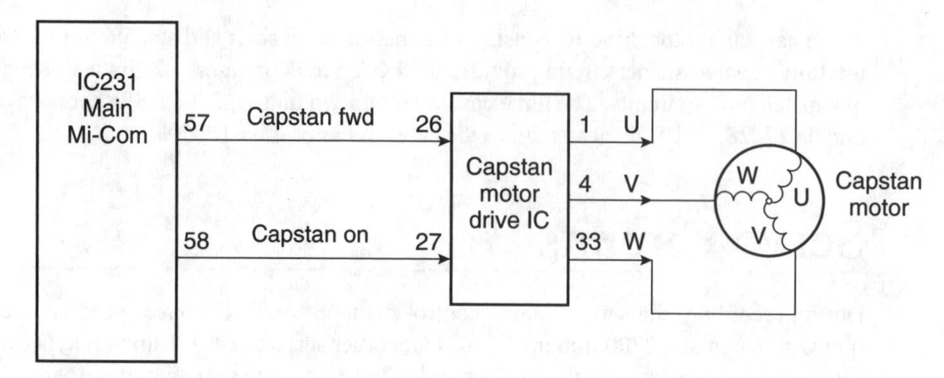

FIGURE 36-17 The typical 8-mm servo circuits.

TYPICAL 8-MM CAPSTAN MOTOR CIRCUITS

Like most camcorder capstan motor circuits, the capstan motor is controlled by the Main Mi-Com IC, to a capstan motor drive IC and to the motor windings. In Canon's ES1000 capstan motor circuits, the capstan forward control comes from pin 57 of IC231 to pin 26 of capstan motor-drive IC (Fig. 36-17). The capstan On signal from pin 58 appears at input pin 27 of the capstan motor-drive IC. The motor drive IC consists of a U-V-W driver internal circuit, which applies voltage to the motor out of pin 1 (U), pin 4 (V), and pin 33 (W), to the capstan motor windings.

TYPICAL 8-MM SERVO CIRCUITS

The main circuits of the 8-mm camcorder servo circuits consists of SS 9-V and EVER 5-V regulators, AFT, reel sensor amp, capstan motor circuits, loading motor circuits, and drum motor circuits. Main Mi-Com IC231 receives signals from the AFT, regulator, reel sensor amp, D - FG, and DPG from the drum motor IC and C.FG from the capstan motor-drive IC (Fig. 36-18). IC231 provides controlling signals and voltages to the capstan motor circuits, loading motor circuits, and drum motor circuits.

The Main Mi-Com IC231 provides a capstan forward (C.FWD), C.ON, and C PWM signal to the capstan motor drive IC to the capstan motor. IC231 provides a load, unload, LMO L.CN, and LMO limit-detect signal to the loading motor drive IC101, which loads and unloads the motor loading mechanism. The Main Mi-Com IC231 provides the DPWM and drum drive signal to drum motor IC180, providing voltage to the W, V, and U motor terminals. An EVER 5-V regulated voltage from IC231 applies voltage to reset and lithium battery detection. SS On provides a drive voltage through Q902 to IC901. The SS 5-V regulator source appears from Main Mi-Com IC231.

TYPICAL VHS-C SERVO CIRCUITS

During recording, the servo system controls the VTR to obtain the VHS track format. The tape speed is controlled at 33.35 mm/s (SP) or 11.1 mm/s (EP) so that the video track pitch is 58 mm (SP) or 19 mm (EP).

The video heads rotate accurately at 2700 rpm so that the length of the video track is fixed at 97.4 mm. The rotating video heads are synchronized with the vertical sync signal of the incoming video signal so that the video track begins at 6.5 H + alpha before the vertical sync signal. During play, continuity and speed (exactly the same used in recording) are maintained by controlling the speed of the video heads accurately at 2700 rpm and by keeping the phase fixed, thereby causing the video heads to trace the video tracks accurately. Actually, the speed and phase of the capstan motor that drives the tape and the cylinder motor that drives the video head are controlled (Table 36-1). Speed control is exercised over the relative speed of the video head and the video track. Phase control is performed for correct tracking.

TYPICAL 8-MM DRUM SERVO CIRCUITS

IC601 Mi-Com controls the servo drum PWM and drum on signals with drum FG and PG signal from the drum motor circuits at terminals 68 and 69. Drum PWM output at pin 76 or IC601 is applied to drum error IC502 and to the drum PWM circuits of dc/dc converter (Fig. 36-19). The drum V_S is applied to pins 13 and 19 of drum motor drive/control IC502. Inside of IC502 are the starter control, timing control, soft switching, back EMF detector, output buffer, and amplifier.

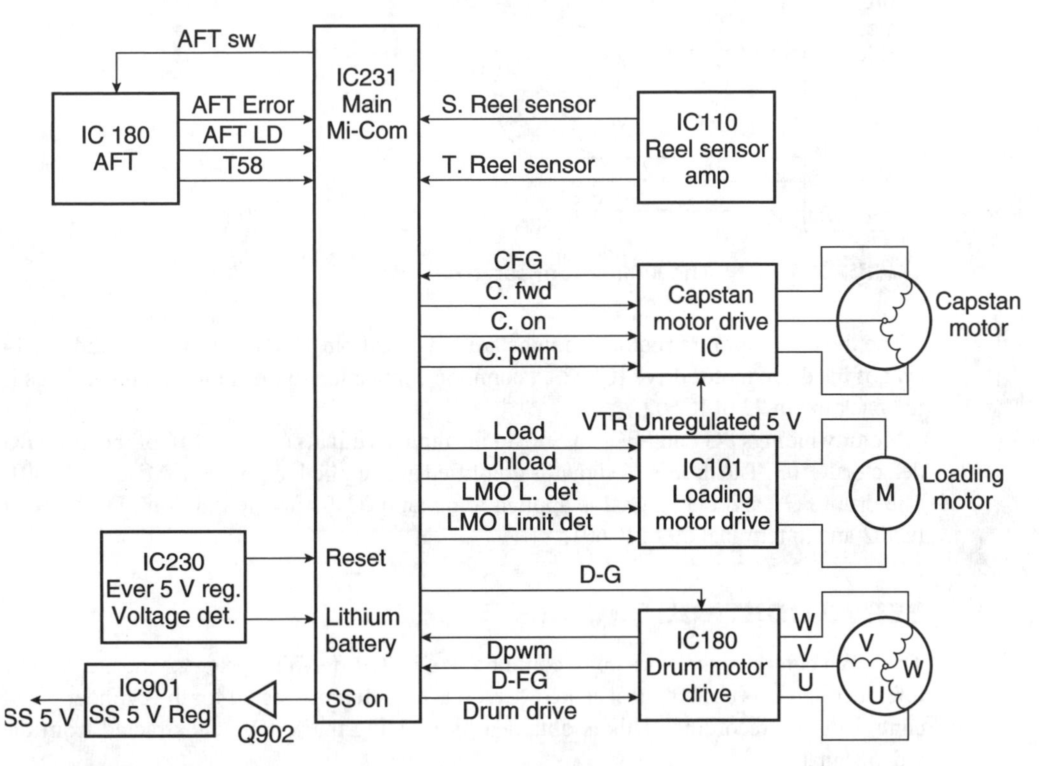

FIGURE 36-18 A block diagram of the servo circuits in the 8-mm camcorder VTR circuits.

TABLE 36-1 SIGNALS USED IN SERVO CONTROL OF VHS-C CAMCORDERS

MOTOR	PHASE/SPEED	MODE	REF. SIGNAL	CONTROL SIGNAL
	Phase	Record	1/2 V SYNC	Tach pulse (45 Hz)
Cylinder		Play	REF 30 Hz	
	Speed	Record/Play	Cylinder FG (CYL.FG: 720 Hz)	
		Record		Flywheel FG
	Phase		REF 30 Hz	(FLW.FG: 360 Hz)
Capstan		Play		Control pulse (CTL30 Hz)
	Speed	Record/Play	Capstan FG (CAPST.FG: 1704 Hz)	

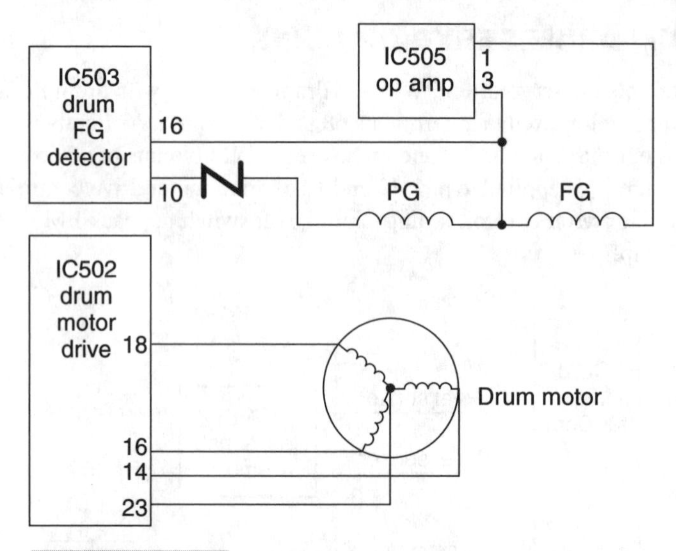

FIGURE 36-19 **The 8-mm drum servo circuits.**

The three motor drum motor windings are fed out of pin 18 (U), pin 16 (V), and pin 14 (W) of the drum motor drive IC502. A common connection to the motor delta windings is fed back to pin 23 of IC502.

Motor windings PG+ and PG– are fed to the input terminals of 10 and 16 of the drum PG detector IC503. The drum PG signal is amplified and applied to pin 68 of Mi-Com IC601. The drum FG+ and FG– signal is applied to op amp IC505 and fed to drum FG detector IC502 and to terminal 69 of IC601.

SERVO CONTROL SIGNALS

The servo control signals are quite common to all VCR or VTR systems. The 1/2-V sync reference signal is used in most camcorders. It is the reference signal for the cylinder phase control during recording. This is obtained by dividing the V SYNC extracted from the video signal.

The play reference signal (REF 30 Hz) is common to all VCR or VTR systems. It is the reference signal for cylinder phase control during play and for the capstan phase during recording. This is obtained by dividing the 3.58-MHz color subcarrier extracted in the chroma-processing circuit down to 30 Hz.

The TACH (PG) pulse is the signal used for cylinder phase control. A magnetic sensor fitted to the lower cylinder generates the signal when it detects the passage of a magnet fitted ahead of the CH1 rotary video head in the upper cylinder Table 36-2.

The TACH (PG) pulse is used also for cylinder phase control in the VHS and VHS-C servo circuits. The magnetic sensor that generates this signal detects the passing of the N-pole of a magnet fitted approximately 6.3 degrees before the CH 3 video head on the upper cylinder. When the video head is rotating at 2700 rpm, the frequency of the TACH (PG) pulse is 45 Hz with its phase advanced by approximately 186.3 degrees from the CH 1 video head position in the VHS-C camcorder. The cylinder phase-control loop times the division of the cylinder FG pulse using this TACH (PG) pulse to convert it to tech (PG) pulses with three frequencies (15 Hz, 30 Hz, and 60 Hz).

Typical 8-mm drum FG pulse signal The Frequency Generator pulse signal is to detect the speed of the drum motor. Usually, the FG pulse controls the speed of the drum motor during record and playback operations. The FG+ and FOG winding from motor circuits is applied to op amp IC505, amplified and passed on to the FG detector IC502 (Fig. 36-20). The drum FG signal is applied to the Mi-Com IC601 at pin 69.

Typical (VHS) cylinder FG (CYL FG) pulse The signal is used to detect the speed of the cylinder motor. Here the FG pulse is a 360-Hz pulse generated by a stator coil and a 16-pole magnet attached to the rotor on the cylinder motor. This FG pulse controls the speed of the cylinder motor during Playback and Record operations.

Typical 8-mm drum FG pulse The Main Mi-Com IC231 controls the speed and phase of the drum motor by comparing the speed and phase of the drum in the Canon ES1000 camcorder. An FG and PG of the FG/PG head in the drum motor assembly is fed to the drum motor drive IC180, which provides a wave-shaper circuit (Fig. 36-21). The shaped wave is fed to drum FG pin terminal 100 of IC231. Likewise, the PG pulse is fed in the same manner. Both of these pulse waveforms can be checked with the scope at terminals 100 and 101 of the Main Mi-Com IC231.

TABLE 36-2 SERVO CONTROL SIGNALS IN THE 8-MM CAMCORDER

MOTOR	PHASE/SPEED	MODE	REFERENCE SIGNAL	FEEDBACK SIGNAL
Cylinder	Phase	Record	½-V SYNC	Tach pulse
	Speed	Play	REF 30 Hz	
	Speed	Both	Cylinder FG (CYL FG 600 Hz)	
Capstan	Phase	Record	REF 30 Hz	Capstan FG (CFG-720 Hz)
	Speed	Both	Capstan FG (CFG-720 Hz)	

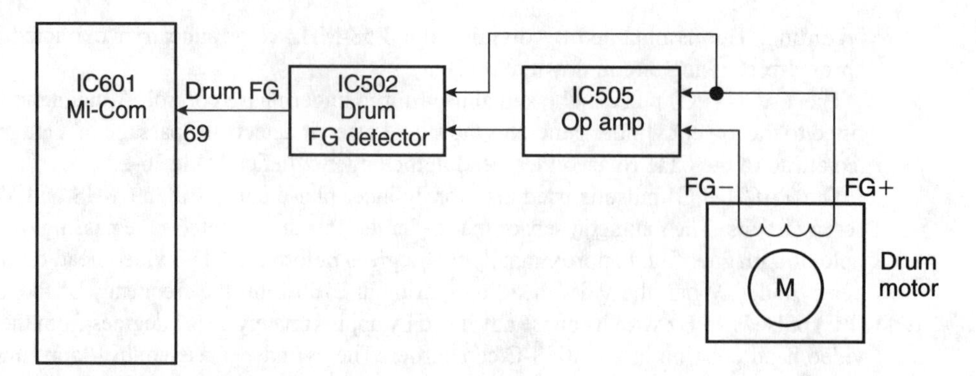

FIGURE 36-20 **A typical 8-mm drum FG pulse from the motor assembly, amplified and detected by IC505 and IC502.**

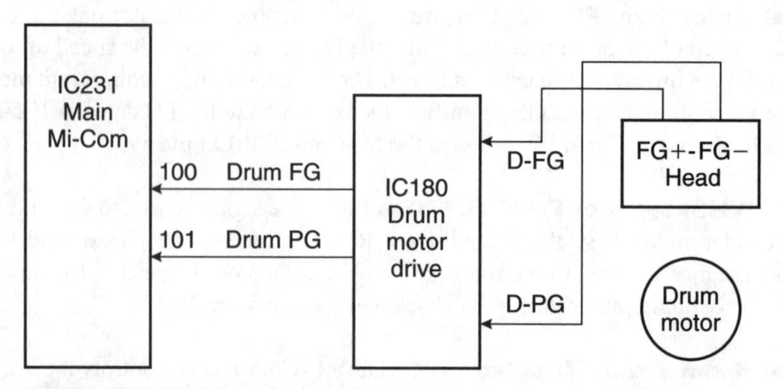

FIGURE 36-21 **The 8-mm drum FG pulse from drum motor, amplified by IC180 and passed on to main MI-COM IC231.**

8-mm lithium battery circuit When the terminal voltage of lithium battery decreases below 2.7 V, the "lithium battery mark" warning indication flashes on the viewfinder screen. Upon receiving the low voltage-level detecting signal at pin 116 of Main Mi-Com IC231, it charges the "chara gene" data to provide the "lithium battery mark" flashing on the viewfinder screen (Fig. 36-22).

Motor Circuits

The large camcorder can contain several small motors, including the drum or cylinder, capstan, loading, autofocus, iris, and zoom motors. The small or inexpensive camcorder might only have a loading, capstan, and drum motor. These small motors operate from a dc source. Most motors are controlled from the main, system-control, servo, and motor-drive IC (Fig. 36-23).

These small motors can be checked with voltage or resistance measurements. A continuity check with the low resistance range of the ohmmeter can determine if the motor winding is open. Measuring the voltage at the motor terminals determines if the motor or drive IC is defective. Applying a small external dc voltage to the motor terminals can determine if the motor is intermittent or slow in rotation.

In this section, a brief description is given on how the motors are controlled and operated. Several brief descriptions are provided for how motors operate in the different camcorders. Removing and replacing the defective motor is given in each motor section. Servicing and troubleshooting the various motors is included in Chapter 40.

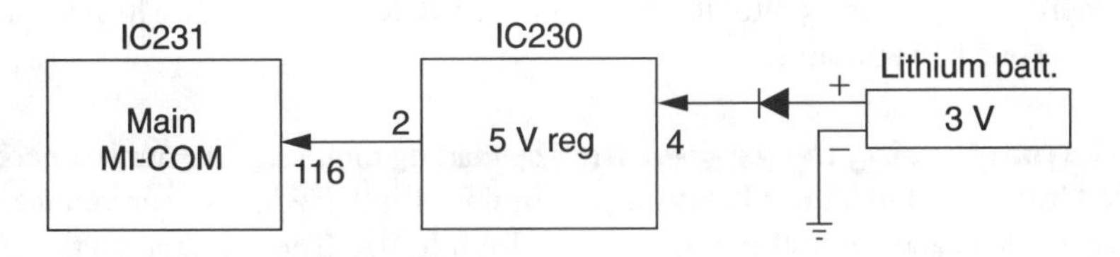

FIGURE 36-22 The lithium battery voltage-detection circuits.

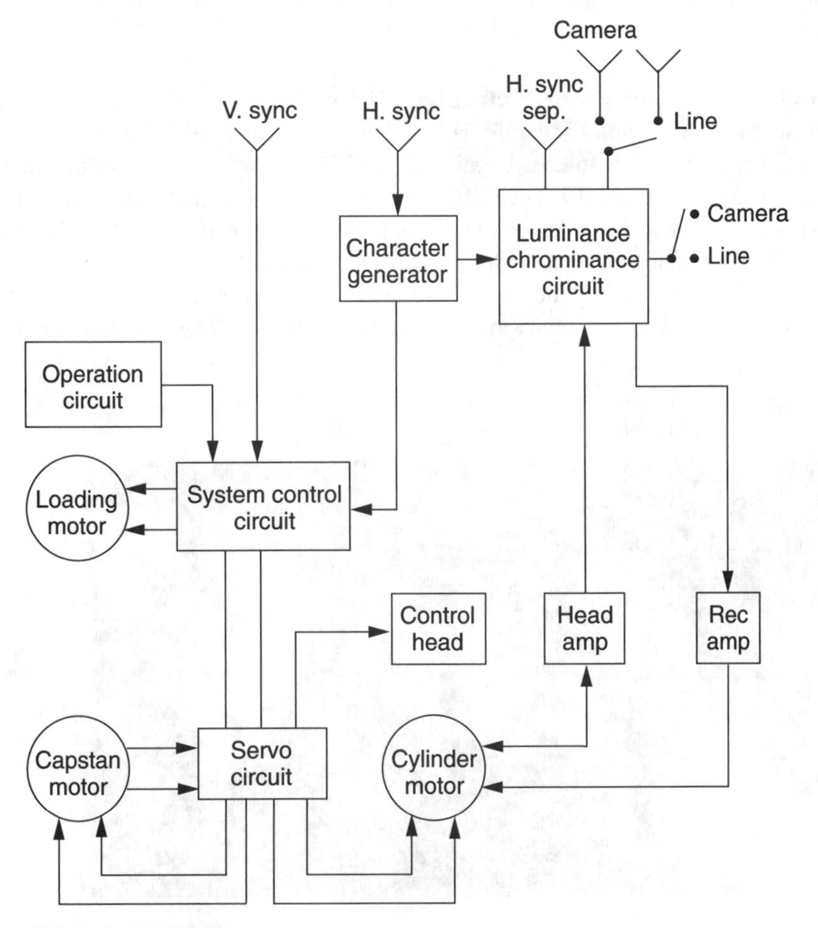

FIGURE 36-23 A block diagram of the different motors and circuits within the system-control circuits.

LOADING MOTORS

The loading motor can eject, load, and unload the video cassette. Its functions include releasing brakes and engaging the Fast Forward/Rewind idler gear and the Playback gear. If you press the Eject button on most camcorders, the loading door opens to receive the cassette. After loading the cassette, the door can be manually or electrically closed. The motor is often located off to the side of the main chassis. Usually, the loading motor is controlled from the system-control IC and a motor-drive IC (Fig. 36-24). The load or mode motor can be one and the same.

Typical 8-mm loading motor circuits The loading motor in the 8-mm camcorder is controlled by the loading motor IC504 at pin terminal 14. A loading motor voltage to unload the cassette is applied to the loading motor from IC504. The LM load voltage for the loading motor is fed from LM REV of pin 74 of IC601 Mi-Com. An LM FWD load is fed to pin 6 of the loading motor IC. Check for dc voltage applied to the motor terminals when the loading mode is entered. Measure the continuity of the dc motor if voltage is found at the motor terminals, but the motor is not rotating.

Canon 8-mm loading motor circuits The loading motor signal from the Main Mi-Com IC231 in the Canon camcorder is found at pin 126 and fed to pin 13 of the loading motor drive IC101 (Fig. 36-25). Terminal 1 of IC231 unloads or reverses the motor rotation and is fed to terminal 14. The LMO limit control signal is sent from pin 127 to pin 15 of drive IC. An LMO limit detector signal is from 128 to pin 12 of IC101. The connector block CN101 contains the load, unload, load and unload voltage that is applied to the motor terminals of the loading motor. Check the dc voltage at the motor terminals if it does not rotate. Also, take a continuity measurement of motor windings to determine if the motor is open.

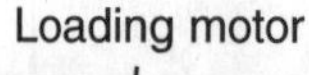

FIGURE 36-24 The loading motor in an RCA VHS camcorder.

CAPSTAN MOTORS

The capstan motor can be belt or gear driven to the various mechanical assemblies. They provide tape movement in the Play, Record, Rewind, Fast Forward, and Search modes. Because the modes operate at different speeds, the dc voltage must be controlled by a servo, capstan speed, and phase-control system. The system or servo IC can control the capstan motor through a capstan motor-drive IC (Fig. 36-26). Often, the capstan and cylinder motors are fed from the same signal source.

Typical 8-mm capstan motor circuits IC231 the main Mi-Com controls the direction of the capstan motor with a capstan on H at pin 58, capstan forward at pin 57, with a C-PG waveform and C-PWM waveform at pin terminals 88 and 2, respectively (Fig. 36-27). The C-PG pulse is received at pin 98 and pin 2 sends out the C-PWM waveform signal to Q151, PWM power drive transistor Q152, and the low-pass filter (LPF) to capstan V_S connector 6.

The capstan FWD signal at pin 2 of CN104 is applied to pin 28 of the capstan motor-drive IC. Capstan on signal is found at pin 3 and connects to pin 27 of the motor-drive IC. Pin 5 of CN104 contains the capstan V_S and is applied to terminal 29 of capstan motor drive IC. A VTR unregulated 6 V is found fed to pin 28 (V_{CC}) of motor drive IC. The internal circuits inside of the motor-drive IC contain the U-V-W driver, start cont, and C-FG wave shaper from the motor FG head.

The capstan motor has a delta winding with applied voltage at terminals 1 (U), terminal 4 (V), and pin 33 (W) to motor terminals. An FG+ and FG– waveform is fed to terminal pins 22 and 23 of capstan motor-drive IC. Check the motor terminals with applied voltage

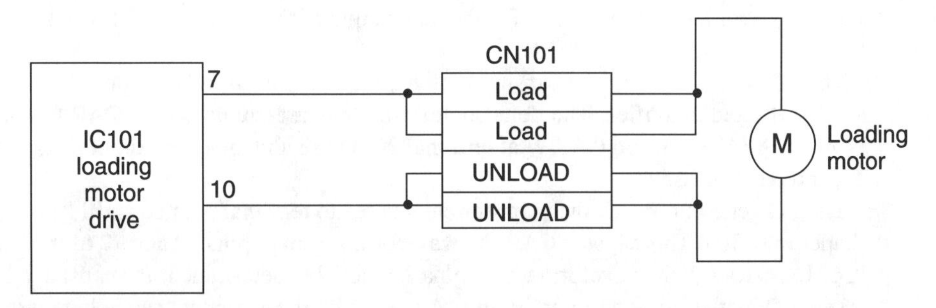

FIGURE 36-25 A typical loading motor drive circuit.

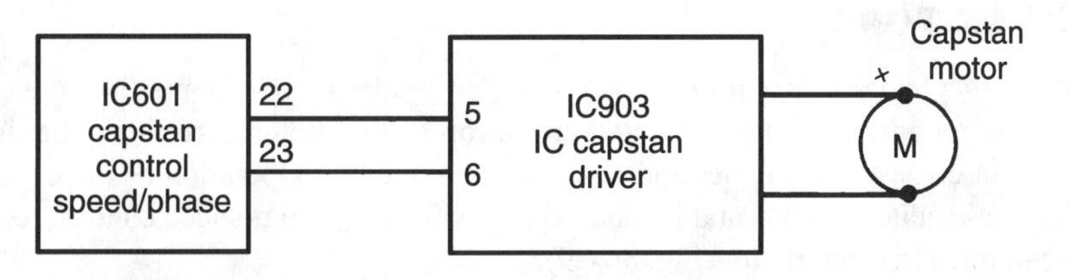

FIGURE 36-26 A block diagram of the capstan motor with control IC601.

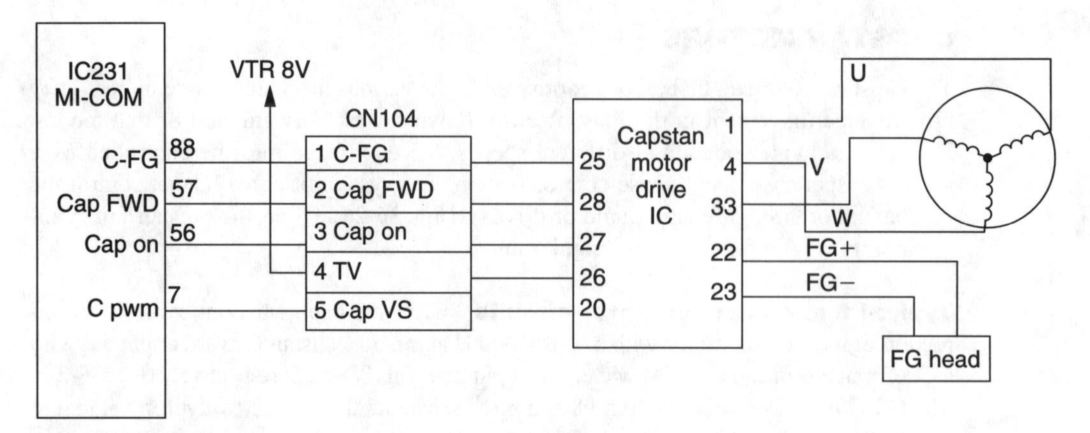

FIGURE 36-27 The 8-mm capstan motor circuit.

and waveforms from the FG head. The capstan FG waveform, in Record mode, can be taken from pin terminal 88 of IC231.

Samsung capstan motor circuits The Samsung 8-mm capstan servo circuits consist of controlling Mi-Com IC601, IC501, capstan PWM, IC503 capstan motor-drive/control IC, and capstan three-phase motor with Hall ICs and FGA and FGB control signals. The capstan On is found at terminal 100 of IC601 and fed to terminal 7 of the capstan motor drive IC503 (Fig. 36-28). A CAP (F) forward and (R) reverse is taken from pin 99 and applied to terminal 18 of IC503. The capstan CAP PWM signal is fed through IC501 with a capstan error applied to pin 8 of IC901. The output of the capstan PWM is fed to pin 4 of IC503.

The internal circuits of the capstan motor drive and control IC503 consist of a matrix, level shift, and amplifier. The delta motor windings are connected to CAP U pin 14 of W501, CAP V at 16, and CAP W at terminal 24. These voltages are fed out of pins 26 and 27, and pin 3 of IC503.

The FG generator inside the motor circuits is fed to terminal pins 10 and 11 of the wave shaper in IC503. This capstan CAP FG waveform is sent to pins 70 and 77 of the Mi-Com IC601. Take a square waveform test at pins 70 and 77 to determine if the capstan FG pulse is sent out by the capstan motor circuits. A continuity measurement can help you to determine if any winding is open inside of the capstan motor.

DRUM MOTORS

Usually, the cylinder or drum motor is controlled by the servo circuits, like the capstan motor. The cylinder or drum is located at the top of the motor assembly. The drum or cylinder can consist of an upper and lower drum assembly. Operation of the cylinder or drum motor is quite complex and it contains the cylinder or drum-speed control, recording phase control, and drum motor (Fig. 36-29).

Typical 8-mm drum motor circuits The 8-mm drum motor circuits consist of Mi-Com IC601 controller, IC502, drum PWM, drum motor drive/control IC502, and drum motor. A PWM drum signal is found at pin 6 of IC601 and fed to IC502 which provides a drum error signal fed to drum PWM IC901 (Fig. 36-30). Drum V_S is fed to terminals 13 and 19 of the drive motor IC502. A starter control, timing control, soft-switching back-EMF detector, output buffer, and amplifier circuits are contained in drum motor driver IC502.

The U winding voltage is fed out of pin 18 to the drum motor. Terminal 16 feeds the V voltage to the V motor winding, and the W winding is tied into terminal 14 of IC502. The center or common terminal of the delta winding motor feeds to pin 23.

Canon 8-mm drum motor circuits A drum forward (FWD) signal is fed from pin 56 of IC231 Mi-Com to pin 26 of drum motor drive IC180. The drum DPWM signal is found at pin 3 and fed to Q161, PWM power drive transistor Q150, and to a low-pass filter network before entering pin 5 of IC180 (Fig. 36-31). Inside IC180, the PWM signal is fed to the U-V-W driver; this, in turn, is fed to the U-V-W windings of the drum motor.

The drum FG/PG pulse is fed from the motor FG/PG head to pins 20 and 18 of IC180. Here, both the FG and PG pulses are wave shaped inside IC180 and the D-FG pulse feeds to pin 100 and D-PG pulse to pin 101 of Main Mi-Com IC231. Both of these waveforms can be scoped at terminals 100 and 101 when the camcorder is in the Record mode. Two

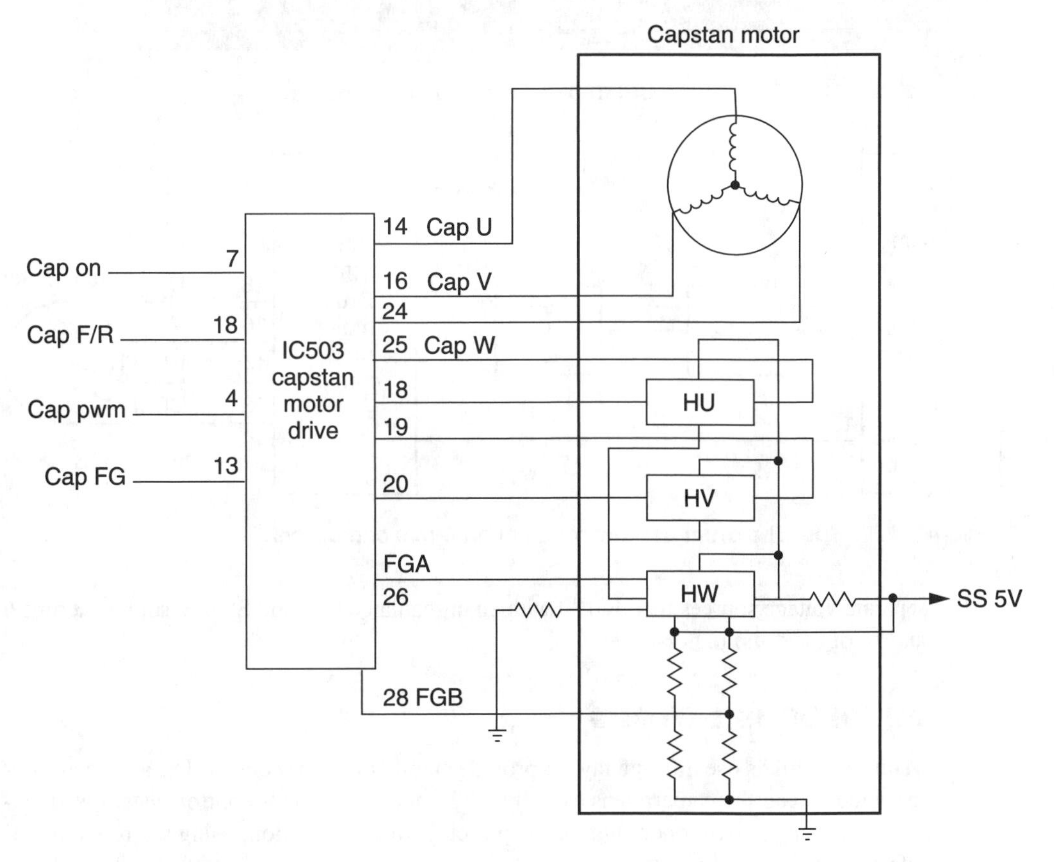

FIGURE 36-28 The Samsung 8-mm capstan motor circuits.

FIGURE 36-29 RCA CPR100 drum servo components.

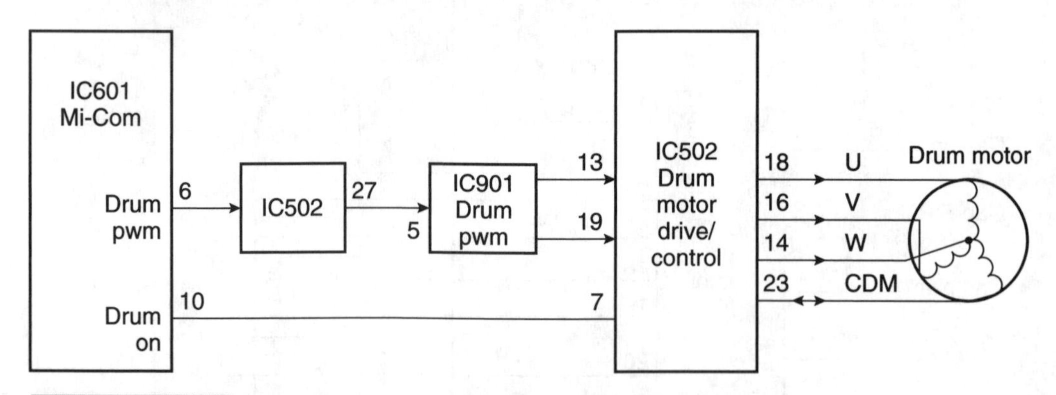

FIGURE 36-30 The drum servo circuits of an 8-mm camcorder.

separate voltage sources feed IC180 with an unregulated 6-V and SS 5-V source, at pins 6 and 14 of the drum motor-drive IC.

AUTOFOCUS MOTORS

Many camcorders use infrared rays to provide automatic focus control. The autofocus motor is located on the camera lens assembly. The automatic focus control system is an external focusing system operating on the principle of triangulation, using the reflection of infrared rays (around 870 nm).

Infrared rays emitted by the infrared LED pass through a projection lens and reach the object. The infrared rays are reflected from the object and back to the sensor via the receiving lens (Fig. 36-32). The receiving lens (sensor) contains two photodiodes. The focus lens is moved until the two photodiodes receive an equal amount of light. The two photodiodes detect the infrared rays reflected back from the object and convert them to current values. The current signals are converted to voltages and are output (Fig. 36-33).

Typical 8-mm autofocus motor circuits The autofocus circuits in the Canon 8-mm camcorder consist of the camera/AF Mi-Com IC1402, the focus drive IC, and the focus motor (Fig. 36-34). The focus CW is at H and CCW rotation is L out of terminal 9 from IC1402 to pin 3 of the focus motor drive IC1336. Pin 12 signal provides focus stop and feed to the terminal of the focus drive IC. The focus power save is at terminal 11, focus monitor at pin 19, focus pulse at pin 35, and focus reset at terminal 10 of IC1402.

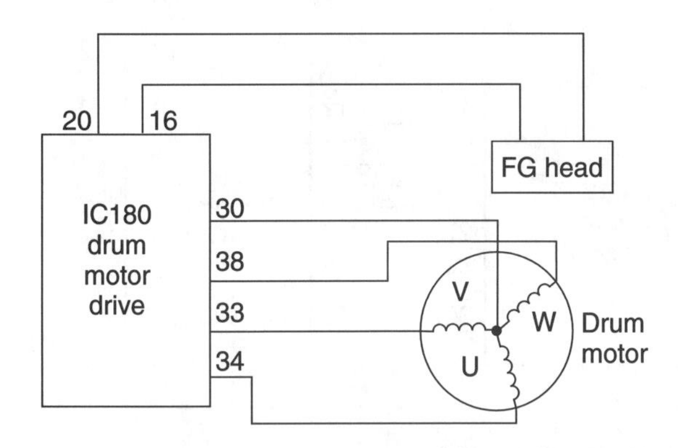

FIGURE 36-31 The drum motor circuits in a Canon 8-mm camcorder.

FIGURE 36-32 The auto focus motor in a VHS camcorder.

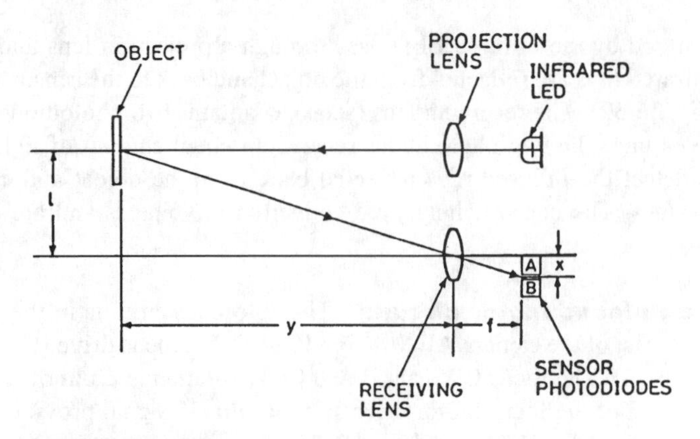

FIGURE 36-33 **Measurement principles of the automatic focusing system.**

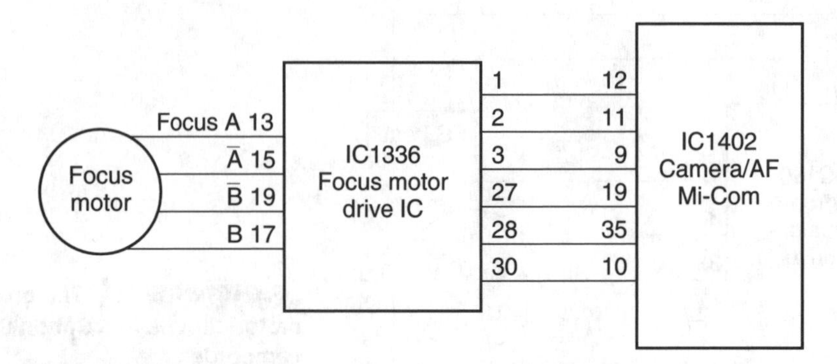

FIGURE 36-34 **The 8-mm focus motor circuits.**

Internal decoder and drive circuits inside of the focus motor drive IC1336 is applied to the connector CN1302 at pins 13, 15, 17, and 19. The focus A, focus Aw, focus Bw, and focus B are applied to the focus motor terminals. Scoping the waveforms on IC1402 and IC1336 with crucial voltage measurement can help you to locate the focus problems.

IRIS MOTOR DRIVES

In the early VHS iris mechanism, a meter system was used to mechanically connect the brake and drive coils. This iris mechanism worked somewhat like the camera aperture. Today, besides this iris mechanical system, you will encounter an iris motor with an AIC (auto iris circuit).

The auto iris circuit operates the lens iris to maintain the optimum average video signal level (Fig. 36-35). The luminance (YE) and wide blanking (W.BLK) signals control the auto iris circuit. The luminance signal level declines with reduced scene brightness. This increases the Q13 bias and the Q13 emitter voltage declines. The reduced Q14 gate voltage reduces the Q14 drain current, decreasing the voltage at pin 6 of IC7.

The op-amp output from pin 7 of IC7 increases and the current flows in the iris driver coil in the direction from pin 2 to pin 4. Now the iris opens to increase the incoming light. Voltage is produced in the iris damper coil with the speed at which the iris operates. As the iris opens, positive voltage appears at pin 3, which increases the voltage at pin 6 of IC7. This damping action provides smooth iris operation.

An increase in scene illumination yields the opposite generation. D7, in the Q14 gate resistance circuit, provides quick iris response to rapid increase in incident light. If the scene brightness declines to where the iris is completely open and pin 13 of IC7 voltage becomes lower than that of pin 12, pin 14 of IC7 comparator output goes high. Then the AGC circuit functions. Spring force is applied to the iris driver in the Enclosed direction. With the power turned off, the 5-V power supply is low and the iris is closed.

ZOOM MOTORS

The zoom motor brings the image close or far away from the lens. In some camcorders, the zoom motor is called a *PZ (power zoom) motor*. The zoom motor is on the lens board assembly. In the early camcorders, the zoom motor was controlled by transistors (Fig. 36-36). The zoom motor drive circuits are controlled by voltage in the zoom motor circuits.

Typical 8-mm zoom motor circuits The zoom motor is controlled from the autofocus Mi-Com ICA07 with a TELE and WIDE signal out of pins 58 and 59. This control signal is sent to zoom motor drive ICA01 at pins 9 and 10 (Fig. 36-37). The zoom motor drive IC provides voltage to open the zoom wide and tele operations. Check the voltage at the dead zoom motor and take continuity measurements. Check ICAO1 for no motor voltage. Do not overlook the possibility that zoom gear train is bent or binding on the lens assembly, to prevent moving lens assembly.

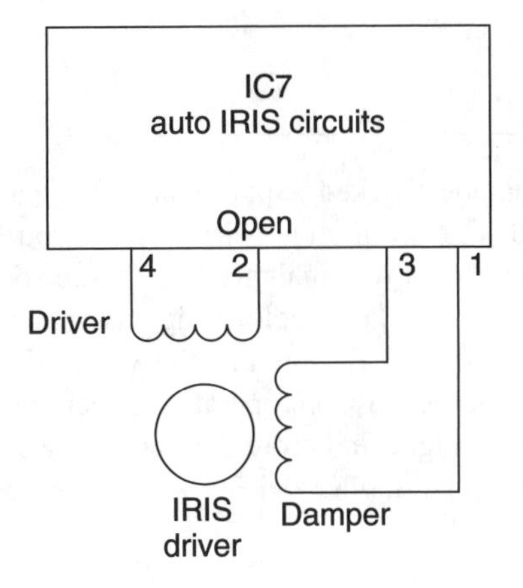

FIGURE 36-35 A block diagram of the auto iris circuits.

FIGURE 36-36 The location of the zoom motor in the VHS-C camcorder.

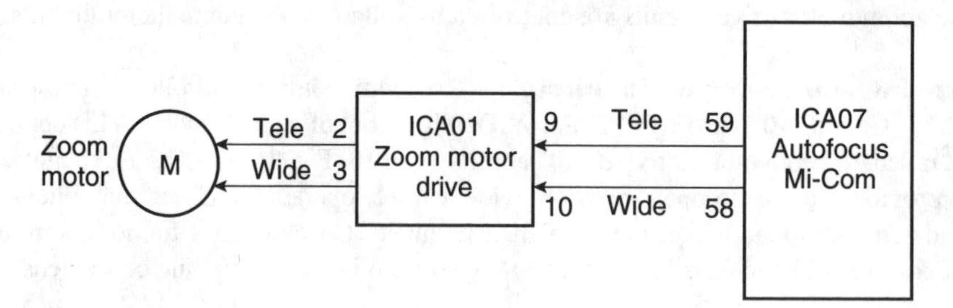

FIGURE 36-37 A typical 8-mm zoom-motor circuits.

Conclusion

The various motors found in the camcorder can be checked with crucial voltage and continuity measurements at each motor terminal. Check the motor-driver IC if you find no voltage at the motor terminals. Test the motor output terminals from the motor-driver IC and also the supply voltage applied to the driver IC. Take waveform signals on the control Mi-Com when the motor and driver circuits are normal. Be sure that the Main Mi-Com IC is defective by checking the various circuits that the Mi-Com operates. Check the supply voltage furnished to the Main Mi-Com IC. You might find more than one voltage supply source fed to the system control Mi-Com IC. Scope the PG and FG waveform from each motor at the Mi-Com control IC.

CAMCORDER AUDIO CIRCUITS

CONTENTS AT A GLANCE

The audio circuits are quite simple when compared to the video or color circuits. The VHS and VHS-C audio circuits are quite similar in operation. The audio control head (A/C) is in both circuits; in the 8-mm camcorders, frequency-modulated audio signals can be recorded by multiplexing them with video signals. PCM recording is also possible, or a conventional fixed head can be used to record audio signals near the edge of the tape. Most 8-mm camcorders use the multiplexing FM audio signals with the video signals, which is called *FM audio signal recording*.

Camcorder Audio Circuits

The audio circuits start at the Play/Record head and amplifies the signal of the REC/Play preamplifier stage in the block diagram of an 8-mm camcorder (Fig. 37-1). The REC/Play signal from the preamp is fed into the AFM audio-processing circuits. The audio signal is fed in and out of the audio-processing circuit to the audio jack.

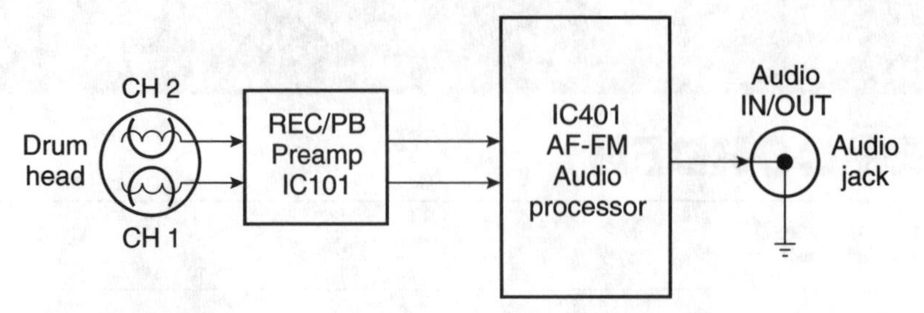

FIGURE 37-1 A block diagram of CH1 and CH2 in Play mode in an 8-mm camcorder.

TYPICAL VHS-C AUDIO CIRCUITS

The VHS audio circuit in the RCA CPR300 camcorder is quite similar to the CPR100 VHS-C model, except for differently numbered components. The audio signal enters the input circuit with the built-in microphone or the external mic (Fig. 37-2). J452 is the external mike jack. Also, external audio can enter the A/V input adapter connector. A circuit in the A/V input adapter reduces the incoming audio signal to a level equal to that produced by either microphone. This audio signal is applied to the audio amp at pin 23 of IC401 (Fig. 37-3). The mic amp (IC401) amplifies the audio signal through the ALC circuit and the output at pin 10 of IC401. The picked-up E-E signal can be heard in the headphones to the A/V output connector.

FIGURE 37-2 The condenser microphone is along side of the front of the camcorder.

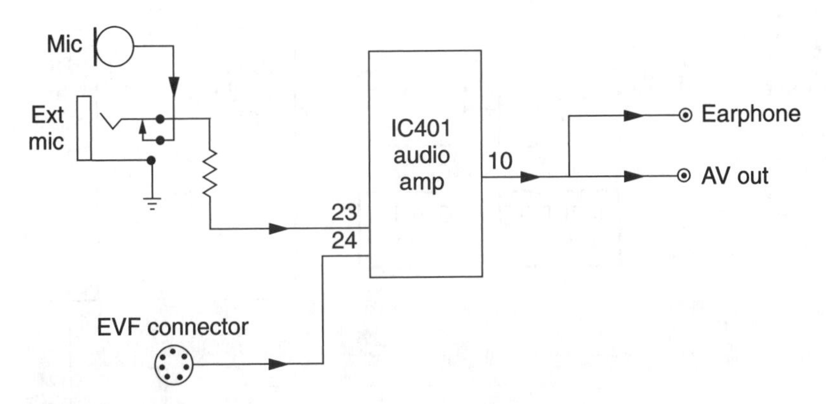

FIGURE 37-3 The audio-input circuits of a VHS-C camcorder.

The Record/Play audio heads are switched by the playback (PB) signal by the system control microprocessor (IC901). When the PB signal is high (Hi), Q401 is turned on and IC402 is low (Lo) at pin 5. SW 1 inside IC402 is turned off. Also, this high (Hi) PB signal is applied to pin 7 of IC402, which turns SW 2 on. This switching grounds the record terminal of the audio head and allows the PB signal to be picked up by the audio head with outputs through IC401 and R401.

TYPICAL 8-MM AUDIO CIRCUITS

The audio circuits in the 8-mm camcorder consist of a microphone, mic board, audio amp IC401, and the audio in/output circuits. The microphone receives the sound of the scene being taken and connects to the amp board and mic input at pin 41 of sound IC401. The audio in and out signal connects to pins 36 and 18, respectively (Fig. 37-4).

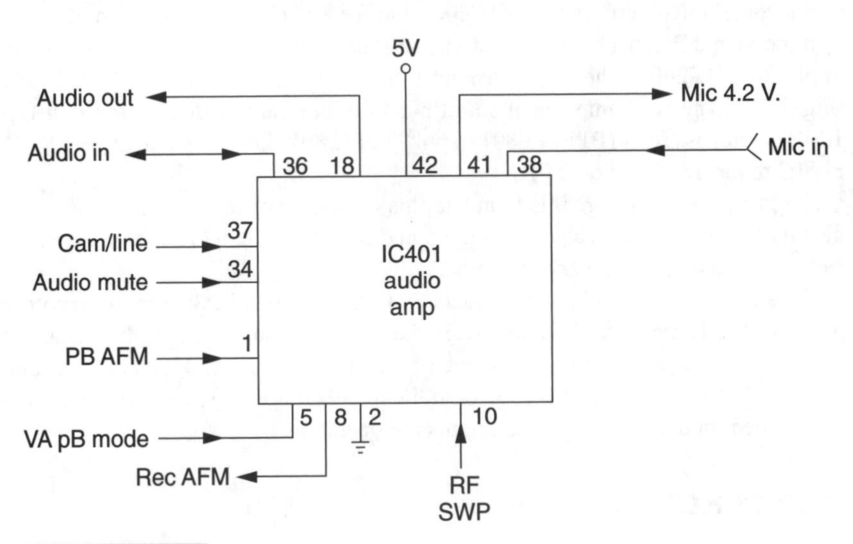

FIGURE 37-4 The 8-mm audio-amp IC circuits.

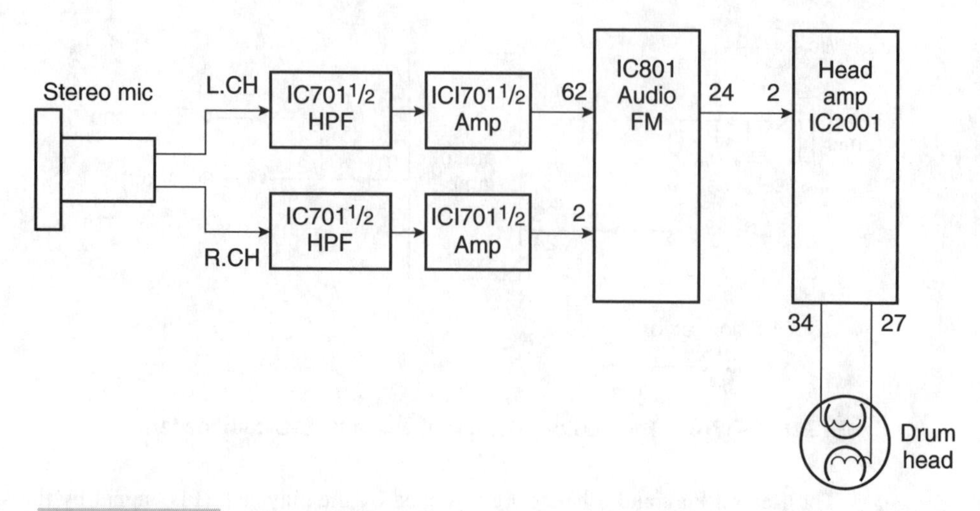

FIGURE 37-5 A block diagram of Canon 8-mm stereo sound circuits.

A 5-V source is found at pin 42 of sound amp, cam/line at pin 37, ALC TC pin 40, and audio mute at pin 34. A PLL LPF input is found on terminal 47, PB AFM at pin 1, REC AFM at pin 8, RF SWP signal at pin 10 and pin 7 is ground at audio amp IC401.

Canon 8-mm audio circuits This 8-mm camcorder has stereo sound. The left stereo mic feeds into an HPF (IC701) and ½ of IC701 where the audio is amplified, and connects to terminal 62 of IC801. Right input microphone audio is applied to Q701 (HPF) and amplified by IC701 to pin 2 of audio-FM IC801 in the Record mode (Fig. 37-5).

The Record signal is at pin 24 of IC801 and fed to head amp IC2001 at pin 2. A CH 1 head signal at pin 34 is applied to connector CN2001 and on to channel 1 drum head. CH 2 signal comes out of terminal 27 of IC2001 and feeds to channel 2 of the drum head circuits.

In the Sound Playback mode, the CH 1 channel picks up audio off of tape and connects to pin 34 of IC2001. The CH 2 channel from head drum goes to pin 28 of head amp IC (Fig. 37-6). The head amp signal out of pin 14 of head amp is fed to amplifier Q801, and 1.5-Hz bandpass filter (BPF) FL801 to pin 27 of IC801. The other signal goes to Q802 and FL802 to pin 18 of audio-FM IC.

The playback audio signal is found at pins 48 and 50 of IC801 fed to IC802 switching, then up to CN301 with L channel at pin 5 and right CH at pin 7. The two stereo in and out jacks are tied to CN301 5 and 7 terminals.

The audio circuits can be signal traced with the scope at head amp, in Record mode at pin 34 of head amp IC2001. Likewise, in the playback audio circuits, the signal can be detected with a scope at terminals 28 and 34. The right (RCH) and left (LCH) channels can be scoped at the connector CN2301 or at the in/out audio jacks, with a dual-trace scope to determine if both audio playback circuits are operating.

MICROPHONES

Most of the microphones used within the camcorder are the unidirectional condenser or electret types. These small mikes must have from 2 to 10 dc voltages applied to the terminals to make them work. In a Pentax PVC850A microphone circuit, dc voltage is applied to pin 2 from R424 of the +5-V source (Fig. 37-7). The audio signal is capacitor-coupled (C420) through the external mike jack, which shorts the terminals to the base of Q403. Q404 and Q403 amplify the audio mike signal and feed it to the line/mic switch (IC402). Some mike circuits contain a low-pass filter to eliminate wind and low-frequency noises.

In the Canon 8-mm camcorder, the mic unit consists of a left and right stereo microphone. The mic input is fed to the recorder PC board. The stereo input signals go through an HPF (high-pass filter) Q701, then to an IC701 amplifier to audio FM IC801 (Fig. 37-8).

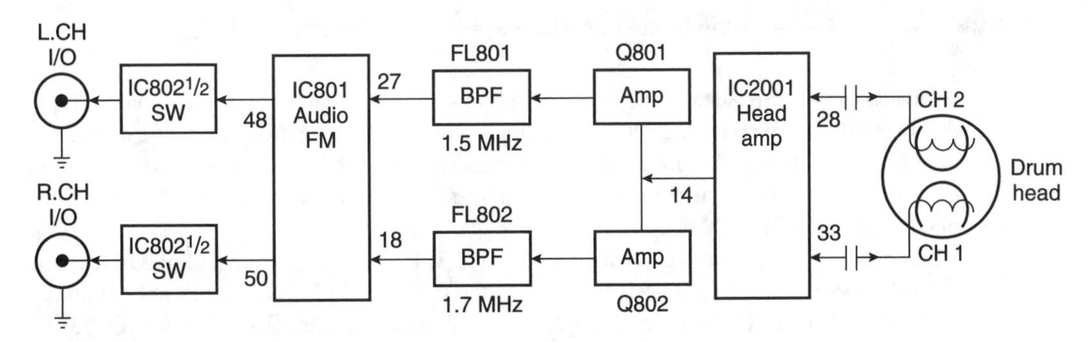

FIGURE 37-6 The audio Playback mode in an 8-mm camcorder.

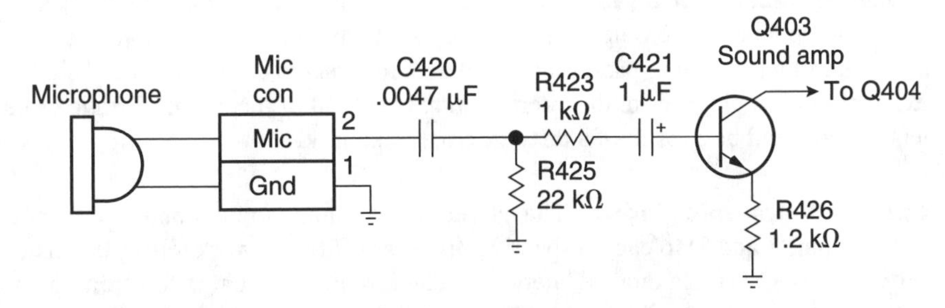

FIGURE 37-7 A typical 8-mm audio microphone preamp circuits.

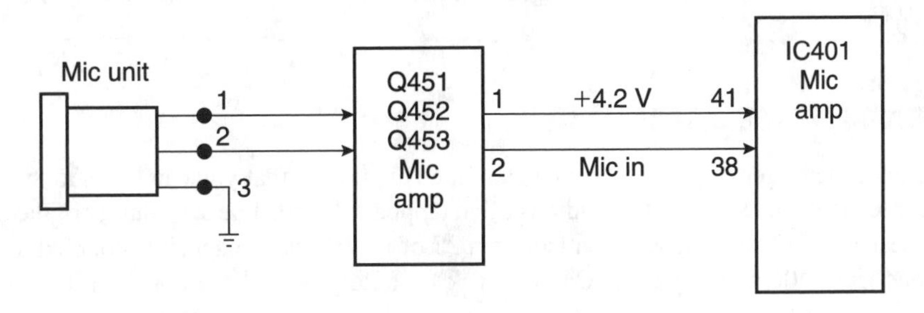

FIGURE 37-8 A Canon 8-mm mic unit and amplifier circuits block diagram.

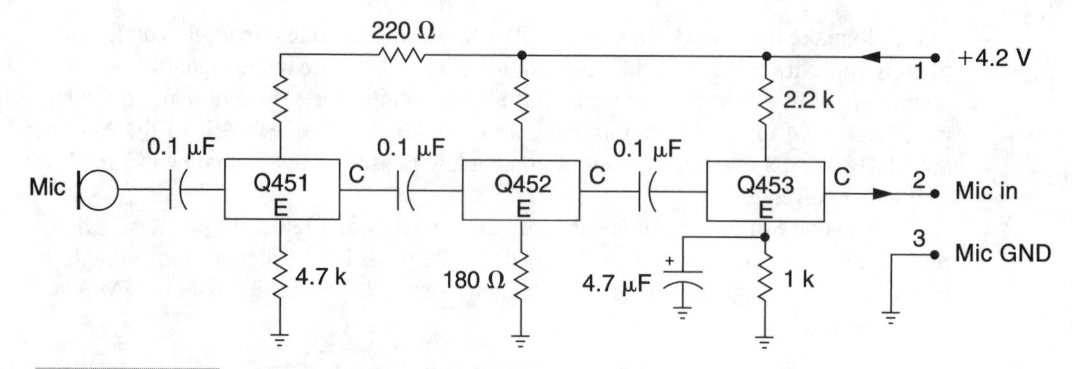

FIGURE 37-9 The Samsung 8-mm microphone amplifier circuits.

Samsung mic circuits The mic unit in Samsung 8-mm camcorder is a monaural type that is fed to the mic board, TP402 connector, and to pin 41 of audio amp IC401 and common ground. When in Record mode, the electret microphone has 4.2 V applied to the mic preamplifier and biased microphone. The ungrounded lead of the mic has a B+ voltage applied through R451 (2.2 kΩ).

The audio from the mic is capacity coupled with C451 (0.1 μF) to the base of buffer transistor Q451 (Fig. 37-9). Audio is taken from the collector of Q451 and capacity coupled with C453 (0.1 μF) to the base of amplifier Q452. The last amplifier transistor (Q453) is directly coupled to Q452; it provides mic input from the collector terminal.

External microphone jack Most camcorders have an external mike jack so that another microphone can be plugged in to get closer to the sound of the subject taken. J401 is a self-shorting jack that bypasses the internal mike signal to the amplifier circuit. When the external mike is plugged in, the internal mike is cut out of the circuit. Erratic or intermittent sound could be a result of a dirty external mike jack.

Canon 8-mm mic input The stereo microphone input connects to connector CN701 (pins 1 and 3) to each respective high-pass filter network (HPF) transistor Q701 (Fig. 37-10). Q701 is a dual RC network. The left audio signal is fed from the network to the IC701 amplifier. The dual IC701 amp operates in both the left and right input circuits. From here, the amplified left channel connects to pin 62 of the audio-FM (IC801). The right stereo channel goes through the same Q701 and IC701 to pin 2 of IC801.

HEADPHONE JACK

Many of the larger camcorders have headphone circuits so that you can listen to the sound recorded and played back. The audio is often tapped off at the line amp output of the audio-processing IC. The audio is taken from pin 28 of IC401 and is capacitor-coupled to headphone jack J402 with capacitor C418. Check for a dirty headphone contact if the sound is erratic or dead.

AUDIO CONTROL HEADS

The audio record/playback (R/P) head might have a bias signal applied from the bias oscillator, which also supplies a bias oscillator frequency to the audio erase and the full erase heads. The audio A/C control head is located after the upper cylinder or drum assembly (Fig. 37-11). The R/P audio head terminals are switched by the head switch IC when either recording or playback operation is used. The audio A/C control head is found in the VHS and VHS-C models. Usually, the audio signal in the 8-mm camcorder is processed to FM and mixed with the video heads.

The video and audio signals are applied to the tape with the REC/PB drum and A/C tape head in the helical scan format. Of course, the video signal is recorded in the center of the tape with the drum or cylinder head and the audio track at the edge of the tape in the VHS

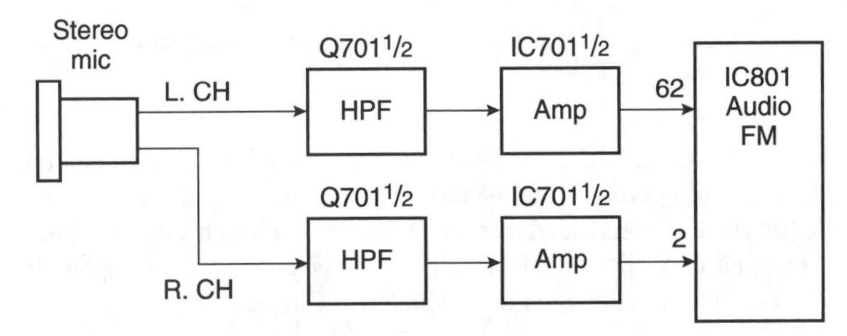

FIGURE 37-10 The 8-mm stereo microphone HPF filter (high-pass filter), amp, and audio FM IC801.

FIGURE 37-11 The audio tape head in a VHS-C VCR transport.

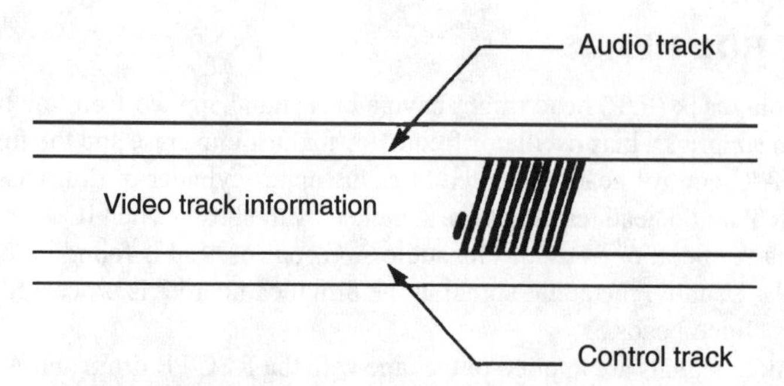

FIGURE 37-12 The audio track is located at the edge of the tape in the VHS and VHS-C formats.

and VHS-C tape formats (Fig. 37-12). The control track keeps the picture and sound synchronized with a CTL head.

Audio erase head Often, the audio head is excited from the bias oscillator circuit and is used to erase the audio portion from the tape. The full-erase head erases the entire tape path. Both the full erase and A/C heads are located on the VTR in the tape path of the VHS and VHS-C machines. Audio tape head alignment is covered in Chapter 38. Troubleshooting the camcorder audio circuits is included in Chapter 40.

Audio 8-mm stereo sound tracks The audio mono stereo sound track is similar to the VHS sound track, except that two tracks are recorded at the bottom side of the tape (Fig. 37-13). The video information is recorded in the center of the track and control data at the top of the tape in the stereo camcorder. In the 8-mm camcorder, stereo sound is recorded in the audio-FM IC. IC801 provides FM stereo modulation in the Record mode of the Canon ES1000 camcorder audio stereo circuits.

AUDIO OUTPUT JACKS

The audio input and output jacks in the Canon 8-mm stereo audio channels are located in the jack unit. The right and left audio output jacks can be cabled to a stereo amplifier with

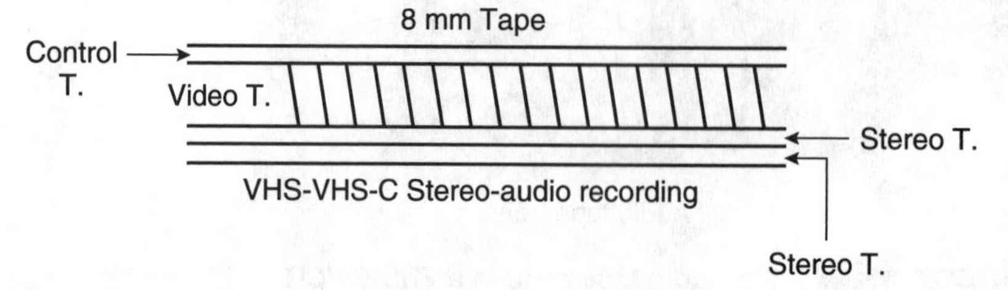

FIGURE 37-13 The VHS-C stereo recording sound tracks on the tape.

FIGURE 37-14 **The audio and video output jacks are located at the back side of this Hitachi camcorder.**

speakers or to the TV stereo audio system (Fig. 37-14). A right stereo output jack is tied to switching IC802 and fed to pin 50 of audio FM IC801. The left stereo channel jack is fed to switching IC802 to pin 48 of IC801. These same stereo jacks are wired for input sound recordings. IC802 is a three-section switching IC.

38

MECHANICAL PROBLEMS AND ADJUSTMENTS

CONTENTS AT A GLANCE

The mechanical VTR or VCR section of the camcorder contains components that provide tape movement. Most of these mechanical components are located in the main chassis. Each part can be outlined or located on the chassis (Fig. 38-1). To service the mechanical section, the VTR main chassis should be removed. The main components are described in this chapter, along with its mode of operation and the tape run path. The audio heads and capstan motor operations are described in Chapters 36 and 37, respectively.

Mechanical movements in the VTR or VCR section consists of any part moving, rotating, and sliding. The most important components are the drum or cylinder, capstan flywheel assembly, take-up and supply reels, pressure roller, impedance roller, guide rollers,

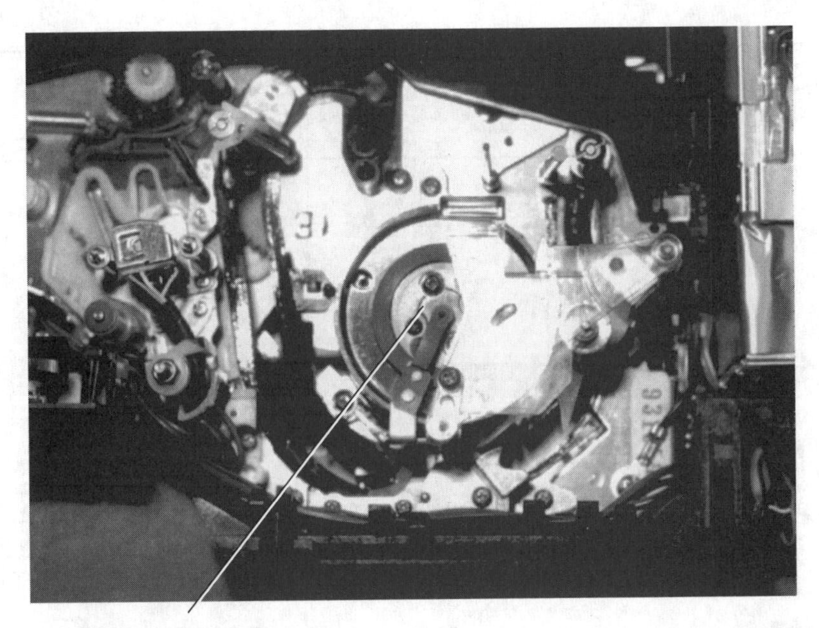

FIGURE 38-1 **The video head can be called a *drum* or *cylinder* in the camcorder.**

gear and pulley assemblies, loading mechanism, tension arm, pressure roller, and cam gear assembly. Most of these components are rotated or moved by a motor, tape, belt or gear driven. Any component in the mechanical section can become worn, dry, and frozen, preventing proper tape movement.

General Head Description

The video tape head, upper cylinder, or drum assembly plays the tape in the Play mode and records on the tape in the Record mode. The conventional tape head (VHS) has two video heads 180 degrees apart on a 62-mm cylinder. The VHS-C camcorder uses the same size tape as the VHS units, except that a smaller video head is used with a 41.33-mm cylinder or drum assembly (Fig. 38-2). The 8-mm camcorder has a smaller tape width (8 mm) with a 40-mm diameter drum or cylinder. The video head assembly is rotated with a cylinder or drum motor.

To reduce the camcorder overall size, the cylinder or drum assembly was reduced with a smaller size of cassette. To maintain compatibility with the standard VHS system, the tape head cylinder diameter was reduced by two-thirds in the VHS-C camcorder. To get the same length of tape around the head surface, a greater tape wrap must be made around the small cylinder or drum assembly. The same tape track length is recorded on the video tape with the same length of tape wrapped around the cylinder or drum. With a smaller-diameter tape head, the speed rotation is increased to maintain relative head-to-tape speed compatibility.

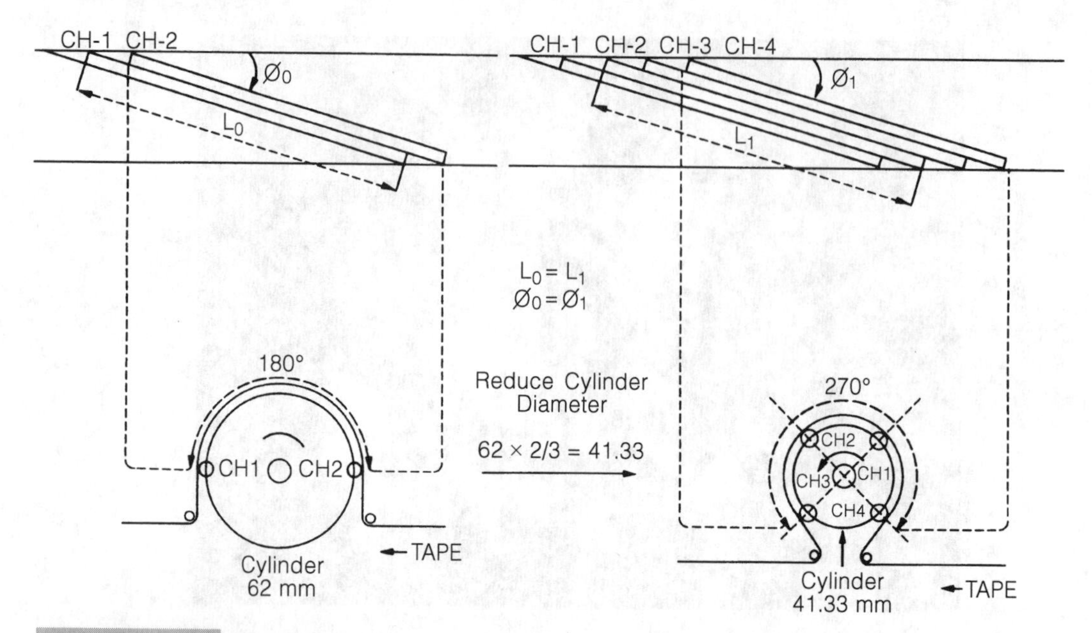

FIGURE 38-2 The VHS cylinder is 62 mm in diameter with a tape wrap of 180 degrees. The VHS-C cylinder is 41.33 mm with a 270-degree tape wrap.

TYPICAL 8-MM VIDEO HEAD CYLINDER

The cylinder is 180 degrees + wind (without PCM recording). The track width is 20 microns (in SP) on a 40-degree cylinder diameter for a tape speed of 14.345 mm/s, which is slower than VHS. The relative speed of the cylinder and tape is 3.8 m/s (Fig. 38-3).

Two video heads (CH 1 and CH 2) are separated and attached to the cylinder, 180 degrees apart. The angle of azimuth for CH 1 is –10 degrees, for CH 2 +10 degrees (as seen from the front surface of the head). The upper cylinder has another CH 3 head, which is

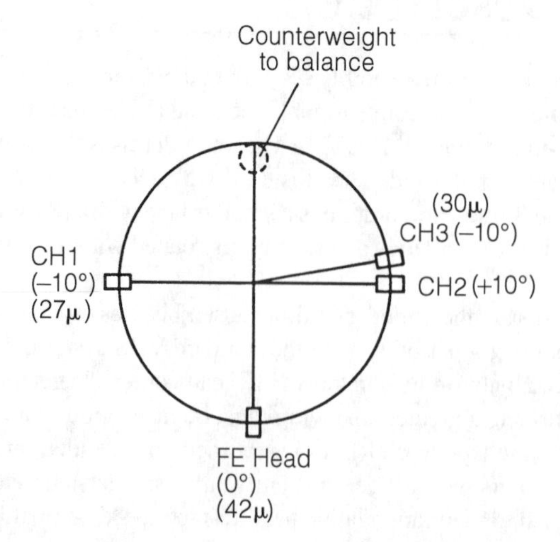

FIGURE 38-3 CH1 and CH2 are 180 degrees apart with the FE head between them in the 8-mm camcorder.

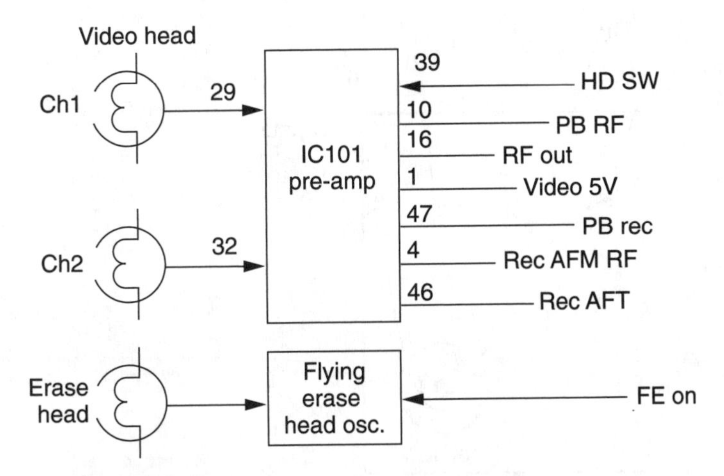

FIGURE 38-4 The CH1, CH2, and FE video heads with preamp IC101 circuits in the 8-mm camcorder.

ahead of the CH 2 head by 1.7 degrees (2.5 H). This head is used for trick play with the azimuth angle at –10 degrees.

Between the CH 1 and CH 2 video heads, 90 degrees counterclockwise from CH 1 head, is a rotating erase head. Opposite to the rotating (flying) erase head is a counterweight to maintain balance. The rotating erase head is 42 microns wide so that it can erase two tracks at the same time. The video tape runs in the same direction as the rotating cylinder rotates.

TYPICAL 8-MM PREAMP CIRCUITS

The video head drum contains CH 1 and CH 2, which are fed into preamp IC101 at pins 29 and 32. CH 1 and CH 2 are mounted 180 degrees apart. Also in the drum is a flying erase head (FE), which is controlled by erase oscillator GM161 from the FE pond circuits. The drum head is rotated by a drum motor (Fig. 38-4).

Mechanical Operations

Most of the mechanical parts contained in the VHS, VHS-C, and 8-mm camcorders are the same, except they might be called by another name. Of course, the VHS camcorder's main tape transport assembly is much larger than the VHS-C or the 8-mm (Fig. 38-5). The bottom side of the VHS-C camcorder is shown.

8-MM STOP-TO-PLAY TAPE PATH

The tape is pulled out of the S-reel. The tape passes the No. 2 guide (impedance roller). The tape passes the No. 3 guide. The angle post (No. 4 guide) positions the tape at the correct angle. The tape wraps on the head drum. The angle post (No. 5 guide) directs the tape

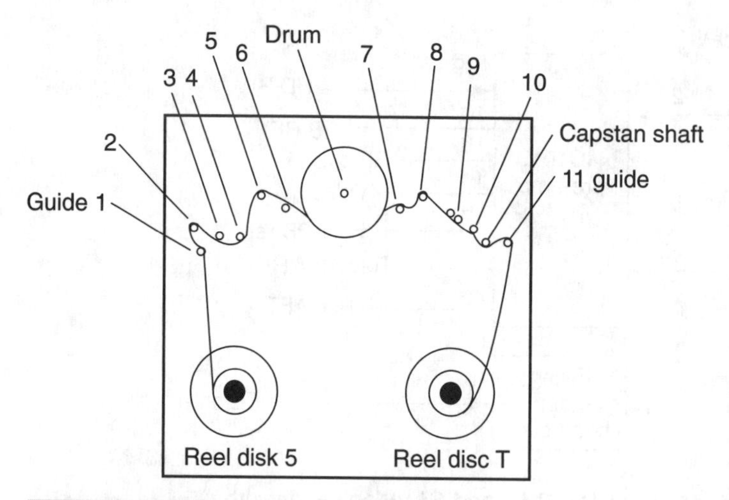

FIGURE 38-5 **The layout of the tape transport with tape guides and reels of an 8-mm camcorder.**

from the head drum to tape guide (No. 6). The tape passes the capstan shaft and the pinch roller to the tape guide (No. 7). The tape is reeled into the T-reel.

TYPICAL VHS-C IMPEDANCE ROLLER

This consists of a body of polyacetal resin and a brass ring fitted around the body. Jitter is reduced by raising the moment of inertia. This is a precision-made roller. Be careful not to damage the surface. If it is damaged, jitter will occur, synchronized with the roller's rotation. Light oil is applied slightly to the bearing. Felt rings (oil barriers) are attached to the top and bottom of the roller to prevent oil from leaking and adhering to the tape. Be sure to replace the felt washers when replacing the impedance roller.

VHS-C GUIDE ROLLERS

These have the same construction as conventional guide rollers and are precision-made rollers molded of polyacetal resin. These rollers change the direction of the tape transport and control the tape height. One is provided on the supply side and the other is on the take-up side (Fig. 38-6). The guide rollers are adjusted so that the bottom edge of the tape runs along a reference line called the *lead* (Fig. 38-7).

TYPICAL 8-MM DRUM ASSEMBLY

The drum assembly is located upon the main mechanical deck assembly. The drum motor is the bottom part of the drum assembly while the top section contains the two CH 1 and CH 2 heads with the flying erase head. The drum head picks up information from the tape and plays it back in the camcorder. Also, the drum applies video and sound recording to the revolving 8-mm tape.

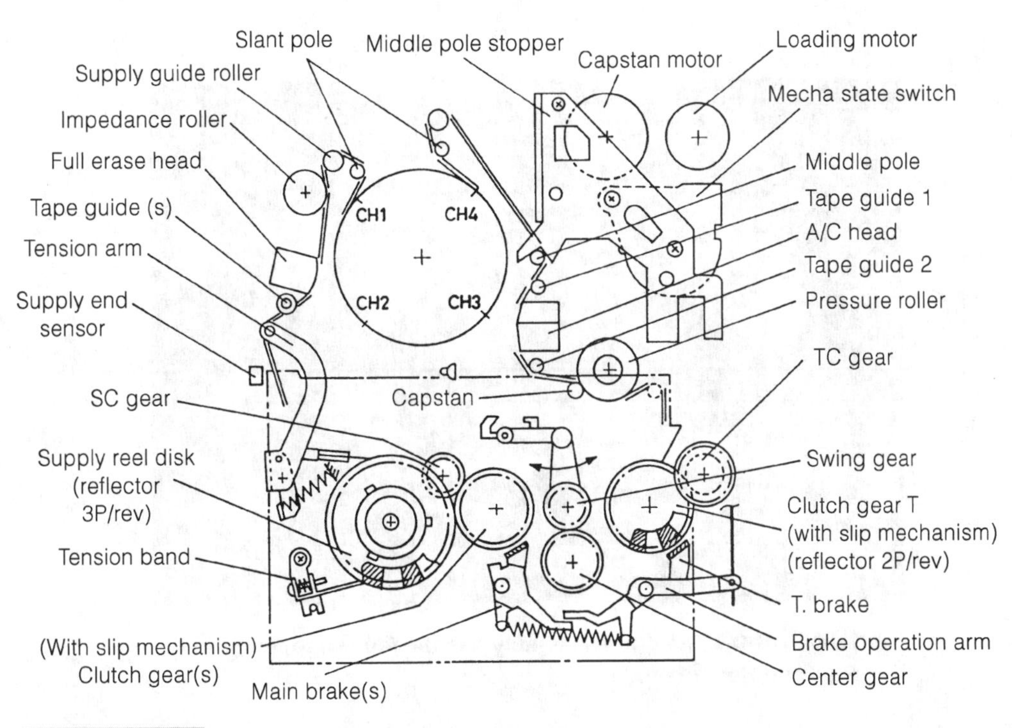

FIGURE 38-6 A VHS-C camcorder with tape head and guide rollers.

FIGURE 38-7 The drum assembly and guide rollers of a VHS-C camcorder.

FIGURE 38-8 **The drum assembly has the tape wrapped around it in Play and Record mode.**

A take-up and supply reel, supplies and takes up the tape after revolving around the drum assembly. In this chassis, the loading motor is at one corner and the capstan motor assembly at the other. The pinch roller is located between capstan and spindle, close to the take-up reel. The capstan and pinch roller move the tape.

Canon 8-mm drum and capstan assembly The drum head in the Canon 8-mm camcorder has a top section (2) and bottom section. The drum head contains CH 1 and CH 2 with the flying erase head (FE). The drum takes off and applies picture and sound to the 8-mm tape. The tape drum is the largest part on the main chassis (Fig. 38-8).

The capstan motor (9) drives the capstan that rotates the tape through pressure or pinch roller around the drum cylinder. A flat loading motor (15) is located at the opposite end of chassis from the capstan motor assembly. The loading motor loads and unloads the 8-mm cassette.

Realistic 150 (VHS-C) capstan assembly The capstan has a diameter of 3 mm, and its speed is 3.5 revolutions per second. The capstan FG detection coil generates 102 pulses per rotation; therefore, the capstan FG is 360 Hz in the SP mode and 120 Hz in the EP mode. The flywheel is supported by two ball bearings (Fig. 38-9).

SLANT AND GUIDE POLES

The slant poles guide the tape so that it winds around the cylinder at an angle. They do not control the tape height. The angle of inclination cannot be varied. Do not try to straighten these slanted poles.

The tape comes from the capstan shaft through the guide pole to be wound by the take-up reel disk. The guide pole allows a neat wind by regulating the tape in the height direction and letting the reel disk wind the tape so that it does not touch the inside walls of the cassette.

8-MM CENTER GEAR AND FRP GEAR

The capstan motor with a square belt drives the center gear. The center gear meshes with the FRP gear and oscillates the FRP gear to left and right in line with the center gear's direction of rotation. If the capstan shaft is rotating counterclockwise (normal winding direction), the center gear rotates clockwise, which oscillates the FRP gear to the take-up side. This allows the take-up reel disk to rotate clockwise, which takes up the tape and enters Play (Record) or Search mode. When the capstan motor is rotating clockwise, all rotation is in reverse, and the supply reel disk rotates counterclockwise to take up the tape in the Rewind or Unloading modes.

Realistic (VHS-C) center gear The torque produced by the capstan motor is transmitted to the center gear through the flywheel. The swing gear is engaged with the center gear, which swings to the left and right, according to the direction of rotation of the center gear, to drive the take-up or supply clutch gear assembly.

Realistic (VHS-C) clutch gear The clutch T gear has a slip mechanism, where the capstan motor rotates the take-up reel hub and permits slipping, to take up the tape. The slip mechanism controls the amount of torque transmitted. This clutch also operates during the Fast-Forward mode. A reflector (two pulses per revolution) fitted on the bottom is used to detect rotation of the take-up reel hub. The clutch T-gear stops rotating when

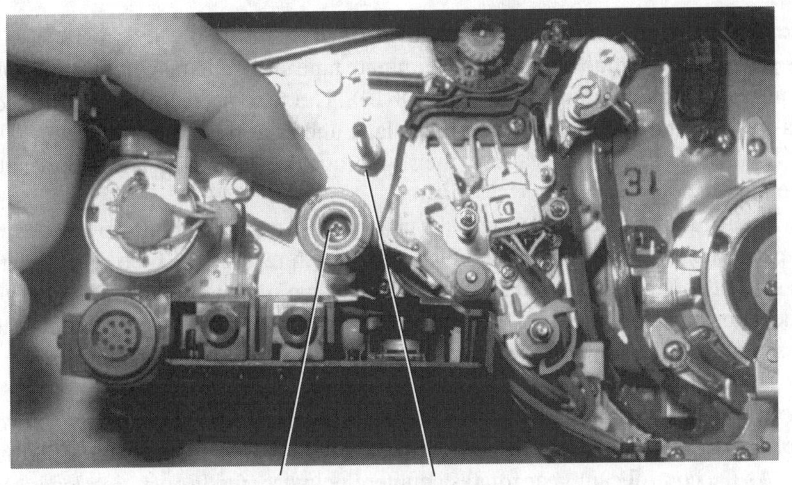

FIGURE 38-9 **The capstan and pinch roller in a VHS camcorder tape path.**

rewinding is complete. The T-brake pressed against the bottom of the clutch T-gear applies reverse torque to the tape at the take-up side during loading and reverse search.

The clutch S gear, like the T-gear, has a slip mechanism that applies capstan motor torque to the supply reel disk, permitting slippage, to take up the loose tape.

TYPICAL 8-MM TAKE-UP REEL DISK

The take-up reel disk operates the same way as the supply reel disk slip mechanism slipping during Take-Up, Fast Forward, and Visual Search modes. At the bottom of the take-up reel disk is the reflecting plate for outputting a sine wave at eight pulses per revolution, which is used to indicate the amount of tape remaining and to detect rpm. The gears on the periphery of the take-up reel disk are the same as those on the supply reel disk. They provide sure transmission of torque.

VHS-C MAIN S BRAKE

The main brake applies braking to clutch gear S when driven by the B-cam lever, which is linked with the groove provided at the top of the control cam. The T-brake (pressed against the bottom of the clutch gear T) applies reverse torque (weak brake) to the take-up reel hub.

LOADING AND DRIVE MECHANISMS

The loading and drive mechanism components are usually located on the bottom side of the main chassis. The top of the main chassis contains the take-up and supply reel, braking mechanism, and tension arms. Under the main chassis are the loading worm and cam gears, lower cylinder motor, loading motor, capstan flywheel, main brake arm, and supply loading and take-up loading cam gears.

Typical 8-mm reel disk drive mechanism The supply reel disks' slip mechanisms generate the torque necessary to eliminate tape slack and take up the tape during reverse visual search and unloading (Fig. 38-10). The take-up reel disk's slip mechanism works in the same way to eliminate tape slack and take up the tape during the take-up modes and visual search. This allows the slip torque on the take-up and supply sides to be independently controlled.

The loading motor force that moves the change base assembly plate up and down passes the capstan motor torque through the FRP and to both reel disks. If the center gear assembly is set in the High position at this time, the capstan motor torque will pass through the slip mechanisms in both reel disks, causing the torque to become weak. Setting the center gear assembly in the Low position transmits torque directly to both reel disks.

Typical (VHS-C) pressure roller drive mechanism The lower groove of the control cam gear is linked with the pin of the P-operation lever (pressure roller operation lever). As the control cam gear rotates counterclockwise, the P-operation lever also turns counterclockwise (Fig. 38-11). The lower end of the P-operation lever has teeth that engage with the teeth in the adjacent gear arm. When the P-operation lever turns counterclockwise, the gear arm rotates clockwise to cause the pressure roller lever to turn

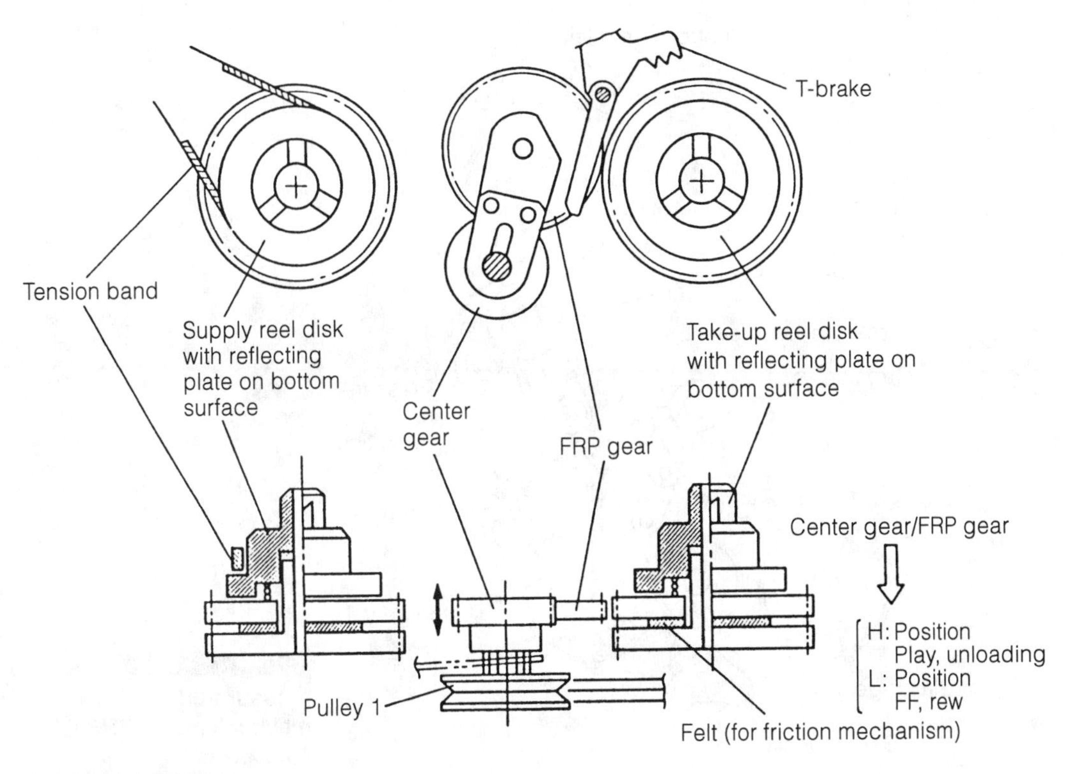

FIGURE 38-10 **A typical 8-mm reel disc-drive mechanism.**

counterclockwise. This moves the pressure roller near the capstan shaft and the toggle mechanism brings the roller into contact with the shaft.

Mechanical Adjustments

From 10 to 20 mechanical adjustments can be made in the camcorder after replacing certain parts and for better Play/Record operation. In the smaller camera, fewer mechanical adjustments can be made because of size. Some of the adjustments can be made with visual means. In others, alignment tapes, operation fixtures, and crucial test equipment must be used. Besides test equipment, several small special hand tools are required.

IMPORTANT PRECAUTIONS

Always disconnect the camcorder from the power (battery or ac adapter) before removing or soldering components. Be careful when removing small screws, bolts, or washers so as not to drop them inside the unit's mechanical areas. Retrieve all loose parts. When removing a component, be careful not to damage or dislodge other parts (Fig. 38-12). Be careful when working around the tape guides and video head drum to prevent breaking or scratching the

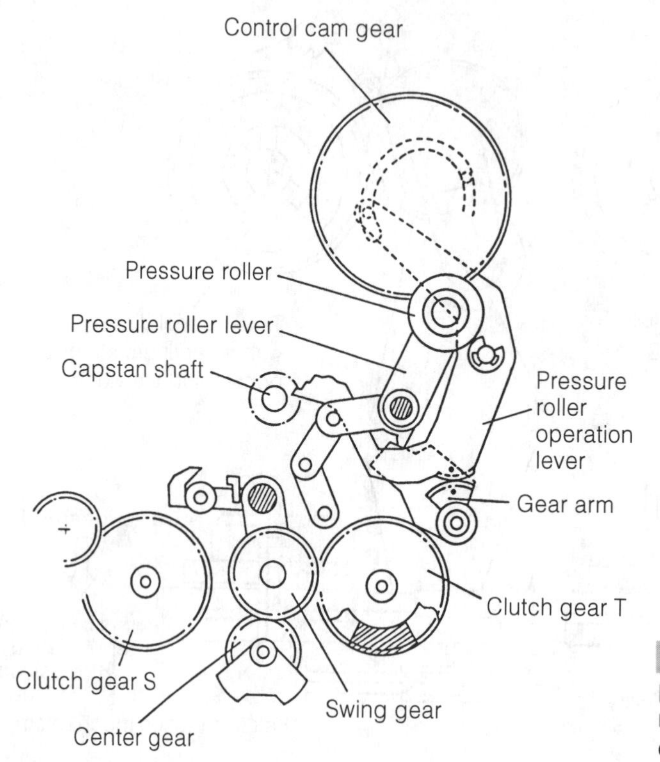

FIGURE 38-11 The pressure-drive mechanism in a VHS-C camcorder.

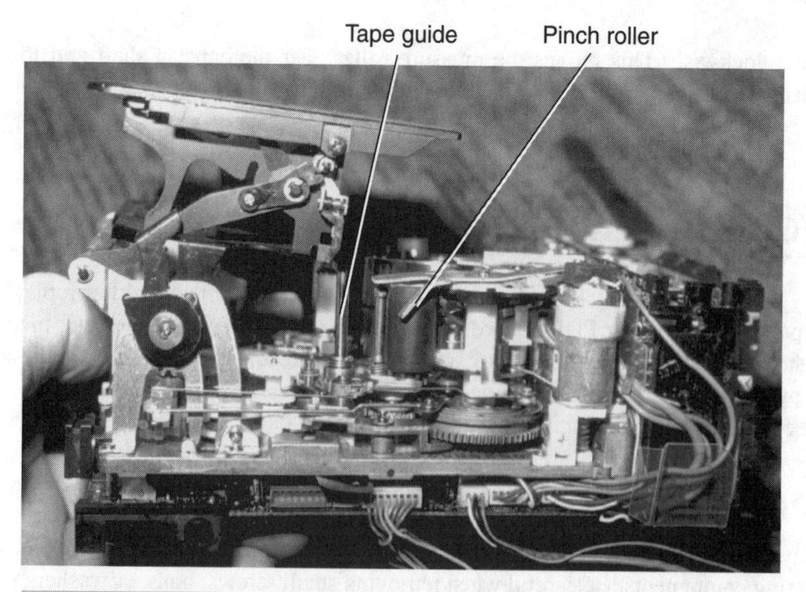

FIGURE 38-12 The mechanical parts in the VTR or VCR transport of a VHS-C camcorder.

video head surface. Do not try to make important mechanical adjustments without the correct test equipment and alignment tapes. Always follow the manufacturer's alignment procedures. Remember, some manufacturers have made precise adjustments at the factory, like the tape transport mechanism, which ordinarily does not require adjustment.

TEST EQUIPMENT

Most manufacturers require the use of alignment and torque-meter cassettes with mechanical adjustments. Some manufacturers require that certain patch cords, adjustment plates, and fixtures be used. Besides the regular bench tools, special head drivers, torque gauge adapter tools, height jigs, and screwdrivers are needed for correct adjustments.

PRELIMINARY ADJUSTMENT STEPS

After replacing or adjusting the tape guide posts, check the coarse adjustment of the guide posts' height to confirm the linearity envelope output on the scope.

When repairing or installing a new upper drum or complete cylinder unit, confirm the envelope output with the scope and make fine adjustments on the A/C head in the horizontal position (VHS and VHS-C models). Do not change the position of tape transport posts P1 and P4 (Fig. 38-13).

After reinstalling all new posts or if any adjustments have been made to the tape guide posts, make coarse adjustments of each tape guide post's height, confirm the tape transport, readjust the pullout post (P5) height, confirm the A/C head tilt (VHS and VHS-C), and confirm the envelope output with the scope.

When installing the A/C head (VHS and VHS-C), confirm the A/C head tilt. Do not change height of the P4 post. Adjust the A/C head coarse height, the A/C head height and azimuth, horizontal course adjustment of the A/C head, and fine horizontal adjustment of the A/C head position.

After installing or adjusting the P4 Post, make the coarse adjustment of the tape guide post heights (only P4 post), confirm the tape transport, confirm the A/C head tilt, and confirm the envelope linearity output on the scope. When installing the pullout post (P5), readjust the pullout post position. Do not readjust any other post after the P5 post. After making several of these post adjustments, you might need to only touch up one or two post heights for accurate running of the tape.

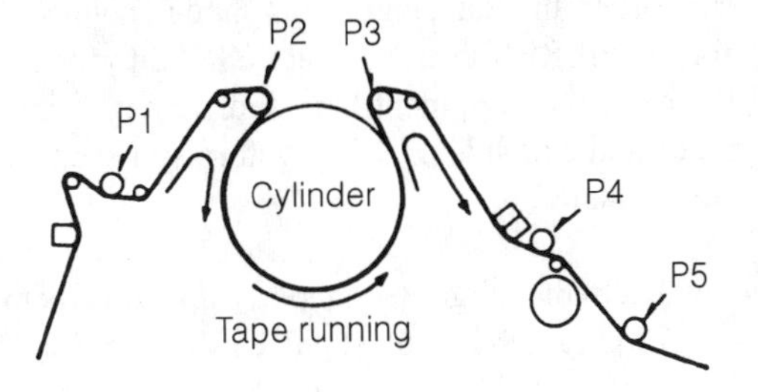

FIGURE 38-13 The loading post location with a VHS cylinder and tape-running path.

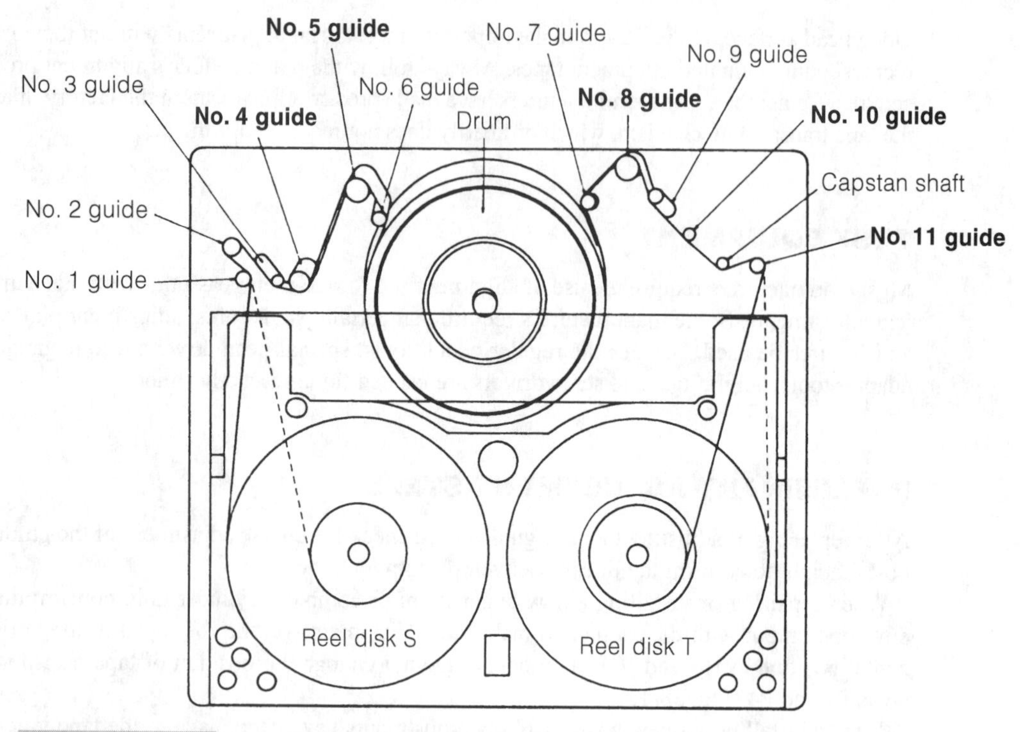

FIGURE 38-14 The tape-path adjustments on the 8-mm camcorder.

Typical 8-mm tape path adjustments The tape path of the 8-mm transport starts at the supply reel to the No. 1 guide, roller lever guide (No. 2), slant pole guide (No. 3), impedance roller (No. 4), guide roller S (No. 5), the pole slant guide (No. 6), pole slant guide (No. 7) of take-up reel T, guide roller T (No. 8), pole slant (No. 9) guide for capstan, shaft pole guide (No. 10), the capstan shaft, and (No. 11) past the review arm to enter into the take-up reel T (Fig. 38-14).

Tape height adjustment is made with the impedance roller, No. 4 (Table 38-1). An envelope waveform is adjusted at guide roller S (No. 5) and guide roller T (No. 8). Tape guide adjustment is made with the shaft pole guide (No. 10) and tape wrinkle check at post review arm (No. 11).

Tracking adjustment Before making any adjustments, clean up the tape paths, guides, rollers and drum or cylinder areas. In the Samsung 8-mm model, connect the oscilloscope connections to CTP101 for the PB RF waveform and CTP102 for head switch and trigger waveforms (Fig. 38-15). Insert the alignment tape and play it to adjust the tracking. Confirm that both the entrance and exit side of RF waveforms of scope are flat. If not, go through the fine-tracking adjustment.

Fine-tracking adjustment Play back the alignment tape. Confirm whether the waveform is flat or not. If it appears as A in Fig. 38-16, adjust guide roller No. 5 clockwise to get the flat envelope waveform. If it appears as B, adjust guide No. 5 counterclockwise to get the flat envelope waveform.

To adjust the drum exit side, confirm if the waveform is flat or not. If waveform shows like C, adjust guide roller T (No. 8) clockwise to get the flat envelope waveform. In case the waveform looks like D, adjust the No. 8 guide counterclockwise to get the flat envelope waveform. Reconfirm on the envelope waveform that the envelope has not changed.

Tape guide adjustment For guide adjustment No. 4, play back the alignment tape. Inspect the tape for a bent area at the lower flange area. Simply, guide the tape to the lower flange. Cue/RPS the tape, then play it back and confirm that the RF waveform rises flat within 2.5 seconds. Repeat this tape adjustment until a normal waveform is obtained.

TABLE 38-1 LAYOUT OF TAPE TRANSPORT ADJUSTMENTS

GUIDE NO.	COMPONENT NAME	ADJUSTMENT PURPOSE
No. 1	Tension pole	—
No. 2	Roller lever/G	—
No. 3	Slant pole SUB	—
No. 4	Impedance roller	Tape height
No. 5	Guide roller "S"	Envelope
No. 6	Pole slant "S"	—
No. 7	Pole slant "T"	—
No. 8	Guide roller "T"	Envelope
No. 9	Pole slant CAP	—
No. 10	Shaft pole guide	Tape guide
No. 11	Post review arm	Tape wrinkle

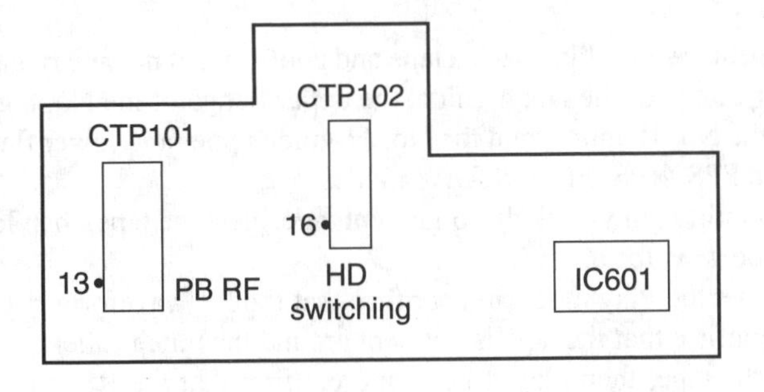

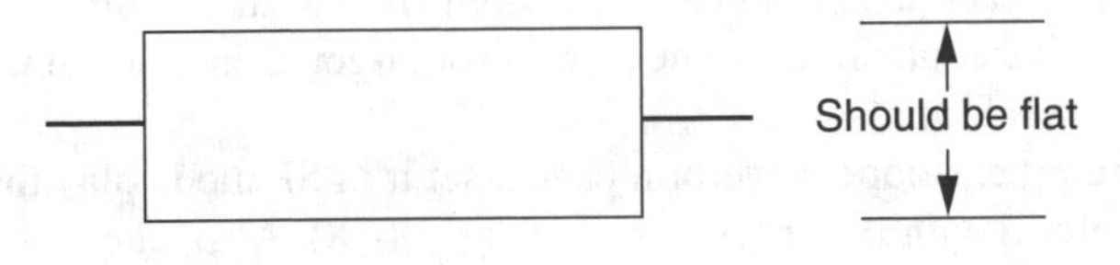

FIGURE 38-15 **The test-point locations and RF waveform in an 8-mm camcorder.**

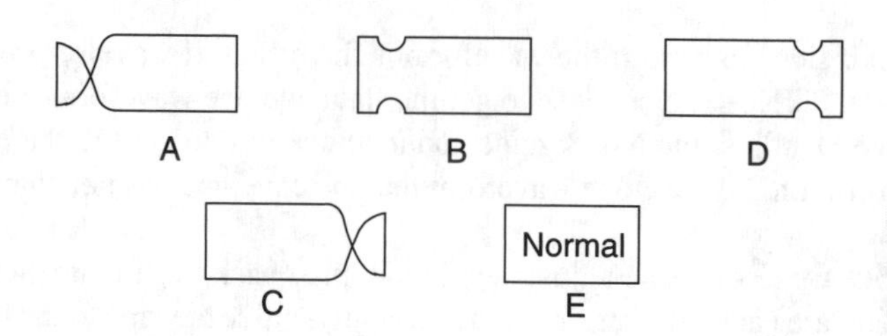

FIGURE 38-16 The 8-mm drum-entrance abnormal and good waveforms.

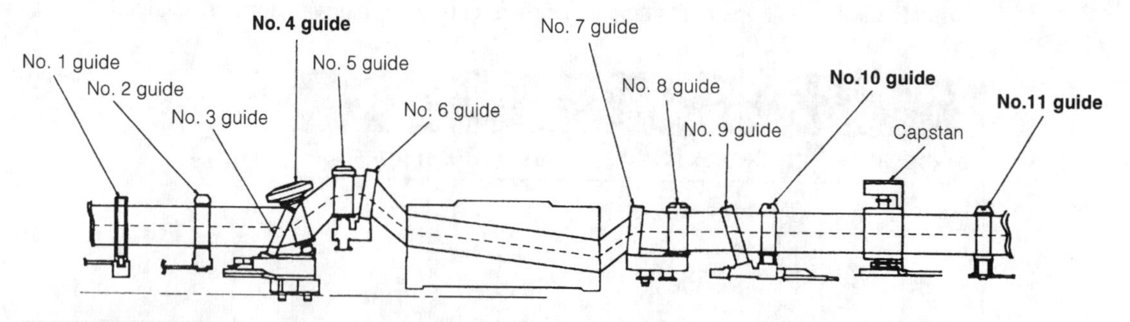

FIGURE 38-17 The tape travel in an 8-mm camcorder.

To adjust tape guide No. 10, play the alignment tape and guide the tape to the lower range. Cue/RPS the tape, then play it back and confirm that the RF waveform rises flat within 2.5 seconds in the Play mode. If the waveform is not normal, repeat the adjustment of the No. 10 guide post (Fig. 38-17). If the tape is bent in the RPS mode, adjust guide No. 11 until the tape is straightened.

To adjust tape guide No. 11, play back the alignment tape. Be sure that the tape is not bent or out of line between guide No. 10 and the capstan. Set the Play mode again and confirm that the tape is not bent. Repeat adjustments No. 10 and 11 if the tape is bent.

Recheck after the adjustment Play back a tape and confirm that no tape rising and curling occurs at the lower flange of the No. 4 guide, the upper flange of the No. 8 guide, the lower flange guide of the No. 10 guide, and the No. 11 guide upper and lower flanges. Check the tape again in the FPS, RPS, FF, and REW modes.

To check the envelope stability, play back the alignment tape, eject the tape, then load it again. Confirm the envelope waveform.

In Stop and Play modes, set the Play mode and confirm that the RF waveform rises flat within 2.5 seconds. Also confirm that the tape is not bent around the pinch roller.

Cue/RPS and FF/REW the tape, then play it back and confirm that the RF waveform rises flat within 2.5 seconds. Also confirm that the tape is not bent around the pinch roller.

For a tracking check, play back the alignment tape. Confirm that the minimum amplitude value (E min) is 65% of the maximum value (E max) or larger. Confirm that no large fluctuations occur on the waveform.

To make the self-recording envelope waveform check, set it in SP mode play the tape, and confirm the RF waveform with the specifications (Fig. 38-18). Also, check the level of amplitude value with the electronic specifications. In each mode, confirm that the difference of the amplitude (CH 1 and CH 2) is 65% or larger. If it does not meet these specifications, replace the drum assembly with a new drum with an original part number.

Typical VHS-C supply guide pole adjustment The guide supply poles serve to improve the tape-transport stability between the cassette output and drum input by maintaining the correct height. Adjust the supply guide roller to obtain smooth tape transport at the lower flange of the guide poles (Fig. 38-19).

Tape interchange ability adjustment A quick check should be made after any service operation is performed that could adversely affect the tape, such as the replacement of the cylinder or drum motor, tape guide, audio/control head, or any component in the tape path. Usually this adjustment is performed only after the tracking preset adjustment is completed. Often, no tape guide adjustments are needed if the unit passes this check.

Audio/control (A/C) head adjustment Incorrect position of the audio/control (A/C) head reduces the playback audio output, has poor signal-to-noise ratio (S/N), and might interfere with servo stability when the head cannot pick up the control signal. The audio/control coarse adjustment should be made before head height, tilt, azimuth, or horizontal position adjustments. The A/C control head is only in the VHS and VHS-C camcorders (Fig. 38-20).

Confirming A/C head tilt Make this adjustment after the height adjustment of P4. Play the tape and be sure that the tape runs between the lower and top limits of -post P4 (Fig. 38-21). Notice the tape movement. If an adjustment is necessary, turn screw (C) clockwise so that the curling is apparent with the lower edge of P4. Now turn the screw (C) counterclockwise so that the curling of tape smooths out.

The tape should be adjusted so that it runs in the center of the guides and tape head. The revolving tape should catch the top part of the control head (C) and bottom of the audio tape (A).

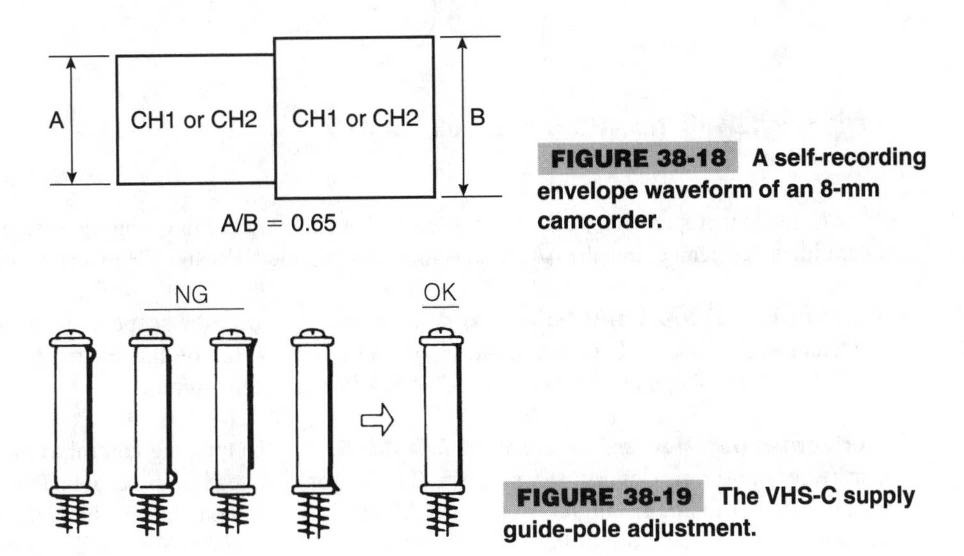

FIGURE 38-18 A self-recording envelope waveform of an 8-mm camcorder.

FIGURE 38-19 The VHS-C supply guide-pole adjustment.

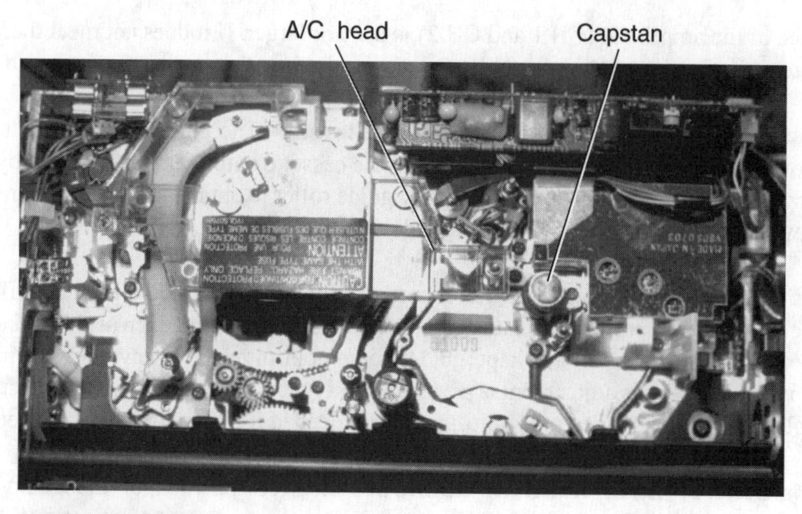

FIGURE 38-20 **The audio control (A/C) head in the RCA VHS camcorder.**

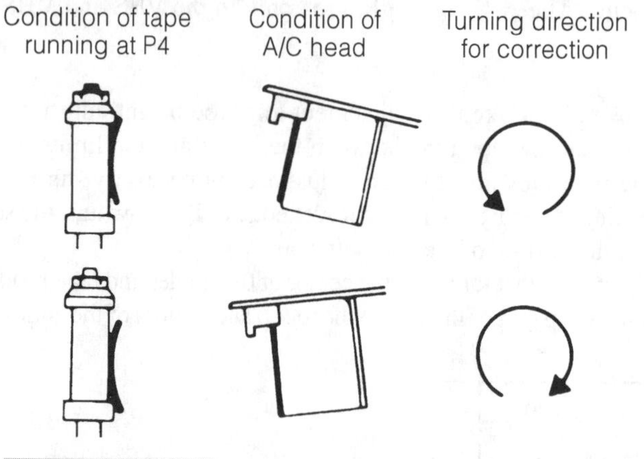

FIGURE 38-21 **The VHS A/C head tilt in a VHS camcorder.**

If the head is not properly adjusted, either the control or audio signals might appear erratic and intermittent. Remember that the A/C head is found only in the VHS and VHS-C camcorders.

Adjustment of A/C head height and azimuth Connect the scope to TP4001 on the audio CBA board. Play the monoscope portion (6 kHz) of the alignment tape (VFM50001H6). Adjust screw B on the A/C head for maximum output.

Horizontal position adjustment of A/C head Set the tracking control to the detent (fixed) position. Connect the scope to TP3501 on the head amp section. Play the monoscope portion of the alignment tape (VFM50001H6) and note the envelope that corresponds to the high period of the head switching signal at TP2005 (Fig. 38-22). Use this envelope for the following adjustments. Slowly turn the azimuth nut so that the envelope is at maximum. Before finding the center of the maximum period of the envelope, rotate the adjustment nut back and forth slightly to confirm the limits on either side of the maximum period. Now determine the center point. Confirmation of the correct adjustment can be made by turning the tracking control to the right and left to correspond with the envelope. The adjustment is okay when the envelope changes symmetrically.

Reel table height adjustments The supply and take-up reel tables should be the same height. Often the reel table heights can be adjusted by changing the stack of washers under the reel assembly within the VHS camcorder (Fig. 38-23). A height fixture or reel

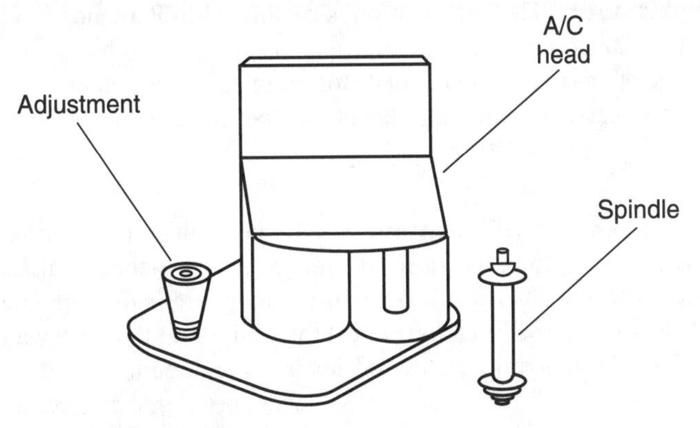

FIGURE 38-22 A close-up view of the A/C head adjustment.

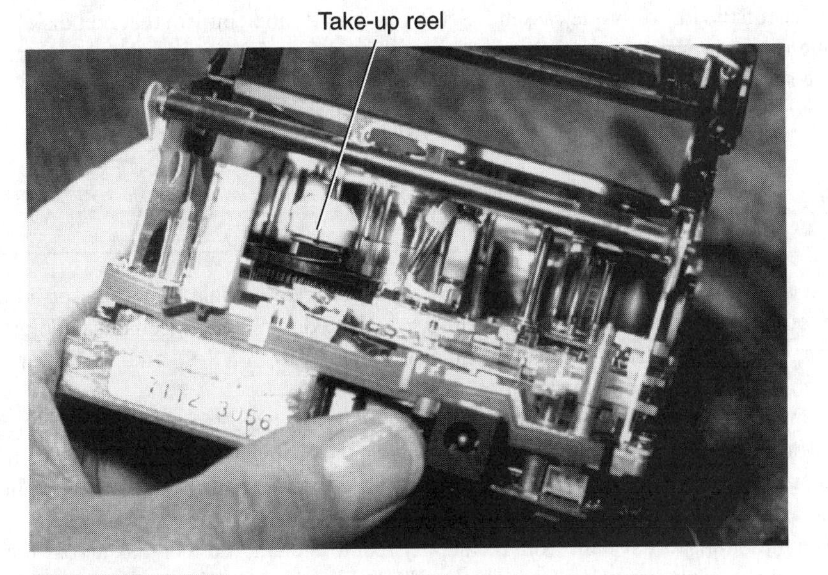

FIGURE 38-23 A close-up of the take-up reel in a VHS-C camcorder.

jig is used for the correct height adjustment. If one of the reels is lower than the other, the adjustment can be made by adding another washer. Usually, reel table height adjustment is only required if a new reel table is installed. If only one reel is replaced, compare the height with the original.

Typical 8-mm reel torque check Insert the torque cassette. Set the camcorder to the Play mode and confirm that the reel disk (T) torque value is 10.5 ± 3.5 g/cm. Set the RPS mode and confirm that reel disk S torque value is within 35 ± 5 g/cm. If the torque value does not meet this specification, replace corresponding reel disk.

Back-tension adjustment The back tension is performed in Play mode to take up tape slack and slippage with correct tension. Some manufacturers use a back tension meter (tentelometer), while others use a video cassette torque meter, which includes back tension and take-up in one cassette. This adjustment can be made with a regular video cassette, small screwdriver, and tension meter.

8-mm forward back-tension adjustment Load the camcorder with the torque cassette. Refer to the replacement parts list for the torque cassette stock number. Place the instrument in the Play mode and confirm that the S-reel base is 6-12 gF/cm. If the torque reading is not within the specification, adjust the position of the B+ lever spring. If the torque is less than 6 gF/cm, relocate the BT lever in the direction of the arrow. If the torque is higher than 12 gF/CM, relocate the BT lever spring in the direction of the black arrow.

8-mm playback tension adjustment Supply an external 7.5-V source to the battery terminals. Insert torque cassette. Set in play mode, and confirm that reel disk S torque value is within 4.5 ± 1 g/cm. Readjust the tension band if the torque value does not meet these specifications.

Conclusion

Most mechanical adjustments are only required after removing, installing, or adjusting. Sometimes when the camcorder is dropped, dented, or bent, mechanical adjustments might be needed. Usually, all three tape formats (VHS, VHS-C, and 8 mm) are given within the same type of adjustment.

Many camcorder manufacturers have their own fixtures, test tapes, and jigs for their own VTR adjustments. You might find that a few special tools are needed for the camcorder you are now repairing. Securing the exact service literature for each camcorder helps to make easier and quick mechanical adjustments.

The tape-transport system from the supply reel to the take-up reel disk across the video heads is the most crucial section in the VTR (Fig. 38-24). The tape-transport compo-

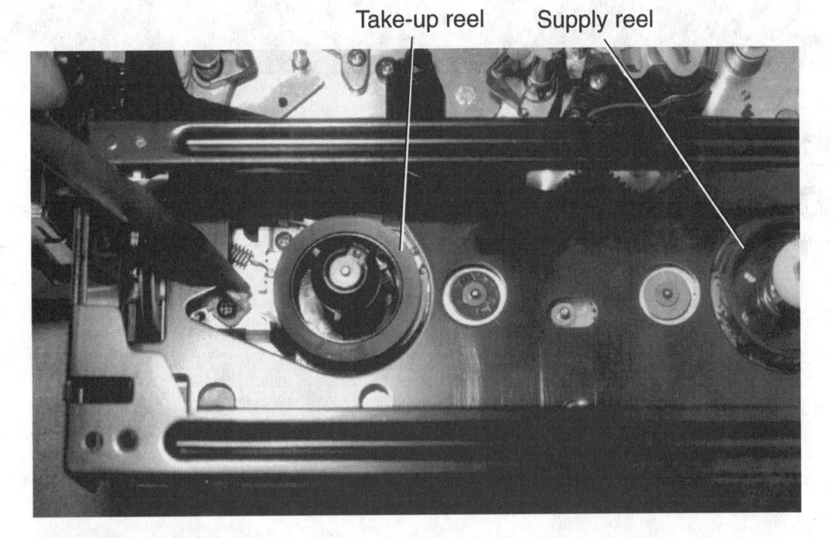

FIGURE 38-24 **The take-up and supply reel tape path in an RCA VHS VCR.**

nents—especially those that contact the tape—should be kept clean. Remove scratches, dust, and oil from these surfaces.

Most tape-transport systems are adjusted before they leave the factory. If the parts are not tampered with or are broken, no adjustments are necessary. Always make the correct mechanical adjustments after replacing a new component to stabilize the transport system. Be sure that exact factory replacements are used.

39

ELECTRICAL ADJUSTMENTS

CONTENTS AT A GLANCE

Electrical adjustments are necessary after you replace crucial components in the camera section. Electronic view-finder (EVF) adjustments might be needed after the viewfinder has been treated roughly or if the camera was accidentally dropped. Sometimes only a touch-up is needed. All electrical adjustments should be followed according to the manufacturer's specifications (Fig. 39-1). Basic adjustment charts are often located in the back of the service manuals.

FIGURE 39-1 Remove the top and bottom covers to access the voltage and camcorder adjustments.

List of Maintenance Tools and Test Equipment

Basically, seven or eight (minimum) test instruments are required for electrical adjustments. Other special tools and servicing jigs might be required by each manufacturer. Many of these test instruments are probably already on your electronic service bench. Basic instruments include:

- Oscilloscope
- Color TV monitor
- Signal generator
- Frequency counter
- Audio tester
- Regulated power supply
- Digital voltmeter (DVM)
- Blank video cassette for recording and playback
- Alignment tape
- Patch cords
- Camera jigs
- Electronic service tools

The 8-mm camcorder maintenance tools and correct test equipment are listed in Table 39-1.

TABLE 39-1 TYPICAL 8-MM CAMCORDER MAINTENANCE AND ADJUSTMENT TOOLS

DESCRIPTION	TOOL NO.	REMARKS
Alignment tape A (color bar)	DY9-1044-001	
Recording current adjusting jig	DY9-1056-000	
Y/C separator	DY9-1093-500	
V sweep master	DY9-1108-100	
Extension cable (12 pins)	DY9-1268-000	
Alignment tape (stereo)	DY9-1291-000	
Extension cable (20 pins)	DY9-1297-000	New
Extension cable (3 pins)	DY9-1298-000	New
Color bar chart	DY9-2002-000	
Gray scale chart	DY9-2005-000	
Color chart viewer (5600 K)	DY9-2039-100 100	
Viewer lamp (5600 K)	DY9-2040-000	
CCA12 filter	DY9-2046-000	

Additional test instruments are:

- Vectorscope
- Light meter
- Tripod
- Light box
- Color viewer—3200 k
- Color bar chart
- Gray-scale chart
- Reflection charts and patterns
- Special screwdrivers
- Camera accessories
- Color temperature-conversion filter
- Remote-control card
- Special camera extension jigs and harnesses

The oscilloscope should be a dual-trace unit with delayed sweep and a minimum bandwidth of 25 MHz. Most electronic technicians servicing the TV chassis already have a dual-trace scope on their work benches. The frequency counter should have a frequency of 20 MHz or greater. Most TV establishments have a digital voltmeter DVM or DMM for crucial voltage adjustments. If not, choose one that can measure up to 1500 Vdc because pickup tubes in the older camcorders have a high sweep voltage.

The vector scope is used for TV repairs and adjustments, such as burst, black balance, white balance, auto white balance, and chroma-balance adjustments. The vector scope can show the color level to be low or off and not show up on the color monitor or charts.

The light meter is to ensure proper light of the camera test patterns while the light box is used instead of the test patterns. The wall charts are inexpensive compared to light boxes. Some manufacturers use both the charts and light boxes in camera adjustments. Secure the wall charts or light box recommended by the camcorder manufacturer on the camcorders that you are servicing or selling. Some camcorders' service manuals contain a back focus and gray-and-white scale chart.

Reflection of Wall Charts

Many wall charts and test patterns are used in different camera tests and alignment. The gray-scale chart, color-bar and back-focus chart are recommended by most. The gray-scale chart is used for troubleshooting, testing, and alignment. The gray-scale chart is used when waveforms are presented by the service manual if by chart or a gray chart on the light box. Besides the gray-scale chart, the several others include the color-bar chart, resolution chart, ball chart, white chart, black-and-white chart, gray-and-white chart, and the back-focus chart. Because the black-and-white chart is reflective, a halogen lamp should be used with it.

The back-focus adjustment using the back-focus chart is to match the distance to the object with the index of the distance on the focus ring. A poor back-focus adjustment can occur when the distance to the object and the index of the distance on the focus ring do not match. The back-focus adjustment chart can be one with lines coming together at the end or a black solid line down the center (Fig. 39-2).

Tools and Fixtures

Most small tools on the TV service bench are all that is required to make electrical adjustments. A blank video cassette is used for recording and playback operations. The cassette alignment tape is used for color, audio, and servo alignment (Table 39-2).

Patch cord, extension cables, and test-jig harnesses can be used in auto and back-focus adjustments, color adjustments, camera dummy setup, iris extension cable, and operation fixtures. Some manufacturers do not have any jig harnesses or external circuit boards for making electrical adjustments. Several camcorders use light balancing and ND and CCD filters over the lens in making electrical adjustments.

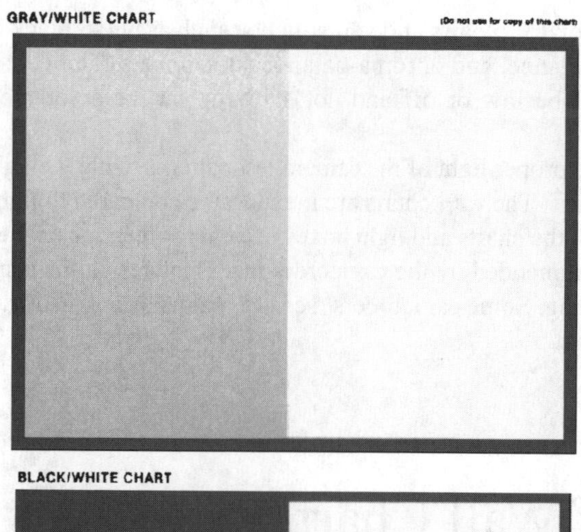

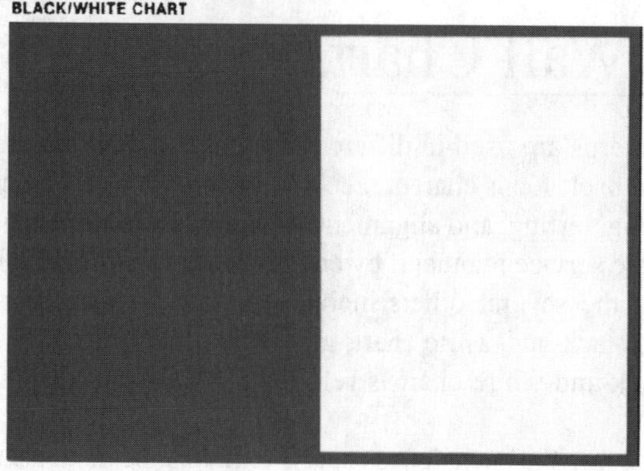

FIGURE 39-2 Black-and-white focus charts for the camcorder back-focus set up.

TABLE 39-2 TEST CASSETTE ALIGNMENT TAPES

MANUFACTURER	TYPE	ALIGNMENT	PART NUMBER
Canon	8 mm	Color bar	DY-1044-001
Minolta	VHS-C	M+C1	7892-8012-01
Mitsubishi	VHS-C	MH–C1	859C35909
Olympus	8 mm	Color bar Monoscope	VFM9010P9D VFM9000P9N
RCA	VHS-C	3 kHz	156502
Pentax	8 mm	NTSC	20HSC-2
NEC	VHS	Alignment	79V40302
Samsung	8 mm	Lion pattern	
Zenith	VHS-C	Color EP mode	MY-C1 CY-CIL

Typical 8-mm Camera Set-Up Procedure

Disconnect the power from the camcorder. Remove the camcorder from the cabinet and separate the camera block assembly from the VCR block assembly. Attach the camera block assembly to the camera fixing tool.

Connect the camera circuit board in the camera section to the main circuit board in the VCR section using extension cable J10. Connect the power circuit board in the VCR section to the camera circuit board in the camera section with extension cable J11. Also, attach the circuit board connecting tool (J9) to J9001 on the camera circuit board using the clamp. Connect the adjustment controller to the circuit board connecting tool, as in drawing.

Apply a regulated dc power source (7.5 to 8.0 Vdc) to power the adjustment controller (black wire to positive). Turn on the dc power supply after powering up the camcorder. When all adjustments have been completed, turn off the adjustment controller first and then turn off the camcorder. Connect the color monitor to the video output.

Camera Setup

For camera adjustments, the camcorder should be mounted on a solid work bench or a heavy-duty tripod for accurate measurements. In the typical VHS-C camcorder, the unit is operated from the ac power supply with the audio/video (A/V) cable between the camcorder and the color video monitor. The video-output cable must be matched with the RCA-BNC adapter to plug into the color monitor.

To make electrical adjustments in the Pentax PV-C850A (8 mm), a light box is used in front of the lens assembly with an RF converter as distribution box. A monitor TV without audio/video jacks can be plugged into the RF converter, or the monitor TV with A/V jacks can also plug into the converter unit. Notice the color-bar generator provides a signal to the RF converter.

TYPICAL 8-MM VAP/AF ADJUSTMENT

Before attempting adjustment, warm up the instruments for at least three minutes. The standard view is obtained when the test chart is displayed over the entire screen of the full-scan monitor. Connect the test equipment as shown in Fig. 39-3. When checking it on an oscilloscope, adjust the angle of view so that the gray-scale section will be 36 μs or the color bar section will be 52 μs. When using the other chart, align the center at the standard angle of view that has been set with the gray-scale or color-bar chart. Shoot the chart from a distance of approximately 1.4 meters, unless otherwise specified.

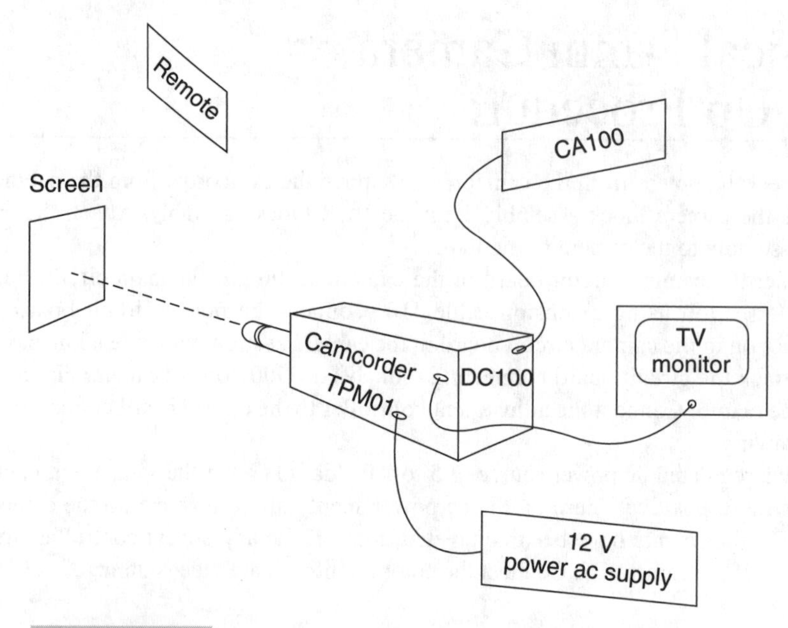

FIGURE 39-3 **A block diagram of the various camcorder adjustment components.**

Camcorder Breakdown

Often, both plastic sides of the camcorder must be removed for electrical adjustments (Fig. 39-4). Some adjustment boards are located on one side while others are on the other side and behind the camera lens assembly. Here, boards 1 through 7 are located on the right side of RCA CPR100 camcorder, and the mike and interface circuit boards at the top of the lens assembly. The process circuit board (10) is at the front and underneath the lens assembly.

Besides containing crucial circuits of the camcorder, the service manuals include test points and the control board layout. Most camcorder service manuals list the various test points and parts for easy camera adjustments. Like on any electronic product, alignment procedures are difficult at first, but after repairing several camcorders, each one becomes easier.

POWER-SUPPLY ADJUSTMENTS

Before attempting to make electrical adjustments, the output of the power-supply voltage must be set correctly (Fig. 39-5). Incomplete adjustments can cause the signal-processing system to operate normally. Most dc power source adjustments are made with a digital voltmeter, blank tape, and dc power supply. These adjustments take only a few minutes, but they are very crucial. Be sure that the lens cap is on.

RCA 8-mm ac adapter adjustment Connect the DVM or DMM at CN521 terminals 1 and 3. Connect the positive lead of the DVM to pin 3 and the ground to pin 1 (Fig.

39-6). Do not connect anything to the ac adapter charger output terminals. Now adjust VR251 so that the meter reads 6.7 Vdc ± 0.1 Vdc.

Canon 8-mm 12-V power source The preparation of the 12-V power supply must be modified. Remove the external housing of the CA-100 power supply. Short-circuit across the negative terminal (–) of the dc jack (JK-1) and control terminal C. Disconnect the battery terminal. Short-circuit across the negative terminal (–) of the battery terminal

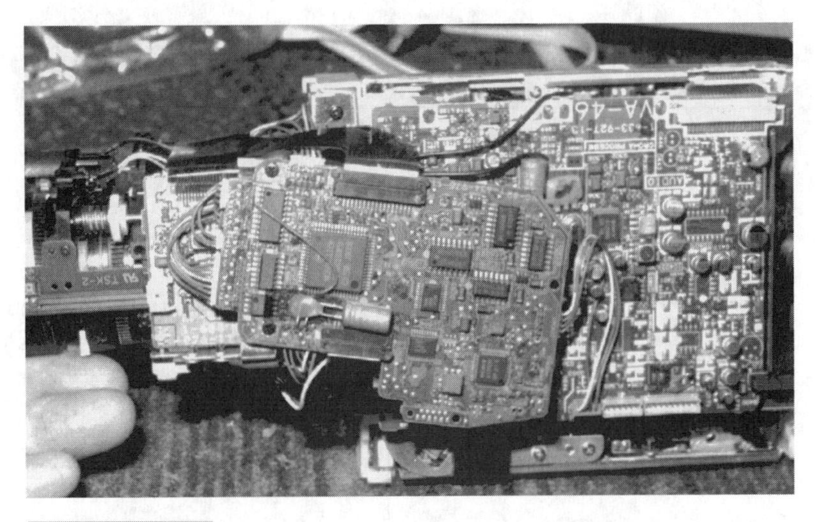

FIGURE 39-4 **Remove both sides to align and adjust the camcorder.**

FIGURE 39-5 **An inside view of the RCA CPR300 power supply.**

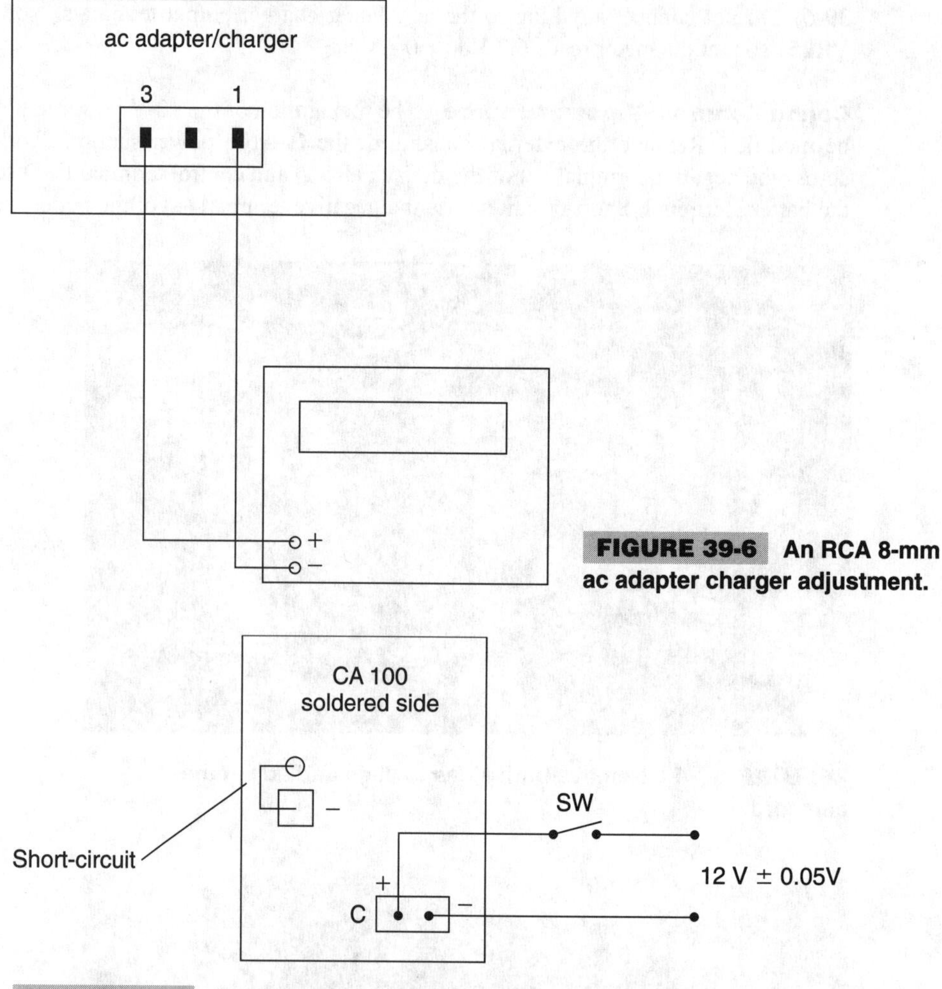

FIGURE 39-6 An RCA 8-mm ac adapter charger adjustment.

FIGURE 39-7 The Canon 8-mm 12-V power source.

land port and the control. Connect a lead wire with the positive terminal (+) of dc jack JK-1, and attach a clip to it with an intermediate switch (Fig. 39-7)

Servo Adjustments

After adjusting the power supply, check the system control, servo system, video, and audio adjustments (in that order). The system-control adjustments can consist of the battery down and tape-end adjustments; the servo-system adjustments can contain drum pulse, capstan sampling, PB switching, SP control delay MMV, EP tracking, and tracking preset.

TYPICAL 8-MM ADJUSTMENT SEQUENCE

The sequence of adjustments should be made in order of the servo/system control circuits, video circuit, and audio circuits within the servo/system circuit to make the OSD character position and the video center adjustments. Next, make the video-circuit adjustments of the playback frequency characteristic confirmation, flying erase-head confirmation, sync AGC adjustment, comb-filter phase adjustment, comb-filter level adjustment, PB line-out Y-level adjustment, PB Y-level confirmation, Y FM carrier frequency adjustment, Y FM deviation adjust, luminance (Y) record current adjustment, chroma (C) record current level adjust, audio (AFM) FM record level confirmation, and ATF record current confirmation (Fig. 39-8).

For the audio circuits, make the deviation confirmation, overall level confirmation, and overall signal-to-noise confirmation.

When performing the VCR section-adjustment procedures, the video signal generator (pattern generator) is used as adjustment signal. It is important that the generator used produces a signal that meets specifications.

Before performing the VCR-adjustment procedures, disassemble the camcorder. Convert the J9 adjustment CBA to J1005 on the (CP1) circuit board. When an external video signal is required, convert the video signal generator.

RCA 8-mm flying-erase-head adjustment Connect the VCR section and test equipment as described in the VCR-adjustment set-up procedure. Short pins 6, 11, and 14 of J1005 together. Place the power switch in the CAM position and turn the standby switch on. Connect the scope to TP181 of CP1CBA and TP143 (ground). Confirm that the frequency is 6.5 to 9.5 MHz and that the oscillation voltage is 7.5 V p-p to 10.5 V p-p (Fig. 39-9). If the required voltage cannot be met, clean the flying erase head. Replace upper cylinder if the voltage cannot be met after cleaning the flying erase head.

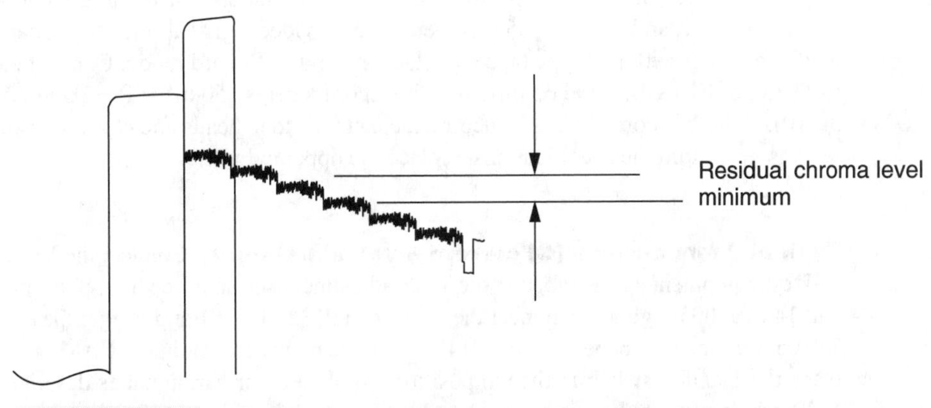

FIGURE 39-8 The typical 8-mm comb filter phase adjustment.

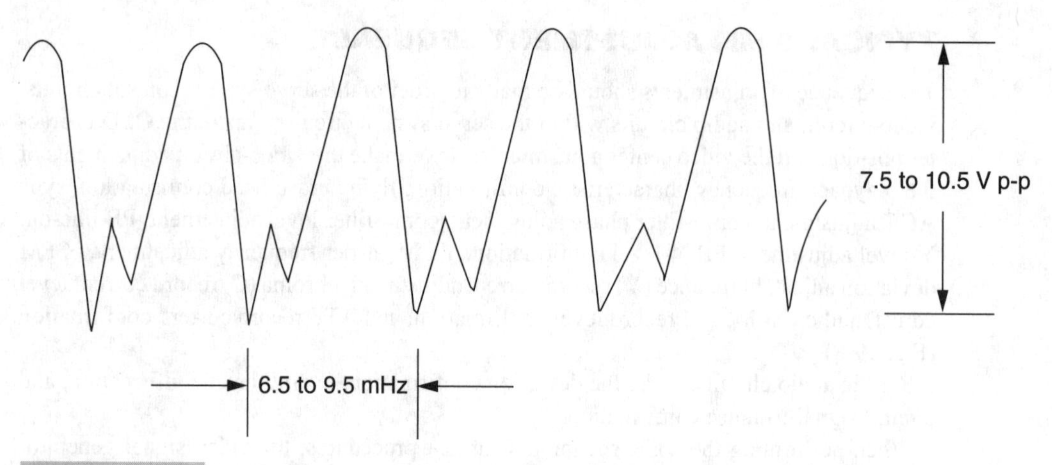

FIGURE 39-9 **An RCA 8-mm flying erase head confirmation.**

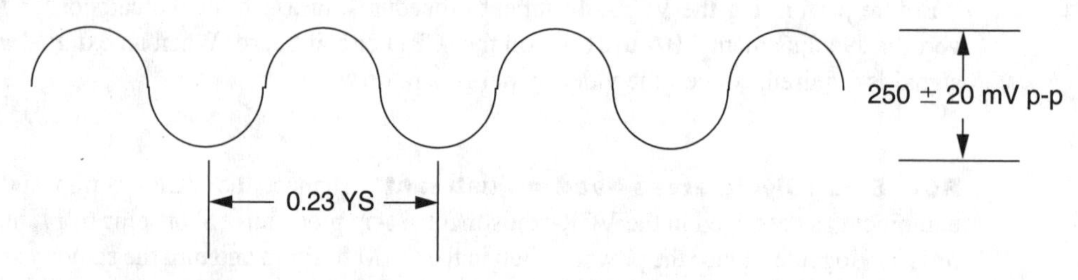

FIGURE 39-10 **A typical 8-mm luminance (Y) record level confirmation.**

Typical 8-mm luminance (Y) record current level adjust Connect the VCR section and test equipment, as described in the VCR adjustment set-up procedure. Place the power switch in the CAM position and place the Standby switch in the On position. Short pins 6, 11, and 14 of J1005 together. With no video signal applied to the camcorder, load camcorder with MP-type tape and place the unit in Record mode. Connect the scope to TP131 of CP1 CBA, and confirm that the record level is 250 mVP-P ± 2U mVP-P (Fig. 39-10). If the Y record current is incorrect, clean the video heads and check it again. If the level is not within the specifications, replace the upper cylinder.

Typical 8-mm chroma (C) current level adjustment Connect the VCR section and test equipment as described in the VCR adjustment set-up procedure. Short pins 6, 11, and 14 of J1005 together. Connect the scope to TP132 of CH 1 and trigger the external to TP105. Ground the scope lead to TP143. Place the power switch in the CAM position and place the Standby switch in the On position. Apply a color-bar signal as described in the VCR adjustment set-up procedure. Load the instrument with an MP-type tape and place the camcorder in Record mode. Adjust VR132 so that the chroma level is 200 V p-p, ±15 V p-p (Fig. 39-11).

CCD Drive Section

Four or five important electrical adjustments are to be made in the CCD drive section, including focus, drive pulse frequency, PLL frequency, V-Sub or OFD, and CCD output adjustments. These adjustments ensure proper focus, set proper drive-pulse frequency, prevent blooming, and provide sufficient CCD or MOS outputs. The lens cap is removed with these tests. Be sure and make the FO/Q, VCO-PLL, AFC, and playback (PB) level adjustment after replacing heads or preamplifier.

BACK-FOCUS ADJUSTMENT

The purpose of the back-focus adjustment is to ensure proper focus tracking throughout the zoom range or to match the distance to the object with the index of the distance on the focus ring. An improper adjustment means that the distance to the object and the index of distance on the focus ring do not match. Many camcorder back-focus adjustments are made with a back-focus chart or light box with test patterns.

General Electric 9-9605 (VHS) focus adjustment Aim the camera at the registration chart with the lens cap off. Look in the viewfinder and adjust VR904 for the best resolution.

Samsung 8-mm iris-level adjustment Set camera E-E, 3100 degrees K on the gray-scale chart. Connect video-output jack waveform-monitor input jack and monitor TV jack, respectively. Press the Play (mode up) Stop (mode down) button so that the OSD state is 02 iris XX XX. Dim the camera at the gray-scale chart evenly illuminated at 1500 to 200 lux (40 US). Adjust the self timer data up int. REC (data down) button so that Y level is 90 IRE +10/–5 IRE. Be sure to set start/stop (confirm) button to memorize setting. The OSD should show OK.

Olympus VX-801U PLL (VCO) frequency adjust Connect the DVM to TP8003. Connect the VF-BA81 to the VX-801 camcorder and set the unit to the VTR Stop mode.

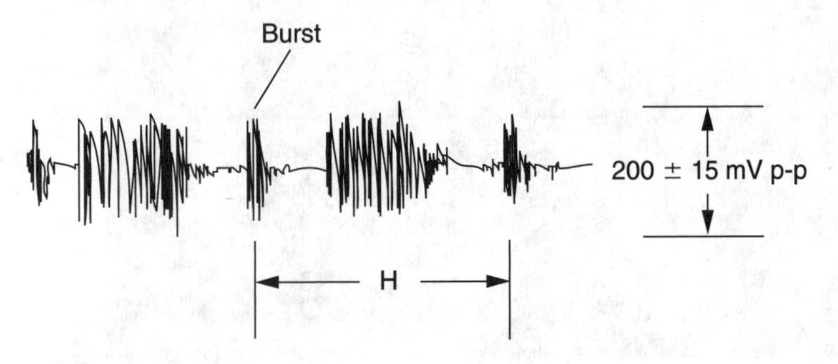

FIGURE 39-11 A typical 8-mm current level adjustment.

Input the color-bar signal through the VF-BA81. Adjust VR8001 to read 2.4 ± 0.1 V on the DVM.

CAMERA ADJUSTMENTS

The process-section adjustments consist of AGC, brightness or luminance, white-balance, auto white-balance, color, and audio-gain adjustments. Within these sections are several other adjustments. The real purpose is to set the brightness of the picture if the picture becomes extremely bright or dim. Set the white balance of the picture so that the color reproduction does not deteriorate. Set the amplitude level of the burst signal so that the color does not become light or dense. Suppress the level differences between the R, G, and B signals in standard illuminations because of color reproduction of poor white in the picture. Set the chroma level so that the color reproduction does not deteriorate. Be sure and make the brightness (Y) adjustment before adjusting the recording system (Fig. 39-12).

Canon 8-mm REC AGC adjustment Set the camcorder to the camera EE mode. The input signal is a white 100% video signal. Connect scope at M.EQ, with a trigger scope external connection to FP2105. Adjust service mode, page 4 - bank 1 - address 01. The specifications should be 500 + or – 10 mV p-p (Fig. 39-13).

Samsung 8-mm auto white balance adjustment Set camera to E-E, 3100-K/5100-K gray-scale chart. Press the play (make up/stop mode-down) button so that the OSD state is 55 AWB XX XX. Connect the vectorscope input jack to the video output jack. For white-balancing indoors, arm camera at a 3100-degree K gray-scale chart, illu-

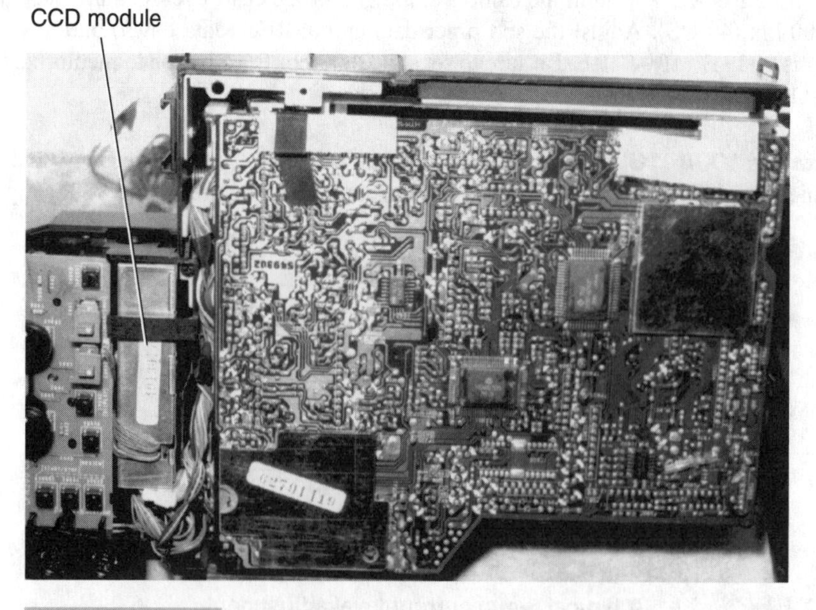

FIGURE 39-12 The CCD module in a VHS camcorder.

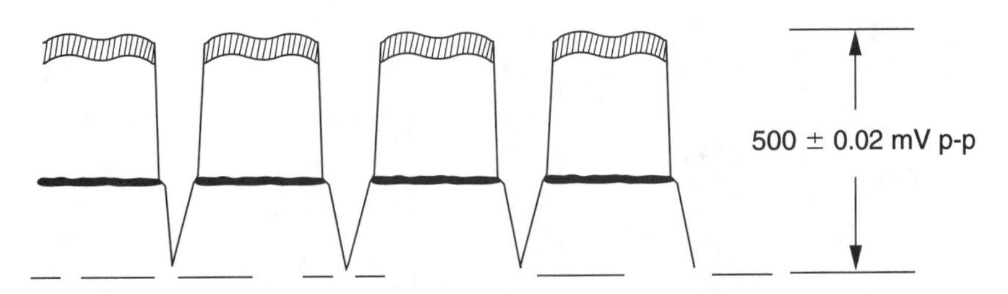

FIGURE 39-13 The Canon 8-mm record AGC adjustment.

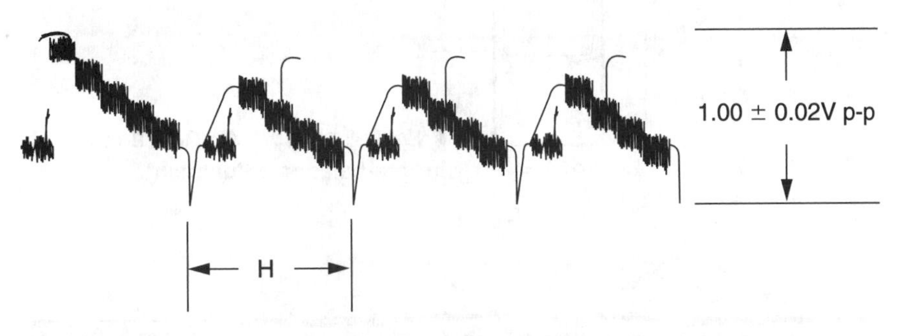

FIGURE 39-14 The 8-mm line-out Y-level adjustment.

minated at 1500 to 2000 lux (40 μs). Press the Start/Stop Confirm button so that the white vector moves to the center of the vectorscope's screen.

To white-balance outdoors, arm the camera at a 5100 OK gray-scale chart, illuminated at 1500 to 2000 lux (40 μs). Press the Start/Stop Confirm button so that the white vector moves to the center of the vectorscope's screen. The OSD should then show OK.

8-mm line out Y level adjust Connect the VCR section and test equipment, as described in the VCR-adjustment set-up procedure. Connect A/V cable and terminate the video output terminal in 75 Ω. Place the power switch in the VCR position. Load the camcorder with the alignment tape and play back the color-bar portion of the tape. Connect the scope to TP141 and external trigger at TP105. Adjust VR108 so that the signal level is 1.0 V p-p ± 0.02 V p-p (Fig. 39-14).

RCA 8-mm chroma balance adjustment Connect the camcorder and equipment as described in the adjustment set-up procedure. Place a jumper between J9101-8 and J9001-15. Attach the C12 color-conversion filter to the camcorder lens. Connect the vectorscope to TP141 and terminate in 15 Ω. Use TP143 as ground (Fig. 39-15).

Set the camcorder to the E-E mode and the focus mode to Manual Focus. Adjust the gain and phase of the vectorscope so that the burst is set to the 75% mark on vectorscope screen. Using the CH up or the CH down buttons, set the channel indication on the adjustment controller to Channel 6. Using the D. CENT, the Data Up and the Data Down button, to focus the luminance points of the color vectors. Pay particular attention to the red vector. Press the Write button to store the data.

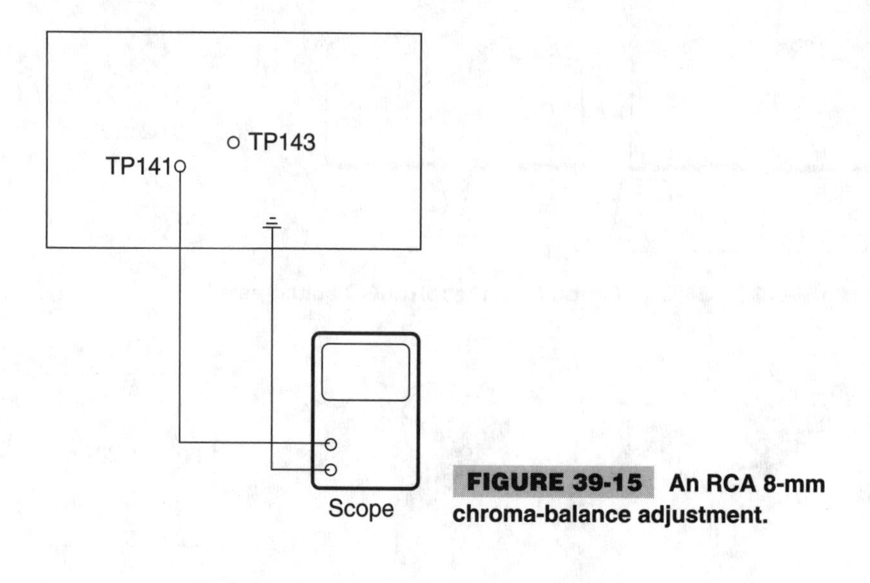

FIGURE 39-15 An RCA 8-mm chroma-balance adjustment.

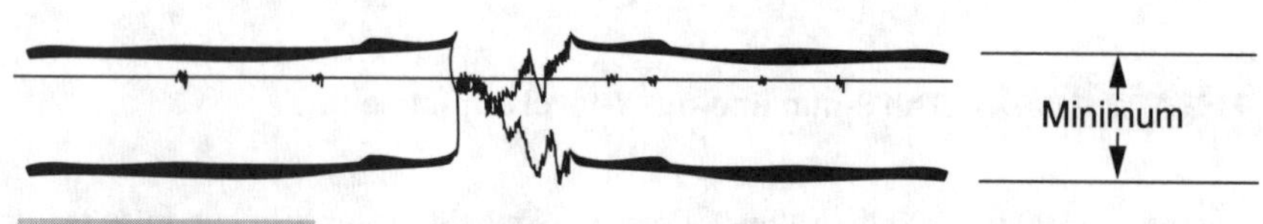

FIGURE 39-16 The Canon 8-mm record C-level adjustment.

Canon 8-mm record C level adjust Connect the red raster signal. Set the camcorder in the camera E-E mode. Connect scope at M. EQ with trigger external terminal of scope at FP2009 test point. Adjust service mode 4 - bank 1 - address OB (Fig. 39-16). Minimize the chroma component to specifications.

Typical (VHS-C) record chroma level adjust This adjustment regulates the recorded color level. Sometimes, if the color level is too high, diamond-shaped beats appear in the picture. The color can be degraded if it is too low. Apply the NTSC color-bar signal (1 / Vp-p) to the EVF jack with the audio/video adapter. Connect the scope probe to TP 201 on the Head Switch/Function Switch board (Fig. 39-17). Check the upper cylinder for an identification number (0 through 4) stamped on top. Rotate the Record Luma Level control (RT 201) fully counterclockwise. Load with blank tape and place in the SP Record mode. Adjust the Record Chroma Level control (RT 202) for correct value.

RCA 8-mm FM audio-level adjust Connect the VCR section and test equipment as described in the VCR adjustment. Place the power switch in the VCR position. Connect the scope to TP201 on the CP1 CBA board. Clip a 22-kΩ resistor from TP201 to ground. Load the camcorder with test tape and play back the FM audio signal. Adjust VR201 so that the amplitude of the lower frequencies and the amplitude of the higher frequencies are equal (Fig. 39-18).

Canon 8-mm Y-recording current adjust No signal at is at the input terminals. Set the camcorder in the camera E-E mode. The scope is connected to the M. EQ and the trigger scope connected to FP2007. Adjust the service mode 4 bank – 1 address OC with specification 210 ± 10 mV p-p.

ELECTRONIC VIEWFINDER (EVF) ADJUSTMENTS

Most of the EVF adjustments are made with the gray and resolution scale for best results. The deflection coil position adjustment is to provide correct tilt of the viewfinder. The centering adjustment is to match the center of the TV and EVF pictures. The vertical-size adjustment is to set the vertical deflection size. The horizontal-hold adjustment sets the

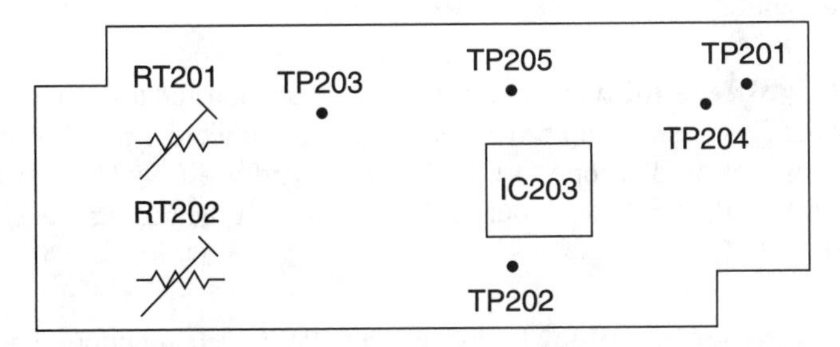

FIGURE 39-17 A typical VHS-C record chroma-level adjustment.

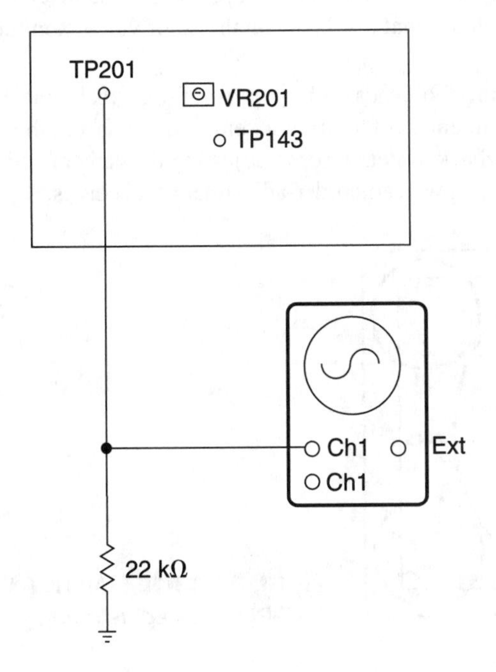

FIGURE 39-18 An RCA 8-mm FM audio level adjustment.

horizontal frequency. Brightness adjustment optimizes the brightness of the picture, and the focus adjustment makes the EVF picture clear and distinct.

RCA 8-mm 848 EVR PLL lock adjust Observe the electronic viewfinder. Connect the VCR section and the test equipment as described in the VCR-adjustment setup procedure and apply the monoscope signal to the camcorder video-output jack. Place the camcorder power switch in the VCR position. Adjust the PLL lock control (VR 985) so that the monoscope pattern is centered on the horizontal scale (Fig. 39-19).

RCA 8-mm EVF contrast adjust Observe the EVF/video monitor. Place the power switch in the CAM position and arm the camcorder at the gray-scale chart. Connect the camcorder video output to a video monitor. Be sure that the camcorder is focused on the chart. Adjust VR980 so that the contrast level displayed on the EVF screen matches the display on the video monitor.

RCA 8-mm EVF brightness adjust Connect the VCR section and test equipment as described in the VCR adjustment set-up procedure. Place the camcorder power switch in the VCR position. Connect oscilloscope to CN973 pin 16 and use CN973 pin 5 as the ground connection. Adjust the brightness control (VR977) so that the pedestal level is 1.8 V p-p ± 0.1 V p-p (Fig. 39-20).

RCA 8-mm EVF color-gain adjust Observe the EVF/video monitor. Place the power switch to the VCR position. Connect the VCR section and test equipment as described in the VCR adjustment set-up procedure. Apply a color-bar video signal to the camcorder video input jack and to the color video monitor. Adjust VR973 so that the color level displayed on the FVF screen matches that displayed on the color video monitor.

Adjustment notes After replacing the heads and the preamplifier, make the FO/Q, VCO-PLL, AFC, and PB level adjustments. If the drum assembly is replaced, the FO/Q adjustment is needed. Adjust the playback system prior to adjusting the camera and deviation. Each manufacturer can have their own camcorder-adjustment procedures.

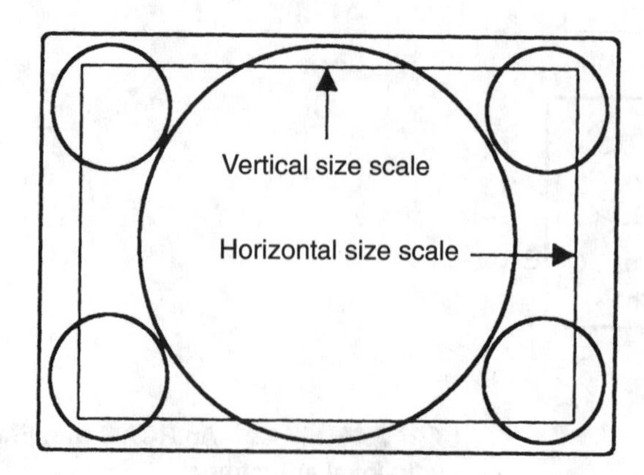

FIGURE 39-19 An RCA 8-mm EVF-PLL lock adjustment.

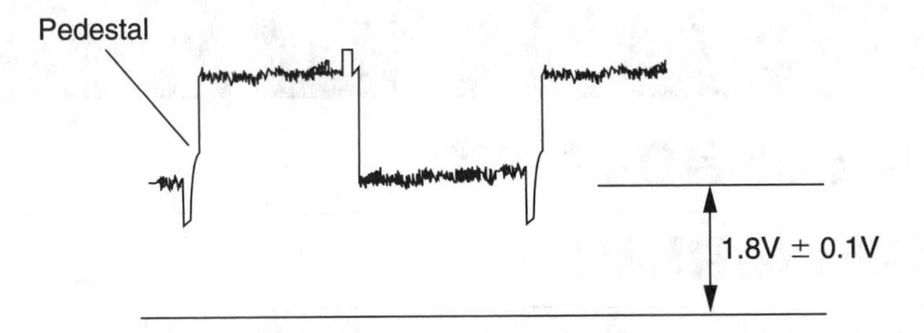

FIGURE 39-20 An RCA 8-mm EVF brightness adjustment.

Use a halogen lamp because most black-and-white charts are the reflective type. Adjust the chart surface temperature to approximately 1500 lux. The chart surface should be lit evenly. When the luminance cannot be reduced down to a specified value, use an ND filter. Wall-chart adjustments are the cheapest and quickest way to go, except that some manufacturers' adjustments are made with the light box.

40

TROUBLESHOOTING AND REPAIRING THE CAMCORDER

CONTENTS AT A GLANCE

Troubleshooting the camcorder might not be as difficult as it seems. Try to match the most likely section on the block diagram with the defective symptom. Then, find the defective section on the regular schematic (Fig. 40-1). Most manufacturers have the schematics broken down into the various sections because one large schematic would be impossible to print.

For instance, if the trouble symptom is no color and the camera section is normal in black-and-white operation, check the chroma circuits. Likewise, if the color picture is normal with no sound, you would check the audio circuits and microphone. If more than one section is defective, check the power source. Although camcorder-servicing methods on the different circuits might be the same as the one that you are working on, they are serviced in the same manner.

Take crucial voltages to locate the defective part. Scope the various circuits to locate missing or improper waveforms. After locating the suspected component, take crucial continuity and resistance measurements. Remember that many of the problems in any electronic product are caused by simple part breakdowns and mechanical operations.

Certain stages and circuits can be tested by simply taking an alignment test. If the circuit or component does not respond, suspect that a stage is defective. For instance, check the flying erase head by making the erase-head adjustment. If voltage is not present, clean the flying erase head. If it is still low in voltage, replace the upper cylinder or drum assembly.

FIGURE 40-1 **Choose small screwdrivers and tools to remove the various screws in the camcorder case.**

Before Troubleshooting

Before troubleshooting the camcorder, obtain the correct schematic, isolate the possible trouble, and list the possible defective components. You can isolate service symptoms with a block diagram and schematic. Be sure that the battery voltage is normal. Connect an external power supply with all switches in the Off position. Dismantle the camcorder to access the defective components.

Voltage Measurements

Crucial voltage measurements with correct waveforms can solve crucial service problems in the camcorder circuits. Only a few camcorder manufacturers print the actual voltage measurements on the schematic. Separate voltage charts are provided in the back of the service manual. Several manufacturers break down the voltage charts with the camcorder in the various modes (Stop, Record, Play, Rewind, Fast Forward, Review Search, and Forward Search).

When the loading motor is not operating properly, check the voltages on the loading-motor (1 and 2) terminals. If the motor does not unload or load, check voltages at pins 14 and 16 of IC391 (Fig. 40-2). Measure the dc voltage at supply pin 15 of IC391. If the voltage is low or missing, check the low voltage power source or dc to dc converter. Remove pin 15 if the voltage is low and determine if the loading motor-drive IC is leaky. Check the load and unload signal at pins 39 and 40 of IC351.

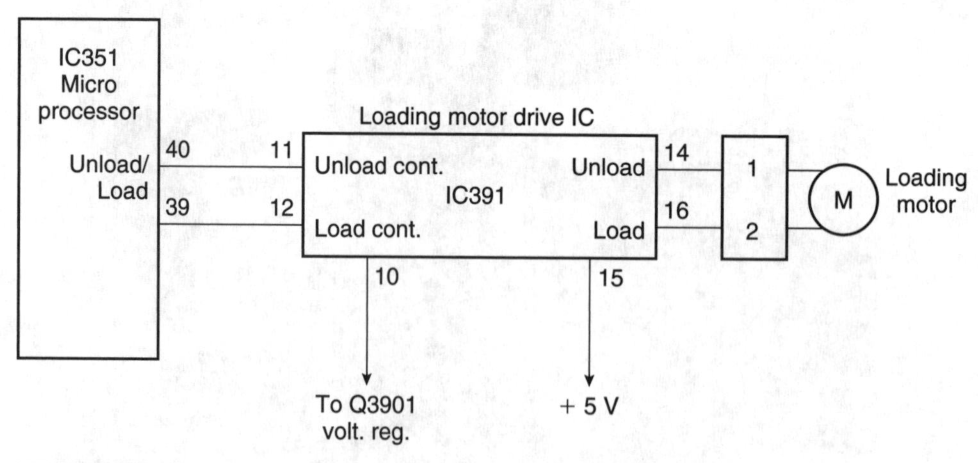

FIGURE 40-2 Check the voltage applied to loading-motor pins 14 and 16 of IC391.

Scope Waveforms

Many crucial waveforms can be taken in the camcorder circuits. Crucial waveforms can determine if the circuit is working or not. Waveforms taken throughout the camcorder signal paths can help you to locate the defective section or stage. Often, signal tracing with the scope and crucial voltage measurements can help you to locate the defective component. Crucial waveforms of the process, servo, sensor, luma/chroma, and the electronic viewfinder (EVF) circuits are listed in the manufacturer's literature.

For instance, to determine if the encoder circuits are defective in the RCA 8-mm camcorder, take crucial waveforms at the color-in (18) and color-out (16) pins. The BFG waveform is taken at pin 36 of IC912 and the sync waveform off of pin 25. If the input waveforms are normal and the output waveforms are missing, suspect that IC912 is defective or that the dc supply voltage is improper. Take a crucial voltage test on the encoder supply voltage, DV_{CC}, pin 3 (4.9 V).

Common Failures

A dead camcorder can result from a blown fuse, poor battery voltage, or a bad ac power supply. Check the battery terminals for poor connections. Test the camcorder for both battery and ac operation to determine if the power source or the camcorder are defective. Sometimes the camera-detection circuits shut the camcorder down if the battery is too low, although it's possible that the tape movement could continue to operate. Monitor the battery voltage with the camcorder operating to determine if the battery breaks down under load. In some models, the cassette must be loaded before any power comes on.

The fuse might be blown or intermittent. Intermittent operation can be caused by a loose fuse in the connector. Sometimes the fuse's metal ends have arced and cause a poor contact, resulting in intermittent operation. When checking the fuse in place, measure the voltage from metal clip to clip to determine if the fuse is open or has poor contact.

NO PICTURE

This can be caused with the lens cap on, showing nothing in the viewfinder. Check the white switch setting and brightness control. Recheck the manual iris setting. Determine if the tape is moving. Connect the camcorder to the TV or monitor to determine if the viewfinder is defective.

No picture in viewfinder This problem can be caused by a defective or improper setting of the Camera/Play switch. Be sure that the switch is in the Play mode. Make sure the

EVF cable is plugged in tight. Check the EVF connection for bent or misaligned prongs. If no picture is in the viewfinder, connect the camcorder to the TV or monitor to be sure that the EVF is not operating.

No picture or sound Be sure that the Camera/Play switch is in the Play mode. Check the power-distribution source for a common defective power source.

NO AUDIO

This problem can result from a defective microphone or cables. Most of these microphones plug directly into the EVF assembly or camera. Clean the mike controls. Be sure that the ring tightener is snug. Try another external microphone. Next, check the VCR sound circuits.

NO OR IMPROPER COLOR

This problem can be caused by poor or insufficient lighting. Check the camera with proper white set-up adjustments. Sometimes bright sun shining through a window causes poor color. Check the color-temperature setting.

FUZZY OR OUT OF FOCUS PICTURE

This can result from improper focus of the zoom lens. When the zoom motor is operating, the picture will naturally be out of focus for a few seconds. Sometimes changing quickly from a scene far away to a close-up shot will cause the picture to be out of focus. Poor lighting can result in fuzzy pictures.

Damaged Parts

Parts damage is usually caused by the operator dropping the camcorder or by general rough treatment. If the camcorder uses a pickup tube as sensor, the glass tube might be broken. You might hear the rattling of glass pieces, indicating pickup tube damage.

CRACKED OR BROKEN BOARDS

Cracked boards can produce dead or intermittent camcorder operation (Fig. 40-3). Be sure that all board sockets are firmly pushed down. Look near heavy objects mounted on the PC board for the cracked areas. Sometimes the PC wiring can be repaired. Try to splice the cracked area. Do not run long wire leads. Be careful around multi-lead microprocessors and ICs because the leads are close together and might be difficult to repair. Replace cracked boards.

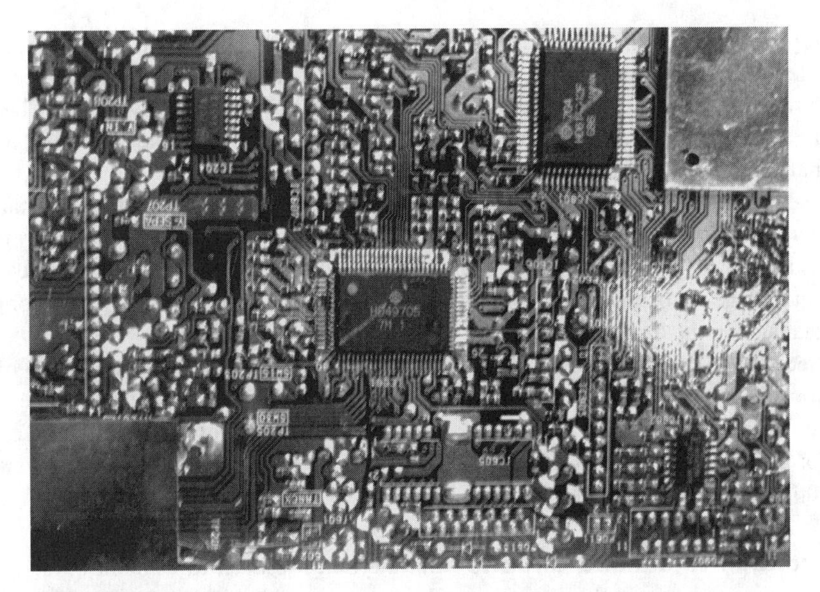

FIGURE 40-3 **Locate the correct surface-mounted (SMD) IC on the PC board for waveform and voltage measurements.**

POORLY SOLDERED JOINTS

Poor soldering can cause intermittent or dead operations. Push down and pry up on a board at different areas to check badly soldered connections. Sometimes poorly soldered connection can be repaired by soldering all of the joints in that area. Be very careful not to apply too much solder so that it doesn't run into another circuit.

You might find a lot of poor joints or broken boards in the camcorder. This is because it is carried around to many places. Check for large parts or shields for poorly soldered connections. Sometimes cooling these joints with cooling solution helps to make it intermittent so that you can pinpoint the problem.

Troubleshooting Various Circuits

Many camcorder manufacturers provide a troubleshooting flowchart of what components to check in the proper order. Besides crucial waveform and voltage measurements, camera electrical adjustments can help to determine if a particular circuit is defective. Do not overlook performing accurate resistance measurements in difficult circuits.

NO POWER

With no VCR or camera operation, go to the power-input circuits. Try the camcorder in both battery and ac operation. A blinking green LED in the viewfinder can indicate that a battery is weak. If nothing, plug in the ac power supply. Check the fuse. Make sure battery or external jack is normal. If okay, measure the voltage at the fuse terminals. Proceed into the circuit and do not overlook an isolation polarity diode in the dc power line. Check the voltage going into the power circuits (Fig. 40-4). If it comes on and then shuts off, it could be that the end-of-reel detection circuit has activated.

Measure the input voltage of –8.5 V at the CN302 pin converter, going to terminal 16 of the dc/dc converter (A3001). Check the 5 V (S) at pin 4 and 15 V at pin 5 that apply to pins 14 and 15 of the converter, respectively (Fig. 40-5). Test the unregulated and regulated output voltage at connector CN301. Check C3003 (220 μF) and C3001 (470 μF) for possible leakage on the input voltages.

Check the VA 5 V and S/S VF 5 V from pins 5 and 6 of A3001 to pins 4 and 5 of connector CN301. If any section of the camcorder does not have adequate voltage, check the voltage at CN301 and then at the converter terminals. For instance, if the capstan motor circuits are not functioning, check the voltage at pins 2 and 4 of the converter IC (A3001).

No voltage to camera circuits Set the Select switch to Camera mode (CAM) and check the picture in the viewfinder. Try the camcorder with the ac power supply and required battery. If the camera or VCR circuits do not operate, suspect that the power-distribution circuit or dc/dc converter are defective. Check the 5-V sources in the Samsung 8-mm camcorder dc/dc converter (Fig. 40-6).

The unregulated voltage comes from pins 1 and 2 of CN901. This voltage is fed through a low-pass filter network (LPF) to the various switching circuits. The unregulated voltage from battery or external power supply is fed to switching transistor Q918 for the camera (CAM 5 V) voltage source. Check the voltages on terminals of Q918 and pin 20 of IC901.

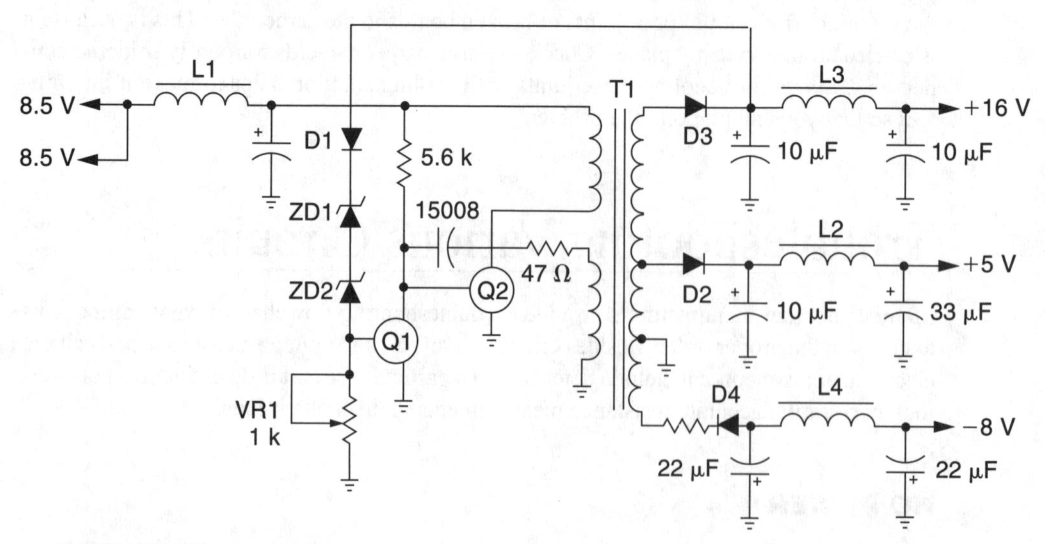

FIGURE 40-4 Typical camcorder power supply circuits.

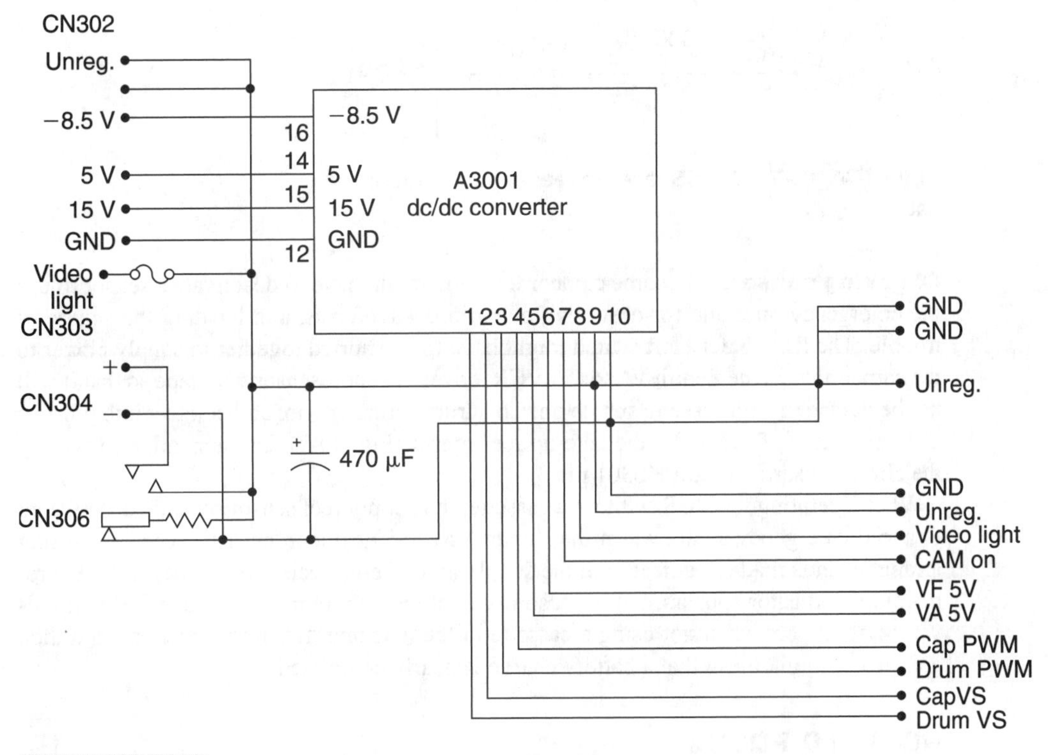

FIGURE 40-5 The 8-mm dc/dc converter input and output voltages.

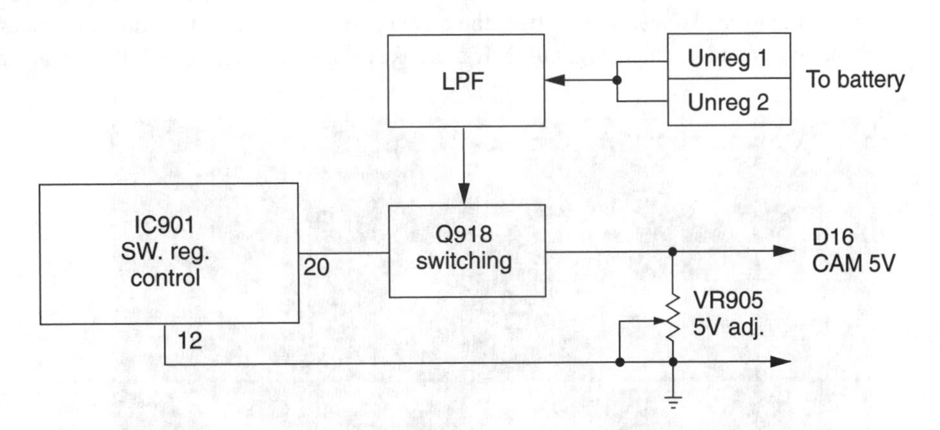

FIGURE 40-6 The 8-mm switching-regulator control circuits.

Test transistor Q918 with a transistor tester or with DMM diode tests. If the voltage will not vary with 5-V ADJ control (VR905), suspect pin 12 of IC901. Check the other voltage sources of EVF 5 V, VTR 5 V, SS 5 V, Drum V_S and Cap V_S, to servo-syscon circuits, 20-V, 15-V, and 9-V sources.

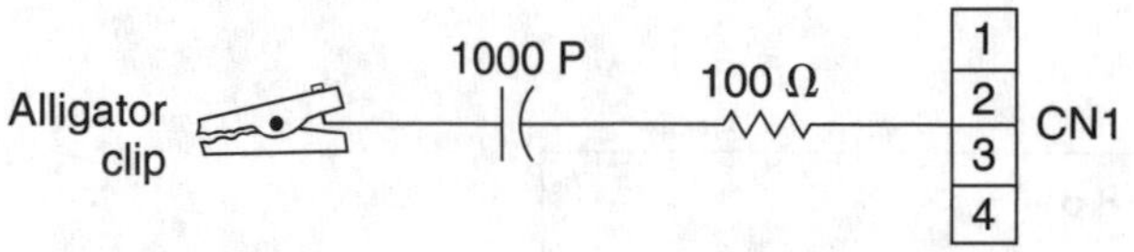

Alligator clip

1000 P

100 Ω

1
2
3
4

CN1

FIGURE 40-7 A VHS take-up reel sensor jumper cable.

Checking sensors In some camcorders, you might have to deactivate a sensor to use the emergency on mode for observing operation, waveforms, and locating the source of trouble. The REC Safety left switch might have to be shorted together to supply power to the camcorder. In the Zenith VM 6150 VCR, cover the photosensor with tape so that it will not be damaged with cassette switch on. Construct a jumper wire and connect it to pin 1 of CN1 and pin 3 of PH S301, the take-up reel sensor (Fig. 40-7). Use a small, stiff wire in the clip for inserting into PHS301 pin 3.

When operating the VCR without a cassette, the supply reel sensor must be deactivated to avoid emergency shutdown in the Zenith VM 6150 camcorder. This type of shorting arrangement is made to defeat both the supply and take-up reels. Sometimes the "tape remaining" indicator appears, but it does not affect the operation. Doublecheck the deactivating of sensors when not using a cassette in the machine. If power comes on and then goes off, it might mean that a battery charge is nearly completed.

NO AUTO FOCUS

Try shifting the camera to different scenes to determine if the lens is changing back and forth in auto focus operation. Often, the camera section must be torn down to access the focus motor and circuits (Fig. 40-8). If there is still no operation, check the voltage at the

FIGURE 40-8 Autofocus assembly on lens assembly of an RCA VHS camcorder.

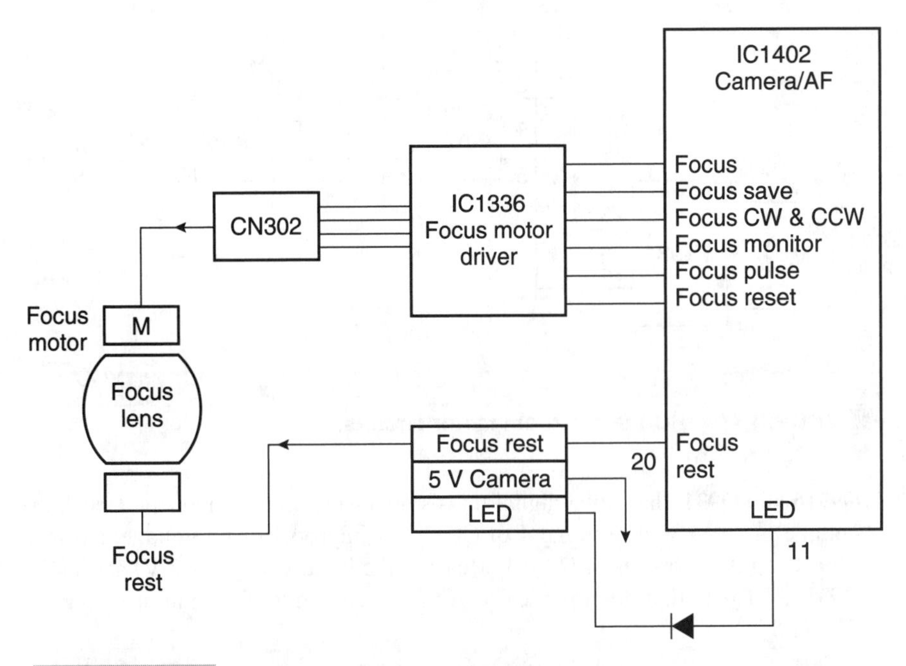

FIGURE 40-9 Typical 8-mm autofocus circuits.

focus motor and focus motor drive IC1336. Check the resistance or continuity of the focus motor in the Canon ES1000 camcorder.

The focus stop, focus power save, focus CW and CCW, focus pulse, and focus reset signals are fed from the camera/AF Mi-Com IC1402 (Fig. 40-9). A focus monitor signal is fed back from pin 27 to pin 19 of IC1402. The camera 5-V source, focus reset, and LED indicator are fed to the focus reset circuits of the camera lens assembly. Take crucial voltages on focus motor-drive IC and the focus motor.

NO POWER ZOOM OPERATION

The iris, auto focus, auto white balance, and power zoom circuits are tied together in some models. If none of these functions are working, suspect that the supply voltage is improper. Press and hold the Tele (T) and Wide (W) switches. Notice if the lens assembly moves out or in. Check the regulated voltage supplied to the zoom drive IC or transistors.

In the RCA 8-mm zoom operation, the zoom circuits operate from a 5-V source fed to a zoom position sensor. The zoom sensor is tied to pins 1, 2, and 3 of CN931, while the negative motor terminal at pin 3 and positive motor terminal at pin 5. The zoom motor is controlled by two zoom drivers, Q9316 and Q9317 (Fig. 40-10).

Check the voltage at the motor terminals and ground with the Zoom Tele or Wide button pressed. Take the continuity at the motor terminals to check for an open motor winding. With no or low voltage, measure the voltage on the zoom drivers. Zoom drivers

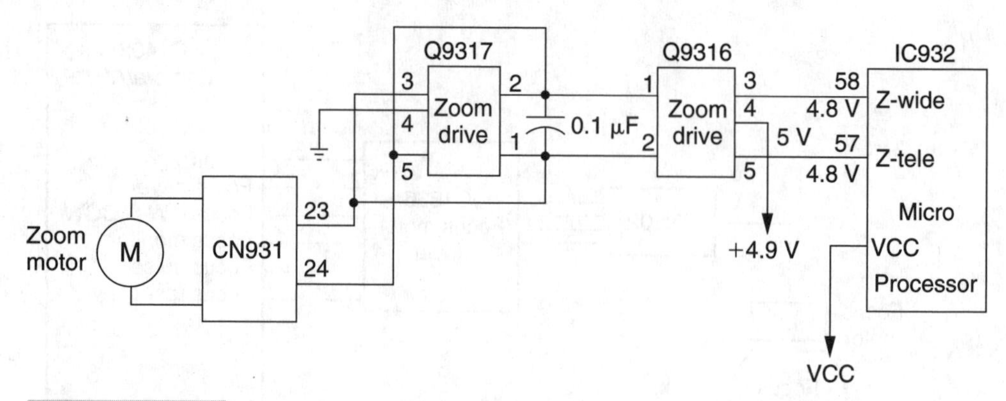

FIGURE 40-10 RCA 8-mm zoom motor circuits.

Q9316 and Q9317 have two digital transistors inside each component. Check the supply voltage pin of 5 V at terminal 4 of Q9316. If improper or no voltage, suspect power-supply regulator transistor Q9313. Measure the input voltage (4.8 V) at pins 3 and 5 of Q9316. Suspect that microprocessor IC932 is defective if no output voltage is at pins 57 and 58.

NO IRIS-CONTROL CIRCUIT

The iris detector and AGC signals are applied at pins 2 and 12 of IC 1404 in the VHS-C iris circuits (Fig. 40-11). First check the B+ voltage (pin 1) with no iris movement. Notice that the drive coil winding of the motor is supplied by +9 V and is tied to pin 9. Measure the damping voltage at pin 7 and 8 of IC 1404. Check the continuity of the damping and driver coils of the iris motor with the low-resistance scale. A low driver voltage can indicate that the iris drive IC is leaky.

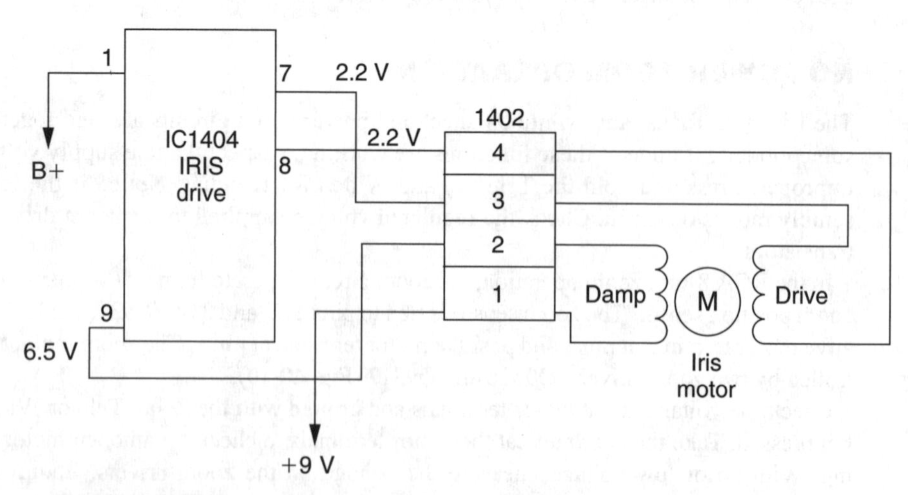

FIGURE 40-11 The VHS-C iris motor circuits.

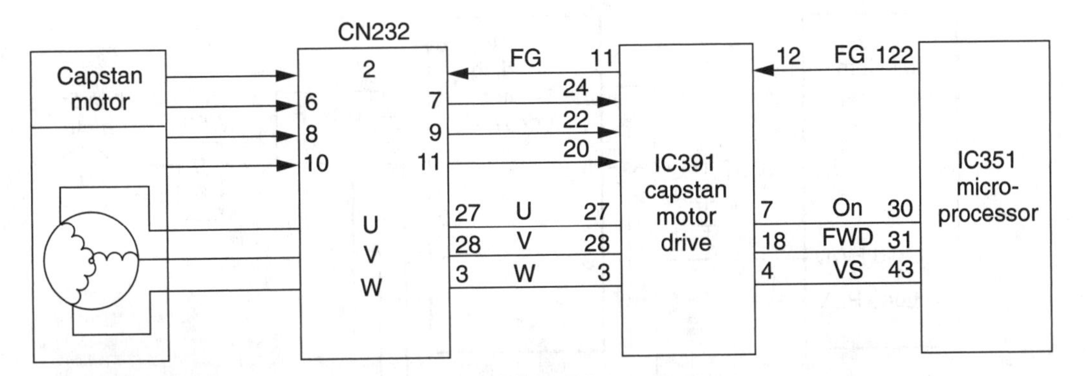

FIGURE 40-12 RCA 8-mm capstan motor circuits.

CAPSTAN DOES NOT ROTATE

The following symptoms might occur in a typical 8-mm camcorder: The tape is inserted and after a couple of seconds, it unloads and indicates drum or cylinder problems. There is no take-up action and the cassette is ejected. These symptoms can indicate that the drum or capstan motor is not rotating. Inspect the motor belt for cracks, breaks, or slippage. If the motor is operating, but the mechanism is not moving, suspect that a capstan motor belt is defective.

Check the voltage applied to the motor terminals in the RCA 8-mm camcorder. No voltage can indicate that a driver is defective or that the drive signal is missing from microprocessor IC351 (Fig. 40-12). Measure the regulated voltage at pin 10 and 5 V at pin 15 of motor driver IC391. Suspect that a voltage source or voltage regulator Q3901 is defective with low or no applied supply voltage. Check the load and unload voltage at pins 11 and 12 of the driver IC. Replace IC391 with correct drive, regulator, and 5-V supply.

DOES NOT EJECT OR LOAD

In the RCA CPP300 camcorder, when loading begins, the main brake releases both reels. The tape from the cassette is loaded by the guide roller. During unloading, the pressure roller is first released. The main brake prevents the tape from fluctuating; the take-up brake is on to prevent the tape from being pulled out of the take-up reel.

The loading motor in the Samsung 8-mm camcorder is operated by a signal voltage from IC601 Mi-Com, loading motor driver IC504, and the loading motor (Fig. 40-13). The Mi-Com IC601 controls the load and unload signal to the loading motor driver IC.

Notice if the loading motor is rotating, when the eject or load circuits are energized. Check the supply voltage at pins 4 and 5, and also pin 12 of IC504. Check the voltages on pin 15 for loading and pin 3 for unloading at the motor terminals. No voltage can indicate that IC504 is defective. If the supply voltage is real low at pin 12, suspect that motor driver IC504 is leaky. Check the door mechanism for bent levers or mechanical problems when the loading motor operates. Check the loading motor flowchart for loading motor operations.

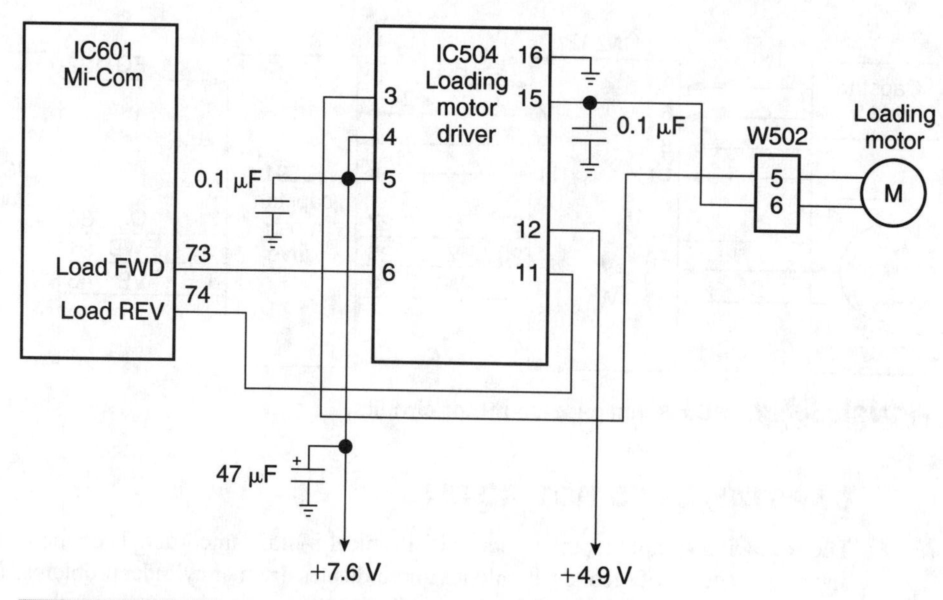

FIGURE 40-13 The 8-mm loading motor circuits.

CYLINDER OF DRUM MOTOR DOES NOT OPERATE

In many camcorders, the drum or cylinder and capstan motors operate from the servo or syscon circuits. The problem might be that when the tape is inserted, it loads for a few seconds and then unloads. There is no drum or cylinder rotation. Check the regulated and unregulated 5 V at pins 5 and 14 of drum motor driver IC180 in the Canon 8-mm (Fig. 40-14). A very low voltage could indicate that a dc/dc connector or battery source are defective, or that IC180 is leaky.

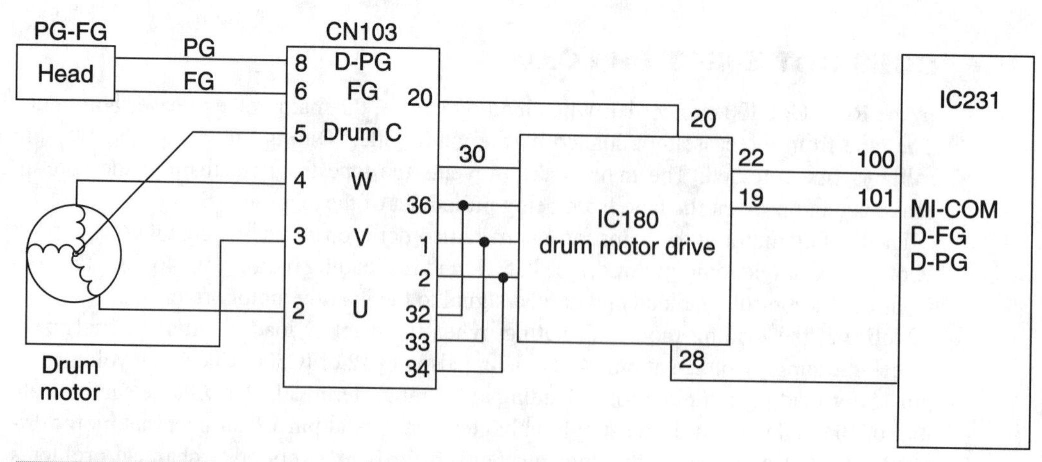

FIGURE 40-14 Canon 8-mm drum-motor circuits.

Check the drum FG and PG waveforms at pins 100 and 101 (IC231). These waveforms are present when the drum motor is operating. Scope the D – PWM signal at pin 3 of servo Mi-Com IC231. Check the continuity of windings V, W, and U of the drum motor when voltage is found at motor terminals. Do not overlook Q151 and Q150 of the PWM power drive applied to pin 5 of IC180.

POOR DRUM MOTOR ROTATION

When the drum or cylinder does not rotate properly during recording in a Samsung 8-mm camcorder, check the servo speed-control circuits. IC502 provides drive control for the drum motor terminals, U, V, and W. Check the drum drive motor IC for correct supply voltage and input drive. Check the input and output terminals of the dc/dc converter of drum PWM circuits. Scope the motor output FG and PG pulses at terminals 68 and 69 of IC601 Mi-Com.

SENSORS NOT WORKING

When none of the trouble detection indicators work, suspect that a system-control IC is defective. Check the voltage supply source of the Mi-Com IC. If the end lamp is not working, check the end lamp voltage and sensor at pin 41 of IC501 (Fig. 40-15). Measure the voltage at the end-sense transistor and LED. Suspect poor cable and socket contacts, or no applied voltage from the sensor detectors of the Samsung 8-mm indicator circuits. Do not overlook the possibility of a defective LED drive transistor.

If there is no supply reel or take-up reel indication, check both reel sensor indicators. Measure the +5 V applied to pin 3 of each device. Check all sensor connections. Do not overlook the possibility that a sensor is defective.

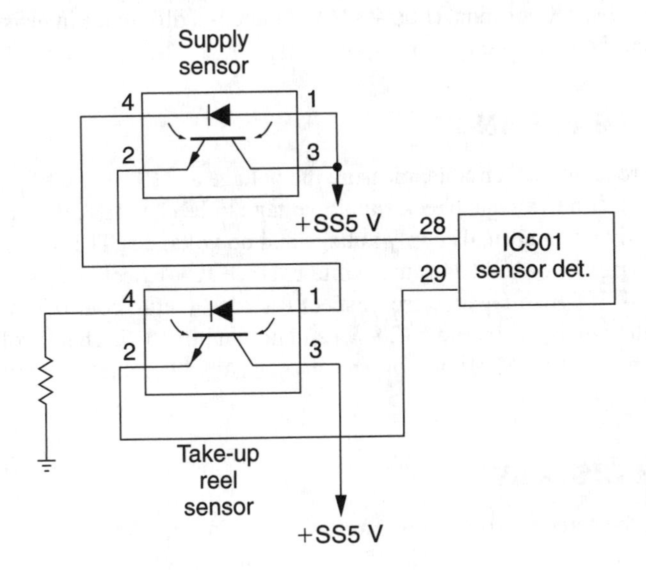

FIGURE 40-15 A typical 8-mm take-up and supply-reel sensor circuit.

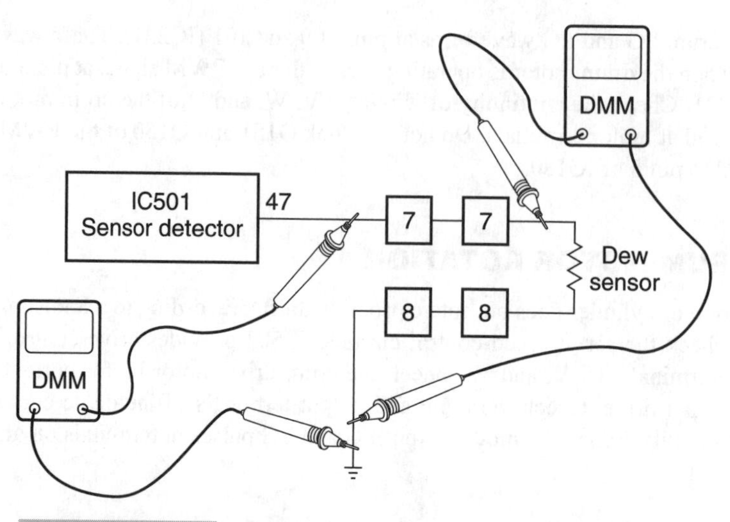

FIGURE 40-16 **Measuring the resistance of the dew sensor and possible poor connections.**

DETECTORS NOT WORKING

The VCR will not operate with a flashing dew test LED. The drum does not rotate during loading. Sometimes in unloading, the system stops for five minutes or so or shuts off. Check for a defective dew sensor or IC controlling the dew sensor circuits. Sometimes, when all LEDs flash, that indicates dew warning. When only some LEDs flash with no operation, that indicates an emergency mode, which requires service.

Intermittent dew sensor operation can result from poor connector or ground contacts. Measure the resistance of the dew sensor. Then measure the resistance where the dew sensor is connected to the sensor IC terminal (Fig. 40-16). If there is a difference in resistance, suspect poor contacts at the socket connector.

IMPROPER WHITE BALANCE

In the RCA CPR100 process video circuits, measure the voltage at TP1107 off of pin 4 of IC1102 (1.4 V to 1.5 V). If no voltage, check supply voltage to IC1102 (B+, 9 V) pin 11. IC1102 might be defective with normal supply voltage and no voltage at TP1107.

Measure the voltage pin 15 of IC1104 (auto white balance IC). Check the voltage at pin 5 (2.9 V) of IC1104. If the voltage is zero, inspect RM1101 (setup control). IC1104 could be defective if the voltage is on pin 5 (2.9 V) and not on pin 15. Recheck voltages on pin 18 (0.3 V) and pin 19 (0.2 V). If no voltages are measured here, test the voltages on IC1107.

NO ON-SCREEN DISPLAY

Check to see if any of the screen display lights are on. If not, inspect the battery display, apply dc voltage to the external battery jack, or exchange the battery. Does the display light (E-F) with any tape rotation? Readjust the battery circuits. Be sure that the camcorder is set to Camera and Record mode. If it has no display, check the 9- or 5-V supply voltage. Measure voltages on power switches, over discharge transistors, and regulators in the power distribution.

You can check the on-screen display, by setting up the on-screen display character position adjustment. Connect the VCR section and test equipment in the VCR set-up adjustment in the RCA 8-mm (Fig. 40-17). Place the Power switch in the CAM and Standby switch in the On position. With the camcorder in Record/Pause mode, check character display adjustment CT321 so that the display is centered.

NO VIDEO MONITOR

To test for lack of video monitor, check the voltages and waveforms on the main video amp, A/V output, and luma process IC. Set the Select switch to Camera mode. Aim the camera at the color-bar chart. If EVF monitor is not good, check IC301 and the EVF circuits. Check video at A/V output, pins 1 and 8. Check pins 10 and 12 of IC301 for the video signal (Fig. 40-18). If there is no signal, measure the B+ voltage on IC301 and suspect IC301.

With no video on TV set or color monitor, check the video at pin 19 of IC201 (1 Vp-p). With no video, measure the voltage at pin 25 of IC201 (5 V). If no voltage, check the 5-V supply line source. If no voltage, check pin 24 of IC201 (video MOM p-1). Proceed to pin 15 of IC201 to look for video signal (1 Vp-p). Suspect a problem with IC201 if the video waveform is at pin 15.

NO VIDEO RECORDING

The Y RF output at pin 39 of Y/C process IC201 in the Samsung 8-mm camcorder is sent to the RF amp IC101. A recording C output is fed from pin 8 to pin 6 of IC101. The chroma (C), luminance (Y), AFM, and AFT are mixed inside the head amp IC101. The output of record amp is fed to CH 1 cylinder head from pin 29 and CH 2 from pin 32

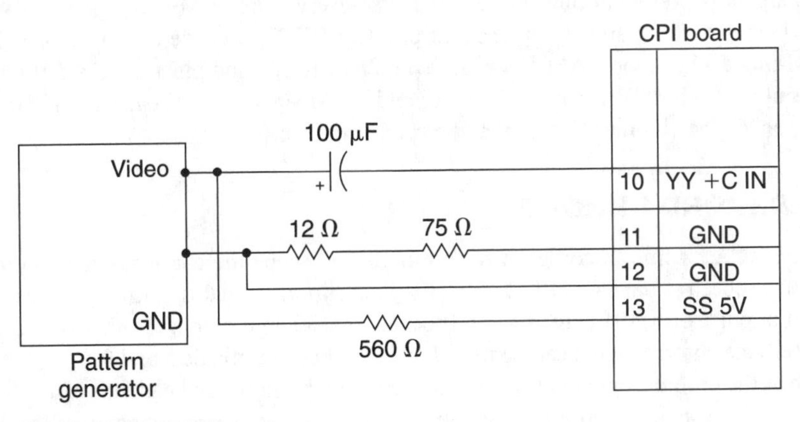

FIGURE 40-17 A typical VCR setup adjustment.

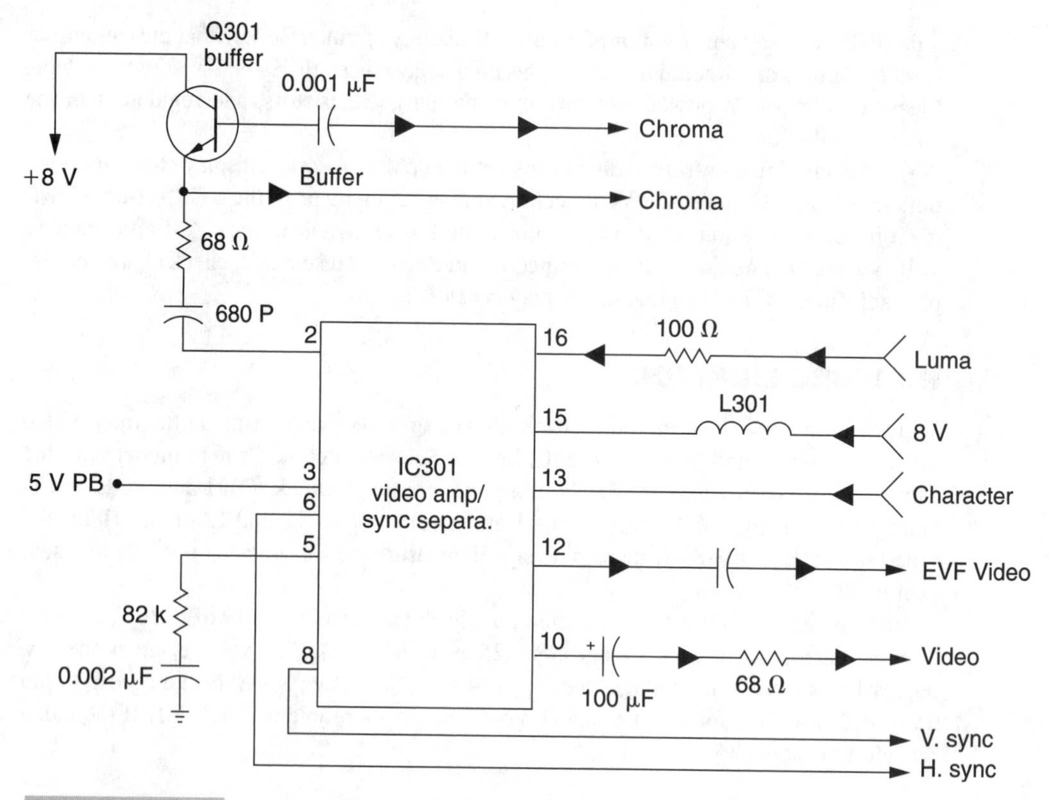

FIGURE 40-18 The VHS-C video signals on the main video IC301.

(Fig. 40-19). The FE head erase oscillator signal is fed through the FE amp to the head cylinder.

Check the video signal at pin 30 of IC201 in the Samsung 8-mm camcorder in VCR mode. If the signal is normal, go to pin 4 of IC301. Proceed to pin 47, 48, and 49 of IC201 if no signal is present at pin 4. Check the sycon/servo circuit Mi-Com, V_{CC} voltage, and IC201 circuit. If the signal is present at pin 4 of IC301, precede to pin 21 of IC201. In CAM mode, check the CAM Y signal at pin 28 of IC201 and pin 1 of IC605 if no Y signal is at pin 28 of IC301. Check the Y RF and Y FM signal to the head amp, IC101. Likewise, check the chroma (C) signal at the head amplifier.

NO RECORD CHROMA

Set the RCA 8-mm camcorder in record mode and scope the chroma signal at pin 53 of IC101. Proceed to pin 8 or TP132 of IC101 if the chroma signal is normal. Check the color signal at pin 8 of IC191 (Fig. 40-20). Proceed to head amp output terminals 28 and 33 of IC191. This chroma signal can be traced to J1003 head terminals 5 and 9.

When the chroma signal is low or missing at any point in the head amp circuits, check that circuit for possible improper supply voltage or defective components. If the chroma signal is not found at pin 8, check the color signal at pin 57 and 59 of IC101. Measure the

V_{CC} supply voltage of IC101. If the voltage is normal with input chroma on IC101 and no output at pin 8, suspect that IC101 is defective.

NO COLOR PLAYBACK

To test for chroma playback, check the signal waveforms and voltages on the luma-process and chroma-process ICs. In the Samsung 8-mm camcorder, check the color-playback signal at pin 45 of IC201 of the Y/C process IC (Fig. 40-21). If the signal is normal, proceed to the chroma signal at pin 64 of IC201. Proceed to the GM141 network and pin 6 of IC101. If the playback chroma signal is still missing, check the color playback at CH 1 and CH 2 heads at pins 29 and 32 of IC101. Do not overlook the possibility that IC101 is defective or supply voltage in IC101 is inaccurate with no playback color recording.

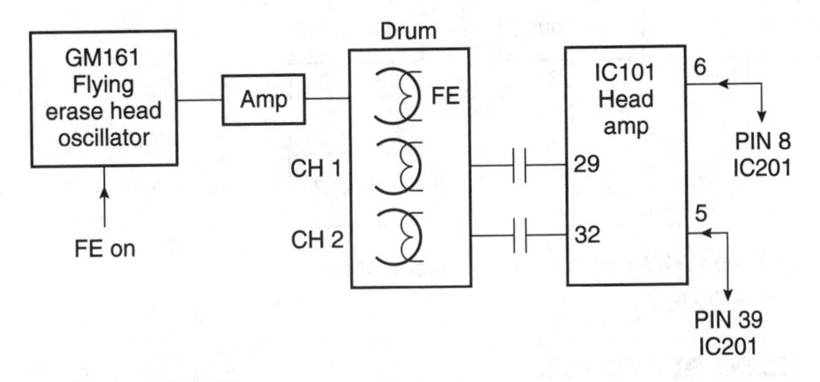

FIGURE 40-19 A block diagram of 8-mm record amp circuits.

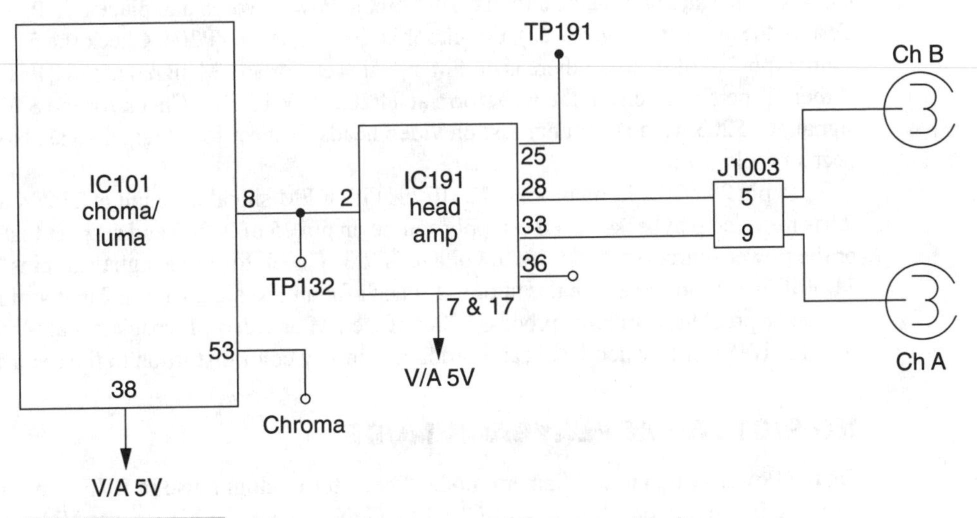

FIGURE 40-20 RCA 8-mm chroma and head-amp circuits.

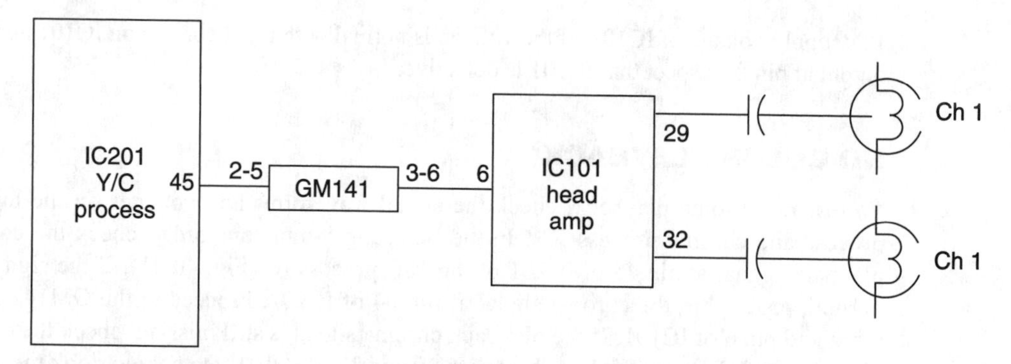

FIGURE 40-21 The 8-mm P/B Y/C process and head-amp circuits.

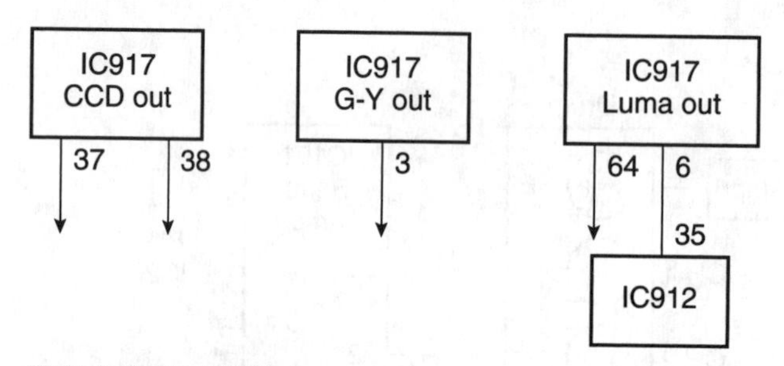

FIGURE 40-22 No picture on P/B mode of an 8-mm camcorder.

NO VIDEO PLAYBACK

The lack of a video playback signal is checked from the tape heads back through luminance/chroma and luminance circuits. Turn on the Power switch and place it in Play mode. Test by the numbers (Fig. 40-22). Check for an FM signal at TP204. Check the 5-V power source (pin 25) of IC203 if there is no FM signal. Test for an FM signal at pin 21 of IC203. If there is no signal, check the waveform at TP202 (SW 15 Hz). Check for the SW 30-Hz signal at TP203. Run continuity test on video heads. Inspect for clogged areas. Now suspect a problem with IC601.

Go to pin 28 of the luma process IC (IC201) if an FM signal is found at TP204 and but it has no video playback. Test the supply source on pin 25 of IC201 and suspect that IC201 or the power source are leaky if the voltage is low. Check for luma signals at pins 23 and 11 of IC201. If no luma signal is at pin 11, check for an FM signal at pin 2 and 4 of IC201. Suspect a problem with luma process IC201 if the FM or video-playback signal is missing. A defective or dirty video head can introduce rainbow color distortion to the recording.

NO PICTURE IN PLAYBACK MODE

Set the Power switch to the Camera mode. Check for random noise in EVF or monitor. If no noise in picture, check pin 14 of CN933 and then check the No Record Video and Incorrect Auto Iris tests. Proceed to the CCD out at pins 37 and 38 of IC917. If it is okay, proceed to IC917 pin 3, 64, 6, and 35, in that order. Then check the supply voltage on IC912. For no signal at pin 3, check IC916 and IC917. Test the LPF block, which consists of Q9165, Q9166, L9190, L9191, C9211, C9256, C9257, C9258, C9261, R9182, R9191, R9192, R9216, R9217, R9218, R9256, and R9257.

NO RECORD CHROMA

If a color flowchart is handy, check the following IC stages to locate the defective component. Check the CAM C at pin 12 of the connector CN933. If it is no good, go to C out at pin 16 of IC912. Proceed to terminals 19 and 20 of IC914 to check R-Y/B-Y out signal. Check IC914 for no color signals and supply voltage (Fig. 40-23).

With normal R-Y and B-Y output signals, check terminals 46 and 48 of IC914. If they are okay, proceed to pins 4 and 5 of IC917. Proceed to pin 4 of IC912 and pins 9, 10, and 11 for BFG, BF, and C BLK signals. If the signals are okay at pins 9, 10, and 11 of IC912, check for a defective IC912 if the signal has no color.

SOUND DOES NOT OPERATE

Check the audio block diagram and schematic for audio problems. Inspect the microphone. Substitute the external microphone to eliminate the regular mic. Most camcorder microphones are condenser types. Can you hear the audio in the headphone monitor?

The audio system can be checked by injecting a 1-kHz audio signal at the microphone jack. Then signal trace the audio record circuits with scope or external power amplifier. In the Samsung 8-mm camcorder, check the audio signal at pin 38 of IC401. Proceed to pin 31 of audio amp IC401 (Fig. 40-24). Check the audio at pin 8 of IC401. Check the microphone if the injected audio signal is okay.

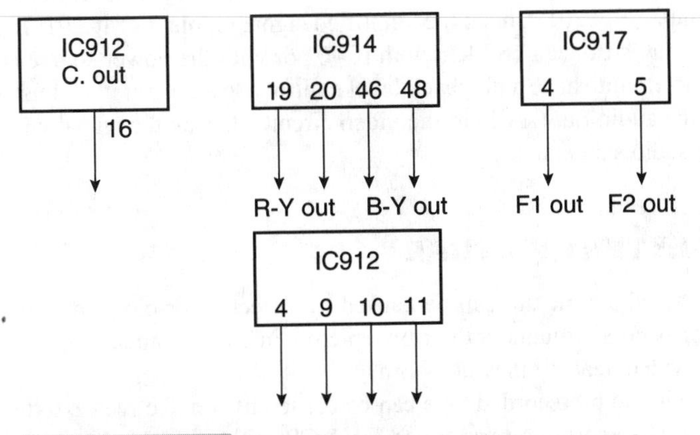

FIGURE 40-23 Check the following IC terminals for if no color is in the picture.

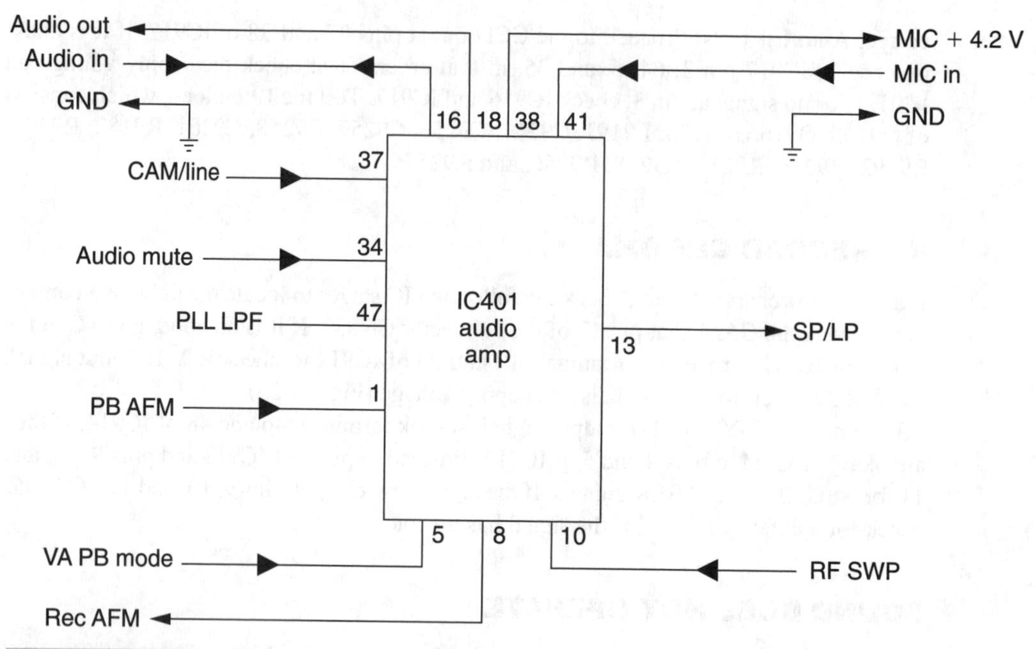

FIGURE 40-24 The 8-mm IC401 audio amp circuits.

If no output AFM signal is at pin 8, check IC401 and the supply voltage at pin 42 (5 V). Proceed to record AFM RF at pin 4 of tape head preamp IC101. The audio signal is amplified by IC101 and is applied to CH 1 and CH 2 of the head cylinder.

NO AUDIO PLAYBACK

Set the Select switch to the Playback mode (PB). The Playback signal from the tape head can be scoped at pin 2 of IC201. Check the amplified signal on pin 8 of IC201. If a signal is going in and not out, suspect a problem with IC401 or with the power source (B+ 5 V, pin 20). Some audio circuits have a playback level control. Be sure that it's adjusted properly. Signal trace the audio out (AV1) to the video circuits. Check the signal with proper waveforms on the audio schematic.

NOISY AND JITTERY PICTURE

Random noise in the whole picture can be caused by defective or clogged video heads. Random noise often occurs throughout the whole picture in both channels. Clean the video heads. Replace the video heads if they are worn.

Intermittent noise in the prerecorded tape can be caused by a defective cassette or poor cylinder sync and CTL recording pulse. Load the VCR with a blank tape and place the VCR in the Record mode. Check for good C sync and CTL-Rec pulse. Intermittent audio can be caused by poor wiring connections around the audio head. Check the lead connections of flying or full erase head—especially where the leads connect to the PC wiring (Fig. 40-25).

Insert a prerecorded tape in the camcorder when a noisy picture is found in playback mode. If the luma level is no good or low, check the signal level at the output of the tape-head preamp IC. Check the preamp IC or head drum for defective components. Do not overlook the possibility that a prerecorded tape is defective.

Horizontal jitter in picture With an intermittent or noisy picture in the Play mode, try another cassette recorded by another VCR. Place the VCR in the Play mode. Adjust the tracking control. Check the waveform at pin 12 of IC601 (Fig. 40-26). Check the CYL phase waveform at pin 17. If no waveform, check the cylinder motor. If there's a normal waveform, suspect IC601. Now check the phase cylinder waveform at pin 64. If there is no waveform, suspect that IC601 is defective. If the waveform is okay, suspect a problem with IC605 (pins 5 and 7).

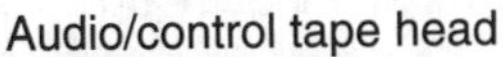

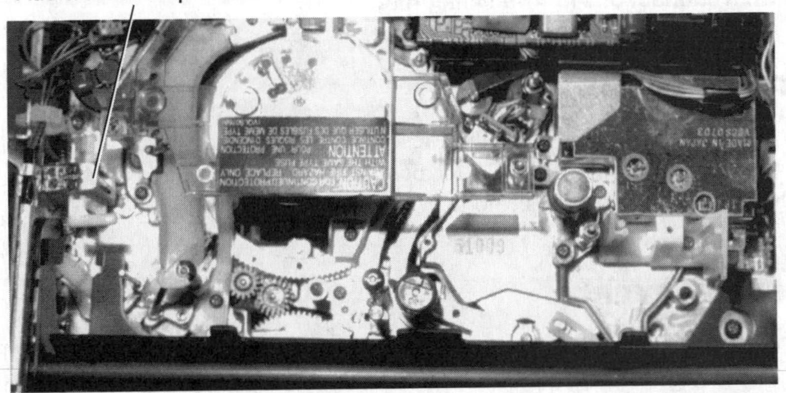

FIGURE 40-25 **The audio tape head in an RCA VHS camcorder.**

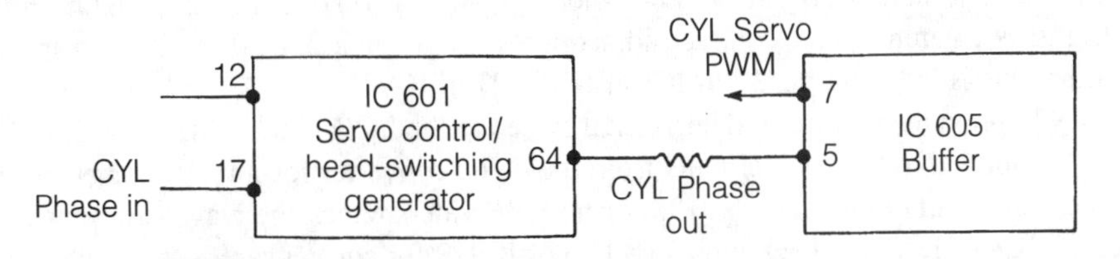

FIGURE 40-26 **Check the tracking control and waveform at pin 12 of IC601 for horizontal jitter noise.**

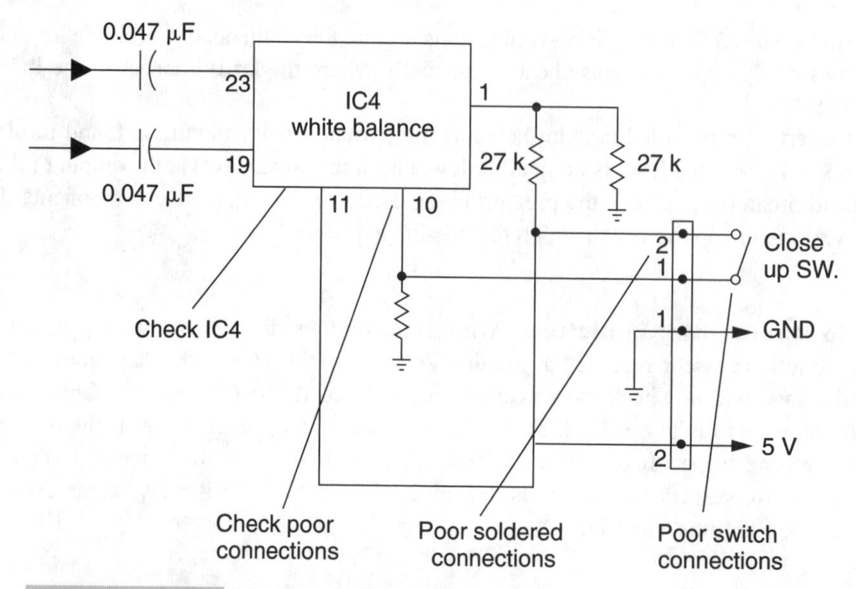

FIGURE 40-27 **Check IC4 if the white balance drifts and closes up the switch connections in VHS camcorder.**

WHITE BALANCE DRIFTS

If the white balance drifts, check the Close-up switch contacts and soldered connections (Fig. 40-27). Check pin 10 of IC4 for poorly soldered connections. Suspect IC4 if the switch and contacts appear to be normal.

NO EVF RASTER

If the electronic viewfinder will not light, check the power source that feeds the EVF circuits. The voltage can be 5, 8, or 9 V. If there is no supply source, check the dc/dc converter or dc circuits. Check the anode voltage on a small CRT from 2 to 3 kV (2.7 kV), pin 2 at 370 V, and pin 3 at 330 V.

The camcorder with color viewfinder includes a back light, LCD, IC972, and IC971 in the RCA 8-mm model (Fig. 40-28). The video input is located at pin 3, where the color, demodulator, matrix, brightness, and colors (A, G, and B) are fed to the LCD from IC971. Brightness, gamma, and contrast adjustments are at pins 32, 30, and 42, respectively. Color sync is fed from pin 37 of IC971 to pin 27 of IC972.

IC972 provides vertical and horizontal sweep to the LCD block. Check the 5, 6, 12, and 7.5 V supplied to the EVF circuits from the dc/dc converter for no EVF raster. Scope the sync at pin 1 of connector CN972. For no EVF video, check the waveform pins at 3, 20, 22, and 24 of IC971. Check pin 27 and 36 of IC972 for correct sync.

NO EVF HORIZONTAL DEFLECTION

Suspect a problem with the horizontal circuit if there is no CRT voltage. Check for input line voltage at pin 5 of output transformer T801 (Fig. 40-29). The heater voltage across pins 3 and 7 is 16.3 V. The horizontal circuit must operate before the voltage is developed. Check for collector voltage on horizontal driver transistor Q803. The fuse could be open (TF801). Suspect a leaky horizontal drive transistor (Q803) or improper drive voltage at the base terminal. Check the horizontal oscillator waveform at pin 2 of deflection IC801. If no deflection waveform, suspect a problem with IC801 or the supply voltage (pin 4, 5.3 V).

NO EVF VERTICAL DEFLECTION

Check the vertical circuits for a horizontal white line or insufficient vertical sweep. The vertical circuits within typical 8-mm model consist of one large vertical and horizontal

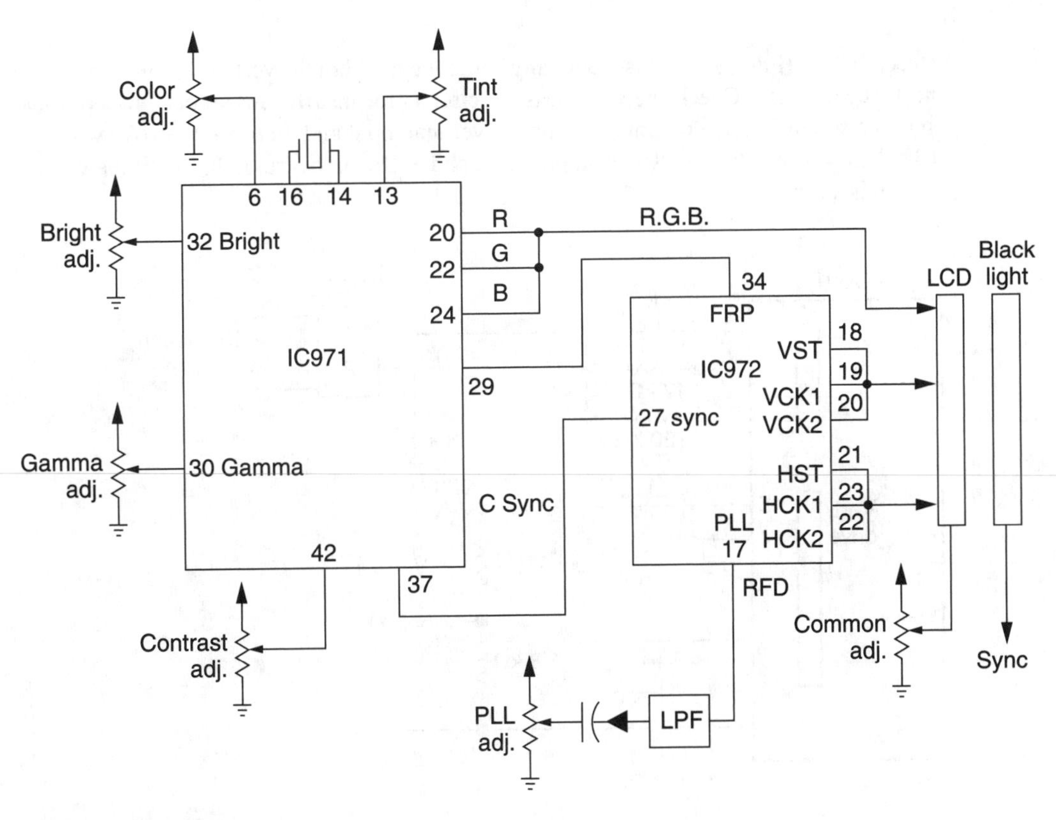

FIGURE 40-28 The RCA 8-mm EVF block diagram.

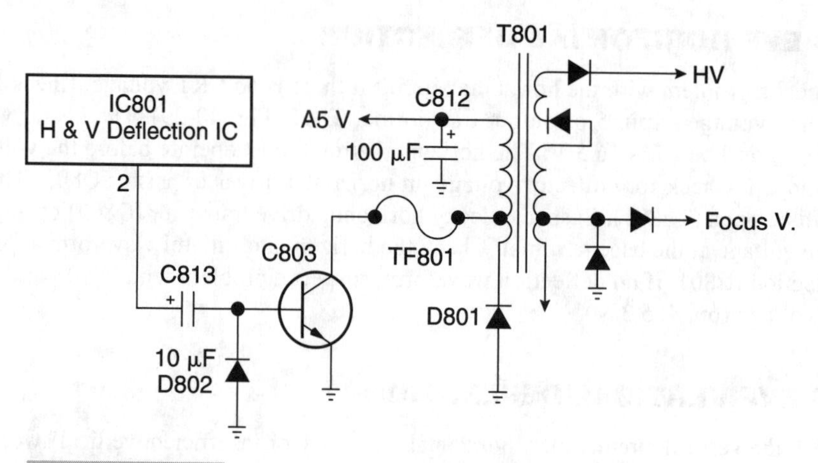

FIGURE 40-29 The black-and-white EVF horizontal flyback circuits.

sweep ICB01 (Fig. 40-30). Just about any component within the vertical circuits can cause no vertical sweep. Check the vertical drive circuits for insufficient vertical sweep. Measure the supply voltage on pin 2. Scope the vertical input and output at the yoke winding. Check all vertical drive voltages on pins 1, 3, and 5. Do not overlook ICB01 if the vertical sweep is absent.

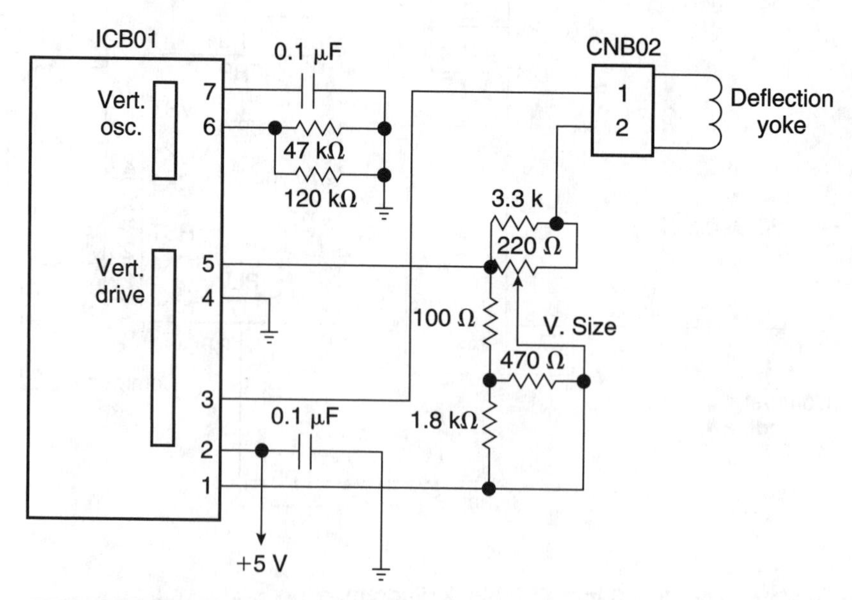

FIGURE 40-30 Typical 8-mm EVF vertical circuits.

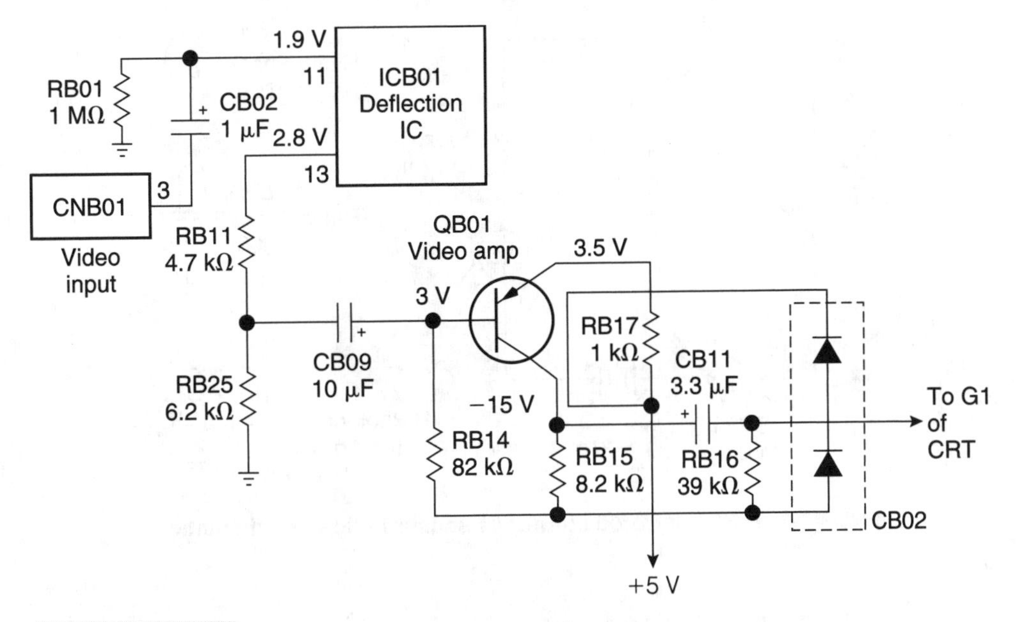

FIGURE 40-31 Samsung 8-mm EVF video circuits.

NO EVF OR WEAK VIDEO

Scope the video waveform at pin 3 of CNB01 and pin 11 of ICB01 (Fig. 40-31). Check the waveform at output pin 13 of ICB01, which is fed to the base of video amp QB01, through 4.7 kΩ and C809 (10 μF). The video amp transistor feeds the amplifier video signal to G1 of the CRT. If video is found at base terminal of QB01 and not at the collector terminal, take voltage measurements at QB01. Suspect trouble with QB01 if the voltages are fairly normal voltages, no video is output. Do not overlook CB02 (1 μF), CB09 (10 μF), and C811 (3.3 μF) for weak video to G1 of B & W CRT in Samsung 8-mm EVF circuits.

INFRARED INDICATOR

A homemade infrared indicator can be made out of only a few parts to indicate that the infrared diode is working in the auto focus circuits (Fig. 40-32). Simply hold the sensor device about one inch away from the infrared diode inside of the camera. This diode is often located at the bottom and front of the lens assembly. When the auto focus and infrared diode is working, a pulsating sound is audible in the piezo buzzer. Moving the infrared indicator away from and then close to the infrared auto focus diode can cause the lens assembly to rotate back and forth. Although the pulsating sound is not very loud, it does indicate that the auto focus infrared diode and driver circuits are ok. No sound indicates that the infrared LED or auto focus circuits are defective.

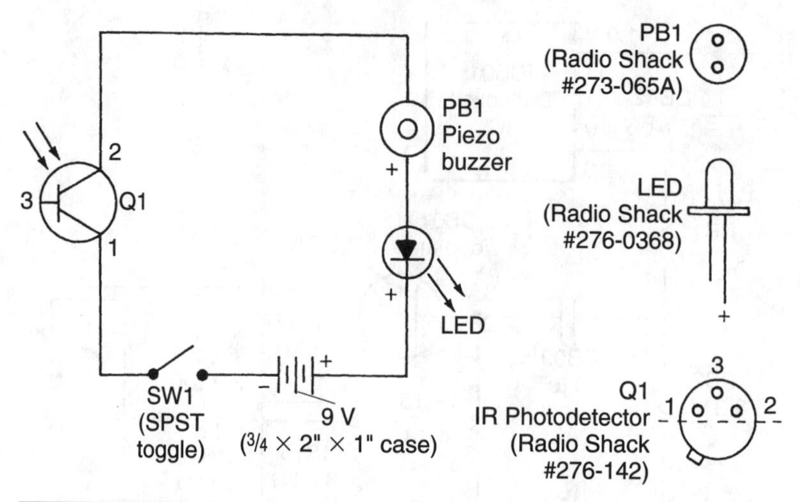

FIGURE 40-32 **Infrared autofocus sensor indicator schematic.**

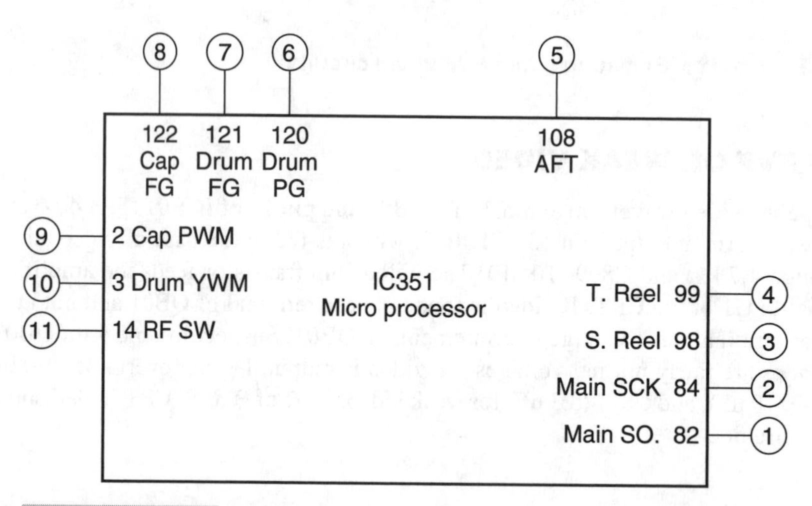

FIGURE 40-33 **Waveforms taken by the number in the RCA 8-mm camcorder.**

SERVO/SYSTEM CONTROL WAVEFORMS

Important waveforms can be found throughout the camcorder circuits. Crucial waveforms are very important in signal tracing or locating defective circuits in the various camcorder circuits. The most important waveforms are those found in the servo system-control circuits. The servo microprocessor (IC351) waveforms in the RCA 8-mm model can determine if the servo circuits are performing correctly (Fig. 40-33).

Waveforms 3 and 4 indicate if the take-up reel and supply reel are rotating at pins 98 and 99 of IC351. The capstan FG, drum FG and drum PG waveforms indicate if the capstan and drum assemblies are rotating at pins 122, 121, and 120. The capstan PWM and drum PWM are found at pins 9 and 10 of IC351 in both Record/Play modes.

THE POWER SUPPLY

The camcorder power supply has much more than just a battery. The camcorder can be operated from a battery or ac power source. The ac power supply can consist of switching, constant voltage, constant current, voltage protection, and voltage-regulation circuits. Besides supplying a dc voltage for operating the camcorder, the batteries can be charged (Fig. 40-34). Also, inside of the camcorder, you can find battery discharge and alarm circuits with power circuit control.

ac power supply Ac power packs or supplies produce 5, 6, 7.2, 9, or 12 V. The Canon ES1000 is supplied with 5 V, RCA PRO845 at 6 V, and Samsung SCX854 at 6 Y with a 7.5-Vac adapter. In the RCA 8-mm camcorder, the ac adapter charger consists of a rectifier circuit that includes ac input, filters, and primary rectifiers. The primary side has a switching element (Q501), PWM-control circuit, and a P/S regulator transformer (Fig. 40-35).

The secondary side circuits consist of a P/S insulator or photocoupler, secondary rectifier (D5201), smoothing circuit, error-amplification circuit (IC521), and VCR output. A charge circuit contains the charge/discharge control circuit (IC523), charge-selector circuit (Q5202), battery output, and discharge circuit (Q5205).

Samsung 8-mm ac adapter The ac input circuits are isolated from the secondary circuits with a photocoupler (IC102). The primary circuits are at ground potential, but the secondary circuits have a floating or hot ground. Note when taking voltages within the ac

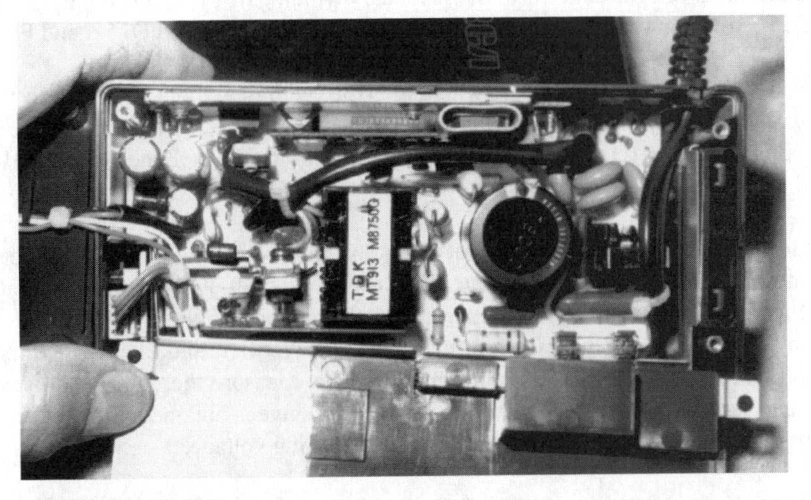

FIGURE 40-34 The ac power supply or adapter provides a charging voltage for the camcorder battery.

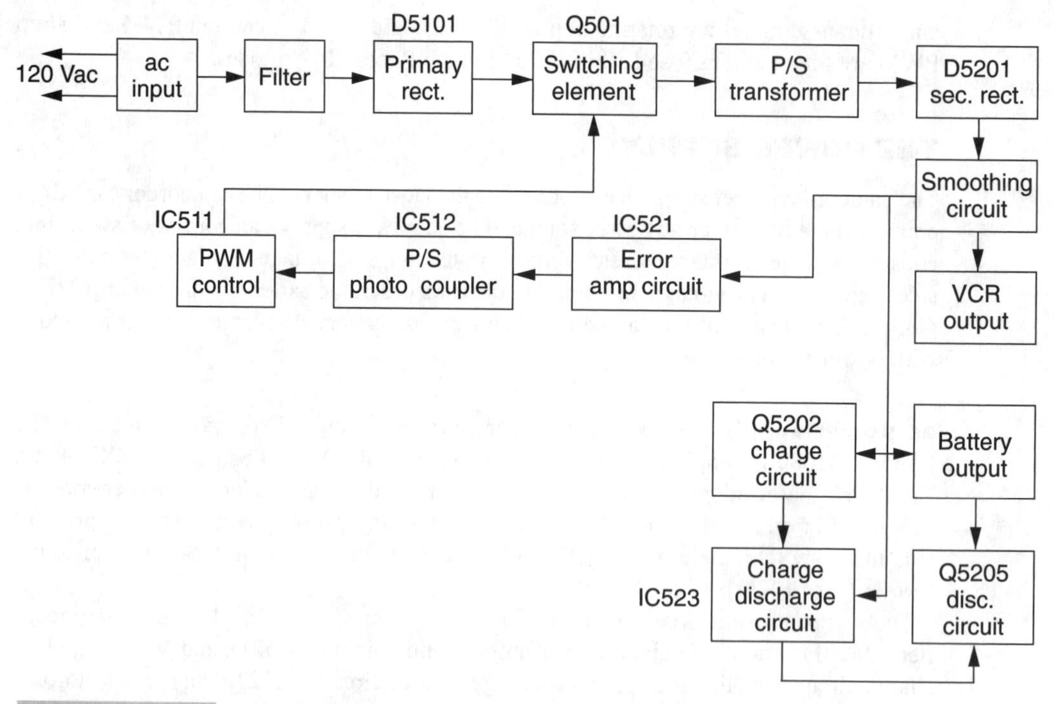

FIGURE 40-35 A block diagram of RCA 8-mm power adapter circuits.

adapter, be sure of the correct ground (Fig. 40-36). The ac circuit consists of full-wave bridge rectifiers, filters, controller IC101, switching Q101 (2SK904), and the primary windings of the power transformer (T101).

The secondary circuit consists of the secondary winding of T101, op amp IC151, switching Q154, digital switching Q153, and dual switching of Q152 and Q151, and F151 (2A) fuse.

A sub 1 chassis contains digital switching Q156, with dual diodes (D152), to the VTR output and charge jack. IC152 regulator is connected to the green LED (D156), digital-switching transistor (Q155), Q157, and switching IC (IC153). Zener diodes D157 and D158 are located in the battery-charging output terminals.

Inside the RCA CPR300 ac adapter A load must be replaced across the ac adapter output terminals, so very little voltage is measured (Fig. 40-37). By placing a battery or 20-Ω load across the battery output terminals, the charger circuits can be serviced. Like most adapter circuits, a power transformer with diode rectifiers and filter capacitors are located. First, inspect and check continuity of ac fuse. Measure the voltage across the primary and secondary circuits to determine if the ac voltage is obtained. Suspect that the IC and transistor regulators are leaky if there is no charging voltage.

Battery charging Many small batteries last an hour or less with constant camcorder operation. The battery can last longer with occasional operation. As the battery

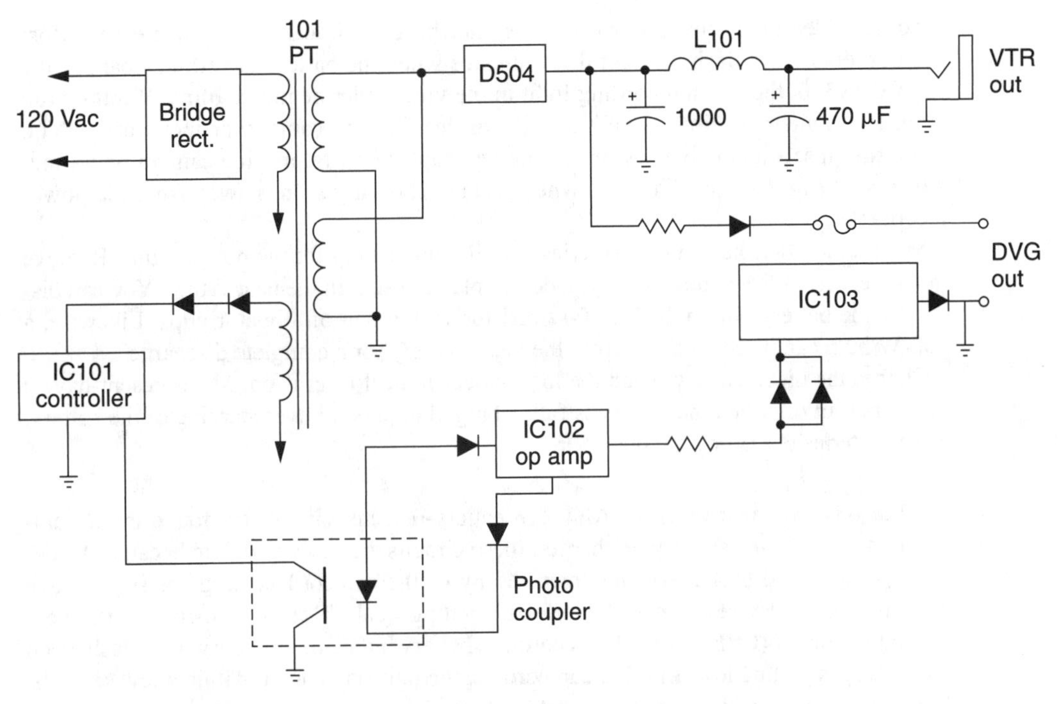

FIGURE 40-36 A block diagram of Samsung ac adapter circuits.

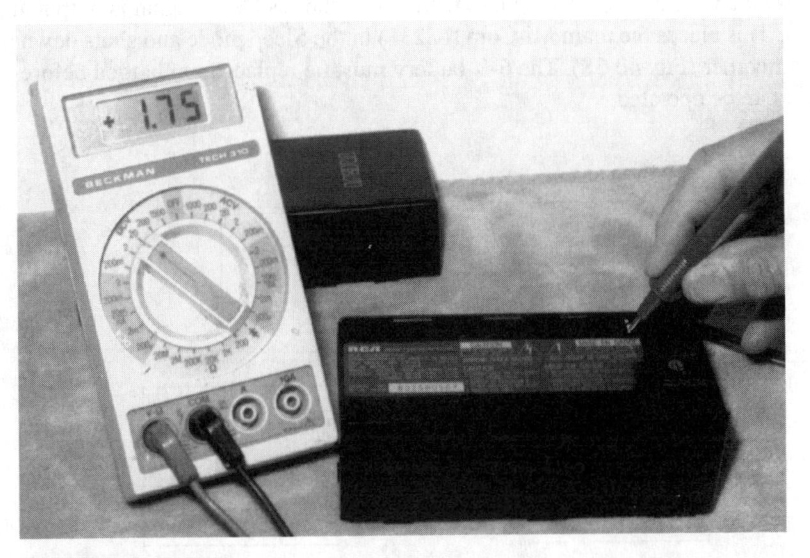

FIGURE 40-37 The RCA ac adapter will not show normal voltage unless a battery load is placed across the charging terminals.

becomes older and is charged many times, the charging time gradually decreases. Most camcorders have a flashing light that indicates when the battery is getting weak. In the RCA CPR300, the green recording light in the viewfinder begins to blink. If left on too long, the camcorder circuits will shut down, leaving the camcorder inoperative. You can either insert a new battery or use the ac adapter unit to provide camera operation. Always take at least two batteries when you are shooting scenes away from the power receptacles.

Most camcorders have nicad batteries with lithium battery for memory circuits. Remove the battery from the camcorder if you do not plan to use it in the near future. You can discharge the battery with a 10-W, 10-Ω resistor with a pan of alligator clips. Likewise, a 100-W bulb can be connected across the nicad battery for a complete discharge.

Charge the nicad battery when the low-battery indicator turns on. Most present-day ac adapters shut off when the battery is fully charged to prevent overcharging of the battery. Most batteries charge within one hour.

Battery-detector circuits Although battery-detector circuits are found in the camcorder and not in the ac adapter/charger, these circuits are discussed here because the detector is activated by battery shutdown. Many of the camcorders (large or small) have some indicating device when the batteries are getting weak. It's usually located in the electronic viewfinder (EVF). Either the recording, battery indicator or battery lines flash when the battery is getting low. In some camcorders, the unit shuts down within a few seconds. Sometimes you can still eject the cassette, but, in some models, you must insert a new battery or apply the ac adapter for any type of operation.

Canon 8-mm battery detection When the unregulated battery of 6 V drops rapidly to cause the EVF 5 V to become less than 4.5 V, the ES Det L signal is output at pin 3 of IC230. This places the main Mi-Com (IC231) in the Sleep mode and shuts down power to the camcorder (Fig. 40-38). The 6-V battery must be replaced or charged before the camcorder can be operated.

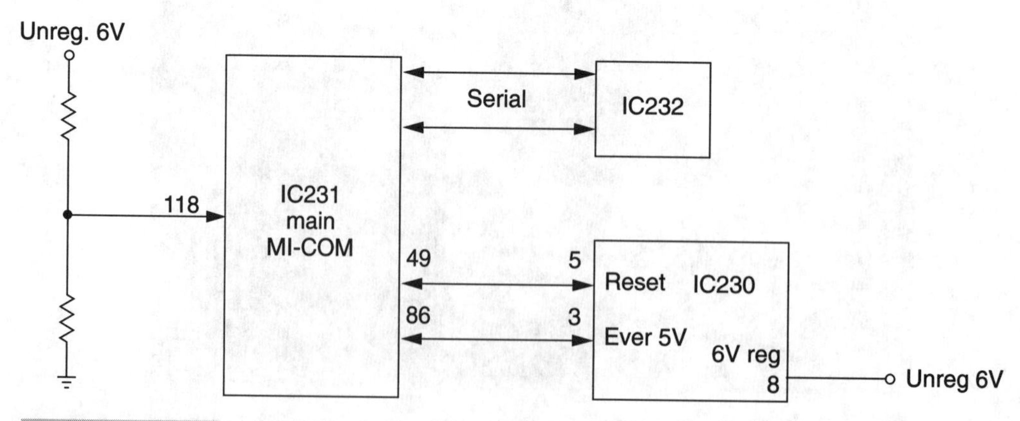

FIGURE 40-38 The battery-detection circuits in an 8-mm camcorder.

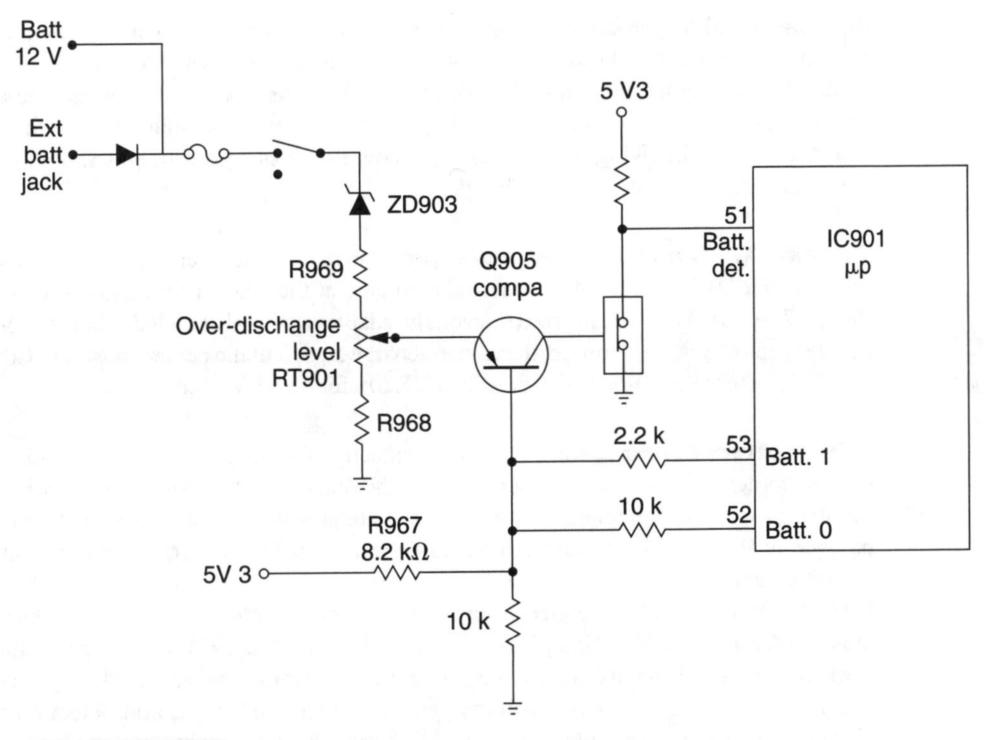

FIGURE 40-39 VHS-C battery overcharge-detection circuit.

Typical VHS-C battery overdischarge-detection circuits The battery-overdischarge detection (ODC) circuit monitors the battery terminal voltage, detects overdischarge, displays characters in the EVF screen to indicate discharge, and, if the voltage is less than normal, it turns the power off. In the VTR mode, "Power Off" corresponds to "Stop." If in the Camera mode, "Power Off" corresponds to "Record Pause."

Battery power (12 V) passes through the fuse (F970) and a latch relay (RL901) to a zener diode (ZD903, 7.5 V), which causes a voltage drop of about 7.5 V (Fig. 40-39). It is then applied through RT901 (overcharge level) to the emitter of the comparator (Q905). Q905 compares the voltage with the reference voltage.

The reference voltage is generated by the digital-to-analog converter controlled by Batt outputs (two bits of data—Batt 1 and Batt 0) from the IC901 and is applied to the base (inverting type) of the comparator (IC905). When the Batt output of the system microprocessor (IC901) is high, the resistors connected to the output pins are connected in parallel with R967 (8.2 kΩ), and the reference voltage is changed, depending on the battery output.

When the battery voltage is applied to the emitter (noninverting input) of the comparator (Q905), it becomes lower than the reference voltage applied to the base of Q905. The output of the switch (Q906) goes high, and IC901 detects battery voltage and judges whether the battery is discharged. The results of this judgment make the output of IC901.

Ac adapter/charger circuit adjustments The ac adapter and charger output voltage must be properly adjusted for accurate and safe voltages when operating the camcorder or charging the batteries. First determine the battery operating voltage and adjust the output of the adapter voltage accordingly. Although each manufacturer has their own adjustments, the following methods indicate how it is done. Adjustments should also be made for required correct current charging.

Typical VHS-C adapter voltage adjustment The adapter voltage adjustment is made with a DVM or DMM. Measure the voltage at the adapter terminals (Fig. 40-40). Place a 20-Ω, 10-W resistor across the voltage adapter terminals as a load. Locate voltage-adapter control VR201. Now set the camera/charge switch to the camera position. Connect the DVM across the 10-W resistor. Adjust VR201 for 13.95 V ±0.05 V.

Other battery connections Besides connecting the small battery or an ac adapter to the camcorder, the reset can be operated with the outside battery pack or auto car battery adapter/charger. Many camcorders have a separate jack where the external power can be plugged in (Fig. 40-41). Some camcorders have a car battery charger that plugs into the cigarette lighter.

The camcorder can be operated from the car-battery adapter or the unit will charge up those batteries while shooting pictures in remote areas. The 12-V car battery voltage is lowered or raised according to the camcorder operating voltage. The car battery charger can have a shutdown, voltage detector, current detector, and detection timer circuits, such as the ac adapter (Fig. 40-42). Follow the manufacturers service literature for correct output voltage and charging current adjustments of the car-battery charger adapter.

Troubleshooting the power adapter Before tearing into the ac adapter, check all cords and plugs. Most ad adapter units fit on the backside of the camcorder in place of the battery. Clean all slip-pressure-type contacts with cleaning fluid. Inspect the dc output cord for breaks or poor connections. Place a load (20-Ω, 10-W resistor) across the charging output terminals and measure the output voltages.

Determine if the ac adapter is dead for both providing voltage and charging batteries. If the unit supplies voltage, but does not charge the batteries, suspect a dirty voltage/charge

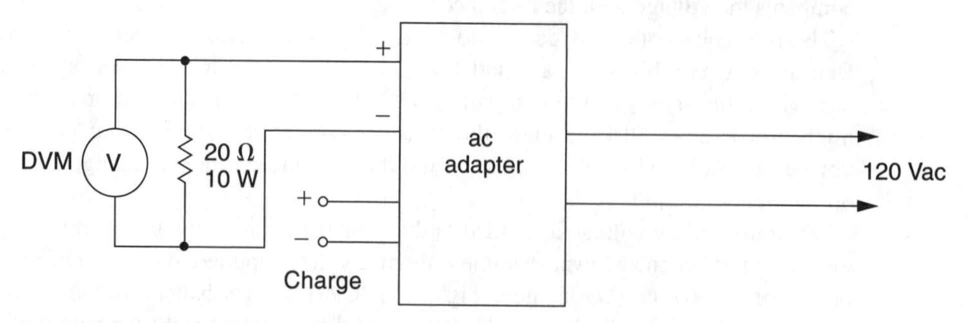

FIGURE 40-40 The ac-adapter voltage-adjustment hookup.

FIGURE 40-41 The battery jack is at the rear of the camcorder.

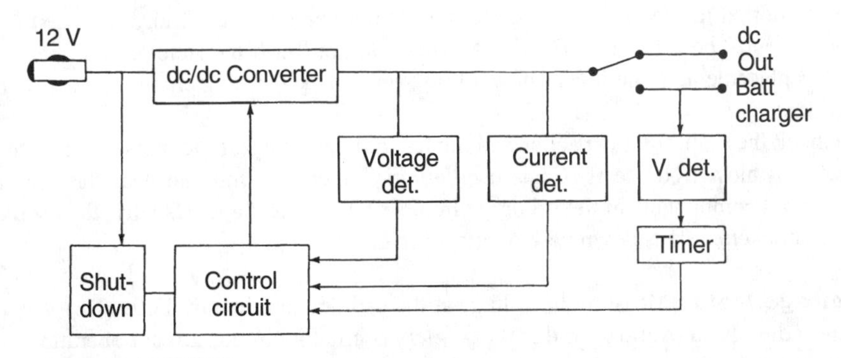

FIGURE 40-42 The block diagram of a car battery voltage/charger adapter.

switch. If the adapter is entirely dead, remove the top cover and inspect the fuse (Fig. 40-43). Replace the blown fuse with exact amperage. Take a close look—sometimes more than one fuse is in these adapters.

To check the output voltage of ac adapter, place a 20- or 10-Ω low resistor across adapter battery terminals or use a 100-W bulb. It's best to solder alligator clips and leads

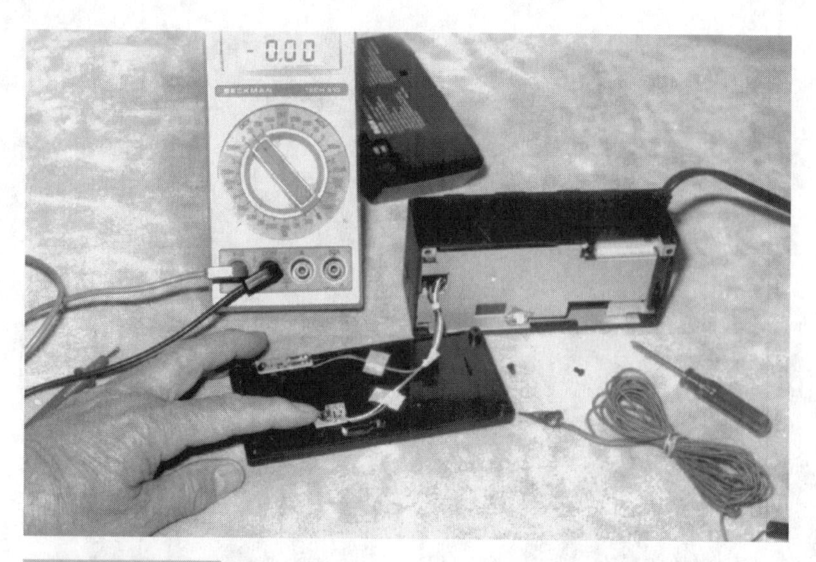

FIGURE 40-43 Remove the cover of an ac adapter to check fuse and components.

to resistor of bulb so that it can be easily clipped across the battery jack terminals. Be sure that no sense or interlock switches must be jumped. Do not forget to remove the jumpers or clips after the adapter has been repaired or tested.

Keeps blowing fuses Suspect that a bridge rectifier is shorted if the main line fuse keeps opening. Check across each diode for leakage. Sometimes one or two diodes become shorted inside of the bridge rectifier. If only one is found leaky, replace the whole component. These fuses are 2- or 3-A types. Do not fret if the adapter unit operates with fuse replacement. Sometimes a power line overload or lightning charge can cause the fuse to open.

Check the main filter capacitor and diodes in the primary of the transformer if the main fuse still blows. Notice if IC1 is oscillating. Sometimes you can hear the oscillations. Check all components in the primary circuits of T1 for leakage. Usually, the oscillator or dc/dc converter shuts down with overloading circuits.

Voltage tests Measure the voltage at the cathode terminal of diode (D8). No voltage here indicates an overload or that the primary oscillator circuits are not operating.

Remove the S2 lead of the secondary from T1 to isolate the secondary circuits. Now take voltage measurements on IC1 and Q1. A very low voltage at pin 3 of IC1 can indicate that an IC is leaky. Higher-than-normal voltage at pin 3 can show that IC1 is open or not oscillating. Proceed to the secondary circuits if the dc voltage is at the collector terminal of D8.

Random transistor and diode tests A quick method of checking for a leaky or shorted diode or transistor is to make a random check with each semiconductor on the board. Place the DMM to diode or transistor test and check the resistance across each diode. All diodes, including zener types, can be checked with this method. Double-check components around the diode if one is found leaky so that you are sure that another component is not indicating the leakage instead of the component that you suspect is faulty.

Check the transistors with a common base check to the collector and then the emitter terminals for leakage or open conditions. Like the diode, a good transistor should only have a low resistance measurement in one direction. The transistor might be open if there is no measurement. Use the regular transistor tester if one is handy. Most transistors in the ac adapters are npn types, but check them to be sure.

Most transistors and diodes can be checked with the transistor tester (Fig. 40-44). Of course, the digital transistors can indicate that resistance joint is open or high. Several digital transistors are used for switching in the ac adapter. If the digital transistor is defective, replace it with an exact-type part.

Checking ICs and other components Often, two to four ICs are in the ac adapter. One is used as a primary control oscillator; the others are found in the voltage control and charging current circuits. Voltage measurements on the IC can help you to find a leaky IC. A low-voltage (V_{cc}) supply pin of the IC might indicate that the IC is leaky. Remove the pin lead from the foil with solder wick and take another voltage test. Measure the pin of the IC to common ground for a low resistance (under 1 kΩ). If the IC has a low-resistance measurement at the supply-voltage pin terminal, it is leaky. Replace it.

The photocoupler enclosed in one component can be checked with the diode test of the DMM (Fig. 40-45). Check the diode like the regular fixed diode with the DMM diode test. Now check the phototransistor side in the same manner. The diode should have a low measurement in one direction and no measurement across the emitter and collector terminals on the transistor side. A very low measurement indicates that the photocoupler is leaky.

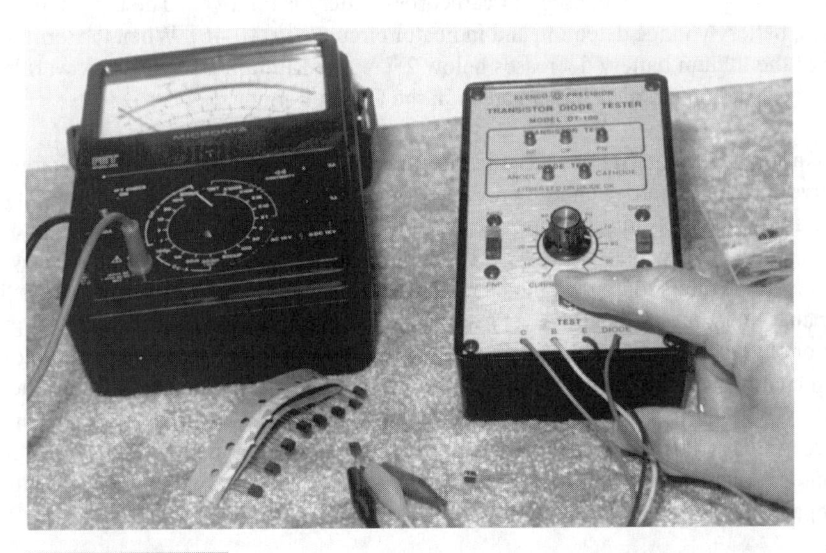

FIGURE 40-44 Check the transistors with the transistor tester.

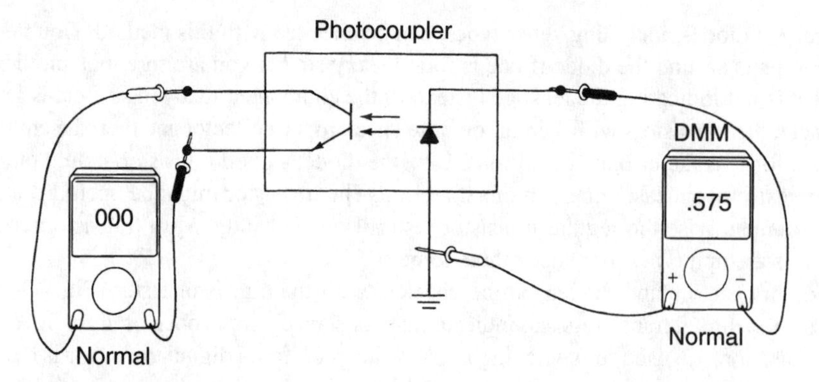

FIGURE 40-45 Checking the photocoupler in the ac adapter isolation circuits.

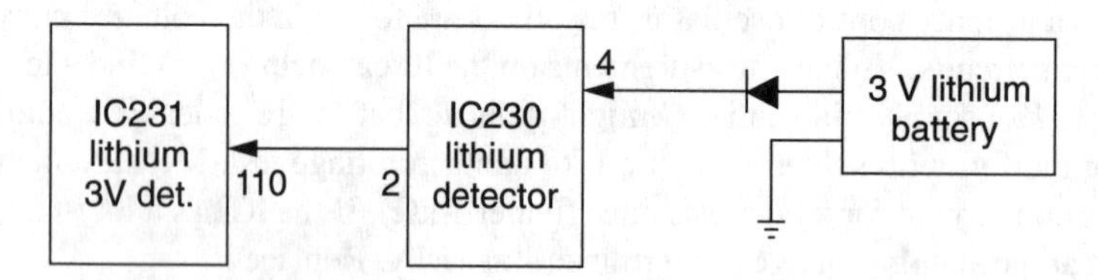

FIGURE 40-46 A Canon 8-mm lithium battery detection and indication circuit.

Lithium batteries The lithium battery is used in some camcorders to backup the memory or quartz circuits. Usually, the lithium battery is a 3-V type. Lithium battery 3-V power is supplied when the regular camcorder battery is removed. The lithium battery can have a battery voltage detection and indicator circuit (Fig. 40-46). When the terminal voltage of the lithium battery decreases below 2.7 V, the Lithium Battery Mark warning indication flashes on the viewfinder screen in the Canon 8-mm.

Battery will not charge The battery might be used up or the battery charger is defective. If other batteries charge on the adapter/charger, suspect that the battery is defective. Try completely discharging the battery. Now try to recharge it. After a couple of hours, if the battery does not charge, discard it. (Do not place it in a fire to destroy it.)

Check the voltage output of the charger with a load made up of resistors across the battery charge output terminals. A very low voltage output can indicate that the charging circuit or charge/voltage switch is defective. Notice if the adapter provides operating voltage for the camcorder. Inspect the Charge/Voltage switch for poor contacts. Shunt the switch contacts with alligator clips. Suspect that the IC charge circuit is defective if it has no charge voltage. Inspect the diode in series with the voltage and charge lead terminal for open or burned conditions. Measure all voltages on the IC (power switch) terminals and compare these voltages with those on the schematic.

INDEX

B

C

R

S

T